PRENTICE HALL

GEOMETRY

Prentice Hall dedicates this mathematics program to all mathematics educators and their students.

PRENTICE HALL
Needham, Massachusetts
Upper Saddle River, New Jersey

AUTHORS

Laurie E. Bass

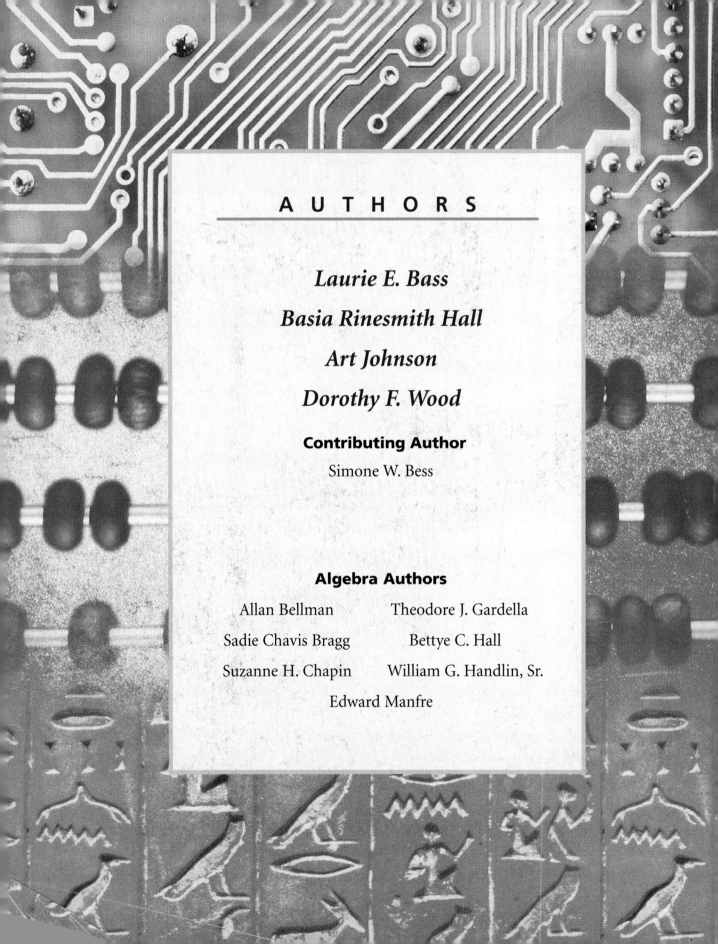

Basia Rinesmith Hall

Art Johnson

Dorothy F. Wood

Contributing Author

Simone W. Bess

Algebra Authors

Allan Bellman

Sadie Chavis Bragg

Suzanne H. Chapin

Theodore J. Gardella

Bettye C. Hall

William G. Handlin, Sr.

Edward Manfre

PRENTICE HALL

GEOMETRY

TOOLS FOR A CHANGING WORLD

Authors, Geometry

Laurie E. Bass
The Fieldston School
Riverdale, New York

Basia Rinesmith Hall
Alief Hastings High School
Alief Independent School District
Alief, Texas

Art Johnson, Ed.D.
Nashua High School
Nashua, New Hampshire

Dorothy F. Wood
Formerly, Kern High School District
Bakersfield, California

Contributing Author
Simone W. Bess, Ed.D.
University of Cincinnati
College of Education
Cincinnati, Ohio

Authors, Algebra & Advanced Algebra

Allan Bellman
Watkins Mill High School
Gaithersburg, Maryland

Sadie Chavis Bragg, Ed.D.
Borough of Manhattan
Community College
The City University of New York
New York, New York

Suzanne H. Chapin, Ed.D.
Boston University
Boston, Massachusetts

Theodore J. Gardella
Formerly, Bloomfield Hills Public Schools
Bloomfield Hills, Michigan

Bettye C. Hall
Mathematics Consultant
Houston, Texas

William G. Handlin, Sr.
Spring Woods High School
Houston, Texas

Edward Manfre
Mathematics Consultant
Albuquerque, New Mexico

PRENTICE HALL
Simon & Schuster Education Group
A VIACOM COMPANY

Printed in the United States of America.

ISBN: 0-13-416785-6

3 4 5 6 7 8 9 10 04 03 02 01 00 99 98 97

REVIEWERS

Series Reviewers

James Gates, Ed.D.
Executive Director Emeritus, National Council of Teachers of Mathematics, Reston, Virginia

Vinetta Jones, Ph.D.
National Director, EQUITY 2000, The College Board, New York, New York

Geometry

Sandra Argüelles Daire
Miami Senior High School
Miami, Florida

Priscilla P. Donkle
South Central High School
Union Mills, Indiana

Tom Muchlinski, Ph.D.
Wayzata High School
Plymouth, Minnesota

Bonnie Walker
Texas ASCD
Houston, Texas

Karen Doyle Walton, Ed.D.
Allentown College of
Saint Francis de Sales
Center Valley, Pennsylvania

Algebra

John J. Brady III
Hume-Fogg High School
Nashville, Tennessee

Elias P. Rodriguez
Leander Junior High School
Leander, Texas

Dorothy S. Strong, Ed.D.
Chicago Public Schools
Chicago, Illinois

Art W. Wilson, Ed.D.
Abraham Lincoln High School
Denver, Colorado

Advanced Algebra

Eleanor Boehner
Methacton High School
Norristown, Pennsylvania

Laura Price Cobb
Dallas Public Schools
Dallas, Texas

William Earl, Ed.D.
Formerly Mathematics
Education Specialist
Utah State Office of Education
Salt Lake City, Utah

Robin Levine Rubinstein
Shorewood High School
Shoreline, Washington

Staff Credits

The people who made up the *Geometry* team — representing editorial, design, marketing, page production, editorial services, production, manufacturing, technology, electronic publishing, and advertising and promotion — and their managers are listed below. Bold type denotes core team members.

Alison Anholt-White, Jackie Zidek Bedoya, Barbara A. Bertell, Bruce Bond, Ellen Brown, Judith D. Buice, Kathy Carter, Kerri Caruso, **Linda M. Coffey, Noralie V. Cox,** Sheila DeFazio, Edward de Leon, Christine Deliee, Gabriella Della Corte, Robert G. Dunn, Barbara Flockhart, Audra Floyd, David Graham, Maria Green, Bridget A. Hadley, Joanne Hudson, Vanessa Hunnibell, Mimi Jigarjian, **Linda D. Johnson,** Elizabeth A. Jordan, Russell Lappa, **Catherine Martin-Hetmansky,** Eve Melnechuk, Cindy A. Noftle, Caroline M. Power, Roger E. Powers, Martha G. Smith, Kira Thaler, Robin Tiano, Christina Trinchero, Stuart Wallace, Cynthia A. Weedel, **Jeff Weidenaar, Pearl B. Weinstein,** Mary Jane Wolfe, Stewart Wood, David Zarowin

We would like to give special thanks to our National Math Consultants, Ann F. Bell, Patricia M. Cominsky, and Brenda Underwood, for all their help in developing this program.

Geometry Contents

CHAPTER 1

Tools of Geometry

Connections and Applications

CHAPTER 2

Investigating Geometric Figures

Connections and Applications

CHAPTER 3

Transformations: Shapes in Motion

Connections and Applications

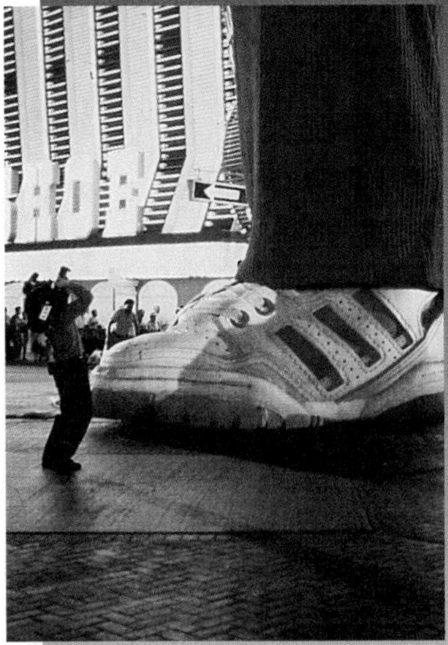

. . . and More!

ASSESSMENT

Chapter Project

Frieze Frames
Analyzing and Creating Frieze Patterns

CHAPTER

4 Triangle Relationships

Connections and Applications

CHAPTER

5 Measuring in the Plane

Connections and Applications

CHAPTER 6

Measuring in Space

Connections and Applications

CHAPTER 7

Reasoning and Parallel Lines

Connections and Applications

CHAPTER 8

Proving Triangles Congruent

Connections and Applications

CHAPTER

9 Quadrilaterals

Connections and Applications

CHAPTER

10 Similarity

Connections and Applications

CHAPTER

11 Right Triangle Trigonometry

Connections and Applications

CHAPTER 12

Chords, Secants, and Tangents

Connections and Applications

To the Student

Students like you helped Prentice Hall develop this program. They identified tools you can use to help you learn now in this course and beyond. In this special "To the Student" section and throughout this program, you will find the **tools you need to help you succeed.** We'd like to hear how these tools work for you. Write us at Prentice Hall Mathematics, 160 Gould Street, Needham, MA 02194 or visit us at http://www.phschool.com.

"... Instead of problem after problem of pointless numbers, we should have a chance to think and to truly understand what we are doing. I personally think that we all should be taught this way."
Chris, Grade 9
Carson City, NV

"... I like to review what I learn as I go, rather than cramming the night before a test."
Ali, Grade 10
St. Paul, MN

"... I learn mathematics best when I draw a diagram or make a graph that helps show what the problem is that I will solve."
Amy, Grade 11
Columbia, SC

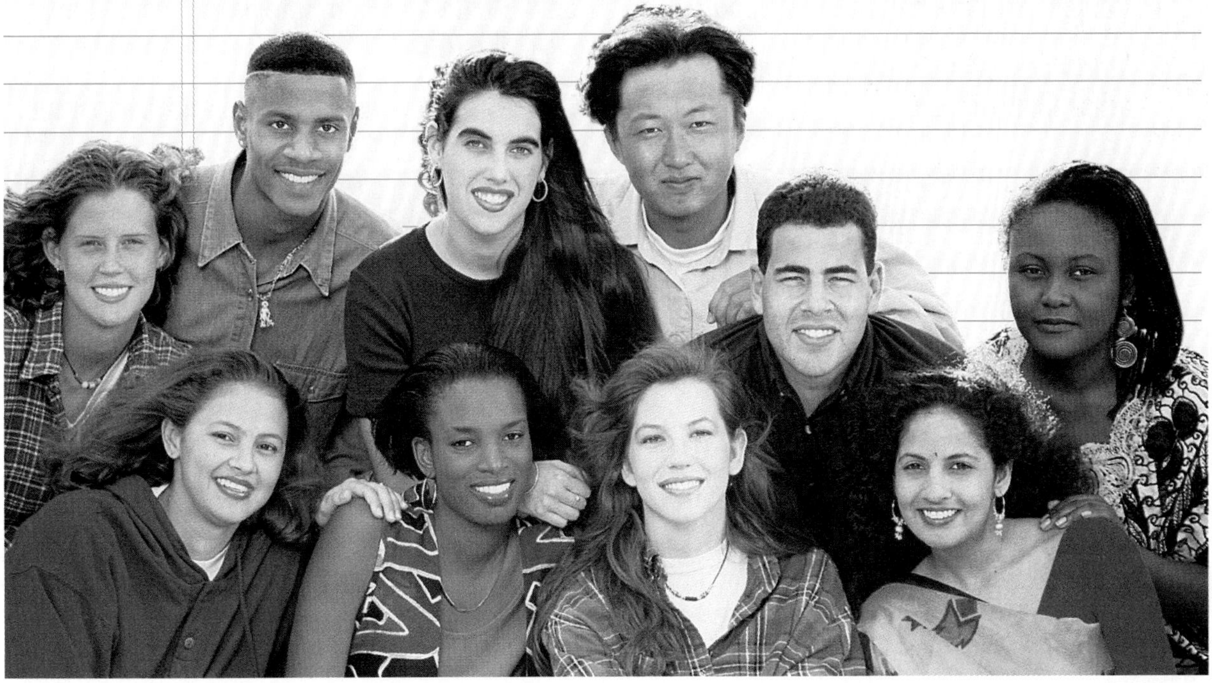

LEARN About Learning!

What comes to your mind when you hear the word **style**? Maybe it's hair style, or style of dress, or walking style. Have you ever thought about your learning style? Just like your hair or your clothes or your walk, everybody has a learning style that they like best because it works best for them. Look around you now. What do you see? Different styles … some like yours, some different from yours. That's the way it is with learning styles, too.

What's Your Best Learning Style?

I understand math concepts best when I...

- ❑ **A. Read about them.**
- ❑ **B. Look at and make illustrations, graphs, and charts that show them.**
- ❑ **C. Draw sketches or handle manipulatives to explore them.**
- ❑ **D. Listen to someone explain them.**

When I study, I learn more when I...

- ❑ **A. Review my notes and the textbook.**
- ❑ **B. Study any graphs, charts, diagrams, or other illustrations.**
- ❑ **C. Write ideas on note cards; then study the ideas.**
- ❑ **D. Explain what I know to another person.**

When I collaborate with a group, I am most comfortable when I...

- ❑ **A. Take notes.**
- ❑ **B. Make visuals for display.**
- ❑ **C. Demonstrate what I know to others.**
- ❑ **D. Give presentations to other groups or the whole class.**

Look for a pattern in your responses.
"A" responses suggest that you learn best by reading;
"B" responses indicate a visual learning style;
"C" responses suggest a tactile, or hands-on, learning style;
"D" responses signal that you probably learn best by listening and talking about what you are learning.

Having a preferred learning style does not limit you to using just that one. Most people learn by using a combination of learning styles. You'll be amazed by the ways that knowing more about yourself and how you learn will help you be successful — not only successful in mathematics, but successful in all your subject areas. When you know how you learn best, you will be well equipped to enter the work place.

Use this chart to help you strengthen your different learning styles.

Learning Style	**Learning Tips**
Learning by ***reading***	✱ Schedule time to read each day. ✱ Carry a book or magazine to read during wait time. ✱ Read what you like to read—it's OK not to finish a book.
Learning by using ***visual*** cues	✱ Visualize a problem situation. ✱ Graph solutions to problems. ✱ Let technology, such as computers and calculators, help you.
Learning by using ***hands-on*** exploration	✱ Make sketches when solving a problem. ✱ Use objects to help you solve problems. ✱ Rely on technology as a tool for exploration and discovery.
Learning by ***listening and talking***	✱ Volunteer to give presentations. ✱ Explain your ideas to a friend. ✱ Listen intently to what others are saying.

Most important, believe in yourself and your ability to learn!

What you do now...	What you do in this course...	What you do in the work place...
Learn best by using a particular learning style . . .	Example 1 Example 2 Relating to the Real World	*Choose a career that you enjoy because it is natural for you.*

Help Teamwork Work for YOU!

Each of us works with other people on teams throughout our lives. What's your job? Your job on a team, that is. Maybe you play center on your basketball team, maybe you count votes for your school elections, perhaps you help decorate the gym for a school function, or maybe you help make scenery for a community play. From relay races to doing your part of the job in the work place, teamwork is required for success.

TEAMWORK CHECKLIST

- **Break apart the large task into smaller tasks, which become the responsibility of individual group members.**
- **Treat the differences in group members as a benefit.**
- **Try to listen attentively when others speak.**
- **Stay focused on the task at hand and the goal to be accomplished.**
- **Vary the tasks you do in each group and participate.**
- **Recognize your own and others' learning styles.**
- **Offer your ideas and suggestions.**
- **Be socially responsible and act in a respectful way.**

What you do now...

Play on a team, decorate the gym, or perform in the band...

What you do in this course...

WORK TOGETHER

What you do in the work place...

Collaborate with coworkers on projects.

Reading + Writing + Talking + Listening:

It's All COMMUNICATION

Ve communicate in songs. We ommunicate in letters. We ommunicate with our body novements. We communicate on he phone. We communicate in yberspace. It's all talking about deas and sharing what you know. t's the same in mathematics — we ommunicate by reading, writing, alking, and listening. Whether we re working together on a project r studying with a friend for a test, ve are communicating.

Ways to Communicate What You Know and Are Able to Do

- ✔ Explain to others how you solve a problem.
- ✔ Listen carefully to others.
- ✔ Use mathematical language in your writing in other subjects.
- ✔ Pay attention to the headings in textbooks — they are signposts that help you.
- ✔ Think about videos and audiotapes as ways to communicate mathematical ideas.
- ✔ Be on the lookout for mathematics when you read, watch television, or see a movie.
- ✔ Communicate with others by using bulletin boards and chat rooms found on the Internet.

What you do now...	What you do in this course...	What you do in the work place...
Teach a young relative a sport...	THINK AND DISCUSS	*Written and verbal communication at work.*

Solving PROBLEMS — *a* SKILL You USE Every DAY

Problem solving is a skill — a skill that you probably use without even knowing it. When you think critically in social studies to draw conclusions about pollution and its stress on the environment, or when a mechanic listens to symptoms of trouble and logically determines the cause, you are both using a mathematical problem-solving skill. Problem solving also involves logical reasoning, wise decision making, and reflecting on our solutions.

Tips for Problem Solving

Recognize that there is more than one way to solve most problems.

When solving a word problem, read it, decide what to do, make a plan, look back at the problem, and revise your answer.

Experiment with various solution methods.

Understand that it is just as important to know how to solve a problem as it is to actually solve it.

Be aware of times you are using mathematics to solve problems that do not involve computation, such as when you reason to make a wise decision.

What you do now...	What you do in this course...	What you do in the work place...
Make decisions based on changing conditions, such as weather...	*PROBLEM SOLVING*	*Synchronize the timing of traffic lights to enhance traffic flow.*

Studying for the TEST Whatever It May Be

SATs, ACTs, chapter tests, and weekly quizzes — they all test what you know and are able to do. Have you ever thought about **how** you can take these tests to your advantage? You are evaluated now in your classes and you will be evaluated when you hold a job.

Pointers for Gaining Points

- Study as you progress through a chapter, instead of cramming for a test.
- Recognize when you are lost and seek help before a test.
- Review important theorems and postulates when studying for a test, and recognize how they relate to each other.
- Study for a test with a friend or study group.
- Take a practice test.
- Think of mnemonic devices to help you, such as Please Excuse My Dear Aunt Sally, which is one way to remember order of operations (parentheses, exponents, multiply, divide, add, subtract).
- Reread test questions before answering them.
- Check to see if your answer is reasonable.
- Think positively and visualize yourself doing well on the test.
- Relax during the test... there is nothing there that you have not seen before.

What you do now...	What you do in this course...	What you do in the work place...
Study notes in preparation for tests and quizzes...	SELF ASSESSMENT **How am I doing?** *Exercises* ON YOUR OWN *Exercises* CHECKPOINT	*Prepare for and participate in a job interview.*

CHAPTER

1 Tools of Geometry

Relating to the Real World

Every day, you trust that people understand the meaning of the words you use. In this chapter, you'll learn some of the basic terms of geometry and become familiar with the tools you'll use.

Lessons	1-1	1-2	1-3	1-4	1-5
	Using Patterns and Inductive Reasoning	*Points, Lines, and Planes*	*Segments, Rays, Parallel Lines and Planes*	*Measuring Angles and Segments*	*Good Definitions*

CHAPTER PROJECT

On Folded Wings

Some people look at a plain sheet of paper and see the hidden form of a swan or a seashell waiting to be revealed. Almost magically, with a few meticulous folds, an origami artist can produce startling replicas of

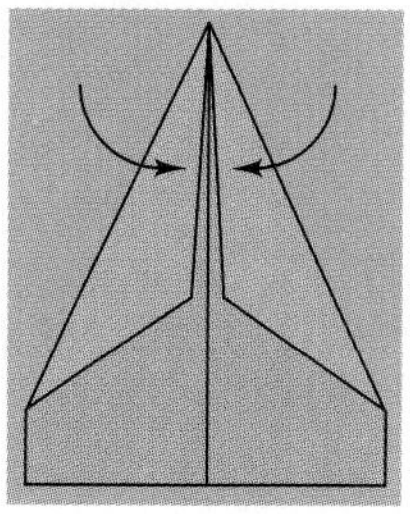

animals, flowers, buildings, vehicles, and even people. The ancient art of paper folding has come to us from Japan, where it has thrived since at least the twelfth century. As a child, every time you made a paper airplane or a paper hat, you were practicing the art of origami.

In this chapter project, you will use paper folding to explore geometric patterns. You will make origami models, and then use the language of geometry to tell others how to make them.

To help you complete the project:

- **p. 10** *Find Out by Doing*
- **p. 22** *Find Out by Creating*
- **p. 31** *Find Out by Researching*
- **p. 44** *Find Out by Writing*
- **p. 58** *Finishing the Project*

Basic Constructions — 1-6

Using Deductive Reasoning — 1-7

The Coordinate Plane — 1-8

1-1 Using Patterns and Inductive Reasoning

What You'll Learn

- Using inductive reasoning to make conjectures

...And Why

To sharpen your ability to reason inductively, a powerful tool used in mathematics and in making everyday decisions

WORK TOGETHER

The shortest path from the school to Longwood Avenue is six blocks long. One path is shown below in red.

PROBLEM SOLVING HINT

Copy the diagram. Find the number of ways to reach each intersection. Look for a number pattern.

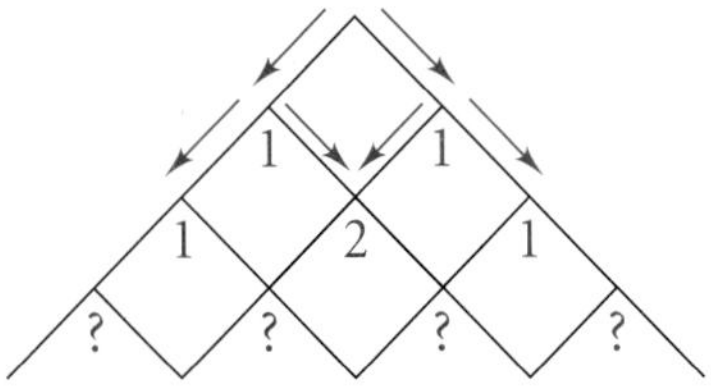

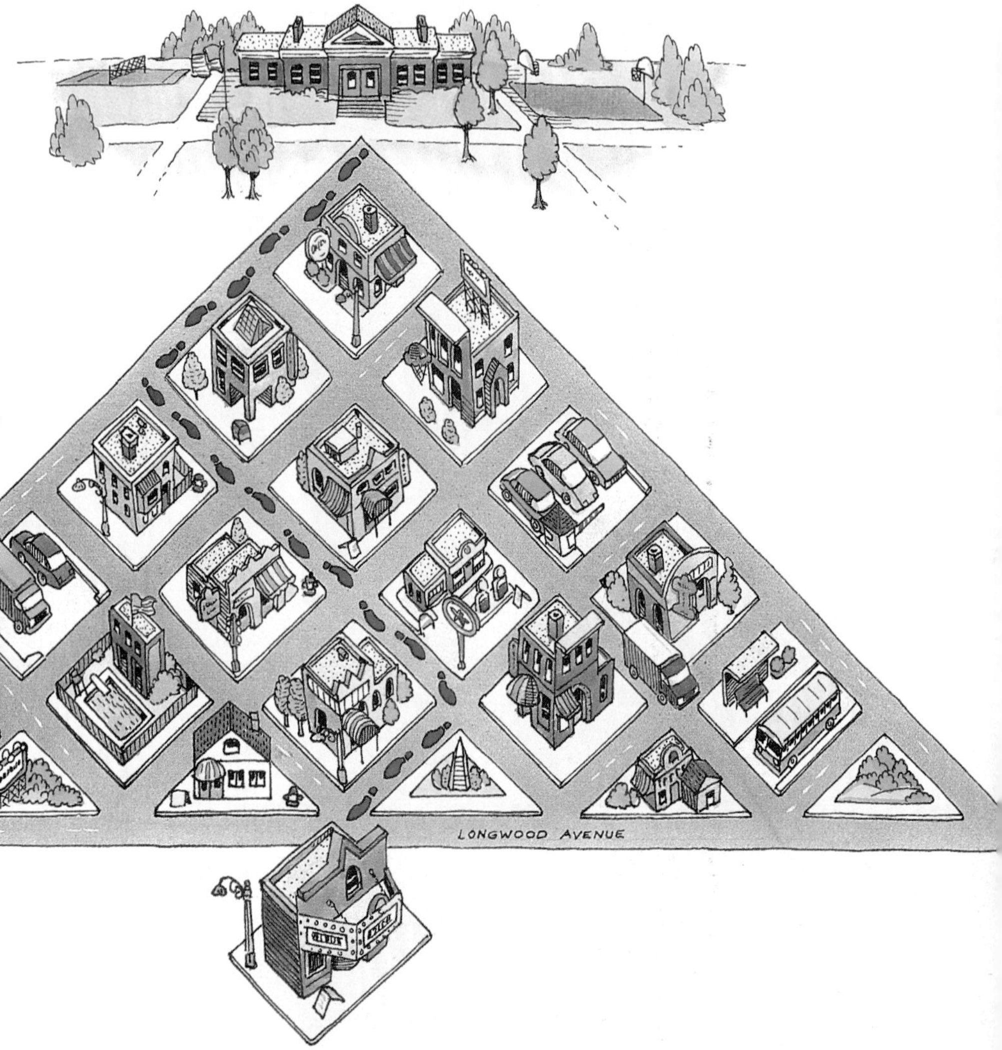

Who? The pattern of numbers you discovered in the Work Together is known as Pascal's Triangle, after the French mathematician Blaise Pascal (1623–1662).

Work with a group to answer the following questions.

1. How many different six-block paths can you take from the school to Longwood Avenue?
2. How many of these paths will end at the corner directly across from the movie theater?

THINK AND DISCUSS

To answer the questions in the Work Together, you used inductive reasoning. **Inductive reasoning** is a type of reasoning that allows you to reach conclusions based on a pattern of specific examples or past events. Mathematicians have made many discoveries using inductive reasoning.

3. **a.** Find the next two terms in this sequence: 2, 4, 6, 8, . . .
b. Describe the pattern you observed.

4. **a.** Find the next two terms in this sequence: 3, 6, 12, 24, . . .
b. Describe the pattern you observed.

5. **a.** Find the next two terms in this sequence: 1, 2, 4, 5, 10, 11, 22, . . .
b. Describe the pattern you observed.

A conclusion reached by using inductive reasoning is sometimes called a **conjecture.** For each sequence above, you found the next term by first finding a pattern, and then using the pattern to make a conjecture about the next term. Inductive reasoning from patterns is a powerful thinking process you will use throughout the year in geometry.

6. **Try This** Describe the next term in this sequence.

Example 1 Relating to the Real World

Manufacturing A skateboard shop finds that for five consecutive months sales of skateboards with small wheels (39 mm to 48 mm in diameter) decreased.

January: 58 February: 55 March: 51 April: 48 May: 45

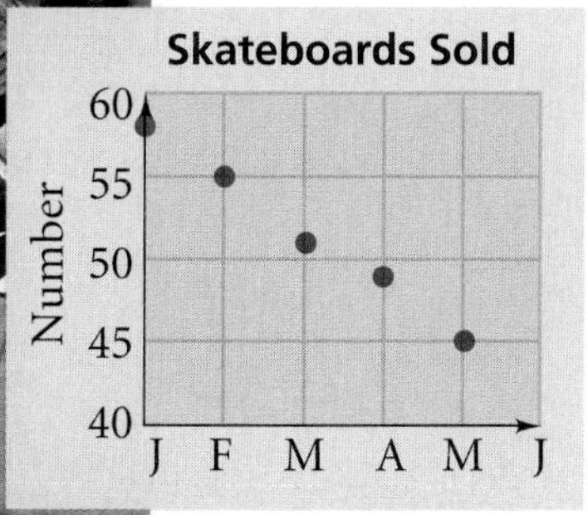

Use inductive reasoning to make a conjecture about the number of small-wheeled skateboards the shop will sell in June.

As the graph at the left shows, the number of small-wheeled skateboards is decreasing by about 3 skateboards each month. The skateboard shop can predict about 42 small-wheeled skateboards will be sold in June.

Not every conjecture or conclusion found by inductive reasoning is correct. The next problem illustrates the limitations of inductive reasoning.

Example 2

If six points on a circle are joined by as many segments as possible, how many nonoverlapping regions will the segments determine?

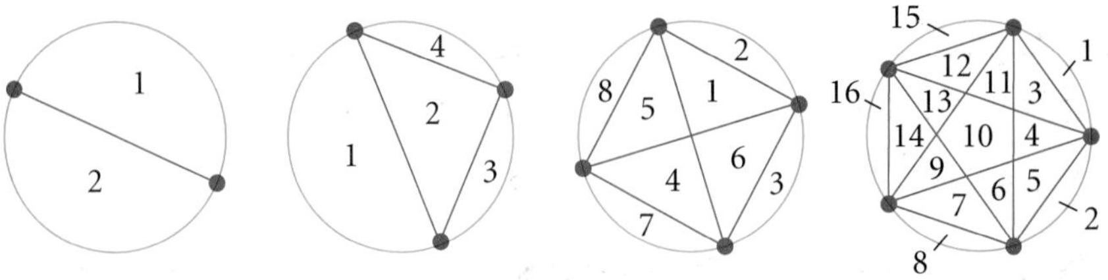

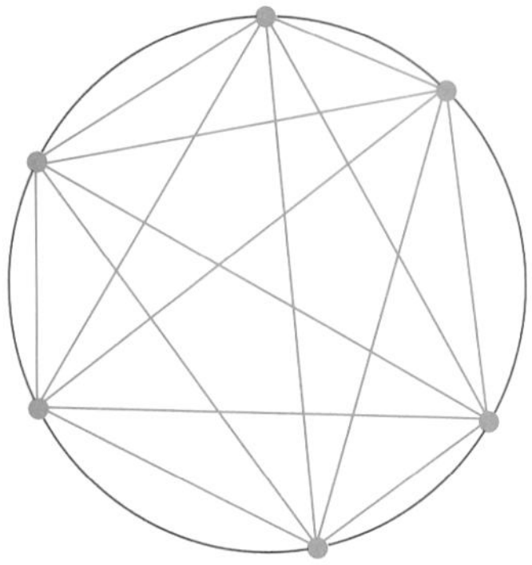

The table at the right shows the data for 2, 3, 4, and 5 points. The number of regions appears to double at each stage. Inductive reasoning would predict that there are 32 regions for 6 points on the circle. And yet, as the diagram at the left shows, there are only 31 regions formed. In this case, the conjecture is incorrect.

Points	Regions
2	2
3	4
4	8
5	16

Because the conjectures arrived at by inductive reasoning are not always true, you should verify them if possible.

7. Try This Candace examined five different examples and came up with this **conjecture:** "If any two positive numbers are multiplied, their product is always greater than either of the two numbers." Is her **conjecture** correct? Explain why or why not.

Sometimes you can use inductive reasoning to solve a problem that at first does not seem to have any pattern.

Example 3

Use inductive reasoning to find the sum of the first 20 odd numbers.

Find the first few sums. Notice that each sum is a perfect square.

1	$=$	1	$=$	1^2
$1 + 3$	$=$	4	$=$	2^2
$1 + 3 + 5$	$=$	9	$=$	3^2
$1 + 3 + 5 + 7$	$=$	16	$=$	4^2

Reasoning inductively, you would expect that the sum of the first 20 odd numbers would be 20^2, or 400.

8. Try This What is your **conjecture** for the sum of the first 30 odd numbers? Use your calculator to verify your **conjecture.**

Exercises ON YOUR OWN

Find the next two terms in each sequence.

1. $5, 10, 20, 40, \ldots$
2. $3, 33, 333, 3333, \ldots$
3. $1, -1, 2, -2, 3, \ldots$
4. $1, \frac{1}{2}, \frac{1}{4}, \frac{1}{8}, \ldots$
5. $15, 12, 9, 6, \ldots$
6. $81, 27, 9, 3, \ldots$
7. O, T, T, F, F, S, S, E, . . .
8. J, F, M, A, M, . . .
9. $1, 2, 6, 24, 120, \ldots$
10. $1, 2, 4, 7, 11, 16, 22, \ldots$
11. $1, \frac{1}{4}, \frac{1}{9}, \frac{1}{16}, \frac{1}{25}, \ldots$
12. $1, \frac{1}{2}, \frac{1}{3}, \frac{1}{4}, \ldots$
13. **Writing** Choose two of the sequences in Exercises 9–12 and describe the pattern.
14. Deano has started working out regularly. When he first started exercising he could do 10 push-ups. After the first month he could do 14 push-ups. After the second month he could do 19, and after the third month he could do 25. How many push-ups would you **predict** he will be able to do after the fifth month of working out? Are you absolutely sure about your prediction? Why or why not?
15. Alexa rides a bus to school. On the first day the trip to school took 25 minutes. On the second day the trip took 24 minutes. On the third day the trip took 26 minutes. On the fourth day the trip took 25 minutes. What **conjecture** would you make?
16. **History** Leonardo of Pisa (c. 1175–c. 1258) was born in Italy and educated in North Africa. He was one of the the first Europeans to use modern numerals instead of Roman numerals. He is also known for the Fibonacci Sequence: 1, 1, 2, 3, 5, 8, 13, ... Find the next three terms.

Draw the next figure in each sequence.

17.

18.
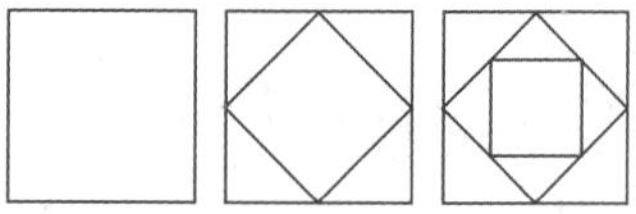

19.
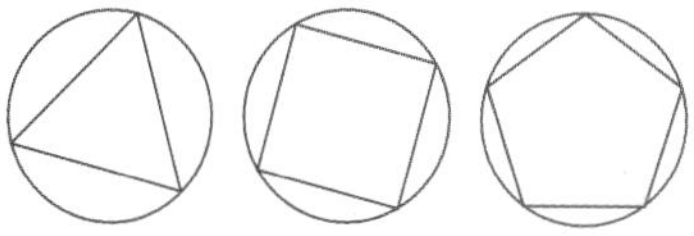

20.
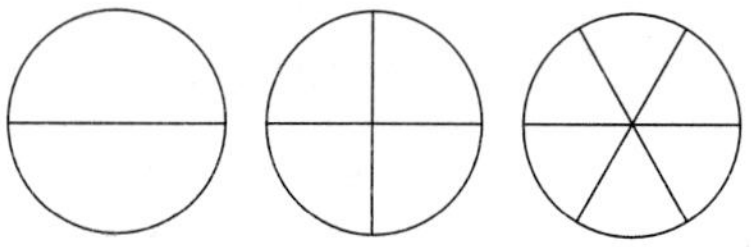

21.

22.
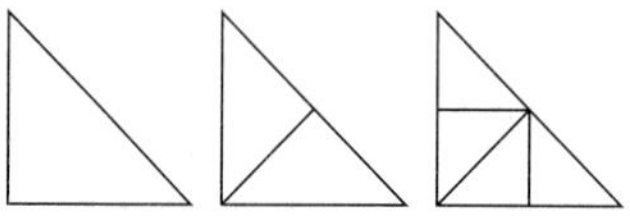

23. Draw two parallel lines on your paper. Locate four points on the paper an equal distance from both lines. Describe the figure you would get if you continued to locate points an equal distance from both lines.
24. Draw a line on your paper. Locate four points on the paper that are each 1 in. from the line. Describe the figure you would get if you continued to locate points that are 1 in. from the line.

TECHNOLOGY HINT

Exercises 23 and 24 could be done using geometry software.

25. For the past four years Paulo has grown 2 in. every year. He is now 16 years old and is 5 ft 10 in. tall. He figures that when he is 22 years old he will be 6 ft 10 in. tall. What would you tell Paulo about his **conjecture**?

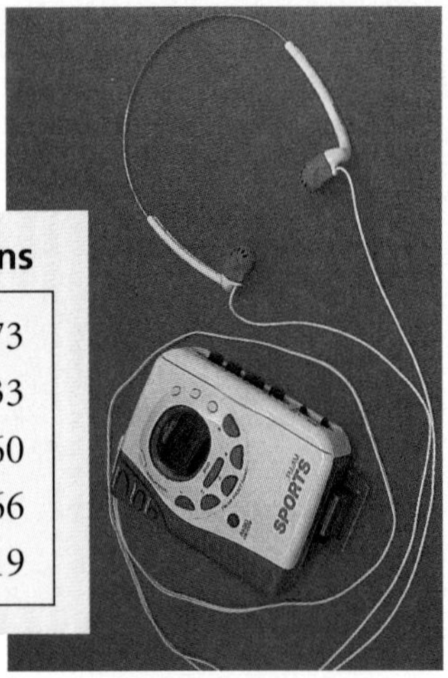

Radio Stations

1950	2,773
1960	4,133
1970	6,760
1980	8,566
1990	10,819

26. After testing her idea with eight different numbers, Jean stated the following **conjecture**: "The square of a number is always greater than the number you started with." What would you tell Jean about her **conjecture**?

27. a. **Communications** The number of radio stations in the United States is increasing. The table shows the number of radio stations for a 40-year period. Make a line graph of the data. Use the graph and inductive reasoning to make a **conjecture** about the number of radio stations in the United States by the year 2010.
    b. How confident are you about your conjecture? Explain.

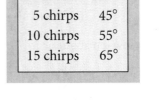

Find the next term in each sequence. Check your answer with a calculator.

28. 12345679 × 9 = 111111111
    12345679 × 18 = 222222222
    12345679 × 27 = 333333333
    12345679 × 36 = 444444444
    12345679 × 45 = ?

29. 1 × 1 = 1
    11 × 11 = 121
    111 × 111 = 12321
    1111 × 1111 = 1234321
    11111 × 11111 = ?

30. **Open-ended** Write two different sequences that begin with the same two numbers.

Chirps per 14 s

5 chirps	45°
10 chirps	55°
15 chirps	65°

31. **Weather** The temperature in degrees Fahrenheit determines how fast a cricket chirps. If you heard 20 cricket chirps in 14 seconds, what do you think the temperature would be?

32. a. A *triangular number* can be represented by a triangular arrangement of dots. The first two triangular numbers are 1 and 3. What are the next three triangular numbers?

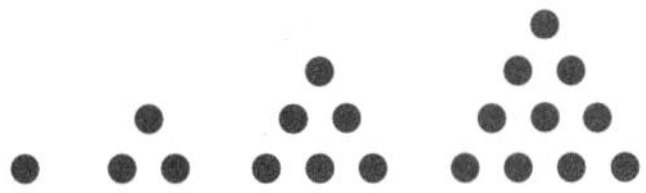

    b. What is the tenth triangular number?
    c. **Algebra** Which of the following expressions represents the nth triangular number?

A. $n(n + 1)$ **B.** $n(n - 2)$ **C.** $\frac{n(n + 1)}{2}$ **D.** $\frac{n(n - 1)}{2}$

33. a. The first two *square numbers* are 1 and 4. Draw diagrams to represent the next two square numbers.

    b. What is the twentieth square number? Describe the pattern.
    c. **Algebra** Write an algebraic expression in terms of n for the nth square number.

34. *History* Nicomachus of Gerasa first described *pentagonal numbers* in *Introductio arithmetica* about A.D. 100. The first three pentagonal numbers are shown. Draw a diagram to represent the next pentagonal number.

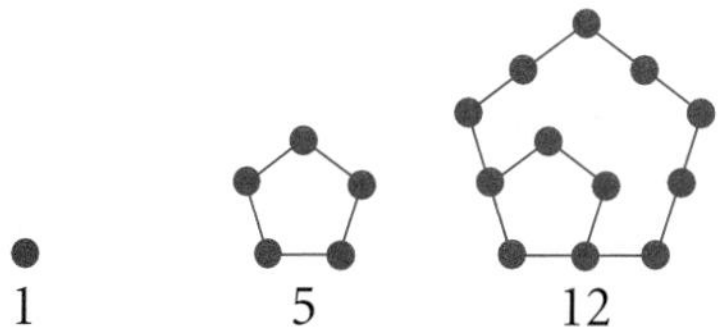

The Race to the Finish Line

Top female runners have been improving about twice as quickly as the fastest men, a new study says. If this pattern continues, women may soon outrun men in competition!

The study is based on world records collected at 10-year intervals, starting in 1905 for men. Reliable women's records were not kept until the 1920s. Women's marathon records date only from 1955.

If the trend continues, the top female and male runners in races ranging from 200 m to 1500 m might attain the same speeds sometime between 2015 and 2055. The rapid improvement in women's marathon records suggests that the marathon record for women will equal that of men even more quickly—perhaps by 2005.

Women's speeds may have improved so quickly because many more women started running competitively in recent decades, according to a professor of anatomy who studies locomotion and gait. This increase in the talent pool of female runners has improved the chance of finding better runners.

35. **a.** What conclusions were reached in the study mentioned in the newspaper clipping?

b. How was inductive reasoning used to reach the conclusions?

c. Explain why the conclusion that women may soon be outrunning men may be incorrect. For which race is the conclusion most suspect? For what reason?

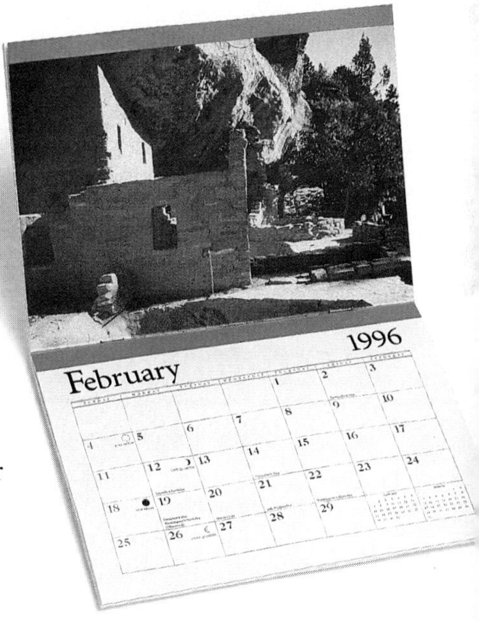

36. *Standardized Test Prep* Which of the following can be a term in the sequence 1, 3, 7, 15, 31, . . . ?

A. 32 **B.** 47 **C.** 55 **D.** 127 **E.** 128

37. **a.** *Leap years* have 366 days. 1984, 1988, 1992, 1996, and 2000 are consecutive leap years. Make a **conjecture** about leap years.

b. Which of the following years do you think will be leap years?

2010, 2020, 2100, 2400

c. *Research* Find out if your **conjecture** for part (a) and your answer for part (b) are correct. How are leap years determined?

38. **a.** *Coordinate Geometry* Graph the following points:

$A(1, 5)$	$B(2, 2)$	$C(2, 8)$	$D(3, 1)$
$E(3, 9)$	$F(6, 0)$	$G(6, 10)$	$H(7, -1)$
$I(7, 11)$	$J(9, 1)$	$K(9, 9)$	$L(10, 2)$
$M(10, 8)$	$N(11, 5)$		

b. Which of the points do not fit the same pattern as the others?

c. Describe the figure you would get if you continued graphing points that fit the pattern.

QUICK REVIEW

The first coordinate is the *x*-coordinate. The second coordinate is the *y*-coordinate.

39. What is the last digit of 2^{85}? Make a table of values and use inductive reasoning.

40. *Patterns* How many different squares are there in this 5-by-5 grid?

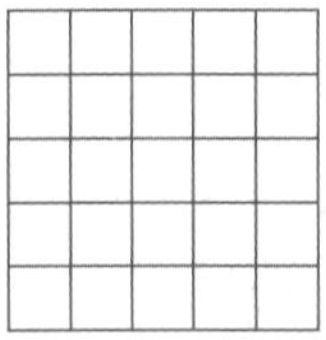

QUICK REVIEW

$2^3 = 2 \times 2 \times 2$

PROBLEM SOLVING HINT

Look at 1-by-1, 2-by-2, and 3-by-3 grids. Count squares and look for a pattern.

Chapter Project

Find Out by Doing

Most origami creations are made by folding square paper. You can create patterns while you practice paper folding.

- Carefully fold a square piece of paper four times as shown.

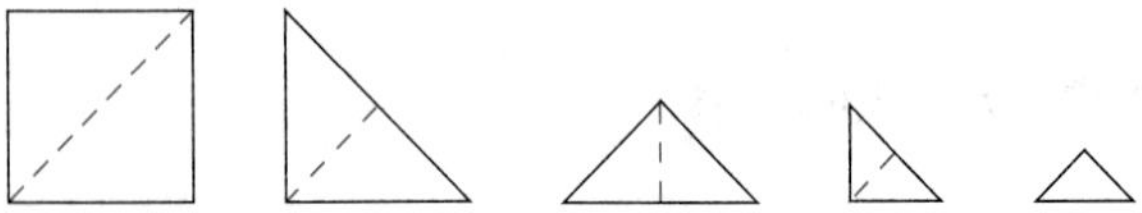

- Unfold the paper after each fold. Count the number of non-overlapping triangles formed. Record your results in a table like the one below.

Fold	1st	2nd	3rd	4th
No. of $\triangle$s	2	■	■	■

- Make a fifth fold. How many triangles are formed? How many triangles do you think will be formed after a sixth fold? Extend the table and describe the number pattern.
- Keep this origami creation for use in upcoming project activities.

Exercises MIXED REVIEW

Graph the following points.

41. $Y(-5, -8)$ **42.** $B(7, -10)$ **43.** $M(9, 12)$ **44.** $Q(-3, 2)$ **45.** $G(-6, 0)$ **46.** $F(-4, -5)$

47. $C(-7, 10)$ **48.** $N(0, -5)$ **49.** $R(4, 8)$ **50.** $H(-4, -9)$ **51.** $W(2, -5)$ **52.** $T(0, 4)$

53. a. In Exercises 41–52, which points are in the fourth quadrant?

b. Which points are on the y-axis?

Getting Ready for Lesson 1-2

54. Copy the diagram at the right. Draw as many different lines as you can to connect pairs of points.

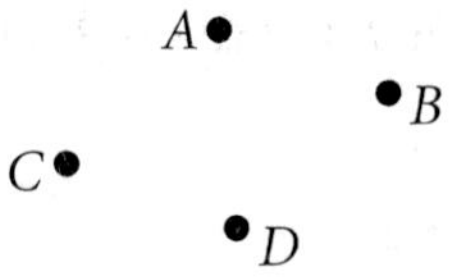

Probability

Before Lesson 1-2

Probability ranges from 0, an impossible event, to 1, a certain event. You can find the probability of an event using this formula.

$$P(\text{event}) = \frac{\text{number of favorable outcomes}}{\text{number of possible outcomes}}$$

Example 1

What is the probability of answering correctly a four-option multiple choice question if you pick an answer at random?

There are 4 possible outcomes. One of them is correct. The probability of getting the correct answer is $\frac{1}{4}$.

Example 2

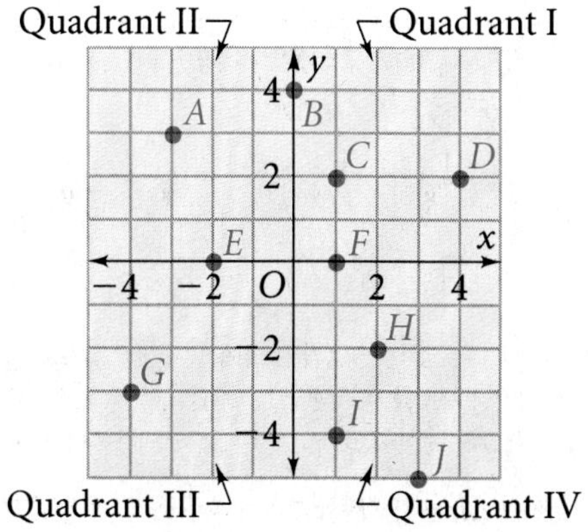

Find the probability that a point picked at random from the graph at the right is in the first quadrant.

List the possible outcomes: *A, B, C, D, E, F, G, H, I, J*.
There are 10 outcomes.

List the favorable outcomes: *C* and *D*.
There are 2 favorable outcomes.

The probability of a point picked at random being in the first quadrant is $\frac{2}{10} = \frac{1}{5}$.

Use the graph from Example 2 to find each probability. Assume points are picked at random.

1. *P*(the point is in the fourth quadrant)

2. *P*(the point is on an axis)

3. *P*(the point is at the origin)

4. *P*(the point has a *y*-coordinate of 2)

5. *P*(the point has an *x*-coordinate less than 4)

6. *P*(the point is on the *x*-axis)

7. *P*(the point is to the right of the *y*-axis)

8. *P*(the point has an *x*-coordinate of 1)

9. *P*(the point is on the *y*-axis)

10. *P*(the point is below the *x*-axis)

11. *P*(the point has a *y*-coordinate greater than 1)

12. *P*(the point is in the third quadrant)

Use the spinner at the right. Find each probability.

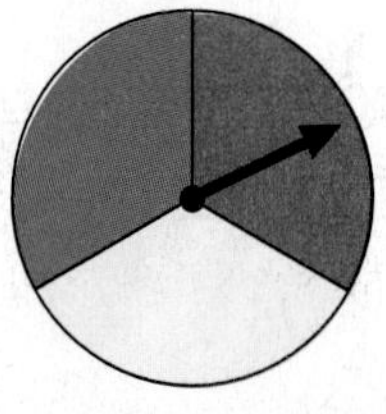

13. *P*(blue)

14. *P*(red)

15. *P*(yellow or red)

16. *P*(purple)

17. *P*(blue or red)

18. *P*(yellow or red or blue)

1-2 Points, Lines, and Planes

What You'll Learn

- Understanding basic terms of geometry
- Understanding basic postulates of geometry

...And Why

To lay the foundation for your study of geometry

What You'll Need

- ruler

WORK TOGETHER

Many constellations are named after animals and mythological figures. It takes some imagination to connect the points representing the stars so that the result is a recognizable figure such as Leo the Lion. There are many different ways to connect the points. How many different lines could be used to connect all ten points?

Ten major stars make up the constellation called Leo the Lion.

Work in groups of three. Make a table and look for a pattern to answer the following questions.

1. Put three points on a circle. Now connect the three points with as many lines as possible. How many lines do you need?

2. Put four points on another circle. How many lines can you draw connecting four points?

3. Repeat for five points on a circle and then for six points. How many lines can you draw to connect the points?

4. Use inductive reasoning to tell how many lines you could draw to connect the ten points of the constellation Leo the Lion.

THINK AND DISCUSS

Basic Terms

Since stars are so far away, they appear quite small to us. We think of them as points even though they are actually quite large. In geometry a **point** has no size. You can think of it as a location. A point is represented by a small dot and is named by a capital letter. All geometric figures are made up of points. **Space** is the set of all points.

5. *Open-ended* Name something in your classroom that is a physical representation of a point.

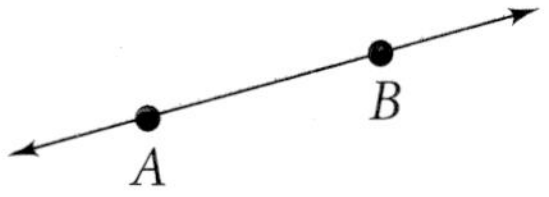

$\overleftrightarrow{AB}$ (or $\overleftrightarrow{BA}$)

You can think of a **line** as a series of points that extends in two opposite directions without end. You can name a line by two points on the line, such as $\overleftrightarrow{AB}$ (read "line AB"). Another way to name a line is with a single lowercase letter, such as line t.

6. Open-ended Describe some physical representations of lines in the real world.

7. Critical Thinking Why do you think arrowheads are used when drawing a line or naming a line such as $\overleftrightarrow{AB}$?

8. Try This Name the line at the left in as many ways as possible.

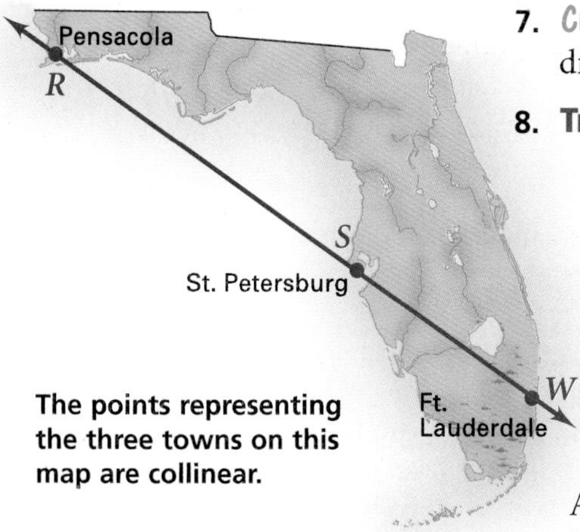

The points representing the three towns on this map are collinear.

Points that lie on the same line are **collinear.**

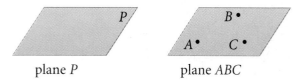

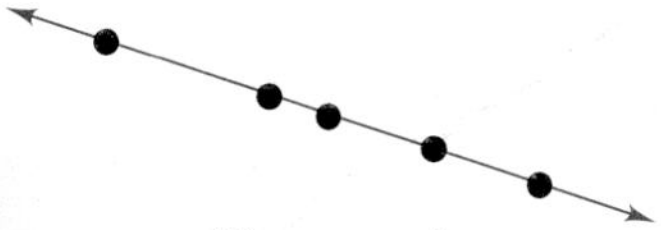

collinear points

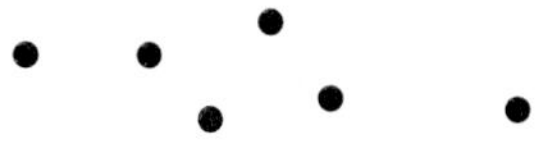

noncollinear points

A **plane** is a flat surface that extends in all directions without end. It has no thickness.

9. Open-ended Name three objects in your classroom that represent planes.

You can name a plane either by a single capital letter or by naming at least three noncollinear points in the plane.

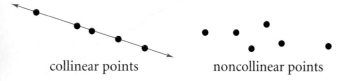

plane *P*

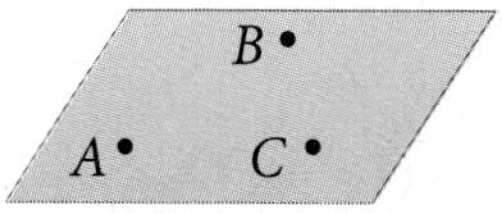

plane *ABC*

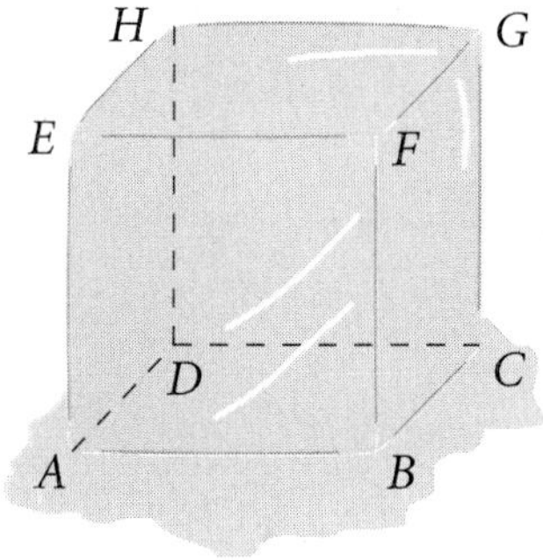

In the diagram, each surface of the ice cube is part of a plane.

10. How many planes are suggested by the surfaces of the ice cube?

11. Try This Name the plane represented by the front of the ice cube in several different ways.

Points and lines in the same plane are **coplanar.**

12. Try This Name a point that is coplanar with the given points.

a. *E, F, G* **b.** *B, C, G*

c. *A, D, E* **d.** *D, C, G*

13. Try This Name two lines that are coplanar with $\overleftrightarrow{AB}$ and $\overleftrightarrow{DC}$.

Basic Postulates

A **postulate** is an accepted statement of fact. You used some of the following geometry postulates in algebra. For example, when you graphed an equation such as $y = -2x + 8$, you began by plotting two points and then you drew the line through those two points.

Postulate 1-1

Through any two points there is exactly one line.

Line t is the only line that passes through points A and B.

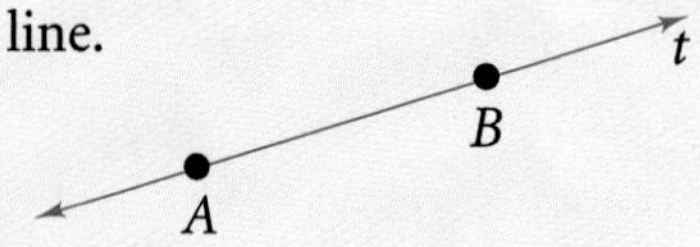

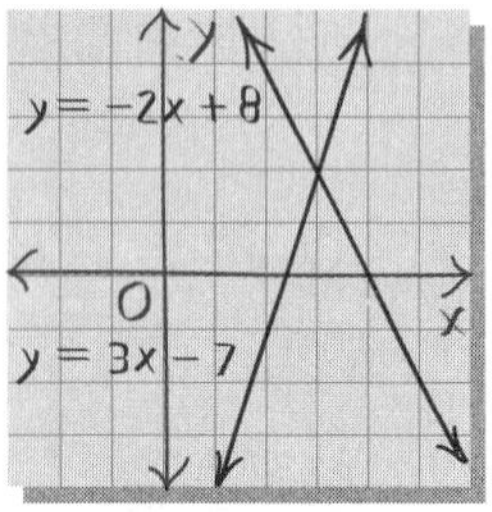

In algebra, one way to solve the following system of equations is to graph the two equations.

$$y = -2x + 8$$
$$y = 3x - 7$$

As the graph shows, the two lines intersect at a single point, (3, 2). The solution to the system of equations is $x = 3$, $y = 2$. This illustrates the following postulate.

Postulate 1-2

If two lines intersect, then they intersect in exactly one point.

14. Open-ended Describe two planes in your classroom that intersect. Also describe the intersection of the planes.

Postulate 1-3

If two planes intersect, then they intersect in a line.

Plane RST and plane STW intersect in $\overleftrightarrow{ST}$.

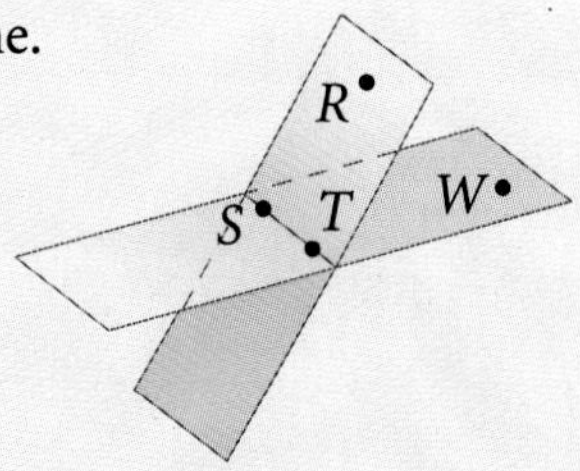

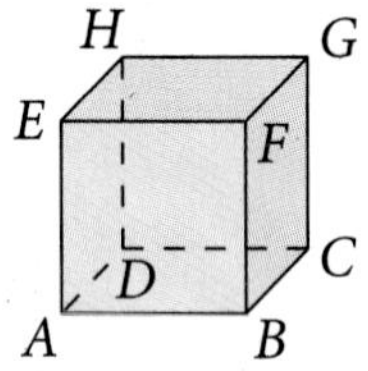

15. a. Try This What is the intersection of plane $HGFE$ and plane $BCGF$?

b. What is the intersection of plane AEF and plane BCG?

A three-legged stool will always be stable, as long as the feet of the stool don't lie on a line. This illustrates the following postulate.

Postulate 1-4

Through any three noncollinear points there is exactly one plane.

Who? Euclid is known for compiling all the geometry of his time into postulates and theorems. His masterwork *The Elements* (about 300 B.C.) is the basis for geometry books today.

Example

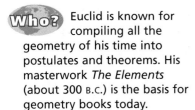

Are points *E, H, B,* and *C* coplanar?
Are points *E, H, F,* and *B* coplanar?

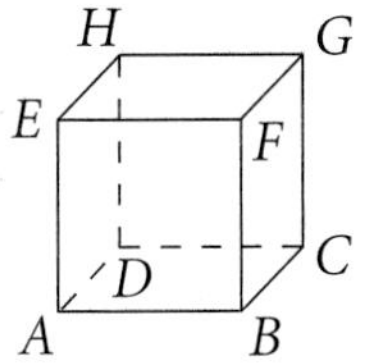

Yes, the plane that contains the three noncollinear points *E, H,* and *B* also contains *C*.

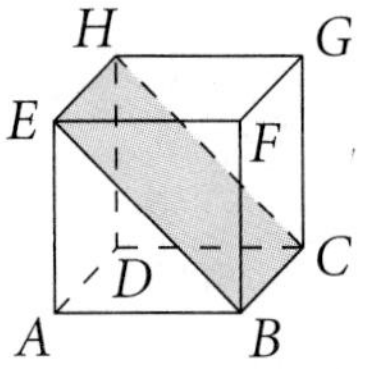

No, points *E, H,* and *F* lie in exactly one plane, which doesn't contain *B*.

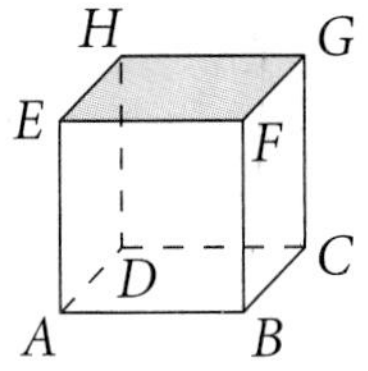

Exercises ON YOUR OWN

Are the points collinear?

1. *A, D, E*
2. *B, C, D*
3. *B, C, F*
4. *A, E, C*
5. *F, B, D*

Are the points coplanar?

6. *B, C, D, F*
7. *A, C, D, F*
8. *B, D, E, F*
9. *A, C, E, F*
10. Name plane *M* in another way.
11. What is the intersection of plane *M* and $\overleftrightarrow{AE}$?
12. What is the intersection of $\overleftrightarrow{AE}$ and $\overleftrightarrow{BD}$?

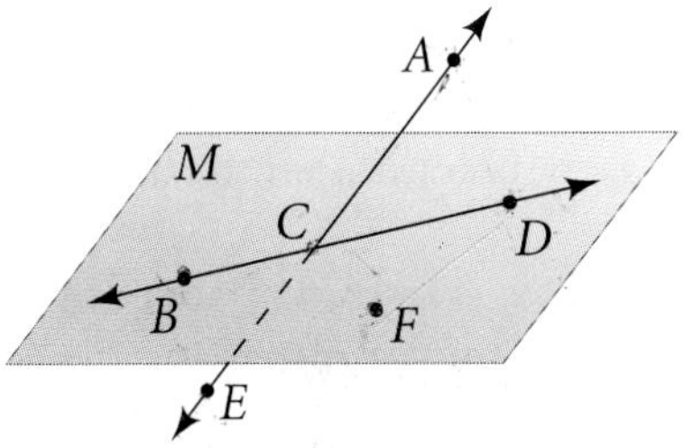

Exs. 1 – 12

Are the following coplanar?

13. *Q, V, R*
14. *X, V, R*
15. *U, V, W, S*
16. *W, V, Q, T*
17. point *X*, $\overleftrightarrow{QT}$
18. $\overleftrightarrow{RS}$, point *X*
19. $\overleftrightarrow{XW}$, $\overleftrightarrow{UV}$
20. $\overleftrightarrow{UX}$, $\overleftrightarrow{WS}$
21. $\overleftrightarrow{UV}$, $\overleftrightarrow{WS}$
22. What is the intersection of plane *QRST* and plane *RSWV*?
23. What is the intersection of $\overleftrightarrow{UV}$ and plane *QTXU*?
24. Name three lines that intersect at point *S*.
25. Name two planes that intersect at $\overleftrightarrow{TS}$.
26. Name another point that is in the same plane as points *Q, T,* and *W*.

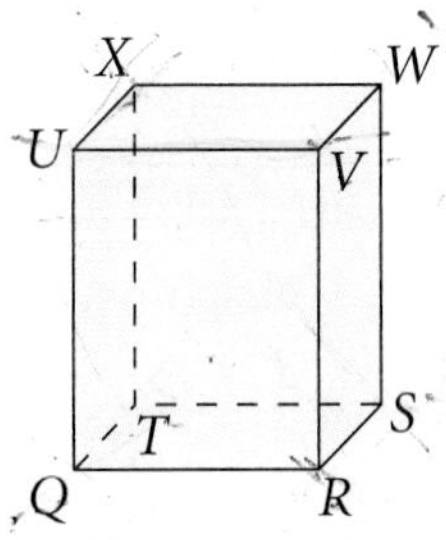

Exs. 13 – 26

27. **Writing** Surveyors and photographers use a *tripod,* or three-legged stand, for their instruments. Use one of the postulates to explain why.

28. **Research** Find out more about Euclid's book *The Elements.* What made it such a significant book? Where did Euclid get his information?

29. How many planes contain three collinear points? Explain.

30. Which postulate is sometimes stated as "Two points determine a line"?

31. **Standardized Test Prep** Which of the following is *not* an acceptable name for the plane shown?
   A. plane RSZ
   B. plane $RSWZ$
   C. plane WSZ
   D. plane $RSTW$
   E. plane STZ

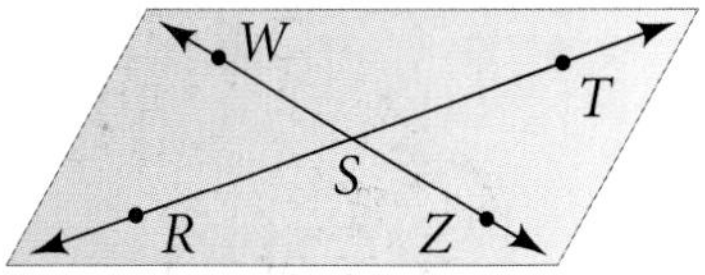

32. How many planes contain each line and point?
   a. $\overleftrightarrow{EF}$ and point Q
   b. $\overleftrightarrow{PH}$ and point E
   c. $\overleftrightarrow{FG}$ and point P
   d. $\overleftrightarrow{EP}$ and point G
   e. Use inductive reasoning. What do you think is true of a line and a point not on the line?

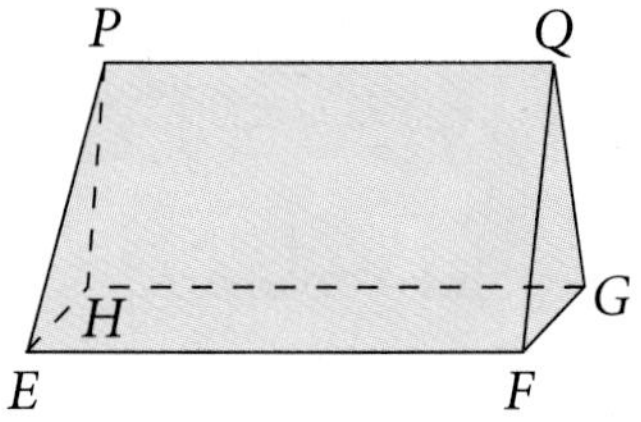

33. **Logical Reasoning** Suppose two lines intersect. How many planes do you think contain both lines? Use the diagram at the right to explain your answer.

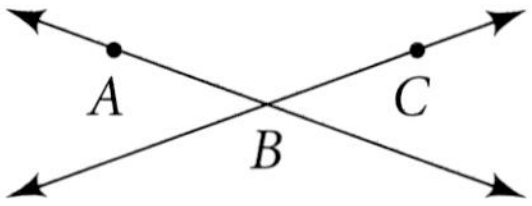

Complete with *always, sometimes,* or *never* to make a true statement.

34. Intersecting lines are _?_ coplanar.

35. Two planes _?_ intersect in exactly one point.

36. Three points are _?_ coplanar.

37. A line and a point not on the line are _?_ coplanar.

38. Four points are _?_ coplanar.

39. Two lines _?_ meet in more than one point.

Probability **Given points *A, B, C,* and *D* as shown, solve each problem.**

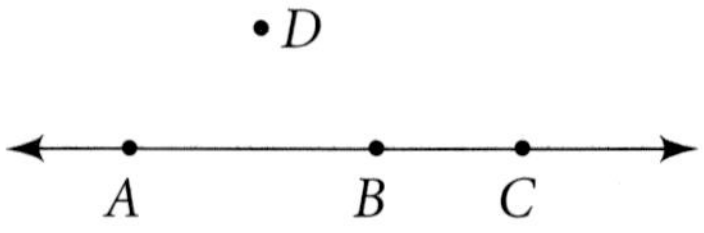

40. Two points are picked at random. Find P(they are collinear).

41. Three points are picked at random.
   a. Find P(they are collinear).
   b. Find P(they are coplanar).

42. **Navigation** Rescue teams use the principles in Postulates 1-1 and 1-2 to determine the location of a distress signal. In the diagram, a ship at point A receives a signal from the northeast. A ship at point B receives the same signal from due west. Trace the diagram and find the location of the distress signal. Explain how the two postulates help to find the location of the distress signal.

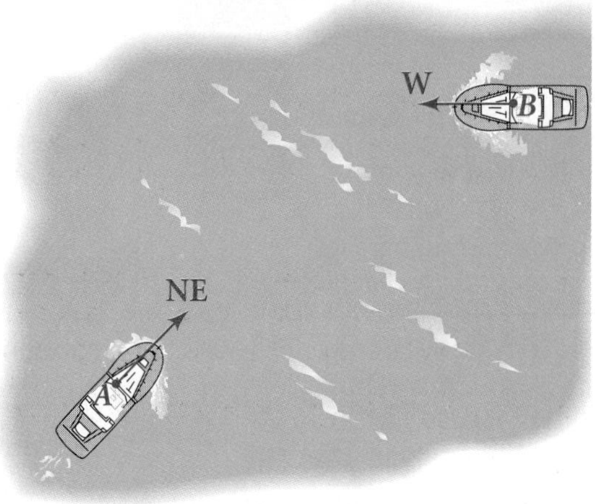

Coordinate Geometry **Are the points collinear? Graph them to find out.**

43. $(1, 1), (4, 4), (-3, -3)$

44. $(2, 4), (4, 6), (0, 2)$

45. $(0, 0), (8, 10), (4, 6)$

46. $(0, 0), (0, 3), (0, -10)$

47. **Open-ended** Give an example from your classroom or your home of three planes intersecting in one line.

48. **Optical Illusions** The diagram at the right is an optical illusion. Which points are collinear, A, B, C or A, B, D? Are you sure? Use a ruler to check your answer.

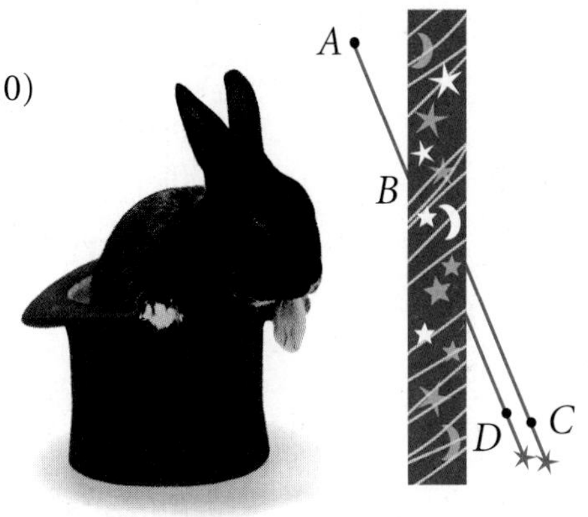

Exercises MIXED REVIEW

Algebra **Evaluate each expression for the given values.**

49. $a^2 + b^2$ for $a = 3$ and $b = -5$

50. $\frac{1}{2}bh$ for $b = 8$ and $h = 11$

51. $2\ell + 2w$ for $\ell = 3$ and $w = 7$

52. $b^2 - 4ac$ for $a = 2$, $b = 5$, and $c = 1$

53. **Patterns** What is the last digit of 3^{45}? Make a table and use inductive reasoning. Explain the pattern.

For more practice with evaluating expressions, see Skills Handbook page 659.

Getting Ready for Lesson 1-3

Will the lines intersect or not?

54.

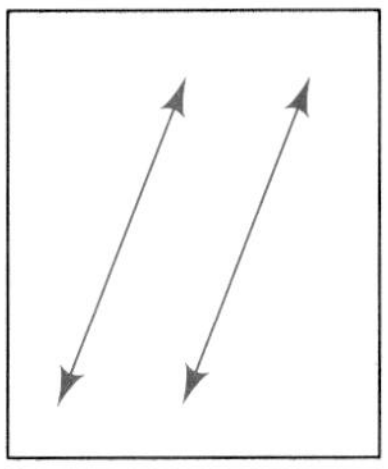

55.

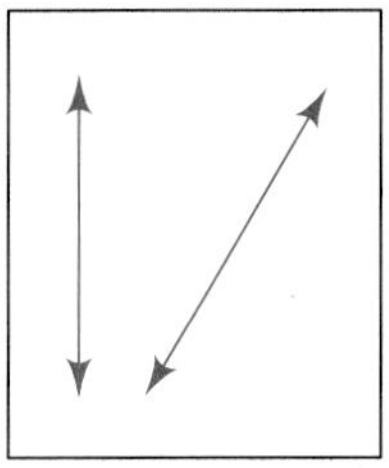

56.

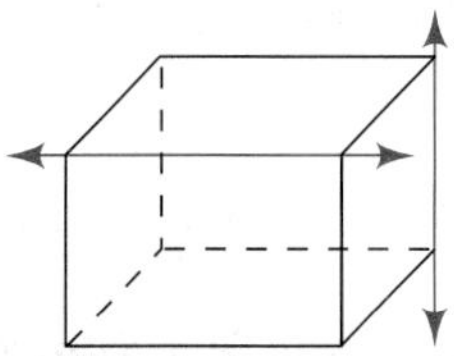

57.

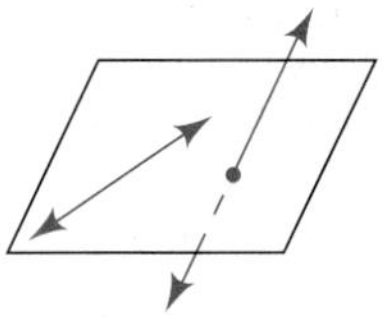

What You'll Learn

- Relating segments and rays to lines
- Recognizing parallel lines and parallel planes

...And Why

To provide a vocabulary of terms needed for communicating in geometry

What You'll Need

graph paper, colored pencils, tape

1-3 Segments, Rays, Parallel Lines and Planes

THINK AND DISCUSS

Many geometric figures, such as squares and angles, use only the parts of lines called segments and rays.

A **segment** is the part of a line consisting of two *endpoints* and all points between them.

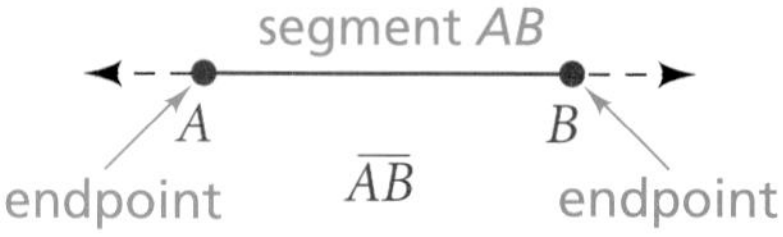

A **ray** is the part of a line consisting of one *endpoint* and all the points of the line on one side of the endpoint.

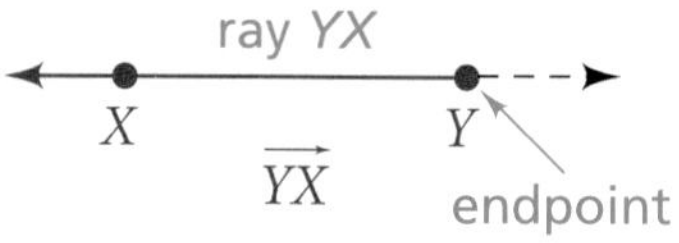

1. Is $\overline{AB}$ the same as $\overline{BA}$? Explain.
2. Is $\overrightarrow{YX}$ the same as $\overrightarrow{XY}$? Explain.
3. How is a ray like a line? How is a ray different from a line?

A ray in geometry is named after the rays of the sun.

Opposite rays are two collinear rays with the same endpoint. Opposite rays always form a line.

4. a. Name four different rays in the figure below.
   b. Name two opposite rays.

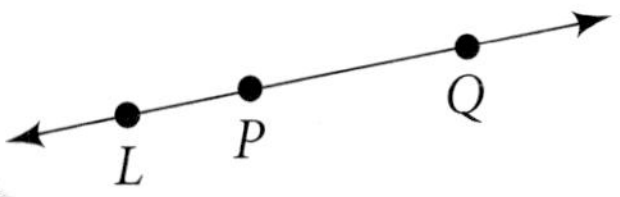

Lines that do not intersect may or may not be coplanar. **Parallel lines** are coplanar lines that do not intersect. Segments and rays are parallel if they lie in parallel lines.

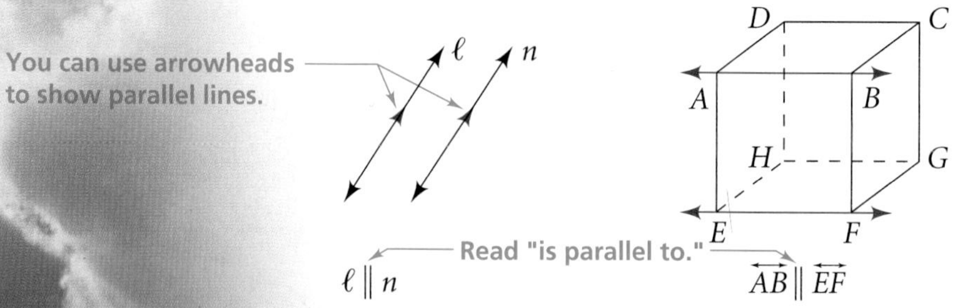

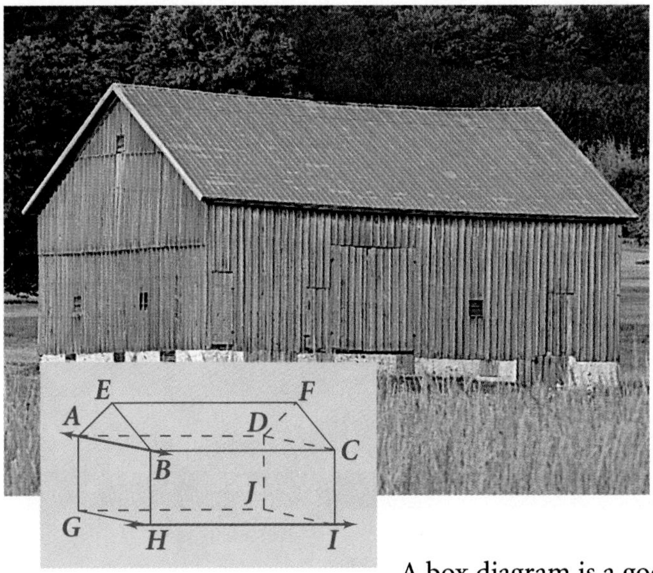

5. Name all the segments shown at the left that are parallel to:
   a. $\overline{DC}$ **b.** $\overline{GJ}$ **c.** $\overline{AE}$

Skew lines do not lie in the same plane. They are neither parallel nor intersecting.

$\overleftrightarrow{AB}$ and $\overleftrightarrow{HI}$ are skew.

Parallel planes are planes that do not intersect.

plane $ABCD \parallel$ plane $GHIJ$

6. Name some other pairs of skew lines in the diagram at the left.
7. Name two more pairs of parallel planes.

A box diagram is a good way to represent parallel lines and segments, skew lines, and parallel planes. Some other ways to draw planes are shown below.

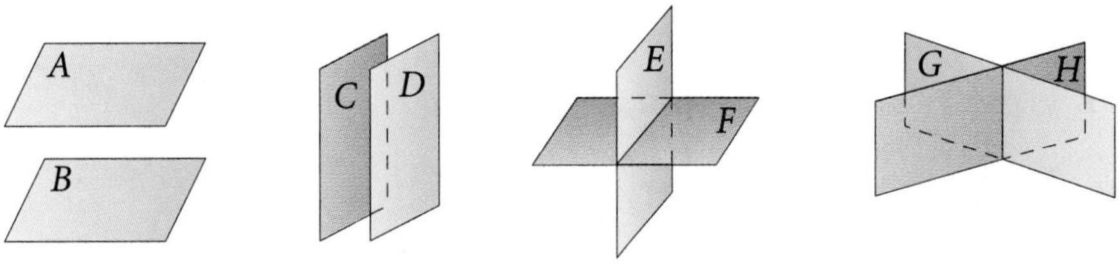

8. Which pairs of planes shown above are parallel? Which are intersecting?

Example

Draw planes A and B intersecting in $\overleftrightarrow{FG}$.

Using graph paper will help you draw parallel lines and representations of planes.

Step 1 Step 2 Step 3

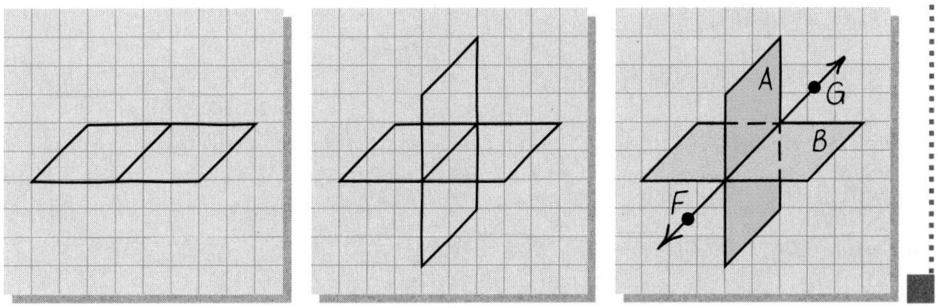

9. **Try This** Use graph paper and colored pencils to draw pairs of parallel and intersecting planes like those above Question 8.

WORK TOGETHER

Work in pairs to answer these questions about the lines and planes determined by the surfaces of a rectangular solid.

Stack your geometry books to form a rectangular solid. Label the vertices *P*, *Q*, *R*, *S*, *T*, *U*, *V*, and *W*. Identify each of the following.

10. three pairs of parallel planes

11. all lines that are parallel to $\overleftrightarrow{PQ}$

12. all lines that are skew to $\overleftrightarrow{PQ}$

Exercises ON YOUR OWN

Name all the segments that are parallel to the given segment.

1. $\overline{AC}$ **2.** $\overline{EF}$ **3.** $\overline{AD}$

4. Name all the lines that form a pair of skew lines with $\overleftrightarrow{AD}$.

5. Name a pair of parallel planes.

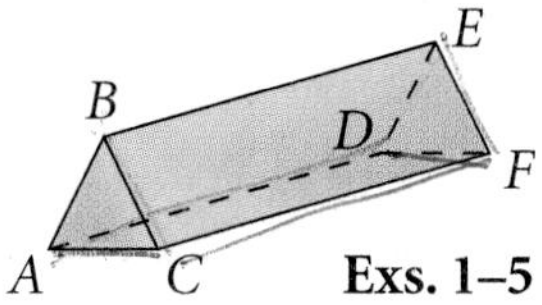

Exs. 1–5

Use the line at the right for Exercises 6–8.

6. a. Name a pair of opposite rays with point *T* as endpoint.
b. Name another pair of opposite rays.

7. Name all the segments shown.

8. Name $\overrightarrow{RT}$ two other ways.

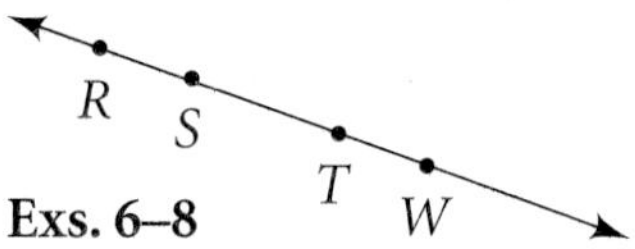

Exs. 6–8

Make a separate sketch for each of the following.

9. Draw three parallel lines *a*, *t*, and *q*.

10. Draw parallel planes *A* and *B*.

11. Draw $\overleftrightarrow{AB}$, $\overleftrightarrow{CD}$, and $\overleftrightarrow{EF}$ so that $\overleftrightarrow{AB} \parallel \overleftrightarrow{CD}$, $\overleftrightarrow{AB}$ and $\overleftrightarrow{EF}$ are skew, and $\overleftrightarrow{CD}$ and $\overleftrightarrow{EF}$ are skew.

12. Draw planes *C* and *D*, intersecting in $\overleftrightarrow{XY}$.

Write *true* or *false*.

13. $\overleftrightarrow{CB} \parallel \overleftrightarrow{GF}$

14. $\overleftrightarrow{ED} \parallel \overleftrightarrow{HG}$

15. plane *AED* $\parallel$ plane *FGH*

16. plane *ABH* $\parallel$ plane *CDF*

17. $\overleftrightarrow{AB}$ and $\overleftrightarrow{HG}$ are skew lines.

18. $\overleftrightarrow{AE}$ and $\overleftrightarrow{BC}$ are skew lines.

19. $\overleftrightarrow{CF}$ and $\overleftrightarrow{AI}$ are skew lines.

20. $\overleftrightarrow{CF}$ and $\overleftrightarrow{AJ}$ are skew lines.

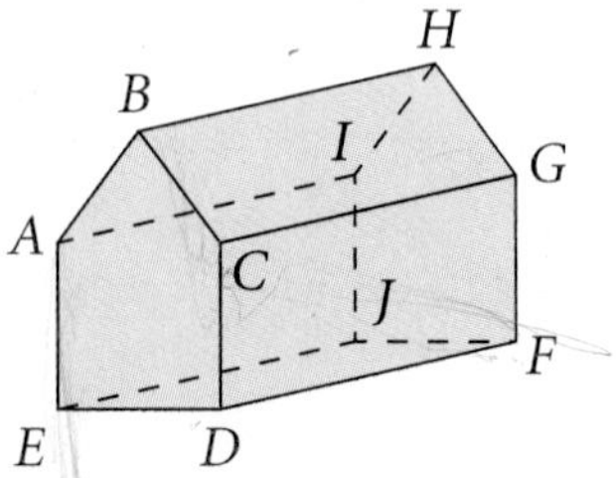

Complete with *always, sometimes,* or *never* to make a true statement.

21. $\overrightarrow{AB}$ and $\overrightarrow{BA}$ are __?__ the same ray.

22. $\overrightarrow{AB}$ and $\overrightarrow{AC}$ are __?__ the same ray.

23. $\overline{AX}$ and $\overline{XA}$ are __?__ the same segment.

24. $\overleftrightarrow{TQ}$ and $\overleftrightarrow{QT}$ are __?__ the same line.

25. Two parallel lines are __?__ coplanar.

26. Skew lines are __?__ coplanar.

27. Opposite rays __?__ form a line.

28. Two lines in the same plane are __?__ parallel.

29. Two planes that do not intersect are __?__ parallel.

30. Two lines that lie in parallel planes are __?__ parallel.

Directions are printed on a compass card, a circle divided into 32 equally-spaced compass points.

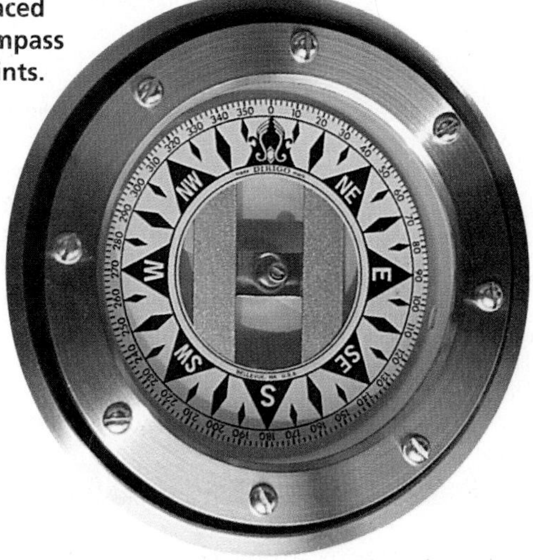

31. Writing **Summarize** the different ways that two lines may be related. Give examples from the real world that illustrate the relationships.

32. Navigation North and south are directions on a compass that are on opposite rays. Name two other pairs of compass directions that are on opposite rays.

33. Coordinate Geometry $\overrightarrow{AB}$ has endpoint $A(2, 3)$ and goes through $B(4, 6)$. Give some possible coordinates for point C so that $\overrightarrow{AB}$ and $\overrightarrow{AC}$ will be opposite rays. Graph your answer.

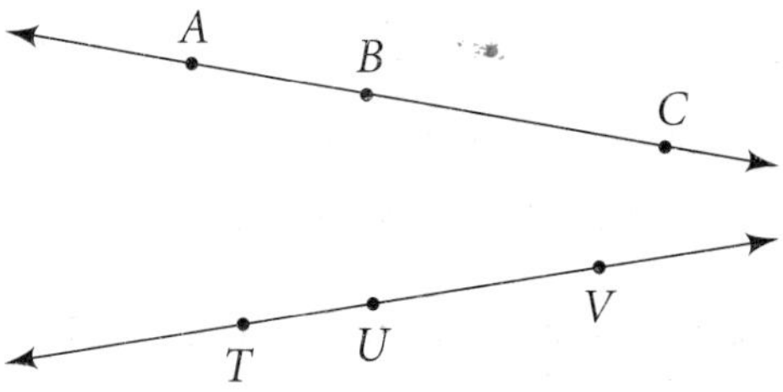

34. Inductive Reasoning Draw a diagram similar to the one shown.
Step 1: Draw $\overline{AU}$ and $\overline{BT}$. Label their intersection point as X.
Step 2: Draw $\overline{AV}$ and $\overline{CT}$. Label their intersection point as Y.
Step 3: Draw $\overline{BV}$ and $\overline{CU}$. Label their intersection point as Z.
Make a **conjecture** about points X, Y, and Z.

35. Critical Thinking Suppose two parallel planes A and B are each intersected by a third plane C. What do you think will be true of the intersection of planes A and C and the intersection of planes B and C? Give an example in your classroom.

36. Research In diamond, each carbon atom bonds to four other carbon atoms in a three-dimensional network. In graphite, each carbon atom bonds to three other carbon atoms in the same plane. The "sheets" or planes of graphite are parallel. Find out how these structures affect the properties of diamond and graphite.

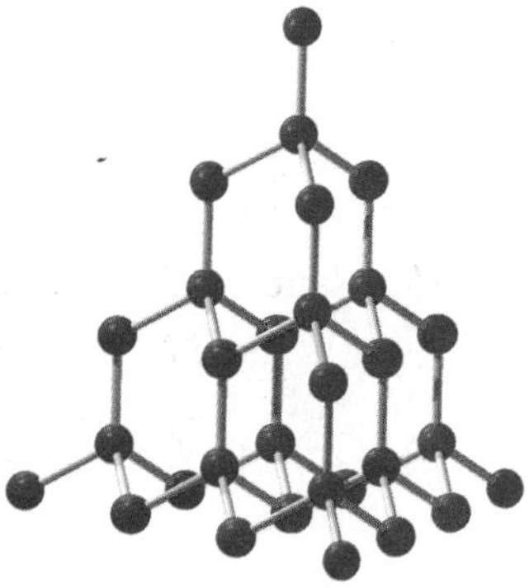

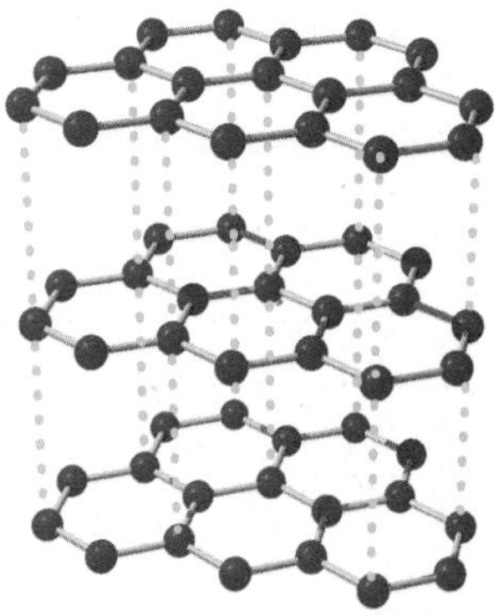

37. *Open-ended* List four pairs of parallel planes in your classroom.

38. *Writing* The term *skew* is from a Middle English word meaning "to escape." Explain why this might be an appropriate origin for the word that names skew lines.

39. *Standardized Test Prep* Which statement(s) can be true about three planes?
I. They intersect in a line. **II.** They intersect in a point. **III.** They have no points in common.
A. I only **B.** II only **C.** I and II only **D.** I and III only **E.** I, II, and III

Chapter Project — *Find Out by Creating*

Some artists create origami by experimenting. They fold and unfold a piece of paper until they see a resemblance to the real world. Take your folded square from the Find Out question on page 10 (or make a new one). Use the existing creases to construct the dog and the flower pictured at the right. Now try to create your own origami, starting with a fresh square of paper.

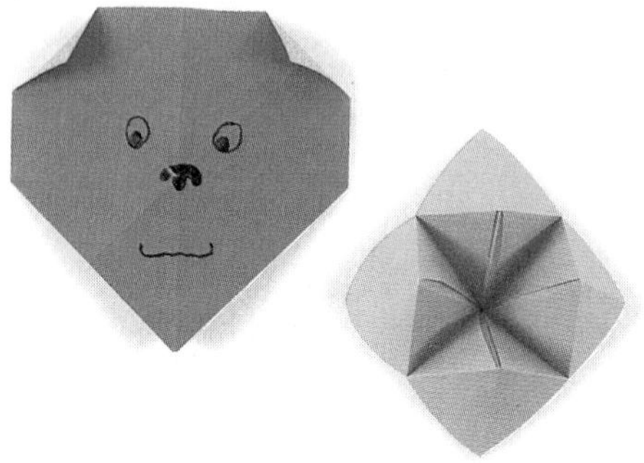

Exercises MIXED REVIEW

Find the next two terms in each sequence.

40. 0.1, 0.12, 0.123, 0.1234, . . .
41. $-1, -2, -4, -7, -11, -16, \ldots$
42. AB, BC, CD, DE, EF, . . .

State the ways you can name each geometric figure.

43. a line
44. a point
45. a plane

46. *Logical Reasoning* Raven made the following **conjecture**: "When you subtract a number from a given number, the answer is always smaller than the given number." Is her **conjecture** correct? Explain.

Getting Ready for Lesson 1-4

Simplify each expression.

47. $\lvert -6 \rvert$	48. $\lvert 3.5 \rvert$	49. $\lvert 7 - 10 \rvert$
50. $\lvert -4 - 2 \rvert$	51. $\lvert 8 - 5 \rvert$	52. $\lvert 4 + 1 \rvert$
53. $\lvert -3 + 12 \rvert$	54. $\lvert -21 + 6 \rvert$	55. $\lvert -11 - (-2) \rvert$

Graph each inequality on a number line.

56. $t > 6$	57. $9 \leq m$	58. $w < -4$
59. $5 > s$	60. $p \leq 7$	61. $-1.5 \geq b$
62. $x \geq -3$	63. $0 < q$	64. $-5 \leq v < -2$

SELF ASSESSMENT

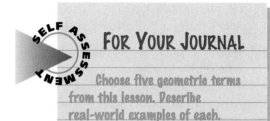

FOR YOUR JOURNAL

Choose five geometric terms from this lesson. Describe real-world examples of each.

Math ToolboX

Algebra Review

Solving Linear Equations

Before Lesson 1-4

Sometimes you need to combine like terms when you are solving linear equations.

Example 1

Solve $(5x + 8) - (2x - 9) = 38$.

$(5x + 8) + (-1)(2x - 9) = 38$	$-(2x - 9) = (-1)(2x - 9)$
$5x + 8 - 2x + 9 = 38$	Use the Distributive Property.
$3x + 17 = 38$	Simplify.
$3x = 21$	Subtract 17 from each side.
$x = 7$	Divide each side by 3.

When you are solving equations with variables on both sides, first get all the variables on the same side of the equation.

Example 2

Solve $4x - 9 = 7x - 15$.

$4x - 9 = 7x - 15$	
$-9 = 3x - 15$	Subtract $4x$ from each side.
$6 = 3x$	Add 15 to each side.
$2 = x$	Divide each side by 3.

Solve.

1. $5x + 10 - 6x + 3 = 6 - 2x - 2$

2. $(6a - 54) - (5a + 27) = 23$

3. $(2 + 4y) - (y + 9) = 26$

4. $7t - 8t + 4 = 5t - 2$

5. $(9k + 30) - (4k + 10) = 100$

6. $6x + 17 = 9x + 2$

7. $(3x + 10) - 5x = 6x - 50$

8. $(3y - 5) + (5y + 20) = 135$

9. $10n + 12 = 14n - 12$

10. $13c + 40 = 9c - 20 + c$

11. $(4w - 28) + (11w + 13) = 180$

12. $7f + 16 = 3f + 48$

13. $(7a + 3) + (-a - 5) = -16$

14. $3x - 35 = 9x - 59$

15. $7y + 44 = 12y + 11$

16. $(11x - 37) + (5x + 59) = 54$

17. $(7t - 21) + (t + 4) = 15$

18. $(5w + 24) + (2w + 13) = 156$

What You'll Learn

- Finding the length of a segment and the measure of an angle

...And Why

To understand the building blocks of many geometric figures

What You'll Need

protractor

1-4 Measuring Angles and Segments

WORK TOGETHER

Your family is traveling on Interstate 80 through Nebraska. You entered the highway at mileage marker 126. You decided to drive as far as you could before stopping for breakfast within $1\frac{1}{2}$ hours. Assume that on the highway you drive at an average speed of 60 mi/h. Work with a partner to answer these questions.

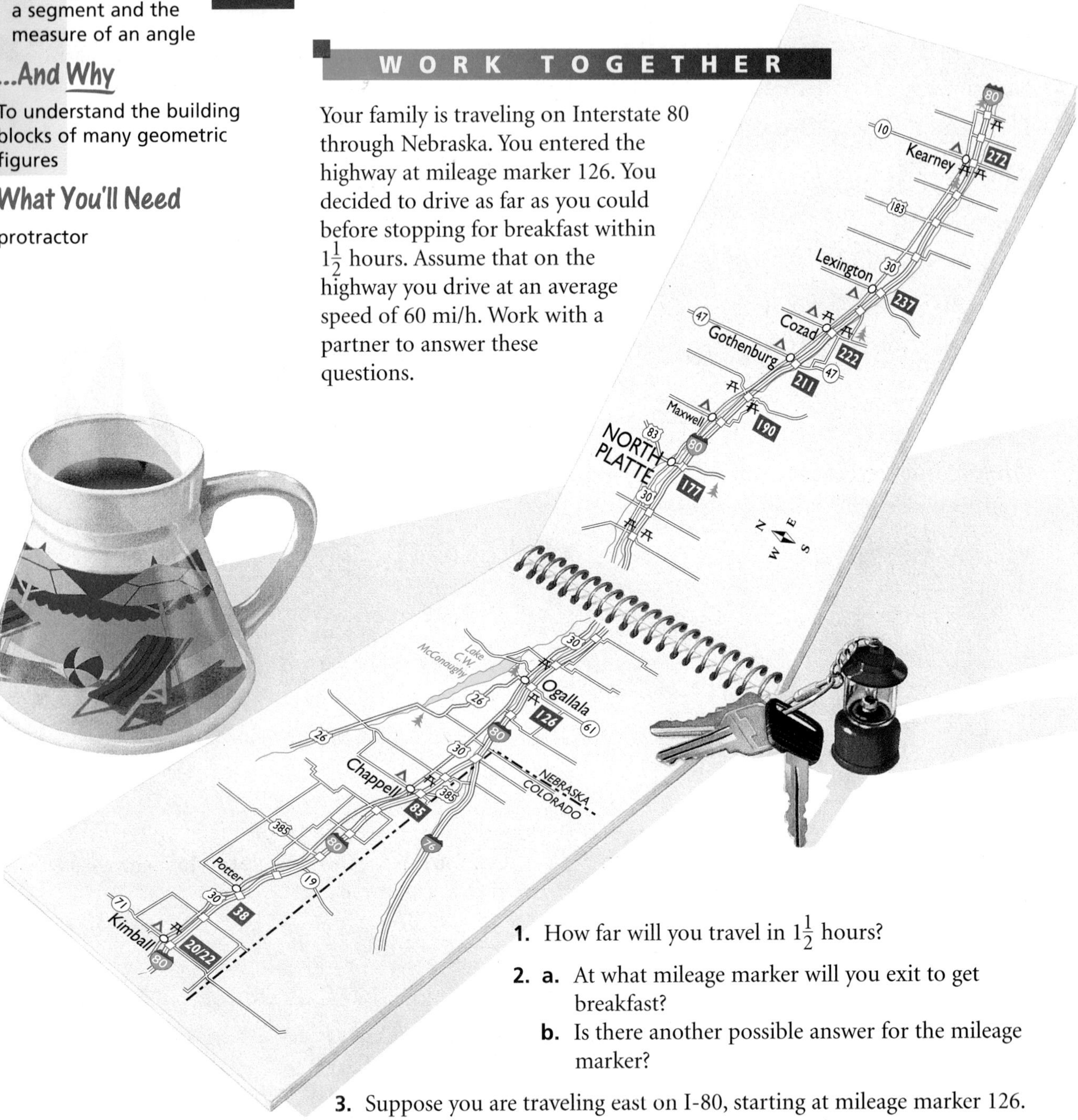

1. How far will you travel in $1\frac{1}{2}$ hours?
2. a. At what mileage marker will you exit to get breakfast?
   b. Is there another possible answer for the mileage marker?
3. Suppose you are traveling east on I-80, starting at mileage marker 126. Not counting side trips for sightseeing, you travel 111 mi on I-80 before you exit for a campground.
   a. Do mileage markers increase or decrease from west to east?
   b. What mileage marker is at your exit?
4. Does the *direction* you travel affect the *distance* you travel?

THINK AND DISCUSS

Measuring Segments

If you picture straightening out the map of Interstate 80, you will have a model for a number line. The mileage markers represent *coordinates*.

Postulate 1-5
Ruler Postulate

The points of a line can be put into a one-to-one correspondence with the real numbers so that the distance between any two points is the absolute value of the difference of the corresponding numbers.

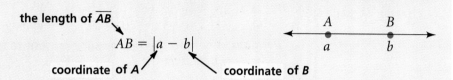

the length of $\overline{AB}$

$$AB = |a - b|$$

coordinate of *A* **coordinate of *B***

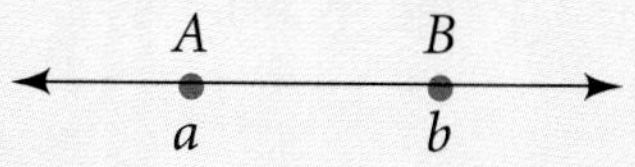

5. **Critical Thinking** Why do you think that absolute value is used to express the distance between two points?

Example 1

PROBLEM SOLVING HINT
Draw a diagram.

Find QS if the coordinate of Q is -3 and the coordinate of S is 21.

$QS = |-3 - 21| = |-24| = 24$

6. Suppose you subtracted -3 from 21 in Example 1. Would you get the same result? Why or why not?

7. **Try This** Find AB if the coordinate of point A is -8, and the coordinate of point B is 11.

Two segments with the same length are **congruent** ($\cong$). In other words, if $AB = CD$, then $\overline{AB} \cong \overline{CD}$. You can use these statements interchangeably. Segments can be marked alike to show that they are congruent.

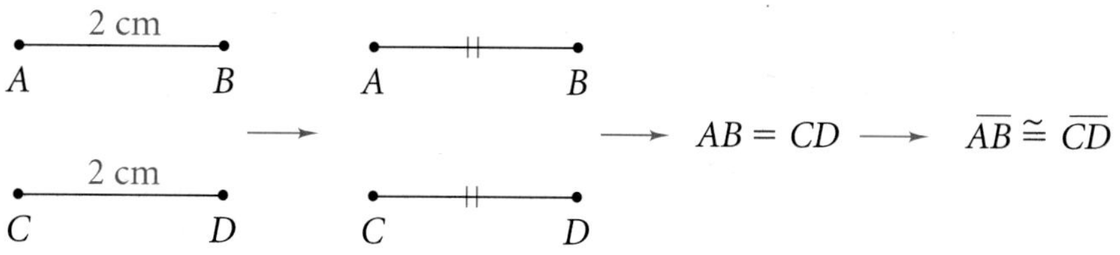

8. a. **Try This** Name two segments that are congruent.
   b. Name a second pair of congruent segments.

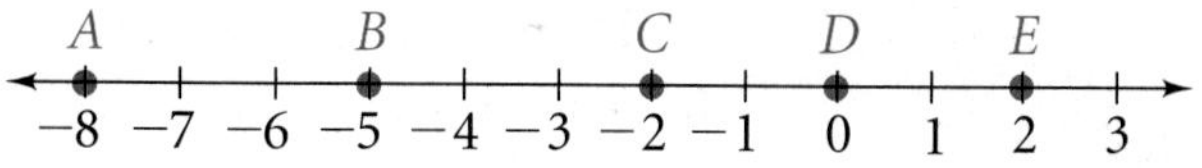

Postulate 1-6
Segment Addition Postulate

If three points A, B, and C are collinear and B is between A and C, then $AB + BC = AC$.

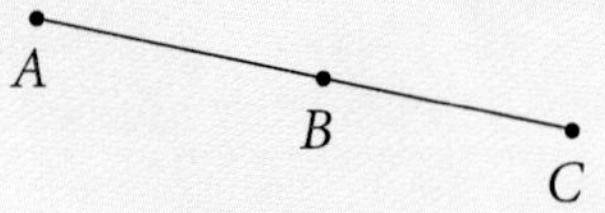

Example 2

Algebra If $DT = 60$, find the value of x. Then find DS and ST.

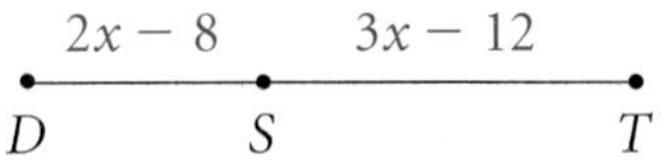

For more practice with solving linear equations, see Skills Handbook, page 663.

$DS + ST = DT$	Segment Addition Postulate
$(2x - 8) + (3x - 12) = 60$	Substitution
$5x - 20 = 60$	Simplify.
$5x = 80$	Add 20 to each side.
$x = 16$	Divide each side by 5.

$DS = 2x - 8 = 2(16) - 8 = 24$

$ST = 3x - 12 = 3(16) - 12 = 36$

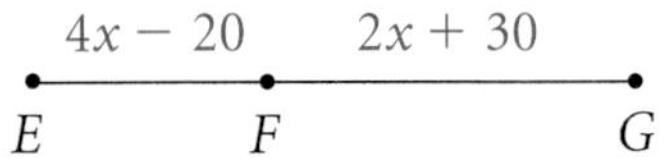

9. Explain how to check the answers in Example 2.

10. a. Try This $EG = 100$. Find the value of x.
b. Find EF and FG.

Measuring Angles

An **angle** (∠) is formed by two rays (called *sides* of the angle) with the same endpoint (called the *vertex* of the angle). You can name an angle several ways.

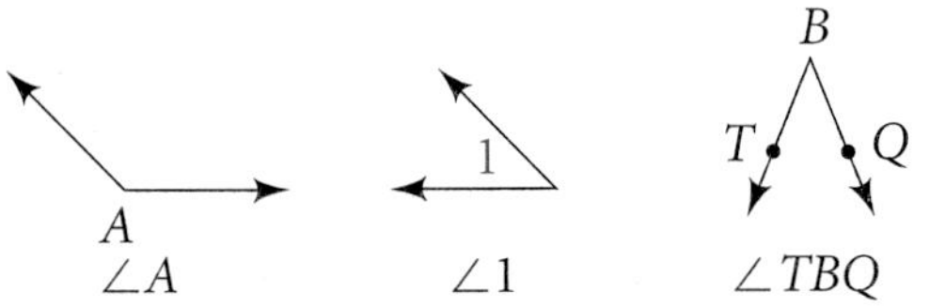

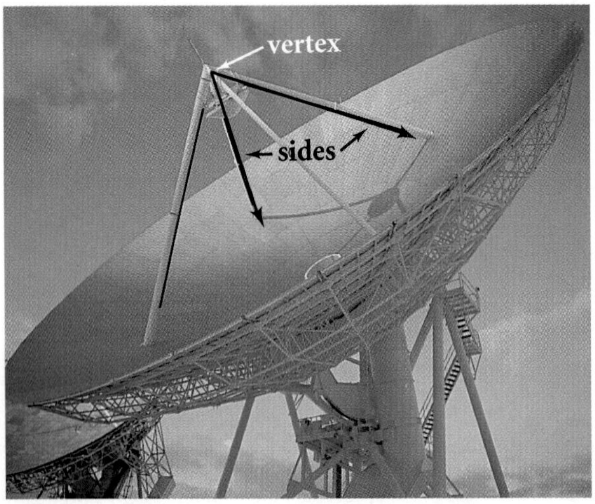

11. a. Name ∠1 two other ways.
b. Name ∠CED two other ways.
c. Would it be correct to refer to any of the angles at the right as ∠E? Why or why not?

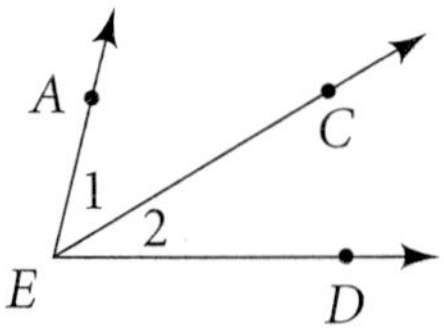

Angles are measured in *degrees*. The *measure* of ∠A is written as $m\angle A$.

$m\angle A = 80$

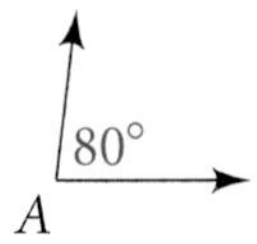

You can use these constructions to construct different geometric figures.

5. **Try This** Draw angles like $\angle X$ and $\angle Y$. Then construct $\angle Z$ so that $m\angle Z = m\angle X + m\angle Y$.

6. Describe how you would construct an angle, $\angle D$, so that $m\angle D = m\angle X - m\angle Y$.

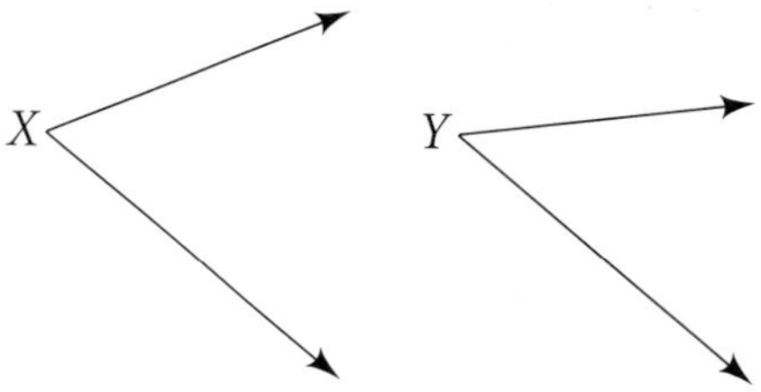

To draw a circle or an arc with a SAFE-T-COMPASS®, use the center hole of the white dial as the center.

You can use a compass and straightedge to construct triangles with sides of specific lengths.

Example

Construct a triangle whose sides have the given lengths.

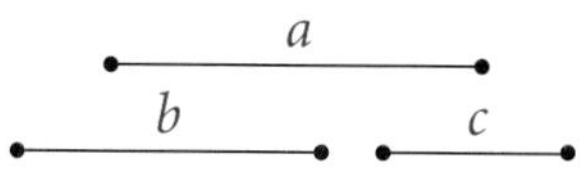

Step 1
Use Construction 1 to construct $\overline{RS}$ with length a.

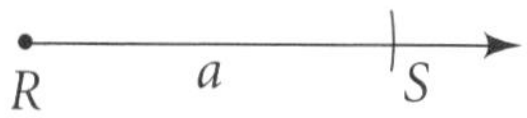

Step 2
Open the compass to the length b. Put the compass point on R and draw an arc.

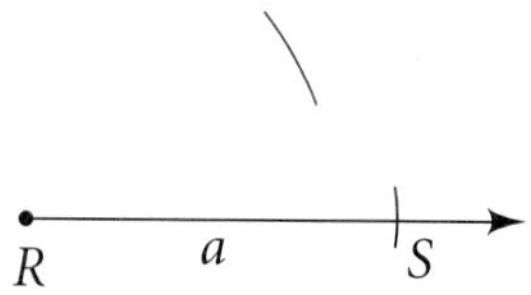

Step 3
Open the compass to the length c. Put the compass point on S and draw an arc. Be sure the two arcs intersect. Label the intersection of the arcs as point T.

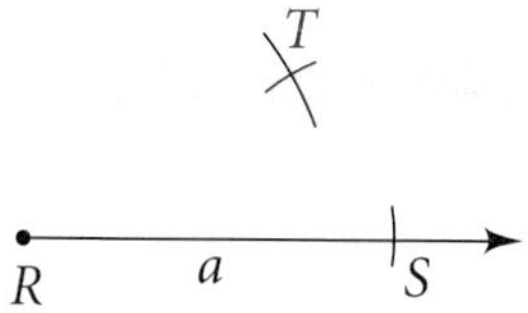

Step 4
Draw segments from point T to both R and S.

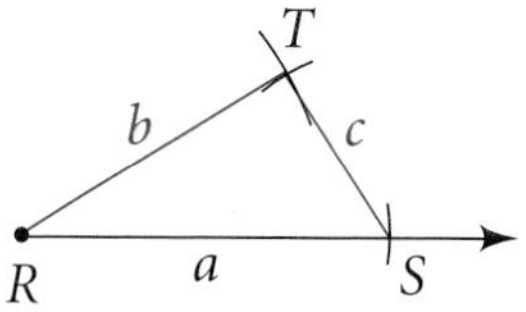

7. **Try This** Draw $\overline{XY}$. Then construct a triangle with three sides the length of $\overline{XY}$.

Constructing Perpendicular Bisectors and Angle Bisectors

The next two constructions will show you how to bisect segments and bisect angles. These constructions and Construction 2 are based on *congruent triangles.* You will see *why* the constructions work in future chapters.

Construction 3
Perpendicular Bisector

Construct the perpendicular bisector of a segment.

Given: $\overline{AB}$

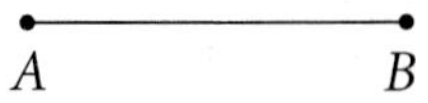

Step 1
Put the compass point on point A and draw an arc. Be sure the opening is greater than $\frac{1}{2}AB$. Keep the same compass setting for Step 2.

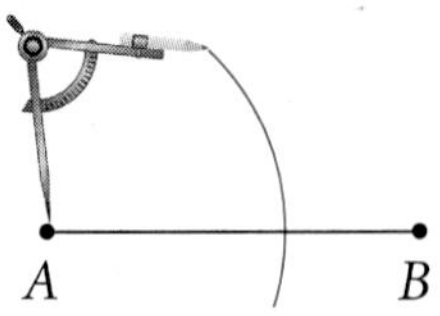

Who? In 1672 Danish mathematician Georg Mohr showed that all constructions that use a compass and a straightedge can be done using only a compass.

Step 2
Put the compass point on point B and draw an arc. Label the points where the two arcs intersect as X and Y.

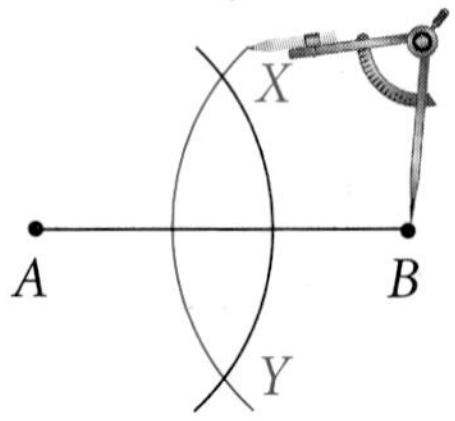

Step 3
Draw $\overleftrightarrow{XY}$. Label the intersection of $\overline{AB}$ and $\overleftrightarrow{XY}$ as point M.

$\overleftrightarrow{XY}$ is the perpendicular bisector of $\overline{AB}$. Point M is the midpoint of $\overline{AB}$.

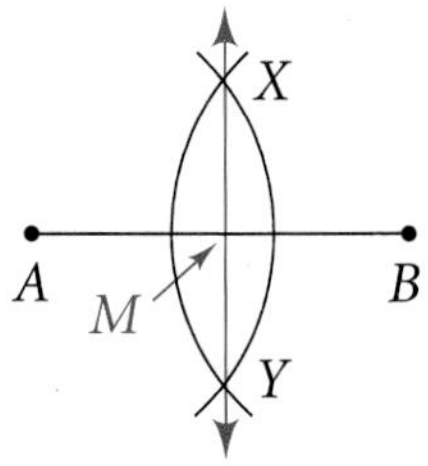

You can use Construction 3 to divide any segment into fourths or eighths.

Construction 4
Angle Bisector

Construct the bisector of an angle.

Given: $\angle A$

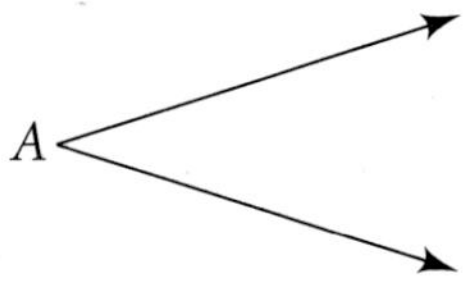

Step 1
Put the compass point on vertex A. Draw an arc that intersects the sides of $\angle A$. Label the points of intersection B and C.

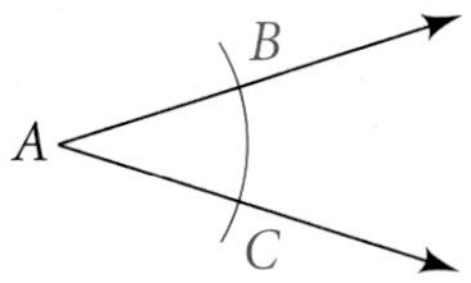

Step 2
Put the compass point on point C and draw an arc. Keep the same compass setting and repeat with point B. Be sure the arcs intersect. Label the point where the two arcs intersect as point X.

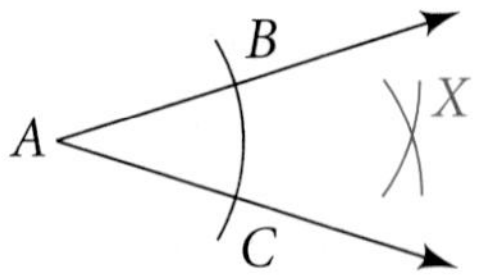

Step 3
Draw $\overrightarrow{AX}$.

$\overrightarrow{AX}$ is the angle bisector of $\angle CAB$.

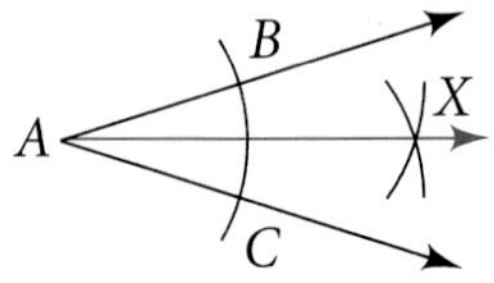

8. Describe how you could construct a 45° angle.

Exercises ON YOUR OWN

Draw a diagram similar to the given one. Then do the construction.

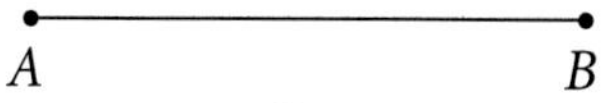

1. Construct $\overline{XY}$ congruent to $\overline{AB}$. Check your work with a ruler.
2. Construct the perpendicular bisector of $\overline{AB}$. Check your work with a ruler and a protractor.

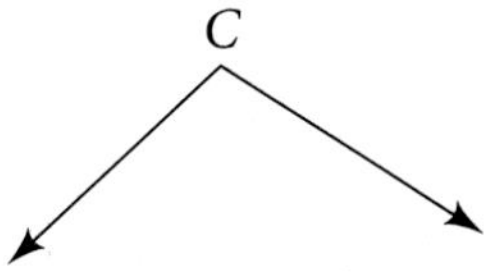

3. Construct the angle bisector of $\angle C$. Check your work with a protractor.

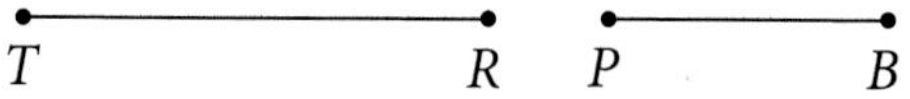

4. Construct $\overline{DE}$ so that $DE = TR + PB$.
5. Construct $\overline{QS}$ so that $QS = TR - PB$.
6. Construct $\overline{XY}$ so that $XY = 2TR$.

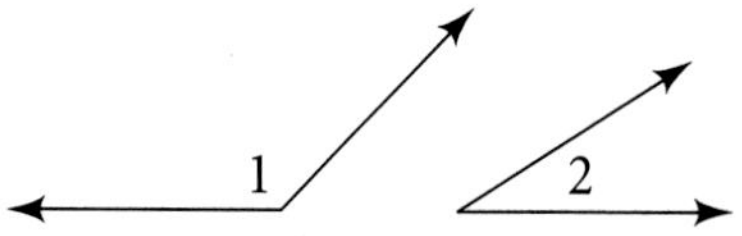

7. Construct $\angle B$ so that $m\angle B = m\angle 1 + m\angle 2$.
8. Construct $\angle C$ so that $m\angle C = m\angle 1 - m\angle 2$.
9. Construct $\angle D$ so that $m\angle D = 2m\angle 2$.

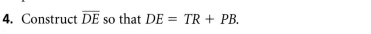

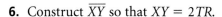

10. Draw an angle that is about 120°. Then construct a congruent angle.
11. Use a ruler to draw two segments that are 4 cm and 5 cm long. Then construct a triangle with sides 4 cm, 4 cm, and 5 cm long.
12. **a.** Construct a 45° angle.
    b. Construct a 135° angle.
13. **Writing** Describe how to construct the midpoint of a segment.
14. **Open-ended** Which method do you prefer for bisecting an angle—paper folding or construction with compass and straightedge? Why?
15. **Patterns** Draw a large triangle with three acute angles. Construct the angle bisectors of all three angles of the triangle. What is true about the intersection of the three angle bisectors? Repeat for another triangle that has an obtuse angle. Make a **conjecture** about the three angle bisectors of any triangle.
16. **Patterns** Draw a large triangle with three acute angles. Construct the perpendicular bisectors of all three sides. What is true about the intersection of the three perpendicular bisectors? Repeat for another triangle that has an obtuse angle. Make a **conjecture** about the three perpendicular bisectors of the sides of any triangle.
17. **a.** Draw a segment, $\overline{AB}$. Construct a triangle whose sides are all congruent to $\overline{AB}$.
    b. Measure the angles of the triangle.
    c. **Writing** Describe how to construct a 60° angle and a 30° angle.

18. Art You can create intricate designs using your compass. Follow these directions to design a *daisy wheel.*

a. Construct a circle. Keeping the same compass setting, put the compass point on the circle and construct an arc. The endpoints of the arc should be on the circle.

b. Keeping the same compass setting, put the compass point on each endpoint of the first arc and draw two new arcs.

c. Continue to make arcs around the circle from the new endpoints of arcs until you get a six-petal daisy wheel.

d. Personalize your daisy wheel by decorating it.

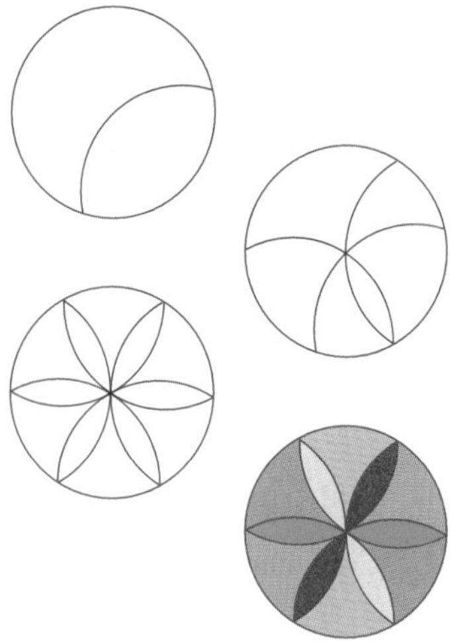

19. a. Use your compass to draw a circle. Locate three points A, B, and C on the circle.

b. Construct the perpendicular bisectors of $\overline{AB}$ and $\overline{BC}$.

c. Critical Thinking Label the intersection of the two perpendicular bisectors as point O. Describe point O.

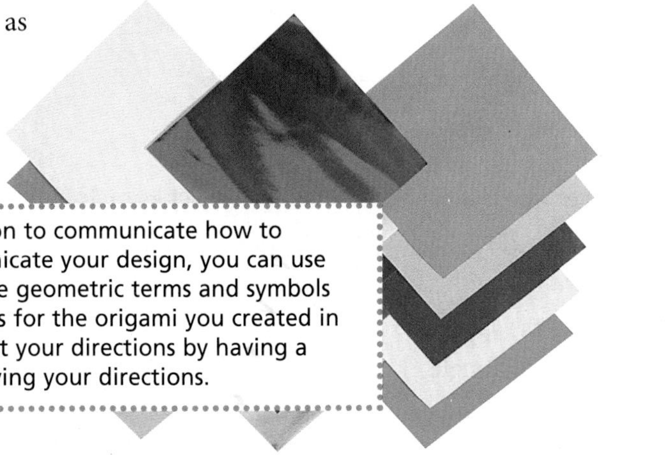

Chapter Project

Find Out by Writing

Origami artists use a special notation to communicate how to construct their creations. To communicate your design, you can use the language of geometry instead. Use geometric terms and symbols along with sketches to write directions for the origami you created in the Find Out question on page 22. Test your directions by having a classmate construct your model following your directions.

Exercises MIXED REVIEW

Use the number line at the right to find the length of each segment.

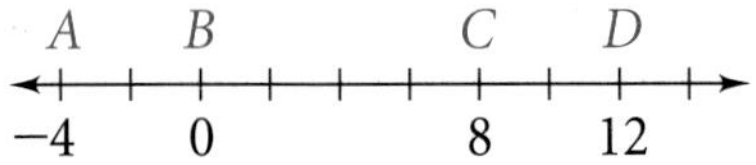

20. $\overline{AC}$ **21.** $\overline{AD}$ **22.** $\overline{CD}$ **23.** $\overline{BC}$

24. Make a sketch of three planes intersecting at one point.

Getting Ready For Lesson 1-7

25. Find the value of x.

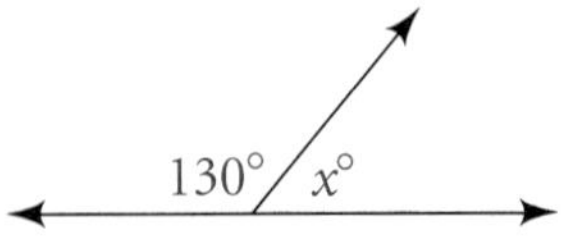

26. Find the value of y.

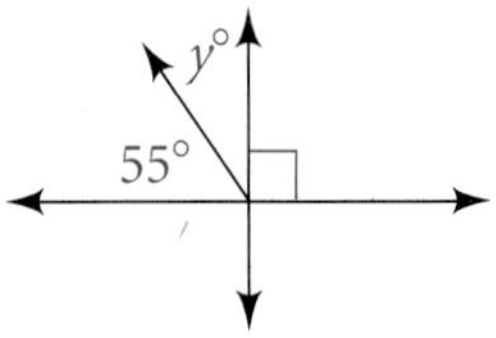

27. Find the value of z.

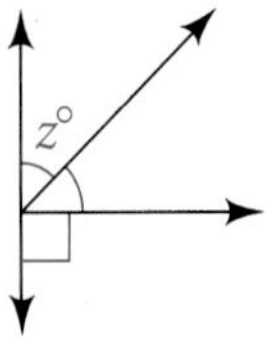

Technology

Exploring Constructions

After Lesson 1-6

Points, lines, and figures are created in geometry software using *draw* or *construct* tools. A figure created by *draw* has no constraints. When the figure is manipulated it moves or changes size freely. A figure created by *construct* is dependent upon an existing object. When you manipulate the existing object, the *construct* object moves or resizes similarly.

In this activity you will explore the difference between *draw* and *construct*. Before you begin, familiarize yourself with your software's tools.

Construct

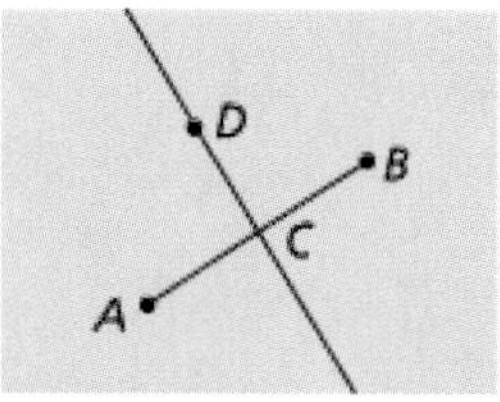

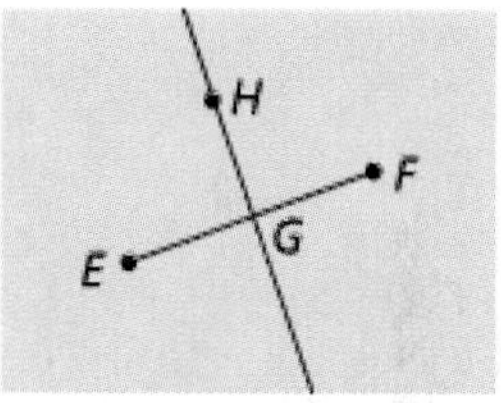

- Draw $\overline{AB}$ and construct perpendicular bisector $\overleftrightarrow{DC}$.
- Draw $\overline{EF}$ and construct G, any point on $\overline{EF}$. Draw $\overleftrightarrow{HG}$. Find EG, GF, and $m\angle HGF$. Attempt to drag G so that $EG = GF$. Attempt to drag H so that $m\angle HGF = 90$. Were you able to draw the perpendicular bisector of $\overline{EF}$? Explain.

Investigate

- Drag A and B. Observe AC, CB, and $m\angle DCB$. Is $\overleftrightarrow{DC}$ always the perpendicular bisector of $\overline{AB}$ no matter how you manipulate the figure?
- Drag E and F. Observe EG, GF, and $m\angle HGF$. How is the relationship between $\overline{EF}$ and $\overleftrightarrow{HG}$ different from the relationship between $\overline{AB}$ and $\overleftrightarrow{DC}$?

Summarize

Write a description of the general difference between *draw* and *construct*. Use your description to explain why the relationship between $\overline{EF}$ and $\overleftrightarrow{HG}$ differs from the relationship between $\overline{AB}$ and $\overleftrightarrow{DC}$.

Extend

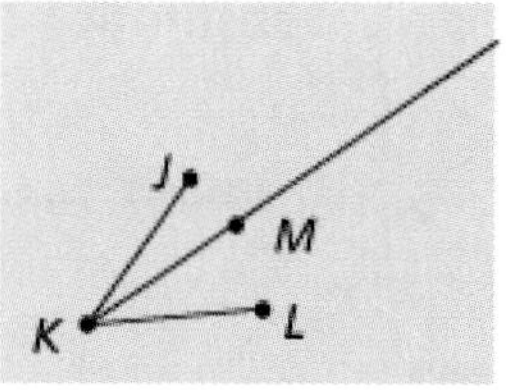

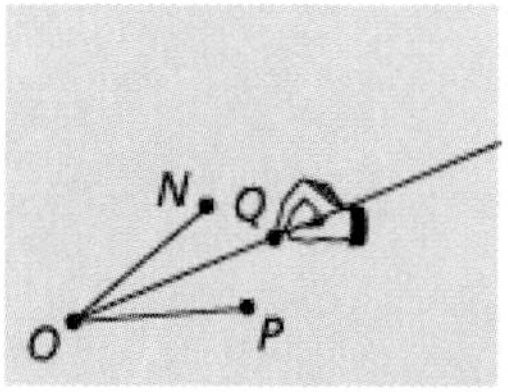

Draw $\angle JKL$ and construct its angle bisector, $\overrightarrow{KM}$. Draw $\angle NOP$. Draw $\overrightarrow{OQ}$ in the interior of $\angle NOP$. Drag Q until $m\angle NOQ = m\angle QOP$. Manipulate both figures and observe the different angle measures. Is $\overrightarrow{KM}$ always the angle bisector of $\angle JKL$? Is $\overrightarrow{OQ}$ always the angle bisector of $\angle NOP$?

1-7 Using Deductive Reasoning

What You'll Learn

- Using deductive reasoning to solve problems and verify conjectures
- Understanding how certain angle pairs are related

...And Why

To reach valid conclusions in geometry and in life

What You'll Need

- tracing paper

THINK AND DISCUSS

Connecting Algebra and Geometry

Deductive reasoning is a process of reasoning logically from given facts to a conclusion. If the given facts are true, deductive reasoning always produces a valid conclusion.

1. **Try This** Maria's parents tell her she can go to the mall with her friends if she finishes her homework. Maria shows her parents her completed homework. What conclusion can you make?

Many people use deductive reasoning in their jobs. A physician diagnosing a patient's illness uses deductive reasoning. A mechanic trying to determine what is wrong with a car uses deductive reasoning. A carpenter uses deductive reasoning to determine what materials are needed at a work site.

Do you remember these Properties of Equality and Real Numbers from algebra?

Properties of Equality and Real Numbers

Addition Property	If $a = b$, then $a + c = b + c$.
Subtraction Property	If $a = b$, then $a - c = b - c$.
Multiplication Property	If $a = b$, then $a \cdot c = b \cdot c$.
Division Property	If $a = b$, then $\frac{a}{c} = \frac{b}{c}$ $(c \neq 0)$.
Substitution Property	If $a = b$, then b can replace a in any expression.
Distributive Property	$a(b + c) = ab + ac$

You may not realize it, but you use deductive reasoning every time you solve an equation.

Example 1

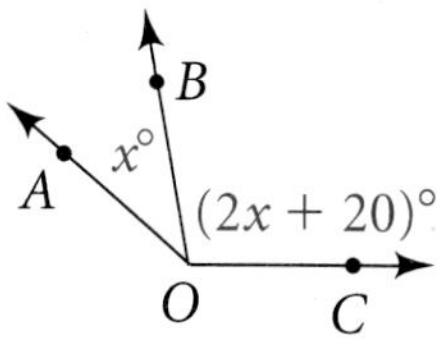

Algebra $m\angle AOC = 140$. Solve for x and justify each step.

$m\angle AOB + m\angle BOC = m\angle AOC$	Angle Addition Postulate
$x + (2x + 20) = 140$	Substitution
$3x = 120$	Subtraction Property of Equality
$x = 40$	Division Property of Equality

These properties of congruence follow from the properties of equality.

QUICK REVIEW

Properties of Equality

Reflexive Property
$a = a$

Symmetric Property
If $a = b$, then $b = a$.

Transitive Property
If $a = b$ and $b = c$, then $a = c$.

Properties of Congruence

Reflexive Property	$\overline{AB} \cong \overline{AB}$
	$\angle A \cong \angle A$
Symmetric Property	If $\overline{AB} \cong \overline{CD}$, then $\overline{CD} \cong \overline{AB}$.
	If $\angle A \cong \angle B$, then $\angle B \cong \angle A$.
Transitive Property	If $\overline{AB} \cong \overline{CD}$ and $\overline{CD} \cong \overline{EF}$, then $\overline{AB} \cong \overline{EF}$.
	If $\angle A \cong \angle B$ and $\angle B \cong \angle C$, then $\angle A \cong \angle C$.

2. Try This Name the property of equality or congruence illustrated.

a. $\angle K \cong \angle K$

b. If $2x - 8 = 10$, then $2x = 18$.

c. If $\overline{RS} \cong \overline{TW}$ and $\overline{TW} \cong \overline{PQ}$, then $\overline{RS} \cong \overline{PQ}$.

d. If $m\angle A = m\angle B$, then $m\angle B = m\angle A$.

WORK TOGETHER

Draw two intersecting lines. Number the angles as shown.

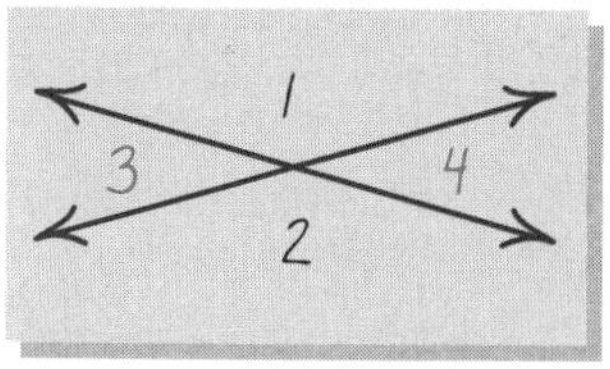

3. Fold the sides of $\angle 1$ onto $\angle 2$. What do you notice?

4. Fold the sides of $\angle 3$ onto $\angle 4$. What do you notice?

5. Compare your results with those of others in your group. Make a **conjecture** about the angles formed by two intersecting lines.

THINK AND DISCUSS

Angle Pairs

In the Work Together you made a conjecture about a pair of angles that has a special name, *vertical angles*. You will learn about several important angle pairs in this lesson.

vertical angles

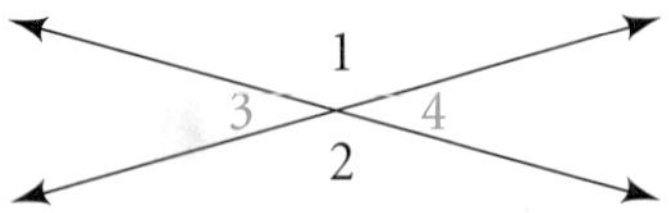

two angles whose sides are opposite rays

adjacent angles

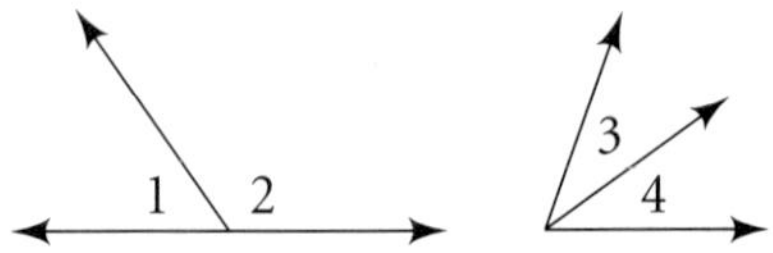

a common side and a common vertex but no common interior points

complementary angles

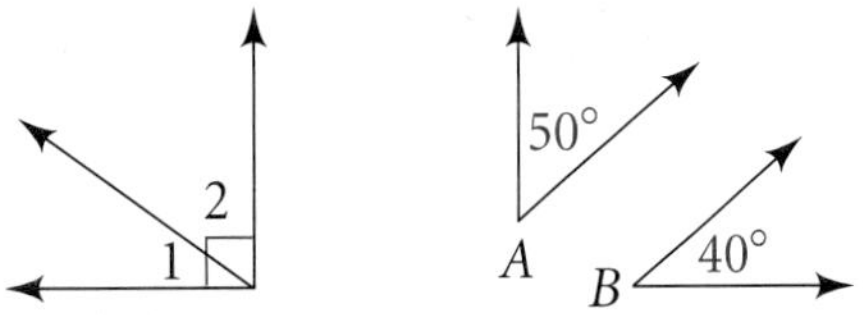

two angles, the sum of whose measures is 90

supplementary angles

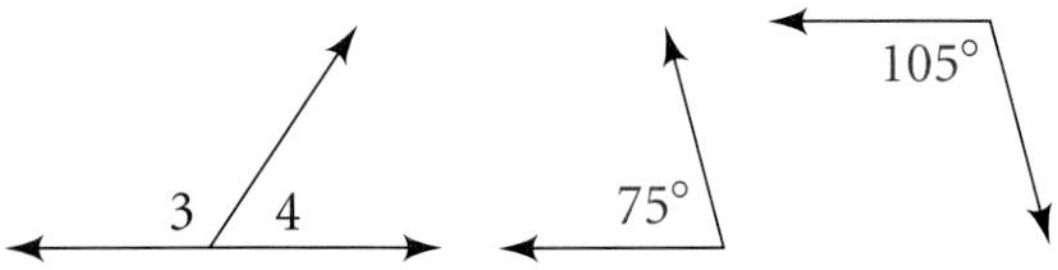

two angles, the sum of whose measures is 180

6. a. Name two pairs of adjacent angles in the photo at the left.

b. Name two pairs of supplementary angles.

In the Work Together you used inductive reasoning to make a conjecture. Now, based on what you know, you can use deductive reasoning to show that your conjecture is always true.

Theorem 1-1
Vertical Angles Theorem

Vertical angles are congruent.

Example 2

Write a convincing argument that the Vertical Angles Theorem is true.

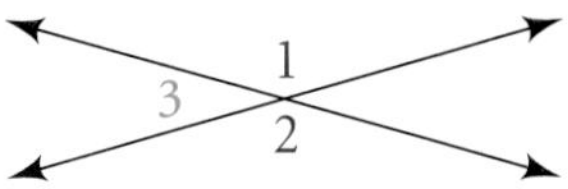

You are given that $\angle 1$ and $\angle 2$ are vertical angles. You must show that $\angle 1 \cong \angle 2$.

By the Angle Addition Postulate,
$m\angle 1 + m\angle 3 = 180$ and $m\angle 2 + m\angle 3 = 180$.
By substitution, $m\angle 1 + m\angle 3 = m\angle 2 + m\angle 3$. Subtract $m\angle 3$ from each side, and you get $m\angle 1 = m\angle 2$, or $\angle 1 \cong \angle 2$.

A convincing argument that uses deductive reasoning is also called a *proof.* A conjecture that is proven is a **theorem.** The Vertical Angles Theorem is actually a special case of the following theorem.

Theorem 1-2
Congruent Supplements Theorem

If two angles are supplements of congruent angles (or of the same angle), then the two angles are congruent.

Example 3

Write a convincing argument that supplements of the same angle are congruent.

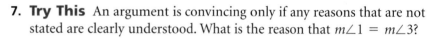

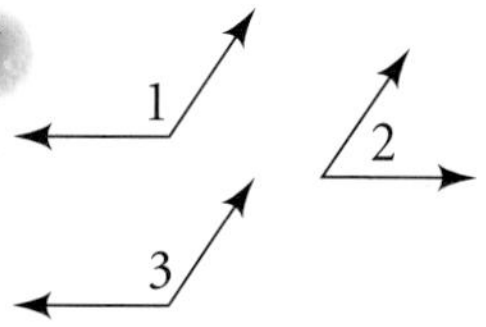

Given: $\angle 1$ and $\angle 2$ are supplementary.
$\angle 3$ and $\angle 2$ are supplementary.

Prove: $\angle 1 \cong \angle 3$

Because $\angle 1$ and $\angle 2$ are supplementary, $m\angle 1 + m\angle 2 = 180$.
Because $\angle 3$ and $\angle 2$ are supplementary, $m\angle 3 + m\angle 2 = 180$.
So $m\angle 1 + m\angle 2 = m\angle 3 + m\angle 2$.
Therefore, $m\angle 1 = m\angle 3$, and $\angle 1 \cong \angle 3$.

7. Try This An argument is convincing only if any reasons that are not stated are clearly understood. What is the reason that $m\angle 1 = m\angle 3$?

The next theorem is much like the Congruent Supplements Theorem.

Theorem 1-3
Congruent Complements Theorem

If two angles are complements of congruent angles (or of the same angle), then the two angles are congruent.

As you will see in Chapter 4, a proof may take many different forms. The format of a proof is not important. Logical use of deductive reasoning is.

You can draw certain conclusions directly from diagrams. You can conclude that angles are
- vertical angles
- adjacent angles
- adjacent supplementary angles.

Unless there are marks that give this information, you cannot assume that
- angles or segments are congruent
- an angle is a right angle
- lines are perpendicular or parallel.

8. What can you conclude? Explain.

a.

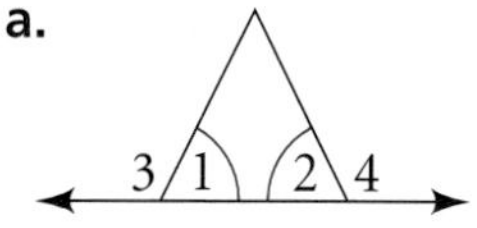

b.

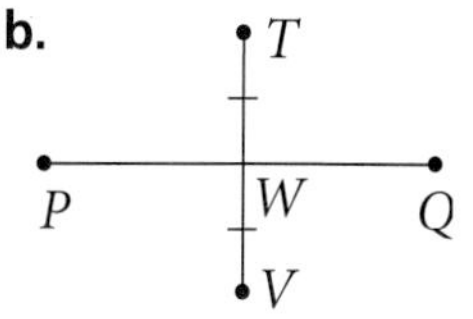

c.

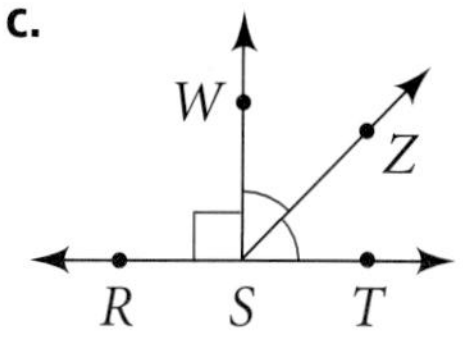

Exercises ON YOUR OWN

Name the property that justifies each statement.

1. $\angle Z \cong \angle Z$
2. If $12x = 84$, then $x = 7$.
3. If $\overline{ST} \cong \overline{QR}$, then $\overline{QR} \cong \overline{ST}$.
4. If $3x + 14 = 80$, then $3x = 66$.
5. If $2x + y = 5$ and $x = y$, then $2x + x = 5$.
6. If $AB - BC = 12$, then $AB = 12 + BC$.
7. If $m\angle A = 15$, then $3m\angle A = 45$.
8. $QR = QR$
9. If $\angle 1 \cong \angle 2$ and $\angle 2 \cong \angle 3$, then $\angle 1 \cong \angle 3$.
10. $2(3x + 5) = 6x + 10$

11. **Writing** How is a theorem different from a postulate?
12. **Open-ended** Give an example of vertical angles in your home.

Algebra **Find the values of the variables.**

13.

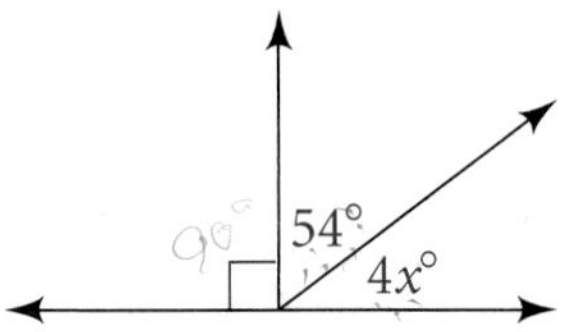

14.

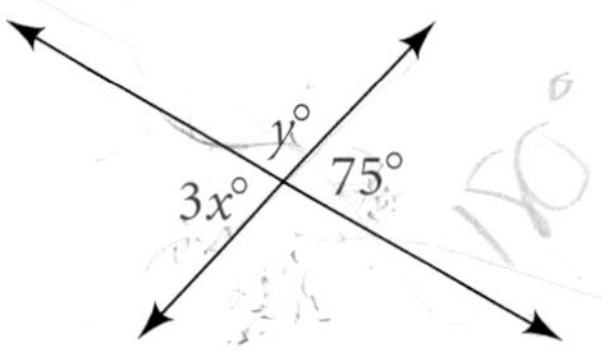

15.

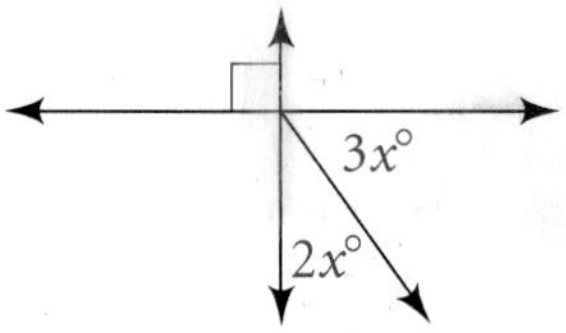

16.

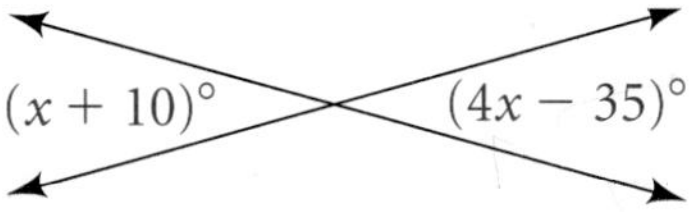

17.

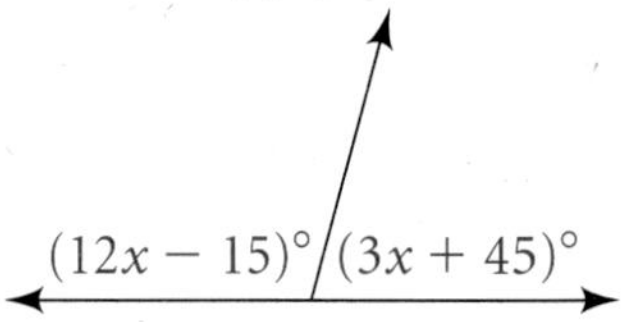

18.

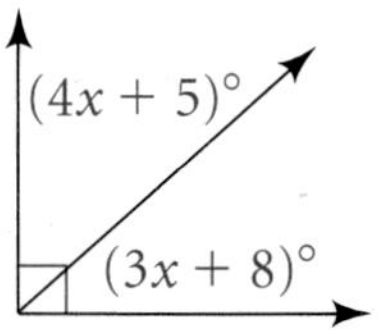

19. a. **Coordinate Geometry** $\angle AOX$ contains points $A(1, 3)$, $O(0, 0)$, and $X(4, 0)$. Find the coordinates of a point B so that $\angle BOA$ and $\angle AOX$ are adjacent complementary angles.
    b. Find the coordinates of a point C so that $\overrightarrow{OC}$ is a side of a different angle that is adjacent to and complementary to $\angle AOX$.
    c. $\angle DOE$ contains points $D(2, 3)$, $O(0, 0)$, and $E(5, 1)$. Find the coordinates of a point F so that $\overrightarrow{OF}$ is a side of an angle that is adjacent to and supplementary to $\angle DOE$.

20. **Algebra** $\angle A$ and $\angle B$ are supplementary angles. $m\angle A = 3x + 12$ and $m\angle B = 2x - 22$. Find the measures of both angles.

21. **Algebra** Solve for x and y.

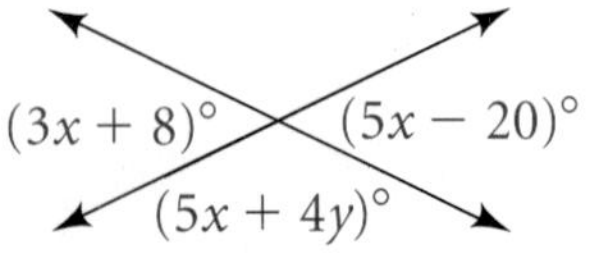

22. *Standardized Test Prep* In the diagrams at the right, which is greater, x or y?

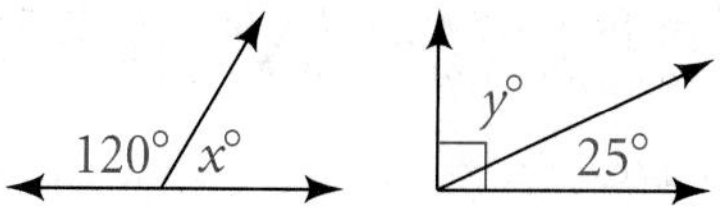

23. *Critical Thinking* If possible, find the measures of the angles described. If it is not possible, explain why.

a. congruent adjacent supplementary angles
b. congruent adjacent complementary angles
c. congruent vertical angles

24. *Algebra* The measure of a supplement of $\angle 1$ is six times the measure of a complement of $\angle 1$. Find the measures of $\angle 1$, its supplement, and its complement.

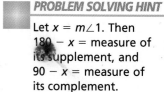

PROBLEM SOLVING HINT

Let $x = m\angle 1$. Then $180 - x$ = measure of its supplement, and $90 - x$ = measure of its complement.

25. One angle is twice as large as its complement. Find the measures of both angles.

What can you conclude about the angles in each diagram? Justify your answers.

26.

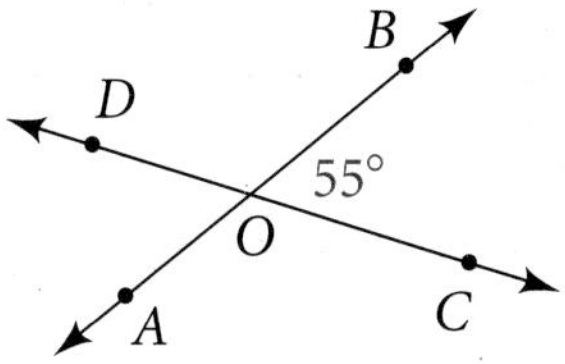

27.

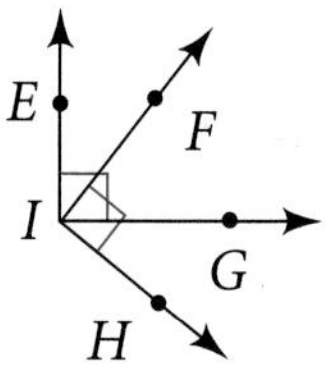

28.

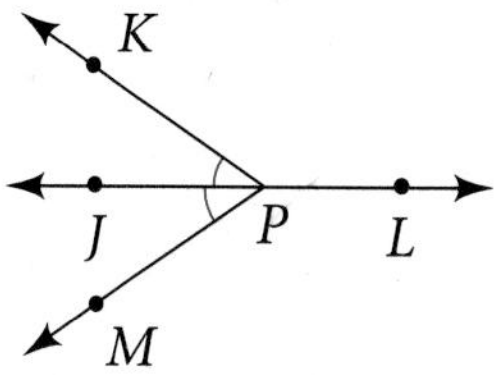

Give a reason for each step.

29.
$$3x - 15 = 105$$
$$3x = 120$$
$$x = 40$$

30.
$$12y + 24 = 96$$
$$12y = 72$$
$$y = 6$$

31.
$$\tfrac{1}{2}x - 5 = 10$$
$$2\left(\tfrac{1}{2}x - 5\right) = 20$$
$$x - 10 = 20$$
$$x = 30$$

32. *Preparing for Proof* Write a convincing argument that complements of the same angle are congruent.

Given: $\angle 1$ and $\angle 2$ are complementary.
$\angle 3$ and $\angle 2$ are complementary.
Prove: $\angle 1 \cong \angle 3$

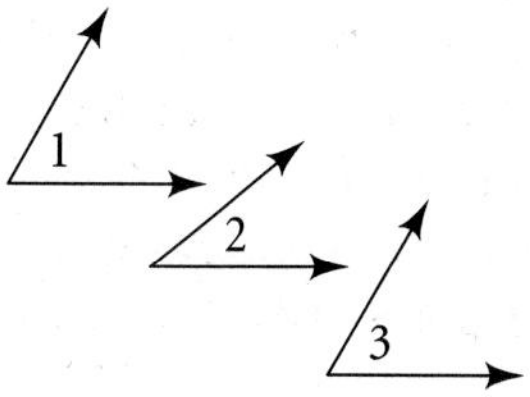

33. *Preparing for Proof* Write a convincing argument that supplements of congruent angles are congruent.

Given: $\angle 1$ and $\angle 2$ are supplementary.
$\angle 3$ and $\angle 4$ are supplementary.
$\angle 2 \cong \angle 4$
Prove: $\angle 1 \cong \angle 3$

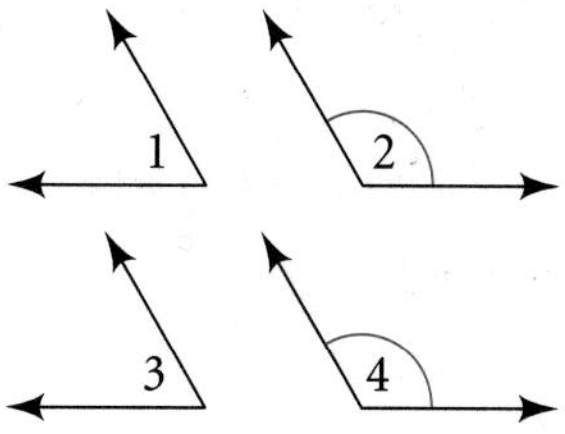

Use the cartoon and deductive reasoning to answer *yes* or *no*. Explain any *no*.

34. Could a person with a red car park here on Tuesday at 10:00 A.M?

35. Could a man with a beard park here on Monday at 10:30 A.M.?

36. Could a woman with a wig park here on Saturday at 10:00 A.M.?

37. Could a person with a blue car park here on Tuesday at 9:05 A.M.?

38. Could a person with a convertible with leather seats park here on Sunday at 6:00 P.M.?

Exercises MIXED REVIEW

Sketch each of the following.

39. three collinear points

40. two intersecting rays

41. two perpendicular lines

42. Find the next three terms in the sequence: $1, 1, \frac{1}{2}, \frac{1}{3}, \frac{1}{5}, \frac{1}{8}, \ldots$

Getting Ready for Lesson 1-8

Calculator **Find the square root of each number to the nearest tenth.**

43. 25

44. 17

45. 123

46. 48

47. 96

48. 1023

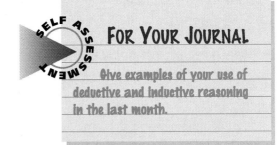

SELF ASSESSMENT

FOR YOUR JOURNAL

Give examples of your use of deductive and inductive reasoning in the last month.

Exercises CHECKPOINT

1. Writing What are the components of a good definition?

2. a. Use a protractor to draw $\angle R$ with measure 78.

b. Construct an angle whose measure is $2 \cdot m\angle R$.

c. Construct an angle whose measure is $\frac{1}{2} \cdot m\angle R$.

3. Open-ended Use a protractor to draw a pair of adjacent complementary angles. Then draw another pair of complementary angles that are *not* adjacent.

4. Standardized Test Prep If $-6 \leq x \leq 1$ is graphed on a number line, its graph is which of the following?

A. 7 points **B.** 8 points **C.** a segment **D.** a line **E.** a ray

What You'll Learn

- Finding the distance between two points in a coordinate plane
- Finding the coordinates of the midpoint of a segment in a coordinate plane

...And Why

To improve skills such as map reading

What You'll Need

tracing paper, ruler, graph paper

1-8 The Coordinate Plane

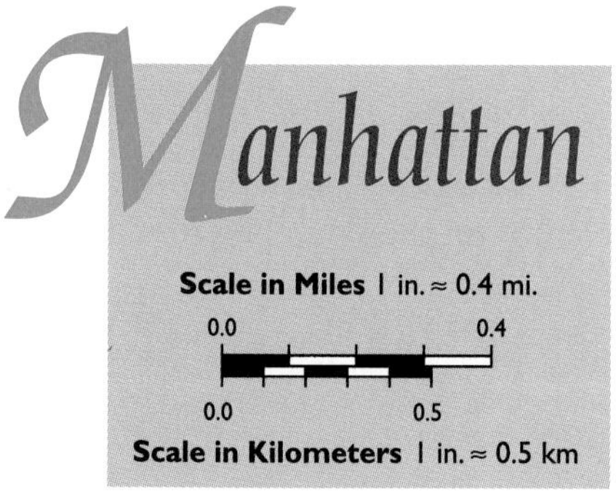

WORK TOGETHER

Much of New York City is laid out in a rectangular grid, as shown in this map. Most of the streets in the grid are either parallel or perpendicular.

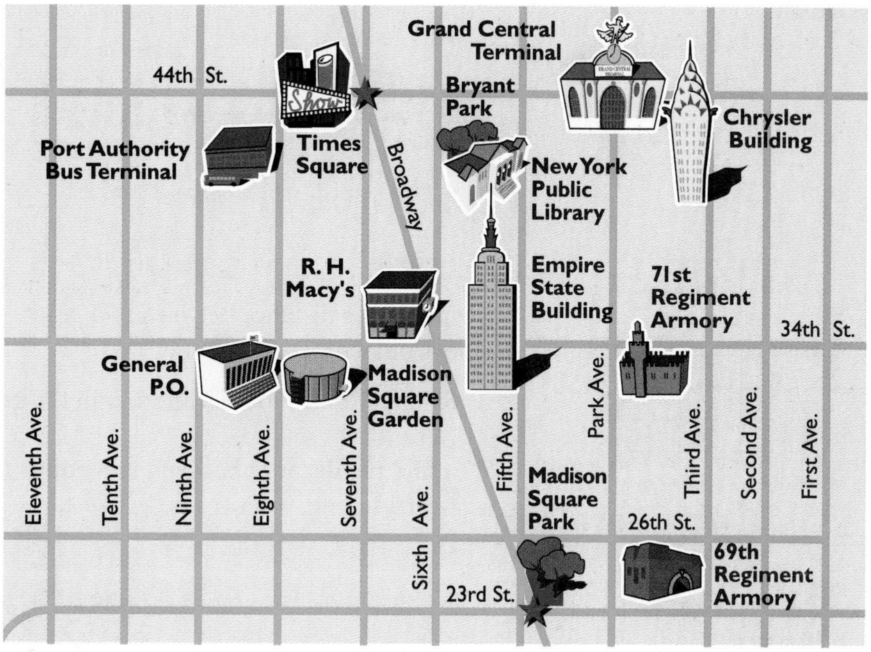

Yvonne's family is on the corner of 44th Street and 7th Avenue. They plan to walk to Madison Square Park at 23rd Street and 5th Avenue. There are several possible routes they can take.

Work with your group. Trace the routes on the map and answer the following questions.

1. Yvonne's father wants to walk east on 44th Street until they reach 5th Avenue. He then plans to walk south on 5th Avenue to Madison Square Park. About how long is his route?
2. Yvonne's mother wants to walk south on 7th Avenue until they reach 23rd Street. She then plans to walk east on 23rd Street to Madison Square Park. About how long is her route?
3. Yvonne notices on the map that Broadway cuts across the grid of streets and leads to Madison Square Park. She suggests walking all the way on Broadway. About how long is her route?
4. Whose route is the shortest? Why?
5. Whose route is the longest? Why?

THINK AND DISCUSS

You can think of a point as a dot, and a line as a series of points. In coordinate geometry you can describe a point with an ordered pair (x, y) and a line with an equation $y = mx + b$.

QUICK REVIEW

The x- and y-axes divide the coordinate plane into four quadrants.

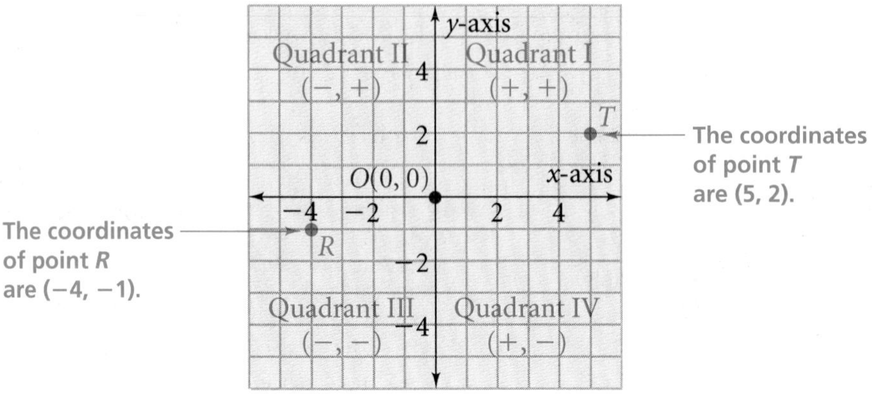

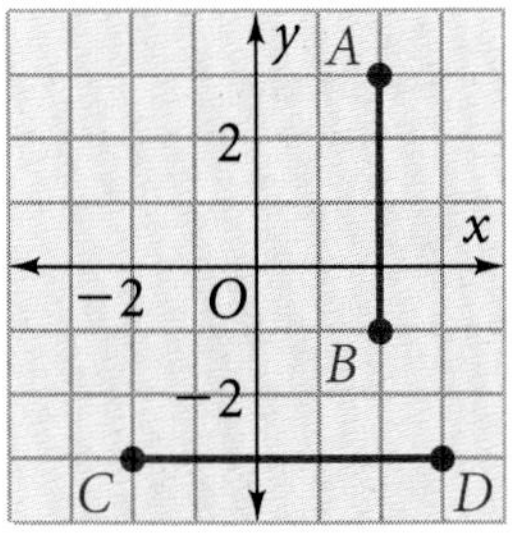

Finding the distance between two points is easy if the points lie on a horizontal or a vertical line.

6. Try This Find AB and CD in the graph at the left.

To find the distance between two points that are not on the same horizontal or vertical line, you can use the Distance Formula.

The Distance Formula

The distance d between two points $A(x_1, y_1)$ and $B(x_2, y_2)$ is

$$d = \sqrt{(x_2 - x_1)^2 + (y_2 - y_1)^2}.$$

Example 1 ***Relating to the Real World***

Transportation Luisa takes the subway from Oak Station to Jackson Station each morning. Oak Station is 1 mi west and 2 mi south of City Plaza. Jackson Station is 2 mi east and 4 mi north of City Plaza. How far does she travel by subway?

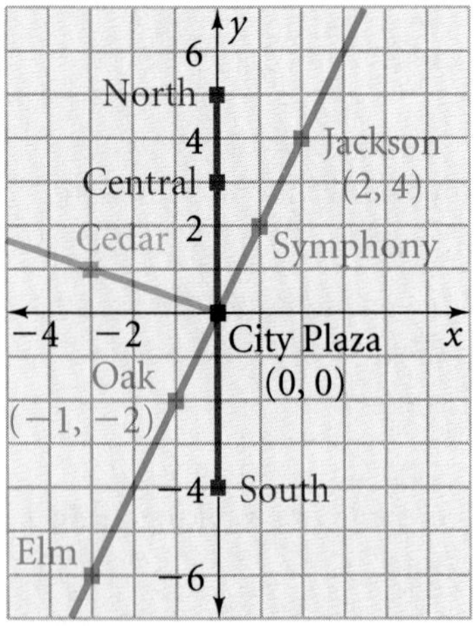

Let (x_1, y_1) and (x_2, y_2) represent Oak and Jackson, respectively. Then, $x_1 = -1$, $x_2 = 2$, $y_1 = -2$, and $y_2 = 4$.

$d = \sqrt{(x_2 - x_1)^2 + (y_2 - y_1)^2}$ — Use the Distance Formula.

$d = \sqrt{(2 - (-1))^2 + (4 - (-2))^2}$ — Substitute.

$d = \sqrt{3^2 + 6^2}$ — Simplify.

$d = \sqrt{9 + 36} = \sqrt{45}$

45 [√] 6.7082039 — Use your calculator.

Luisa travels about 6.7 mi by subway.

The system of longitude and latitude lines on Earth is a type of coordinate system.

7. In Example 1 suppose you let (x_1, y_1) be (2, 4) and (x_2, y_2) be (−1, −2). Would you get the same result? Why or why not?

8. **Calculator** Find the length of $\overline{AB}$ with endpoints $A(1, -3)$ and $B(-4, 4)$ to the nearest tenth.

You know how to find the midpoint of a segment on a number line. Now you will use a similar process to find the midpoint of a segment in the coordinate plane.

Graph $\overline{TS}$ with endpoints $T(4, 3)$ and $S(8, 5)$. Let $\overline{TR}$ be a horizontal segment and $\overline{SR}$ be a vertical segment. Add the coordinates to your drawing as you answer these questions.

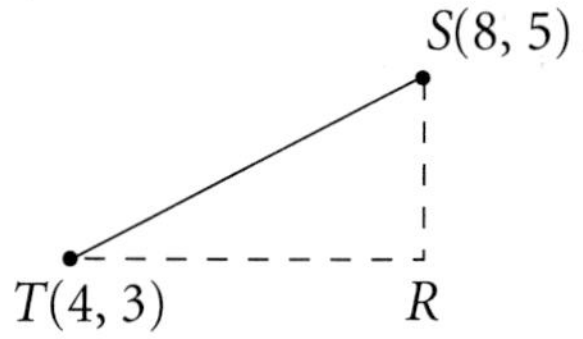

9. What are the coordinates of point R?

10. a. What are the coordinates of M_1, the midpoint of $\overline{TR}$?
    b. What are the coordinates of M_2, the midpoint of $\overline{SR}$?

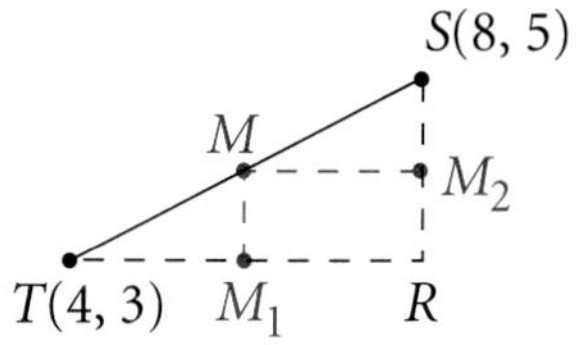

11. What are the coordinates of M, the midpoint of $\overline{TS}$?

This specific case shows you how the Midpoint Formula was developed.

The Midpoint Formula

The coordinates of the midpoint M of $\overline{AB}$ with endpoints $A(x_1, y_1)$ and $B(x_2, y_2)$ are the following:

$$M = \left(\frac{x_1 + x_2}{2}, \frac{y_1 + y_2}{2}\right)$$

QUICK REVIEW

$\frac{x_1 + x_2}{2}$ is the *arithmetic mean* of x_1 and x_2. Another word for arithmetic mean is *average*.

Example 2

Find the coordinates of the midpoint M of $\overline{QS}$ with endpoints $Q(3, 5)$ and $S(7, -9)$.

Let $x_1 = 3, x_2 = 7, y_1 = 5$, and $y_2 = -9$.

$$x\text{-coordinate of } M = \frac{x_1 + x_2}{2} = \frac{3 + 7}{2} = \frac{10}{2} = 5$$

$$y\text{-coordinate of } M = \frac{y_1 + y_2}{2} = \frac{5 + (-9)}{2} = \frac{-4}{2} = -2$$

The coordinates of point M are $(5, -2)$.

12. **Try This** Find the coordinates of the midpoint of $\overline{AB}$ with endpoints $A(2, -5)$ and $B(6, 13)$.

Exercises ON YOUR OWN

Graph each point in the same coordinate plane.

1. $A(5, -3)$ 2. $B(-3, 0)$ 3. $C(6, 2)$ 4. $D(-6, 2)$ 5. $E(-4, 3)$ 6. $F(0, 5)$

Choose **Use mental math, pencil and paper, or a calculator to find the distance between the points to the nearest tenth.**

7. $J(2, -1), K(2, 5)$
8. $L(10, 14), M(-8, 14)$
9. $N(-11, -11), P(-11, -3)$
10. $A(0, 3), B(0, 12)$
11. $C(12, 6), D(-8, 18)$
12. $E(6, -2), F(-2, 4)$
13. $Q(12, -12), T(5, 12)$
14. $R(0, 5), S(12, 3)$
15. $X(-3, -4), Y(5, 5)$

Find the coordinates of the midpoint of $\overline{HX}$.

16. $H(0, 0), X(8, 4)$
17. $H(-1, 3), X(7, -1)$
18. $H(13, 8), X(-6, -6)$
19. $H(7, 10), X(5, -8)$
20. $H(-6.3, 5.2), X(1.8, -1)$
21. $H\left(5\frac{1}{2}, -4\frac{3}{4}\right), X\left(2\frac{1}{4}, -1\frac{1}{4}\right)$

22. The midpoint of $\overline{QS}$ is the origin. Point Q is located in Quadrant II. What quadrant contains point S?

23. $M(5, 12)$ is the midpoint of $\overline{AB}$. The coordinates of point A are $(2, 6)$. What are the coordinates of point B?

24. The midpoint of $\overline{QT}$ has coordinates $(3, -4)$. The coordinates of point Q are $(2, 3)$. What are the coordinates of point T?

25. The coordinates of A, B, C, and D are given at the right. Graph the points and draw the segments connecting them in order. Are the lengths of the sides of $ABCD$ the same? Explain.

$A(-6, 2)$ $B(-3, 5)$
$C(-6, 6)$ $D(-9, 5)$

26. **Open-ended** Graph $A(-2, 1)$ and $B(2, 3)$. Draw $\overleftrightarrow{AB}$. For each point described, give two sets of possible coordinates if they exist. Otherwise write "exactly one point" and give the coordinates.
    a. point D so that $\overleftrightarrow{CD}$ contains $C(-1, 4)$ and is parallel to $\overleftrightarrow{AB}$
    b. point E so that $\overleftrightarrow{AE}$ is parallel to $\overleftrightarrow{BC}$, $\overleftrightarrow{BC}$ contains $C(-1, 4)$, and $\overleftrightarrow{EC}$ is parallel to $\overleftrightarrow{AB}$
    c. point G so that $\overleftrightarrow{FG}$ contains $F(0, 2)$ and is perpendicular to $\overleftrightarrow{AB}$
    d. point H so that $\overleftrightarrow{HJ}$ contains $J(4, 2)$ and is perpendicular to $\overleftrightarrow{AB}$

27. **Writing** Why do you think that some cities are designed with a rectangular grid instead of a triangular grid or some other shape?

28. Graph the points $A(2, 1)$, $B(6, -1)$, $C(8, 7)$, and $D(4, 9)$. Draw quadrilateral $ABCD$. Use the Midpoint Formula to determine the midpoints of $\overline{AC}$ and $\overline{BD}$. What do you notice?

For each graph, find (a) the length of $\overline{AB}$ to the nearest tenth and (b) the coordinates of the midpoint of $\overline{AB}$.

29.

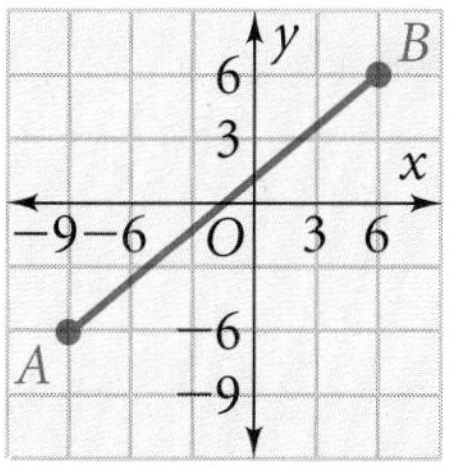

30.

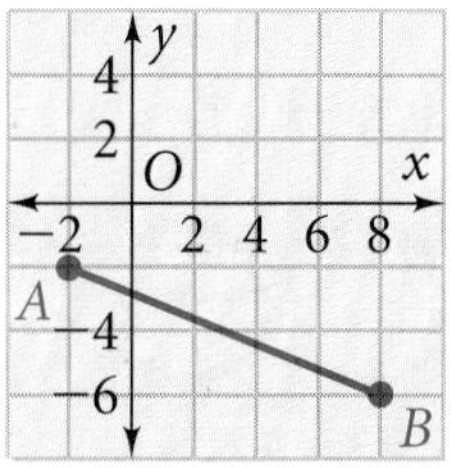

31.

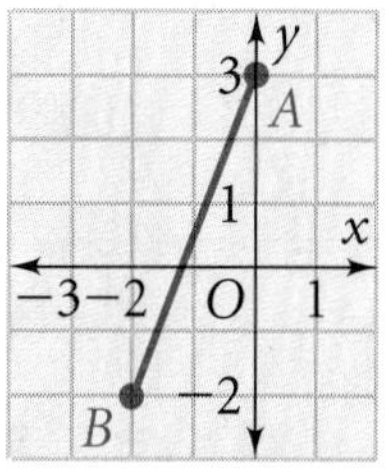

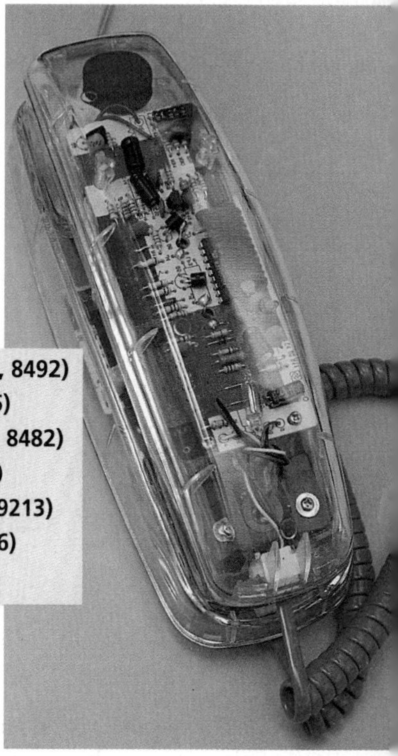

32. *Communications* Long-distance rates for telephone calls are determined mainly by the distance between the two ends of the call. To determine these distances, long-distance telephone companies have divided North America into a grid, with each unit equaling $\sqrt{0.1}$ mile. The distance between two customers is then determined by using the Distance Formula. The coordinates for certain customers in several cities are listed. Find the distance between customers in the following cities:

a. Boston and San Francisco
b. Houston and Chicago
c. Denver and New Orleans

San Francisco (8719, 8492)
Chicago (3439, 5985)
New Orleans (2637, 8482)
Denver (5899, 7501)
Los Angeles (7878, 9213)
Houston (3537, 8936)
Boston (1248, 4422)

33. *Geometry in 3 Dimensions* You can use three coordinates (x, y, z) to locate points in three dimensions. Point P has coordinates $(3, -3, 5)$.

a. Give the coordinates of points $A, B, C, D, E, F,$ and G.
b. Draw three axes like those shown. Then graph $R(4, 5, 9)$.

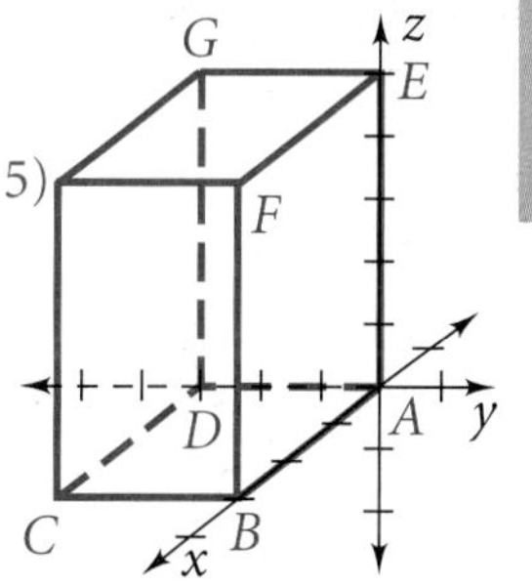

Exercises MIXED REVIEW

Find the measure of the complement and supplement of each angle, if possible.

34. $m\angle B = 56$ **35.** $m\angle R = 18$ **36.** $m\angle D = 179$

37. $m\angle P = 23.5$ **38.** $m\angle T = 108$ **39.** $m\angle E = 78$

40. The length of $\overline{AC}$ is 45. If $AB = x + 8$ and $BC = 3x - 3$, find the value of x.

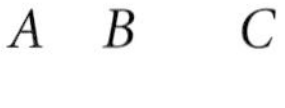

SELF ASSESSMENT

PORTFOLIO

For your portfolio, select one or two items from your work for this chapter. Here are some possibilities.

- corrected work
- an origami creation
- a journal entry

Explain why you have included each selection you make.

CHAPTER PROJECT

Finishing the Chapter Project

On Folded Wings

Find Out questions on pages 10, 22, 31, and 44 should help you to complete your project. Prepare a *Geometry in Origami* display. Include the models you made, instructions for making them, and the geometric patterns you discovered. You may want to display origami creations in a hallway display case. Consider adding more origami creations or writing a short report on the history of origami. You can use the books listed below to find out more about origami.

Reflect and Revise

Ask a classmate to review your display with you. Together, check that your models are well constructed, your directions are clear and correct, and your explanations are sensible. Have you used geometric terms correctly? Is the display attractive as well as informative?

Follow Up

Use paper folding to illustrate some of the geometric terms, such as *midpoint, angle bisector,* and *perpendicular bisector*, that you learned in this chapter.

For More Information

Gray, Alice and Kunihiko Kasahara. *The Magic of Origami*. Tokyo: Japan Publications, Inc., 1977.

Jackson, Paul. *Step-By-Step Origami*. London: Anness Publishing, 1995.

Kenneway, Eric. *Complete Origami.* New York: St. Martin's Press, 1987.

Montroll, John. *Easy Origami*. New York: Dover Publications, 1992.

Weiss, Stephen. *Origami That Flies*. New York: St. Martin's Press, 1984.

CHAPTER 1

Wrap Up

Key Terms

acute angle (p. 27)
adjacent angles (p. 48)
angle (p. 26)
angle bisector (p. 34)
collinear (p. 13)
compass (p. 39)
complementary angles (p. 48)
congruent angles (p. 28)
congruent segments (p. 25)
conjecture (p. 5)
construction (p. 39)
coordinate (p. 25)
coplanar (p. 13)
deductive reasoning (p. 46)
inductive reasoning (p. 5)
line (p. 13)
measure of an angle (p. 26)
midpoint (p. 33)
obtuse angle (p. 27)
opposite rays (p. 18)
parallel lines (p. 18)
parallel planes (p. 19)
perpendicular bisector (p. 34)
perpendicular lines (p. 33)
plane (p. 13)
point (p. 12)
postulate (p. 14)
proof (p. 49)
quadrant (p. 54)
ray (p. 18)
right angle (p. 27)
segment (p. 18)
segment bisector (p. 33)
skew lines (p. 19)
space (p. 12)
straight angle (p. 27)
straightedge (p. 39)
supplementary angles (p. 48)
theorem (p. 49)
vertical angles (p. 48)

How am I doing?

SELF ASSESSMENT

- State three ideas from this chapter that you think are important. Explain your choices.
- Describe three types of mathematical statements.

Using Patterns and Inductive Reasoning 1-1

You use **inductive reasoning** when you make conclusions from specific examples or patterns. You can use inductive reasoning to make conjectures. A **conjecture** describes a conclusion reached from observations or inductive reasoning. Because conjectures are not always valid, you should verify them if possible.

Find the next two terms of each sequence and describe the pattern.

1. 1, 5, 9, 13, . . .

2. 1, 3, 7, 15, 31, . . .

3. $\frac{1}{2}, \frac{2}{3}, \frac{3}{4}, \frac{4}{5}, \ldots$

4. 0, 1, −2, 3, −4, . . .

5. Draw the next figure in the sequence.

6. a. Calculator Find the last two digits of 76^2, 76^4, 276^2, and 376^3.

b. Make a **conjecture** about powers of numbers whose last two digits are 76.

Basic Concepts 1-2, 1-3

Points that lie on a line are **collinear.** Points and lines in the same plane are **coplanar.**

Two coplanar lines that do not intersect are **parallel.** Two lines in space that are not parallel and do not intersect are **skew.** Two planes that do not intersect are **parallel.**

Segments and **rays** are parts of lines.

7. Critical Thinking Explain why the postulate "Through any three noncollinear points there is exactly one plane" applies only to noncollinear points.

Use the figure to answer Exercises 8–13.

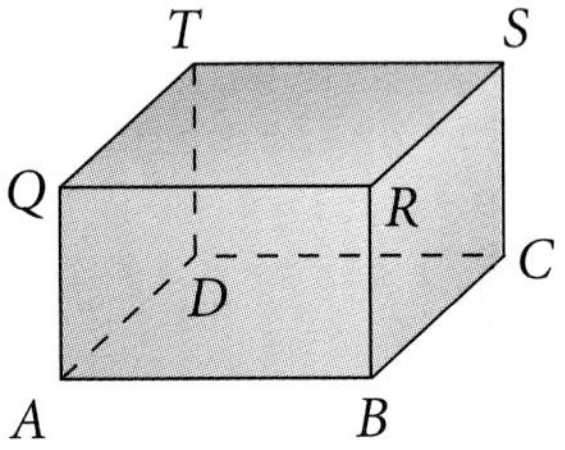

8. Name two intersecting lines.

9. Name a pair of skew lines.

10. Name three noncollinear points.

11. Name four noncoplanar points.

12. Name a pair of parallel planes.

13. Name three lines that intersect at D.

Complete with *always, sometimes,* or *never* to make a true statement

14. A line and a point are __?__ coplanar.

15. Two segments are __?__ coplanar.

16. Skew lines are __?__ coplanar.

17. Opposite rays __?__ have the same endpoint.

18. Two points are __?__ collinear.

19. Parallel lines are __?__ skew.

Angles, Segments, and Deductive Reasoning 1-4, 1-7

Segments with the same length are **congruent.** An **angle** is formed by two rays with the same endpoint. Angles are measured in degrees. Angles with the same measure are **congruent.**

Deductive reasoning is the process of reasoning logically from given facts to a conclusion. If the given facts are true, deductive reasoning always produces a valid conclusion.

Special relationships exist between certain angle pairs. For example, vertical angles are congruent. The sum of the measures of a pair of **complementary angles** is 90. The sum of the measures of a pair of **supplementary angles** is 180.

20. Open-ended Describe a real-world situation where you use deductive reasoning.

21. Find possible coordinates of point Q on the number line so that $PQ = 5$.

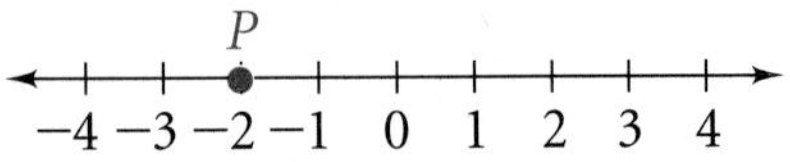

Algebra **Find the value of each variable in the diagrams below.**

22.

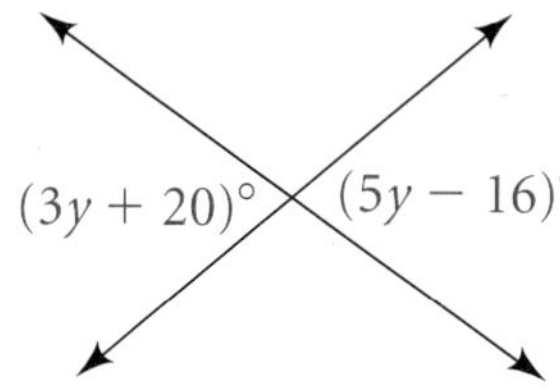

23.

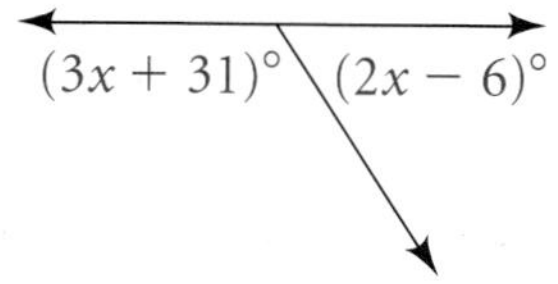

24.

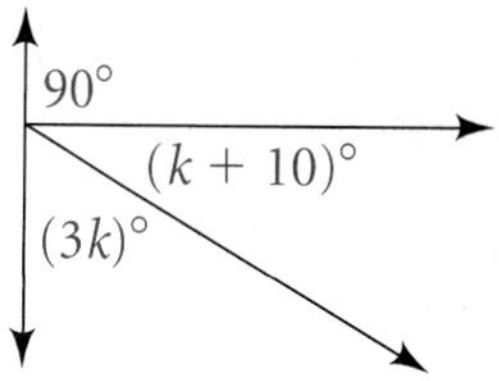

What can you conclude from each diagram? Justify your answers.

25.

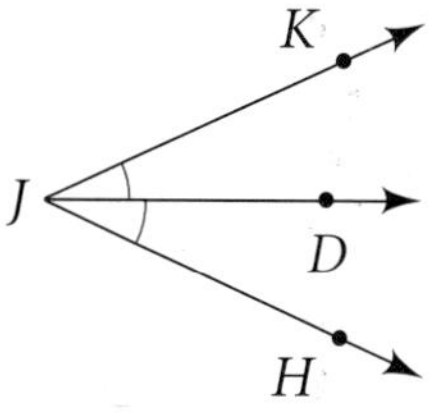

26.

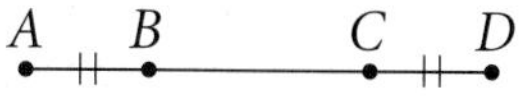

27.

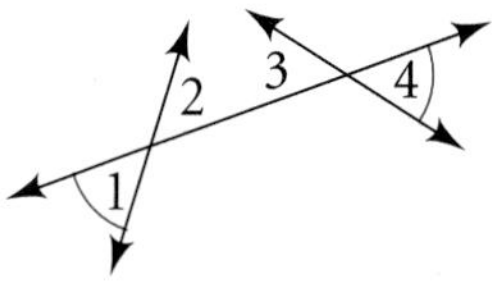

Good Definitions

1-5

A good definition is precise. A good definition uses terms that have been previously defined or are commonly accepted.

The **midpoint** of a segment divides a segment into two congruent segments. **Perpendicular lines** intersect at right angles. A **perpendicular bisector** of a segment is perpendicular to a segment at its midpoint. An **angle bisector** divides an angle into two congruent angles.

28. **Writing** Rico defines a book as something you read. Explain what's wrong with this definition. Write a good definition for the word *book*.

29. **Critical Thinking** In the diagram at the right, $\overleftrightarrow{LJ}$ is the perpendicular bisector of $\overline{BK}$. Is $\overleftrightarrow{BK}$ necessarily the perpendicular bisector of $\overline{LJ}$? Explain.

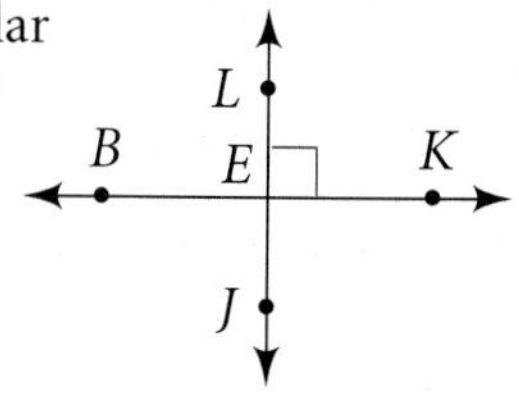

30. **Algebra** In the diagram at right, point E is the midpoint of $\overline{BK}$, $BE = 3y - 2$, and $EK = 2y + 1$. Find the value of y.

31. Find the coordinate of the midpoint of $\overline{GH}$.

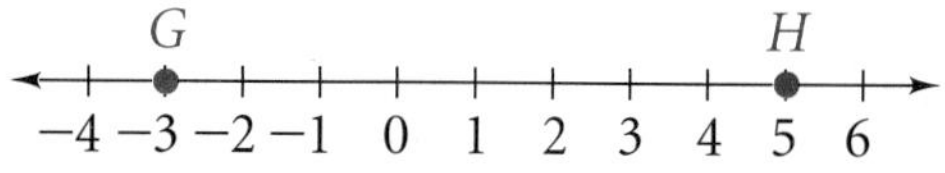

32. **Standardized Test Prep** In which figure is the $m\angle 1$ *not* equal to 60?

A.

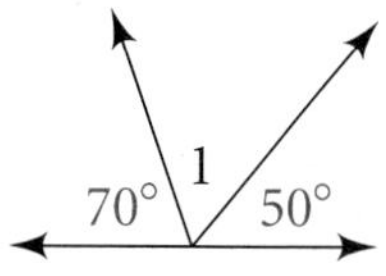

B.

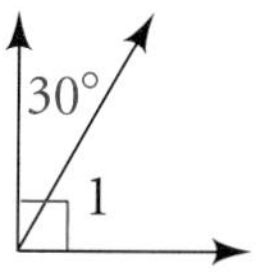

C.

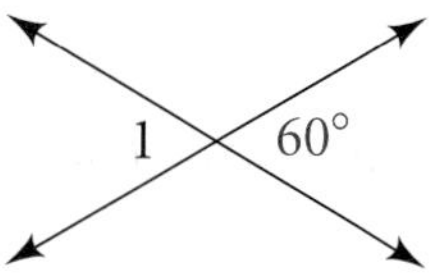

D.

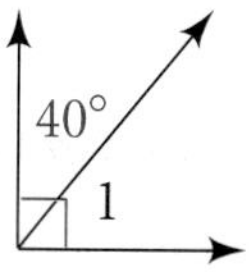

E.

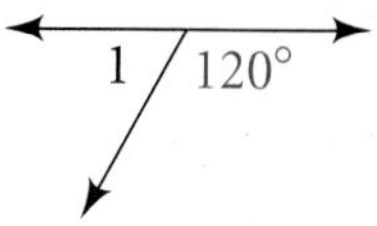

Basic Constructions 1-6

A **construction** is the process of making geometric figures using a **straightedge** and **compass.**

33. Use a protractor to draw a 72° angle. Then construct a congruent angle.

34. a. Construct $\overline{AB} \cong \overline{PQ}$.

b. Construct the perpendicular bisector of $\overline{AB}$.

35. **Writing** Describe how to construct a 45° angle.

The Coordinate Plane 1-8

The **x-axis** and the **y-axis** intersect at the **origin** (0, 0) and determine a **coordinate plane.** You can find the coordinates of the midpoint M of $\overline{AB}$ with endpoints $A(x_1, y_1)$ and $B(x_2, y_2)$ using the **Midpoint Formula.**

$$M = \left(\frac{x_1 + x_2}{2}, \frac{y_1 + y_2}{2}\right)$$

You can find the distance d between points $A(x_1, y_1)$ and $B(x_2, y_2)$ using the **Distance Formula.**

$$d = \sqrt{(x_2 - x_1)^2 + (y_2 - y_1)^2}$$

Graph each point in the same coordinate plane.

36. $A(-1, 5)$ **37.** $B(0, 4)$ **38.** $C(-1, -1)$ **39.** $D(6, 2)$ **40.** $E(-7, 0)$ **41.** $F(3, -4)$

42. $\overline{GH}$ has endpoints $G(-3, 2)$ and $H(3, -2)$. Find the coordinates of the midpoint of $\overline{GH}$.

43. **Calculator** Find the distance between points $B(4, -3)$ and $D(3, 0)$ to the nearest tenth.

Getting Ready for... CHAPTER 2

44. a. Graph points $A(-1, 3)$, $B(-1, 6)$, and $C(4, 3)$ in the same coordinate plane and connect them to form a triangle.

b. **Choose** Use pencil and paper, mental math, or a calculator to find the lengths of the three sides of the triangle.

c. List the sides in order from longest to shortest.

Name a real-world object that has the given geometric shape.

45. triangle **46.** circle **47.** rhombus **48.** cylinder **49.** cube

CHAPTER 1 Assessment

Find the next two terms in each sequence.

1. 8, -4, 2, -1, . . .
2. 0, 2, 4, 6, 8, . . .
3. 

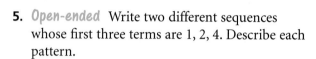

4. 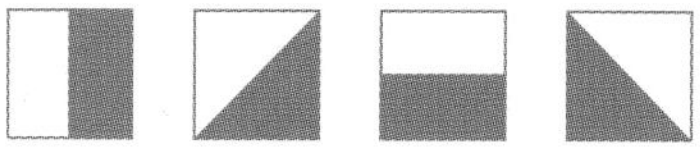
5. **Open-ended** Write two different sequences whose first three terms are 1, 2, 4. Describe each pattern.

Use the figure to answer Exercises 6–10.

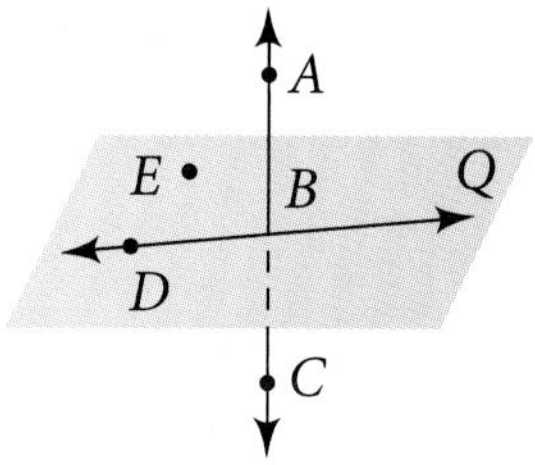

6. Name three collinear points.
7. Name four coplanar points.
8. Name four noncoplanar points.
9. What is the intersection of $\overleftrightarrow{AC}$ and plane Q?
10. How many planes contain each line and point?
    a. $\overleftrightarrow{BD}$ and point A b. $\overleftrightarrow{AB}$ and point C
    c. $\overleftrightarrow{BE}$ and point C d. $\overleftrightarrow{BD}$ and point E

Is the definition of the term in red acceptable? If not, write a good definition.

11. A pencil is a writing instrument.
12. Vertical angles are angles that are congruent.
13. Complementary angles are angles that form a right angle.

Complete with *always, sometimes,* or *never* to make each statement true.

14. $\overrightarrow{LJ}$ and $\overrightarrow{TJ}$ are __?__ opposite rays.
15. Four points are __?__ coplanar.
16. Skew lines are __?__ noncoplanar.
17. Two lines that lie in parallel planes are __?__ parallel.
18. The intersection of two planes is __?__ a point.
19. **Algebra** $JK = 48$. Find the value of x.

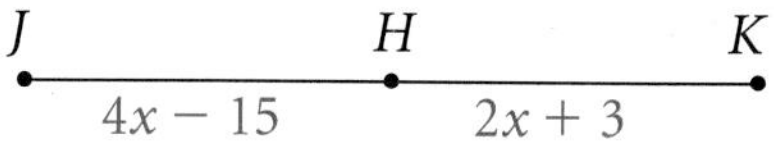

Algebra Use the figure to find the values of the variables in Exercises 20 and 21.

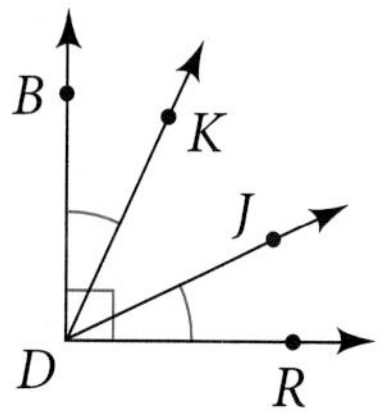

20. $m\angle BDK = 3x + 4$, $m\angle JDR = 5x - 10$
21. $m\angle BDJ = 7y + 2$, $m\angle JDR = 2y + 7$
22. **Writing** Why is it useful to have more than one way of naming an angle?

Name the property of equality or congruence that justifies each statement.

23. If $UV = KL$ and $KL = 6$, then $UV = 6$.
24. If $m\angle 1 + m\angle 2 = m\angle 4 + m\angle 2$, then $m\angle 1 = m\angle 4$.
25. $\angle ABC \cong \angle ABC$
26. If $\frac{1}{2}m\angle D = 45$, then $m\angle D = 90$.
27. If $\angle DEF \cong \angle HJK$, then $\angle HJK \cong \angle DEF$.

Use the figure to complete Exercises 28–32.

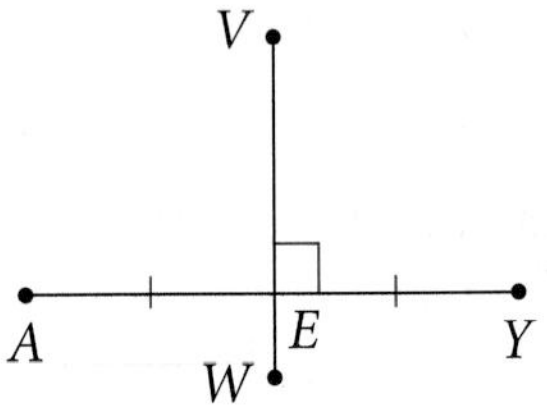

28. $\overline{VW}$ is the __?__ of $\overline{AY}$.

29. $EW + EV =$ ■

30. If $EY = 3.5$, then $AY =$ ■.

31. $\frac{1}{2}$■ $= AE$

32. ■ is the midpoint of ■.

33. Construct a triangle whose sides have the given lengths.

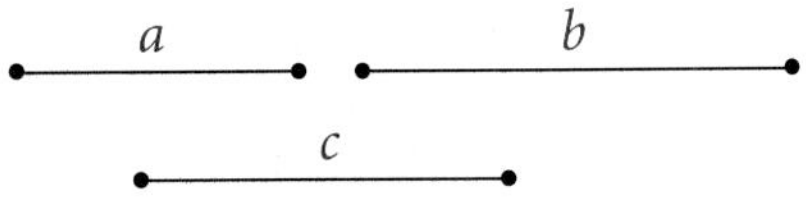

Algebra **Use the figure to complete Exercises 34 and 35. $\overrightarrow{AE}$ bisects $\angle DAC$. Find the values of the variables.**

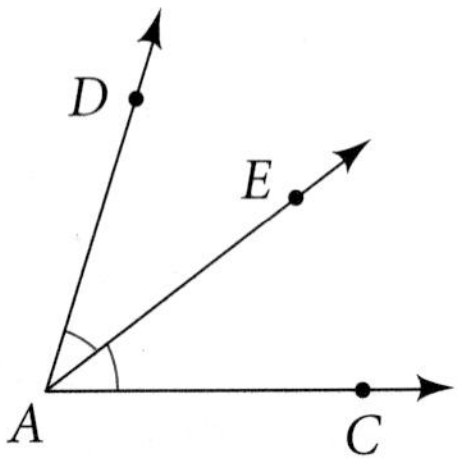

34. $m\angle DAE = 4y + 4$, $m\angle CAE = 6y - 12$

35. $m\angle CAE = 6c - 12$, $m\angle DAC = 72$

36. **Standardized Test Prep** The measure of an angle is $2z$. What is the measure of its supplement?

A. $90 - 2z$
B. 180
C. $2z$
D. $2z - 180$
E. $180 - 2z$

37. Find the measure of each angle.

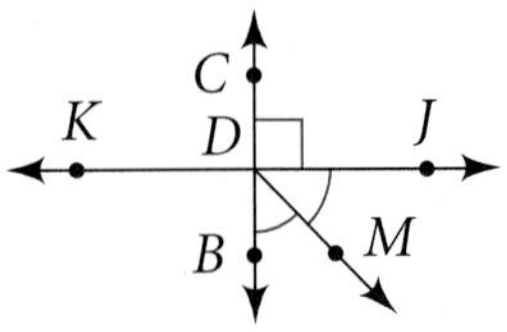

a. $\angle CDM$ **b.** $\angle KDM$
c. $\angle JDK$ **d.** $\angle JDM$
e. $\angle CDB$ **f.** $\angle CDK$

Use the graph to complete Exercises 38–41.

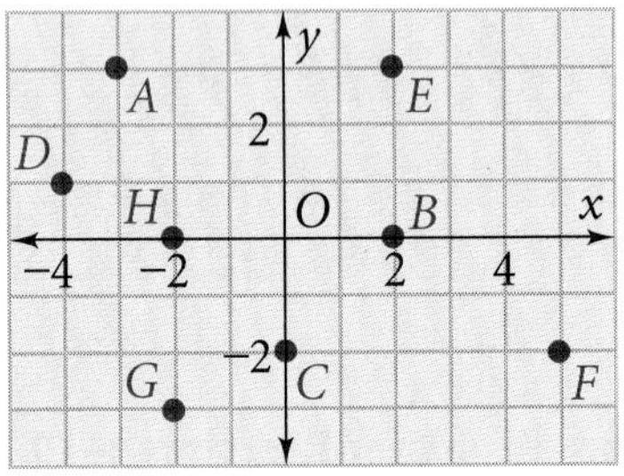

38. Find the coordinates of each labeled point.

39. Find the coordinates of the midpoint of $\overline{EH}$.

40. Find the coordinates of the midpoint of $\overline{AF}$.

41. Find the length of each segment to the nearest tenth.

a. $\overline{AC}$ **b.** $\overline{BH}$
c. $\overline{GB}$ **d.** $\overline{AF}$

What can you conclude from each diagram? Justify your answers.

42.

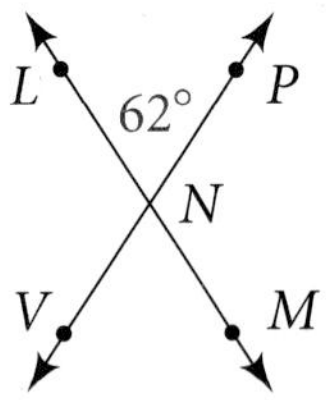

43.

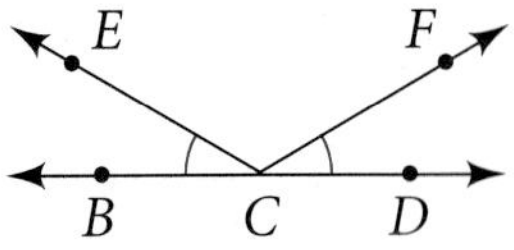

44. The coordinates of the midpoint of $\overline{RS}$ are $(-1, 1)$. The coordinates of R are $(3, 3)$. What are the coordinates of S?

45. Use a protractor to draw a 60° angle. Then construct its bisector.

Preparing for Standardized Tests

For Exercises 1–7, choose the correct letter.

1. The intersection of two planes is
   A. a point. **B.** a ray. **C.** a segment.
   D. a line. **E.** none of the above

2. Estimate $m\angle BCD$.

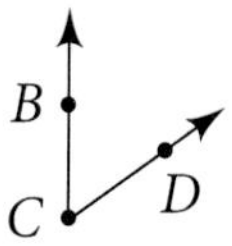

   A. 25 **B.** 55 **C.** 85
   D. 105 **E.** 125

3. Which is the next figure in the sequence?

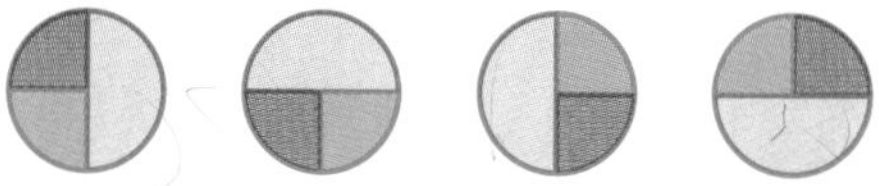

   A. 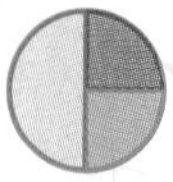 **B.** 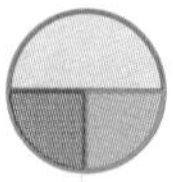 **C.** 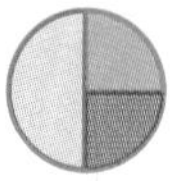

   D. 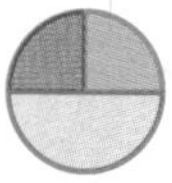 **E.** none of the above

4. Find the values of x and y.

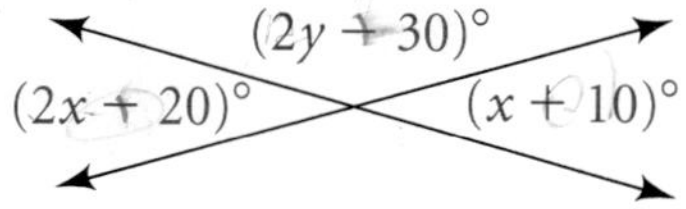

   A. $x = 40, y = 140$ **B.** $x = 10, y = 170$
   C. $x = 30, y = 85$ **D.** $x = 30, y = 90$
   E. none of the above

5. Two right angles can be
   I. vertical. **II.** adjacent.
   III. complementary. **IV.** supplementary.

   A. I and II **B.** II and III **C.** III and IV
   D. I and IV **E.** I, II, and IV

6. Find the distance between $A(4, -5)$ and $B(-2, 1)$ to the nearest tenth.
   A. 8.5 **B.** 7.2 **C.** 6.7 **D.** 6.3 **E.** 4.5

7. Which point lies the farthest from the origin?
   A. $(0, -7)$ **B.** $(-3, 8)$ **C.** $(-4, -3)$
   D. $(5, 1)$ **E.** $(-6, 0)$

For Exercises 8–11, compare the boxed quantity in Column A with the boxed quantity in Column B. Choose the best answer.

A. The quantity in Column A is greater.
B. The quantity in Column B is greater.
C. The two quantities are equal.
D. The relationship cannot be determined on the basis of the information supplied.

$\overline{MV}$ bisects $\overline{NW}$.

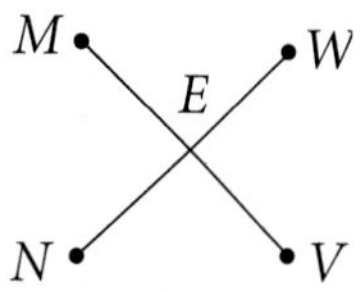

	Column A	Column B
8.	$m\angle MEN$	$m\angle VEW$
9.	$m\angle MEW$	$m\angle VEW$
10.	EW	EN
11.	EM	EV

12. **Open-ended** Write the measures of three noncongruent acute angles. Find the complement and supplement of each.

13. Find the coordinates of the midpoint of $\overline{CD}$ with endpoints $C(5, 7)$ and $D(10, -3)$.

14. **Writing** Explain the difference between inductive and deductive reasoning.

CHAPTER

2 Investigating Geometric Figures

Relating to the Real World

Take a look around you. Chances are that the objects you see are made of simple geometric figures. The door is rectangular; the light fixture is circular; the windowpanes are square. In order to make sense of the world, you need to start with these basic figures. And that's what you'll learn in this chapter—the basic characteristics of the geometric figures that you see every day.

Lessons	2-1	2-2	2-3	2-4	2-5
	Triangles	*Polygons*	*Parallel and Perpendicular Lines in the Coordinate Plane*	*Classifying Quadrilaterals*	*Circles*

CHAPTER PROJECT

AMAZING SPACE

What is it about puzzles that makes them so popular? Is it that they always have a clear solution, in contrast to real-life problems? Or is it the way simple tasks can become complex and complex ones simple? Whatever the reasons, people have been solving puzzles since at least 2400 B.C. when magic squares were popular in China.

In this chapter project, you will explore shape puzzles, such as tangrams, which also originated in China. As you solve puzzles and create your own, you will see how simple shapes can mystify and enlighten—revealing secrets of how our world fits together.

To help you complete the project:

- p. 74 *Find Out by Doing*
- p. 81 *Find Out by Exploring*
- p. 95 *Find Out by Analyzing*
- p. 107 *Find Out by Investigating*
- p. 115 *Find Out by Modeling*
- p. 116 *Finishing the Project*

Congruent and Similar Figures 2-6

Isometric and Orthographic Drawings 2-7

What You'll Learn

- Finding the measures of angles of a triangle
- Classifying triangles

...And Why

To increase your knowledge of triangles, the simplest polygons used in the design of furniture, buildings, and bridges

What You'll Need

- scissors
- protractor

TECHNOLOGY HINT

The Work Together could be done using geometry software.

2-1 Triangles

WORK TOGETHER

Work in a group to explore the angle measures of triangles.

- Have each member of your group draw and cut out a large triangle.
- Number the angles and tear them off.
- Place the three angles adjacent to each other to form one angle.

1. Compare your results with group members. Describe your observations.

2. Make a **conjecture** about the sum of the measures of the angles of a triangle.

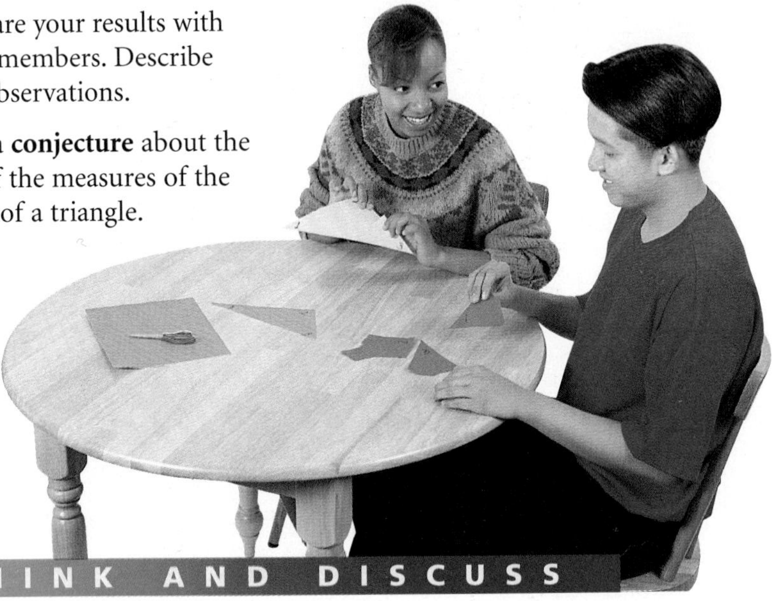

THINK AND DISCUSS

The Triangle Angle-Sum Theorem

The Work Together demonstrates the following theorem that you will prove in Chapter 7.

Theorem 2-1
Triangle Angle-Sum Theorem

The sum of the measures of the angles of a triangle is 180.

$m\angle A + m\angle B + m\angle C = 180$

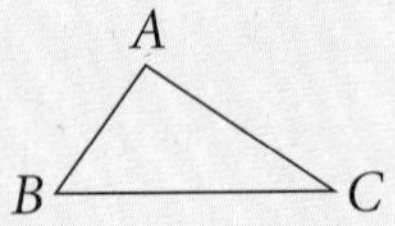

3. Try This Find the measure of each numbered angle.

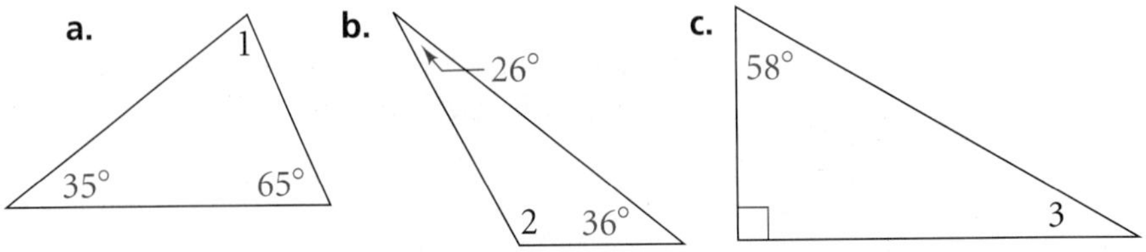

Example 1

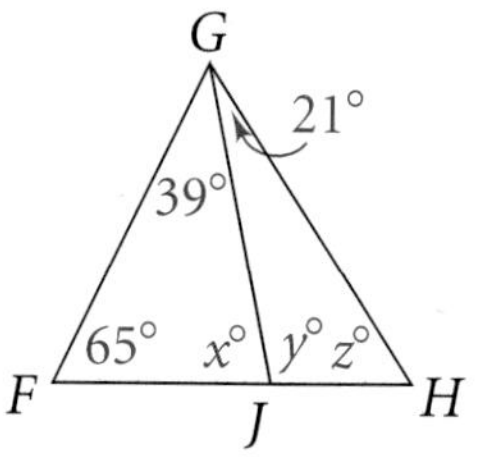

Find the values of x, y, and z.

To find the value of x, use $\triangle FJG$.

$65 + 39 + x = 180$	Triangle Angle-Sum Theorem
$104 + x = 180$	Simplify.
$x = 76$	Subtract 104 from each side.

To find the value of y, look at $\angle FJH$. It is a straight angle.

$m\angle GJF + m\angle GJH = 180$	Angle Addition Postulate
$76 + y = 180$	Substitution
$y = 104$	Subtract 76 from each side.

To find the value of z, use $\triangle GJH$.

$104 + 21 + z = 180$	Triangle Angle-Sum Theorem
$125 + z = 180$	Simplify the left side.
$z = 55$	Subtract 125 from each side.

PROBLEM SOLVING

Look Back Check your answers by finding the sum of the measures of the angles in each triangle.

4. **Critical Thinking** Describe how you could use $\triangle FGH$ instead of $\triangle GJH$ to find the value of z.
5. **Try This** Find the values of x and y.

a.

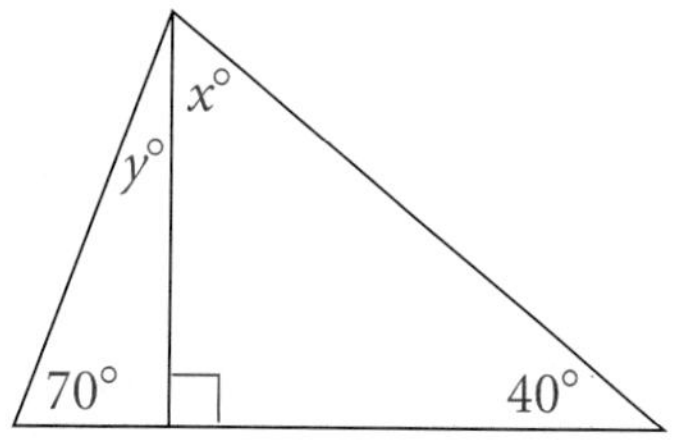

b.

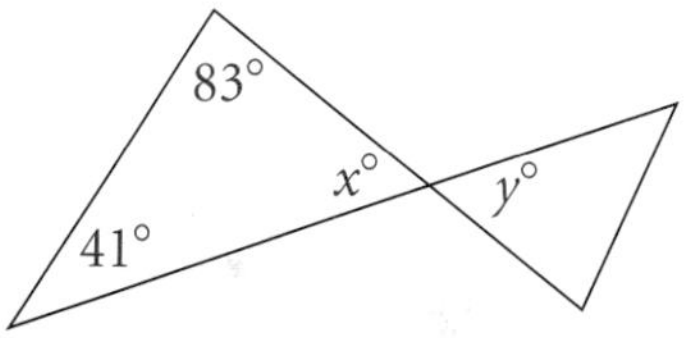

Exterior Angles of a Triangle

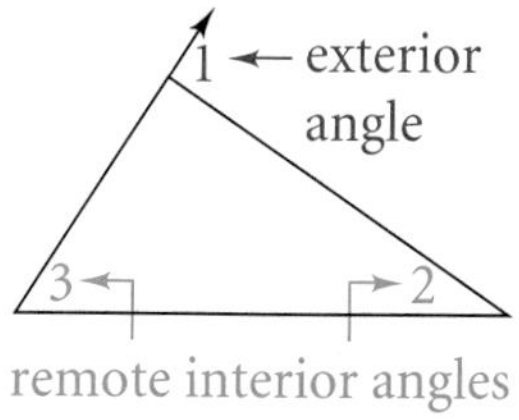

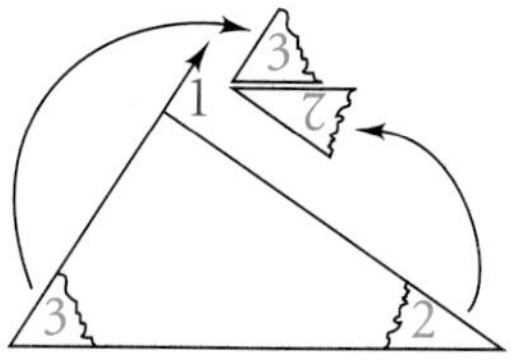

An **exterior angle** of a polygon is an angle formed by a side and an extension of a side. For each exterior angle of a triangle, the two non-adjacent interior angles are called its **remote interior angles.**

6. a. **Manipulatives** Draw and label a large triangle like the one shown at the left. Cut out the two remote interior angles and place them on the exterior angle as shown. What do you observe?
   b. Make a **conjecture** about the measure of an exterior angle of a triangle.

Your **conjecture** in Question 6 is the basis of the following theorem, which you will justify in Exercise 32.

Theorem 2-2
Exterior Angle Theorem

The measure of each exterior angle of a triangle equals the sum of the measures of its two remote interior angles.

$m\angle 1 = m\angle 2 + m\angle 3$

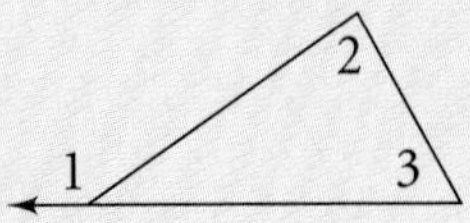

7. Try This Find the measure of each numbered exterior angle.

a.

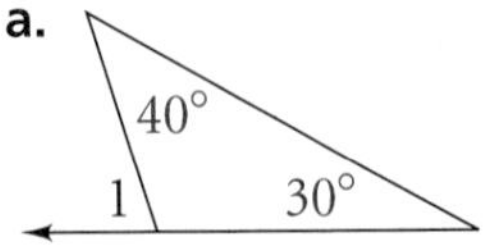

b.

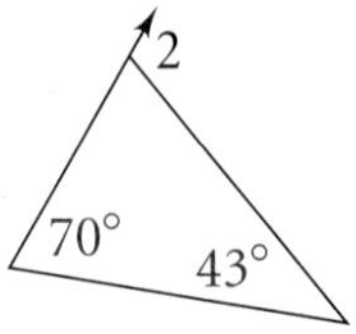

c.

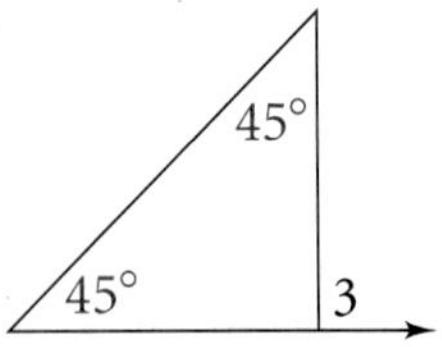

A **corollary** is a statement that follows directly from a theorem. The following statement is a corollary to the Exterior Angle Theorem.

Corollary to Theorem 2-2

The measure of an exterior angle of a triangle is greater than the measure of either of its remote interior angles.

$m\angle 1 > m\angle 2$ and $m\angle 1 > m\angle 3$

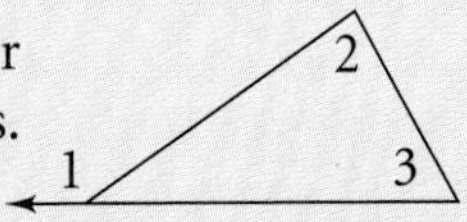

8. Use what you know from the Exterior Angle Theorem to explain why this corollary makes sense.

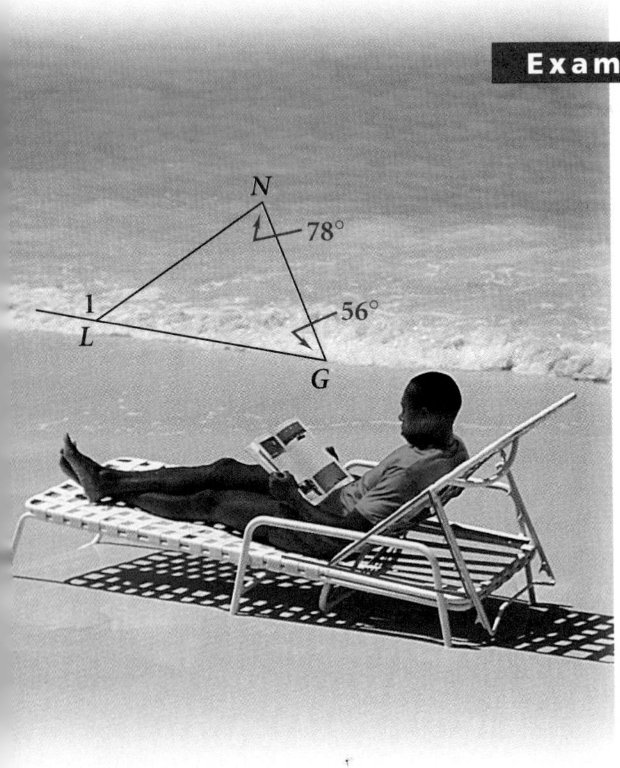

Example 2 ***Relating to the Real World***

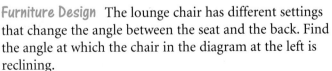

Furniture Design The lounge chair has different settings that change the angle between the seat and the back. Find the angle at which the chair in the diagram at the left is reclining.

$m\angle 1 = m\angle LGN + m\angle GNL$	**Exterior Angle Theorem**
$m\angle 1 = 56 + 78$	**Substitution**
$m\angle 1 = 134$	**Simplify.**

The chair is reclining at a 134° angle.

9. Explain how you could have found $m\angle 1$ without using the Exterior Angle Theorem.

10. Try This You change the setting on the chair so that $m\angle NGL = 90$ and $m\angle GNL = 45$. Find the angle at which the chair is reclining.

Classifying Triangles

In Chapter 1, you classified an angle by its measure. You can classify a triangle by its sides and by its angles.

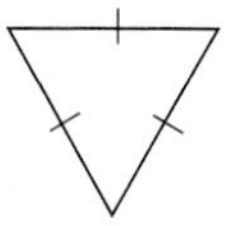

equilateral
all sides congruent

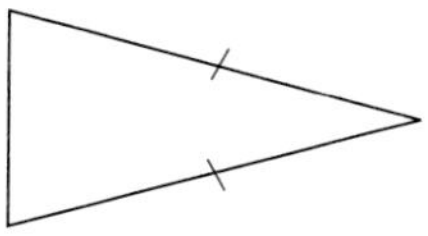

isosceles
at least two sides congruent

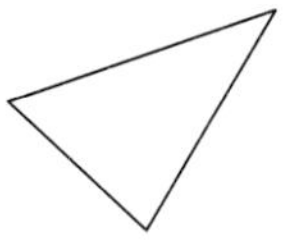

scalene
no sides congruent

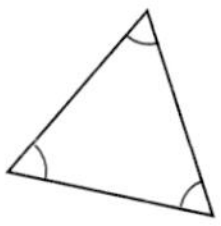

equiangular
all angles congruent

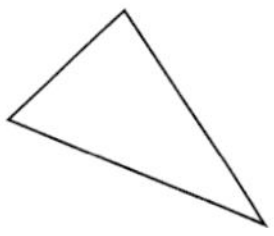

acute
all angles acute

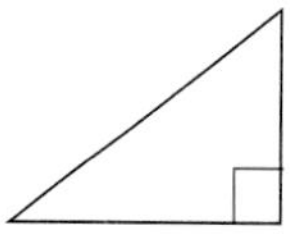

right
one right angle

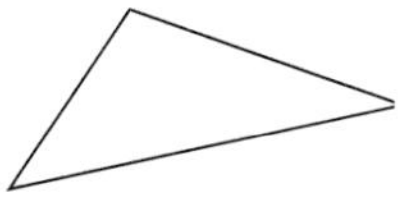

obtuse
one obtuse angle

11. Try This Draw and label a triangle to fit each description. If no triangle can be drawn, write *not possible* and explain.

a. acute scalene
b. right isosceles
c. obtuse equilateral
d. acute isosceles

Exercises ON YOUR OWN

Use a protractor and a centimenter ruler to measure the sides and angles of each triangle. Classify each triangle by its sides and angles.

1.

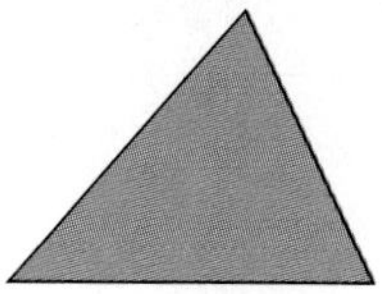

2.

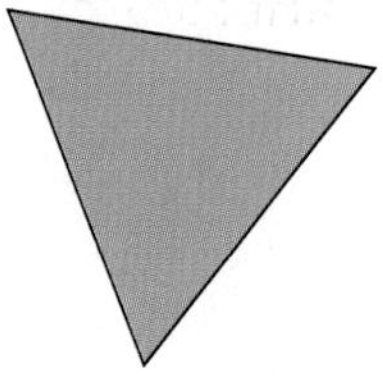

3.

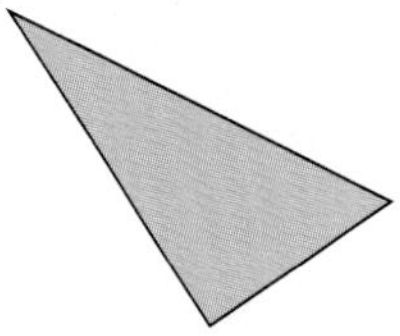

4.

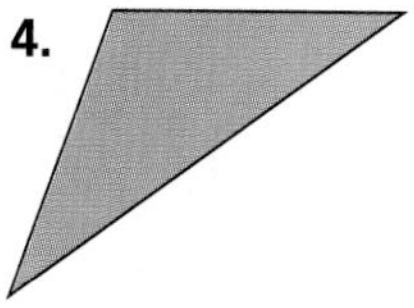

Sketch a triangle to fit each description. If no triangle can be drawn, write *not possible* and explain.

5. obtuse scalene
6. right equilateral
7. acute equilateral
8. obtuse isosceles

9. a. Logical Reasoning What is the measure of each angle of an equiangular triangle? **Justify** your reasoning.

b. What is the sum of the measures of the acute angles of a right triangle? **Justify** your reasoning.

Choose **Use paper and pencil, mental math, or a calculator to find the values of the variables.**

10.
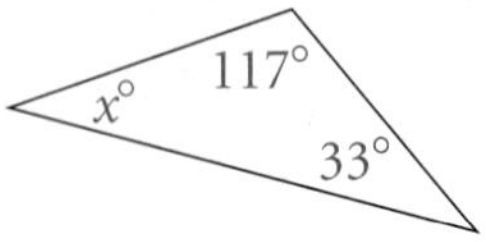

11.
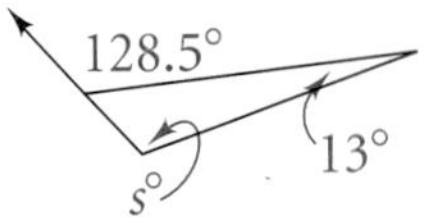

12.
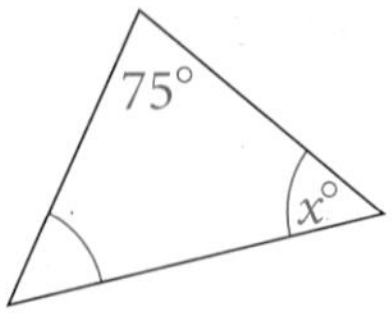

13.
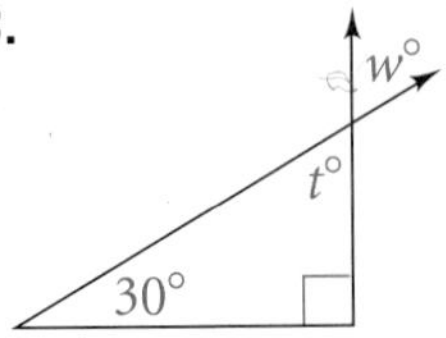

14.
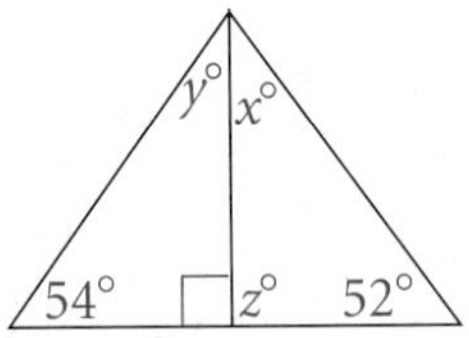

15.
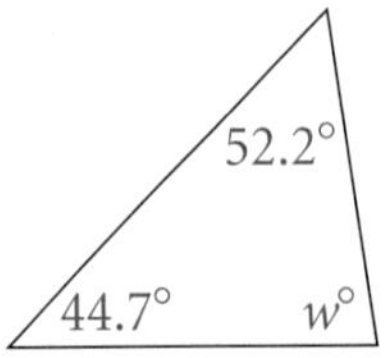

16.
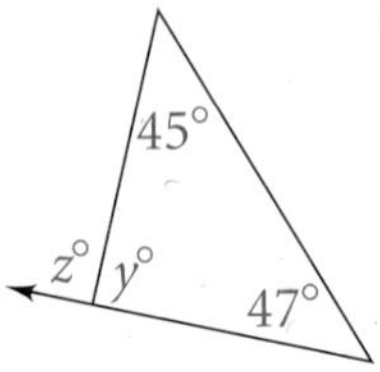

17.
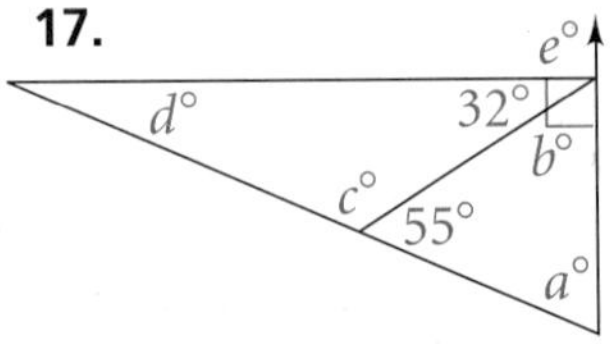

Use the figure at the right for Exercises 18–20.

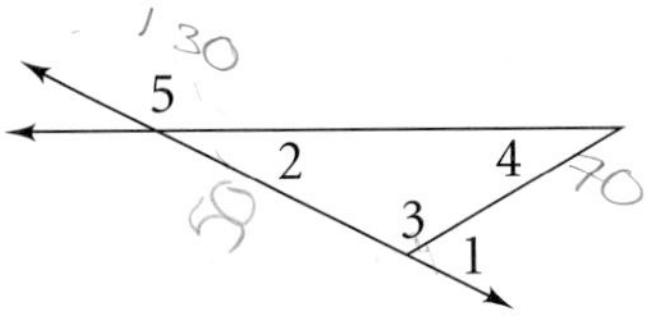

18. Find $m\angle 3$ if $m\angle 5 = 130$ and $m\angle 4 = 70$.

19. Find $m\angle 1$ if $m\angle 5 = 142$ and $m\angle 4 = 65$.

20. Find $m\angle 2$ if $m\angle 3 = 125$ and $m\angle 4 = 23$.

21. **Standardized Test Prep** The measures of the angles of a triangle are shown below. In which case is x *not* an integer?
A. $x, 2x, 3x$ **B.** $x, 3x, 5x$ **C.** $x, 3x, 4x$ **D.** $x, 4x, 7x$ **E.** $2x, 3x, 4x$

22. The measure of one angle of a triangle is 115. The other two angles are congruent. Find their measures.

23. **Writing** Is every equilateral triangle isosceles? Is every isosceles triangle equilateral? Explain.

24. **Algebra** A right triangle has acute angles whose measures are in the ratio 1 : 2. Find the measures of these angles. (*Hint:* Let x and $2x$ represent the angle measures.)

25. **Music** The top of a grand piano is held open by props of varying lengths, depending upon the desired volume of the music. The longest prop makes an angle of 57° with the piano. What is the angle of opening between the piano and its top?

26. **Critical Thinking** Rosa makes the following claim: The ratio of the lengths of the three sides of a triangle equals the ratio of the three angle measures. Give examples to support her claim or one counterexample to disprove it.

Algebra **Find the measures of the angles of each triangle. Classify each triangle by its angles.**

27.

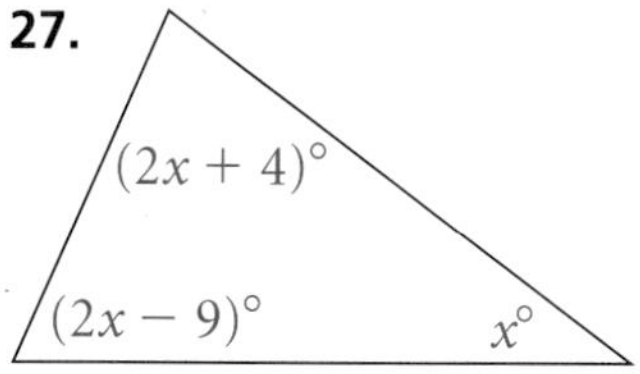

28.

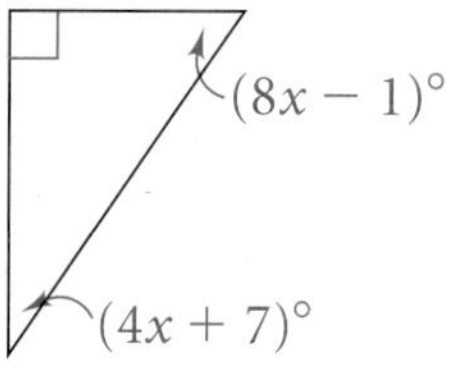

29. **Geometry on a Sphere** Suppose you were measuring the angles of a triangle on a globe. The meridians of longitude pass through both poles and are perpendicular to the equator. Will the sum of the measures of the angles of this triangle be equal to, greater than, or less than 180? Explain.

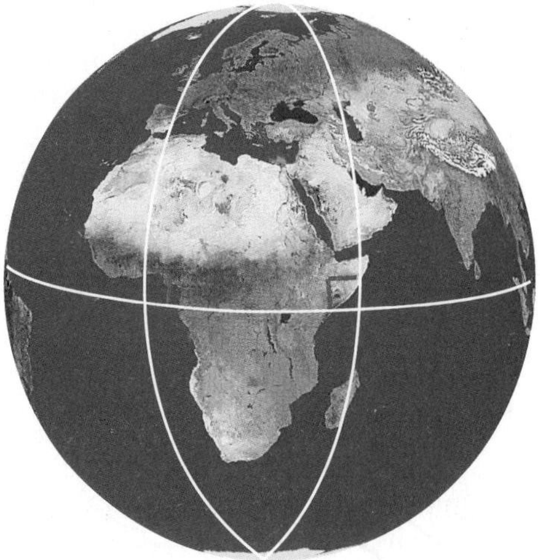

30. **a.** **Algebra** The ratio of the angle measures in $\triangle BCR$ is 2 : 3 : 4. Find the angle measures.

b. What type of triangle is $\triangle BCR$?

31. **a.** **Probability** Find the probability that a triangle is equiangular if the measure of each of its angles is a multiple of 30.

b. Find the probability that a triangle is equiangular if the measure of each of its angles is a multiple of 20.

PROBLEM SOLVING HINT

Make a table showing all possibilities.

32. **Logical Reasoning** Complete the following statements to **justify** the Exterior Angle Theorem.

a. By the Angle Addition Postulate, $m\angle 1 + m\angle 4 = \blacksquare$.

b. By the Triangle Angle-Sum Theorem, $m\angle 2 + m\angle 3 + m\angle 4 = \blacksquare$.

c. By the __?__ Property, $m\angle 1 + m\angle 4 = m\angle 2 + m\angle 3 + m\angle 4$.

d. By the __?__ Property of Equality, $m\angle 1 = m\angle 2 + m\angle 3$.

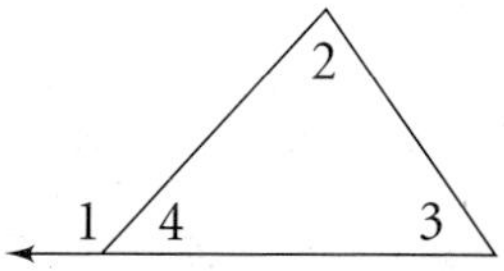

33. The measures of the angles of $\triangle RST$ are $\sqrt{x}$, $2\sqrt{x}$, and $3\sqrt{x}$.

a. Find the value of x and the measures of the angles.

b. What type of triangle is $\triangle RST$?

34. **Weaving** Patricia Tsinnie, a Navajo weaver, often uses isosceles triangles in her designs.

a. Trace the design shown below. Then use a colored pencil to outline isosceles triangles used in the design.

b. **Open-ended** Make a repeated design of your own that uses isosceles triangles.

Chapter Project ***Find Out by Doing***

The *tangram,* known in China as the *ch'i-ch'iao t'u,* meaning "ingenious seven-piece plan," is one of the oldest manipulative puzzles. You can use paper folding to make your own tangram.

- Fold a square sheet of paper in half four times and then unfold it. Draw the segments shown to form seven tangram pieces, called *tans.*

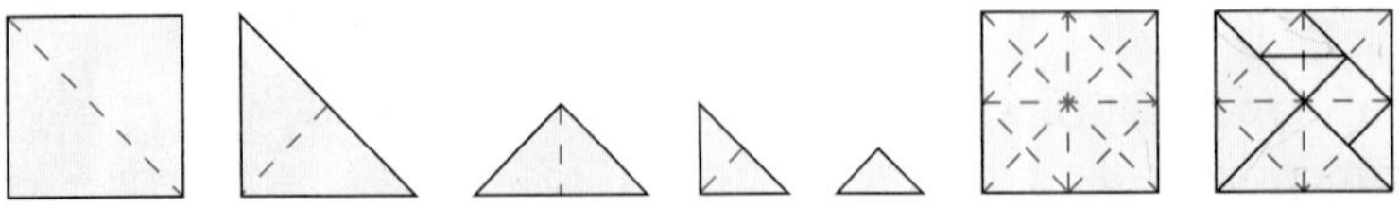

- Cut out the seven tans.

How many of the tans are triangles? Form other triangles by placing tans together. Make a sketch of each. Classify each triangle by its sides and angles. Can you make one triangle using all seven tans?

Exercises MIXED REVIEW

Draw the next figure in each sequence.

35.

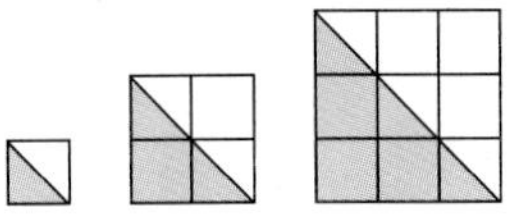

36.

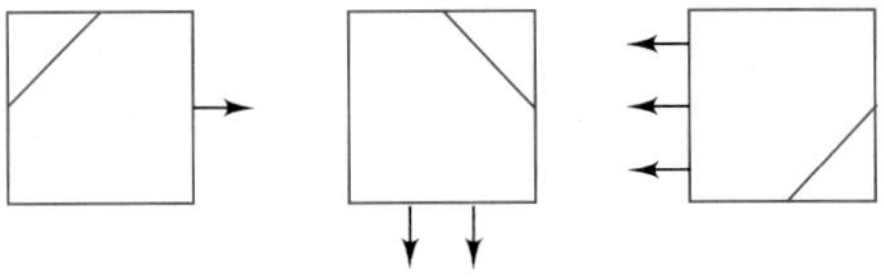

Find the length of $\overline{WZ}$ to the nearest tenth.

37. $W(8, -2)$ and $Z(2, 6)$

38. $W(-4.5, 1.2)$ and $Z(3.5, -2.8)$

39. In the figure, $m\angle AOB = 3x + 20$, $m\angle BOC = x + 32$, and $m\angle AOC = 80$. Find the value of x.

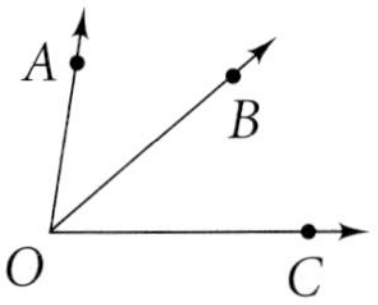

Getting Ready for Lesson 2-2

Find (a) the measure of each angle of quadrilateral *ABCD,* and (b) the sum of the measures of the angles of quadrilateral *ABCD.*

40.

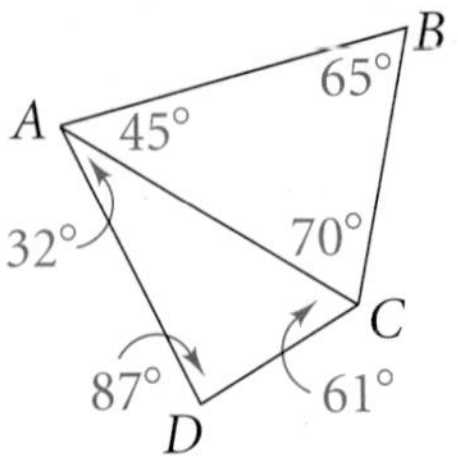

41.

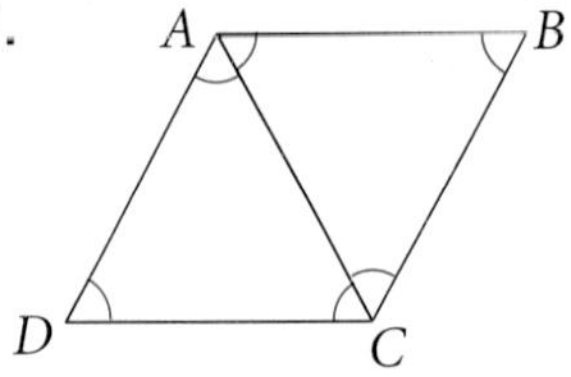

42.

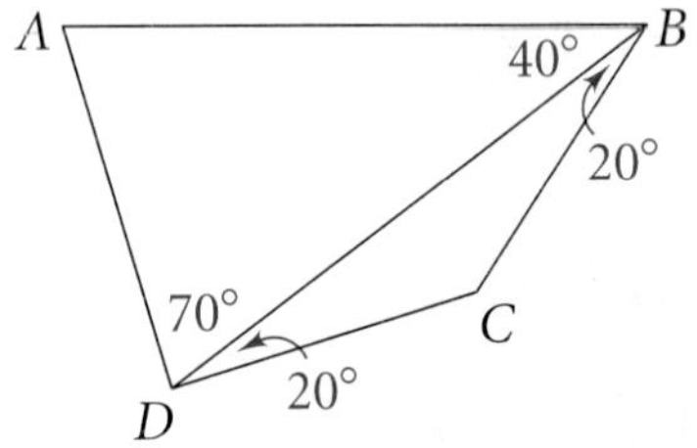

Exploring the Exterior Angles of a Polygon

With Lesson 2-2

Work in pairs or small groups.

Construct

Use geometry software. Construct a convex polygon. Extend each side as shown. To measure the exterior angles you will need to mark a point on each ray.

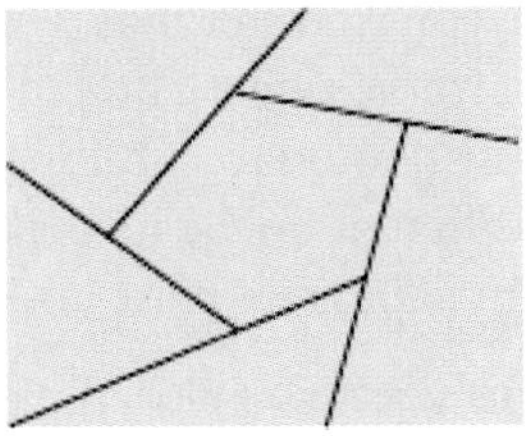

Investigate

- Measure each exterior angle.
- Calculate the sum of the measures of the exterior angles.
- Manipulate the polygon, making sure it remains convex. Observe the sum of the measures of the exterior angles.

Conjecture

- Write a conjecture about the sum of the measures of the exterior angles (one at each vertex) of a convex polygon.
- Test your conjecture with another polygon.

Extend

The figures below show a polygon that is decreasing in size until finally it "disappears." Describe how these figures could be used as a justification for your conjecture.

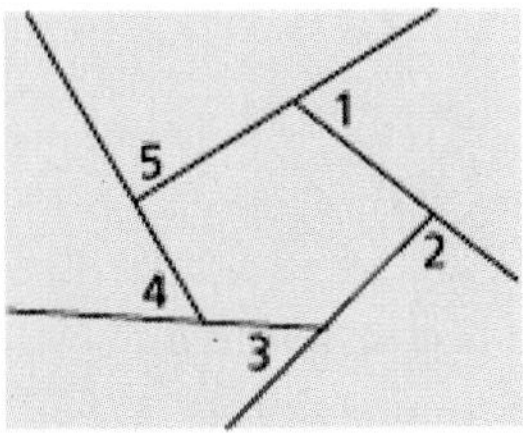

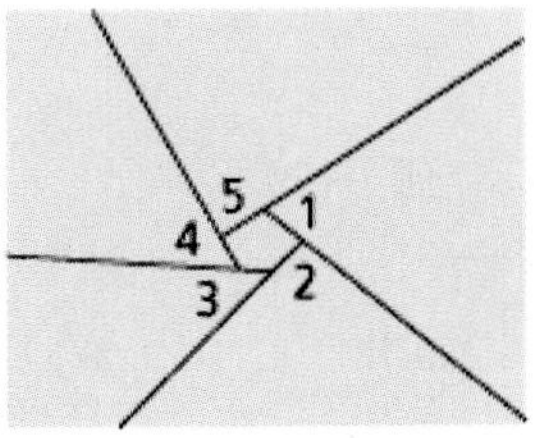

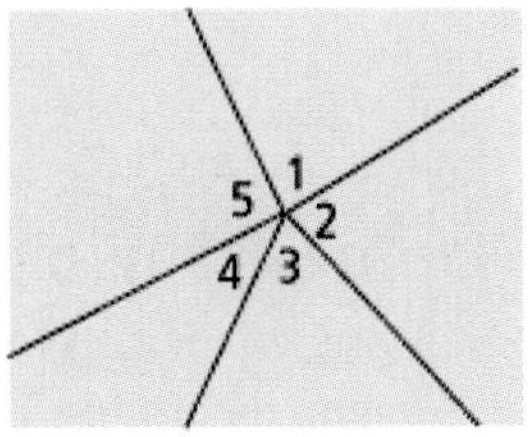

2-2 Polygons

What You'll Learn

- Classifying polygons
- Finding the sum of the measures of the interior and exterior angles of polygons

...And Why

To increase your understanding of polygons, which are found in art and nature as well as in manufactured products

THINK AND DISCUSS

Polygons and Interior Angles

Walking around a city, you can see polygons in buildings, windows, and traffic signs. The grillwork in the photo is a combination of different polygons that form a pleasing pattern.

A **polygon** is a closed plane figure with at least three sides. The sides intersect only at their endpoints and no adjacent sides are collinear. To identify a polygon, start at any vertex and list the other vertices consecutively.

Polygon	Number of Sides
triangle	3
quadrilateral	4
pentagon	5
hexagon	6
heptagon	7
octagon	8
nonagon	9
decagon	10
dodecagon	12
n-gon	n

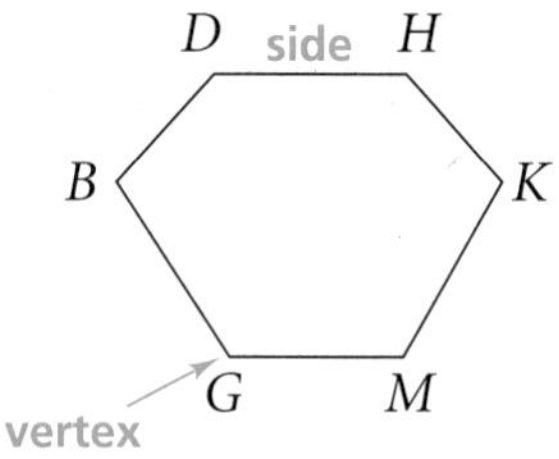

polygon $DHKMGB$

sides: $\overline{DH}, \overline{HK}, \overline{KM}, \overline{MG}, \overline{GB}, \overline{BD}$

vertices: D, H, K, M, G, B

angles: $\angle D, \angle H, \angle K, \angle M, \angle G, \angle B$

You can classify polygons by the number of sides. The most common polygons are listed at the left.

A polygon is **convex** if no diagonal contains points outside the polygon.

A polygon is **concave** if a diagonal contains points outside the polygon.

QUICK REVIEW

A diagonal of a polygon is a segment that connects two nonconsecutive vertices.

In this textbook, the term *polygon* refers to a convex polygon unless otherwise stated.

1. **Open-ended** Draw a convex and a concave octagon. Draw diagonals and explain why one octagon is convex and the other is concave.

WORK TOGETHER

You can use triangles and the Triangle Angle-Sum Theorem to find the measures of the interior angles of a polygon. Work with a partner. Record your data in a table like the one shown below.

- Sketch polygons with 4, 5, 6, 7, and 8 sides.
- Divide each polygon into triangles by drawing all the diagonals from one vertex.
- Multiply the number of triangles by 180 to find the sum of the measures of the interior angles of each polygon.

2. *Inductive Reasoning* Look for a pattern in the table. Write a rule for finding the sum of the measures of the interior angles of a polygon with n sides.

Polygon	Number of Sides	Number of Triangles Formed	Sum of the Interior Angle Measures
	4	■	■ • 180 = ■

THINK AND DISCUSS

The results of the Work Together suggest the following theorem.

Theorem 2-3
Polygon Interior Angle-Sum Theorem

The sum of the measures of the interior angles of an n-gon is $(n - 2)180$.

3. **Try This** Find the sum of the measures of the interior angles of each polygon.

a. 15-gon **b.** 20-gon
c. decagon **d.** dodecagon

Example 1

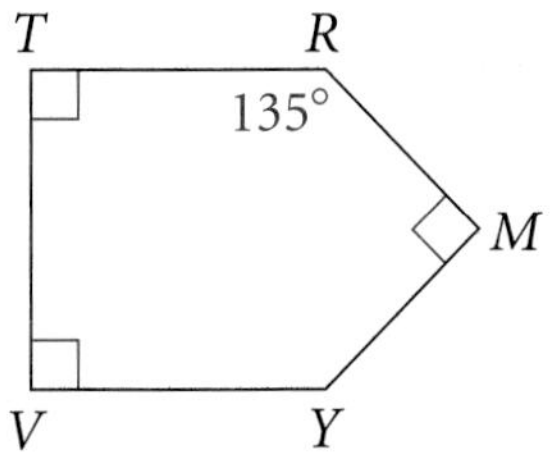

Find $m\angle Y$ in $TVYMR$.

Use the Polygon Interior Angle-Sum Theorem for $n = 5$.

$$m\angle T + m\angle V + m\angle Y + m\angle M + m\angle R = (5 - 2)180$$
$$90 + 90 + m\angle Y + 90 + 135 = 540$$
$$m\angle Y + 405 = 540$$
$$m\angle Y = 135$$

Exterior Angles

Technology The figures below show one exterior angle drawn at each vertex of each polygon.

4. a. Find the sum of the measures of the exterior angles (one at each vertex) of each polygon.

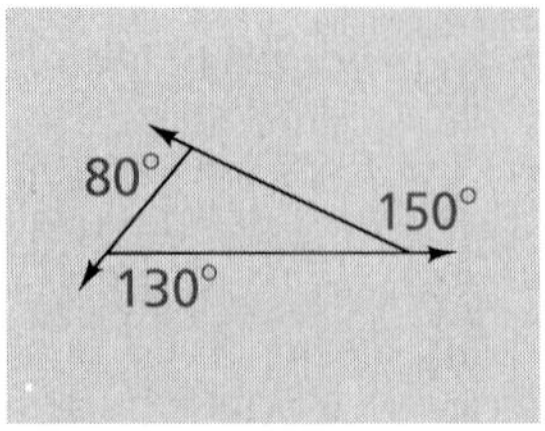

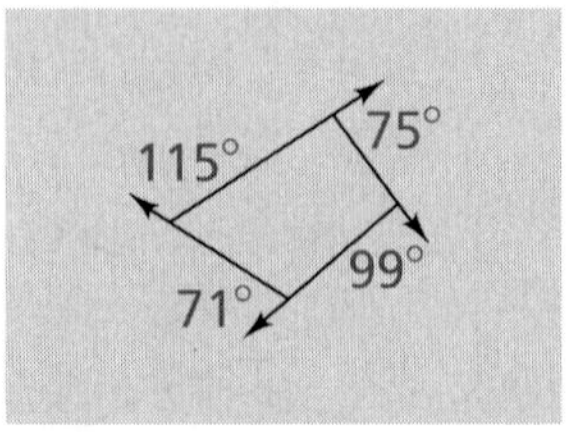

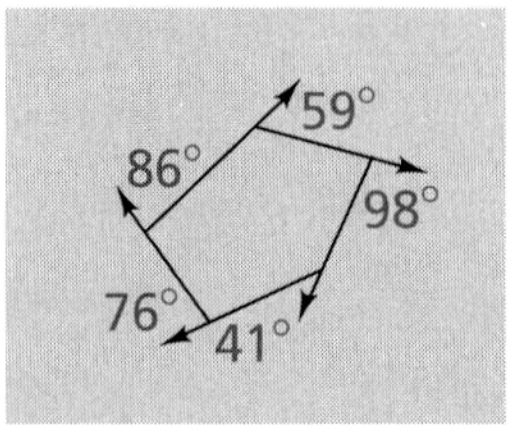

b. Make a conjecture about the sum of the measures of the exterior angles (one at each vertex) of a polygon.

This inductive reasoning leads to the following theorem.

Theorem 2-4
Polygon Exterior Angle-Sum Theorem

The sum of the measures of the exterior angles of a polygon, one at each vertex, is 360.

An **equilateral polygon** has all sides congruent. An **equiangular polygon** has all angles congruent. A **regular polygon** is equilateral and equiangular.

Regular Octagon

Example 2

Find the measure of an interior angle and an exterior angle of a regular octagon.

Method 1: Find the measure of an interior angle first.

Sum of the measures of the interior angles = $(8 - 2)180 = 1080$

- Measure of one interior angle $= \frac{1080}{8} = 135$
- Measure of its adjacent exterior angle $= 180 - 135 = 45$

Method 2: Find the measure of an exterior angle first.

Sum of the measures of the exterior angles = 360

- Measure of one exterior angle $= \frac{360}{8} = 45$
- Measure of its adjacent interior angle = $180 - 45 = 135$

5. Try This Find the measure of an interior angle and an exterior angle of a regular dodecagon.

Example 3 Relating to the Real World

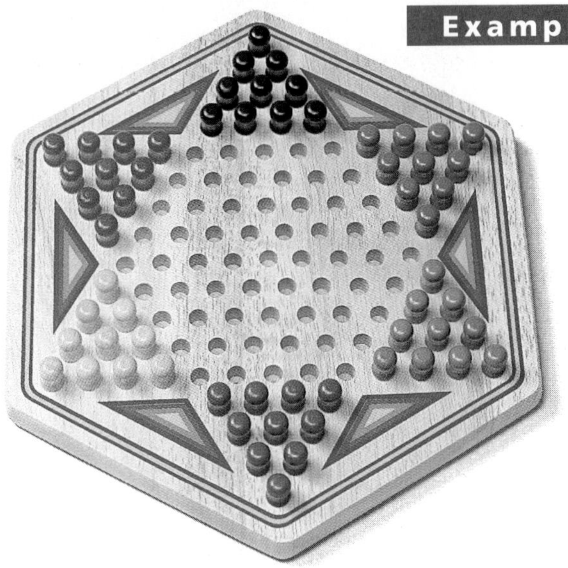

Manufacturing Mindco, a toy manufacturer, is packaging a Chinese checkers game. The game consists of colored pegs and a regular hexagonal wooden board. Mindco is packaging the game in a rectangular box using four right triangles made of foam. Find the measures of the acute angles of each foam triangle.

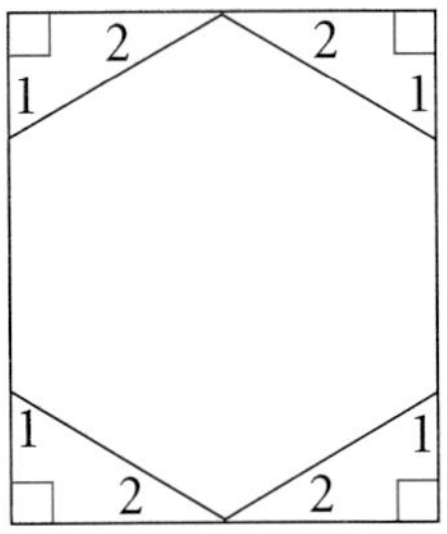

$\angle 1$ is an exterior angle of a regular hexagon.

Find $m\angle 1$ and then use the Triangle Angle-Sum Theorem to find $m\angle 2$.

$6 \cdot m\angle 1 = 360$	Sum of the measures of the exterior ⦣s of a polygon = 360.
$m\angle 1 = 60$	Divide each side by 6.
$m\angle 1 + m\angle 2 + 90 = 180$	Triangle Angle-Sum Thm.
$60 + m\angle 2 + 90 = 180$	Substitution
$m\angle 2 + 150 = 180$	Simplify.
$m\angle 2 = 30$	Subtract 150 from each side.

The measures of the acute angles of each foam triangle are 60 and 30.

Exercises ON YOUR OWN

Classify each polygon by its number of sides. Identify which polygons are convex and which are concave.

1. 

2. 

3. 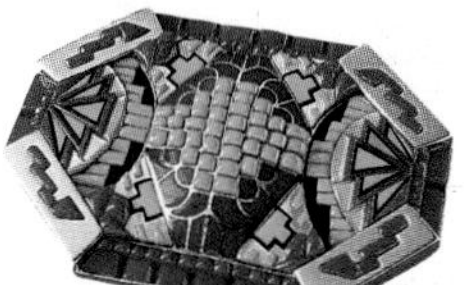

4. 

Draw each regular polygon.

Sample: dodecagon

Draw a circle. Use a protractor to locate 12 points equidistant around a circle. (These points will be located every 30 degrees around a circle, since $360° \div 12 = 30°$.) Connect these points to form a regular dodecagon.

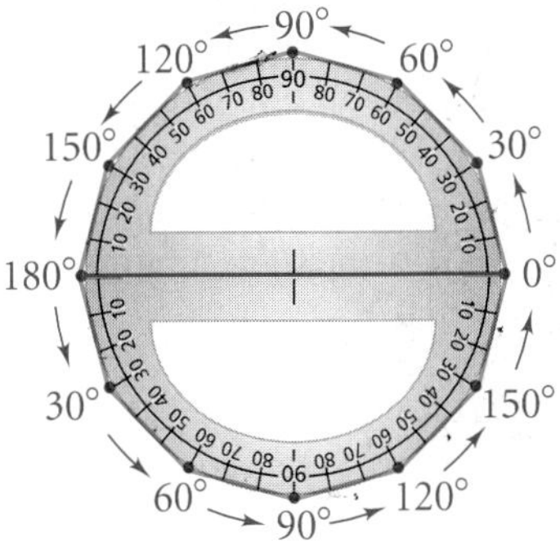

5. triangle

6. quadrilateral

7. pentagon

8. hexagon

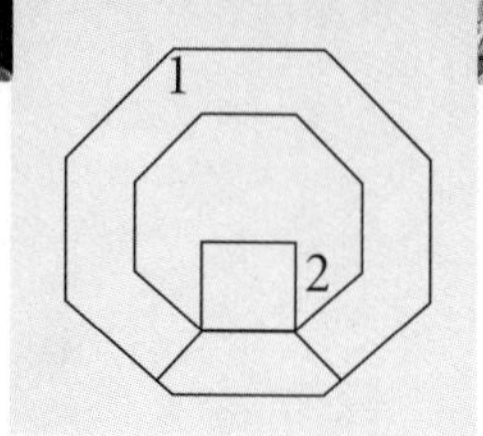

9. **Design** A theater-in-the-round is constructed so that the audience surrounds the stage. Such theaters are not always circular in shape, however. Classify the theater-in-the-round shown in the diagram below the photo, by the number of sides. Find the measure of each numbered angle.

Find the measure of an interior angle and an exterior angle for each regular polygon.

10. pentagon **11.** nonagon **12.** 18-gon **13.** y-gon

The measure of an exterior angle of a regular polygon is given. Find the number of sides.

14. 72 **15.** 36 **16.** 18 **17.** x

18. **Writing** Keach said that he drew a regular polygon and measured one of its interior angles. He got 130°. Explain to him why this is impossible.

Choose **Use pencil and paper, mental math, or a calculator to find the values of the variables.**

19.

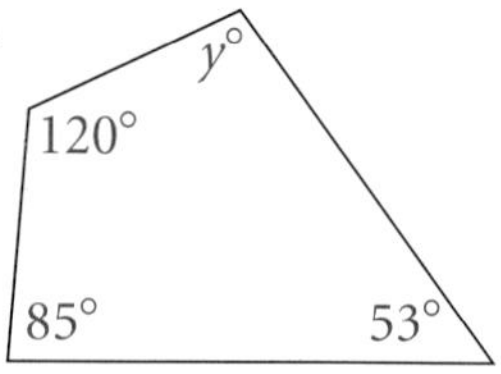

20.

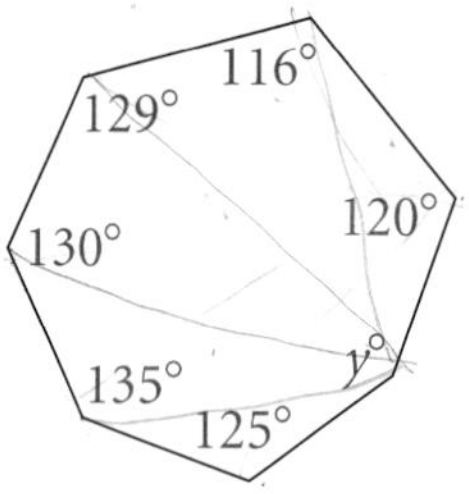

21.

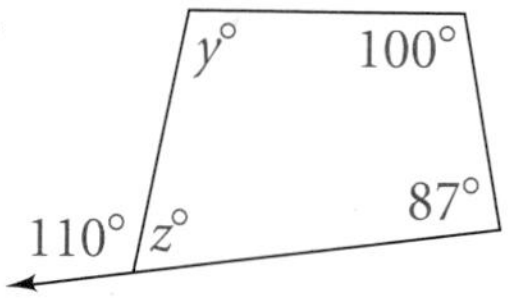

22.

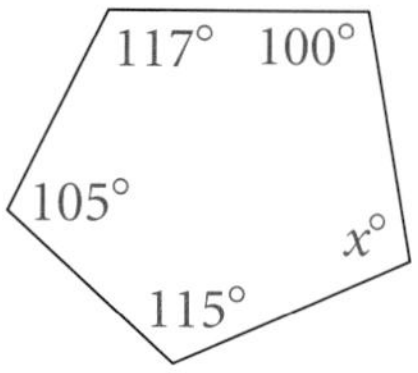

23.

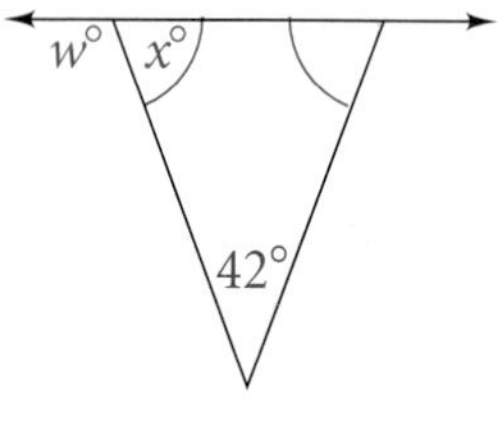

24.

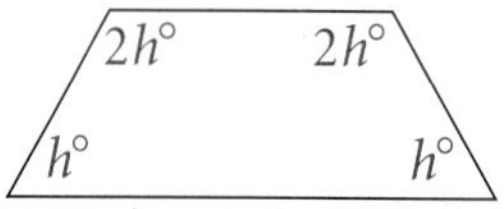

25.

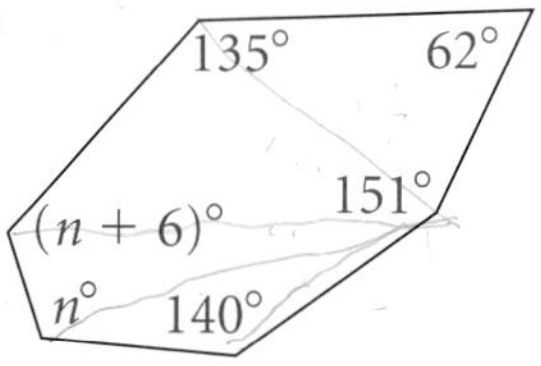

26.

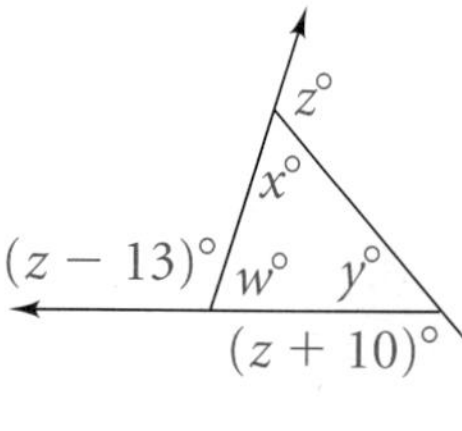

27. a. **Calculator** Find the measure of an interior angle of a regular n-gon for $n = 20, 40, 60, 80, \ldots 200$. Record your results as ordered pairs (n, measure of each interior angle).
   b. Show your results on a graph like the one at the right.
   c. **Data Analysis** What predictions would you make about the measure of an interior angle of a regular 1000-gon?
   d. Is there a regular n-gon with an interior angle of 180°? Explain.

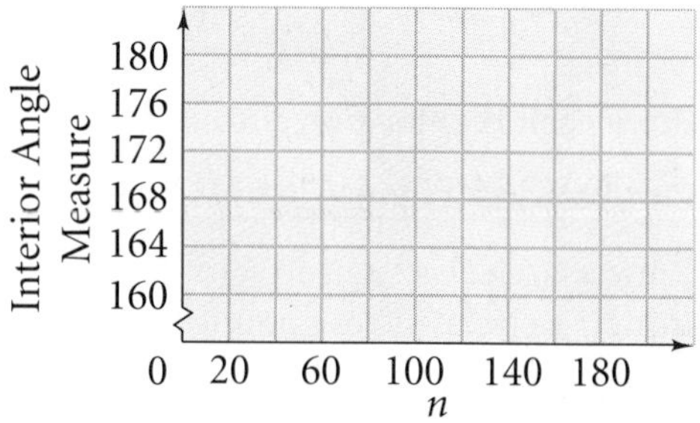

28. **Probability** Find the probability that the measure of an interior angle of a regular n-gon is a positive integer if n is an integer and $3 \leq n \leq 12$.

29. a. *Open-ended* Sketch a quadrilateral that is not equiangular.
b. Sketch an equiangular quadrilateral that is not regular.

30. *Critical Thinking* Laura suggests another way to find the sum of the measures of the interior angles of an n-gon. She picks an interior point of the figure, draws segments to each vertex, counts the number of triangles, multiplies by 180, then subtracts 360. Does her method work? Explain.

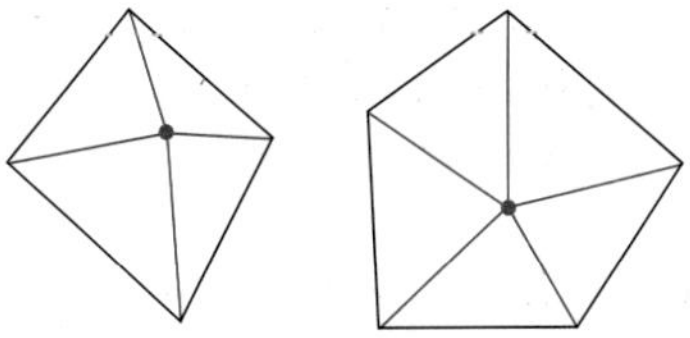

Chapter Project ***Find Out by Exploring***

In 1942, two mathematicians at the University of Chekiang in China proved that only 13 convex polygons could be formed by using all 7 tans. They were able to form 1 triangle, 6 quadrilaterals, 2 pentagons, and 4 hexagons. Try to make these using your set of tans. Make a sketch of each figure.

Exercises MIXED REVIEW

Identify the following.

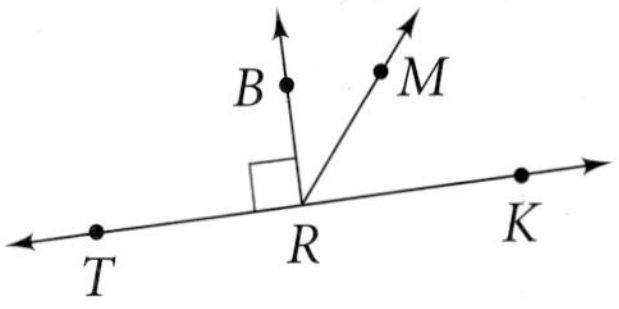

31. a pair of opposite rays

32. two right angles

33. a pair of supplementary angles

34. a pair of complementary angles

Choose **Use paper and pencil, mental math, or a calculator to find the values of the variables.**

35. 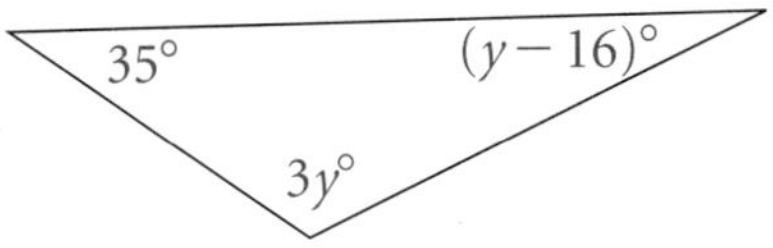

36. 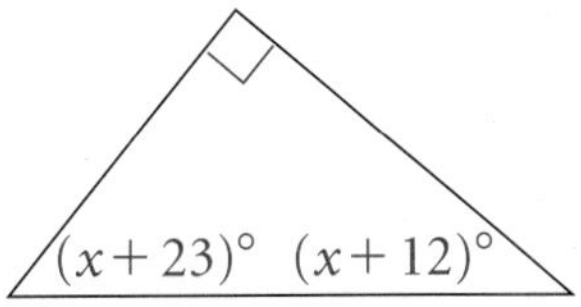

37. 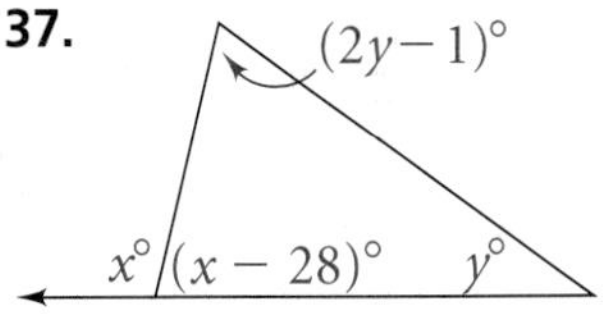

38. The measure of an angle is one-third the measure of its supplement. Find the measures of the angles.

39. The measure of an angle is four times the measure of its complement. Find the measures of the angles.

Getting Ready for Lesson 2-3

Find $\frac{y_2 - y_1}{x_2 - x_1}$.

40. $(x_1, y_1) = (3, 5)$ and $(x_2, y_2) = (1, 4)$

41. $(x_1, y_1) = (-2, 6)$ and $(x_2, y_2) = (3, 1)$

42. $(x_1, y_1) = (1, -8)$ and $(x_2, y_2) = (1, 2)$

43. $(x_1, y_1) = (-5, 3)$ and $(x_2, y_2) = (1, 3)$

Math ToolboX

Technology

Exploring Equations of Lines

With Lesson 2-3

Work in pairs or small groups. To make sure that the viewing grid on your graphing calculator screen is square, press ZOOM 5.

Investigate

- Use your graphing calculator to graph the lines $y = x$, $y = 2x$, and $y = 3x$. Experiment with other equations of lines in the form $y = mx$. Substitute fractions and negative numbers for m. How does the value of m affect the graph of a line?
- Graph the lines $y = x + 1$, $y = x + 2$, and $y = x + 3$. Experiment with other equations of lines in the form $y = x + b$. Substitute fractions and negative numbers for b. How does the value of b affect the graph of a line?

Conjecture

When you graph an equation in the form $y = mx + b$, how do the values of m and b affect the graph of the line? List all your **conjectures.**

Extend

- Graph the lines $y = -\frac{2}{3}x$, $y = -\frac{2}{3}x + 3$, and $y = -\frac{2}{3}x - 2$. What appears to be true of these lines? How are the values of m related? Make a **conjecture** about the graphs of equations of lines where the values of m are the same. Test your **conjecture** by graphing several equations of lines where the values of m are the same.
- Graph each pair of lines:

  a. $y = \frac{1}{2}x$, $y = -2x$

  b. $y = \frac{3}{4}x$, $y = -\frac{4}{3}x$

  c. $y = 5x$, $y = -\frac{1}{5}x$

  What appears to be true of these lines? For each pair, how are the values of m related? Make a **conjecture** about the graphs of equations of lines where the product of the values of m is -1. Test your **conjecture** by graphing several pairs of equations of lines where the product of the values of m is -1.

What You'll Learn

- Graphing lines in the coordinate plane
- Recognizing parallel and perpendicular lines by their slopes

...And Why

To familiarize yourself with parallel and perpendicular lines, which are essential to the construction of houses, furniture, and machinery

What You'll Need

graph paper, ruler

2-3 Parallel and Perpendicular Lines in the Coordinate Plane

THINK AND DISCUSS

Slope and Graphing Lines

In everyday life the word *slope* refers to the steepness of a mountain, the grade of a road, or the pitch of a roof. In algebra, the *slope* of a line is the ratio of the vertical change to the horizontal change between any two points on a line.

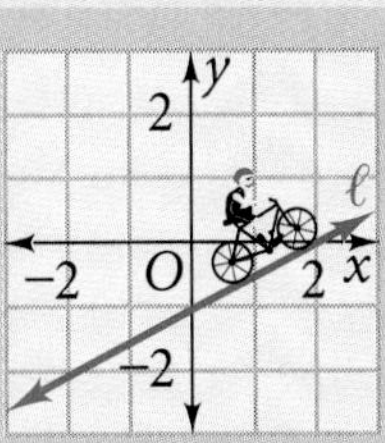

positive slope

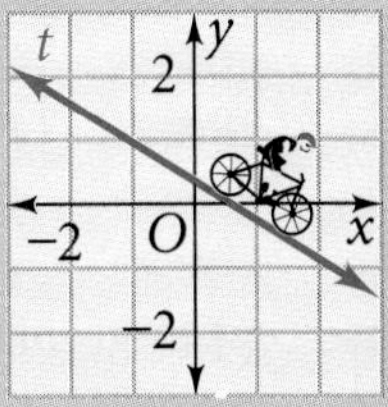

negative slope

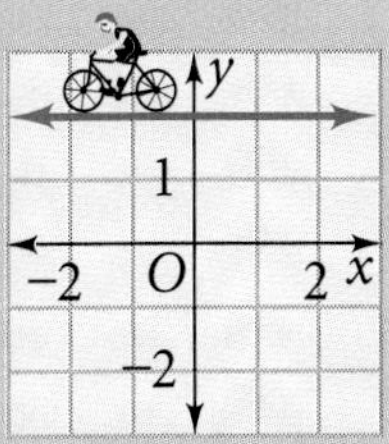

zero slope

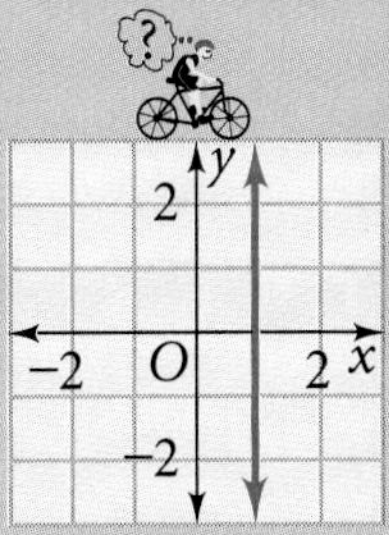

undefined slope

QUICK REVIEW

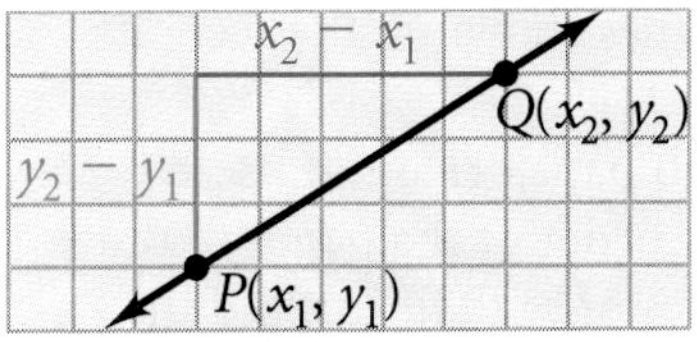

$$\text{Slope} = \frac{\text{vertical change (rise)}}{\text{horizontal change (run)}}$$

$$m = \frac{y_2 - y_1}{x_2 - x_1}$$

1. a. Use $\frac{\text{vertical change}}{\text{horizontal change}}$ to find the slopes of line ℓ and line t shown above.
   b. Explain why the slope of a horizontal line is zero.
   c. Explain why the slope of a vertical line is undefined.

2. **Try This** Use these pairs of points to answer parts (a)–(c).

   $R(-3, -4)$, $S(5, -4)$ $\quad$ $C(-2, 2)$, $D(4, -2)$
   $K(-3, 3)$, $T(-3, 1)$ $\quad$ $P(3, 0)$, $Y(0, -5)$

   a. Graph and label a line that contains each pair of points.
   b. **Mental Math** Decide whether the slope of each line is positive, negative, zero, or undefined.
   c. Find the slope of each line.

You can graph a line by starting at a given point and using the slope of the line to plot another point.

Quick Review

The slope-intercept form of a linear equation is

$y = mx + b$.

slope y-intercept

Example 1

Graph the line $y = \frac{3}{4}x + 2$.

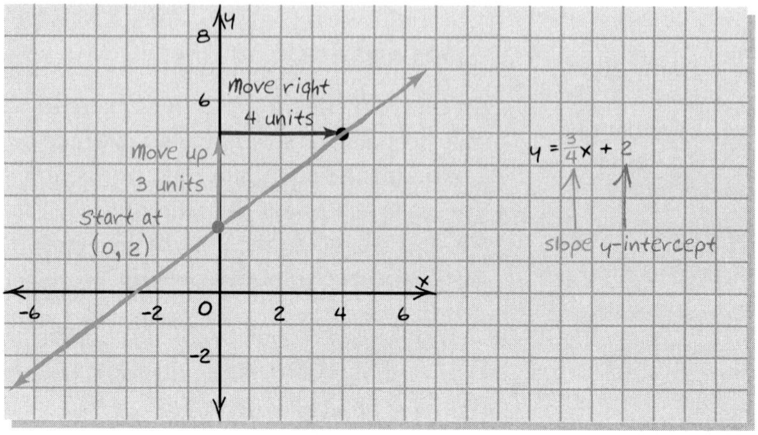

3. **Try This** Use the slope and y-intercept to graph each line.
   a. $y = 3x - 4$ b. $y = \frac{1}{2}x + 3$ c. $y = -2x - 1$.
4. Describe how you would graph a line with a slope of $-\frac{1}{2}$ and a y-intercept of 0.

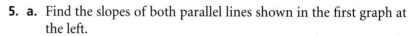

WORK TOGETHER

Work in a group to explore the slopes of parallel and perpendicular lines in a coordinate plane.

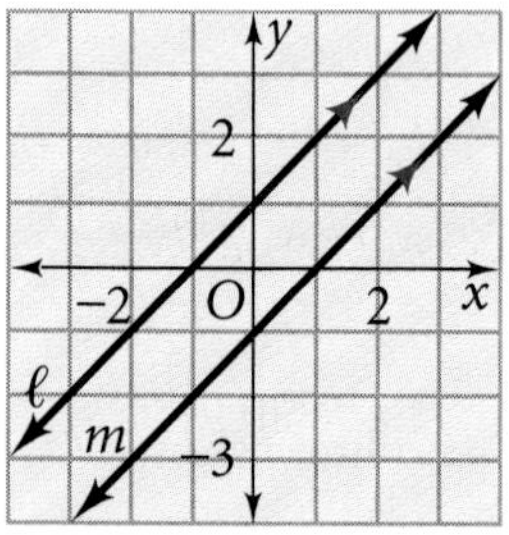

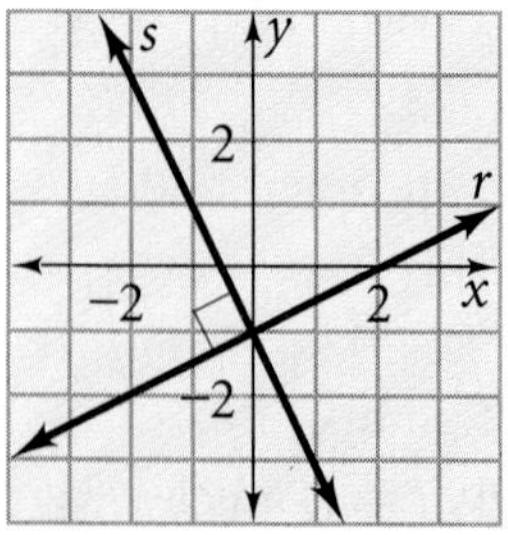

5. a. Find the slopes of both parallel lines shown in the first graph at the left.
   b. Have each member draw a pair of nonvertical parallel lines on graph paper and find the slopes of both lines.
   c. Compare your results with group members. Make a **conjecture** about the slopes of parallel lines.
6. a. Find the slopes of both perpendicular lines shown in the second graph at the left.
   b. Have each member draw a line on graph paper. (Do not use a vertical or horizontal line.) Then use a corner of the paper to draw a line perpendicular to it. Find the slopes of both lines. Find the product of the slopes.
   c. Compare your results with group members. Make a **conjecture** about the product of the slopes of two perpendicular lines.

THINK AND DISCUSS

Parallel and Perpendicular Lines

Your observations in the Work Together are summarized below.

> **Slopes of Parallel and Perpendicular Lines**
>
> The slopes of nonvertical parallel lines are equal. Two lines with the same slope are parallel. Vertical lines are parallel.
>
> The product of the slopes of two perpendicular lines, neither of which is vertical, is -1. If the product of the slopes of the two lines is -1, then the lines are perpendicular. A horizontal and a vertical line are perpendicular.

7. Try This Which of these lines are parallel? Which are perpendicular?

a. $y = 2x + 1$ **b.** $y = -x$ **c.** $y = x - 4$

d. $y = \frac{1}{2}$ **e.** $y = -2x + 3$

8. a. Graph $y = 5$, $y = -1$, and $x = -4$.

b. Which lines are parallel? Which are perpendicular?

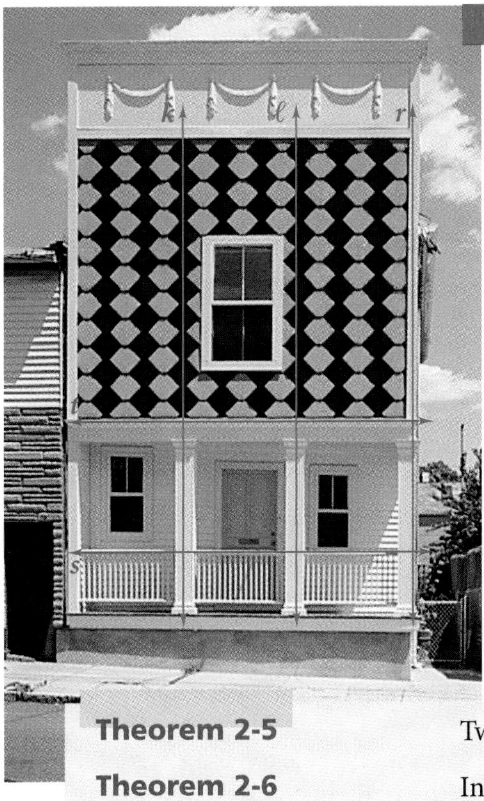

Example 2

Find the slope of a line perpendicular to $y = -3x + 4$.

The slope of line $y = -3x + 4$ is -3.
Let m be the slope of the perpendicular line.

$-3 \cdot m = -1$ The product of the slopes is -1.

$m = \frac{1}{3}$ Divide each side by -3.

9. *Open-ended* Give an equation of a line parallel to $y = -3x + 4$.

Architecture There are parallel and perpendicular lines in the photo at the left.

10. If line $k \parallel$ line ℓ and line $r \parallel$ line ℓ, what is the relationship between lines k and r?

11. If line $t \perp$ line k and line $s \perp$ line k, what is the relationship between lines t and s?

Questions 10 and 11 demonstrate the following theorems.

Theorem 2-5 Two lines parallel to a third line are parallel to each other.

Theorem 2-6 In a plane, two lines perpendicular to a third line are parallel to each other.

Exercises ON YOUR OWN

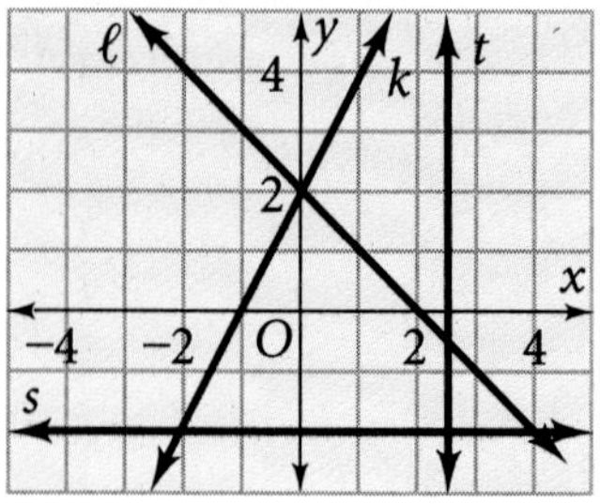

1. Identify the slope of each line in the graph as positive, negative, zero, or undefined.

2. a. What is the slope of the x-axis? Explain.
   b. What is the slope of the y-axis? Explain.

3. Writing A classmate claims that having no slope and having a slope of 0 are the same. Is your classmate right? Explain.

Choose **Use pencil and paper, mental math, or a calculator to find the slopes of $\overleftrightarrow{AB}$ and $\overleftrightarrow{CD}$. Then determine if the lines are parallel, perpendicular, or neither.**

4. $A(-1, \frac{1}{2})$, $B(-1, 2)$, $C(3, 7)$, $D(3, -1)$
5. $A(-2, 3)$, $B(-2, 5)$, $C(1, 4)$, $D(2, 4)$
6. $A(2, 4)$, $B(5, 4)$, $C(3, 2)$, $D(0, 8)$
7. $A(-3, 2)$, $B(5, 1)$, $C(2, 7)$, $D(1, -1)$
8. $A(1, -3)$, $B(3, 2)$, $C(4, 5)$, $D(2, 0)$
9. $A(4.5, 5)$, $B(2, 5)$, $C(1.5, -2)$, $D(3, -2)$

10. Use slope to show that the opposite sides of hexagon $RSTUVW$ are parallel.

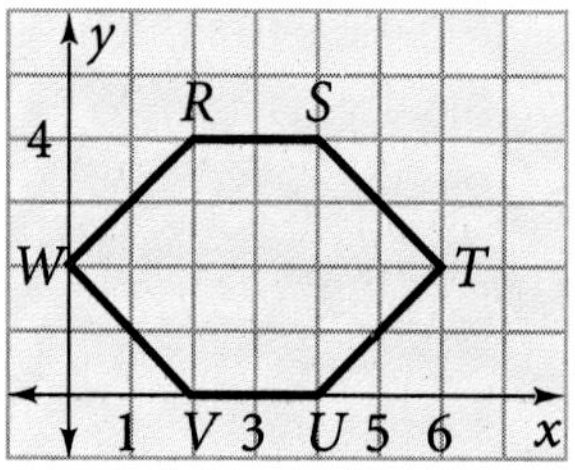

11. Use slope to determine whether a triangle with vertices (3, 2), (8, 5), and (0, 10) is a right triangle. Explain.

Sketch each pair of lines. Tell whether they are parallel, perpendicular, or neither.

12. $y = 5x - 4$
    $x = 2$
13. $y = \frac{1}{3}x - 1$
    $y = -3x + 5$
14. $x = -2$
    $x = 0$
15. $y = -4$
    $y = \frac{1}{4}$
16. $y = 2.5x + 1$
    $y = 2.5$
17. $y = x - \frac{1}{2}$
    $y = x + 1$
18. $y = \frac{3}{4}x - 2$
    $y = -2$
19. $y = -7$
    $x = -5$

20. Standardized Test Prep Which of the lines is *not* parallel to the line $y = -\frac{2}{3}x + 8$?
    A. $2x + 3y = 1$ B. $4x = 3 - 6y$ C. $x = -1.5y$ D. $24 = 2x - 3y$ E. $9y = -6x - 2$

21. a. Sketch vertical line t containing $(-5, 2)$.
    b. Write an equation for line t.
    c. On the same graph, sketch horizontal line s containing $(-5, 2)$.
    d. Write an equation for line s.
    e. What is the relationship between line t and line s? Explain.

22. a. Sketch line w perpendicular to $y = 5$, and containing $(1, 4)$.
    b. Write an equation for line w.
    c. On the same graph, sketch line r parallel to $y = 5$, and containing $(1, 4)$.
    d. Write an equation for line r.
    e. What is the relationship between line r and line w? Explain.

23. **Building** A law concerning wheelchair accessibility states that the slope of a ramp must be no greater than $\frac{1}{12}$. A local civic center plans to install a ramp. The height from the pavement to the main entrance is 3 ft and the distance from the sidewalk to the building is 10 ft. Is it possible for the center to design a ramp that complies with this law? Explain.

24. a. **Open-ended** Find the equations of two lines perpendicular to line $y = 3x - 2$.
    b. Find the equations of two lines parallel to line $y = 3x - 2$.

25. **Manipulatives** Use straws, toothpicks, or pencils.
    a. **Geometry in 3 Dimensions** Show how two lines that are each perpendicular to a third line can be perpendicular to each other.
    b. Show how two lines that are each perpendicular to a third line can be skew to each other.

26. a. Graph the points $A(1, 7)$, $B(2, 5)$, and $C(5, -1)$.
    b. What appears to be the relationship between these three points?
    c. Find the slopes of $\overleftrightarrow{AB}$, $\overleftrightarrow{BC}$, and $\overleftrightarrow{AC}$.
    d. **Logical Reasoning** Use part (c) to **justify** your answer to part (b).

27. **Geometry on a Sphere** Suppose you are investigating "lines" on a globe. The lines pass through both poles and are perpendicular to the equator. Are the lines of longitude "parallel" to each other? Explain.

28. a. Sketch line c with a slope -5 and line f with slope $\frac{1}{5}$.
    b. On the same graph, sketch line w perpendicular to line f.
    c. What is the relationship between line c and line w? Explain.

29. a. **Calculator** Find the slopes of each line containing the origin and $(\frac{1}{n}, 10)$ for $n = 1, 2, 3, \ldots 10$. Record your results as ordered pairs (n, slope).
    b. Show your results on a graph like the one shown.
    c. **Data Analysis** What predictions can you make about the slope of the line when $n = 100$?
    d. **Critical Thinking** Do you think that the slope will get infinitely large? Explain your reasoning.

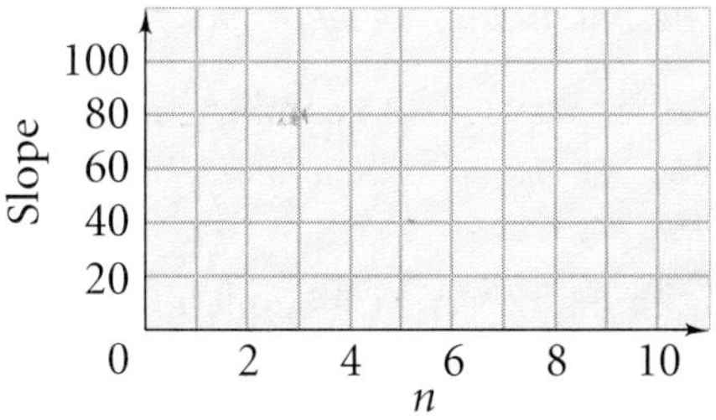

Exercises MIXED REVIEW

Use the diagram for the Exercises 30–32.

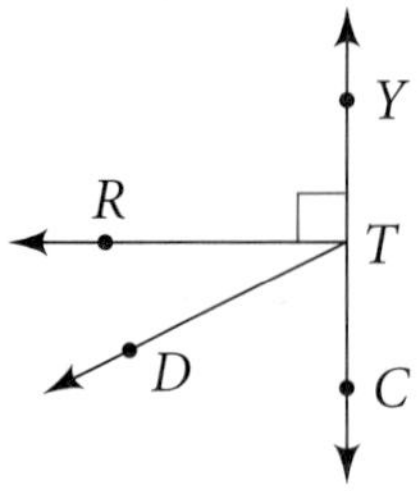

30. Name an acute angle, an obtuse angle, a right angle, and a straight angle.

31. Find $m\angle DTR$ when $m\angle CTD = 64.5$.

32. Find the value of x when $m\angle CTD = x + 32$ and $m\angle DTY = 3x + 20$. Then find the measures of both angles.

33. The measure of one acute angle of a right triangle is three times the measure of the other acute angle. Find the measures of the angles.

Getting Ready for Lesson 2-4

In each quadrilateral below, name any sides that appear to be parallel.

34.

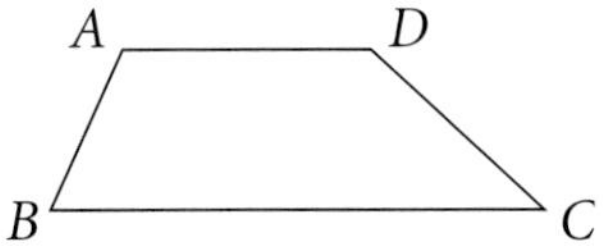

35.

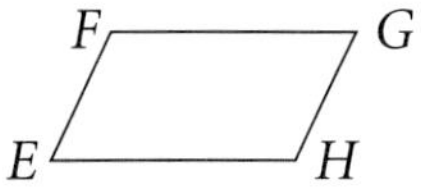

SELF ASSESSMENT

FOR YOUR JOURNAL

Describe real-world examples of parallel and perpendicular lines that you see on the way to school.

Exercises CHECKPOINT

Draw each figure. Mark congruent sides and angles.

1. equilateral triangle **2.** isosceles triangle **3.** right triangle **4.** acute triangle

5. obtuse triangle **6.** regular octagon **7.** concave decagon **8.** convex nonagon

9. Writing Describe two methods you can use to find the measure of each interior angle of a regular polygon.

Find the slope of lines $\overleftrightarrow{RS}$ and $\overleftrightarrow{TV}$. Then determine if $\overleftrightarrow{RS}$ and $\overleftrightarrow{TV}$ are *parallel, perpendicular,* or *neither.*

10. $R(-2, 6)$, $S(3, 4)$, $T(2, 5)$, $V(0, 0)$

11. $R(6, -1)$, $S(7, 0)$, $T(3, -4)$, $V(0, -1)$

12. $R(9, 1)$, $S(5, 6)$, $T(3, 8)$, $V(-2, 4)$

13. $R(5, -7)$, $S(-4, -9)$, $T(6, 2)$, $V(-3, 0)$

14. a. Open-ended Write an equation of a nonvertical line with a y-intercept of 2.

b. Write an equation of a line perpendicular to this line having the same y-intercept.

Math ToolboX

Algebra Review

Writing Linear Equations

After Lesson 2-3

When you know the slope of a line and a point on it, you can use the slope-intercept form of a line to write a linear equation.

Example 1

Write an equation of a line that has a slope of $\frac{1}{4}$ and contains the point $R(8, -3)$.

$y = mx + b$	Use the slope-intercept form.
$-3 = \frac{1}{4}(8) + b$	Substitute the slope and the x- and y-coordinates of the point.
$-3 = 2 + b$	Simplify.
$-5 = b$	Solve for b.

The equation of the line is $y = \frac{1}{4}x - 5$.

You can also use the slope-intercept form of a linear equation when you know the coordinates of two points on the line.

Example 2

Write an equation of a line containing $A(9, -2)$ and $B(3, 4)$.

$m = \frac{y_2 - y_1}{x_2 - x_1}$	Use the formula to find the slope.
$m = \frac{4 - (-2)}{3 - 9} = -1$	Substitute the coordinates of both points.
$y = -1x + b$	Substitute -1 for m in $y = mx + b$.
$4 = -1(3) + b$	Substitute the coordinates of one of the points and solve for b.
$7 = b$	

The equation of the line is $y = -1x + 7$ or $y = -x + 7$.

Write an equation of the line with the given slope, and containing point *T*.

1. $m = 3, T(0, 5)$
2. $m = \frac{2}{3}, T(-6, -1)$
3. $m = -\frac{1}{2}, T(4, -8)$
4. $m = 1, T(-1, 3)$
5. $m = -\frac{5}{4}, T(4, 3)$
6. $m = -1, T(-1, -8)$
7. $m = \frac{3}{2}, T(4, 4)$
8. $m = \frac{3}{4}, T(-12, -9)$

Write an equation of the line containing points *C* and *D*.

9. $C(9, -2), D(3, 4)$
10. $C(2, 1), D(-2, 3)$
11. $C(0, 3), D(-5, 0)$
12. $C(-5, 0), D(-2, 1)$
13. $C(2, 0), D(3, 5)$
14. $C(3, -1), D(2, -3)$
15. $C(0, 0), D(8, -2)$
16. $C(-8, 3), D(4, -6)$

What You'll Learn

- Defining and classifying special types of quadrilaterals

...And Why

To learn about the most commonly used polygons in buildings, architecture, and design

What You'll Need

- toothpicks

2-4 Classifying Quadrilaterals

WORK TOGETHER

Some quadrilaterals have special names.

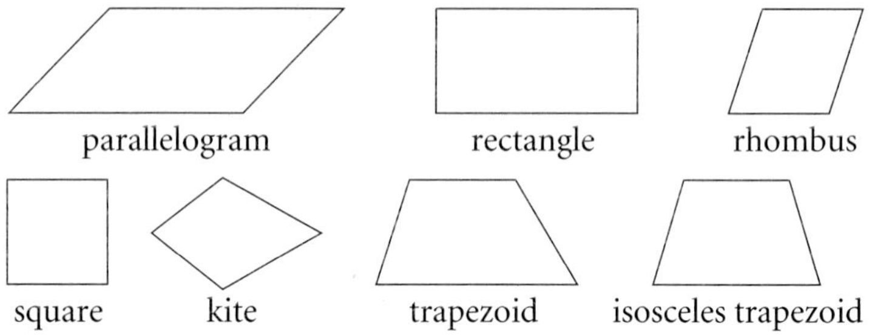

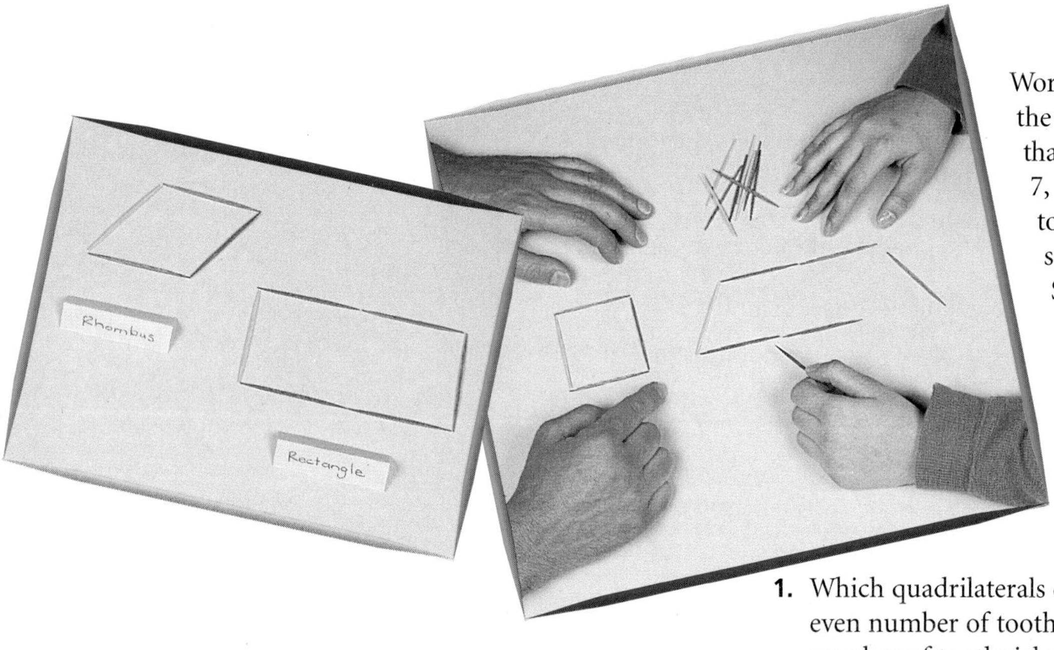

Work in a group to build all the different quadrilaterals that you can using 4, 5, 6, 7, and 8 toothpicks. Each toothpick represents a side or a part of a side. Sketch and name each quadrilateral you form. Two examples are shown at the left.

1. Which quadrilaterals can you build with an even number of toothpicks? with an odd number of toothpicks? Explain.

THINK AND DISCUSS

As you made quadrilaterals out of toothpicks, you probably noticed some of the properties of special quadrilaterals.

2. What appears to be true about the sides of the following quadrilaterals?
   a. rhombus **b.** trapezoid **c.** parallelogram **d.** kite
3. What appears to be true about the angles of a rectangle and a square?

You can use characteristics of special quadrilaterals to define them.

Special Quadrilaterals

A **parallelogram** is a quadrilateral with both pairs of opposite sides parallel. In Chapter 9, you will prove that both pairs of opposite sides are also congruent. The symbol for a parallelogram is ▱.

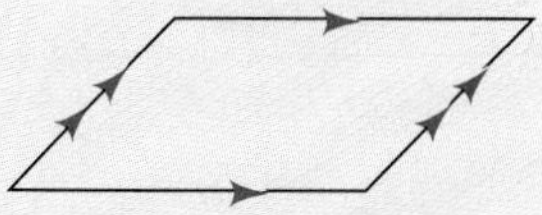

A **rhombus** is a parallelogram with four congruent sides.

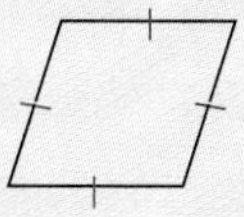

A **rectangle** is a parallelogram with four right angles.

A **square** is a parallelogram with four congruent sides and four right angles.

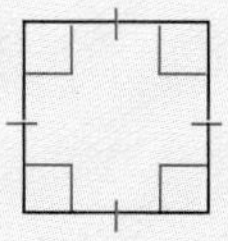

A **kite** is a quadrilateral with two pairs of adjacent sides congruent and no opposite sides congruent.

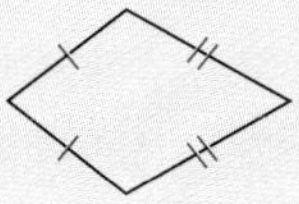

A **trapezoid** is a quadrilateral with exactly one pair of parallel sides.

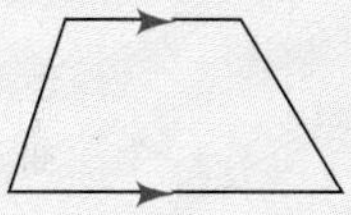

An **isosceles trapezoid** is a trapezoid whose nonparallel sides are congruent.

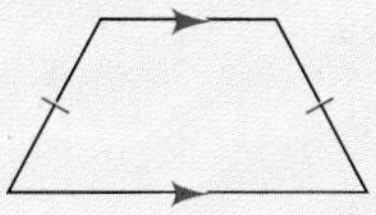

Example 1

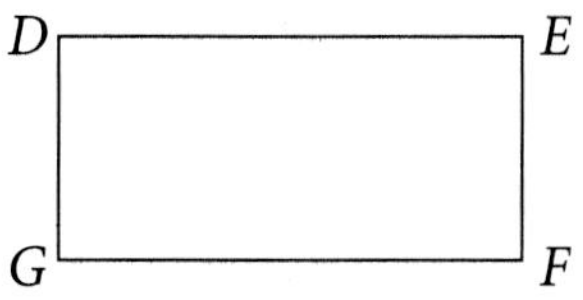

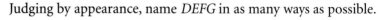

Judging by appearance, name *DEFG* in as many ways as possible.

It is a quadrilateral because it has four sides.

It is a parallelogram because both pairs of opposite sides are parallel.

It is a rectangle because it has four right angles. ■

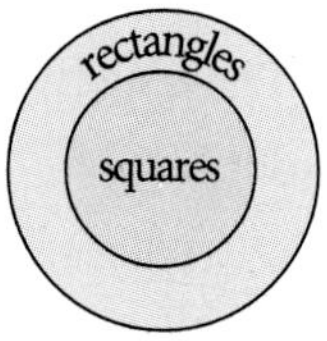

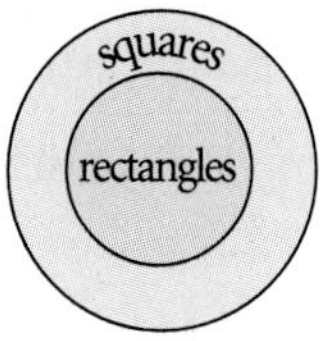

4. Which name do you think gives the most information about *DEFG*? Explain.

5. *Logical Reasoning* Which Venn diagram at the left shows the relationship between rectangles and squares? Explain.

The diagram below shows the relationships among special quadrilaterals.

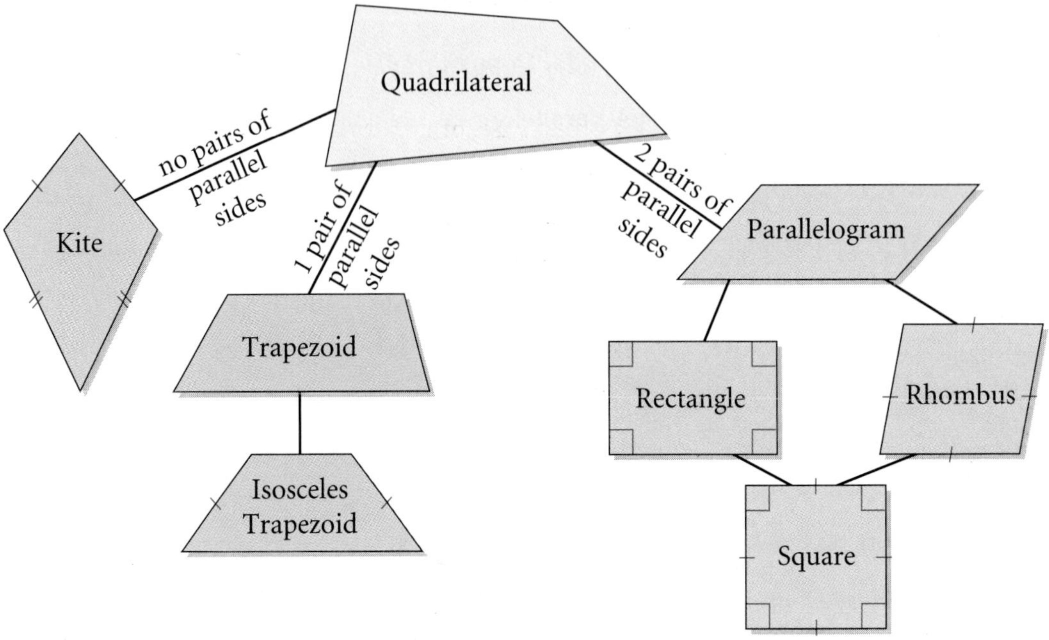

You can use the definitions of special quadrilaterals and what you know about slope and distance to classify a quadrilateral.

Example 2

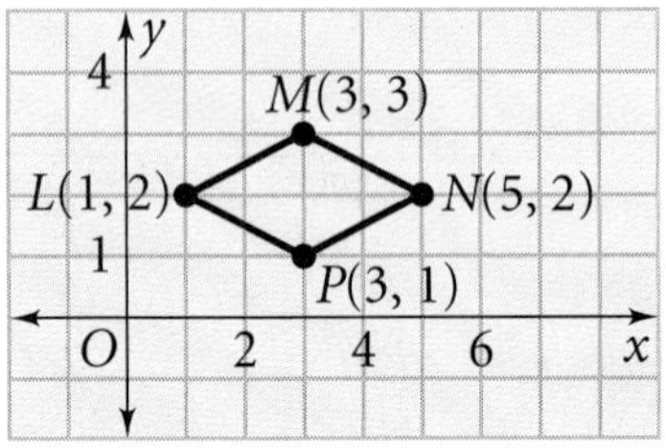

Determine the most precise name for quadrilateral *LMNP*.

Find the slope of each side.

slope of $\overline{LM} = \frac{3-2}{3-1} = \frac{1}{2}$ slope of $\overline{NP} = \frac{2-1}{5-3} = \frac{1}{2}$

slope of $\overline{MN} = \frac{3-2}{3-5} = -\frac{1}{2}$ slope of $\overline{LP} = \frac{2-1}{1-3} = -\frac{1}{2}$

Both pairs of opposite sides are parallel, so *LMNP* is a parallelogram.

Now use the Distance Formula to see if any of the sides are congruent.

$LM = \sqrt{(3-1)^2 + (3-2)^2} = \sqrt{5}$

$MN = \sqrt{(3-5)^2 + (3-2)^2} = \sqrt{5}$

$NP = \sqrt{(5-3)^2 + (2-1)^2} = \sqrt{5}$

$LP = \sqrt{(1-3)^2 + (2-1)^2} = \sqrt{5}$

All sides are congruent, so *LMNP* is a rhombus.

For practice with radicals, see Skills Handbook, page 660.

6. Explain how you know that *LMNP* is *not* a square.
7. **Try This** Determine the most precise name for quadrilateral *ABCD* with vertices $A(0, 4)$, $B(3, 0)$, $C(0, -4)$, $D(-3, 0)$.

You can use the definitions of special quadrilaterals to find lengths of sides of objects like kites.

Example 3

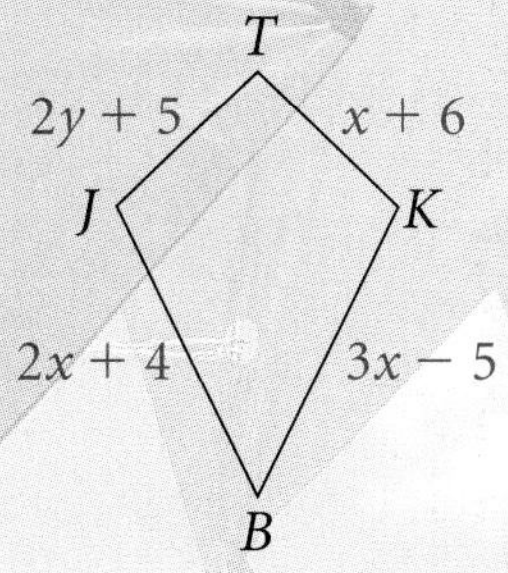

Find the values of the variables in the kite at the left.

$KB = JB$	Definition of kite
$3x - 5 = 2x + 4$	Substitution
$x - 5 = 4$	Subtract $2x$ from each side.
$x = 9$	Add 5 to each side.
$KT = x + 6$, $KT = 15$	Substitute 9 for x.
$KT = JT$	Definition of kite
$15 = 2y + 5$	Substitution
$10 = 2y$	Subtract 5 from each side.
$5 = y$	Divide each side by 2.

8. What are the lengths of the longer sides of the kite?

9. What types of triangles are formed when you draw diagonal $\overline{KJ}$?

10. What types of triangles are formed when you draw diagonal $\overline{BT}$?

Exercises ON YOUR OWN

Judging by appearance, name each quadrilateral in as many ways as possible.

1.

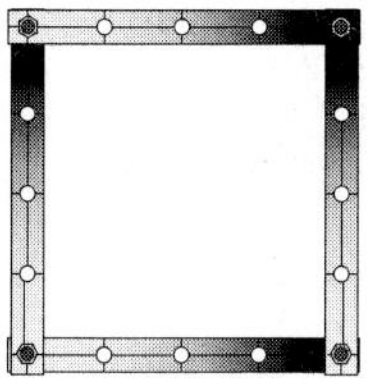

2.

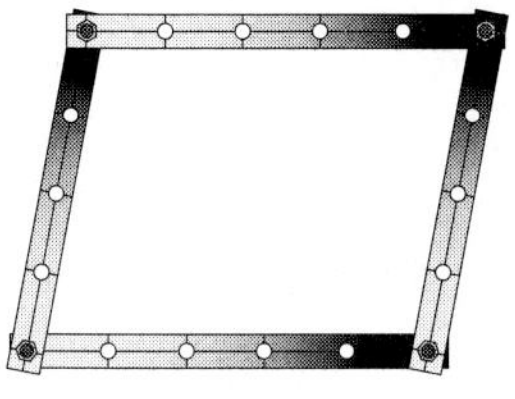

3.

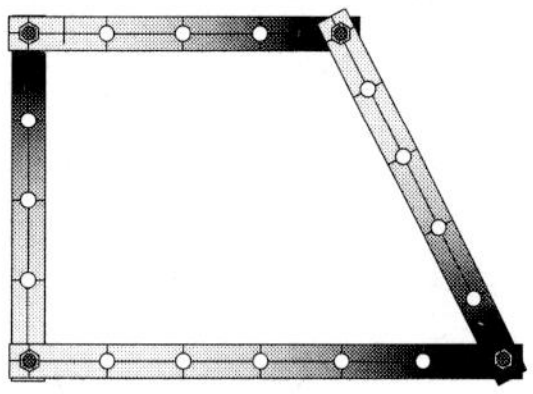

4.

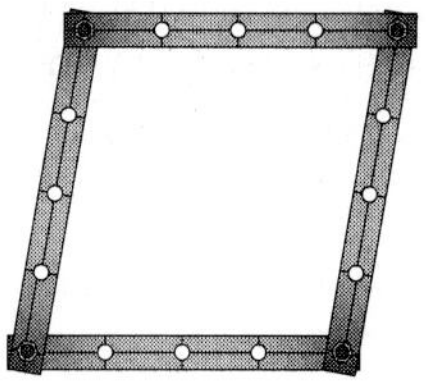

Copy the Venn diagram. Show the relationships of special quadrilaterals by adding the labels *Rectangles, Rhombuses,* and *Trapezoids.* Then use the diagram to decide whether each statement is true or false.

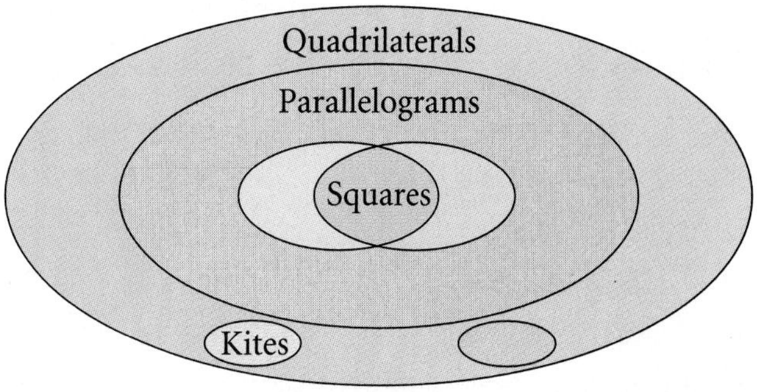

5. All squares are rectangles.

6. A trapezoid is a parallelogram.

7. A rhombus can be a kite.

8. Some parallelograms are squares.

9. A quadrilateral is a parallelogram.

10. Art Inspired by Cubists like Pablo Picasso, American artist Charles Demuth created *My Egypt.* Identify the geometric figures in this oil painting. List all the special quadrilaterals you see.

Draw each figure on graph paper.

11. parallelogram that is neither a rectangle nor a rhombus

12. trapezoid with a right angle

13. rhombus that is not a square

14. Writing Describe the difference between a rhombus and a kite.

15. **a.** Open-ended Graph and label points $K(-3, 0)$, $L(0, 2)$, and $M(3, 0)$. Find possible coordinates for point N so that $KLMN$ is a kite.

b. Explain why there is more than one possible fourth vertex.

Coordinate Geometry **Graph and label each quadrilateral with the given vertices. Use slope and/or the Distance Formula to determine the most precise name for each figure.**

16. $A(3, 5)$, $B(7, 6)$, $C(6, 2)$, $D(2, 1)$

17. $W(-1, 1)$, $X(0, 2)$, $Y(1, 1)$, $Z(0, -2)$

18. $J(2, 1)$, $K(5, 4)$, $L(7, 2)$, $M(2, -3)$

19. $R(-2, -3)$, $S(4, 0)$, $T(3, 2)$, $V(-3, -1)$

20. Paper Folding Fold a rectangular piece of paper in half horizontally and then vertically. Draw and then cut along the line connecting the two corners containing a fold. What quadrilateral do you find when you unfold the paper? Explain.

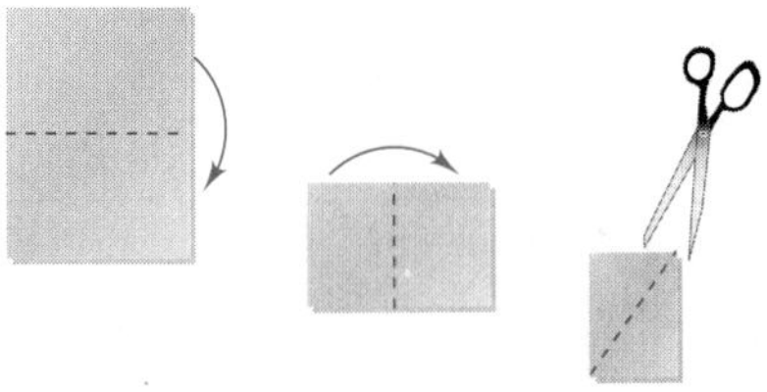

Name all special quadrilaterals that satisfy the given conditions. Make a sketch to support your answer.

21. exactly one pair of congruent sides

22. two pairs of parallel sides

23. four right angles

24. adjacent sides that are congruent

Algebra **Find the values of the variables and the lengths of the sides.**

25. kite $ABCD$

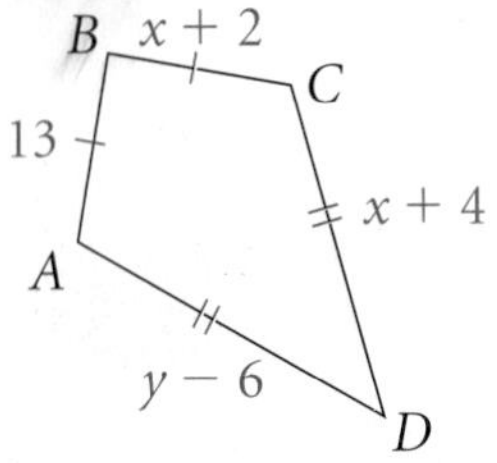

26. isosceles trapezoid $DEFG$

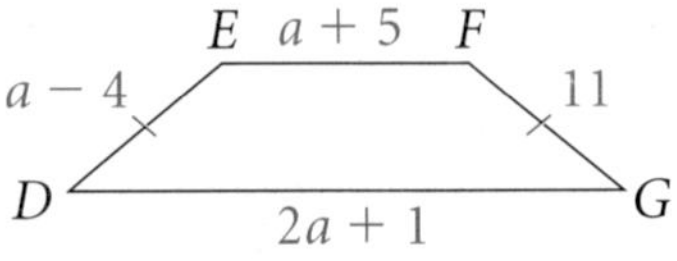

27. rhombus $HIJK$

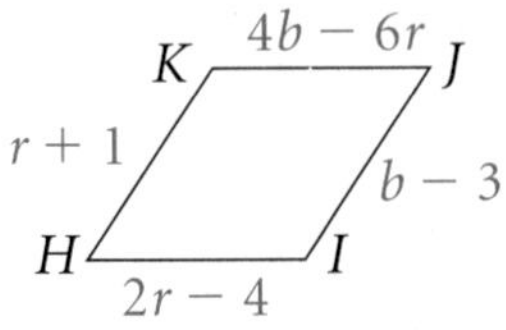

Part of each quadrilateral is covered. Name all special quadrilaterals that each could be. Explain each choice.

28.
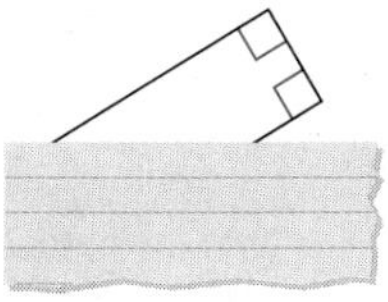

29.
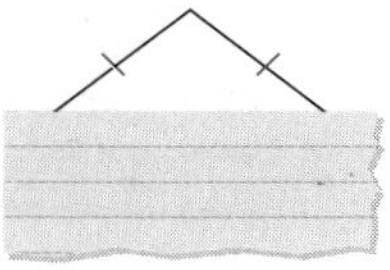

30.
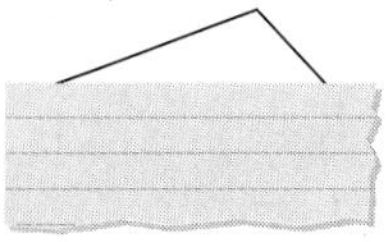

31.
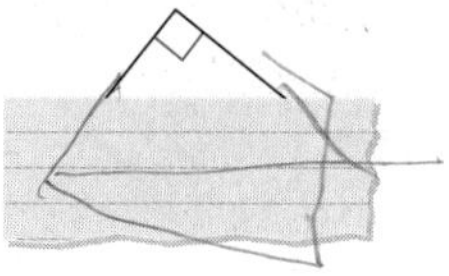

Chapter Project — *Find Out by Analyzing*

You can create another geometric puzzle, called *pentominoes,* by joining five unit squares. Each square shares a side with at least one other square.

These are pentominoes.

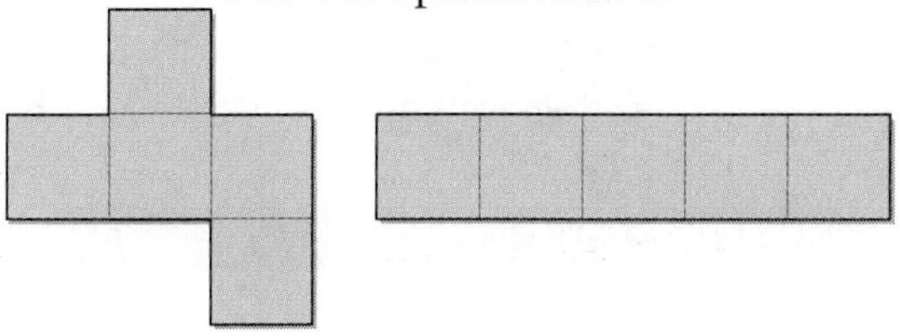

These are not pentominoes.

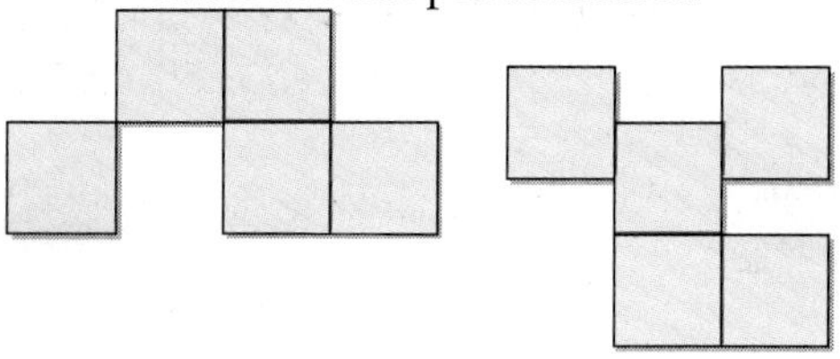

- There are twelve different pentominoes. Sketch the other ten pentominoes on graph paper.
- Make a set of pentominoes out of cardboard. Use any three pentominoes to form a 3 x 5 rectangle. Find and record as many solutions as you can.

Exercises — MIXED REVIEW

Use a calculator to find TR to the nearest tenth.

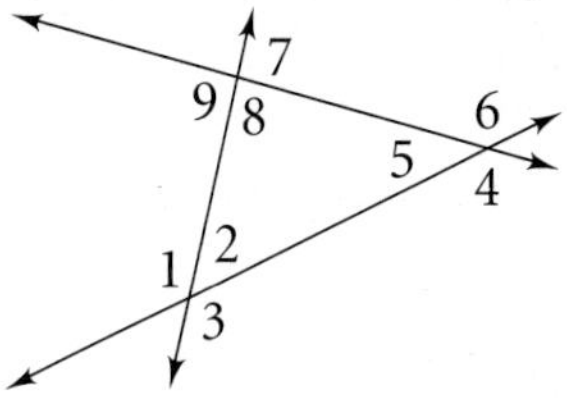

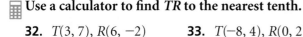

32. $T(3, 7), R(6, -2)$ **33.** $T(-8, 4), R(0, 2)$

34. a. Find $m\angle 5$ if $m\angle 7 = 103$ and $m\angle 2 = 48$.
b. Find $m\angle 1$ if $m\angle 7 = 110$ and $m\angle 6 = 153$.
c. Find $m\angle 9$ if $m\angle 3 = 138$ and $m\angle 5 = 51$.

35. a. Sketch the line perpendicular to $x = 2$ containing the point $(3, -1)$.
b. Sketch the line parallel to $x = 2$ containing the point $(3, -1)$.

Getting Ready for Lesson 2-5

Estimate the percent of the circle that is shaded.

36.
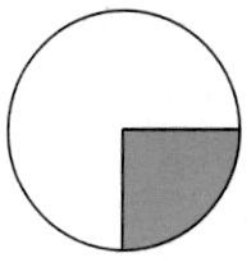

37.
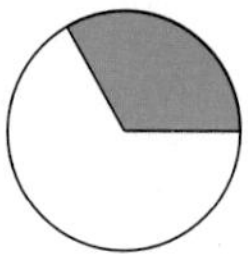

38.
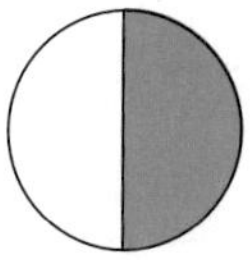

39.
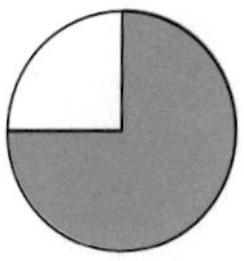

40.
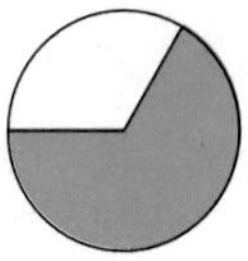

2-5 Circles

What You'll Learn

- Measuring central angles and arcs of circles.
- Displaying data in a circle graph.

...And Why

To learn about a geometric figure at the heart of the design of many things, including amusement park rides, satellite orbits, and circle graphs

What You'll Need

- compass
- protractor

WORK TOGETHER

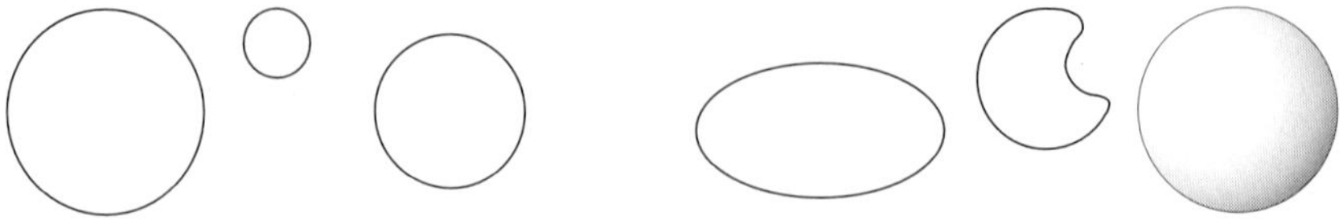

Have each person in your group write a definition of a circle. Exchange papers. Examine the definition you receive. Try to find a counterexample. Continue until your group has agreed on a definition of a circle.

THINK AND DISCUSS

Parts of a Circle

A **circle** is the set of all points in a plane *equidistant* from a given point, called the center. You name a circle by its center. Circle P ($\odot P$) is shown.

A **radius** is a segment that has one endpoint at the center and the other endpoint on the circle. $\overline{PC}$ is a radius. $\overline{PA}$ and $\overline{PB}$ are also radii.

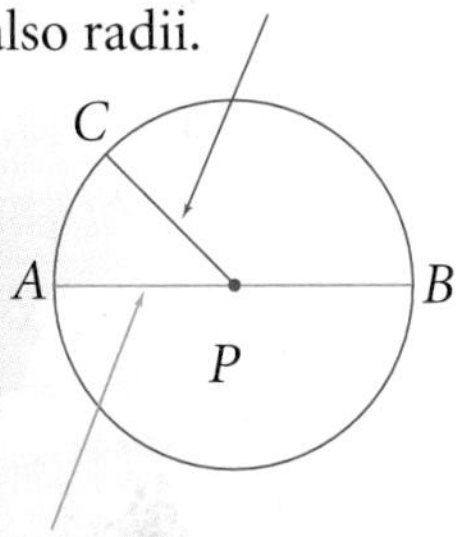

A **diameter** is a segment that contains the center of the circle and has both endpoints on the circle. $\overline{AB}$ is a diameter.

QUICK REVIEW

The length r of a radius is *the* radius of a circle. The length d of a diameter is *the* diameter of a circle. $d = 2r$

1. Amusement Parks The Scream Weaver at Paramount's Carowinds, North Carolina, is a gondola ride with a diameter of 13.1 m. Find the length of each arm.
2. Complete: The center of a circle is the __?__ of a diameter.

Example 1

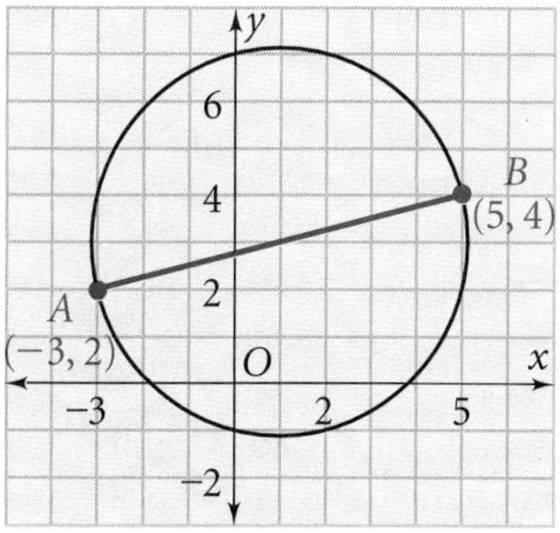

Coordinate Geometry $\overline{AB}$ is a diameter of the circle. Find the coordinates of the center and the radius of the circle.

The center P of the circle is the midpoint of the diameter.

$$P = \left(\frac{-3 + 5}{2}, \frac{2 + 4}{2}\right) \qquad \text{Use the Midpoint Formula.}$$

$$P = (1, 3) \qquad \text{Simplify.}$$

The radius is the distance from the center to any point on the circle.

$$PB = \sqrt{(1 - 5)^2 + (3 - 4)^2} \qquad \text{Use the Distance Formula.}$$

$$PB = \sqrt{(-4)^2 + (-1)^2} \qquad \text{Simplify.}$$

$$PB = \sqrt{16 + 1} = \sqrt{17}$$

The center of the circle is (1, 3). The radius is $\sqrt{17}$.

For practice with radicals, see Skills Handbook, page 660.

3. **Try This** Find the coordinates of the center and the radius of a circle with diameter $\overline{AB}$ whose endpoints are $A(1, 3)$ and $B(7, -5)$.

Central Angles and Arcs

You often see circle graphs in newspapers and magazines. When you make a circle graph you have to find the measure of each wedge, or central angle. A **central angle** is an angle whose vertex is the center of the circle.

Example 2 Relating to the Real World

Source: *The New York Times*

How Do You Spend Your Day?

To learn how people really spend their time, a market research firm studied the hour-by-hour activities of more than 3,000 people. The participants were between 18 and 90 years old. Each participant was sent a 24-h recording sheet every March for 3 years from 1992 to 1994. The study found that people spend most of their time sleeping, working, and watching television.

Statistics Find the measure of each central angle in the circle graph.

Since there are 360 degrees in a circle, multiply each percent by 360 to find the measure of each central angle in the circle graph.

Sleep: 31% of 360 = 111.6
Food: 9% of 360 = 32.4
Work: 20% of 360 = 72
Must Do: 7% of 360 = 25.2
Entertainment: 18% of 360 = 64.8
Other: 15% of 360 = 54

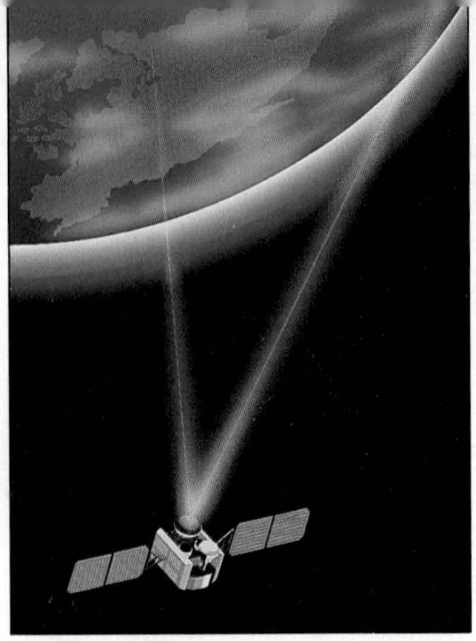

An *arc* is a part of a circle. There are three types of arcs: a *semicircle*, a *minor arc*, and a *major arc*.

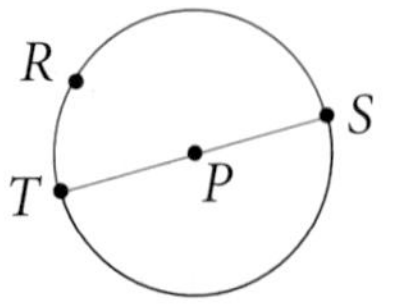

$\widehat{TRS}$ is a semicircle.
$m\widehat{TRS} = 180$

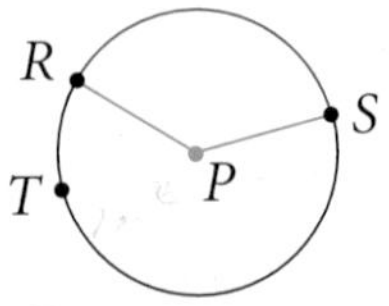

$\widehat{RS}$ is a minor arc.
$m\widehat{RS} = m\angle RPS$

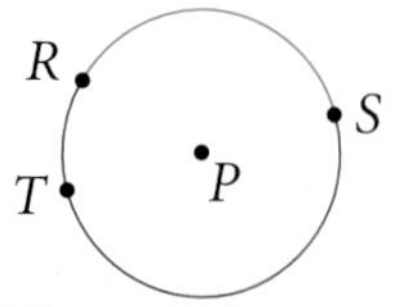

$\widehat{RTS}$ is a major arc.
$m\widehat{RTS} = 360 - m\widehat{RS}$

A **semicircle** is half a circle. The measure of a semicircle is 180. A **minor arc** is shorter than a semicircle. The measure of a minor arc is the measure of its corresponding central angle. A **major arc** is longer than a semicircle. The measure of a major arc is 360 minus the measure of its related minor arc.

What? Geosynchronous satellites stay above the same point on Earth's equator. They orbit Earth in a near-circular path every 24 hours at a distance 42,000 km from the Earth's center. The radius of Earth is about 6000 km. How far above Earth's surface are these satellites?

4. **Critical Thinking** What kind of arcs can you name with only two points? What kind of arcs must you name with three points? Why?

Adjacent arcs are two arcs in the same circle that have exactly one point in common. You can add the measures of adjacent arcs just as you can add the measures of adjacent angles.

Arc Addition Postulate

The measure of the arc formed by two adjacent arcs is the sum of the measures of the two arcs.

$m\widehat{ABC} = m\widehat{AB} + m\widehat{BC}$

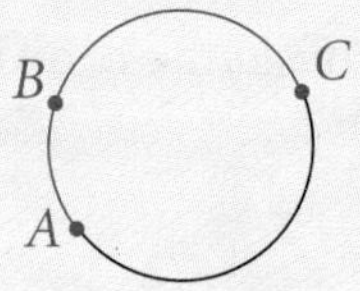

Example 3

Find the measure of each arc.

a. $\widehat{BC}$ **b.** $\widehat{BD}$ **c.** $\widehat{ABC}$ **d.** $\widehat{AB}$ **e.** $\widehat{BAD}$

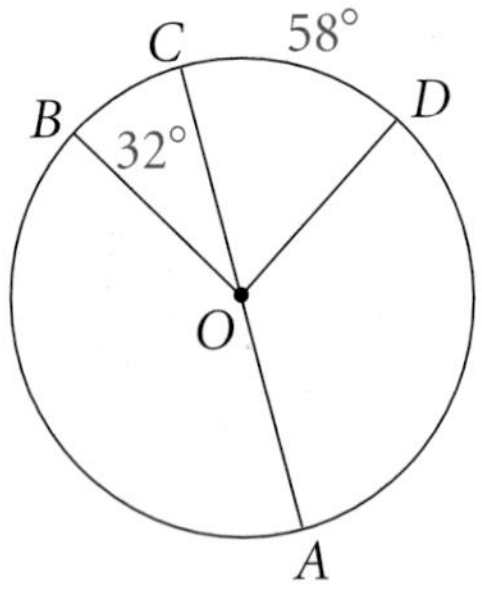

a. $m\widehat{BC} = m\angle BOC = 32$

b. $m\widehat{BD} = 32 + 58 = 90$ — $m\widehat{BD} = m\widehat{BC} + m\widehat{CD}$

c. $m\widehat{ABC} = 180$ — $\widehat{ABC}$ is a semicircle.

d. $m\widehat{AB} = 180 - 32 = 148$

e. $m\widehat{BAD} = 360 - m\widehat{BD}$
$= 360 - 90 = 270$

5. **Try This** Find $m\angle COD$, $m\widehat{CDA}$, and $m\widehat{AD}$.

6. **a.** Use a compass to draw a circle. Then use a protractor to draw $\widehat{JK}$ with measure 75.
   b. Explain how to draw a longer arc that has a measure of 75.

Exercises ON YOUR OWN

Identify the following in $\odot O$.

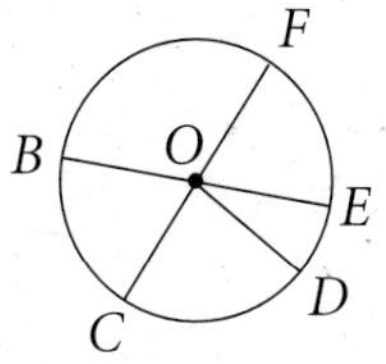

1. two minor arcs **2.** two major arcs **3.** two semicircles

4. three radii **5.** two diameters **6.** a pair of adjacent arcs

7. an acute central angle **8.** an obtuse central angle **9.** a pair of congruent angles

Find the diameter of a circle with the given radius.

10. 20 ft **11.** 5 cm **12.** $3\frac{1}{2}$ m **13.** $6\sqrt{2}$ in. **14.** r mi

Find the radius of a circle with the given diameter.

15. 13 cm **16.** 10.5 m **17.** $5\sqrt{3}$ in. **18.** $\frac{1}{3}$ ft **19.** d km

20. *Printing* Newspaper companies use offset presses to print. The paper passes between cylinders. The radius of each large cylinder is 7 in. Find PR in the diagram.

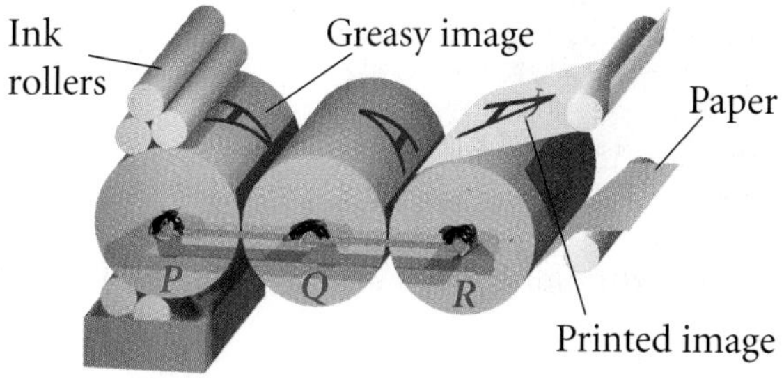

21. *Paper Folding* Use a compass to make a circle. Cut out the circle and use paper folding to form 90°, 45°, and 135° central angles.

Coordinate Geometry **Find the coordinates of the center and the radius of each circle with diameter $\overline{AB}$.**

22. $A(3, 4)$, $B(-3, -4)$ **23.** $A(0, 4)$, $B(-4, 6)$ **24.** $A(-2, -2)$, $B(3, 10)$

25. $A(2, 3)$, $B(-4, 5)$ **26.** $A(-6, -2)$, $B(0, 6)$ **27.** $A(-1, -12)$, $B(7, 3)$

Find the measure of each arc in $\odot P$.

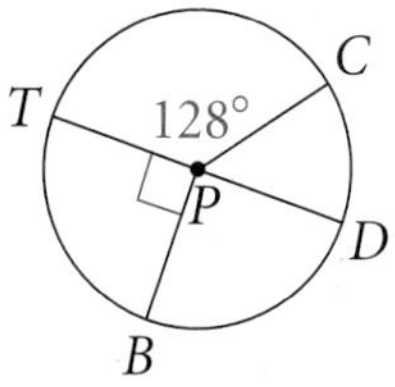

28. $\overparen{TC}$ **29.** $\overparen{TBD}$ **30.** $\overparen{BTC}$

31. $\overparen{CD}$ **32.** $\overparen{CBD}$ **33.** $\overparen{TCD}$

34. $\overparen{TDC}$ **35.** $\overparen{TB}$ **36.** $\overparen{BC}$

37. Use a compass to draw $\odot A$. Then use a protractor to draw $\overparen{XY}$ with measure 105.

38. *Statistics* Americans throw out more than 150 million tons of garbage each year. The circle graph shows the percent of different materials found in a typical city trash collection.

a. Find the measure of the central angle for each category (rounded to the nearest whole number).

b. Find the sum of the measures of these angles.

c. *Writing* Explain why the sum might not equal 360.

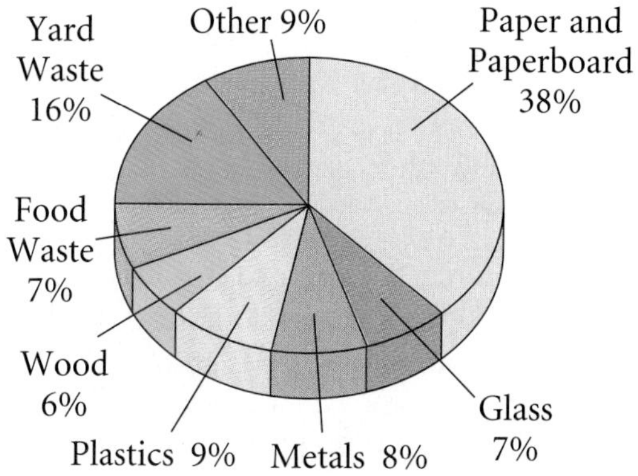

Source: *The American Almanac*

39. **a.** How many degrees does a minute hand move in 1 minute? in 5 minutes? in 20 minutes?

b. How many degrees does an hour hand move in 5 minutes? in 10 minutes? in 20 minutes?

c. What is the measure of the angle formed by the hands of the clock at 8:25?

Find each indicated measure for ⊙*O*.

40. **a.** $m\angle EOF$

b. $m\widehat{EJH}$

c. $m\widehat{FH}$

d. $m\angle FOG$

e. $m\widehat{JEG}$

f. $m\widehat{HFJ}$

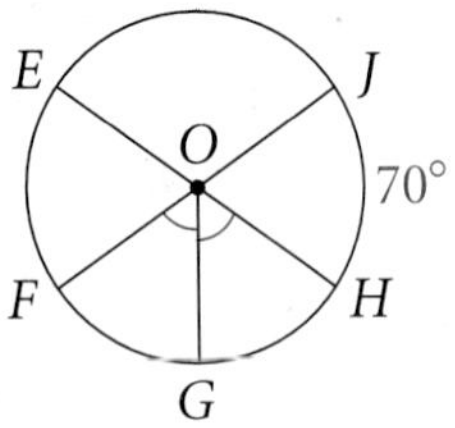

41. **a.** $m\widehat{TR}$

b. $m\angle COD$

c. $m\widehat{BT}$

d. $m\widehat{BR}$

e. $m\widehat{BTR}$

f. $m\widehat{TRB}$

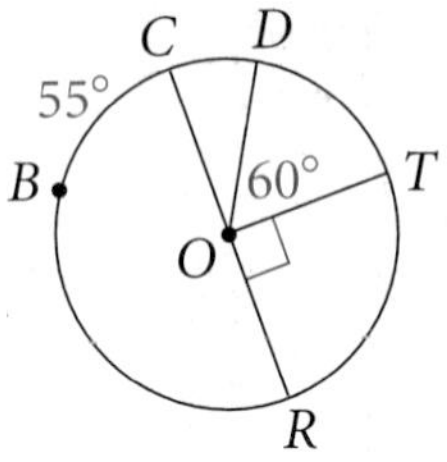

42. **a.** $m\angle LOM$

b. $m\widehat{QP}$

c. $m\widehat{PMQ}$

d. $m\angle QOL$

e. $m\widehat{QLP}$

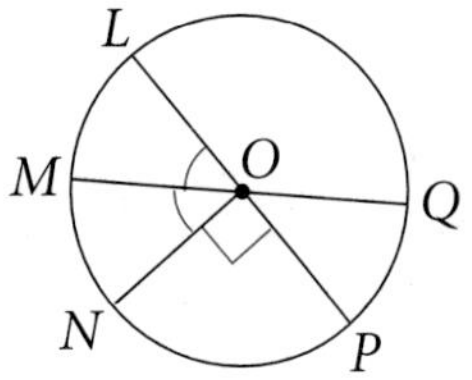

43. **a.** $m\angle KOV$

b. $m\widehat{KZ}$

c. $m\angle SOW$

d. $m\widehat{YVK}$

e. $m\widehat{WSZ}$

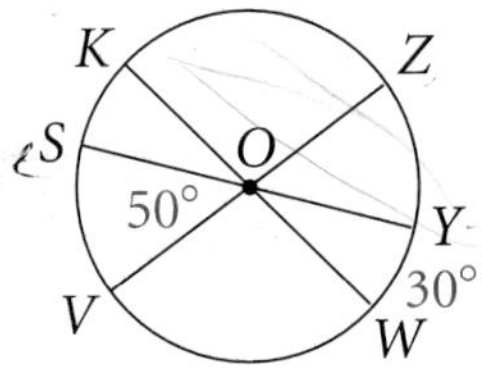

Algebra **Find $m\widehat{PQ}$ in ⊙*A*.**

44.

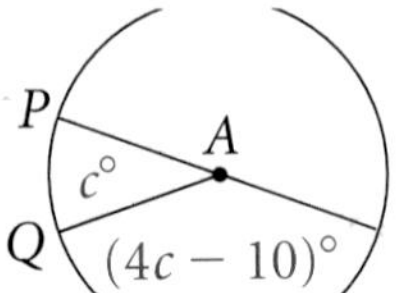

45.

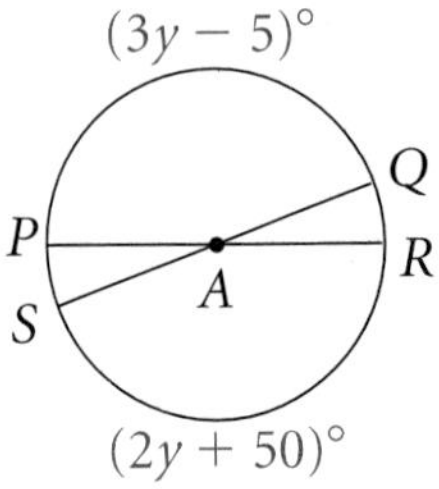

46.

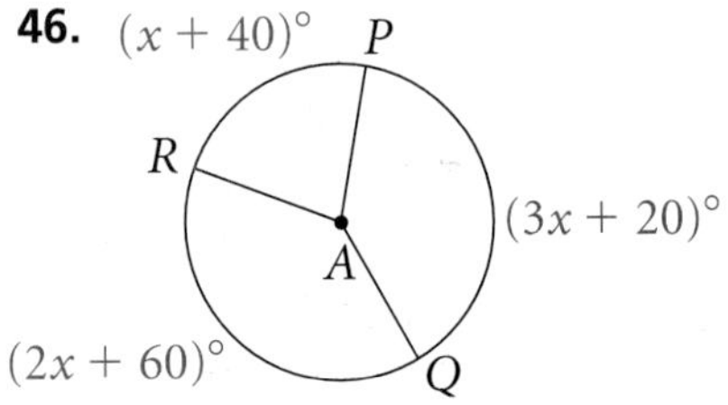

47. Travel Five streets come together at a traffic circle. Vehicles travel counterclockwise around the circle. Use arc measure to give directions to someone who wants to get to East Street from Neponset Street.

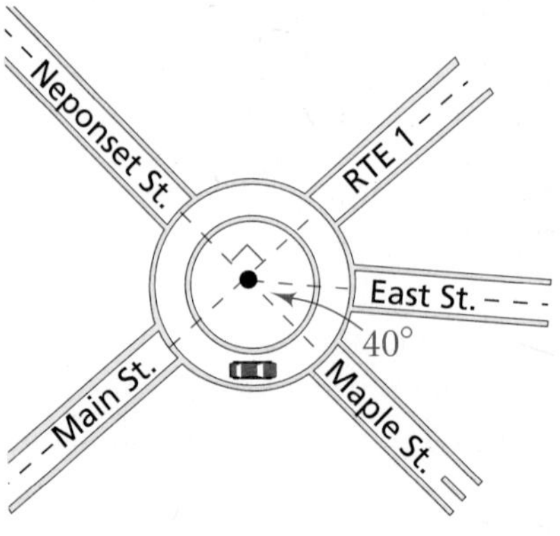

48. Research Describe the traditional life of the Zulus of South Africa. Find out how they use circles when building and designing their villages.

49. *Open-ended* Make a circle graph showing how you spend a 24-hour weekday.

50. *Statistics* In 1995, O'Reilly & Associates surveyed people 18 and older in the United States to find out "Who's Using the Internet?" The table shows Internet users by age groups. (Of those surveyed, 2% refused to answer.) Display the data in a circle graph.

Who's Using the Internet?

Age	% of Surveyed Internet Users	Age	% of Surveyed Internet Users
18–24 years	20%	35–44 years	25%
25–29 years	20%	44–54 years	15%
30–34 years	15%	55+ years	3%

Source: *U.S. News & World Report*

Exercises MIXED REVIEW

Find the values of the variables.

51.

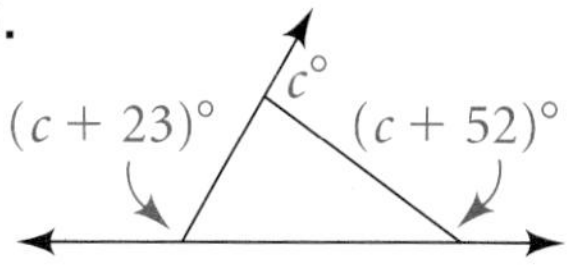

52.

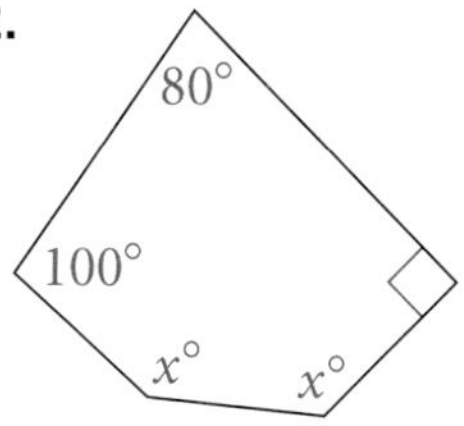

53.

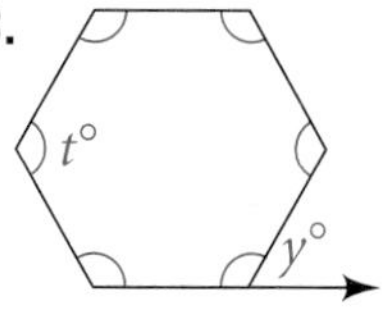

54.

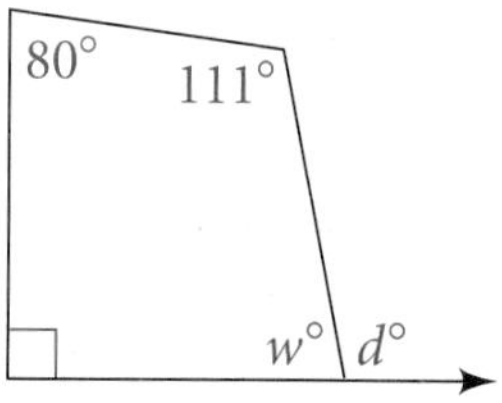

55. Find the measure of an angle that is 16° less than its complement.

Getting Ready for Lesson 2-6

56. The diagram below at the left shows a 4-by-4 square on graph paper. You can divide the square into two identical pieces by cutting along grid lines. One way to do this is to make a vertical or a horizontal line. Another way is shown below at the right. Find the four other ways of doing this.

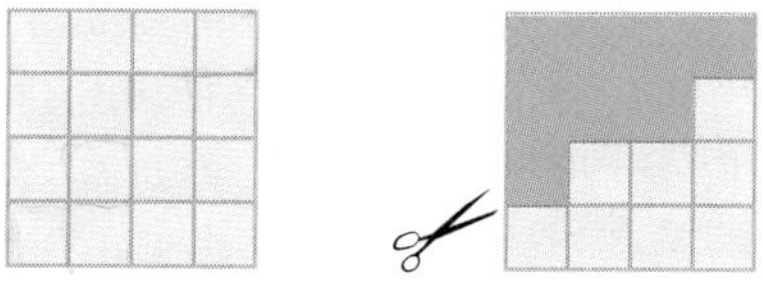

SELF ASSESSMENT

FOR YOUR JOURNAL

Imagine a world without circles. Describe what it would be like to live in this world for a day.

2-6 Congruent and Similar Figures

What You'll Learn

- Measuring congruent and similar figures
- Using properties of congruence and similarity

...And Why

To model real-world situations, such as mass production and photography

What You'll Need

- scissors
- protractor
- centimeter ruler

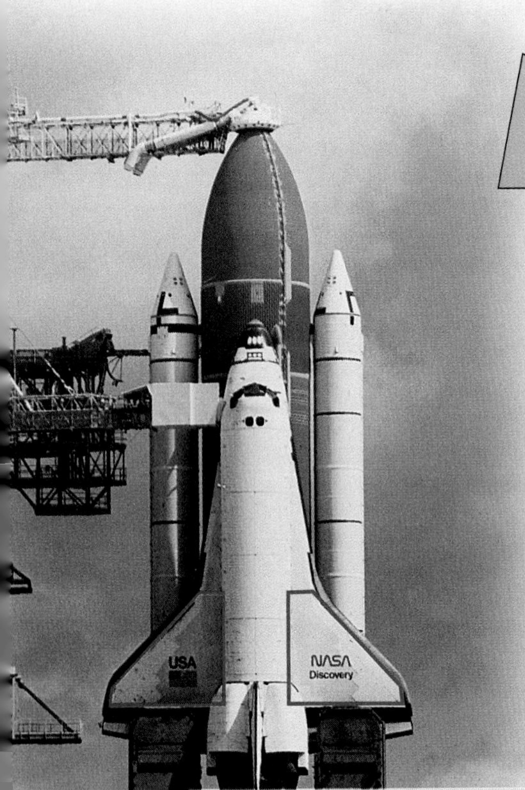

THINK AND DISCUSS

Congruent Figures

Congruent figures have exactly the same size and shape. When two figures are congruent you can slide, flip, or turn one so that it fits exactly on the other one.

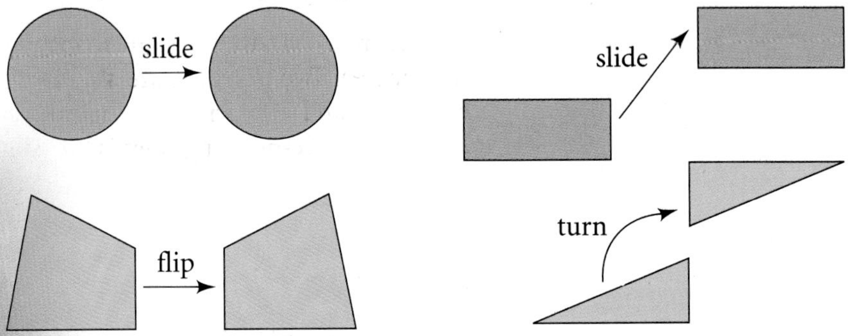

Congruent circles have congruent radii. **Congruent polygons** have congruent corresponding parts. The matching angles and sides of congruent polygons are called *corresponding parts.* Matching vertices are *corresponding vertices.* When you name congruent polygons, always list corresponding vertices in the same order.

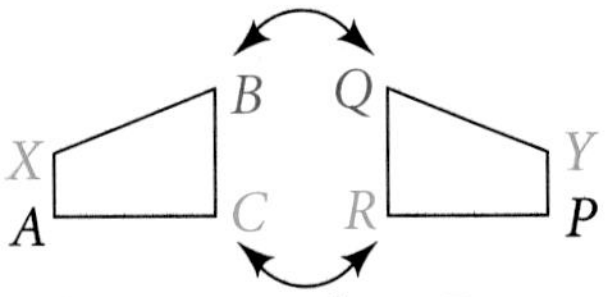

C corresponds to R.
$\angle B$ corresponds to $\angle Q$.
$\overline{AX}$ corresponds to $\overline{PY}$.
$ACBX \cong PRQY$

Example 1 **Relating to the Real World**

Rocketry The fins of the rocket are congruent pentagons. Find $m\angle B$.

Because the fins are congruent, $\angle B \cong \angle E$. So you can find $m\angle B$ by first finding $m\angle E$.

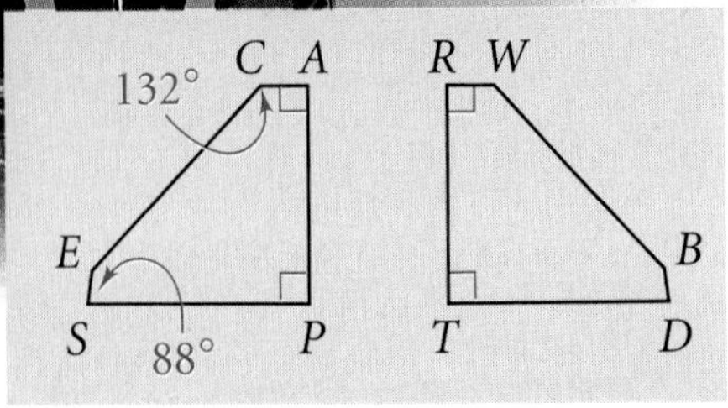

By the Polygon Interior Angle-Sum Theorem, you know that the sum of the measures of the interior angles of pentagon *SPACE* is $(5 - 2)180$, or 540.

$$m\angle S + m\angle P + m\angle A + m\angle C + m\angle E = 540$$
$$88 + 90 + 90 + 132 + m\angle E = 540$$
$$400 + m\angle E = 540$$
$$m\angle E = 140$$

So $m\angle E = m\angle B = 140$.

Example 2

$\triangle TJD \cong \triangle RCF$. List congruent corresponding parts.

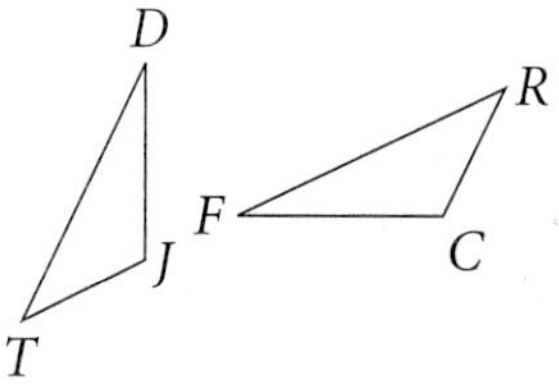

$\overline{TJ} \cong \overline{RC}$ $\overline{JD} \cong \overline{CF}$ $\overline{DT} \cong \overline{FR}$

$\angle T \cong \angle R$ $\angle J \cong \angle C$ $\angle D \cong \angle F$

1. **Try This** $\triangle WYS \cong \triangle MKV$. List congruent corresponding parts.

Similar Polygons

QUICK REVIEW

Corresponding sides are *proportional* if the ratios of their lengths are equal.

Two figures that have the same shape but not necessarily the same size are similar ($\sim$). Two polygons are **similar** if (1) corresponding angles are congruent and (2) corresponding sides are proportional. The ratio of the lengths of corresponding sides is the **similarity ratio.**

Example 3

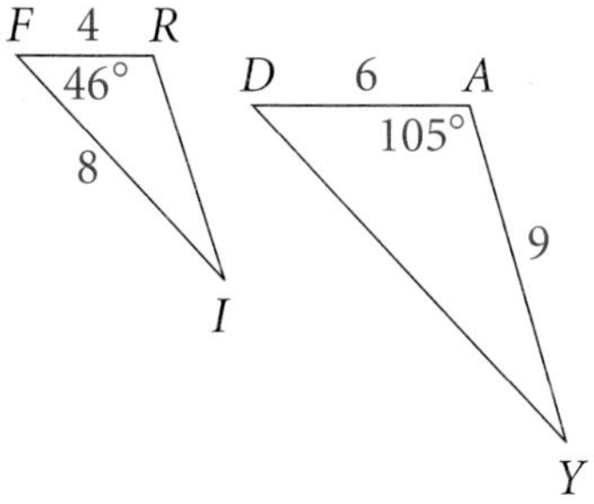

$\triangle FRI \sim \triangle DAY$. Find:

a. the similarity ratio **b.** $m\angle R$ **c.** RI

a. Since $\overline{FR}$ and $\overline{DA}$ are corresponding sides, the similarity ratio is $\frac{FR}{DA} = \frac{4}{6} = \frac{2}{3}$.

b. $\angle R$ corresponds to $\angle A$, so $m\angle R = m\angle A$. $m\angle A = 105$, so $m\angle R = 105$.

c. Write a proportion to solve for RI.

$\frac{FR}{DA} = \frac{RI}{AY}$

$\frac{4}{6} = \frac{RI}{9}$ Substitution

$36 = 6 \cdot RI$ Use cross-products.

$6 = RI$ Divide each side by 6.

PROBLEM SOLVING

Look Back Describe how you could use the similarity ratio to solve for *RI*.

2. **Try This** Find DY and $m\angle D$.
3. **a.** Find the perimeter of $\triangle FRI$.
   b. Find the perimeter of $\triangle DAY$.
   c. What is the ratio of the perimeter of $\triangle FRI$ to the perimeter of $\triangle DAY$?
   d. Compare your answer to part (c) to the similarity ratio. Make a **conjecture** about the ratio of the perimeters of similar figures.
   e. Test your **conjecture** by drawing and measuring other pairs of similar polygons.
4. **Critical Thinking** What type of similar figures have a similarity ratio of 1?

Example 4 Relating to the Real World

Photography You want to enlarge a photo that is 4 in. tall and 6 in. wide into a poster. The poster will be 24 in. wide. How tall will it be?

$\frac{6}{24} = \frac{4}{x}$ Write a proportion.

$6 \cdot x = 24 \cdot 4$ Use cross-products.

$x = 16$ Divide each side by 6.

The poster will be 16 in. tall.

5. What is the similarity ratio of the photo to the poster?

6. *Critical Thinking* A school photo package comes with an 8 in.-by-10 in. photo and a 5 in.-by-7 in. photo. Are the photos similar? Explain.

WORK TOGETHER

Trace the diagram and cut out the seven pieces. (Or use the tans you made in Lesson 2-1.)

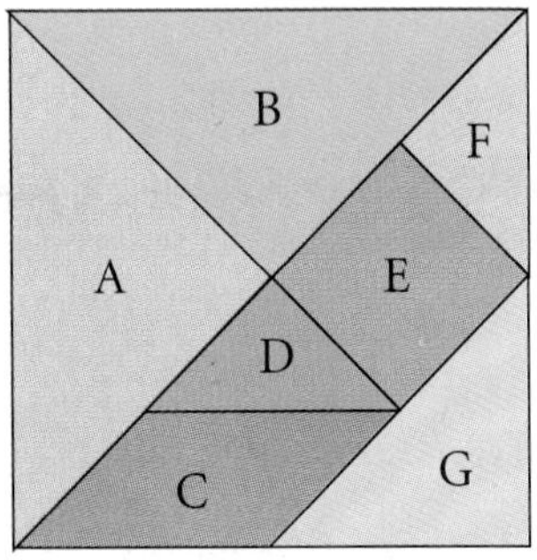

7. Which pieces are congruent?

8. Which pieces are similar? (Check that all pairs of corresponding angles are congruent and that corresponding sides are proportional.)

9. Find other congruent and similar pairs by placing pieces together. Record your answers.

Here are some examples:

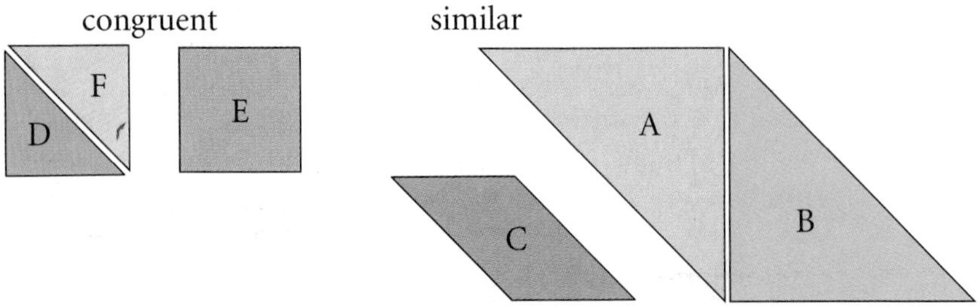

Exercises ON YOUR OWN

1. Identify the pairs of triangles that appear to be congruent.

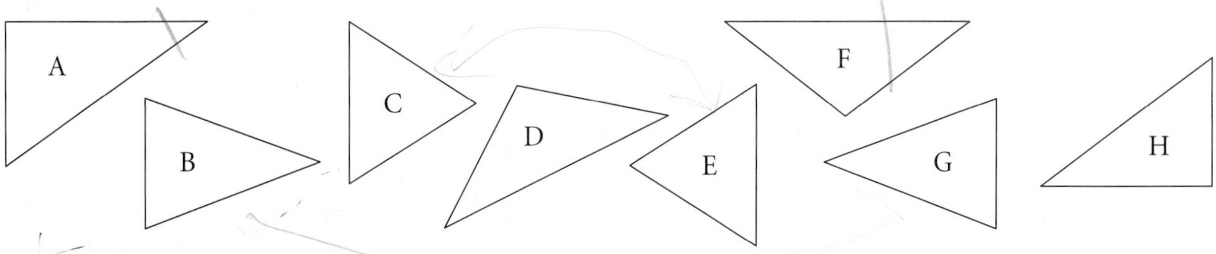

$\triangle LMC \cong \triangle BJK$. Complete the congruence statements.

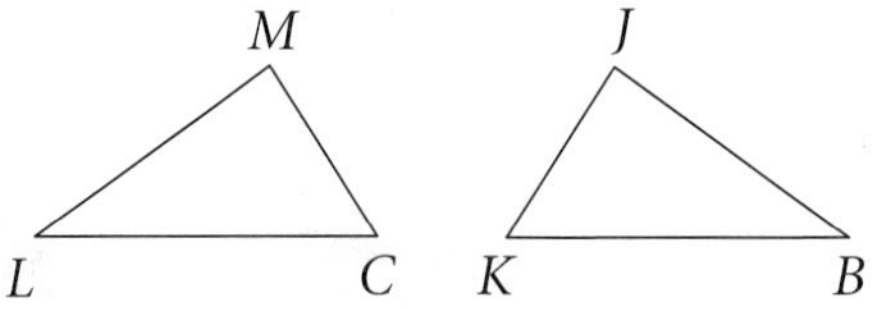

2. $\overline{LC} \cong$ ■
3. $\overline{KJ} \cong$ ■
4. $\overline{JB} \cong$ ■
5. $\angle L \cong$ ■
6. $\angle K \cong$ ■
7. $\angle M \cong$ ■
8. $\triangle CML \cong$ ■
9. $\triangle KBJ \cong$ ■
10. $\triangle MLC \cong$ ■

$JDRT \sim JHYX$. Complete the proportions and congruence statements.

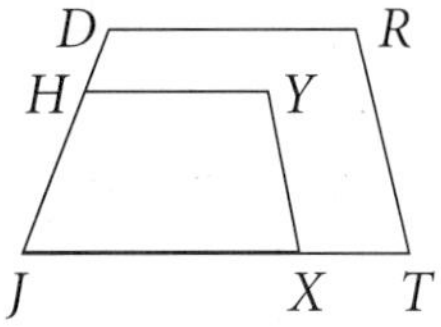

11. $\frac{JD}{JH} = \frac{DR}{■}$
12. $\frac{RT}{YX} = \frac{■}{JX}$
13. $\frac{■}{DR} = \frac{YX}{RT}$
14. $\angle D \cong$ ■
15. $\angle Y \cong$ ■
16. $\angle T \cong$ ■

$\triangle DFG \sim \triangle HKM$. Use the diagram to find the following.

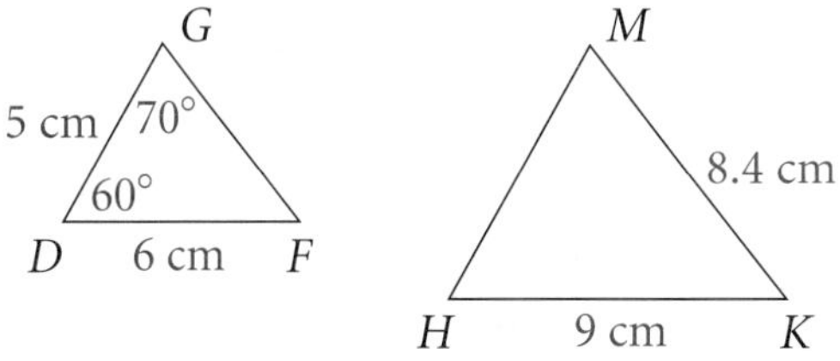

17. the similarity ratio of $\triangle DFG$ to $\triangle HKM$
18. the similarity ratio of $\triangle HKM$ to $\triangle DFG$
19. $m\angle F$
20. $m\angle K$
21. $m\angle M$
22. $\frac{DF}{HK}$
23. HM
24. GF

In Exercises 25–27, $POLY \cong SIDE$.

25. List four pairs of congruent angles.
26. List four pairs of congruent sides.
27. Complete the congruence statements.
   a. $OLYP \cong$ ■ b. $DESI \cong$ ■

28. **Art** An art class is painting a mural for a spring festival. The students are working from a diagram that is 48 in. long and 36 in. high. Find the length of the mural if its height is to be 12 ft.

29. **Research** A *fractal* is a self-similar geometric pattern. It is made up of parts that are similar to the object itself. A good example of this is a fern frond. Break off any leaflet and the leaflet looks like a small fern frond. Investigate fractals and **summarize** your findings.

30. **a.** Draw two different-sized squares on graph paper. Find the ratio of each pair of corresponding sides.
    b. Are these squares similar? Explain.
    c. Draw two different-sized rectangles on graph paper. Find the ratio of each pair of corresponding sides.
    d. Are these rectangles similar? Explain.
    e. **Writing** Are all squares similar? Are all rectangles similar? Explain. Sketch pictures to support your conclusion.

31. **Critical Thinking** Kimi claims that all circles are similar. Is she right? Explain.

Write a congruence statement for each pair of triangles.

32.
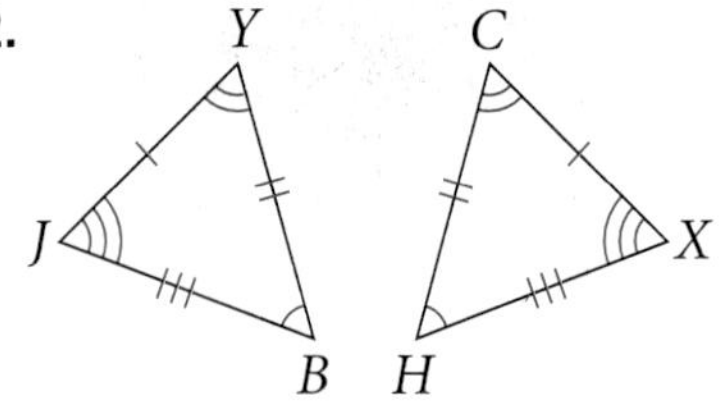

33.
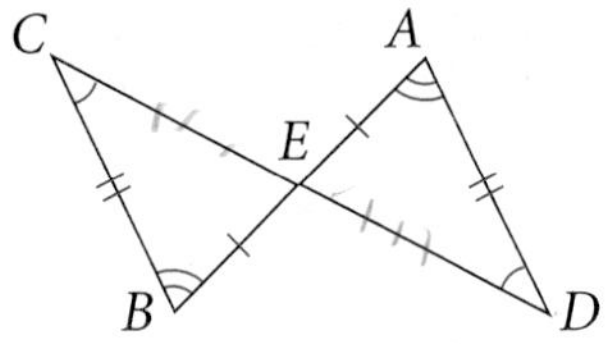

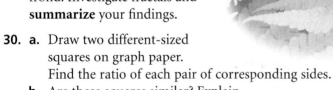

E is the midpoint of $\overline{CD}$.

34.
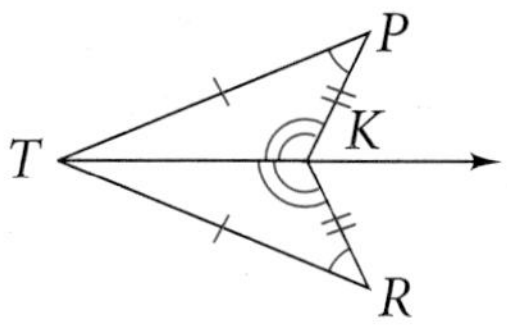

$\overrightarrow{TK}$ bisects $\angle PTR$.

Choose **Use pencil and paper, mental math, or a calculator to find the values of the variables.**

35.
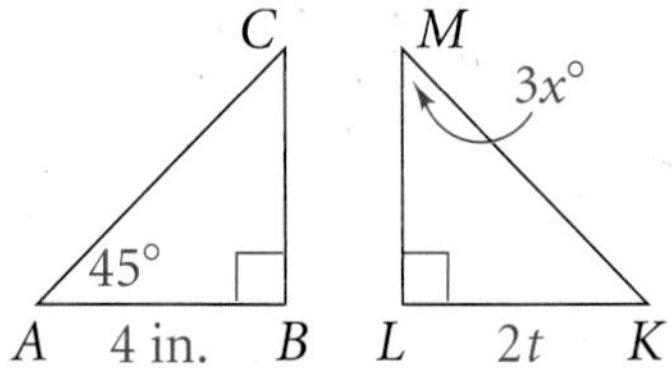

$\triangle ABC \cong \triangle KLM$

36.
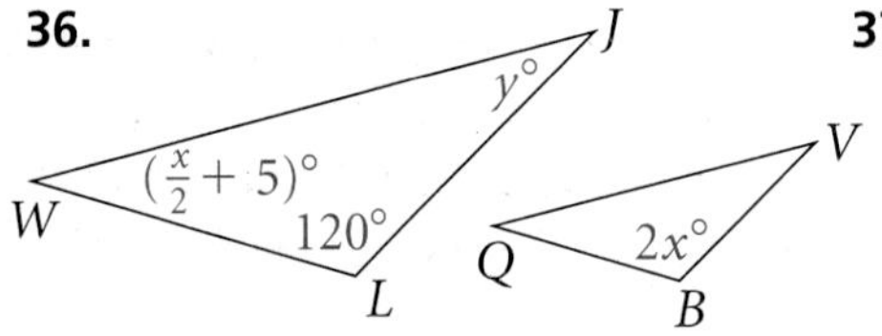

$\triangle WLJ \sim \triangle QBV$

37.
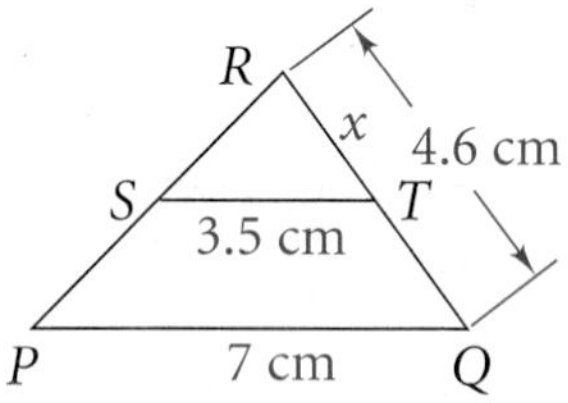

$\triangle PRQ \sim \triangle SRT$

38. **Standardized Test Prep** $\triangle KJH$ is congruent to the triangle shown. Which of these *cannot* be the coordinates of point H?
    A. (5, 0) **B.** (5, 4) **C.** (6, 4)
    D. (6, 0) **E.** (7, 0)

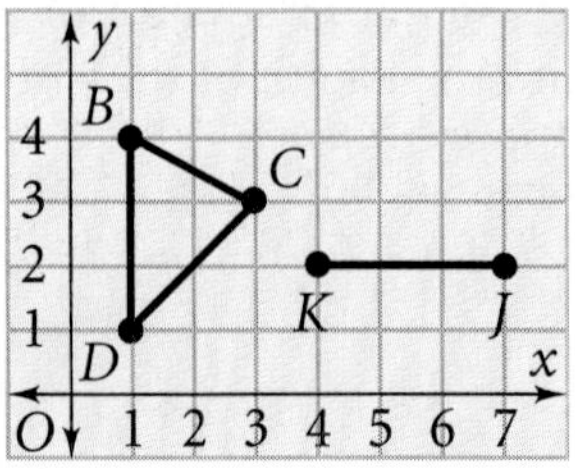

39. When you make an enlargement on a copy machine, the enlargement is similar to the original. You are making an enlargement of $\triangle ABC$. Suppose you choose the 120% enlargement setting. This means that the similarity ratio of the enlargement to the original is $\frac{120}{100}$, or $\frac{6}{5}$.

a. Find the lengths of the sides of the enlargement.

b. Find the measures of the angles of the enlargement.

c. Suppose you use the 150% setting. Find the lengths of the sides of the triangle.

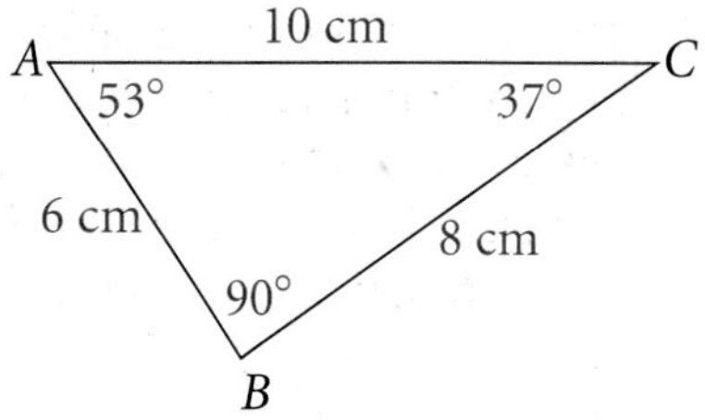

The triangles are similar. Find the similarity ratio of the first to the second.

40.

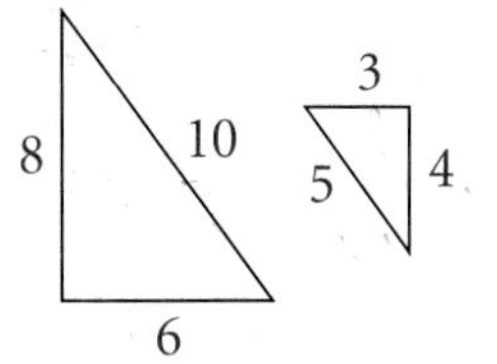

41.

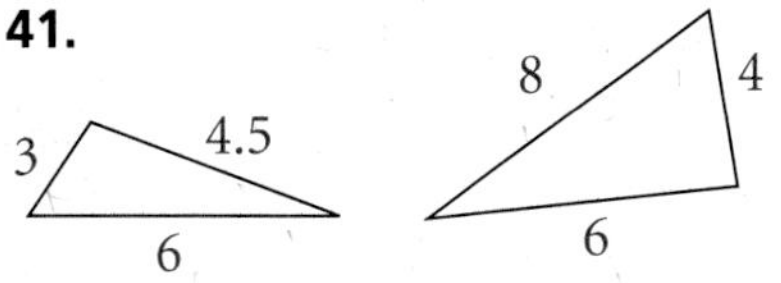

42.

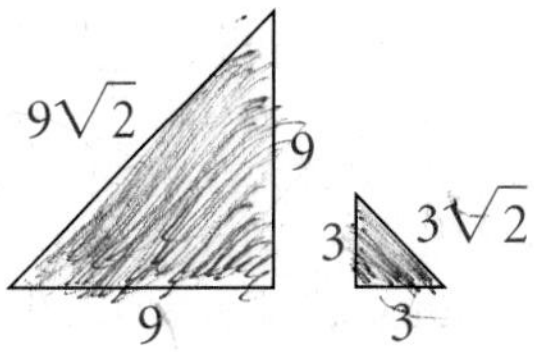

43. **a.** ***Photography*** Is an 8 in.-by-10 in. photograph similar to a 16 in.-by 20 in. photograph? Explain.

b. Is a 4 in.-by-5 in. photograph similar to a 5 in.-by-7 in. photograph? Explain.

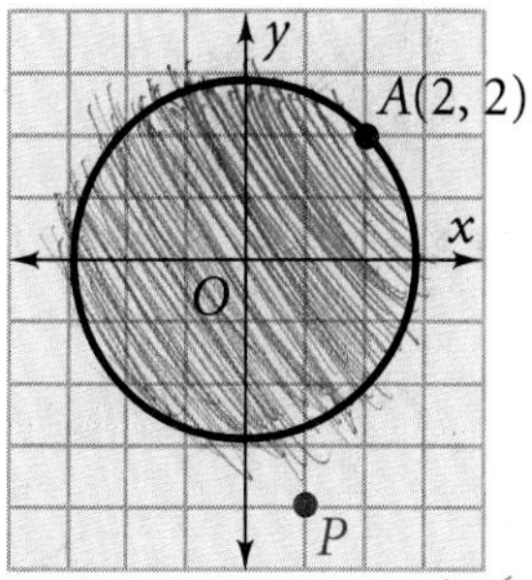

44. ***Standardized Test Prep*** The circle shown is congruent to a circle with center $P(1, -4)$. Which of these *cannot* be the coordinates of a point on $\odot P$?

A. $(-1, -2)$ **B.** $(-1, -6)$ **C.** $(3, -2)$ **D.** $(3, -4)$ **E.** $(3, -6)$

45. ***Open-ended*** Draw two quadrilaterals that have sides in the ratio 2 : 1 and yet are not similar.

Chapter Project

Find Out by Investigating

Use the pentominoes you made in the Find Out exercise on page 95 to investigate similarity.

- How many more pentominoes will it take to complete the 3-by-15 rectangle below? How do you know? Complete the rectangle and record your solution on graph paper.

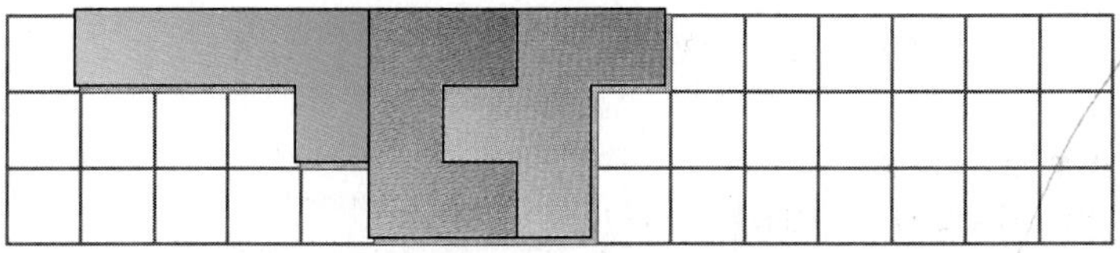

- Is this 3-by-15 rectangle *similar* to the rectangular pentomino piece? Explain. If it is, what is the similarity ratio?

Exercises MIXED REVIEW

46. *Constructions* Draw a 3-in. segment. Construct the perpendicular bisector of that segment. Check your work with a ruler and a protractor.

47. *Coordinate Geometry* Connect $A(3, 3)$, $B(5, 5)$, $T(9, 1)$, and $S(9, -3)$ in order. What type of quadrilateral is $ABTS$? Explain.

48. $M(-1, 0)$ is the midpoint of $\overline{AB}$. The coordinates of A are $(5, 1)$. Find the coordinates of B.

49. What is the measure of each interior angle of a regular 18-gon?

SELF ASSESSMENT

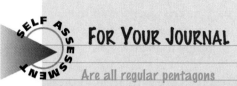

FOR YOUR JOURNAL

Are all regular pentagons similar? Are all regular hexagons similar? Make a conjecture about regular polygons and similarity. Justify your reasoning.

Getting Ready for Lesson 2-7

Imagine you are looking down at each of these figures from above. Describe the geometric figure you see.

50.

51.

52.

Exercises CHECKPOINT

Find the center and radius of each circle with diameter $\overline{AB}$.

1. $A(4, 1), B(7, 5)$ **2.** $A(0, 8), B(3, 6)$ **3.** $A(-3, 9), B(4, -2)$ **4.** $A(-2, -5), B(-8, 4)$

Find the measure of each arc or angle in $\odot A$.

5. $\angle WAX$ **6.** $\overset{\frown}{RX}$ **7.** $\angle SAR$

8. $\overset{\frown}{TRW}$ **9.** $\angle TAW$ **10.** $\overset{\frown}{RT}$

11. $\overset{\frown}{SR}$ **12.** $\overset{\frown}{XST}$ **13.** $\overset{\frown}{STW}$

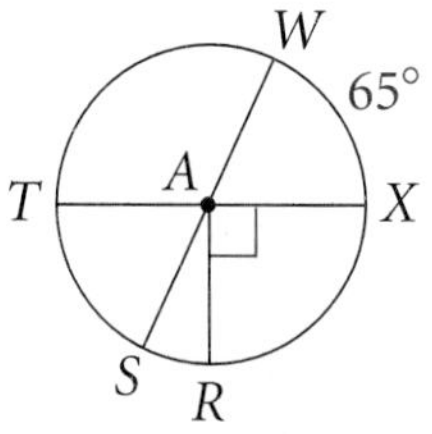

Draw a pair of figures to fit each description.

14. congruent right triangles **15.** similar pentagons **16.** similar rectangles

17. What is the most precise name for a quadrilateral with vertices at (3, 5), (−1, 4), (7, 4), and (3, −5)?

A. kite **B.** rectangle **C.** parallelogram **D.** rhombus **E.** trapezoid

What You'll Learn

- Drawing isometric and orthographic views of objects

...And Why

To practice a skill used in industries of all sorts

What You'll Need

- isometric dot paper
- graph paper
- straightedge
- cubes

2-7 Isometric and Orthographic Drawings

THINK AND DISCUSS

Isometric Drawings

The figures that you've studied so far in this chapter have been two-dimensional. The *faces* of three-dimensional figures are two-dimensional. Many types of industries use two-dimensional drawings of three-dimensional objects. The makers of some animated cartoons for example, use computers to create *wire-frame models* of three-dimensional characters. These wire-frame images—basically complicated stick figures—are made up of polygons.

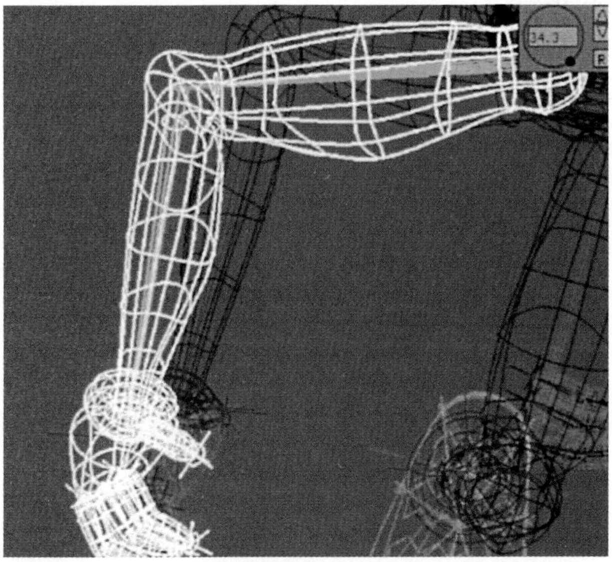

The wire frame model of the insect's leg is made up of polygons. Notice how they are joined to give the illusion of a curved surface.

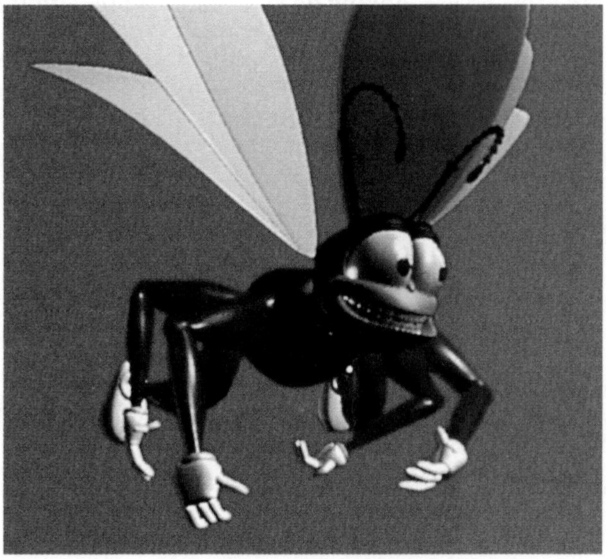

The computer forms the final image by adding a "skin" to the wire-frame and then lighting the figure.

One way to show a three-dimensional object is with an isometric drawing. An **isometric drawing** shows a corner view. It shows three sides of an object in a single drawing. Here are two examples.

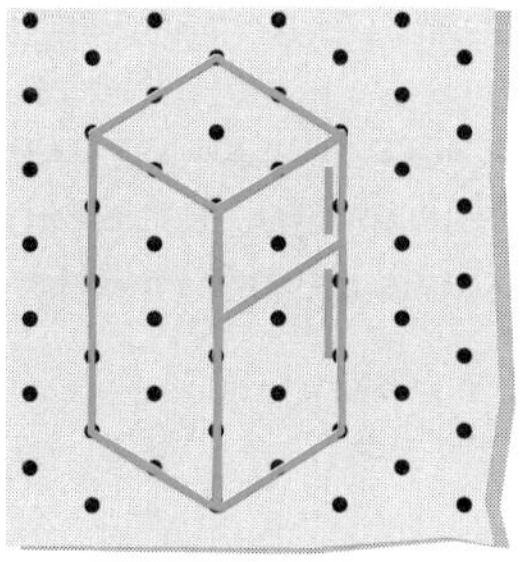

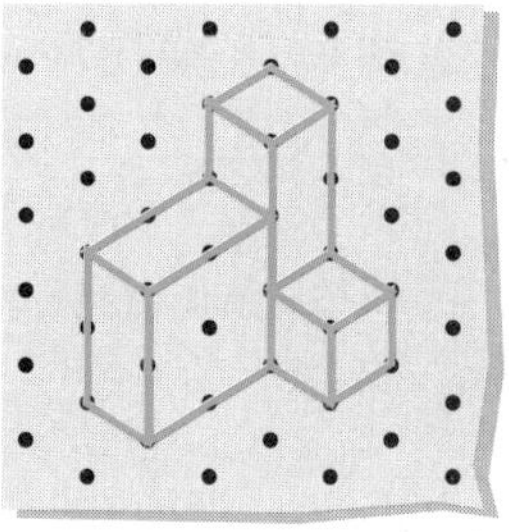

You can use isometric dot paper to draw cube structures.

Example 1

Create an isometric drawing of the cube structure at the left.

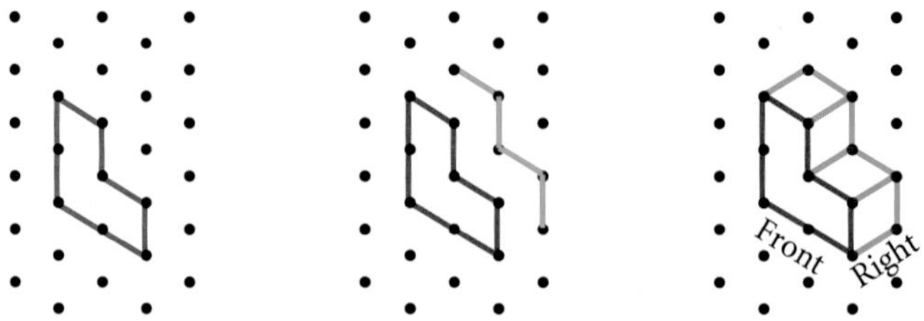

Step 1 Step 2 Step 3

1. **Try This** Build a structure using 4 cubes. Create an isometric drawing of your structure.

Another way to show some types of three-dimensional objects is with a foundation drawing. A **foundation drawing** shows the base of a structure and the height of each part. The diagram at the right is a foundation drawing of the Sears Tower in Chicago.

54	67	54
67	98	67
41	98	41

The Sears Tower is made up of nine sections. The numbers tell how many stories tall each section is.

Example 2

Create a foundation drawing for the isometric drawing below.

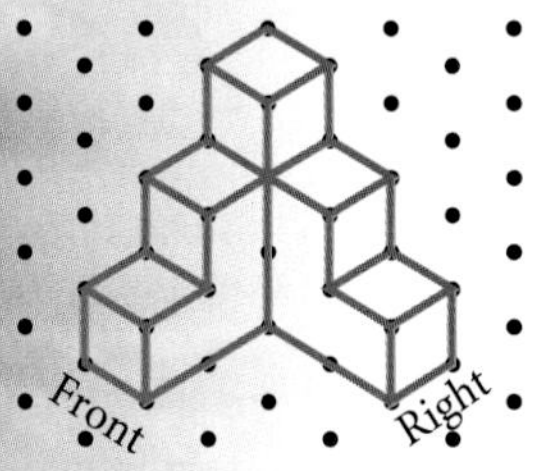

Isometric Drawing

Foundation Drawing

2. a. How many cubes are needed to make the structure in Example 2?
   b. Which drawing did you use to answer part (a), the foundation drawing or the isometric drawing? Why?

Orthographic Drawings

A third way to show three-dimensional figures is with an orthographic drawing. An **orthographic drawing** shows a top view, front view, and right-side view. Here is an example.

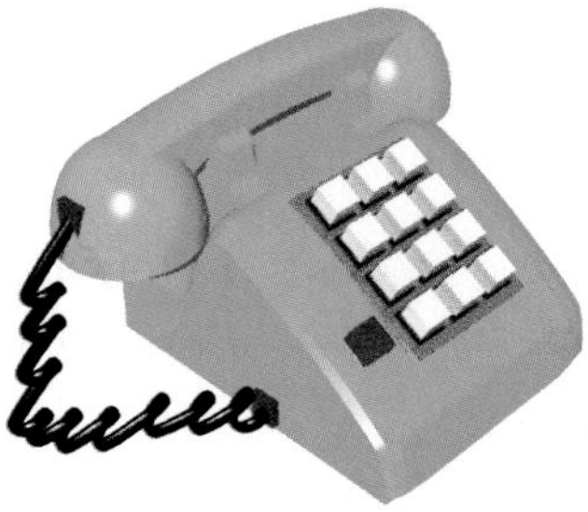

Isometric Drawing

Front

Top

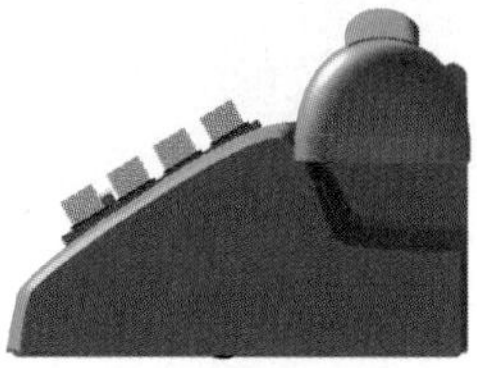

Right

Example 3

Create an orthographic drawing for each isometric drawing at the left.

Isometric Drawing

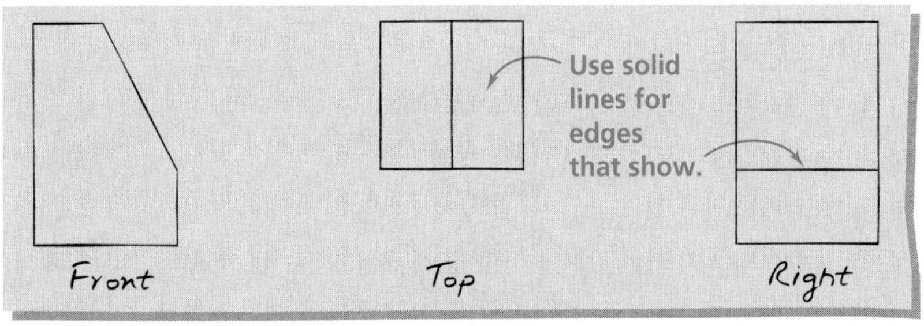

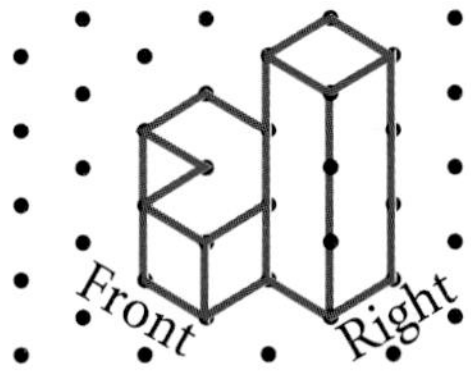

Isometric Drawing

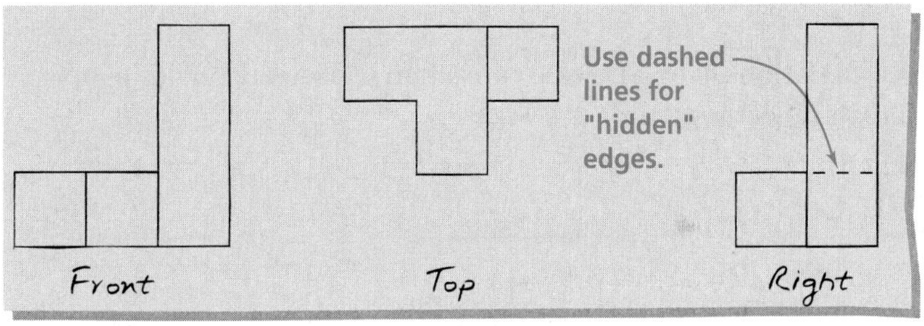

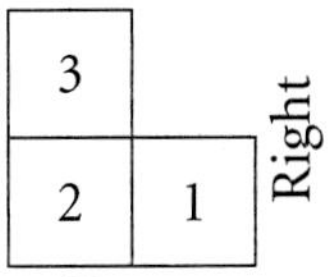

Front

3. *Open-ended* Choose a simple object in your classroom. Make an orthographic drawing showing three views.

4. a. *Manipulatives* Build a cube structure for the foundation drawing at the left. Then create an isometric drawing for the structure.

   b. Create an orthographic drawing showing a top view, front view, and right-side view.

WORK TOGETHER

Step 1: Work in a group to build a structure using 12 cubes. Label the sides of the structure *Front, Back, Right,* and *Left.*

Step 2: Have each member of your group use graph paper to create a foundation drawing of the structure. Then use isometric dot paper to create an isometric drawing. In both drawings, label the sides.

Step 3: Trade isometric drawings with another group. Use their drawings to re-create their structure. Use the other group's foundation drawings to check that you built their structure correctly.

Step 4: Repeat Steps 1–3, substituting orthographic drawings for the isometric drawings in Steps 2 and 3.

Exercises ON YOUR OWN

For each figure, (a) create a foundation drawing, and (b) create an orthographic drawing.

1.

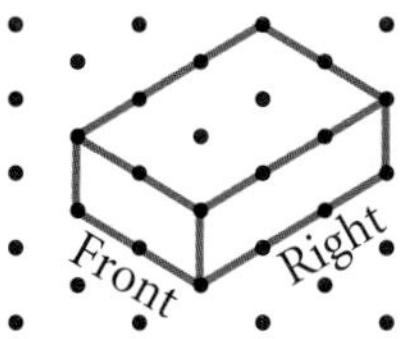

2.

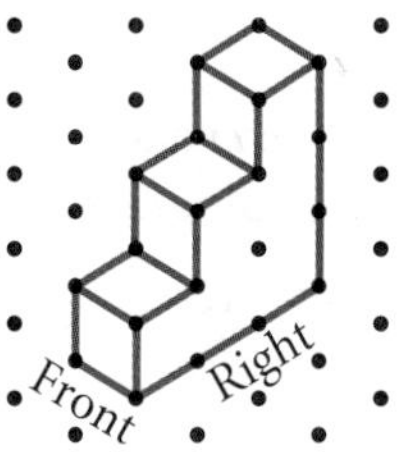

3.

4.

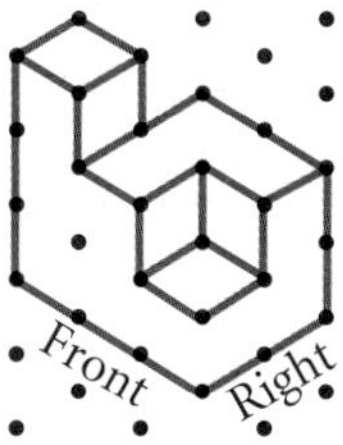

For each foundation plan, (a) create an isometric drawing on dot paper, and (b) create an orthographic drawing.

5.

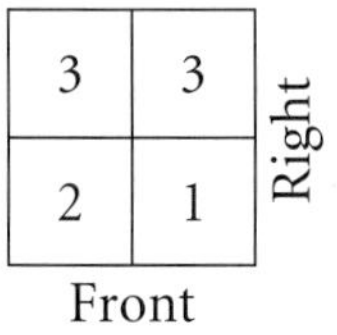

6.

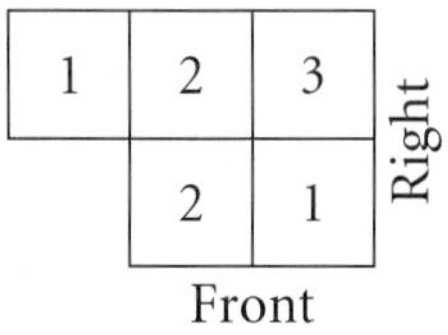

7.

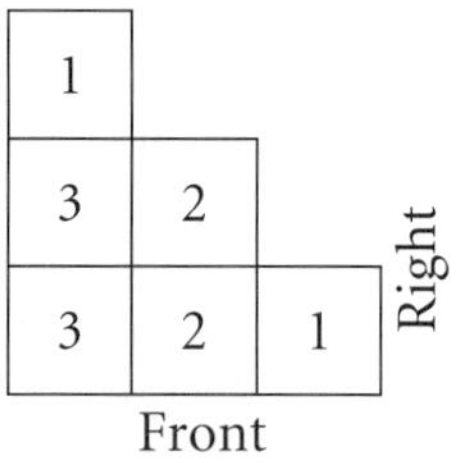

8.

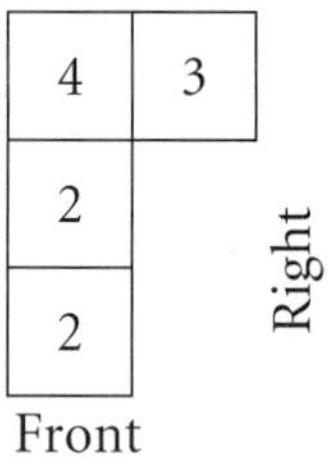

9. a. *Open-ended* Create an isometric drawing of a figure that can be constructed using 8 cubes.
   b. Create an orthographic drawing of this structure.
   c. Create a foundation plan for the figure.

Read the comic strip and complete Exercises 10 and 11.

SHOE by Jeff MacNelly

10. What type of drawing that you've studied in this lesson is a "bird's-eye view"?

11. *Writing* Photographs of the Washington Monument are typically not taken from a bird's-eye view. Describe a situation in which you would want a photo showing a bird's-eye view.

Cross Sections **Imagine cutting straight through an orange. The *cross section* will always be a circle. Describe the cross section in each diagram.**

Sample:

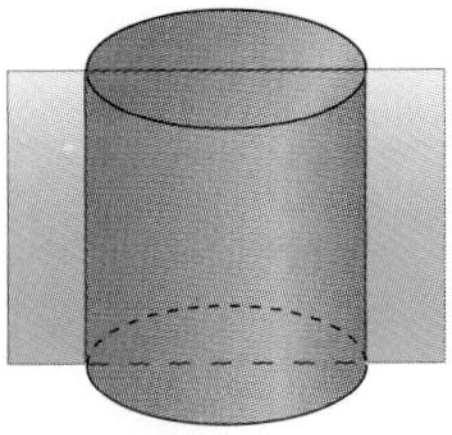

This cross section of a cylinder is a rectangle. The width of the rectangle is the diameter of the circular base of the cylinder. The length of the rectangle is the height of the cylinder.

12.

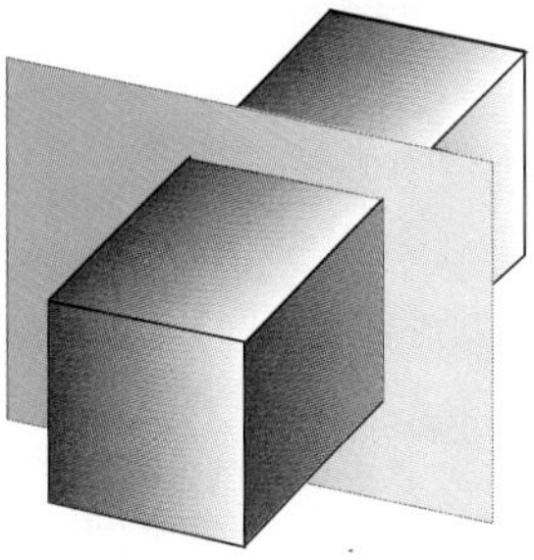

13.

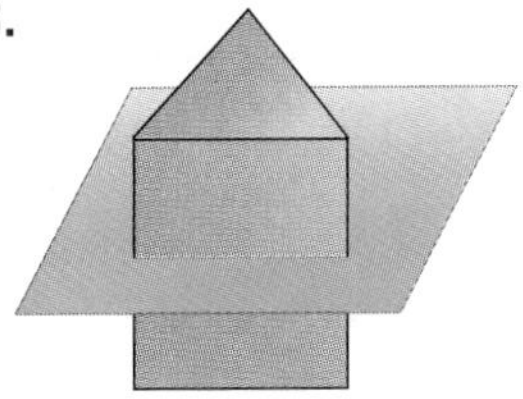

14.

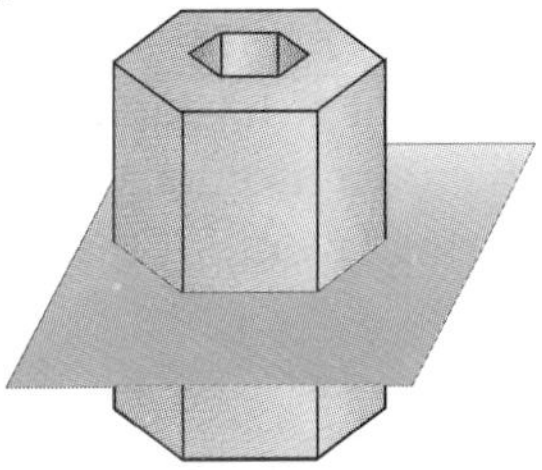

15.

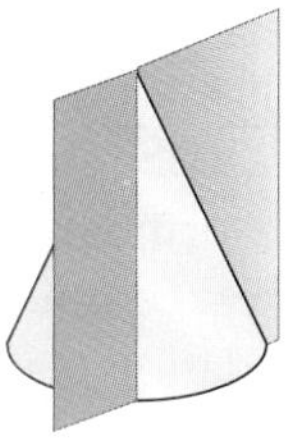

16. *Probability* Jake makes a structure using 27 wooden cubes. He paints four of the faces blue and leaves the top and bottom unpainted. Then he takes the structure apart and places the cubes in a bag. Leah closes her eyes, reaches into the bag, and pulls out a cube.

a. What is the probability that the cube is unpainted?

b. What is the probability that two of its faces are blue?

c. What is the probability that only one of its faces is blue?

Match each isometric drawing with the correct orthographic drawing.

17.
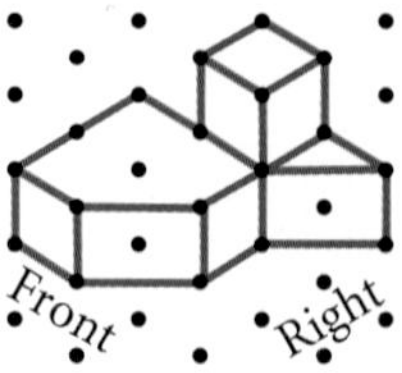

18.
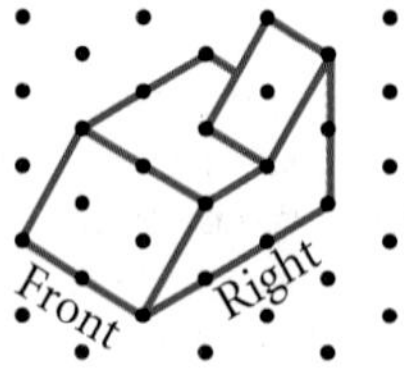

19.

20.
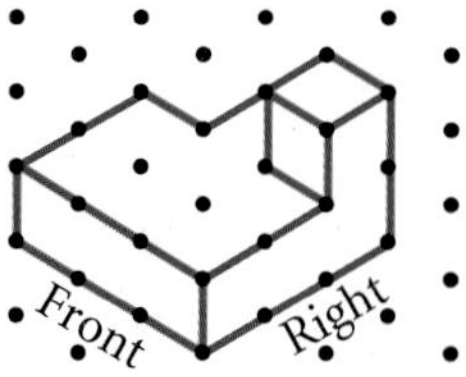

A.
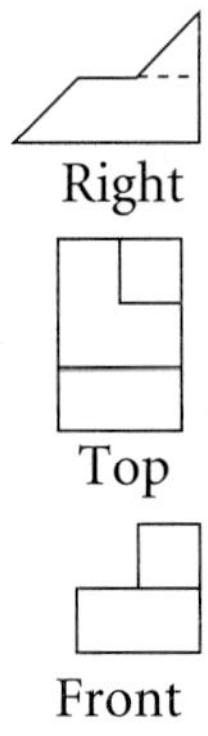

B.
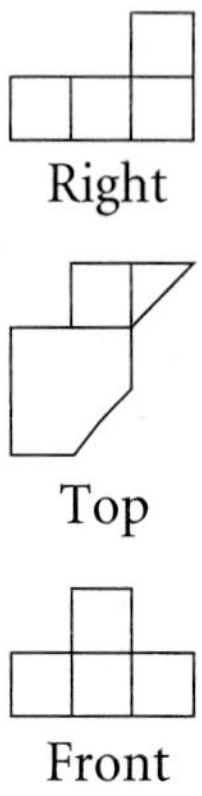

C.
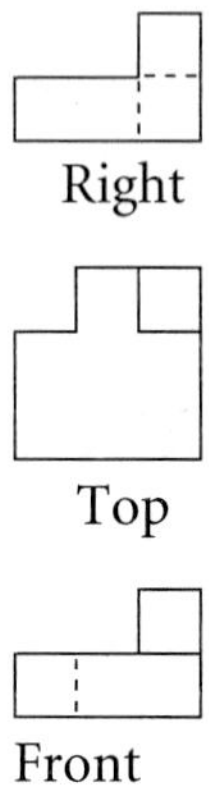

D.

Create an orthographic drawing for each isometric drawing.

21.
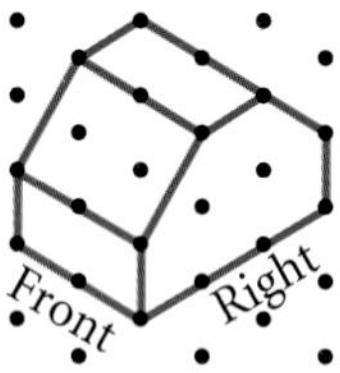

22.
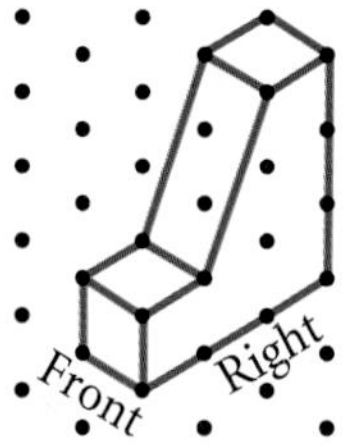

23.
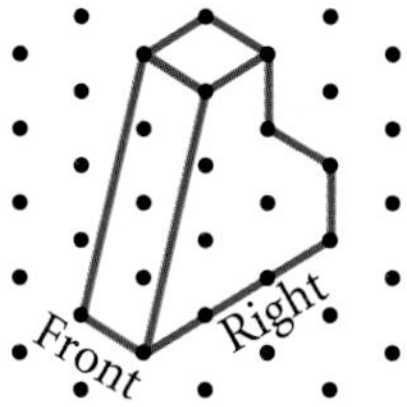

24.
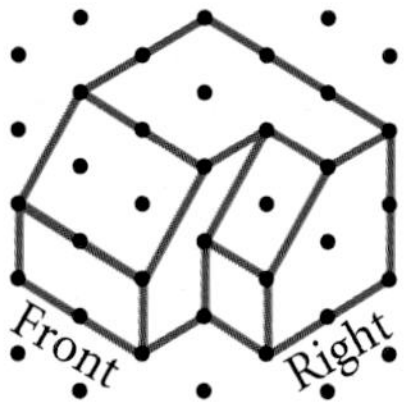

25. **Engineering** Engineers use an *engineering layout* to describe structures. A complete layout includes three orthographic views and an isometric view. Make a complete engineering layout for the foundation plan.

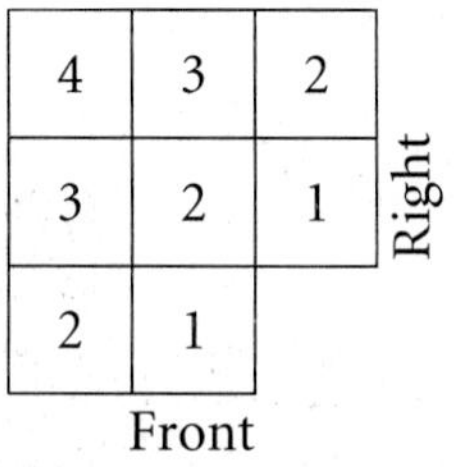

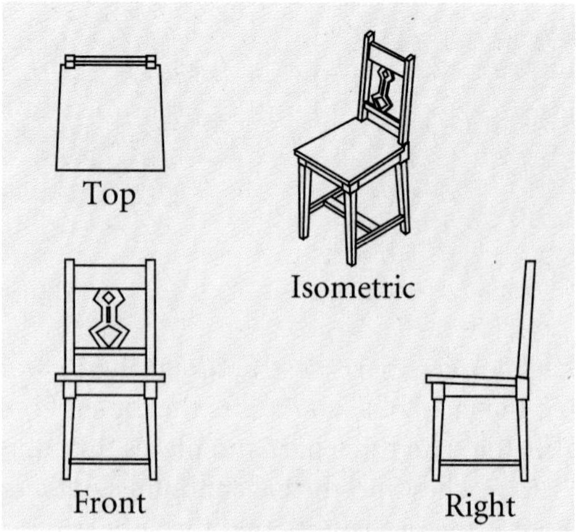

Engineering Layout

Chapter Project

Find Out by Modeling

Soma is a three-dimensional puzzle made of unit cubes. All Soma pieces consist of 3 or 4 unit cubes. Each cube shares a face with at least one other cube. Every Soma piece must contain at least two cubes, which meet to form a 90° dihedral angle. (A *dihedral angle* is an angle formed by two planes.) There are exactly seven Soma pieces, and all are different. The photograph shows two Soma pieces.

- Build a set of Soma pieces by gluing cubes together.
- Draw the other five Soma pieces on isometric dot paper.
- Use all seven Soma pieces to build a cube.

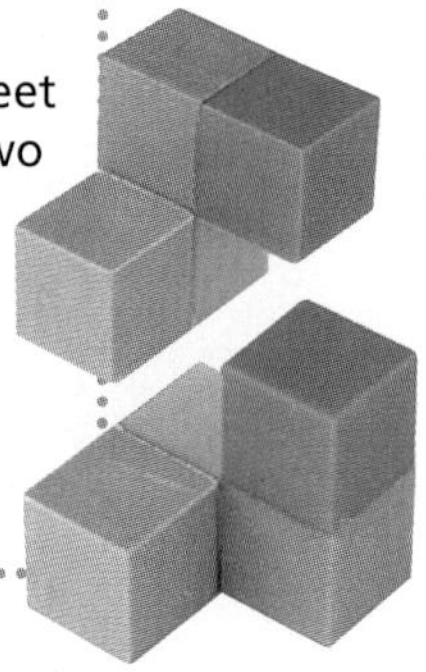

Exercises MIXED REVIEW

Find the length of each segment with the given endpoints. Then find the coordinates of the midpoint of each segment.

26. $A(4, 6)$ and $B(2, 9)$ **27.** $G(-3, -1)$ and $H(1, 2)$

28. Use slope to determine whether points $A(0, 1)$, $B(-3, -2)$, and $C(4, 5)$ are collinear.

Draw a diagram that illustrates each concept.

29. adjacent angles **30.** perpendicular bisector

SELF ASSESSMENT

PORTFOLIO

For your portfolio, select one or two items from your work for this chapter. Here are some possibilities:
- most improved work
- a Journal entry

Explain why you have included each selection.

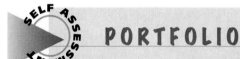

Geometry at Work

Industrial Designer

Industrial designers work on two-dimensional surfaces to develop products that have three-dimensional appeal to consumers. They use computer-aided design (CAD) software to create two-dimensional screen images and manipulate them for three-dimensional effects. Fashion designers use CAD to study their creations on electronic human forms from various angles and distances.

Mini Project: Design an athletic shoe. Use an engineering layout to display your design. Include three orthographic views and an isometric view.

CHAPTER PROJECT

Finishing the Chapter Project

AMAZING SPACE

Find Out exercises on pages 74, 81, 95, 107, and 115 should help you to complete your project. Prepare a *Geometric Diversions* display. Include the models you made, instructions for making them, and the geometric patterns you discovered.

Reflect and Revise

Ask a classmate to review your display with you. Together, check that your solutions are correct, your diagrams clear, and your explanations sensible. Have you used geometric terms correctly? Is the display organized, comprehensive, and visually appealing? Consider doing more research (using some of the books listed below) on other popular puzzles and writing a short report.

Follow Up

Now that you've had some experience exploring geometric puzzles, create your own puzzle. Start with any two- or three-dimensional figure and go from there. Challenge your classmates!

For More Information

Costello, Matthew J. *The Greatest Puzzles of All Time.* New York: Prentice Hall Press, 1988.

Gardner, Martin. *The Scientific American Mathematical Puzzles & Diversions.* New York: Simon & Schuster, 1959.

Gardner, Martin. *The 2nd Scientific American Book of Mathematical Puzzles & Diversions.* New York: University of Chicago Press, 1987.

Kenney, Margaret J., Stanley J. Bezuszka, and Joan D. Martin. *Informal Geometry Explorations.* Palo Alto, California: Dale Seymour Publications, 1992.

Reid, Ronald C. *Tangrams-330 Puzzles.* New York: Dover Publications, Inc., 1965.

CHAPTER 2

Wrap Up

Key Terms

acute triangle (p. 71)
adjacent arcs (p. 98)
circle (p. 96)
central angle (p. 97)
concave (p. 76)
congruent circles (p. 102)
congruent polygons (p. 102)
convex (p. 76)
corollary (p. 70)
diameter (p. 96)
equiangular polygon (p. 78)
equiangular triangle (p. 71)
equilateral polygon (p. 78)
equilateral triangle (p. 71)
exterior angle (p. 69)
foundation drawing (p. 110)
isometric drawing (p. 109)
isosceles trapezoid (p. 91)
isosceles triangle (p. 71)
kite (p. 91)
major arc (p. 98)
minor arc (p. 98)
obtuse triangle (p. 71)
orthographic drawing (p. 111)
parallelogram (p. 91)
polygon (p. 76)
radius (p. 96)
rectangle (p. 91)
regular polygon (p. 78)
remote interior angle (p. 69)
rhombus (p. 91)
right triangle (p. 71)
scalene triangle (p. 71)
semicircle (p. 98)
similar polygons (p. 103)
similarity ratio (p. 103)
square (p. 91)
trapezoid (p. 91)

How am I doing?

SELF ASSESSMENT

- State three ideas from this chapter that you think are important. Explain your choices.
- Describe different ways of classifying triangles and quadrilaterals.

Triangles 2-1

The sum of the measures of the angles of a triangle is 180. The measure of each **exterior angle** of a triangle equals the sum of the measures of its two **remote interior angles,** and is therefore greater than the measure of either remote interior angle.

You can classify triangles according to their sides and angles.

Find the values of the variables. Then classify each triangle by its sides and angles.

1.

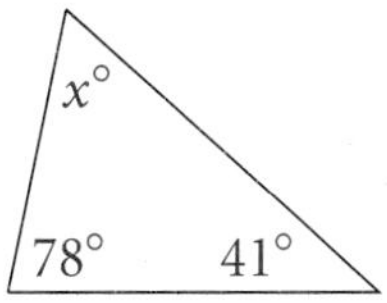

2.

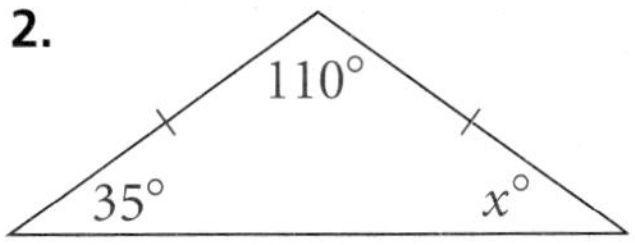

3.

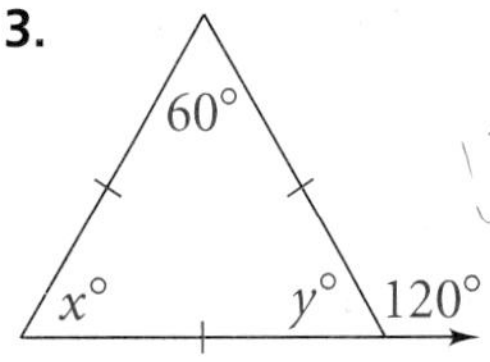

4.

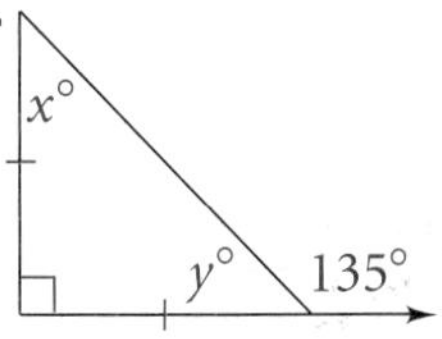

5. *Standardized Test Prep* The measures of the angles of different triangles are shown below. Which triangle is obtuse?

A. $x + 10, x - 20, x + 25$ **B.** $x, 2x, 3x$
C. $20x + 10, 30x - 2, 7x + 1$ **D.** $10x - 3, 14x - 20, x + 3$
E. none of the above

Polygons and Classifying Quadrilaterals — 2-2, 2-4

A **polygon** is a closed plane figure with at least three sides. A polygon is **convex** if no diagonal contains points outside the polygon. Otherwise, it is **concave.** A **regular polygon** is equilateral and equiangular.

The sum of the measures of the interior angles of an n-gon is $(n - 2)180$. The sum of the measures of the exterior angles (one at each vertex) of an n-gon is 360.

Quadrilaterals have four sides. Some quadrilaterals have special names.

Find the measure of each interior angle and exterior angle for each regular polygon.

6. a hexagon **7.** an octagon **8.** a decagon **9.** a 24-gon

Algebra **Find the values of the variables and the lengths of the sides.**

10. isosceles trapezoid $ABCD$

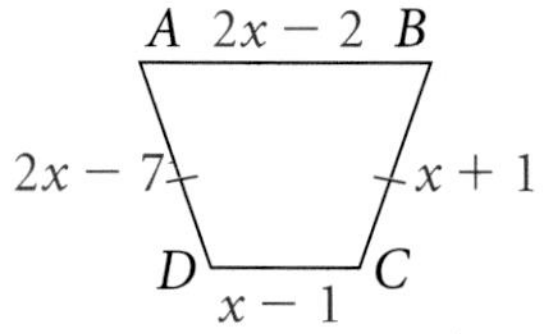

11. kite $KLMN$

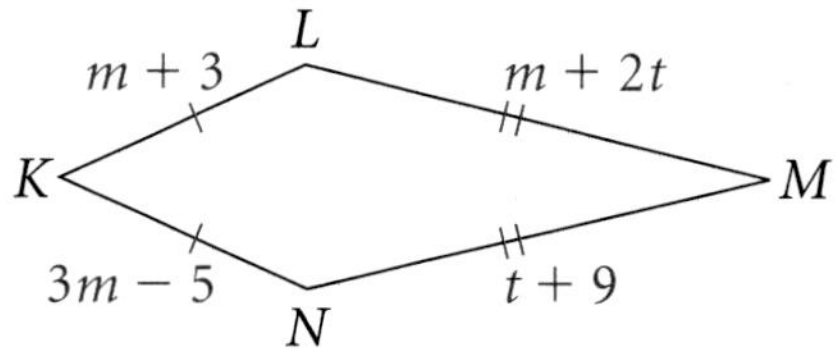

12. rhombus $PQRS$

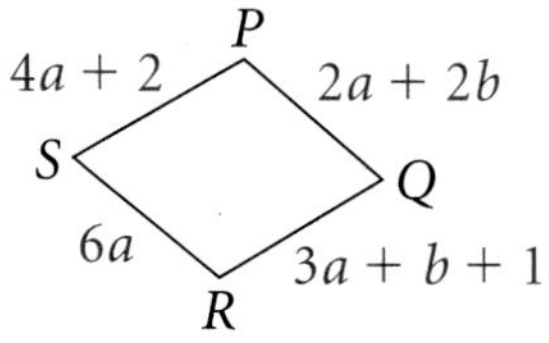

Parallel and Perpendicular Lines in the Coordinate Plane — 2-3

The slopes of two nonvertical parallel lines are equal. Two lines parallel to a third line are parallel to each other.

The product of the slopes of two nonvertical perpendicular lines is -1. In a plane, two lines perpendicular to a third line are parallel to each other.

Find the slopes of $\overleftrightarrow{AB}$ and $\overleftrightarrow{CD}$. Then determine if the lines are parallel, perpendicular, or neither.

13. $A(-1, -4), B(2, 11), C(1, 1), D(4, 10)$

14. $A(2, 10), B(-1, -2), C(3, 7), D(0, -5)$

15. $A(-3, 3), B(0, 2), C(1, 3), D(-2, -6)$

16. $A(-1, 3), B(6, 10), C(-6, 0), D(4, 10)$

Circles — 2-5

A **circle** is the set of all points in a plane equidistant from one point called the center. The measure of a minor arc is the measure of its corresponding central angle. The measure of a major arc is 360 minus the measure of its related minor arc. **Adjacent arcs** have exactly one point in common.

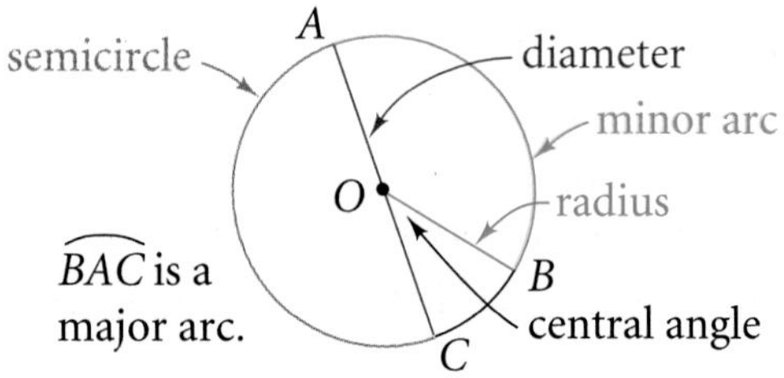

Find the coordinates of the center and the radius of each circle with diameter $\overline{AB}$.

17. $A(4, 0), B(4, 6)$ **18.** $A(2, -3), B(0, 1)$ **19.** $A(-4, -5), B(2, -1)$ **20.** $A(7, 2), B(4, 8)$

Find each measure.

21. $m\angle APD$ **22.** $m\overset{\frown}{AC}$

23. $m\overset{\frown}{ABD}$ **24.** $m\angle CPA$

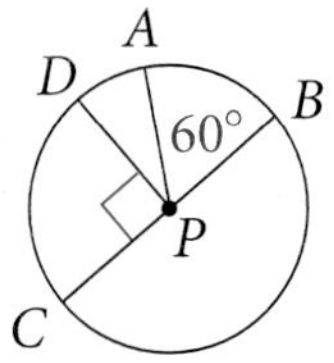

Congruent and Similar Figures 2-6

Congruent polygons have congruent corresponding parts. **Similar polygons** have congruent corresponding angles and proportional corresponding sides. The ratio of the lengths of corresponding sides is the **similarity ratio.**

***RSTUV* ≅ *KLMNO*. Complete each congruence statement.**

25. $\overline{TS} \cong$ ■ **26.** $\angle N \cong$ ■ **27.** $\overline{LM} \cong$ ■ **28.** $VUTSR \cong$ ■

29. Open-ended Sketch a pair of similar hexagons.

Isometric and Orthographic Drawings 2-7

There are different ways to make two-dimensional drawings of three-dimensional objects. An **isometric drawing** shows three sides of an object in one drawing. A **foundation drawing** shows the bottom of a structure and the height of each part. An **orthographic drawing** shows the top, front, and right-side view of an object.

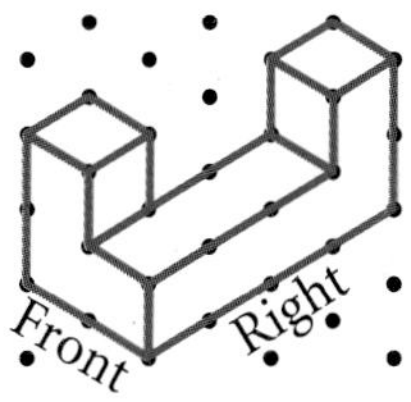

30. Use the isometric drawing at the right.
  a. Make an orthographic drawing.
  b. Make a foundation drawing.

31. Writing Describe a situation in which an orthographic drawing is useful.

Getting Ready for CHAPTER 3

Draw a pair of figures to fit each description.

32. similar acute triangles with similarity ratio $\frac{1}{2}$

33. similar quadrilaterals with similarity ratio $\frac{3}{4}$

Draw each figure. Then draw a line dividing the figure in half.

34. regular octagon **35.** rectangle **36.** isosceles trapezoid **37.** kite

CHAPTER 2 Assessment

Use a protractor and a centimeter ruler to classify each triangle by its angles and sides.

1. 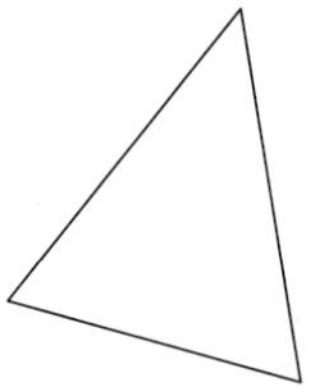

2. 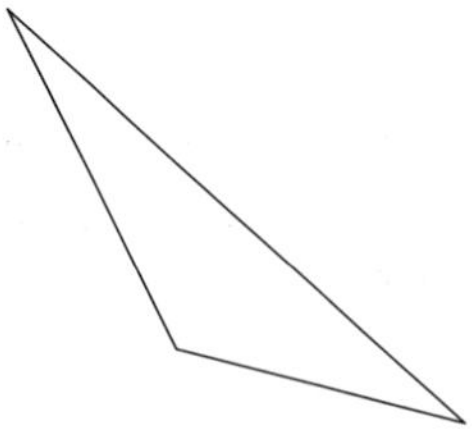

Algebra **Find the values of the variables.**

3. 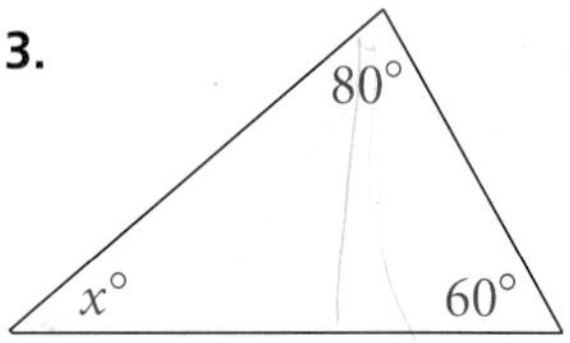

4. 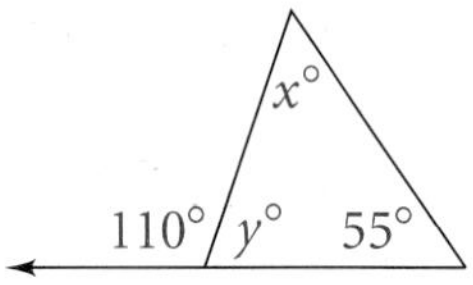

5. 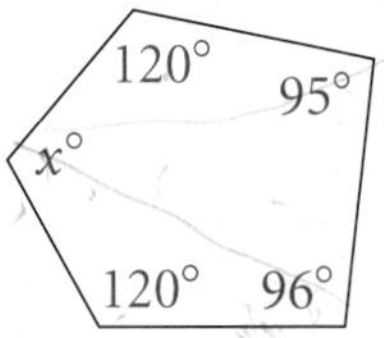

6. 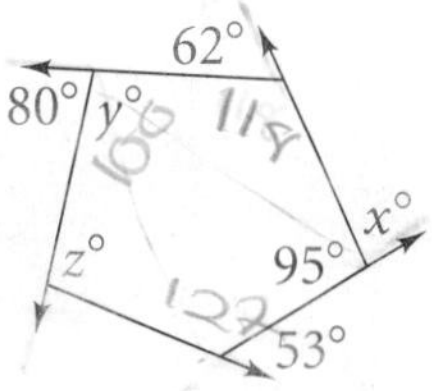

7. Writing Explain how you can determine if a polygon is concave or convex.

Sketch each pair of lines. Tell whether they are *parallel, perpendicular,* or *neither.*

8. $y = 4x + 7$
   $y = -\frac{1}{4}x - 3$

9. $y = 3x - 4$
   $y = 3x + 1$

10. $y = x + 5$
    $y = -5x - 1$

11. $y = x - 6$
    $y = -x + 2$

Coordinate Geometry **Graph quadrilateral *ABCD*. Then determine the most precise name for each figure.**

12. $A(1, 2), B(11, 2), C(7, 5), D(4, 5)$

13. $A(3, -2), B(5, 4), C(3, 6), D(1, 4)$

14. $A(1, -4), B(1, 1), C(-2, 2), D(-2, -3)$

15. Open-ended Write the coordinates of four points that determine each figure.
    a. square b. parallelogram
    c. rectangle d. trapezoid

Find the radius of a circle with the given diameter.

16. 6 ft 17. 5.1 m 18. $4\sqrt{5}$ in.

Find each measure for $\odot P$.

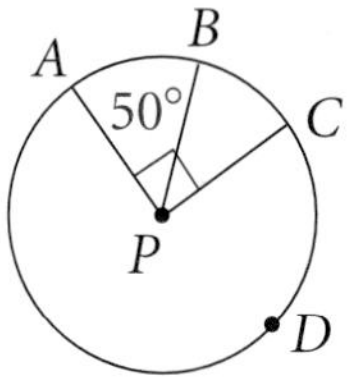

19. $m\angle BPC$

20. $m\widehat{AB}$

21. $m\widehat{ADC}$

22. $m\widehat{ADB}$

Find the values of the variables for each pair of similar figures.

23. 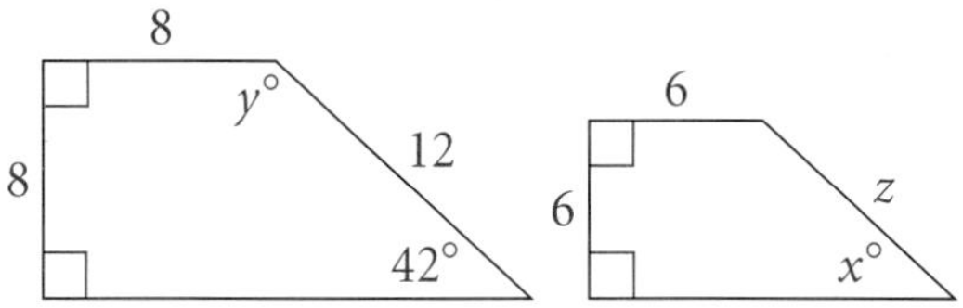

24. 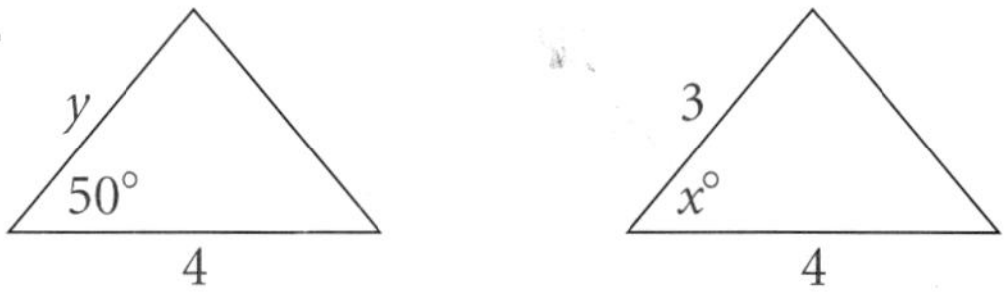

25. Standardized Test Prep What is the measure of each exterior angle of a regular 12-gon?

    A. 150 B. 30 C. 300 D. 210 E. 36

Use the figure below for Exercises 26 and 27.

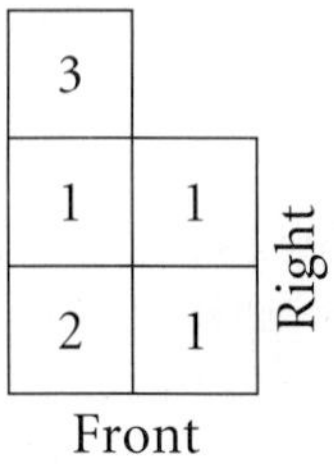

26. Create an isometric drawing.

27. Create an orthographic drawing.

Cumulative Review

For Exercises 1–10, choose the correct letter.

1. What is a name for the quadrilateral below?

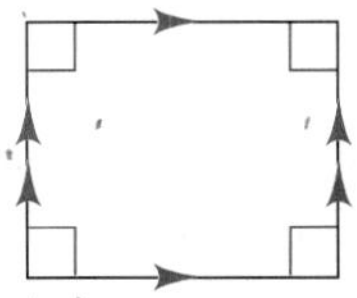

I. square **II.** rectangle
III. rhombus **IV.** parallelogram
A. I only **B.** IV only **C.** I and II
D. II and IV **E.** I and III

2. Which can be the intersection of three distinct planes?
I. a point **II.** a line
III. a plane **IV.** a ray
A. I only **B.** II only **C.** I and II
D. II and IV **E.** I, II, and III

3. Which line is parallel to $y = 3x - 2$?
A. $y = \frac{1}{3}x + 5$ **B.** $y = 3$
C. $y = -3x + 1$ **D.** $y = -\frac{1}{3}x - 4$
E. none of the above

4. Find the diameter of a circle with radius $6\sqrt{5}$.
A. $12\sqrt{5}$ **B.** $6\sqrt{10}$ **C.** $3\sqrt{5}$
D. $6\sqrt{2.5}$ **E.** none of the above

5. What is $m\angle CDF$?

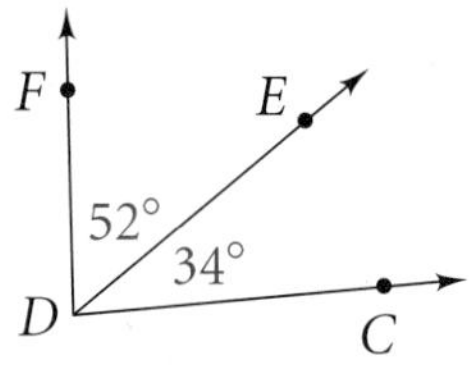

A. 18 **B.** 274 **C.** 86 **D.** 94
E. cannot be determined from the information given

6. $\triangle ABC$ is obtuse. Two vertices of the triangle are $A(3, 4)$ and $B(-1, 1)$. What could be the coordinates of point C?
A. $(-1, -3)$ **B.** $(3, 1)$ **C.** $(0, 0)$
D. $(-1, 4)$ **E.** none of the above

7. What is the length of the segment with endpoints at $A(1, 7)$ and $B(-3, -1)$?
A. 8 **B.** $4\sqrt{5}$ **C.** 40 **D.** $2\sqrt{10}$ **E.** $\sqrt{5}$

8. What is the next number in the pattern 1, −4, 9, −16 . . . ?
A. −25 **B.** −5 **C.** 5 **D.** 25 **E.** 35

Compare the boxed quantity in Column A with the boxed quantity in Column B. Choose the best answer.

A. The quantity in Column A is greater.
B. The quantity in Column B is greater.
C. The two quantities are equal.
D. The relationship cannot be determined on the basis of the information supplied.

	Column A	Column B
	$\angle A$ is a supplement of $\angle B$.	
9.	$m\angle A$	$m\angle B$
10.	the measure of each interior angle of a regular hexagon	the measure of each exterior angle of a regular octagon

Find each answer.

11. What are the coordinates of the midpoint of $\overline{CD}$ with endpoints $C(5, 3)$ and $D(0, -7)$?

12. *Open-ended* Write an equation of a line perpendicular to the line $y = 6x + 4$.

13. Create an isometric drawing for this foundation plan.

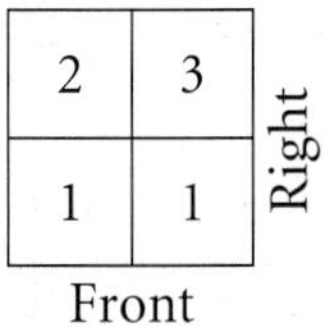

14. Draw an angle. Then construct another angle congruent to the first.

CHAPTER

3 Transformations: Shapes in Motion

Relating to the Real World

Many geometric figures in the real world do not sit still. They move; they change shape; they change size. You can describe these changes with transformational geometry. The principles you will study here have applications in many diverse fields—from science and architecture to music and history.

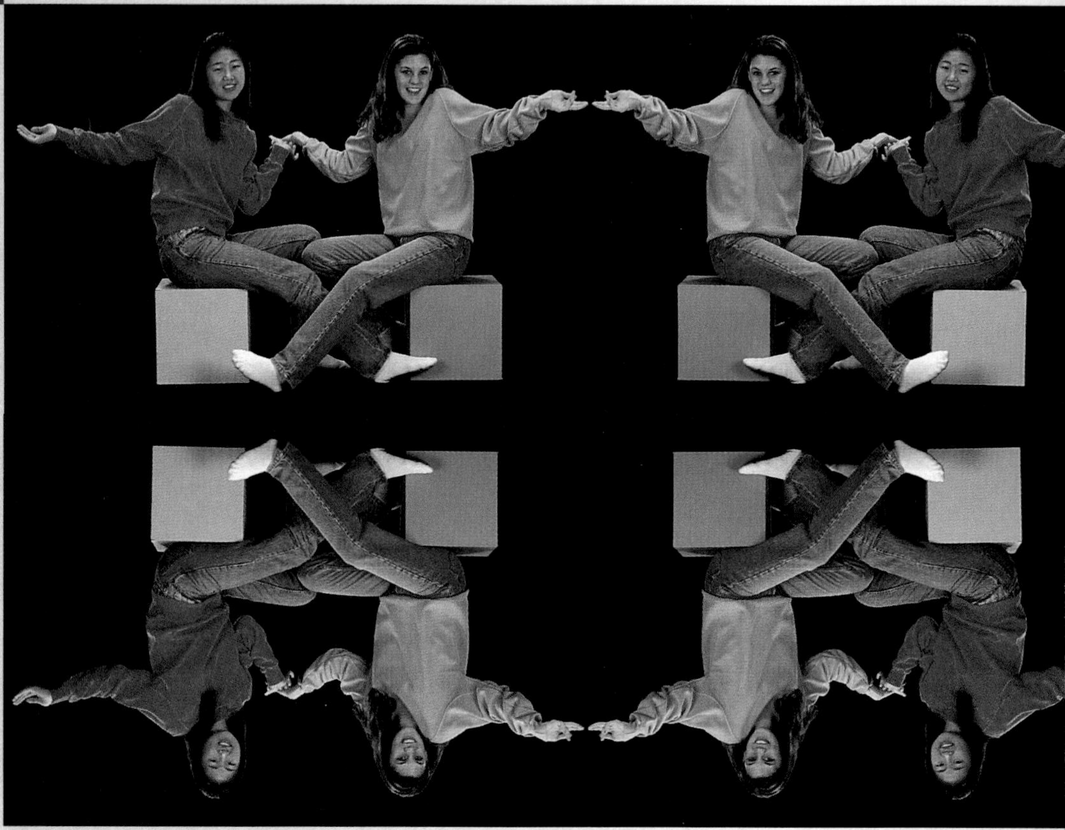

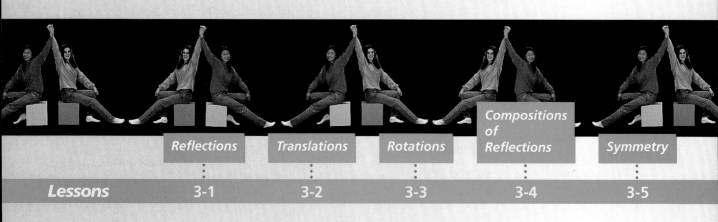

Tessellations

3-6

Dilations

3-7

CHAPTER PROJECT

FRIEZE FRAMES

Ukrainian painted eggs, dollar bills, Native American pottery, Japanese kimonos, automobile tire treads, and African cloth are products of vastly diverse cultures, but these things all have something in common. They contain strips of repeating patterns, called *frieze patterns.*

In this chapter project, you will explore the underlying relationships among frieze patterns from around the world. You will also create your own designs. You will see how distinct civilizations—separated by oceans and centuries—are linked by their use of geometry to express themselves and to beautify their world.

To help you complete the project:

- **p. 137** *Find Out by Investigating*
- **p. 143** *Find Out by Modeling*
- **p. 150** *Find Out by Investigating*
- **p. 158** *Find Out by Classifying*
- **p. 171** *Find Out by Creating*
- **p. 173** *Finishing the Project*

What You'll Learn

- Identifying isometries
- Locating reflection images of figures

...And Why

To describe and explain the kinds of motions you encounter every day

What You'll Need

- straightedge
- protractor
- ruler
- graph paper
- MIRA™ (optional)

3-1 Reflections

THINK AND DISCUSS

An Introduction to Transformations

Have you ever put together a jigsaw puzzle? Think about opening the box and emptying all of the puzzle pieces onto a table.

1. Describe the kinds of motions you use to put the pieces together.

You probably didn't realize it, but you use transformations when you assemble a puzzle. A **transformation** is a change in position, shape, or size of a figure. The photos below illustrate four basic transformations that you will study. Each transformed figure is the **image** of the original figure. The original figure is called the **preimage.**

The figure flips.

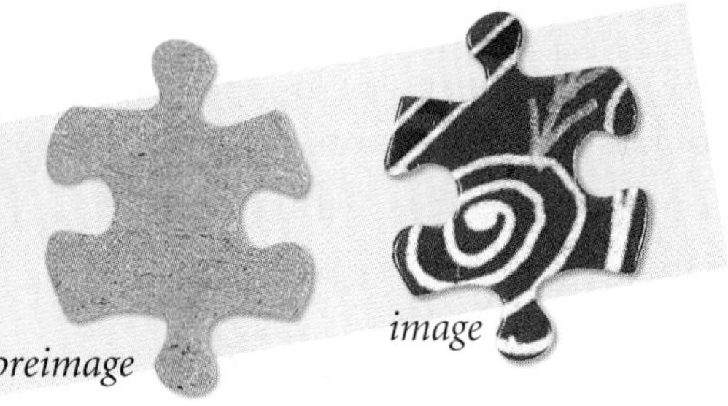

The figure slides.

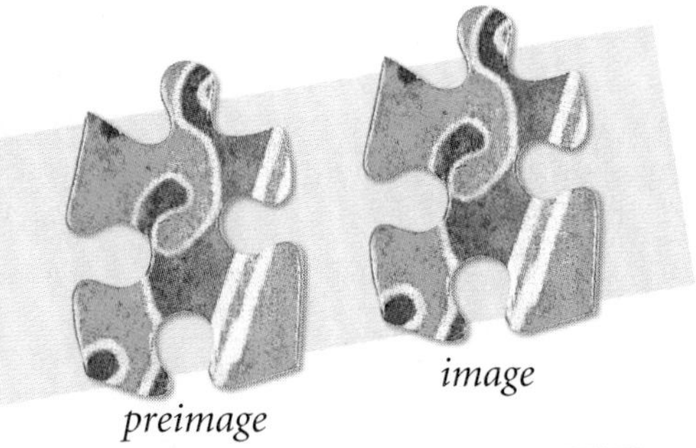

The figure turns.

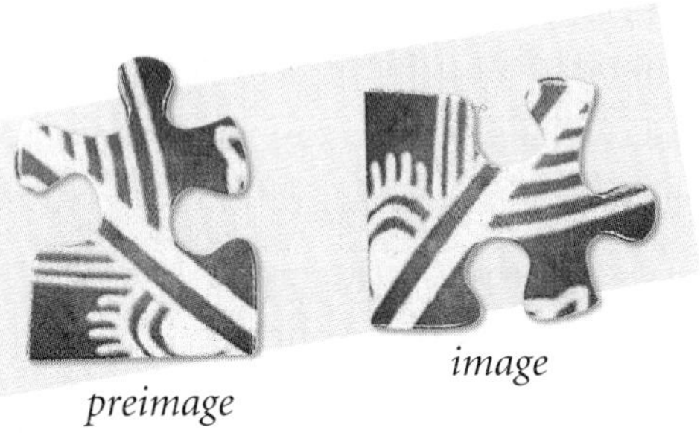

The figure changes size.

An **isometry** is a transformation in which the original figure and its image are congruent.

2. Which of the transformations shown above appear to be isometries?

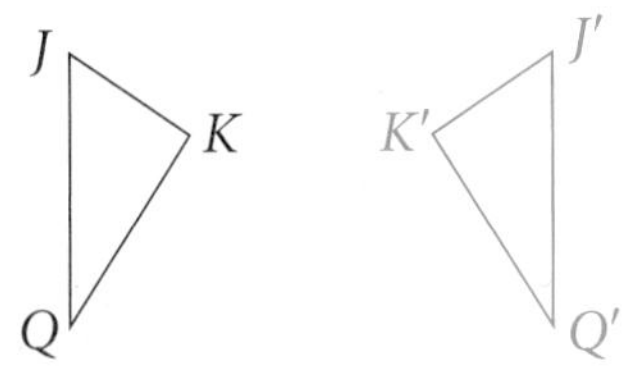

$\triangle JKQ \rightarrow \triangle J'K'Q'$

$\triangle JKQ$ maps to $\triangle J'K'Q'$.

A transformation **maps** a figure onto its image. An arrow (⟶) indicates a mapping. **Prime notation** is sometimes used to identify image points. In the diagram, K' (read "K prime") is the image of K. Notice that corresponding points of the original figure and its image are listed in the same order, just as corresponding points of congruent and similar figures are listed in the same order.

Example 1

In the diagram, $E'F'G'H'$ is the image of $EFGH$.

a. Name the images of $\angle F$ and $\angle H$.

b. List all pairs of corresponding sides.

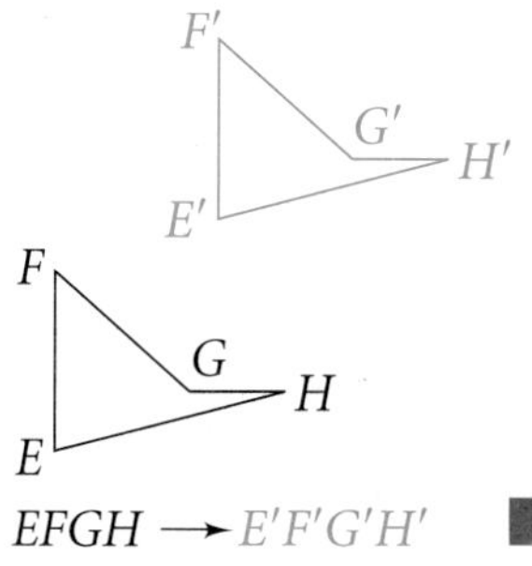

a. $\angle F'$ is the image of $\angle F$.
$\angle H'$ is the image of $\angle H$.

b. $\overline{EF}$ and $\overline{E'F'}$; $\overline{FG}$ and $\overline{F'G'}$;
$\overline{EH}$ and $\overline{E'H'}$; $\overline{GH}$ and $\overline{G'H'}$

3. Which of the four types of transformations shown on the previous page is illustrated in Example 1?

4. Try This List the corresponding segments and angles for the transformation $TORN \longrightarrow SAKE$.

Reflections

A flip is also known as a *reflection.* You see reflections almost every day. This morning, for example, you probably looked in the mirror before you headed out the door. In the following activity, you will investigate some properties of reflections.

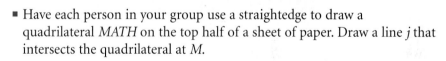

WORK TOGETHER

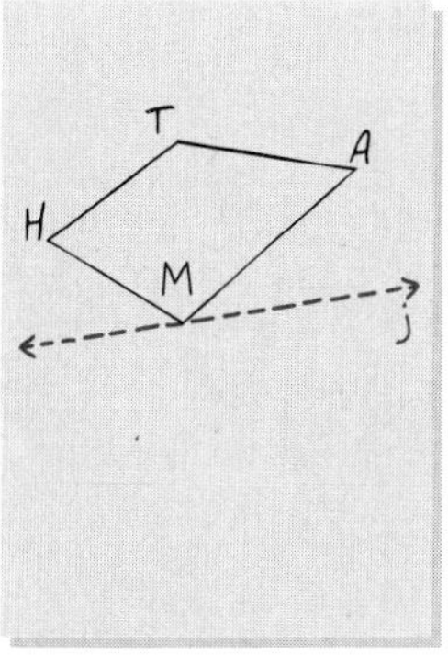

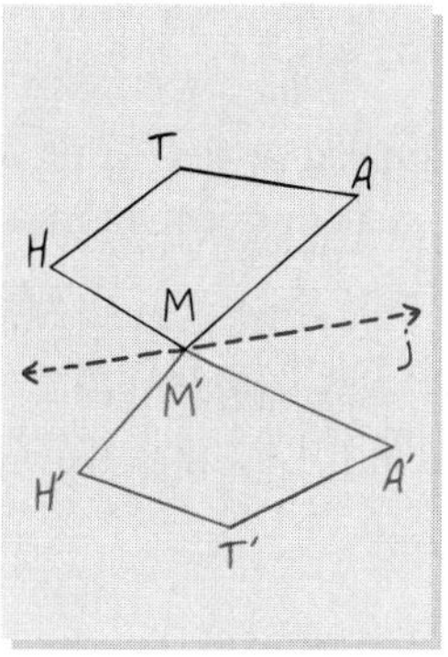

- Have each person in your group use a straightedge to draw a quadrilateral *MATH* on the top half of a sheet of paper. Draw a line *j* that intersects the quadrilateral at *M.*
- Fold the paper along line *j*, then use a straightedge to trace the reflection image of *MATH* onto the bottom portion of your paper. (You could also create the image by using a MIRA™.) Label the corresponding vertices of the image *M', A', T',* and *H'.*

5. Measure corresponding angles and segments in *MATH* and *M'A'T'H'.* Is a reflection an isometry? Explain.

6. Make a **conjecture** about the image of a point that lies on the line of reflection. (*Hint:* Consider point *M* and its image.)

7. a. In your original figure, did you write the labels *M, A, T,* and *H* in clockwise or counterclockwise order around the quadrilateral?

b. In the reflection image, do the labels *M', A', T',* and *H'* appear in clockwise or counterclockwise order around the quadrilateral?

c. What property of reflections do parts (a) and (b) suggest?

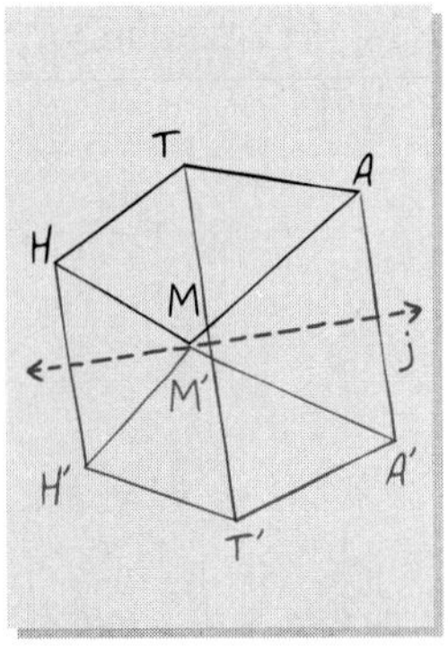

- Use a straightedge to draw segments $\overline{AA'}$, $\overline{TT'}$, and $\overline{HH'}$.

8. a. Line *j* divides each segment you drew into two parts. Compare the lengths of the two parts of each of the segments.

b. Line *j* forms four angles with each segment you drew. Use a protractor to find the measures of each of the angles.

c. Use your answers to parts (a) and (b) to complete the statement: Line *j* is the __?__ of the segment that connects a point and its image.

THINK AND DISCUSS

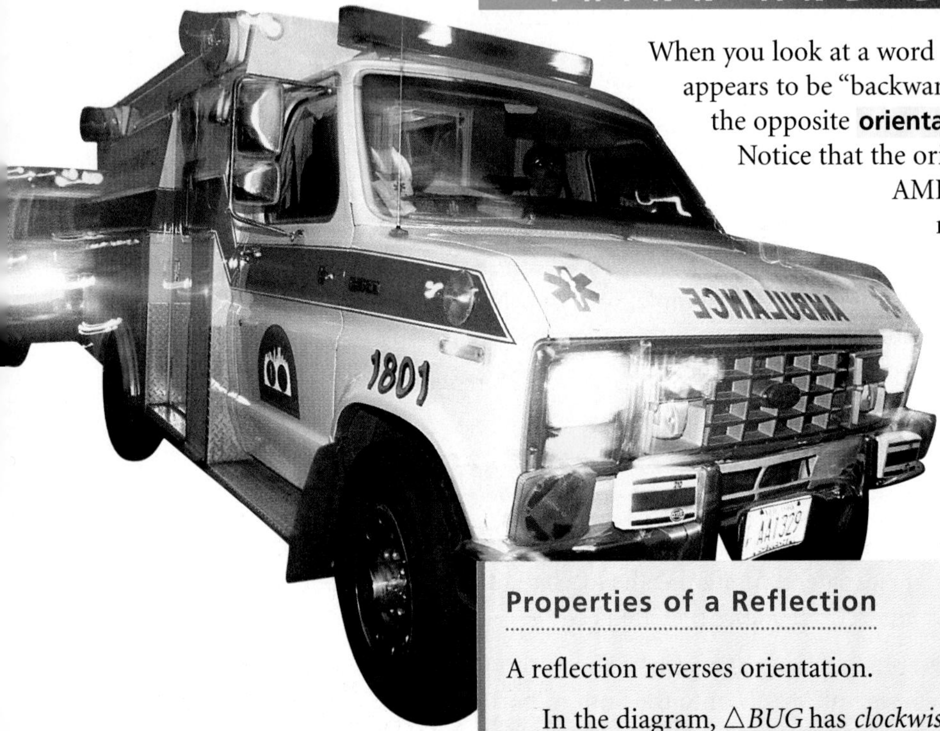

When you look at a word in a mirror, the image appears to be "backwards." The reflected word has the opposite **orientation** of the original word. Notice that the orientation of the word AMBULANCE in the photograph is reversed. The fronts of emergency vehicles often have mirror-image words on them so that drivers looking through rear-view mirrors can easily read them.

You discovered both of the following properties of reflections in the Work Together.

Properties of a Reflection

A reflection reverses orientation.

In the diagram, $\triangle BUG$ has *clockwise* orientation, so its image $\triangle B'U'G'$ has *counterclockwise* orientation.

A reflection is an isometry.

In the diagram, $\triangle BUG \cong \triangle B'U'G'$.

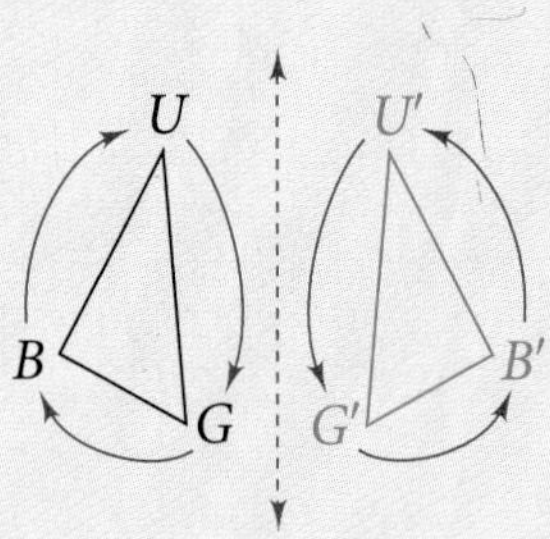

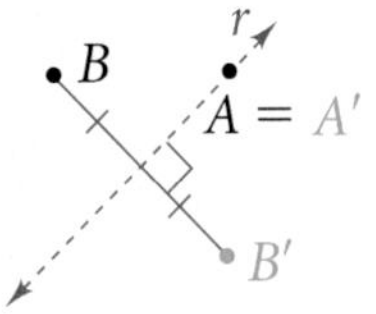

The other properties of reflections that you explored in the Work Together form the basis of the definition of a reflection. A **reflection** in line *r* is a transformation for which the following are true.

- If a point *A* is on line *r*, then the image of *A* is itself (that is, $A = A'$).
- If a point *B* is not on line *r*, then *r* is the perpendicular bisector of $\overline{BB'}$.

9. *Critical Thinking* Suppose you are given a point R and its reflection image R'. How could you find the line of reflection?

Example 2

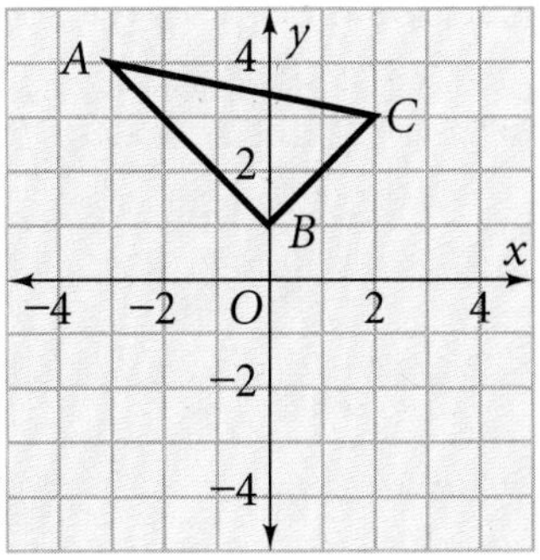

Coordinate Geometry Copy $\triangle ABC$ and draw its reflection image in each line.

a. the x-axis **b.** the y-axis

You can find A', B', and C' by paper folding or by locating points such that the line of reflection is the perpendicular bisector of $\overline{AA'}$, $\overline{BB'}$, and $\overline{CC'}$.

a.

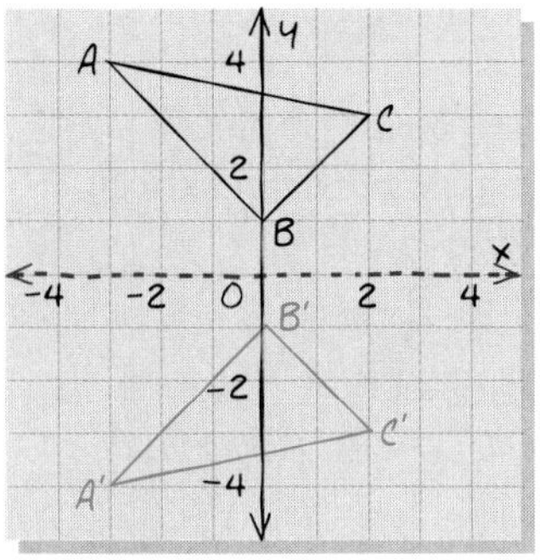

b.

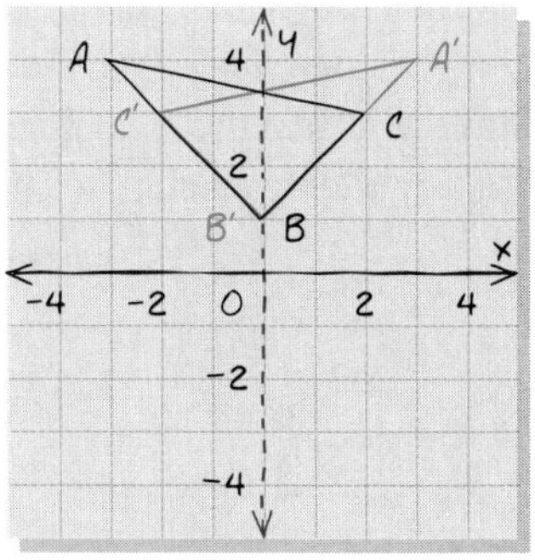

10. **Try This** Copy $\triangle ABC$ and draw its reflection image in $x = 3$.

Example 3 Relating to the Real World

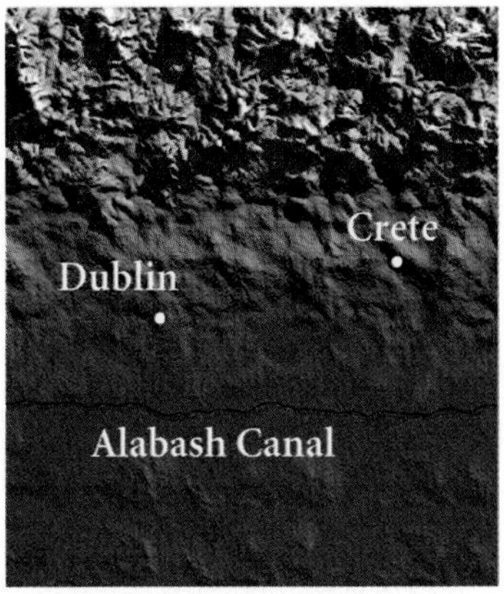

Engineering The state government wants to build a pumping station along the Alabash Canal to serve the towns of Crete and Dublin. Where along the canal should the pumping station be built to minimize the amount of pipe needed to connect the towns to the pump?

You need to find the point P on ℓ such that $DP + PC$ is as small as possible. Locate C', the reflection image of C in ℓ. Because a reflection is an isometry, $PC = PC'$, and $DP + PC = DP + PC'$. The sum $DP + PC'$ is smallest when D, P, and C' are collinear. So the pump should be located at the point P where $\overline{DC'}$ intersects ℓ.

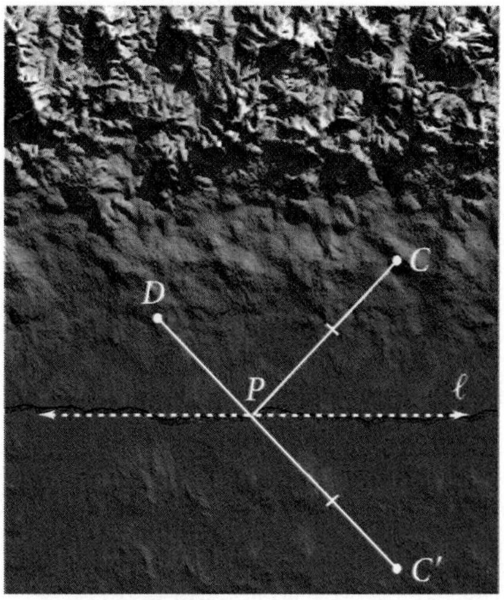

11. *Critical Thinking* Ursula began to solve Example 3 by reflecting point D in line ℓ. Will her method work? Explain.

Exercises ON YOUR OWN

Copy each diagram, then find the reflection image of the figure in line ℓ.

1.

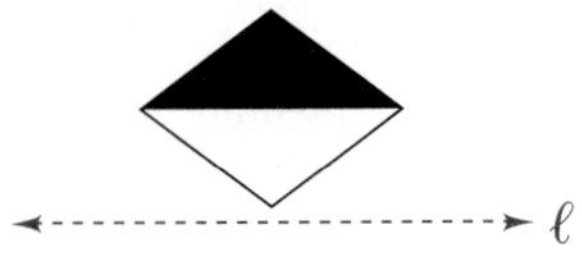

2.

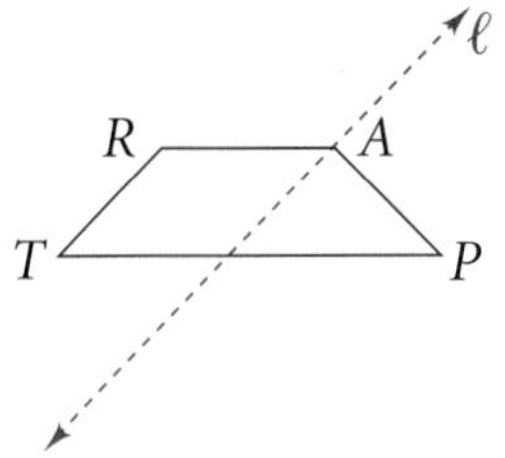

3.

In each diagram, the blue figure is the image of the black figure.
(a) List the corresponding sides.
(b) State whether the transformation appears to be an isometry.
(c) State whether the figures have the *same* or *opposite* orientation.

4.

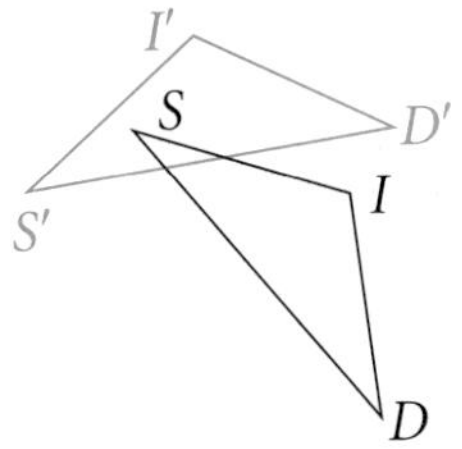

5.

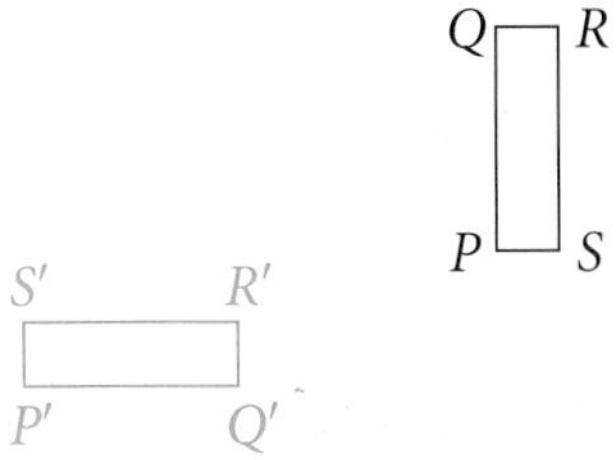

6.

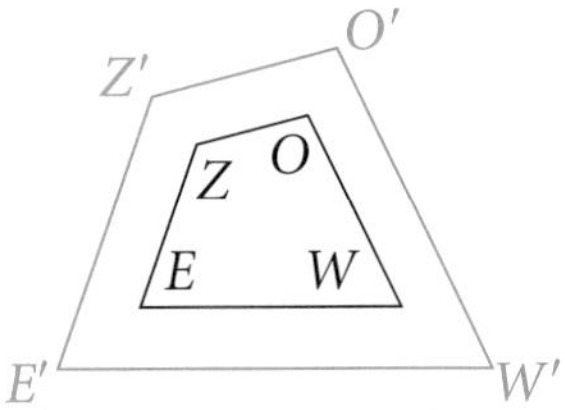

7.

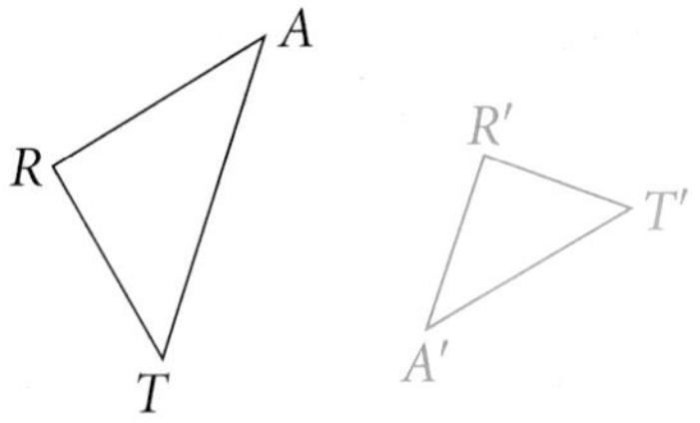

8.

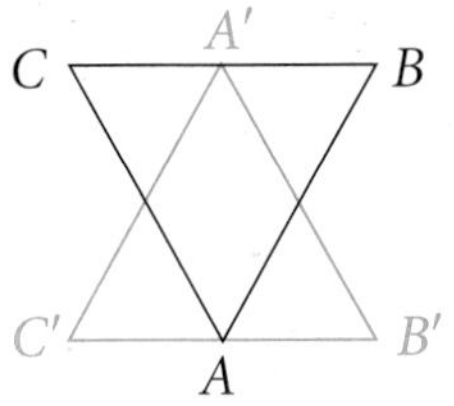

9.

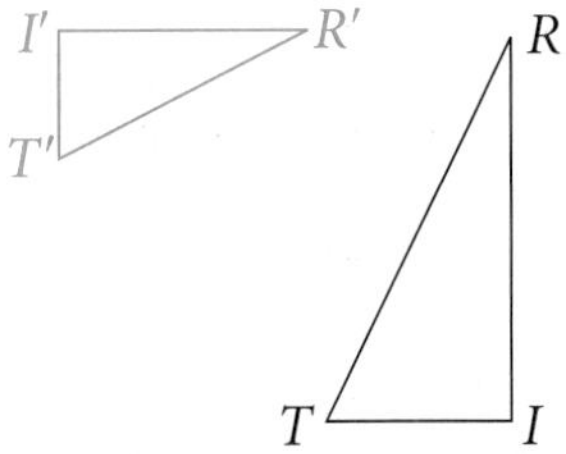

Coordinate Geometry **Given points $J(1, 4)$, $A(3, 5)$, and $R(2, 1)$, draw $\triangle JAR$ and its reflection image in the given line.**

10. the x-axis

11. $y = 2$

12. the y-axis

13. $x = -1$

14. $y = 5$

15. $x = 2$

16. $y = -x$

17. $y = x - 3$

18. **Critical Thinking** Given that the transformation $\triangle ABC \longrightarrow \triangle A'B'C'$ is an isometry, list everything you know about the two figures.

19. **Writing** Describe an example from everyday life of a flip, a slide, a turn, and a size change.

20. **Surveillance** SafeCo specializes in installing security cameras in department stores. Copy the diagram onto your paper. At what point on the mirrored wall should camera *C* be aimed in order to photograph door *D*?

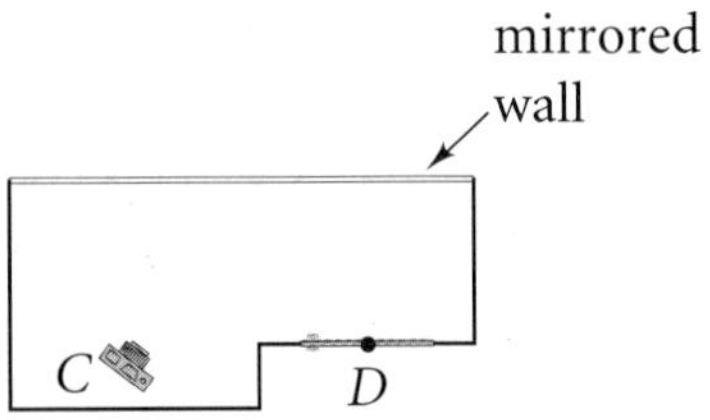

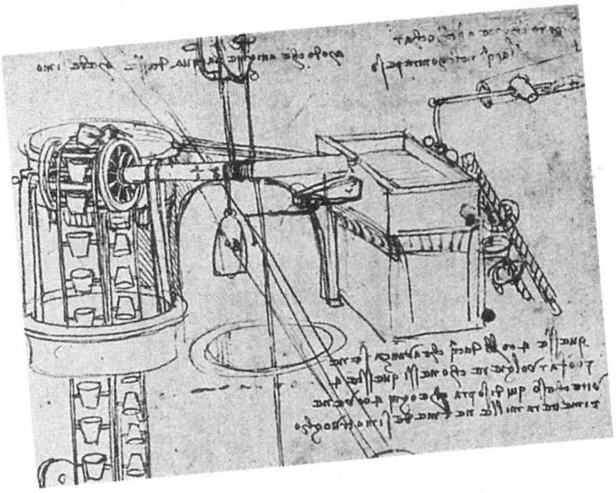

21. *History* An unusual characteristic of the work of 15th-century artist and scientist Leonardo da Vinci is his handwriting, which is a mirror image of normal handwriting.

a. Explain why the fact that da Vinci was left-handed might have made it seem normal for him to write in this manner.

b. Write the mirror image of this sentence. Use a mirror to check that your image is correct.

Copy each pair of figures, then find the line of reflection in which one figure is mapped onto the other.

22.

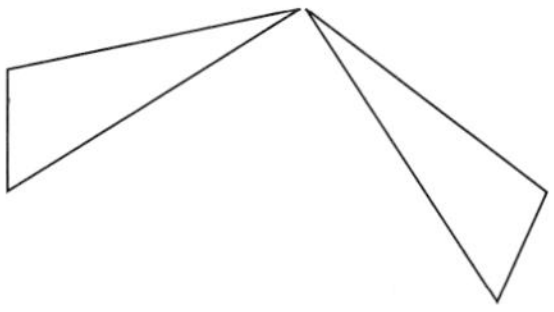

23.

24.

Some Drugs Aren't Ambidextrous

Your head aches. You take a pain reliever that cures this ache, but then your stomach starts to hurt. As you've no doubt noticed, many drugs have unwanted side-effects.

Most drugs are really made up of two versions of the same molecule—each a mirror image of the other. One version is known as an *R-isomer* and the other, an *S-isomer*. The isomers can have different healing properties.

Researchers have recently learned how to create pure batches of each isomer. They then run tests to determine which one produces the fewest side-effects. Drug companies can then produce drugs that contain only the "good" isomer.

For example, the R-isomer of the drug albuterol relieves asthma, while its twin has been shown to increase the chances of having future attacks.

Albuterol's Structure

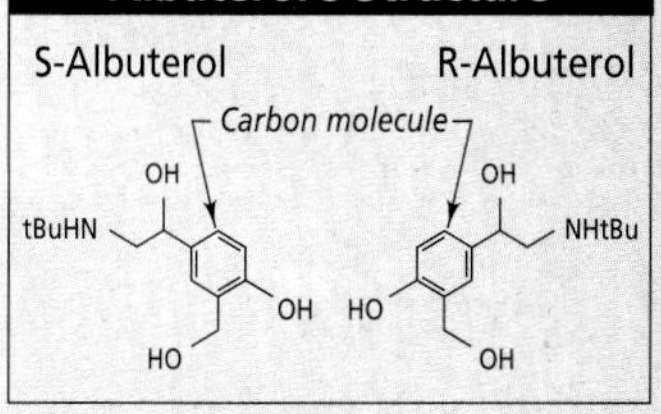

Source: *Wall Street Journal*

25. *Pharmacy* Consider the two "isomers" shown. In order for this drug to cure an illness, it needs to fit into the "receptor molecule" shown.

a. Which isomer will cure the illness?

b. *Open-ended* Give three examples from everyday life of objects that come in a left-handed version and a right-handed version.

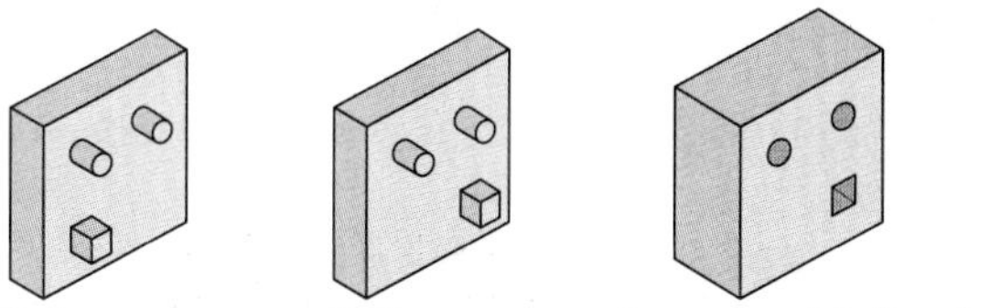

26. *Standardized Test Prep* What is the image of $(-4, 5)$ under a reflection in the line $y = x$?

A. $(5, -4)$ **B.** $(4, 5)$ **C.** $(-4, -5)$ **D.** $(4, -5)$ **E.** $(-5, 4)$

27. Which panels in the comic strip show the kind of reflection you studied in this lesson? Explain your answer.

28. a. Paper Folding Plot the points $Y(-2, 5)$, $A(5, 0)$, and $K(-1, -2)$. Draw $\triangle YAK$, then use paper folding to find its image in $y = x$. Label the image $\triangle Y'A'K'$.

   b. Patterns Look for a pattern in the coordinates of $\triangle YAK$ and $\triangle Y'A'K'$. Write a general rule for finding the image of any point under a reflection in $y = x$.

29. Critical Thinking Under a reflection, do all points move the same distance? If not, which points move the farthest?

Source: *K-Hito (Ricardo Garcia López). "Macaco."* © K-Hito

Exercises MIXED REVIEW

What type of quadrilateral is $ABB'A'$?

30.

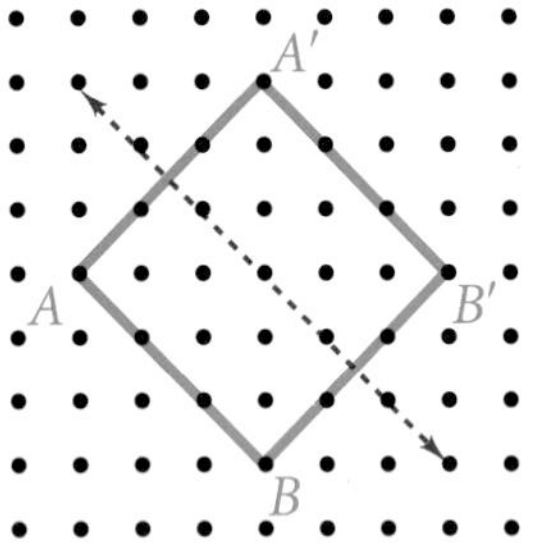

31.

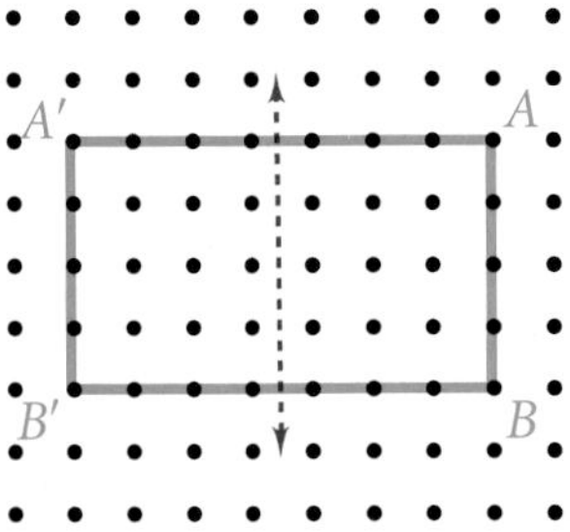

32.

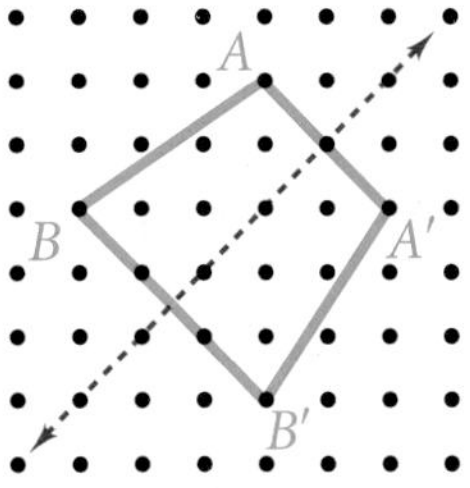

33. The points $P(2, 5)$ and $Q(-4, -7)$ are reflection images of one another.
   a. Find the coordinates of the midpoint of $\overline{PQ}$.
   b. Find the slope of $\overline{PQ}$.
   c. Find the slope of a line perpendicular to $\overline{PQ}$.
   d. Use your answers to parts (a) and (c) to write the equation of the line in which P is reflected to Q.

SELF ASSESSMENT

FOR YOUR JOURNAL

List the properties of reflections. Then draw a polygon, a line of reflection, and the reflection image of your polygon. Explain how the figures illustrate each property listed.

Getting Ready for Lesson 3-2

34. In the diagram, an isometry maps $\triangle HUB$ to $\triangle H'U'B'$.
   a. Is the transformation a flip, slide, or turn?
   b. Compare the slopes of $\overline{HH'}$, $\overline{UU'}$, and $\overline{BB'}$.
   c. Compare the lengths of $\overline{HH'}$, $\overline{UU'}$, and $\overline{BB'}$.

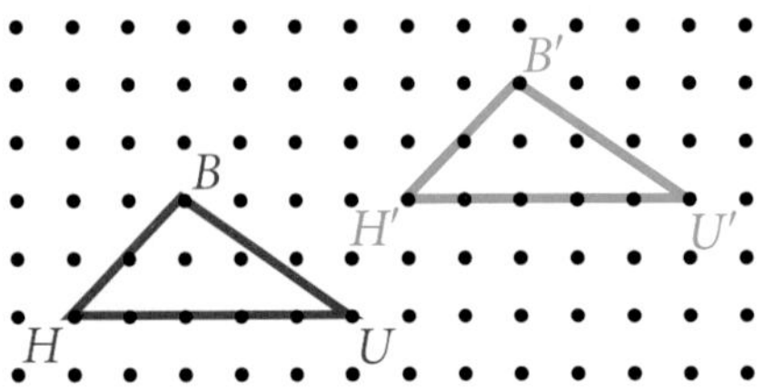

Math ToolboX

Algebra Review

Matrices

Before Lesson 3-2

A **matrix** is a rectangular arrangement of numbers. A matrix is usually written inside brackets. You identify the size of a matrix by the number of rows and columns. This matrix has two rows and three columns, so it is a 2 × 3 matrix. Each item in a matrix is called an **entry.**

$$\begin{bmatrix} -1 & 4 & 9 \\ -5 & 3 & 7 \end{bmatrix}$$

You can add or subtract matrices if they are the same size. You do this by adding or subtracting corresponding entries.

Example

Add $\begin{bmatrix} 6 & -5 \\ 0 & 3 \end{bmatrix} + \begin{bmatrix} 1 & 8 \\ -2 & 10 \end{bmatrix}$.

$$\begin{bmatrix} 6 & -5 \\ 0 & 3 \end{bmatrix} + \begin{bmatrix} 1 & 8 \\ -2 & 10 \end{bmatrix} = \begin{bmatrix} 6+1 & -5+8 \\ 0+(-2) & 3+10 \end{bmatrix}$$ Add corresponding entries.

$$= \begin{bmatrix} 7 & 3 \\ -2 & 13 \end{bmatrix}$$

Add or subtract each pair of matrices.

1. $\begin{bmatrix} 3 & 8 \\ 1 & 5 \end{bmatrix} + \begin{bmatrix} 8 & 2 \\ 0 & 7 \end{bmatrix}$

2. $\begin{bmatrix} -6 & 3 \\ -8 & 1 \end{bmatrix} - \begin{bmatrix} -4 & -9 \\ 3 & 5 \end{bmatrix}$

3. $\begin{bmatrix} 1 & -6 \\ 2 & -7 \end{bmatrix} + \begin{bmatrix} \frac{1}{2} & -1 \\ \frac{2}{3} & -2 \end{bmatrix}$

4. $\begin{bmatrix} \frac{1}{3} & \frac{3}{4} \\ \frac{1}{2} & \frac{2}{5} \end{bmatrix} - \begin{bmatrix} -\frac{1}{6} & \frac{1}{4} \\ -\frac{3}{5} & \frac{2}{3} \end{bmatrix}$

5. $\begin{bmatrix} 2 & 9 \\ 6 & 7 \end{bmatrix} + \begin{bmatrix} 6 & 2.3 \\ 9 & 4.1 \end{bmatrix}$

6. $\begin{bmatrix} 3 & -7 & 4 \\ 0 & -4 & 9 \end{bmatrix} + \begin{bmatrix} -9 & 4 & 10 \\ 3 & -11 & 2 \end{bmatrix}$

7. $\begin{bmatrix} 5 & -3.5 \\ 10 & 14 \\ -5 & 4.7 \end{bmatrix} + \begin{bmatrix} -6.1 & 0.8 \\ 7 & -5 \\ 8.3 & 9 \end{bmatrix}$

8. $\begin{bmatrix} 4 & 2 & 9 \\ -11 & 20 & 5 \\ -18 & 21 & -2 \end{bmatrix} - \begin{bmatrix} 8 & 17 & 4 \\ -34 & 26 & -9 \\ 3 & 0 & 17 \end{bmatrix}$

9. Use the matrices to find the total number of students per grade involved in each activity.

Greenfield High School North

	Sports	Drama	Debate
9th	146	5	11
10th	201	15	4
11th	205	11	7
12th	176	19	13

Greenfield High School South

	Sports	Drama	Debate
9th	301	13	9
10th	345	8	6
11th	245	11	11
12th	220	11	9

3-2 Translations

What You'll Learn

- Finding translation images of figures
- Using vectors and matrix addition to represent translations

...And Why

To use translations in the arts, computer graphics, navigation, manufacturing, music, and other fields

What You'll Need

- centimeter ruler
- scissors
- graph paper

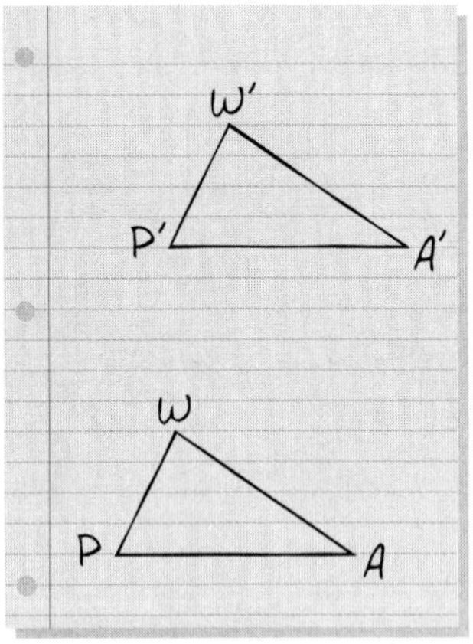

WORK TOGETHER

- Have each member of your group draw a triangle, cut it out, and label it $\triangle PAW$.
- Place your triangle on a sheet of lined paper so that $\overline{PA}$ lies on a horizontal line. Trace the triangle, and label it $\triangle PAW$.
- Slide the cutout to another location on your paper so that $\overline{PA}$ again lies on a horizontal line. Trace the triangle, and label it $\triangle P'A'W'$.

1. Does the transformation $\triangle PAW \longrightarrow \triangle P'A'W'$ appear to be an isometry? Explain.
2. Does the transformation $\triangle PAW \longrightarrow \triangle P'A'W'$ change the orientation of the triangle? Explain.

- Use a straightedge to draw $\overline{PP'}$, $\overline{AA'}$, and $\overline{WW'}$. Measure each segment with a ruler.

3. What do you notice about the lengths of the segments?
4. Notice the positions of $\overline{PP'}$, $\overline{AA'}$, and $\overline{WW'}$ in relation to one another. What appears to be true about them? Compare your answer with others in your group.

THINK AND DISCUSS

The sliding motion that maps $\triangle PAW$ to $\triangle P'A'W'$ in the Work Together is an example of a translation. A **translation** is a transformation that moves points the same distance and in the same direction. In the Work Together, you discovered the following properties of a translation.

Properties of a Translation

A translation is an isometry.

A translation does not change orientation.

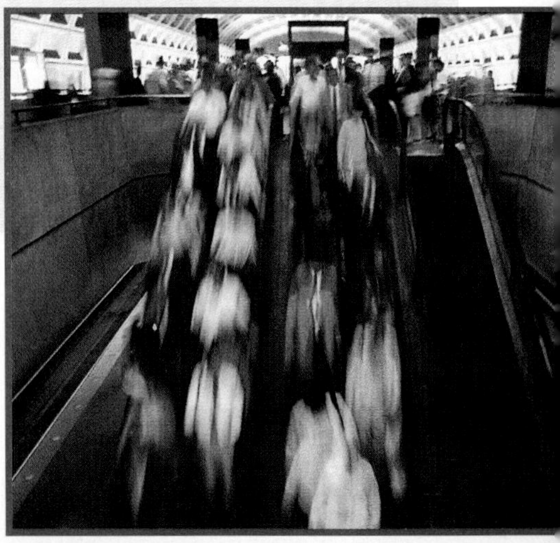

5. Elevators, escalators, and people movers all suggest translations. Name some other examples of translations from the real world.

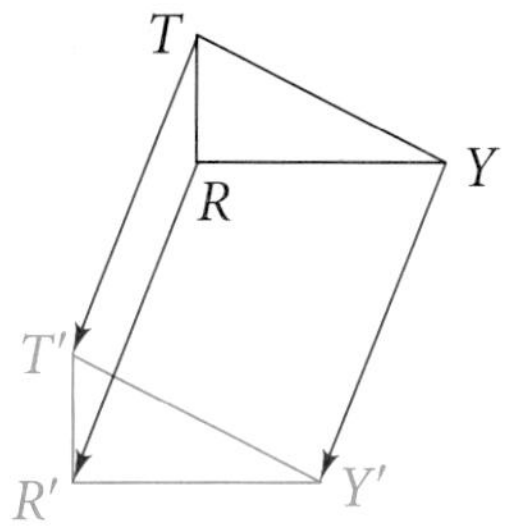

The distance and direction of a translation can be expressed as a *vector*. In the diagram, $\overrightarrow{TT'}$, $\overrightarrow{RR'}$, and $\overrightarrow{YY'}$ are vectors. Vectors have an *initial point* and a *terminal point*. T, R, and Y are initial points, and T', R', and Y' are terminal points. Note that although diagrams of vectors look identical to diagrams of rays, vectors do not go on forever in the indicated direction—they have a fixed length.

Example 1

Use the given vector and rectangle to create a sketch of a box.

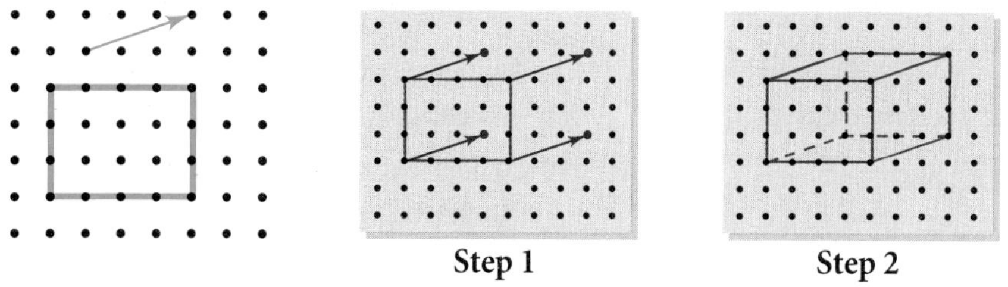

Copy the rectangle, then translate each of its vertices 3 units to the right and 1 unit up. Next, connect points to form the box. Use dashed lines for parts of the figure that are hidden from view.

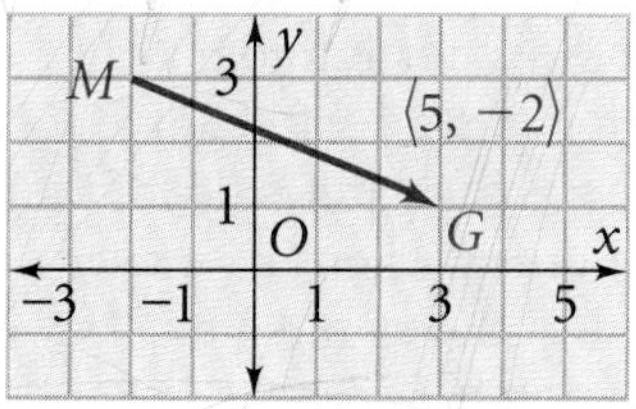

You can use *ordered pair notation*, $\langle x, y \rangle$, to represent a vector on the coordinate plane. In the notation, x represents horizontal change from the initial point to the terminal point and y represents vertical change from the initial point to the terminal point. The notation for vector $\overrightarrow{MG}$ is $\langle 5, -2 \rangle$.

6. **Try This** Describe the vector in Example 1 by using ordered pair notation.

7. Use vector notation to describe the vector with initial point (1, 3) and terminal point (6, 1).

Example 2

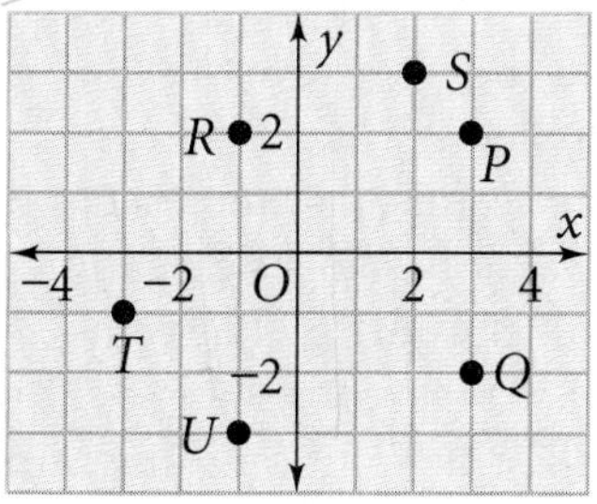

a. What is the image of P under the translation $\langle 0, -4 \rangle$?
b. What vector describes the translation $S \longrightarrow U$?

a. The vector $\langle 0, -4 \rangle$ represents a translation of 4 units down. The image of P is Q.
b. To get from S to U, you move 3 units left and 6 units down. The vector that describes this translation is $\langle -3, -6 \rangle$.

8. **Try This** Refer to the diagram in Example 2.
   a. What is the image of S under the translation $\langle -3, -1 \rangle$?
   b. What vector describes the translation $T \longrightarrow P$?

9. Describe in words the distance and direction of the translation represented by the vector $\langle 18, 0 \rangle$.

$$\begin{array}{lccc} & S & U & D \\ x\text{-coordinate} & & & \\ y\text{-coordinate} & & & \end{array} \begin{bmatrix} -1 & 2 & 3 \\ -1 & -5 & 2 \end{bmatrix}$$

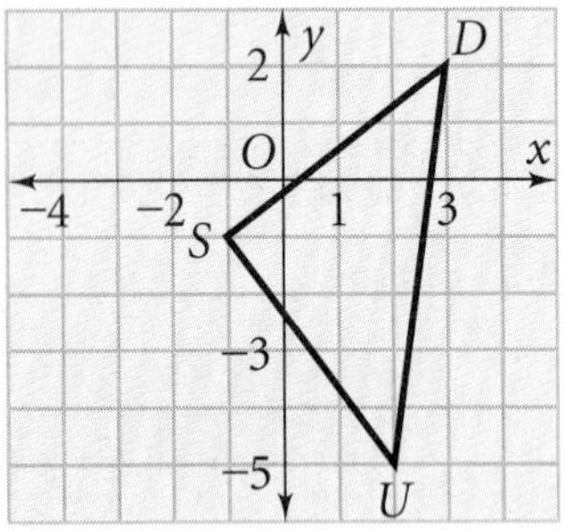

You can use matrices to help you translate figures in the coordinate plane. To do so, start by creating a matrix for the figure, as shown at the left.

Example 3

Use matrices to find the image of $\triangle SUD$ under the translation $\langle 4, -5 \rangle$.

To find the image of $\triangle SUD$, you add 4 to all of the x-coordinates and -5 to all of the y-coordinates.

Vertices of Preimage — Translation Matrix — Vertices of Image

$$\begin{array}{c} S \quad U \quad D \\ \begin{bmatrix} -1 & 2 & 3 \\ -1 & -5 & 2 \end{bmatrix} \end{array} + \begin{bmatrix} 4 & 4 & 4 \\ -5 & -5 & -5 \end{bmatrix} = \begin{array}{c} S' \quad U' \quad D' \\ \begin{bmatrix} 3 & 6 & 7 \\ -6 & -10 & -3 \end{bmatrix} \end{array}$$

10. Check the answer to Example 3 by sketching $\triangle SUD$ and $\triangle S'U'D'$ on the same set of axes.

Example 4 — Relating to the Real World

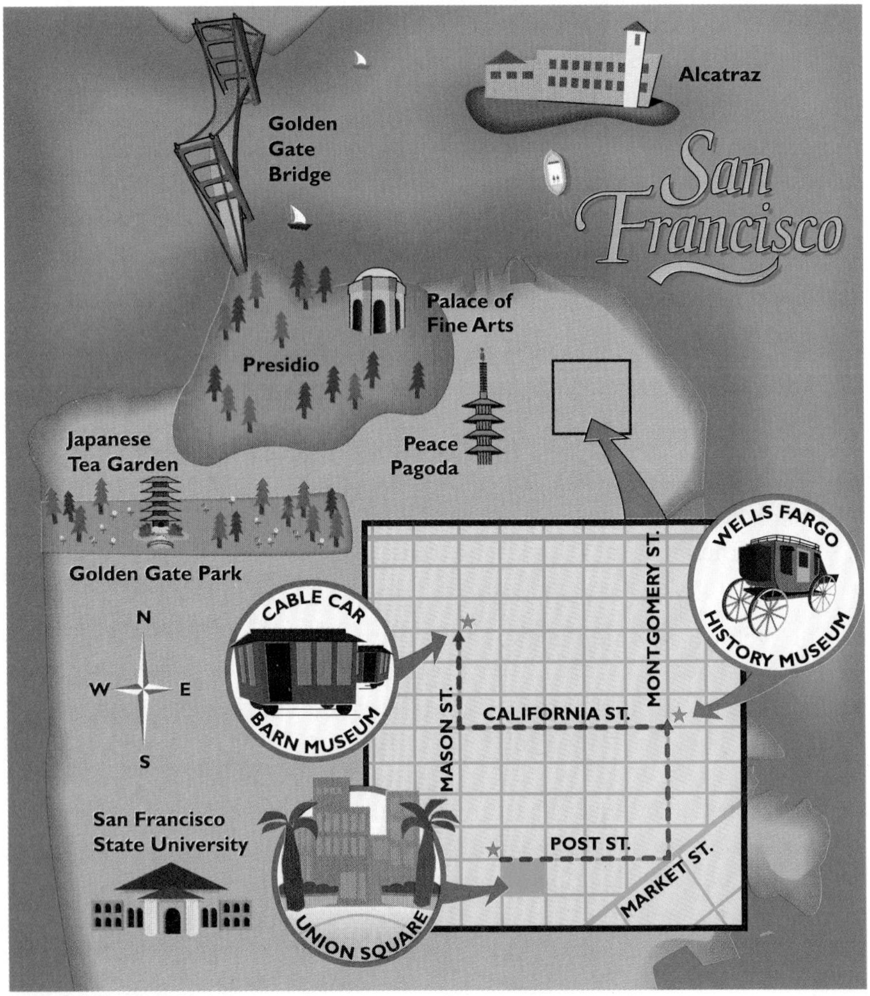

Travel Yolanda Pérez is visiting San Francisco. From her hotel near Union Square, she walked 4 blocks east and 4 blocks north to the Wells Fargo History Museum to see a stagecoach and relics of the gold rush. Then she walked 5 blocks west and 3 blocks north to the Cable Car Barn Museum. How many blocks from her hotel is she now?

As shown in the diagram, she is 1 block west and 7 blocks north of her hotel.

You can also solve this problem by using vectors. The vector $\langle 4, 4 \rangle$ represents a walk of 4 blocks east and 4 blocks north. The vector $\langle -5, 3 \rangle$ represents her second walk. The solution is the sum of the x- and y-coordinates of each vector:

$\langle 4, 4 \rangle + \langle -5, 3 \rangle = \langle -1, 7 \rangle$.

Example 4 shows the composition of two translations. The term **composition** describes any two transformations in which the second transformation is performed on the image of the first transformation. As the solution to Example 4 suggests, a composition of translations can be rewritten as a single translation.

Exercises ON YOUR OWN

In each diagram, the blue figure is the image of the red figure. Use ordered pair notation to represent each translation.

1.
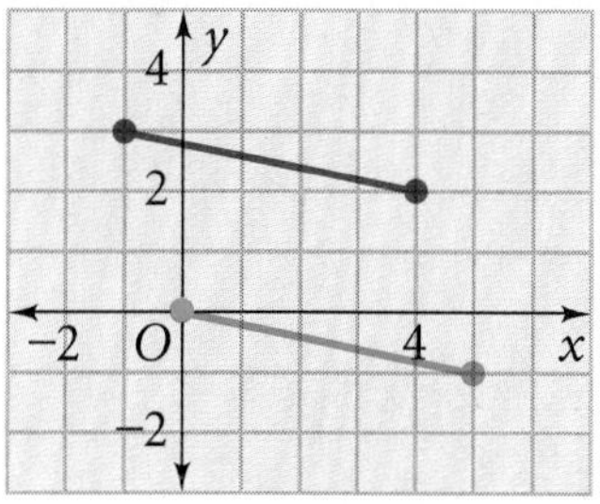

2.
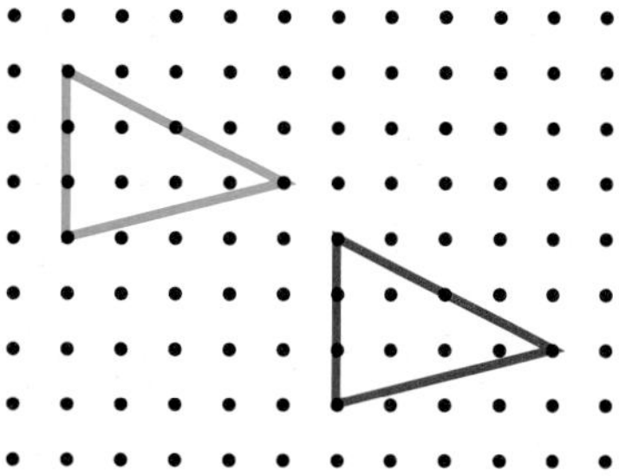

3.
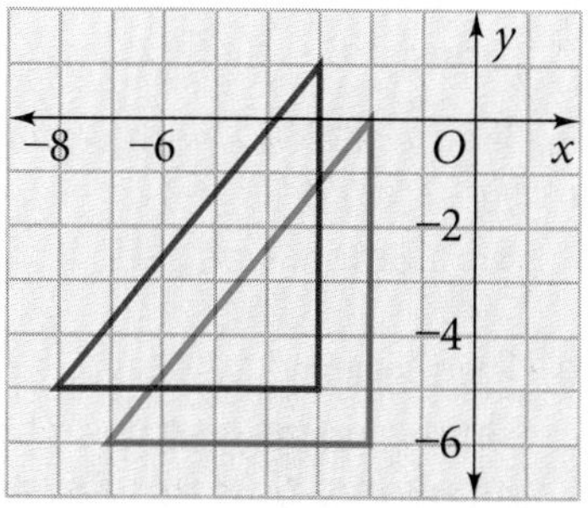

4.
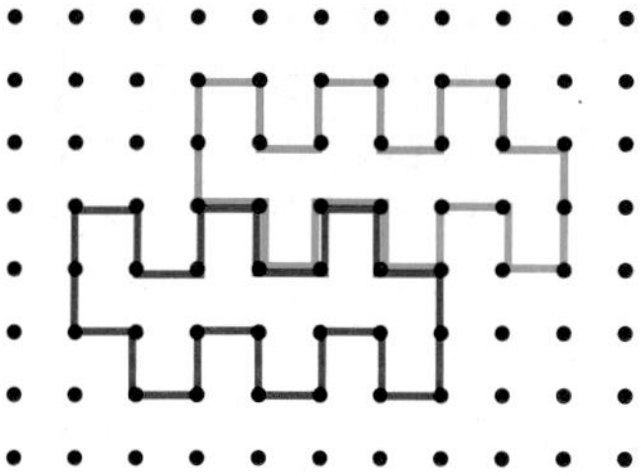

5.
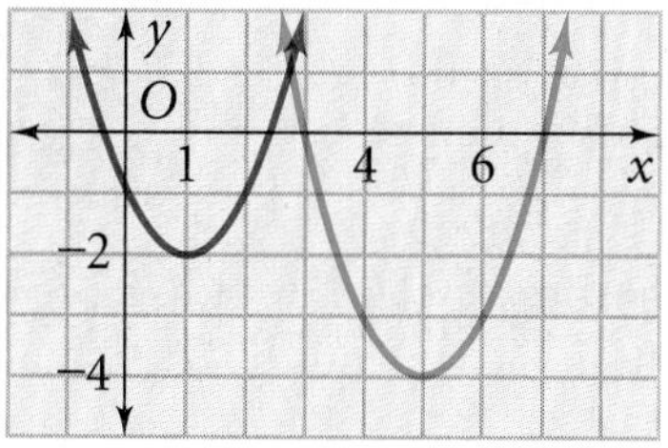

6.
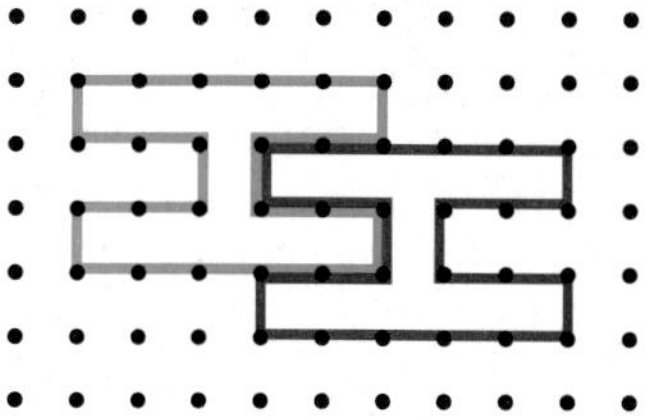

In Exercises 7–12, refer to the figure at the right.

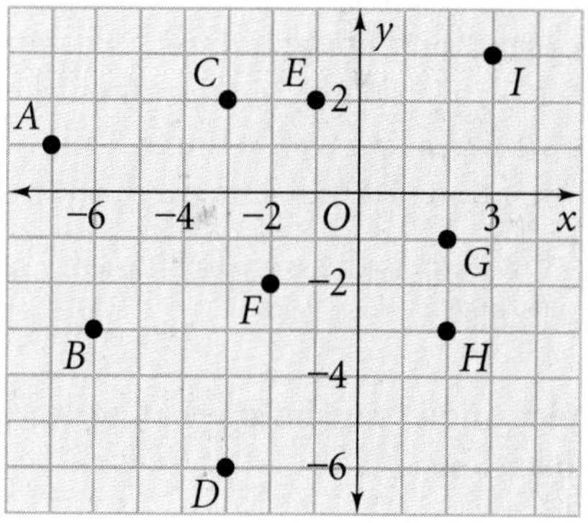

7. What is the image of F under the translation $\langle -1, 4\rangle$?

8. What vector describes the translation $G \longrightarrow H$?

9. What is the image of E under the translation $\langle 4, 1\rangle$?

10. What vector describes the translation $B \longrightarrow E$?

11. What is the image of F under the translation $\langle 4, -1\rangle$?

12. What vector describes the translation $I \longrightarrow C$?

Use each figure and vector to sketch a three-dimensional figure.

13.
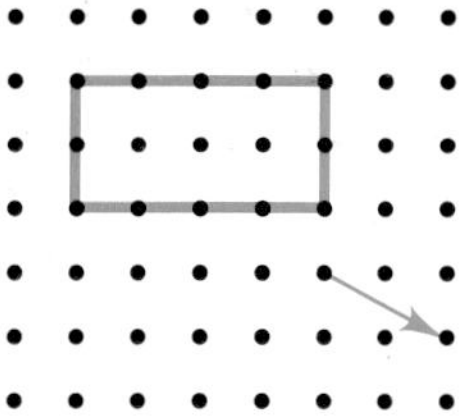

14.
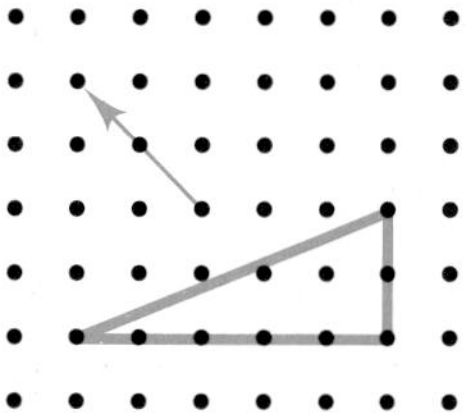

15.
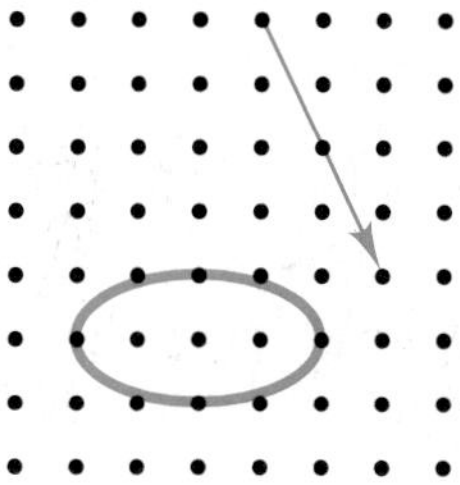

16. **Sailing** Emily left Galveston Bay at the east jetty and sailed 4 km north to an oil rig. She then sailed 5 km west to Redfish Island. Finally, she sailed 3 km southwest to the Spinnaker Restaurant. Draw vectors on graph paper that show her journey.

In Exercises 17–19, use matrix addition to find the image of each figure under the given translation.

17. Figure: $\triangle ACE$ with vertices $A(7, 2)$, $C(-8, 5)$, $E(0, -6)$
Translation: $\langle -9, 4 \rangle$

18. Figure: $\triangle PUN$ with vertices $P(1, 0)$, $U(4, 6)$, $N(-5, 8)$
Translation: $\langle 11, -13 \rangle$

19. Figure: $\square NILE$ with vertices $N(2, -5)$, $I(2, 2)$, $L(-3, 4)$, $E(-3, -3)$
Translation: $\langle -3, -4 \rangle$

20. **Photography** When you snap a photograph, a shutter opens to expose the film to light. The amount of time that the shutter remains open is known as the *shutter speed.* The photographer of the train used a long shutter speed to create an image that suggests a translation. Sketch a picture of your own that suggests a translation.

21. **Coordinate Geometry** $\triangle MUG$ has coordinates $M(2, -4)$, $U(6, 6)$ and $G(7, 2)$. A translation maps point M to $(-3, 6)$. Find the coordinates of U' and G' under this translation.

22. **Visiting Colleges** Nakesha and her parents are visiting colleges. They leave their home in Enid, Oklahoma, and head for Tulsa, which is 107 mi east and 18 mi south of Enid. From Tulsa, they head to Norman, which is 83 mi west and 63 mi south of Tulsa. Where is Norman in relation to Enid? Draw a diagram to show your solution.

23. **Writing** Is the transformation $\triangle HYP \longrightarrow \triangle H'Y'P'$ a translation? Explain.

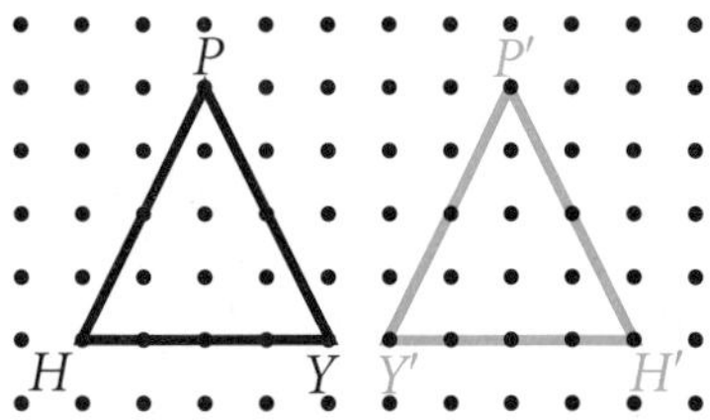

Find a single translation that has the same effect as each composition of translations.

24. $\langle 2, 5 \rangle$ followed by $\langle -4, 9 \rangle$

25. $\langle -3, 7 \rangle$ followed by $\langle 3, -7 \rangle$

26. $\langle 12, 0.5 \rangle$ followed by $\langle 1, -3 \rangle$

27. **Coordinate Geometry** $\square ABCD$ has vertices $A(3, 6)$, $B(5, 5)$, $C(4, 2)$, and $D(2, 3)$. The figure is translated so that the image of point C is the origin.
   a. Find the vector that describes the translation.
   b. Graph $\square ABCD$ and its image.

28. *Open-ended* You work for a company that specializes in creating unique, artistic designs for business stationery. One of your clients is Totter Toys. You have been assigned to create a design that forms a border at the top of their stationery. Create a design that involves translations to present to your client.

Chapter Project

Find Out by Investigating

A **frieze pattern,** or **strip pattern,** is a design that repeats itself along a straight line. Every frieze pattern can be mapped onto itself by a translation. Some can also be mapped onto themselves by other transformations, such as reflections.

- Decide whether each frieze pattern can be mapped onto itself by a reflection in a horizontal line, a vertical line, or both.

a. Navaho Design

b. Design from Sandwich Islands

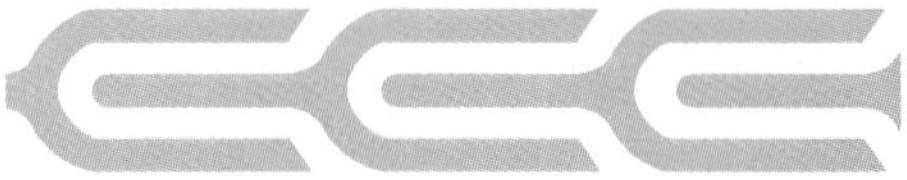

c. Medieval Ornament

d. Arabian Design

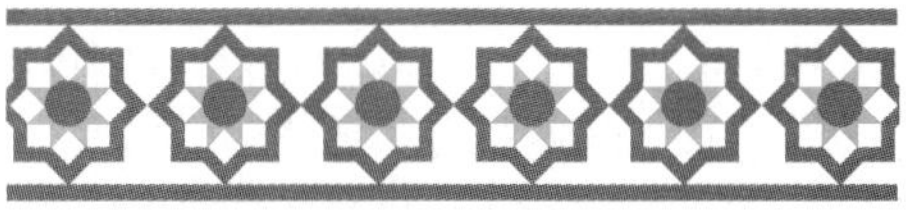

Exercises MIXED REVIEW

For Exercises 29 and 30, refer to the diagram.

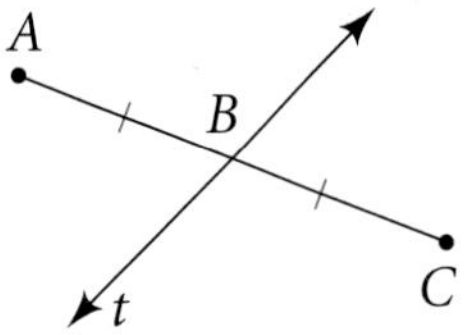

29. Line t is a __?__ of $\overline{AC}$.

30. $AB = 3x - 8$ and $BC = 5x - 36$. Find AC.

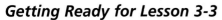

31. a. *Algebra* Graph $y = 2x - 3$, then draw its image under the translation $\langle 0, 5 \rangle$.

b. Find the slope and y-intercept of the preimage and the image.

c. How are the two lines related?

Getting Ready for Lesson 3-3

Cooking **What temperature will the oven be if the knob is turned the given number of degrees in a clockwise direction?**

32. 120° **33.** 180° **34.** 210° **35.** 270°

3-3 Rotations

What You'll Learn

- Identifying and locating rotation images of figures

...And Why

To understand real-life objects that involve rotation, such as clocks, combination locks, and laser disc players

What You'll Need

- straightedge
- colored pencils (optional)
- protractor
- compass

WORK TOGETHER

Before beginning the activity, have each member of your group fold a piece of paper in half lengthwise and widthwise and then cut it into fourths.

Step 1: Place a piece of the paper over the figure below. Trace the six points on the circle, the center of the circle, and the triangle.

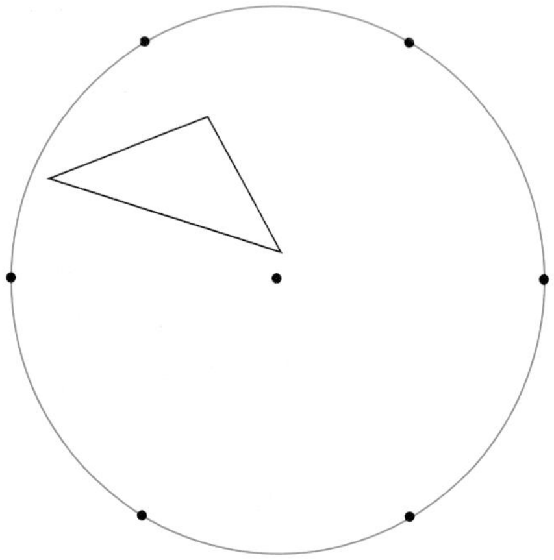

If you start with this in Step 4 . . .

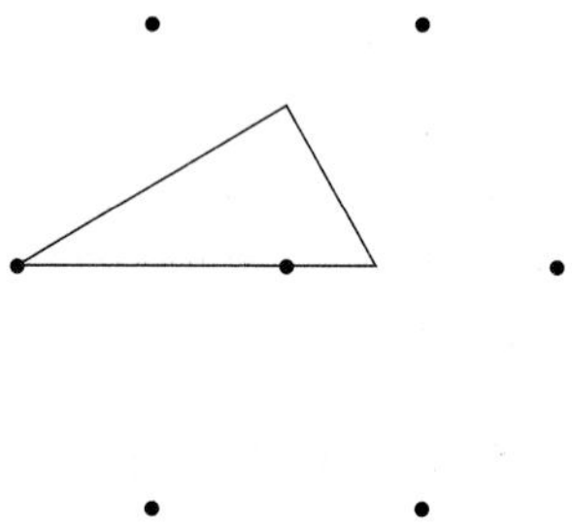

You could end up with this . . .

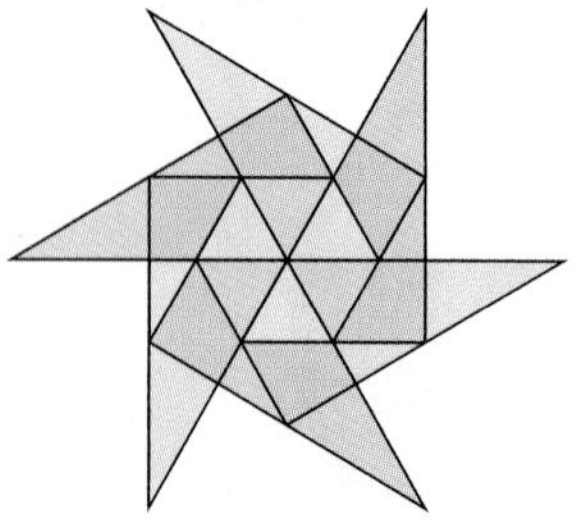

Step 2: Place the point of your pencil on the center of the circle and then rotate the paper until the six points again overlap. Trace the triangle in its new location.

Step 3: Repeat Step 2 until there are six triangles on your paper. Compare drawings within your group to be sure that your results look the same.

Step 4: Now it's your turn to be creative. Place a piece of paper over the figure above, trace the six points on the circle and the center of the circle, and then draw your own triangle on the paper.

Step 5: Place the paper from Step 4 on your desktop, and then use a blank piece of paper to repeat the process in Steps 1–3. Color your design, and then create a display of your group's designs.

THINK AND DISCUSS

In the Work Together, you used rotations to create a design. In order to describe a rotation, you need to know the center of rotation, the angle of rotation, and the direction of the rotation.

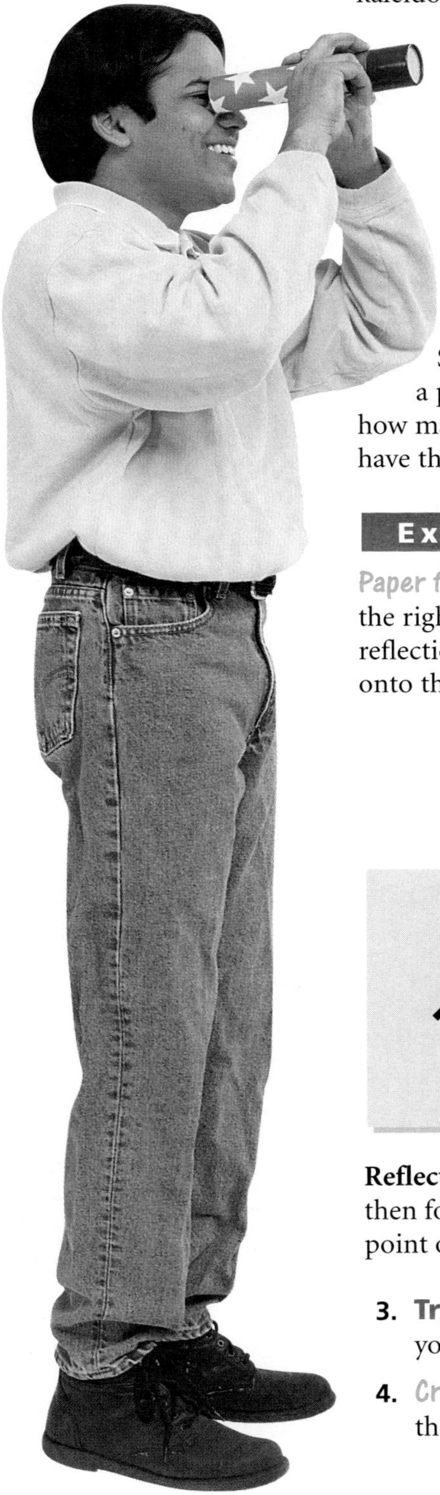

Example 1 **Relating to the Real World**

Creative Art Determine the angle between the mirrors in this kaleidoscope image.

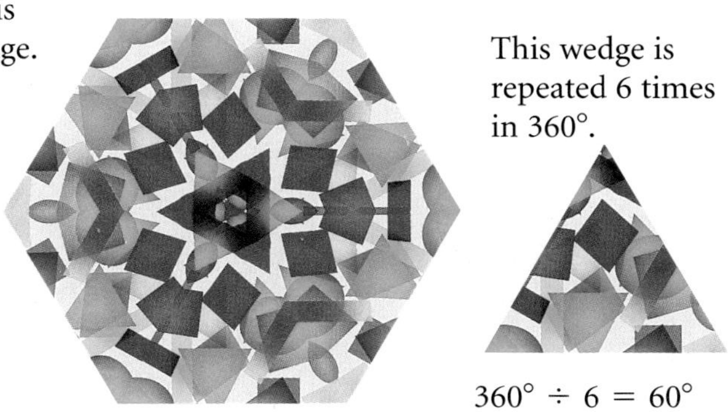

This wedge is repeated 6 times in 360°.

$360° \div 6 = 60°$

Suppose you are given two congruent figures in random positions in a plane. Could you use reflections to map one onto the other? If so, how many reflections would be needed? Example 2 shows that if the figures have the same orientation, only two reflections are needed.

Example 2

Paper folding The two P's at the right are congruent. Use two reflections to map one figure onto the other.

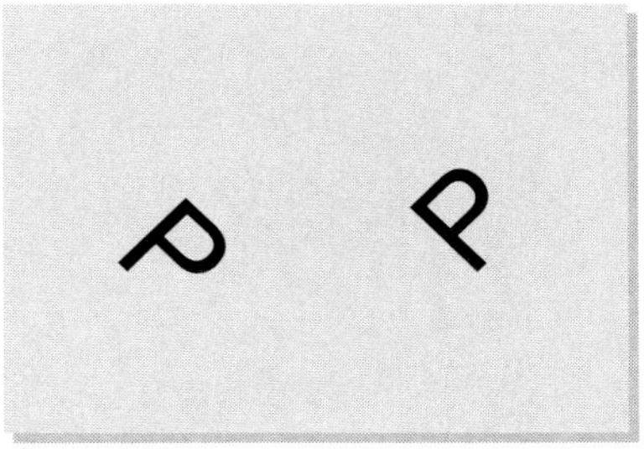

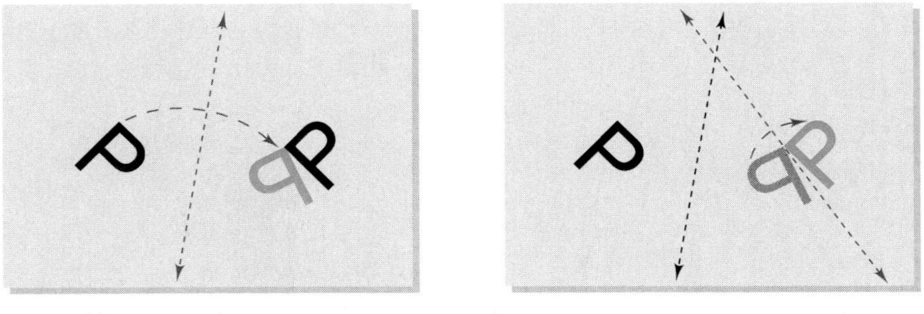

Reflection 1: Pick a point on one P, then fold it onto the corresponding point on the other P.

Reflection 2: Fold the two P's that share a point onto one another.

3. **Try This** Trace the two P's in Example 2 in different locations on your paper; then use paper folding to map one figure onto the other.
4. *Critical Thinking* What single isometry could be used to map one of the P's onto the other? Explain.

Compositions of Three Reflections

Example 2 shows that if two congruent figures have the *same* orientation, you can map one onto the other by exactly two reflections. If two congruent figures have *opposite* orientation, you may need to use three reflections.

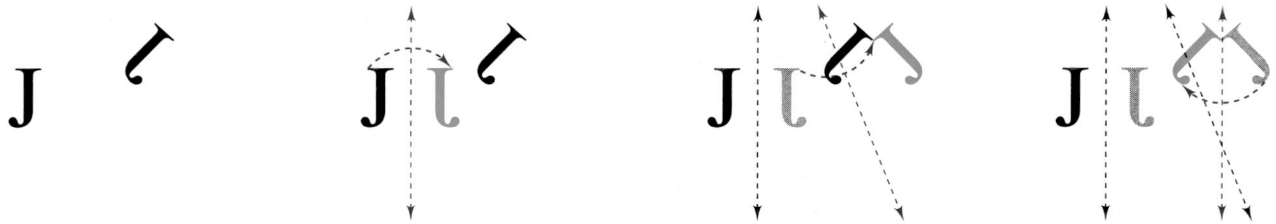

Given: two figures with opposite orientation

Reflect one figure in any line to change its orientation.

Then reflect the image twice, as shown in Example 2.

These paper-folding techniques illustrate the following theorem.

Theorem 3-3 In a plane, two congruent figures can be mapped onto one another by a composition of at most three reflections.

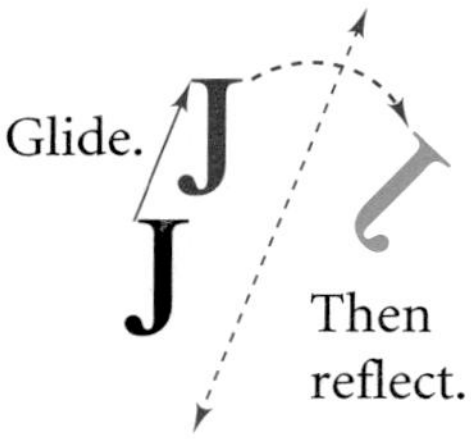

A composition of three reflections in lines that intersect in more than one point is called a **glide reflection.** It is called a glide reflection because any such composition of reflections can be rewritten as a translation (or glide) followed by a reflection in a line parallel to the translation vector.

5. *Critical Thinking* Explain why a glide reflection changes orientation.

Example 3

Coordinate Geometry Find the image of $\triangle TEX$ under a glide reflection where the glide is given by the vector $\langle 0, -5 \rangle$ and the reflection is in $x = 0$.

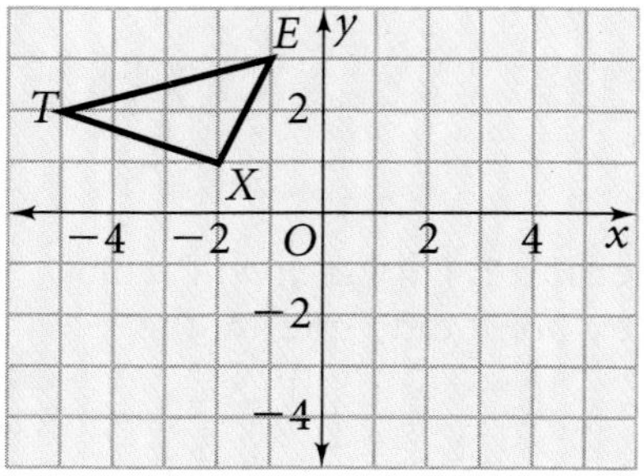

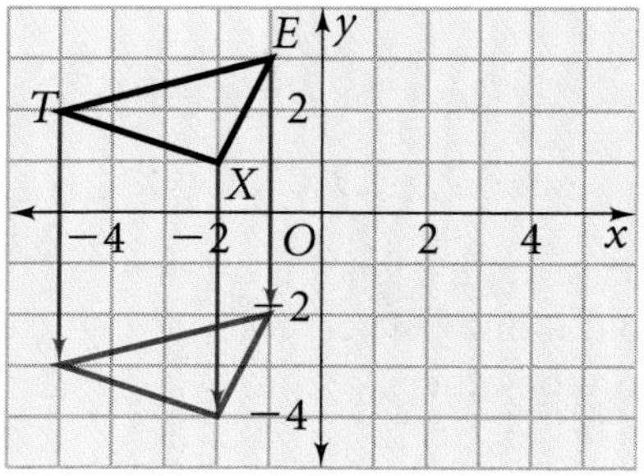

Translate $\triangle TEX$ by the vector $\langle 0, -5 \rangle$.

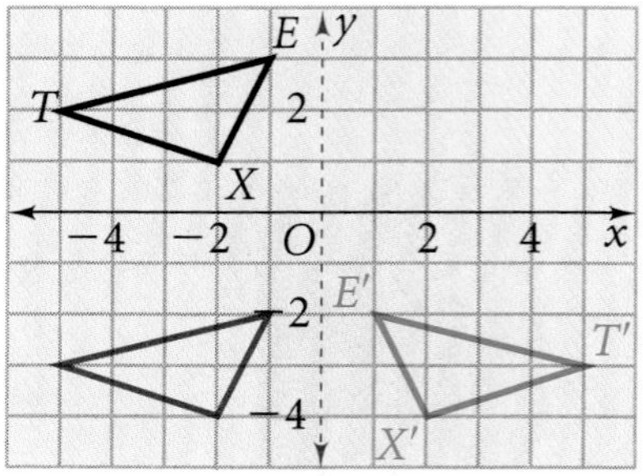

Reflect the image in $x = 0$.

6. Would the result of Example 3 be the same if you reflected $\triangle TEX$ first, then translated it?

7. **Try This** Find the image of $\triangle TEX$ under a glide reflection given by the vector $\langle 1, 0 \rangle$ and a reflection in $y = -2$.

You can map any two congruent figures onto one another by a single reflection, translation, rotation, or glide reflection. These four transformations are the only isometries.

Theorem 3-4
Isometry Classification Theorem

There are only four isometries. They are the following.

reflection — translation — rotation — glide reflection

Example 4

Each pair of figures is congruent. What isometry maps one to the other?

a. 

b. 

PROBLEM SOLVING HINT
Use Logical Reasoning.

a. These figures have the same orientation, so the transformation must be either a translation or a rotation. It's obviously not a translation, so it must be a rotation.

b. These figures have opposite orientation, so the transformation must be either a reflection or a glide reflection. Since it's not a reflection, it must be a glide reflection.

Exercises ON YOUR OWN

Creative Art **What is the angle between the mirrors for each kaleidoscope image?**

1. 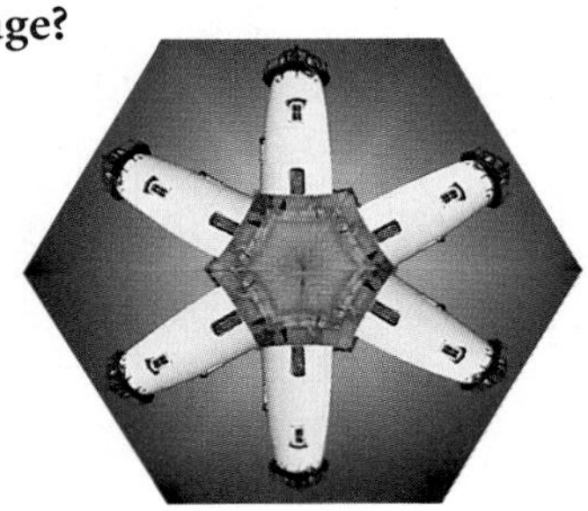

2. 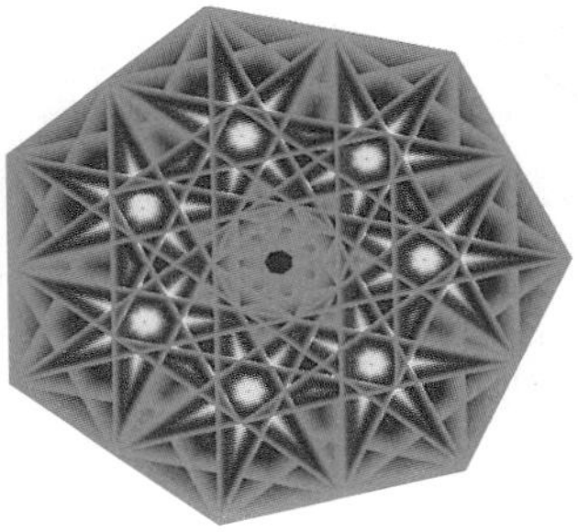

3. 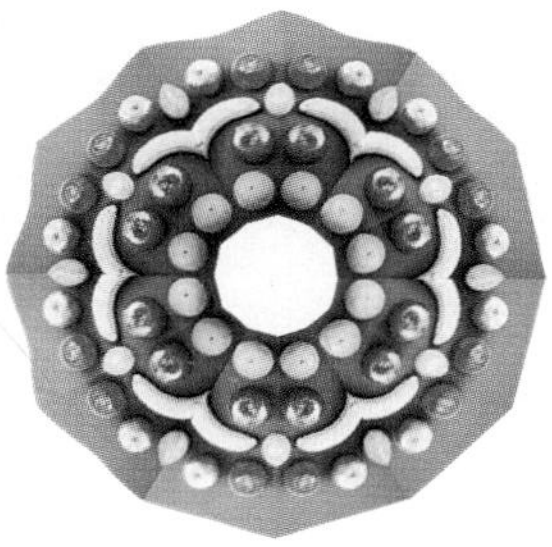

Match each image of the figure at the left with one of the following isometries: I. reflection II. rotation III. translation IV. glide reflection

4. 

a. 
b. 
c. 

d. 

5. 

a. 

b. 
c. 
d. 

6. 
a. 
b. 
c. 
d. 

Coordinate Geometry **Find the image of $\triangle PNB$ under each glide reflection.**

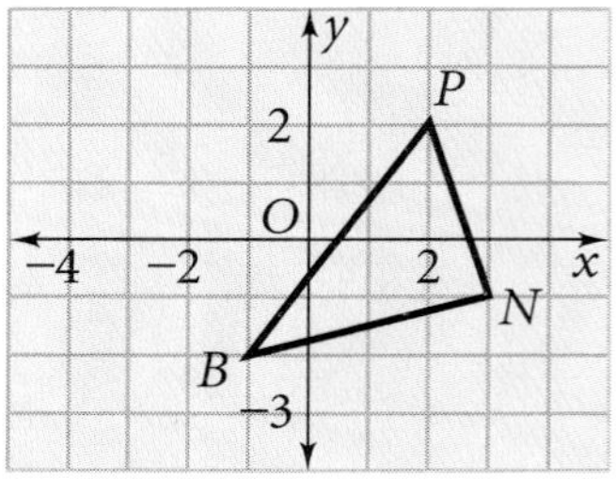

7. $\langle 2, 0\rangle$ and $y = 3$

8. $\langle 0, -3\rangle$ and $x = 0$

9. $\langle 0, 3\rangle$ and $x = -2$

10. $\langle -2, 0\rangle$ and $y = -1$

11. $\langle 2, 2\rangle$ and $y = x$

12. $\langle -1, 1\rangle$ and $y = -x$

Is the isometry that maps the black figure to the blue figure a translation, reflection, rotation, or glide reflection?

13. 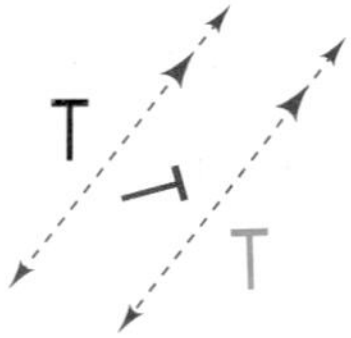

14. 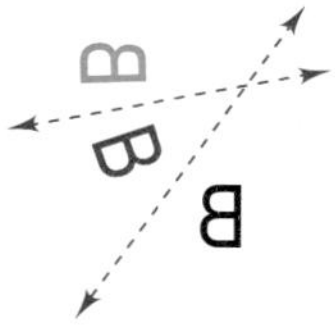

15. 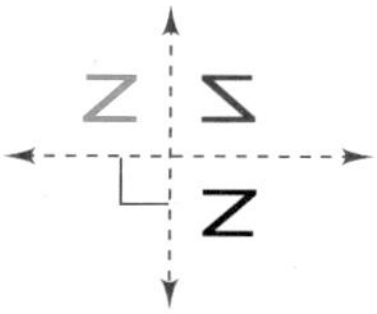

16. 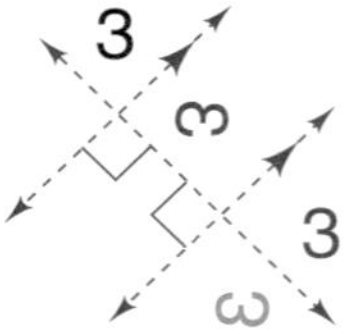

17. 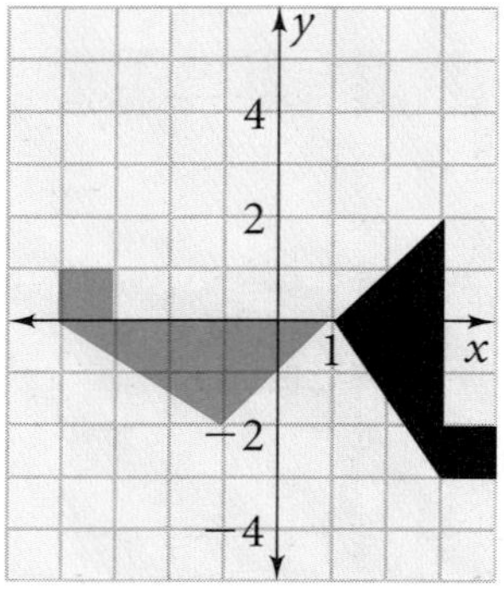

18. 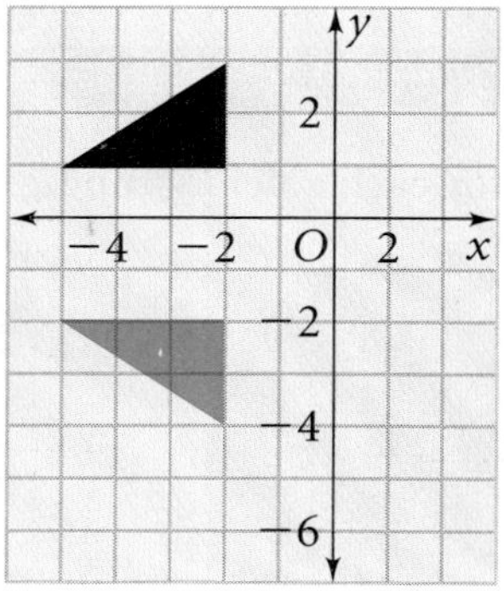

19. 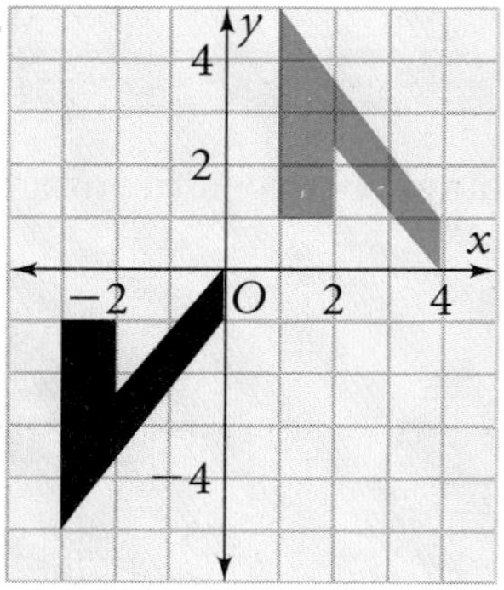

20. 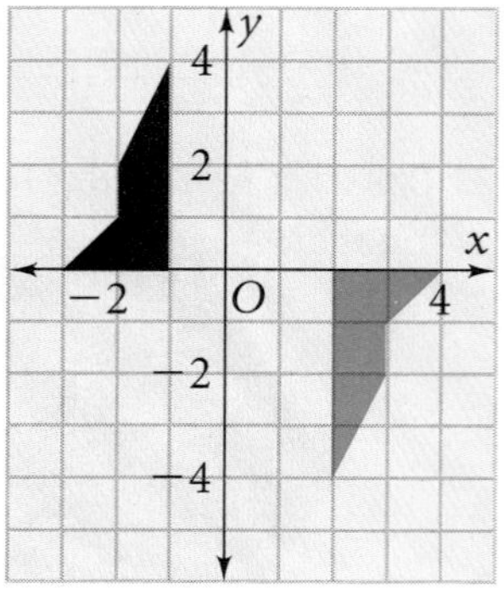

State whether each mapping is a reflection, rotation, translation, or glide reflection.

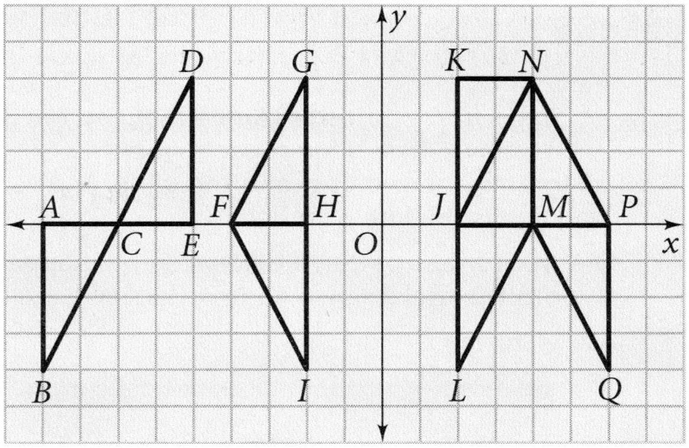

21. $\triangle ABC \longrightarrow \triangle EDC$

22. $\triangle EDC \longrightarrow \triangle PQM$

23. $\triangle MNJ \longrightarrow \triangle EDC$

24. $\triangle HIF \longrightarrow \triangle HGF$

25. $\triangle PQM \longrightarrow \triangle JLM$

26. $\triangle MNP \longrightarrow \triangle EDC$

27. $\triangle JLM \longrightarrow \triangle MNJ$

28. $\triangle PQM \longrightarrow \triangle KJN$

29. **Writing** Reflections and glide reflections are *odd isometries*, while translations and rotations are *even isometries*. Use what you learned in this lesson to explain why these categories make sense.

30. **Paper folding** Fold a rectangular piece of paper into sixths as shown. Then use scissors to cut a nonregular polygon into the folded paper. Unfold the paper and number each of the six figures represented by the holes. What isometries map Figure 1 onto Figures 2, 3, 4, 5, and 6?

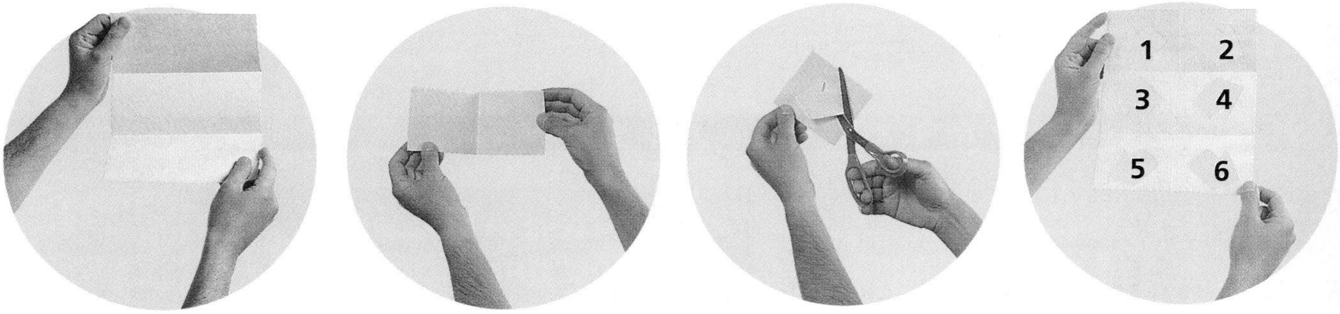

31. **Probability** Suppose you toss two cardboard cutouts of congruent figures into the air so that they land in random positions on the floor. Consider the isometry that maps one of the figures onto the other. Which of the four isometries, if any, are most likely to occur?

PROBLEM SOLVING HINT

Cut out two congruent figures and experiment. Look for a pattern.

32. **Architecture** These housing plans were created by the Swiss architect Le Corbusier for a development in Pessac, France. They illustrate each of the four isometries. Name the isometry illustrated by each design.

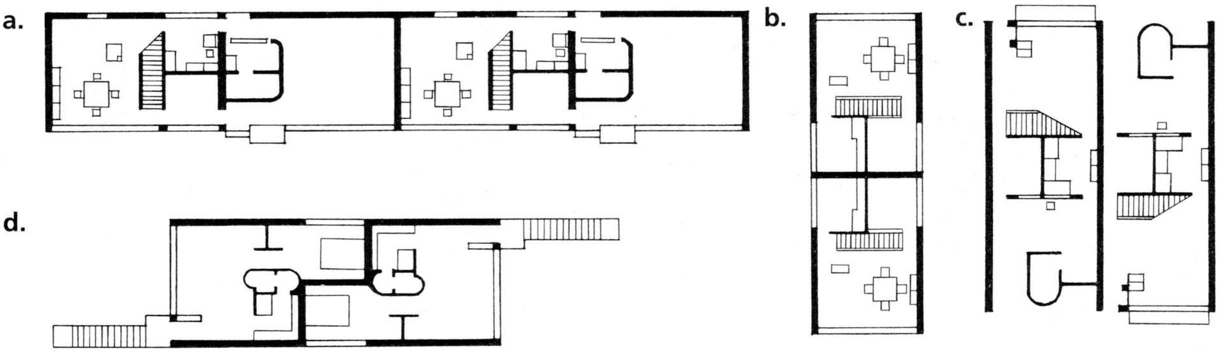

Chapter Project *Find Out by Investigating*

Some frieze patterns can be mapped onto themselves by a glide reflection. In the diagrams, the vector shows the glide. The red line is the line of reflection.

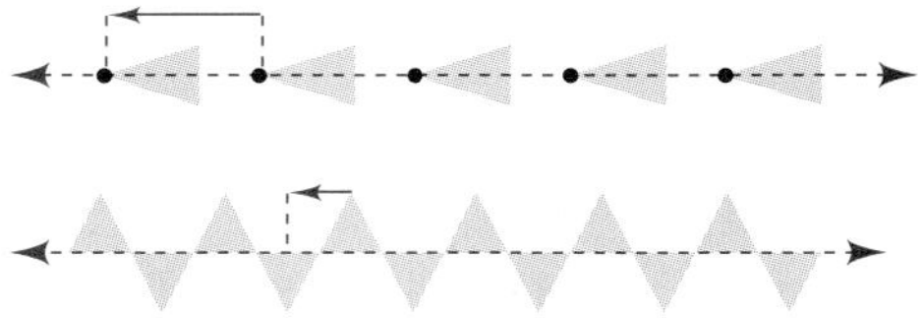

- Which frieze patterns below can be mapped onto themselves by a glide reflection?

a. Victorian Design

b. Chinese Design

c. Nigerian Design

d. Turkish Design

Exercises MIXED REVIEW

Data Analysis **Refer to the circle graph.**

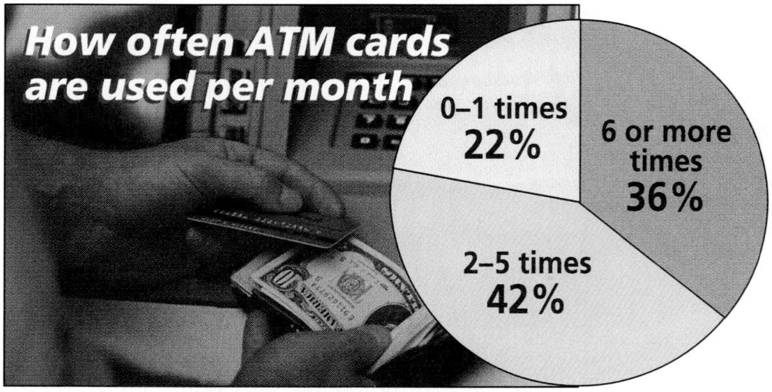

Source: *Research Partnership survey for Cirrus Systems*

33. Find the measure of each central angle.

34. What percent of people use their ATM cards fewer than six times per month?

35. Which statements are true, based on the graph?

a. Most people use their ATM cards two or more times per month.

b. Most people have ATM cards.

c. At least 36% of people use their ATM cards one or more times per week.

Getting Ready for Lesson 3-5

Find the image of each figure under the given transformation.

36.

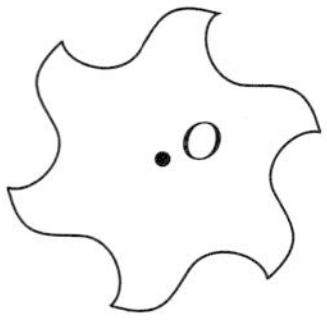

Rotation of 60° about O

37.

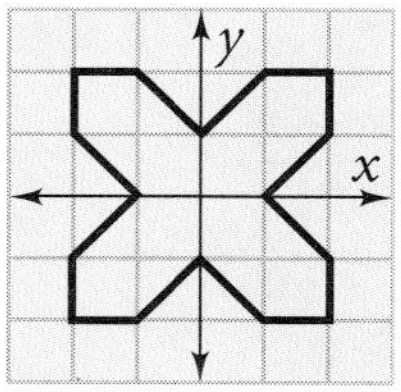

Reflection in y-axis

38.

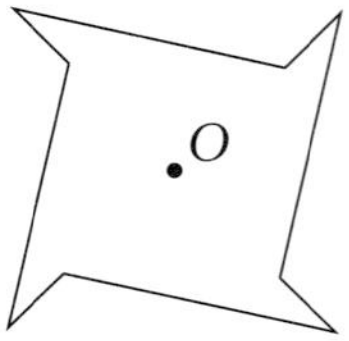

Rotation of 90° about O

Kaleidoscopes

Before Lesson 3-5

The mirrors in a kaleidoscope reflect objects to create a *symmetrical* design. Work in pairs or small groups to create your own kaleidoscope.

Construct

- Use geometry software. Draw a line and construct a point on the line. Rotate the line 60° about the point. Rotate the image 60° about the point.
- Construct a polygon in the interior of an angle, as shown. Reflect the polygon in the closest line in a clockwise direction. Reflect the image in the next closest line in a clockwise direction. Continue reflecting until the kaleidoscope is filled. Then hide the lines of reflection.

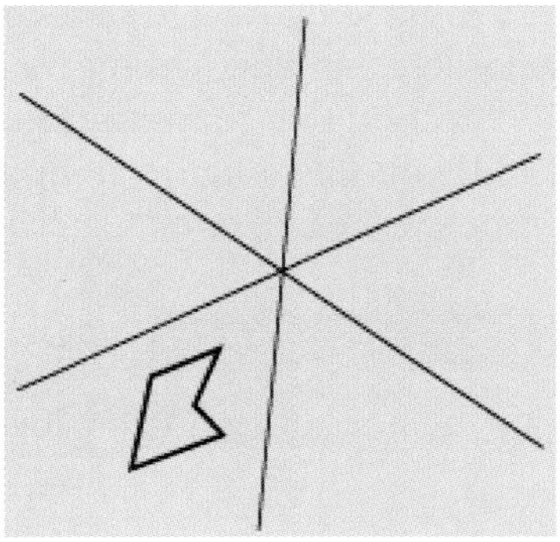

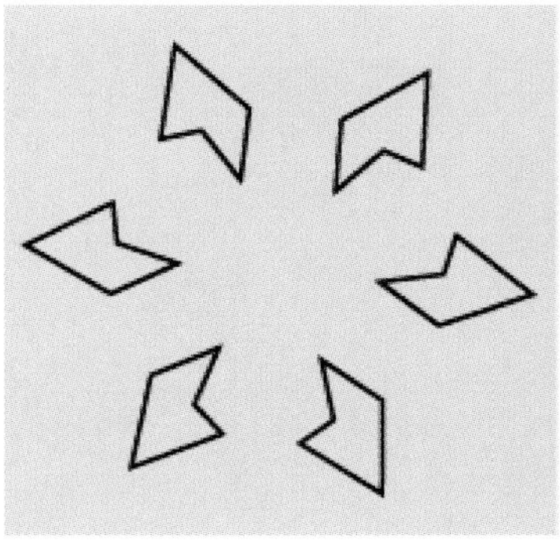

Investigate

- Manipulate the original figure by dragging any of its vertices or selecting and moving it. As the figure is manipulated, what happens to the images? Does the design remain symmetrical? Continue manipulating the original figure until you are satisfied with your design. Print the design and color it.
- Now add other figures in addition to the original polygon. Reflect these figures to create a more interesting design, as shown at the right. (You may need to temporarily show the hidden lines of reflection.) Print your design and color it.

Extend

Create a kaleidoscope with four lines of reflection. Draw a line and construct a point on the line. Rotate the line 45° about the point and repeat until you have four lines. Add a figure to the interior of an angle and reflect it as described above.

What You'll Learn

- Identifying types of symmetry in figures

...And Why

To understand a topic that influences art, dance, and poetry, and is an important tool of scientists

3-5 Symmetry

THINK AND DISCUSS

Reflectional Symmetry

A figure has **symmetry** if there is an isometry that maps the figure onto itself. A plane figure has **reflectional symmetry,** or **line symmetry,** if there is a reflection that maps the figure onto itself. If you fold a figure along a line of symmetry, the halves match exactly.

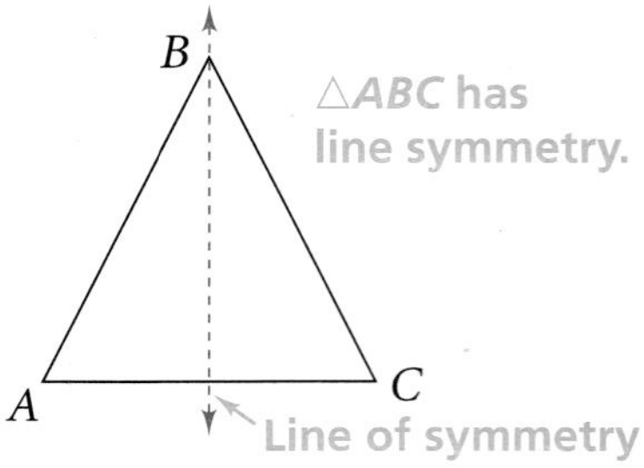

Example 1

Draw the lines of symmetry for each figure.

a.

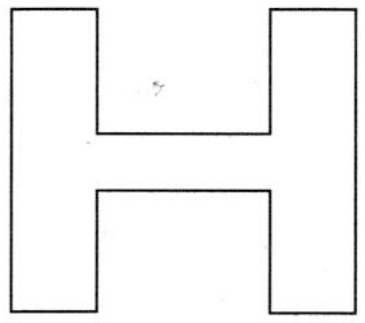

b.

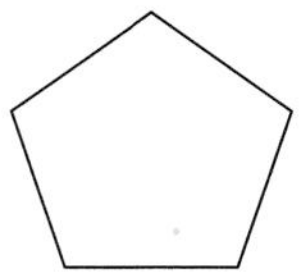

c.

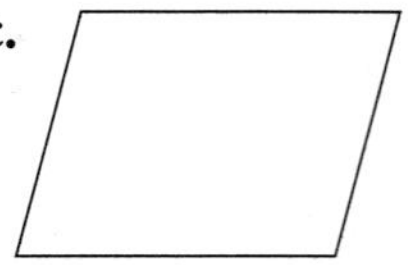

a.

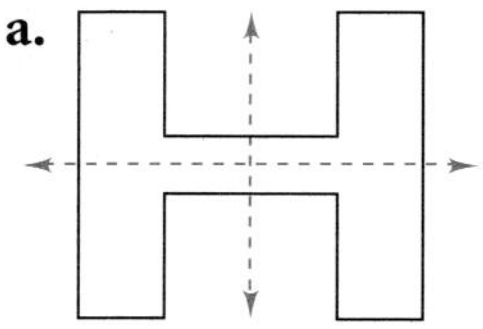

b.

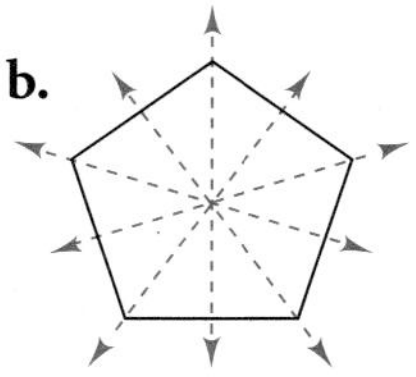

c. This figure has no lines of symmetry.

Symmetry is especially important to left-handers. Because about 95% of people are right-handed, right-handed versions of objects are easier to find (and often less expensive!) than left-handed versions of the same objects.

Paul McCartney of the Beatles used converted right-handed bass guitars until he was given a left-handed Rickenbacker bass in the mid-1960s.

Left-handed people appreciate reflectional symmetry because objects that have it can be used by both left-handers and right-handers.

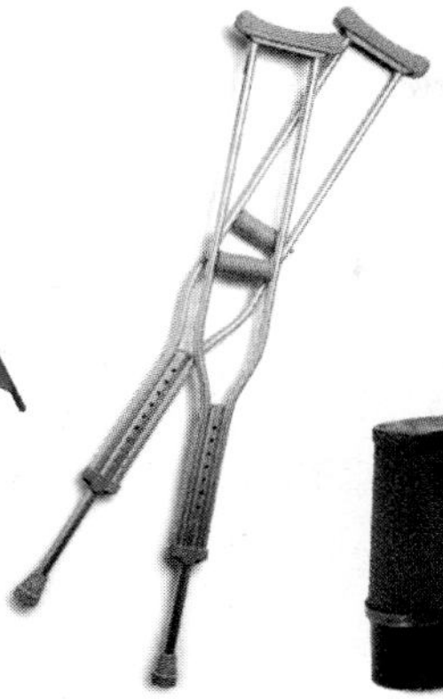

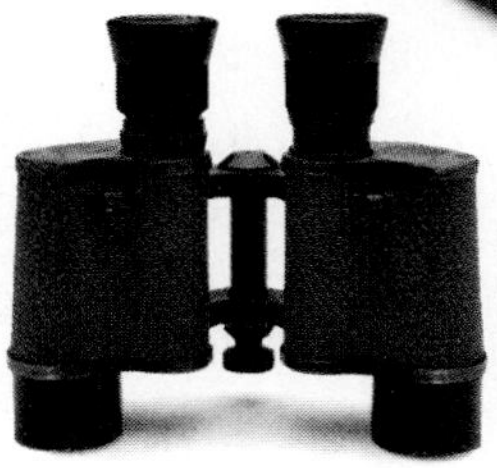

Three-dimensional objects with reflectional symmetry can be divided into two congruent parts by a plane. You can sketch these symmetries in two dimensions by using orthographic views (top, front, or right side).

Example 2 **Relating to the Real World**

Technical Drawing Show the reflectional symmetries of each object by sketching an orthographic view.

a.

b.

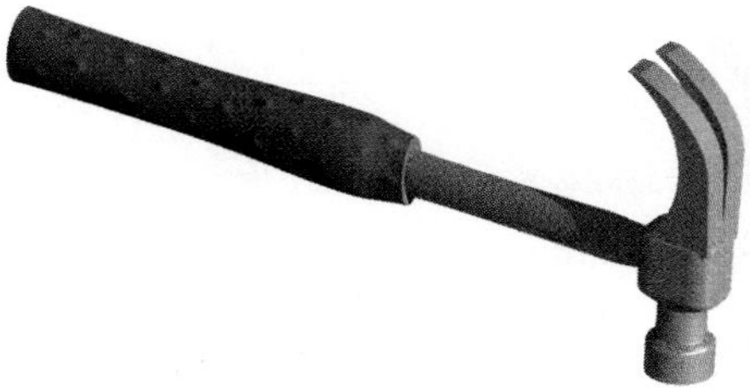

a.

Top View

b.

Front View

1. **Critical Thinking** Name an object that has more than one plane of symmetry.

Rotational Symmetry

A figure has **rotational symmetry** if there is a rotation of 180° or less that maps the figure onto itself.

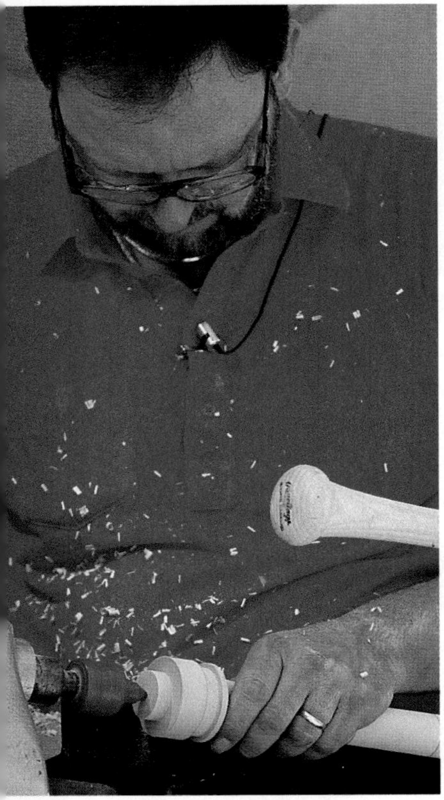

The spinning motion of a lathe ensures that objects created on it have rotational symmetry.

Example 3

Which figures have rotational symmetry? For those that do, give the angle of rotation.

a. 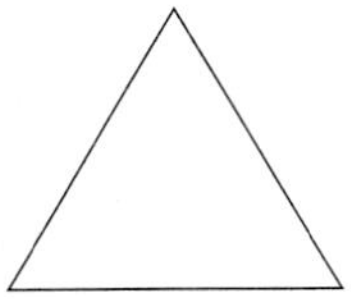

b. 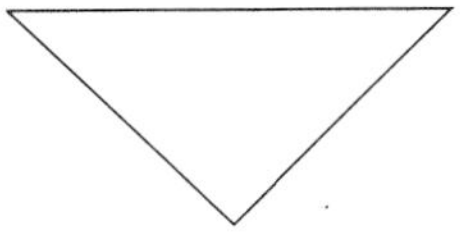

c. 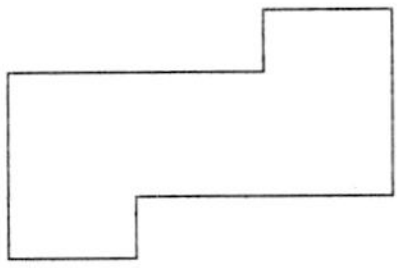

a. 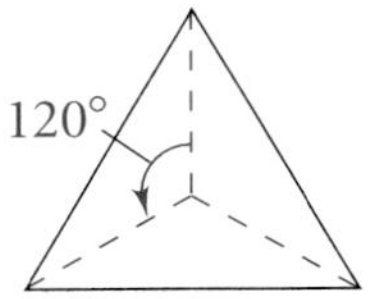

b. This figure does not have rotational symmetry.

c. 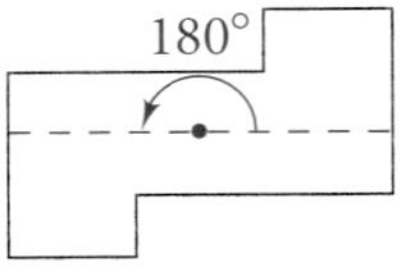

2. If a figure has rotational symmetry, must it also have line symmetry? Explain your answer.

A rotation of 180° is known as a **half-turn.** If a half-turn maps a plane figure onto itself, the figure has **point symmetry.**

3. Try This Which figures have point symmetry?

a. 

b. 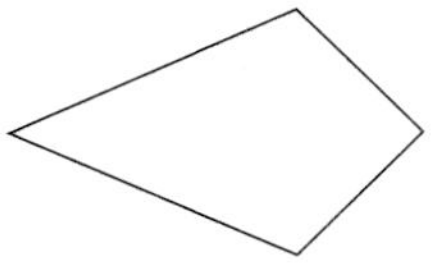

c. 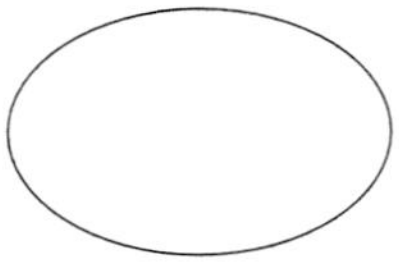

So far, you've looked at figures with reflectional and rotational symmetry. As you may have guessed, figures may also have *translational* or *glide reflectional symmetry*. You will discuss these symmetries in the next lesson.

Type of Symmetry	Points
Reflectional Symmetry	1
Rotational Symmetry of 180°	2
Rotational Symmetry other than 180°	3

WORK TOGETHER

- Work in groups to find examples of symmetrical objects in your classroom. For each object that you find, sketch an orthographic view and list its symmetries. You will have only ten minutes in which to search, so plan your time wisely!
- Determine your group's score by using the chart at the left.

Exercises ON YOUR OWN

What types of symmetry does each figure have? If it has reflectional symmetry, sketch the figure and the line(s) of symmetry. If it has rotational symmetry, state the angle of rotation.

1. 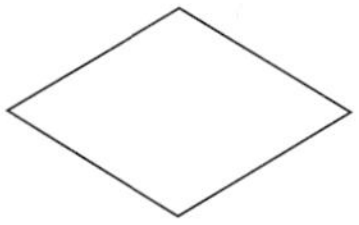

2. 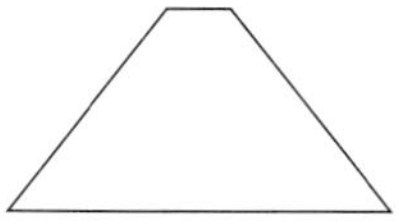

3. 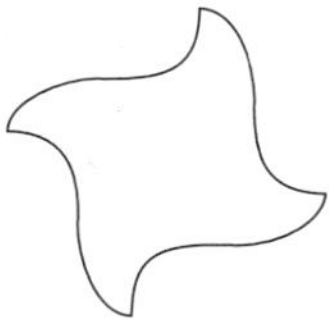

4. 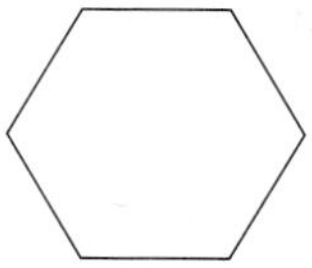

5. 

6. 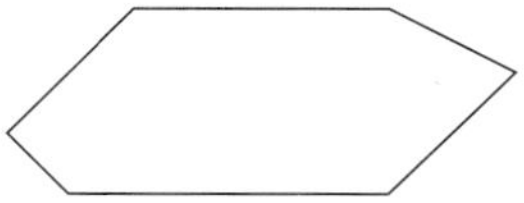

7. 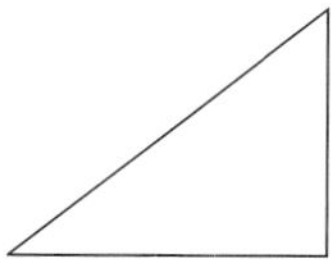

8. 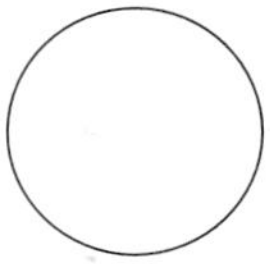

What types of symmetry are shown in each photograph?

9. 

10. 

11. 

12. 

13. Sketch a triangle that has reflectional symmetry but not rotational symmetry.

14. **a.** Copy the tree diagram of quadrilaterals on page 92. Then draw each figure's lines of symmetry.

b. *Patterns* How do the symmetries of the figures in the top portion of the tree diagram compare with the symmetries of those lower in the diagram?

Each diagram shows a shape folded along a red line of symmetry. Sketch the unfolded figure.

15. 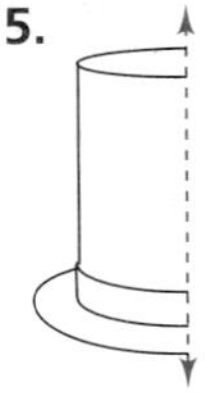

16. 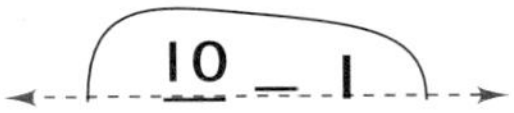

17. 

18. a. The word CODE has a horizontal line of symmetry through its center. Find three other words that have this type of symmetry.
    b. The word WAXY, when printed vertically, has a vertical line of symmetry. Find three other words that have this type of symmetry.

Advertising **Many automobile manufacturers have symmetrical logos. Describe the symmetry, if any, in each logo.**

19. 20. 21. 22. 23.

24. 25. 26. 27. 28.

29. Research Many company logos are symmetrical. Find three symmetrical logos in the Yellow Pages of your local phone book. Copy each logo, identify the name of the business, and describe the type(s) of symmetry illustrated.

Geometry in 3 Dimensions **Show the reflectional symmetries of each object by sketching an orthographic view.**

30. 

31. 

32. 

33. **a.** *Languages* Copy the chart at the right. Then use the alphabets below to list the letters in each category. Some letters will appear in more than one category.

b. Which alphabet is more symmetrical? Explain your reasoning.

A B C D E F G H I J K L M N O P Q R S T U V W X Y Z

Α Β Γ Δ Ε Ζ Η Θ Ι Κ Λ Μ Ν Ξ Ο Π Ρ Σ Τ Υ Φ Χ Ψ Ω

Type of Symmetry

Language	Horizontal Line	Vertical Line	Point
English			
Greek			

34. *Writing* Use what you learned in Lesson 3-4 to explain why a figure that has two or more lines of symmetry must also have rotational symmetry.

Algebra **Sketch the graph of each equation. Describe the symmetry of each graph.**

35. $y = x^2$ **36.** $y = (x - 2)^2$ **37.** $y = x^3$

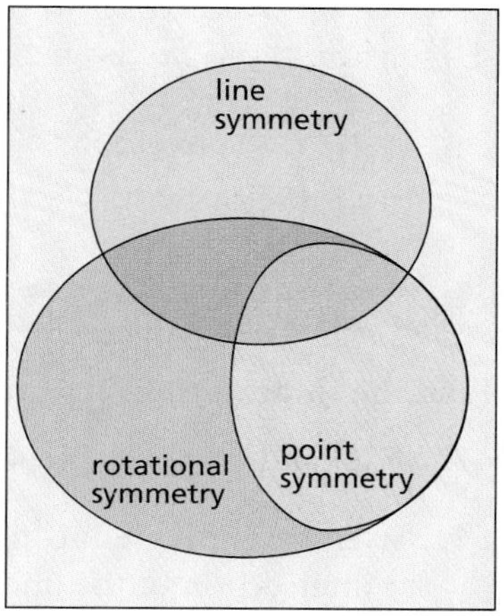

38. *Open-ended* Copy the Venn diagram; then draw a figure in each of its six regions that shows that type of symmetry.

39. *Open-ended* The equation $\frac{10}{10} - 1 = 0 \div \frac{83}{83}$ is not only true, but also symmetrical. Write four other equations or inequalities that are both true and symmetrical.

40. **a.** Is the line that contains the bisector of an angle a line of symmetry of the angle? Explain.

b. Is a bisector of a segment a line of symmetry of the segment? Explain.

Each of the logos below is nearly symmetrical. For each logo, describe how you could alter it to make it symmetrical. Then, describe the symmetries of your altered logo.

41.

42.

43.

Coordinate Geometry **A figure has a vertex at (3, 4). If the figure has the given type of symmetry, state the coordinates of another vertex of the figure.**

44. line symmetry in the y-axis **45.** line symmetry in the x-axis **46.** point symmetry in the origin

Chapter Project — Find Out by Classifying

It may surprise you to find out that when you classify frieze patterns by their symmetries, there turn out to be only seven different types. Each pattern is identified by a different two-character code: 11, 1*g*, *m*1, 12, *mg*, 1*m*, or *mm*. Use the flow chart at the right to classify each frieze pattern below.

a. Caucasian Rug Design, Kazak

b. French, Empire Motif

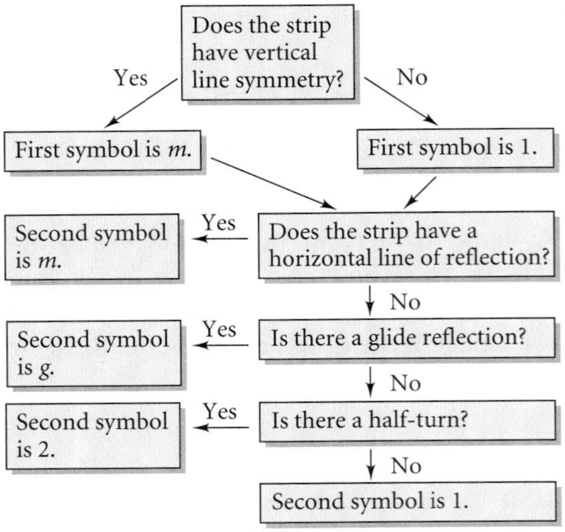

Exercises MIXED REVIEW

Find the slope of the line through the given points.

47. (5, 2) and (3, 0) **48.** (−2, 4) and (3, −7) **49.** (−5, −8) and (1, 6) **50.** (9, −1) and (−2, −7)

51. Given three different coplanar lines, what is the least number of points of intersection of the lines? the greatest number?

Getting Ready for Lesson 3-6

52. Refer to the figure. What is $m\angle 1 + m\angle 2 + m\angle 3 + m\angle 4$?

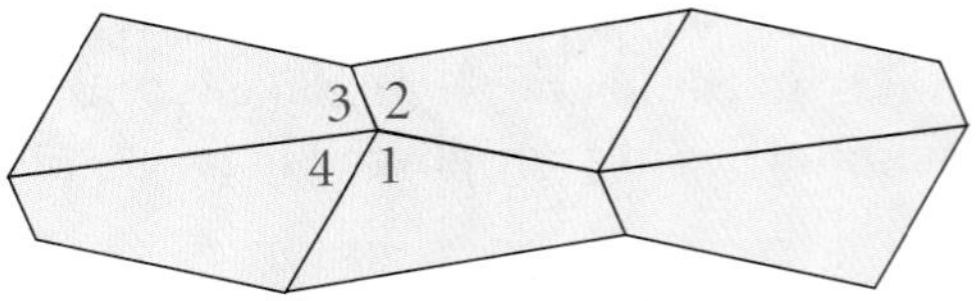

Exercises CHECKPOINT

What types of symmetry does each figure have?

1.

2.

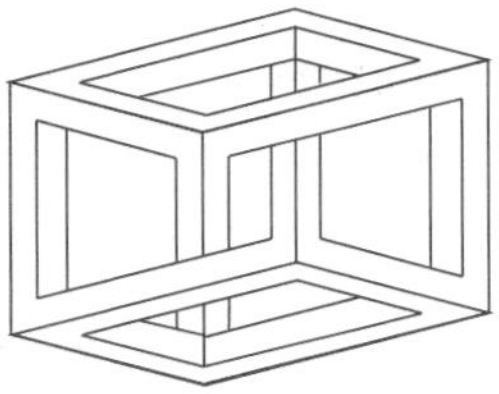

3.

4. $\triangle CAL$ has vertices $C(0, -1)$, $A(-3, 2)$, and $L(-1, -2)$. Find the image of $\triangle CAL$ under a glide reflection in $\langle 0, 4\rangle$ and $x = -2$.

5. **Open-ended** Sketch lines ℓ_1 and ℓ_2 so that the composition of reflections in the two lines is a translation.

What You'll Learn

- Identifying figures that tessellate
- Identifying symmetries of tessellations

...And Why

To recognize tessellations in nature, architecture, art, and other areas of life

What You'll Need

- scissors
- clear tape
- ruler

3-6 Tessellations

THINK AND DISCUSS

Identifying Figures that Tessellate

A **tessellation** is a repeating pattern of figures that completely covers a plane without gaps or overlaps. Tessellations are also called **tilings.** A set of figures that can be used to create a tessellation is said to *tessellate.* You can find tessellations in art, nature, and everyday life.

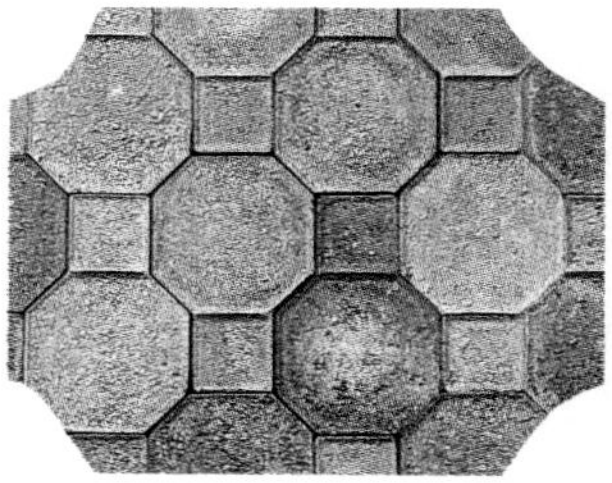

Example 1 Relating to the Real World

Source: © 1996 M. C. Escher Heirs/Cordon Art—Baam—Holland. All rights reserved.

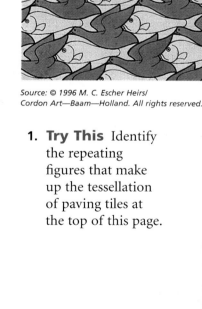

Art Identify the repeating figures that make up this tessellation.

Repeating figures

Who? The Dutch artist Maurits Cornelis Escher (1898–1972) used transformational geometry in intriguing ways in his work. His work is very popular with the public, scientists, and mathematicians.

Source: © 1996 M. C. Escher Heirs/Cordon Art—Baam—Holland. All rights reserved.

1. **Try This** Identify the repeating figures that make up the tessellation of paving tiles at the top of this page.

Source: © 1996 M. C. Escher Heirs/Cordon Art—Baam—Holland. All rights reserved.

Because the figures in a tessellation do not overlap or leave gaps, the sum of the measures of the angles around any vertex must be 360.

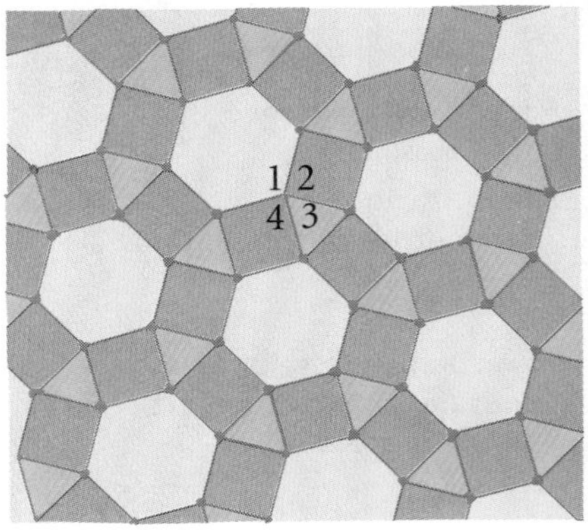

2. a. The tessellation shown consists of regular polygons. Find the measures of angles 1, 2, 3, and 4.

b. Check your answer to part (a) by making sure that the sum of the measures is 360.

A **pure tessellation** is a tessellation that consists of congruent copies of one figure. It may surprise you that there are only three pure tessellations made up of regular polygons. To see why this is the case, consider the following diagrams.

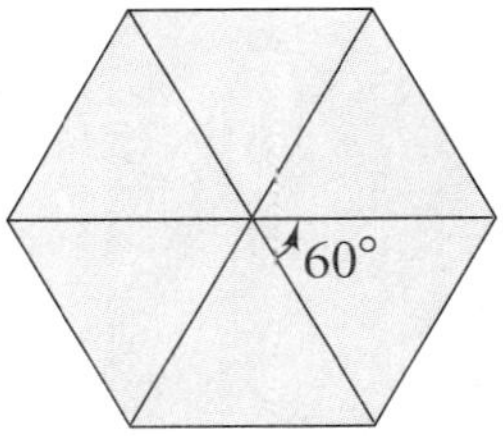

$360° \div 60° = 6$

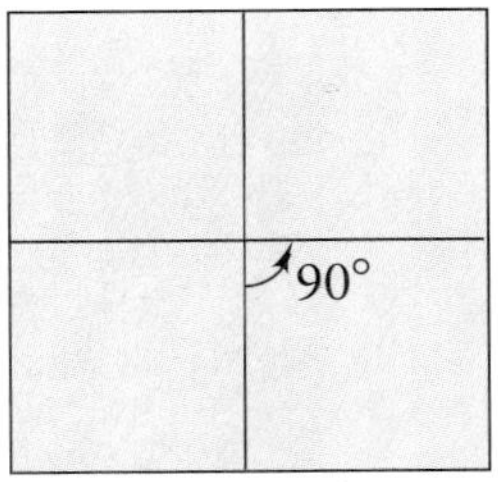

$360° \div 90° = 4$

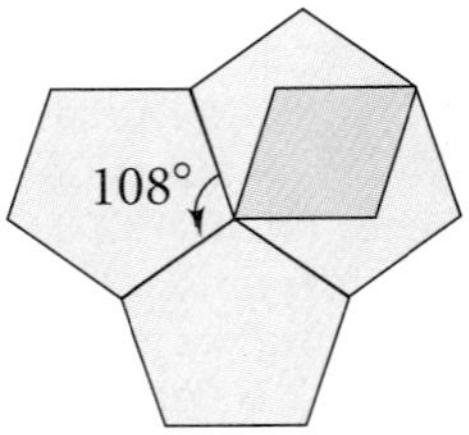

$360° \div 108° \approx 3.3$

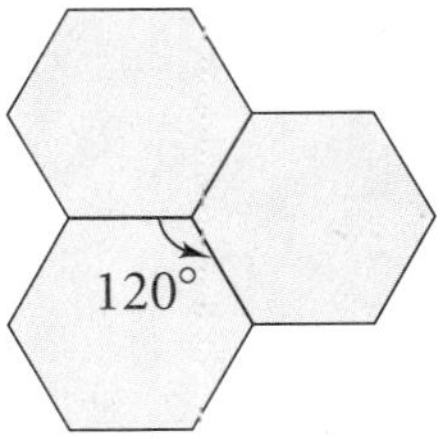

$360° \div 120° = 3$

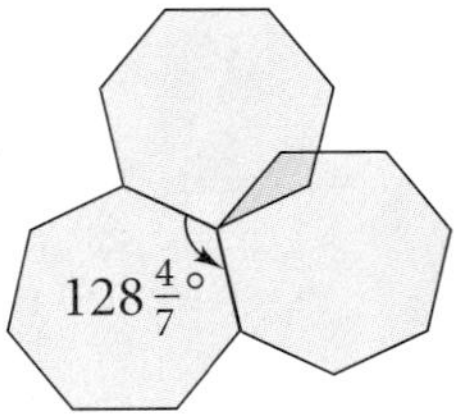

$360° \div 128\frac{4}{7}° \approx 2.8$

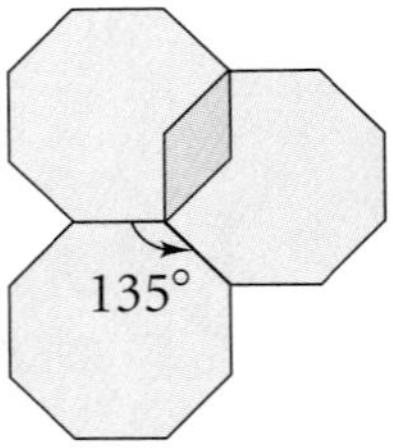

$360° \div 135° \approx 2.6$

3. *Critical Thinking* Explain why no regular polygons with more than six sides tessellate.

Regular triangles and quadrilaterals tessellate, but what about other triangles and quadrilaterals? Work in groups to explore this problem.

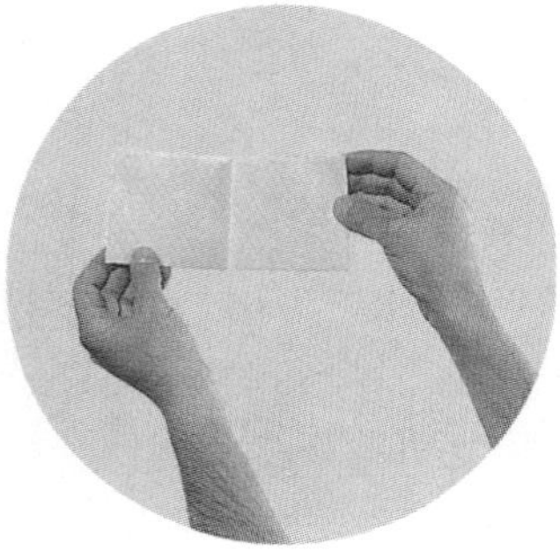

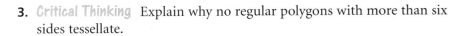

WORK TOGETHER

Step 1: Have each member of your group fold four pieces of paper into sixths, as shown in the photos.

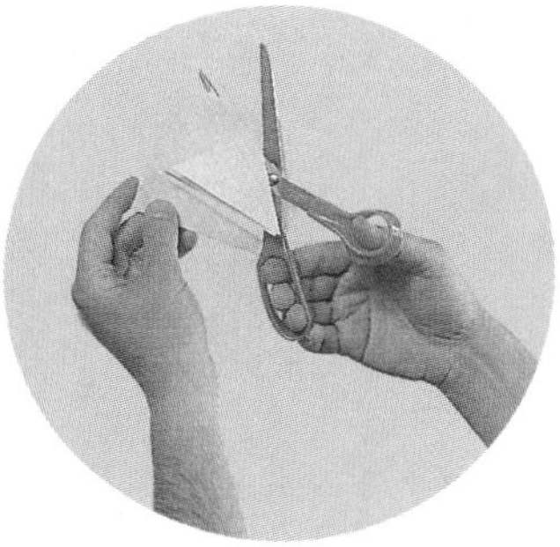

Step 2: Draw a triangle on one of the four folded pieces of paper; then cut through all six sheets of paper to create six congruent triangles. Create six more congruent triangles by tracing one of your cutouts onto another folded piece of paper and then cutting out six triangles.

4. a. Try to arrange your twelve congruent triangles into a tessellation. (*Hint:* Arrange vertices so that they meet.) Compare results within your group.

b. Use what you know about the sum of the measures of the angles of a triangle to explain your results from part (a).

Step 3: Set your triangles aside. Use the method in Step 2 to create twelve congruent quadrilaterals.

5. a. Try to arrange your twelve congruent quadrilaterals into a tessellation. Compare results within your group.

b. Use what you know about the sum of the measures of the angles of a quadrilateral to explain your results from part (a).

THINK AND DISCUSS

In the Work Together, you discovered the following properties of tessellations.

Theorem 3-5 Every triangle tessellates.

Theorem 3-6 Every quadrilateral tessellates.

Tessellations and Symmetry

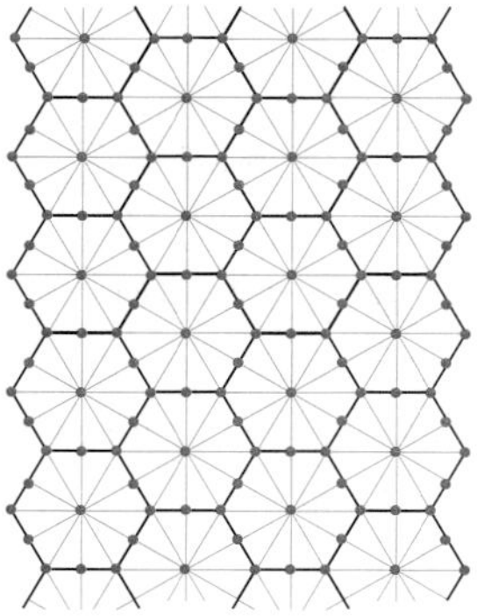

This pure tessellation of regular hexagons has reflectional symmetry in each of the blue lines. It has rotational symmetry centered at each of the red points. The tessellation also has two other types of symmetry—translational symmetry and glide reflectional symmetry.

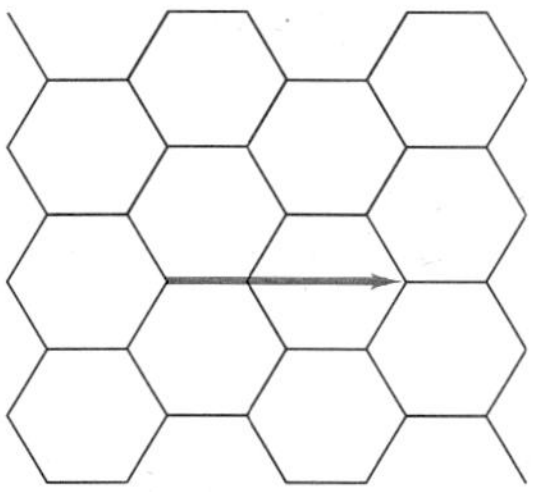

Translational Symmetry

A translation maps the tessellation onto itself.

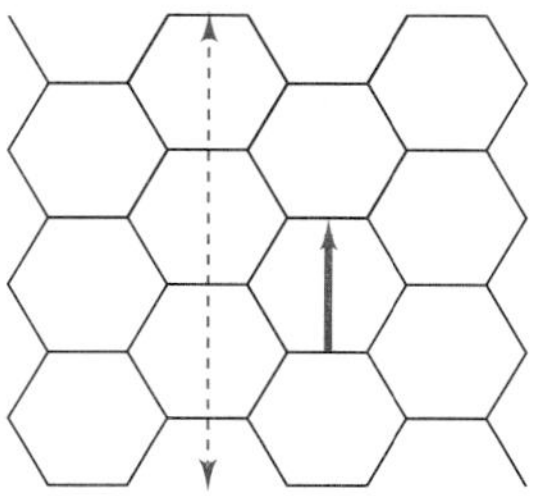

Glide Reflectional Symmetry

A glide reflection maps the tessellation onto itself.

Example 2

List the symmetries of each tessellation.

a.

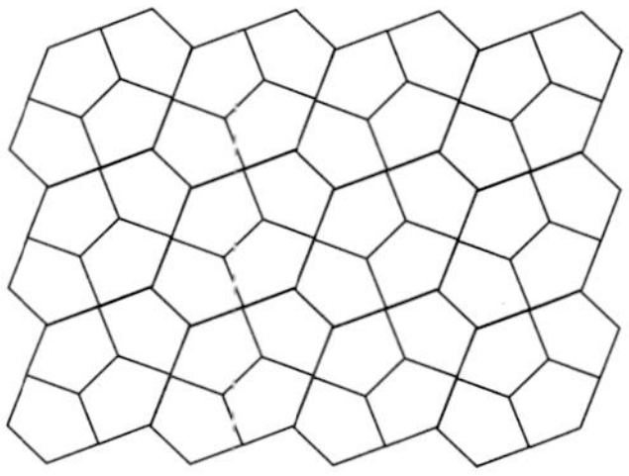

b.

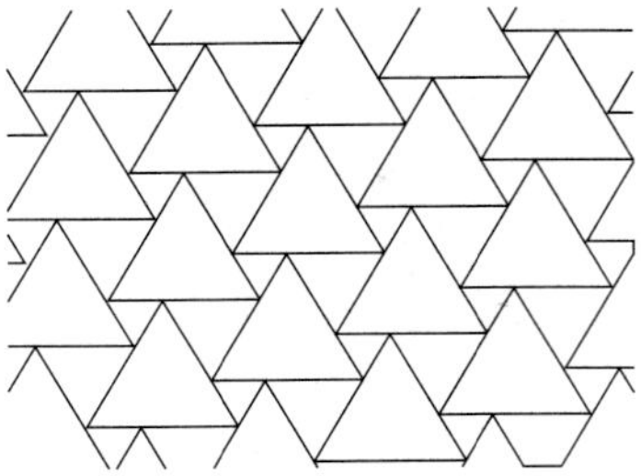

a.

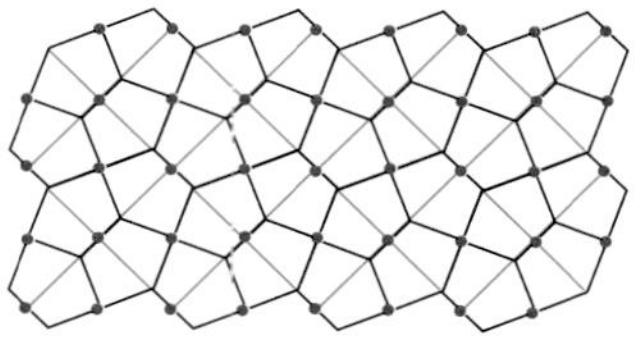

b.

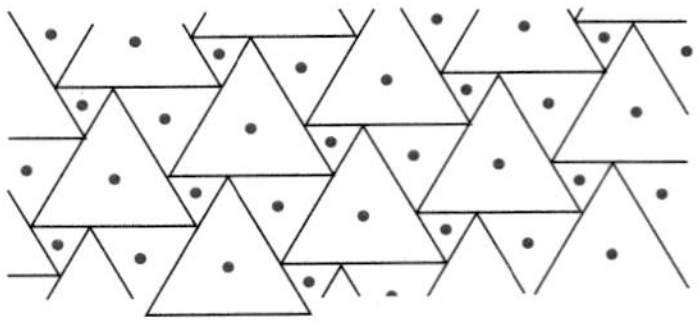

a.
- Line symmetry in the blue lines
- Rotational symmetry in the red points
- Translational symmetry
- Glide reflectional symmetry

b.
- Rotational symmetry in the red points
- Translational symmetry

WORK TOGETHER

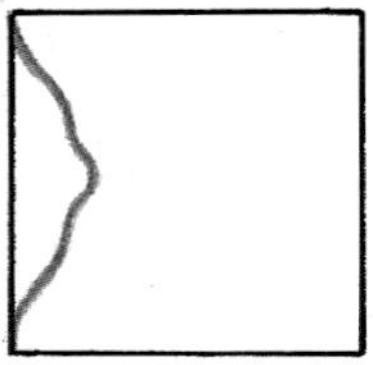

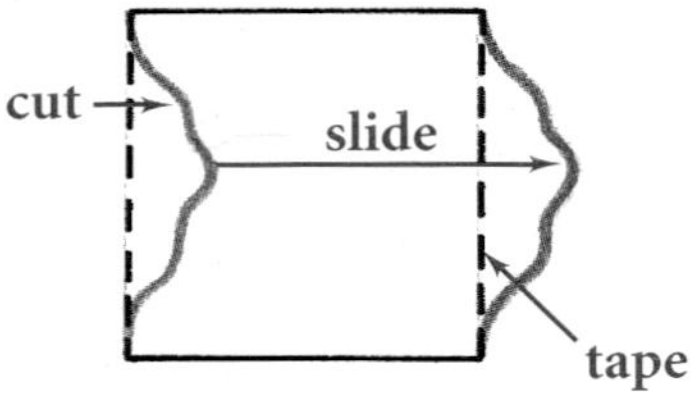

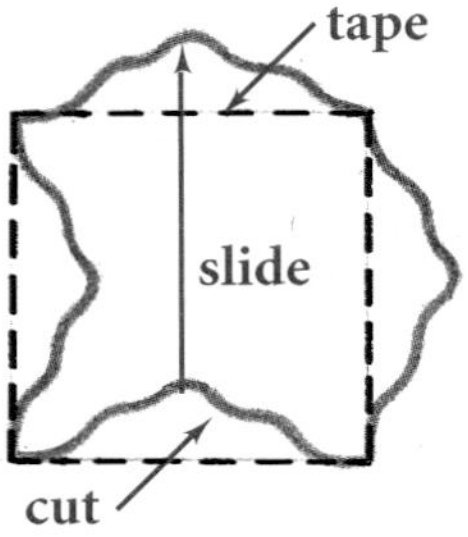

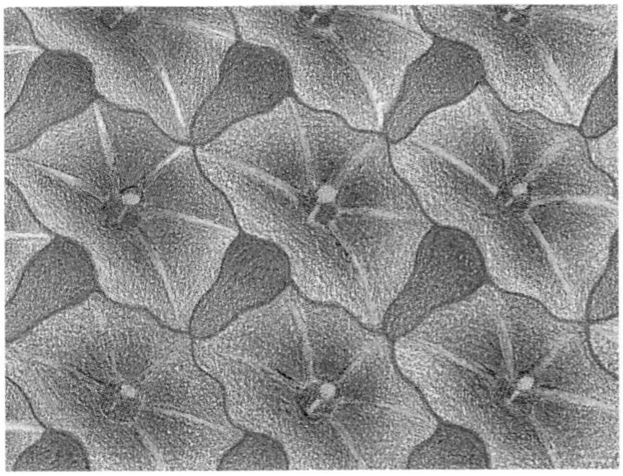

Create your own "Escher-like" tessellation! Start by having each member of your group draw a 1.5-inch square on a blank piece of paper and cut it out.

- Sketch a curve from one vertex to a consecutive vertex.
- Cut along the curve that you sketched and slide the resulting cutout to the opposite side of the square. Tape it in place using clear tape.
- Repeat this process using the remaining two sides of the square.
- Rotate the figure you end up with. What does it look like? A penguin with a hat on? A knight on horseback? A dog with floppy ears? Sketch whatever you come up with on your figure.
- Create a tessellation using your figure.

Exercises ON YOUR OWN

Identify the repeating figure or figures that make up each tessellation.

1. 

Fabric by *Fabric Traditions*

2. 

Arabian design

3. 

Honeycomb

Describe the symmetries of each tessellation. Copy a portion of the tessellation and draw any centers of rotational symmetry or lines of symmetry.

4. 

5. 

6. 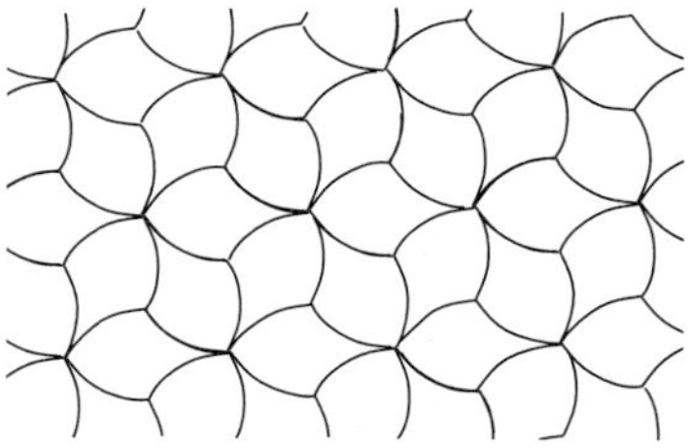

7. 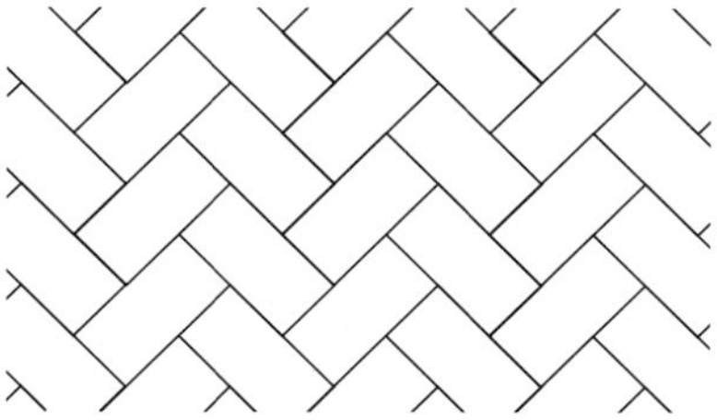

8. 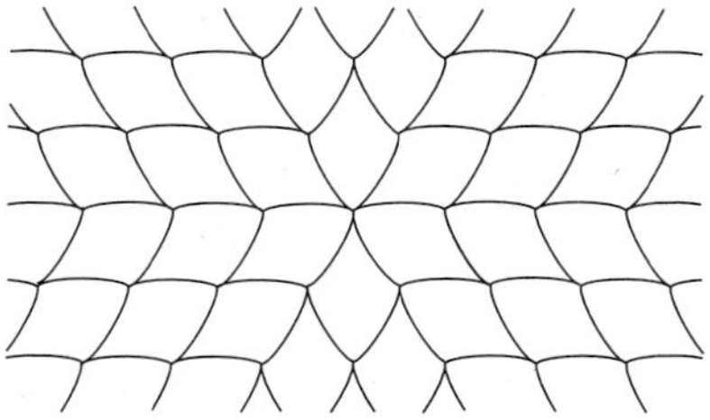

9. 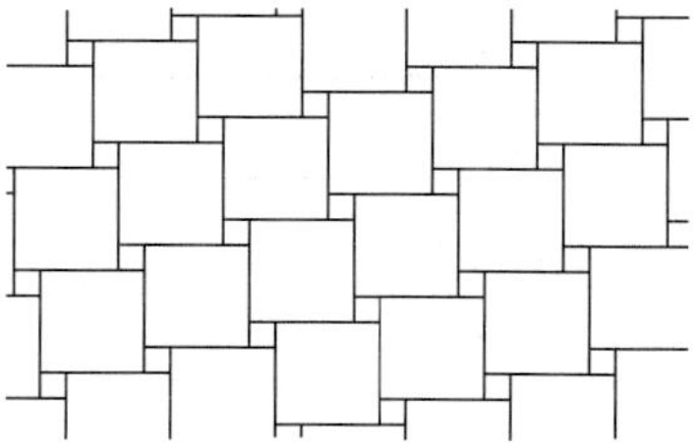

10. The figure shown at the right can be used to tile the plane in several different ways. Make copies of the figure and sketch two different tessellations.

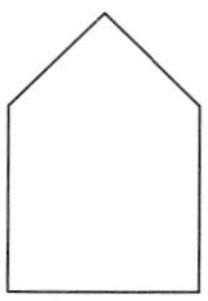

11. **Open-ended** Find and sketch two examples of tessellations found at home or at school.

12. **Writing** Is it possible to tile the plane with regular decagons? Explain why or why not.

Use each figure to create a tessellation on dot paper.

13.

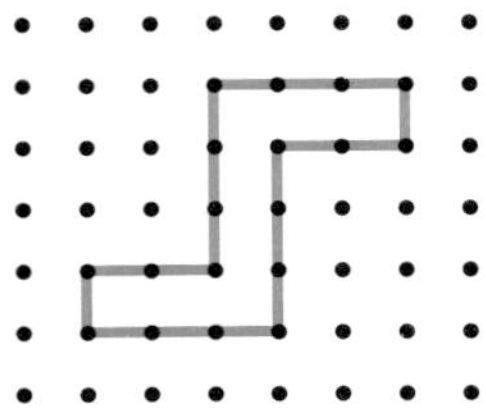

14.

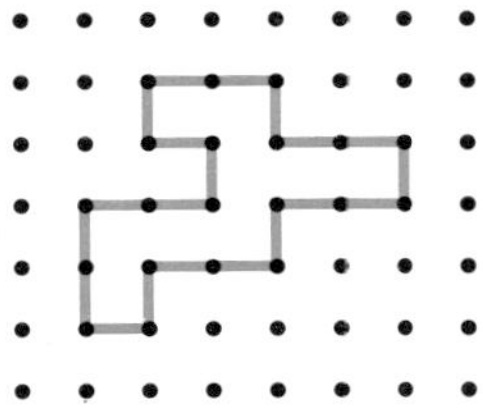

15.

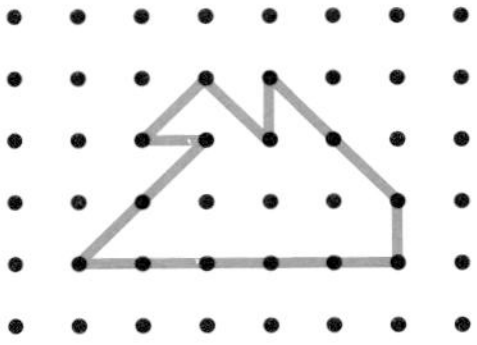

16.

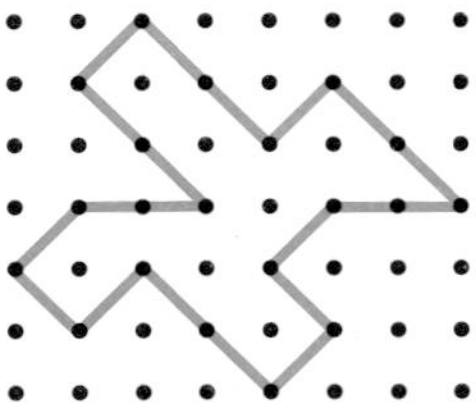

Identify the repeating figure or figures that make up each tessellation.

17.

Mongolian design

18.

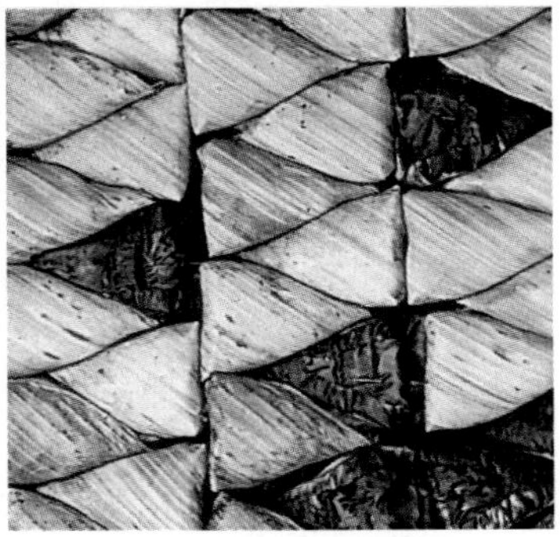

Rice wrapped in banana leaves

19.

Design by François Brisse

20. Follow the steps below to create an "Escher-like" tessellation based on a regular hexagon.

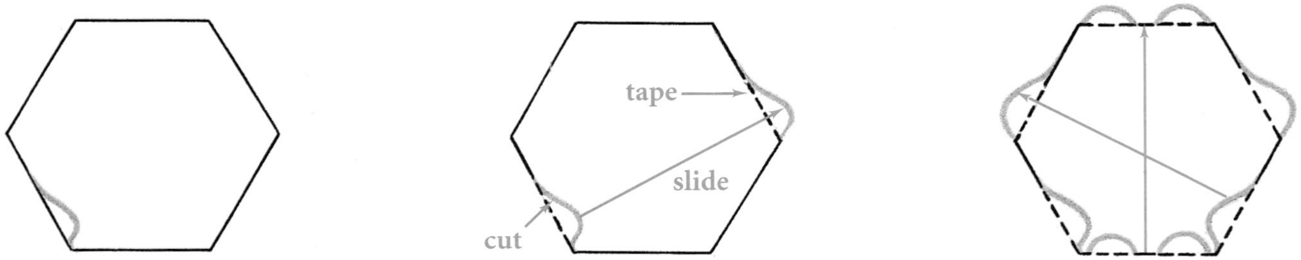

- Trace and cut out the regular hexagon above.
- Sketch a curve from one vertex to an adjacent vertex.
- Cut along the curve that you sketched. Slide the resulting cutout to the opposite side of the hexagon. Tape it in place using clear tape.
- Repeat this process on the remaining two pairs of opposite sides.
- Decorate the figure that you end up with and use it to make a tessellation.

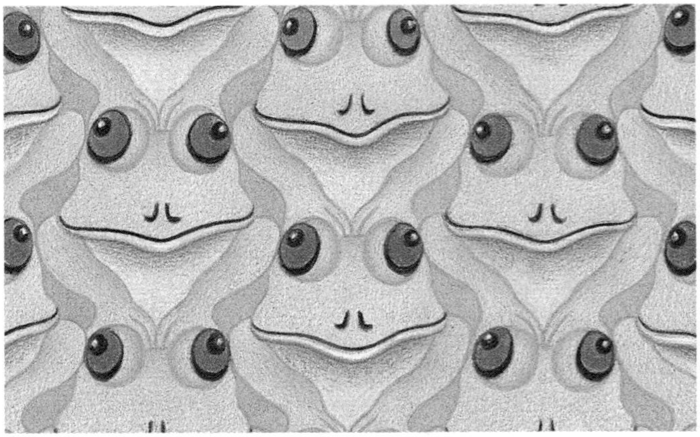

Exercises MIXED REVIEW

Classify each triangle by its sides and angles.

21.

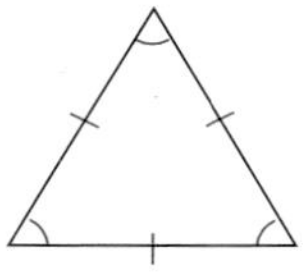

22.

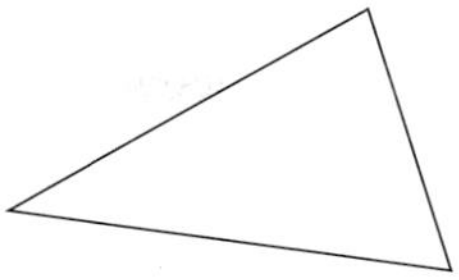

23.

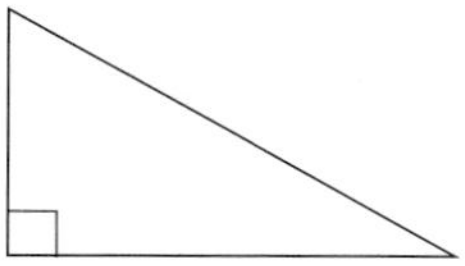

24.

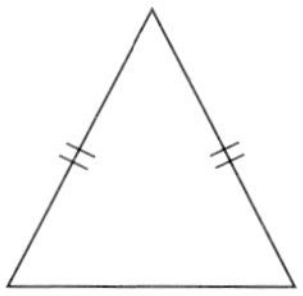

25. Given $\odot A$, identify each of the following.

a. a diameter **b.** a major arc
c. a minor arc **d.** a radius
e. a central angle **f.** a pair of adjacent arcs

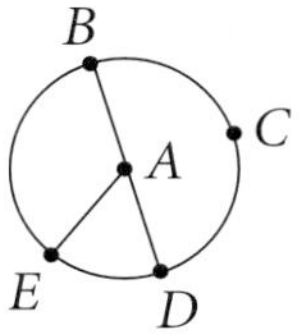

26. a. Find the measure of an interior angle of a regular 3-gon, 7-gon, and 42-gon.

b. Find the sum of the measures of the three angles from part (a).

SELF ASSESSMENT

FOR YOUR JOURNAL

Think about different products that you can buy at a grocery store. How is the idea of tessellation important in the design of containers for products?

Getting Ready for Lesson 3-7

27. In the diagram, $\triangle ABC \sim \triangle DEF$.

a. Find AB and EF.

b. Find the scale factor of the enlargement.

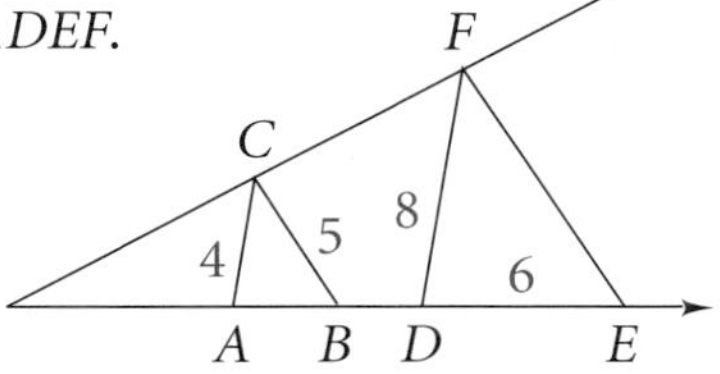

not drawn to scale

A Point in Time

1500 1600 2000

The Grand Mosaic of Mexico

A mosaic is a picture or decorative design made by setting tiny pieces of glass, stone, or other materials in clay or plaster. A mosaic may be a tessellation. Most mosaics, however, do not have a repeating pattern of figures. The art of constructing mosaics goes back at least 6,000 years to the Sumerians, who used tiles to both decorate and reinforce walls.

The Romans gave us the word *tessellate,* from the Latin *tessellare,* "to pave with tiles." During the second century A.D., Roman architects used 2 million tiles to create the magnificent mosaic of Dionysus in Germany. Large as it was, the Dionysus was less than a third the size of the work created in **1950** by the Mexican artist Juan O'Gorman. O'Gorman's mosaic, which depicts the cultural history of Mexico, is ten stories high and covers all four sides of the library of the National University of Mexico. Constructed of some 7.5 million stones, it is the largest mosaic in the world.

Connections: Human Development . . . and more

What You'll Learn

- Locating dilation images of figures

...And Why

To recognize applications of dilations in maps, photographs, scale models, and architectural blueprints

What You'll Need

- centimeter ruler
- calculator
- graph paper

3-7 Dilations

WORK TOGETHER

- Have each member of your group draw a triangle and a point outside the triangle. Draw your figures in roughly the same positions as shown in the diagram. Label the triangle $\triangle RST$ and the point C.

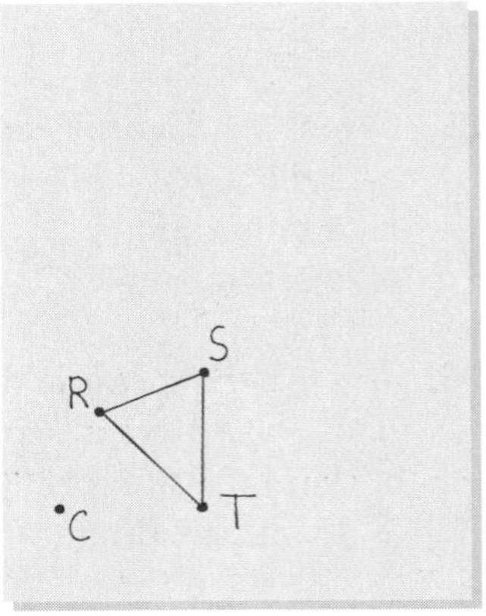

- Draw $\overrightarrow{CR}$, $\overrightarrow{CS}$, and $\overrightarrow{CT}$.
- Have each member of your group select a different number n from the set $\{\frac{1}{3}, \frac{1}{2}, 1\frac{1}{2}, 2, 3\}$.
- Measure $\overline{CR}$, $\overline{CS}$, and $\overline{CT}$ to the nearest millimeter. Then calculate $n \cdot CR$, $n \cdot CS$, and $n \cdot CT$.
- Use a ruler to locate point R' on $\overrightarrow{CR}$ so that $CR' = n \cdot CR$. Locate points S' and T' in the same manner. Draw $\triangle R'S'T'$.

1. Compare corresponding angle measures in triangles $\triangle RST$ and $\triangle R'S'T'$. What do you notice? Compare results within your group.

2. Measure the lengths of the sides of $\triangle RST$ and $\triangle R'S'T'$ to the nearest millimeter. Then use a calculator to find the values of the ratios $\frac{R'S'}{RS}$, $\frac{S'T'}{ST}$, and $\frac{T'R'}{TR}$ to the nearest hundredth. What do you notice?

3. Use your results from Questions 1 and 2 to complete the statement: $\triangle RST$ is __?__ to $\triangle R'S'T'$.

THINK AND DISCUSS

The transformation that you performed in the Work Together is known as a *dilation.* Every dilation has a center and a scale factor. In the Work Together, the center of the dilation was C. The **scale factor** n described the size change from the original figure to the image. The dimensions of the image were n times that of the preimage. A **dilation** with center C and scale factor n, where $n > 0$, maps a point R to R' in such a way that R' is on $\overrightarrow{CR}$ and $CR' = n \cdot CR$. The center of dilation C is its own image (that is, $C' = C$).

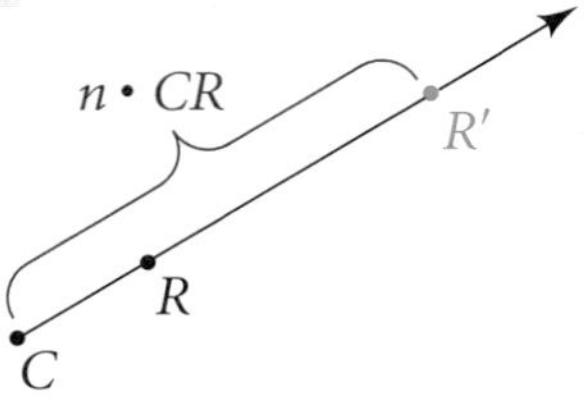

This is not a similarity transformation. Do you see why?

As you noticed in the Work Together, a dilation maps a figure to a similar figure. A dilation is a **similarity transformation.**

4. **Critical Thinking** Describe the dilation image of a figure when the scale factor is 1.

Example 1

Find the scale factor for the dilation that maps the red figure onto the blue figure.

a.

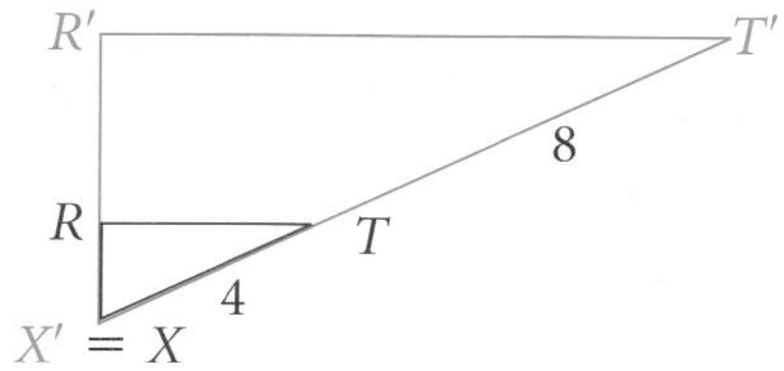

b.

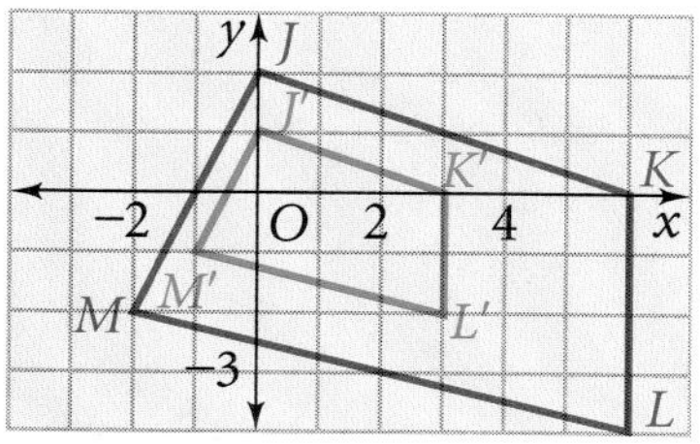

a. $\frac{T'X'}{TX} = \frac{8 + 4}{4} = 3$ **b.** $\frac{K'L'}{KL} = \frac{2}{4} = \frac{1}{2}$

There are two types of dilations. If the image is larger than the original figure, the dilation is an **enlargement.** If the image is smaller than the original figure, the dilation is a **reduction.**

5. What scale factors produce enlargements? reductions?

Example 2 Relating to the Real World

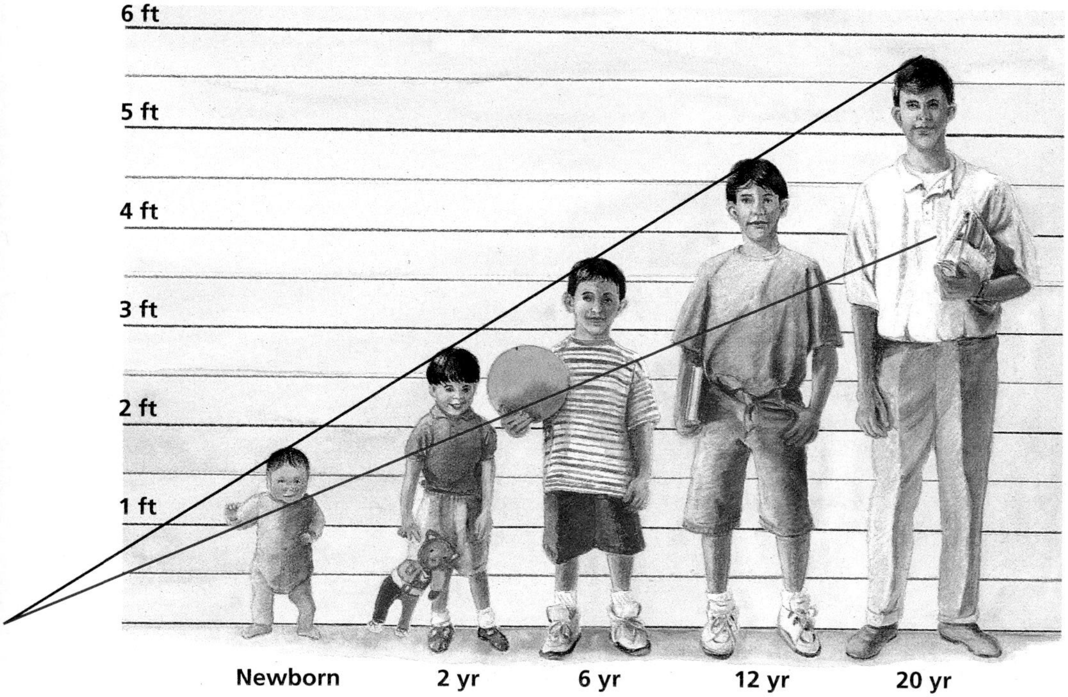

Human Development The diagram shows the growth of a human male from infancy through adulthood. Can human development be modeled by a dilation? Explain.

No. If human development could be modeled by a dilation, the red line would align with the chin in each figure. The 5 ft 9 in. adult figure shown would have a head about $1\frac{3}{4}$ ft long!

To find the image of a point on the coordinate plane under a dilation with center (0, 0), you multiply the x-coordinate and y-coordinate by the scale factor. Here are two examples.

Scale factor 4
$(x, y) \longrightarrow (4x, 4y)$

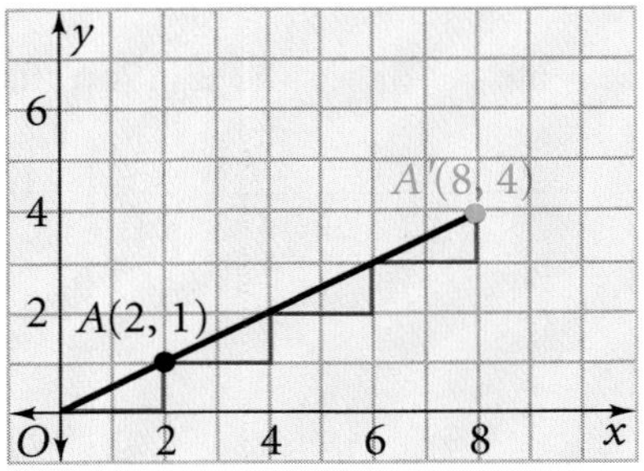

Scale factor $\frac{1}{3}$
$(x, y) \longrightarrow (\frac{1}{3}x, \frac{1}{3}y)$

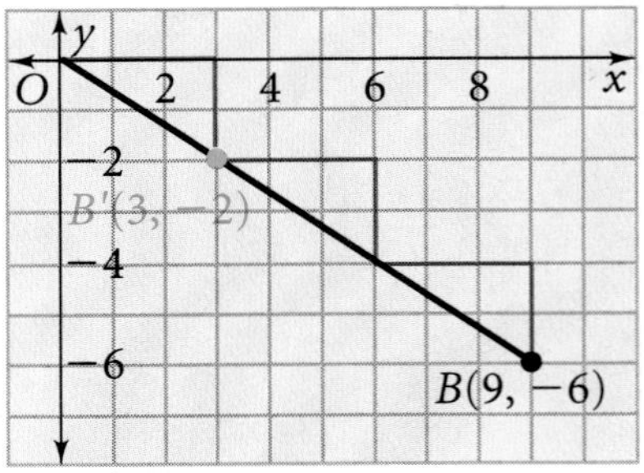

You can use matrices to perform dilations that are centered at the origin.

Example 3

Use matrices to find the image of $\triangle PZG$ under a dilation centered at the origin with scale factor 3.

$$\begin{array}{l} \\ x\text{-coordinate} \\ y\text{-coordinate} \end{array} \begin{array}{c} \begin{array}{ccc} P & Z & G \end{array} \\ \begin{bmatrix} 2 & -1 & 1 \\ 0 & \frac{1}{2} & -2 \end{bmatrix} \end{array}$$

To find the dilation image of $\triangle PZG$, you multiply all the x-coordinates and y-coordinates by 3.

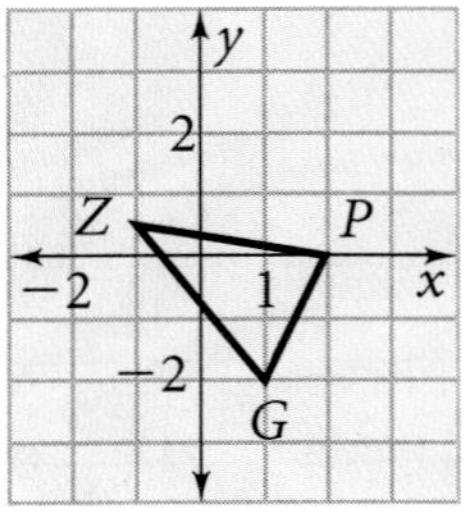

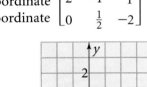

Vertices of Preimage — Vertices of Image

$$3 \cdot \begin{array}{c} \begin{array}{ccc} P & Z & G \end{array} \\ \begin{bmatrix} 2 & -1 & 1 \\ 0 & \frac{1}{2} & -2 \end{bmatrix} \end{array} = \begin{array}{c} \begin{array}{ccc} P' & Z' & G' \end{array} \\ \begin{bmatrix} 6 & -3 & 3 \\ 0 & \frac{3}{2} & -6 \end{bmatrix} \end{array}$$

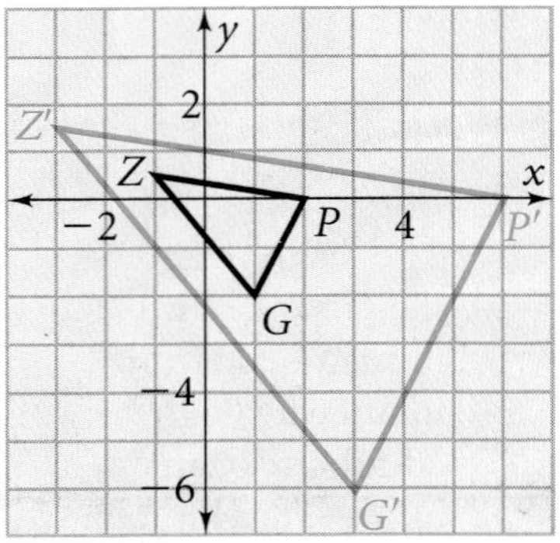

6. Try This Use matrices to find the image of $\triangle PZG$ under a dilation centered at the origin with scale factor $\frac{1}{2}$.

The type of multiplication shown in Example 3, in which each entry of a matrix is multiplied by the same number, is called **scalar multiplication.**

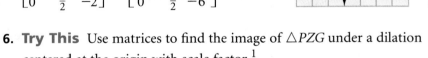

Exercises ON YOUR OWN

Copy $\triangle TBA$ and point O. Draw $\triangle T'B'A'$ under the dilation with the given center and scale factor.

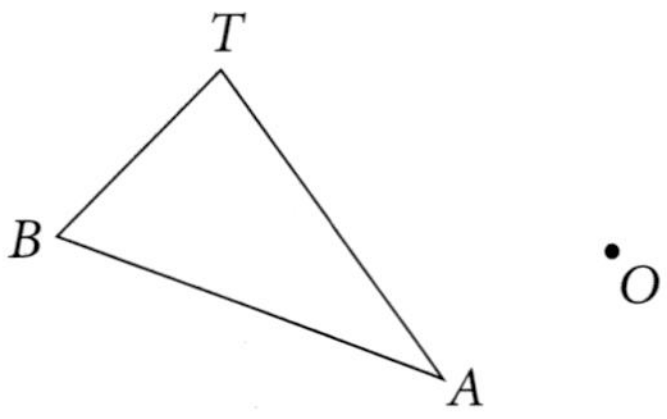

1. Center O, scale factor $\frac{1}{2}$
2. Center B, scale factor 3
3. Center T, scale factor $\frac{1}{3}$
4. Center O, scale factor 2

The blue figure is a dilation image of the red figure.
(a) Determine whether the dilation is a reduction or an enlargement.
(b) Find the scale factor.

5. 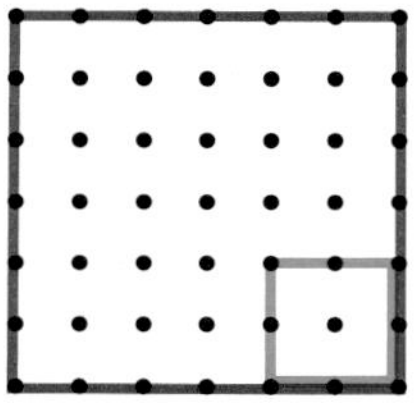

6. 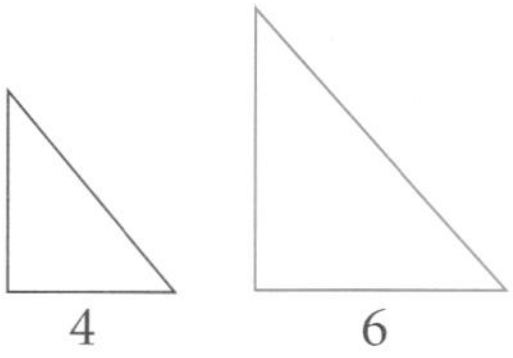

7. 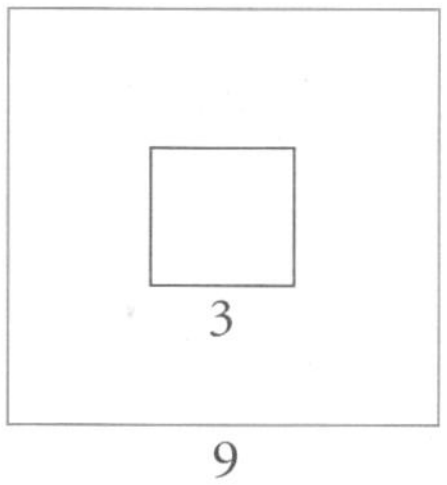

8. 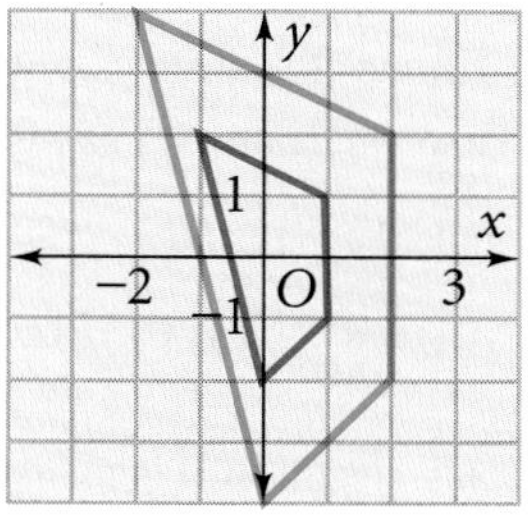

9. 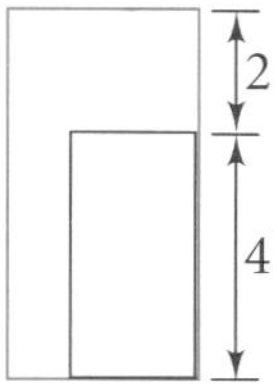

10. 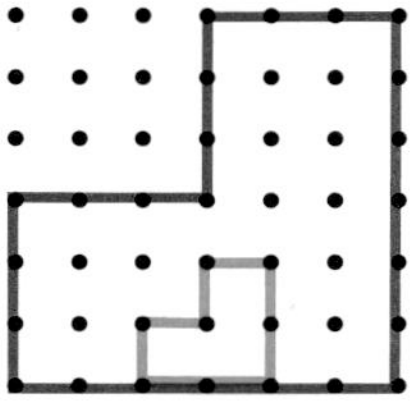

11. 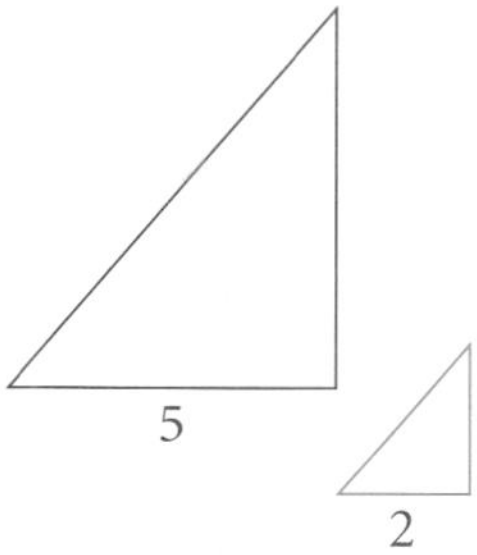

12. 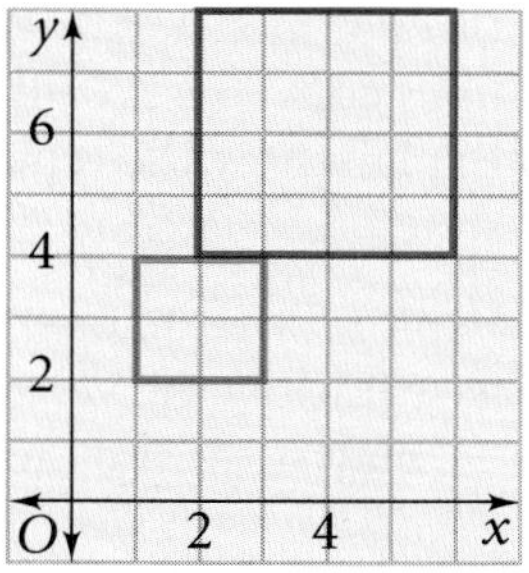

Use scalar multiplication to find the vertices of $\triangle A'B'C'$ under a dilation with center (0, 0) and the given scale factor.

13. $\begin{array}{r} \\ x\text{-coordinate} \\ y\text{-coordinate} \end{array} \begin{array}{c} \begin{array}{ccc} A & B & C \end{array} \\ \begin{bmatrix} 1 & 3 & 5 \\ 0 & 2 & 1 \end{bmatrix} \end{array}$
scale factor 3

14. $\begin{array}{r} \\ x\text{-coordinate} \\ y\text{-coordinate} \end{array} \begin{array}{c} \begin{array}{ccc} A & B & C \end{array} \\ \begin{bmatrix} -2 & 1 & 1 \\ -2 & 1 & -1 \end{bmatrix} \end{array}$
scale factor $\frac{1}{4}$

15. $\begin{array}{r} \\ x\text{-coordinate} \\ y\text{-coordinate} \end{array} \begin{array}{c} \begin{array}{ccc} A & B & C \end{array} \\ \begin{bmatrix} -2 & -4 & -3 \\ 0 & -3 & 0 \end{bmatrix} \end{array}$
scale factor 2

16. **Entertainment** In the film *Honey, I Blew Up the Kid*, a botched scientific experiment causes a two-year-old boy to grow to a height of 112 ft. If the average height of a two-year-old boy is 3 ft, what is the scale factor of this enlargement?

17. **Movies** The projection of a film onto a movie screen is an example of a dilation. Most movies are shot on film that is 35 mm wide. If the width of a movie screen is 12 m, what is the scale factor of this enlargement?

18. A regular triangle with 4-in. sides undergoes a dilation with scale factor 2.5.
a. What are the side lengths of the image?
b. What are the angle measures of the image?

A giant two-year-old walks through the streets of Las Vegas in a scene from *Honey, I Blew Up the Kid.*

19. *Constructions* Copy $\triangle GHI$ and point X onto your paper. Use a compass and straightedge to construct the image of $\triangle GHI$ under a dilation with center X and scale factor 2.

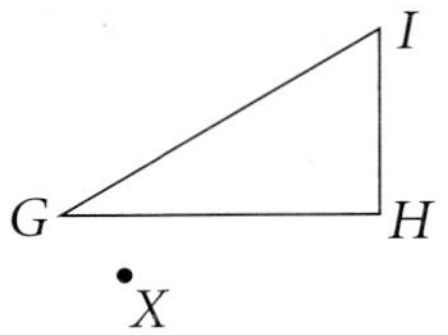

Coordinate Geometry **Graph $MNPQ$ and its image $M'N'P'Q'$ under a dilation with center (0, 0) and the given scale factor.**

20. $M(-1, -1)$, $N(1, -2)$, $P(1, 2)$, $Q(-1, 3)$; scale factor 2

21. $M(0, 0)$, $N(4, 0)$, $P(6, -2)$, $Q(-2, -2)$; scale factor $\frac{1}{2}$

22. *Art* Perspective drawing uses converging lines to give the illusion that an object is three-dimensional. The point at which the lines converge is called the vanishing point. Explain how the type of perspective drawing shown is related to dilations.

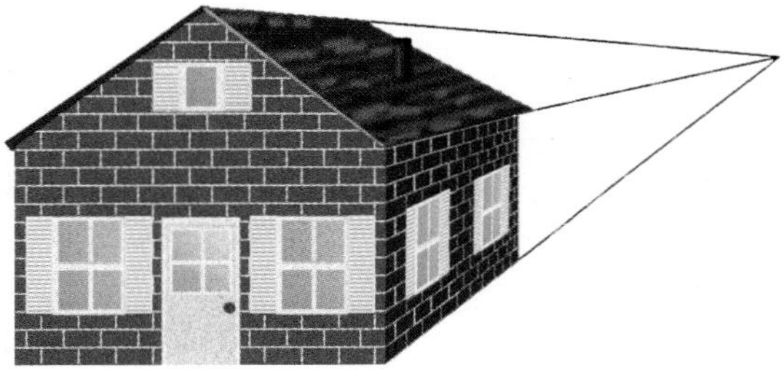

23. Explore what happens if you use a negative scale factor.

a. Multiply the vertex matrix for $ABCD$ by -3. The result is a vertex matrix for the image of $ABCD$ under a dilation centered at the origin with scale factor -3.

$$\begin{matrix} A & B & C & D \end{matrix}$$
$$\begin{bmatrix} 2 & -2 & -2 & 2 \\ 2 & 2 & -2 & -2 \end{bmatrix}$$

b. Graph $ABCD$ and $A'B'C'D'$ on the same set of axes.

c. *Critical Thinking* Compare each point to its image. What conclusion can you draw about the effect of a negative scale factor?

24. *Critical Thinking* Given $\overline{AB}$ and its dilation image $\overline{A'B'}$, explain how to find the center of dilation. Assume $\overline{AB}$ and $\overline{A'B'}$ are not collinear.

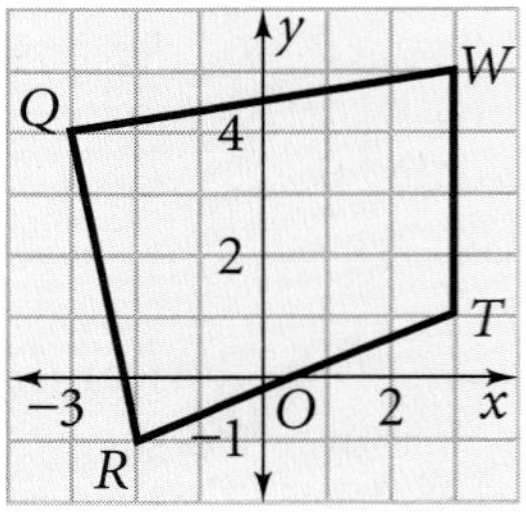

Use scalar multiplication to find the vertices of the image of $QRTW$ under a dilation with center (0, 0) and the given scale factor.

25. scale factor 3

26. scale factor $\frac{1}{2}$

27. scale factor 2

28. scale factor 0.9

Graphic Design **The designs at the right were created by graphic artist Scott Kim. For each design, (a) describe the location of the center of dilation and (b) find the scale factor for the repeated reductions.**

29.

30.

31. **Technology** Use geometry software or drawing software to create a design that involves repeated dilations. Print your design and color it. Feel free to use other transformations along with dilations.

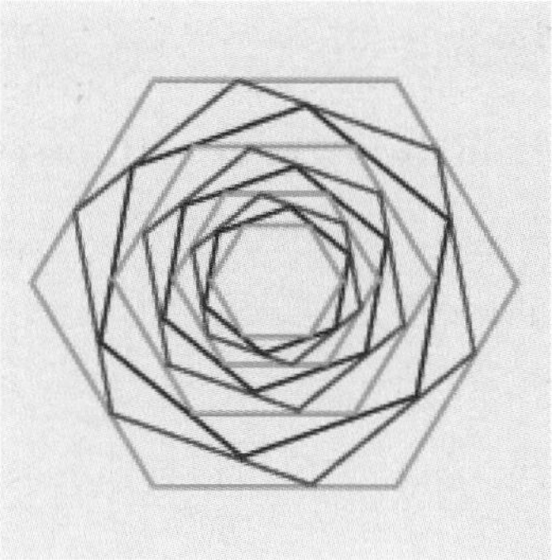

32. **Standardized Test Prep** $\triangle P'L'H'$ is the image of $\triangle PLH$ under a dilation with center X and scale factor 3. If P and L lie on $\overline{P'L'}$, what must be true of X?

A. X is in the exterior of $\triangle PLH$. **B.** X is in the interior of $\triangle PLH$.
C. X is on $\overline{L'H'}$. **D.** X is on $\overline{PL}$. **E.** $X = H$

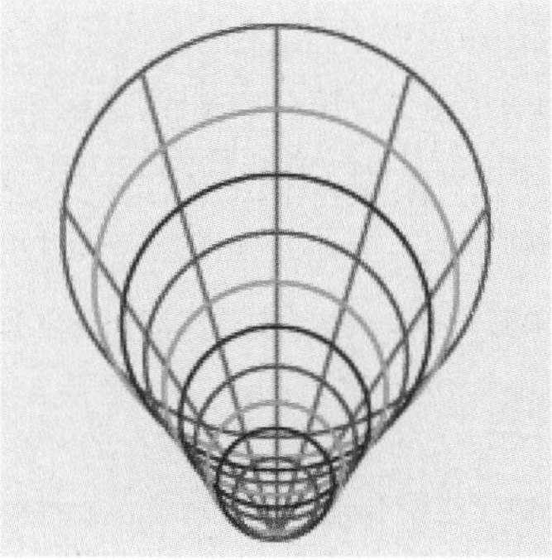

A dilation maps $\triangle HIJ$ to $\triangle H'I'J'$. Find the missing values.

33. HI = 8 in.
IJ = 5 in.
HJ = 6 in.
$H'I'$ = 16 in.
$I'J'$ = ■ in.
$H'J'$ = ■ in.

34. HI = 7 cm
IJ = 7 cm
HJ = ■ cm
$H'I'$ = 5.25 cm
$I'J'$ = ■ cm
$H'J'$ = 12 cm

35. HI = ■ ft
IJ = 30 ft
HJ = 24 ft
$H'I'$ = 8 ft
$I'J'$ = ■ ft
$H'J'$ = 6 ft

Write *true* or *false*. Explain your answers.

36. A dilation with a scale factor greater than 1 is a reduction.

37. Under a dilation, corresponding angles of the image and preimage are congruent.

38. A dilation is an isometry.

39. A dilation changes orientation.

40. A dilation image cannot have any points in common with its preimage.

Chapter Project

Find Out by Creating

In previous Find Out questions, you investigated and classified frieze patterns from a variety of cultures. Now you can make your own. Use graph paper, dot paper, geometry or drawing software, or cutouts (such as your pentominoes from the Chapter 2 project). Make at least one frieze pattern for each of the seven types summarized below.

The Seven Types of Frieze Patterns

Type	Symmetries	Example
11	T	P P P P P P P
12	T, H	Z Z Z Z Z Z Z
$m1$	T, RV	Y Y Y Y Y Y Y
1g	T, G	D W D M D W D
1m	T, RH, G	D D D D D D D
mg	T, H, RV, G	M W M W M W M
mm	T, H, RV, RH, G	I I I I I I I

Key to Symmetries:
T = **T**ranslation
H = **H**alf-turn
RV = **R**eflection in **V**ertical Line
RH = **R**eflection in **H**orizontal Line
G = **G**lide reflection

Exercises MIXED REVIEW

Find the value of each variable.

41.

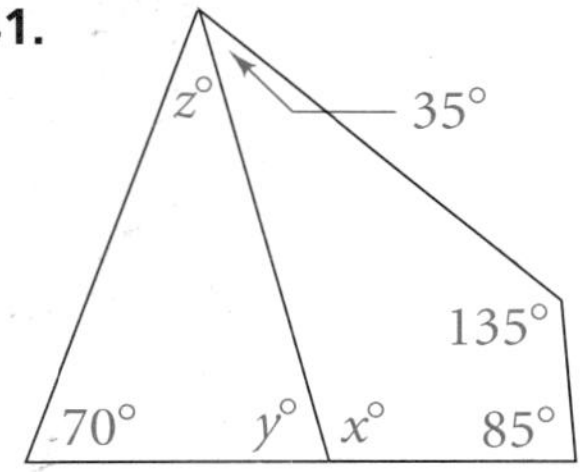

42.

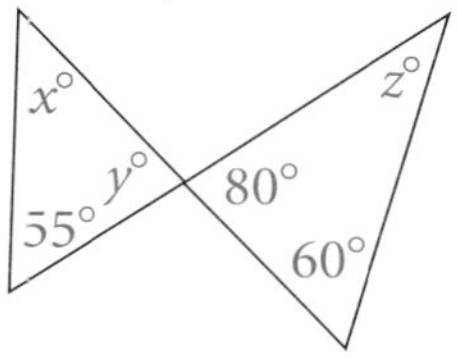

43.

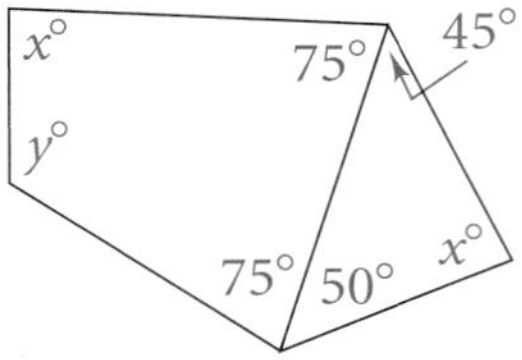

Graph each set of points and state which type of quadrilateral it determines.

44. $(-1, -2), (1, 4), (-3, 4), (3, -2)$

45. $(2, -1), (6, 2), (10, -1), (8, 2)$

46. $(-7, 1), (-5, 3), (-2, -4), (0, -2)$

47. *Geometry in 3 Dimensions* Planes T and G are parallel, and plane M is perpendicular to plane G. Sketch the three planes.

48. Find the coordinates of the midpoint of the segment with endpoints $A(4, 8)$ and $B(-3, -6)$.

SELF ASSESSMENT

PORTFOLIO

For your portfolio, select one or two items from your work for this chapter. Here are some possibilities:

- tessellations
- kaleidoscope patterns
- computer-generated designs involving transformations

Explain why you have included each selection.

Geometry at Work

Graphic Artist

Not too long ago, the graphic artist's main tools were the pen and paintbrush. Today, graphic artists are just as likely to have a computer mouse in hand as either of these. Computers have helped graphic artists produce effects that could only be imagined previously. Some of the "special effects" on the cover of this book, for instance, were produced using design software. Computers have also helped lower the cost of producing graphic art, because they allow the artist to work more quickly and to store and transport images easily.

You can find each of the transformations that you've studied in this chapter in software written for graphic artists. At the click of a mouse, the graphic artist rotates, reflects, translates, dilates, and creates!

Mini Project: Use a computer drawing program to create a new logo for your school's stationery. Try to use transformations in your logo.

Finishing the Chapter Project

CHAPTER PROJECT

FRIEZE FRAMES

Find Out questions and activities on pages 137, 143, 150, 158, and 171 will help you complete your project. Prepare a frieze pattern display. Include a brief explanation of frieze patterns as well as your original designs for each of the seven types of frieze patterns. Find more examples of frieze patterns from various cultures, as well as examples from buildings, clothing, and other places in your home, school, and community. Classify each example into one of the seven categories, and include them in your display.

Reflect and Revise

Ask a classmate to review your display with you. Together, check that your diagrams and explanations are clear and your information accurate. Have you used geometric terms correctly? Is the display attractive, organized, and complete? Have you included material that no one else has included? Revise your work as needed. Consider doing more research.

Follow Up

Use logical reasoning and what you've learned about transformations to explain why there are no more than seven different frieze patterns. List all other possible combinations of symmetries and show how each can be ruled out.

For More Information

Appleton, Le Roy H. *American Indian Design and Decoration*. New York: Dover, 1971.

Hargittai, István and Magdolna. *Symmetry: A Unifying Concept*. Berkeley, California: Ten Speed Press, 1994.

Schuman, Jo Miles. *Art from Many Hands: Multicultural Art Projects for Home and School*. Englewood Cliffs, New Jersey: Prentice Hall, 1981.

Stevens, Peter S. *Handbook of Regular Patterns: An Introduction to Symmetry in Two Dimensions*. Cambridge, Massachusetts: MIT, 1981.

CHAPTER 3

Wrap Up

Key Terms

composition (p. 135)
dilation (p. 166)
enlargement (p. 167)
entry (p. 131)
frieze pattern (p. 137)
glide reflection (p. 146)
glide reflectional symmetry (p. 161)
half-turn (p. 154)
image (p. 124)
isometry (p. 124)
line symmetry (p. 152)
map (p. 124)
matrix (p. 131)
orientation (p. 126)
point symmetry (p. 154)
preimage (p. 124)
prime notation (p. 124)
pure tessellation (p. 160)
reduction (p. 167)
reflection (p. 126)
reflectional symmetry (p. 152)
rotation (p. 139)
rotational symmetry (p. 154)
scalar multiplication (p. 168)
scale factor (p. 166)
similarity transformation (p. 167)
strip pattern (p. 137)
symmetry (p. 152)
tessellation (p. 159)
tiling (p. 159)
transformation (p. 124)
translation (p. 132)
translational symmetry (p. 161)

How am I doing?

SELF ASSESSMENT

- Describe the different transformations and their properties.
- State the different types of symmetry and draw examples of each.

Reflections 3-1

A **transformation** is a change made to the position, shape, or size of a figure. An **isometry** is a transformation in which the original figure and its image are congruent. The diagram shows a **reflection** of B to B' in line r. A reflection is an isometry that changes a figure's orientation.

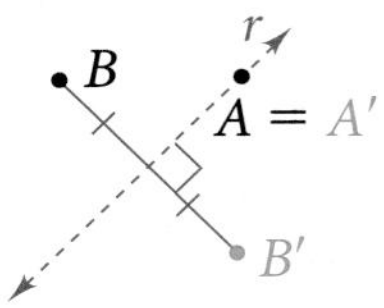

Coordinate Geometry **Given points $A(6, 4)$, $B(-2, 1)$, and $C(5, 0)$, draw $\triangle ABC$ and its reflection image in the given line.**

1. the x-axis
2. $x = 4$
3. $x = -3$
4. $y = x$

Trace each figure, then find its reflection image in line j.

5.

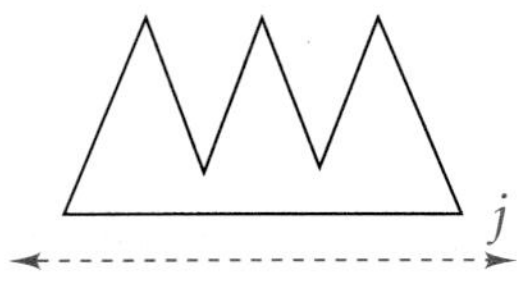

6.

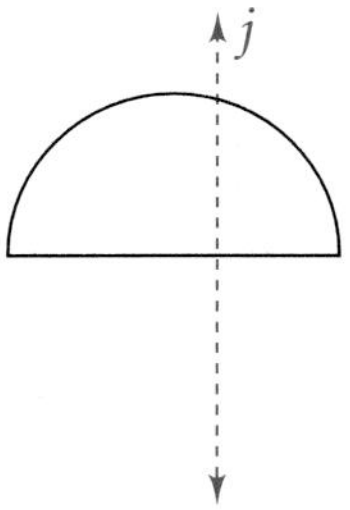

7.

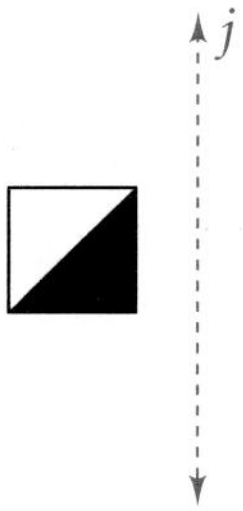

8.

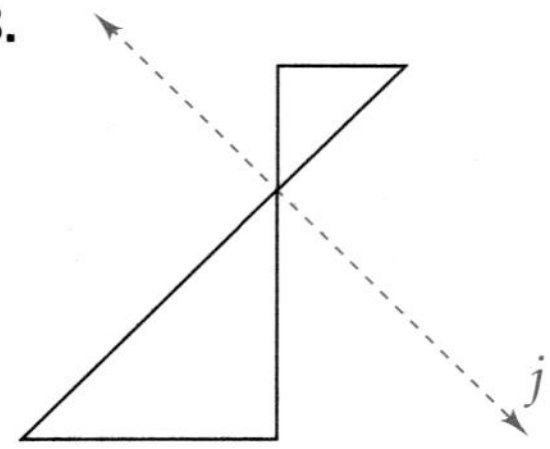

9. Standardized Test Prep What is the image of $(5, 4)$ under a reflection in $y = 2$?

A. $(-5, 4)$ **B.** $(5, -4)$ **C.** $(-1, 4)$ **D.** $(-5, -4)$ **E.** $(5, 0)$

Translations 3-2

A **translation** is a transformation that moves points the same distance and in the same direction. A translation is an isometry that does not change orientation. You can use vectors and matrices to describe a translation.

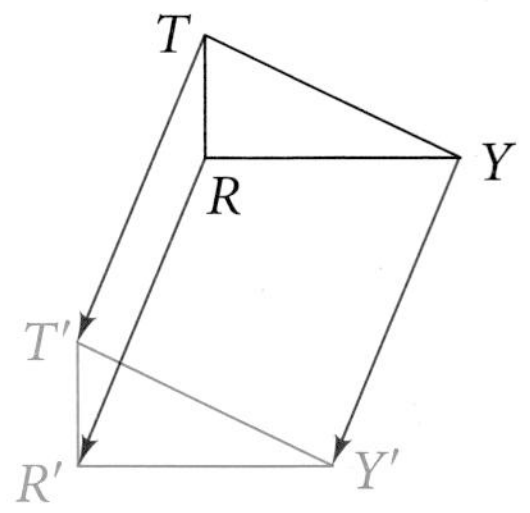

Under a **composition** of two transformations, the second transformation is performed on the image of the first.

10. *Open-ended* Draw a figure in the fourth quadrant. Draw a translation vector under which the figure would move to the second quadrant.

Use matrix addition to find the image of each figure under the given translation.

11. $\triangle ABC$ with vertices $A(5, 9)$, $B(4, 3)$, $C(1, 2)$
Translation: $\langle 2, 3 \rangle$

12. $\triangle RST$ with vertices $R(0, -4)$, $S(-2, -1)$, $T(-6, 1)$
Translation: $\langle -4, 7 \rangle$

Find a single translation that has the same effect as each composition of translations.

13. $\langle 4, 8 \rangle$ followed by $\langle -2, 0 \rangle$

14. $\langle -5, -7 \rangle$ followed by $\langle 3, 6 \rangle$

15. $\langle 10, -9 \rangle$ followed by $\langle 1, 5 \rangle$

Rotations 3-3

The diagram shows a **rotation** of point V about point R through $x°$. A rotation is an isometry that does not change orientation.

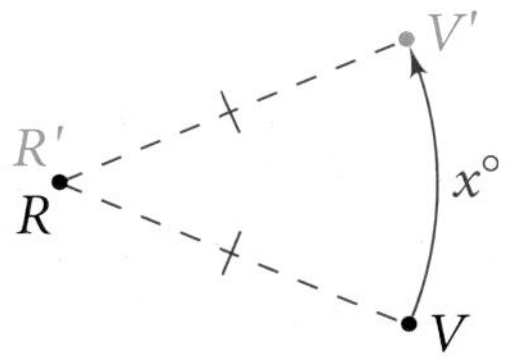

Copy each figure and point *P*. Rotate the figure the given number of degrees about *P*. Label the vertices of the image.

16. 180°

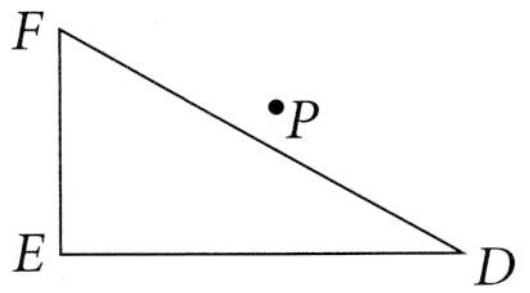

17. 135°

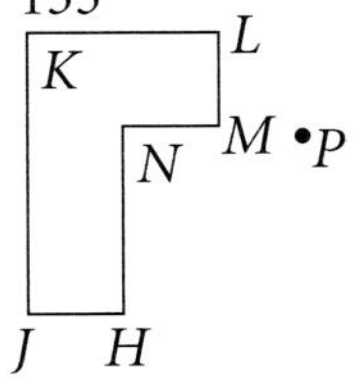

18. 60°

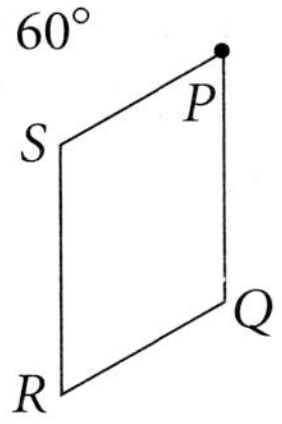

19. 90°

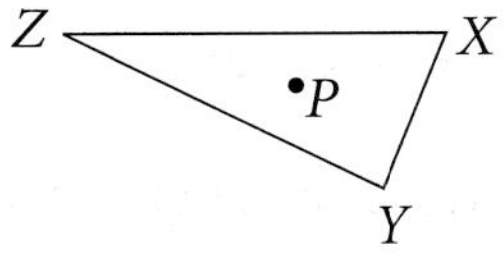

Find the image of each point under a 90° rotation about the origin.

20. $(5, 2)$

21. $(0, 3)$

22. $(-4, 1)$

23. $(7, 0)$

24. $(-2, -8)$

25. $(0, 0)$

Compositions of Reflections 3-4

A composition of reflections in two parallel lines is a translation. A composition of reflections in two intersecting lines is a rotation. The diagram shows a **glide reflection.** The only four isometries are reflection, translation, rotation, and glide reflection.

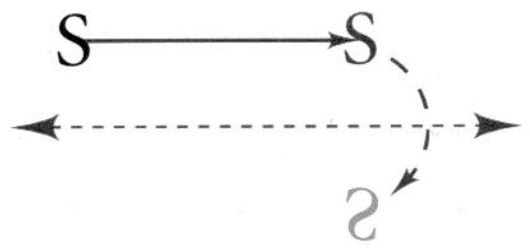

Match each image of the figure at the left with one of the following isometries. I. reflection II. rotation III. translation IV. glide reflection

26. 

a. 
b. 
c. 

d. 

27. 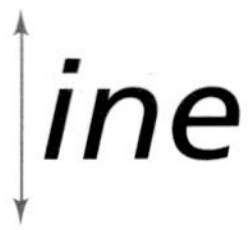

a. 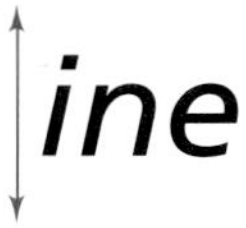

b. 
c. 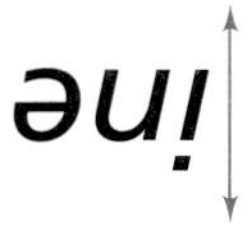
d. 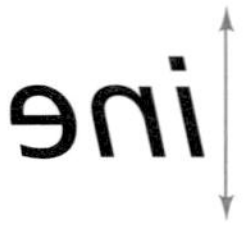

28. $\triangle TAM$ has vertices $T(0, 5)$, $A(4, 1)$, and $M(3, 6)$. Find the image of $\triangle TAM$ under a glide reflection in $\langle -4, 0 \rangle$ and $y = -2$.

Symmetry 3-5

A figure has **symmetry** if there is an isometry that maps the figure onto itself. A plane figure has **reflectional symmetry,** or **line symmetry,** if there is a line in which the figure is reflected onto itself. A figure has **rotational symmetry** if there is a rotation of 180° or less that maps the figure onto itself. If a plane figure can be mapped onto itself by a rotation of 180° (a **half-turn**), it has **point symmetry.**

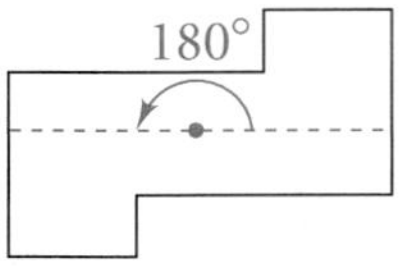

Point Symmetry

What type(s) of symmetry does each figure have? If it has rotational symmetry, state the angle of rotation.

29. 

30. 

31. 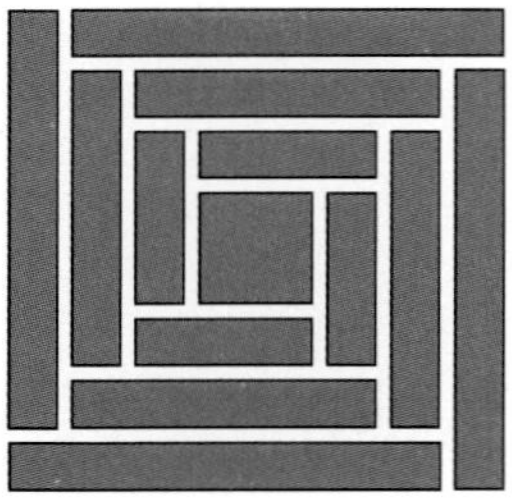

Tessellations 3-6

A **tessellation**, or **tiling**, is a repeating pattern of figures that completely covers a plane without gaps or overlaps. Every triangle and quadrilateral tessellates. Tessellations can have many kinds of symmetries, including **translational symmetry** and **glide reflectional symmetry.**

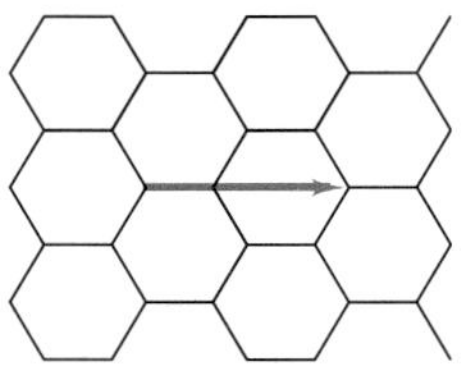
Translational Symmetry
A translation maps the tessellation onto itself.

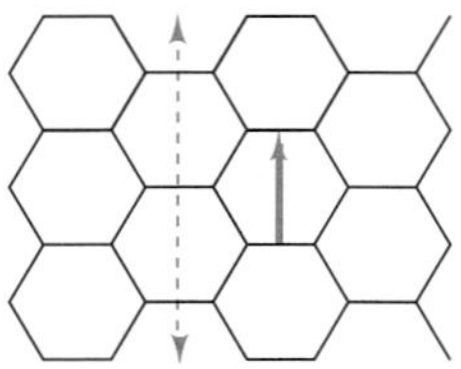
Glide Reflectional Symmetry
A glide reflection maps the tessellation onto itself.

In Exercises 32–34, (a) identify the repeating figure(s) that make up each tessellation, and (b) describe the symmetries of each tessellation.

32.

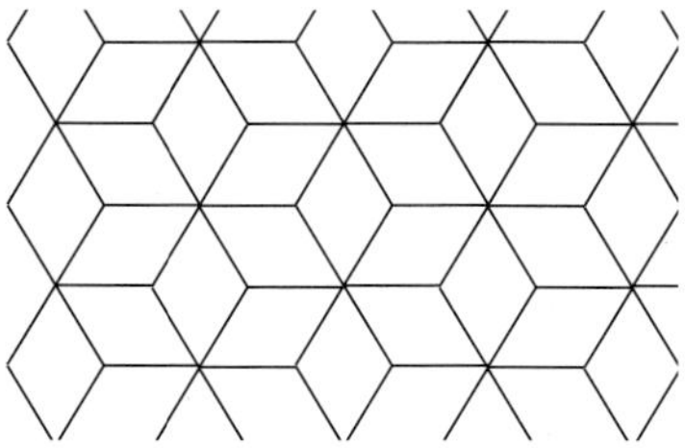

33.

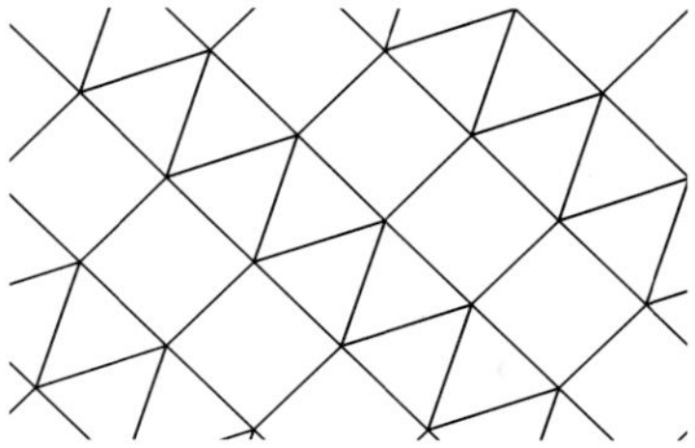

34.

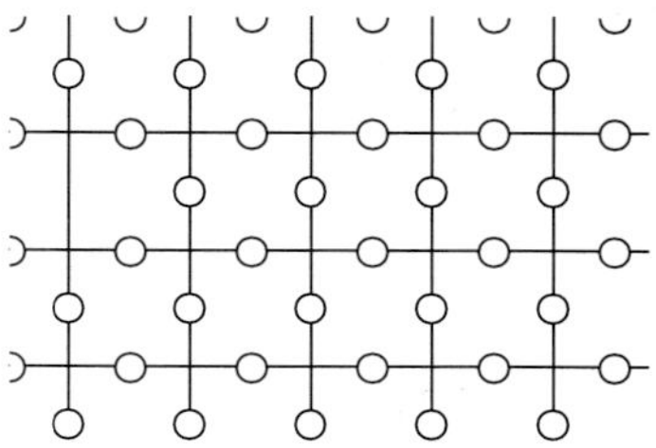

Dilations 3-7

The diagram shows a **dilation** with center C and **scale factor** n. A dilation is a **similarity transformation** because it maps figures to similar figures. When the scale factor is greater than 1, the dilation is an **enlargement.** When the scale factor is less than 1, the dilation is a **reduction.** In the coordinate plane, you can use **scalar multiplication** to find the image of a figure under a dilation centered at the origin.

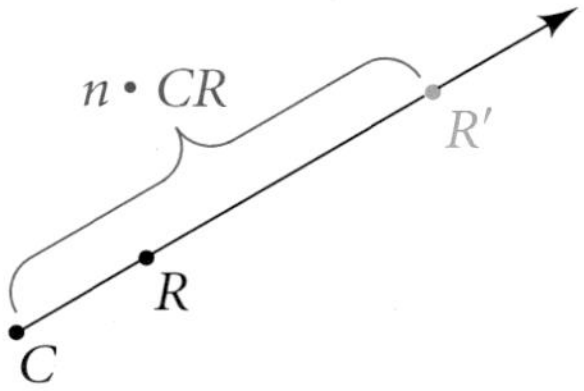

Use matrices to find the image of each set of points under a dilation centered at the origin with the given scale factor.

35. $M(-3, 4)$, $A(-6, -1)$, $T(0, 0)$, $H(3, 2)$; scale factor 5

36. $A(7, -1)$, $N(-4, -3)$, $D(0, 2)$; scale factor 2

37. $W(4, 5)$, $I(2, 6)$, $T(3, 8)$, $H(0, 7)$; scale factor 3

38. $F(-4, 0)$, $U(5, 0)$, $N(-2, -5)$; scale factor $\frac{1}{2}$

39. *Writing* Explain how each of the five transformations you studied in this chapter affects the orientation of a figure and its image.

Getting Ready for CHAPTER 4

Find the coordinates of the midpoint of the segment with the given endpoints.

40. $C(3, 5)$ and $D(1, 11)$ **41.** $E(6, 7)$ and $F(-4, 7)$ **42.** $T(5, 8)$ and $W(5, 0)$

Classify each triangle by its sides and angles.

43.

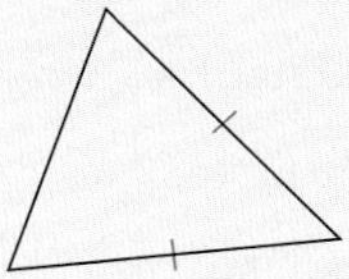

44.

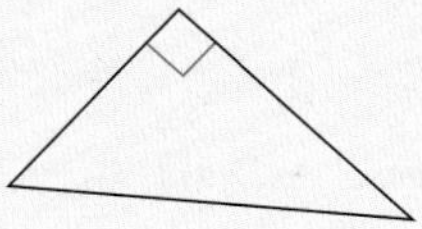

45.

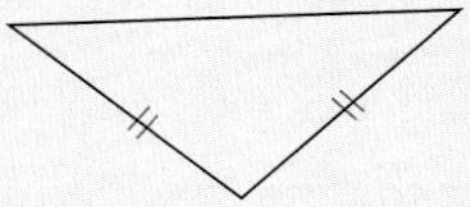

46.

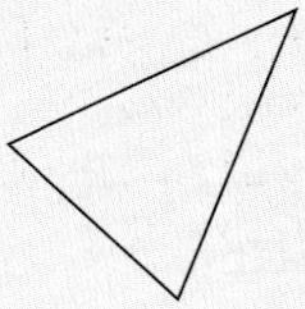

CHAPTER 3 Assessment

Coordinate Geometry **Find the coordinates of the vertices of the image of *ABCD* under each transformation.**

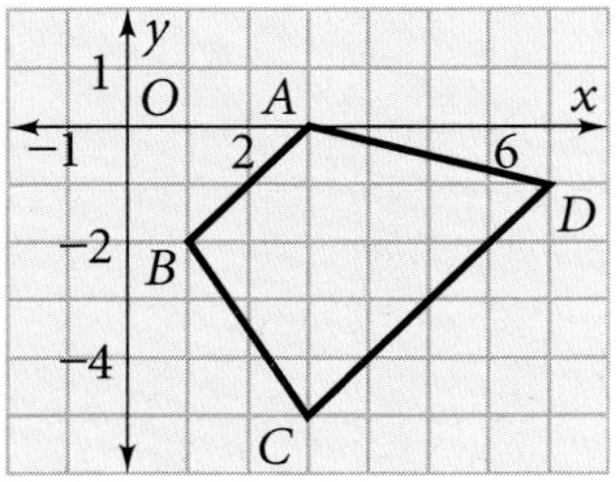

1. reflection in $x = -4$
2. translation $\langle -6, 8 \rangle$
3. rotation of 90° about the point (0, 0)
4. dilation with scale factor $\frac{2}{3}$ centered at the origin
5. glide reflection in $\langle 0, 3 \rangle$ and $x = 0$
6. reflection in $y = x$
7. rotation of 270° about the point (0, 0)
8. dilation with scale factor 5 centered at (0, 0)
9. glide reflection in $\langle -2, 0 \rangle$ and $y = 5$

What type of transformation has the same effect as each composition of transformations?

10. translation $\langle 4, 0 \rangle$ followed by reflection in $y = -4$
11. translation $\langle 4, 8 \rangle$ followed by $\langle -2, 9 \rangle$
12. reflection in $y = 7$, then in $y = 3$
13. reflection in $y = x$, then in $y = 2x + 5$

14. **Writing** Line m intersects $\overline{UH}$ at N, and $UN = NH$. Must H be the reflection image of U in line m? Explain why or why not.

Open-ended **Sketch a figure that has each type of symmetry.**

15. reflectional
16. rotational
17. point

18. Describe the symmetries of this tessellation. Copy a portion of the tessellation and draw any centers of rotational symmetry or lines of symmetry.

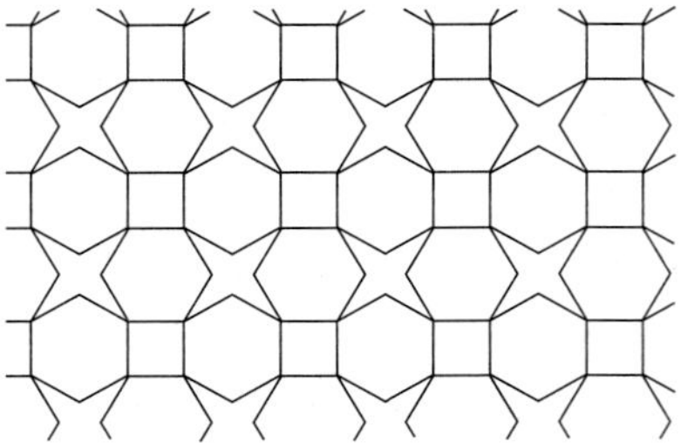

What type(s) of symmetry does each figure have?

19.

20.

21. **Standardized Test Prep** Which of the following letters does *not* tessellate?

A. 
B. 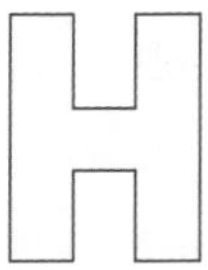
C. 
D. 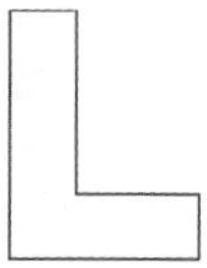
E. 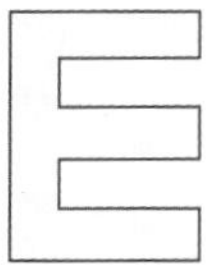

Find the image of $\triangle ABC$ under a dilation with center (0, 0) and the given scale factor.

22. $A(2, 4)$, $B(3, 7)$, $C(5, 1)$; scale factor 4
23. $A(0, 0)$, $B(-3, 2)$, $C(1, 7)$; scale factor $\frac{1}{2}$
24. $A(-2, 2)$, $B(2, -2)$, $C(3, 4)$; scale factor 3

CHAPTER 3

Preparing for Standardized Tests

For Exercises 1–10, choose the correct letter.

1. What is the image of (−3, 8) reflected in $y = 4$?
A. (3, 8) **B.** (3, −8) **C.** (−3, −8)
D. (11, 8) **E.** (−3, 0)

2. Which angles could an obtuse triangle have?
I. a right angle **II.** two acute angles
III. an obtuse angle **IV.** two vertical angles

A. I and II **B.** II and III **C.** III and IV
D. I and IV **E.** none of the above

3. What is the image of (4, 5) rotated 270° about the origin?
A. (−5, 4) **B.** (−4, −5) **C.** (−5, −4)
D. (5, −4) **E.** none of the above

4. Which lines are perpendicular to $y = 2x - 4$?
I. $y = \frac{1}{2}x + 6$ **II.** $y = 0.5x + 11$
III. $y = 2x + 7$ **IV.** $y = -0.5x - 1$

A. I and II **B.** III only **C.** IV only
D. II and IV **E.** I, II, and IV

5. The point (2, 7) is reflected in the *y*-axis, then translated by ⟨0, −7⟩. Where is its image?
A. *x*-axis **B.** third quadrant
C. fourth quadrant **D.** origin
E. first quadrant

6. Which figure does *not* tessellate?

A.
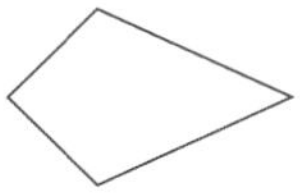

B.
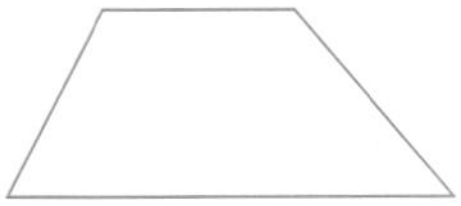

C.

D.
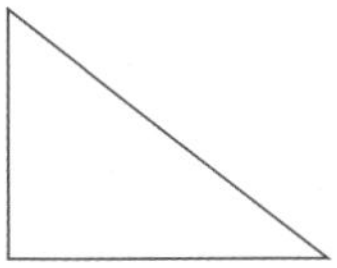

E. none of the above

7. Calculator What is the distance between the points (3, 5) and (−4, −9)?
A. 15.7 **B.** 4.1 **C.** 6.1
D. 15 **E.** 21

8. What is the midpoint of the segment with endpoints (0, −4) and (−4, 7)?
A. $(-4, 3)$ **B.** $(-2, 3)$ **C.** $(-2, \frac{3}{2})$
D. $(-4, \frac{3}{2})$ **E.** $(2, -3)$

Compare the boxed quantity in Column A with the boxed quantity in Column B. Choose the best answer.

A. The quantity in Column A is greater.
B. The quantity in Column B is greater.
C. The two quantities are equal.
D. The relationship cannot be determined on the basis of the information supplied.

	Column A	Column B

$\triangle ABC \rightarrow \triangle A'B'C'$

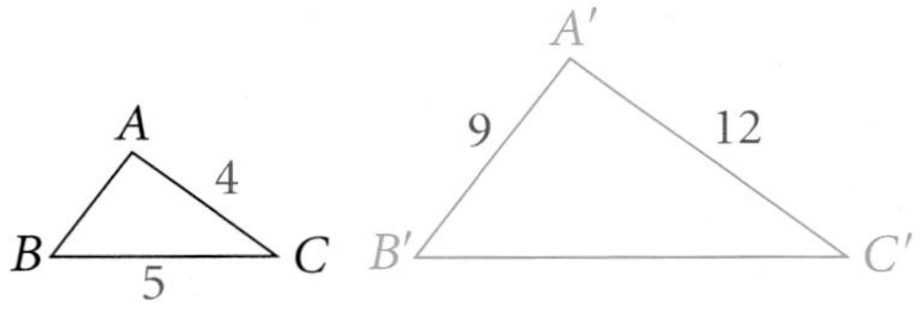

9.	the scale factor	*AB*

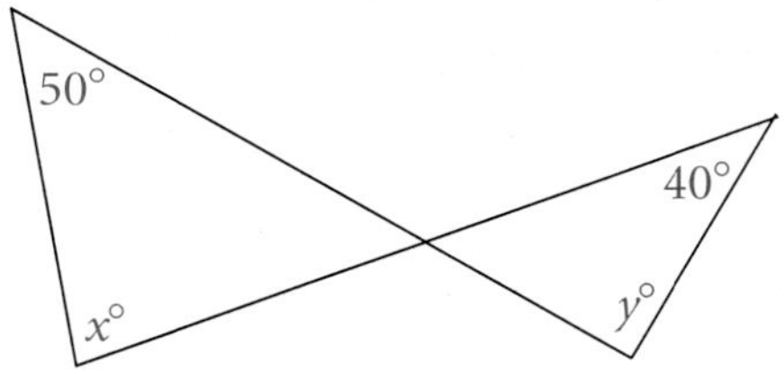

10.	*x*	*y*

Find each answer.

11. Which transformations are isometries?

12. You are building a scale model of a building. The front of the actual building will be 60 ft wide and 100 ft tall. The front of your model is 3 ft by 5 ft. What is the scale factor of the reduction?

CHAPTER

4 Triangle Relationships

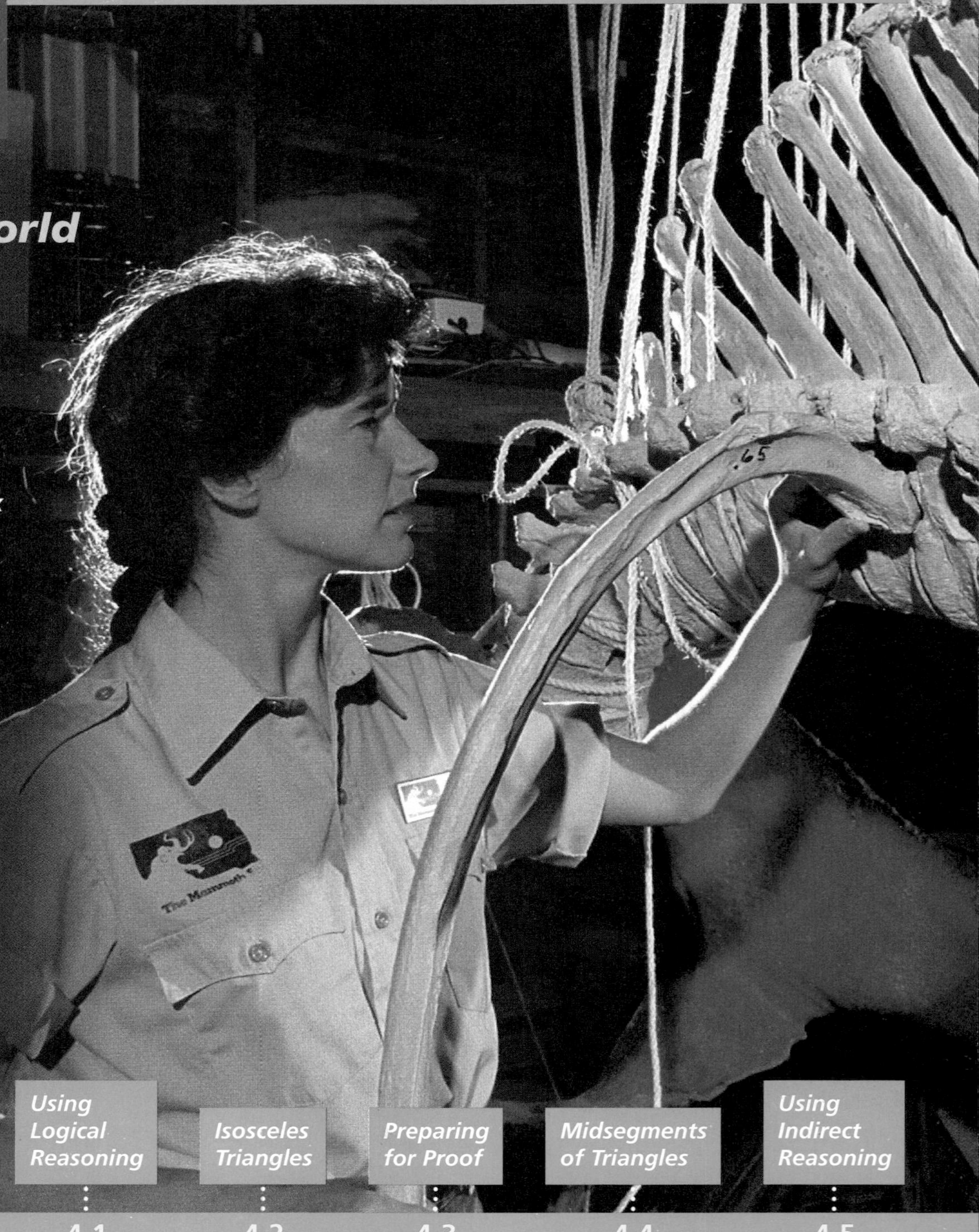

Relating to the Real World

You are sitting in a meeting around a large table. An important decision needs to be made. Your boss turns to you and asks, "So, what do you think we should do?" Your response is clear and well-reasoned, and your co-workers nod in approval. The reasoning skills that you learn now will help you to succeed in life. In this chapter you will study the tools needed to become a better problem solver and logical thinker.

Lessons					
	Using Logical Reasoning	*Isosceles Triangles*	*Preparing for Proof*	*Midsegments of Triangles*	*Using Indirect Reasoning*
	4-1	4-2	4-3	4-4	4-5

CHAPTER PROJECT

PUZZLING PIECES

A paleontologist makes sense of the past by piecing together fossils. By using logical reasoning, she makes a few bones tell the story of an entire species. That feeling of discovery—that sense of "aha!"—has driven people throughout history to use logic to solve puzzles.

In your project for this chapter, you will explore ways of solving logic puzzles. You will also create your own puzzles. You will see how logic, though used for amusement, is the backbone of mathematics—a powerful tool for determining truth.

To help you complete the project:

- **p. 187** *Find Out by Listing*
- **p. 198** *Find Out by Organizing*
- **p. 205** *Find Out by Analyzing*
- **p. 225** *Find Out by Writing*
- **p. 233** *Finishing the Project*

Triangle Inequalities — 4-6

Bisectors and Locus — 4-7

Concurrent Lines — 4-8

What You'll Learn

- Writing and interpreting different types of conditional statements

...And Why

To help you analyze situations and become a better problem solver

4-1 Using Logical Reasoning

THINK AND DISCUSS

Conditionals and Converses

- If Raúl's major is bagpipe, then he attends Carnegie-Mellon University.
- If a second goes by, then Earth has moved another $18\frac{1}{2}$ mi along its orbit.
- If a movie is scary, then the concession stands sell more popcorn.
- If you are not completely satisfied, then your money will be refunded.

1. You make and hear *if-then* statements like these many times each day. What are some *if-then* statements you have heard?

Another name for an *if-then statement* is a **conditional.** Every conditional has two parts. The part following *if* is the **hypothesis,** and the part following *then* is the **conclusion.**

Example 1

Identify the hypothesis and conclusion in this statement:

If it is February, then there are only 28 days in the month.

Hypothesis: It is February.
Conclusion: There are only 28 days in the month.

If you want double the love, then buy a pair.

2. Try This Identify the hypothesis and conclusion in the photo caption at the left.

When you determine whether a conditional is true or false, you determine its **truth value.** To show that a conditional is false, you need to find only one *counterexample* for which the hypothesis is true and the conclusion is false. The conditional in Example 1 is false because during a leap year February has 29 days.

3. Find a counterexample for this conditional: If the name of a state contains the word *New*, then the state borders an ocean.

Many sentences can be written as conditionals. For example, the sentence

Quadrilaterals have four sides.

can be rewritten in if-then form as

If a polygon is a quadrilateral, then it has four sides.

The **converse** of a conditional interchanges the hypothesis and conclusion.

Example 2

Write the converse of this statement: If a polygon is a quadrilateral, then it has four sides.

Conditional: If a polygon is a quadrilateral, then it has four sides.

Converse: If a polygon has four sides, then it is a quadrilateral. ■

Literature Notice that both statements in Example 2 have the same truth value. This is *not* true of all conditionals and their converses, as Alice discovers in this passage from Lewis Carroll's *Alice's Adventures in Wonderland.*

The Hatter opened his eyes very wide on hearing this; but all he *said* was "Why is a raven like a writing-desk?"

"Come, we shall have some fun now!" thought Alice. "I'm glad they've begun asking riddles—I believe I can guess that," she added aloud.

"Do you mean that you think you can find out the answer to it?" said the March Hare.

"Exactly so," said Alice.

"Then you should say what you mean," the March Hare went on.

"I do," Alice hastily replied; "at least—at least I mean what I say—that's the same thing, you know."

"Not the same thing a bit!" said the Hatter. "Why, you might just as well say that 'I see what I eat' is the same thing as 'I eat what I see'!"

The Hatter's statement "I see what I eat" can be rewritten in if-then form as "If I eat it, then I see it." The converse, "I eat what I see," can be rewritten as "If I see it, then I eat it."

4. Are the truth values for the Hatter's statement and its converse the same? Explain.

Biconditionals, Inverses, and Contrapositives

When a conditional and its converse are true, you can combine them as a **biconditional.** The conditionals from Example 2 can be combined as

If a polygon is a quadrilateral, then it has four sides, *and* if a polygon has four sides, then it is a quadrilateral.

This long sentence can be shortened by using the phrase *if and only if:*

A polygon is a quadrilateral *if and only if* it has four sides.

You learned in Chapter 1 that any good definition is "reversible." This means that any good definition can be written as a biconditional.

Example 3

Write the two statements that make up this definition: A right angle has measure 90. Then write the definition as a biconditional.

Conditional: If an angle is a right angle, then its measure is 90.

Converse: If the measure of an angle is 90, then it is a right angle.

Biconditional: An angle is a right angle if and only if its measure is 90.

The **negation** of a statement has the opposite meaning. For example, the negation of "An angle is acute" is "An angle is *not* acute."

5. Write the negation of each statement.
a. Two angles are vertical. **b.** Two lines are not parallel.

The **inverse** of a conditional negates both the hypothesis and the conclusion. The **contrapositive** of a conditional interchanges and negates both the hypothesis and the conclusion.

Example 4

Write the inverse and the contrapositive of this statement: If a figure is a square, then it is a rectangle.

Conditional: If a figure is a square, then it is a rectangle.

Negate both.

Inverse: If a figure is *not* a square, then it is *not* a rectangle.

Conditional: If a figure is a square, then it is a rectangle.

Negate both.

Contrapositive: If a figure is *not* a rectangle, then it is *not* a square.

6. Find the truth values of the three statements in Example 4.

7. Try This Write the inverse and contrapositive of this statement: If a quadrilateral is a parallelogram, then it has two pairs of parallel sides.

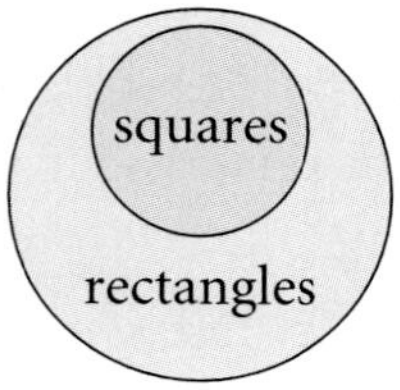

This Venn diagram shows the conditional "If a figure is a square, then it is a rectangle." It also shows its contrapositive "If a figure is *not* a rectangle, then it is *not* a square" because any point *not* in the larger circle is *not* in the smaller circle. Since the same diagram shows both statements, a conditional and its contrapositive always have the same truth value.

Summary of Conditionals

Statement	Form	Example
conditional	If ■, then □.	If an angle is a straight angle, then its measure is 180.
converse	If □, then ■.	If the measure of an angle is 180, then it is a straight angle.
inverse	If *not* ■, then *not* □.	If an angle is *not* a straight angle, then its measure is *not* 180.
contrapositive	If *not* □, then *not* ■.	If the measure of an angle is *not* 180, then it is *not* a straight angle.
biconditional	■ if and only if □.	An angle is a straight angle if and only if its measure is 180.

Exercises ON YOUR OWN

1. Identify the hypothesis and conclusion in the cartoon.

FRANK AND ERNEST By BOB THAVES

For Exercises 2–5: (a) Rewrite each statement in if-then form. (b) Underline the hypothesis once and the conclusion twice.

2. Glass objects are fragile.

3. $3x - 7 = 14$ implies that $3x = 21$.

4. Numbers that have 2 as a factor are even.

5. An isosceles triangle has two congruent sides.

Find a counterexample for each statement.

6. If it is not a weekday, then it is Saturday.

7. Odd integers less than 10 are prime.

8. If you live in a country that borders the United States, then you live in Canada.

9. If you play a sport with a ball and bat, then you play baseball.

10. a. *Open-ended* Write a conditional with the same truth value as its converse.

b. Write a conditional whose converse has the opposite truth value.

For each statement, write (a) the converse, (b) the inverse, and (c) the contrapositive.

11. If you eat all of your vegetables, then you will grow.

12. Transformations If a figure has point symmetry, then it has rotational symmetry.

13. If a triangle is a right triangle, then it has a 90° angle.

14. If a quadrilateral has exactly two congruent sides, then it is not a rhombus.

15. If two segments are congruent, then they have the same length.

16. If you do not work, you will not get paid.

17. If a polygon is a pentagon, then the sum of the measures of its angles is 540.

18. If a conditional statement is false, then its contrapositive is false.

For Exercises 19–26: (a) Write the converse of each statement. (b) Determine the truth value of the statement and its converse. (c) If both statements are true, write a biconditional.

19. If you travel from the United States to Kenya, then you have a passport.

20. Coordinate Geometry If a point is in the first quadrant, then its coordinates are positive.

21. Chemistry If a substance is water, then its chemical formula is H_2O.

22. Transformations If a figure has two lines of symmetry, then it has rotational symmetry.

23. Coordinate Geometry If two nonvertical lines are parallel, then their slopes are equal.

24. If two angles are complementary, then the sum of their measures is 90.

25. If you are in Indiana, then you are in Indianapolis.

26. Probability If the probability that an event will occur is 1, then the event is certain to occur.

27. a. Consumer Issues Advertisements often suggest conditional statements. For example, an ad might imply that if you don't buy a product, you won't be popular. What conditional is implied in the ad at the right?

b. Research Find magazine ads that use conditionals effectively. Make a poster to display these ads.

Write the two conditionals that make up each biconditional.

28. A swimmer wins a race if and only if she swims the fastest.

29. A number is divisible by 3 if and only if the sum of its digits is divisible by 3.

30. Two angles are congruent if and only if they have the same measure.

31. Writing Jynona knows that vertical angles are congruent. She thinks the converse is also true. Is she correct? Explain.

For each Venn diagram: (a) Write a conditional statement. (b) Write the contrapositive of the conditional.

32. 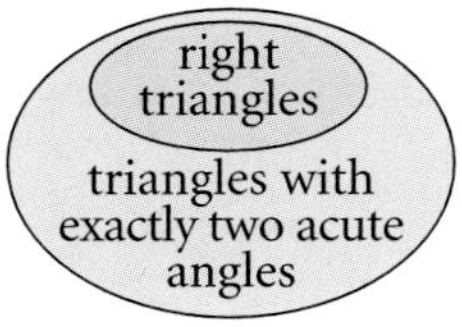

33. 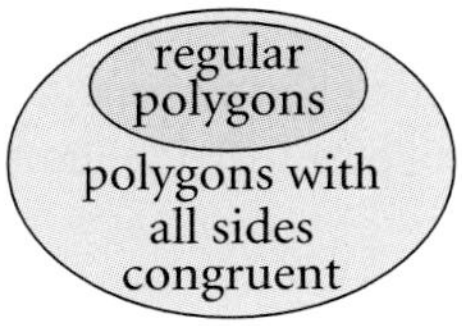

34. 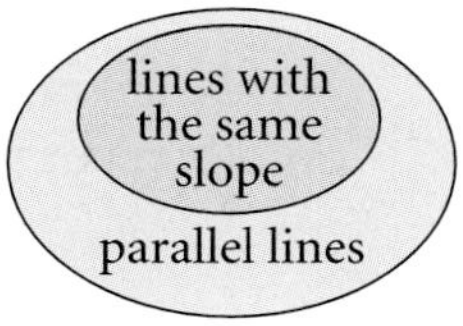

35. 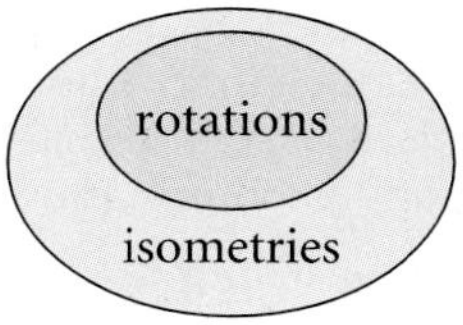

Chapter Project — *Find Out by Listing*

Three red hats and three blue hats are packed in three boxes, with two hats to a box. The boxes are all labeled incorrectly. To determine what each box actually contains, you may select one hat from one box, without looking at the contents of that box. Explain how this will allow you to determine the contents of each box. (*Hint:* List all possible solutions; then use logic to solve.)

Exercises MIXED REVIEW

Coordinate Geometry **Sketch a line with the given slope containing the given point.**

36. $m = \frac{1}{2}$, (2, −6)

37. $m = 1$, (0, 5)

38. $m = -2$, (−3, 6)

39. $m = \frac{1}{3}$, (0, 0)

40. An angle's measure is 10 more than its supplement. Find the measures of both angles.

41. Transformational Geometry Locate the coordinates of the image of $\triangle ABC$ with vertices $A(0, 3)$, $B(-4, -6)$, $C(6, 1)$ under a 180° rotation about the origin.

Getting Ready for Lesson 4-2

Find the value of *x*.

42. 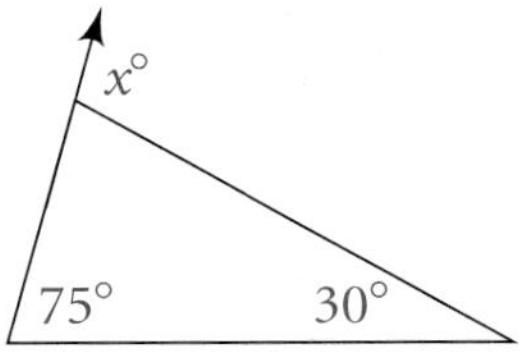

43. 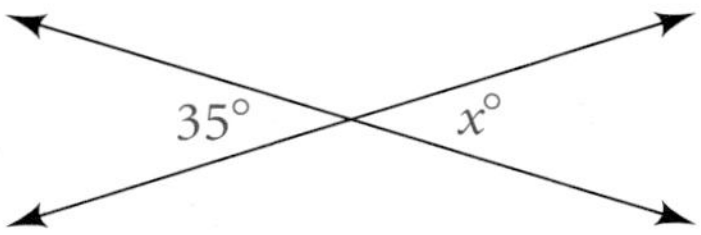

44. 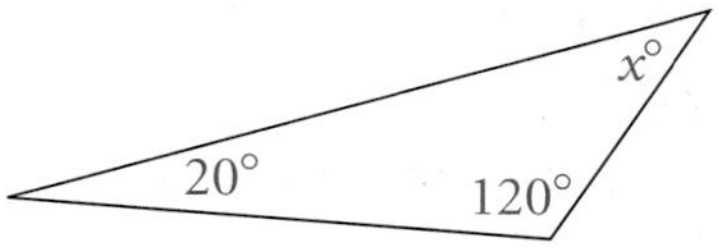

4-2 Isosceles Triangles

What You'll Learn

- Using and applying properties of isosceles triangles

...And Why

To understand a geometric figure used in the designs of many buildings and bridges

What You'll Need

- straightedge
- compass
- scissors

WORK TOGETHER

Have each member of your group construct a different isosceles triangle and then cut it out. Be sure to include acute and obtuse triangles.

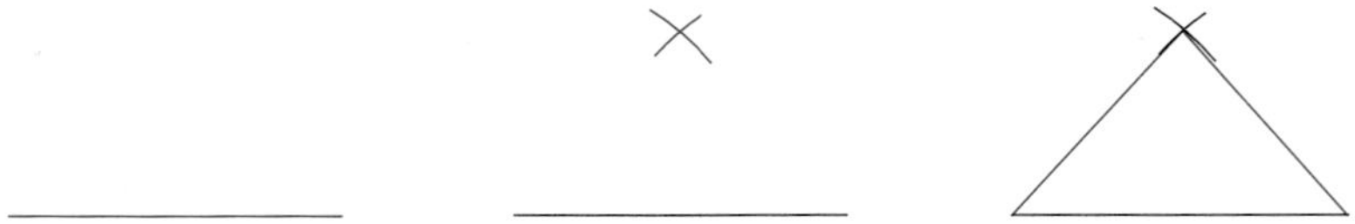

- Label the triangle $\triangle ABC$, with A and B opposite the congruent sides.
- Bisect $\angle C$ by folding the triangle so that the congruent sides overlap. Label the intersection of the fold line and $\overline{AB}$ as point D.

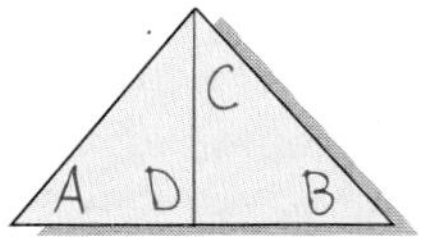

TECHNOLOGY HINT

The Work Together could be done using geometry software.

1. What do you notice about $\angle A$ and $\angle B$? Compare results within your group.

2. **a.** What types of angles do $\angle CDA$ and $\angle CDB$ appear to be?
b. What do you notice about $\overline{AD}$ and $\overline{DB}$?
c. Use your answers to parts (a) and (b) to complete the statement: $\overline{CD}$ is the __?__ of $\overline{AB}$.

- Construct a new triangle that has two congruent angles. Cut out the triangle and label it $\triangle EFG$, where $\angle E \cong \angle F$.

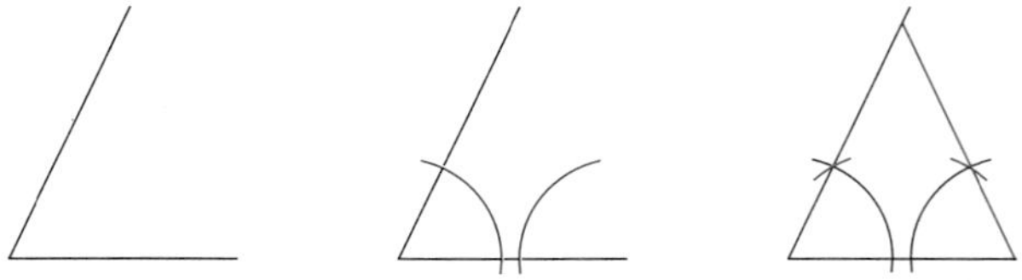

3. Fold the triangle so that the congruent angles overlap.
a. What do you notice about $\overline{EG}$ and $\overline{FG}$? Compare results within your group.
b. What type of triangle is $\triangle EFG$?

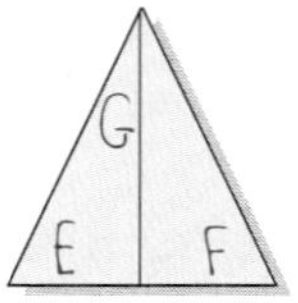

THINK AND DISCUSS

Isosceles triangles are common in the real world. You can find them in structures such as bridges and buildings. The congruent sides of an isosceles triangle are the **legs.** The third side is the **base.** The two congruent sides form the **vertex angle.** The other two angles are the **base angles.**

Your observations from the Work Together suggest the following theorems. The proofs of these theorems involve properties of congruent triangles that you will study in Chapter 8.

Theorem 4-1
Isosceles Triangle Theorem

If two sides of a triangle are congruent, then the angles opposite those sides are also congruent.

If $\overline{AC} \cong \overline{BC}$, then $\angle A \cong \angle B$.

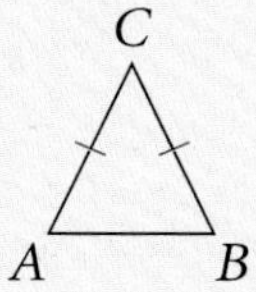

Theorem 4-2

The bisector of the vertex angle of an isosceles triangle is the perpendicular bisector of the base.

If $\overline{AC} \cong \overline{BC}$ and $\overline{CD}$ bisects $\angle ACB$, then $\overline{CD} \perp \overline{AB}$ and $\overline{CD}$ bisects $\overline{AB}$.

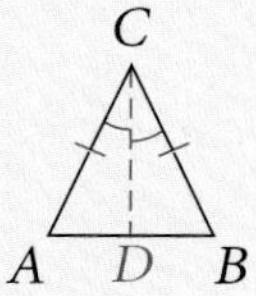

Theorem 4-3
Converse of Isosceles Triangle Theorem

If two angles of a triangle are congruent, then the sides opposite the angles are congruent.

If $\angle A \cong \angle B$, then $\overline{AC} \cong \overline{BC}$.

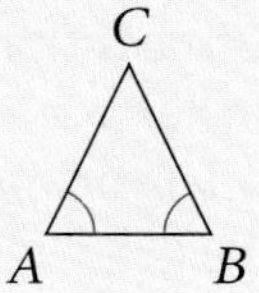

4. Write the Isosceles Triangle Theorem and its converse as a biconditional.

Example 1

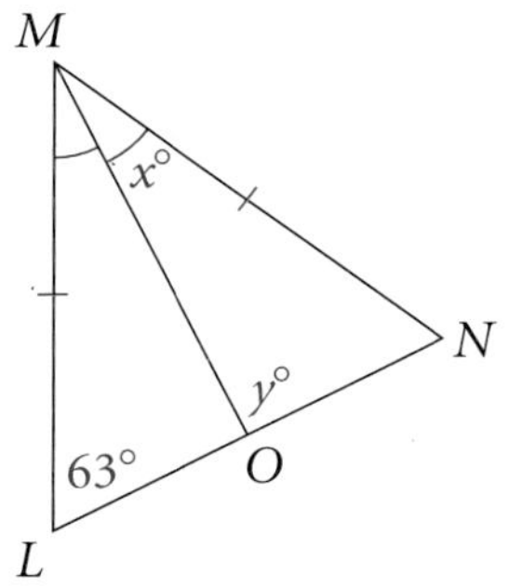

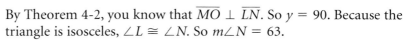

Algebra Find the values of x and y.

By Theorem 4-2, you know that $\overline{MO} \perp \overline{LN}$. So $y = 90$. Because the triangle is isosceles, $\angle L \cong \angle N$. So $m\angle N = 63$.

$m\angle N + x + y = 180$	Triangle Angle-Sum Theorem
$63 + x + 90 = 180$	Substitution
$x = 27$	Subtract 153 from each side.

So $x = 27$ and $y = 90$.

5. Try This Suppose $m\angle L = 43$. Find the values of x and y.

Example 2

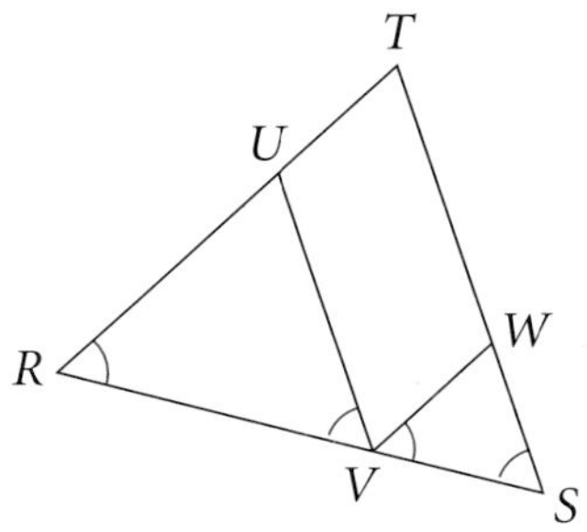

Complete each statement. Explain your answers.

a. $\overline{RT} \cong$ __?__ **b.** $\overline{RU} \cong$ __?__ **c.** $\overline{VW} \cong$ __?__

a. $\overline{RT} \cong \overline{ST}$ because $\angle R \cong \angle S$.

b. $\overline{RU} \cong \overline{VU}$ because $\angle R \cong \angle RVU$.

c. $\overline{VW} \cong \overline{SW}$ because $\angle WVS \cong \angle S$.

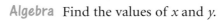

QUICK REVIEW

An equilateral triangle has three congruent sides. An equiangular triangle has three congruent angles.

6. a. Choose two sides of $\triangle EFG$. What must be true about the angles opposite these sides? Why?

b. Repeat part (a) with a different pair of sides.

c. What is true about the angles of an equilateral triangle?

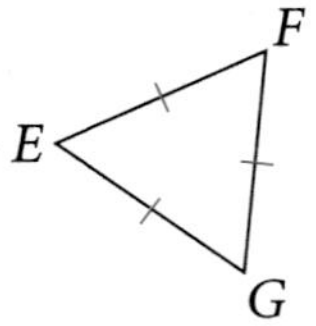

7. a. Choose two angles of $\triangle MNO$. What must be true about the sides opposite these angles? Why?

b. Repeat part (a) with a different pair of angles.

c. What is true about the sides of an equiangular triangle?

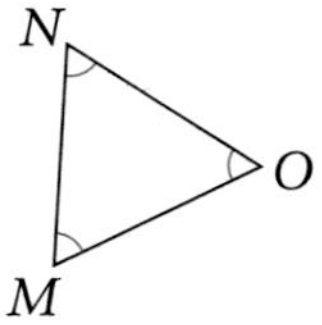

Your observations from Questions 6 and 7 are summarized below.

Corollary to Isosceles Triangle Theorem

If a triangle is equilateral, then it is equiangular.

If $\overline{XY} \cong \overline{YZ} \cong \overline{ZX}$, then $\angle X \cong \angle Y \cong \angle Z$.

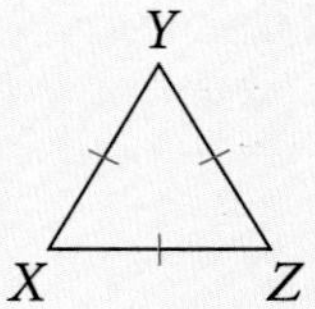

Corollary to Converse of Isosceles Triangle Theorem

If a triangle is equiangular, then it is equilateral.

If $\angle X \cong \angle Y \cong \angle Z$, then $\overline{XY} \cong \overline{YZ} \cong \overline{ZX}$.

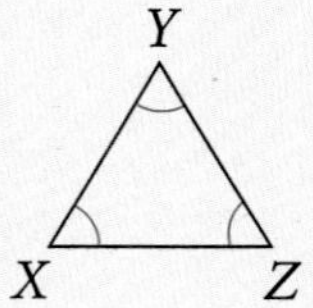

8. Use the corollaries above to write a biconditional.

Example 3

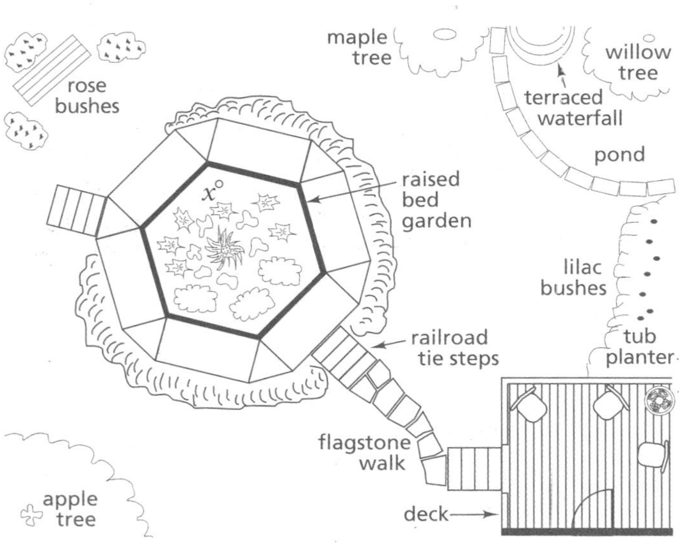

Relating to the Real World

Landscaping A landscaper is building a raised bed garden to fit in the hexagonal space in the diagram. The path around the garden consists of rectangles and equilateral triangles. What is the measure of the angle marked x?

Each angle of a rectangle measures 90. Each angle of an equilateral triangle measures 60. (Why?)

$$x + 90 + 60 + 90 = 360$$
$$x = 120$$

The measure of the angle is 120.

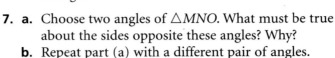

Exercises ON YOUR OWN

Algebra **Find the values of x and y.**

1.

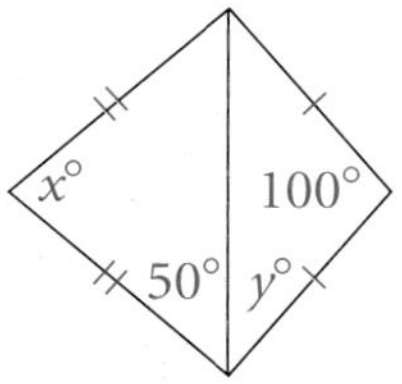

2.

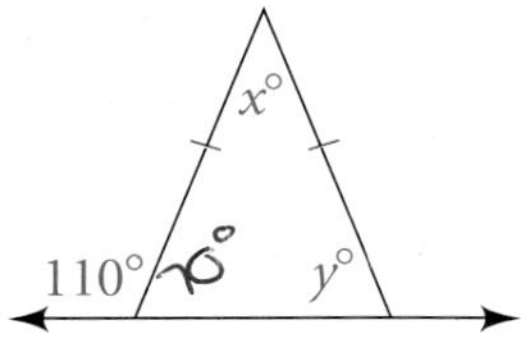

3.

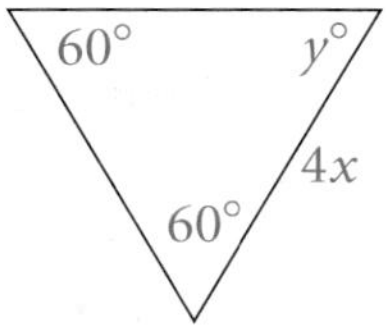

Perimeter is 54.

4.

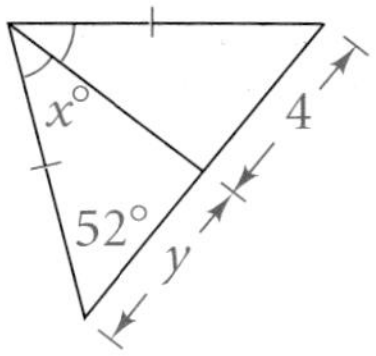

5.

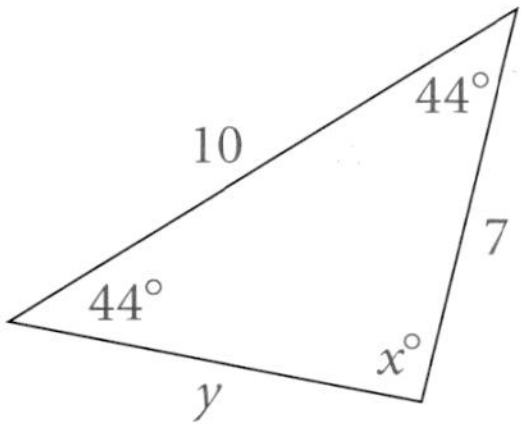

6.

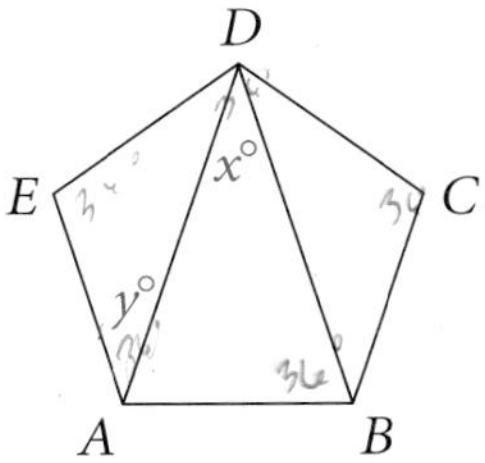

ABCDE is a regular pentagon.

7.

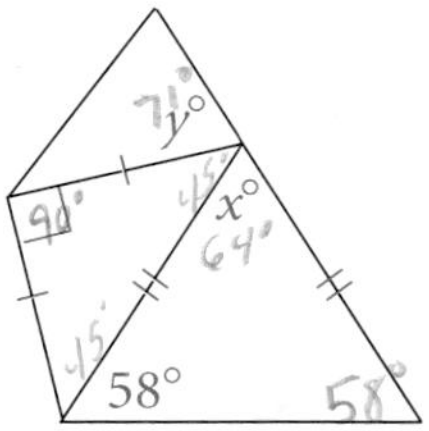

8.

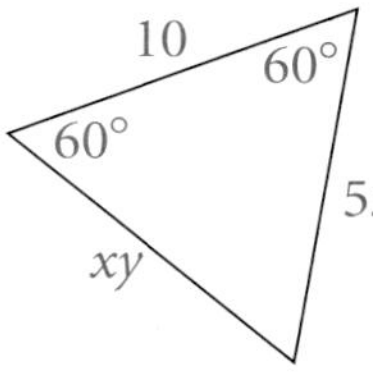

9. **Architecture** The Air Force Academy Cadet Chapel has 17 spires that point to the sky. Each spire is an isosceles triangle with a 40° vertex angle. Find the measures of the base angles.

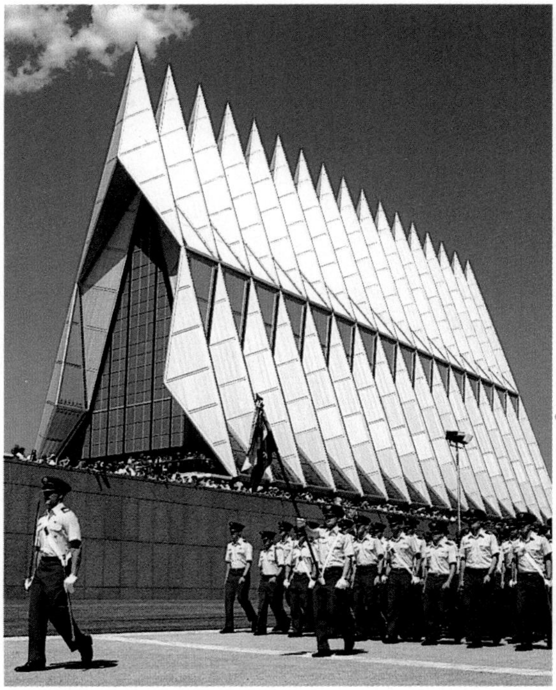

10. **Critical Thinking** What are the measures of the base angles of an isosceles right triangle? Explain.

11. **Coordinate Geometry** The vertices of the base angles of an isosceles triangle are at (0, 0) and (6, 0). Describe the possible locations of the third vertex.

Logical Reasoning **Determine whether each statement is true or false. If it is false, provide a counterexample.**

12. If a quadrilateral is equilateral, then it is equiangular.

13. If a quadrilateral is equiangular, then it is equilateral.

14. Every isosceles triangle has at least one line of symmetry.

15. Every equilateral triangle has exactly three lines of symmetry.

16. **Graphic Arts** The logo of the National Council of Teachers of Mathematics is shown at the right.

a. Trace the logo onto your paper. Highlight an obtuse isosceles triangle in the design and then find its angle measures.

b. **Open-ended** Repeat part (a) for each of the following figures: kite, pentagon, hexagon.

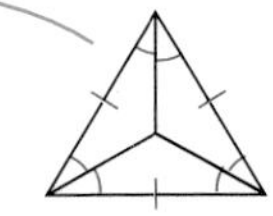

The triangles in the logo have these congruent sides and angles.

Coordinate Geometry **For each pair of points, there are six points that could be the third vertex of an isosceles right triangle. Find the coordinates of each point.**

17. (0, 0) and (5, 5) **18.** (2, 3) and (5, 6)

19. *Algebra* A triangle has angle measures $x + 15$, $3x - 35$, and $4x$.

a. Find the value of x.

b. Find the measure of each angle.

c. What type of triangle is it? Why?

20. *Writing* If a triangle is equiangular, is it also isosceles? Explain.

21. a. *Communications* In the diagram, what type of triangle is formed by the cable pairs and the ground?

b. What are the two different base lengths of the triangles?

c. How is the tower related to each of the triangles?

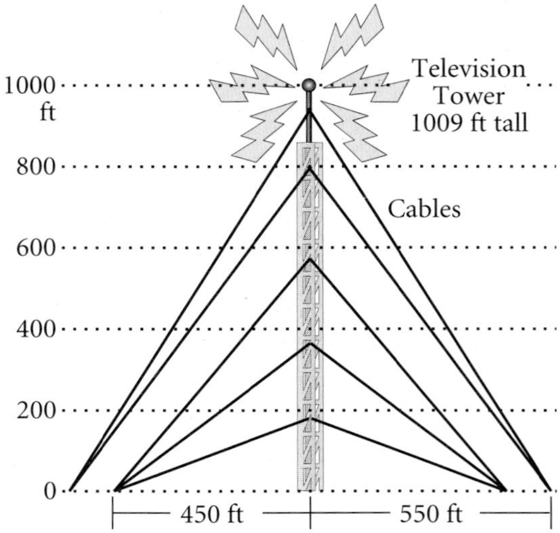

Tower cables extend to both widths.

Find each value.

22. If $m\angle L = 58$, then $m\angle LKJ = \blacksquare$.

23. If $JL = 5$, then $ML = \blacksquare$.

24. If $m\angle JKM = 48$, then $m\angle J = \blacksquare$.

25. If $m\angle J = 55$, then $m\angle JKM = \blacksquare$.

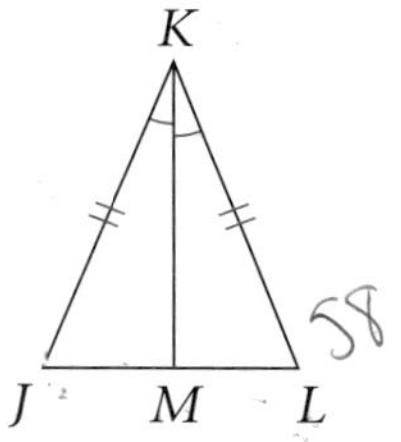

Choose **Use mental math, pencil and paper, or a calculator to find the values of the variables.**

26.

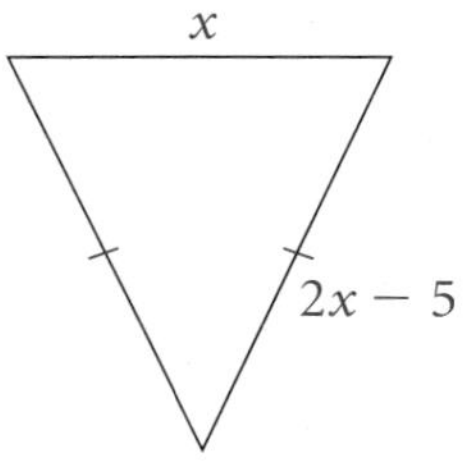

Perimeter is 20.

27.

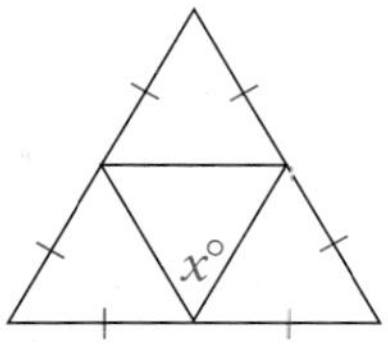

28.

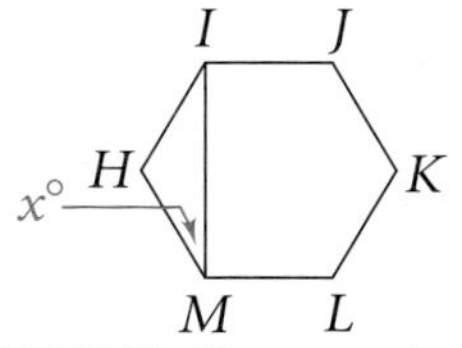

HIJKLM is a regular hexagon.

29.

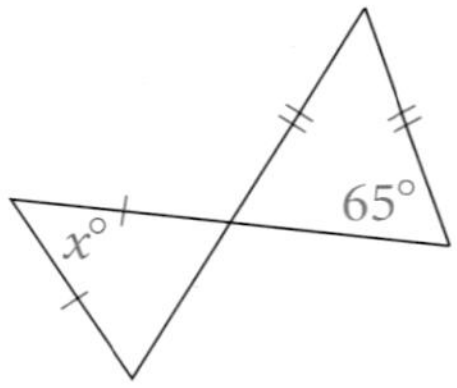

30.

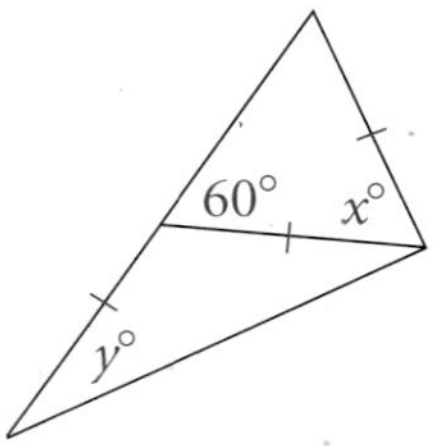

31.

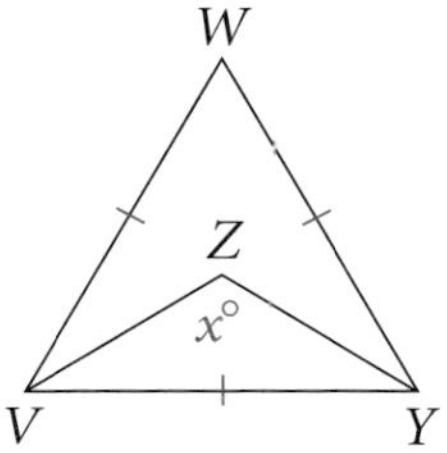

$\overline{VZ}$ and $\overline{YZ}$ are angle bisectors.

32.

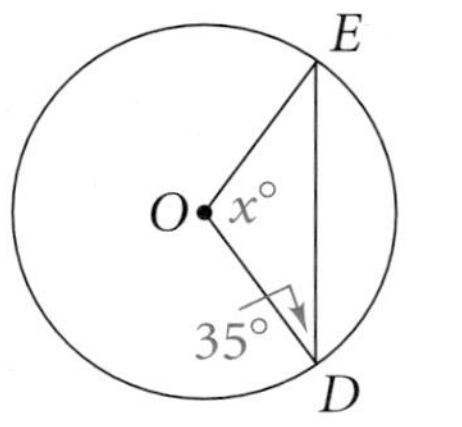

$\overline{OD}$ and $\overline{OE}$ are radii.

33.

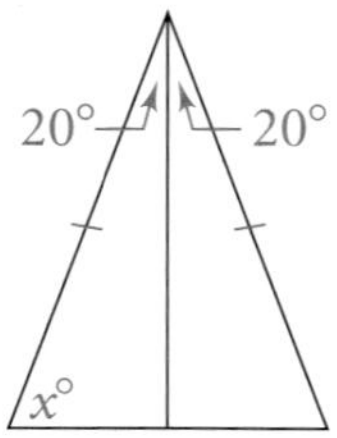

34. *Crafts* This design is used in Hmong crafts and in Islamic and Mexican tiles. To create it, the artist starts by drawing a circle and four equally spaced diameters.

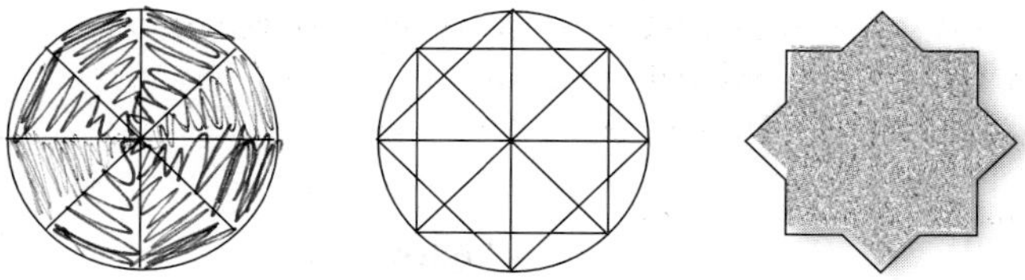

Step 1 **Step 2** **Step 3**

a. How many different sizes of isosceles right triangles can you find in Step 2? Trace an example of each onto your paper.

b. For each size of triangle that you traced, count the number of times it appears in the diagram.

35. *Critical Thinking* Patrick defines the base of an isosceles triangle as "the bottom side of an isosceles triangle." Is his definition a good one? Explain why or why not.

36. *Standardized Test Prep* A square and a regular hexagon are placed so that they have a common side. Find $m\angle HAS$.

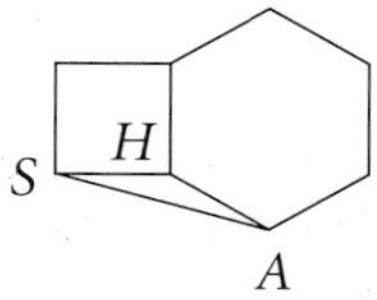

A. 9 **B.** 10 **C.** 15 **D.** 20 **E.** 30

Algebra **Find the values of *m* and *n*.**

37.

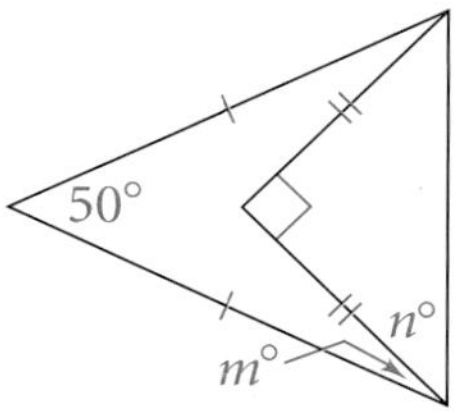

38.

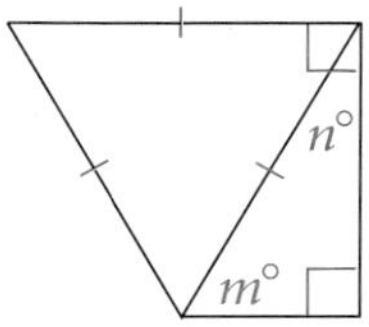

39.

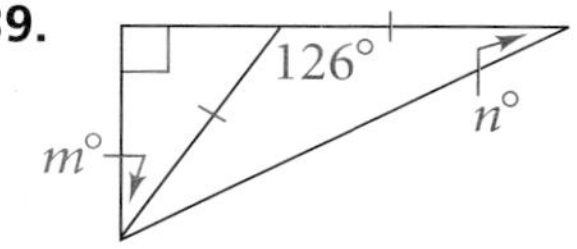

40.

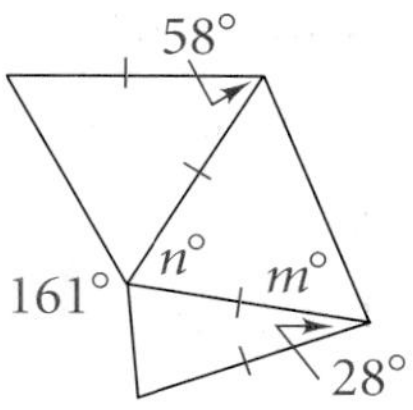

Exercises MIXED REVIEW

Coordinate Geometry **The endpoints of a diameter of a circle are given. Find the coordinates of the center and the length of a radius.**

41. (3, 8), (−1, 2) **42.** (−2, 5), (−5, 2) **43.** (3, 7), (−2, 6)

44. *Coordinate Geometry* Find the equation of the line that passes through (0, 4) and is parallel to $y = -3x - 5$.

45. Find the number of sides of a regular polygon whose exterior angles measure 15°.

Getting Ready for Lesson 4-3

What can you conclude from each diagram? Justify your answers.

46.

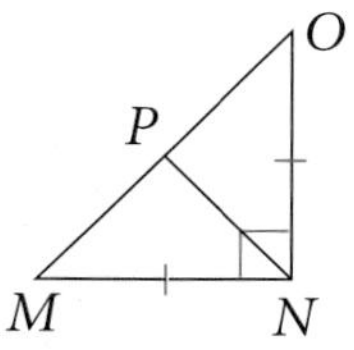

$\angle MNP \cong \angle ONP$

47.

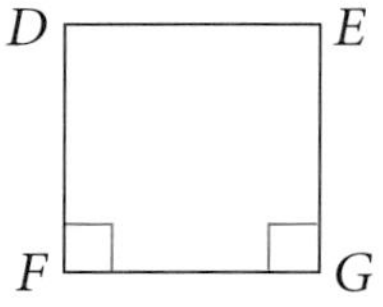

48.

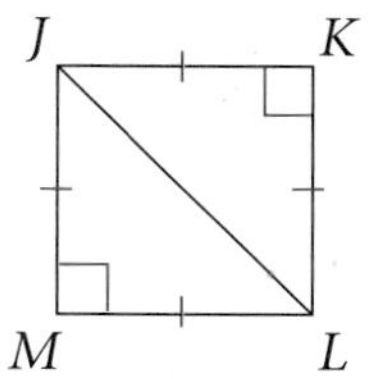

What You'll Learn

- Using different styles of proofs to write convincing arguments

...And Why

To help you think logically

4-3 Preparing for Proof

THINK AND DISCUSS

In everyday life, proof takes many forms.

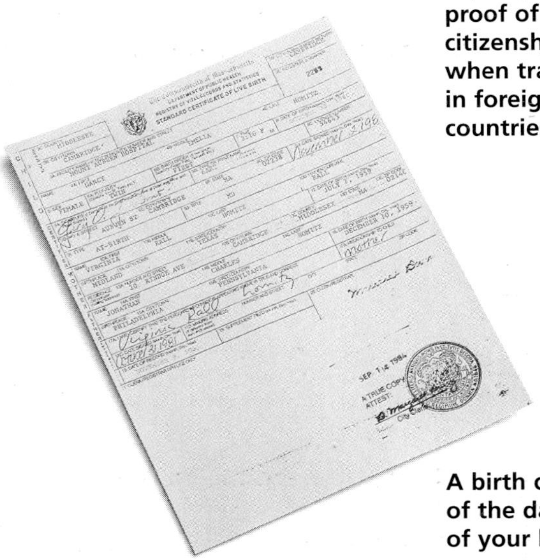

A birth certificate is proof of the date and location of your birth.

A passport is proof of your citizenship when traveling in foreign countries.

A punch clock record is proof of the number of hours you have worked.

1. Give some other examples of proof from everyday life.

Proofs in geometry can take several forms. Each form has these basic parts:

A list of the given information.

A list of what is to be proved.

A logical series of statements that lead from the given information to what is to be proved.

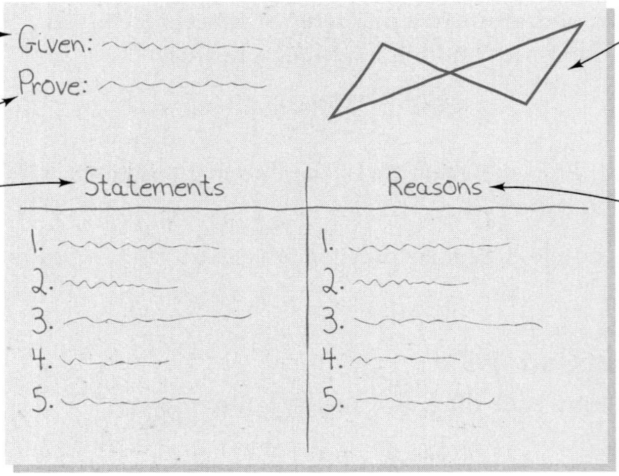

A diagram showing the given information.

The reasons why each statement is true. Reasons can include postulates, definitions, properties, previously stated theorems, or given facts.

The diagram shows a **two-column proof.** In this style of proof, statements appear in one column and reasons in another.

In this lesson, you will study the proofs of the four theorems at the top of the next page.

Theorem 4-4	If a triangle is a right triangle, then the acute angles are complementary.
Theorem 4-5	If two angles of one triangle are congruent to two angles of another triangle, then the third angles are congruent.
Theorem 4-6	All right angles are congruent.
Theorem 4-7	If two angles are congruent and supplementary, then each is a right angle.

In a proof of a theorem, the *Given* information and the figure relate to the hypothesis of the theorem. You *Prove* the conclusion of the theorem.

Theorem 4-4
If a triangle is a right triangle, then the acute angles are complementary.

Hypothesis

Given: $\triangle EFG$ with right angle $\angle F$
Prove: $\angle E$ and $\angle G$ are complementary.

Conclusion

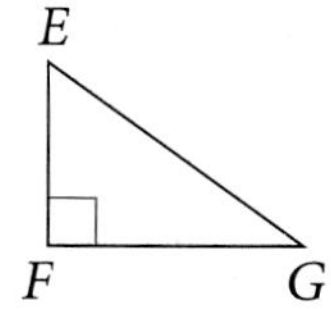

Two-Column Proof of Theorem 4-4

Statements	Reasons
1. $\angle F$ is a right angle.	**1.** Given
2. $m\angle F = 90$	**2.** Def. of right angle
3. $m\angle E + m\angle F + m\angle G = 180$	**3.** Triangle Angle-Sum Thm.
4. $m\angle E + 90 + m\angle G = 180$	**4.** Substitution
5. $m\angle E + m\angle G = 90$	**5.** Subtraction Prop. of Equality
6. $\angle E$ and $\angle G$ are complementary.	**6.** Def. of complementary angles

In a **paragraph proof,** the statements and reasons appear in sentences within a paragraph.

Example

Write a paragraph proof for Theorem 4-5.

Given: $\angle X \cong \angle Q$ and $\angle Y \cong \angle R$
Prove: $\angle Z \cong \angle S$

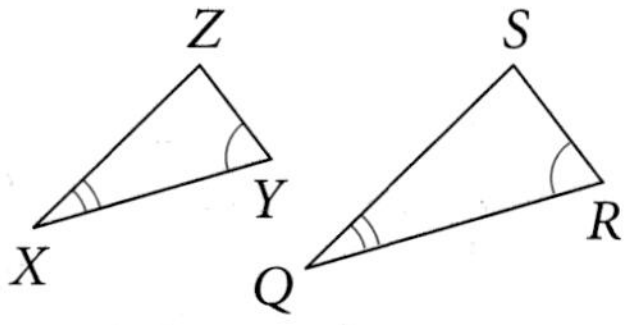

Paragraph Proof
By the Triangle Angle-Sum Theorem, $m\angle X + m\angle Y + m\angle Z = 180$ and $m\angle Q + m\angle R + m\angle S = 180$. By substitution, $m\angle X + m\angle Y + m\angle Z = m\angle Q + m\angle R + m\angle S$. We are given that $\angle X \cong \angle Q$ and $\angle Y \cong \angle R$ (or $m\angle X = m\angle Q$ and $m\angle Y = m\angle R$). Subtracting equal quantities from both sides of the equation leaves $m\angle Z = m\angle S$, so $\angle Z \cong \angle S$.

WORK TOGETHER

Work in pairs.

2. Rewrite this two-column proof of Theorem 4-6 as a paragraph proof.

Given: $\angle X$ and $\angle Y$ are right angles.

Prove: $\angle X \cong \angle Y$

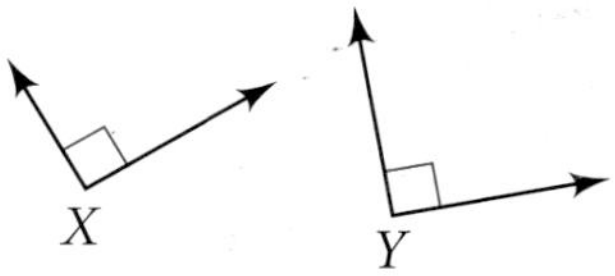

Two-Column Proof

Statements	Reasons
1. $\angle X$ and $\angle Y$ are right angles.	**1.** Given
2. $m\angle X = 90$, $m\angle Y = 90$	**2.** Def. of right angles
3. $m\angle X = m\angle Y$, or $\angle X \cong \angle Y$	**3.** Substitution

3. Rewrite this paragraph proof of Theorem 4-7 as a two-column proof.

Given: $\angle W$ and $\angle V$ are congruent and supplementary.

Prove: $\angle W$ and $\angle V$ are right angles.

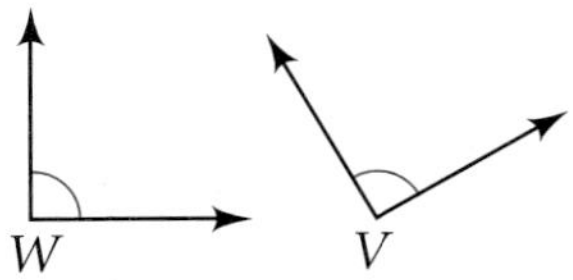

Paragraph Proof

$\angle W$ and $\angle V$ are congruent and supplementary, so $m\angle W = m\angle V$ and $m\angle W + m\angle V = 180$. Substituting $m\angle W$ for $m\angle V$ gives $m\angle W + m\angle W = 180$. Therefore, $m\angle W = 90$. Since $\angle W \cong \angle V$, $m\angle V = 90$, too. Thus both angles are right angles.

Exercises ON YOUR OWN

What can you conclude from each diagram? Justify your answers.

1.

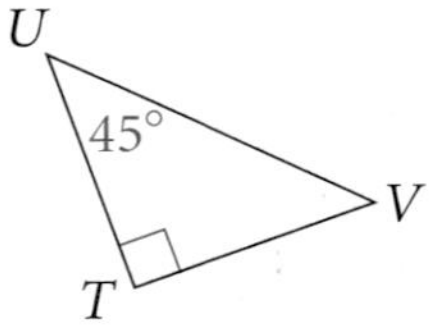

2.

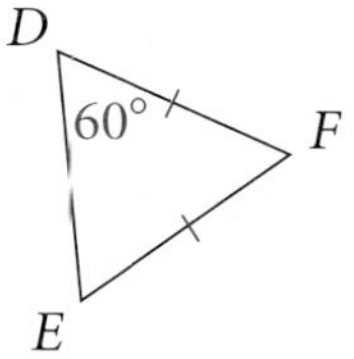

3.

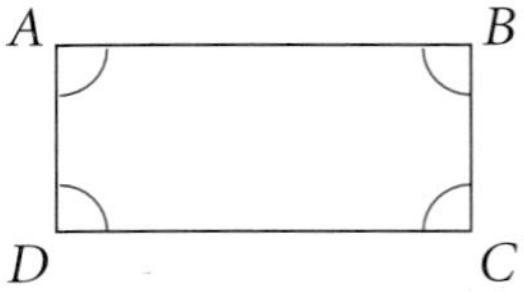

4.

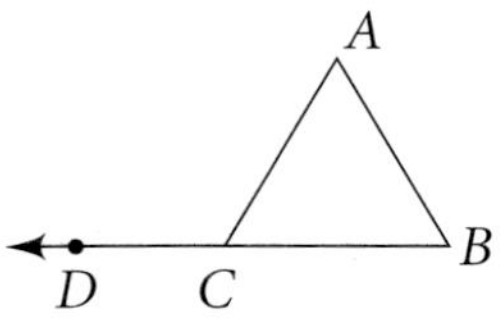

5.

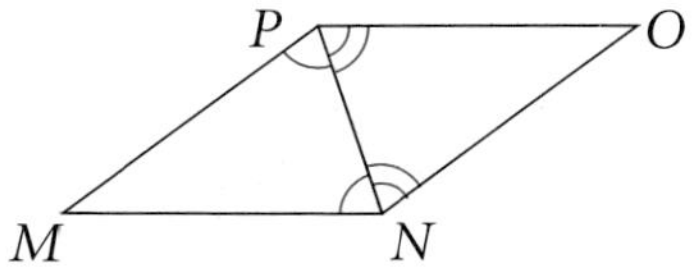

6.

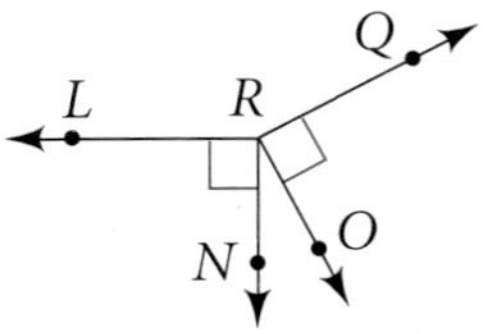

7. *Standardized Test Prep* Find the value of x in the figure.

A. 20 **B.** 30 **C.** 45 **D.** 60 **E.** none of these

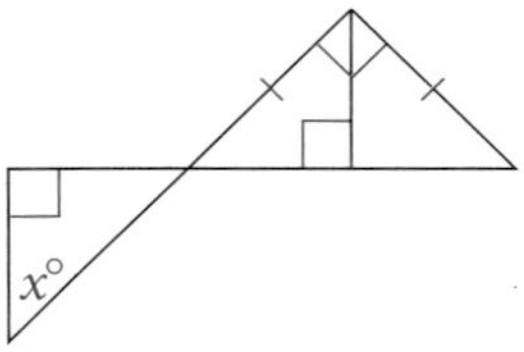

8. *Open-ended* Explain the similarities and differences between paragraph and two-column proofs. Which do you prefer? Why?

9. **Writing** Russell Black Elk found this multiple-choice question in a puzzle book. Solve the puzzle and explain your solution.

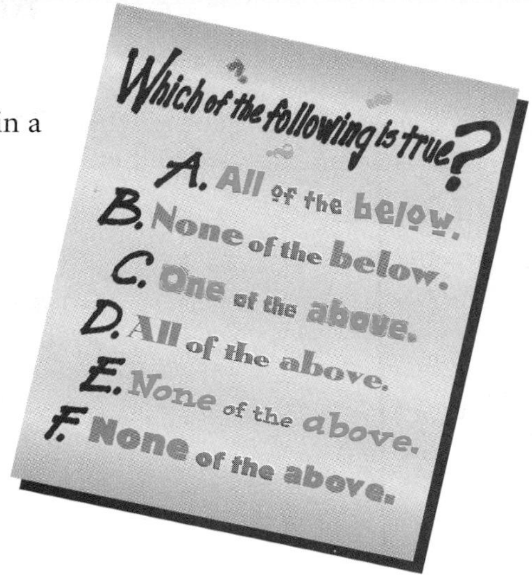

10. Rewrite this paragraph proof as a two-column proof.

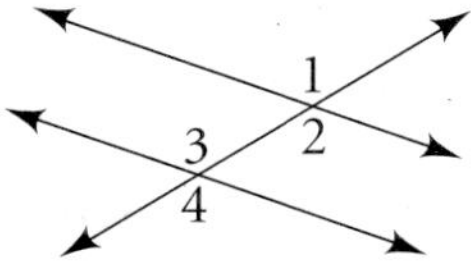

Given: $\angle 1 \cong \angle 4$

Prove: $\angle 2 \cong \angle 3$

$\angle 1$ and $\angle 2$ are vertical angles, as are $\angle 3$ and $\angle 4$. Vertical angles are congruent, so $\angle 1 \cong \angle 2$ and $\angle 3 \cong \angle 4$. We are given that $\angle 1 \cong \angle 4$. By the Transitive Property of Congruence, $\angle 2 \cong \angle 4$ and $\angle 2 \cong \angle 3$.

Refer to the diagrams to complete each statement.

11. a. $\angle OKN \cong$ ___?___
   b. $\angle LKO \cong$ ___?___
   c. $\angle LOK \cong$ ___?___

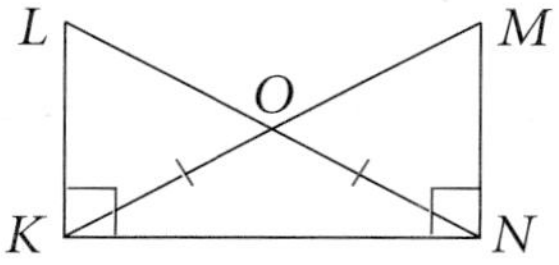

12. a. $m\angle USR =$ ___?___
   b. $m\angle RUS =$ ___?___
   c. $m\angle SUQ =$ ___?___
   d. $m\angle USQ =$ ___?___
   e. $m\angle QST =$ ___?___

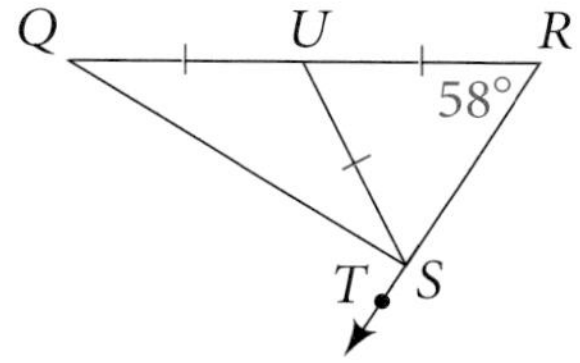

13. **Preparing for Proof** The reasons given in this proof are correct, but they are in the wrong order. List them in the correct order.

Given: $m\angle AOB = m\angle BOC$

Prove: $\overleftrightarrow{OA} \perp \overleftrightarrow{OB}$

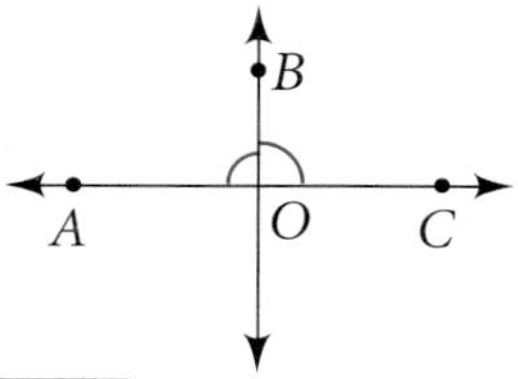

Statements	Reasons
1. $m\angle AOB = m\angle BOC$	a. Division Prop. of Equality
2. $m\angle AOB + m\angle BOC = 180$	b. Substitution
3. $m\angle AOB + m\angle AOB = 180$ or $2(m\angle AOB) = 180$	c. Def. of right angle
4. $m\angle AOB = 90$	d. Given
5. $\angle AOB$ is a right angle.	e. Def. of perpendicular lines
6. $\overleftrightarrow{OA} \perp \overleftrightarrow{OB}$	f. Angle Addition Postulate

14. **Preparing for Proof** Rewrite this proof as a paragraph proof.

Given: $\overline{XZ} \cong \overline{YZ}$ and $\overline{XW} \cong \overline{YW}$

Prove: $m\angle 1 = m\angle 2$

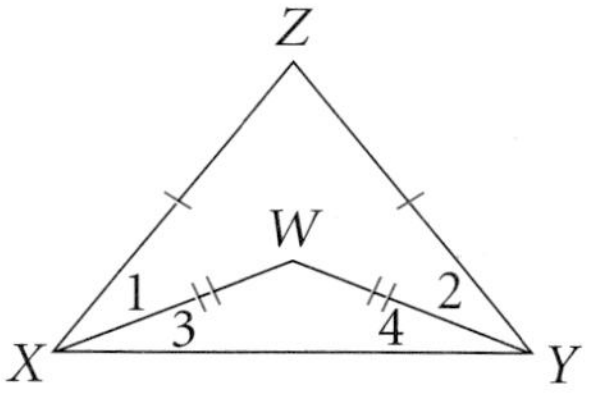

Statements	Reasons
1. $\overline{XZ} \cong \overline{YZ}$	1. Given
2. $\angle ZXY \cong \angle ZYX$ or $m\angle ZXY = m\angle ZYX$	2. Base $\angle$s of an isosceles $\triangle$ are $\cong$.
3. $\overline{XW} \cong \overline{YW}$	3. Given
4. $\angle 3 \cong \angle 4$ or $m\angle 3 = m\angle 4$	4. Base $\angle$s of an isosceles $\triangle$ are $\cong$.
5. $m\angle ZXY = m\angle 1 + m\angle 3$; $m\angle ZYX = m\angle 2 + m\angle 4$	5. Angle Addition Postulate
6. $m\angle 1 + m\angle 3 = m\angle 2 + m\angle 4$	6. Substitution
7. $m\angle 1 = m\angle 2$	7. Subtraction Prop. of Equality

15. **Logical Reasoning** Explain why this statement is true:
If $m\angle 1 + m\angle 2 = 180$ and $m\angle 2 + m\angle 3 = 180$, then $\angle 1 \cong \angle 3$.

16. **Logical Reasoning** Explain why this proof is invalid.

Given: $a = b$

Prove: $1 = 2$

Statements	Reasons
1. $a = b$	**1.** Given
2. $ab = b^2$	**2.** Multiplication Prop. of Equality
3. $ab - a^2 = b^2 - a^2$	**3.** Subtraction Prop. of Equality
4. $a(b - a) = (b + a)(b - a)$	**4.** Distributive Property
5. $a = b + a$	**5.** Division Prop. of Equality
6. $a = a + a$	**6.** Substitution
7. $a = 2a$	**7.** Distributive Property
8. $1 = 2$	**8.** Division Prop. of Equality

17. Supply the reasons to complete this proof of the Corollary to the Isosceles Triangle Theorem stated on page 190.

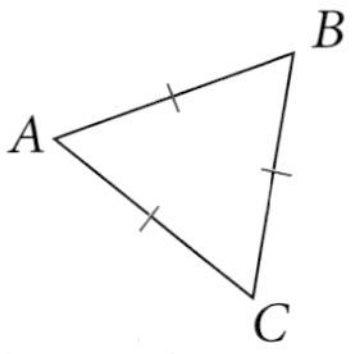

Given: $\triangle ABC$ with $\overline{AB} \cong \overline{BC} \cong \overline{CA}$

Prove: $\angle A \cong \angle B$, $\angle B \cong \angle C$, and $\angle A \cong \angle C$

Statements	Reasons
1. $\overline{AB} \cong \overline{BC} \cong \overline{CA}$	**a.** ?
2. $\angle A \cong \angle C$	**b.** ?
3. $\angle C \cong \angle B$	**c.** ?
4. $\angle A \cong \angle B$	**d.** ?

18. Complete this paragraph proof.

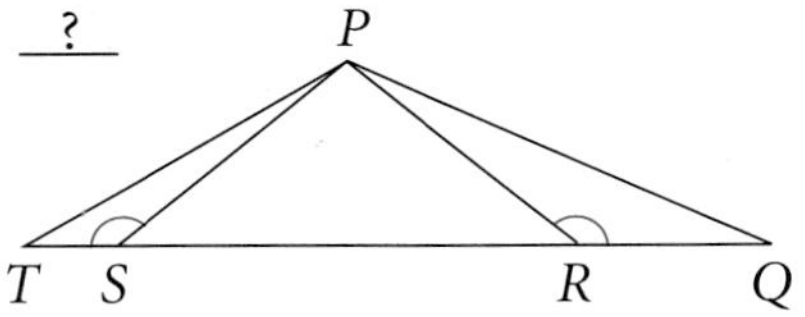

Given: $\angle PST \cong \angle PRQ$

Prove: $\triangle PSR$ is isosceles.

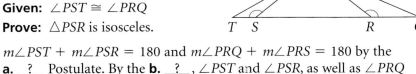

$m\angle PST + m\angle PSR = 180$ and $m\angle PRQ + m\angle PRS = 180$ by the **a.** ? Postulate. By the **b.** ? , $\angle PST$ and $\angle PSR$, as well as $\angle PRQ$ and $\angle PRS$, are supplementary. We are given that **c.** ? . By the Congruent Supplements Thm., **d.** ? . If two $\angle$s of a $\triangle$ are $\cong$, the sides opposite them are $\cong$, so **e.** ? . By **f.** ? , $\triangle PSR$ is isosceles.

Chapter Project

Find Out by Organizing

A drummer, guitarist, and keyboard player named Amy, Bob, and Carla are in a band. Use the clues to determine which instrument each plays.

Carla and the drummer wear different-colored shirts.
The keyboard player is older than Bob.
Amy, the youngest band member, lives next door to the guitarist.

You can solve this type of logic puzzle by eliminating possibilities. Make a grid. Put an X in a box once you eliminate it as a possibility.

Instrument	Amy	Bob	Carla
Drums			
Guitar			
Keyboard			

Exercises MIXED REVIEW

Statistics **For Exercises 19 and 20, refer to the circle graph.**

Source: *Louis Harris & Associates survey*

19. Find the measure of each central angle.

20. Are the statements *true* or *false?* Explain.

- **a.** Most adults want students who work to save or invest money.
- **b.** All of those surveyed are parents of high-school students.
- **c.** Most adults aren't in favor of allowing students who work to spend their earnings as they please.

21. **a.** Transformational Geometry Graph points $A(4, -8)$ and $A'(-6, 1)$. Sketch $\overrightarrow{AA'}$.

- **b.** Describe the translation of A to A' in words and using ordered pair notation.

Getting Ready for Lesson 4-4

22. **a.** Coordinate Geometry Find the slope of the line containing $A(-2, 3)$ and $B(3, 1)$.

- **b.** Find the coordinates of the midpoint of $\overline{AB}$.

A Point in Time

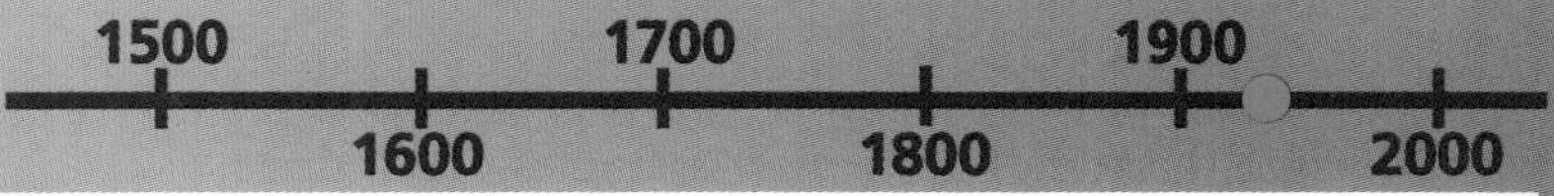

The Logic of Agatha Christie

Most people are not detectives, but as a young woman, the English writer Agatha Christie (1890–1976) correctly deduced that many people would *like* to be. In **1920** she published her first book, a detective novel entitled *The Mysterious Affair at Styles* in which she introduced the eccentric and ultra-logical Belgian detective Hercule Poirot. In this and in many subsequent novels, Poirot solved mysteries not with guns or car chases but with logical reasoning. He and Agatha Christie's other principal detective character, Jane Marple, always won in the end because they were more patient, more diligent, and, above all, more *clever* than the crooks. Readers made best-sellers of many of Christie's 78 detective novels. Many of her novels, including *Murder on the Orient Express* and *Death on the Nile,* have been adapted into popular films.

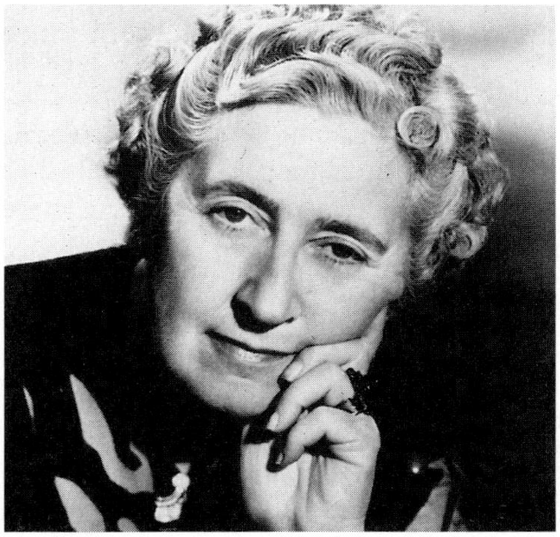

To date, her books have sold some 2 billion copies in 44 languages, making Agatha Christie by far the world's top-selling writer of fiction.

Investigating Midsegments

Before Lesson 4-4

Work in pairs or small groups.

Construct

Use geometry software to draw a triangle. Label it $\triangle ABC$. Construct the midpoints of $\overline{AB}$ and $\overline{AC}$, label them D and E, respectively, and then connect them with a *midsegment.*

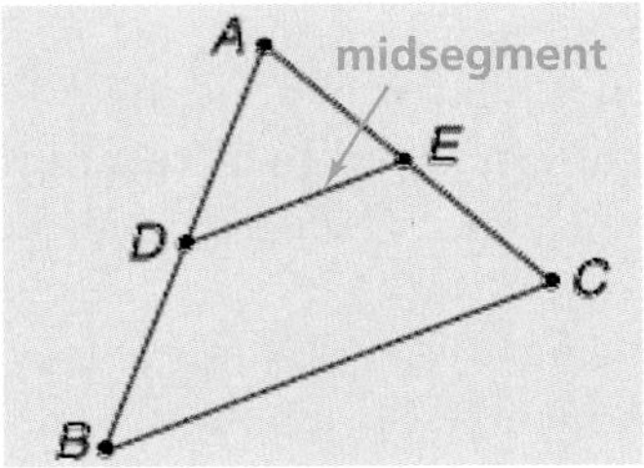

Investigate

- Measure the lengths of $\overline{DE}$ and $\overline{BC}$. Calculate $\frac{DE}{BC}$.
- Measure the slopes of $\overline{DE}$ and $\overline{BC}$.
- Manipulate the triangle and observe the lengths and slopes of the segments.

Conjecture

List all your **conjectures** about a midsegment.

Extend

- Construct the other two midsegments in $\triangle ABC$. Measure the angles of the four triangles determined by the three midsegments. How do they compare to the angles of the original triangle? Make **conjectures.**
- Measure the sides of the four triangles. Make a **conjecture** about the four triangles determined by the midsegments of a triangle. Make a **conjecture** about the relationship between the original triangle and the four triangles.
- Measure the areas of the four triangles formed by the midsegments of a triangle. Measure the area of the original triangle. Make a **conjecture** about the areas. Do the same for perimeter.

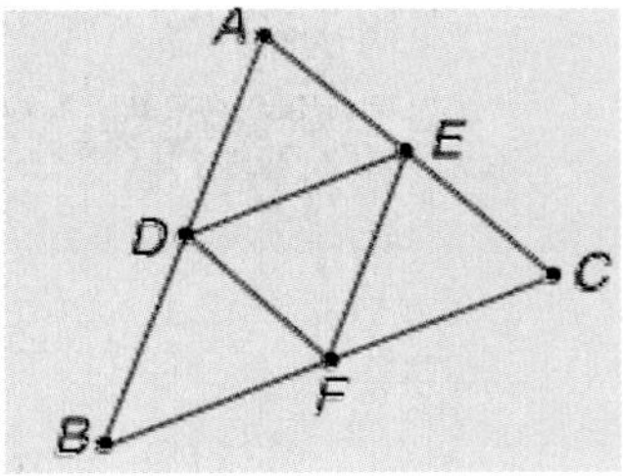

Connections Architecture . . . and more

4-4 Midsegments of Triangles

What You'll Learn

- Using properties of midsegments to solve problems

...And Why

- To help you find lengths and distances indirectly

What You'll Need

- scissors
- straightedge

WORK TOGETHER

Have each member of your group draw and cut out a large scalene triangle. Be sure to include right, acute, and obtuse triangles.

- Label the vertices of your triangle A, B, and C.

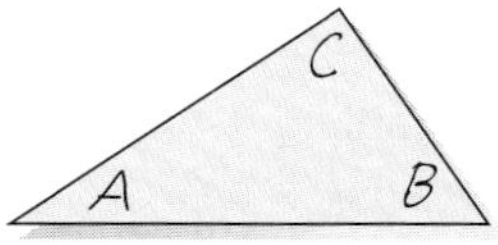

- Fold A onto C to find the midpoint of $\overline{AC}$. Do the same for $\overline{BC}$. Label the midpoints L and N, then draw $\overline{LN}$.

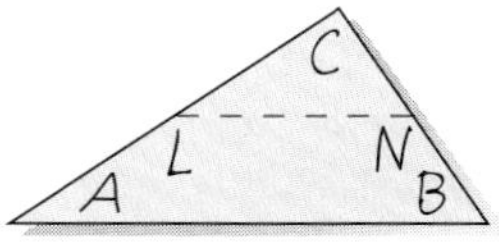

- Fold your triangle on $\overline{LN}$.

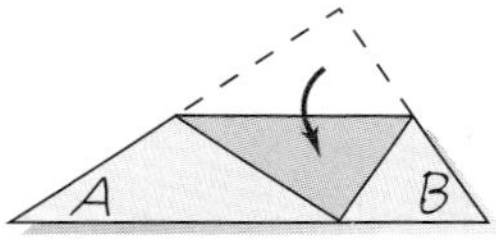

- Fold A to C. Do the same for B.

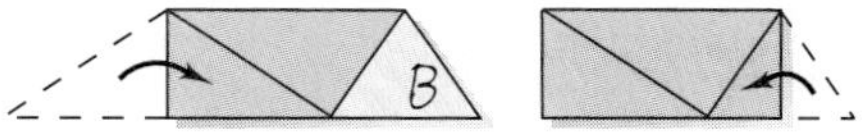

1. a. What type of quadrilateral does the folded triangle appear to form?
   b. What does this tell you about $\overline{LN}$ and $\overline{AB}$?
2. How does LN compare to AB? Explain.
3. Make a **conjecture** about how the segment joining the midpoints of two sides of a triangle is related to the third side of the triangle.

THINK AND DISCUSS

The segment you constructed in the Work Together is a midsegment. A **midsegment** of a triangle is a segment connecting the midpoints of two of its sides.

Theorem 4-8
Triangle Midsegment Theorem

If a segment joins the midpoints of two sides of a triangle, then the segment is parallel to the third side and half its length.

You can prove the Triangle Midsegment Theorem by using coordinate geometry and algebra. This style of proof is called a *coordinate proof.* You begin the proof by placing a triangle in a convenient spot on the coordinate plane. You then choose variables for the coordinates of the vertices.

Coordinate Proof of Theorem 4-8

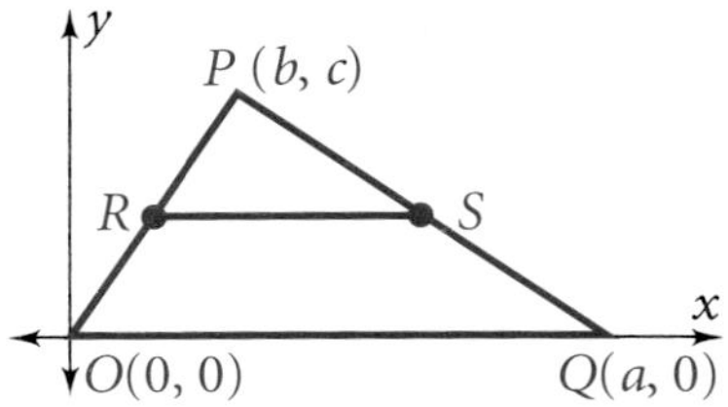

Given: R is the midpoint of $\overline{OP}$.
S is the midpoint of $\overline{QP}$.

Prove: $\overline{RS} \parallel \overline{OQ}$ and $RS = \frac{1}{2}OQ$

- Use the Midpoint Formula to find the coordinates of R and S.

$$R: \left(\frac{0+b}{2}, \frac{0+c}{2}\right) = \left(\frac{b}{2}, \frac{c}{2}\right) \qquad S: \left(\frac{a+b}{2}, \frac{0+c}{2}\right) = \left(\frac{a+b}{2}, \frac{c}{2}\right)$$

- To prove that $\overline{RS}$ and $\overline{OQ}$ are parallel, show that their slopes are equal. Because the y-coordinates of R and S are the same, the slope of $\overline{RS}$ is zero. The same is true for $\overline{OQ}$. Therefore, $\overline{RS} \parallel \overline{OQ}$.
- Use the Distance Formula to find RS and OQ.

$$RS = \sqrt{\left(\frac{a+b}{2} - \frac{b}{2}\right)^2 + \left(\frac{c}{2} - \frac{c}{2}\right)^2}$$

$$= \sqrt{\left(\frac{a}{2} + \frac{b}{2} - \frac{b}{2}\right)^2 + 0^2} = \sqrt{\left(\frac{a}{2}\right)^2} = \frac{a}{2} = \frac{1}{2}a$$

$$OQ = \sqrt{(a-0)^2 + (0-0)^2}$$

$$= \sqrt{a^2 + 0^2} = a$$

Therefore, $RS = \frac{1}{2}OQ$.

QUICK REVIEW

Midpoint Formula:
$\left(\frac{x_1 + x_2}{2}, \frac{y_1 + y_2}{2}\right)$

Distance Formula:
$\sqrt{(x_2 - x_1)^2 + (y_2 - y_1)^2}$

For practice with simplifying radicals, see Skills Handbook page 660.

4. Select numerical values for the coordinates of P and Q.
  a. Use the Midpoint Formula to find the coordinates of R and S.
  b. **Verify** that $RS = \frac{1}{2}OQ$ for the values you chose.
  c. **Verify** that $\overline{RS} \parallel \overline{OQ}$ for the values you chose.

Example 1

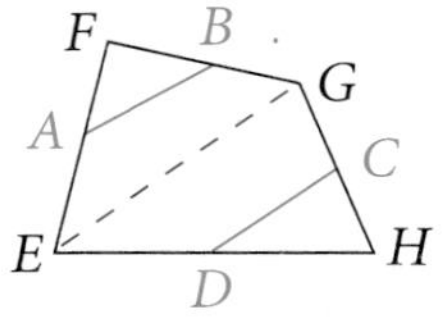

In quadrilateral $EFGH$, the points A, B, C, and D are midpoints and $EG = 18$ cm. Find AB and CD.

Consider $\triangle EFG$. By the Triangle Midsegment Theorem, $AB = \frac{1}{2}EG$. Using the same reasoning for $\triangle EHG$, you get $CD = \frac{1}{2}EG$. Because $EG = 18$ cm, $\frac{1}{2}EG = 9$ cm. Therefore, $AB = CD = 9$ cm.

5. *Critical Thinking* Is $\overline{AB} \parallel \overline{CD}$? **Justify** your answer.

6. **Try This** $FH = 25$ cm. Find BC and AD.

7. *Critical Thinking* What type of quadrilateral is $ABCD$? Explain.

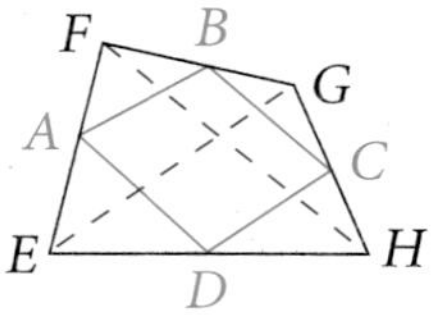

You can use the Triangle Midsegment Theorem to find lengths of segments that might otherwise be difficult to measure.

Example 2 ***Relating to the Real World***

Indirect Measurement DeAndre swims the length of a lake and wants to know the distance he swam. Here is what he does to find out.

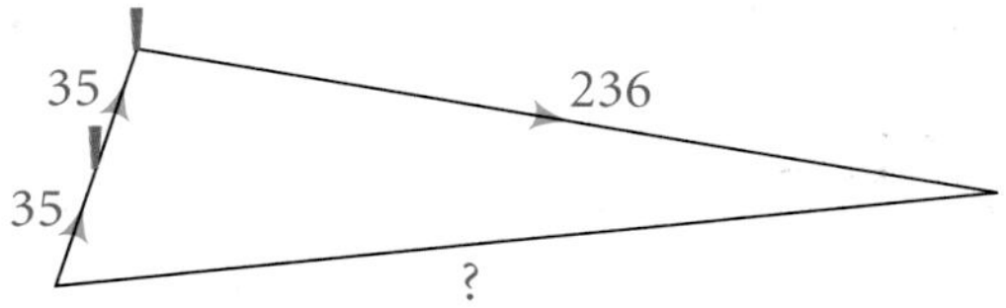

Step 1: From the edge of the lake, he paces 35 strides and sets a stake.

Step 2: He paces 35 more strides in the same direction and sets another stake.

Step 3: He paces to the other end of the lake, counting 236 strides.

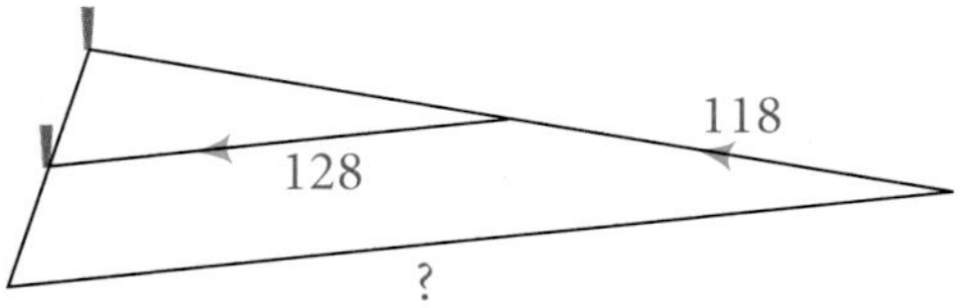

Step 4: He paces half the distance to the second stake (118 strides).

Step 5: He paces to the first stake, counting 128 strides.

If DeAndre's stride averages 3 ft, about how far did he swim?

2(128 strides) = 256 strides — Triangle Midsegment Theorem

$256 \text{ } \cancel{\text{strides}} \times \frac{3 \text{ ft}}{1 \text{ } \cancel{\text{stride}}} = 768 \text{ ft}$ — Convert strides to feet.

DeAndre swam approximately 768 ft.

Exercises ON YOUR OWN

Mental Math **Find the value of x.**

1.

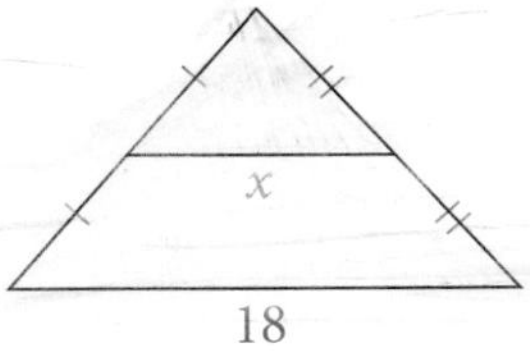

2.

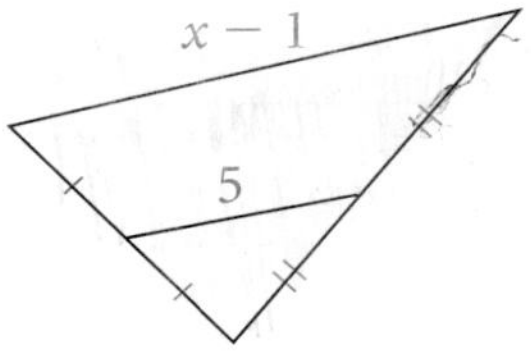

3.

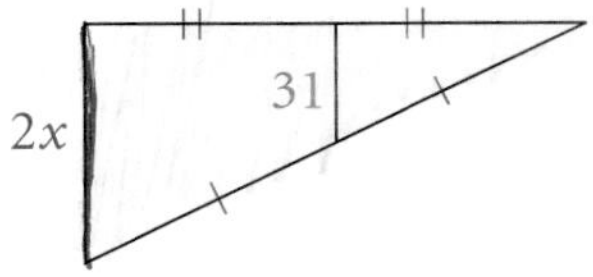

4. $\overline{IJ}$ is a midsegment of $\triangle FGH$.
   a. $IJ = 7$. Find FG.
   b. $FH = 13$ and $GH = 10$. Find the perimeter of $\triangle HIJ$.

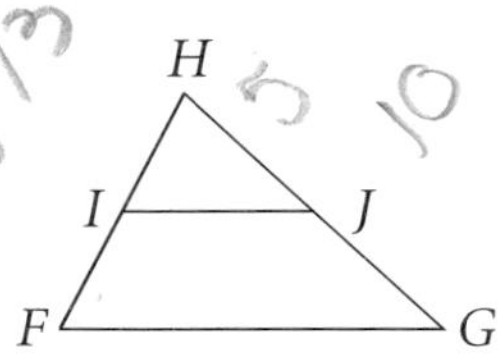

5. **Coordinate Geometry** The coordinates of the vertices of a triangle are $E(1, 2)$, $F(5, 6)$, and $G(3, -2)$.
   a. Find the coordinates of H, the midpoint of $\overline{EG}$, and J, the midpoint of $\overline{FG}$.
   b. **Verify** that $\overline{HJ} \parallel \overline{EF}$.
   c. **Verify** that $HJ = \frac{1}{2}EF$.

6. **Architecture** The triangular face of the Rock and Roll Hall of Fame in Cleveland, Ohio, is isosceles. The length of the base is 229 ft 6 in. The face consists of smaller triangles determined by the midsegments. What is the length of the base of the highlighted triangle?

7. a. Draw a triangle and label it $\triangle FST$. Then draw the midsegment opposite $\overline{ST}$.
   b. **Transformational Geometry** Describe a transformation of $\triangle FST$ that produces the same diagram as in part (a).
   c. What does your answer to part (b) tell you about the triangles formed in part (a)?

8. **Open-ended** Draw a triangle and its three midsegments. Compare the four triangles determined by the midsegments. Repeat the experiment with a different triangle. Make a **conjecture** about your observations.

Choose Use mental math, pencil and paper, or a calculator to find the values of the variables.

9.

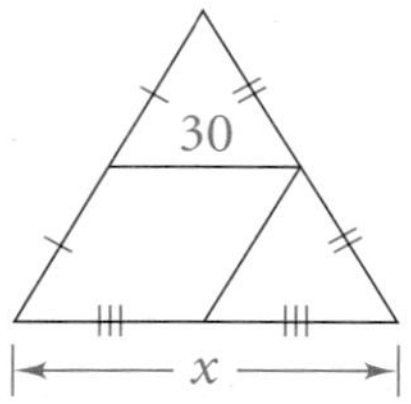

10.

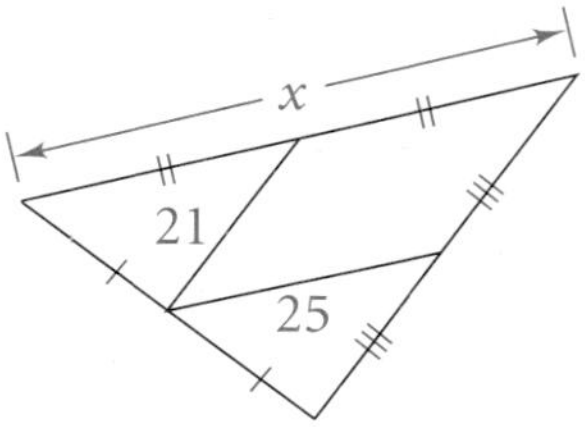

11.

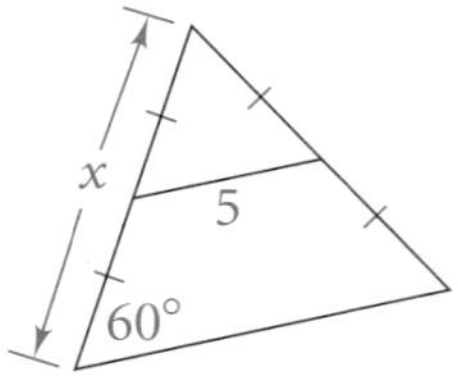

12.

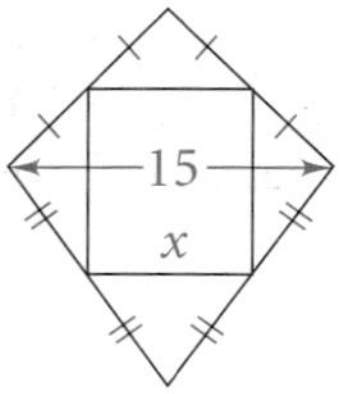

13.

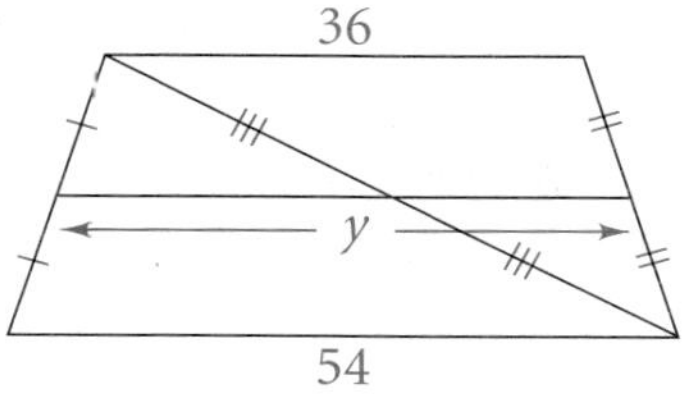

14.

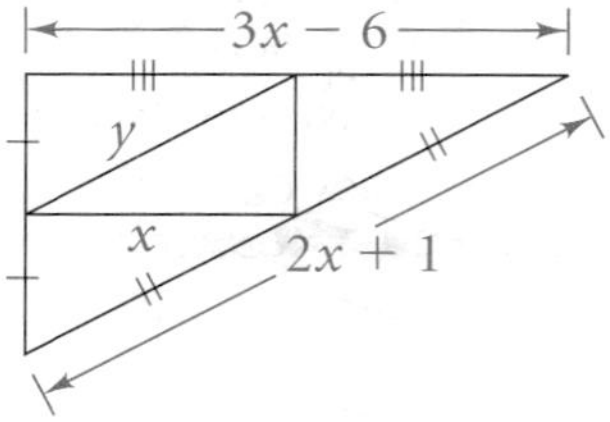

15. **Creative Arts** Marita is designing a kite for a competition. She plans to use a decorative ribbon to connect the midpoints of the sides of the kite. The diagonals of the kite measure 64 cm and 90 cm. Find the amount of ribbon she will need.

16. **Preparing for Proof** This is a proof that the midsegments of an equilateral triangle form an equilateral triangle. The reasons given in the proof are correct, but they are in the wrong order. List them in the correct order.

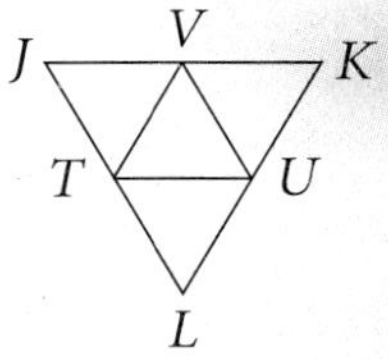

Given: Equilateral $\triangle JKL$ with midpoints T, U, and V

Prove: $\triangle TUV$ is equilateral.

Statements	Reasons
1. $\triangle JKL$ is equilateral.	a. Def. of equilateral triangle
2. $JK = KL = JL$	b. Multiplication Prop. of Equality
3. $\frac{1}{2}JK = \frac{1}{2}KL = \frac{1}{2}JL$	c. Substitution
4. T, U, and V are midpoints.	d. Def. of equilateral triangle
5. $TU = \frac{1}{2}JK$; $UV = \frac{1}{2}JL$; $TV = \frac{1}{2}KL$	e. Given
6. $TU = UV = TV$	f. Triangle Midsegment Thm.
7. $\triangle TUV$ is equilateral.	g. Given

17. **Patterns** The vertices of the smallest square are the midpoints of the sides of the larger square. The vertices of that square are the midpoints of the largest square. Find the length of the sides of the largest square. Write a sentence or two explaining the pattern.

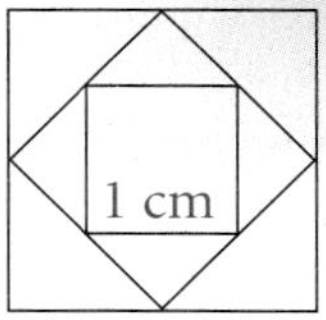

Chapter Project

Find Out by Analyzing

Try your powers of logic on this new version of an old puzzle.

Alan, Ben, and Cal are seated as shown with their eyes closed. Three hats are placed on their heads from a box they know contains 3 red and 2 blue hats. They open their eyes and look forward.

Alan says, "I cannot deduce what color hat I'm wearing."

Hearing that, Ben says, "I cannot deduce what color I'm wearing, either."

Cal then says, "I know what color I'm wearing!"

How does Cal know what color his hat is? (*Hint*: Use one of the strategies you used to solve the previous two puzzles on pages 187 and 198.)

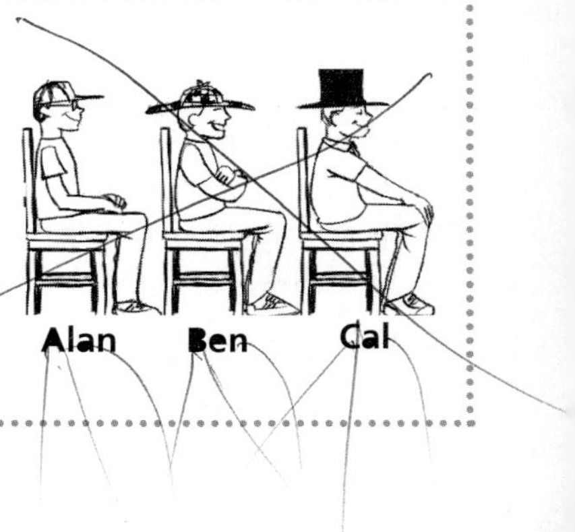

Exercises MIXED REVIEW

Coordinate Geometry **Find the distance between the given points.**

18. $B(0, 9)$ and $E(4, 9)$ **19.** $V(8, -7)$ and $W(4, -4)$ **20.** $A(3, 2)$ and $N(-1, 0)$

21. **Transformational Geometry** State the coordinates of the point (6, 2) after a rotation of 90° about the origin.

22. **Standardized Test Prep** A circle and a square lie in a plane. What is the maximum number of points of intersection of the figures?

A. one **B.** two **C.** four **D.** six **E.** eight

SELF ASSESSMENT

FOR YOUR JOURNAL

Make a list of the properties of triangles you've learned so far in this chapter. Include a diagram to illustrate each property. Add to your list as you continue with the chapter.

Getting Ready for Lesson 4-5

Write the negation of each statement.

23. Lines m and n intersect. **24.** The integer x is odd.

25. $\triangle ABC$ is scalene. **26.** $\overline{AB} \parallel \overline{KM}$

27. A given angle is neither acute, right, nor obtuse. What is its measure? Explain how you know.

Exercises CHECKPOINT

For Exercises 1 and 2: (a) Write the converse of each conditional. (b) Determine the truth value of the conditional and its converse. (c) If both statements are true, write a biconditional.

1. If a triangle is equilateral, then the measure of at least one of its angles is 60.

2. If a polygon is regular, then its angles are congruent.

3. **Writing** Explain the difference between a converse and a contrapositive.

4. **Standardized Test Prep** $\triangle ABC$ is isosceles, and $AC = 3x$, $AB = 2x + 30$, and $BC = 3x + 40$. Which of the following is true?

A. $\angle A \cong \angle B$ **B.** $\angle A \cong \angle C$ **C.** $\angle B \cong \angle C$

D. $\angle A \cong \angle B \cong \angle C$ **E.** none of the above

5. The measures of the angles of a triangle are $6x$, $7x - 9$, and $8x$.

a. **Algebra** Find the value of x.

b. What are the measures of the angles?

c. What type of triangle is this? Explain.

6. Determine whether this statement is true or false: If a triangle is obtuse, then it is not isosceles. **Justify** your answer.

7. M, N, O, and P are midpoints of the sides of trapezoid $ABCD$. $AC = BD = 18$. Find the perimeter of $MNOP$.

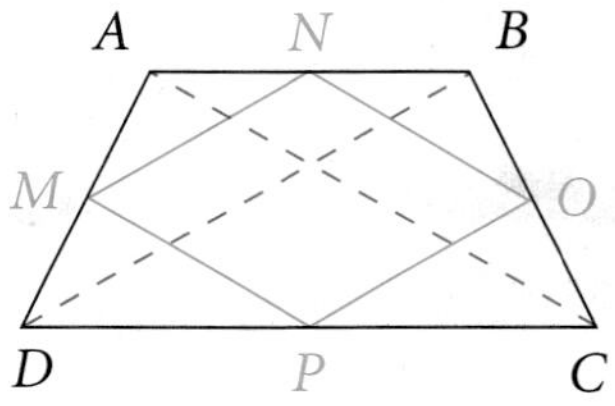

4-5 Using Indirect Reasoning

What You'll Learn

- Writing convincing arguments by using indirect reasoning

...And Why

To help you in real-life situations in which proving something directly isn't possible

THINK AND DISCUSS

Suppose that your brother tells you, "Susan called a few minutes ago." You have two friends named Susan, and you know that one of them is at play rehearsal. You deduce that the other Susan must be the caller.

This type of reasoning is called indirect reasoning. In **indirect reasoning,** all possibilities are considered and then all but one are proved false. The remaining possibility must be true. Mathematical proofs involving indirect reasoning usually follow the pattern in the following example.

Example 1 Relating to the Real World

Consumer Issues Use indirect reasoning to prove this statement: If Jaeleen spends more than $50 to buy two items at a bicycle shop, then at least one of the items costs more than $25.

Given: The cost of two items is more than $50.
Prove: At least one of the items costs more than $25.

- Begin by assuming that the opposite of what you want to prove is true. That is, assume that neither item costs more than $25.
- This means that both items cost $25 or less. This, in turn, means that the two items together cost $50 or less. This contradicts the given information that the amount spent is more than $50. So, the assumption that neither item costs more than $25 must be incorrect.
- Therefore, at least one of the items costs more than $25.

The three parts of the proof in Example 1 are summarized below.

Writing an Indirect Proof

Step 1: Assume that the opposite of what you want to prove is true.

Step 2: Use logical reasoning to reach a contradiction of an earlier statement, such as the given information or a theorem. Then state that the assumption you made was false.

Step 3: State that what you wanted to prove must be true.

1. Write the first step of an indirect proof of each statement.
  a. Quadrilateral *TRWX* does not have four acute angles.
  b. An integer *n* is divisible by 5.
  c. The shoes cost no more than $20.

2. Identify the pair of statements that form a contradiction.
  a. **I.** $\triangle ABC$ is acute.
   II. $\triangle ABC$ is scalene.
   III. $\triangle ABC$ is equiangular.
  b. **I.** $m\angle 1 + m\angle 2 = 180$
   II. $m\angle 1 - m\angle 2 = m\angle 2$
   III. $m\angle 1 \leq m\angle 2$
  c. **I.** Both items that Val bought cost more than $10.
   II. Val spent $34 for the two items.
   III. Neither of the two items that Val bought cost more than $15.

Example 2

Write an indirect proof.

Given: $\triangle LMN$
Prove: $\triangle LMN$ has at most one right angle.

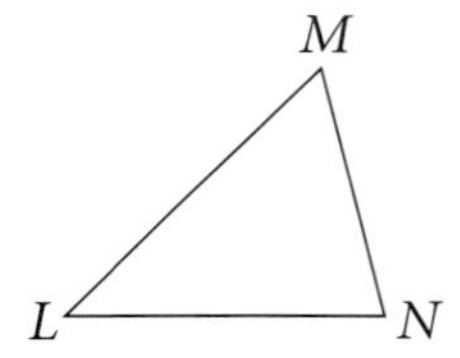

Indirect Proof

Step 1: Assume $\triangle LMN$ has more than one right angle. That is, assume that $\angle L$ and $\angle M$ are both right angles.

Step 2: If $\angle L$ and $\angle M$ are both right angles, then $m\angle L = m\angle M = 90$. According to the Triangle Angle-Sum Theorem, $m\angle L + m\angle M + m\angle N = 180$. Substitution gives $90 + 90 + m\angle N = 180$. Solving leaves $m\angle N = 0$. This means that there is no $\triangle LMN$, which contradicts the given statement. So the assumption that $\angle L$ and $\angle M$ are both right angles must be false.

Step 3: Therefore, $\triangle LMN$ has at most one right angle. ■

3. **Try This** Use indirect reasoning to show that a parallelogram with three right angles is a rectangle.

WORK TOGETHER

Play *What's My Number* in groups of three or four. Here's how to play.

- One member of the group chooses a number from 1 to 20.
- The remaining members ask yes-or-no questions about the number until they know what the number is.
- The person answering the questions records the number of questions required to guess the number.

Play the game at least three times, rotating roles with each game.

4. **Critical Thinking** Describe the best strategy for playing *What's My Number* to a friend who is just learning to play it.

Exercises ON YOUR OWN

Write the first step of an indirect proof of each statement.

1. It is raining outside.

2. $\angle J$ is not a right angle.

3. $\triangle PEN$ is isosceles.

4. At least one angle is obtuse.

5. $\overline{XY} \cong \overline{AB}$

6. $m\angle 2 > 90$

Identify the pair of statements that forms a contradiction.

7.
I. $\triangle PQR$ is equilateral.
II. $\triangle PQR$ is a right triangle.
III. $\triangle PQR$ is isosceles.

8.
I. *ABCD* is a parallelogram.
II. *ABCD* is a trapezoid.
III. *ABCD* has two acute angles.

9.
I. ℓ and m are skew.
II. ℓ and m do not intersect.
III. ℓ is parallel to m.

10.
I. $\overline{FG} \parallel \overline{KL}$
II. $\overline{FG} \perp \overline{KL}$
III. $\overline{FG} \cong \overline{KL}$

What conclusion follows from each pair of statements?

11. There are three types of drawbridges: bascule, lift, and swing. This drawbridge does not swing or lift.

12. If this were the day of the party, our friends would be home. No one is home.

13. Every air traffic controller in the world speaks English on the job. Sumiko does not speak English.

14. If two nonvertical lines are perpendicular, then the product of their slopes is -1. The product of the slopes of nonvertical lines ℓ and n is not -1.

15. Given $\triangle ABC$ with $BC > AC$, use indirect reasoning to show that $\angle A \not\cong \angle B$.

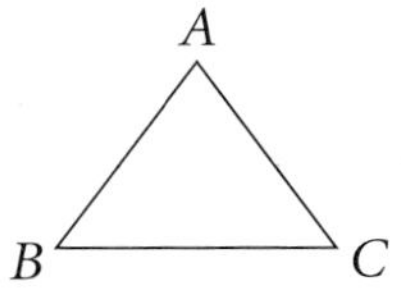

16. *Preparing for Proof* Complete this indirect proof that every quadrilateral contains at least one right or acute angle.

Assume that a quadrilateral does not contain **a.** _?_. That is, assume that the measure of each of the angles is greater than **b.** _?_. The sum of the measures of the four angles, then, is greater than **c.** _?_. By the **d.** _?_ Theorem, however, the sum of the measures of the interior angles of a quadrilateral is **e.** _?_. This contradicts **f.** _?_, so the assumption that **g.** _?_ is false. Therefore, **h.** _?_.

17. *Standardized Test Prep* Which of the following represents the measure of an interior angle of a regular polygon?

A. 30 **B.** 50 **C.** 70 **D.** 100 **E.** 140

PROBLEM SOLVING HINT

Use indirect reasoning. Eliminate all incorrect answer choices. The remaining choice must be correct.

18. Earlene lives near a noisy construction site at which work ends promptly at 5:00 each weekday. Earlene thinks, "Today is Tuesday. If it were before 5:00, I would hear construction noise, but I don't hear any. So it must be later than 5:00."

a. What does Earlene prove?
b. What assumption does she make?
c. What fact contradicts the assumption?

For Exercises 19–25, write a convincing argument that uses indirect reasoning.

19. Fresh skid marks appear behind a green car at the scene of an accident. Show that the driver of the green car applied the brakes.

20. Ice is forming on the sidewalk in front of Toni's house. Show that the temperature outside must be 32°F or less.

21. The sum of the measures of the interior angles of a polygon is 900. Show that the polygon is not a hexagon.

22. Show that a quadrilateral can have at most three acute angles.

23. An obtuse triangle cannot contain a right angle.

Given: $\triangle PQR$ with obtuse $\angle Q$

Prove: $m\angle P \neq 90$

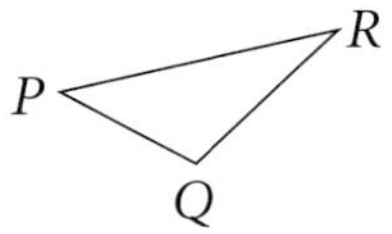

24. If a triangle is isosceles, then a base angle is not a right angle.

Given: $\overline{BC} \cong \overline{AC}$

Prove: $\angle B$ is not a right angle.

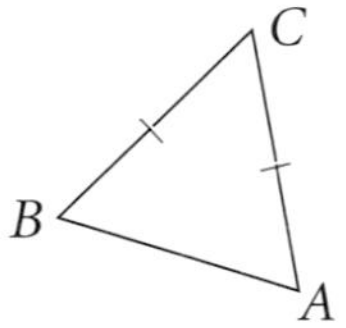

25. Mr. Pitt is a suspect in a robbery. Here are the facts:

- A robbery occurred in Charlotte, North Carolina, at 1:00 A.M. on April 9.
- A hotel clerk in Maine saw Mr. Pitt at 12:05 A.M. on April 9.
- Mr. Pitt has an identical twin brother who was in Chicago during the time the robbery took place.
- Mr. Pitt has receipts for purchases made in Maine on April 8.

Show that Mr. Pitt was not the robber.

26. Open-ended Describe a real-life situation in which you used an indirect argument to convince someone of your point of view. Outline your argument.

27. Mysteries In Arthur Conan Doyle's story "The Sign of the Four," Sherlock Holmes talks to his sidekick Watson about how a culprit enters a room that has only four entrances: a door, a window, a chimney, and a hole in the roof.

"You will not apply my precept," he said, shaking his head. "How often have I said to you that when you have eliminated the impossible, whatever remains, *however improbable,* must be the truth? We know that he did not come through the door, the window, or the chimney. We also know that he could not have been concealed in the room, as there is no concealment possible. Whence, then, did he come?"

How did the culprit enter the room? Explain.

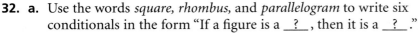

Exercises MIXED REVIEW

Geometry in 3 Dimensions **Refer to the diagram for Exercises 28–31.**

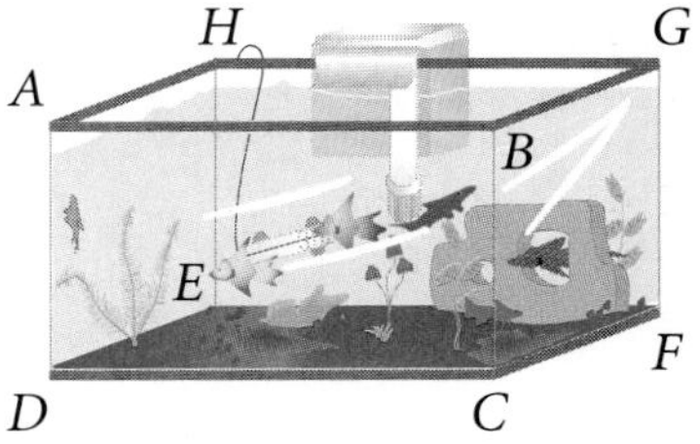

28. Name two parallel planes.

29. Name two intersecting planes.

30. Name two skew lines.

31. Name four coplanar points.

32. a. Use the words *square, rhombus,* and *parallelogram* to write six conditionals in the form "If a figure is a __?__, then it is a __?__."
   b. Probability What is the probability that a conditional chosen at random from among the six is true?
   c. Probability What is the probability that the converse of a randomly chosen statement from part (a) is true?
   d. Probability What is the probability that the contrapositive of a randomly chosen statement from part (a) is true?

33. Transformational Geometry An isometry maps $LEFT \longrightarrow BURN$. Which statement is *not* necessarily true?
   A. $EF = UR$
   B. $\angle TLE \cong \angle BUR$
   C. $\angle F \cong \angle R$
   D. $LF = BR$
   E. $FT = RN$

Getting Ready for Lesson 4-6

Coordinate Geometry **Graph the triangles whose vertices are given. List the sides in order from shortest to longest.**

34. $A(5, 0), B(0, 8), C(0, 0)$

35. $P(2, 4), Q(-5, 1), R(0, 0)$

36. $G(3, 0), H(4, 3), J(8, 0)$

Algebra Review

Solving Inequalities

Before Lesson 4-6

The solutions of an inequality are all the numbers that make the inequality true. Below is a review of the Properties of Inequality. To solve inequalities you will use the Addition and Multiplication Properties of Inequality.

Properties of Inequality

For all real numbers a, b, c, and d:

Addition	If $a > b$ and $c \geq d$, then $a + c > b + d$.
Multiplication	If $a > b$ and $c > 0$, then $ac > bc$. If $a > b$ and $c < 0$, then $ac < bc$.
Transitive	If $a > b$ and $b > c$, then $a > c$.
Comparison	If $a = b + c$ and $c > 0$, then $a > b$.

Example

Solve $-6x + 7 > 25$.

$-6x + 7 - 7 > 25 - 7$ ← Add −7 to each side (or subtract 7 from each side).

$\frac{-6x}{-6} < \frac{18}{-6}$ ← Multiply each side by $-\frac{1}{6}$ (or divide each side by −6). Remember to reverse the order of the inequality.

$x < -3$ ← Simplify.

Solve each inequality.

1. $7x - 13 \leq -20$
2. $3z + 8 > 16$
3. $-2x + 5 < 16$
4. $8y + 2 \geq -14$
5. $5a + 1 \leq 91$
6. $-x - 2 > 17$
7. $-4z - 10 < -12$
8. $9x - 8 \geq 82$
9. $6n + 3 \leq -18$
10. $c + 13 > 34$
11. $3x - 5x + 2 < 12$
12. $x - 19 < -78$
13. $-n - 27 \leq 92$
14. $-9t + 47 < 101$
15. $8x - 4 + x > -76$
16. $2(y - 5) > -24$
17. $8b + 3 \geq 67$
18. $-3(4x - 1) \geq 15$
19. $r - 9 \leq -67$
20. $\frac{1}{2}(4x - 7) \geq 19$
21. $5x - 3x + 2x < -20$
22. $9x - 10x + 4 < 12$
23. $-3x - 7x \leq 97$
24. $8y - 33 > -1$
25. $4a + 17 \geq 13$
26. $-4(5z + 2) > 20$
27. $x + 78 \geq -284$

What You'll Learn

- Using inequalities involving triangle side lengths and angle measures to solve problems

...And Why

To use triangle inequalities in solving real-world problems when only some measures of a triangle are known

What You'll Need

scissors, straws, ruler, protractor

4-6 Triangle Inequalities

WORK TOGETHER

Work in groups of three. Have each member of your group cut straws into 2-, 3-, 4-, 5-, and 6-in. segments.

- Have each member pick three segments at random and test whether they form a triangle. Continue picking segments until you find three sets of segments that form a triangle and three that do not form a triangle. Record your results in a table like the one shown.

Lengths of Segments			Triangle Formed?
No. 1	No. 2	No. 3	
2	3	4	Yes
2	3	5	No
5	4	3	Yes

1. Pick a row of data that yields a triangle. Compare each quantity.
   a. Segment 1 + Segment 2 __?__ Segment 3
   b. Segment 1 + Segment 3 __?__ Segment 2
   c. Segment 2 + Segment 3 __?__ Segment 1
2. Pick a row of data that does *not* yield a triangle. Use it to complete parts (a)–(c) of Question 1.
3. Patterns Compare the results of Questions 1 and 2 within your group. Look for a pattern. Write a **conjecture** about the sum of the lengths of two sides of a triangle compared to the length of the third side.

THINK AND DISCUSS

Triangle Inequality Theorem

Your observations from the Work Together suggest the following theorem.

Theorem 4-9
Triangle Inequality Theorem

The sum of the lengths of any two sides of a triangle is greater than the length of the third side.

$XY + YZ > XZ$
$YZ + XZ > XY$
$XZ + XY > YZ$

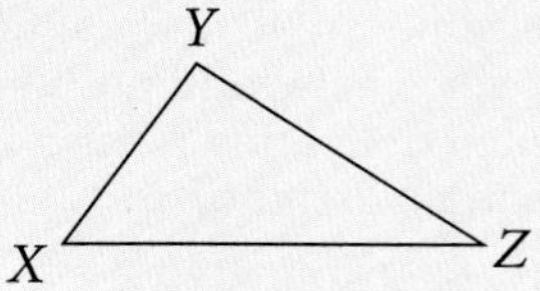

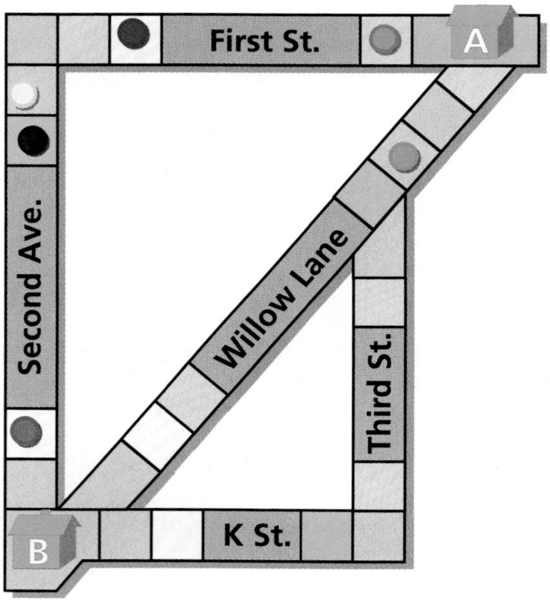

4. Use the Triangle Inequality Theorem to explain which route is the shortest distance from House *A* to House *B*.

Example 1

Is it possible for a triangle to have sides with the given lengths? Explain.

a. 3 cm, 7 cm, and 8 cm
$3 + 7 > 8$
$8 + 7 > 3$
$3 + 8 > 7$ **Yes**

b. 3 ft, 6 ft, and 10 ft
$3 + 6 \not> 10$ **No**

The sum of any two numbers in (a) is greater than the third number. In part (b), the sum of 3 and 6 is less than 10.

5. Try This Is it possible to form a triangle with side lengths 4 cm, 6 cm, and 10 cm? Explain.

WORK TOGETHER

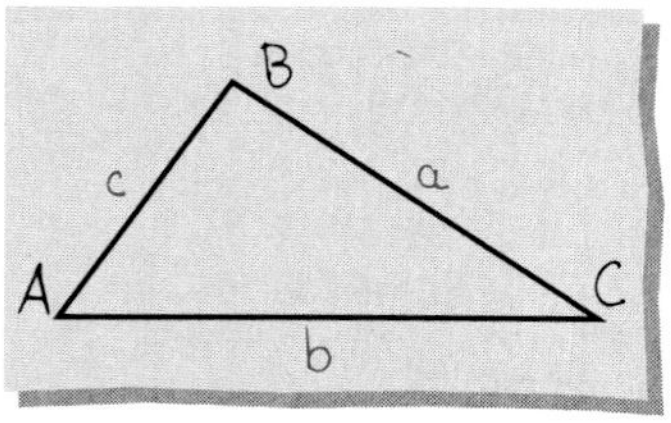

Work in groups of three. Have each member of your group draw a large scalene triangle. Label each triangle as shown. Include right, obtuse, and acute triangles.

- Measure the sides and angles of each triangle.

6. a. Use the letters *A*, *B*, and *C* to complete:
$m\angle \blacksquare < m\angle \blacksquare < m\angle \blacksquare$.

b. Use the letters *a*, *b*, and *c* to complete:
$\blacksquare < \blacksquare < \blacksquare$.

7. Compare your answer to Question 6 with others in your group. Make a **conjecture** about the longest side and the largest angle in a triangle.

THINK AND DISCUSS

Inequalities Relating Sides and Angles

Your observations in the Work Together lead to Theorems 4-10 and 4-11.

Theorem 4-10

If two sides of a triangle are not congruent, then the larger angle lies opposite the longer side.

If $XZ > XY$, then $m\angle Y > m\angle Z$.

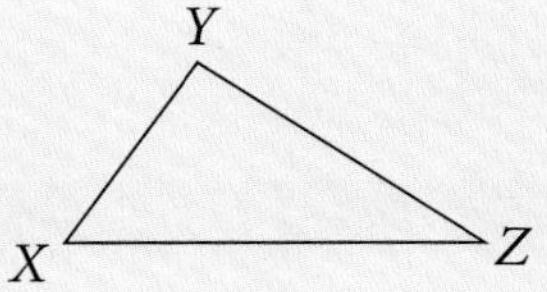

You will justify Theorem 4-10 in Exercise 35.

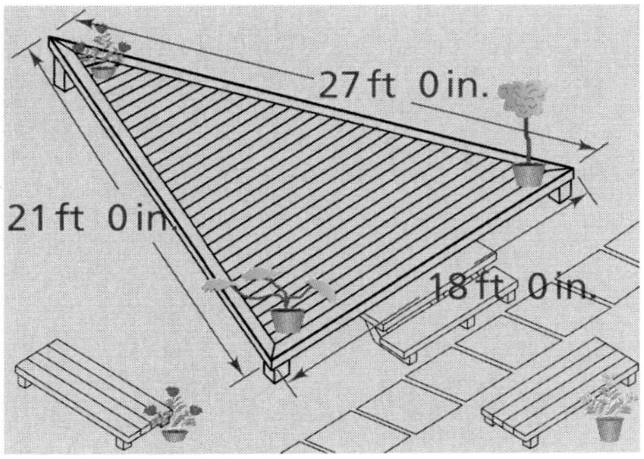

Example 2 **Relating to the Real World**

Architecture A landscape architect is designing a triangular deck. She wants to place benches in the two largest corners. In which corners should she place the benches?

The two largest corners are opposite the two longest sides, 27 ft and 21 ft.

8. Try This The sides of a triangle are 14 in., 7 in., and 8 in. long. The smallest angle is opposite which side?

Theorem 4-11

If two angles of a triangle are not congruent, then the longer side lies opposite the larger angle.

Indirect Proof of Theorem 4-11

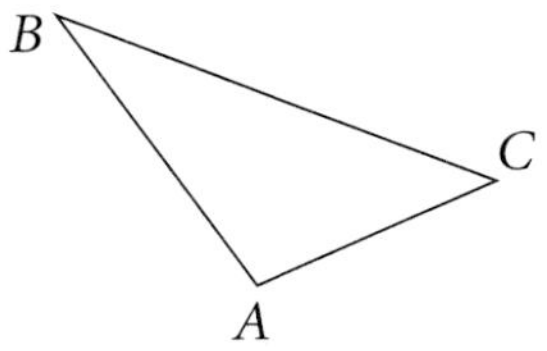

Given: $m\angle A > m\angle B$
Prove: $BC > AC$

Step 1: Assume $BC \not> AC$. That is, assume $BC < AC$ or $BC = AC$.

Step 2: If $BC < AC$, then $m\angle A < m\angle B$ (Theorem 4-10). That contradicts the given fact that $m\angle A > m\angle B$. Therefore, the assumption that $BC < AC$ must be false.

If $BC = AC$, then $m\angle A = m\angle B$ (Isosceles Triangle Theorem). This also contradicts the given fact that $m\angle A > m\angle B$. Therefore the assumption that $BC = AC$ must be false.

Step 3: Therefore, $BC > AC$ must be true.

Example 3

In $\triangle TUV$, which side is shortest?

By the Triangle Angle-Sum Theorem, $m\angle T = 60$. The smallest angle in $\triangle TUV$ is $\angle U$. Therefore, by Theorem 4-11, the shortest side is $\overline{TV}$.

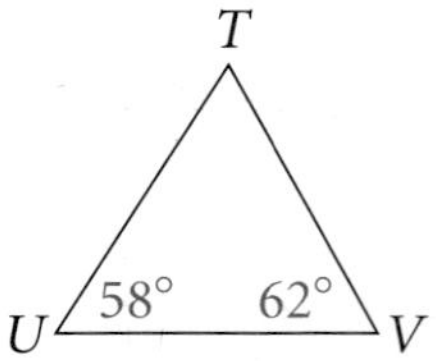

Exercises ON YOUR OWN

Is it possible for a triangle to have sides with the given lengths? Explain.

1. 2 in., 3 in., 6 in.
2. 11 cm, 12 cm, 15 cm
3. 6 ft, 10 ft, 13 ft
4. 8 m, 10 m, 19 m
5. 2 yd, 9 yd, 10 yd
6. 4 m, 7 m, 9 m
7. 5 in., 5 in., 5 in.
8. 1 cm, 15 cm, 15 cm

List the sides of each triangle in order from shortest to longest.

9.
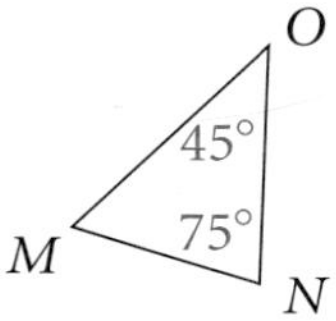

10.
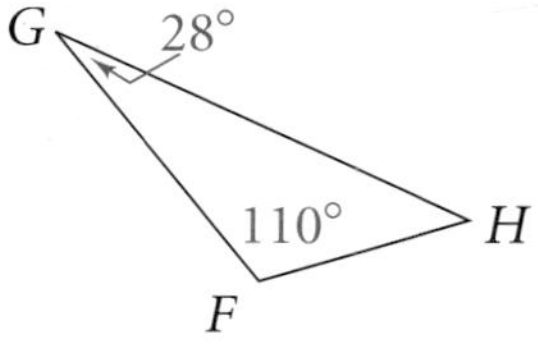

11.
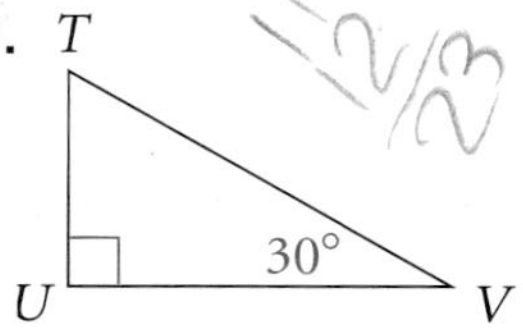

12.
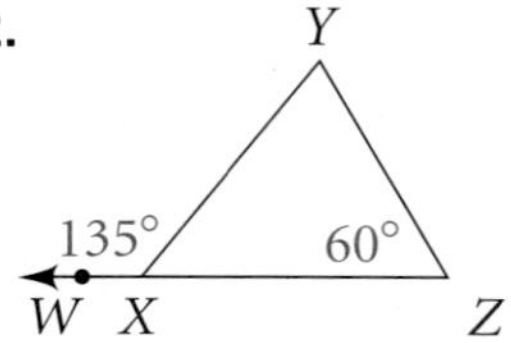

13. $\triangle ARK$, where $m\angle A = 90$, $m\angle R = 40$, and $m\angle K = 50$

14. $\triangle INK$, where $m\angle I = 20$, $m\angle N = 120$, and $m\angle K = 40$

List the angles of each triangle in order from smallest to largest.

15.
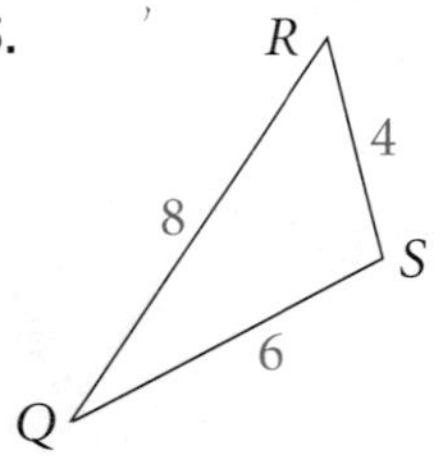

16.
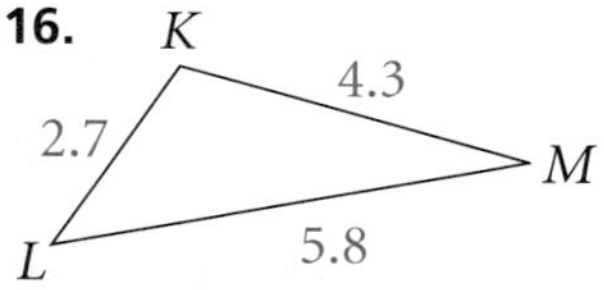

17.
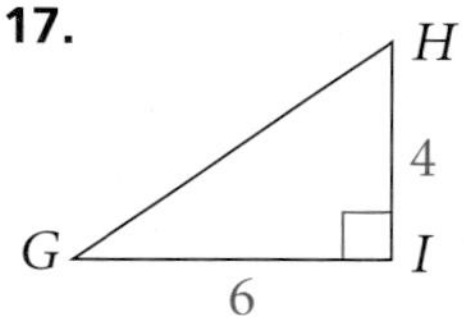

18.
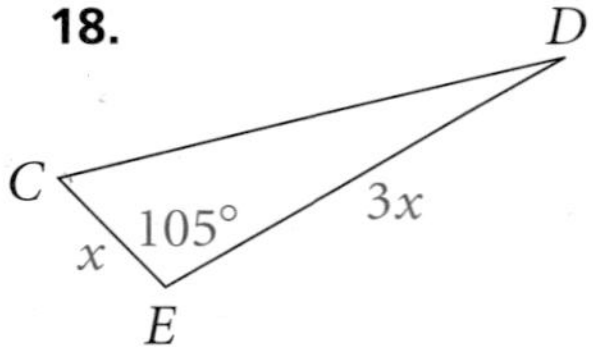

19. $\triangle ABC$, where $AB = 8$, $BC = 5$, and $AC = 7$

20. $\triangle DEF$, where $DE = 15$, $EF = 18$, and $DF = 5$

21. $\triangle XYZ$, where $XY = 2$, $YZ = 4$, and $XZ = 3$

22. **Geometry in 3 Dimensions** Refer to the pyramid shown. Explain why the altitude, $\overline{PT}$, must be shorter than edges $\overline{PA}$, $\overline{PB}$, $\overline{PC}$, and $\overline{PD}$.

23. **Probability** A student picks two straws, one 6 cm long and the other 9 cm long. She picks another straw at random from a group of four straws whose lengths are 3 cm, 5 cm, 11 cm, and 15 cm. What is the probability that the straw she picks will allow her to form a triangle?

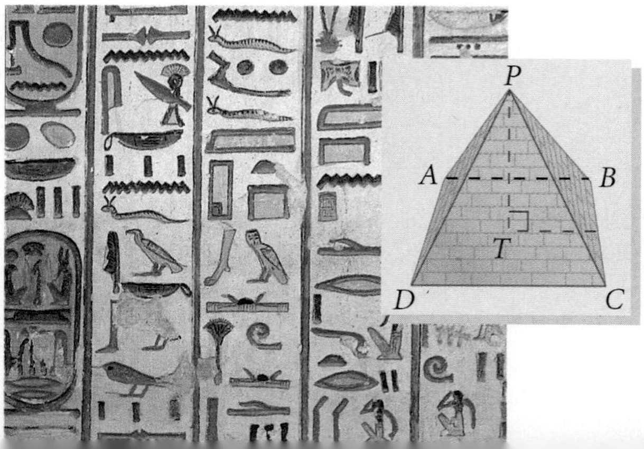

24. *Standardized Test Prep* Compare the quantities in Columns A and B. Select the best answer.

Column A	Column B
a	b

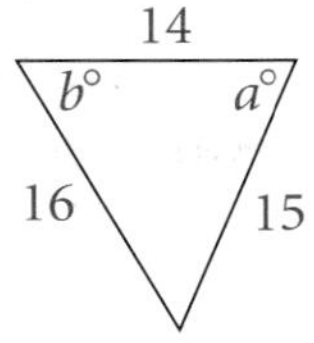

A. The quantity in Column A is greater.
B. The quantity in Column B is greater.
C. The quantities are equal.
D. The relationship cannot be determined from the information given.

25. *Technology* Darren used geometry software to draw $\triangle ACF$ and $\triangle DEH$ so that $\overline{AC} \cong \overline{DE}$ and $\overline{AF} \cong \overline{DH}$. He then manipulated point H to collect the data in the spreadsheet.

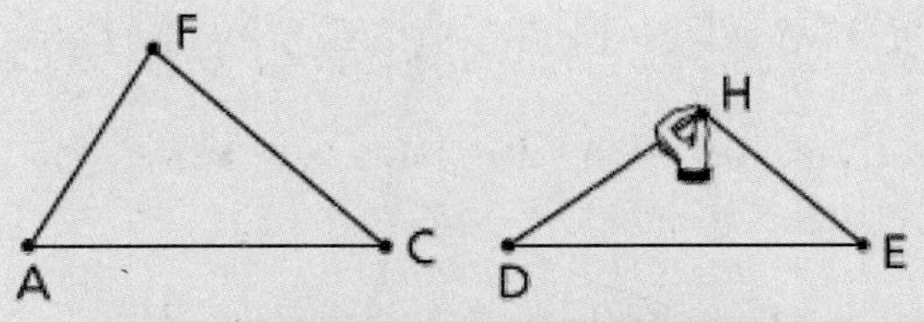

	∠FAC	∠HDE	FC	HE
1	58.3	104.3	3.21	4.98
2	58.3	94.2	3.21	4.64
3	58.3	65.2	3.21	3.50
4	58.3	50.7	3.21	2.88
5	58.3	34.7	3.21	2.18

a. Compare these values for each row of the spreadsheet.
$m\angle FAC$ __?__ $m\angle HDE$ FC __?__ HE

b. *Patterns* What pattern do you notice in part (a)?

c. The pattern that you noticed is stated as the Hinge Theorem. Complete the theorem:
Hinge Theorem *If two sides of one triangle are congruent to two sides of another triangle, and the included angle of the first triangle is greater than the included angle of the second triangle, then* __?__.

26. *Critical Thinking* Pliers, scissors, and many other tools are hinged. When the angle between the blades of a pair of scissors increases, what happens to the distance between the tips of the blades? Explain how this relates to the Hinge Theorem in Exercise 25.

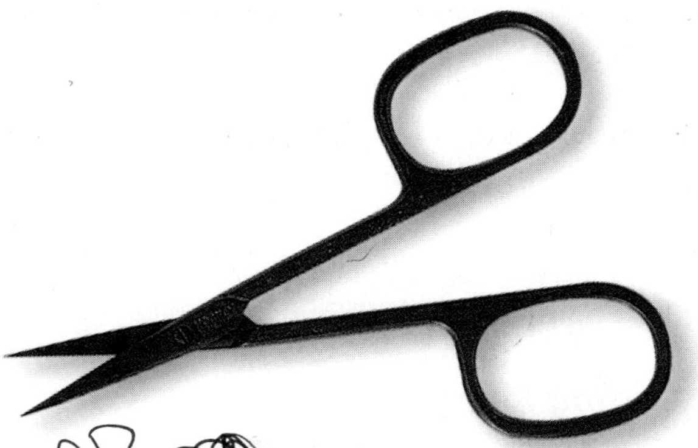

27. *Critical Thinking* The Shau family is crossing Kansas on Highway 70. A sign reads "Topeka 110 miles, Wichita 90 miles." Avi says, "I didn't know that it was only 20 miles from Topeka to Wichita." Explain to Avi why the distance between the two cities doesn't have to be 20 miles.

Critical Thinking **Which segment is the shortest?**

28.

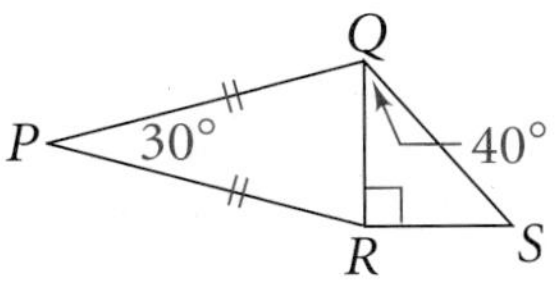

29.

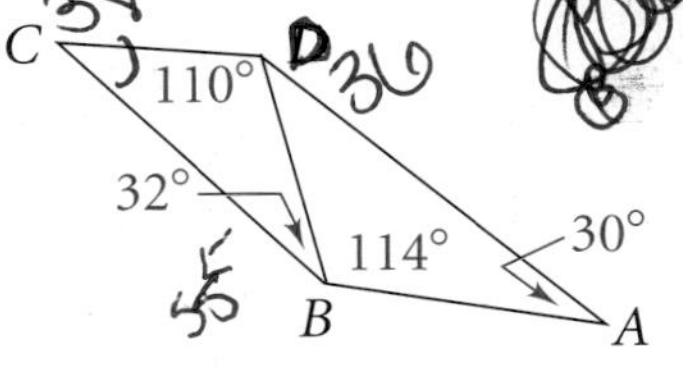

30.

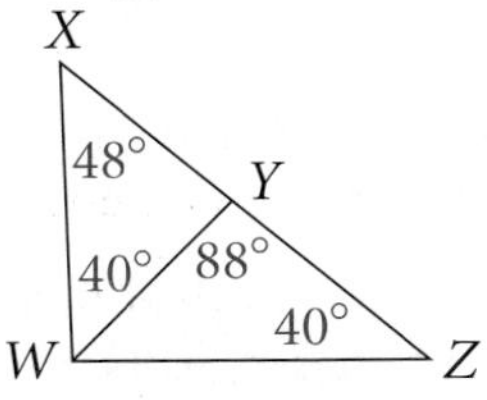

The lengths of two sides of a triangle are given. Write an inequality to represent the range of values for *z*, the length of the third side.

31. 8 ft, 12 ft **32.** 5 in., 16 in. **33.** 6 cm, 6 cm **34.** x, y, where $x \geq y$

35. *Preparing for Proof* Theorem 4-10 states: If two sides of a triangle are not congruent, then the larger angle lies opposite the longer side. To prove this theorem, begin with $\triangle TOY$, with $OY > TY$. Find P on $\overline{OY}$ so that $\overline{TY} \cong \overline{PY}$. Draw $\overline{TP}$. Supply a reason for each statement.

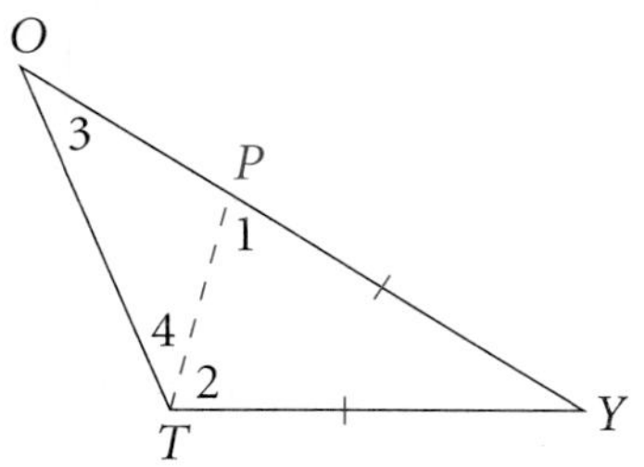

Given: $OY > TY$ and $\overline{TY} \cong \overline{PY}$

Prove: $m\angle OTY > m\angle 3$

Statements	Reasons
1. $\overline{YP} \cong \overline{YT}$	**a.** ?
2. $m\angle 1 = m\angle 2$	**b.** ?
3. $m\angle OTY = m\angle 4 + m\angle 2$	**c.** ?
4. $m\angle OTY > m\angle 2$	**4.** Comparison Prop. of Ineq. (p. 212)
5. $m\angle OTY > m\angle 1$	**d.** ?
6. $m\angle 1 = m\angle 3 + m\angle 4$	**e.** ?
7. $m\angle 1 > m\angle 3$	**7.** Comparison Prop. of Ineq.
8. $m\angle OTY > m\angle 3$	**f.** ?

36. *Logical Reasoning* A corollary to Theorem 4-11 states: The perpendicular segment from a point to a line is the shortest segment from the point to the line. Show that $PA > PT$, given that $\overline{PT} \perp \overline{TA}$.

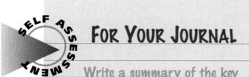

Exercises MIXED REVIEW

37. *Transformational Geometry* $\triangle SKY$ has vertices $S(-1, 0)$, $K(3, 8)$, and $Y(5, 4)$. A translation maps K to $K'(2, 1)$. Find the coordinates of the images of S and Y under this translation.

38. *Open-ended* Use a rhombus to create a tessellation.

39. Find the measure of an interior angle and an exterior angle of a regular 15-gon.

SELF ASSESSMENT

FOR YOUR JOURNAL

Write a summary of the key points of this lesson. Include diagrams with your summary where appropriate.

Getting Ready for Lesson 4-7

Describe the set of red points in terms of their distance from the set of blue points.

40.

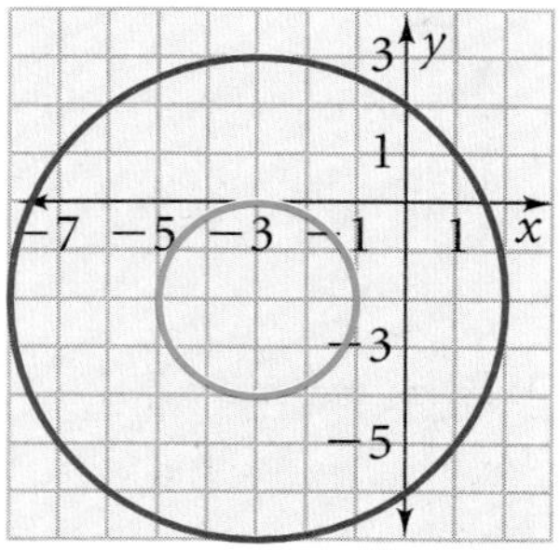

41.

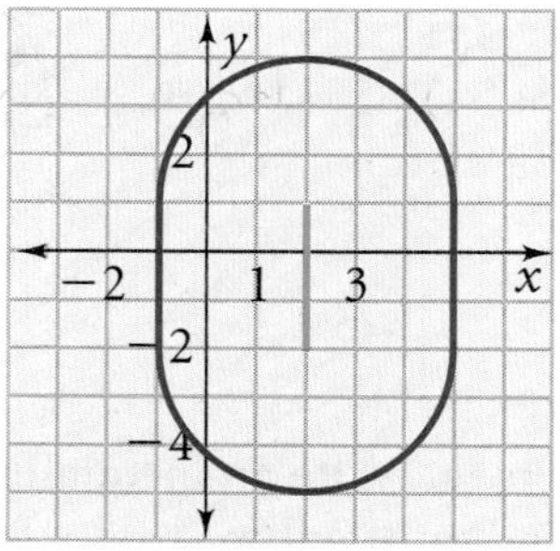

4-7 Bisectors and Locus

What You'll Learn

- Using properties of angle bisectors and perpendicular bisectors
- Solving locus problems

...And Why

To learn methods used by circuit designers and many other professionals who solve problems in which stated conditions must be met

What You'll Need

straightedge, compass, ruler, MIRA™

TECHNOLOGY HINT

The Work Together could be done using geometry software.

WORK TOGETHER

- Draw a segment and construct its perpendicular bisector by using a straightedge and a compass or a MIRA™.
- Draw a point on the perpendicular bisector and then draw segments connecting this point to the endpoints of the segment.

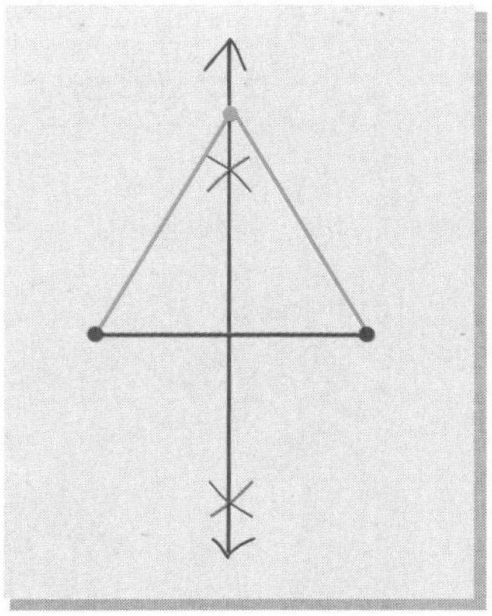

1. Compare the lengths of the segments. What do you notice?
2. Repeat this process with some other points. Does what you noticed still hold true?
3. Use what you discovered to complete this statement: If a point is on the perpendicular bisector of a segment, then __?__ .
4. **a.** Write the converse of the statement you completed in Question 3.
   b. Explain how the converse is related to the method you use to construct a perpendicular bisector.

THINK AND DISCUSS

Perpendicular Bisectors and Locus

The properties you discovered in the Work Together are summarized in the following theorems.

Theorem 4-12 **Perpendicular Bisector Theorem**

If a point is on the perpendicular bisector of a segment, then it is equidistant from the endpoints of the segment.

Theorem 4-13 **Converse of Perpendicular Bisector Theorem**

If a point is equidistant from the endpoints of a segment, then it is on the perpendicular bisector of the segment.

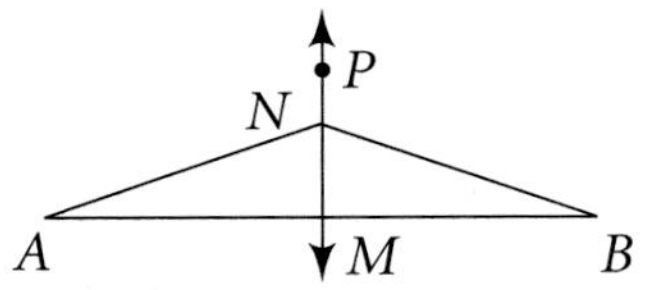

5. **a.** Given that $AN = BN$, what can you conclude about point N?
   b. Given that $\overline{MP}$ is the perpendicular bisector of $\overline{AB}$, what can you conclude about AN and BN? about $\triangle ABN$?
6. Rewrite the two theorems as a single biconditional statement.

7. Given isosceles triangle $\triangle URI$ with vertex angle $\angle R$, what does Theorem 4-13 tell you about point R?

Example 1 — Relating to the Real World

National Landmarks Find the set of points in Washington, D.C., that are equidistant from the Jefferson Memorial and the White House.

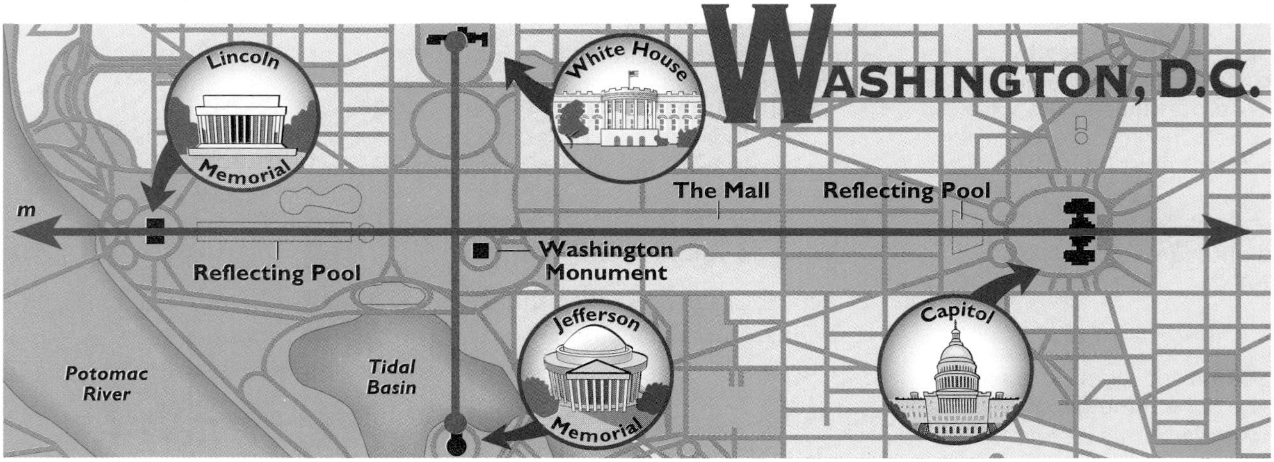

The red segment connects the Jefferson Memorial and the White House. All points on the perpendicular bisector m of this segment are equidistant from its endpoints. The perpendicular bisector passes through some of Washington's most famous landmarks.

8. For which landmarks does line m appear to be a line of symmetry?

Example 1 involves the concept of locus. A **locus** is a set of points that meets a stated condition. In Example 1, the condition is "equidistant from the Jefferson Memorial and the White House." The locus is the perpendicular bisector of the segment connecting the two landmarks.

Who? Benjamin Banneker (1731–1806) was an American astronomer, farmer, mathematician, and surveyor. In 1791, Banneker assisted in laying out the boundaries of Washington, D.C.

Example 2

Sketch the locus of points in a plane that are 1 cm from point C.

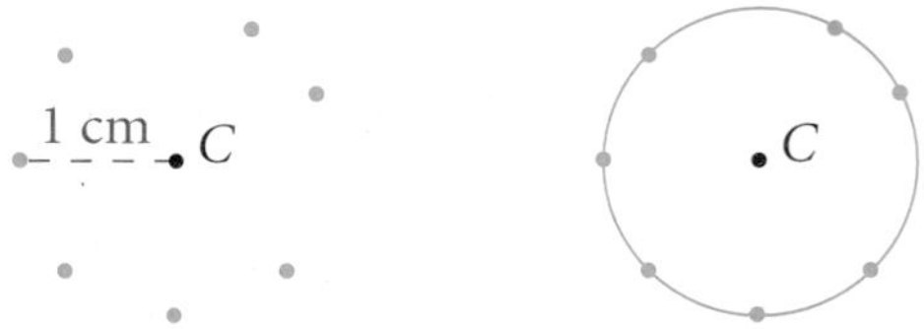

Locate several points 1 cm from C. Keep doing so until you see a pattern. The locus is a circle with center C and radius 1 cm.

9. Try This Sketch the locus of points in a plane that are 1 cm from a segment $\overline{VC}$.

10. What would the locus be in Example 2 if the words "in a plane" were replaced with "in space"?

As the answer to Question 10 implies, a locus of points in a plane and a locus of points in space can be quite different. For example, consider this condition: all points 3 cm from line ℓ.

The locus in a plane looks like this . . .

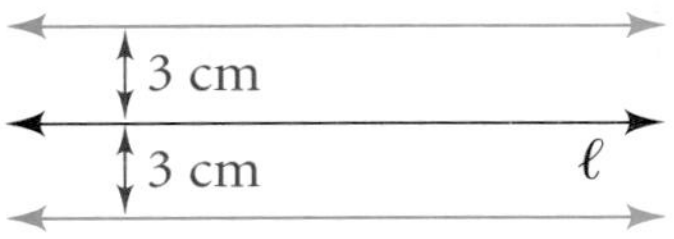

The locus is two parallel lines, each 3 cm from line ℓ.

The locus in space looks like this . . .

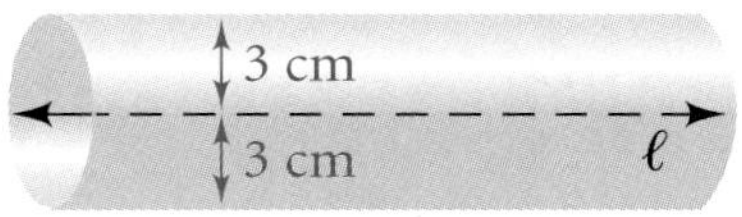

The locus is an endless cylinder with radius 3 cm and center-line ℓ.

QUICK REVIEW

To bisect an angle using paper folding, fold the paper so that the sides of the angle overlap.

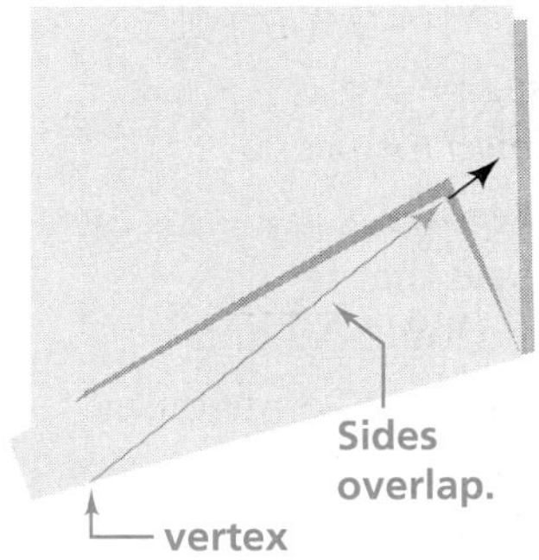

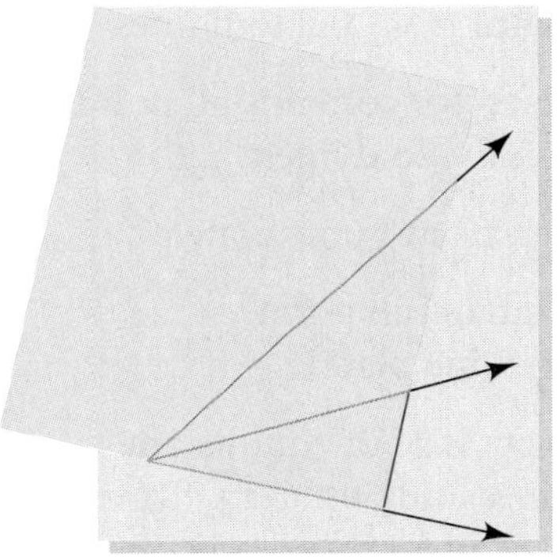

Angle Bisectors and Locus

The Work Together below involves the distance from a point to a line. The **distance from a point to a line** is the length of the perpendicular segment from the point to the line.

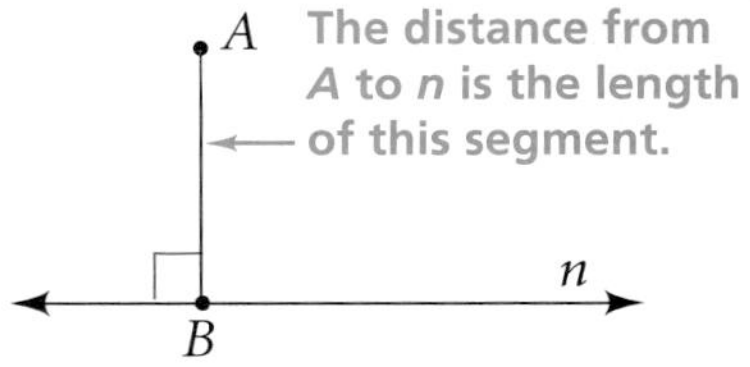

11. Let C be any point on line n other than B. Show that $AB < AC$.

WORK TOGETHER

- Draw an angle and use paper folding to create its bisector.
- Draw a point on the bisector. Use the corner of a piece of paper to create perpendicular segments from both sides of the angle to the point.

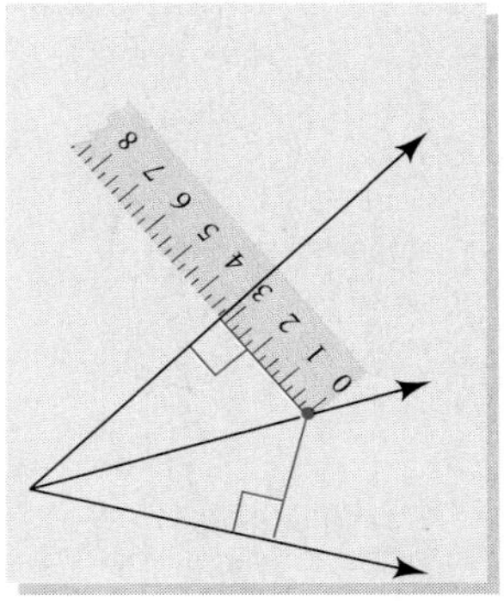

12. Measure the lengths of the two segments you drew. What do you notice?

13. Repeat with two other points. Does what you noticed still hold true?

14. Use what you discovered to complete the statement: If a point is on the bisector of an angle, then __?__.

THINK AND DISCUSS

The property that you discovered in the Work Together is stated below. Its converse, which also is true, follows it.

Theorem 4-14
Angle Bisector Theorem

If a point is on the bisector of an angle, then it is equidistant from the sides of the angle.

Theorem 4-15
Converse of Angle Bisector Theorem

If a point in the interior of an angle is equidistant from the sides of the angle, then it is on the angle bisector.

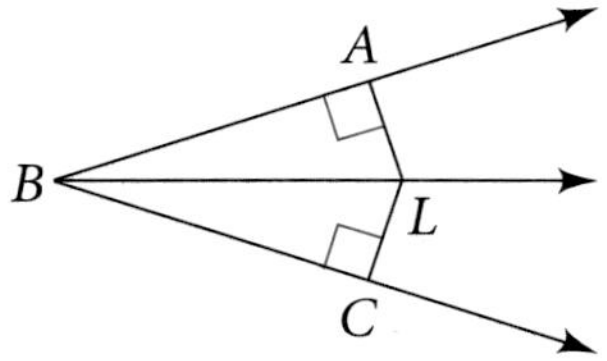

15. a. Given that $m\angle ABL = m\angle CBL$, what can you conclude about AL and CL?

b. Given that $AL = CL$, what can you conclude about $m\angle ABL$ and $m\angle CBL$?

16. What is the locus of all points in the interior of an angle that are equidistant from the sides of the angle?

Sometimes the condition in a locus problem has more than one part.

Example 3 Relating to the Real World

Sports What is a common name for the part of a baseball field that is equidistant from the foul lines and 60 ft 6 in. from home plate?

There are two parts to the condition stated in the problem.

Part 1: Equidistant from the foul lines

The red angle bisector contains all points equidistant from the foul lines.

Part 2: 60 ft 6 in. from home plate

The blue arc contains all points 60 ft 6 in. from home plate.

The pitcher's plate is both equidistant from the foul lines and 60 ft 6 in. from home plate.

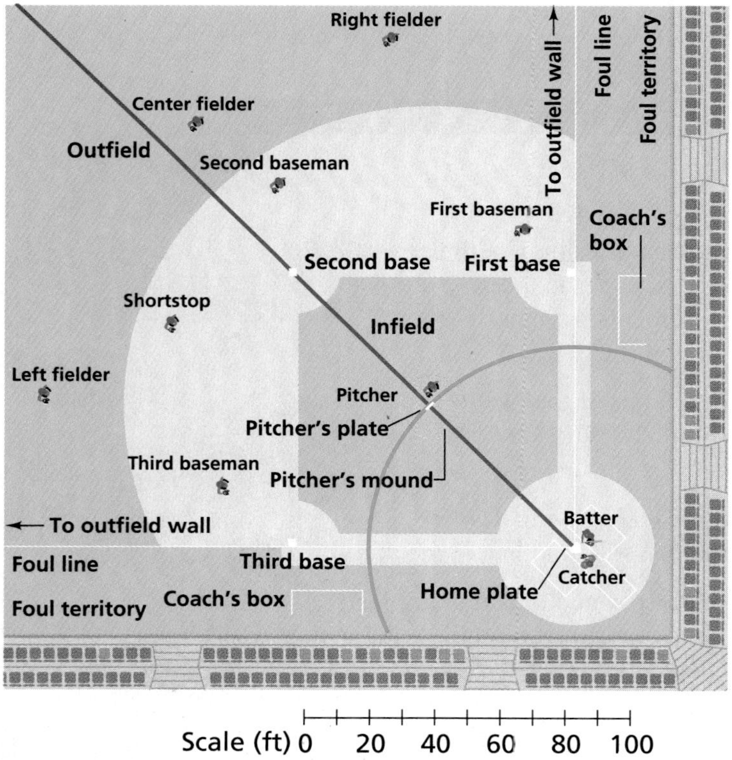

Exercises ON YOUR OWN

Sketch and label each locus.

1. all points in a plane 4 cm from a point X
2. all points in a plane 1 in. from a line $\overleftrightarrow{UV}$
3. all points in a plane 1 in. from a segment $\overline{UV}$
4. all points in space 3 cm from a point F
5. all points in space a distance a from a line $\overleftrightarrow{DE}$
6. all points in a plane 1 in. from a circle with radius 0.5 in.
7. all points in a plane equidistant from two parallel lines
8. all points in a plane equidistant from the endpoints of $\overline{PQ}$
9. all points in space equidistant from two parallel lines
10. all points in a plane equidistant from $\overleftrightarrow{MN}$ and $\overleftrightarrow{OP}$ where $\overleftrightarrow{MN} \perp \overleftrightarrow{OP}$
11. all points in a plane 3 cm from $\overline{GH}$ and 5 cm from G, where $GH = 4.5$ cm
12. all points equidistant from two parallel planes
13. Find the locus of points equidistant from the sides of $\angle JKL$ and on $\odot C$.

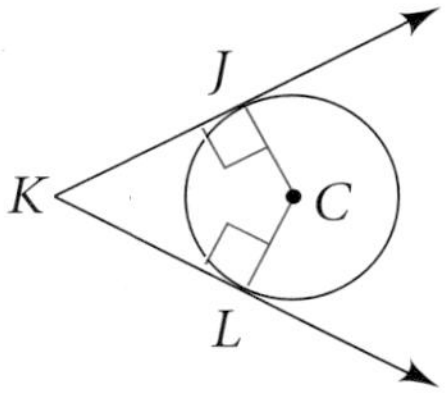

14. **Coordinate Geometry** Write an equation for the locus of points equidistant from (2, −3) and (6, 1).
15. **Sports** What is the common name for the part of a baseball field that is equidistant from first and third bases and 127 ft from home plate?
16. **Standardized Test Prep** The highlighted points are the locus of points on the coordinate plane that are
    - **A.** 2 units from the origin and 1 unit from the y-axis.
    - **B.** 1 unit from the origin and 2 units from the y-axis.
    - **C.** 2 units from the origin and 1 unit from the x-axis.
    - **D.** 1 unit from the origin and 2 units from the x-axis.
    - **E.** none of the above

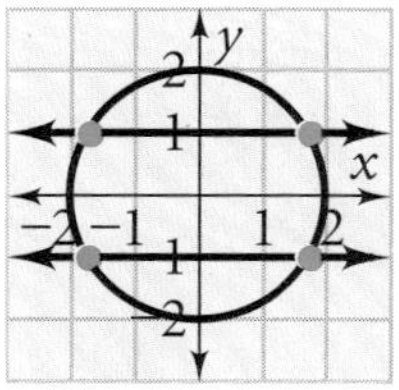

17. **a.** What is the locus of the tip of the minute hand on a clock?
    b. What is the locus of the tip of the hour hand on a clock?
18. **Open-ended** The smoke trails of the jets show the paths that they travel. Find a picture from a book or magazine or draw a sketch that shows an object and its path.
19. **Open-ended** Give two examples of locus from everyday life, one in a plane and one in space.
20. **Logical Reasoning** Rosie says that it is impossible to find a point equidistant from three collinear points. Is she correct? Explain your thinking.

21. Paul and Priscilla Wilson take new jobs in Shrevetown and need to find a place to live. Paul says "Let's try to move somewhere equidistant from both of our offices." Priscilla says, "Let's try to stay within three miles of downtown." Trace the map and suggest some locations for the Wilsons.

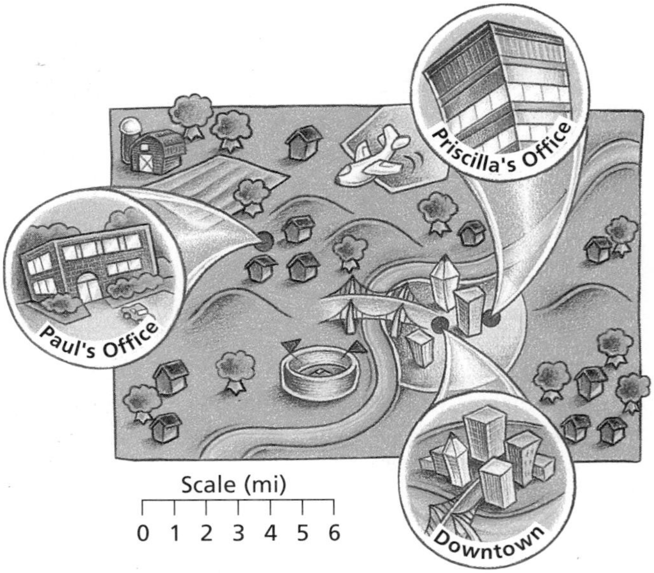

22. *Open-ended* Name a food with the shape described by each locus.
   a. all points in space no more than $1\frac{1}{2}$ in. from a point
   b. all points in space no more than $\frac{1}{2}$ in. from a circle with radius 2 in.

Coordinate Geometry **Sketch each locus on the coordinate plane.**

23. all points 3 units from the origin

24. all points 5 units from $x = 2$

25. all points equidistant from $A(0, 2)$ and $B(2, 0)$

26. all points equidistant from $P(1, 3)$ and $Q(5, 1)$

27. all points 4 units from the y-axis

28. all points equidistant from $y = 3$ and $y = -1$

Describe the locus that each blue figure represents.

29.

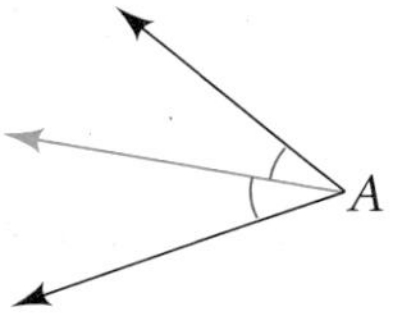

30.

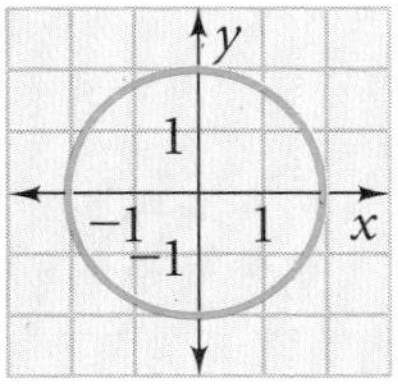

31.

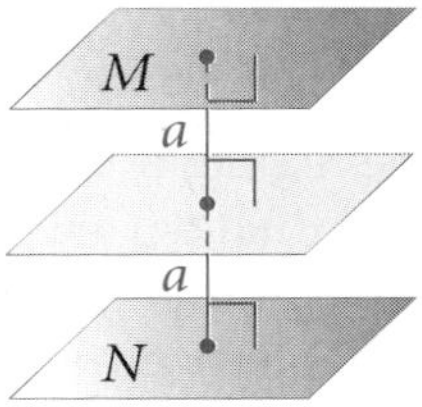

32.

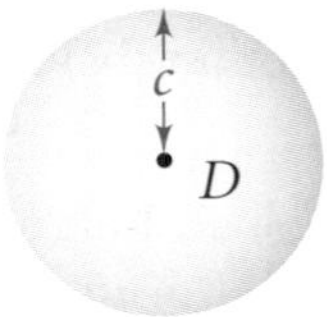

33. Think about the path of a child on a swing.
   a. Draw a side view of this path.
   b. Draw a top view of this path.
   c. Draw a front view of this path.

34. Alejandro draws a segment to use as the base of an isosceles triangle.
   a. Draw a segment to represent Alejandro's base. Locate three points that could be the vertex of the isosceles triangle.
   b. Describe the locus of points for the vertex of Alejandro's isosceles triangle.
   c. *Writing* Explain why the locus you described is the only possibility for the vertex of the triangle.

Sketch each path.

35. the swimmer's left foot

36. a doorknob as the door opens

37. a knot in the middle of a jump rope as the rope is being used

Chapter Project ***Find Out by Writing***

Make up a logic puzzle in which the solver must use clues to match the people, places, or things you describe. Have a classmate solve your puzzle. If necessary, refine your puzzle based on input from your classmate.

Exercises MIXED REVIEW

Identify the following in $\odot O$.

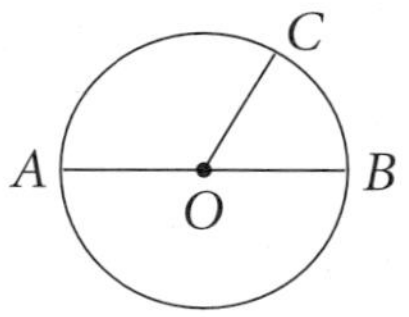

38. a central angle and its intercepted arc

39. two major arcs

40. a pair of adjacent arcs

Getting Ready for Lesson 4-8

Constructions **Draw a large triangle. Construct each figure.**

41. an angle bisector

42. a midpoint of a side

43. a perpendicular bisector of a side

Exercises CHECKPOINT

List the sides from shortest to longest.

1.

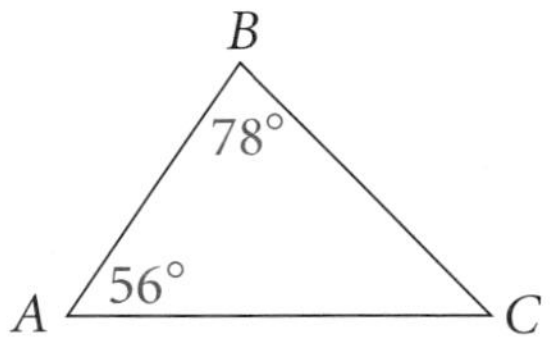

2.

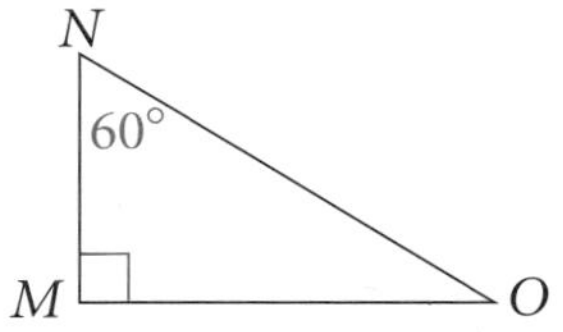

3.

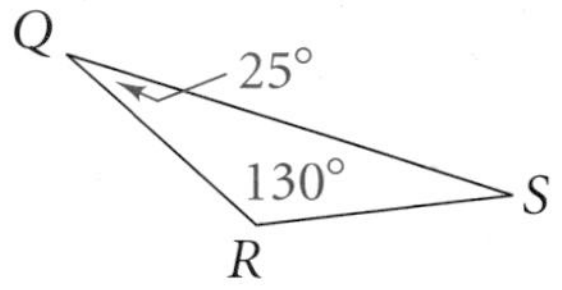

4. Open-ended Write a conditional that is true. Then write the first step of an indirect proof for your conditional.

Write the first step of an indirect proof of each statement.

5. February has fewer than 30 days.

6. A pentagon has at most three right angles.

Sketch and label the locus of points.

7. all points in a plane a distance d from a line $\overleftrightarrow{PR}$

8. all points in space 1 in. from a plane

Math ToolboX — Technology

Exploring Special Segments in Triangles

Before Lesson 4-8

Work in pairs or small groups.

Construct

Use geometry software.

1. Construct a triangle and the three perpendicular bisectors of its sides.
2. Construct a triangle and its three angle bisectors.
3. Construct a triangle. Construct a line through a vertex of the triangle that is perpendicular to the line containing the side opposite that vertex. Similarly, draw the perpendiculars from the other two vertices. Since an *altitude* of a triangle is the segment from a vertex to the line containing the opposite side, the three lines you have drawn contain the three altitudes of the triangle.

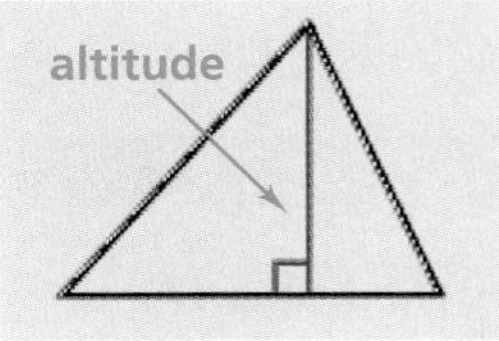

4. Construct a triangle. Construct the midpoint of a side of the triangle and draw the segment from the midpoint to the opposite vertex. This segment is called a *median* of the triangle. Construct the other two medians.

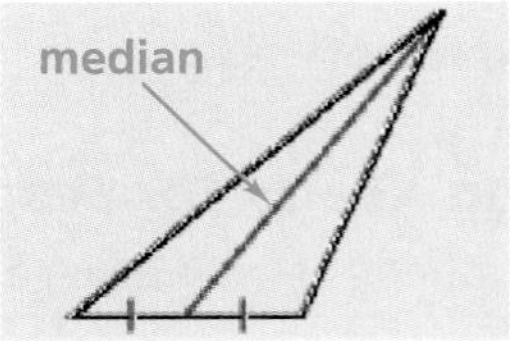

Investigate

- What do you notice about the lines, rays, or segments that you constructed in each triangle?
- Manipulate the triangles. Does the property still hold in the manipulated triangles?

Conjecture

List your **conjectures** about the angle bisectors, the perpendicular bisectors, the lines containing the altitudes, and the medians of a triangle.

Extend

- In what types of triangles do the perpendicular bisectors of the sides intersect inside the triangle? on the triangle? outside the triangle?
- Describe the triangles for which the lines containing the altitudes, the angle bisectors, or the medians intersect inside, on, or outside the triangles.

4-8 Concurrent Lines

What You'll Learn

- Identifying properties of perpendicular bisectors, angle bisectors, altitudes, and medians of a triangle

...And Why

To understand the points of concurrency that are used in architecture, construction, and transportation

What You'll Need

- scissors

WORK TOGETHER

Step 1: Each member of your group should draw and cut out five triangles: two acute, two right, and one obtuse. Make them big enough so that they are easy to fold.

Step 2: Use paper folding to create the angle bisectors of each angle of an acute triangle. What do you notice about the angle bisectors?

Step 3: Repeat Step 2 with a right and an obtuse triangle. Does what you discovered still hold true?

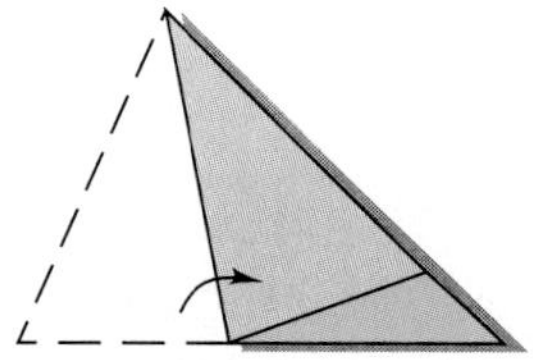

Folding an Angle Bisector

1. Make a **conjecture** about the bisectors of the angles of a triangle.

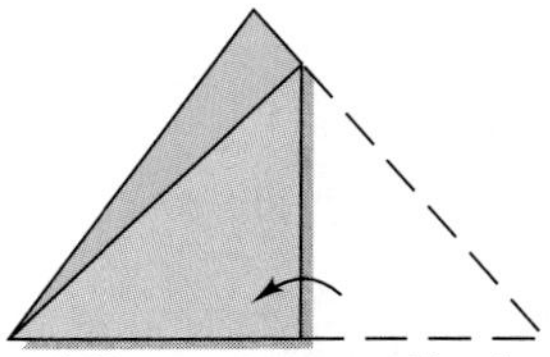

Folding a Perpendicular Bisector

Step 4: Use paper folding to create the perpendicular bisectors of each of the sides of an acute triangle. What do you notice?

Step 5: Repeat Step 4 with a right triangle. What do you notice?

Step 6: Draw an obtuse triangle in the middle of a piece of paper (do not cut it out!), then repeat Step 4 with this triangle. What do you notice?

2. Make a **conjecture** about the perpendicular bisectors of the sides of a triangle.

THINK AND DISCUSS

Perpendicular Bisectors and Angle Bisectors

When three or more lines intersect in one point, they are **concurrent.** The point at which they are concurrent is the **point of concurrency.** Your explorations in the Work Together lead to the following theorems.

Theorem 4-16 The perpendicular bisectors of the sides of a triangle are concurrent at a point equidistant from the vertices.

Theorem 4-17 The bisectors of the angles of a triangle are concurrent at a point equidistant from the sides.

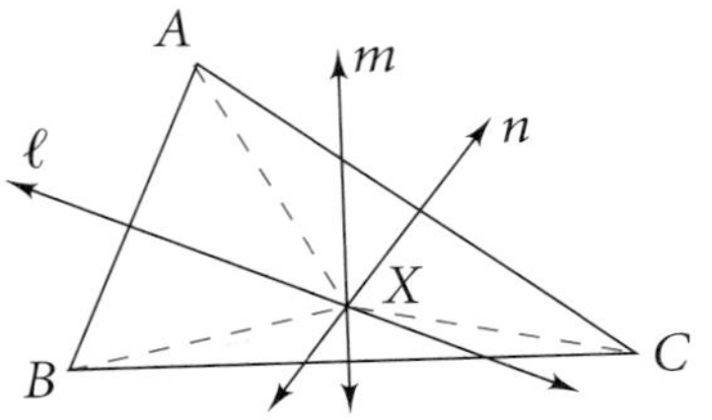

Paragraph Proof of Theorem 4-16

Given: Lines ℓ, m, and n are perpendicular bisectors of the sides of $\triangle ABC$. X is the intersection of lines ℓ and m.

Prove: Line n contains point X, and X is equidistant from A, B, and C.

Since m is the perpendicular bisector of $\overline{BC}$, $BX = CX$. Similarly, since ℓ is the perpendicular bisector of $\overline{AB}$, $AX = BX$. By substitution, $AX = CX$. So by the Converse of the Perpendicular Bisector Theorem, X is on line n. Since $AX = BX = CX$, X is equidistant from A, B, and C.

Example 1 **Relating to the Real World**

Recreation The Jacksons want to install a circular pool in their backyard. They want the pool to be as large as possible. Where would the largest possible pool be located?

The point of concurrency of the angle bisectors is equidistant from the sides of the triangular yard (Theorem 4-17). If any other point were chosen as the center of the pool, it would be closer to at least one of the sides of the yard, and the pool would have to be smaller. ■

As Example 1 suggests, the points of concurrency in Theorems 4-16 and 4-17 have some interesting properties related to circles.

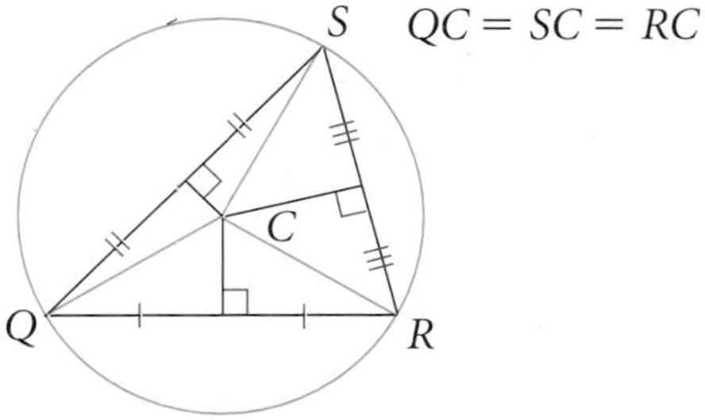

Point C, called the *circumcenter*, is equidistant from the vertices of $\triangle QRS$. The circle is *circumscribed* around the triangle.

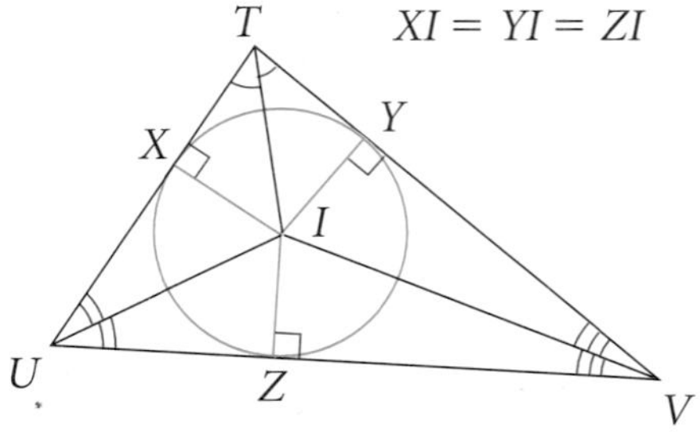

Point I, called the *incenter*, is equidistant from the sides of $\triangle TUV$. The circle is *inscribed* in the triangle.

Example 2

Coordinate Geometry Find the center of the circle that circumscribes $\triangle OPS$.

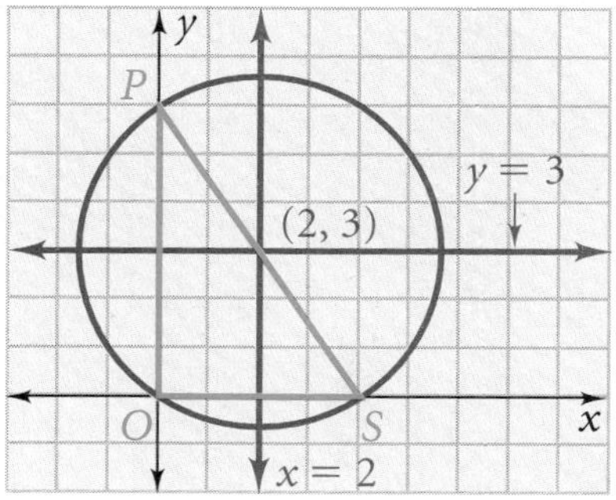

Two of the perpendicular bisectors of the sides of the triangle are $x = 2$ and $y = 3$. These lines intersect at $(2, 3)$. This point is the center of the circle.

3. **Critical Thinking** Explain why it was not necessary to find the third perpendicular bisector in Example 2.

4. **Try This** Find the center of the circle that circumscribes the triangle with vertices $(0, 0)$, $(-8, 0)$, and $(0, 6)$.

Medians and Altitudes

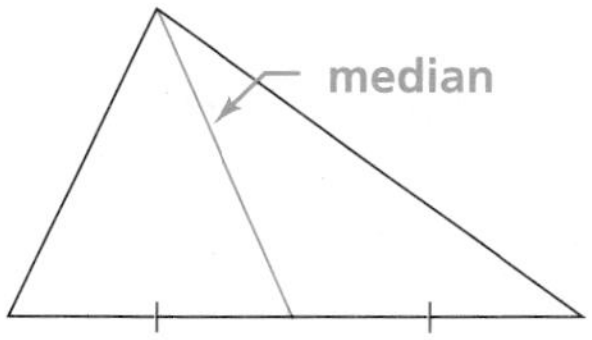

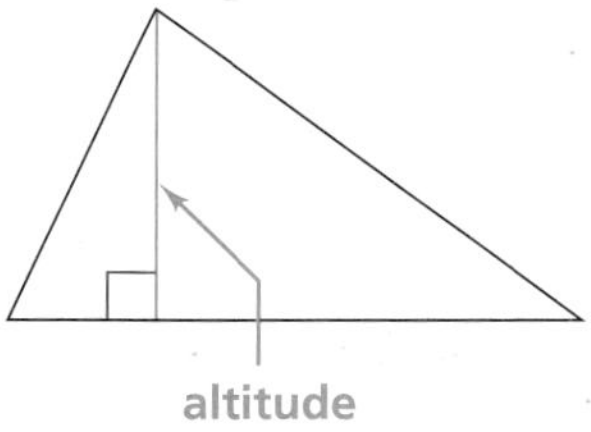

Two other special segments of a triangle are medians and altitudes. The **median of a triangle** is a segment whose endpoints are a vertex and the midpoint of the side opposite the vertex. The **altitude of a triangle** is a perpendicular segment from a vertex to the line containing the side opposite the vertex.

Unlike angle bisectors and medians, an altitude of a triangle may lie outside the triangle. Consider the following diagrams.

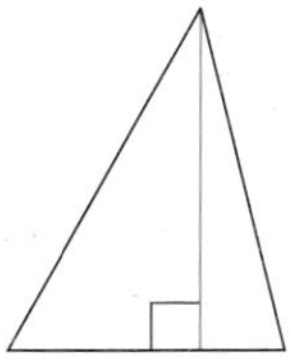

Acute Triangle:
Altitude is inside.

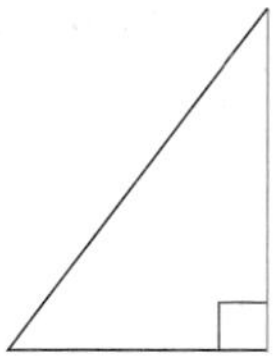

Right Triangle:
Altitude is a side.

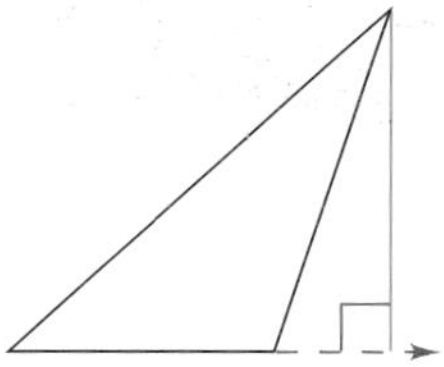

Obtuse Triangle:
Altitude is outside.

You can use paper folding to find altitudes and medians.

To find an altitude . . .

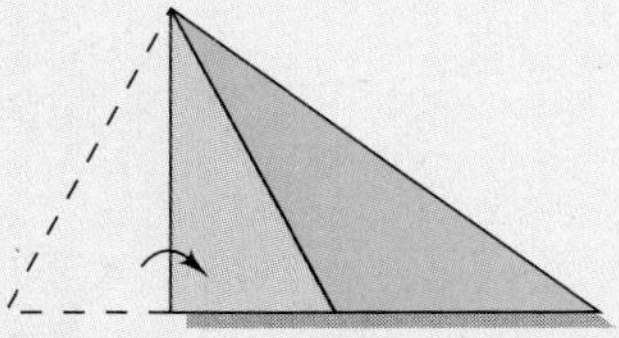

Fold so that a side overlaps itself and the fold contains a vertex.

To find a median . . .

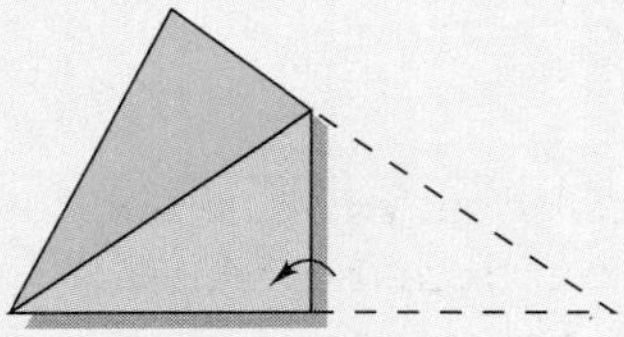

Fold one vertex to another to find the midpoint of a side,

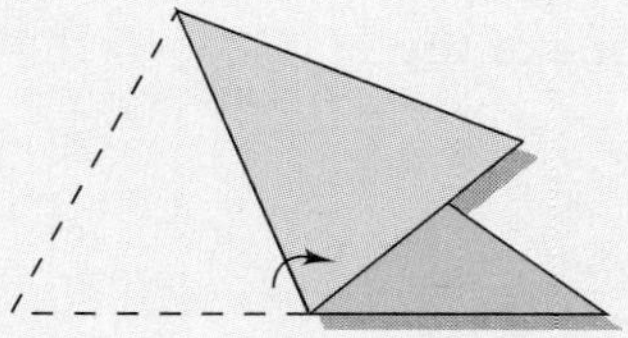

then fold from the midpoint to the opposite vertex.

Example 3

a. *Paper Folding* Use paper folding to create the altitudes from each of the three vertices of an acute triangle.

b. Use paper folding to create the three medians of an acute triangle.

The orange triangle shows the paper-folded medians.
The blue triangle shows the paper-folded altitudes.

Notice in Example 3 that the three altitudes and three medians are concurrent. This property holds true for any triangle.

Theorem 4-18 The lines that contain the altitudes of a triangle are concurrent.

Theorem 4-19 The medians of a triangle are concurrent.

5. *Critical Thinking* For what type of triangle is the point of concurrency of the altitudes outside the triangle?

Exercises ON YOUR OWN

Is $\overline{AB}$ a perpendicular bisector, an angle bisector, an altitude, a median, or none of these?

1.
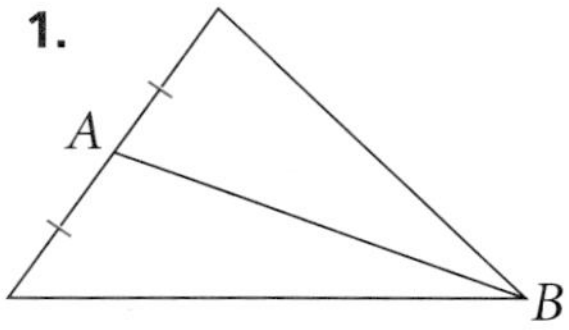

2.
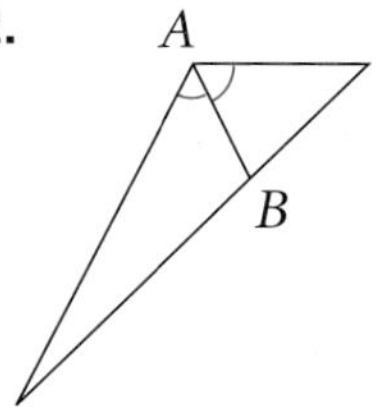

3.
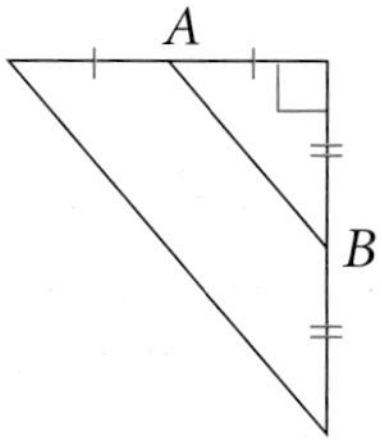

4.
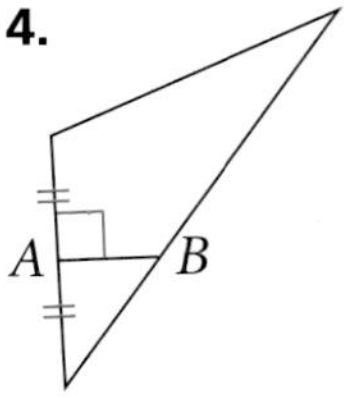

5.
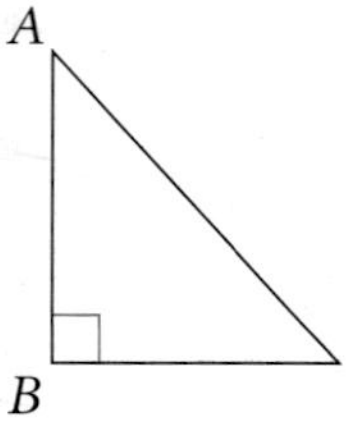

For each triangle, give the coordinates of the point of concurrency of (a) the perpendicular bisectors of the sides and (b) the altitudes.

6.

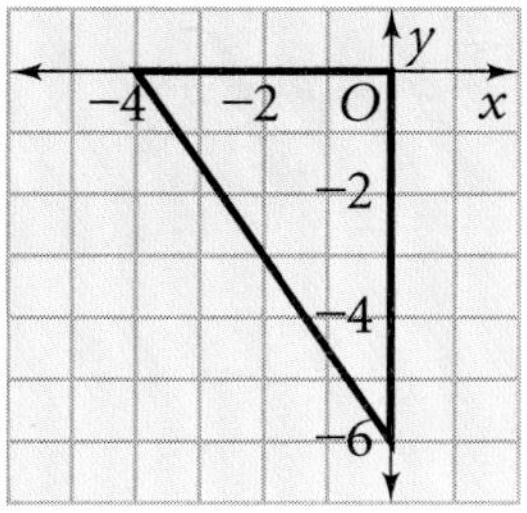

7.

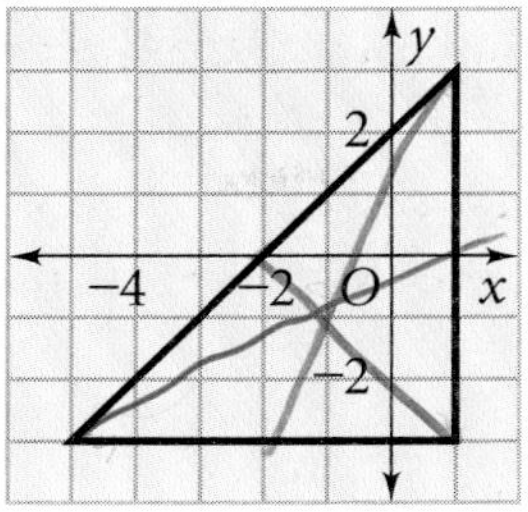

8.

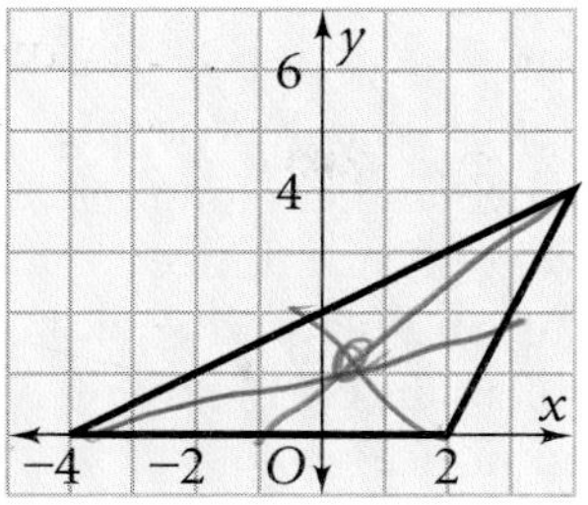

9.

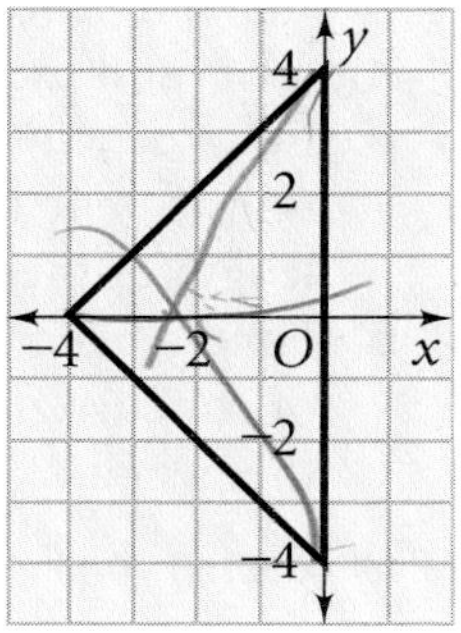

The points of concurrency for the lines and segments listed in I–IV have been drawn on the triangles. Match the points with the lines and segments.

I. perpendicular bisectors of sides

II. angle bisectors

III. medians

IV. altitudes

10.

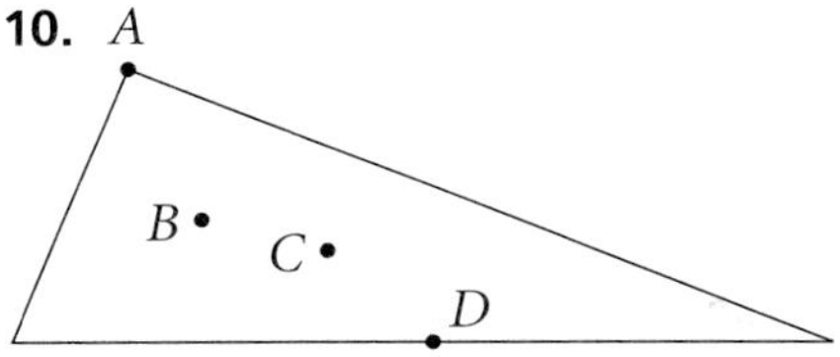

11.

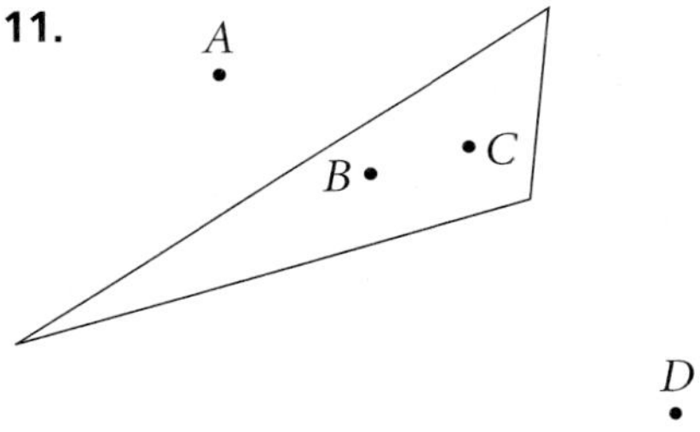

12. **History of Mathematics** Leonard Euler proved in 1765 that for any triangle, three of the four points of concurrency are collinear. The line that contains these three points is known as *Euler's Line*. Refer to Exercises 10 and 11 to determine which point of concurrency does *not* lie on Euler's Line.

13. **Park Design** Where should park officials place a drinking fountain in Altgeld Park so that it is equidistant from the tennis court, the playground, and the volleyball court?

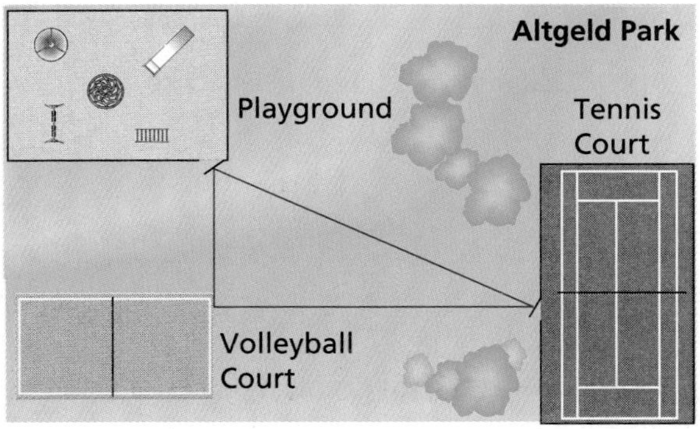

14. **Open-ended** Draw a triangle and construct the perpendicular bisectors of two of its sides. Then construct the circle that circumscribes the triangle.

15. **Coordinate Geometry** $\triangle DEF$ has vertices $D(0, 0)$, $E(12, 0)$, and $F(0, 12)$.

a. Find the equations of the lines that contain the three altitudes.

b. Find the equations of the three perpendicular bisectors of the sides.

c. **Writing** Are any of the lines in parts (a) and (b) the same? Explain.

16. **Locus** Three students are seated at uniform distances around a circular table. Copy the diagram and shade the points on the table that are closer to Moesha than to Jan or Chandra.

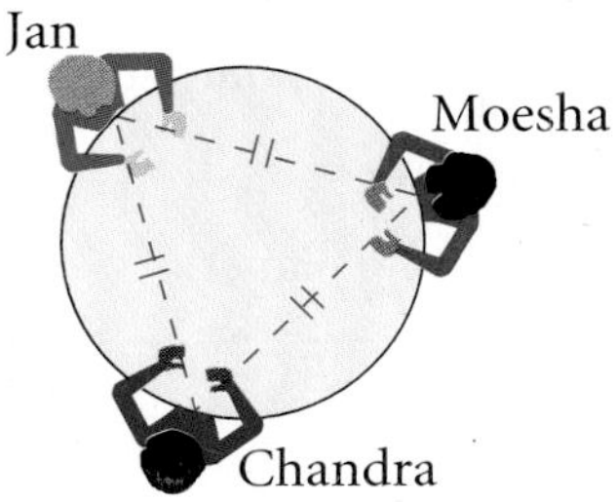

17. **Manipulatives** Medians of triangles have special physical properties related to balance. Draw a triangle on heavy cardboard, construct its medians and then cut it out.

a. Put a pencil on a table and place the triangle on the pencil so that a median lies along the length of the pencil. What do you notice?

b. Hold a pencil straight up and place the point of concurrency of the medians on the tip of the pencil. What do you notice?

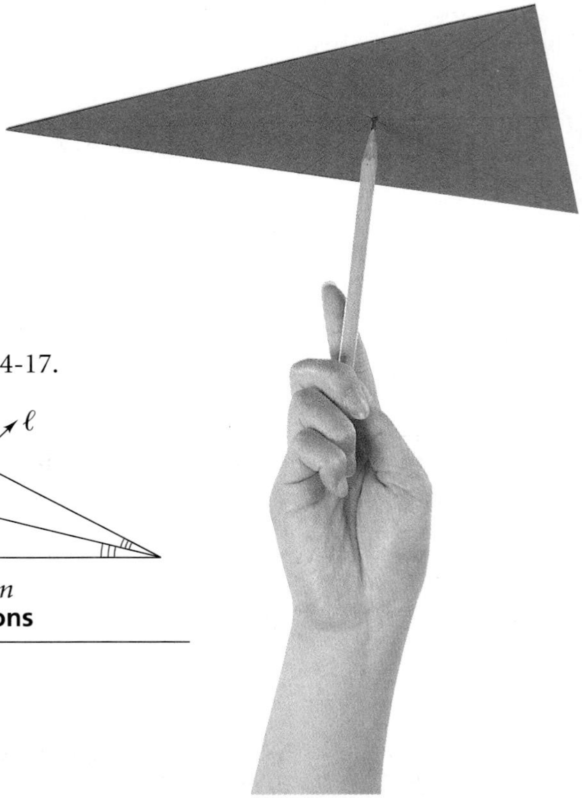

The point on which a figure balances is called the *center of gravity.*

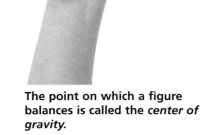

18. **Preparing for Proof** Complete the proof of Theorem 4-17.

Given: Rays ℓ, m, and n are bisectors of the angles of $\triangle ABC$. X is the intersection of rays ℓ and m.

Prove: Ray n contains point X; $DX = EX = FX$

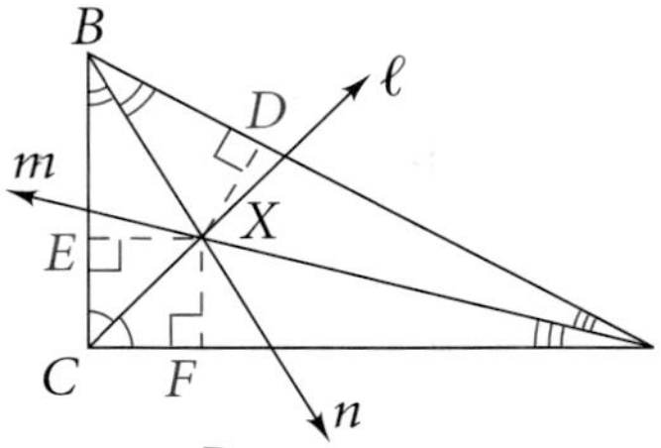

Statements	Reasons
1. Rays ℓ, m, and n are bisectors of the angles of $\triangle ABC$. X is the intersection of rays ℓ and m.	**a.** ?
2. $FX = EX$ and $DX = FX$	**b.** ?
3. $DX = EX$	**c.** ?
4. Ray n contains X.	**d.** ?
5. $DX = EX = FX$	**5.** Transitive Prop. of Equality

Exercises MIXED REVIEW

Coordinate Geometry **Classify each triangle as acute, obtuse, or right.**

19. $M(-3, -4)$, $N(2, 5)$, $L(2, -4)$

20. $Q(-6, 1)$, $T(0, 0)$, $V(4, 3)$

21. $B(3, -4)$, $P(3, 5)$, $J(6, 5)$

Geometry in 3 Dimensions **In the diagram, *ABCD* is a square. Give an example of each figure or pair of figures.**

22. a plane and a point not on the plane

23. skew lines

24. intersecting planes

25. concurrent lines **26.** parallel lines

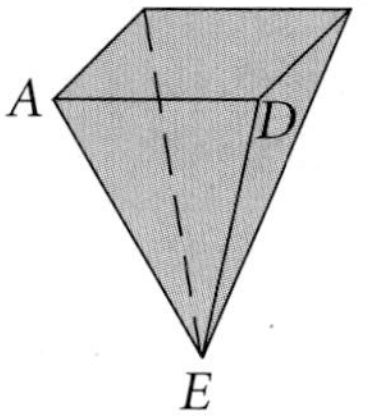

SELF ASSESSMENT

PORTFOLIO

For your portfolio, select one or two items from your work for this chapter. Here are some possibilities.
- corrected work
- work that you think best shows the things you learned or did in this chapter

Explain why you have included each selection.

Finishing the Chapter Project

PUZZLING PIECES

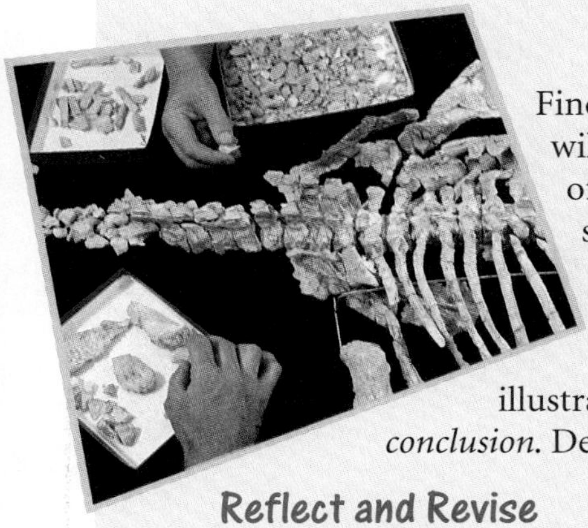

Find Out questions on pages 187, 198, 205, and 225 will help you complete your project. Prepare a display of logic puzzles. Include your own puzzles and solutions, those from the Find Out questions, or those found through research. Look at *Alice's Adventures in Wonderland* as well as the books listed below. Use puzzles and solutions to illustrate terms of logic, such as *conditional* and *conclusion.* Decide how best to organize your display.

Reflect and Revise

Review your display with a classmate. Together, check that your solutions are correct and that your diagrams and explanations are clear. Have you used logic terms correctly? Is the display attractive, organized, and comprehensive? Revise your work as needed. Consider doing more research.

Follow Up

A paradox is self-contradictory. Study these three paradoxes:

1. In a town, there is a librarian who reads to every person who does not read to himself or herself.
2. **A.** Statement (B) is false. **B.** Statement (A) is true.
3. This sentence contains two errers.

What can you conclude from each? Does the librarian read to himself? Is statement (A) true? How many errors does the sentence have? Study other paradoxes and create your own. Explain why each is a paradox.

For More Information

Gardner, Martin. *Aha! Gotcha: Paradoxes to Puzzle and Delight.* New York: W.H. Freeman and Company, 1982.

Shannon, George. *Stories to Solve: Folktales from Around the World.* New York: Greenwillow Books, 1985.

Smullyan, Raymond M. *Alice in Puzzle-Land: A Carrollian Tale for Children Under Eighty.* New York: William Morrow, 1982.

CHAPTER 4

Wrap Up

Key Terms

altitude of a triangle (p. 229)
base (p. 188)
base angle (p. 188)
biconditional (p. 183)
conclusion (p. 182)
concurrent (p. 227)
conditional (p. 182)
contrapositive (p. 184)
converse (p. 183)
distance from a point to a line (p. 221)
hypothesis (p. 182)
indirect reasoning (p. 207)
inverse (p. 184)
legs (p. 188)
locus (p. 220)
median of a triangle (p. 229)
midsegment (p. 201)
negation (p.184)
point of concurrency (p. 227)
paragraph proof (p. 195)
truth value (p. 182)
two-column proof (p. 194)
vertex angle (p. 188)

How am I doing?

SELF ASSESSMENT

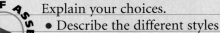

- State three ideas from this chapter that you think are important. Explain your choices.
- Describe the different styles of proof.

Using Logical Reasoning 4-1

An *if-then statement* is a **conditional.** The part following *if* is the **hypothesis,** and the part following *then* is the **conclusion.** You find its **truth value** when you determine if a conditional is true or false.

Statement	Form	Example
conditional	If ■, then □.	If a polygon is a triangle, then it has three sides.
converse	If □, then ■.	If a polygon has three sides, then it is a triangle.
inverse	If *not* ■, then *not* □.	If a polygon is *not* a triangle, then it does *not* have three sides.
contrapositive	If *not* □, then *not* ■.	If a polygon does *not* have three sides, then it is *not* a triangle.
biconditional	■ if and only if □.	A polygon is a triangle if and only if it has three sides.

For Exercises 1–4: (a) Write the converse. (b) Determine the truth value of the conditional and its converse. (c) If both statements are true, write a biconditional.

1. If you are in Australia, then you are south of the equator.
2. If an angle is obtuse, then its measure is greater than 90 and less than 180.
3. If it is snowing, then it is cold outside.
4. If a figure is a square, then its sides are congruent.

5. *Open-ended* Write a conditional and then write its contrapositive.

Isosceles Triangles 4-2

If two sides of a triangle are congruent, then the angles opposite those sides are also congruent. The bisector of the vertex angle of an isosceles triangle is the perpendicular bisector of the base. If two angles of a triangle are congruent, then the sides opposite the angles are congruent.

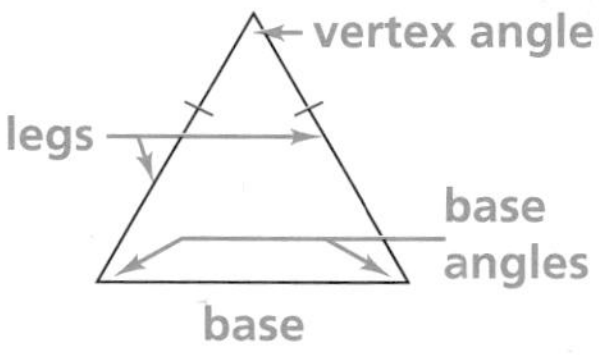

Find the values of *x* and *y*.

6.

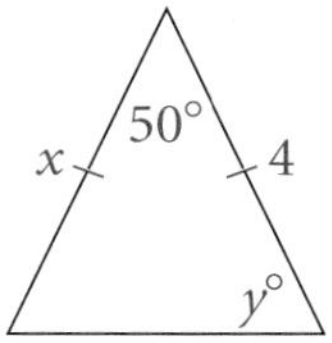

7.

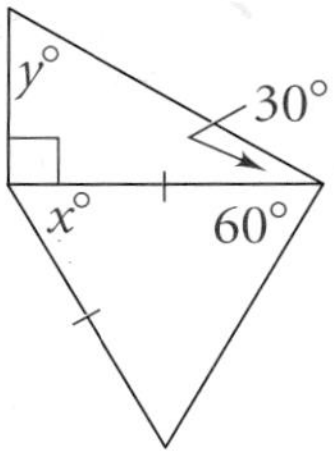

8.

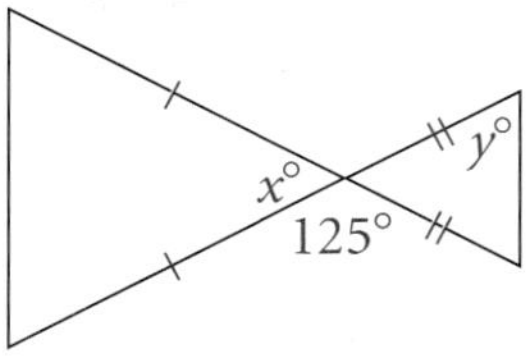

9.

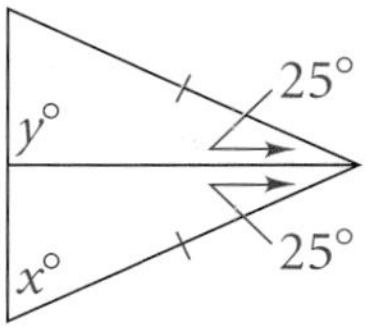

Midsegments of Triangles 4-3, 4-4

There are different types of proofs, including the **paragraph proof** and the **two-column proof.**

A **midsegment** of a triangle is a segment that connects the midpoints of its sides. The midsegment connecting two sides of a triangle is parallel to the third side and half its length.

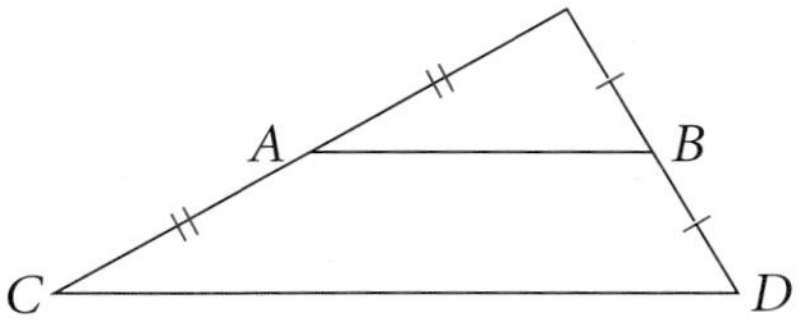

$\overline{AB} \parallel \overline{CD}$

$AB = \frac{1}{2}CD$

10. Preparing for Proof The reasons given in this proof are correct, but they are in the wrong order. List them in the correct order.

Given: $\triangle BCA \cong \triangle CDE$
$\triangle CDE \cong \triangle EGF$

Prove: $\overline{AF} \parallel \overline{BG}$

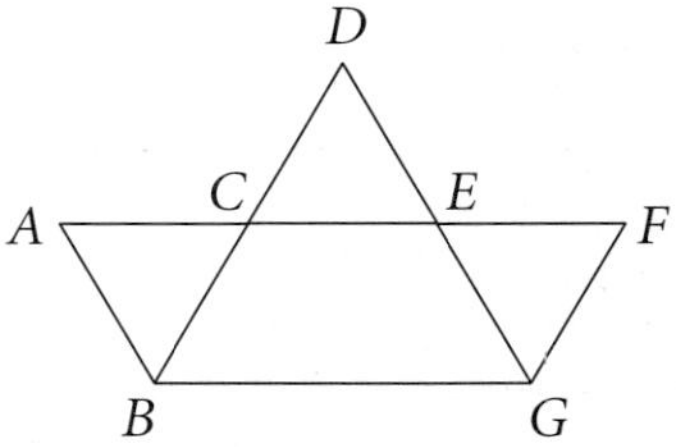

Statements	Reasons
1. $\triangle BCA \cong \triangle CDE$ $\triangle CDE \cong \triangle EGF$	a. Def. of midpoint
2. $\overline{BC} \cong \overline{CD}$, $\overline{DE} \cong \overline{EG}$	b. Triangle Midsegment Thm.
3. C is the midpoint of $\overline{BD}$. E is the midpoint of $\overline{DG}$.	c. Def. of congruent polygons
4. $\overline{AF} \parallel \overline{BG}$	d. Given

11. Writing Rewrite the proof above as a paragraph proof.

What can you conclude from each diagram? Justify your answers.

12.

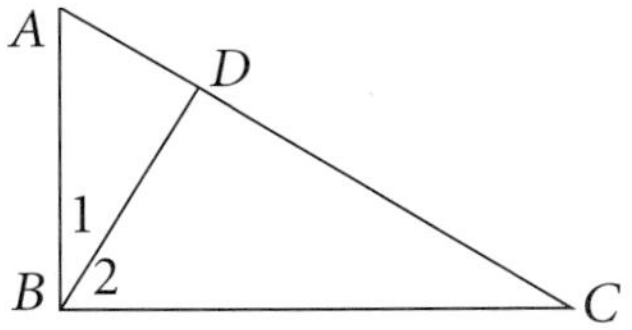

$\angle 1$ and $\angle 2$ are complementary.

13.

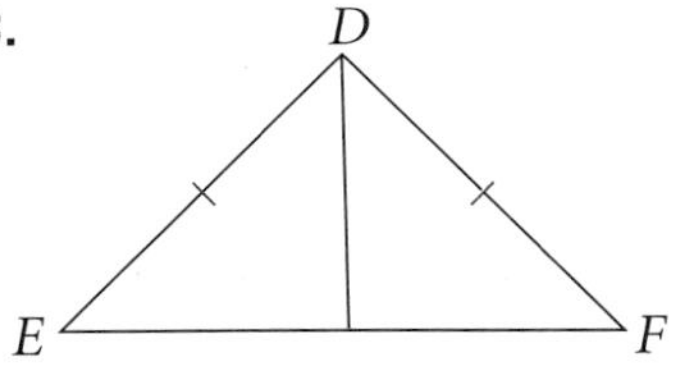

14.

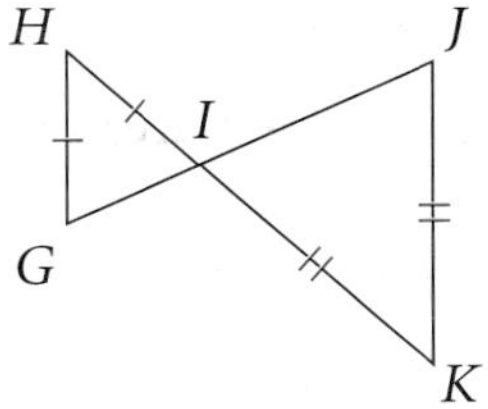

Algebra **Find the value of x.**

15. 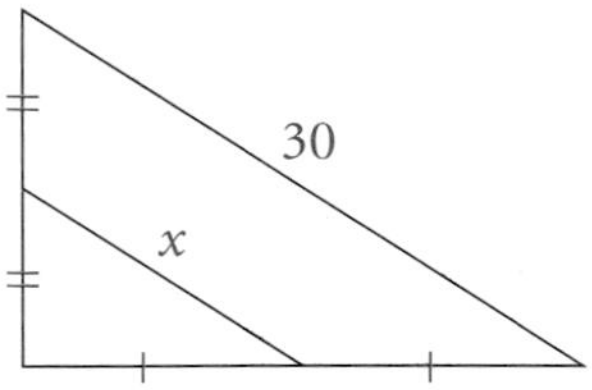

16. 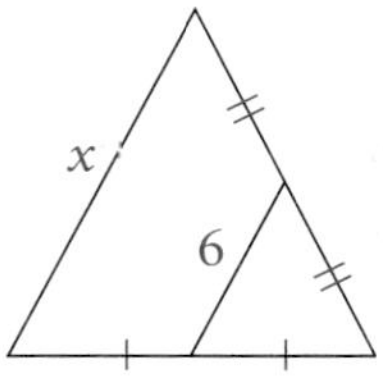

17. 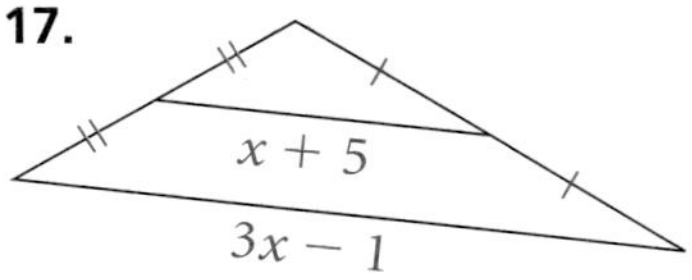

Using Indirect Reasoning 4-5

To use **indirect reasoning**, consider all possibilities and then prove all but one false. The remaining possibility must be true.

There are three steps in an indirect proof.

Step 1: Assume the opposite of what you want to prove is true.
Step 2: Use logical reasoning to reach a contradiction of an earlier statement, such as the given information or a theorem. Then state that the assumption you made was false.
Step 3: State that what you wanted to prove must be true.

Write a convincing argument that uses indirect reasoning.

18. Mary walks into a newly-painted room and finds 2 paint brushes rinsing in water. Show that the room was not painted with oil-based paint.

19. The product of two numbers is even. Show that at least one of the two numbers must be even.

20. Show that a triangle can have at most one obtuse angle.

21. Show that an equilateral triangle cannot have an obtuse angle.

Triangle Inequalities 4-6

The sum of the lengths of any two sides of a triangle is greater than the length of the third side. If two sides of a triangle are not congruent, the larger angle lies opposite the larger side.

List the angles and sides in order from smallest to largest.

22. 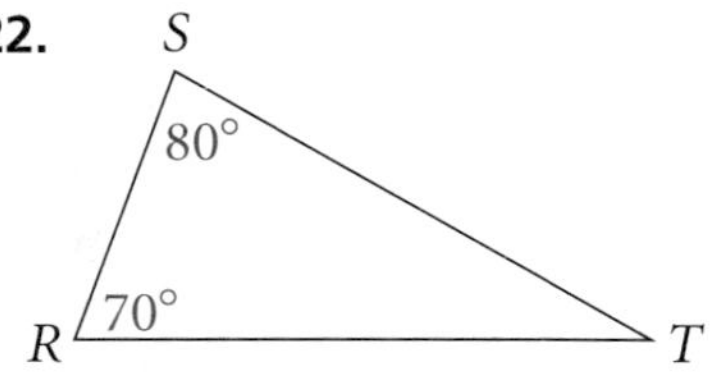

23. 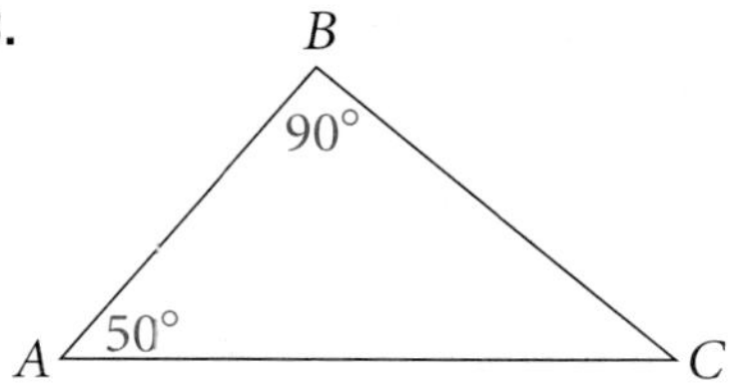

24. 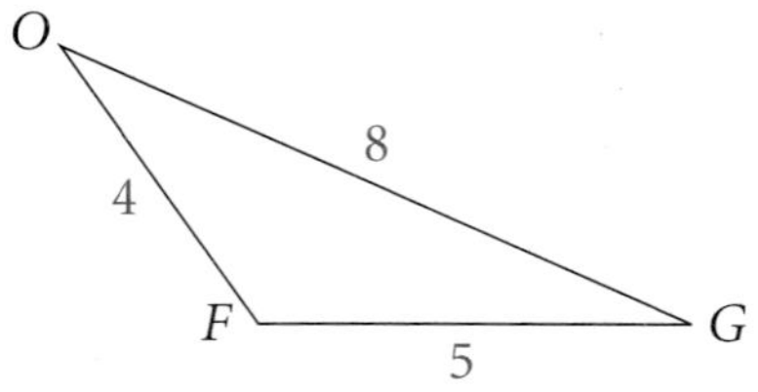

25. *Standardized Test Prep* Two sides of a triangle have lengths 4 and 7. Which could *not* be the length of the third side?
A. 5 **B.** 2 **C.** 7 **D.** 10 **E.** 8

Bisectors and Locus 4-7

A point is on the perpendicular bisector of a segment if and only if it is equidistant from the endpoints of the segment. A point is on the bisector of an angle if and only if it is equidistant from the sides of the angle. A set of points that meet a stated condition is a **locus.**

Sketch and label the locus of points.

26. all points in a plane 2 cm from a circle with radius 1 cm

27. all points in a plane equidistant from two points

28. all points in space a distance a from $\overline{DS}$

29. all points in space a distance b from a point P

Concurrent Lines 4-8

When three or more lines intersect in one point, they are **concurrent.**

The **median of a triangle** is a segment joining a vertex and the midpoint of the side opposite the vertex. The **altitude of a triangle** is a perpendicular segment from a vertex to the line containing the side opposite the vertex.

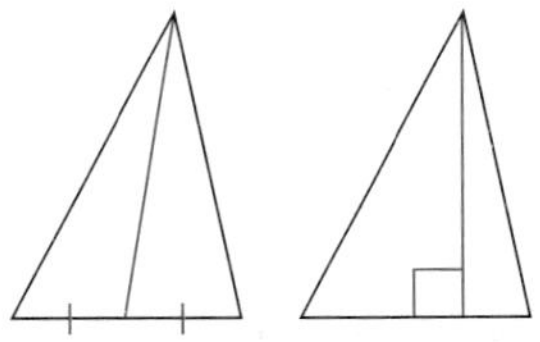

For any given triangle, each of the following are concurrent:

- The perpendicular bisectors of the sides
- The bisectors of the angles
- The medians
- The lines containing the altitudes

Coordinate Geometry **Graph $\triangle ABC$ with vertices $A(2, 3)$, $B(-4, -3)$, and $C(2, -3)$. Find the coordinates of each point of concurrency.**

30. perpendicular bisectors **31.** medians **32.** altitudes

Getting Ready for CHAPTER 5

Find the perimeter and area of a rectangle with the given dimensions.

33. $\ell = 5$ cm, $w = 3$ cm **34.** $\ell = 6.2$ ft, $w = 9.0$ ft **35.** $\ell = 0.5$ m, $w = 1.5$ m

Find the value of $\sqrt{a^2 + b^2}$ for the given values of a and b.

36. $a = 4, b = 3$ **37.** $a = 5, b = 12$ **38.** $a = 9, b = 12$

CHAPTER 4 Assessment

For each statement, write (a) the converse, (b) the inverse, and (c) the contrapositive.

1. If a polygon has eight sides, then it is an octagon.

2. If it is a leap year, then it is an even-numbered year.

3. If it is snowing, then it is not summer.

Algebra **Find the values of *x* and *y*.**

4.

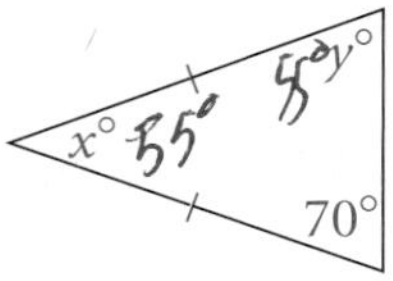

5.

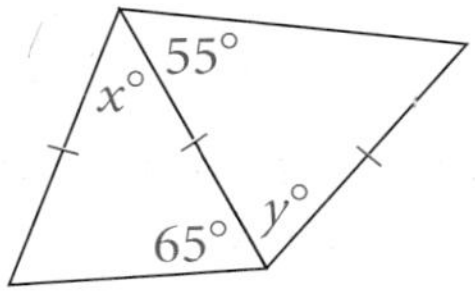

6. What can you conclude from the diagram? Justify your answer.

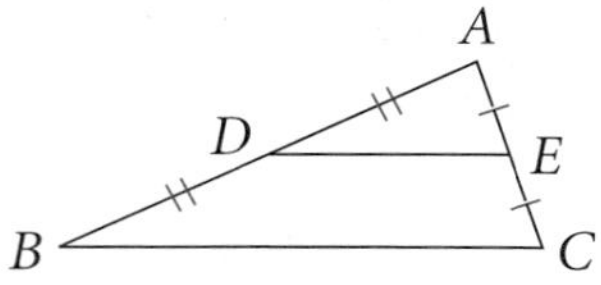

Identify the pair of statements that forms a contradiction.

7. **I.** $\triangle PQR$ is a right triangle.
II. $\triangle PQR$ is an obtuse triangle.
III. $\triangle PQR$ is scalene.

8. **I.** $\angle DAS \cong \angle CAT$
II. $\angle DAS$ and $\angle CAT$ are vertical.
III. $\angle DAS$ and $\angle CAT$ are adjacent.

List the angles of $\triangle ABC$ from smallest to largest.

9. $AB = 9$, $BC = 4$, $AC = 12$

10. $AB = 10$, $BC = 11$, $AC = 9$

11. $AB = 3$, $BC = 9$, $AC = 7$

12. Open-ended Write three lengths that cannot be the lengths of sides of a triangle. Explain your answer.

List the sides of each triangle in order from smallest to largest.

13.

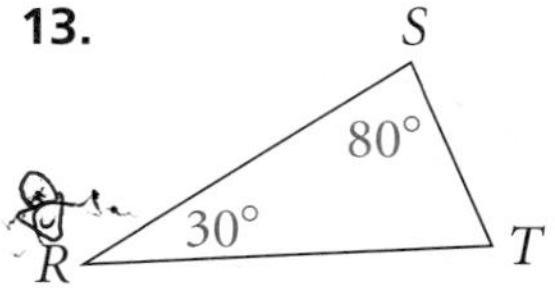

14.

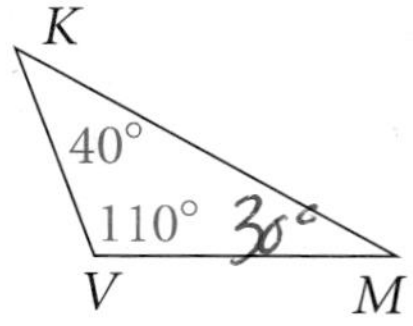

Find the value of *x*.

15.

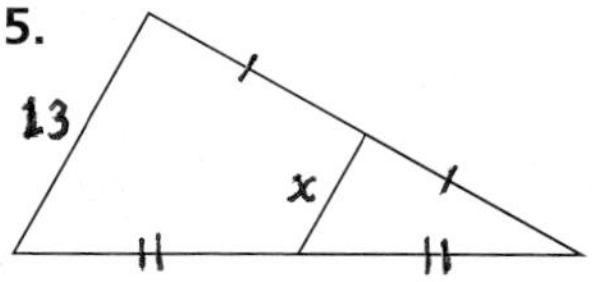

16.

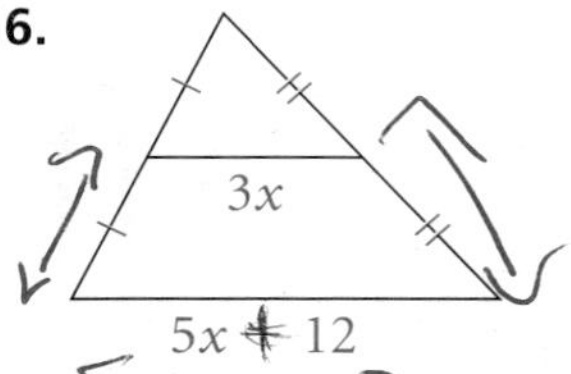

17. Writing Use indirect reasoning to explain why the following statement is true: If an isosceles triangle is obtuse, then the obtuse angle is the vertex angle.

Coordinate Geometry **Sketch each locus on a coordinate plane.**

18. all points 6 units from the origin

19. all points 3 units from the line $y = -2$

20. all points equidistant from points (2, 4) and (0, 0)

21. all points equidistant from the axes

Sketch each figure. Determine whether the point of concurrency is in the interior, exterior, or on the triangle.

22. acute triangle, perpendicular bisectors

23. obtuse triangle, medians

24. right triangle, altitudes

25. Standardized Test Prep $\triangle ABC$ has vertices $A(2, 5)$, $B(2, -3)$, and $C(10, -3)$. Which point of concurrency is at (6, 1)?

A. angle bisectors **B.** altitudes
C. perpendicular bisectors **D.** medians
E. none of the above

CHAPTER 4

Cumulative Review

For Exercises 1–11, choose the correct letter.

1. Which could be the measures of the angles of a triangle?

I. 37, 89, 54	**II.** 125, 45, 10
III. 100, 75, 15	**IV.** 60, 60, 60

A. I only **B.** IV only **C.** I and II
D. II and IV **E.** I, II, and IV

2. $\triangle DEB$ has vertices $D(3, 7)$, $E(1, 4)$, and $B(-1, 5)$. In which quadrants is the image of $\triangle DEB$ under a 90° rotation about the origin?
A. I and II **B.** II and III **C.** III and IV
D. I and IV **E.** none of the above

3. What is the next term in the sequence 1000, 200, 40, 8, . . . ?
A. 1.6 **B.** 3 **C.** 4 **D.** 0.16
E. none of the above

4. What is the converse of the statement "If a strawberry is red, then it is ripe"?
A. If a strawberry is not red, then it is not ripe.
B. If a strawberry is ripe, then it is red.
C. A strawberry is ripe if and only if it is red.
D. If a strawberry is not ripe, then it is not red.
E. A strawberry is not ripe if and only if it is not red.

5. Which quadrilateral cannot contain four right angles?
A. square **B.** trapezoid **C.** rectangle
D. rhombus **E.** parallelogram

6. A dilation centered at the origin maps $(-3, 6)$ to $(-9, 18)$. Which of the following does *not* represent the same dilation?
A. $(0, 5) \longrightarrow (0, 15)$ **B.** $(1, -4) \longrightarrow (3, -12)$
C. $(4, 3) \longrightarrow (9, 12)$ **D.** $(6, 2) \longrightarrow (18, 6)$
E. $(-2, -7) \longrightarrow (-6, -21)$

7. What is the length of a midsegment parallel to the side of a triangle 6 cm long?
A. 3 cm **B.** 9 cm **C.** 12 cm
D. 15 cm **E.** none of the above

8. What is the image of $(6, -9)$ under the translation $\langle 5, -2 \rangle$?
A. $(11, -11)$ **B.** $(11, -7)$ **C.** $(1, 11)$
D. $(-1, -11)$ **E.** none of the above

9. What type of polygon is shown?

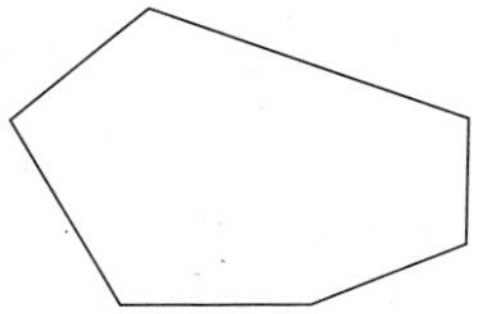

A. triangle **B.** quadrilateral **C.** hexagon
D. octagon **E.** none of the above

Compare the boxed quantity in Column A with the boxed quantity in Column B. Choose the best answer.

A. The quantity in Column A is greater.
B. The quantity in Column B is greater.
C. The two quantities are equal.
D. The relationship cannot be determined from the information given.

	Column A	Column B
	$\angle CDB$ and $\angle RDM$ are vertical angles.	
10.	$m\angle CDB$	$m\angle BDR$
	Lines ℓ and t are perpendicular.	
11.	the slope of ℓ	the slope of t

Find each answer.

12. **Open-ended** Write a conditional that has the same truth value as its converse.

13. Construct a right triangle. Then construct the bisectors of two of its angles.

14. **Coordinate Geometry** Find the locus of points in a plane equidistant from the lines $y = x$ and $y = -x$.

CHAPTER

5 Measuring in the Plane

Relating to the Real World

From architects to race car designers, many careers make use of the concepts you will learn in this chapter. You will find the perimeters and areas of polygons and circles. You will also study one of the most famous theorems in geometry: the Pythagorean Theorem.

CHAPTER PROJECT

And Sew On

Throughout history, people in all corners of the world have used patterns in their clothing, rugs, wall hangings, and blankets. Some of these articles were symbols of wealth and power. Today, textiles and fabrics still reflect social identity and cultural expression.

In your project for this chapter, you will explore patchwork techniques used by Native Americans and American pioneers. You will use these techniques to design your own quilt.

To help you complete the project:

- **p. 248** *Find Out by Modeling*
- **p. 254** *Find Out by Creating*
- **p. 273** *Find Out by Researching*
- **p. 290** *Find Out by Calculating*
- **p. 292** *Finishing the Project*

Areas of Regular Polygons — 5-6

Circles: Circumference and Arc Length — 5-7

Areas of Circles, Sectors, and Segments of Circles — 5-8

5-1 Understanding Perimeter and Area

What You'll Learn

- Finding area and perimeter of squares and rectangles

...And Why

To find the perimeters of banners, animal pens, and gardens

To find the surface area to be covered by carpet or by tiles

What You'll Need

- centimeter grid paper

WORK TOGETHER

In your group, draw each figure on centimeter grid paper.

- a rectangle with length 5 cm and width 3 cm
- a rectangle with base 8 cm and height 2 cm
- a rectangle with each side 4 cm

1. Record the perimeter of each rectangle.
2. Are the rectangles with equal perimeters congruent, similar, or neither?
3. To find the area of each rectangle, count the number of square centimeters in its interior. Record the area of each rectangle.
4. Are the rectangles with the same area congruent, similar, or neither?
5. Do the rectangles with equal perimeters have the same area?

THINK AND DISCUSS

Finding Perimeter

In the Work Together, you found the **perimeter of a polygon** by finding the sum of the lengths of the sides. In special cases such as squares and rectangles, you may use formulas for perimeter.

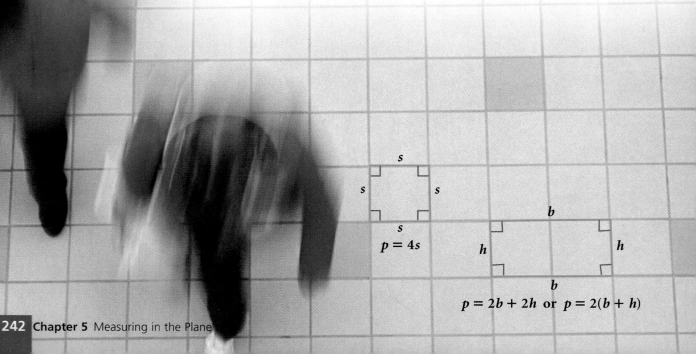

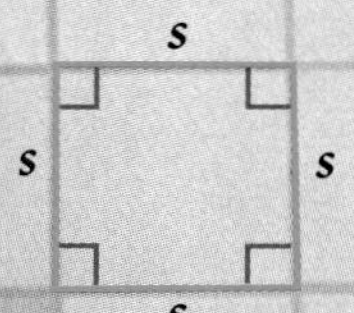

$p = 4s$

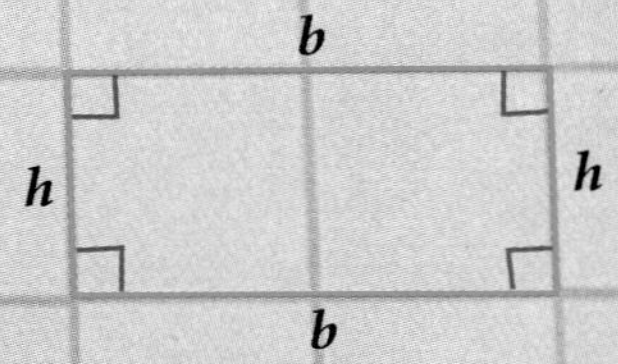

$p = 2b + 2h$ or $p = 2(b + h)$

6. Can you use the formula for the perimeter of a square to find the perimeter of any rectangle? Explain.

You can use the Distance Formula to find perimeter in the coordinate plane.

Example 1

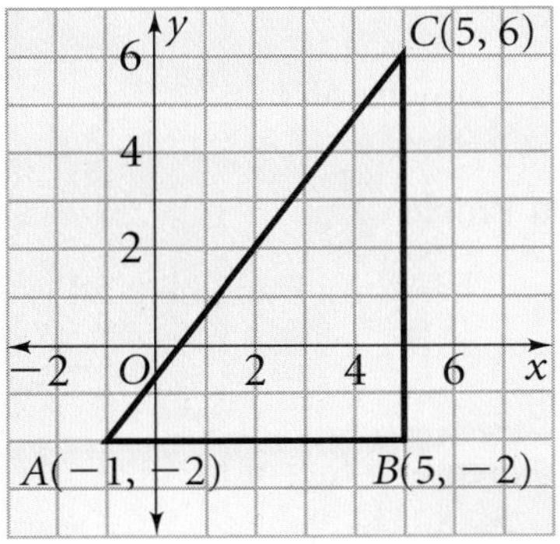

Coordinate Geometry Find the perimeter of $\triangle ABC$.

Find the length of each side. Add the lengths to find the perimeter.

$AB = 5 - (-1) = 6$

$BC = 6 - (-2) = 8$

$AC = \sqrt{(5 - (-1))^2 + (6 - (-2))^2}$ Use the Distance Formula.

$= \sqrt{6^2 + 8^2} = \sqrt{100} = 10$

$AB + BC + AC = 6 + 8 + 10 = 24$

The perimeter of $\triangle ABC$ is 24 units.

7. **Try This** Graph the quadrilateral with vertices $K(-3, -3)$, $L(1, -3)$, $M(1, 4)$, and $N(-3, 1)$. Find the perimeter of $KLMN$.

Finding Area

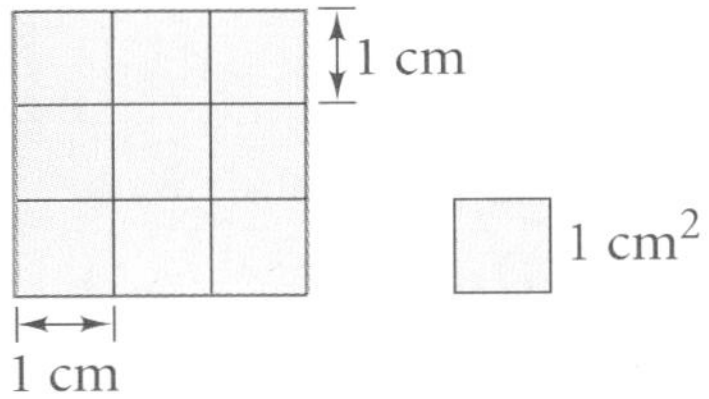

The **area of a polygon** is the number of square units enclosed by the polygon. The blue square at the left encloses nine smaller red squares. Each red square has sides 1 cm long and is called a square centimeter. By counting square centimeters, you see that the blue square has area 9 cm².

Postulate 5-1 The area of a square is the square of the length of a side.

$A = s^2$

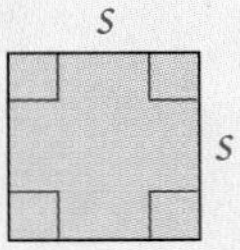

Postulate 5-2 If two figures are congruent, their areas are equal.

Postulate 5-3 The area of a region is the sum of the areas of its nonoverlapping parts.

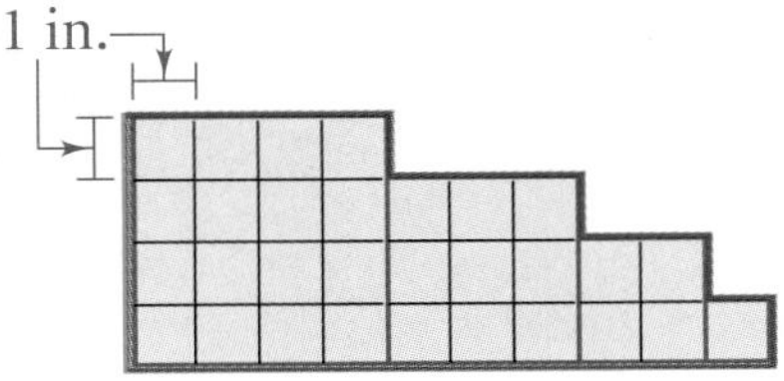

8. a. **Try This** What is the area of a square whose sides are 12 in. long?
   b. What is the area of a square whose sides are 1 ft long?
   c. How many square inches are in a square foot?

9. a. By counting squares, find the area of the polygon outlined in blue.
   b. Use Postulate 5-1 to find the area of each square outlined in red.
   c. How does the sum of your answers to part (b) compare to your answer to part (a)? Which postulate does this **verify?**

You can select any side of a rectangle to be the base. Because adjacent sides are perpendicular, the length of a side adjacent to the base is the height.

Theorem 5-1
Area of a Rectangle

The area of a rectangle is the product of its base and height.

$$A = bh$$

To find area, you must use the same units for all dimensions.

Example 2 Relating to the Real World

Design You are designing a rectangular banner. The banner will be 2 yd long and 4 ft wide. How much material will you need?

$2 \text{ yd} = 6 \text{ ft}$	Change units using 1 yd = 3 ft.
$A = bh$	Use the formula for the area of a rectangle.
$= 6(4) = 24$	Substitute 6 for *b* and 4 for *h*.

The area of the banner is 24 ft^2. You will need at least 24 ft^2 of material.

PROBLEM SOLVING

Look Back Find the area of the banner in Example 2 by first changing all units to yards.

10. **Try This** Find the area of a rectangle with length 75 cm and width 2 m.

You can use a graphing calculator or spreadsheet technology to find maximum and minimum values for area and perimeter problems.

Example 3 Relating to the Real World

Animal Science You have 32 yd of fencing. You want to make a rectangular pen for a calf you are raising for a 4-H project. What are the dimensions of the rectangle that will result in the maximum area? What is the maximum area?

Draw some possible rectangular pens and find their areas.

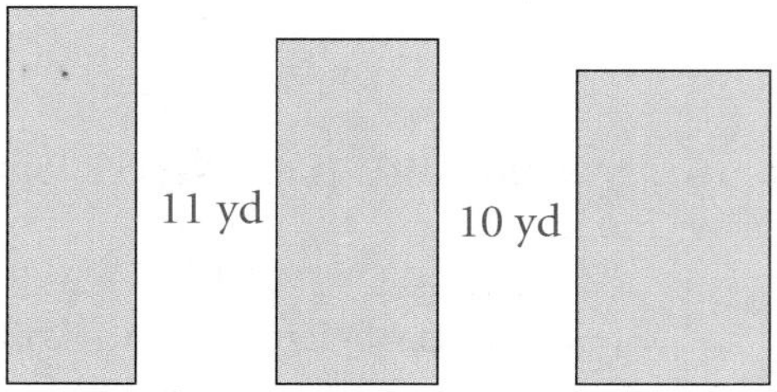

$A = 48 \text{ yd}^2$ $\quad A = 55 \text{ yd}^2$ $\quad A = 60 \text{ yd}^2$

	A	B	C
1	b	h = 16 − b	A = bh
2	1	15	15
3	2	14	28
4	3	13	39
5	...	...	...

Create a spreadsheet to find area. Choose values for b from 0 to 15.

Make a graph of the spreadsheet values. Graph values of b on the horizontal axis and values of A on the vertical axis. Connect the points with a smooth curve.

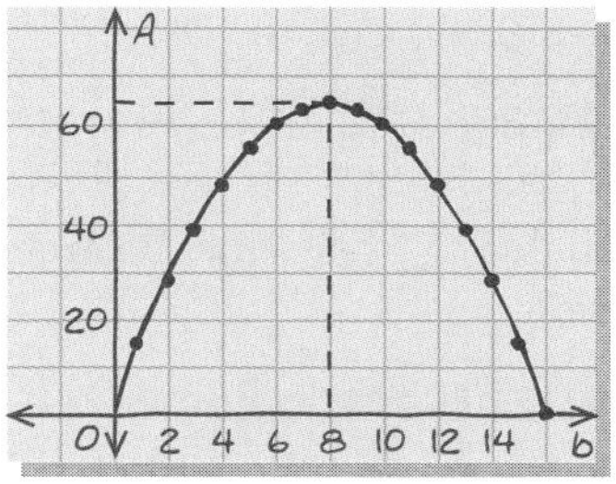

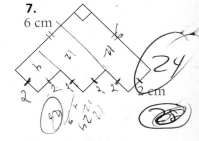

The maximum value occurs at $b = 8$. When $b = 8$, $h = 8$ and $A = 64$.

To have the maximum area for your calf, you should fence a square with sides 8 yd long. The maximum area is 64 yd^2.

11. Critical Thinking Show how the equation $h = 16 - b$ was derived from the formula $2b + 2h = 32$.

12. Use the answer to Example 3. Make a **conjecture** by completing this statement: If you have a fixed amount of fencing to enclose a rectangle, you can get the maximum area by enclosing a ___?___.

Exercises ON YOUR OWN

Estimation **Estimate the perimeter of each item.**

1. the cover of this book **2.** the cover of your notebook **3.** a classroom bulletin board

Mental Math **Find the perimeter of each figure.**

4.

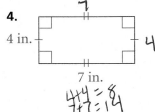

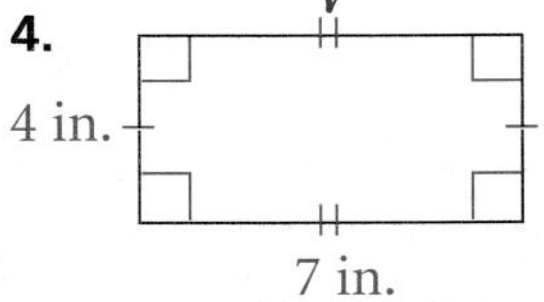

5.

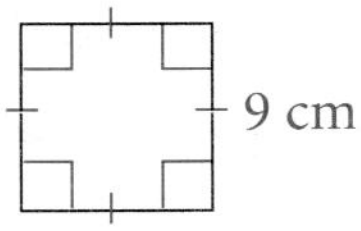

6.

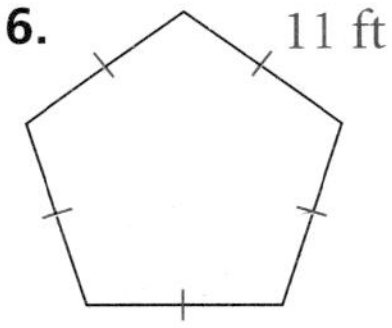

7.

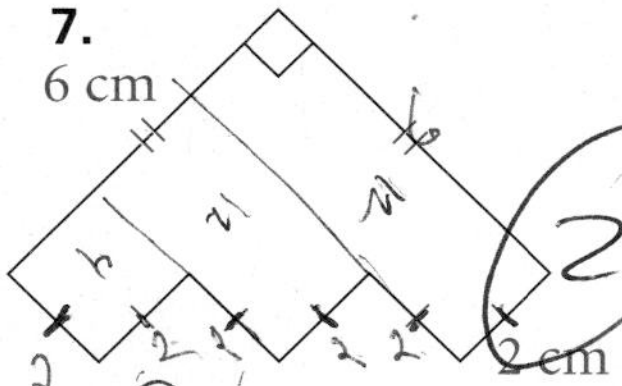

Find the perimeter of each rectangle with the given base and height.

8. 21 in., 7 in. **9.** 16 cm, 23 cm **10.** 24 m, 36 m **11.** 14 ft, 23 ft

The figures below are drawn on centimeter graph paper. Find the area of the shaded portion of each figure.

12.

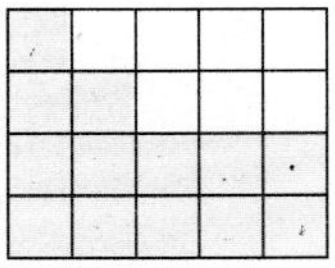

13.

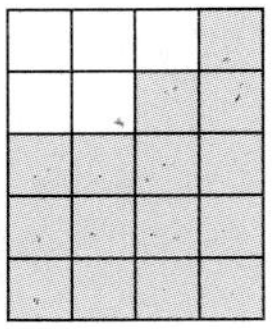

14.

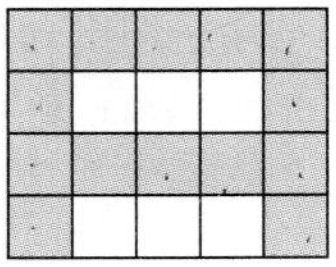

15.

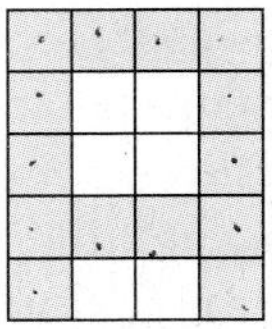

Coordinate Geometry **Graph each rectangle *ABCD* and find its area.**

16. $A(0, 0)$, $B(0, 4)$, $C(5, 4)$, $D(5, 0)$

17. $A(1, 4)$, $B(1, 7)$, $C(5, 7)$, $D(5, 4)$

18. $A(-3, 2)$, $B(-2, 2)$, $C(-2, -2)$, $D(-3, -2)$

19. $A(-2, -6)$, $B(-2, -3)$, $C(3, -3)$, $D(3, -6)$

20. A rectangle is 11 cm wide. Its area is 176 cm^2. What is the length of the rectangle?

21. **Coordinate Geometry** Points $A(1, 1)$, $B(10, 1)$, $C(10, 8)$, $D(7, 8)$, $E(7, 5)$, $F(4, 5)$, $G(4, 8)$, and $H(1, 8)$ are the coordinates of the vertices of polygon *ABCDEFGH*.

a. Draw the polygon on graph paper.
b. Find the perimeter of the polygon.
c. Divide the polygon into rectangles.
d. Find the area of the polygon.

22. The perimeter of a rectangle is 40 cm and the base is 12 cm. What is the area?

23. A square and a rectangle have equal areas. The rectangle is 64 cm by 81 cm. What is the perimeter of the square?

Find the area of each rectangle with the given base and height.

24. 4 ft 6 in., 4 in.

25. 1 yd 18 in., 4 yd

26. 2 ft 3 in., 6 in.

Building Safe Stairs

Since falls are a major cause of injury, it makes sense to be concerned about the safety of stairs. According to John Templer, the world's foremost authority on stairs, steps with a 7-in. riser and an 11-in. tread form the safest possible stairs.

Prior to his investigations, Francois Blondel's formula, dated 1675, had recommended that stair measurements conform to the formula 2(riser) + tread = 25.5 in.

Source: *Smithsonian*

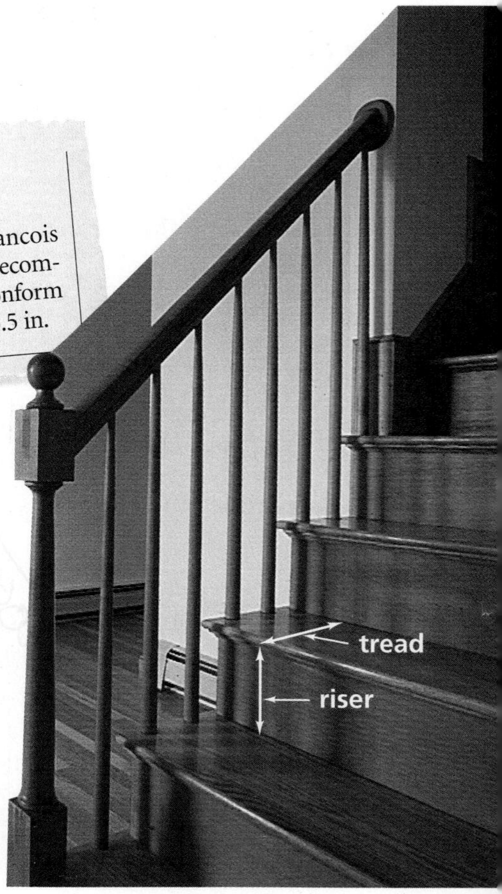

27. **Carpeting** You use John Templer's dimensions to build a stairway with six steps. You want to carpet the stairs with a 3-ft wide runner from the bottom of the first riser to the top of the sixth riser.

a. Find the area of the runner.
b. Since a roll of carpet is 12 ft across, a rectangle of carpet that measures 3 ft by 12 ft is cut from the roll to make the runner. How many square feet of the material will be wasted?
c. The carpet costs $17.95/$yd^2$. You must pay for the entire piece that is cut. Find the cost of the carpet.
d. Binding for the edge of the runner costs $1.75/yd. How much will the binding cost if the two long edges of the runner are bound?

Writing Tell whether you need to know area or perimeter in order to determine how much of each item to buy. Explain your choice.

28. edging for a garden

29. paint for a basement floor

30. wallpaper for a bedroom

31. weatherstripping for a door

32. **Tiling** The Art Club is tiling an 8 ft-by-16 ft wall at the entrance to the school. They are creating a design by using different colors of 4 in.-by-4 in. tiles. How many tiles do the students need?

33. **Gardening** You want to make a 900-ft^2 rectangular garden to grow corn. In order to keep raccoons out of your corn, you must fence the garden. You want to use the minimum amount of fencing so that your costs will be as low as possible.

- **a.** List some possible dimensions for the rectangular garden. Find the perimeter of each rectangle.
- **b.** **Technology** Create a spreadsheet listing integer values of b and the corresponding values of h and P. What dimensions will give you a garden with the minimum perimeter?

34. You want to build a rectangular corral by using one side of a barn and fencing the other three sides. You have enough material to build 100 ft of fence.

- **a.** **Technology** Create a spreadsheet listing integer values of b and the corresponding values of h and A.
- **b.** **Coordinate Geometry** Make a graph using your spreadsheet values. Graph b on the horizontal axis and A on the vertical axis.
- **c.** Describe the dimensions of the corral with the greatest area.

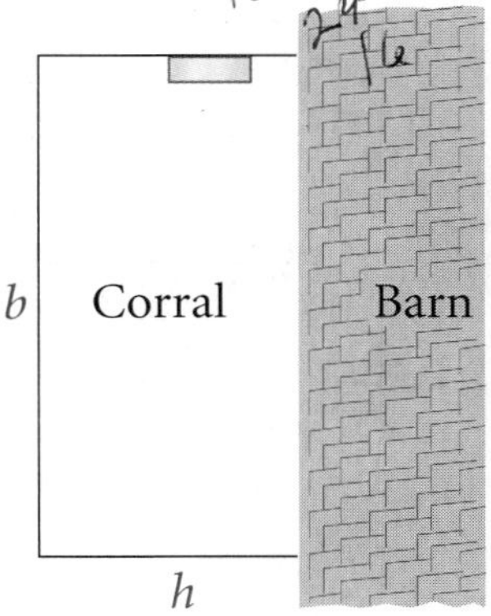

Find the area of the shaded portion of each figure. All angles in the figures are right angles.

35.

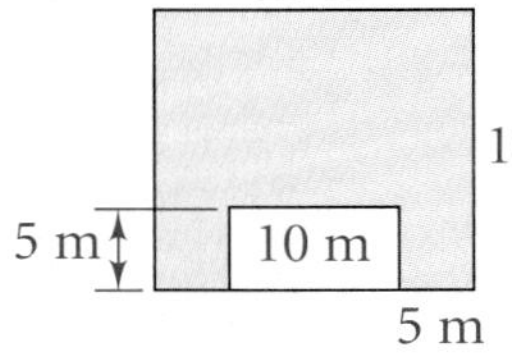

36.

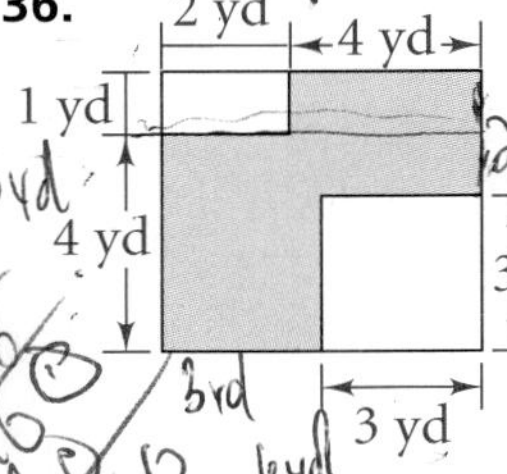

37.

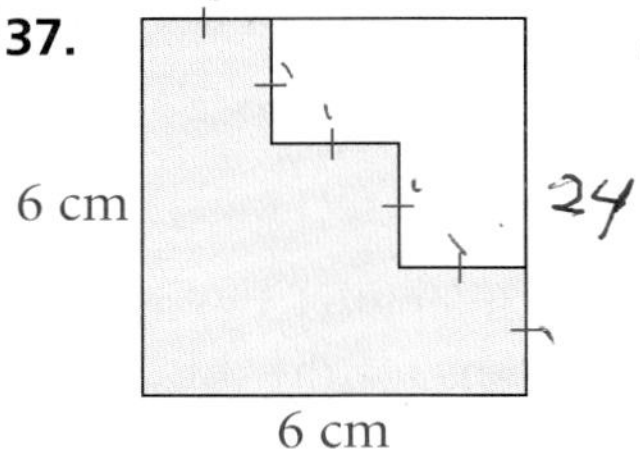

38.

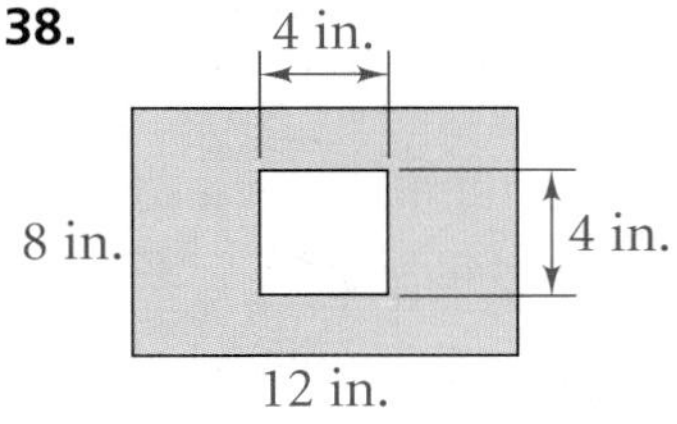

Coordinate Geometry Graph each quadrilateral *ABCD*. Find its perimeter.

39. $A(-2, 2), B(0, 2), C(4, -1), D(-2, -1)$

40. $A(-4, -1), B(4, 5), C(4, -2), D(-4, -2)$

41. $A(0, 1), B(3, 5), C(5, 5), D(5, 1)$

42. $A(-5, 3), B(7, -2), C(7, -6), D(-5, -6)$

43. **Open-ended** The area of a 5 in.-by-5 in. square is the same as the sum of the areas of a 3 in.-by-3 in. square and a 4 in.-by-4 in. square. Find two or more squares that have the same total area as an 11 in.-by-11 in. square.

PROBLEM SOLVING HINT

Make a list of perfect squares.

44. Standardized Test Prep The length of a rectangle is increased by 50% and the width is decreased by 50%. How is the area affected?

A. increased by 25% **B.** decreased by 25% **C.** increased by 50%
D. decreased by 50% **E.** unchanged

Chapter Project Find Out by Modeling

You can create a quilt by sewing together congruent squares to form blocks. To model a quilt block, cut four 3 in.-by-3 in. squares out of $\frac{1}{4}$-in. graph paper. Place one square on top of another and make a seam by stapling the two squares together $\frac{1}{4}$ in. from one of the edges. Unfold the squares and press the seam flat in the back.
Repeat this with the two other squares. Then place the two sections on top of each other. Staple a $\frac{1}{4}$-in. seam from one end to the other. Unfold and press the seams back.

- What is the total area of the four paper squares that you started with?
- What is the area of your finished quilt block?

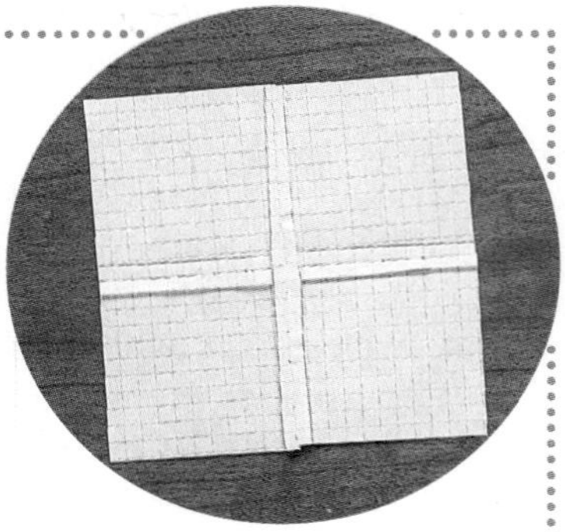

Exercises MIXED REVIEW

Coordinate Geometry **Find the coordinates of the midpoint of a segment with the given endpoints.**

45. $A(4, 1), B(7, 9)$ **46.** $G(0, 3), H(3, 8)$

47. $R(-2, 7), S(-6, -1)$

Write the converse of each conditional.

48. If you make a touchdown, then you score six points.

49. If it is Thanksgiving, then it is November.

50. If a figure is a square, then it is a rectangle.

51. A triangle has two sides with lengths 3 m and 5 m. What is the range of possible lengths for the third side?

SELF ASSESSMENT

FOR YOUR JOURNAL

Explain the difference between area and perimeter. Use examples to show how and when each type of measurement might be used.

Getting Ready for Lesson 5-2

Each rectangle is divided into two congruent triangles. Find the area of each triangle.

52. 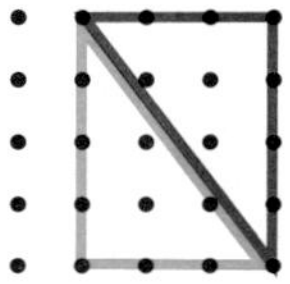

53. 

54. 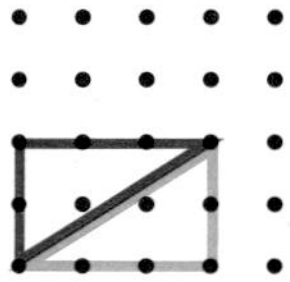

55. 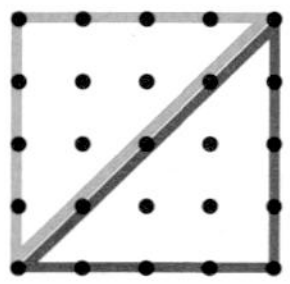

What You'll Learn

- Finding areas of parallelograms and triangles

...And Why

To solve design problems in architecture and landscaping

What You'll Need

- centimeter grid paper
- straightedge
- scissors
- tape

5-2 Areas of Parallelograms and Triangles

WORK TOGETHER

Have each member of your group cut out a different rectangle from centimeter grid paper.

- Record the base, height, and area of each rectangle.
- Cut out a triangle from one side of the rectangle as shown below. Tape it to the opposite side to form a parallelogram.

1. Compare each original rectangle with the parallelogram formed. With your group, list all the ways the rectangle and the parallelogram are the same and all the ways they are different.

THINK AND DISCUSS

Areas of Parallelograms

In the Work Together, you cut a rectangle into two pieces and used the pieces to form another parallelogram. The area of the parallelogram was the same as the area of the rectangle. This suggests the following theorem.

Theorem 5-2
Area of a Parallelogram

The area of a parallelogram is the product of any base and the corresponding height.

$$A = bh$$

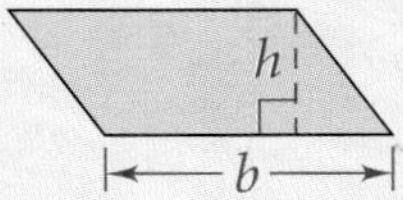

You can choose any side to be a **base** of a parallelogram. An **altitude** is any segment perpendicular to the line containing the base drawn from the side opposite the base. The **height** is the length of the altitude.

2. Draw any parallelogram and draw altitudes to two adjacent sides.

Example 1

Coordinate Geometry What is the area of ▱*PQRS* with vertices *P*(1, 2), *Q*(6, 2), *R*(8, 5), and *S*(3, 5)?

Graph ▱*PQRS*. If you choose $\overline{PQ}$ as the base, then the height is 3.

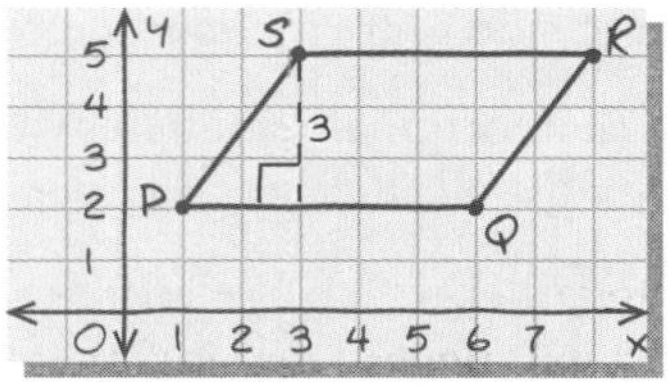

$$b = PQ = 5$$
$$h = 3$$
$$A = bh = 5(3)$$
$$= 15$$

▱*PQRS* has area 15 square units.

3. Try This What is the area of ▱*EFGH* with vertices *E*(−4, 3), *F*(0, 3), *G*(1, −2), and *H*(−3, −2)?

You can use the area formula to find missing dimensions in a parallelogram.

Example 2

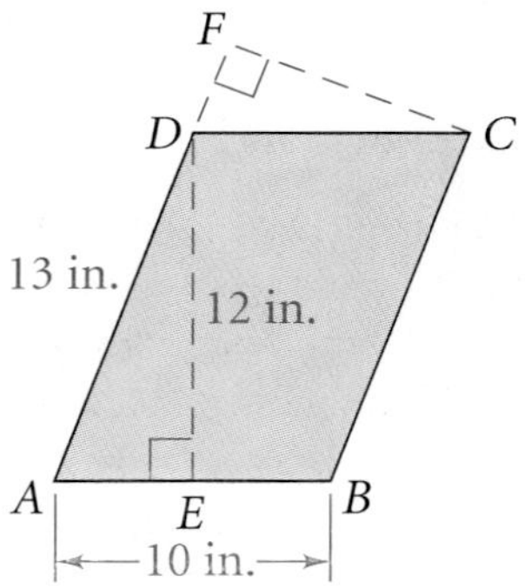

In ▱*ABCD*, $\overline{DE}$ and $\overline{CF}$ are altitudes. Find *CF* to the nearest tenth.

Find the area of ▱*ABCD*. Then use the area formula to find *CF*.

$A = bh$

$= 10(12)$ Use base *AB* and height *DE*.

$= 120$

The area of ▱*ABCD* is 120 in.^{2}.

$A = bh$

$120 = 13(CF)$ Use base *AD* and height *CF*.

$CF = \frac{120}{13}$ Divide each side by 13.

≈ 9.2

$\overline{CF}$ is about 9.2 in. long.

4. Try This A parallelogram has sides 15 cm and 18 cm. The altitude perpendicular to the line containing the 15 cm side is 9 cm long. Sketch the parallelogram. Then find the length of the altitude perpendicular to the line containing the 18-cm side.

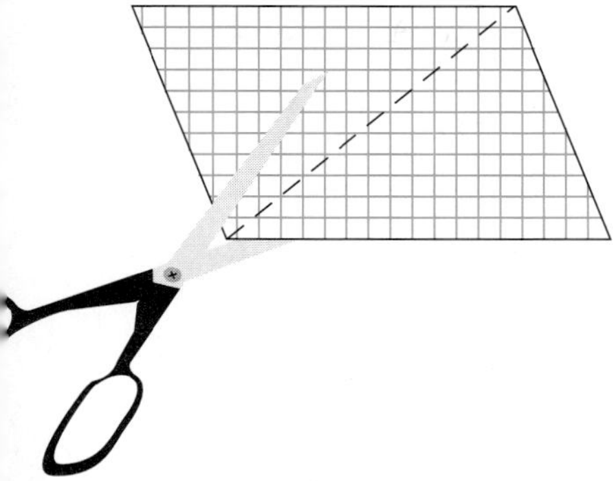

WORK TOGETHER

Work in groups. Have each member of your group cut out a different parallelogram from centimeter grid paper.

- Record the base, height, and area of each parallelogram.
- Cut each parallelogram along a diagonal as shown, forming two triangles.

5. How does the area of each triangle compare to the area of the parallelogram?

THINK AND DISCUSS

Areas of Triangles

In the Work Together, you cut a parallelogram into two congruent triangles of equal area. This suggests the following theorem.

Theorem 5-3
Area of a Triangle

The area of a triangle is half the product of any base and the corresponding height.

$A = \frac{1}{2}bh$

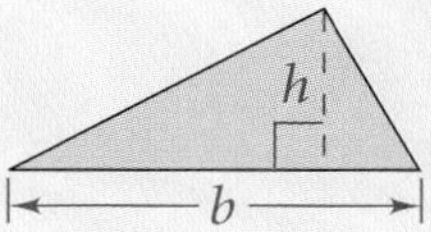

You can choose any side to be a **base** of a triangle. The corresponding **height** is the length of an altitude drawn to the line containing that base.

Example 3 ***Relating to the Real World***

Architecture When designing a building, an architect must be sure that the building can stand up to hurricane force winds, which have a velocity of 73 mi/h or more. The formula $F = 0.004Av^2$ gives the force F in pounds exerted by a wind blowing against a flat surface. A is the area of the surface in square feet, and v is the wind velocity in miles per hour. How much force is exerted by a 73 mi/h wind blowing directly against the side of this building?

Find the area of the side of the building.

triangle area $= \frac{1}{2}bh = \frac{1}{2}(20)6 = 60 \text{ ft}^2$

rectangle area $= bh = 20(12) = 240 \text{ ft}^2$

area of end of building $= 60 + 240 = 300 \text{ ft}^2$

$F = 0.004Av^2$ — **Use the formula for force.**

$= 0.004(300)(73)^2$ — **Substitute 300 for A and 73 for v.**

$= 6394.8$

The force is about 6400 lb.

Exercises ON YOUR OWN

Find the area of each figure.

1. $\square ABJF$
2. $\triangle BDJ$
3. $\triangle DKJ$
4. $\square BDKJ$
5. $\square ADKF$
6. $\triangle BCJ$

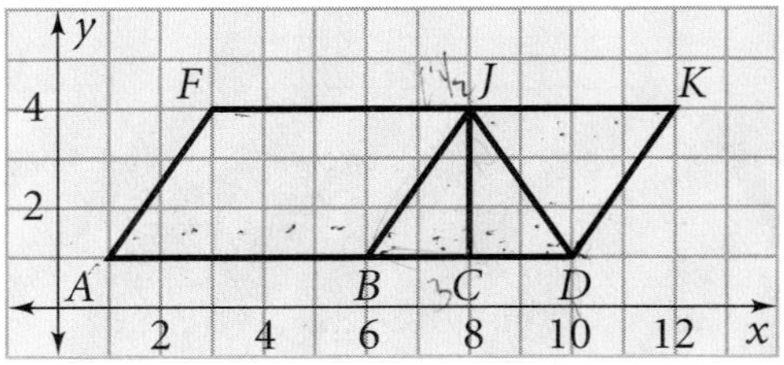

7. The area of a parallelogram is 24 in.^{2} and the height is 6 in. Find the length of the base.
8. An isosceles right triangle has area of 98 cm^2. Find the length of each leg.

Find the area of each shaded region.

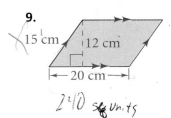

9.

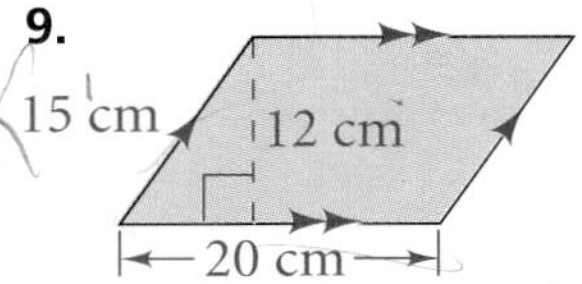

10.

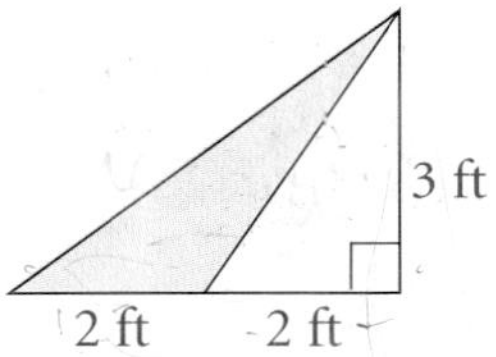

11.

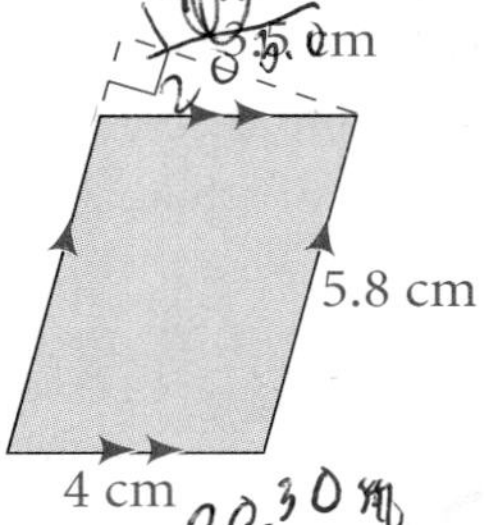

12.

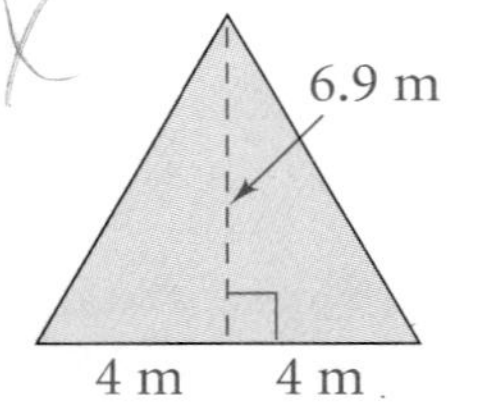

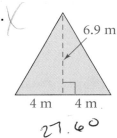

Coordinate Geometry **(a) Graph the lines. (b) Find the area of the triangle enclosed by the lines.**

13. $y = x, x = 0$, and $y = 7$
14. $y = x + 2, y = 2, x = 6$
15. $y = -\frac{1}{2}x + 3, y = 0, x = -2$

Find the value of *h* in each parallelogram.

16.

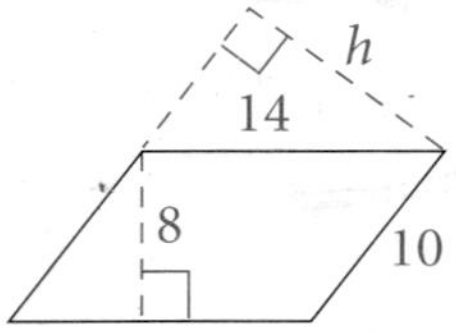

17.

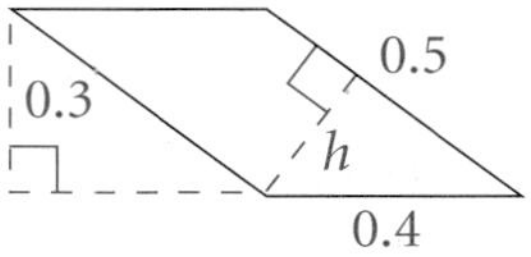

18.

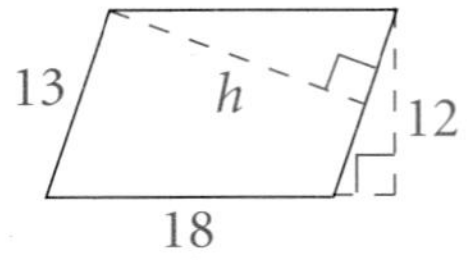

19. **Landscaping** Taisha's Bakery has a plan for a 50 ft-by-31 ft parking lot. The four parking spaces are congruent parallelograms, the driving area is a rectangle, and the two unpaved areas for flowers are congruent triangles.
   a. **Writing** Explain two different ways to find the area of the region that must be paved.
   b. **Verify** your answer to part (a) by using each method to find the area.

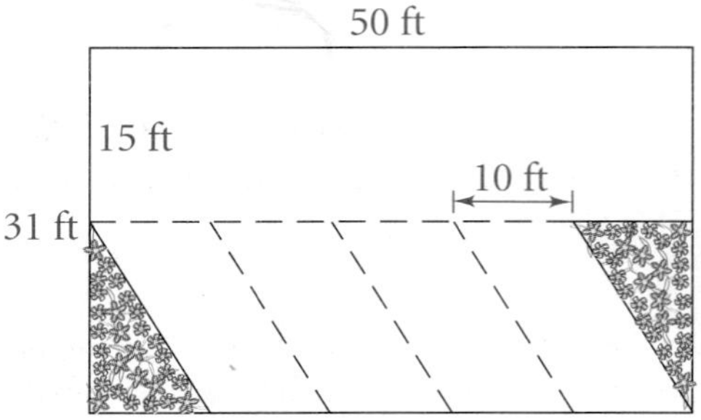

20. **Algebra** In a triangle, a base and the corresponding height are in the ratio 3 : 2. The area is 108 in.^{2}. Find the base and the corresponding height.

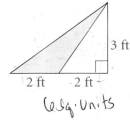

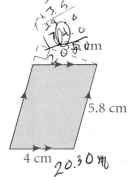

Find the area of each figure.

21. 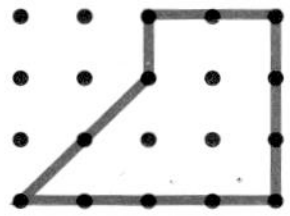

22. 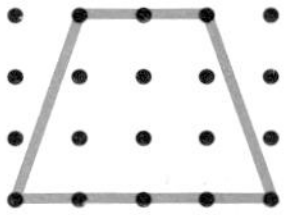

23. 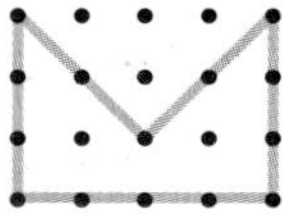

24. **Probability** Ann drew these three figures on a grid. A fly landed at random at a point on the grid.

- **a.** Is the fly more likely to have landed on one of the figures or on the blank grid? Explain.
- **b.** Suppose you know the fly landed on one of the figures. Is the fly more likely to have landed on one figure than on another? Explain.

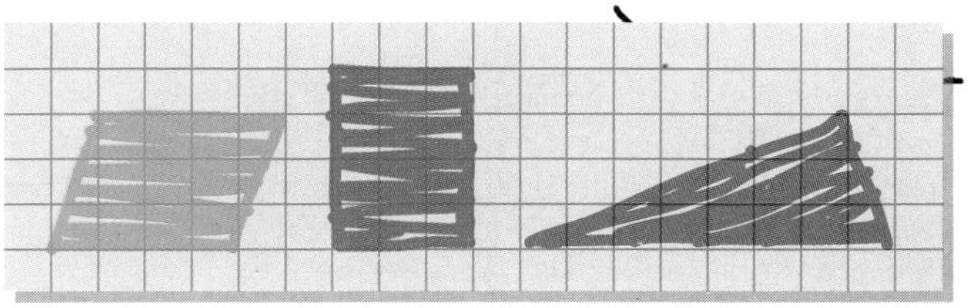

Find the area of each figure.

25. 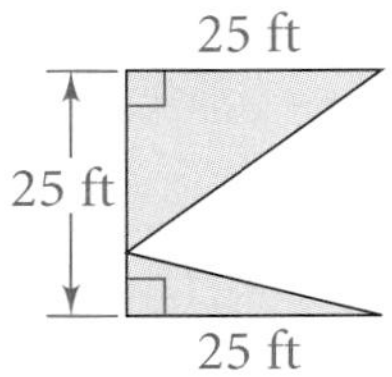

26. 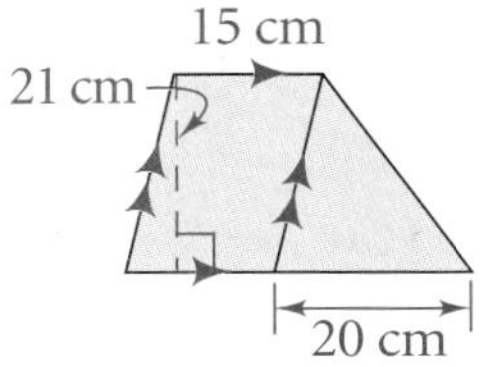

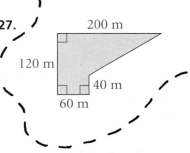

27. 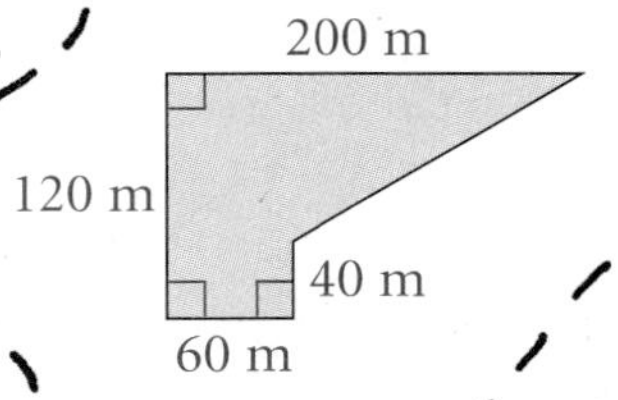

28. **Open-ended** Using graph paper, draw an acute triangle, an obtuse triangle, and a right triangle, each with area 12 units^2.

29. **Technology** Juanita used geometry software to create the figure at the right. She drew segment $\overline{AB}$, chose point C, and constructed line k parallel to $\overline{AB}$ through point C. Then Juanita chose point D on line k. Next she dragged point D along line k to form different triangles. How do the areas of the triangles compare? Explain.

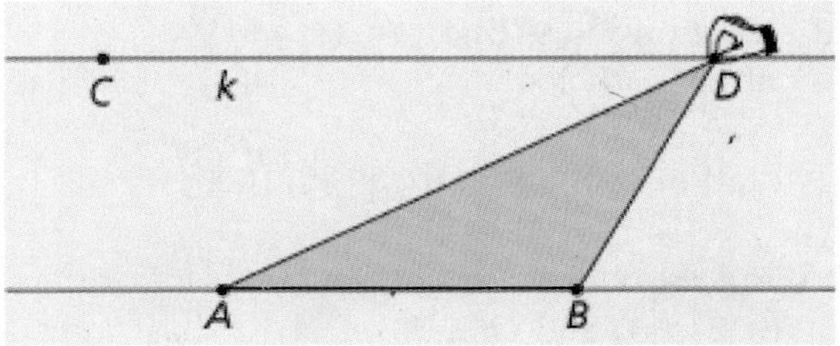

The ancient Greek mathematician Heron is most famous for his formula for the area of a triangle in terms of its sides *a*, *b*, and *c*.

$$A = \sqrt{s(s-a)(s-b)(s-c)}, \text{ where } s = \tfrac{1}{2}(a+b+c)$$

Use Heron's formula and a calculator to find the area of each triangle. Round your answer to the nearest whole number.

30. $a = 8$ in., $b = 9$ in., $c = 10$ in.

31. $a = 15$ m, $b = 17$ m, $c = 21$ m

32.
- **a.** Use Heron's formula to find the area of the triangle at the right.
- **b.** **Verify** your answer to part (a) by using the formula $A = \frac{1}{2}bh$.

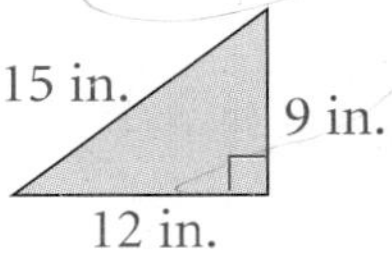

Coordinate Geometry **The vertices of a polygon are given. Graph each polygon and find its area.**

33. $A(3, 9), B(8, 9), C(2, -3), D(-3, -3)$

34. $E(1, 1), F(4, 5), G(11, 5), H(8, 1)$

35. $M(-2, -5), L(1, -5), N(2, -2)$

36. $R(1, 2), S(1, 6), T(4, 1)$

Chapter Project

Find Out by Creating

Your class can model a quilt by using the quilt blocks your classmates created in the *Find Out* activity on page 248. Here is one suggestion for a design.

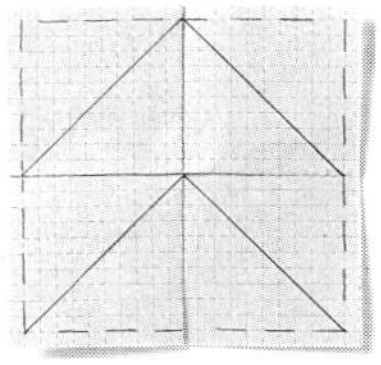

On each block, mark off a $\frac{1}{4}$-in. border for seams. Draw the four diagonals pictured.

Staple four blocks together in a row, keeping the orientation shown at the left throughout the row. Do this until you have four rows.

Staple the rows together, turning the second and fourth row upside down. Color the blocks to create a three-dimensional illusion.

Exercises MIXED REVIEW

Transformations **Find the coordinates of the images of *A, B, C,* and *D* after each transformation.**

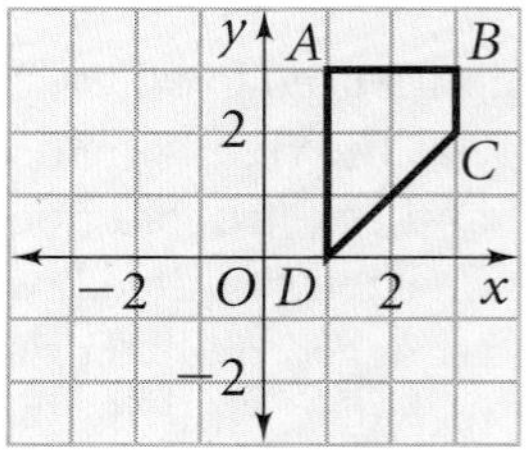

37. reflection in the line $x = 1$

38. translation $\langle -4, -7 \rangle$

39. rotation 180° about the point (0, 0)

40. Find the coordinates of the midpoint of the segment joining $P(-2, -3)$ and $Q(9, 12)$.

Getting Ready for Lesson 5-3

Square the lengths of the sides of each triangle. What do you notice?

41.

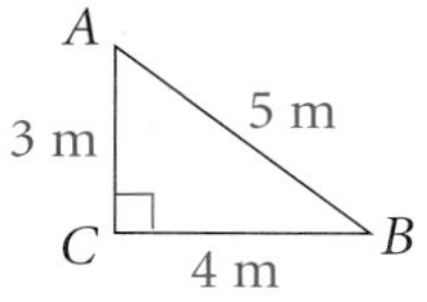

42.

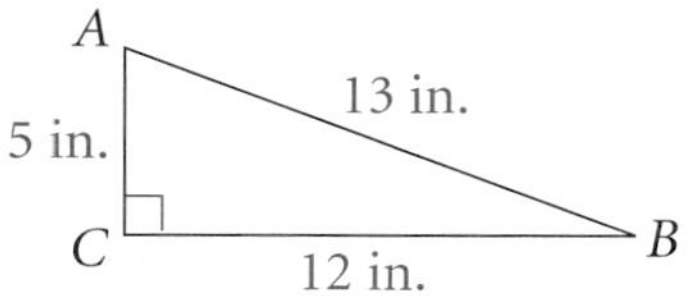

43.

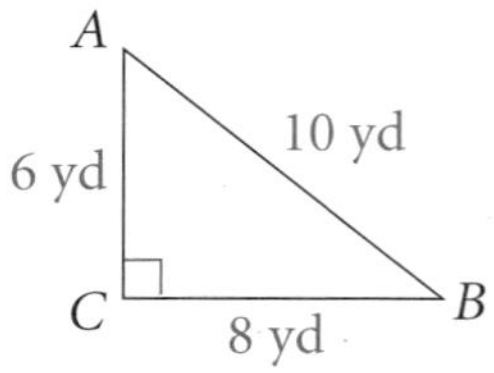

Math ToolboX

Algebra Review

Simplifying Radicals

Before Lesson 5-3

You can multiply and divide numbers that are under radical signs.

Example 1

Simplify the expressions $\sqrt{2} \cdot \sqrt{8}$ and $\sqrt{294} \div \sqrt{3}$.

$\sqrt{2} \cdot \sqrt{8} = \sqrt{2 \cdot 8}$ Write both numbers under one radical.

$= \sqrt{16}$ Simplify the expression under the radical.

$= 4$ Factor out perfect squares and simplify.

$$\sqrt{294} \div \sqrt{3} = \sqrt{\frac{294}{3}} = \sqrt{98} = \sqrt{49 \cdot 2} = 7\sqrt{2}$$

A radical expression is in simplest radical form when all the following are true.

- The number under the radical sign has no perfect square factors other than 1.
- The number under the radical sign does not contain a fraction.
- The denominator does not contain a radical expression.

Example 2

Write $\sqrt{\frac{4}{3}}$ in simplest form.

$\sqrt{\frac{4}{3}} = \frac{\sqrt{4}}{\sqrt{3}}$ Rewrite the single radical as the quotient of two radicals.

$= \frac{2}{\sqrt{3}}$ Simplify.

$= \frac{2}{\sqrt{3}} \cdot \frac{\sqrt{3}}{\sqrt{3}}$ Multiply by a form of 1 to rationalize the denominator.

$= \frac{2\sqrt{3}}{3}$

Simplify each expression.

1. $\sqrt{5} \cdot \sqrt{10}$ **2.** $\sqrt{243}$ **3.** $\sqrt{128} \div \sqrt{2}$ **4.** $\sqrt{\frac{125}{4}}$

5. $\sqrt{6} \cdot \sqrt{8}$ **6.** $\frac{\sqrt{36}}{\sqrt{3}}$ **7.** $\frac{\sqrt{144}}{\sqrt{2}}$ **8.** $\sqrt{3} \cdot \sqrt{12}$

9. $\sqrt{72} \div \sqrt{2}$ **10.** $\sqrt{169}$ **11.** $24 \div \sqrt{8}$ **12.** $\sqrt{300} \div \sqrt{5}$

Connections Recreation . . . and more

5-3 The Pythagorean Theorem and Its Converse

What You'll Learn

- Using the Pythagorean Theorem and its converse

...And Why

To solve problems involving boundaries, packaging, and satellites

What You'll Need

scissors, graph paper, colored paper, straightedge

WORK TOGETHER

Work in groups. Using graph paper, draw any rectangle. Label the sides *a* and *b*. Cut four rectangles with length *a* and width *b* from the graph paper. Then cut each rectangle on its diagonal, *c*, forming eight congruent triangles.

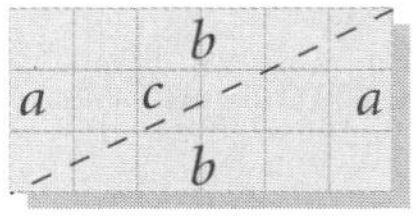

Cut three squares from the colored paper, one with sides of length *a*, one with sides of length *b*, and one with sides of length *c*.

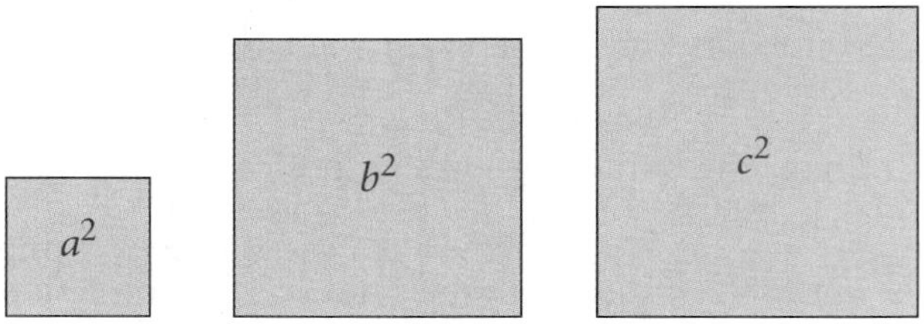

Separate the pieces into groups.

Group 1: four triangles and the two smaller squares

Group 2: four triangles and the largest square

Arrange the pieces of each group to form a square.

1. Write an algebraic expression for the area of each of the squares you formed.
2. How do the areas of the two squares you formed compare?
3. What can you conclude about the areas of the squares you cut from colored paper?
4. Express your conclusion as an algebraic equation.

THINK AND DISCUSS

The Pythagorean Theorem

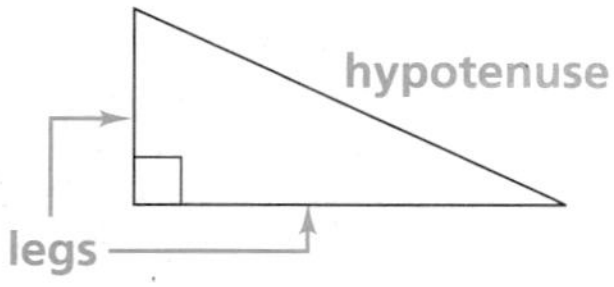

In a right triangle, the side opposite the right angle is the longest side. It is the **hypotenuse.** The other two sides are the **legs of a right triangle.**

The Work Together presents a justification of the well-known right triangle relationship called the Pythagorean Theorem.

Theorem 5-4
Pythagorean Theorem

In a right triangle, the sum of the squares of the lengths of the legs is equal to the square of the length of the hypotenuse.

$$a^2 + b^2 = c^2$$

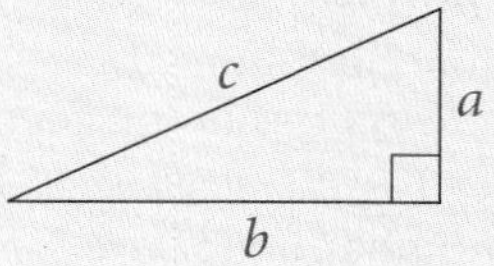

5. a. A right triangle has sides of lengths 20, 29, and 21. What is the length of the hypotenuse?

b. Verify that the Pythagorean Theorem is true for the right triangle in part (a).

Example 1 ***Relating to the Real World***

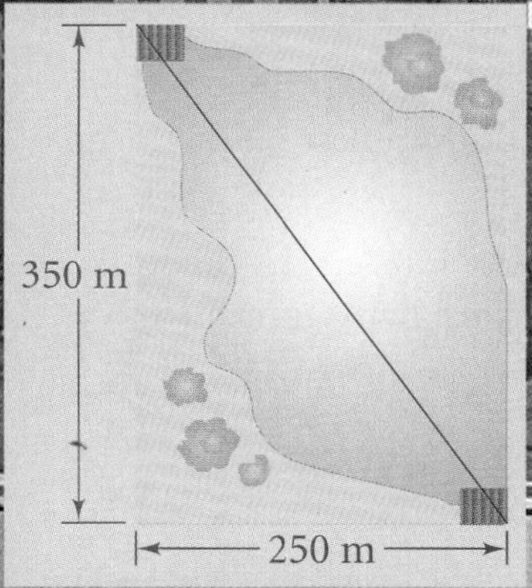

Recreation A city park department rents paddle boats at docks near each entrance to the park. About how far is it to paddle from one dock to the other?

You can find the distance between the two docks by finding the hypotenuse of the right triangle.

$a^2 + b^2 = c^2$	Use the Pythagorean Theorem.
$250^2 + 350^2 = c^2$	Substitute 250 for *a* and 350 for *b*.
$62{,}500 + 122{,}500 = c^2$	Simplify.
$185{,}000 = c^2$	
$c = \sqrt{185{,}000}$	Find the square root.

185,000 [√] [=] 430.11626

It is about 430 m from one dock to the other.

6. Try This Find the length of the hypotenuse of a right triangle with legs of lengths 7 and 24.

QUICK REVIEW

A radical expression is in simplest radical form when all the following are true.

- The number under the radical sign has no perfect square factors other than 1.
- The number under the radical sign does not contain a fraction.
- The denominator does not contain a radical expression.

For practice with radical expressions, see Skills Handbook page 660.

Sometimes you will leave your answer in simplest radical form.

Example 2

Find the value of x. Leave your answer in simplest radical form.

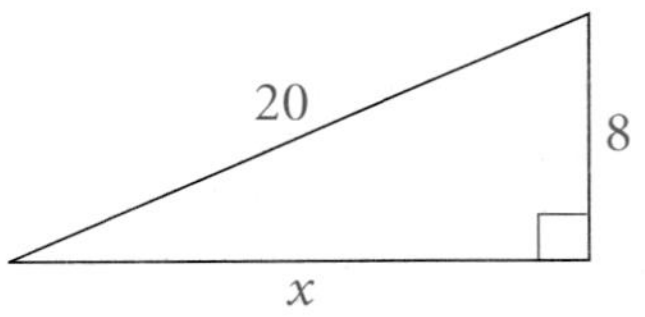

Use the Pythagorean Theorem.

$a^2 + b^2 = c^2$

$8^2 + x^2 = 20^2$ — Substitute.

$64 + x^2 = 400$ — Simplify.

$x^2 = 336$ — Subtract 64 from each side.

$x = \sqrt{336}$ — Find the square root.

$x = \sqrt{16(21)}$ — Simplify.

$x = 4\sqrt{21}$

7. **Try This** The hypotenuse of a right triangle has length 12. One leg has length 6. Find the length of the other leg in simplest radical form.

When the lengths of the sides of a right triangle are integers, the integers form a **Pythagorean triple.** Here are some common Pythagorean triples.

3, 4, 5   5, 12, 13   8, 15, 17   7, 24, 25

8. *Open-ended* Choose an integer. Multiply each number of a Pythagorean triple by that integer. **Verify** that the result is a Pythagorean triple.

The Converse of the Pythagorean Theorem

You can use the Converse of the Pythagorean Theorem to determine whether a triangle is a right triangle.

Theorem 5-5
Converse of the Pythagorean Theorem

If the square of the length of one side of a triangle is equal to the sum of the squares of the lengths of the other two sides, then the triangle is a right triangle.

The Converse of the Pythagorean Theorem leads to the inequalities below. You can use them to determine whether a triangle is obtuse or acute.

In $\triangle ABC$ with longest side c,
if $c^2 > a^2 + b^2$, then the triangle is obtuse, and
if $c^2 < a^2 + b^2$, then the triangle is acute.

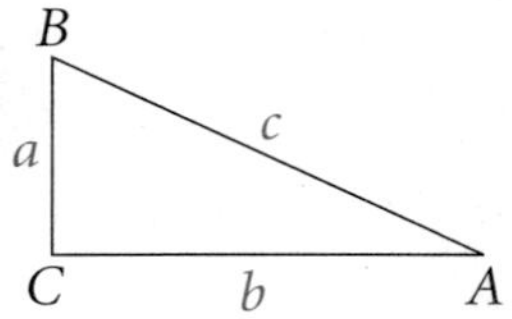

Who? Czech-American mathematician Olga Taussky-Todd (1906–1995) studied Pythagorean triangles. In 1970, she won the Ford Prize for her research.

When? The Pythagorean Theorem is named for Pythagoras, a Greek mathematician who lived in the sixth century B.C. We now know that the Babylonians, Egyptians, and Chinese were aware of this relationship before Pythagoras.

The diagram below illustrates an ancient Greek proof of the Pythagorean Theorem for an isosceles right triangle.

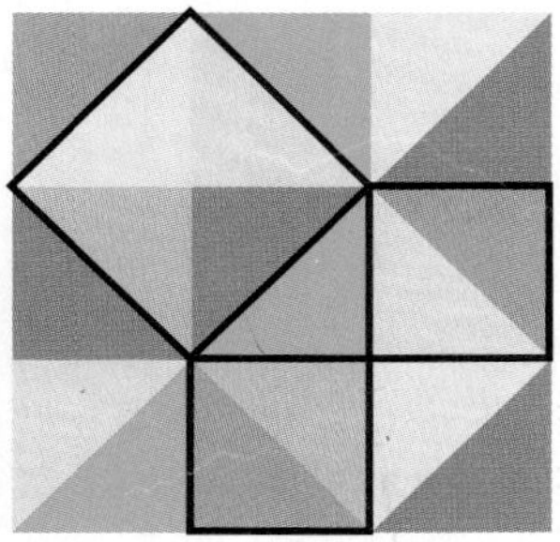

Example 3

The numbers represent the lengths of the sides of a triangle. Classify each triangle as acute, obtuse, or right.

a. 13, 84, 85

$$85^2 \stackrel{?}{=} 13^2 + 84^2$$
$$7225 \stackrel{?}{=} 169 + 7056$$
$$7225 = 7225$$

Compare c^2 to $a^2 + b^2$. Substitute the length of the longest side for *c*. $c^2 = a^2 + b^2$

The triangle is a right triangle.

b. 6, 11, 14

$$14^2 \stackrel{?}{=} 6^2 + 11^2$$
$$196 \stackrel{?}{=} 36 + 121$$
$$196 > 157$$

Compare c^2 to $a^2 + b^2$. Substitute the length of the longest side for *c*. $c^2 > a^2 + b^2$

The triangle is an obtuse triangle.

9. Try This A triangle has sides of lengths 7, 8, and 9. Classify the triangle as acute, obtuse, or right.

Exercises ON YOUR OWN

Algebra **Find the value of *x*. Leave your answer in simplest radical form.**

1.

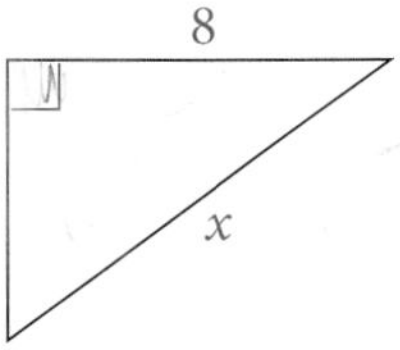

2.

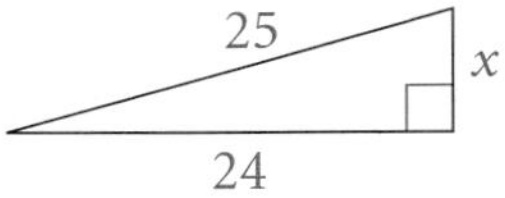

3.

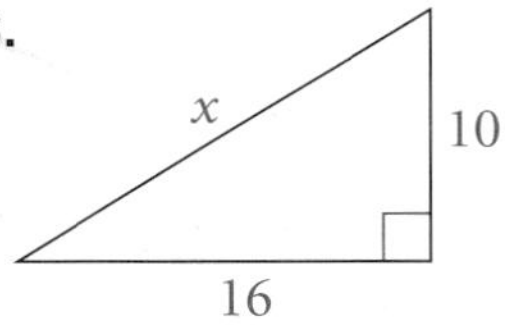

4.

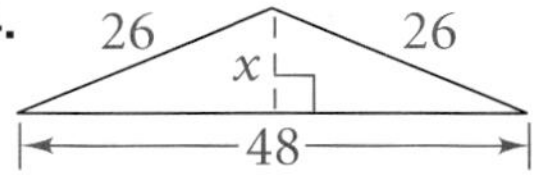

5.

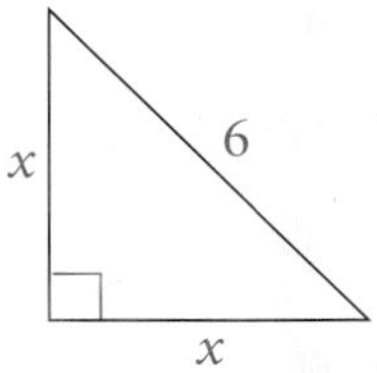

6.

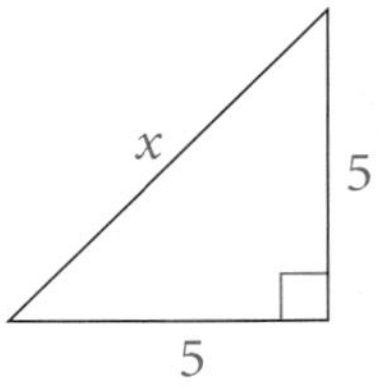

7.

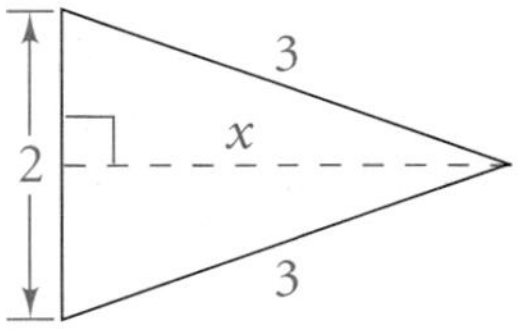

8.

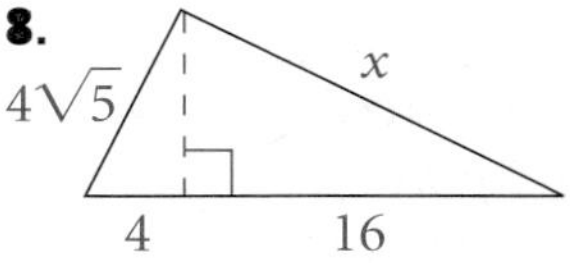

9. A 15-ft ladder is leaning against a building. The base of the ladder is 5 ft from the building. To the nearest foot, how high up the building does the ladder reach?

PROBLEM SOLVING HINT
Draw a diagram.

10. A brick walkway forms the diagonal of a square playground. The walkway is 24 m long. To the nearest tenth of a meter, how long is a side of the playground?

Choose **Use mental math, paper and pencil, or a calculator. The lengths of the sides of a triangle are given. Classify each triangle as acute, right, or obtuse.**

11. 15, 8, 21 **12.** 12, 16, 20 **13.** $2, 2\frac{1}{2}, 3$ **14.** 30, 34, 16

15. 0.3, 0.4, 0.6 **16.** 11, 12, 15 **17.** $\sqrt{3}, 2, 3$ **18.** 1.8, 8, 8.2

19. 20, 21, 28 **20.** 31, 23, 12 **21.** 30, 40, 50 **22.** $\sqrt{11}, \sqrt{7}, 4$

23. **Ancient Egypt** Each year the Nile River overflowed its banks and deposited fertile silt on the valley farmlands. Although the flood was helpful to farmers, it often destroyed boundary markers. Egyptian surveyors used a rope with knots at 12 equal intervals to help reconstruct boundaries.

a. **Writing** Explain how a surveyor could use this rope to form a right angle.

b. **Research** Find out why the Nile no longer floods as it did in ancient Egypt.

24. **Open-ended** Draw a right triangle with three sides that are integers. Draw the altitude to the hypotenuse. Label the lengths of the three sides and the altitude.

Calculator **Use the triangle at the right. Find the missing length to the nearest tenth.**

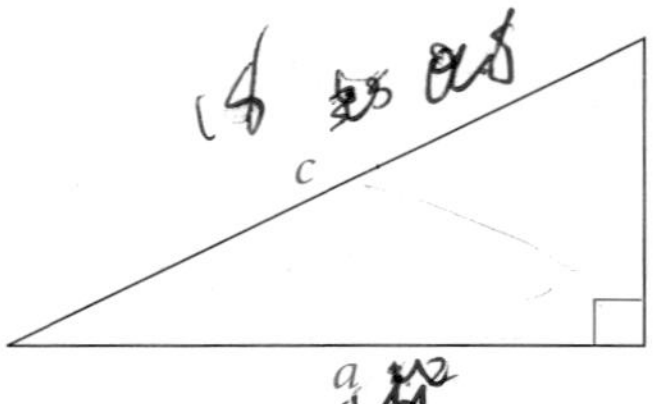

25. $a = 3, b = 7, c =$ ■ **26.** $a = 1.2, b =$ ■$, c = 3.5$

27. $a =$ ■$, b = 23, c = 30$ **28.** $a = 0.7, b =$ ■$, c = 0.8$

29. $a = 8, b = 8, c =$ ■ **30.** $a =$ ■$, b = 9, c = 18$

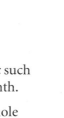

31. **Sewing** You want to embroider a square design. You have an embroidery hoop with a 6-in. diameter. Find the largest value of x such that the entire square will fit in the hoop. Round to the nearest tenth.

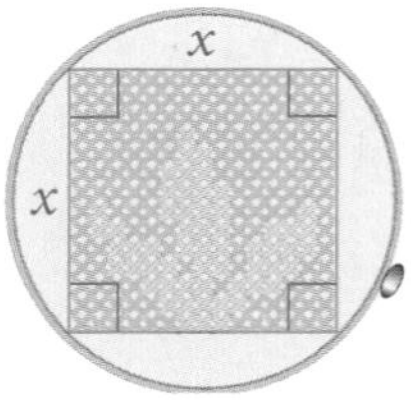

32. A rectangle has 10-in. diagonals and the lengths of its sides are whole numbers. Use the problem-solving strategy *Guess and Test* to find the perimeter of the rectangle.

Find the area of each figure. Leave your answer in simplest radical form.

33.

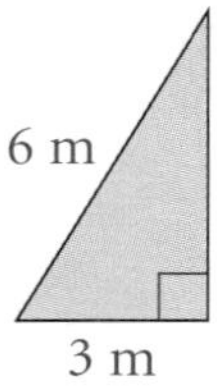

34.

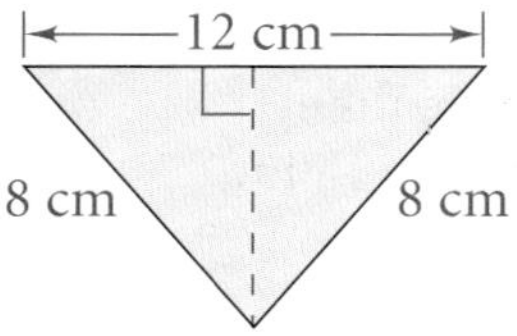

35.

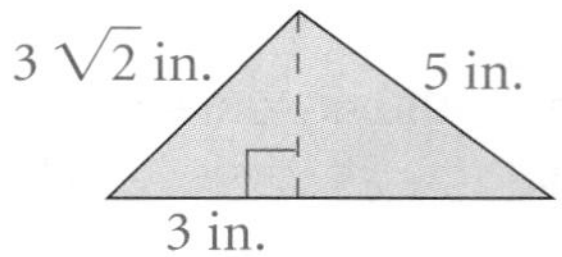

36.

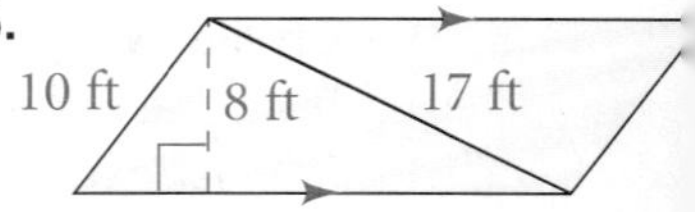

37. *Coordinate Geometry* You can use the Pythagorean Theorem to prove the Distance Formula. Let points $P(x_1, y_1)$ and $Q(x_2, y_2)$ be the endpoints of the hypotenuse of a right triangle.

a. Write an algebraic expression to complete each of the following: $PR = ▒$ and $QR = ▒$.

b. By the Pythagorean Theorem, $PQ^2 = PR^2 + QR^2$. Rewrite this statement, substituting the algebraic expressions you found for PR and QR in part (a).

c. Complete the proof by finding the square root of each side of the equation that you wrote in part (b).

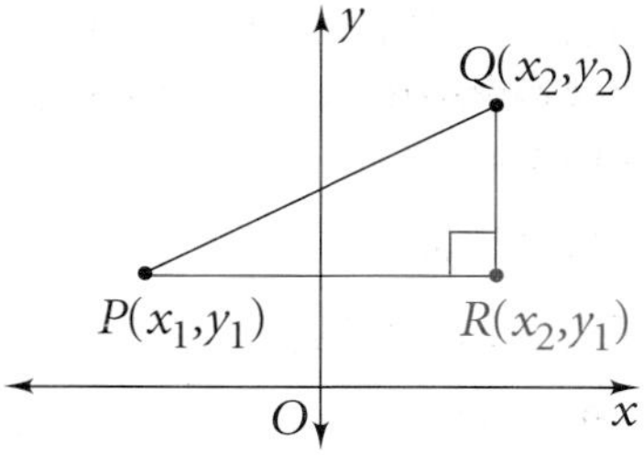

Find a third number so that the three numbers form a Pythagorean triple.

38. 9, 41	**39.** 14, 48	**40.** 60, 61	**41.** 8, 17
42. 20, 21	**43.** 13, 85	**44.** 12, 37	**45.** 63, 65

46. *Logical Reasoning* You can use the diagram at the right to prove the Pythagorean Theorem.

a. Find the area of the large square in terms of c.

b. Find the area of the large square in terms of a and b by finding the area of the four triangles and the small square.

c. Write an equation setting your answers to part (a) and part (b) equal to each other. Complete the proof by simplifying the equation.

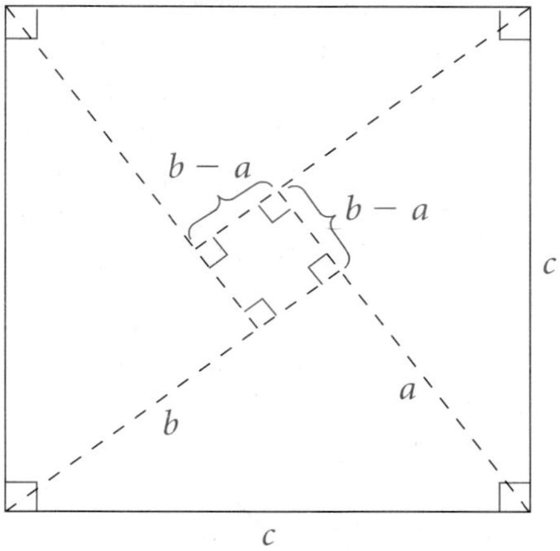

Calculator **The figures below are drawn on centimeter graph paper. Find the perimeter of each shaded figure to the nearest tenth.**

47.

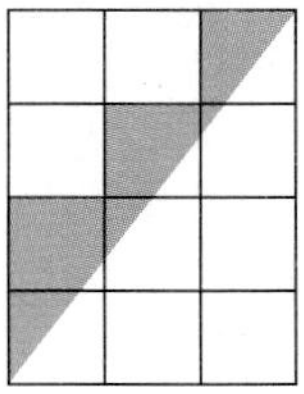

48.

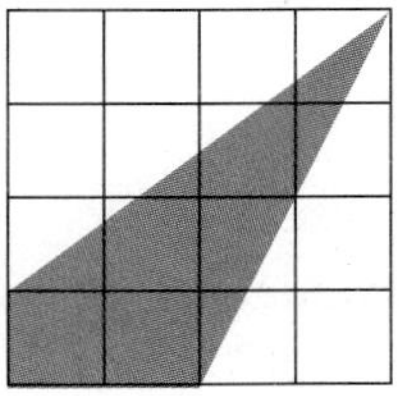

49.

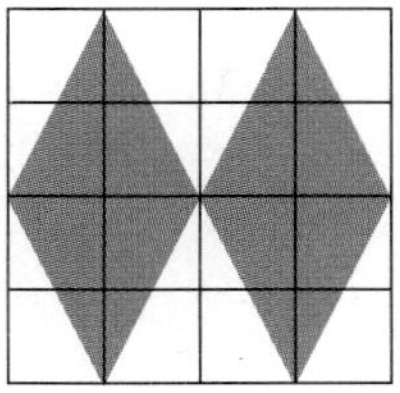

50.

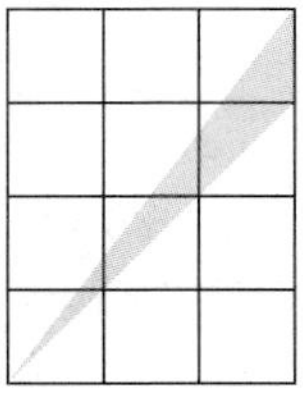

51. *Geometry in 3 Dimensions* The box at the right is a rectangular solid.

a. Use $\triangle ABC$ to find the length of the diagonal of the base, d_1.

b. Use $\triangle ABD$ to find the length of the diagonal of the box, d_2.

c. You can **generalize** steps in parts (a) and (b). Use the fact that $AC^2 + BC^2 = d_1^2$ and $d_1^2 + BD^2 = d_2^2$ to write a one-step formula to find d_2.

d. *Calculator* Use the formula you wrote to find the length of the longest fishing pole you can pack in a box with dimensions 18 in., 24 in., and 16 in.

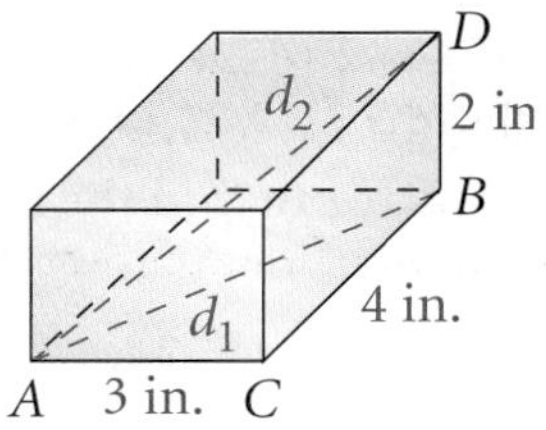

Geometry in 3 Dimensions **Points $P(x_1, y_1, z_1)$ and $Q(x_2, y_2, z_2)$ are points in a three-dimensional coordinate system. Use the following formula to find PQ. Leave your answer in simplest radical form.**

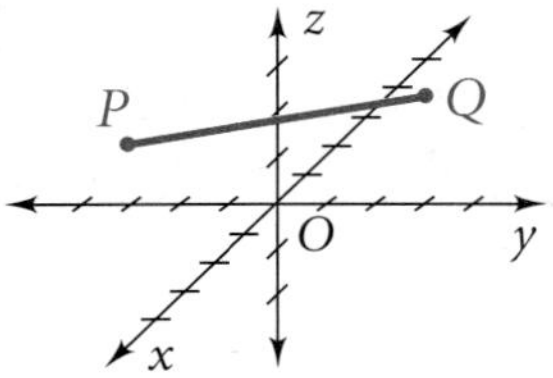

$$d = \sqrt{(x_2 - x_1)^2 + (y_2 - y_1)^2 + (z_2 - z_1)^2}$$

52. $P(0, 0, 0)$, $Q(1, 2, 3)$

53. $P(0, 0, 0)$, $Q(-3, 4, -6)$

54. $P(-1, 3, 5)$, $Q(2, 1, 7)$

55. $P(3, -4, 8)$, $Q(-1, 6, 2)$

56. **Space** The Hubble Space Telescope is orbiting Earth 600 km above Earth's surface. Earth's radius is about 6370 km. Use the Pythagorean Theorem to find the distance, x, from the telescope to Earth's horizon. Round your answer to the nearest ten kilometers.

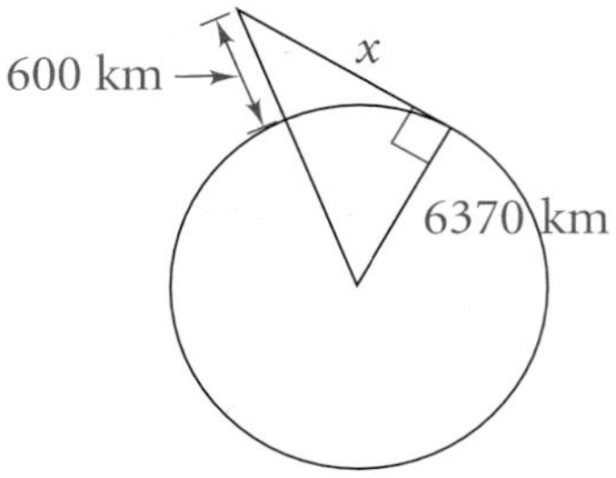

not drawn to scale

57. **a.** The ancient Greek philosopher Plato used the expressions $2n$, $n^2 - 1$, and $n^2 + 1$ to produce Pythagorean triples. Choose any integer greater than 1. Substitute for n and evaluate the three expressions.

b. **Verify** that your answers to part (a) form a Pythagorean triple.

58. **Standardized Test Prep** $\triangle ABC$ has perimeter 20 in. What is its area?

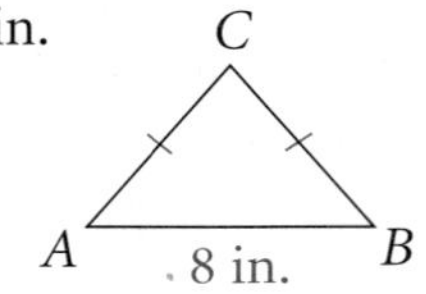

A. 12 in.^{2} **B.** 16 in.^{2} **C.** 24 in.^{2}

D. $8\sqrt{5}$ in.^{2} **E.** $16\sqrt{5}$ in.^{2}

Exercises MIXED REVIEW

Sketch each figure after a counterclockwise rotation of 90° about C.

59.

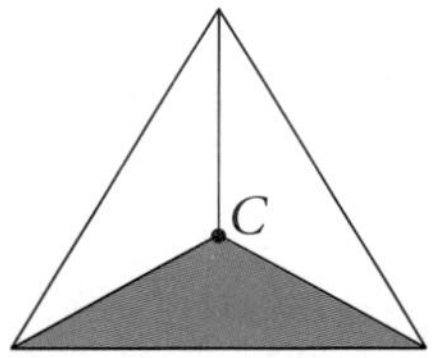

60.

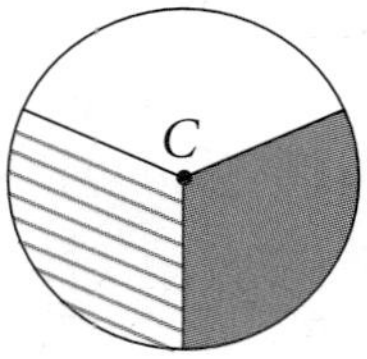

61.

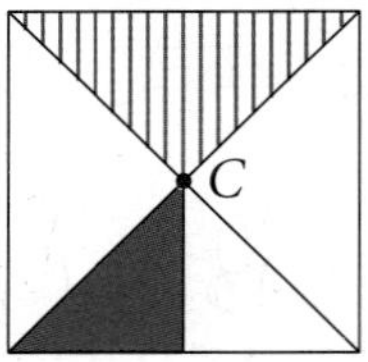

62. An angle is 87°. What is the measure of its complement?

Getting Ready for Lesson 5-4

Use a protractor to find the measures of the angles of each triangle.

63.

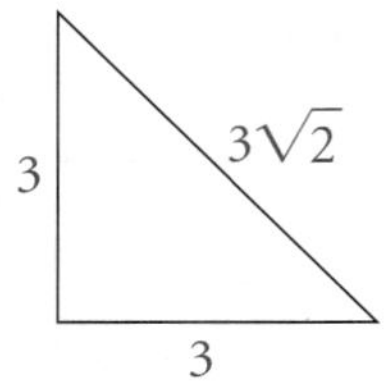

64.

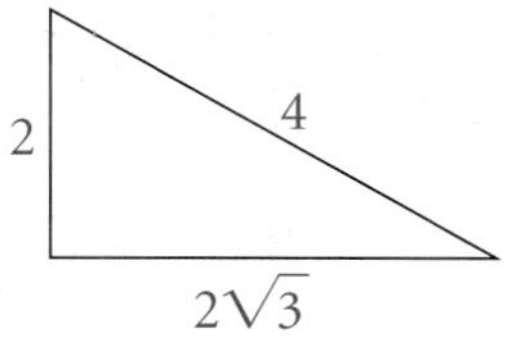

65.

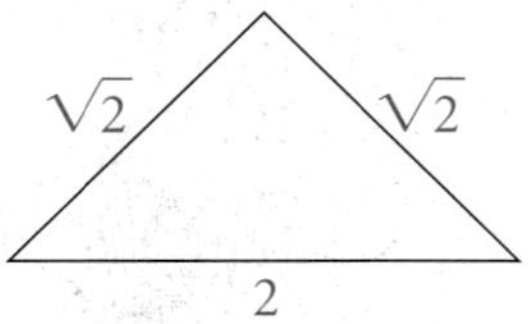

What You'll Learn

- Using the properties of 45°-45°-90° and 30°-60°-90° triangles

...And Why

To study figures in real life, including baseball diamonds and helicopter blades, which use special right triangles

What You'll Need

- centimeter grid paper
- metric ruler
- calculator
- protractor

5-4 Special Right Triangles

WORK TOGETHER

Work in a group. Have each person draw a different isosceles right triangle on centimeter grid paper. Choose integer values for the lengths of the legs.

- Record the length of each leg. Then use the Pythagorean Theorem to find the length of the hypotenuse. Leave your answers in simplest radical form.
- Organize your group's data in a table like the one below. Look for a pattern relating the side lengths of each triangle.

Triangle	Leg Length	Hypotenuse Length
Triangle 1	■	■
Triangle 2	■	■

- Make a **conjecture** about the relationship between the lengths of the legs and the length of the hypotenuse of an isosceles right triangle.

THINK AND DISCUSS

45°-45°-90° Triangles

1. What do you know about the measures of the acute angles of an isosceles right triangle?
2. If the measures of the angles of a triangle are 45, 45, and 90, why are the legs of the triangle congruent?

Another name for an isosceles right triangle is a 45°-45°-90° triangle.

3. a. Use the Pythagorean Theorem to solve for y in terms of x. Leave your answer in simplest radical form.
   b. Do the results of part (a) agree with the pattern you found in the Work Together?

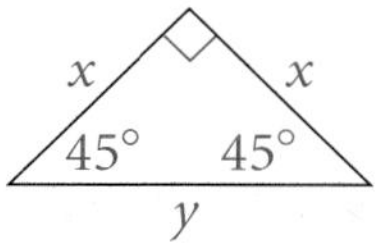

4. Find the value of each variable *without* using the Pythagorean Theorem.

a.

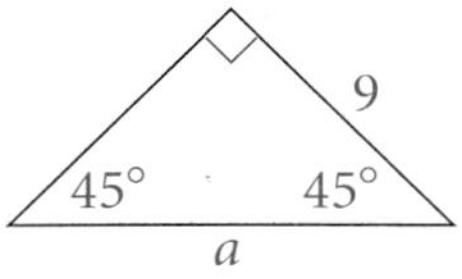

b.

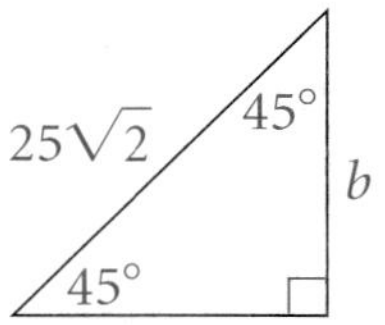

c.

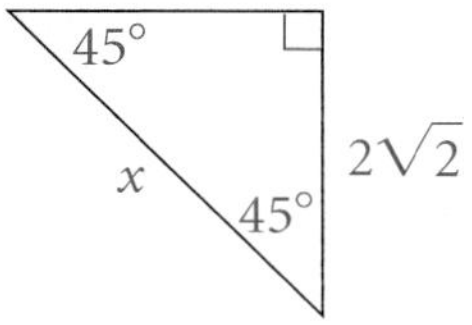

The pattern you observed in the Work Together (and generalized in Question 3) is the basis of the following theorem.

Theorem 5-6
45°-45°-90° Triangle Theorem

In a 45°-45°-90° triangle, both legs are congruent and the length of the hypotenuse is $\sqrt{2}$ times the length of a leg.

$$\text{hypotenuse} = \sqrt{2} \cdot \text{leg}$$

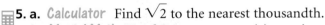

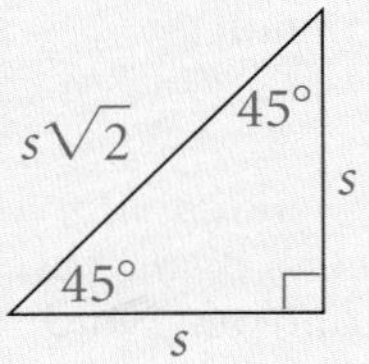

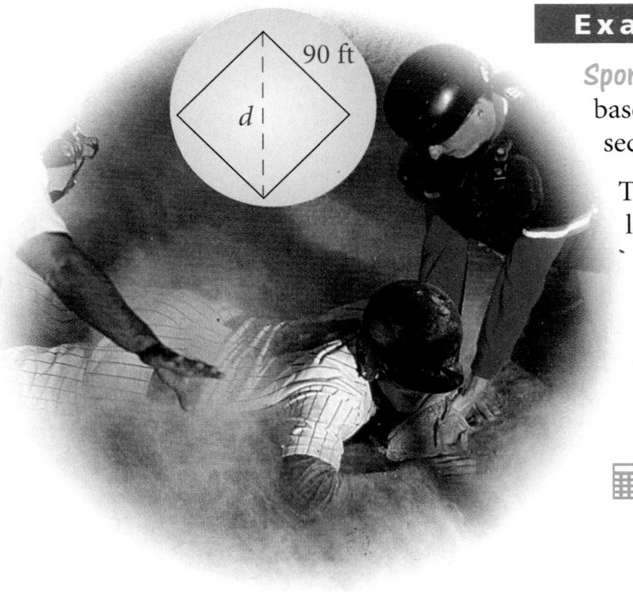

Example 1 — Relating to the Real World

Sports A baseball diamond is a square. The distance from base to base is 90 ft. To the nearest foot, how far does the second baseman throw a ball to home plate?

The distance d from second base to home plate is the length of the hypotenuse of a 45°-45°-90° triangle.

$d = 90\sqrt{2}$ — hypotenuse $= \sqrt{2} \cdot$ leg

90 [×] 2 [√] [=] 127.27922

The distance from second base to home plate is about 127 ft.

5. a. *Calculator* Find $\sqrt{2}$ to the nearest thousandth.
b. *Mental Math* Use the answer to part (a) to estimate the length of a diagonal of a square with sides 100 ft long.

WORK TOGETHER

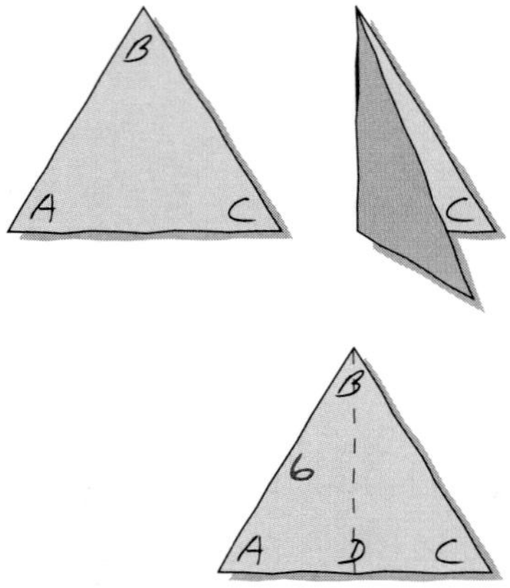

Work with a group.

- Draw an equilateral triangle with sides 6 cm long and cut it out. Label the vertices A, B, and C. Fold vertex A onto vertex C as shown at the left. Unfold the triangle and label the fold-line $\overline{BD}$.
- With your group, make a list of everything you know about $\triangle ABC$, $\triangle ABD$, and $\triangle CBD$, their angles and their sides.

6. Name a pair of congruent triangles.

7. a. Find $m\angle A$, $m\angle ADB$, and $m\angle ABD$.
b. Name $\triangle ABD$ using its angle measures.

8. a. Complete: $\overline{DB}$ is the _?_ of $\overline{AC}$.
b. If $AB = 6$, what is AD?
c. Use the Pythagorean Theorem to find BD in simplest radical form.
d. Find the ratios $\frac{AB}{AD}$ and $\frac{BD}{AD}$.

THINK AND DISCUSS

30°-60°-90° Triangles

The ratios you found in Question 8 part (d) suggest the following theorem about 30°-60°-90° triangles.

Theorem 5-7
30°-60°-90° Triangle Theorem

In a 30°-60°-90° triangle, the length of the hypotenuse is twice the length of the shorter leg. The length of the longer leg is $\sqrt{3}$ times the length of the shorter leg.

hypotenuse = 2 · shorter leg

longer leg = $\sqrt{3}$ · shorter leg

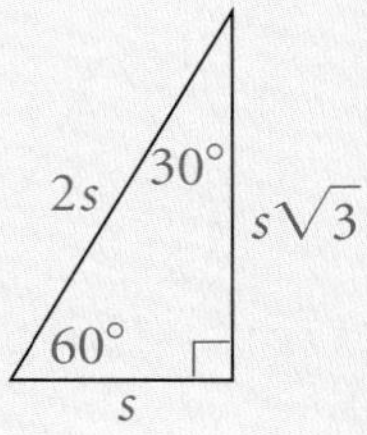

Justification:

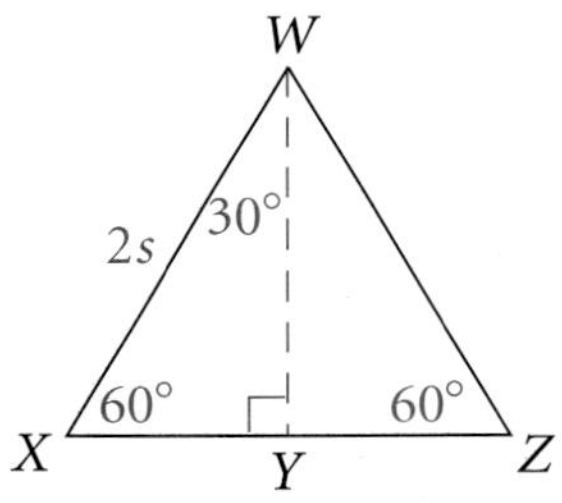

Refer to $\triangle WXZ$ at the left. Since $\overline{WY}$ is the perpendicular bisector of $\overline{XZ}$, $XY = \frac{1}{2}XZ$. That means that if $XW = 2s$, then $XY = s$.

$XY^2 + YW^2 = XW^2$	Use the Pythagorean Theorem.
$s^2 + YW^2 = (2s)^2$	Substitute s for XY and $2s$ for XW.
$YW^2 = 4s^2 - s^2$	Subtract s^2 from each side.
$YW^2 = 3s^2$	
$YW = s\sqrt{3}$	Find the square root of each side.

Example 2

Algebra Find the value of each variable.

a.

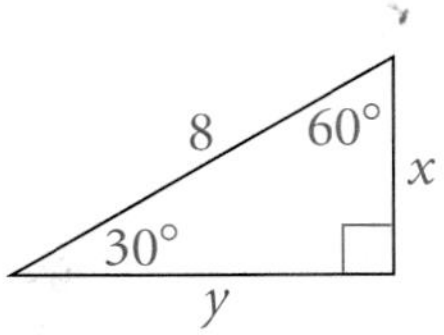

a. $8 = 2x$	hypotenuse = 2 · shorter leg
$x = 4$	
$y = x\sqrt{3}$	longer leg = $\sqrt{3}$ · shorter leg
$y = 4\sqrt{3}$	Substitute 4 for x.

b.

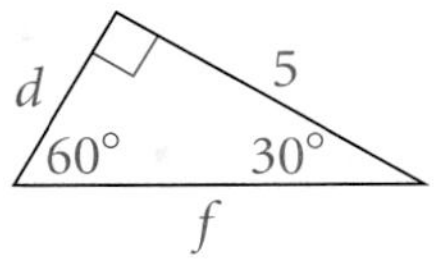

b. $5 = d\sqrt{3}$	longer leg = $\sqrt{3}$ · shorter leg
$d = \frac{5}{\sqrt{3}} \cdot \frac{\sqrt{3}}{\sqrt{3}} = \frac{5\sqrt{3}}{3}$	Simplify.
$f = 2d$	hypotenuse = 2 · shorter leg
$f = 2 \cdot \frac{5\sqrt{3}}{3} = \frac{10\sqrt{3}}{3}$	Substitute $\frac{5\sqrt{3}}{3}$ for d.

9. **Try This** The shorter leg of a 30°-60°-90° triangle has length $\sqrt{6}$. What are the lengths of the other two sides? Leave your answers in simplest radical form.

You can use the properties of 30°-60°-90° triangles to find the dimensions you need to calculate area.

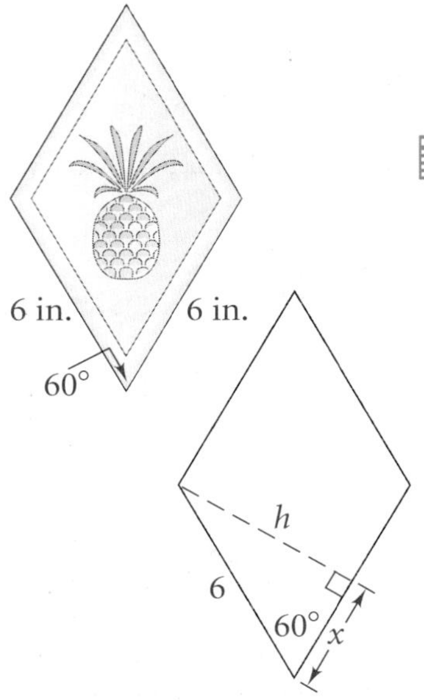

Example 3 Relating to the Real World

Design The rhombus at the left is a glass panel for a door. How many square inches of colored glass will you need for the panel?

Draw an altitude of the rhombus. Label x and h as shown.

$6 = 2x$	hypotenuse = 2 · shorter leg
$x = 3$	
$h = 3\sqrt{3}$	longer leg = $\sqrt{3}$ · shorter leg

Use the value of h to find the area.

$A = bh$	Use the formula for area of a parallelogram.
$= 6(3\sqrt{3})$	Substitute 6 for b and $3\sqrt{3}$ for h.

6 [×] 3 [×] 3 [√] [=] 31.176915

You will need about 31.2 in.^{2} of colored glass.

Exercises ON YOUR OWN

Find the value of each variable. Leave your answer in simplest radical form.

1.
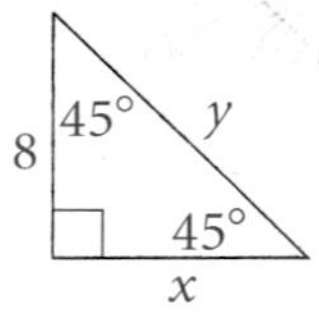

2.
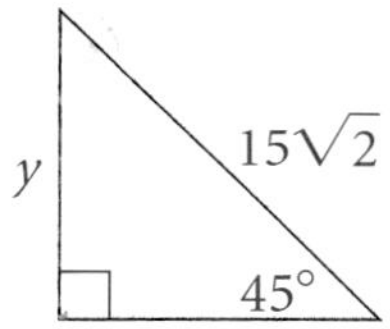

3.
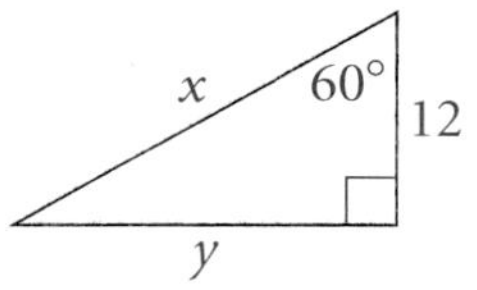

4.
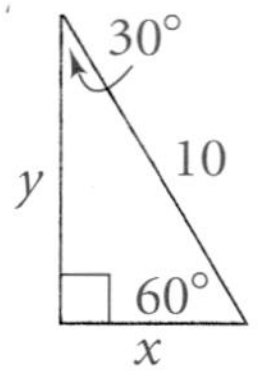

5.
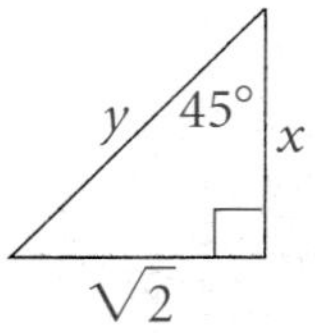

6.
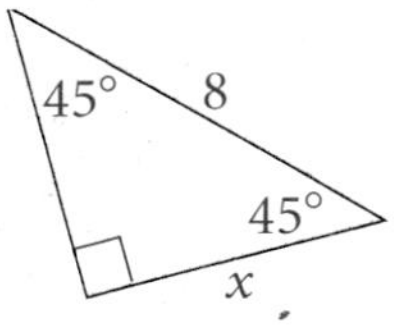

7.
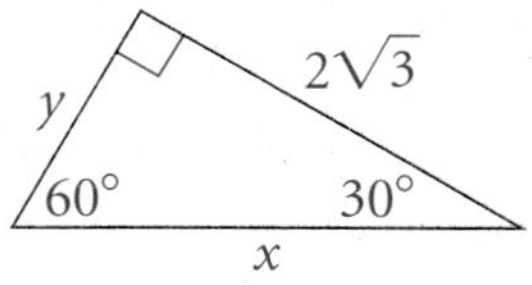

8.
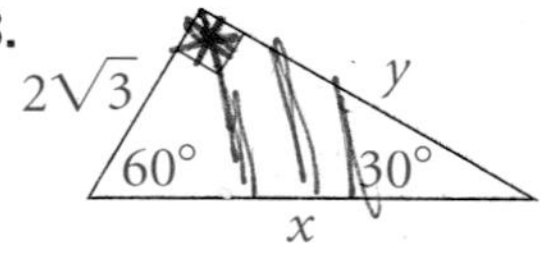

9. a. **Farming** A conveyor belt carries bales of hay from the ground to the loft of a barn 27.5 ft above ground. The belt makes a 30° angle with the ground. How far does a bale of hay travel on the conveyor belt?

b. The conveyor belt moves at 100 ft/min. How long does it take for a bale of hay to go from the ground to the barn loft?

Find the value of each variable. Leave your answer in simplest radical form.

10.

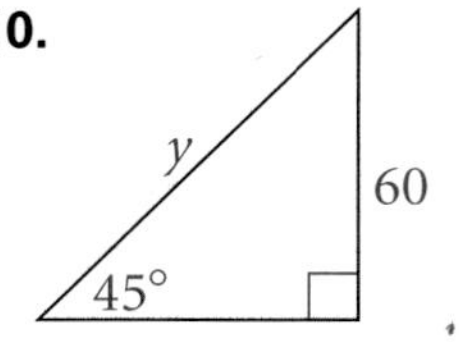

11.

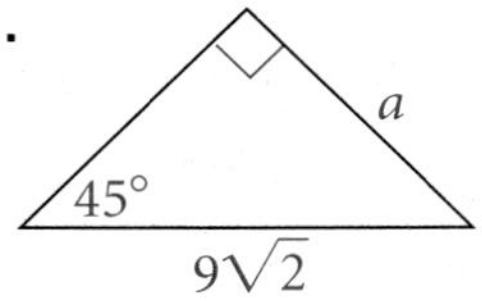

12.

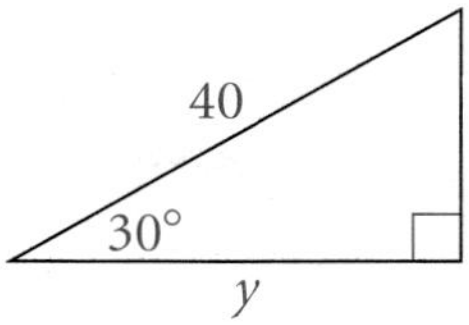

13.

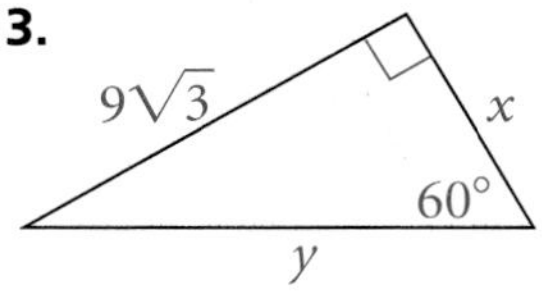

14.

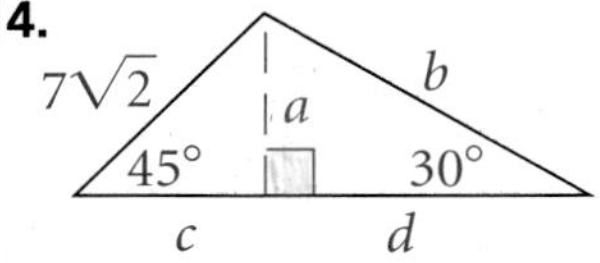

15.

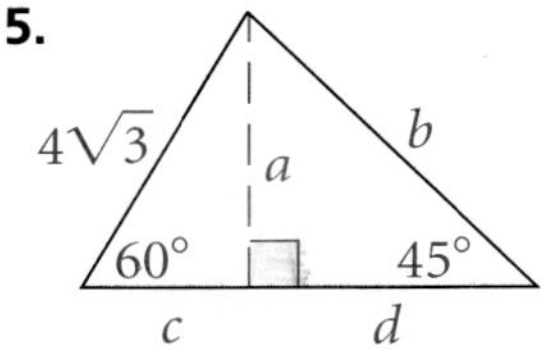

16.

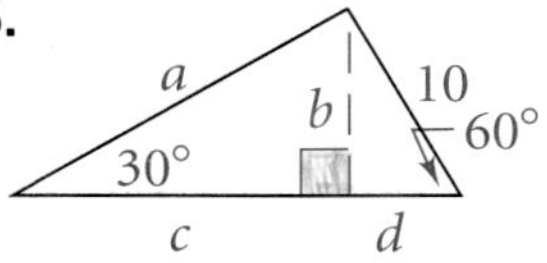

17.

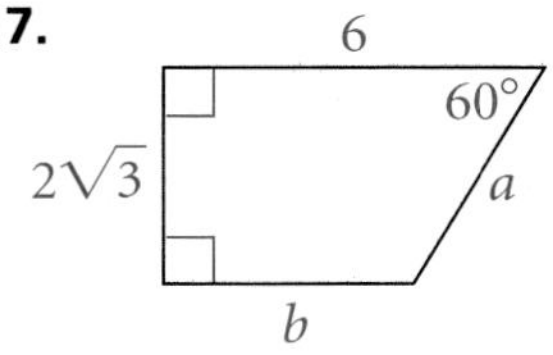

18. **Writing** Sandra drew this triangle. Rika said that the lengths couldn't be correct. With which student do you agree? Explain.

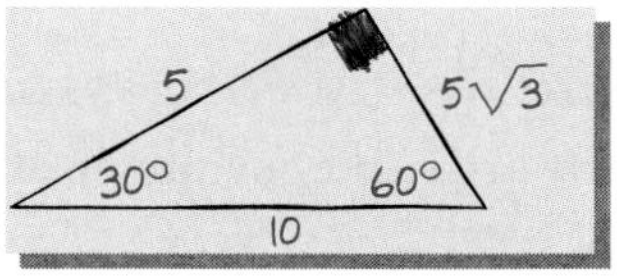

19. **Standardized Test Prep** Which of the following *cannot* be the lengths of sides of a 30°-60°-90° triangle?

A. $\frac{1}{2}, 1, \frac{\sqrt{3}}{2}$ **B.** $\sqrt{3}, 2\sqrt{3}, 3$ **C.** $1, \frac{1}{2}, \sqrt{3}$

D. $2\sqrt{2}, \sqrt{2}, \sqrt{6}$ **E.** $2, 4, 2\sqrt{3}$

Calculator **Find the area of each figure. When an answer is not a whole number, round to the nearest tenth.**

20.

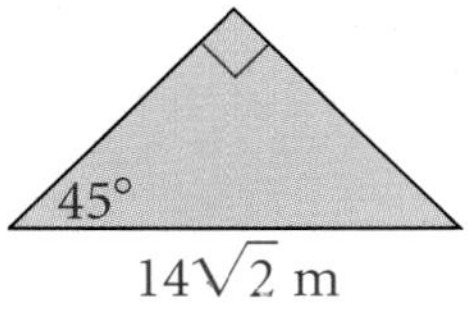

21.

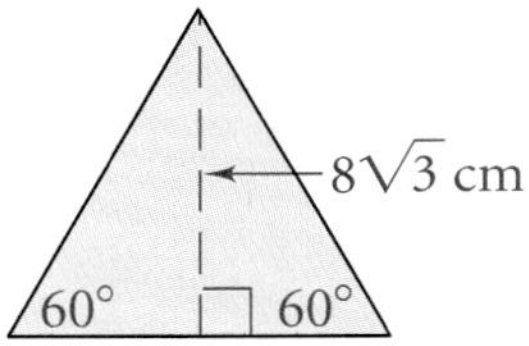

22.

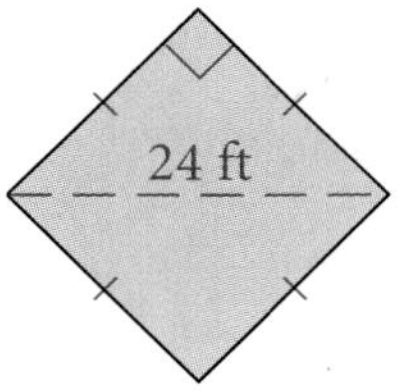

23.

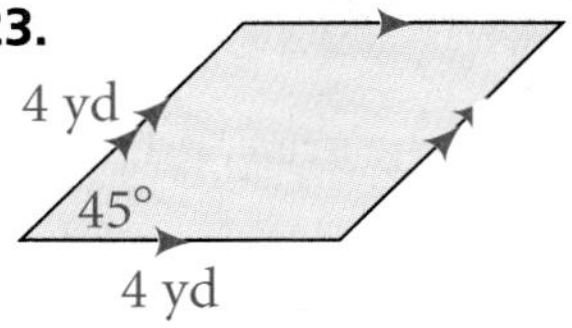

24. **Helicopters** The blades of a helicopter meet at right angles and are all the same length. The distance between the tips of two consecutive blades is 36 ft. How long is each blade? Round your answer to the nearest tenth.

25. **Open-ended** The hypotenuse of a 30°-60°-90° triangle is 12 ft long. Write a real-life problem that you can solve using this triangle. Show your solution.

26. A rhombus has a 60° angle and sides 5 cm long. What is its area? Round your answer to the nearest tenth.

27. a. *Geometry in 3 Dimensions* Find the length d, in simplest radical form, of the diagonal of a cube with sides 1 unit long.
b. Find the length d of the diagonal of a cube with sides 2 units long.
c. *Generalize* Find the length d of the diagonal of a cube with sides s units long.

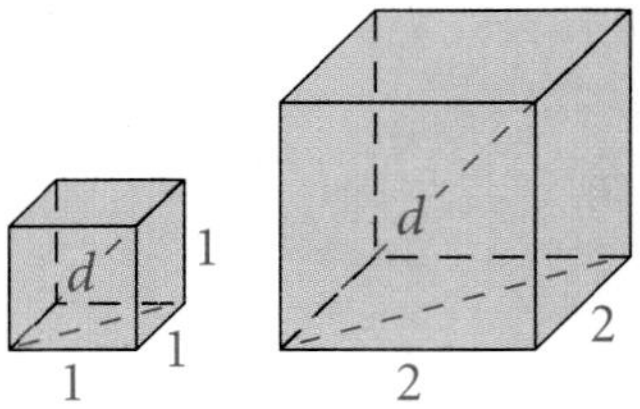

Exercises MIXED REVIEW

Find the slope of $\overline{AB}$.

28. $A(1, 0), B(-2, 3)$

29. $A(-5, 4), B(-1, 8)$

30. Find the equation of the line with slope $\frac{1}{2}$ containing the point $(-2, 5)$.

SELF ASSESSMENT

For Your Journal

Summarize what you know about special right triangles. Give two real-life examples of special right triangles.

Getting Ready For Lesson 5-5

Find the area and perimeter of each trapezoid to the nearest tenth.

31.

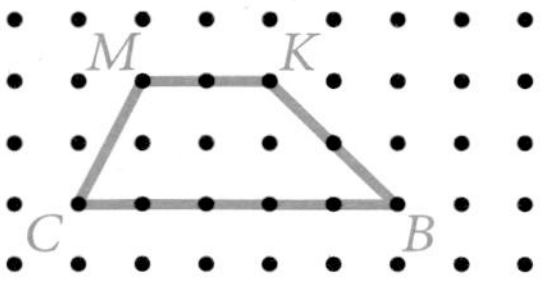

32.

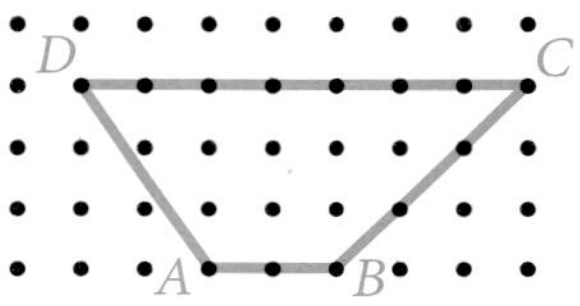

Exercises CHECKPOINT

Find the area and perimeter of each figure.

1.

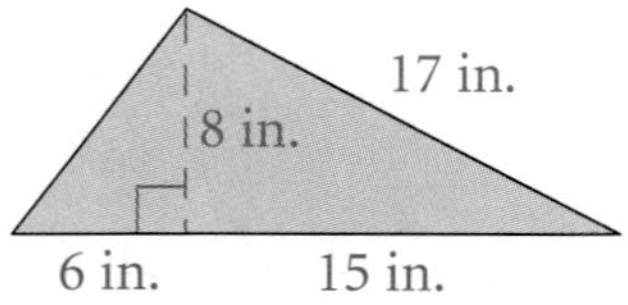

2.

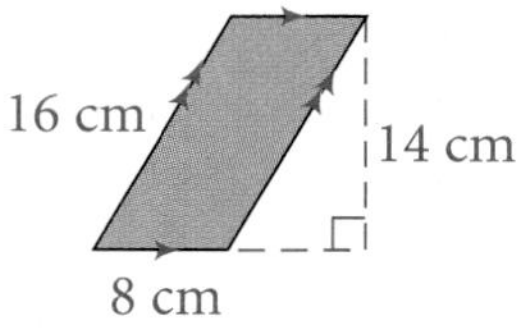

3.

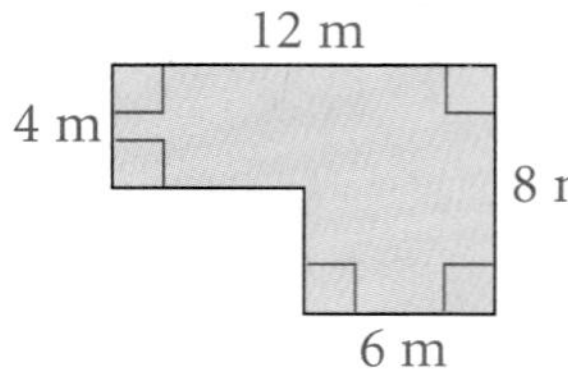

Algebra **Find the value of each variable. Leave your answer in simplest radical form.**

4.

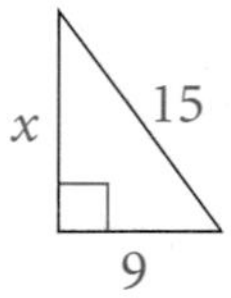

5.

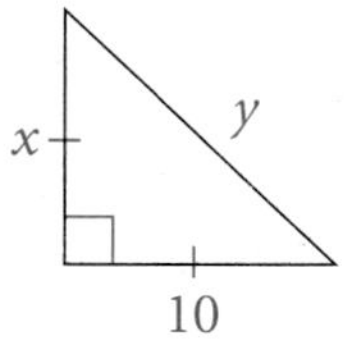

6.

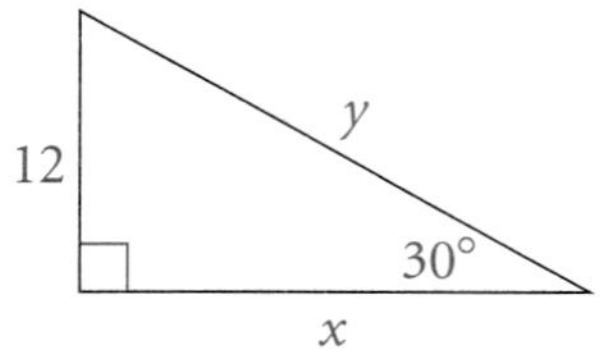

7. *Standardized Test Prep* Which numbers could represent the lengths of the sides of an acute triangle?

A. 3, 4, 5 **B.** 6, 8, 9 **C.** 14, 45, 50 **D.** 5, 12, 13 **E.** 5, 9, 13

8. *Open-ended* Sketch a rectangle and a triangle with the same perimeter. Label the lengths of the sides of the figures.

5-5 Areas of Trapezoids

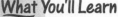

What You'll Learn

- Finding the areas of trapezoids

...And Why

To approximate the areas of irregular figures

What You'll Need

- lined paper
- scissors

WORK TOGETHER

Work in groups. Fold a piece of lined paper in half along one of the lines. On two lines of the paper, draw parallel segments of different lengths. Connect the endpoints of the segments to form a trapezoid. Cut through both layers of the folded paper, so that you will have two congruent trapezoids. Label b_1, b_2, and h for each trapezoid.

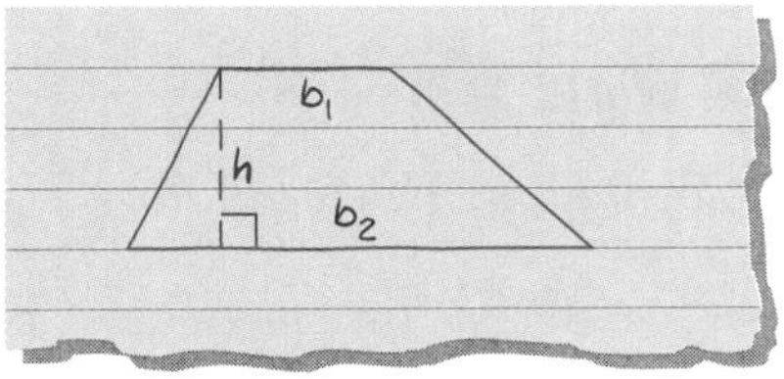

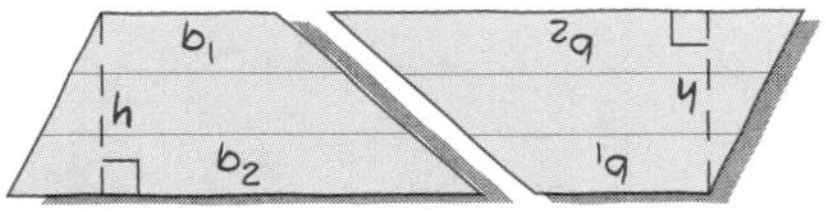

- Arrange the congruent trapezoids to form a parallelogram as shown at the left below.

1. a. Write an expression for the length of the base of the parallelogram.

b. Write an expression for the area of the parallelogram using b_1, b_2, and h.

2. How does the area of each trapezoid compare to the area of the parallelogram?

3. Use your answers to Questions 1 and 2 to write a formula for the area of each trapezoid.

THINK AND DISCUSS

In a trapezoid, the parallel sides are the **bases.** The nonparallel sides are the **legs.** The **height** h is the perpendicular distance between the two parallel bases.

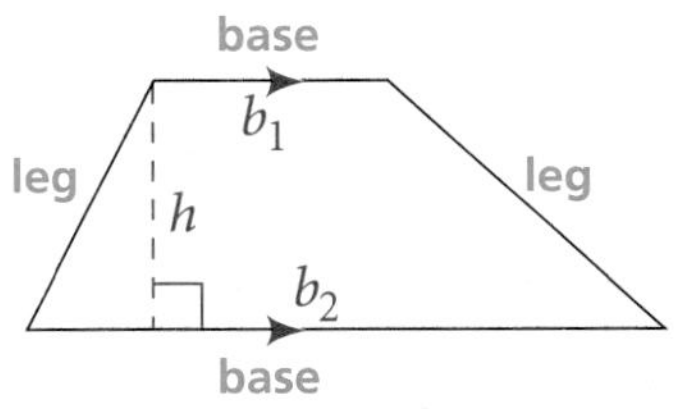

Your observations in the Work Together suggest the following theorem.

Theorem 5-8
Area of a Trapezoid

The area of a trapezoid is half the product of the height and the sum of the lengths of the bases.

$$A = \frac{1}{2}h(b_1 + b_2)$$

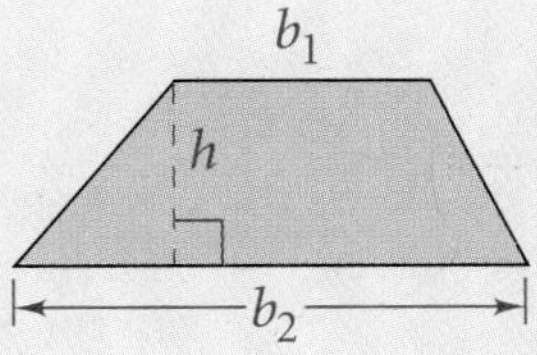

4. *Critical Thinking* When finding the area of a trapezoid, does it make a difference which base is labeled b_1 and which base is labeled b_2? Explain.

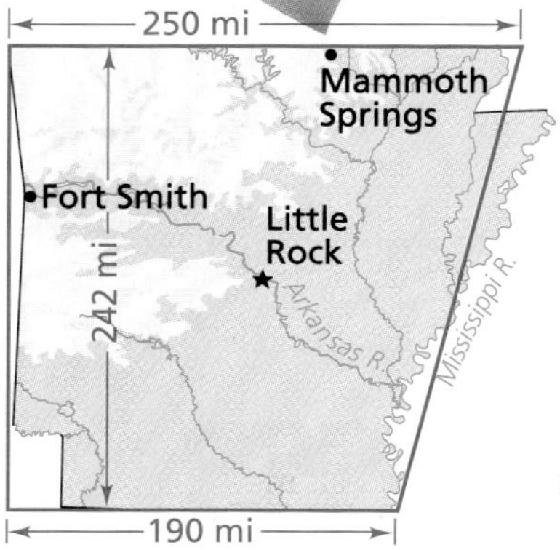

Example 1 Relating to the Real World

Geography Approximate the area of Arkansas by finding the area of the trapezoid shown.

$$A = \frac{1}{2}h(b_1 + b_2) \qquad \text{Use the area formula for a trapezoid.}$$
$$= \frac{1}{2}(242)(190 + 250) \qquad \text{Substitute.}$$
$$= 53{,}240$$

The area of Arkansas is about 53,240 mi^2.

5. Try This Find the area of a trapezoid with height 7 cm and bases 12 cm and 15 cm.

Sometimes properties of special right triangles can help you find the area of a trapezoid.

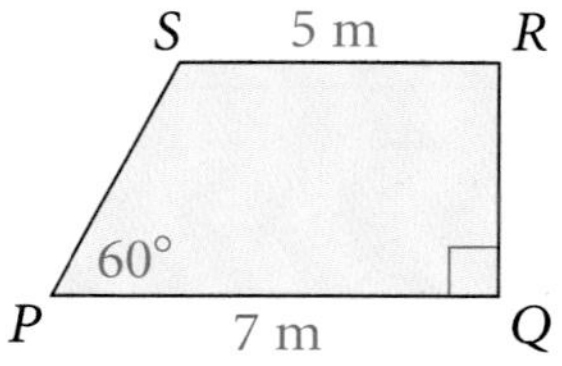

Example 2

Find the area of trapezoid *PQRS*. Leave your answer in simplest radical form.

You can draw an altitude that divides the trapezoid into a rectangle and a 30°-60°-90° triangle. Find *h*.

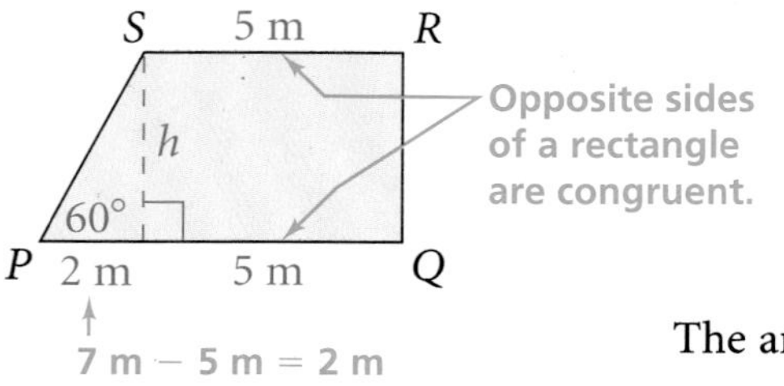

$$h = 2\sqrt{3} \qquad \text{longer leg} = \text{shorter leg} \cdot \sqrt{3}$$
$$A = \frac{1}{2}h(b_1 + b_2) \qquad \text{Use the area formula for a trapezoid.}$$
$$= \frac{1}{2}(2\sqrt{3})(5 + 7) \qquad \text{Substitute.}$$
$$= 12\sqrt{3} \qquad \text{Simplify.}$$

The area of trapezoid *PQRS* is $12\sqrt{3}$ m^2.

6. Suppose $m\angle P = 45$. Find the area of trapezoid *PQRS*.

Find the area of each trapezoid.

1.

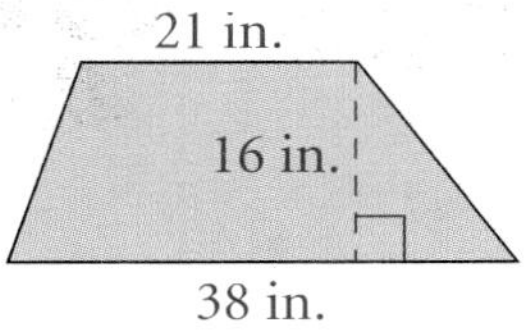

2.

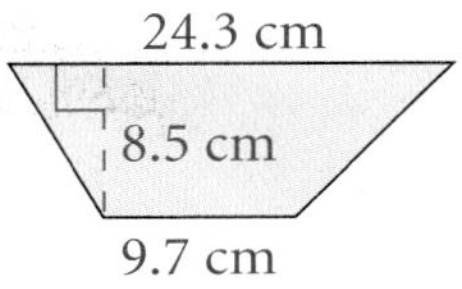

3.

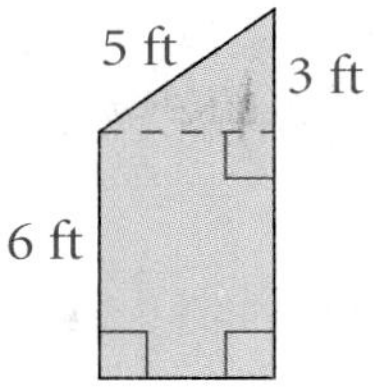

4.

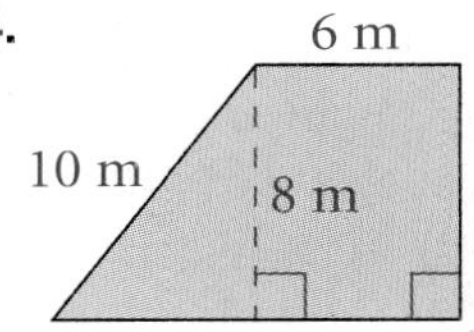

5. **Geography** Approximate the area of Nevada by finding the area of the trapezoid shown.

6. **Research** On a state map, select a town or county that is shaped like a trapezoid. Use the scale of the map to find values for b_1, b_2, and h. Then approximate the area.

7. The area of a trapezoid is 80 ft^2. Its bases have lengths 26 ft and 14 ft. Find its height.

8. a. A trapezoid has two right angles, bases of lengths 12 m and 18 m, and a height of 8 m. Sketch the trapezoid.
   b. What is the perimeter?
   c. What is the area?

Find the area of each trapezoid. Leave your answer in simplest radical form.

9.

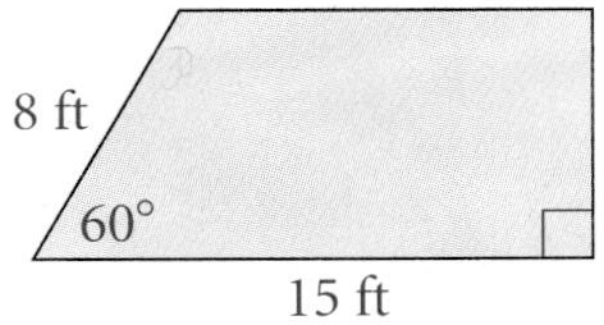

10.

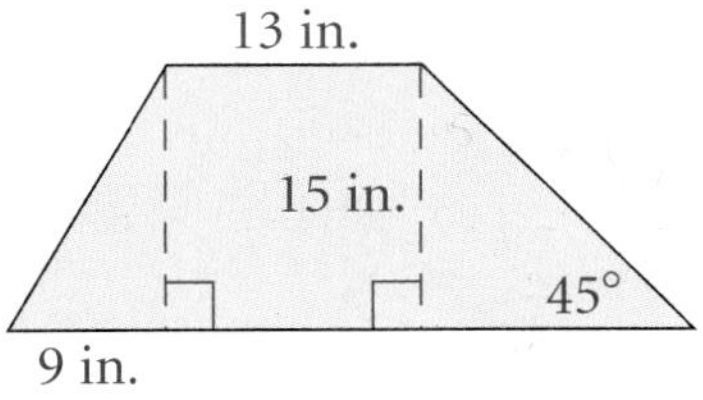

11.

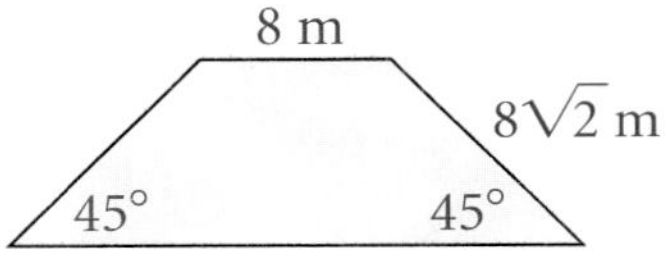

12. **Geometry in 3 Dimensions** A rain gutter has a trapezoidal cross section. The bottom is 4 in. wide, the top is 6 in. wide, and the gutter is 4 in. deep. What is the area of an end-piece?

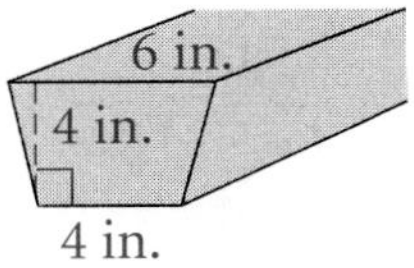

13. Draw a trapezoid. Label its bases and height b_1, b_2, and h. Then draw a diagonal of the trapezoid.
   a. Write an expression for the area of each triangle determined by the diagonal.
   b. **Writing** Explain how you can justify the trapezoid area formula using the areas of the two triangles.

14. **Open-ended** Draw a trapezoid. Measure its height and the lengths of its bases. Find its area.

15. **Crafts** You plan to lace together four isosceles trapezoids and a square to make the trash basket shown. How much material will you need?

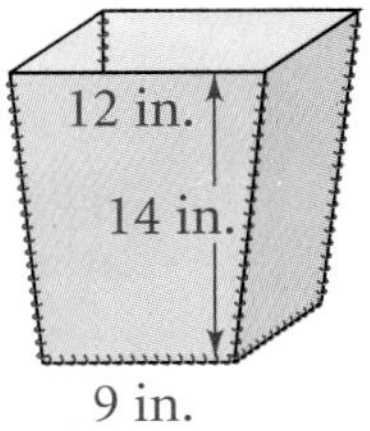

16. The area of an isosceles trapezoid is 160 cm^2. Its height is 8 cm and the length of its shorter base is 14 cm. Find the length of the longer base.

17. **Algebra** One base of a trapezoid is twice as long as the other. The height is the average of the two bases. The area is 324 cm^2. Find the height and the lengths of the bases. (*Hint:* Let the lengths of the bases be $2x$ and $4x$.)

Calculator **Find the area of each trapezoid to the nearest tenth.**

18.

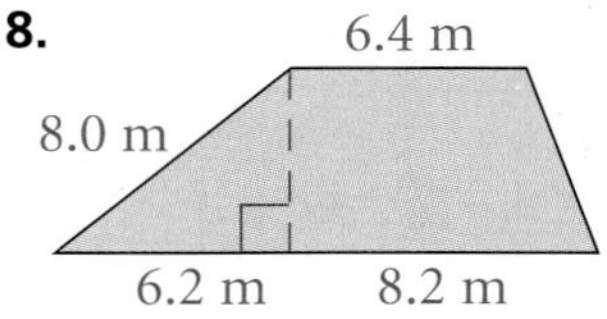

19.

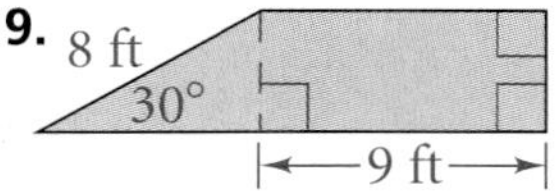

20.

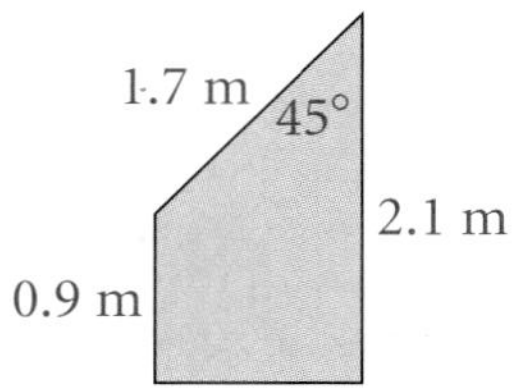

21.

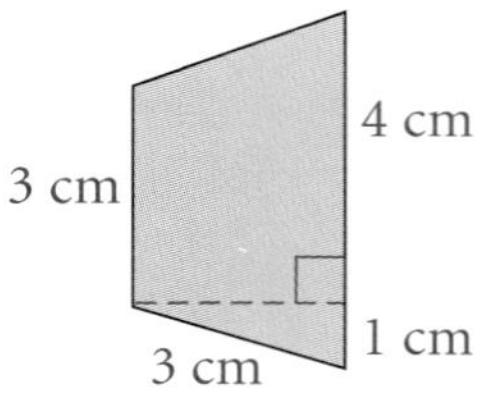

22. a. **Coordinate Geometry** Graph the lines $x = 0$, $x = 6$, $y = 0$, and $y = x + 4$.
   b. What quadrilateral do the lines form?
   c. Find the area of the quadrilateral.

23. **Recreation** A town youth center is building a skateboarding ramp. The ramp is 4 m wide, and the surface of the ramp is modeled by the equation $y = 0.25x^2$. You want to paint the front face of the ramp. Use the triangles and trapezoids shown to approximate the area of the face.

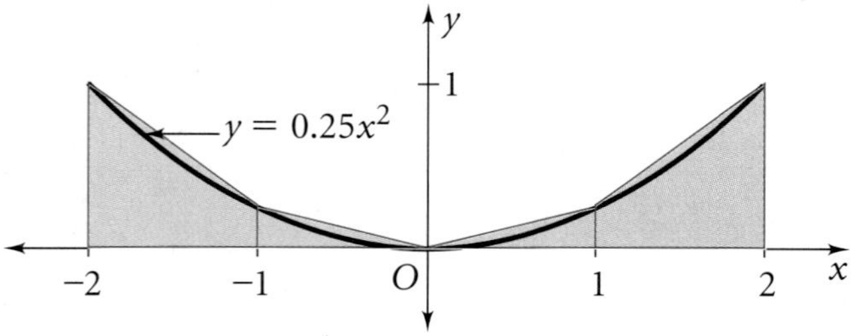

Exercises MIXED REVIEW

Open-ended **Find a possible length for the third side of a triangle that has two sides with the given lengths.**

24. 7 cm, 10 cm
25. 2 in., 8 in.
26. 13 mm, 6 mm
27. 4 ft, 9 ft

28. **Locus** Describe the locus of points in a plane equidistant from the sides of an angle. What is another name for this locus?

Getting Ready for Lesson 5-6

Find the area of each regular polygon. If your answer is not an integer, leave it in simplest radical form.

29.

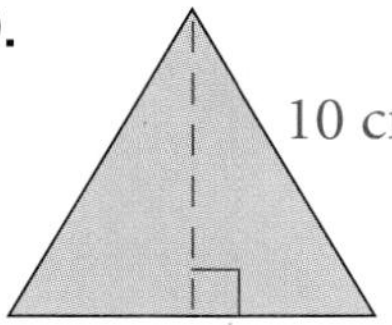

30.

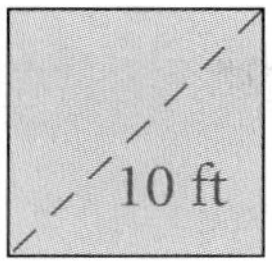

31.

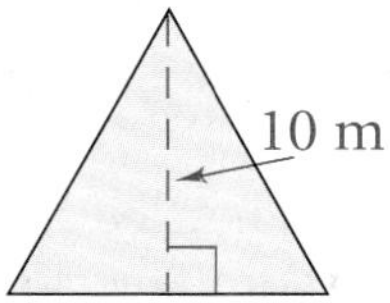

Chapter Project

Find Out by Researching

In the early 1900s, Seminoles of southern Florida developed a method of arranging strips of fabric to create geometric designs. These patchwork patterns sometimes serve to advertise the clan to which the wearer belongs.

- Research the patchwork techniques used by the Seminoles.
- Create your own Seminole patchwork design with colored paper, graphics software, or fabric.
- In the photo, notice the angled strips with trapezoids at each end. Explain, with diagrams or models, how this effect was created.

A Point in Time

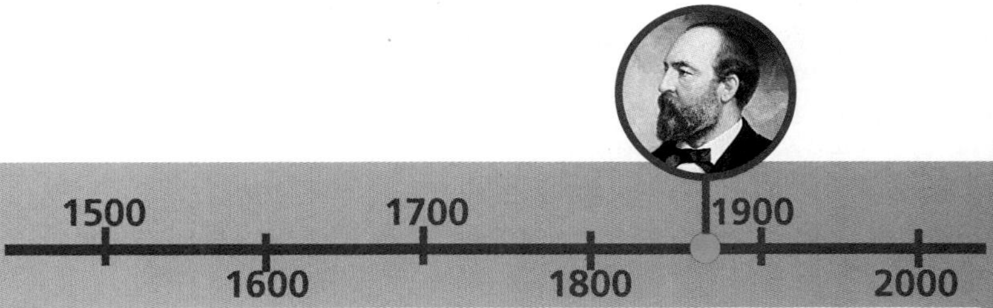

Presidential Proof

Presidents are known more often for their foreign policy than for their mathematical creativity. James Garfield, the 20th President of the United States, was an exception. After serving as a general in the Civil War, in **1876** Garfield demonstrated this proof of the Pythagorean Theorem.

In the diagram, $\triangle NRM$ and $\triangle RPQ$ are congruent right triangles with sides of length *a, b,* and *c.* The legs of isosceles right triangle *NRP* have length *c.* The three triangles form trapezoid *MNPQ.* The sum of the areas of the three right triangles equals the area of trapezoid *MNPQ.*

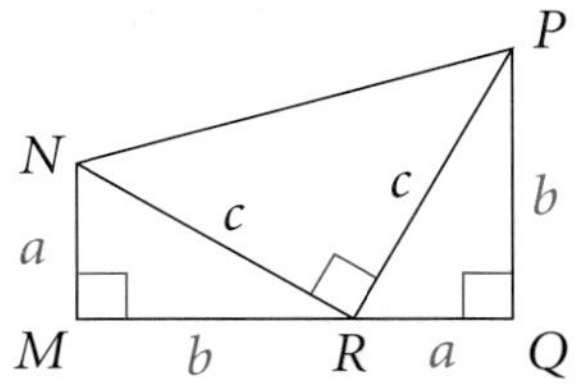

Areas of Triangles	=	Area of Trapezoid
$\frac{1}{2}ab + \frac{1}{2}ab + \frac{1}{2}c^2$	=	$\frac{1}{2}(a + b)(a + b)$
$ab + \frac{1}{2}c^2$	=	$\frac{1}{2}a^2 + ab + \frac{1}{2}b^2$
$\frac{1}{2}c^2$	=	$\frac{1}{2}a^2 + \frac{1}{2}b^2$
c^2	=	$a^2 + b^2$

Connections Racing Cars and Boats . . . and more

5-6 Areas of Regular Polygons

What You'll Learn

- Finding areas of regular polygons

...And Why

To find amounts of materials used in manufacturing and in architecture

QUICK REVIEW

A regular polygon is any polygon that is both equilateral and equiangular.

THINK AND DISCUSS

You can circumscribe a circle about any regular polygon. The **center** of a regular polygon is the center of the circumscribed circle. The **radius** of a regular polygon is the distance from the center to a vertex. The **apothem** of a regular polygon is the perpendicular distance from the center to a side.

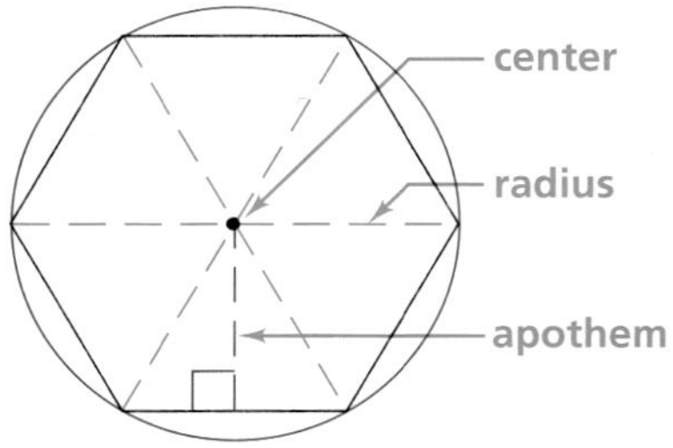

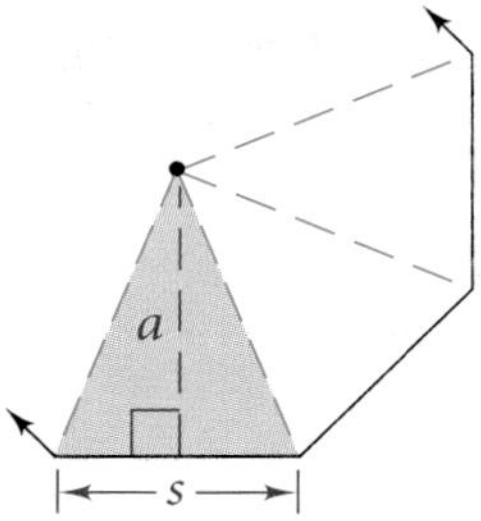

1. Suppose you have a regular n-gon with side s. The radii divide the figure into n congruent isosceles triangles.
   a. Write an expression for the area of each isosceles triangle in terms of the apothem a and the length of each side s.
   b. There are n congruent triangles. Use your answer to part (a) to write an expression for the area of the n-gon.
   c. The perimeter p of the n-gon is ns. Substitute p for ns in your answer to part (b) to find a formula for area in terms of a and p.

Your answers to Question 1 suggest the following theorem.

Theorem 5-9
Area of a Regular Polygon

The area of a regular polygon is half the product of the apothem and the perimeter.

$A = \frac{1}{2}ap$

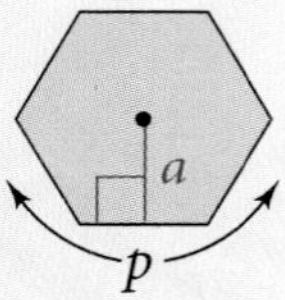

Example 1

Find the area of a regular decagon with a 12.3-in. apothem and 8-in. sides.

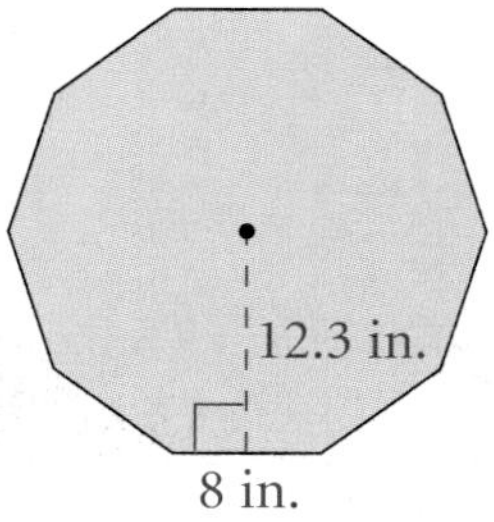

$p = ns$ — Find the perimeter.

$= 10(8) = 80$ — A decagon has 10 sides; $n = 10$.

$A = \frac{1}{2}ap$ — Use the formula for the area of a regular polygon.

$= \frac{1}{2}(12.3)(80) = 492$

The regular decagon has area 492 in.^{2}.

2. **Try This** Find the area of a regular pentagon with sides of length 11.6 cm and apothem 8 cm.

QUICK REVIEW

Regular polygons that tessellate are triangles, squares, and hexagons.

Engineers use regular polygons that tessellate because they fill the plane without wasting space. You can use special right triangles to find their areas.

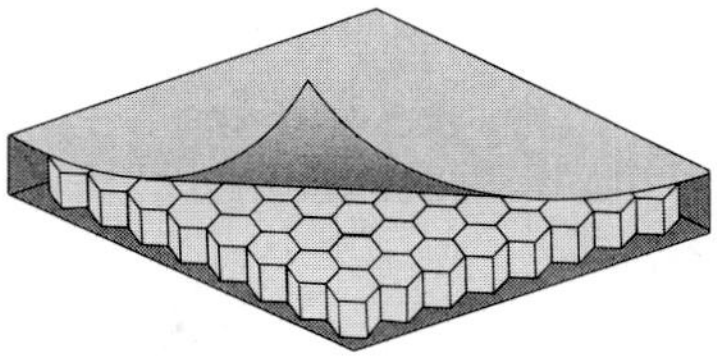

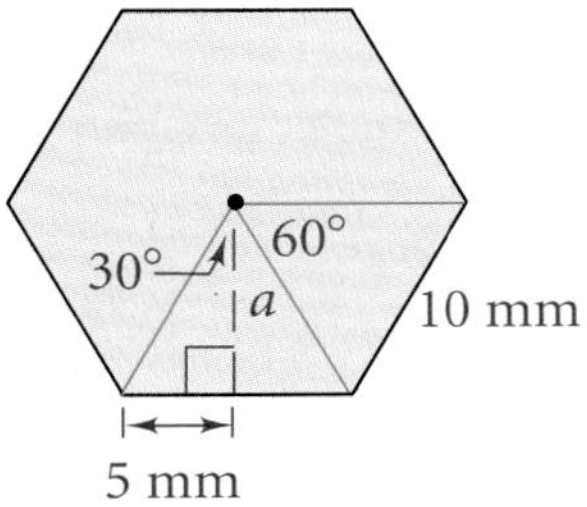

Example 2 **Relating to the Real World**

Racing Cars and boats used for racing need to be strong and durable, yet lightweight. One material that designers use to build body shells is a honeycomb of regular hexagonal prisms sandwiched between two layers of outer material. The honeycomb is plastic and provides strength and resilience without adding a lot of weight. The figure at the left is a cross section of one hexagonal cell.

The radii of a regular hexagon form six 60° angles at the center. So, you can use a 30°-60°-90° triangle to find the apothem *a*.

$a = 5\sqrt{3}$	longer leg = $\sqrt{3}$ · shorter leg
$p = ns$	Find the perimeter of the hexagon.
$= 6(10) = 60$	Substitute 6 for *n* and 10 for *s*.
$A = \frac{1}{2}ap$	Find the area.
$= \frac{1}{2}(5\sqrt{3})(60)$	Substitute.

0.5 [×] 5 [×] 3 [√] [×] 60 [=] 259.80762

The area is about 260 mm^2.

3. *Estimation* About how many hexagonal cells are in a 10 cm by 10 cm square panel?

4. **Try This** The apothem of a regular hexagon is 15 ft. Find the area of the hexagon.

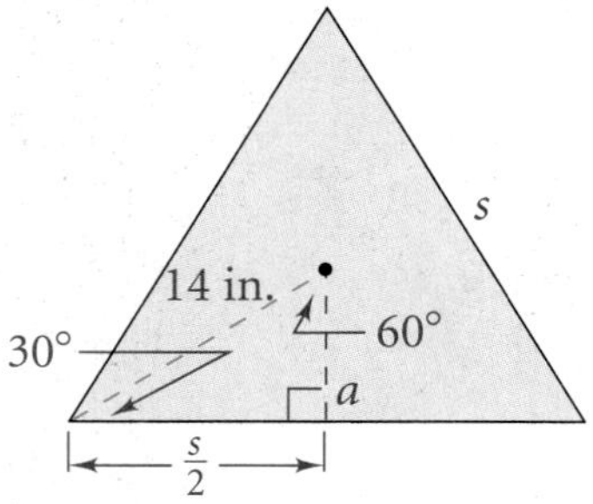

Example 3

Find the area of an equilateral triangle with radius 14 in. Leave your answer in simplest radical form.

You can use a 30°-60°-90° triangle to find the apothem *a* and the length of a side *s*.

$a = 7$	hypotenuse = 2 · shorter leg
$\frac{s}{2} = 7\sqrt{3}$	longer leg = $\sqrt{3}$ · shorter leg
$s = 14\sqrt{3}$	
$p = ns$	Find the perimeter.
$p = 3(14\sqrt{3})$	Substitute 3 for *n* and $14\sqrt{3}$ for *s*.
$= 42\sqrt{3}$	
$A = \frac{1}{2}ap$	Use the formula for area of a regular polygon.
$= \frac{1}{2}(7)(42\sqrt{3})$	Substitute 7 for *a* and $42\sqrt{3}$ for *p*.
$= 147\sqrt{3}$	

The area of the triangle is $147\sqrt{3}$ in.^{2}.

PROBLEM SOLVING

Look Back Check the solution to Example 3 by finding the area of the triangle using a different formula.

5. **Try This** Find the area of a square with radius 4 in.

Exercises ON YOUR OWN

Each regular polygon has radii and an apothem as shown. Find the measure of each numbered angle.

1.

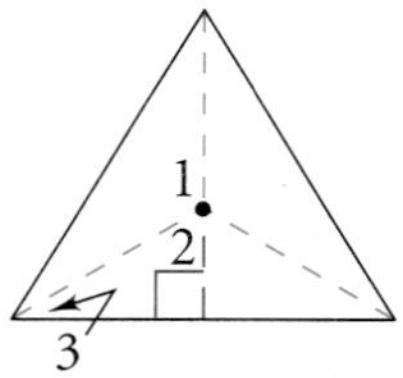

2.

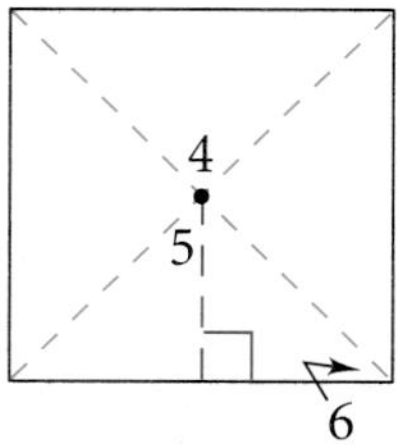

3.

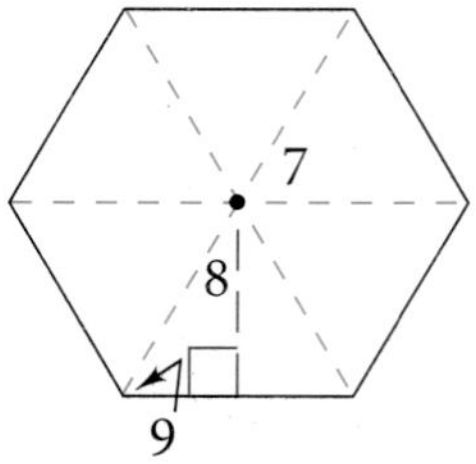

4. A regular pentagon has apothem 24.3 cm and side 35.4 cm. Find its area to the nearest tenth.
5. A regular octagon has apothem 60.5 in. and side 50 in. Find its area.
6. The apothem of a regular decagon is 19 m. Each side is 12.4 m. Find its area.

Calculator **Find the area of each regular polygon to the nearest tenth.**

7.

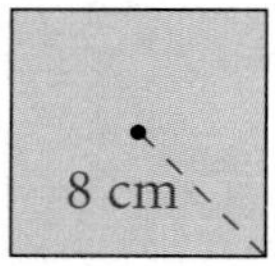

8.

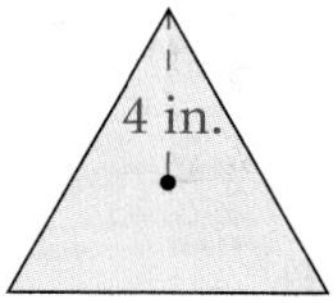

9.

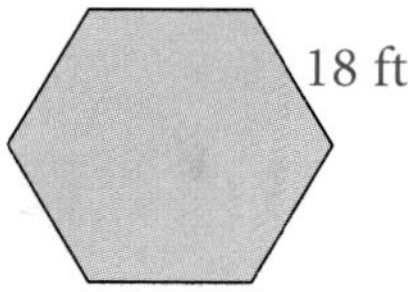

10.

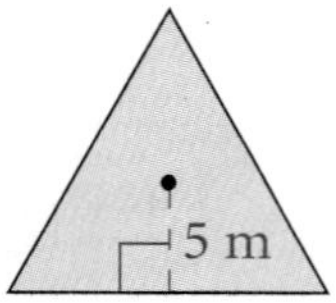

11. *Architecture* The floor of this gazebo is a regular octagon. Each side is 8 ft long, and its apothem is 9.7 ft. To the nearest tenth, find the area of the floor.

12. *Calculator* The area of a regular polygon is 36 in.^{2}. Find the length of a side if the polygon has the given number of sides. Round your answer to the nearest tenth.

a. 3 **b.** 4 **c.** 6

d. *Estimation* Suppose the polygon is a pentagon. What would you expect the length of its side to be? Explain.

13. *Writing* Explain why the radius of a regular polygon cannot be less than the apothem.

14. *Open-ended* Create a design using equilateral triangles and regular hexagons that have sides of the same length. Find the area of the completed design.

Find the area of each regular polygon with the given radius or apothem. Leave your answer in simplest radical form.

15.

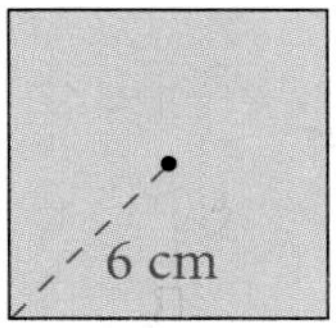

16.

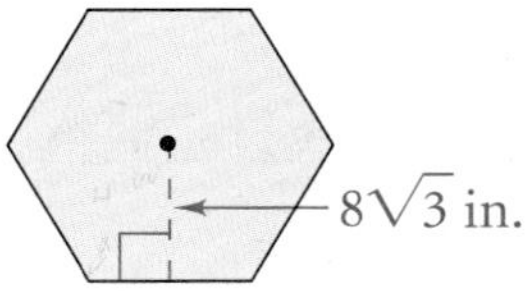

17.

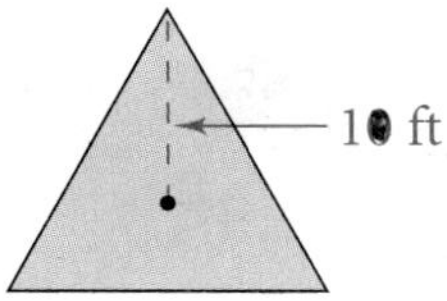

18.

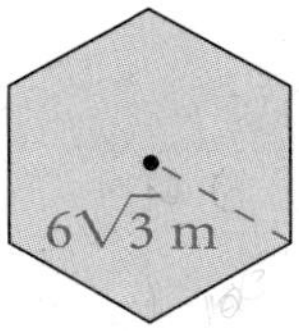

19. *Critical Thinking* To find the area of an equilateral triangle, you can use the formula $A = \frac{1}{2}bh$ or $A = \frac{1}{2}ap$. A third way to find the area of an equilateral triangle is to use the formula $A = \frac{1}{4}s^2\sqrt{3}$.

a. Verify the formula $A = \frac{1}{4}s^2\sqrt{3}$ by finding the area of Figure 1 using the formula $A = \frac{1}{2}bh$.

b. Verify the formula $A = \frac{1}{4}s^2\sqrt{3}$ by finding the area of Figure 2 using the formula $A = \frac{1}{2}ap$.

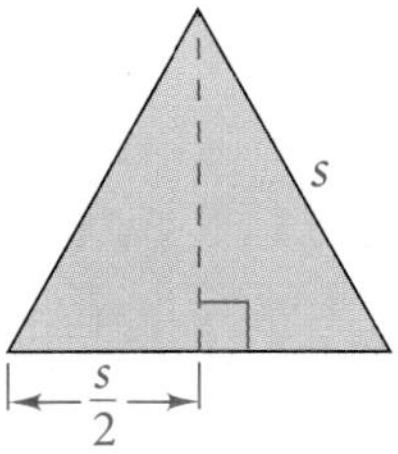

Figure 1

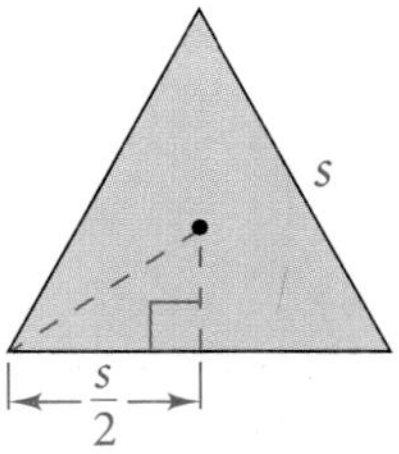

Figure 2

20. *Standardized Test Prep* A square and an equilateral triangle share a common side. What is the ratio of the area of the triangle to the area of the square?

A. $1 : \sqrt{2}$ **B.** $\sqrt{2} : \sqrt{3}$ **C.** $\sqrt{3} : 4$ **D.** $\sqrt{2} : 1$ **E.** $\sqrt{3} : 1$

21. **Coordinate Geometry** A regular octagon with center at the origin and radius 4 is graphed in the coordinate plane.

a. Since V_2 lies on the line $y = x$, its x- and y-coordinates are equal. Use the Distance Formula to find the coordinates of V_2 to the nearest tenth.

b. Use the coordinates of V_2 and the formula $A = \frac{1}{2}bh$ to find the area of $\triangle V_1OV_2$ to the nearest tenth.

c. Use your answer to part (b) to find the area of the octagon to the nearest tenth.

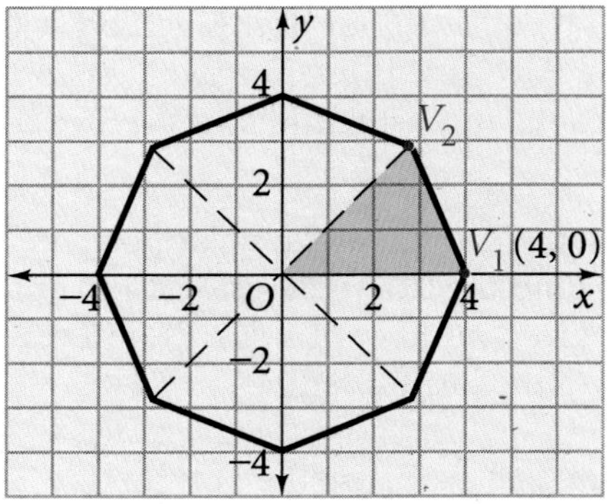

22. **Satellites** One of the smallest satellites ever developed is in the shape of a pyramid. Each of the four faces of the pyramid is an equilateral triangle with sides about 13 cm long. What is the area of one equilateral triangular face of the satellite? Round your answer to the nearest tenth.

Find the area of each regular polygon. If your answer is not an integer, you may leave it in simplest radical form.

23.

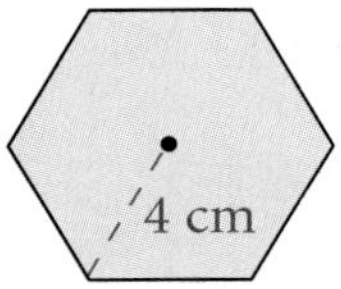

24.

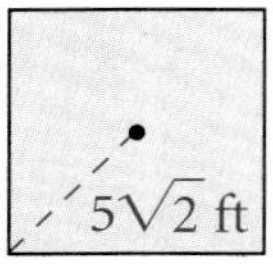

25.

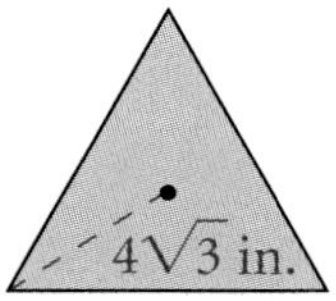

26.

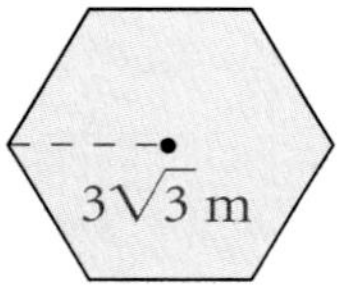

Exercises MIXED REVIEW

Create an isometric drawing for each foundation drawing.

27.

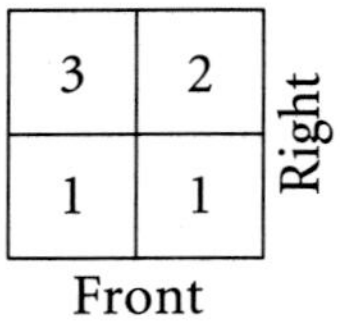

28.

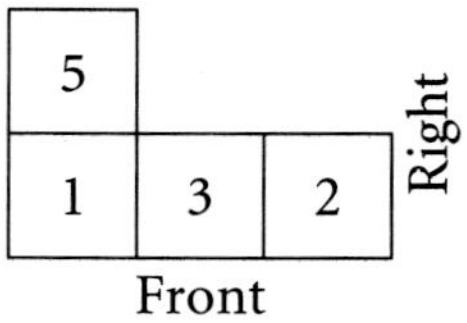

29.

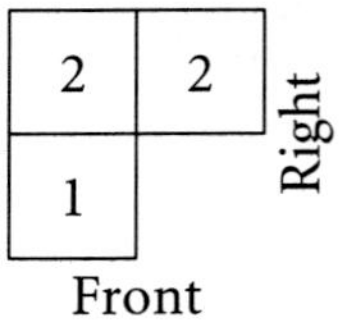

30. The measure of an angle is three more than twice the measure of its supplement. Find the measure of the angle.

Getting Ready for Lesson 5-7

Calculator **Evaluate each expression. Round your answer to the nearest hundredth.**

31. $2\pi r$ for $r = 4$

32. πd for $d = 7.3$

33. $2\pi r$ for $r = 5$

34. πd for $d = 11.8$

5-7 Circles: Circumference and Arc Length

What You'll Learn

- Finding the circumference of a circle and the length of an arc

...And Why

To solve problems involving auto safety, metalworking, and amusement park rides

What You'll Need

- circular objects
- string
- metric ruler
- calculator

WORK TOGETHER

Work in groups. Each member of your group should have one circular object such as a juice can or a jar lid.

- Measure the diameter of each circle to the nearest millimeter.
- Find the circumference of each circle by wrapping a string around each object. Straighten the string and measure its length to the nearest tenth of a centimeter.
- Organize your group's data in a table like the one below. Calculate the ratio $\frac{\text{circumference}}{\text{diameter}}$ to the nearest hundredth.

Name of Object	Circumference (C)	Diameter (d)	$\frac{C}{d}$
jelly-jar lid	19.6 cm	6.2 cm	3.16

1. Make a **conjecture** about the relationship between the circumference and the diameter of a circle.

THINK AND DISCUSS

Circumference

The ratios you found in the Work Together are estimates of the number **pi** (π), the ratio of the circumference of a circle to its diameter.

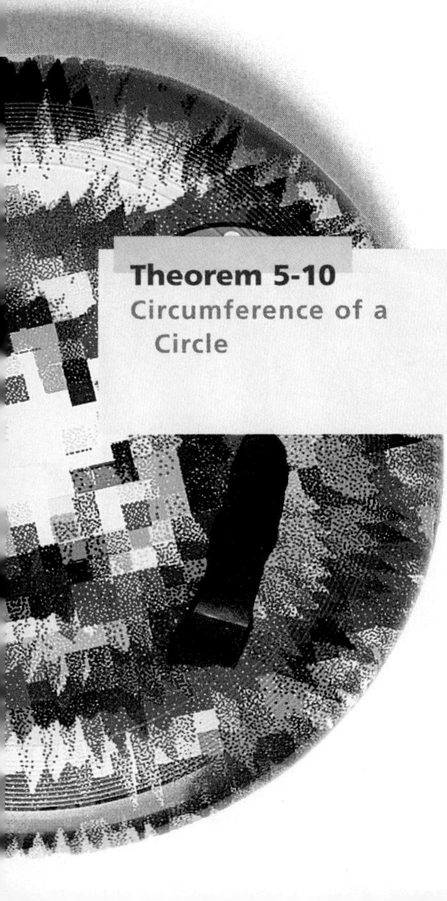

Theorem 5-10
Circumference of a Circle

The circumference of a circle is π times the diameter.

$C = \pi d$ or $C = 2\pi r$

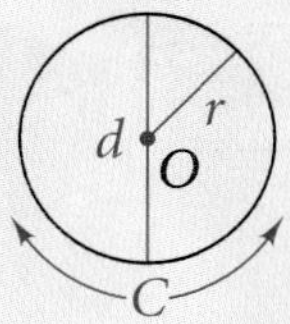

Example 1

Find the circumference of $\odot A$ and $\odot B$. Leave your answer in terms of π.

a.

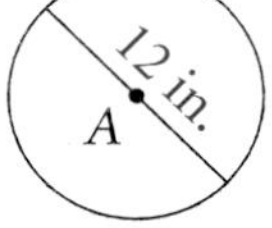

$C = \pi d$
$C = 12\pi$ in.

b.

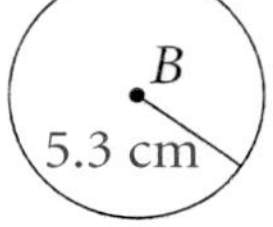

$C = 2\pi r$
$C = 2 \cdot \pi \cdot 5.3$
$C = 10.6\pi$ cm

2. **Try This** What is the radius of a circle with circumference 18π m?

Since the number π is irrational, you cannot write it as a decimal. You can use 3.14, $\frac{22}{7}$, or the [π] key on your calculator as approximations for π.

Two circles that lie in the same plane and have the same center are **concentric circles.**

Example 2 Relating to the Real World

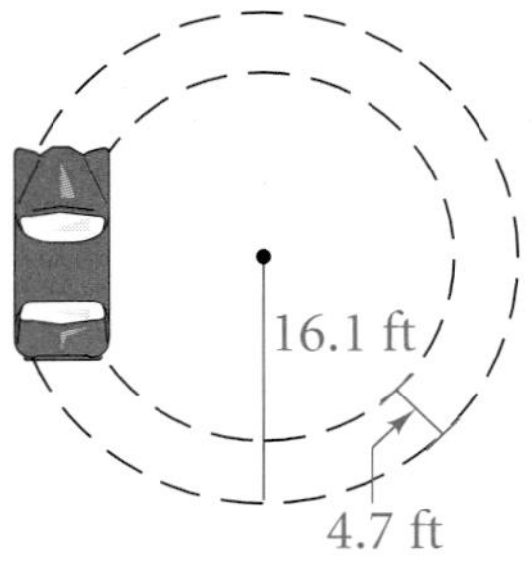

Automobiles A manufacturer advertises that a new car has a turning radius of only 16.1 ft. The distance between the two front tires is 4.7 ft. How much farther do the outside tires have to travel in making a complete circle than the tires on the inside?

The outside and inside tires travel on concentric circles. The radius of the outer circle is 16.1 ft. To find the radius of the inner circle, you must subtract 4.7 ft.

circumference of outer circle $= 2\pi(16.1) = 32.2\pi$

radius of the inner circle $= 16.1 - 4.7 = 11.4$ ft

circumference of inner circle $= 2\pi(11.4) = 22.8\pi$

The difference in the two distances is $32.2\pi - 22.8\pi = 9.4\pi$.

9.4 [×] [π] [=] 29.530971

The outside tires travel about 29.5 ft farther than the inside tires.

3. Try This The diameter of a bicycle wheel is 26 in. To the nearest whole number, how many revolutions does the wheel make when the bicycle travels 100 yd?

Arc Length

In Chapter 2 you found the measure of an arc in degrees. You can also find the **arc length,** which is a fraction of a circle's circumference.

An arc of 60° represents $\frac{60}{360}$ or $\frac{1}{6}$ of the circle. Its arc length is $\frac{1}{6}$ the circumference of the circle.

This observation suggests the following generalization.

Theorem 5-11
Arc Length

The length of an arc of a circle is the product of the ratio $\frac{\text{measure of the arc}}{360}$ and the circumference of the circle.

length of $\widehat{AB} = \frac{m\widehat{AB}}{360} \cdot 2\pi r$

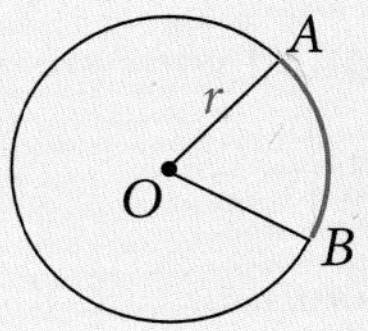

Example 3

Find the length of the arc shown in red on each circle. Leave your answer in terms of π.

a.

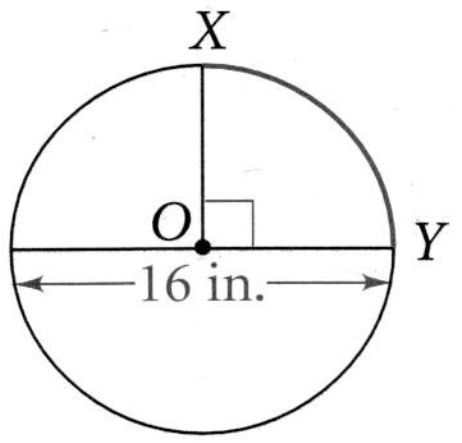

b.

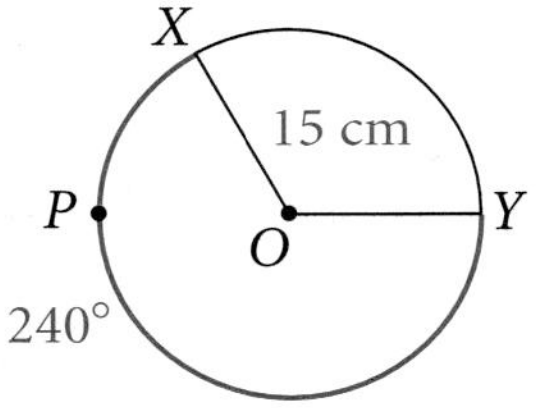

length of $\widehat{XY} = \frac{m\widehat{XY}}{360} \cdot \pi d$

length of $\widehat{XY} = \frac{90}{360} \cdot \pi(16)$

$= 4\pi$ in.

length of $\widehat{XPY} = \frac{m\widehat{XPY}}{360} \cdot 2\pi r$

length of $\widehat{XPY} = \frac{240}{360} \cdot 2\pi(15)$

$= 20\pi$ cm

4. **Try This** Find the length of a semicircle with radius 1.3 m. Leave your answer in terms of π.

5. **Critical Thinking** Is it possible for two arcs of different circles to have the same measure, but different lengths? Support your answer with an example.

6. **Critical Thinking** Is it possible for two arcs of different circles to have the same length but have different measures? Support your answer with an example.

Your answers to Questions 5 and 6 illustrate that for arcs to be congruent two things must be true. **Congruent arcs** are arcs that have the same measure and are in the same circle or in congruent circles.

Exercises ON YOUR OWN

Find the circumference of $\odot O$. Leave your answer in terms of π.

1.

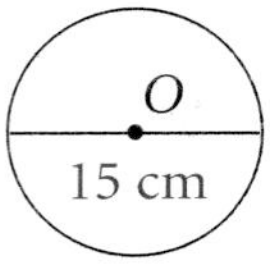

2.

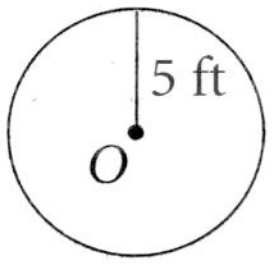

3.

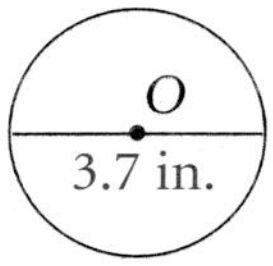

4.

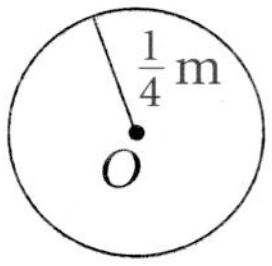

Calculator **Find the circumference of each circle with the given radius or diameter. Round your answer to the nearest hundredth.**

5. $r = 9$ in. **6.** $d = 7.3$ m **7.** $d = \frac{1}{2}$ yd **8.** $r = 0.13$ cm

Find the circumference of each circle. Then find the length of the arc shown in red on each circle. Leave your answer in terms of π.

9.

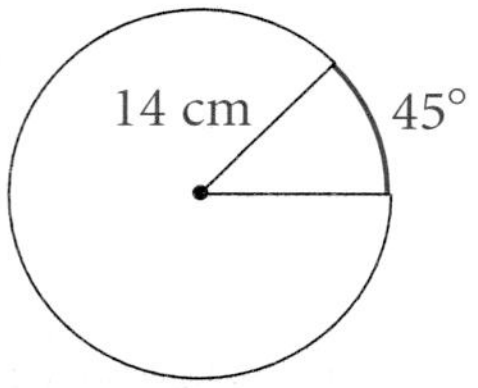

10.

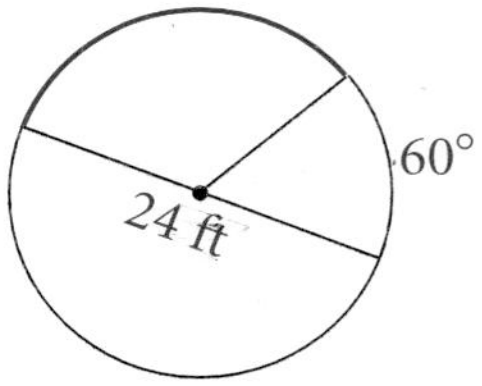

11.

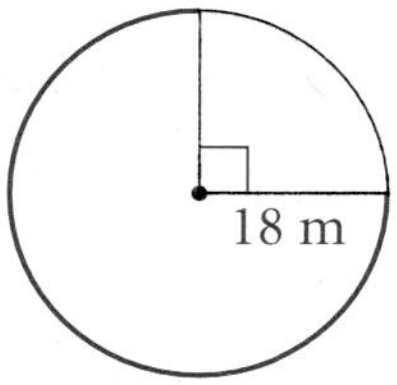

12.

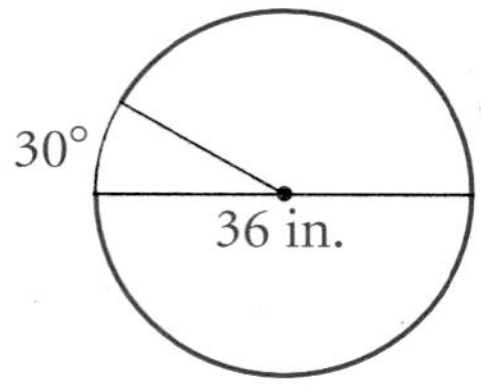

13. The circumference of a circle is 100π in. Find each of the following.
a. the diameter **b.** the radius **c.** the length of an arc of 120°

14. **Coordinate Geometry** The endpoints of a diameter of a circle are $A(1, 3)$ and $B(4, 7)$. Find each of the following.
a. the coordinates of the center **b.** the diameter **c.** the circumference

15. **Metalworking** Miya constructed a wrought-iron arch to top the entrance to a mall. The 11 bars between the two concentric semicircles are each 3 ft long. Find the length of the wrought iron used to make this structure. Round your answer to the nearest foot.

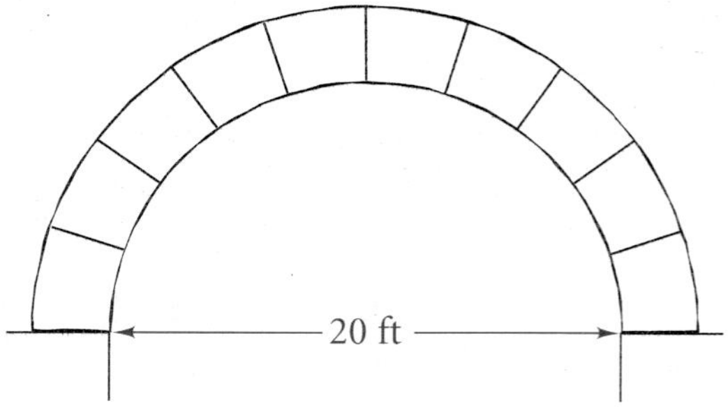

16. A 60° arc of $\odot A$ has the same length as a 45° arc of $\odot B$. Find the ratio of the radius of $\odot A$ to the radius of $\odot B$.

17. **Space Travel** The orbit of the space station *Mir* is 245 mi above Earth. How much greater is the circumference of *Mir's* orbit than the circumference of Earth? Earth's radius is about 3960 mi. Leave your answer in terms of π.

18. **Open-ended** Use a compass and protractor to draw two noncongruent arcs with the same measure.

Calculator **Find the length of the arc shown in red on each circle. Round your answer to the nearest hundredth.**

19.
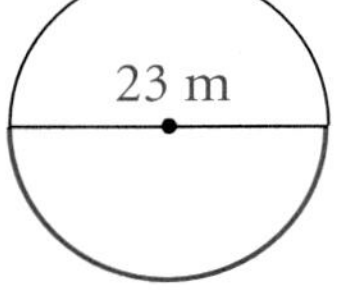

20.
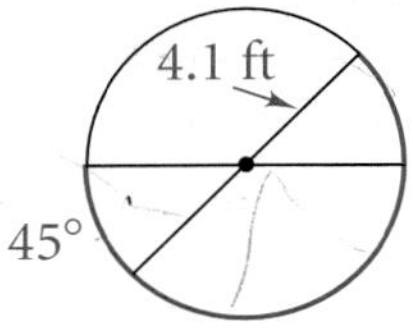

21.
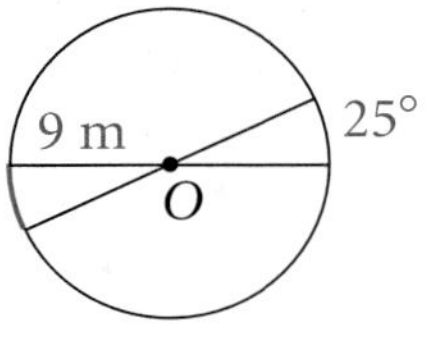

22.
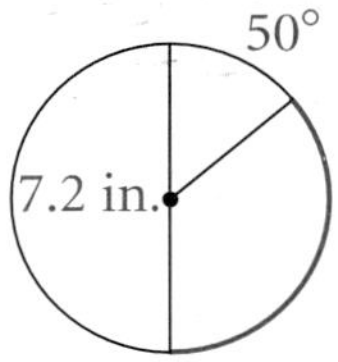

23. Find the perimeter of the shaded portion of the figure at the right. Leave your answer in terms of π.

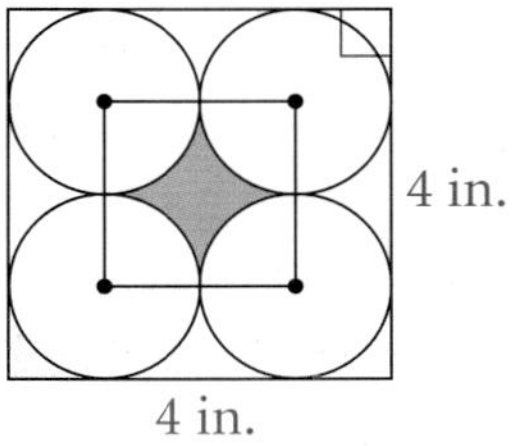

24. *Standardized Test Prep* The length of $\widehat{AB}$ is 6π cm and $m\widehat{AB} = 120$. What is the diameter of the circle?

A. 2 cm **B.** 6 cm **C.** 9 cm
D. 18 cm **E.** 24 cm

25. *Coordinate Geometry* Find the length of a semicircle with endpoints (3, 7) and (3, −1). Round your answer to the nearest tenth.

CALVIN AND HOBBES by Bill Watterson

Cartoon **Use what you learned from Calvin's father to answer the following questions.**

26. In one revolution, how much farther does a point 10 cm from the center of the record travel than a point 3 cm from the center? Round your answer to the nearest hundredth.

27. *Writing* Kendra and her mother plan to ride the merry-go-round. Two horses on the merry-go-round are side by side. For a more exciting ride, should Kendra sit on the inside or the outside? Explain your reasoning.

Exercises MIXED REVIEW

Locus **Sketch and label each locus.**

28. all points in a plane equidistant from points A and B

29. all points in space equidistant from parallel lines m and n

30. **Transformations** A triangle has vertices $A(3, 2)$, $B(4, 1)$, and $C(4, 3)$. Find the coordinates of the image of the triangle under a glide reflection in $\langle 0, 1 \rangle$ and $x = 0$.

Getting Ready for Lesson 5-8

Estimation **A circle is drawn on three different grids. Use the scale of each grid to estimate the area of each circle in square inches.**

31.

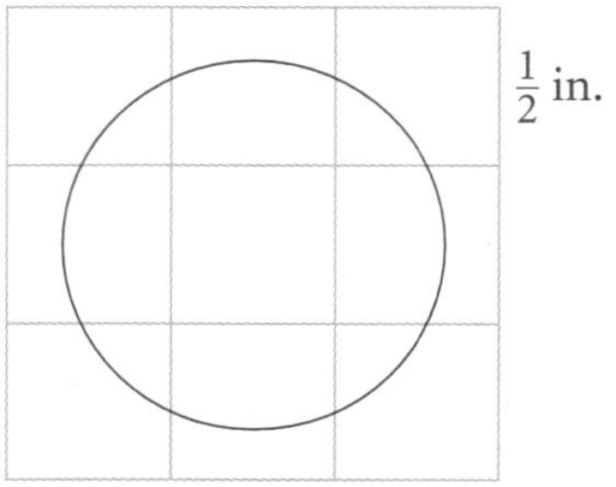

32.

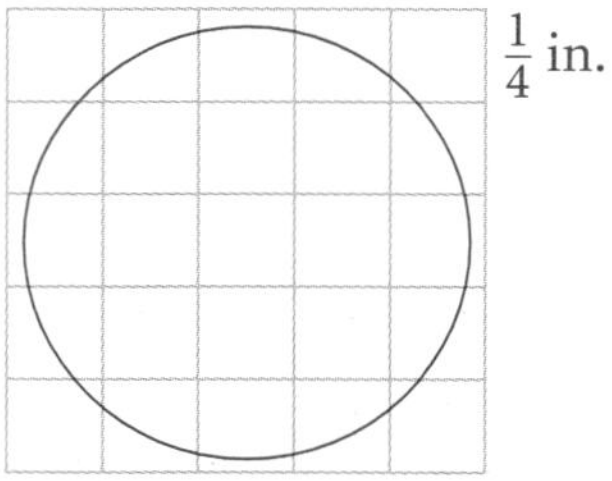

33.

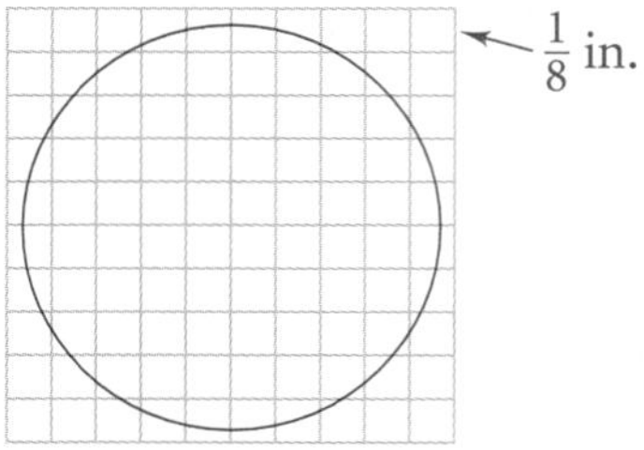

Exercises CHECKPOINT

Find the area of each trapezoid or regular polygon. Leave your answer in simplest radical form.

1.

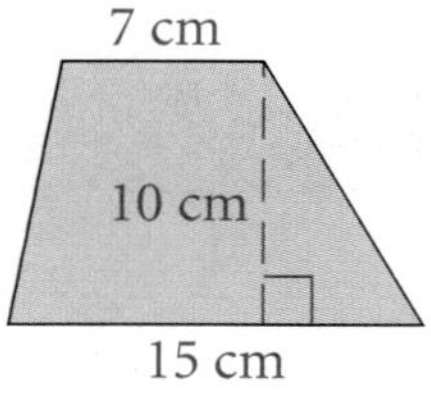

2.

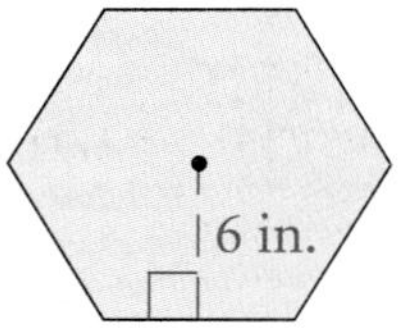

3.

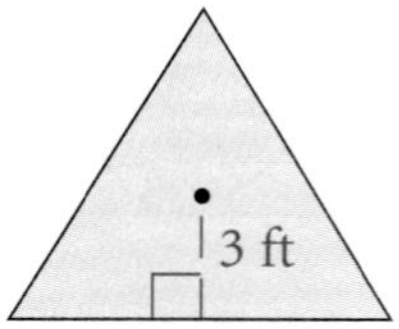

Calculator **Find the circumference of a circle with the given radius. Round your answer to the nearest hundredth.**

4. 8 in. **5.** 2 m **6.** 5 ft **7.** 1.4 km **8.** 9 mm

9. In a circle of radius 18 mm, $m\widehat{AB} = 45$. Find the length of $\widehat{AB}$. Leave your answer in terms of π.

10. **Writing** Explain at least two ways to find the area of an equilateral triangle. Use an example to illustrate your explanation.

5-8 Areas of Circles, Sectors, and Segments of Circles

What You'll Learn

- Computing the areas of circles, sectors, and segments of circles

...And Why

To solve real-world problems in food preparation, archaeology, and biology

What You'll Need

- compass
- scissors
- tape

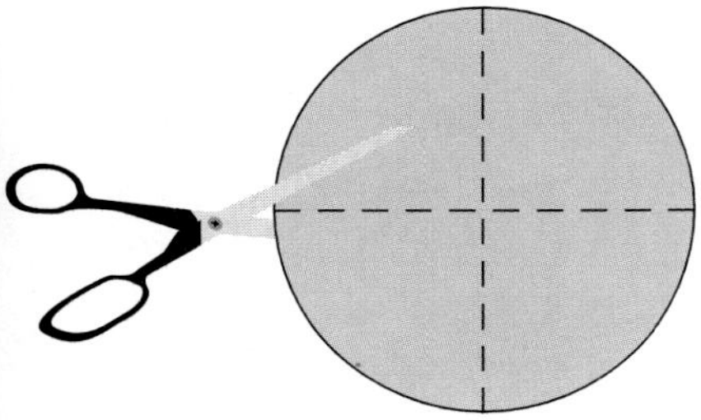

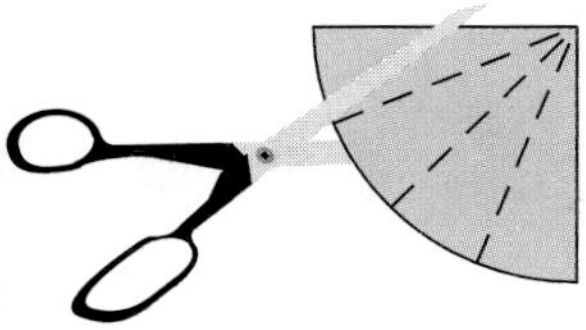

WORK TOGETHER

Work in groups. Have each member of your group use a compass to draw a large circle. Fold the circle in half horizontally and vertically. Cut the circle into four wedges on the fold lines. Then fold each wedge into quarters. Cut each wedge on the fold lines. You will have 16 wedges. Tape the wedges to a piece of paper to form the figure below.

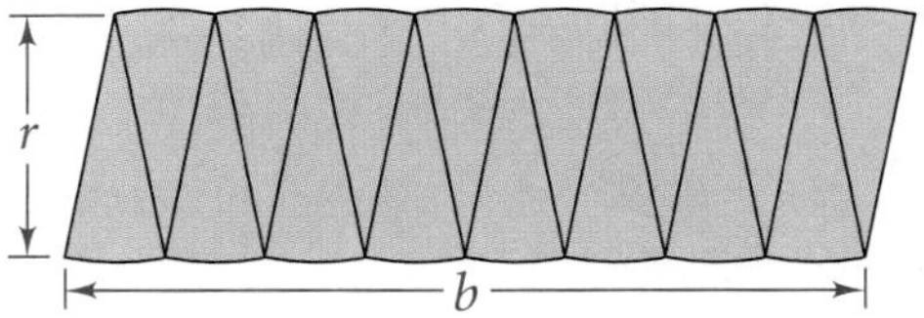

1. How does the area of the figure compare with the area of the circle?
2. The base of the figure is formed by arcs of the circle. Write an equation relating the length of the base b to the circumference C of the circle.
3. Write an equation for the length of the base b in terms of the radius r of the circle.
4. If you increase the number of wedges, the figure you create becomes more and more like a rectangle with base b and height r. Write an expression for the area of the rectangle in terms of r.

THINK AND DISCUSS

Areas of Circles

Your observations in the Work Together suggest the following theorem.

Theorem 5-12
Area of a Circle

The area of a circle is the product of π and the square of the radius.

$$A = \pi r^2$$

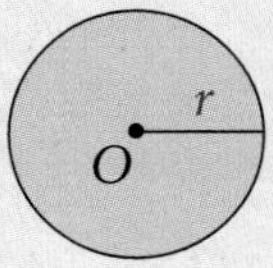

5. **Try This** What is the area of a circle with radius 15 cm? Leave your answer in terms of π.

Example 1 — Relating to the Real World

Food The diameter of a small pizza is 10 in. How much more pizza do you get if you order a medium pizza with diameter 12 in.?

$$\text{radius of small pizza} = \frac{10}{2} = 5 \qquad r = \frac{d}{2}$$

$$\text{radius of medium pizza} = \frac{12}{2} = 6 \qquad r = \frac{d}{2}$$

Use the formula for the area of a circle.

$$\text{area of small pizza} = \pi(5)^2 = 25\pi \qquad A = \pi r^2$$

$$\text{area of medium pizza} = \pi(6)^2 = 36\pi \qquad A = \pi r^2$$

$$\text{difference in area} = 36\pi - 25\pi = 11\pi$$

11 [×] [π] [=] 34.557519

The medium pizza has about 35 in.^{2} more pizza than the small pizza.

6. Suppose the small pizza in Example 1 costs \$5.00 and the medium pizza costs \$6.00. Which pizza is a better buy? Explain your answer.

Sectors and Segments

A **sector of a circle** is the region bounded by two radii and their intercepted arc. A slice of pizza is an example of a sector of a circle. You name a sector using one endpoint of the arc, the center of the circle, and the other endpoint of the arc. Sector *XOY* is at the left.

The area of a sector is a fractional part of the area of a circle. The ratio of a sector's area to a circle's area is $\frac{\text{measure of the arc}}{360}$.

Theorem 5-13
Area of a Sector of a Circle

The area of a sector of a circle is the product of the ratio $\frac{\text{measure of the arc}}{360}$ and the area of the circle.

Area of sector $AOB = \frac{m\widehat{AB}}{360} \cdot \pi r^2$

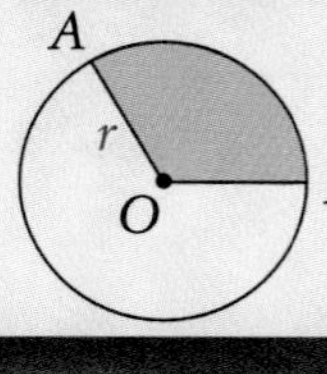

Example 2

Find the area of sector ZOM. Leave your answer in terms of π.

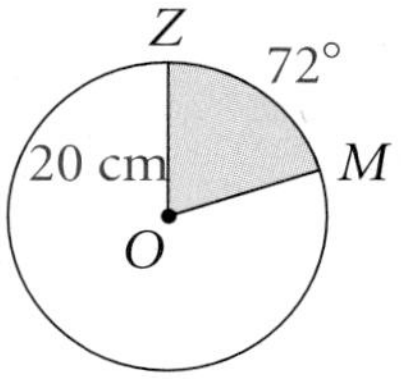

$$\text{area of sector } ZOM = \frac{m\overset{\frown}{ZM}}{360} \cdot \pi r^2$$
$$= \frac{72}{360} \cdot \pi(20)^2$$
$$= 80\pi$$

The area of sector ZOM is 80π cm^2.

7. Calculator A circle has diameter 8.2 m. What is the area of a sector with a 125° arc? Round your answer to the nearest tenth.

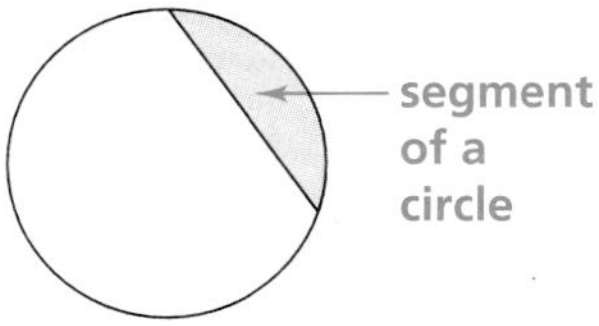

The part of a circle bounded by an arc and the segment joining its endpoints is a **segment of the circle.** To find the area of a segment, draw radii to form a sector. The area of the segment equals the area of the sector minus the area of the triangle formed.

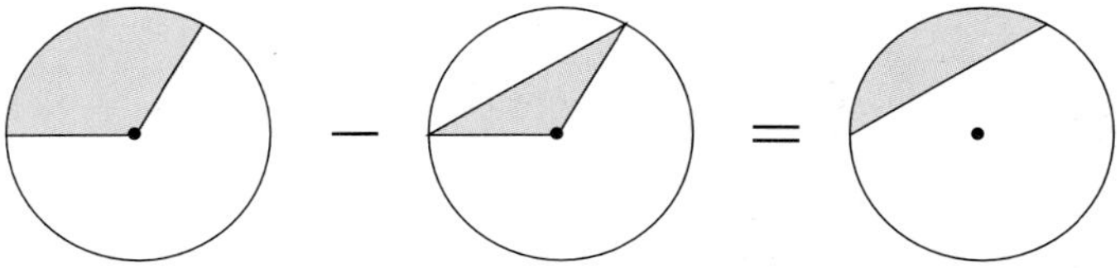

Area of sector   Area of triangle   Area of segment

Example 3

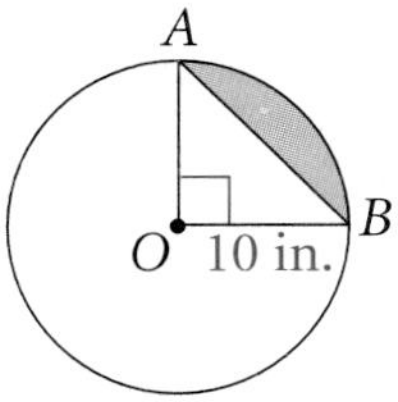

Find the area of the shaded segment.

$$\text{area of sector } AOB = \frac{m\overset{\frown}{AB}}{360} \cdot \pi r^2 \qquad \text{Formula for area of a sector}$$
$$= \frac{90}{360} \cdot \pi(10)^2 \qquad \text{Substitute.}$$
$$= \frac{1}{4} \cdot 100\pi$$
$$= 25\pi$$

$$\text{area of } \triangle AOB = \frac{1}{2}bh \qquad \text{Formula for area of a triangle}$$
$$= \frac{1}{2}(10)(10) \qquad \text{Substitute.}$$
$$= 50$$

area of segment $= 25\pi - 50$

25 [×] [π] [−] 50 [=] 28.539816

The area of the segment is about 28.5 in.^{2}

8. Try This A circle has radius 12 cm. Find the area of a segment of the circle bounded by a 60° arc and the segment joining its endpoints. Round your answer to the nearest tenth.

Exercises ON YOUR OWN

Find the area of each circle. Leave your answers in terms of π.

1.

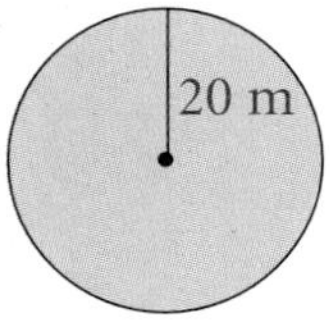

2.

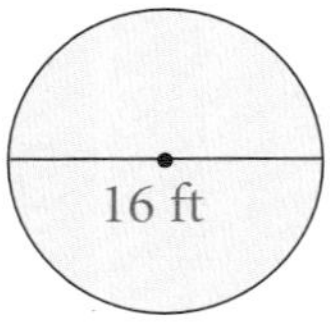

3.

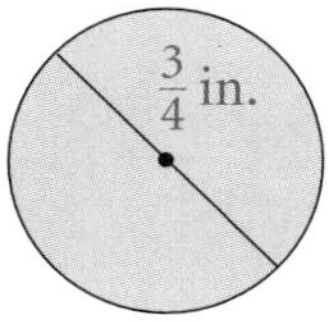

4.

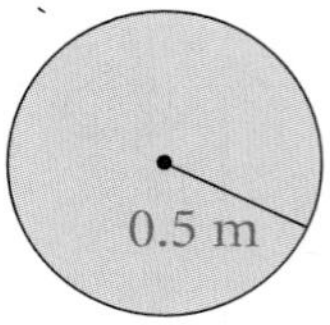

5. A circle has area 225π m^2. What is its diameter?

6. How many circles with radius 4 in. will have the same total area as a circle with radius 12 in.?

7. *Coordinate Geometry* The endpoints of a diameter of $\odot A$ are (2, 1) and (5, 5). Find the area of $\odot A$. Leave your answer in terms of π.

Calculator **Find the area of each circle. Round your answer to the nearest hundredth.**

8. $r = 7$ ft **9.** $d = 8.3$ m **10.** $d = 0.24$ cm

11. *Archaeology* Off the coast of Sweden, divers are working to bring up artifacts from a ship that sank several hundred years ago. The line to a diver is 100 ft long, and the diver is working at a depth of 80 ft. What is the area of the circle that the diver can cover? Round your answer to the nearest square foot.

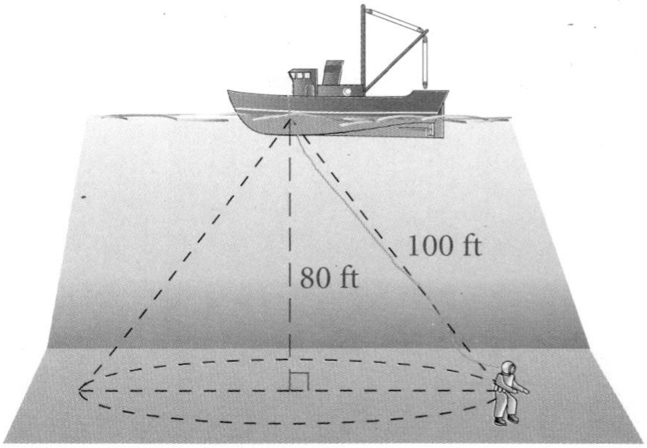

Find the area of each shaded sector of a circle. Leave your answer in terms of π.

12.

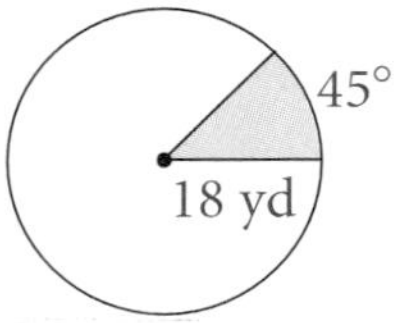

13.

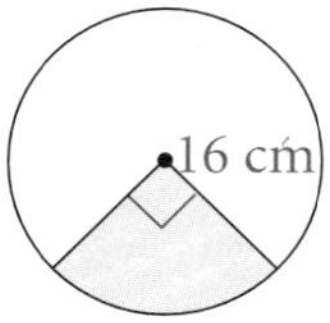

14.

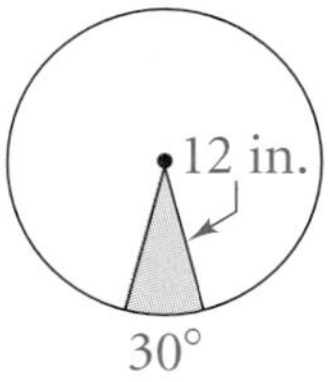

15.

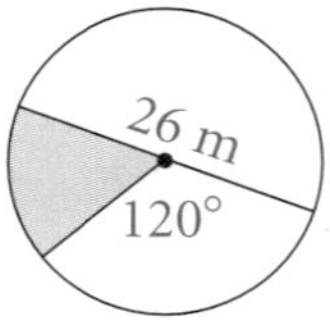

16. *Games* A dartboard has a diameter of 20 in. and is divided into 20 congruent sectors. Find the area of one sector. Round your answer to the nearest tenth.

17. *Animal Habitats* In the Pacific Northwest, a red fox has a circular home range with a radius of about 718 m. To the nearest thousand, about how many square meters are in a red fox's home range?

Calculator **Find the area of each shaded segment of a circle. Round your answer to the nearest hundredth.**

18.
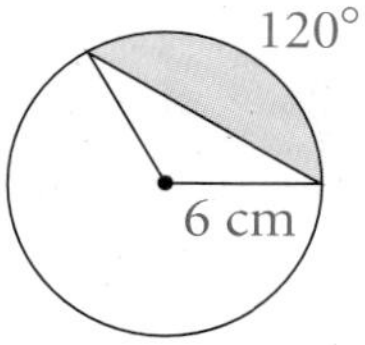

19.
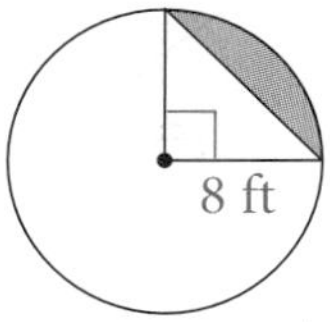

20.
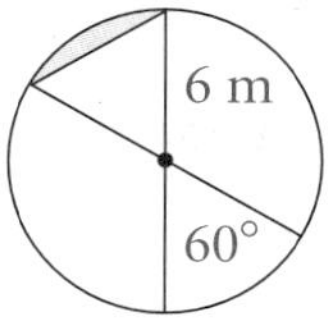

21.
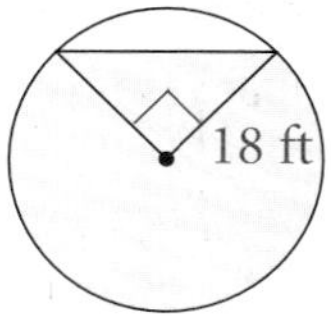

22. **Writing** The American Institute of Baking suggests a technique for cutting and serving a tiered cake. The tiers of a cake have the same height and have diameters 8 in. and 13 in. The top layer and the circle directly under it are cut into 8 pieces and the exterior ring of the 13-inch layer is cut into 12 pieces. Which piece would be biggest, a top, bottom-inside, or bottom-outside piece? Explain your answer.

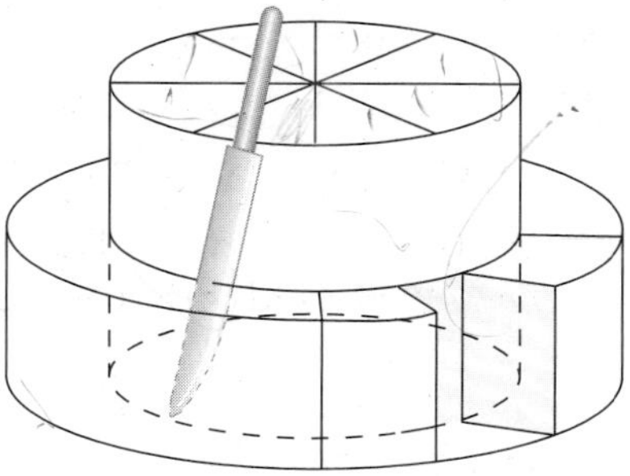

23. A sector of a circle with a 90° arc has area 36π in.^{2}. What is the radius of the circle?

Find the area of the shaded figure. Leave your answer in terms of π.

24.
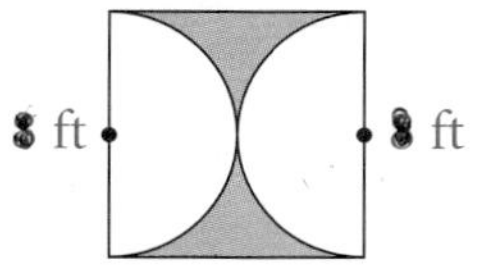

25.
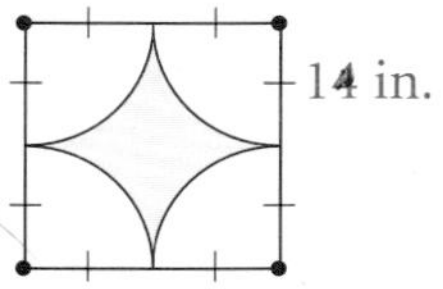

26.
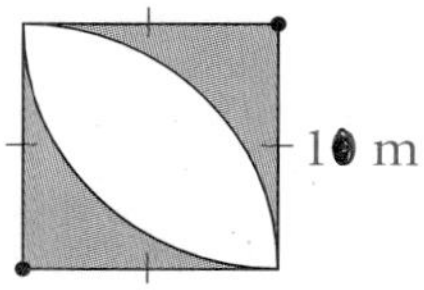

27.
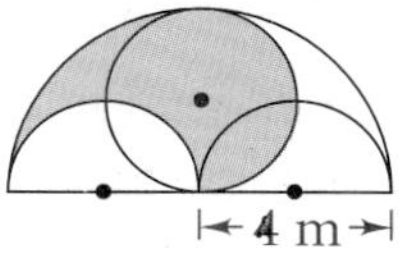

28. **Open-ended** Draw a diagram for a sector of a circle such that the sector has area 16π cm^2. Label the radius of the circle and the measure of the arc of the sector.

29. An 8 ft-by-10 ft floating dock is anchored in the middle of a pond. The bow of a canoe is tied to one corner of the dock with a 10-ft rope.

a. Sketch a diagram of the area in which the bow of the canoe can travel.

b. Write a plan for finding the area.

c. Find the area. Round your answer to the nearest square foot.

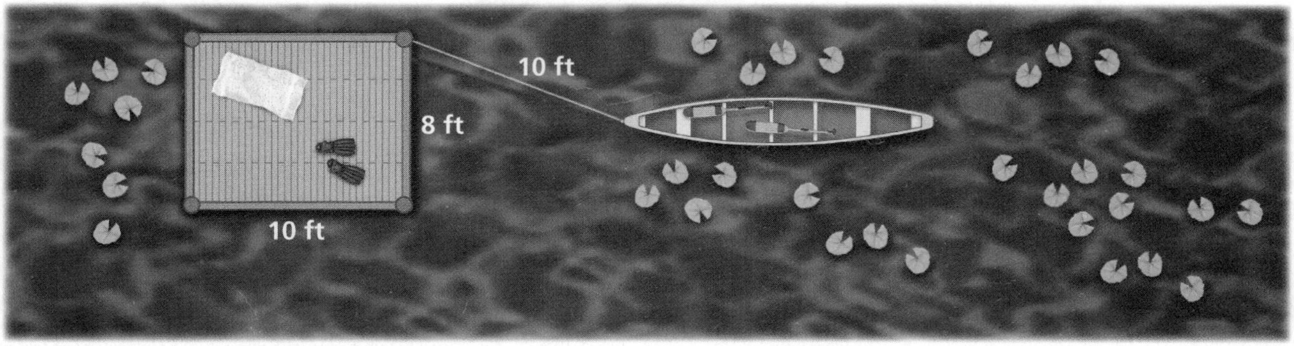

Chapter Project — Find Out by Calculating

The circles on this quilt were sewn onto the background cloth. Cutting circles from rectangles leaves some waste. Explore whether you can reduce waste by using smaller circles.

- Compare the percent of material wasted when the shaded circles are cut from squares A, B, and C below.

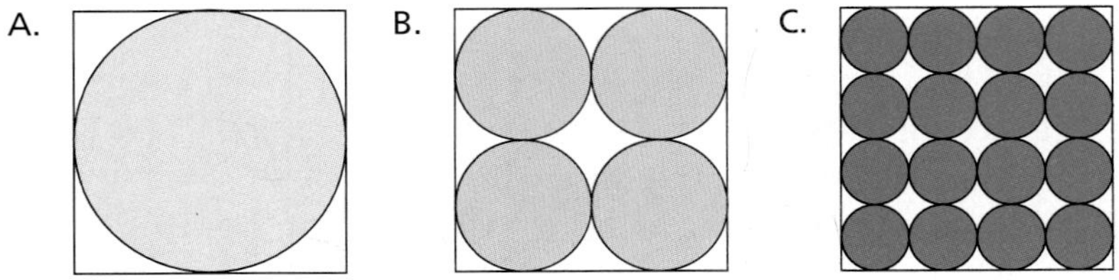

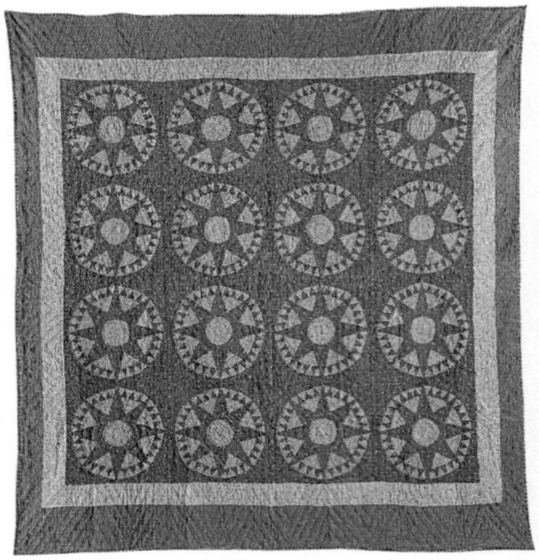

- Estimate how many times longer it would take to cut out the 16 circles from square C than the 1 circle from square A. Support your estimate with calculations.

Exercises MIXED REVIEW

Data Analysis **Use the line graph for Exercises 30–32.**

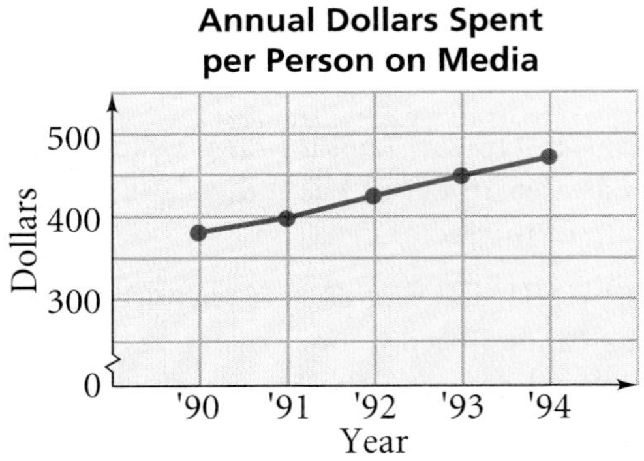

30. In 1990, how much did the average person spend on media such as printed material, videos, and recordings?

31. How much has spending increased from 1990 to 1994?

32. **Predict** how much the average person will spend on media in the year 2000.

33. What is the area of a 30°-60°-90° triangle with hypotenuse 12 cm? Leave your answer in simplest radical form.

The measures of two angles of a triangle are given. Find the measure of the third angle. Then classify the triangle by its sides and angles.

34. 54°, 108°

35. 72°, 36°

36. 36°, 54°

37. 78°, 34°

38. 60°, 60°

39. 90°, 45°

SELF ASSESSMENT — PORTFOLIO

For your portfolio, select one or two items from your work for this chapter. Here are some possibilities.
- corrected work
- a journal entry

Explain why you have included each selection.

Math ToolboX

Technology

Exploring Area and Circumference

After Lesson 5-8

A polygon that is *inscribed* in a circle has all its vertices on the circle. Work in pairs or small groups. Investigate the ratios of the perimeters and areas of inscribed regular polygons to the circumference and area of the circle in which they are inscribed. Begin by making a table like this.

Regular Polygon			Circle		Ratios	
Sides	Perimeter	Area	Circumference	Area	$\frac{\text{Perimeter}}{\text{Circumference}}$	$\frac{\text{Polygon Area}}{\text{Circle Area}}$
3						

Construct

Use geometry software to construct a circle. Find its circumference and area and record them in your table. Inscribe an equilateral triangle in the circle. Your software may be able to do this for you automatically, or you can construct three points on the circle and move them so they are approximately evenly spaced on the circle. Then draw the triangle.

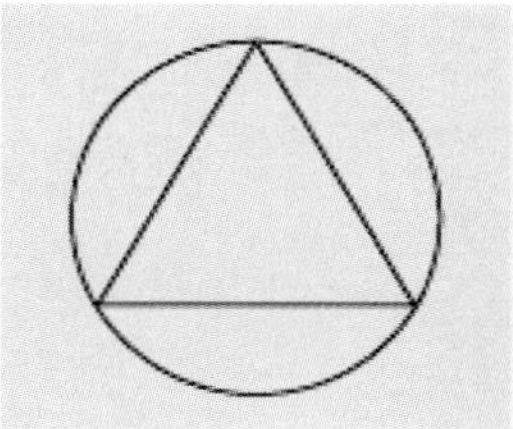

Investigate

Use your geometry software to measure the perimeter and area of the triangle and to calculate the ratios $\frac{\text{triangle perimeter}}{\text{circle circumference}}$ and $\frac{\text{triangle area}}{\text{circle area}}$. Record the results.

Manipulate the circle to change its size. Do the ratios you calculated stay the same or change?

Now inscribe a square in a circle and fill in your table for a polygon of four sides. Do the same for a regular pentagon.

Conjecture

What will happen to the ratios $\frac{\text{perimeter}}{\text{circumference}}$ and $\frac{\text{polygon area}}{\text{circle area}}$ as you increase the number of sides of the polygon?

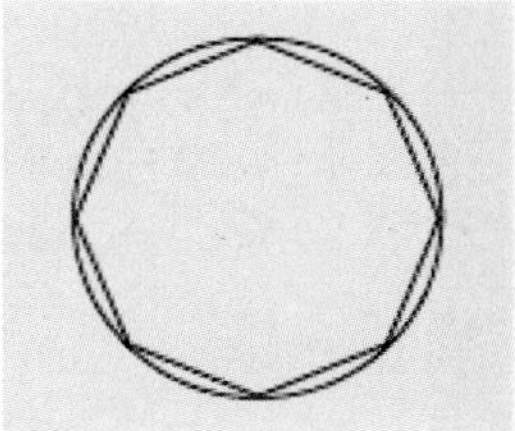

Extend

- Extend your table to include polygons of 12 sides. Does your conjecture still hold? Compare the two columns of ratios in your table. How do they differ?
- Estimate the perimeter and area of a polygon of 100 sides that is inscribed in a circle with a radius of 10 cm.

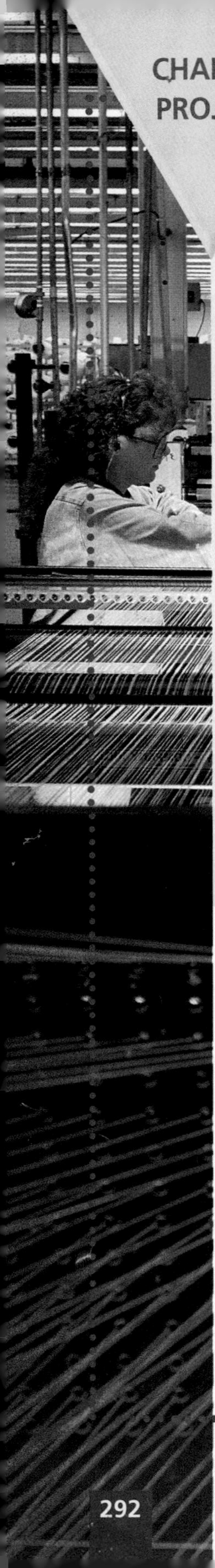

CHAPTER PROJECT

Finishing the Chapter Project

And Sew On

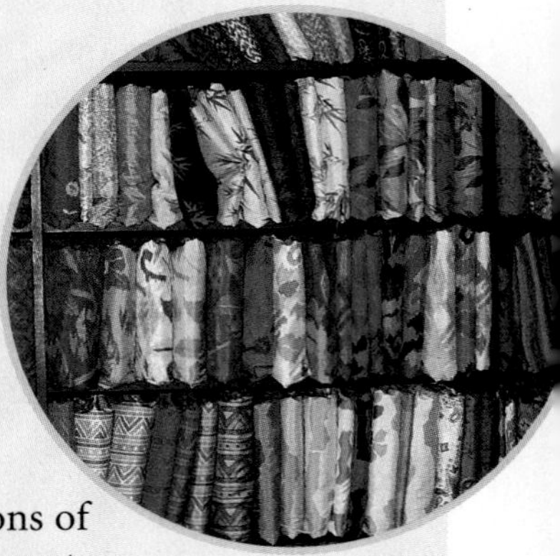

Find Out exercises on pages 248, 254, 273, and 290 should help you complete your project. Design a quilt for your bed. Use one of the techniques you learned. Include the dimensions of the quilt. List the size, shape, color, and number of each different piece. If available, use iron-on patches to color several small blocks that establish the design. Then iron the design on a T-shirt or a piece of fabric.

Reflect and Revise

Ask a classmate to review your project with you. Together, check that your quilt design is complete, your diagrams are clear, and your explanations and information are accurate. Is the display attractive, organized, and comprehensive? Consider doing more research (using some of the books listed below) on textiles of different cultures.

Follow Up

Go to a fabric store to find the different widths in which fabrics are sold. Determine the amount of each fabric you need, and estimate the cost.

For More Information

Bradkin, Cheryl G. *Basic Seminole Patchwork.* Mountain View, California: Leone Publications, 1990.

Fisher, Laura. *Quilts of Illusion.* Pittstown, New Jersey: The Main Street Press, 1988.

Kapoun, Robert W. *Language of the Robe: American Indian Trade Blankets.* Salt Lake City, Utah: Peregrine Smith Books, 1948.

Norden, Mary. *Ethnic Needlepoint Designs from Asia, Africa, and the Americas.* New York: Watson-Guptill Publications, 1993.

Schevill, Margot Blum. *Maya Textiles of Guatemala.* Austin, Texas: University of Texas Press, 1993.

CHAPTER 5

Wrap Up

Key Terms

altitude (p. 250)
apothem (p. 274)
arc length (p. 281)
area of a circle (p. 285)
area of a parallelogram (p. 249)
area of a polygon (p. 243)
area of a rectangle (p. 244)
area of a square (p. 243)
area of a trapezoid (p. 269)
area of a triangle (p. 251)
base (pp. 250, 251, 269)
center (p. 274)
circumference of a circle (p. 279)
concentric circles (p. 280)
congruent arcs (p. 281)
Converse of the Pythagorean Theorem (p. 258)
45°-45°-90° triangle (p. 264)
height (pp. 250, 251, 269)
hypotenuse (p. 256)
legs of a right triangle (p. 256)
legs of a trapezoid (p. 269)
perimeter of a polygon (p. 242)
pi (p. 279)
Pythagorean Theorem (p. 257)
Pythagorean triple (p. 258)
radius (p. 274)
sector of a circle (p. 286)
segment of a circle (p. 287)
30°-60°-90° triangle (p. 265)

How am I doing?

SELF ASSESSMENT

- State three ideas from this chapter that you think are important. Explain your choices.
- Explain how to find the area of different polygons.

Understanding Perimeter and Area — 5-1

The **perimeter of a polygon** is the sum of the lengths of its sides.

The formula for the perimeter of a square is $P = 4s$. The formula for the perimeter of a rectangle is $P = 2b + 2h$.

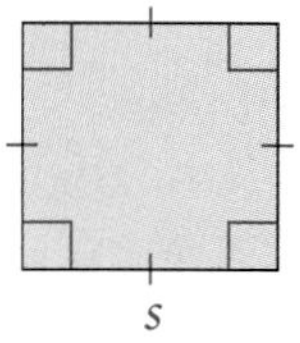

The **area of a polygon** is the number of square units it encloses. The area of a region is the sum of the area of its nonoverlapping parts. If two figures are congruent, their areas are equal.

The formula for the **area of a square** is $A = s^2$. The formula for the **area of a rectangle** is $A = bh$.

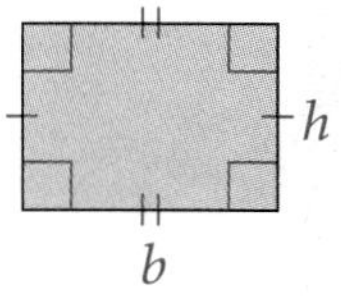

Find the perimeter and area of each figure.

1.

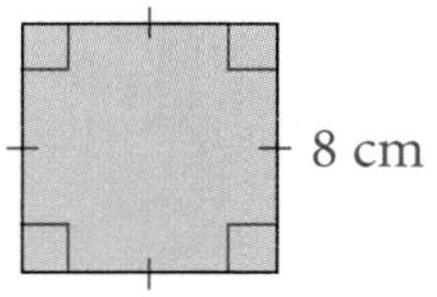

2.

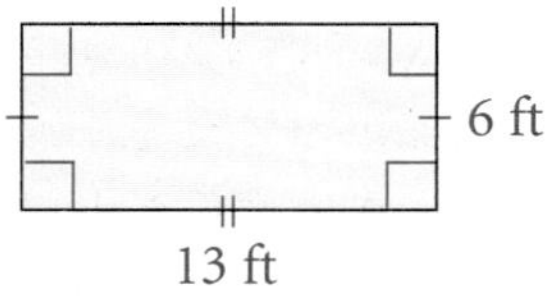

3.

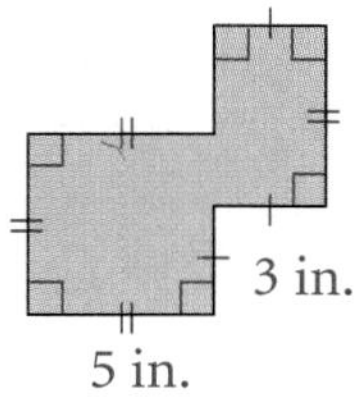

4. *Open-ended* Draw a polygon with perimeter 33 cm.

Areas of Parallelograms and Triangles 5-2

You can find the area of a parallelogram or a triangle if you know the **base** and **height**. The **area of a parallelogram** is $A = bh$. The **area of a triangle** is $A = \frac{1}{2}bh$.

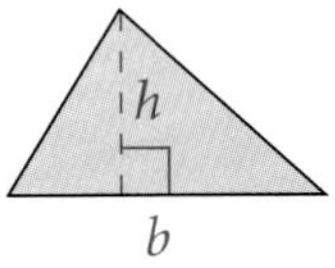

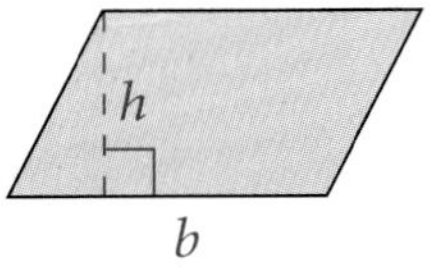

Find the area of each figure.

5.

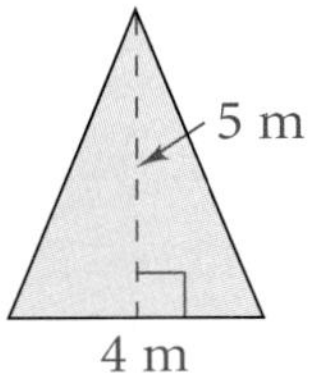

6.

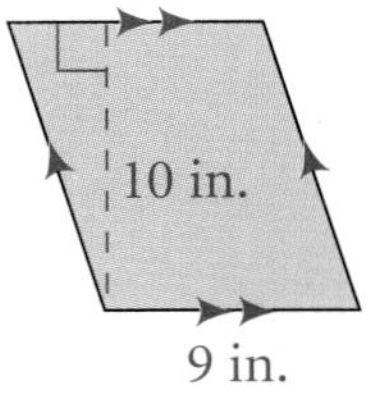

7.

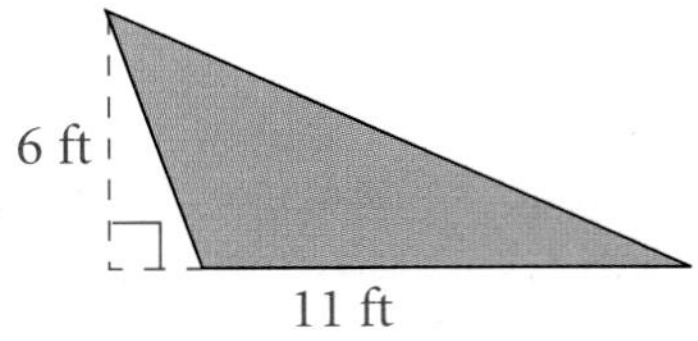

The Pythagorean Theorem and Its Converse 5-3

The **Pythagorean Theorem** states that in a right triangle the sum of the squares of the lengths of the legs equals the square of the length of the hypotenuse, or $a^2 + b^2 = c^2$.

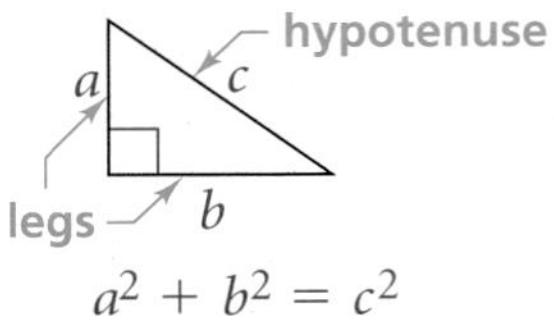

Positive integers a, b, and c form a **Pythagorean triple** if $a^2 + b^2 = c^2$.

The **Converse of the Pythagorean Theorem** states that if the square of the length of one side of a triangle is equal to the sum of the squares of the lengths of the other two sides, then the triangle is a right triangle.

In a triangle with longest side c, if $c^2 > a^2 + b^2$, the triangle is obtuse, and if $c^2 < a^2 + b^2$, the triangle is acute.

Find each value of x. If your answer is not an integer, you may leave it in simplest radical form.

8.

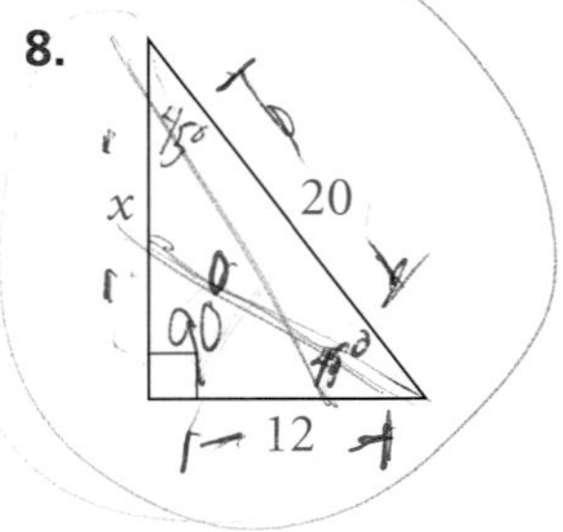

9.

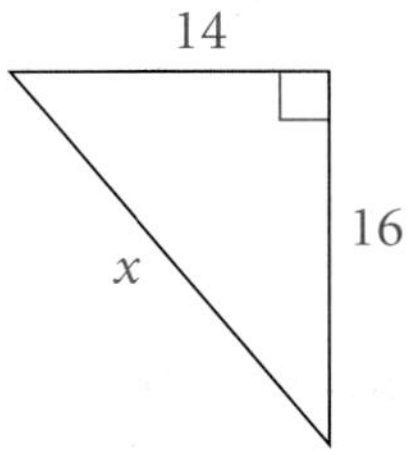

10.

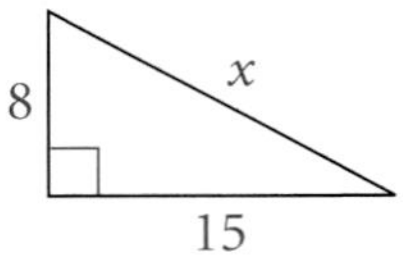

Special Right Triangles 5-4

In a **45°-45°-90° triangle,** both legs are congruent and the length of the hypotenuse is $\sqrt{2}$ times the length of a leg.

In a **30°-60°-90° triangle,** the length of the hypotenuse is twice the length of the shorter leg. The length of the longer leg is $\sqrt{3}$ times the length of the shorter leg.

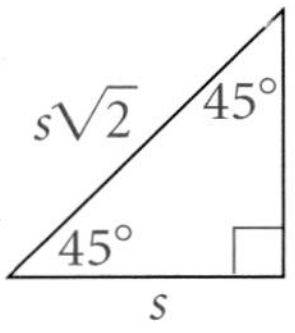

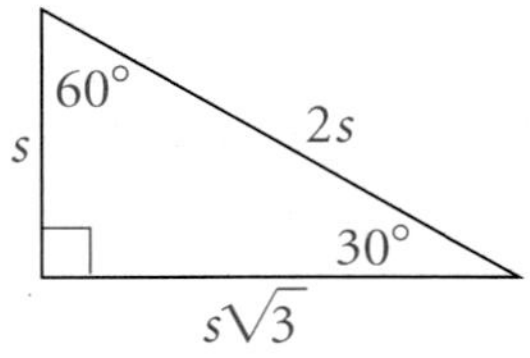

Find the value of each variable. If your answer is not an integer, you may leave it in simplest radical form.

11.

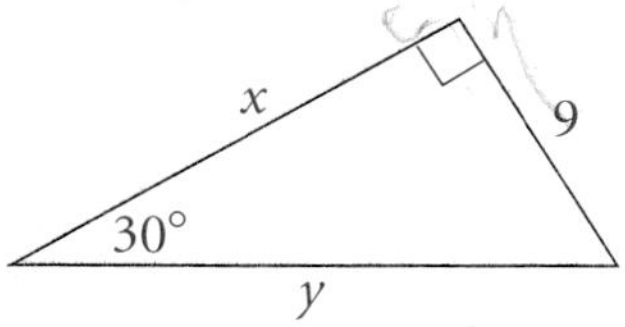

12.

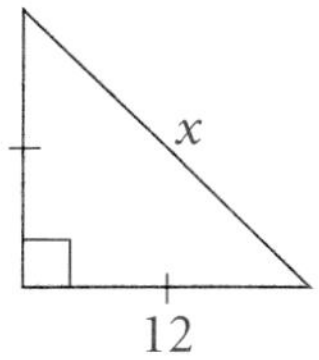

13.

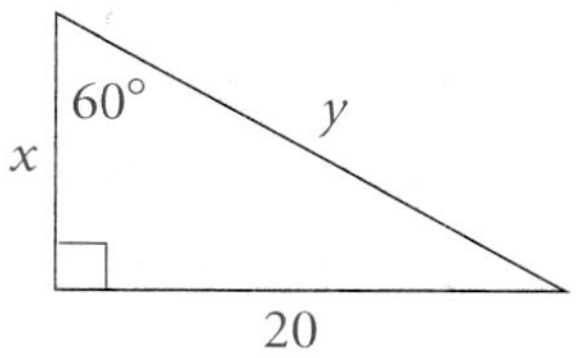

14. Standardized Test Prep A triangle has sides with lengths 4, 4, and $4\sqrt{2}$. What kind of triangle is it?

A. acute isosceles **B.** scalene right **C.** equilateral
D. obtuse right **E.** isosceles right

Areas of Trapezoids 5-5

The two parallel sides of a trapezoid are **bases.** The nonparallel sides are **legs.** The **height** is the perpendicular distance between the two bases. The **area of a trapezoid** is $A = \frac{1}{2}h(b_1 + b_2)$.

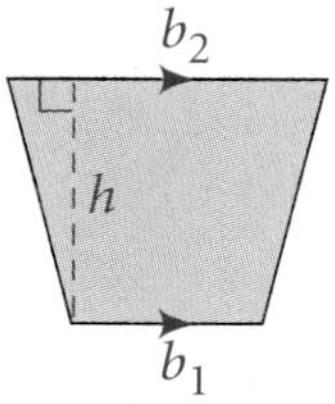

Find the area of each trapezoid. If your answer is not an integer, you may leave it in simplest radical form.

15.

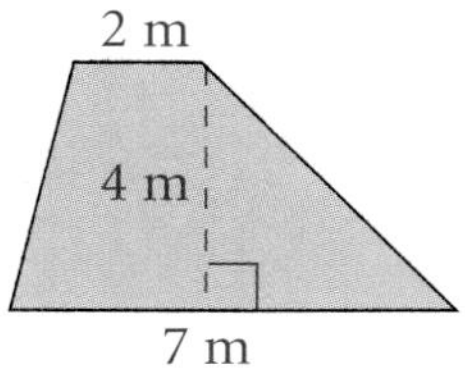

16.

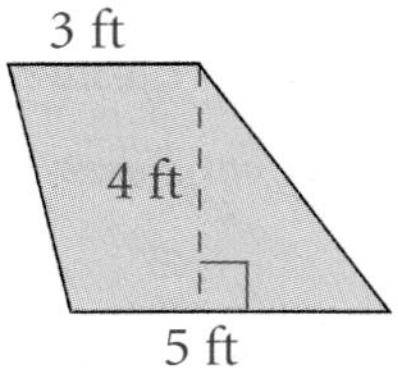

17.

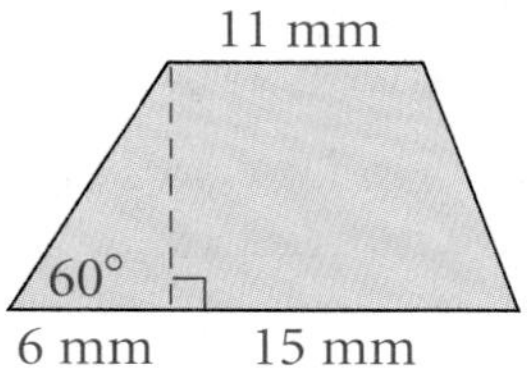

18. Writing Explain how the formula for the area of a trapezoid is related to the formula for the area of a triangle.

Areas of Regular Polygons 5-6

The **center** of a regular polygon is the center of its circumscribed circle. The **radius** is the distance from the center to a vertex. The **apothem** is the perpendicular distance from the center to a side. The area of a regular polygon with apothem a and perimeter p is $A = \frac{1}{2}ap$.

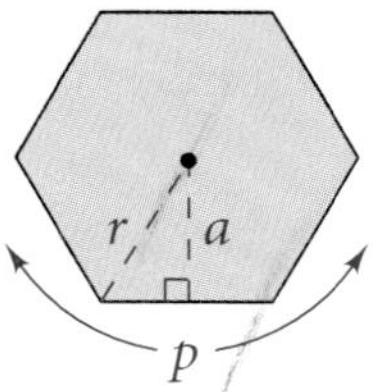

Calculator **Sketch each regular polygon with the given radius. Then find its area. Round your answer to the nearest tenth.**

19. triangle; radius 4 in. **20.** square; radius 8 mm **21.** hexagon; radius 7 cm

Circles: Circumference and Arc Length 5-7

The **circumference of a circle** is $C = \pi d$ or $C = 2\pi r$. The **length of an arc** is a fraction of a circle's circumference.

The length of $\widehat{AB} = \frac{m\widehat{AB}}{360} \cdot 2\pi r$.

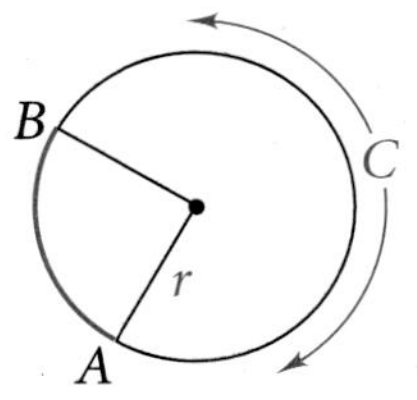

Find the circumference of each circle and the length of each arc shown in red. Leave your answer in terms of π.

22.

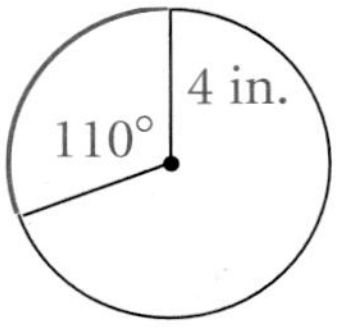

23.

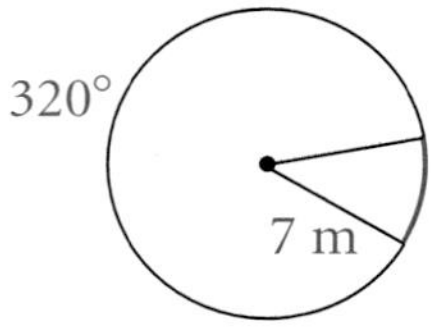

24.

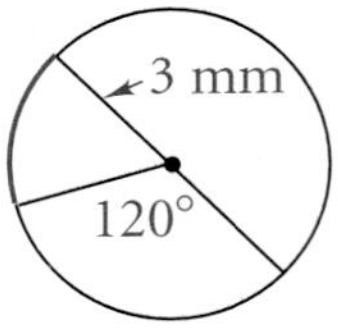

Areas of Circles, Sectors, and Segments of Circles 5-8

The **area of a circle** is $A = \pi r^2$. The part of a circle bounded by two radii and their intercepted arc is a **sector of a circle.**

The area of sector $APB = \frac{m\widehat{AB}}{360} \cdot \pi r^2$.

The part of a circle bounded by an arc and the segment joining its endpoints is a **segment of a circle.** The area of a segment of a circle is the difference between the areas of the related sector and triangle.

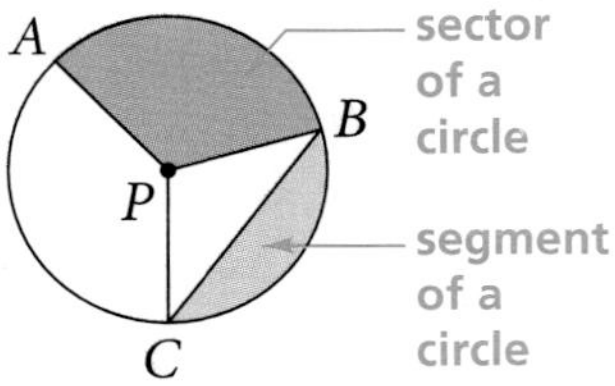

Calculator **Find the area of each shaded region. Round your answer to the nearest hundredth.**

25.

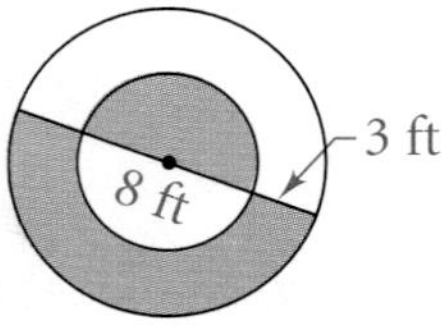

26.

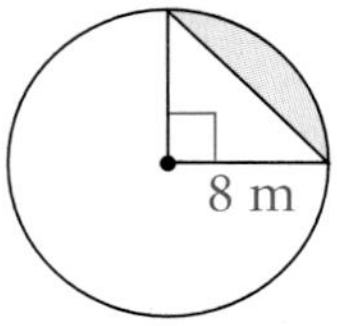

27.

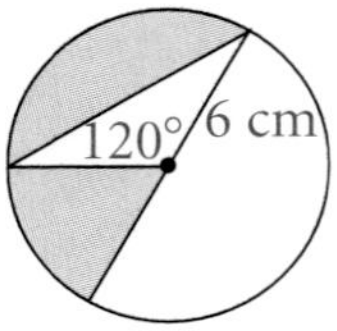

Getting Ready for... CHAPTER 6

Find the area of each figure. If your answer is not an integer, round to the nearest tenth.

28.

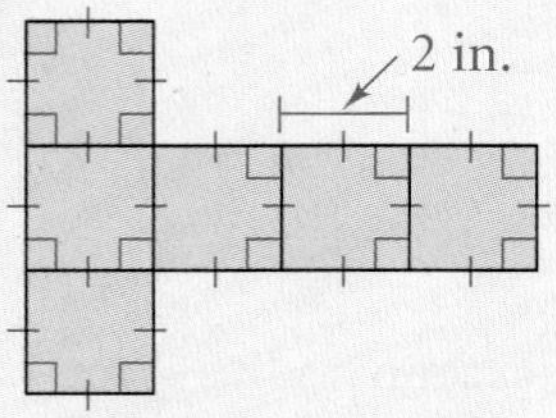

29.

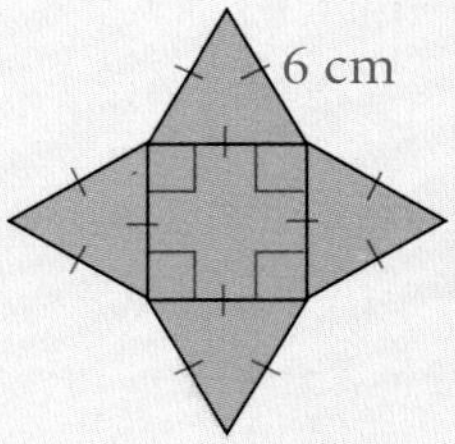

30.

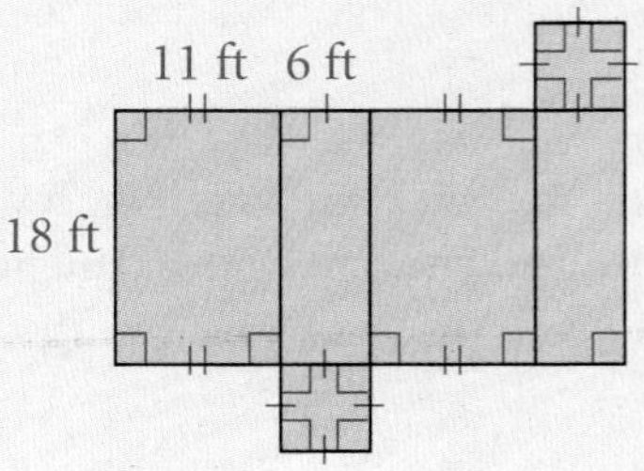

Assessment

Find the perimeter of each figure.

1.

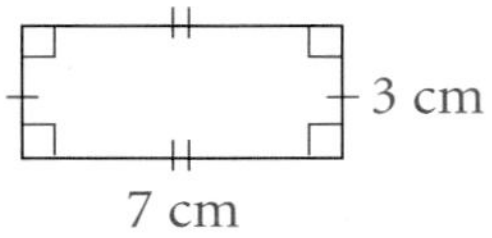

2.

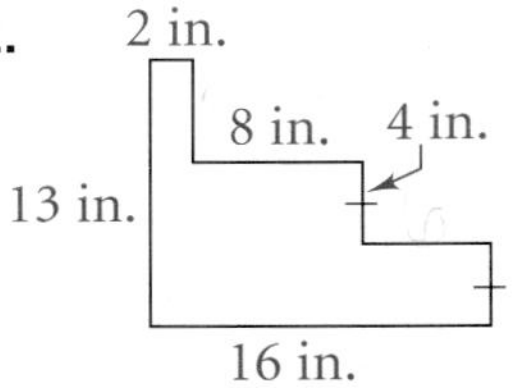

3. You have 64 ft of fencing. What are the dimensions of the rectangle of greatest area you could enclose?

Find the area of each figure. If your answer is not an integer, round to the nearest tenth.

4.

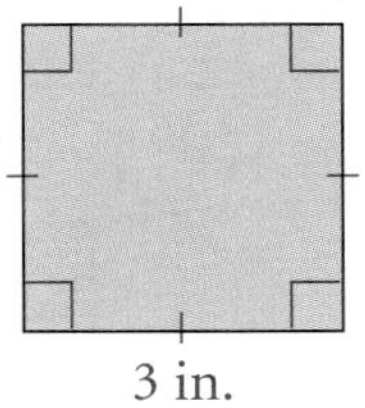

5.

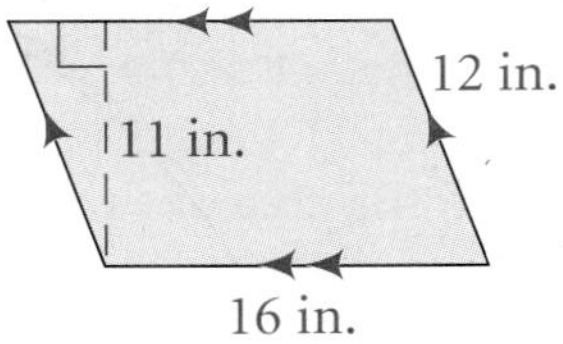

6.

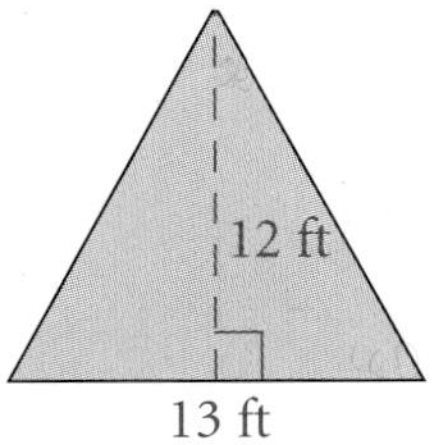

7.

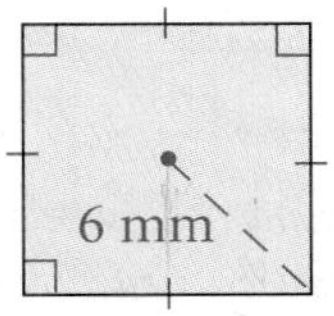

8.

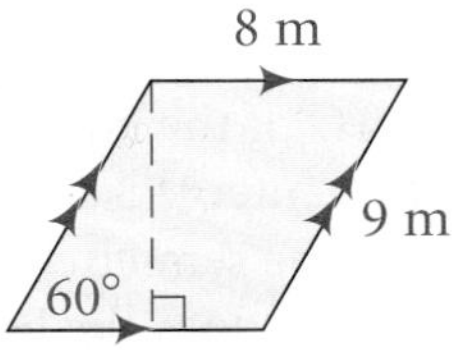

9.

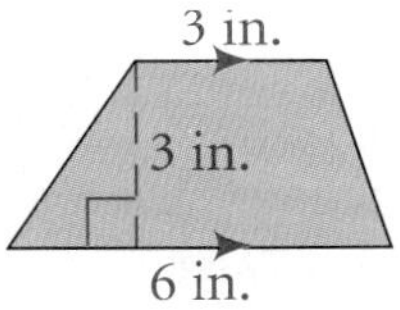

10. *Coordinate Geometry* A quadrilateral has vertices at $A(0, 7)$, $B(-2, 7)$, $C(-2, 0)$, and $D(0, 0)$. Find the area and perimeter of $ABCD$.

11. *Open-ended* An equilateral triangle, a square, and a regular pentagon all have the same perimeter. What can this perimeter be if all figures have sides that are integers?

Find the area of each regular polygon. Round to the nearest tenth.

12.

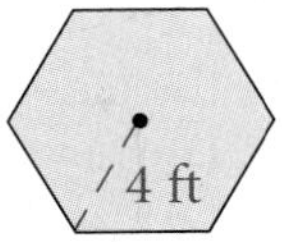

13.

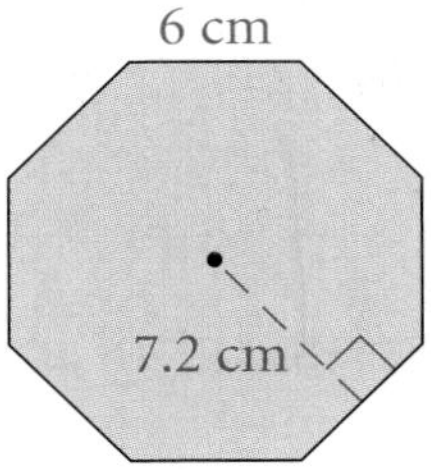

The lengths of two sides of a right triangle are given. Find the length of the third side. Leave your answers in simplest radical form.

14. one leg 9, other leg 6

15. one leg 12, hypotenuse 17

16. hypotenuse 20, leg 10

17. *Standardized Test Prep* Which integers form Pythagorean triples?

I. 15, 36, 39
II. 6, 8, 10
III. 16, 30, 34
IV. 10, 12, 14

A. I only **B.** IV only **C.** II and III
D. III and IV **E.** I, II, and III

Find the values of the variables. Leave your answers in simplest radical form.

18.

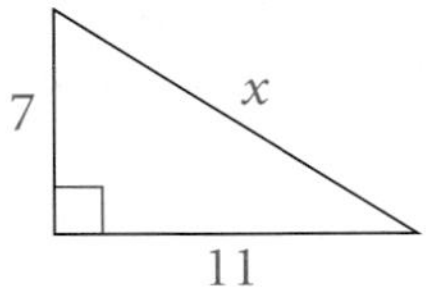

19.

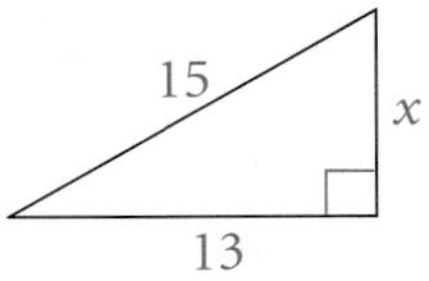

20.

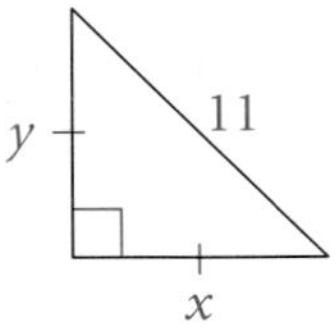

21.

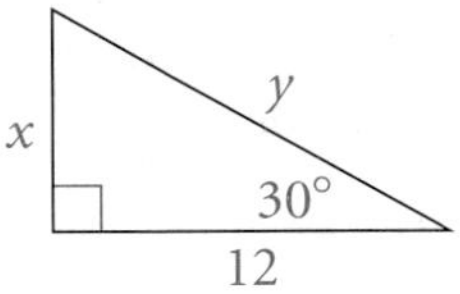

The lengths of three sides of a triangle are given. Describe each triangle as acute, right, or obtuse.

22. 9 cm, 10 cm, 12 cm

23. 8 m, 15 m, 17 m

24. 5 in., 6 in., 10 in.

25. **Writing** Explain how you can use the length of the shorter side of a 30°-60°-90° triangle to find the lengths of the other two sides.

Find the length of each arc shown in red. Leave your answers in terms of π.

26.

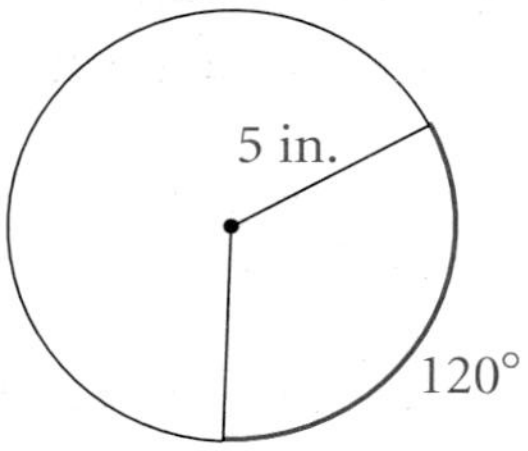

27.

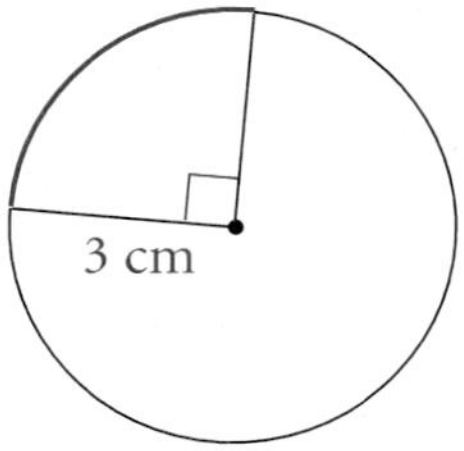

28.

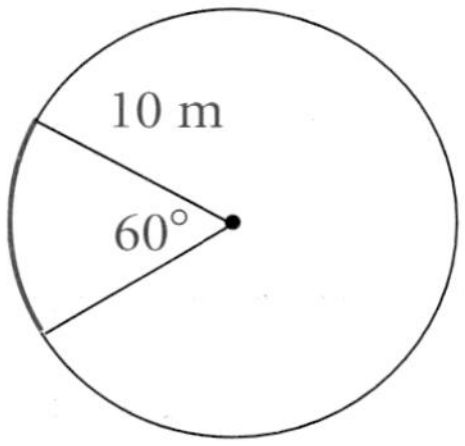

29.

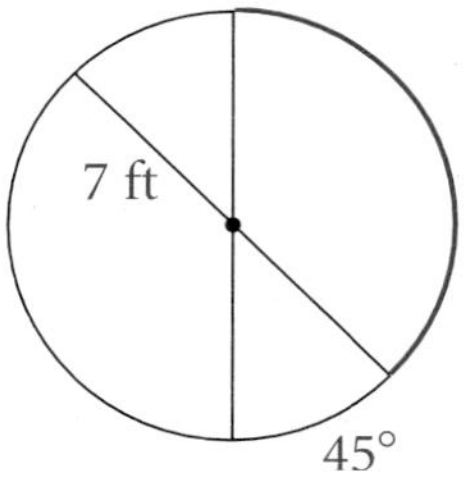

30. **Standardized Test Prep** Which length is greatest?

A. the diagonal of a 4-in. square
B. the diameter of a circle with 3-in. radius
C. the circumference of a circle with 2-in. radius
D. the length of a semicircle of a circle with diameter 6 in.
E. the perimeter of a 2 in.-by-3 in. rectangle

Find the area and circumference of a circle with the given radius or diameter. Leave your answers in terms of π.

31. $r = 4$ cm

32. $d = 10$ in.

33. $d = 7$ ft

34. $r = 12$ m

Calculator **Find the area of each shaded region. Round your answer to the nearest hundredth.**

35.

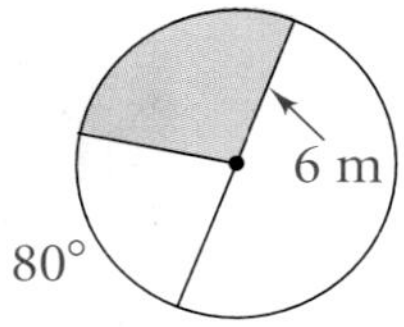

36.

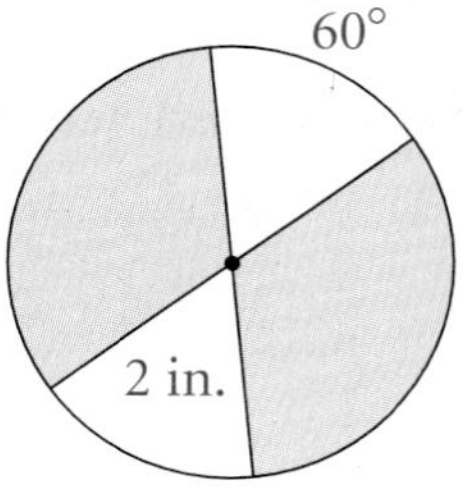

37.

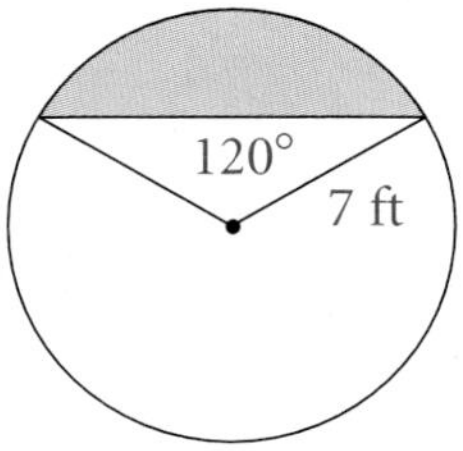

38.

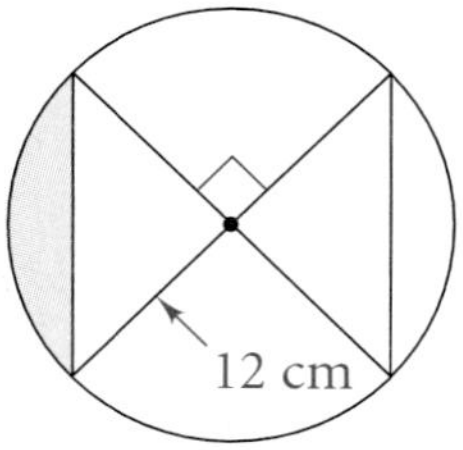

39. **Open-ended** Use a compass to draw a circle. Shade a sector of the circle and find its area.

Find the area of each figure. Leave your answer in terms of π.

40.

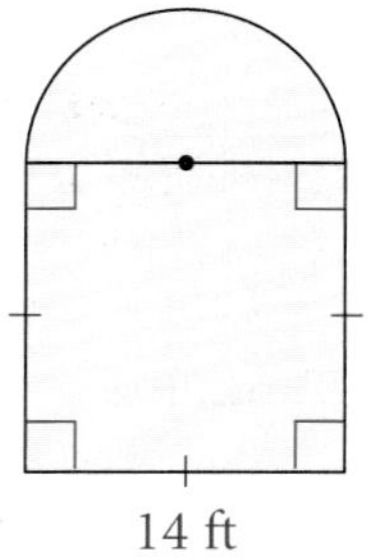

41.

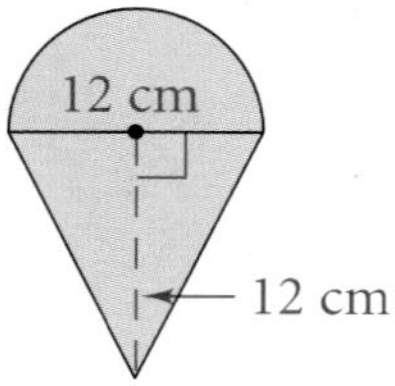

42. **Sports** Netball players in different positions are restricted to different parts of the court. A wing defense can play in the center third and in her own goal third, except in the semicircle around the net. How much area does she have to play in? Round your answer to the nearest tenth.

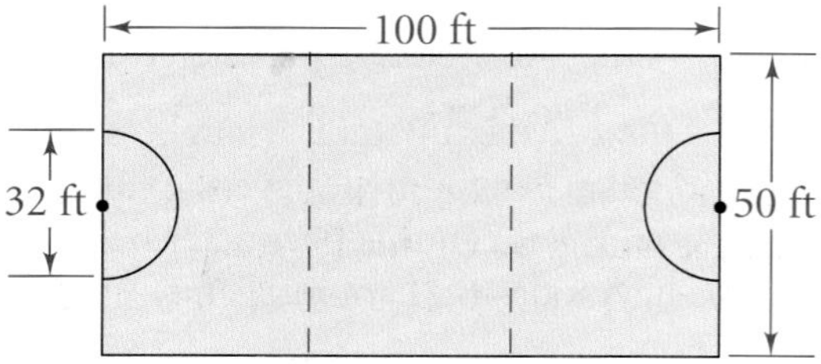

CHAPTER 5

Preparing for Standardized Tests

For Exercises 1–10, choose the correct letter.

1. What is the area of a rectangle with vertices at $(-2, 5)$, $(3, 5)$, $(3, -1)$, and $(-2, -1)$?
A. 30 **B.** 25 **C.** 4 **D.** 24 **E.** 56

2. An isosceles triangle has two angles measuring 49 and 82. What is the measure of the third angle?
A. 82 **B.** 49 **C.** 51
D. 8 **E.** none of the above

3. The lengths of two sides of a triangle are 6 cm and 3 cm. What *cannot* be the length of the third side?
I. 12 cm **II.** 9 cm **III.** 5 cm **IV.** 1 cm
A. I and II **B.** II and III **C.** III and IV
D. I and IV **E.** I, II, and IV

4. What is the circumference of a circle with radius 5 ft?
A. 5π ft **B.** 25π ft **C.** 10π ft
D. 125π ft **E.** 15π ft

5. The length of the hypotenuse of an isosceles right triangle is 8 in. What is the length of one leg?
A. 4 in. **B.** $4\sqrt{2}$ in. **C.** $8\sqrt{2}$ in.
D. 2 in. **E.** none of the above

6. Which figure has area 30 ft^2?

A.
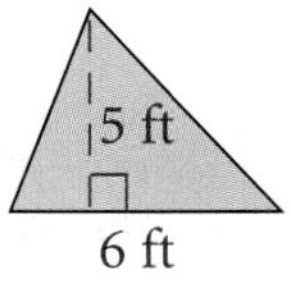

B.
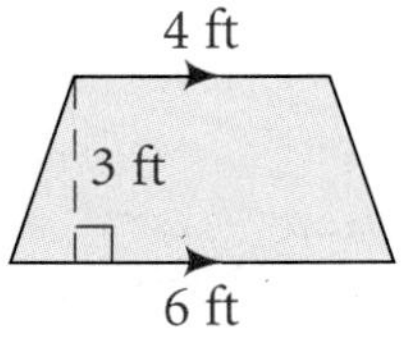

C.
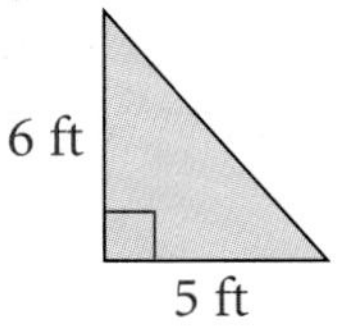

D.
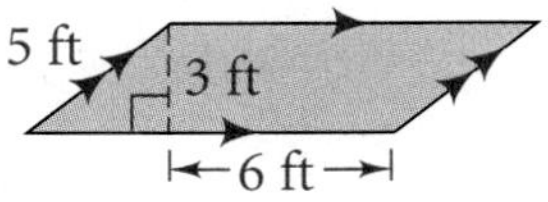

E. none of the above

7. Which *cannot* be the perimeter of an equilateral triangle with integer side lengths?
A. 36 m **B.** 48 m **C.** 16 m **D.** 24 m **E.** 9 m

8. $\overline{DE}$ is a midsegment of $\triangle ABC$. Find DE.

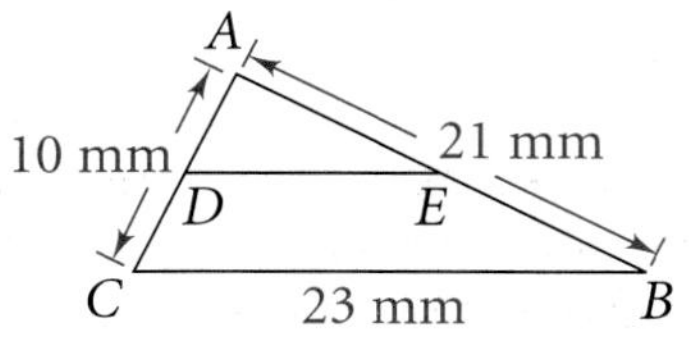

A. 11.5 mm **B.** 11 mm **C.** 10.5 mm
D. 10 mm **E.** 5 mm

Compare the boxed quantity in Column A with the boxed quantity in Column B. Choose the best answer.

A. The quantity in Column A is greater.
B. The quantity in Column B is greater.
C. The two quantities are equal.
D. The relationship cannot be determined on the basis of the information supplied.

	Column A	Column B
	perimeter of square $RSTV = 12x$	
9.	side of $RSTV$	$3x$
10.	RT	RV

11. Critical Thinking Is it possible to sketch a triangle in which exactly one of the altitudes is outside the triangle? Explain.

12. The area of a circle is 144π cm^2. What is the area of a sector with a 90° arc?

13. Writing Explain how to use the Converse of the Pythagorean Theorem to determine if three numbers can represent the lengths of the sides of a right triangle. Include an example.

14. Open-ended Draw a circle. Then draw a second circle with a sector with a 90° arc so that the area of the sector equals the area of the first circle.

CHAPTER

6 Measuring in Space

Relating to the Real World

What types of space objects do you observe each day? As you ride down the street, you might pass by an office building, a water tower, or a house with a dormer window. At the grocery store, you see a variety of boxes, containers, cans, and bottles on the shelves. You can describe many space objects as prisms, cylinders, cones, pyramids, or combinations of these. In this chapter you will learn how to find the surface areas and volumes of space objects.

Lessons

Lessons	6-1	6-2	6-3	6-4	6-5
	Space Figures and Nets	*Surface Areas of Prisms and Cylinders*	*Surface Areas of Pyramids and Cones*	*Volumes of Prisms and Cylinders*	*Volumes of Pyramids and Cones*

CHAPTER PROJECT

The Place is Packed

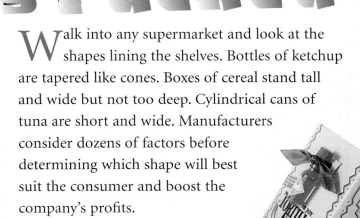

Walk into any supermarket and look at the shapes lining the shelves. Bottles of ketchup are tapered like cones. Boxes of cereal stand tall and wide but not too deep. Cylindrical cans of tuna are short and wide. Manufacturers consider dozens of factors before determining which shape will best suit the consumer and boost the company's profits.

In this chapter project, you will explore package design and uncover some of the reasons for the shapes that manufacturers have chosen. You will also design and construct your own package. You will see how spatial sense and business sense go hand in hand to determine the shapes of things you use every day.

To help you complete the project:

- **p. 306** *Find Out by Doing*
- **p. 314** *Find Out by Measuring*
- **p. 329** *Find Out by Analyzing*
- **p. 335** *Find Out by Investigating*
- **p. 353** *Finishing the Project*

What You'll Learn

- Recognizing nets of various space figures

...And Why

To help you visualize the faces of space figures that you encounter in everyday life

What You'll Need

straightedge, centimeter grid paper, scissors, tape

6-1 Space Figures and Nets

THINK AND DISCUSS

As you can see in the photo below, most buildings are polyhedrons. A **polyhedron** is a three-dimensional figure whose surfaces are polygons. The polygons are the **faces** of the polyhedron. An **edge** is a segment that is the intersection of two faces. A **vertex** is a point where edges intersect.

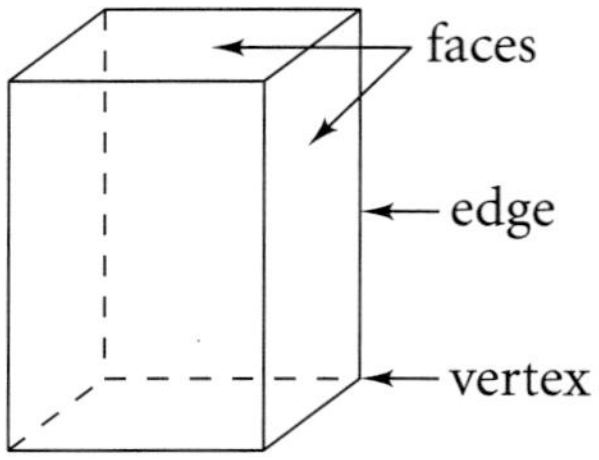

1. a. **Try This** How many faces does the polyhedron below have?
   b. How many edges does it have?
   c. How many vertices does it have?

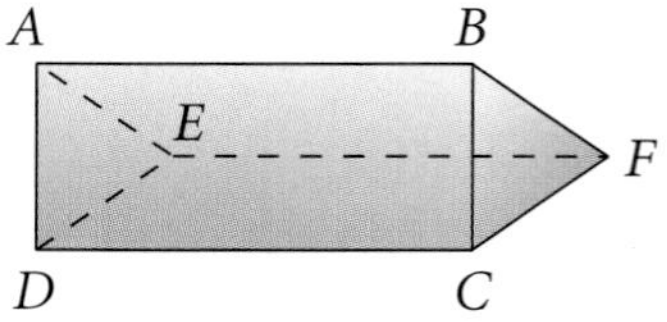

A **net** is a two-dimensional pattern that you can fold to form a three-dimensional figure. Packagers use nets to design boxes.

Who? Grace Chisholm Young (1868–1944) was the first woman to officially receive a doctorate in Germany. In her *First Book of Geometry,* she provides nets to cut out and fold to help students visualize theorems in solid geometry.

WORK TOGETHER

- Working in a group, draw a larger copy of this net on grid paper. Cut it out and fold it to make a cube.

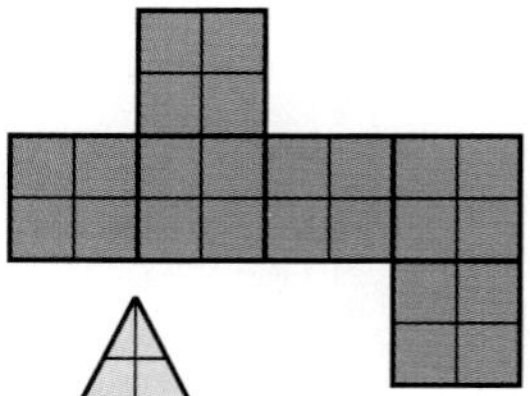

2. a. Describe the faces of the cube.
   b. Write a definition of cube.

- Draw a larger copy of this net on grid paper. Cut it out and fold it to make a pyramid.

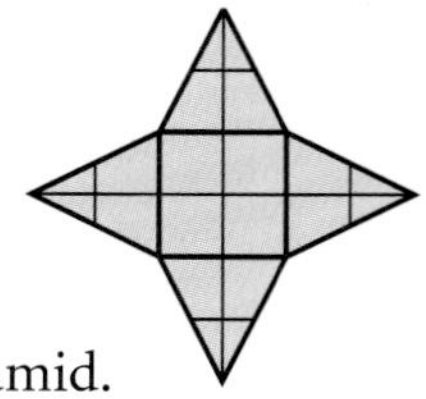

3. a. Describe the faces of the pyramid.
   b. Describe the faces of other pyramids you have seen.
   c. Complete this definition: A pyramid is a polyhedron whose base is a polygon and whose other faces are __?__.

THINK AND DISCUSS

Example ***Relating to the Real World***

Packaging Draw a net for the graham cracker box. Label the net with the appropriate dimensions.

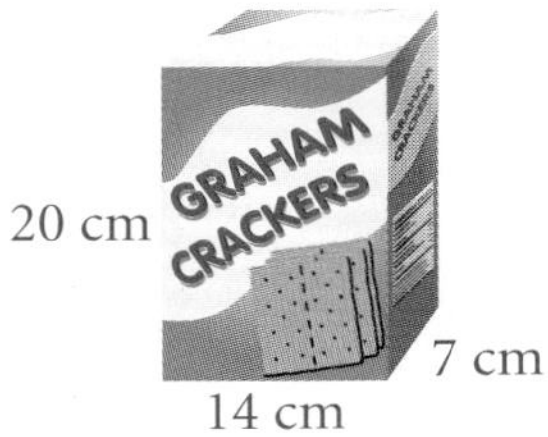

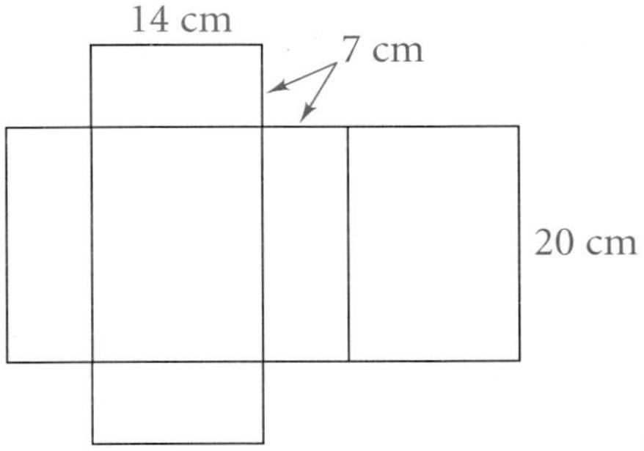

4. Draw a different net for this box. Include the dimensions.

WORK TOGETHER

Work in a group to draw larger copies of the nets below. Use the nets to make three-dimensional models.

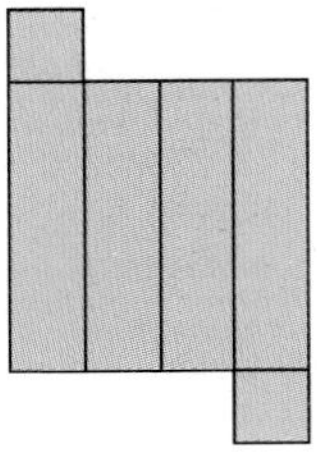

Figure 1

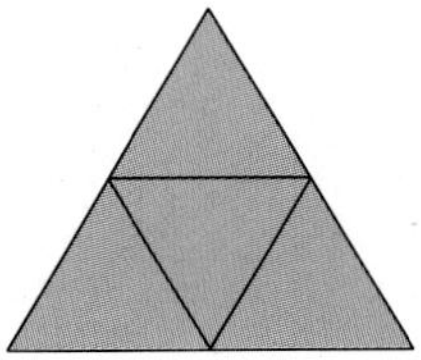

Figure 2

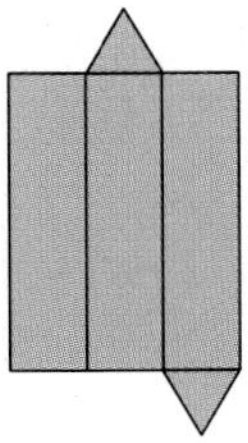

Figure 3

5. Complete the table below, using these three models and the cube and pyramid you made for the Work Together on page 302.

Polyhedron	Number of Faces (F)	Number of Vertices (V)	Number of Edges (E)
Cube	■	■	■
Pyramid	■	■	■
Figure 1	■	■	■
Figure 2	■	■	■
Figure 3	■	■	■

6. Look for a pattern in your table. Write a formula for E in terms of F and V. Then, compare your results with those of other groups. Euler discovered that this relationship is true for any polyhedron. This formula is known as Euler's Formula.

Exercises ON YOUR OWN

1. Which nets will fold to make a cube?

A. 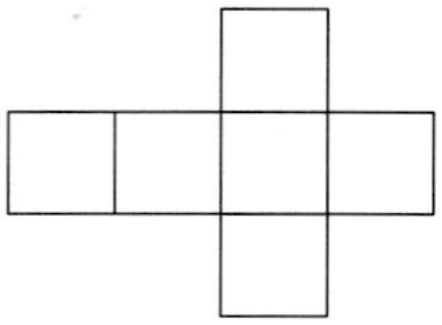

B. 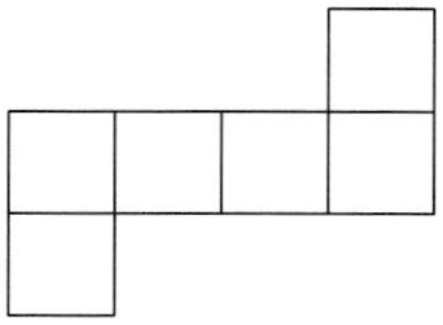

C. 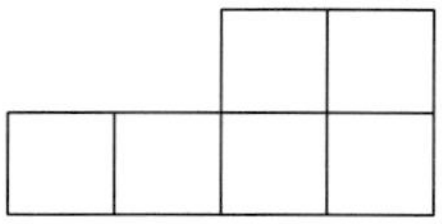

D. 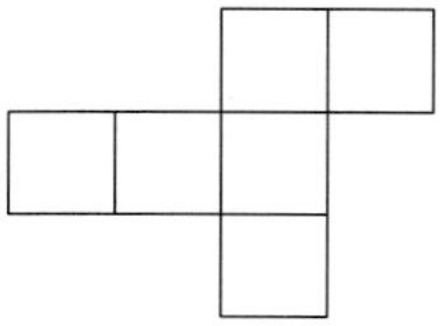

2. Which nets will fold to make a pyramid with a square base?

A. 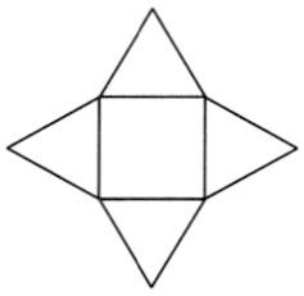

B. 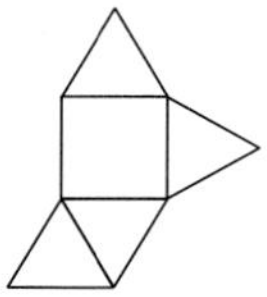

C. 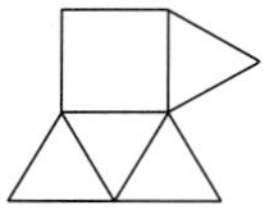

D. 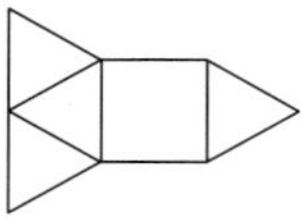

Match each three-dimensional figure with its net.

3. 

4. 

5. 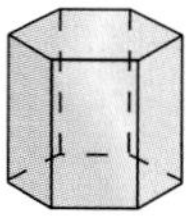

6. 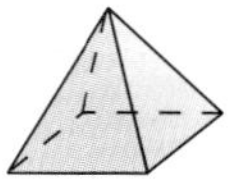

7. 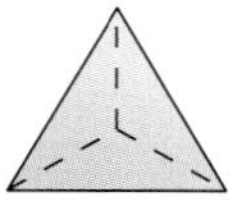

A. 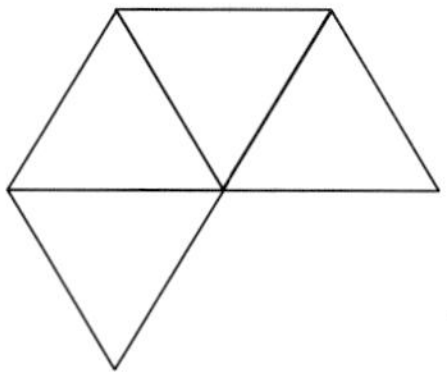

B. 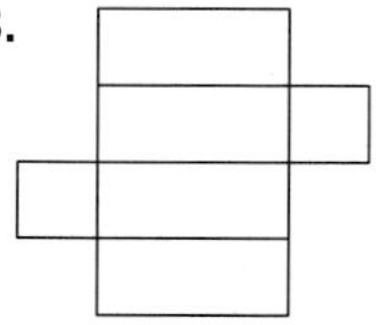

C. 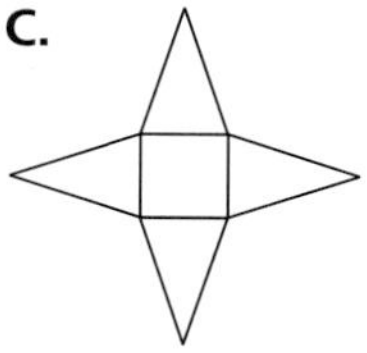

D. 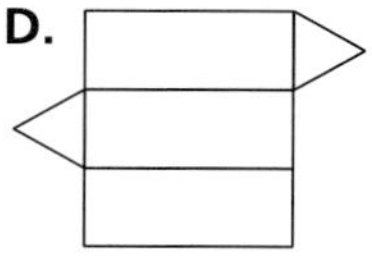

E. 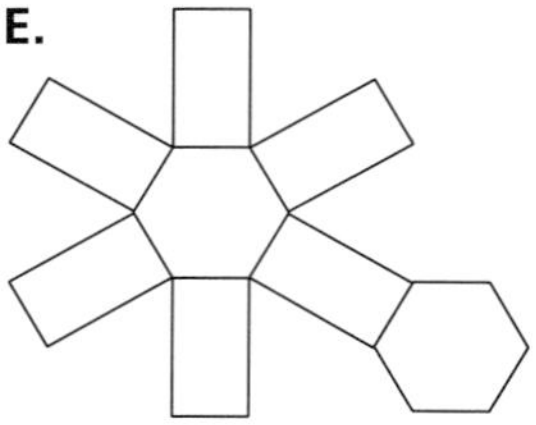

8. The fourth-century Andean textile at the right is now on display at the Museum of Fine Arts, Boston, Massachusetts.
   a. Which of the three outlined figures could be nets for the same polyhedron?
   b. Describe this polyhedron.

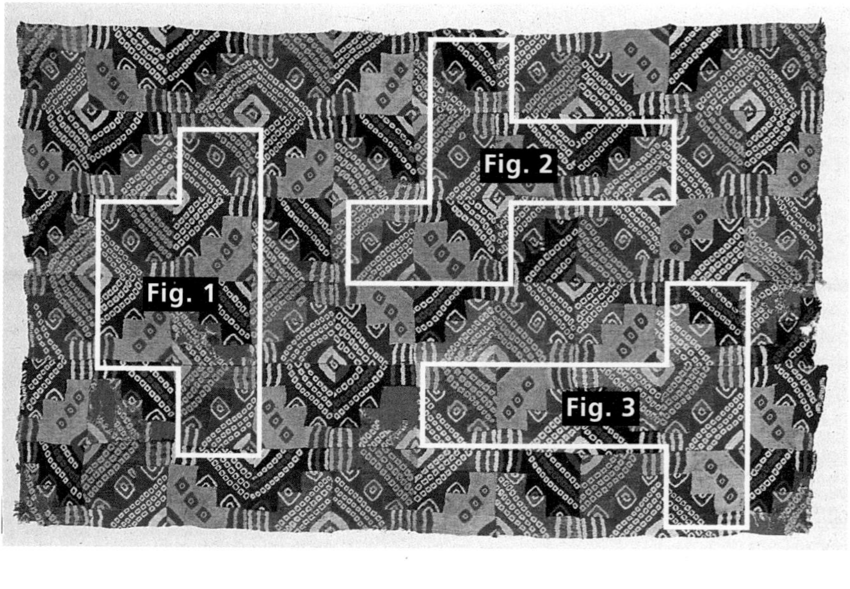

Spatial Visualization **Think about how each net can be folded to form a cube. What is the color of the face that will be opposite the red face?**

9. 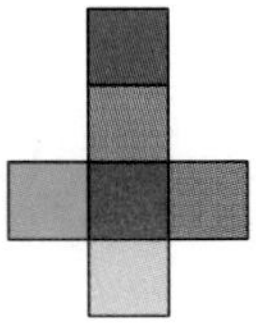

10. 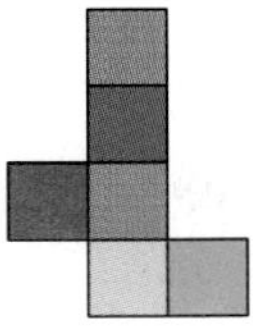

11. 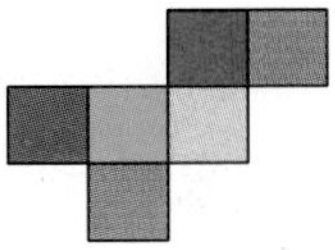

12. 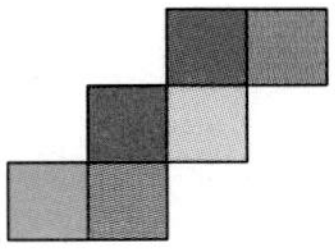

13. a. There are eleven different nets for a cube. Four of them are shown above. Draw as many of the other seven as you can.
    b. **Writing** If you were going to make 100 cubes for a mobile, which of the eleven nets would you use? Explain why.

PROBLEM SOLVING HINT

Two nets are the same if one is a rotation or reflection of the other.

14. There are seven different nets for a pyramid with a square base. Draw as many of them as you can.

15. The total area of a net for a cube is 216 in.^{2}. What is the length of an edge of the cube?

16. a. **Open-ended** Draw a polyhedron whose faces are all rectangles. Label the lengths of its edges.
    b. Use graph paper to draw two different nets for the polyhedron.

17. There are five regular polyhedrons. They are called *regular* because all their faces are congruent regular polygons, and the same number of faces meet at each vertex. They are also called Platonic Solids after the Greek philosopher Plato (427–347 B.C.)

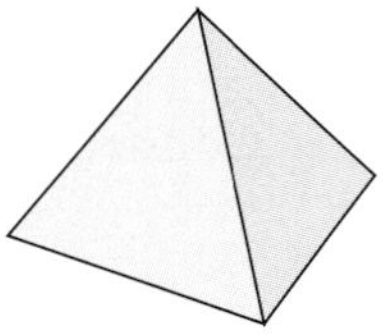

Tetrahedron

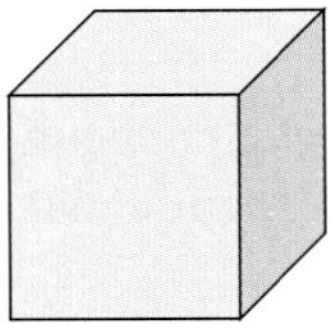

Hexahedron

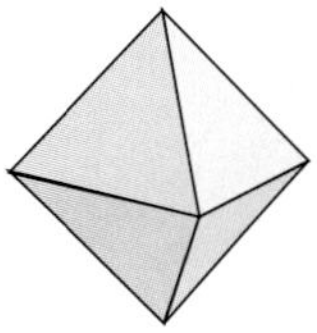

Octahedron

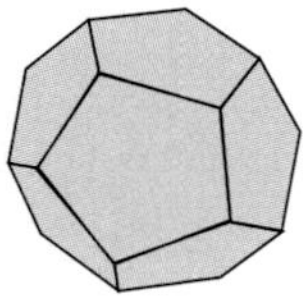

Dodecahedron

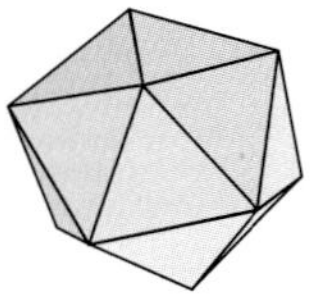

Icosahedron

a. Match each of the nets below with a Platonic Solid.

A. 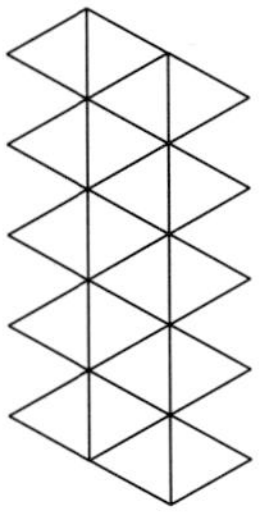

B. 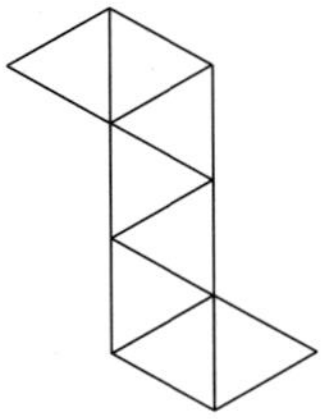

C. 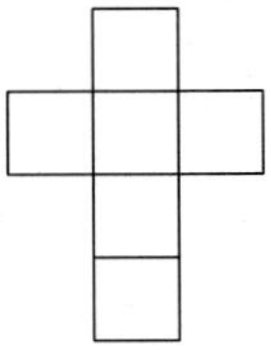

D. 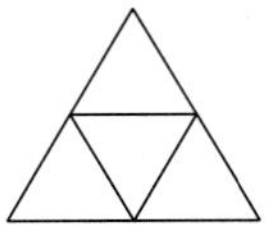

E. 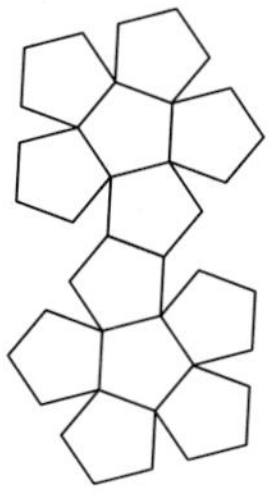

b. Which two Platonic Solids have common names? What are those names?

c. **Verify** that Euler's Formula, which you discovered in the Work Together on page 303, is true for an octahedron.

Chapter Project

Find Out by Doing

With one or two sheets of cardboard and a rubber band, you can make a pop-up version of one of the Platonic Solids.

- Draw and cut out two larger copies of this net. All of the faces are congruent regular pentagons.
- Fold along the dotted lines and then flatten again.
- Place one net on top of the other as shown. (The flaps of the top pattern should fold downward and the flaps of the bottom upward.) Hold the patterns flat and stretch a rubber band (shown in red) over and under the protruding vertices. Watch the flat patterns pop into a dodecahedron.

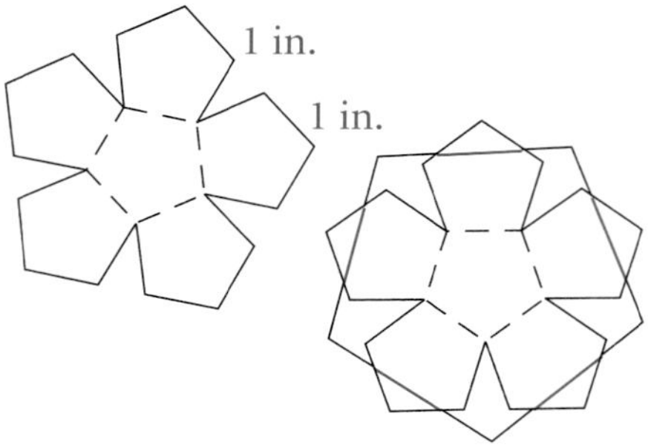

Exercises MIXED REVIEW

18. a. Biology The size of a jaguar's territory depends on how much food is available. Where there is a lot of food, such as in a forest, jaguars have circular territories about 3 mi in diameter. Use $\pi \approx 3$ to **estimate** the area of such a region.

b. Where food is less available, a jaguar may need up to 200 mi^2. **Estimate** the radius of this circular territory.

19. Open-ended Draw a triangle that has no lines of symmetry.

20. Constructions Copy $\overline{TR}$. Construct its perpendicular bisector.

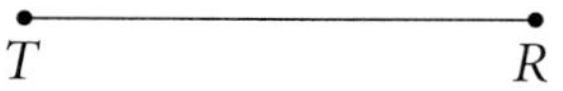

21. In $\triangle ABC$, $AC = 20$, $AB = 18$, and $BC = 13$. Which angle of the triangle has the greatest measure?

Getting Ready for Lesson 6-2

Find the area of each net.

22.

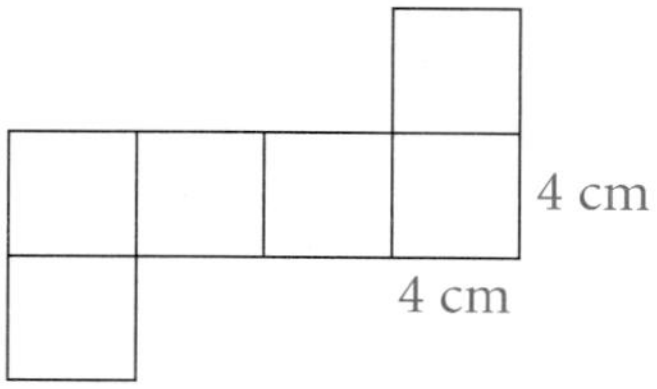

23.

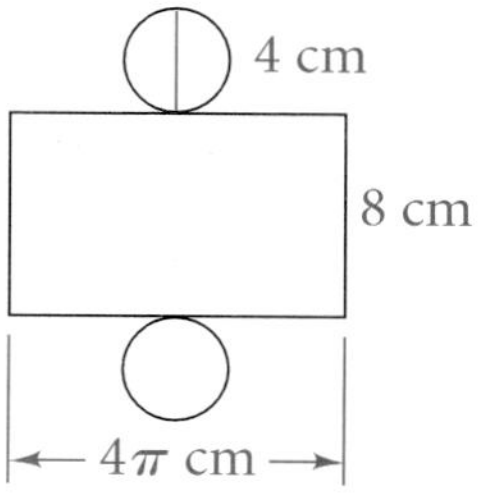

24.

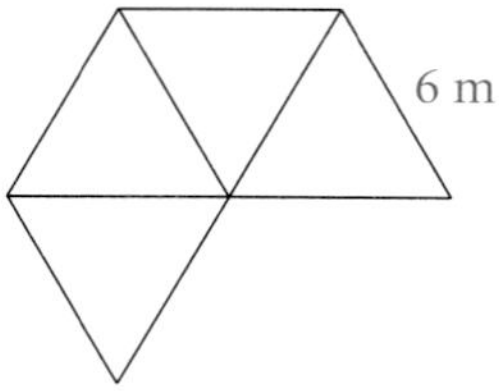

Math ToolboX

Algebra Review

Dimensional Analysis

Before Lesson 6-2

You can use conversion factors to change from one unit of measure to another. The process of analyzing units to decide which conversion factors to use is called **dimensional analysis.**

Since 60 min = 1 h, $\frac{60\text{ min}}{1\text{ h}}$ equals 1. You can use $\frac{60\text{ min}}{1\text{ h}}$ to convert hours to minutes.

$$7\cancel{\text{h}} \cdot \frac{60\text{ min}}{1\cancel{\text{h}}} = 420\text{ min}$$

The hour units cancel, and the result is minutes.

Sometimes you need to use a conversion factor more than once.

Example

The area of the top of a desk is 8 ft^2. Convert the area to square inches.

You need to convert feet to inches.

feet to inches ... feet to inches

$$8\text{ ft} \cdot \text{ft} \cdot \frac{12\text{ in.}}{1\text{ ft}} \cdot \frac{12\text{ in.}}{1\text{ ft}} = 8\cancel{\text{ft}} \cdot \cancel{\text{ft}} \cdot \frac{12\text{ in.}}{1\cancel{\text{ft}}} \cdot \frac{12\text{ in.}}{1\cancel{\text{ft}}}$$ ← The feet units cancel. The result is square inches.

$$= 8 \cdot 12 \cdot 12\text{ in.}^2$$ ← Simplify.

$$= 1152\text{ in.}^2$$

The area of the desktop is 1152 in.^{2}.

Choose the correct conversion factor for changing the units.

1. centimeters to meters
   A. $\frac{100\text{ cm}}{1\text{ m}}$ **B.** $\frac{1\text{ m}}{100\text{ cm}}$
2. yards to feet
   A. $\frac{3\text{ ft}}{1\text{ yd}}$ **B.** $\frac{1\text{ yd}}{3\text{ ft}}$
3. inches to yards
   A. $\frac{36\text{ in.}}{1\text{ yd}}$ **B.** $\frac{1\text{ yd}}{36\text{ in.}}$

Write each quantity in the given unit.

4. 4 m = ■ cm
5. 360 in. = ■ yd
6. 9 mm = ■ cm
7. 17 yd = ■ ft
8. 2.5 ft = ■ in.
9. 35 m = ■ km
10. 2 yd = ■ in.
11. 23 cm = ■ mm
12. 2 ft^2 = ■ in.^{2}
13. 500 mm^2 = ■ cm^2
14. 840 in.^{2} = ■ ft^2
15. 3 km^2 = ■ cm^2
16. 7 m^2 = ■ km^2
17. 360 in.^{2} = ■ yd^2
18. 900 cm^3 = ■ m^3
19. 4 yd^3 = ■ ft^3

What You'll Learn

- Investigating the surface areas and lateral areas of prisms and cylinders

...And Why

To use surface areas of objects in daily life, from CD cases and videocassette boxes to buildings and storage tanks

What You'll Need

- straightedge
- $\frac{1}{4}$-in. graph paper

6-2 Surface Areas of Prisms and Cylinders

WORK TOGETHER

In 1993 the music recording industry changed the size of compact disc packaging. The change to smaller packaging was made in response to the concerns of major recording artists who were worried about the effects of wasted packaging on the environment. Nets for the two packages are shown.

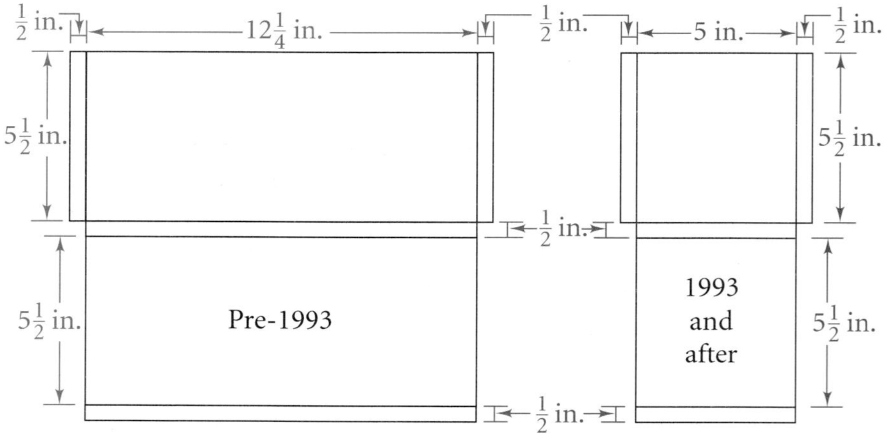

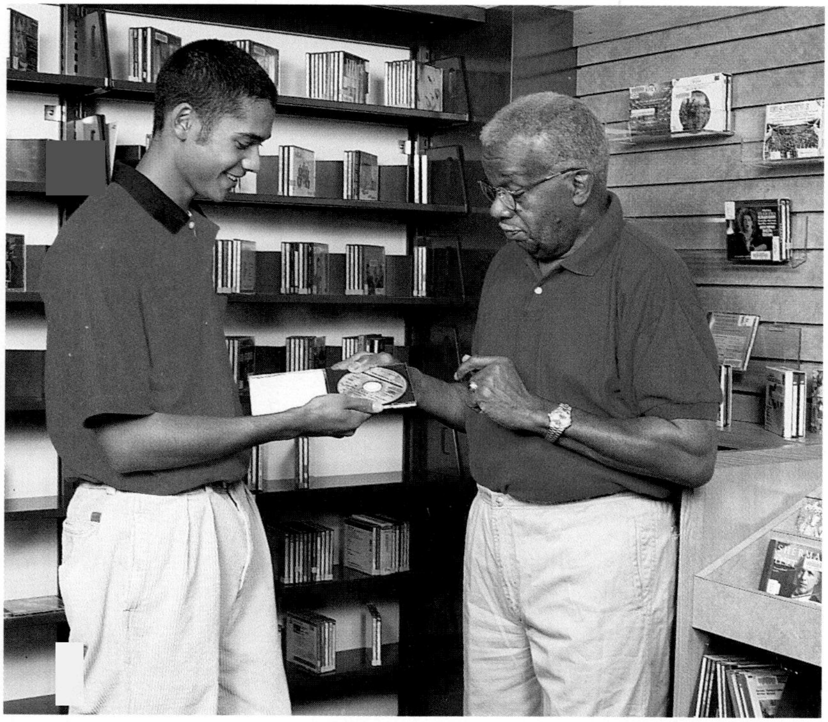

Work with a group. Draw nets on graph paper for each of the two packages.

1. a. What is the area of the net for the pre-1993 CD packaging?
   b. What is the area of the net for the new, smaller CD packaging?
   c. How many square inches of packaging are saved by using the smaller packaging?
2. How many pairs of congruent rectangles are in each net?
3. *Critical Thinking* Why do you think the earlier packaging was so large?

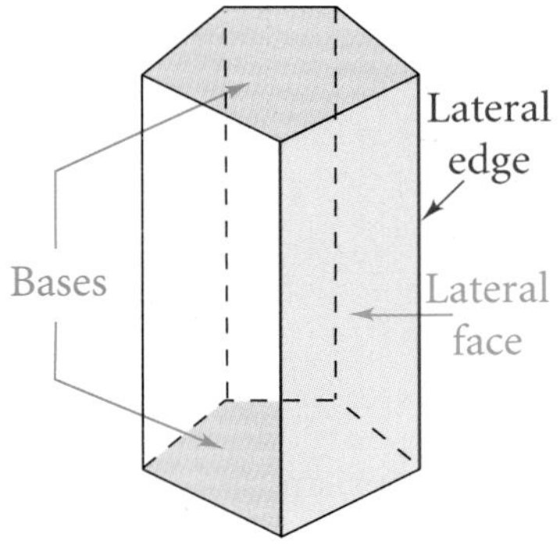

pentagonal prism

THINK AND DISCUSS

Lateral Areas and Surface Areas of Prisms

The CD packages in the Work Together are examples of rectangular prisms. A **prism** is a polyhedron with two congruent, parallel **bases.** The other faces are **lateral faces.** A prism is named for the shape of its bases.

4. Match each prism with one of the following names: triangular prism, rectangular prism, hexagonal prism, octagonal prism.

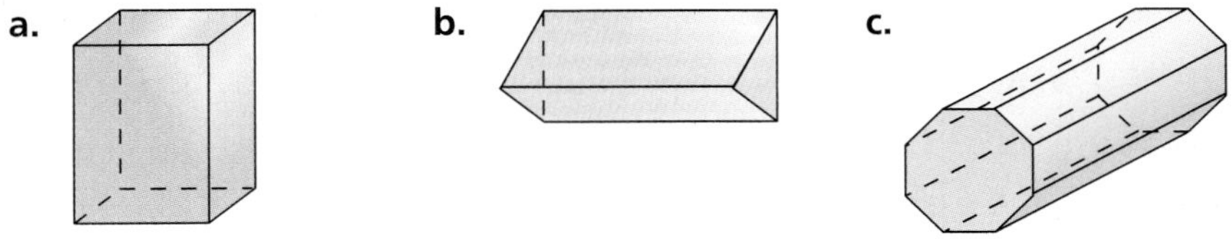

An **altitude** of a prism is a perpendicular segment that joins the planes of the bases. The **height** h of the prism is the length of an altitude. A prism may be either *right* or *oblique.* In a **right prism** the lateral faces are rectangles and a lateral edge is an altitude. In this book you may assume that a prism is a right prism unless you are told otherwise.

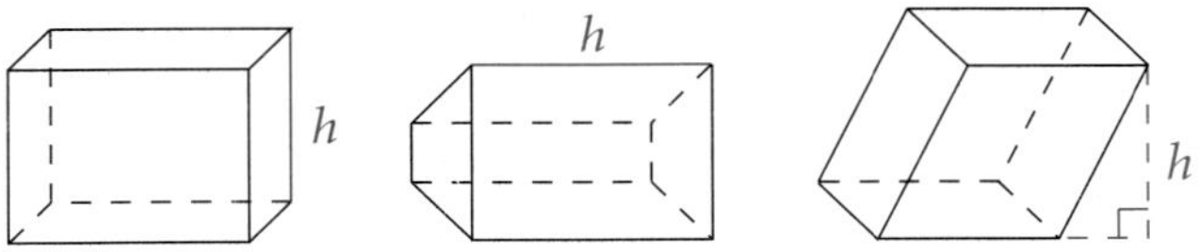

right prisms **oblique prism**

5. *Open-ended* Draw a right triangular prism and an oblique triangular prism. Draw and label an altitude in each figure.

The **lateral area** of a prism is the sum of the areas of the lateral faces. The **surface area** is the sum of areas of the lateral faces and the two bases. You can also find these areas by using formulas.

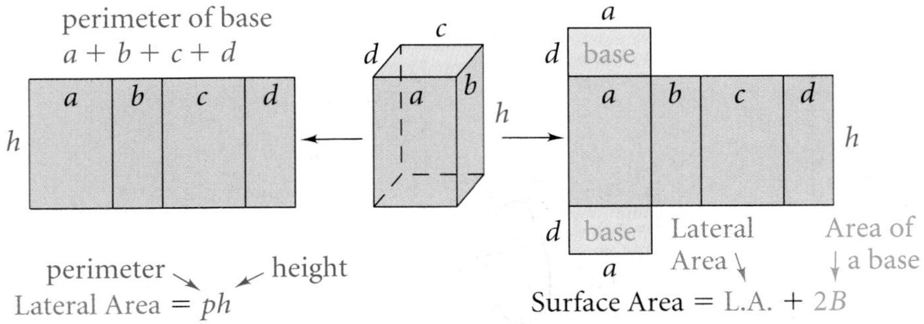

Theorem 6-1
Lateral and Surface Areas of a Right Prism

The lateral area of a right prism is the product of the perimeter of the base and the height.

$$L.A. = ph$$

The surface area of a right prism is the sum of the lateral area and the areas of the two bases.

$$S.A. = L.A. + 2B$$

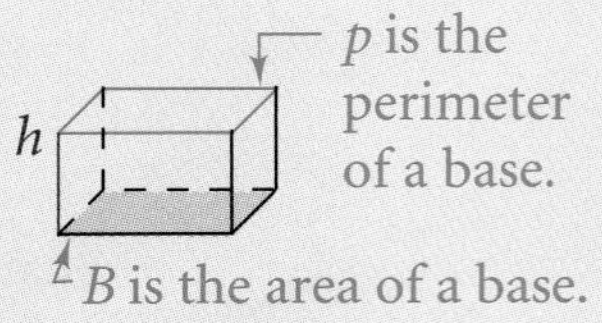

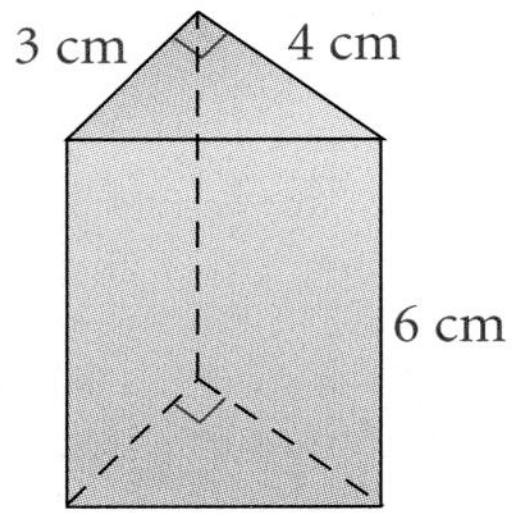

Example 1

Find (a) the lateral area and (b) the surface area of the prism.

The hypotenuse of the triangular base is 5 cm, because the sides form a Pythagorean triple.

a. $L.A. = ph$ — Use the formula for Lateral Area.
$= 12 \cdot 6$ — $p = 3 + 4 + 5 = 12$ cm
$= 72$

The lateral area of the prism is 72 cm^2.

b. $S.A. = L.A. + 2B$ — Use the formula for Surface Area.
$= 72 + 2(6)$ — $B = \frac{1}{2}(3 \cdot 4) = 6$ cm^2
$= 84$

The surface area of the prism is 84 cm^2.

6. **Try This** A **cube** is a prism with square faces. Suppose a cube has edges 5 in. long. What is its lateral area? its surface area?

Lateral Areas and Surface Areas of Cylinders

Like a prism, a **cylinder** has two congruent parallel bases. However, the bases of a cylinder are circles. An **altitude** of a cylinder is a perpendicular segment that joins the planes of the bases. The **height** *h* of a cylinder is the length of an altitude.

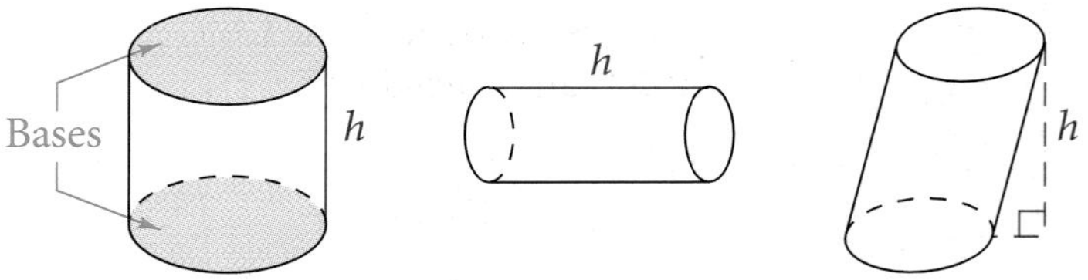

right cylinders **oblique cylinder**

In this book you may assume that a cylinder is a *right* cylinder, like the first two cylinders above, unless you are told otherwise.

To find the lateral area of a cylinder, visualize "unrolling" it. The area of the resulting rectangle is the **lateral area** of the cylinder. The **surface area** of a cylinder is the sum of the lateral area and the areas of the two circular bases. You can find formulas for these areas by looking at a net for a cylinder.

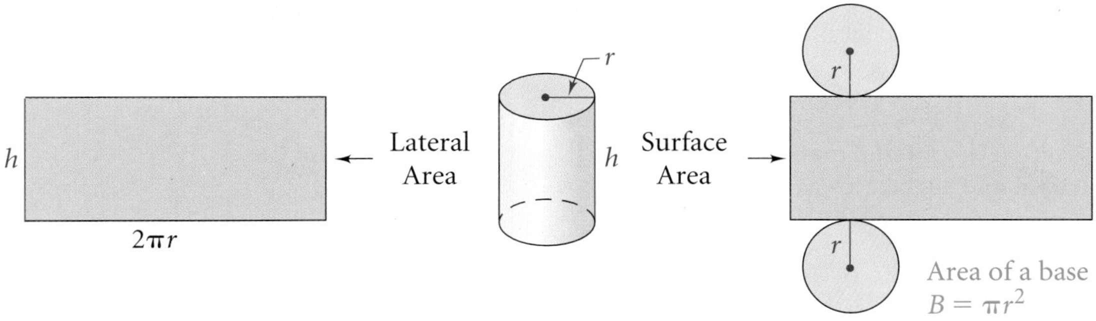

Theorem 6-2
Lateral and Surface Areas of a Right Cylinder

The lateral area of a right cylinder is the product of the circumference of the base and the height of the cylinder.

$$\text{L.A.} = 2\pi rh, \text{ or L.A.} = \pi dh$$

The surface area of a right cylinder is the sum of the lateral area and the areas of the two bases.

$$\text{S.A.} = \text{L.A.} + 2B$$

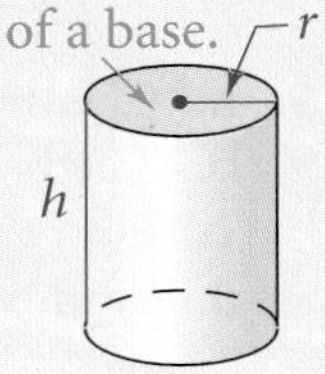

7. **Try This** The radius of the base of a cylinder is 4 in. and its height is 6 in.
   a. Find the lateral area of the cylinder in terms of π.
   b. Find the surface area of the cylinder in terms of π.

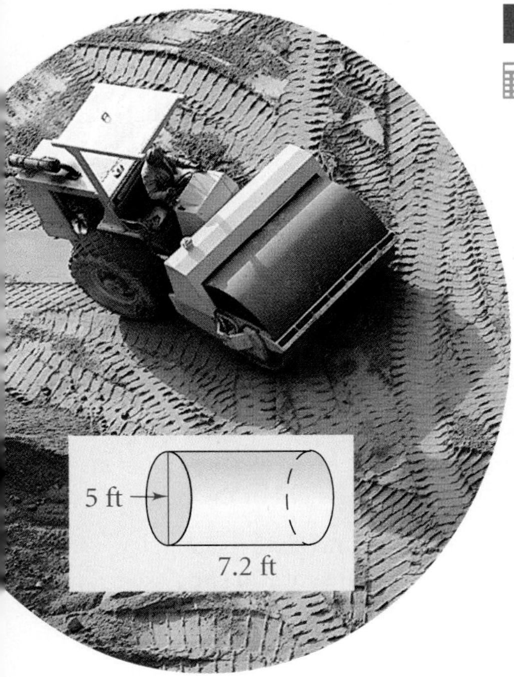

Example 2 **Relating to the Real World**

Machinery The wheel of the steamroller at the left is a cylinder. How many square feet does a single revolution of the wheel cover? Round your answer to the nearest square foot.

The area covered is the lateral area of a cylinder that has a diameter of 5 ft and a height of 7.2 ft.

$\text{L.A.} = \pi dh$	Use the formula for Lateral Area of a cylinder.
$= \pi(5)(7.2)$	Substitute.
$= 36\pi$	Simplify.

36 [×] [π] [=] 113.09734

A single revolution of this steamroller wheel covers about 113 ft^2.

8. **Estimation** Use $\pi \approx 3$ to estimate the lateral area and surface area of a cylinder with height 10 cm and radius 10 cm.

Exercises ON YOUR OWN

Each structure is made of 12 unit cubes. What is the surface area of each figure?

1. 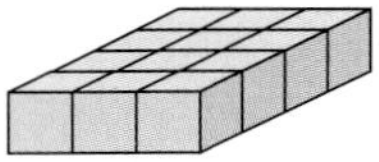

2. 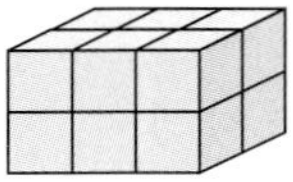

3. 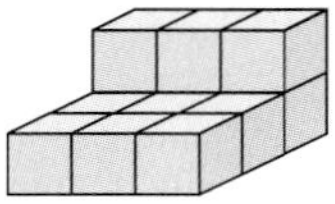

4. 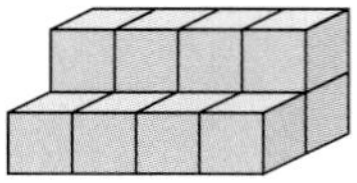

Choose **Use mental math, paper and pencil, or a calculator to find the lateral and surface areas of each figure.**

5. 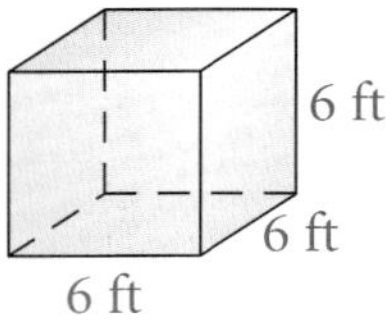

6. 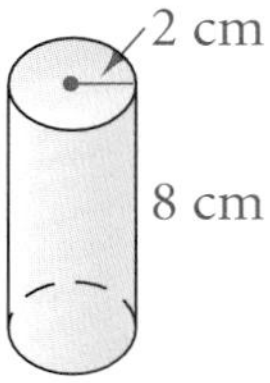

7. 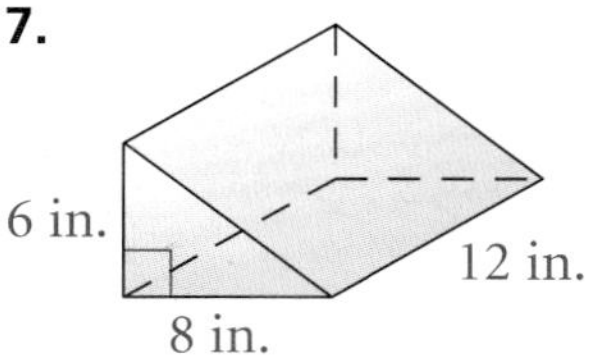

8. 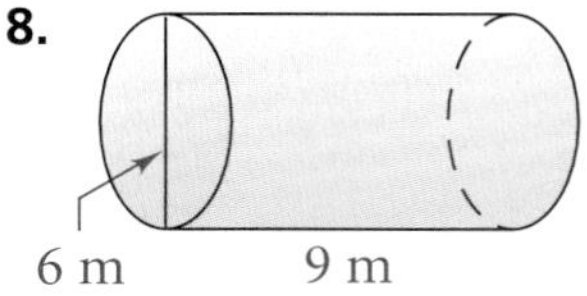

9. **a.** Classify the prism.
b. The bases are regular hexagons. Find the sum of their areas.
c. Find the lateral area of the prism.
d. Find the surface area of the prism.

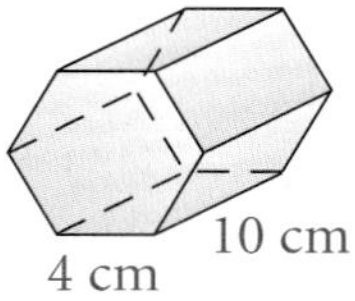

Calculator **Find the lateral area of each object. When an answer is not a whole number, round to the nearest tenth.**

10. 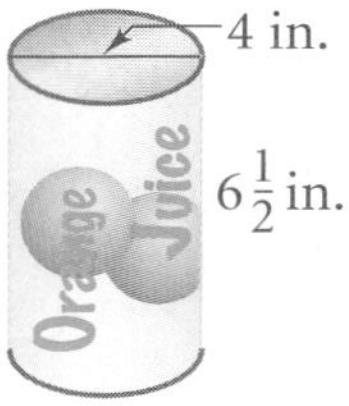

11. 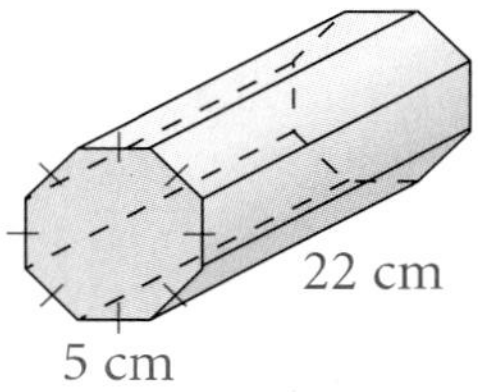

12. 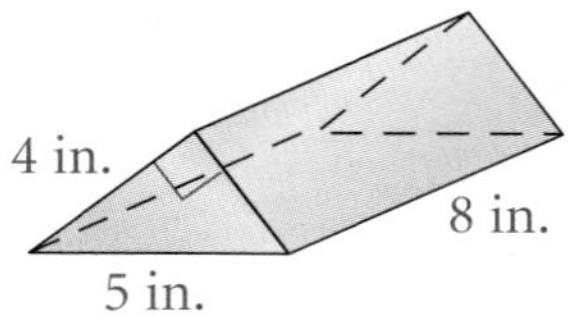

13. *Manufacturing* A standard drinking straw is 195 mm long and has a diameter of 6 mm. How many square centimeters of plastic are needed to make 1000 straws? Round your answer to the nearest hundred.

14. *Packaging* A typical video cassette tape box is open on one side. How many square inches of cardboard are in a typical video-cassette tape box?

15. *Algebra* A triangular prism has base edges 4 cm, 5 cm, and 6 cm long. Its lateral area is 300 cm^2. What is the height of the prism?

16. *Open-ended* Draw a net for a rectangular prism with a surface area of 220 cm^2.

17. a. Geometry in 3 Dimensions List the three coordinates (x, y, z) for vertices A, B, C, and D of the rectangular prism.
b. What is AB? c. What is BC? d. What is CD?
e. What is the surface area of the prism?

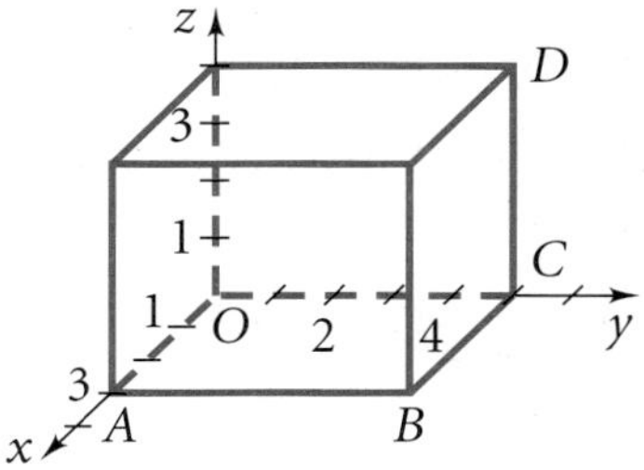

18. Estimation Estimate the surface area of a cube with edges 4.95 cm long.

19. Writing Explain how a cylinder and a prism are alike and how they are different.

Calculator **Find the surface area of each figure. When an answer is not a whole number, round to the nearest tenth.**

20.

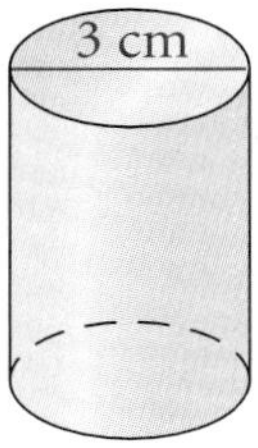

21.

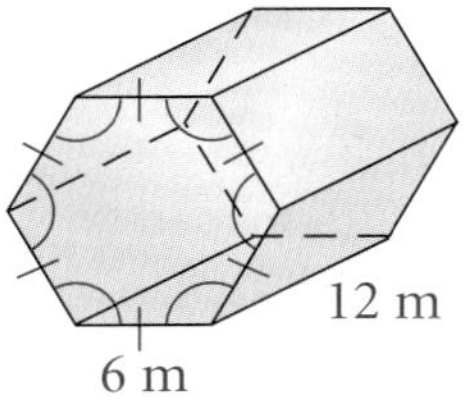

22.

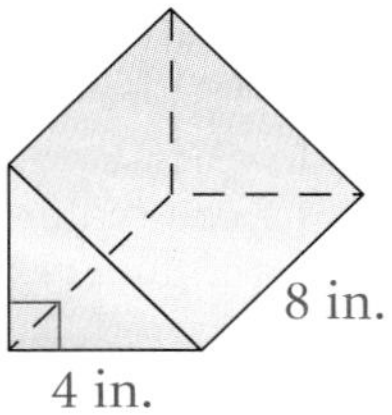

23.

24. a. Coordinate Geometry Suppose the rectangle shown at the right is rotated 360° about the y-axis. What space figure will the rotating rectangle generate?
b. Find the surface area of this figure in terms of π.
c. What will be the surface area in terms of π if the rectangle is rotated 360° about the x-axis?

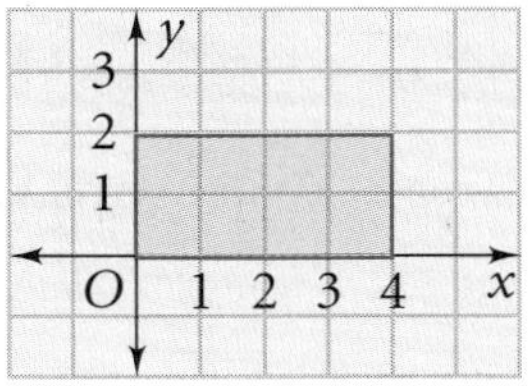

25. Standardized Test Prep If the radius and height of a cylinder are both doubled, then the surface area is ▇.
A. the same B. doubled C. tripled D. quadrupled
E. not enough information given to determine the amount of change in the surface area

26. Algebra The sum of the height and radius of a cylinder is 9 m. The surface area of the cylinder is $54\pi \text{ m}^2$. Find the height and radius.

27. Frieze Patterns From about 3500 B.C. to 2500 B.C., Sumerians etched cylindrical stones to form seals. They used the imprint from rotating the seal to make an official signature. These seals make interesting frieze patterns.
a. Two and one-quarter revolutions of a cylinder created the frieze pattern at the right. What are the dimensions of the cylinder? Round to the nearest tenth.
b. What type of transformation appears in the frieze pattern?

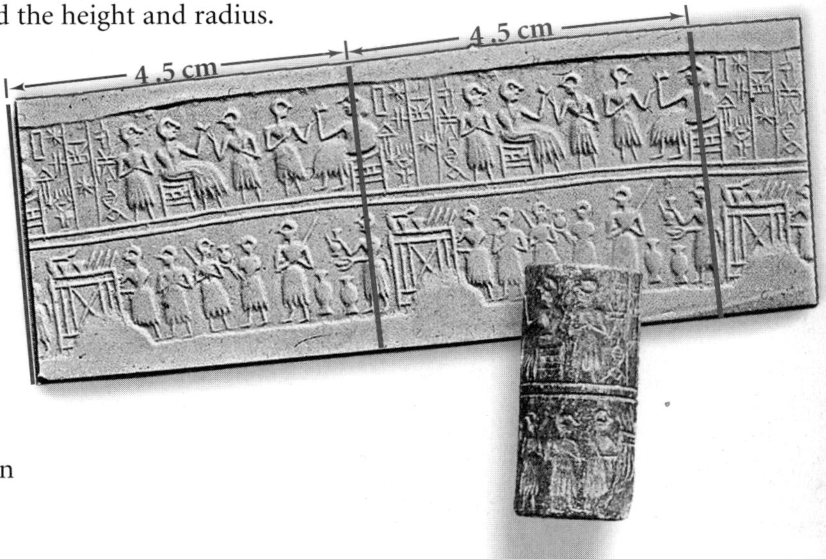

Chapter Project — Find Out by Measuring

Collect some empty cardboard containers shaped like prisms and cylinders.

- Measure each container and calculate its surface area.
- Flatten each container by carefully separating the places where it has been glued together. Find the total area of the packaging material used.
- For each container, find the percent by which the area of the packaging material exceeds the surface area of the container.

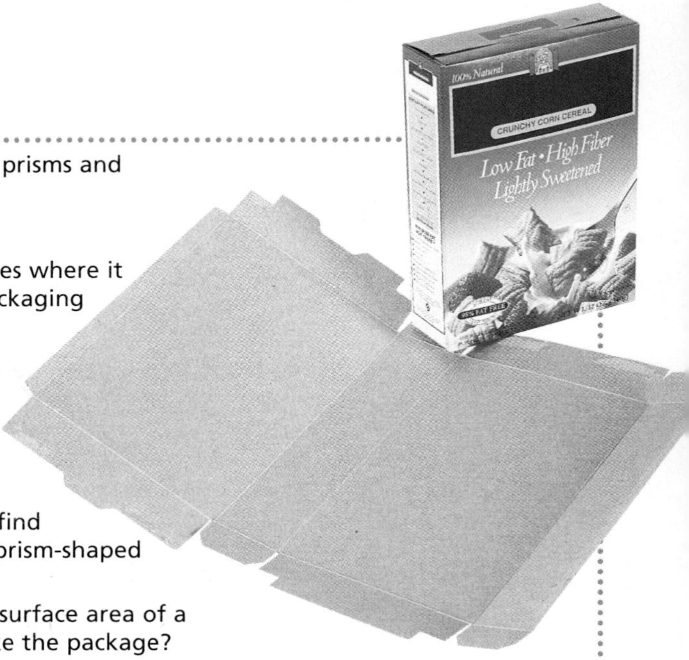

1. How does an unfolded and flattened prism-shaped package differ from a net for a prism?
2. Compare the percents you calculated. What did you find out about the amount of extra material needed for prism-shaped containers? for cylindrical containers?
3. Why would a manufacturer be concerned about the surface area of a package? about the amount of material used to make the package?

Exercises MIXED REVIEW

Find the area of each figure. You may leave answers in simplest radical form.

28.

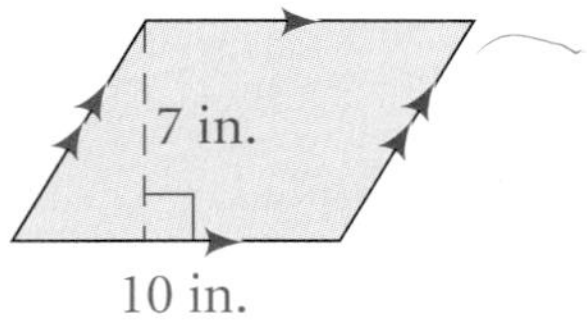

29.

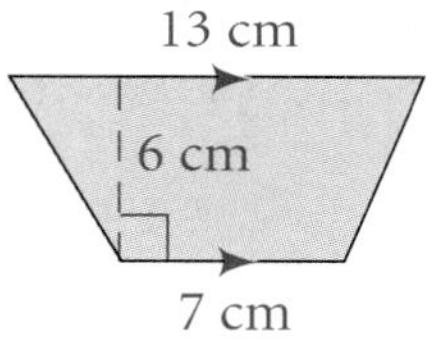

30.

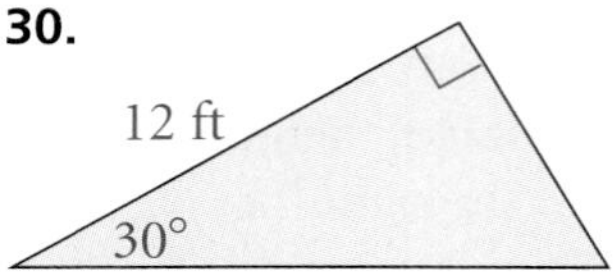

Transformations **Find the coordinates of the images of points $B(-4, 2)$, $I(0, -3)$, and $G(1, 0)$ under each reflection.**

31. in the y-axis **32.** in the x-axis **33.** in the line $x = 4$

34. Write a definition for a trapezoid. Write your definition as a biconditional.

Getting Ready for Lesson 6-3

Calculator **Use a calculator to find the length of each hypotenuse to the nearest tenth.**

35.

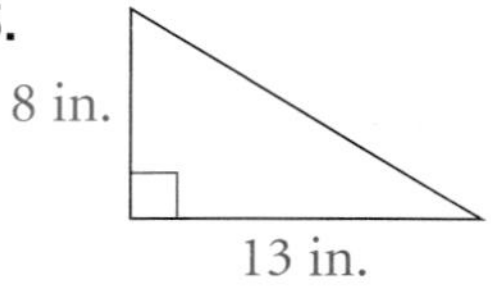

36.

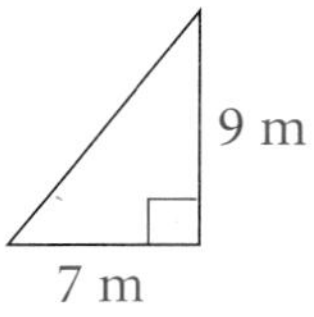

37.

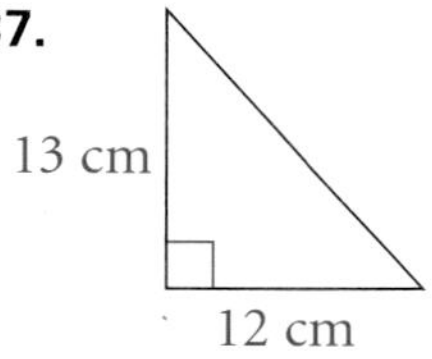

SELF ASSESSMENT

For Your Journal

Give examples of products that are usually packaged in cylindrical containers. Then give examples of products that are usually packaged in boxes that are prisms. Explain why you think manufacturers choose particular container shapes.

Exploring Surface Area

After Lesson 6-2

Work in pairs or small groups.

Input

Use your spreadsheet program to investigate the surface areas of rectangular prisms with square bases. Consider square prisms with a volume of 100 cm^3 and see how the surface area (S.A.) changes as the length of the side (s) of a base changes.

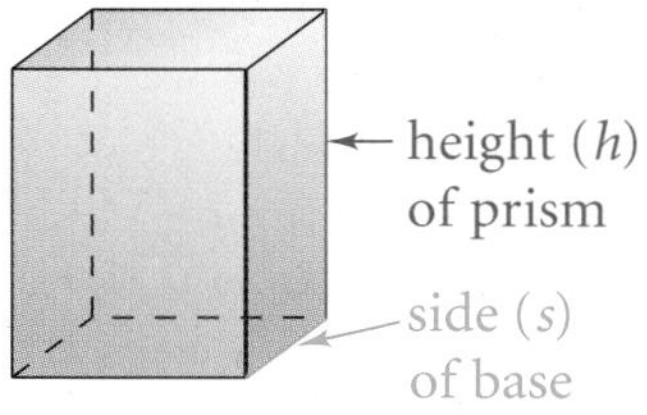

The volume of a prism equals the area of a base times its height ($V = Bh$). Therefore, the height of the prism equals the volume divided by the area of the base ($h = \frac{V}{B} = \frac{V}{s^2}$). The surface area of a square prism equals two times the area of the base plus four times the area of a face (S.A. $= 2B + 4sh = 2s^2 + 4sh$). Set up your spreadsheet as follows.

	A	B	C
1	Square Prisms with a Volume of 100 cm^3		
2			
3	sides (s) of base	height (h) of prism	Surface Area (S.A.)
4		= 100/A4^2	= 2*A4^2+4*A4*B4

Investigate

Copy the formulas down several rows and enter different values for the length of the side of the base in Column A. How small can the surface area be? How large can it be?

Conjecture

Which dimensions give a very large surface area? Which dimensions give the smallest surface area? How do the side of the base and the height compare in the prism of smallest surface area? What is the shape of the square prism that has the smallest surface area?

Extend

- If a square prism has a volume of 1000 cm^3, what dimensions would give the smallest surface area?

6-3 Surface Areas of Pyramids and Cones

What You'll Learn

- Finding the lateral areas and surface areas of pyramids and cones

...And Why

To find the lateral areas of pyramids, such as the Great Pyramid at Giza, and of conical tower roofs

What You'll Need

calculator, metric ruler, compass, protractor, scissors, tape

THINK AND DISCUSS

Lateral Areas and Surface Areas of Pyramids

Many Egyptian pharaohs built pyramids as burial tombs. The Fourth Dynasty, 2615–2494 B.C., was the age of the great pyramids. The builders knew and understood the mathematical properties of a pyramid.

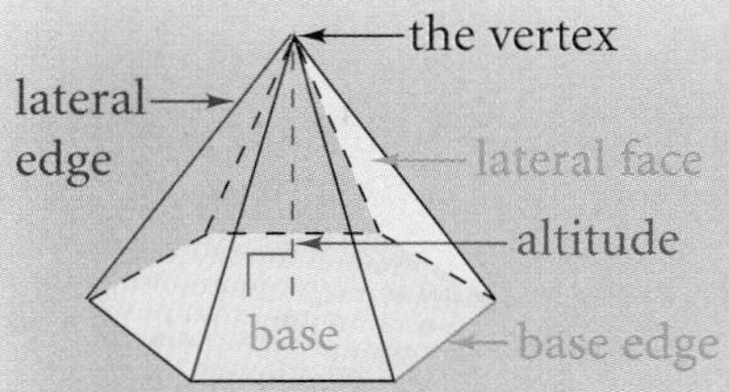

A **pyramid** is a polyhedron in which one face (the **base**) can be any polygon and the other faces (the **lateral faces**) are triangles that meet at a common vertex (called the **vertex** of the pyramid). You can name a pyramid by the shape of its base. The **altitude** of a pyramid is the perpendicular segment from the vertex to the plane of the base. The length of the altitude is the **height** h of the pyramid.

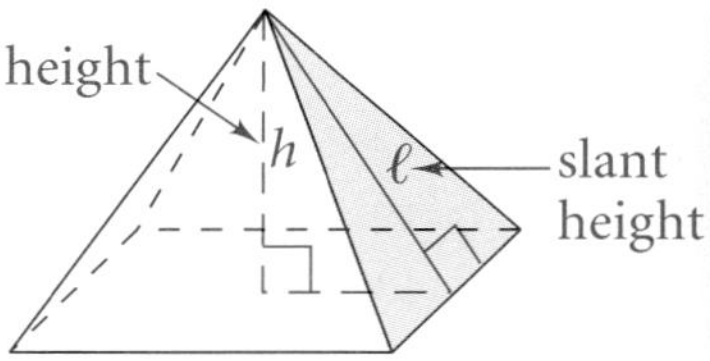

A **regular pyramid** is a pyramid whose base is a regular polygon. The lateral faces are congruent isosceles triangles. The **slant height** ℓ is the length of the altitude of a lateral face of the pyramid. In this book, you can assume a pyramid is regular unless you are told otherwise.

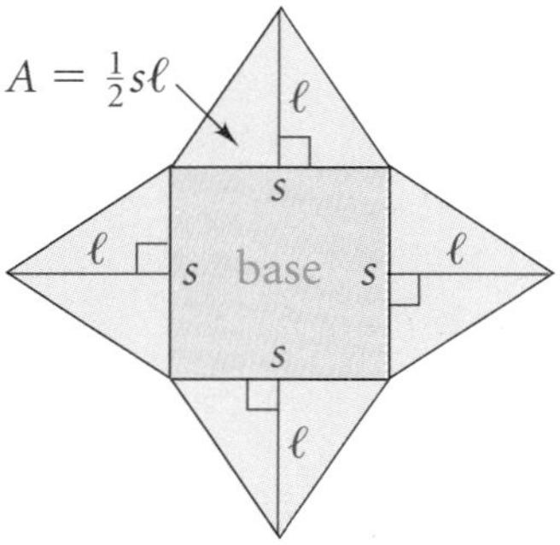

You can find a formula for the lateral area of a pyramid by looking at its net. The **lateral area** is the sum of the areas of the congruent lateral faces.

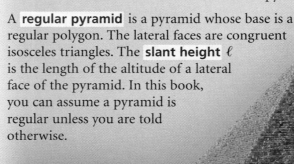

L.A.	$= 4(\frac{1}{2}s\ell)$	The area of each lateral face is $\frac{1}{2}s\ell$.
	$= \frac{1}{2}(4s)\ell$	Commutative Property of Multiplication
	$= \frac{1}{2}p\ell$	The perimeter p of the base is $4s$.

To find the **surface area** of a pyramid, add the area of its base to its lateral area.

Theorem 6-3
Lateral and Surface Areas of a Regular Pyramid

The lateral area of a regular pyramid is half the product of the perimeter of the base and the slant height.

$$\text{L.A.} = \frac{1}{2}p\ell$$

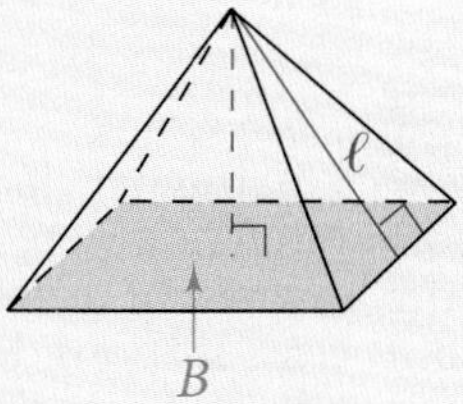

The surface area of a regular pyramid is the sum of the lateral area and the area of the base.

$$\text{S.A.} = \text{L.A.} + B$$

1. Try This A regular hexagonal pyramid has base edges 60 m long and slant height 35 m. Find the perimeter of its base and its lateral area.

Sometimes the slant height of a pyramid is not given. You must calculate it before you can find the lateral or surface area.

Example 1 — Relating to the Real World

Social Studies The Great Pyramid at Giza, Egypt, was built about 2580 B.C. as a final resting place for Pharaoh Khufu. At the time it was built, its height was about 481 ft. Each edge of the square base was about 756 ft long. What was the lateral area of the pyramid?

- The legs of right $\triangle ABC$ are the height of the pyramid and the apothem of the base. The height of the pyramid is 481 ft. The apothem of the base is $\frac{756}{2}$, or 378 ft. You can use the Pythagorean Theorem to find the slant height ℓ.

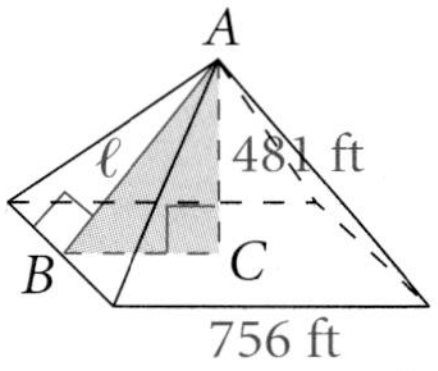

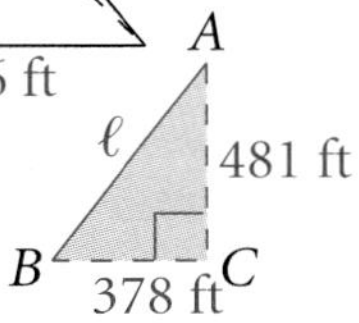

$\ell^2 = AC^2 + BC^2$
$\ell^2 = 481^2 + 378^2$
$\ell = \sqrt{481^2 + 378^2}$ Find the square root of each side.

481 [x^2] [+] 378 [x^2] [=] [$\sqrt{\ }$] 611.75567

- Now use the formula for the lateral area of a pyramid. The perimeter of the base is approximately $4 \cdot 756$, or 3024 ft.

$\text{L.A.} = \frac{1}{2}p\ell$
$= \frac{1}{2}(3024)(611.75567)$ Substitute.

0.5 [×] 3024 [×] 611.75567 [=] 924974.57

The lateral area of the Great Pyramid at Giza was about 925,000 ft^2.

2. Try This Find the surface area of the Great Pyramid at Giza.

Lateral Areas and Surface Areas of Cones

Who? The Greek mathematician Apollonius (about 225 B.C.) wrote a work about the cone and its cross sections. His *Conic Sections* earned him the title "The Great Geometer."

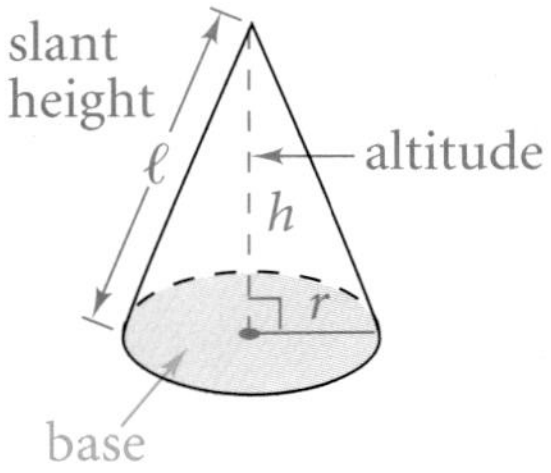

A **cone** is like a pyramid, but its base is a circle. In a **right cone** the **altitude** is a perpendicular segment from the vertex to the center of the base. The **height** h is the length of the altitude. The **slant height** ℓ is the distance from the vertex to a point on the edge of the base. In this book, all the cones discussed will be right cones.

The formulas for the lateral area and surface area of a cone are similar to those for a pyramid.

Theorem 6-4
Lateral and Surface Areas of a Right Cone

The lateral area of a right cone is half the product of the circumference of the base and the slant height.

$$\text{L.A.} = \frac{1}{2} \cdot 2\pi r \cdot \ell, \text{ or L.A.} = \pi r \ell$$

The surface area of a right cone is the sum of the lateral area and the area of a base.

$$\text{S.A.} = \text{L.A.} + B$$

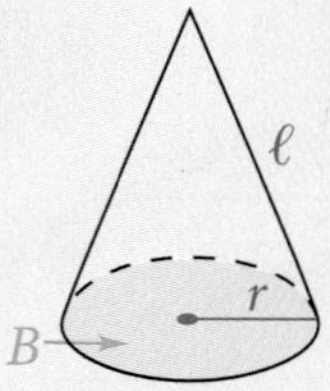

Example 2

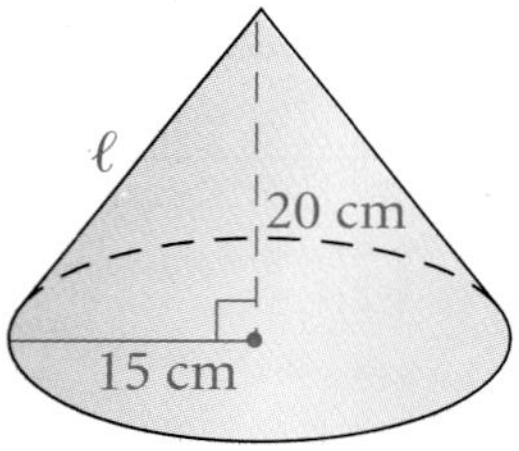

The radius of the base of a cone is 15 cm. Its height is 20 cm. Find its lateral area in terms of π.

- To determine the lateral area, you first must find the slant height ℓ.

$\ell^2 = 20^2 + 15^2$ Use the Pythagorean Theorem.
$= 400 + 225$ Simplify.
$= 625$
$\ell = 25$ Find the square root of each side.

The slant height of the cone is 25 cm.

- Now you can use the formula for the lateral area of a cone.

L.A. $= \pi r \ell$
$= \pi(15)25$ Substitute.
$= 375\pi$ Simplify.

The lateral area of the cone is $375\pi \text{ cm}^2$.

3. a. Calculator Find the lateral area of the cone in Example 2 to the nearest square centimeter.
b. Find the surface area of the cone to the nearest square centimeter.

WORK TOGETHER

- Work in a group. Use a compass to draw three congruent circles with radii 8 cm. Use a protractor to draw the shaded sectors shown below.

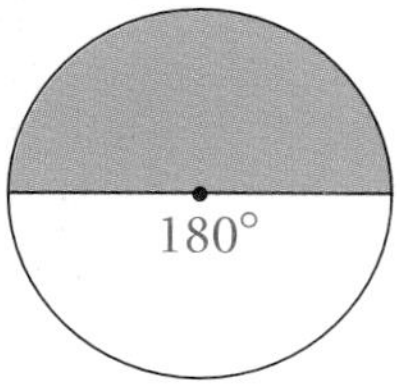

Figure 1

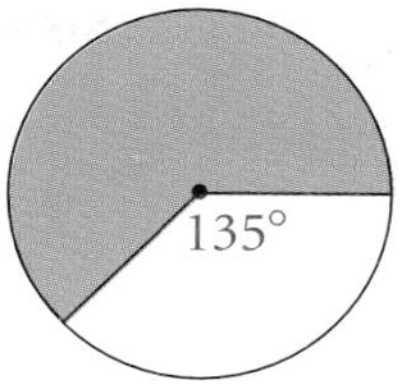

Figure 2

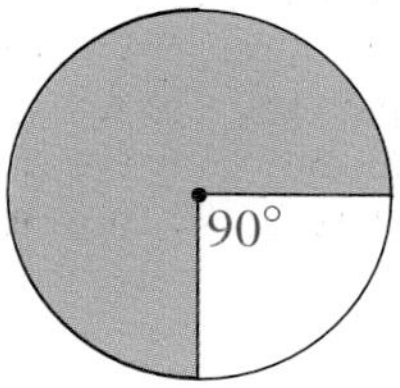

Figure 3

Cone	C	r	L.A.
1	■	■	■
2	■	■	■
3	■	■	■

- Cut out the sectors. Tape the radii of each sector together, without overlapping, to form a cone without a base.

4. Find the slant height of each cone.

5. Find the circumference C and radius r of each cone. Then find the lateral area of each cone. Record your results in a table like the one shown at the left.

Exercises ON YOUR OWN

Choose **Use mental math, pencil and paper, or a calculator to find the slant height of each figure.**

1.

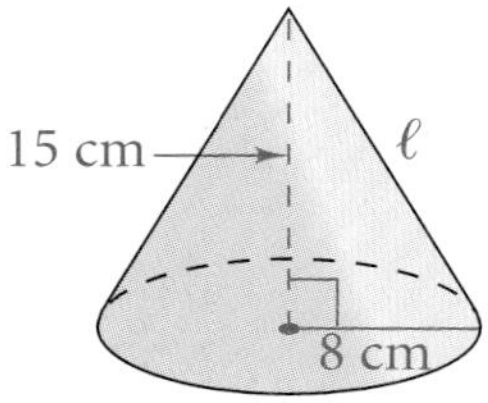

2.

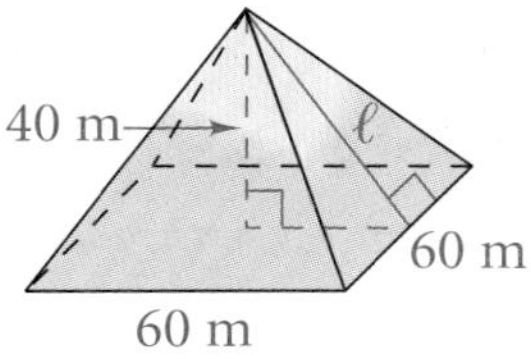

3.

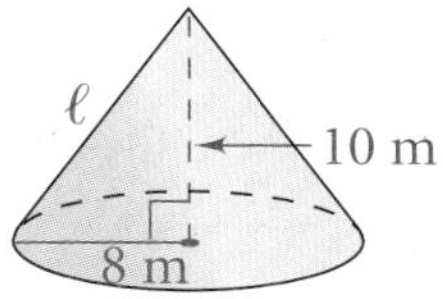

4.

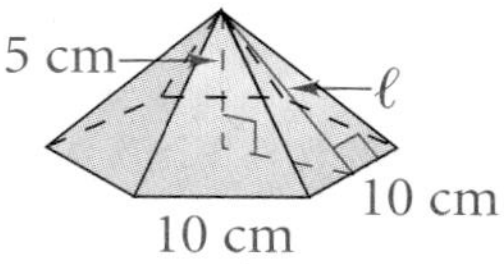

Find the lateral area of each figure. You may leave answers in terms of π.

5.

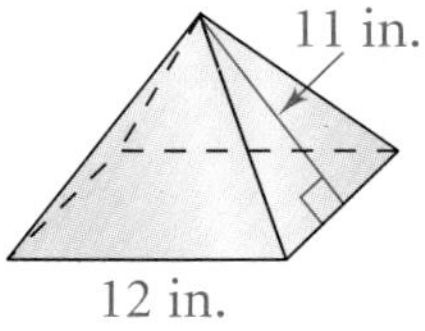

6.

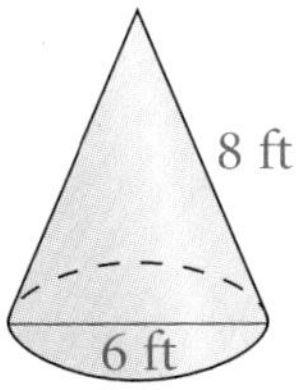

7.

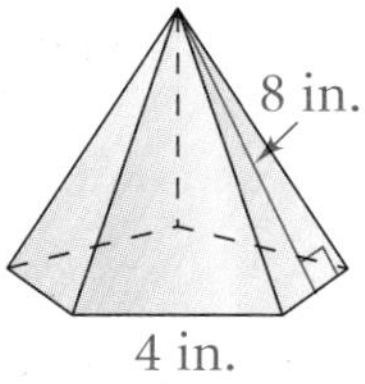

8.

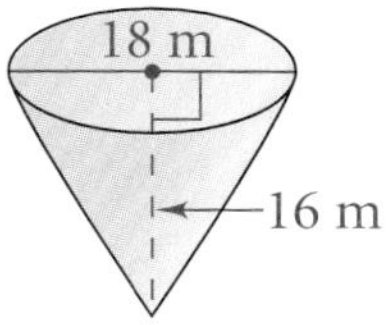

9. *Writing* How are a cone and a pyramid alike? How are they different?

10. *Architecture* The roof of the tower in a castle is shaped like a cone. The height of the roof is 30 ft and the radius of the base is 15 ft. What is the area of the roof? Round your answer to the nearest tenth.

11. **Critical Thinking** Anita says that when she uses her calculator to find the surface area of a cone she uses the formula S.A. $= (\ell + r)r\pi$. Explain why this formula works. Why do you think Anita uses this formula?

12. **Manufacturing** The hourglass shown at the right is made by connecting two glass cones inside a glass cylinder. Which has more glass, the two cones or the cylinder? Explain.

13. A regular square pyramid has base edges 10 in. long and height 4 in. Sketch the pyramid and find its surface area. Round your answer to the nearest tenth.

14. **Algebra** The lateral area of a cone is 48π in.^{2}. The radius is 12 in. Find the slant height.

15. **Open-ended** Draw a pyramid with a lateral area of 48 cm^2. Label its dimensions. Then find its surface area.

16. **a.** **Coordinate Geometry** Describe the figure that is formed when the right triangle at the right is rotated 360° about the x-axis. Find its lateral area. Leave your answer in terms of π.

b. What is the lateral area of the figure formed when the triangle is rotated 360° about the y-axis? Leave your answer in terms of π.

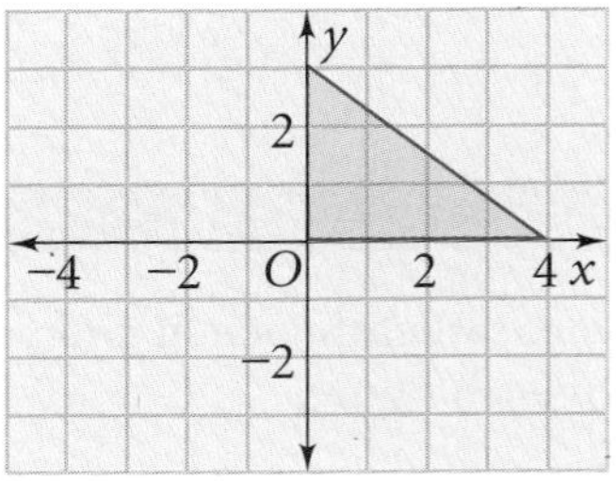

Calculator **Find the lateral area of each figure to the nearest tenth.**

17.

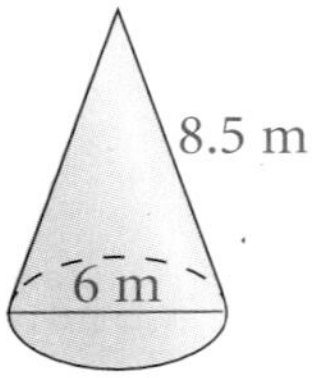

18.

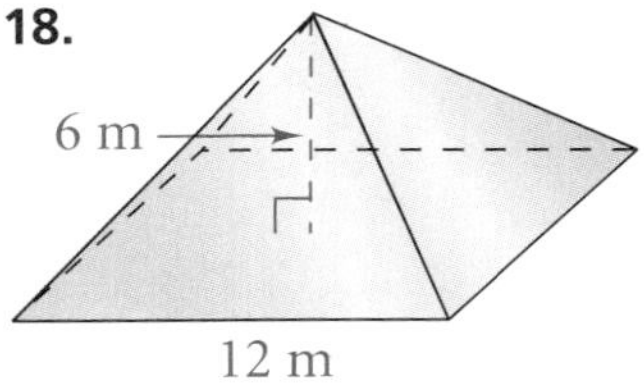

19.

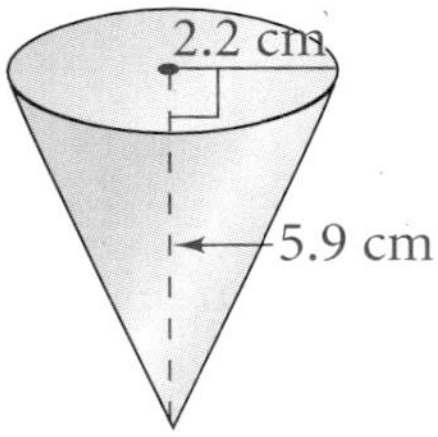

20.

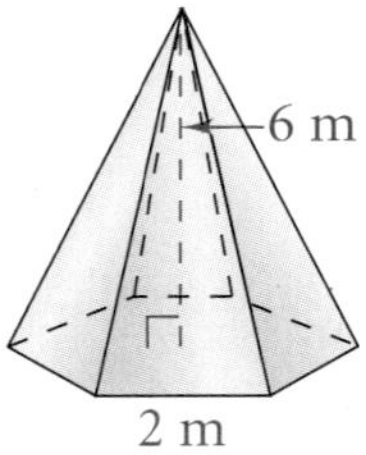

Calculator **Find the surface area of each figure to the nearest tenth.**

21.

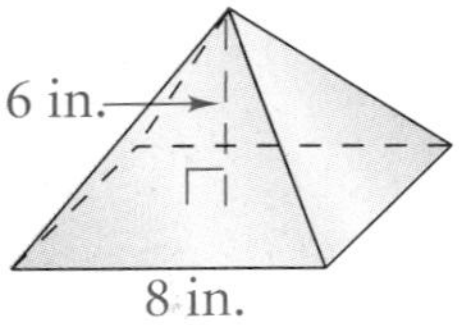

22.

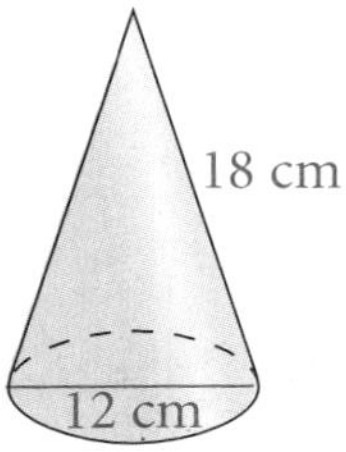

23.

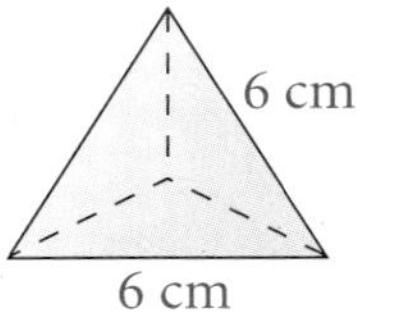

24.

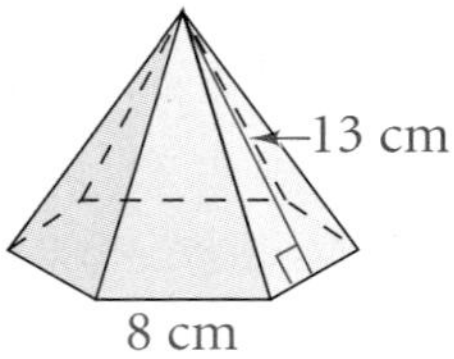

Exercises ON YOUR OWN

Choose Use mental math, paper and pencil, or a calculator to find the volume of each figure. When an answer is not an integer, round to the nearest tenth.

1.

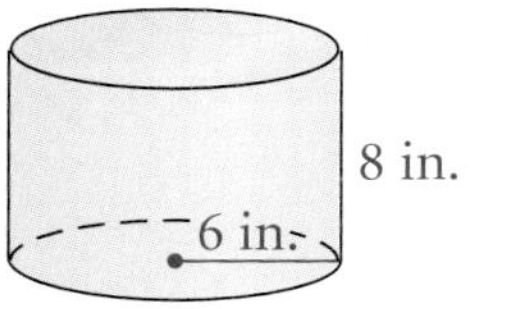

2.

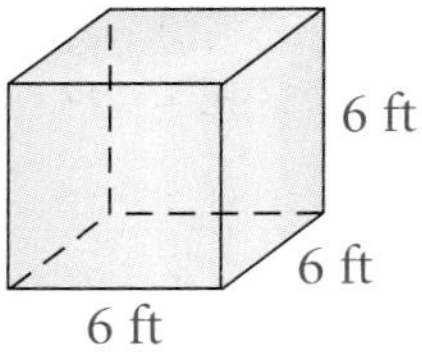

3.

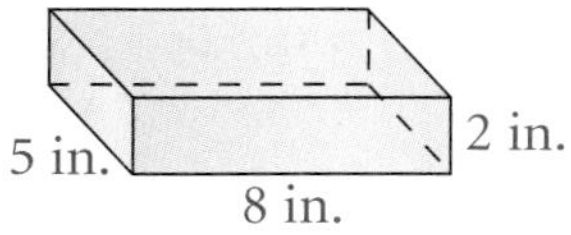

4.

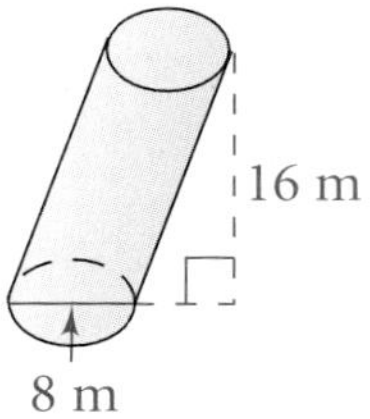

Calculator Find the volume of each figure. When an answer is not a whole number, round to the nearest tenth.

5.

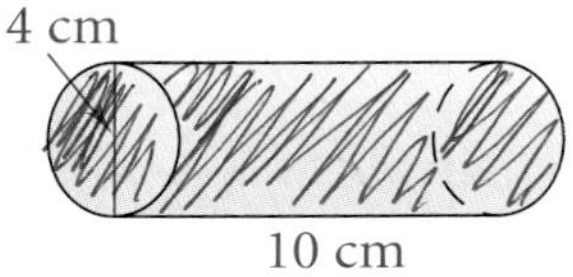

6.

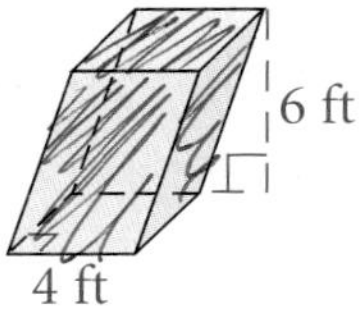

square base

7.

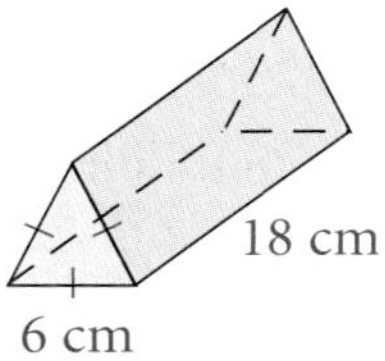

8.

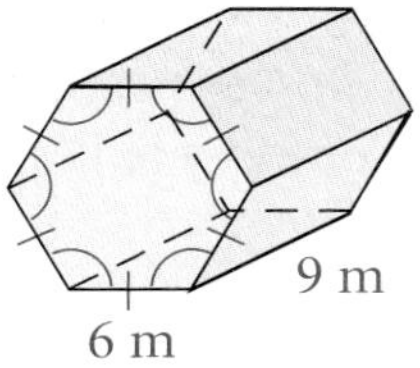

9. a. What is the volume of a 7 ft-by-4 ft-by-1 ft waterbed mattress?

b. To the nearest pound, what is the weight of the water in a full mattress? (Water weighs 62.4 lb/ft^3.)

10. *Geometry in 3 Dimensions* Find the volume of the rectangular prism at the right.

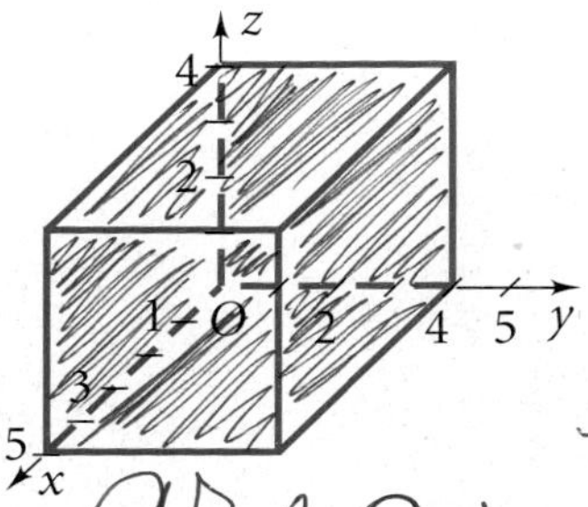

11. *Open-ended* Give the dimensions of two rectangular prisms that each have a volume of 80 cm^3 but have different surface areas.

12. *Landscaping* Zia is planning to landscape her backyard. The yard is a 70 ft-by-60 ft rectangle. She plans to put down a 4-in. layer of topsoil. She can buy bags of topsoil at \$2.50 per 3-ft^3 bag, with free delivery. Or, she can buy bulk topsoil for \$25.00 per yd^3, plus a \$20 delivery fee. Which option is less expensive? Explain.

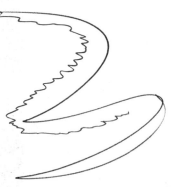

13. *Water Resources* One of the West Delaware water-supply tunnels is a 105-mi long cylinder with a diameter of 13.5 ft. To the nearest million cubic feet, how much earth was removed when the tunnel was built?

14. *Standardized Test Prep* The volume of a cube is 1000 cm^3. What is its surface area?

A. 60 cm^2
B. 600 cm^2
C. 100 cm^2
D. about 4630 cm^2
E. cannot be determined

Find the height of each prism or cylinder with the given volume.

15.

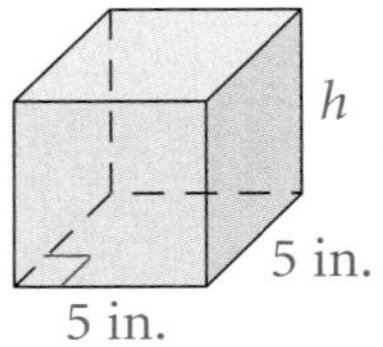

$V = 125$ in.^{3}

16.

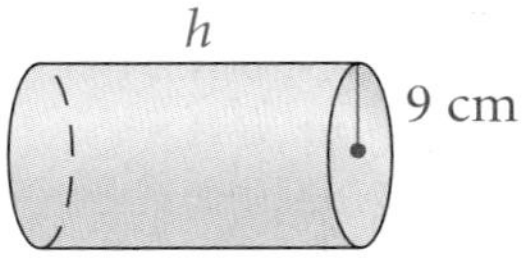

$V = 243\pi$ cm^3

17.

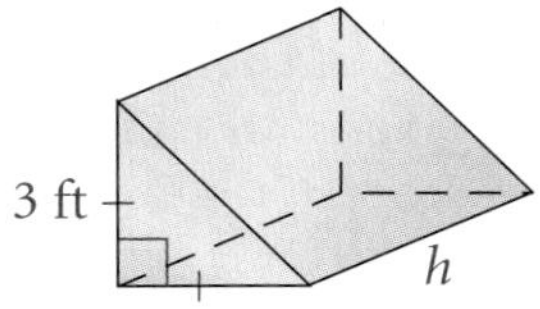

$V = 27$ ft^3

18.

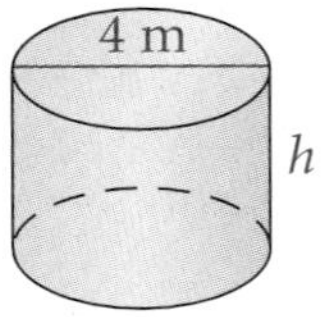

$V = 12\pi$ m^3

19. **a.** What is the volume of the "ordinary" cube in the cartoon if each edge is 18 in. long?

b. What is the volume of the "improved" cube if each edge is half as long as an edge of the "ordinary" cube?

c. *Writing* Do you agree with the cartoon statement that the "improved" cube is half the size of the "ordinary" cube? Explain.

GUINDON © News America Syndicate, 1985

An improved cube (right), half the size of ordinary cubes. It has a convenient carrying handle.

20. *Environmental Engineering* A scientist has suggested that one way to keep indoor air relatively pollution free is to provide two or three pots of flowers such as daisies for every 100 ft^2 of floor space with an 8-ft ceiling. How many pots of daisies would a 35 ft-by-45 ft-by-8 ft classroom need?

21. **a.** The volume of a cylinder is 600π cm^3. The radius of a base of the cylinder is 5 cm. What is the height of the cylinder?

b. The volume of a cylinder is 135π cm^3. The height of the cylinder is 15 cm. What is the radius of a base of the cylinder?

A cylinder has been cut out of each figure. Find the volume of the remaining figure. Round your answer to the nearest tenth.

22.

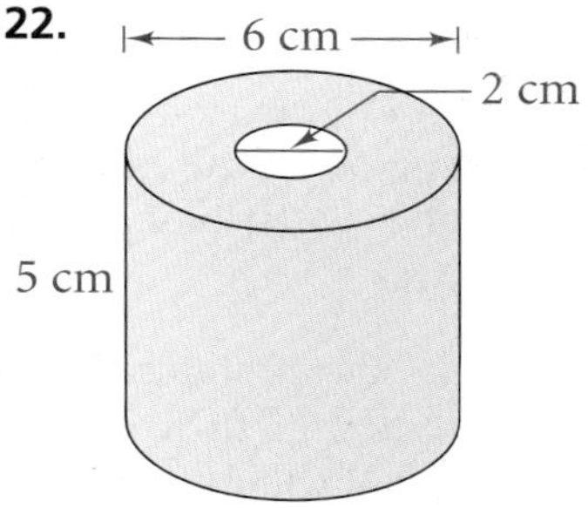

23.

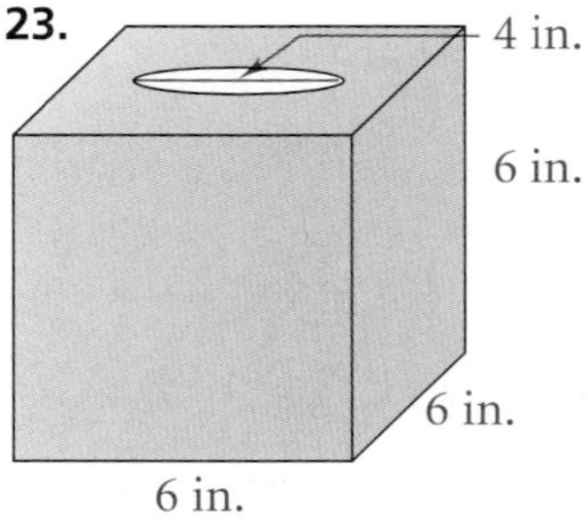

24. *Plumbing* The outside diameter of a pipe is 5 cm. The inside diameter is 4 cm. If the pipe is 4 m long, what is the volume of the metal used for this length of pipe? Round your answer to the nearest whole number.

25. a. *Geometry in 3 Dimensions* What is the volume, in terms of π, of the cylinder formed if the rectangle at the right is rotated 360° about the x-axis?

b. What is the volume, in terms of π, if the rectangle is rotated 360° about the y-axis?

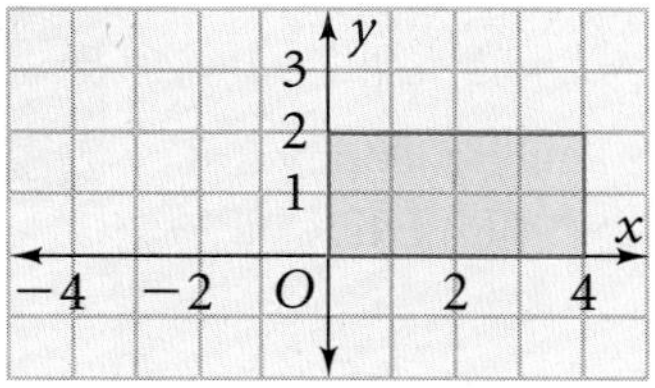

Chapter Project — *Find Out by Analyzing*

Copy and complete the table below for four *different* rectangular prisms that each have a volume of 216 cm^3.

Length (cm)	Width (cm)	Depth (cm)	Volume (V) (cm^3)	Surface Area (S.A.) (cm^2)	Ratio V : S.A.
6	6	■	216	■	■
■	■	■	216	■	■

1. Which of the prisms uses the container material most efficiently? least efficiently? Explain.
2. Why would a manufacturer be concerned with the ratio of volume to surface area?
3. Why do you think cereal boxes are not shaped to give the greatest ratio of volume to surface area?

Exercises MIXED REVIEW

Draw and label △*ABC*. List the sides from shortest to longest.

26. $m\angle A = 67, m\angle B = 34, m\angle C = 79$

27. $m\angle A = 101, m\angle B = 13, m\angle C = 66$

28. $m\angle A = 98, m\angle B = 73, m\angle C = 9$

29. $m\angle A = 28, m\angle B = 81, m\angle C = 71$

30. *Critical Thinking* The area of a triangle is 24 cm^2. The longest side of the triangle is 13 cm long. The second longest side is 12 cm long. Is the triangle a right triangle? Explain.

Getting Ready for Lesson 6-5

Use the Pythagorean Theorem to find the height *h* of each space figure.

31.

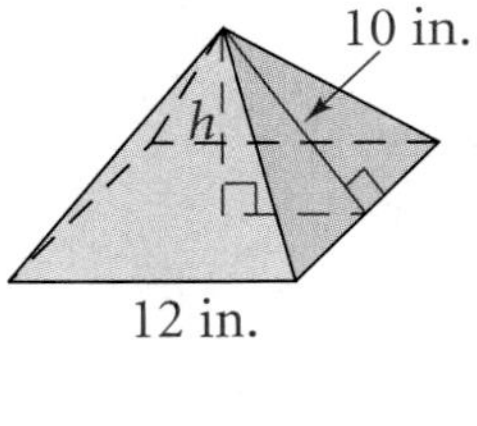

32.

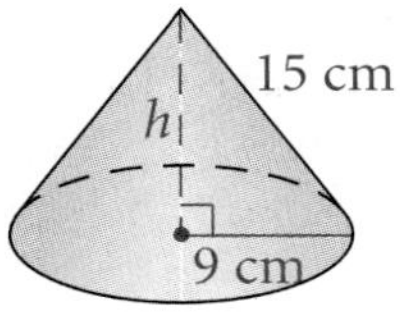

SELF ASSESSMENT

FOR YOUR JOURNAL

Describe some everyday situations in which you would want to know the volumes of prisms.

What You'll Learn

- Finding volumes of pyramids and cones

...And Why

To solve problems concerning the amount of space in things such as a convention center or a popcorn box

What You'll Need

- cardboard
- ruler
- scissors
- tape
- rice
- calculator

6-5 Volumes of Pyramids and Cones

WORK TOGETHER

You know how to find the volume of a prism. Work in a group to explore the volume of a pyramid.

- Draw the nets shown below on cardboard.
- Cut out the nets and tape them together to make a cube and a regular square pyramid. Each model will have one open face.

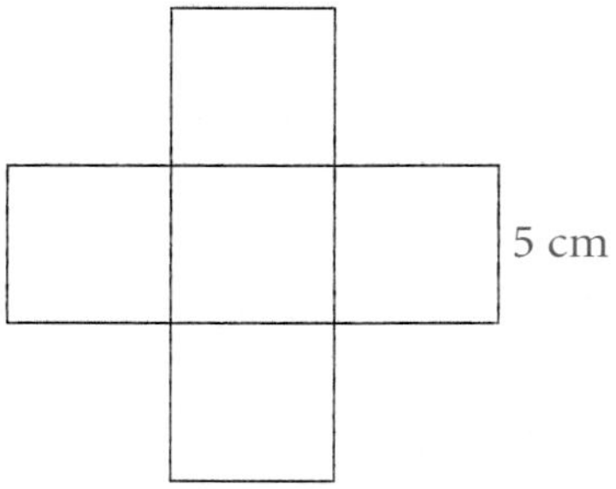

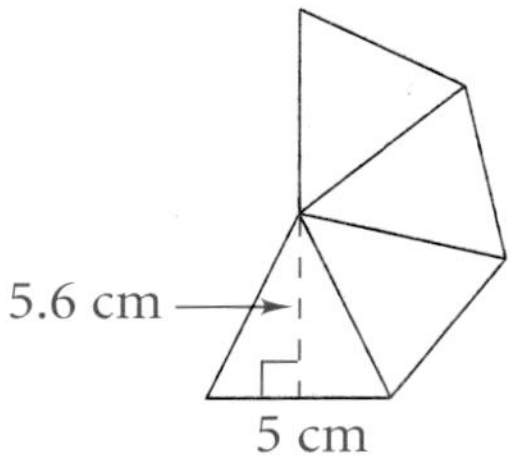

1. Compare the areas of the bases of the cube and the pyramid.
2. Compare the heights of the cube and the pyramid.
3. Fill the pyramid with rice. Then pour the rice from the pyramid into the cube. How many pyramids full of rice does the cube hold?
4. The volume of the pyramid is what fractional part of the volume of the cube?

THINK AND DISCUSS

Volumes of Pyramids

The Work Together demonstrates the following theorem.

Theorem 6-8
Volume of a Pyramid

The volume of a pyramid is one third the product of the area of the base and the height of the pyramid.

$$V = \frac{1}{3}Bh$$

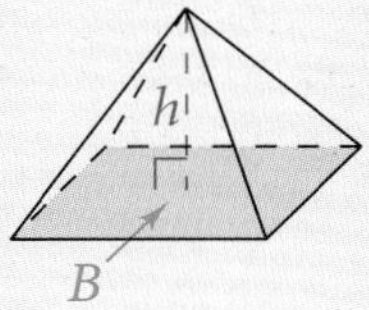

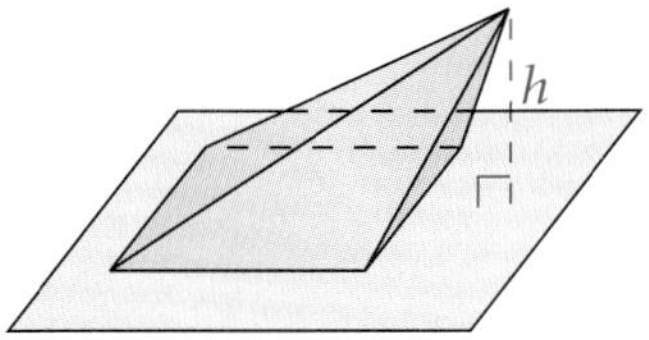

Oblique Pyramid

Because of Cavalieri's Principle, the volume formula is true for all pyramids, including *oblique* pyramids. The **height** h of an oblique pyramid is the length of the perpendicular segment from the vertex to the plane of the base.

Example 1 **Relating to the Real World**

Architecture The Pyramid is an arena in Memphis, Tennessee. The area of the base of the Pyramid is about 300,000 ft^2. Its height is 321 ft. What is the volume of the Pyramid?

$V = \frac{1}{3}Bh$ — Use the formula for the volume of a pyramid.

$= \frac{1}{3}(300{,}000)(321)$ — Substitute.

$= 32{,}100{,}000$ — Simplify.

The volume is about 32,100,000 ft^3.

5. **Try This** Find the volume of a regular square pyramid with base edges 12 in. long and height 8 in.

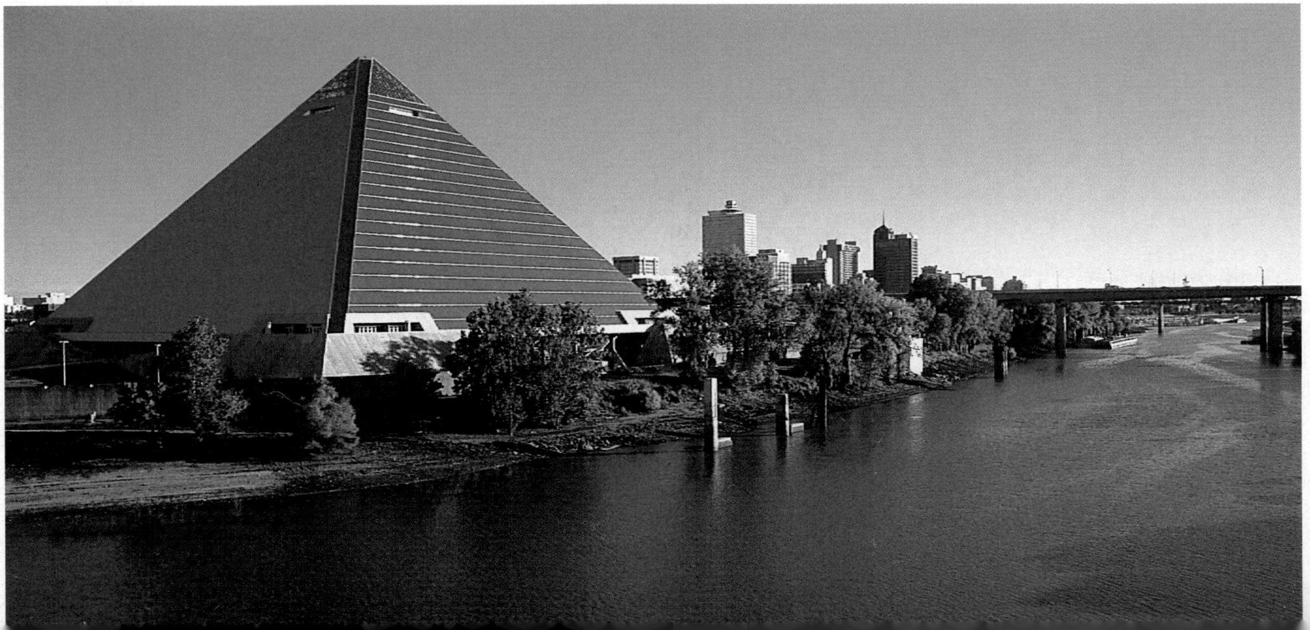

To find the volume of a pyramid you need to know its height.

Example 2

Find the volume of a regular square pyramid with base edges 40 ft long and slant height 25 ft.

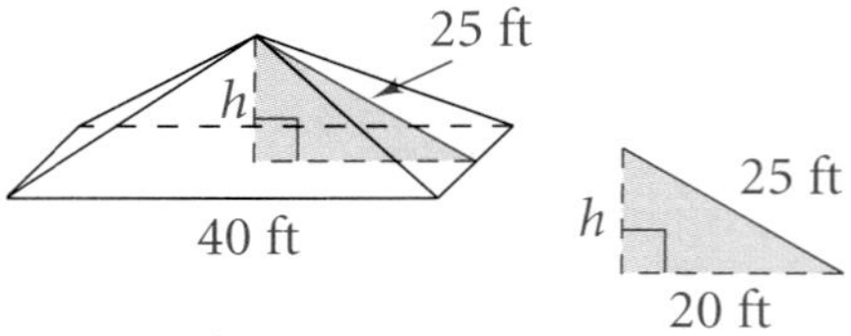

- Find the height of the pyramid.

$25^2 = h^2 + 20^2$ Use the Pythagorean Theorem.
$625 = h^2 + 400$ Simplify.
$h^2 = 225$ Subtract 400 from each side.
$h = 15$ Find the square root of each side.

- Find the volume of the pyramid.

$V = \frac{1}{3}Bh$ Use the formula for volume of a pyramid.
$= \frac{1}{3}(40 \cdot 40)15$ Substitute.
$= 8000$ Simplify.

The volume of the pyramid is 8000 ft^3.

6. Try This Find the volume of a regular square pyramid with base edges 24 m long and slant height 13 m.

Volumes of Cones

In the Work Together, you discovered that the volume of a pyramid is one third the volume of a prism with the same base and height. You can also find that the volume of a cone is one third the volume of a cylinder with the same base and height.

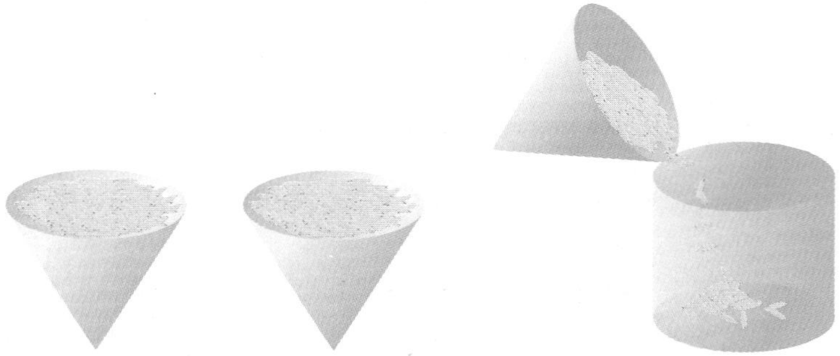

Theorem 6-9
Volume of a Cone

The volume of a cone is one third the product of the area of the base and the height.

$$V = \frac{1}{3}Bh, \text{ or } V = \frac{1}{3}\pi r^2 h$$

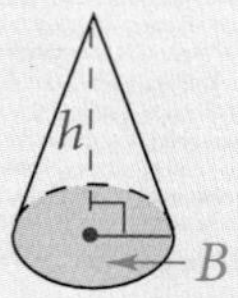

This volume formula applies to all cones, including *oblique* cones.

Example 3

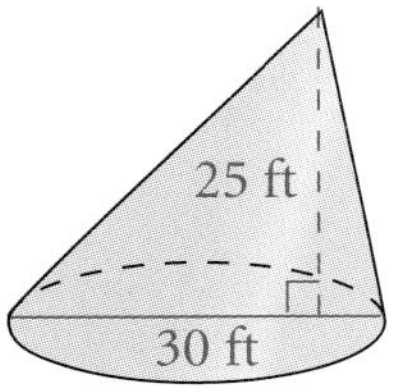

Find the volume of the oblique cone with diameter 30 ft and height 25 ft. Round to the nearest whole number.

$r = \frac{30}{2} = 15$ — The radius is half the diameter.

$V = \frac{1}{3}\pi r^2 h$ — Use the formula for the volume of a cone.

$= \frac{1}{3}\pi(15)^2 25$ — Substitute.

1 [÷] 3 [×] [π] [×] 15 [x^2] [×] 25 [=] 5890.4862

The volume of the cone is about 5890 ft^3.

7. Try This Find the volume of a cone with radius 3 in. and height 8 in. Round to the nearest tenth.

Exercises ON YOUR OWN

Calculator **Find the volume of each figure. When an answer is not a whole number, round to the nearest tenth.**

1.

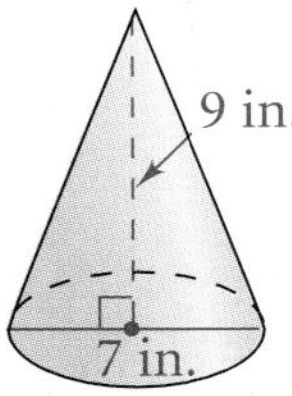

2.

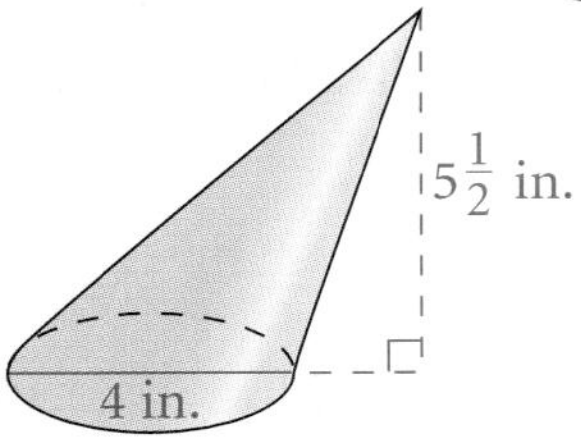

3.

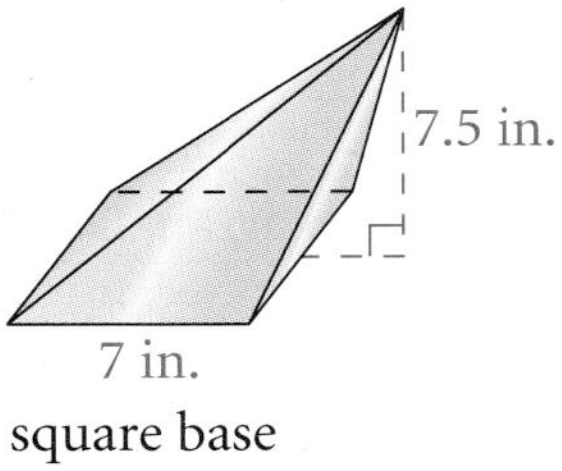

4.

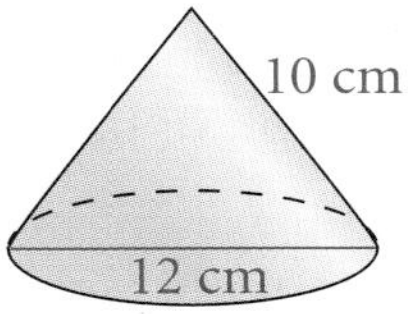

5. a. **Mental Math** A cone with radius 3 ft and height 10 ft has a volume of 30π ft^3. What is its volume when the radius is doubled?

b. What is the volume when the height of the original cone is doubled?

c. What is the volume when both the radius and the height of the original cone are doubled?

6. a. The largest tepee in the United States belongs to a member of the Crow (Native Americans of the Great Plains). It is 43 ft high and 42 ft in diameter. Find its volume to the nearest cubic foot.

b. How does this compare with the volume of your classroom?

Find the volume of each figure. You may leave answers in terms of π or in simplest radical form.

7.

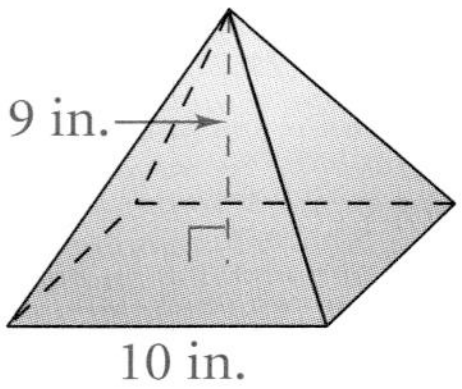

8.

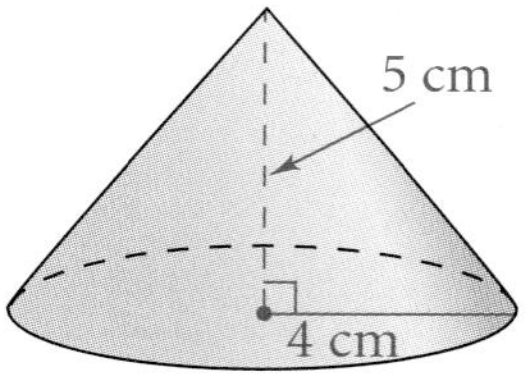

9.

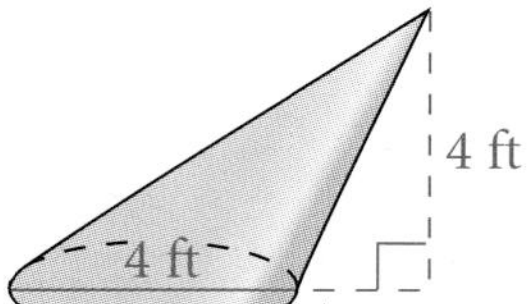

10.

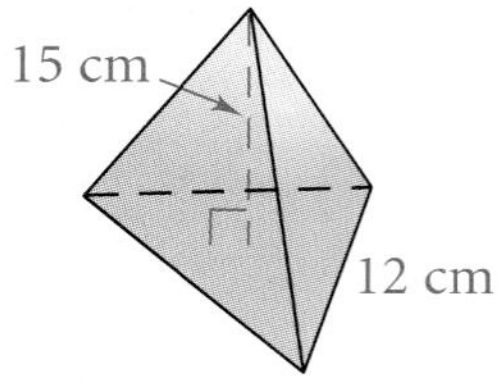

11. *Open-ended* A cone has a volume of 600π in.^{3}. Find two possible sets of dimensions for the height and radius of the cone.

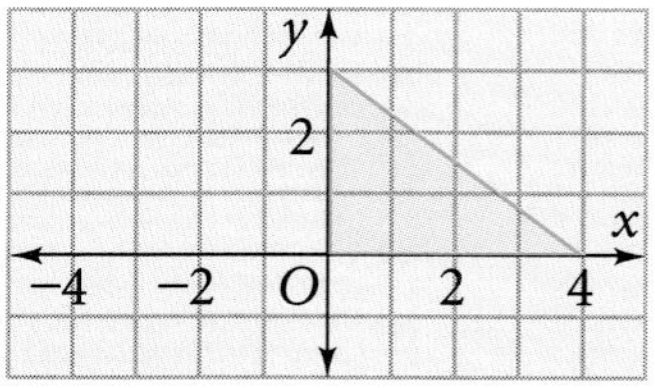

12. a. *Coordinate Geometry* Suppose you rotate the right triangle shown 360° about the x-axis. What is the volume of the resulting cone in terms of π?

b. Suppose you rotate the triangle 360° about the y-axis. What is the volume in terms of π?

13. The two cylinders pictured at the right are congruent. How does the volume of the larger cone compare to the total volume of the two smaller cones? Explain.

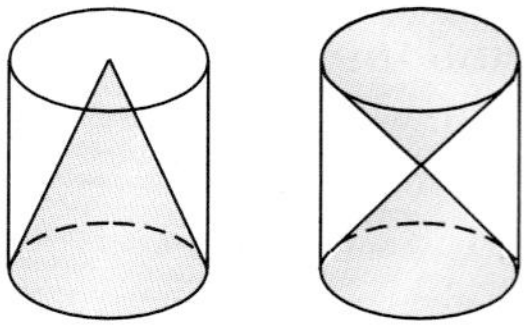

14. The volume of a cone is 36π cm^3. If the radius is 6 cm, what is the height of the cone?

15. The volume of a regular square pyramid is 600 in.^{3}. The height is 8 in. What is the length of each base edge?

16. To the nearest tenth, find the volume of a regular hexagonal pyramid with base edges 12 cm long and height 15 cm.

17. a. *Architecture* The Transamerica Building in San Francisco is a pyramid 800 ft tall with a square base 149 ft on each side. What is its volume to the nearest thousand cubic feet?

b. If the Transamerica Building had been built as a *prism* with the same square base, how tall to the nearest foot would it have to be to have the same volume as the existing building?

18. **Critical Thinking** A movie theater sells popcorn in cylindrical containers that are 4 in. in diameter and 10 in. high. As a special promotion, the theater plans to sell popcorn in a cone-shaped container for the same price. The diameter of the container will remain the same. The promotional cone will use the same amount of cardboard as the cylindrical container. Do you think this promotional cone is a good value? Why or why not?

19. Which container has the greatest volume?

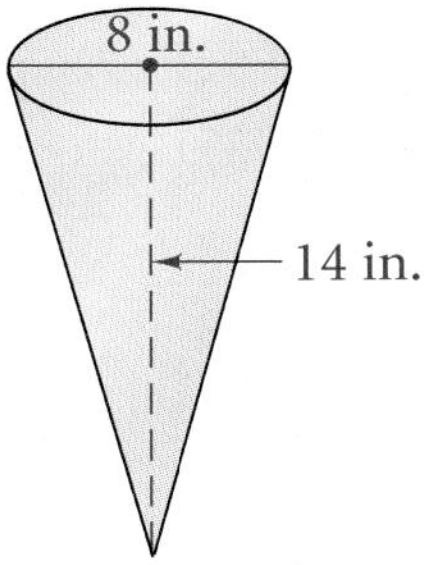

Figure 1

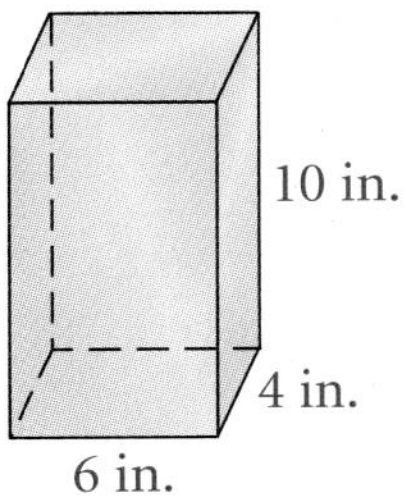

Figure 2

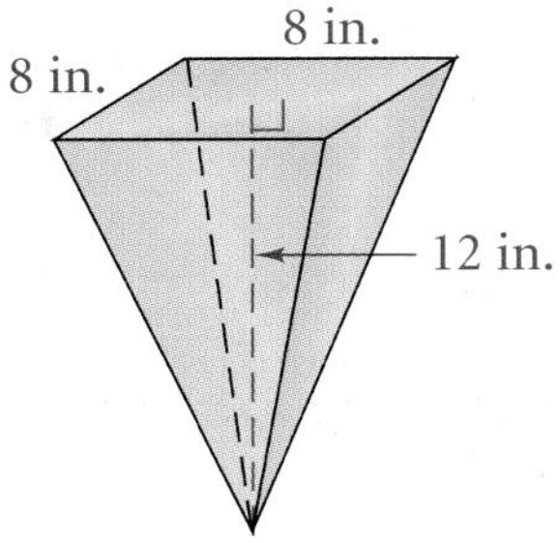

Figure 3

Algebra **Find the value of the variable in each figure. You may leave answers in simplest radical form.**

20.

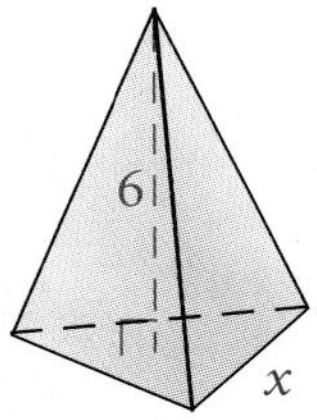

Volume = $18\sqrt{3}$

21.

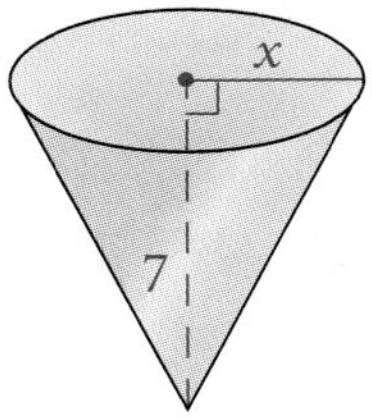

Volume = 21π

22.

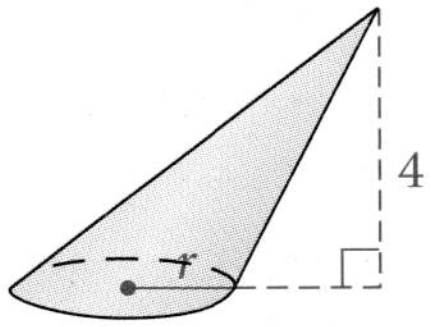

Volume = 24π

23.

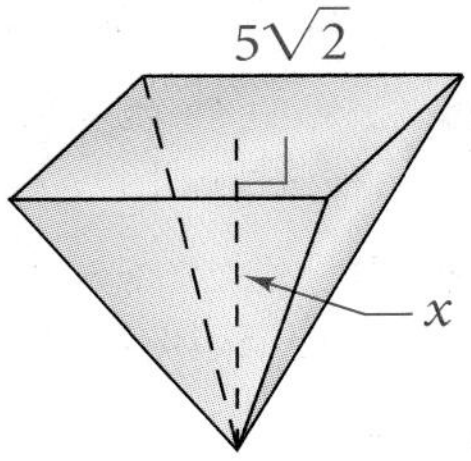

Volume = 150

24. **Writing** The figures at the right can be covered by an equal number of straws that are the same length. Describe how Cavalieri's Principle could be adapted to compare the areas of these figures.

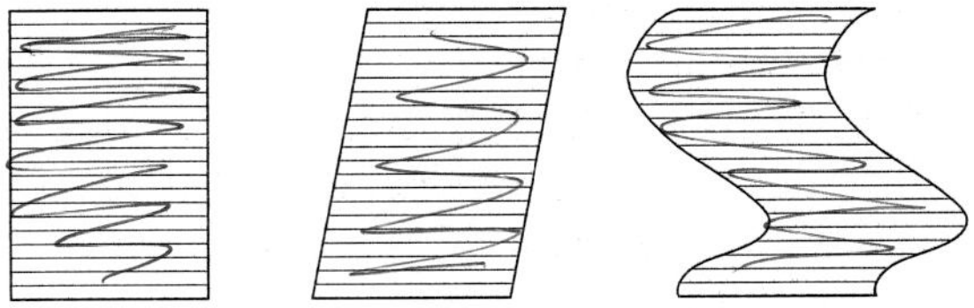

Chapter Project **Find Out by Investigating**

Do some container shapes seem to give more product for the same money? Go to a supermarket and identify a variety of container shapes. Do some shapes make you think there's more in them than there actually is? What factors do you think a manufacturer considers in deciding the shape of a container? Write a report about your findings.

Exercises MIXED REVIEW

25. The vertices of quadrilateral $ABCD$ are $A(-4, -1)$, $B(-1, 3)$, $C(7, -3)$, and $D(4, -7)$.

a. Use the Distance Formula to find the length of each side.

b. Find the slope of each side.

c. Determine the most accurate name for quadrilateral $ABCD$.

26. Movies The largest permanent movie screen in the world is in Jakarta, Indonesia. It is 96 ft by 70.5 ft. What is its area?

SELF ASSESSMENT

FOR YOUR JOURNAL

Explain why finding the volume of a cylinder is like finding the volume of a prism. Then explain why finding the volume of a cone is like finding the volume of a pyramid.

Getting Ready for Lesson 6-6

Calculator **Find the area and circumference of a circle with the given radius. Round answers to the nearest tenth.**

27. 6 in. **28.** 5 cm **29.** 2.5 ft **30.** 1.2 m

Geometry at Work

Packaging Engineer

Each year, more than one trillion dollars in manufactured goods are packaged in some kind of a container. To create each new box, bag, or carton, packaging engineers must balance such factors as safety, environmental impact, and attractiveness against cost of production.

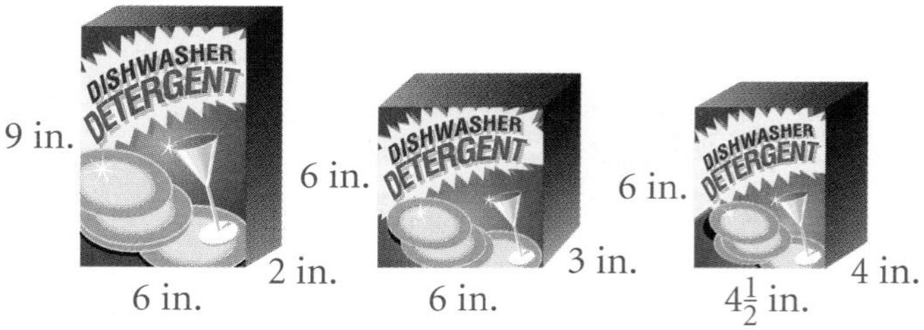

Consider the three boxes of dishwasher detergent. All three boxes have volumes of 108 in.^{3}, the volume of a standard box of automatic dishwasher detergent. The boxes have different shapes, however, and different surface areas. The box on the left has the greatest surface area and therefore costs the most to produce. Nevertheless, despite the higher cost, the box on the left has become standard. There are many reasons why a company might choose more expensive packaging. In this case, the least expensive package on the right is too difficult for a consumer to pick up and pour. To offset the extra cost, packaging engineers try to choose a package design that will attract enough additional buyers to outweigh the higher cost of production.

Mini Project: Choose a product that you enjoy using. Design and make a new package for the product. Calculate the surface area and volume of your package and describe advantages it has over the current package.

Connections Manufacturing . . . **and more**

6-6 Surface Areas and Volumes of Spheres

What You'll Learn

- Calculating the surface areas and volumes of spheres

...And Why

To solve real-world problems such as finding the surface area of a soccer ball and the volume of a scoop of ice cream

What You'll Need

- ruler
- scissors
- tacks
- foam balls (cut in half)
- string
- calculator

WORK TOGETHER

How do you compute the surface area of Earth or the amount of leather covering a baseball? Work in a group to explore the surface area of a sphere.

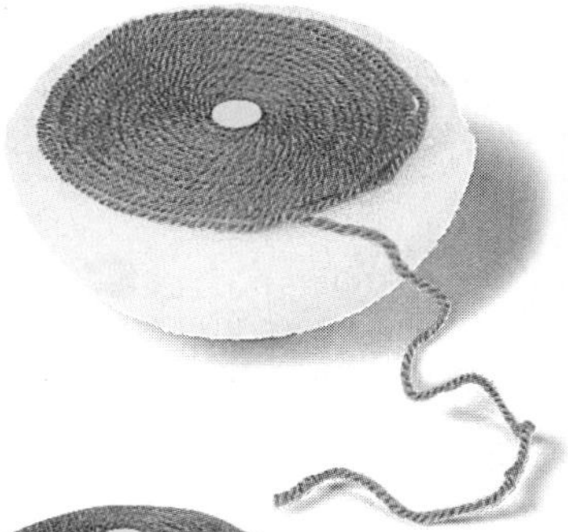

- The surface of half a foam ball consists of two parts—a plane circular region and a curved surface. Place a tack in the center of the circular region. Wind string around the tack covering the entire circular region.
- Once the circular region is covered, cut off any excess string. Measure the length x of string that covered the circular region.

- Place a tack in the center of the curved surface. Wind another string around the tack covering the entire curved surface.
- Once the curved surface is covered, cut off any excess string. Measure the length y of the string that covered the curved surface.

1. **a.** How do x and y compare?
b. Express y as a multiple of x.

2. **a.** How much string would you need to cover the entire surface of an uncut foam ball? Express your answer as a multiple of y.
b. Express your answer to part (a) as a multiple of x.

3. A string of length x covers an area of πr^2 where r is the radius of the foam ball. Substitute πr^2 for x in the expression you wrote for Question 2(b) to find a formula for the surface area of a sphere.

THINK AND DISCUSS

Finding the Surface Area of a Sphere

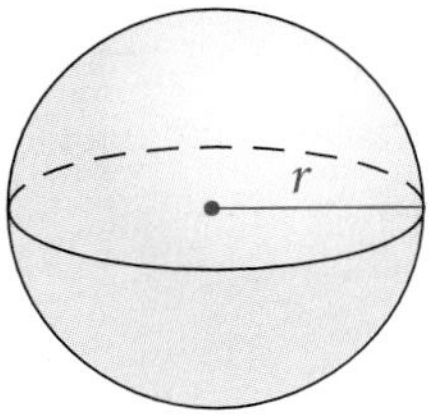

A **sphere** is the set of all points in space equidistant from a given point called the **center.**

4. How would you define a *radius* of a sphere?

5. How would you define a *diameter* of a sphere?

Your discovery in the Work Together leads to the following theorem.

Theorem 6-10
Surface Area of a Sphere

The surface area of a sphere is four times the product of π and the square of the radius of the sphere.

$$\text{S.A.} = 4\pi r^2$$

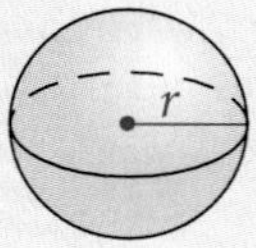

Example 1 Relating to the Real World

Manufacturing Manufacturers make soccer balls with a radius of 11 cm by sewing together 20 regular hexagons and 12 regular pentagons. Templates for guiding the stitching are shown at the left. Approximate the surface area of a soccer ball to the nearest square centimeter using the following two methods.

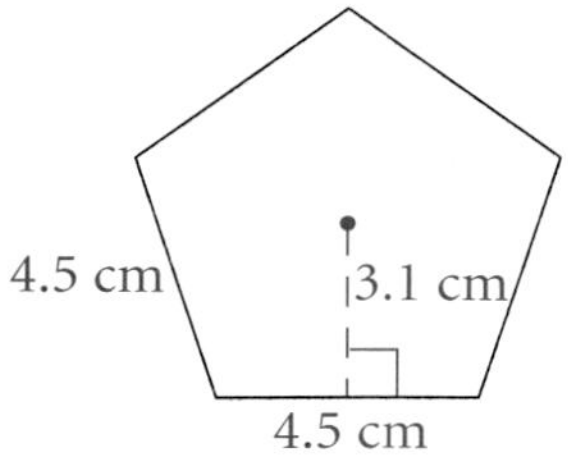

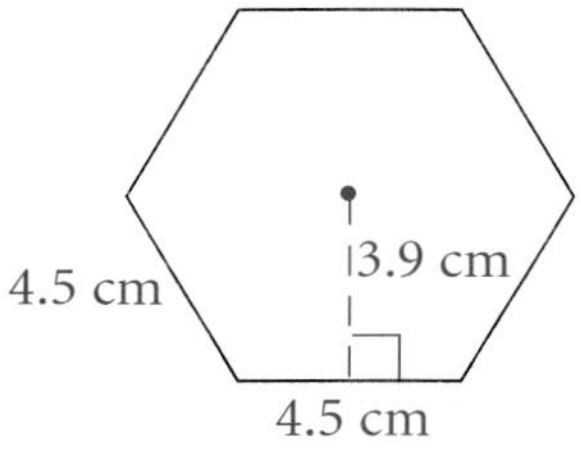

Method 1:
Find the sum of the areas of the pentagons and hexagons.

regular pentagon

$$\begin{aligned} A &= \tfrac{1}{2}ap \\ &= \tfrac{1}{2}(3.1)(5 \cdot 4.5) \\ &= 34.875 \end{aligned}$$

regular hexagon

$$\begin{aligned} A &= \tfrac{1}{2}ap \\ &= \tfrac{1}{2}(3.9)(6 \cdot 4.5) \\ &= 52.65 \end{aligned}$$

Area of the 12 regular pentagons $= (12)(34.875) = 418.5$

Area of the 20 regular hexagons $= (20)(52.65) = 1053$

The sum of the areas of the pentagons and the hexagons is about 1472 cm^2.

Method 2:
Use the formula for the surface area of a sphere.

$\text{S.A.} = 4\pi r^2$ — Use the formula for surface area.

$= 4 \cdot \pi \cdot 11^2$ — Substitute.

4 [×] [π] [×] 11 [x^2] [=] 1520.5308

The surface area of the soccer ball is about 1521 cm^2.

6. Compare the answers for the two methods. Explain why they differ.

Finding the Volume of a Sphere

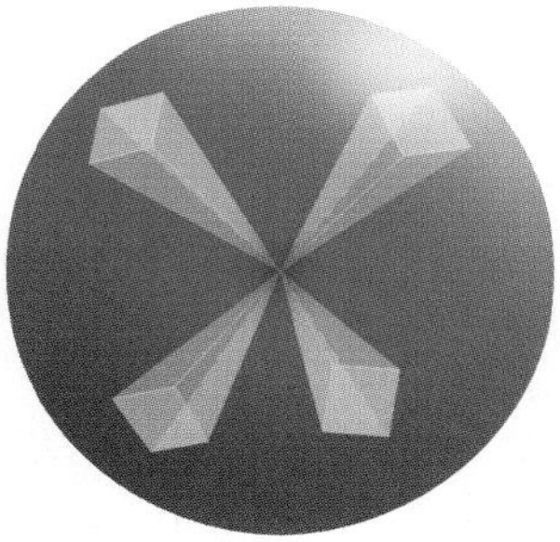

You can fill a sphere with a large number n of small pyramids. The vertex of each pyramid is the center of the sphere. The height of each pyramid is approximately the radius r of the sphere. The sum of the areas of all the bases approximates the surface area of the sphere. You can use this model to derive a formula for the volume of a sphere.

Volume of each pyramid $= \frac{1}{3}Bh$

Sum of the volumes of n pyramids $= n \cdot \frac{1}{3}Br$ — Substitute r for h.

$= \frac{1}{3} \cdot (nB) \cdot r$

$= \frac{1}{3} \cdot (4\pi r^2) \cdot r$ — Replace nB with the surface area of a sphere.

$= \frac{4}{3}\pi r^3$

The volume of a sphere is $\frac{4}{3}\pi r^3$.

Theorem 6-11
Volume of a Sphere

The volume of a sphere is four thirds the product of π and the cube of the radius of the sphere.

$$V = \frac{4}{3}\pi r^3$$

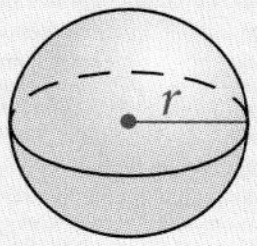

Example 2

The volume of a sphere is 4849.05 m^3. What is the surface area of the sphere? Round your answer to the nearest tenth.

- Find the radius r.

$V = \frac{4}{3}\pi r^3$	Use the formula for the volume of a sphere.
$4849.05 = \frac{4}{3}\pi r^3$	Substitute.
$4849.05\left(\frac{3}{4\pi}\right) = r^3$	Multiply both sides by $\frac{3}{4\pi}$.
$\sqrt[3]{4849.05\left(\frac{3}{4\pi}\right)} = r$	Find the cube root of each side.

4849.05 [×] 3 [÷] 4 [÷] [π] [=] [$\sqrt[x]{y}$] 3 [=] *10.500001*

The radius of the sphere is about 10.5 m.

- Find the surface area of the sphere.

S.A. $= 4\pi r^2$	Use the formula for the surface area of a sphere.
$= 4\pi(10.5)^2$	Substitute.

4 [×] [π] [×] 10.5 [x^2] [=] *1385.4424*

The surface area of the sphere is about 1385.4 m^2.

QUICK REVIEW

The *cube root* of x is the number that when *cubed* is x.

7. **Try This** The volume of a sphere is 20,579 in.^{3}. What is the radius of the sphere to the nearest whole number?

8. **Try This** The radius of a sphere is 15 m. What is the volume to the nearest hundred?

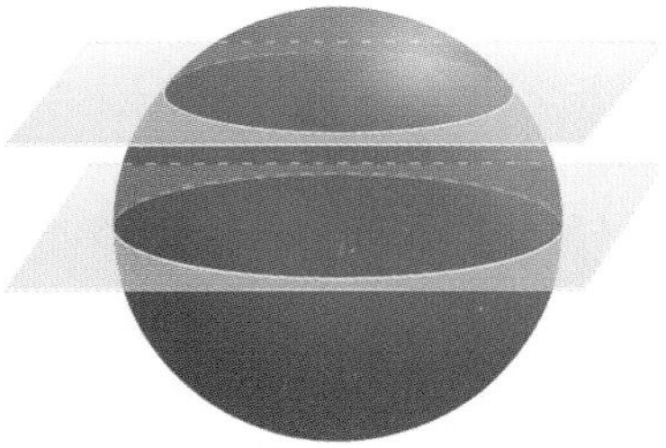

When a plane and a sphere intersect in more than one point, the intersection is a circle. If the center of the circle is also the center of the sphere, the circle is called a **great circle** of the sphere. The circumference of a great circle is the **circumference of the sphere.** A great circle divides a sphere into two **hemispheres.**

9. **Geography** What is the name of the best-known great circle on Earth?

10. **Geography** Describe the Northern Hemisphere of Earth.

Exercises ON YOUR OWN

Calculator **Find the surface area of each ball to the nearest tenth.**

1. 

$d = 23.9$ cm

2. 

$d = 68$ mm

3. 

$d = 2\frac{3}{4}$ in.

4. 

$d = 1.68$ in.

5. **Coordinate Geometry** Find the surface area and volume of the sphere formed by rotating the semicircle at the right 360° about the x-axis. Leave your answers in terms of π.

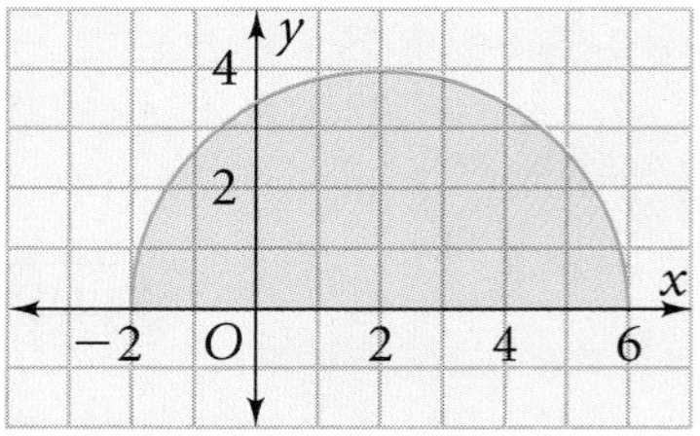

6. A balloon has a 14-in. diameter when it is fully inflated. Half the air is let out of the balloon. Assuming the balloon is a sphere, what is the new diameter? Round your answer to the nearest inch.

7. **Meteorology** On July 16, 1882, a massive thunderstorm over Dubuque, Iowa, produced huge hailstones. The diameter of some of the hailstones was 17 in. Ice weighs about 0.033 lb/in.^{3}. What was the approximate weight of these hailstones to the nearest pound?

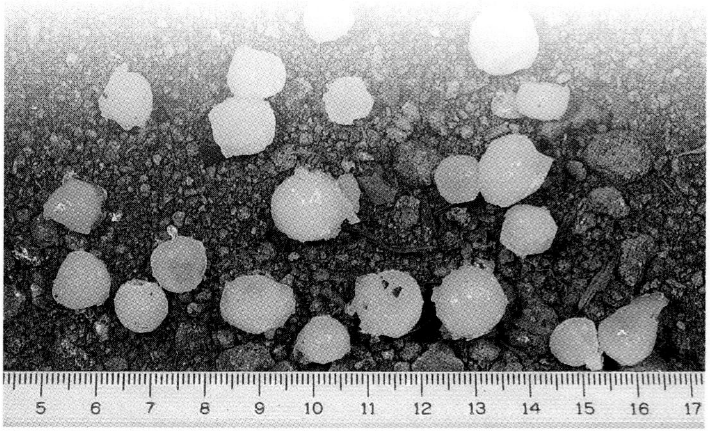

Find the volume of each sphere. Leave your answers in terms of π.

8.
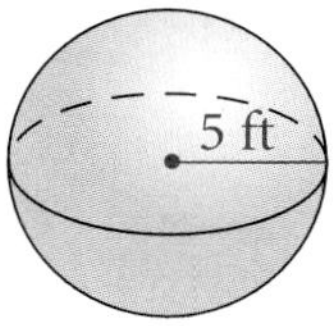

9.
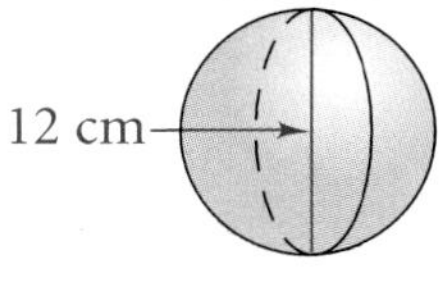

10.
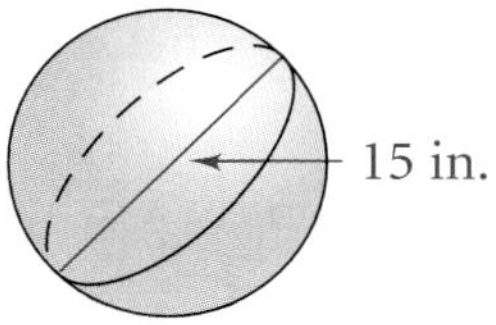

11.
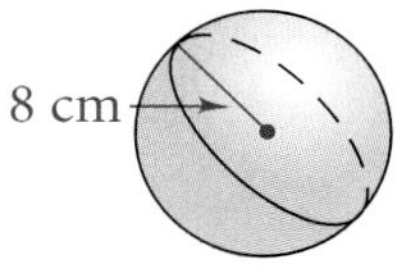

12. *Sports* The circumference of a bowling ball is about 27 in. Find its volume to the nearest tenth.

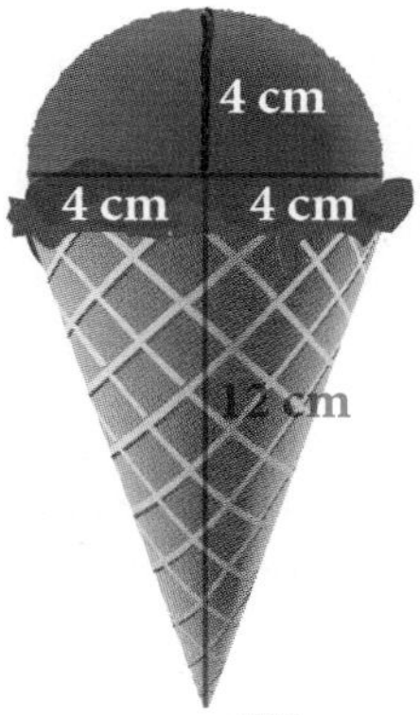

13. If the sphere of ice cream shown melts, is the cone large enough to hold the melted ice cream? Explain.

14. *Geometry in 3 Dimensions* The center of a sphere has coordinates (0, 0, 0). The radius of the sphere is 5.

a. Name the coordinates of six points on the sphere.

b. Tell whether each of the following points is inside, outside, or on the sphere. (*Hint:* Use the formula from page 262.)
$A(0, -3, 4)$, $B(1, -1, -1)$, $C(4, -6, -10)$

Believe It Or Not

J.C. Payne, a Texas farmer, is the world champion string collector. The ball of string he wound over a three-year period has a circumference of 41.5 ft. It weighed 13,000 lb.

Listed in the Guinness Book of Records, the ball of string is now on display in a museum devoted to oddities. It took almost a dozen men with forklift trucks to load the ball onto a truck to move it to the museum.

15. a. What is the volume of the ball of string to the nearest cubic foot?

b. What is the weight of the string per cubic foot?

c. If the diameter of the string is 0.1 in., what is the approximate length of the string to the nearest mile? (*Hint:* Think of the unwound string as a long cylinder.)

16. The radius of Earth is approximately 3960 mi. The area of Australia is about 2,940,000 mi^2.

a. Find the surface area of the Southern Hemisphere.

b. *Probability* If a meteorite falls randomly in the Southern Hemisphere, what is the probability that it will fall in Australia?

CALCULATOR HINT

When answers will be used in later calculations, keep or store them in unrounded form so that rounding errors will not be introduced into the final answer.

17. The sphere just fits in a cube with edges 6 in. long.
   a. What is the radius of the sphere?
   b. What is the volume of the space between the sphere and cube to the nearest tenth?

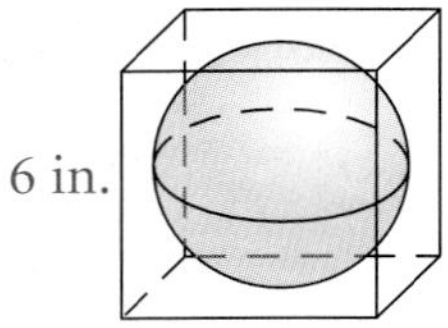

18. *Open-ended* Give the dimensions of a cylinder and a sphere that have the same volume.

19. The sphere just fits in the cylinder. Archimedes (about 287–212 B.C.) asked to have this figure engraved on his tombstone because he was the first to find the ratio of the volume of the sphere to the volume of the cylinder. What is the ratio that Archimedes discovered?

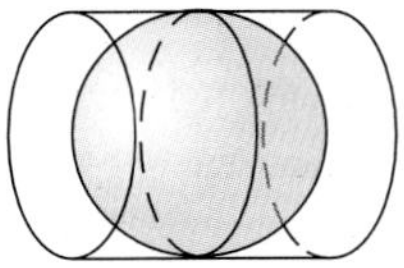

20. Which is greater, the total volume of three balls that each have diameter of 3 in. or the volume of one ball that has a diameter of 8 in.? Explain your answer.

21. A cube with edges 6 in. long just fits in the sphere. The diagonal of the cube is the diameter of the sphere.
   a. Find the length of the diagonal of the cube and the radius of the sphere. Leave your answers in simplest radical form.
   b. What is the volume of the space between the sphere and the cube to the nearest tenth?

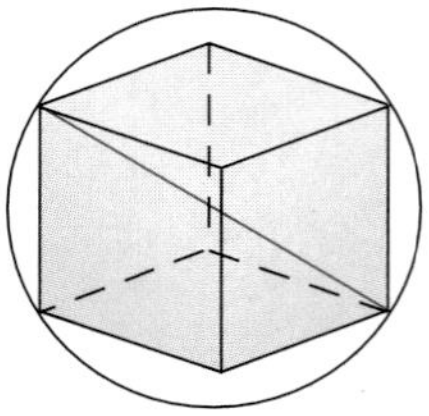

22. *Estimation* Use $\pi \approx 3$ to estimate the surface area and volume of a sphere with radius 30 cm.

23. *Standardized Test Prep* The cone and sphere just fit in cubes that are the same size. What is the ratio of the volume of the cone to the volume of the sphere?
   A. 1 : 2 B. 1 : 3 C. 1 : 4
   D. 2 : 3 E. cannot be determined

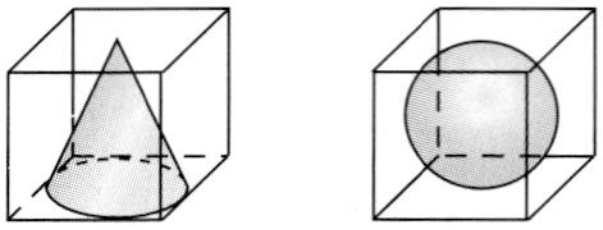

24. a. The number of square meters of surface area of a sphere equals the number of cubic meters of volume. What is the radius of the sphere?
   b. *Algebra* The ratio of the surface area of a sphere in square meters to its volume in cubic meters is 1 : 5. What is the radius of the sphere?

25. *Science* The density of steel is about 0.28 lb/in.^{3}. Could you lift a steel ball with radius 4 in.? with radius 6 in.? Explain.

26. A plane that intersects a sphere is 8 cm from the center of the sphere. The radius of the sphere is 17 cm. What is the area of the cross section to the nearest whole number?

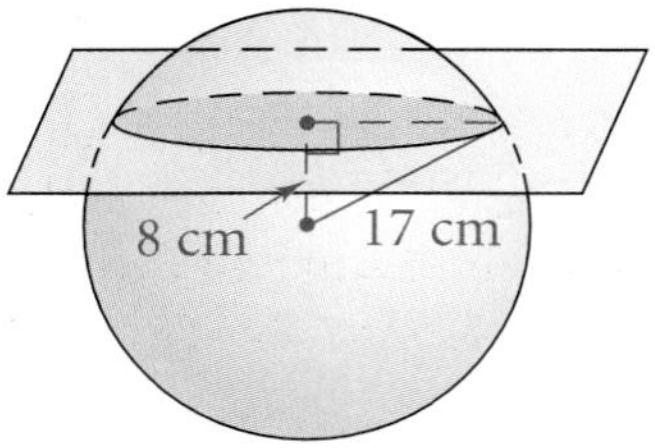

27. a. *Algebra* If a cube and a sphere have the same volume, which has the greater surface area? Explain.
   b. *Writing* Explain why spheres are rarely used for packaging.

Exercises MIXED REVIEW

Make a sketch for each description.

28. a pair of supplementary angles

29. an obtuse triangle and one midsegment

30. a polygon with three lines of symmetry

31. a line segment and its perpendicular bisector

32. *Open-ended* Binary numbers are written with only the digits 0 and 1. These digits can be written so they have both vertical and horizontal lines of symmetry. Using such digits, write two binary numbers that do not have the same symmetries and describe the differences.

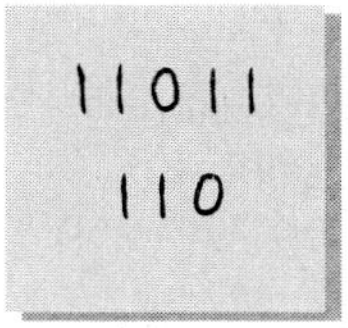

Getting Ready for Lesson 6-7

Calculator **Find the total volume of each pair of figures to the nearest tenth.**

33.

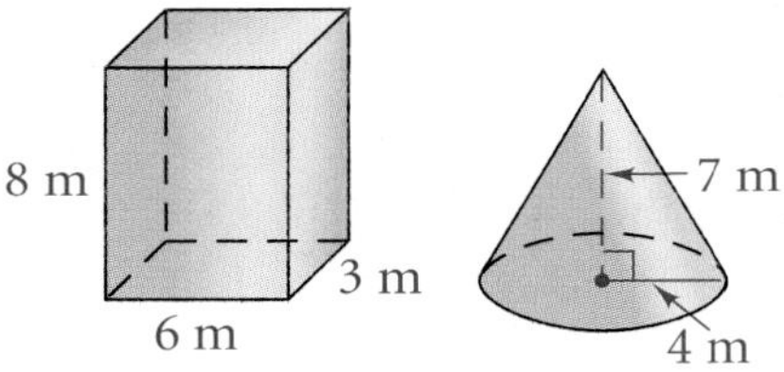

34.

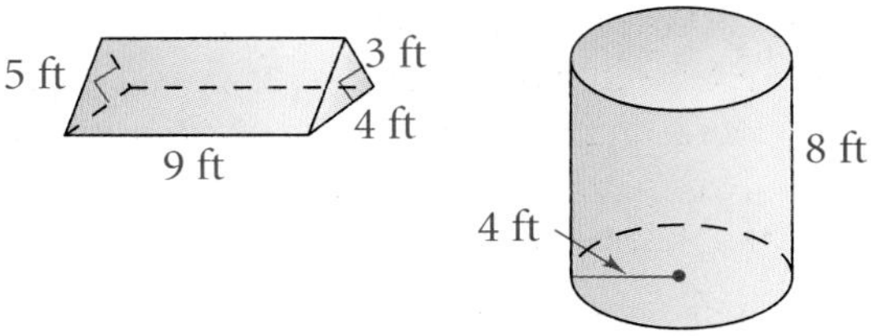

Exercises CHECKPOINT

Calculator **Find the surface area and the volume of each figure to the nearest tenth.**

1.

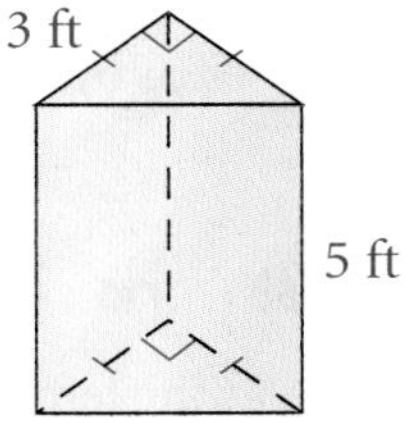

2.

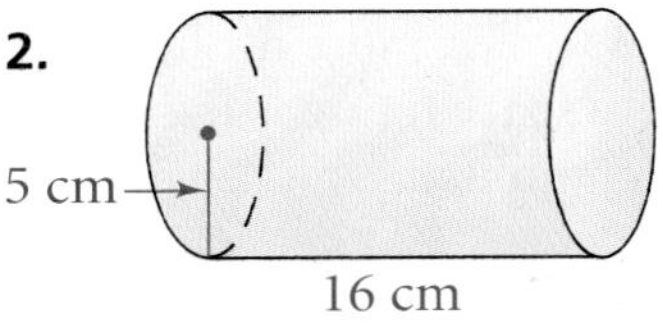

3.

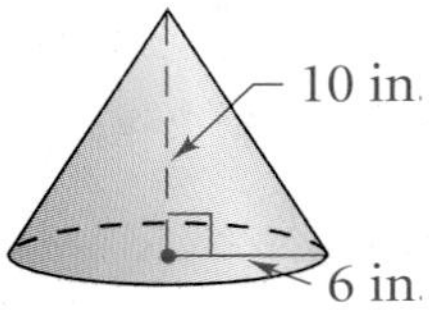

4.

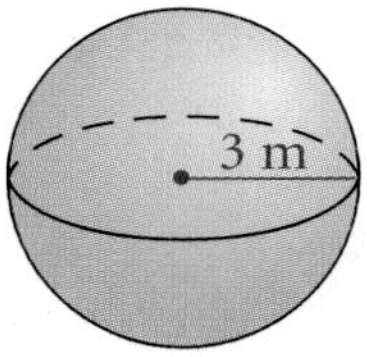

5. *Standardized Test Prep* What is the surface area of a sphere with radius 5 in.?

A. 5π in.^{2} **B.** 20π in.^{2} **C.** 100π in.^{2} **D.** 125π in.^{2} **E.** $166\frac{2}{3}\pi$ in.^{2}

6. *Critical Thinking* Tennis balls are packaged as shown at the right. Which is greater, the volume of a tennis ball or the space around the three balls? Explain.

6-7 Composite Space Figures

What You'll Learn

- Recognizing composite space figures, which combine two or more simple figures

...And Why

To find the volumes and surface areas of composite space figures such as silos and backpacks

What You'll Need

- unit cubes

WORK TOGETHER

1. Work in a group to build the prisms shown in Figures 1 and 2 using unit cubes. Find the surface area and volume of each prism.
2. What is the sum of the surface areas of the two prisms? What is the sum of their volumes?
3. Place one prism on top of the other as in Figure 3. Find the surface area and volume of the resulting space figure.
4. Compare your answers to Questions 2 and 3. How does the relationship of the volumes differ from that of the surface areas?

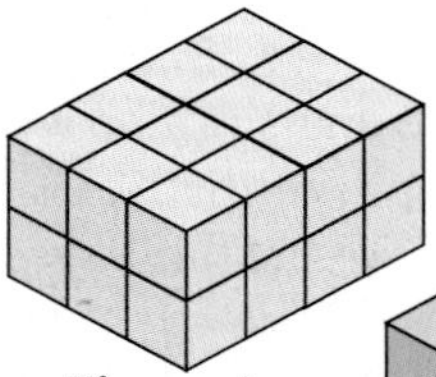
Figure 1

Figure 2

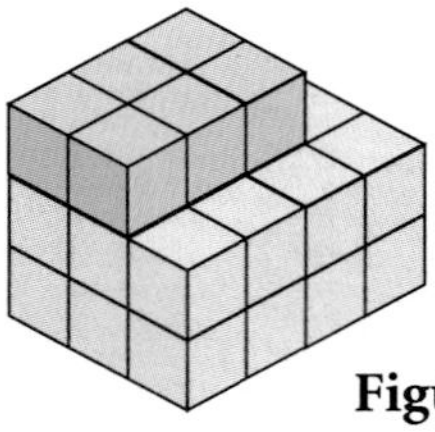
Figure 3

THINK AND DISCUSS

You can use what you know about the volumes of three-dimensional figures such as prisms, pyramids, cones, cylinders, and spheres to find the volume of a composite space figure. A **composite space figure** combines two or more of these figures. The volume of a composite space figure is the sum of the volumes of the figures that are combined.

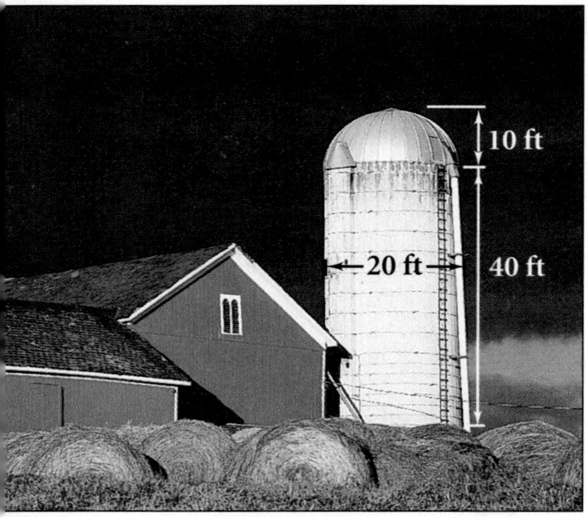

Example 1 *Relating to the Real World*

Agriculture Find the volume of the grain silo at the left.

The silo combines a cylinder and a hemisphere.

- Volume of the cylinder $= \pi r^2 h = \pi(10)^2(40) = 4000\pi$
- Volume of the hemisphere $= \frac{1}{2}(\frac{4}{3}\pi r^3) = \frac{2}{3}\pi(10)^3 = \frac{2000\pi}{3}$
- Volume of the composite figure $= 4000\pi + \frac{2000\pi}{3}$

4000 ⊠ π ⊞ 2000 ⊠ π ÷ 3 ⊟ 14660.766

The volume of the silo is about 14,700 ft^3.

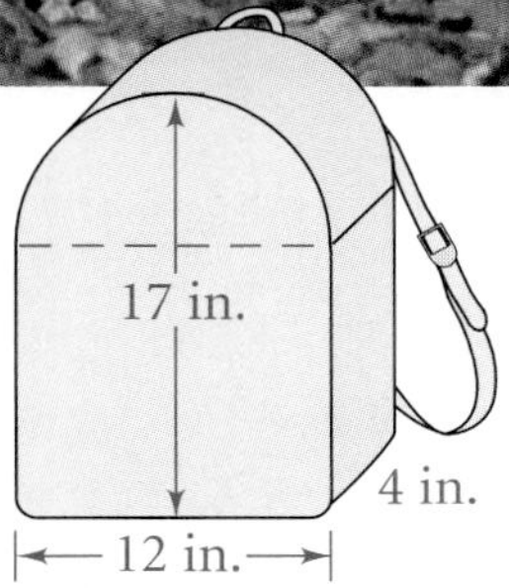

5. Try This Find the volume of each composite space figure to the nearest whole number.

a. 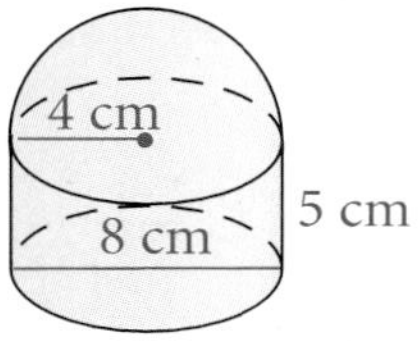

b. 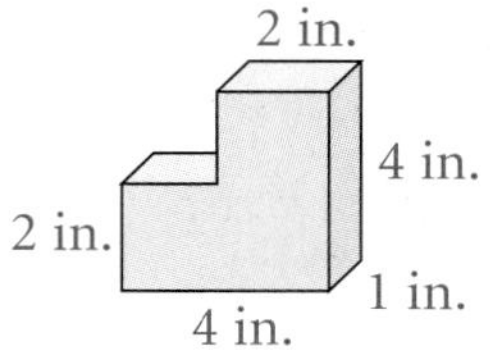

You can use geometric figures to approximate the shape of a real-world object. Then you can estimate the volume and surface area of the object.

Example 2 — Relating to the Real World

Estimation What space figures can you use to approximate the shape of the backpack? Use these space figures to estimate the volume of the backpack.

- You can use a prism and half of a cylinder to approximate the shape of the backpack.

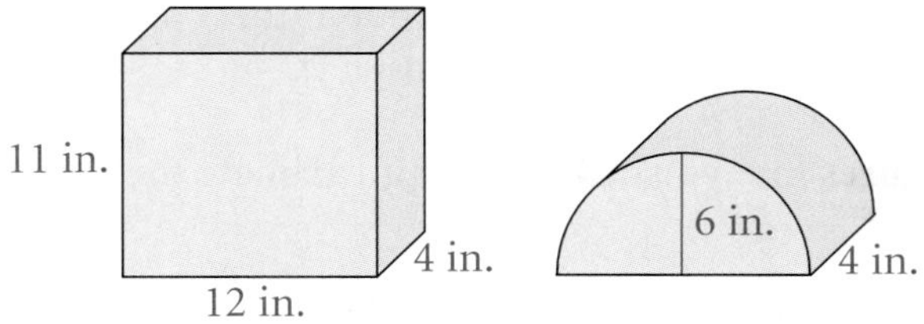

- Volume of the prism $= Bh = (12 \cdot 4)11 = 528$
- Volume of the half cylinder $= \frac{1}{2}(\pi r^2 h) = \frac{1}{2}\pi(6)^2(4)$
  $= \frac{1}{2}\pi(36)(4) \approx 226$
- Sum of the two volumes $= 528 + 226 = 754$

The approximate volume of the backpack is 754 in.^{3}.

6. What is the approximate surface area of this backpack?

7. Try This Describe the space figures that you can use to approximate the shape of each object.

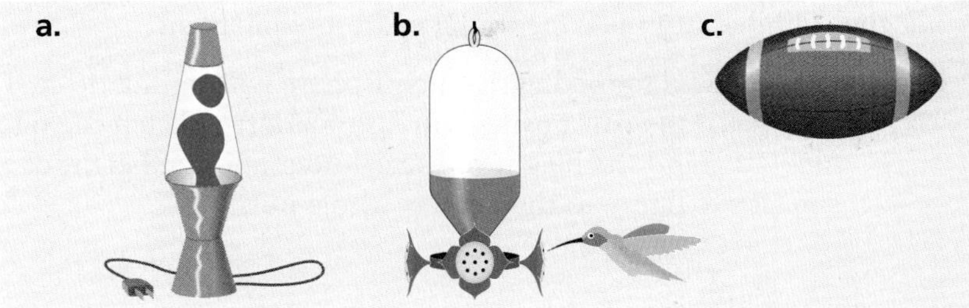

Exercises ON YOUR OWN

Find the volume of each composite space figure. You may leave answers in terms of π.

1.

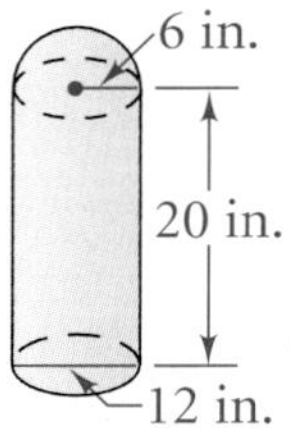

2.

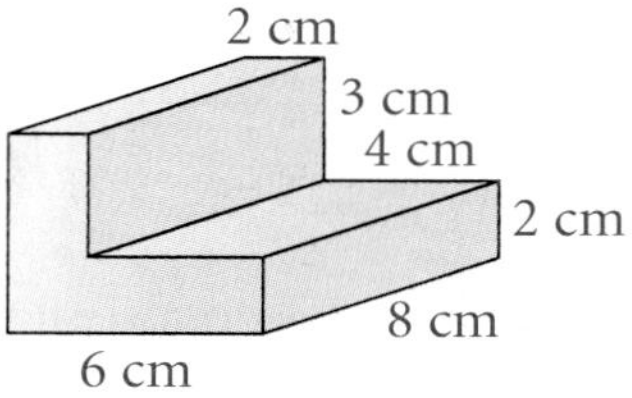

3.

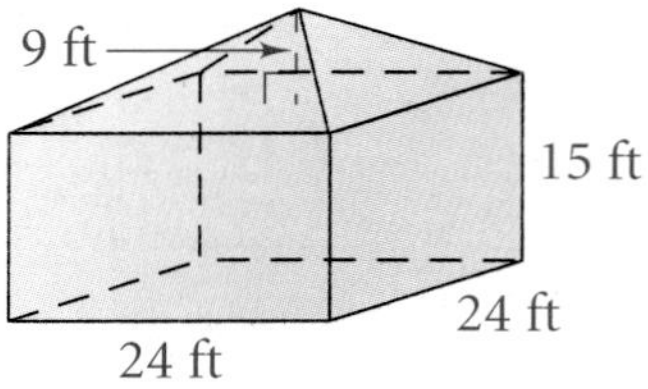

4. **Open-ended** Draw a composite three-dimensional figure, label its dimensions, and find either its surface area or its volume.

5. **Manufacturing** Find the volume of the lunch box shown at the right to the nearest cubic inch.

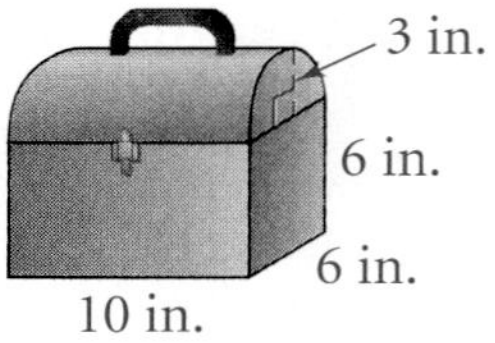

6. **Writing** Describe your home, school, or some other building that is a composite space figure. Explain why it is a composite space figure.

7. a. **Coordinate Geometry** Draw a sketch of the composite space figure formed by rotating this triangle 360° about the x-axis.
   b. Find the volume of the figure in terms of π.

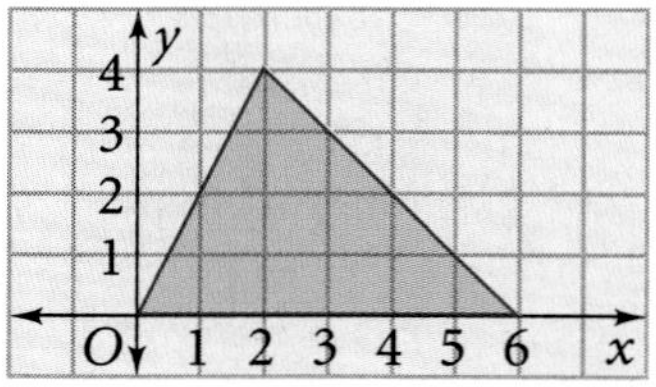

Describe space figures that you can use to approximate the shape of each object.

8.

9.

10.

11. **Writing** Describe how you would find the volume of an octahedron if you know the length of an edge. (See the picture on page 305.)

12. In the diagram at the right, what is the volume of the space between the cylinder and the cone to the nearest cubic inch?

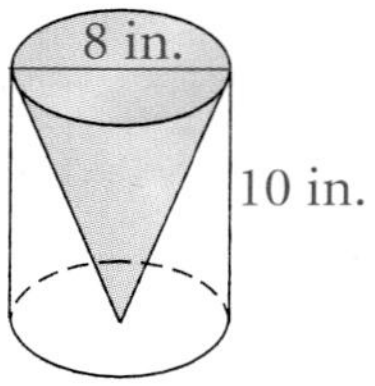

13. a. A cylinder is topped with a hemisphere with radius 15 ft. The total volume of the composite figure is 6525π ft^3. Sketch the figure.
   b. **Algebra** Find the height of the cylinder and the height of the composite figure to the nearest whole number.

14. *Coordinate Geometry* Find the surface area of the figure formed by rotating the figure at the right 360° about the y-axis. Leave your answer in terms of π.

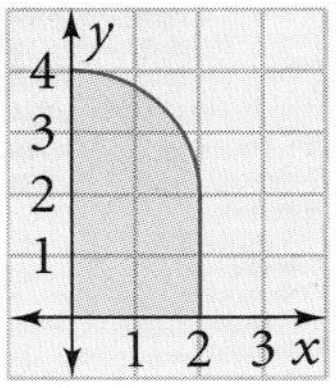

15. *Carpentry* Builders use a plumb bob to establish a vertical line. To the nearest cubic centimeter, find the volume of the plumb bob shown at the right. It combines a hexagonal prism with a pyramid.

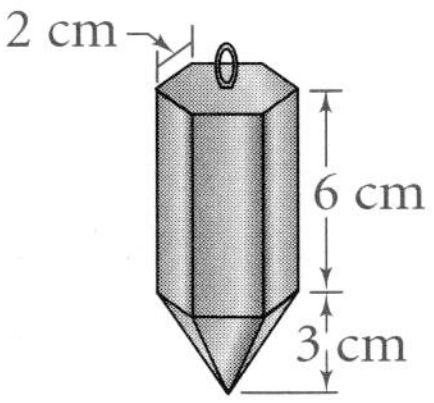

Engineering Steelworkers use I-beams like the one shown to build bridges and overpasses.

16. To the nearest cubic foot, find the volume of steel needed to make the beam.

17. a. To the nearest square foot, what is the surface area of the beam?
    b. One gallon of paint covers about 450 square feet. How many 1-gallon cans of paint do you need to paint the beam?

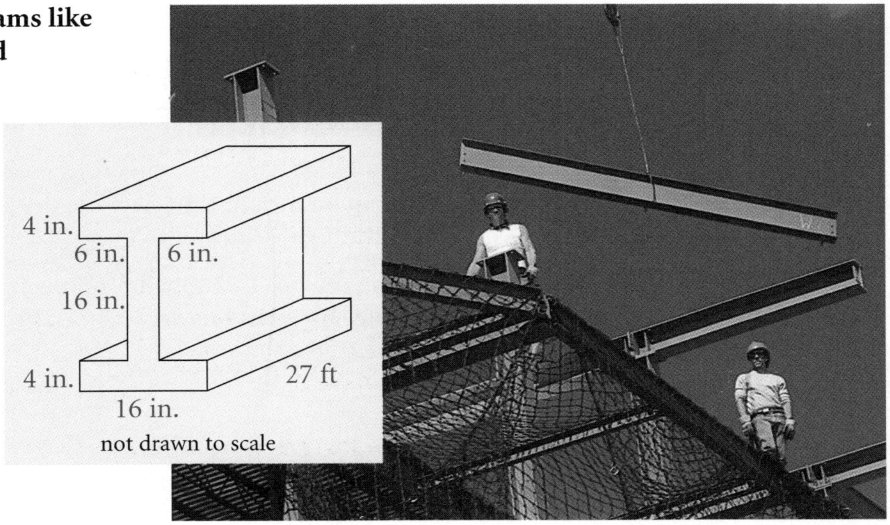

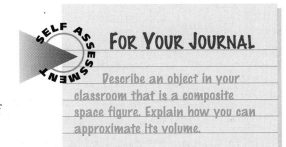

Exercises MIXED REVIEW

Write an indirect proof for each statement.

18. If a triangle is isosceles, then a base angle is not a right angle.

19. In $\triangle MNP$, if $MP < MN$, then $\angle P \ncong \angle N$.

20. *Coordinate Geometry* Find the circumference of a circle if the endpoints of a diameter are (3, 7) and (3, −1). Leave your answer in terms of π.

SELF ASSESSMENT

FOR YOUR JOURNAL

Describe an object in your classroom that is a composite space figure. Explain how you can approximate its volume.

Getting Ready for Lesson 6-8

You roll a number cube. Find each probability.

21. $P(4)$

22. $P(\text{odd number})$

23. $P(\text{prime number})$

24. $P(2 \text{ or } 5)$

Connections — Commuting . . . and more

6-8 Geometric Probability

What You'll Learn

- Using geometric models to find the probability of events

...And Why

To solve probability problems about games, waiting times, and other random events

THINK AND DISCUSS

Using a Segment Model

Mathematicians use models to represent the real world. **Geometric probability** uses geometric figures to represent occurrences of events. Then the occurrences can be compared by comparing measurements of the figures.

QUICK REVIEW

The probability of an event is the ratio of the number of favorable outcomes to the number of possible outcomes.

$P(\text{event}) = \dfrac{\text{favorable outcomes}}{\text{possible outcomes}}$

Example 1 — Relating to the Real World

Commuting Mr. Hedrick's bus runs every 25 minutes. If he arrives at his bus stop at a random time, what is the probability that he will have to wait 10 minutes or more?

Assuming the bus is stopped for a negligible length of time, the 25 minutes between bus arrivals can be represented by $\overline{AB}$.

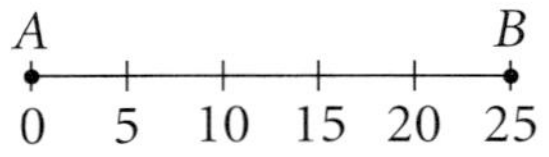

If Mr. Hedrick arrives 5 minutes after a bus has left (represented by point C on the segment below), he has to wait 20 minutes for a bus to arrive at time B. If he arrives 20 minutes after a bus has left (represented by point D), he has to wait only 5 minutes. If he arrives at any time represented by a point on $\overline{AR}$, he has to wait 10 minutes or more.

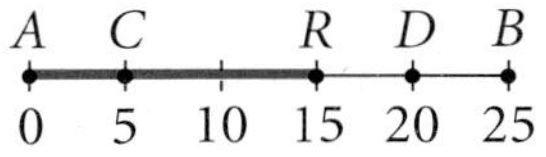

$$P(\text{waiting 10 min or more}) = \frac{\text{length of } \overline{AR}}{\text{length of } \overline{AB}} = \frac{15}{25} = \frac{3}{5}$$

The probability that Mr. Hedrick waits 10 minutes or more is $\frac{3}{5}$, or 60%.

1. **Try This** What is the probability that Mr. Hedrick has to wait 20 minutes or more?
2. *Critical Thinking* Suppose a bus arrives at Mr. Hedrick's bus stop every 25 minutes and waits 5 minutes before leaving. Draw a diagram and find the probability that Mr. Hedrick has to wait 10 minutes or more to get on the bus.

Using an Area Model

For some probability situations, you can use an area model.

Example 2 **Relating to the Real World**

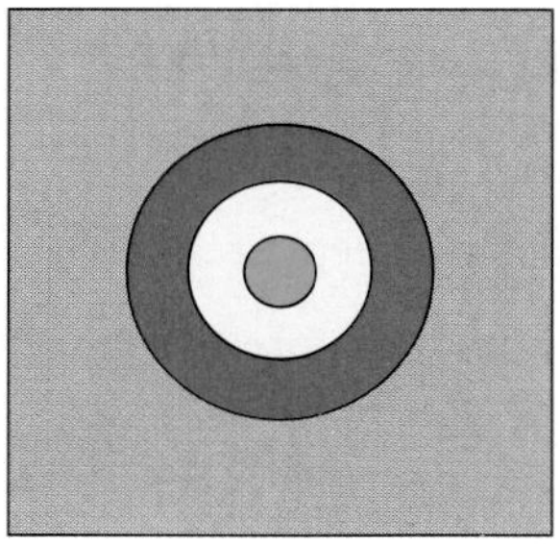

12 in.

12 in.

Dart Game At a carnival, you can win prizes if you throw a dart into the blue, yellow, or red regions of a 12 in.-by-12 in. dartboard. The radii of the concentric circles are 1, 2, and 3 inches. Assume that your dart lands on the board and that it is equally likely to land on any point. Find the probability of hitting each colored region.

$$P(\text{blue}) = \frac{\text{blue area}}{\text{area of square}} = \frac{\pi(1)^2}{12^2} = \frac{\pi}{144} \approx 2.2\%$$

Use a calculator.

$$P(\text{yellow}) = \frac{\text{yellow area}}{\text{area of square}} = \frac{\pi(2^2) - \pi(1^2)}{12^2} = \frac{3\pi}{144} \approx 6.5\%$$

$$P(\text{red}) = \frac{\text{red area}}{\text{area of square}} = \frac{\pi(3^2) - \pi(2^2)}{144} = \frac{5\pi}{144} \approx 10.9\%$$

The probability of hitting the blue region is about 2.2%, of hitting the yellow region is about 6.5%, and of hitting the red region is about 10.9%.

3. **Try This** What is the probability of winning some prize?
4. **Try This** What is the probability of winning no prize?
5. **Critical Thinking** If the radius of the blue circle were doubled, would the probability of hitting blue be doubled? Explain.

Some carnival games are more difficult than they appear. Consider the following coin toss game.

Example 3 Relating to the Real World

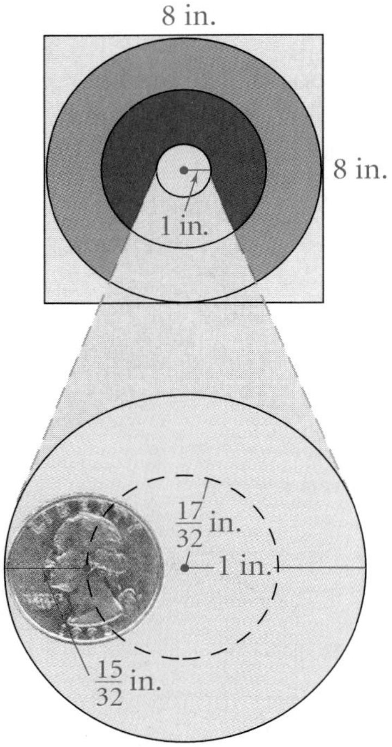

Coin Toss To win a prize, you must toss a quarter so that it lands entirely within a green circle of radius 1 in. The radius of a quarter is $\frac{15}{32}$ in. Suppose that the center of a tossed quarter is equally likely to land at any point within the 8-in. square shown at the left. What is the probability that a quarter will land completely within the green circle?

When a quarter is within the green circle, its center must be more than $\frac{15}{32}$ in. from the edge of the circle. So, the center of the quarter must be less than $\frac{17}{32}$ in. from the center of the green circle. This means that the center of the quarter must be within the dashed circle of radius $\frac{17}{32}$ in. as shown in the diagram.

$$P(\text{quarter within green circle}) = \frac{\text{area of dashed circle}}{\text{area of square}}$$

$$= \frac{\pi\left(\frac{17}{32}\right)^2}{(8)^2} \approx 1.4\%$$

The probability of the quarter landing in the green circle is only about 1.4%.

6. **Try This** Suppose you toss 100 quarters. Would you expect to win a prize?

7. a. *Critical Thinking* For every 1000 quarters tossed, about how many prizes would be won?
   b. Suppose the prize is \$10. About how much profit would the game operator expect for every 1000 quarters tossed?

Exercises ON YOUR OWN

Darts are thrown at random at each of the boards shown. If a dart hits the board, find the probability that it will land in the shaded area.

1. 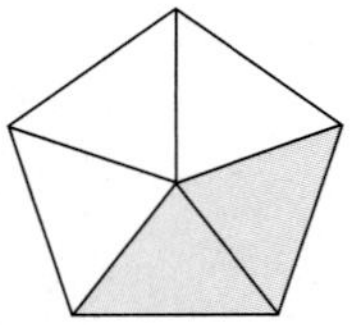

2. 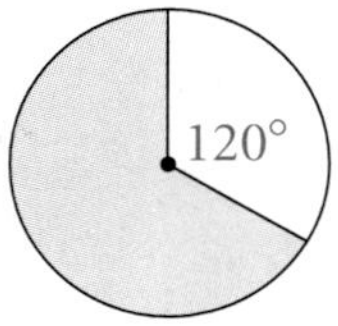

3. 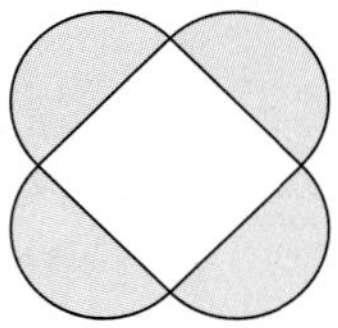

4. 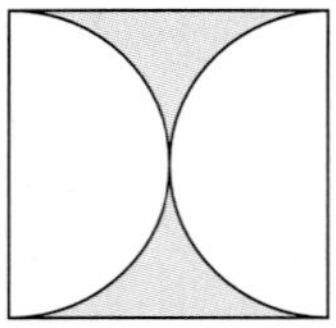

5. *Transportation* A rapid transit line runs trains every ten minutes. Draw a geometric model and find the probability that randomly arriving passengers will not have to wait more than four minutes.

Sketch a geometric model for each exercise and solve.

6. **Ducking Stool** At a fund raiser, volunteers sit above a tank of water hoping that they won't get wet. You can trip the ducking mechanism by throwing a baseball at a metal disk mounted in front of a 1-m square backboard. The volunteer gets wet even if only the edge of the baseball hits the disk. The radius of a baseball is 3.6 cm and the radius of the disk is 8 cm. What is the probability that a baseball that hits the backboard at a random point will hit the disk?

7. Amy made a tape recording of a choir rehearsal. The recording began 21 minutes into the 60-minute tape and lasted 8 minutes. Later she inadvertently erased a 15-minute segment somewhere on the tape.
   a. Make a drawing showing the possible starting times of the erasure. Explain how Amy knows that the erasure did not start after the 45-minute mark.
   b. Make a drawing showing the starting times of the erasure that would erase the entire choir rehearsal. Find the probability that the entire rehearsal was erased.

8. Kimi has a 4-in. straw and a 6-in. straw. She wants to cut the 6-in. straw into two pieces so that the three pieces form a triangle.
   a. If she cuts the straw to get two 3-in. pieces, can she form a triangle?
   b. If the two pieces are 1 in. and 5 in., can she form a triangle?
   c. If Kimi cuts the straw at a random point, what is the probability that she can form a triangle?

9. **Archery** An archery target with a radius of 61 cm has five scoring zones of equal widths. The colors of the zones are gold, red, blue, black, and white. The width of each colored zone is 12.2 cm and the radius of the gold circle is also 12.2 cm. If an arrow hits the target at a random point, what is the probability that it hits the gold region?

10. During the summer, the drawbridge over the Quisquam River is raised every half hour to allow boats to pass. It remains open for 5 min. What is the probability that a motorist arriving at the bridge will find it raised?

11. **Astronomy** Meteorites (mostly dust-particle size) are continually bombarding Earth. The radius of Earth is about 3960 mi. The area of the United States is about 3,679,245 mi^2. What is the probability that a meteorite landing on Earth will land in the United States?

12. a. *Open-ended* Design a dartboard game to be used at a charity fair. Specify the size and shape of the regions of the board.

b. *Writing* Describe the rules for using your dartboard and the prizes that winners receive. Explain how much money you would expect to raise if the game were played 100 times.

13. *Sonar Sub Search* A ship is trying to locate a disabled remote-controlled research submarine. The captain has determined that the submarine is within a 2000 m-by-2000 m square region. The depth of the ocean in this region is about 6000 m. The ship moves to a position in the center of the square region. Its sonar sends out a signal that covers a conical region. What is the probability that the ship will locate the submarine with the first pulse?

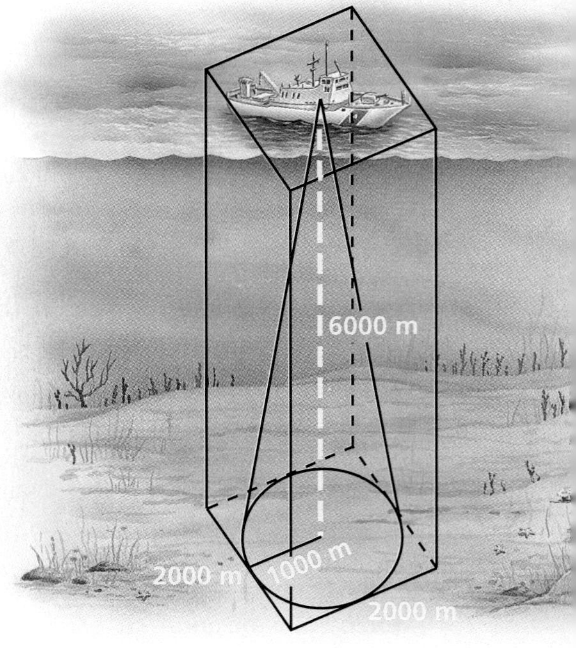

Exercises MIXED REVIEW

Data Analysis **Use the triple bar graph for Exercises 14–16.**

14. Which type of TV show is more popular with Latin American teens than with other teens?

15. In which region is stand-up comedy most popular?

16. *Critical Thinking* How can you tell that each teen who was polled was allowed to choose more than one type of TV show?

What Teens 15–18 Like to Watch on TV by Region of the World

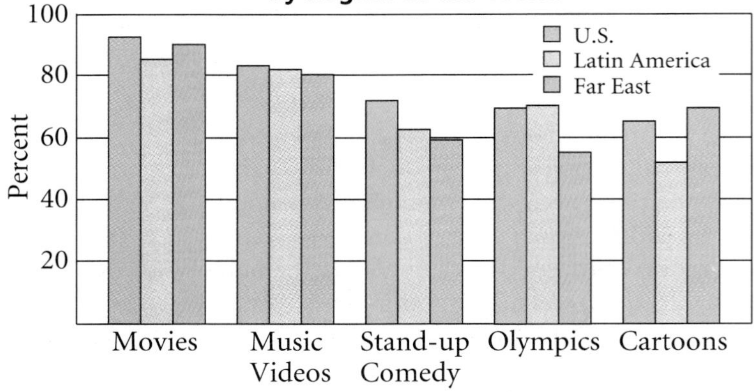

Find the perimeter of each figure.

17.

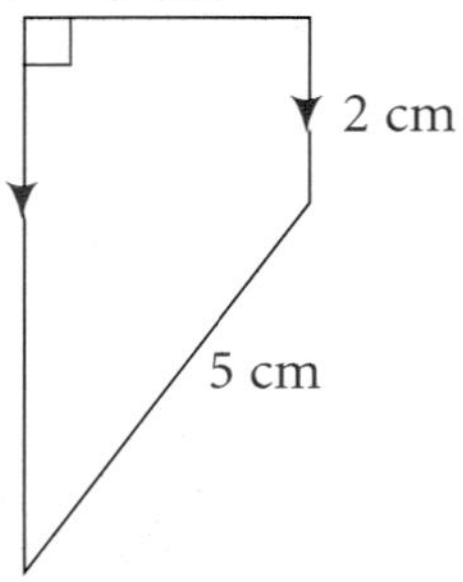

18.

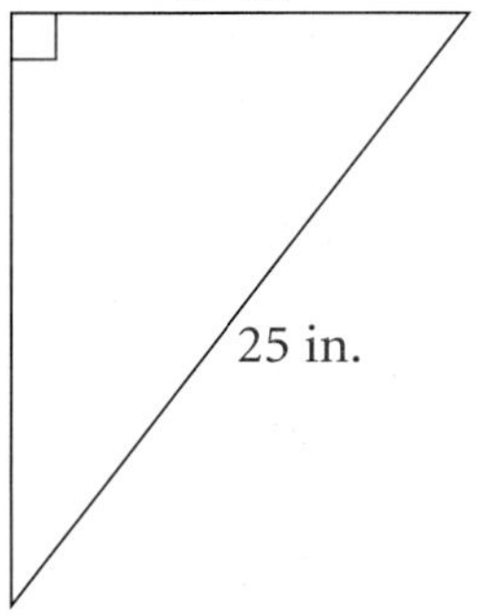

SELF ASSESSMENT

PORTFOLIO

For your portfolio, select one or two items from your work for this chapter. Some possibilities:

- corrected work
- work that you think best shows the things you did

Explain why you included each selection.

Finishing the Chapter Project

The Place is Packed

Find Out questions on pages 306, 314, 329, and 335 should help you complete your project. Design and construct your own package for a product. Specify the dimensions, surface area, amount and type of packaging material used, and volume of the package. Justify your design with mathematical and economic arguments.

Reflect and Revise

Ask a classmate to review your project with you. Together, check that your package design is complete, your diagrams and explanations clear, and your information accurate. Have you used geometric terms correctly? Have you considered other possible designs? Is the display attractive, organized, and comprehensive? Revise your work as needed.

Follow Up

Find pictures of packaged products in newspaper or magazine advertisements. Identify the shape of each package. Give possible reasons why the manufacturer chose each package design. Display your work on a poster.

For More Information

Botersman, Jack. *Paper Capers*. New York: Henry Holt and Company, 1986.

Davidson, Patricia, and Robert Willcutt. *Spatial Problem Solving with Paper Folding and Cutting*. New Rochelle, New York: Cuisenaire Company of America, 1984.

Pearce, Peter. *Structure in Nature Is a Strategy for Design*. Cambridge, Massachusetts: MIT Press, 1978.

Shell Centre for Mathematical Education. *Be a Paper Engineer*. Essex, England: Longman Group UK Limited, 1988.

CHAPTER 6

Wrap Up

Key Terms

altitude (pp. 309, 310, 316, 318)
bases (pp. 309, 316)
center (p. 337)
circumference of a sphere (p. 340)
composite space figure (p. 344)
cone (p. 318)
cube (p. 310)
cylinder (p. 310)
edge (p. 302)
faces (p. 302)
geometric probability (p. 348)
great circle (p. 340)
height (pp. 309, 310, 316, 318, 331)
hemisphere (p. 340)
lateral area (pp. 309, 311, 316)
lateral face (pp. 309, 316)
net (p. 302)
oblique cylinder (p. 310)
oblique prism (p. 309)
polyhedron (p. 302)
prism (p. 309)
pyramid (p. 316)
regular pyramid (p. 316)
right cone (p. 318)
right cylinder (p. 310)
right prism (p. 309)
slant height (pp. 316, 318)
sphere (p. 337)
surface area (pp. 309, 311, 316)
vertex (pp. 302, 316)
volume (p. 324)

How am I doing?

- State three ideas from this chapter that you think are important. Explain your choices.
- Describe how to find the volumes and surface areas of different space figures.

Space Figures and Nets — 6-1

A **polyhedron** is a three-dimensional figure whose surfaces are polygons. The polygons are **faces** of the polyhedron. An **edge** is a segment that is the intersection of two faces. A **vertex** is a point where edges intersect. A **net** is a two-dimensional pattern that folds to form a three-dimensional figure.

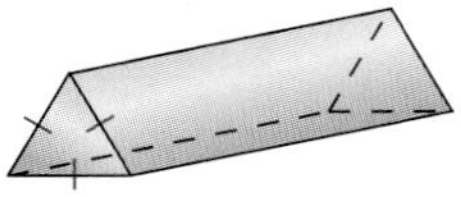

1. *Open-ended* Draw a net for the space figure at the right.
2. *Standardized Test Prep* Which figure has the most edges?
   - **A.** rectangular prism
   - **B.** hexagonal prism
   - **C.** pentagonal prism
   - **D.** octagonal prism
   - **E.** cannot be determined from the information given

Match each three-dimensional figure with its net.

3. 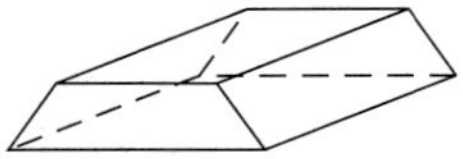

4. 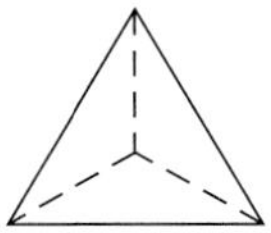

5. 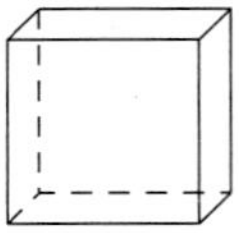

A. 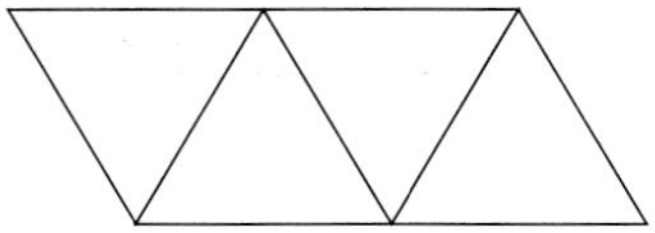

B. 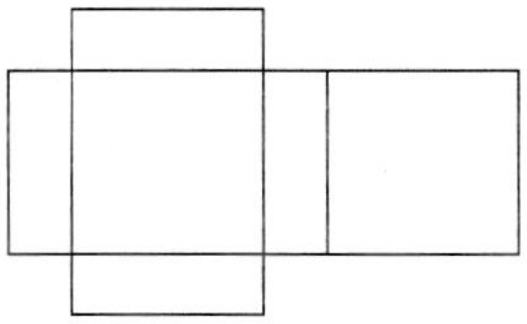

C. 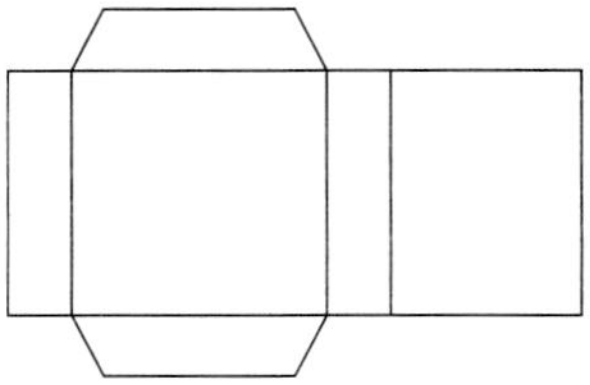

Surface Areas of Prisms and Cylinders 6-2

The **lateral area** of a **right prism** is the product of the perimeter of the base and the height. The **surface area** of a prism is the sum of the lateral area and the areas of the two bases.

L.A. = ph
S.A. = L.A. + $2B$

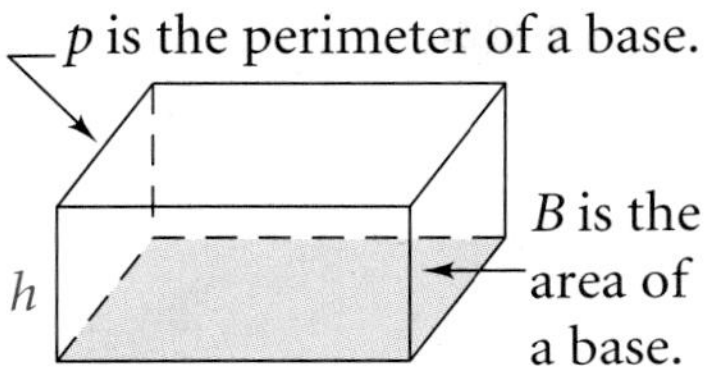

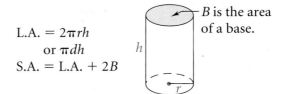

The **lateral area** of a **right cylinder** is the product of the circumference of the base and the height of the cylinder. The **surface area** is the sum of the lateral area and the areas of the two bases.

L.A. = $2\pi rh$
or πdh
S.A. = L.A. + $2B$

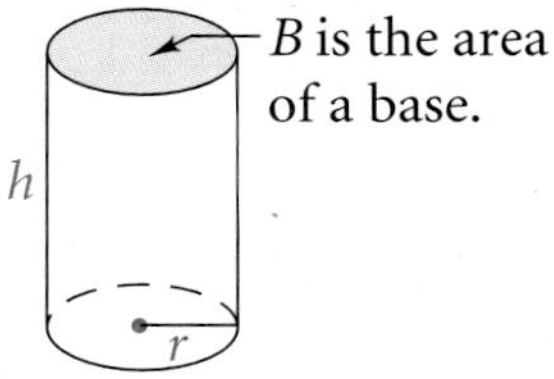

Find the surface area of each figure. You may leave answers in terms of π.

6.

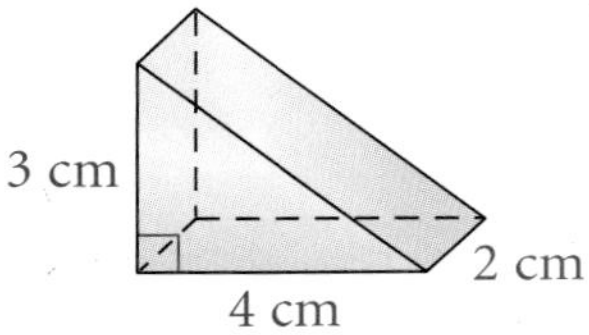

7.

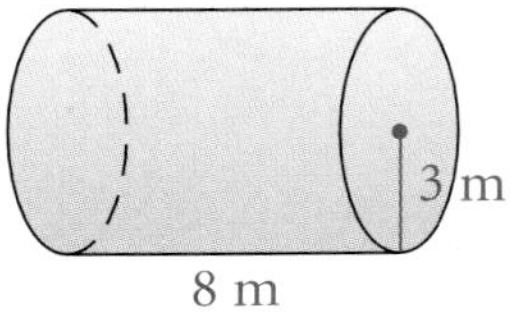

8.

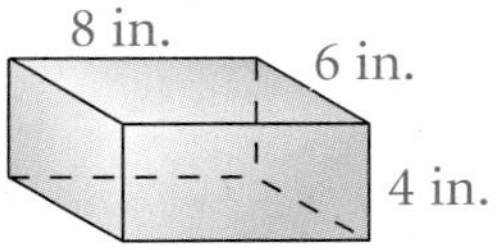

9.

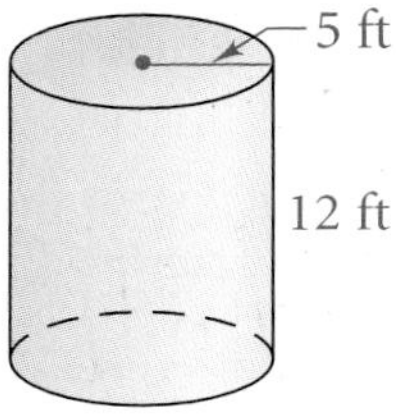

Surface Areas of Pyramids and Cones 6-3

The **lateral area** of a **regular pyramid** is half the product of the perimeter of the base and the slant height. The **surface area** is the sum of the lateral area and the area of the base.

L.A. = $\frac{1}{2}p\ell$
S.A. = L.A. + B

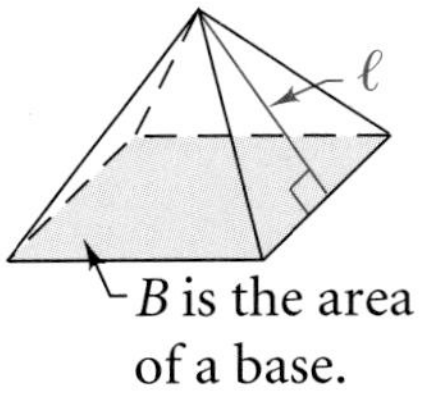

The **lateral area** of a **right cone** is half the product of the circumference of the base and the slant height. The **surface area** is the sum of the lateral area and the area of the base.

L.A. = $\pi r\ell$
S.A. = L.A. + B

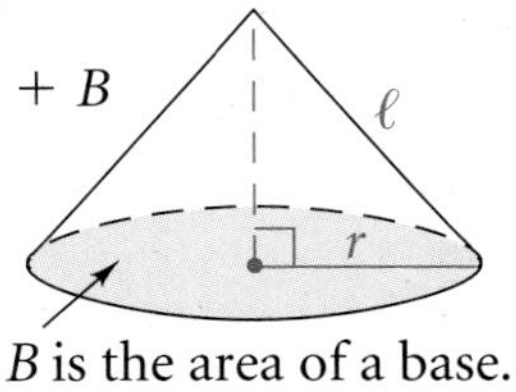

Calculator **Find the lateral area and surface area of each figure. When an answer is not a whole number, round to the nearest tenth.**

10.

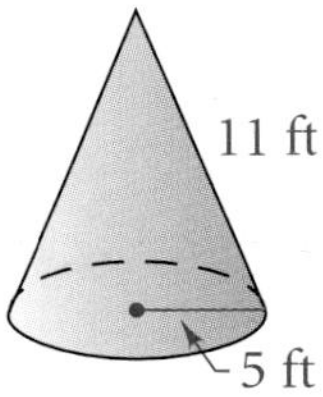

11.

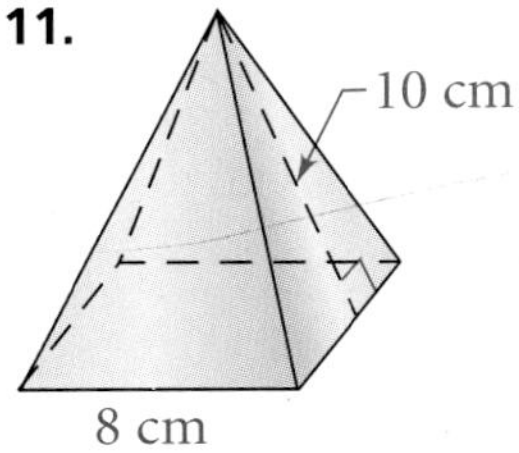

12.

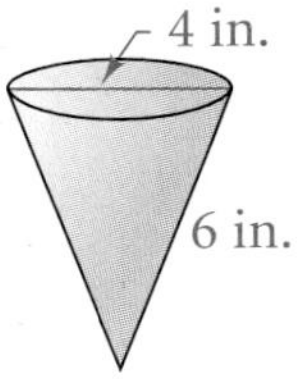

13.

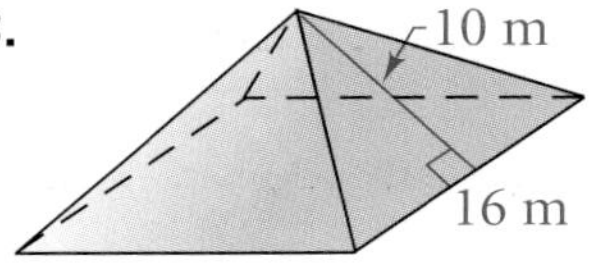

Volumes of Prisms and Cylinders 6-4

The **volume** of a space figure is the space that the figure occupies. Volume is measured in cubic units.

The **volume** of a **prism** is the product of the area of a base and the height of the prism.

The **volume** of a **cylinder** is the product of the area of a base and the height of the cylinder.

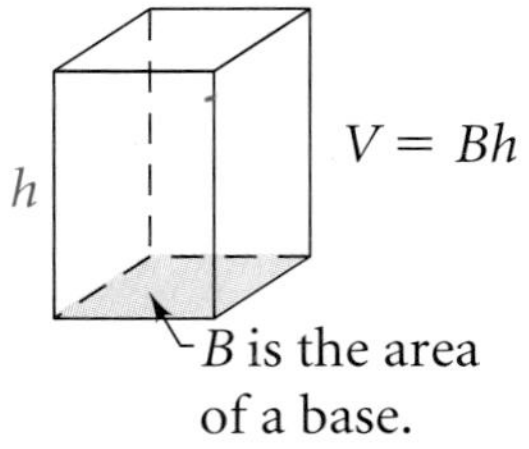

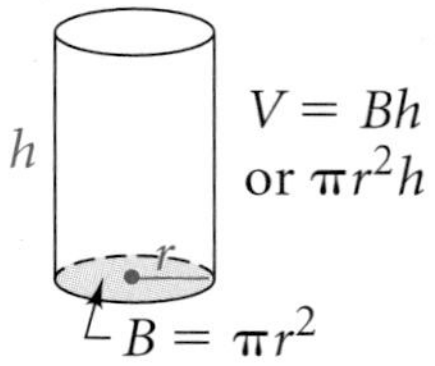

Find the volume of each figure. You may leave answers in terms of π.

14.
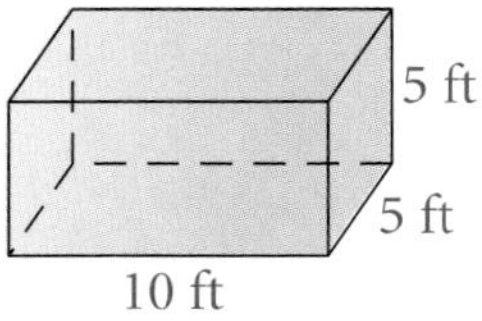

15.
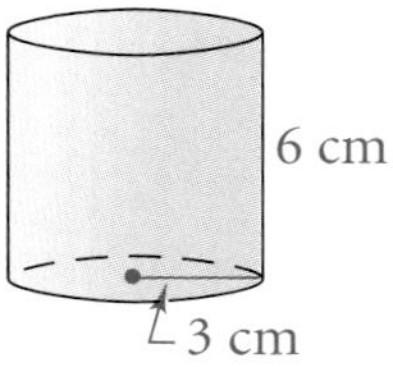

16.
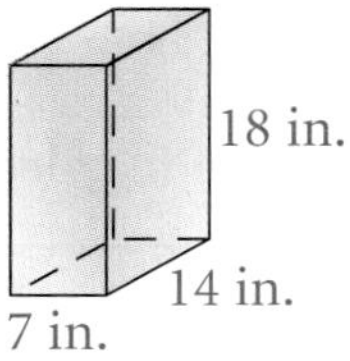

17.
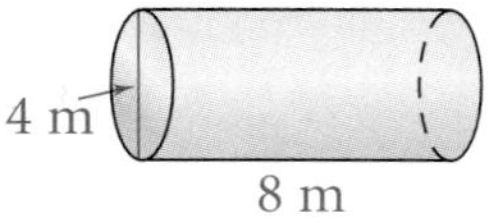

18. Writing Compare finding the volume and the surface area of a prism. What are the similarities and differences?

Volumes of Pyramids and Cones 6-5

The **volume** of a **pyramid** is one third the product of the area of the base and the height of the pyramid.

The **volume** of a **cone** is one third the product of the area of the base and the height of the cone.

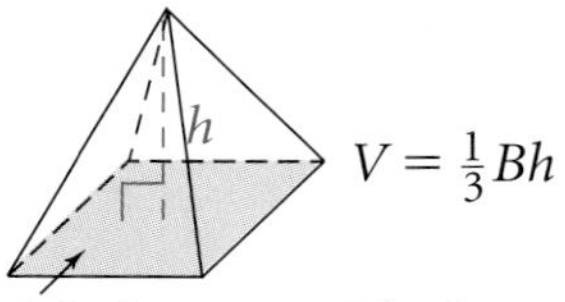

B is the area of the base.

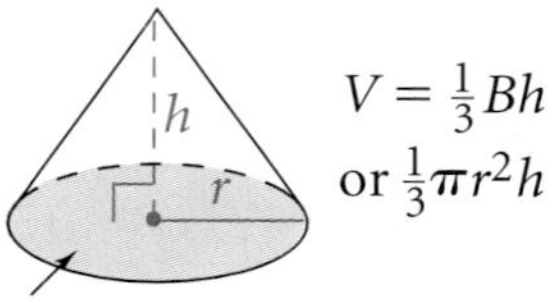

B is the area of the base.

Calculator **Find the volume of each figure. Round to the nearest tenth.**

19.
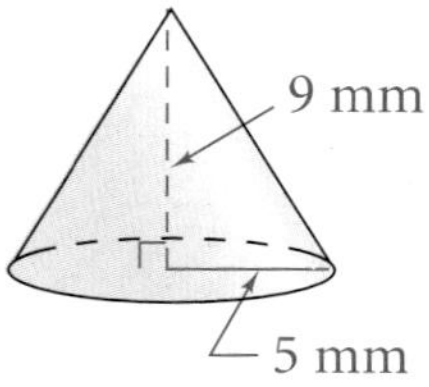

20.
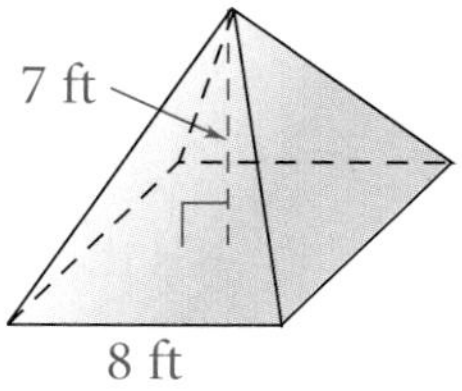

21.
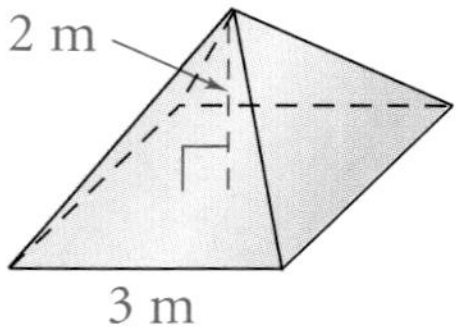

22.
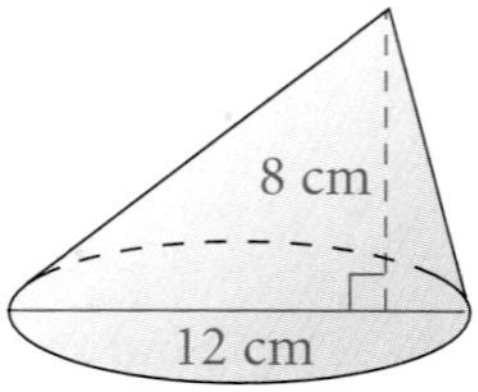

Surface Areas and Volumes of Spheres 6-6

A **sphere** is the set of points in space equidistant from a given point called the **center**.

The **surface area of a sphere** is four times the product of π and the square of the radius of the sphere. The **volume of a sphere** is $\frac{4}{3}$ the product of π and the cube of the radius of the sphere.

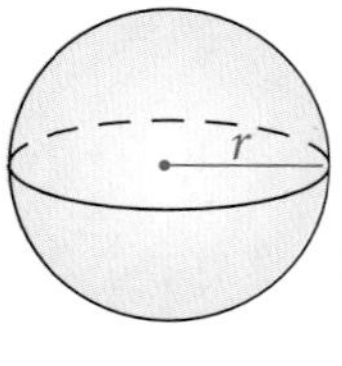

S.A. $= 4\pi r^2$

$V = \frac{4}{3}\pi r^3$

Calculator **Find the surface area and volume of a sphere with the given radius or diameter. Round answers to the nearest tenth.**

23. $r = 5$ in. **24.** $d = 7$ cm **25.** $d = 4$ ft **26.** $r = 0.8$ ft

27. *Sports* The circumference of a lacrosse ball is 8 in. Find its volume to the nearest tenth of a cubic inch.

Composite Space Figures 6-7

A **composite space figure** combines two or more space figures. The volume of a composite figure is the sum of the volumes of the combined figures.

Calculator **Find the volume of each figure. When an answer is not a whole number, round to the nearest tenth.**

28.

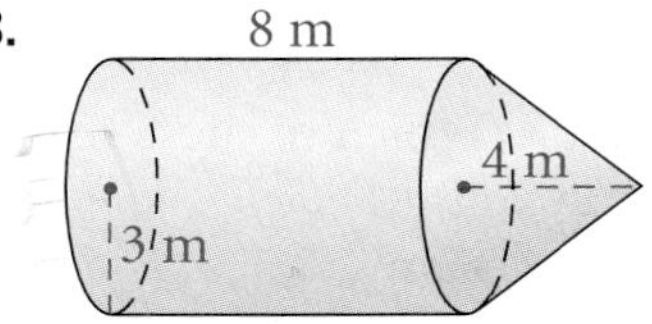

29.

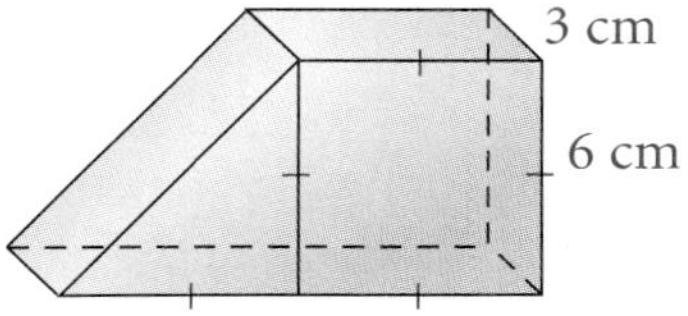

30.

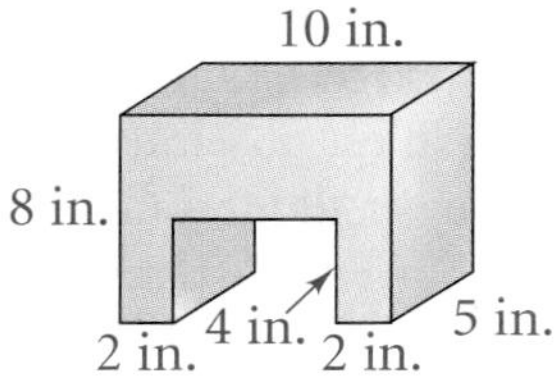

Geometric Probability 6-8

Geometric probability uses geometric figures to represent occurrences of events. You can use a segment model or an area model. Compare the part that represents favorable outcomes to the whole, which represents all outcomes.

Darts are thrown at random at each of the boards shown. If a dart hits the board, find the probability that it will land in the shaded area.

31.

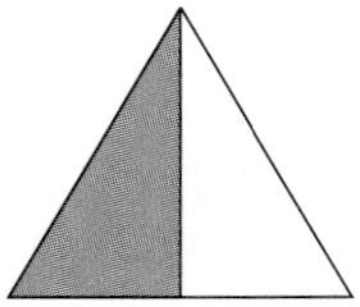

32.

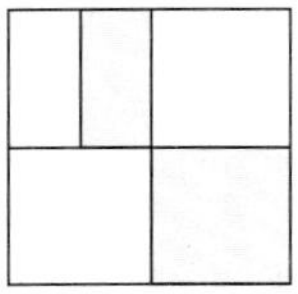

33.

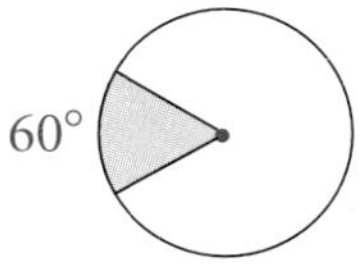

34. *Critical Thinking* If you are modeling probability with a segment, does the length of the segment matter? Explain.

Getting Ready for... CHAPTER 7

Use the rectangular prism at the right.

35. List three pairs of parallel segments.

36. List three pairs of perpendicular segments.

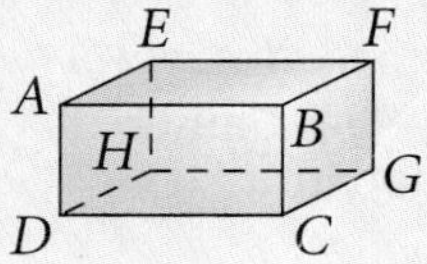

CHAPTER 6

Assessment

Draw a net for each figure. Label the net with appropriate dimensions.

1.

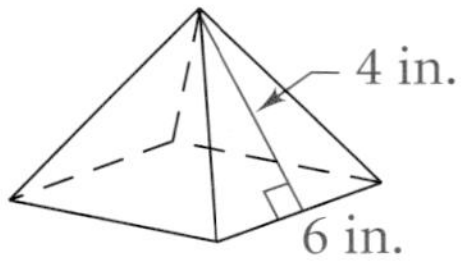

2.

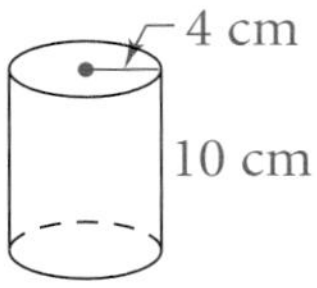

3. a. *Aviation* The "black box" information recorder on an airplane is a rectangular prism. The base is 15 in. by 8 in., and it is 15 in. to 22 in. tall. What are the largest and smallest possible volumes for the box?

b. Newer flight data recorders are smaller and record more data. A new recorder is 8 in. by 8 in. by 13 in. What is its volume?

Calculator **Find the volume and surface area of each figure. When an answer is not a whole number, round to the nearest tenth.**

4.

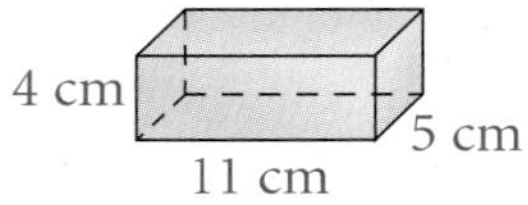

5.

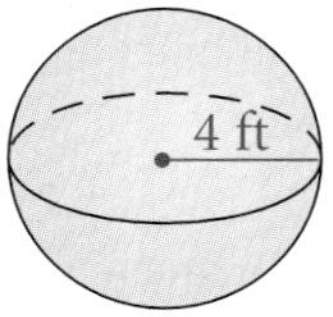

6.

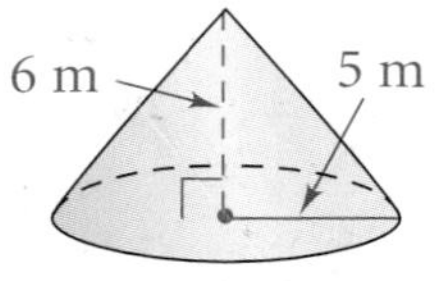

7.

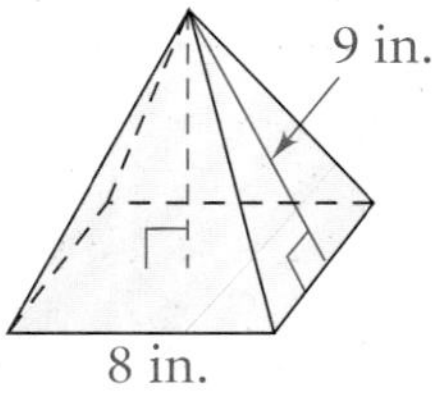

8.

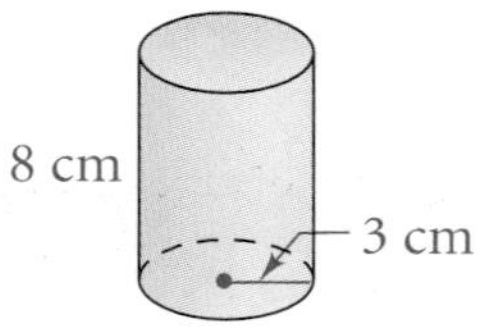

9.

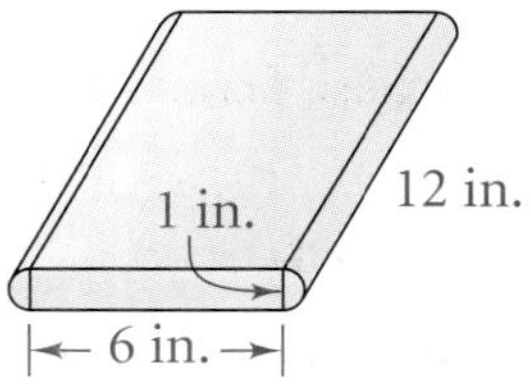

10. *Open-ended* Draw two different space figures that have a volume of 100 in.^{3}. Label the dimensions of each figure.

11. a. Describe the figure that is formed when the triangle is rotated 360° about the y-axis.

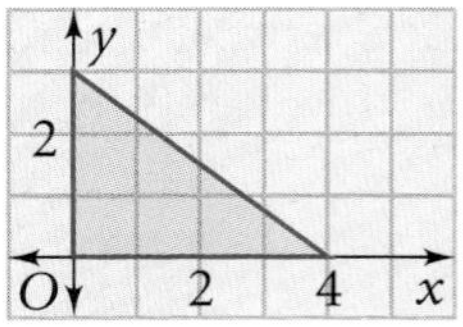

b. What is its volume and lateral area in terms of π?

12. How many gallons of paint do you need to paint the walls of a bedroom? The floor is 12 ft by 15 ft and the walls are 7 ft high. One gallon of paint covers about 450 square feet.

13. *Standardized Test Prep* Which of these space figures has the greatest volume?

A. cube with an edge of 5 cm
B. cylinder with radius 4 cm and height 4 cm
C. pyramid with a square base with sides of 6 cm and height 6 cm
D. cone with radius 4 cm and height 9 cm
E. rectangular prism with a 5 cm-by-5 cm base and height 6 cm

14. *Writing* Describe a real-world situation in which you would use the volume of an object. Then describe another situation in which you would use the lateral area of an object.

15. a. *Estimation* What space figures can you use to approximate the shape of the spring water bottle?

b. Estimate the volume of the bottle.

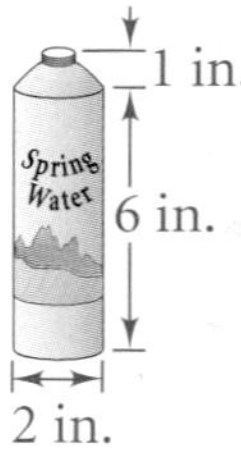

16. *Probability* Every 20 minutes from 4:00 P.M. to 7:00 P.M., a commuter train crosses Main Street. For three minutes a gate stops cars from passing as the train goes by. What is the probability that a motorist driving by during this time will have to stop at the train crossing?

17. *Probability* What is the probability that a dart tossed randomly at this board will hit the shaded area?

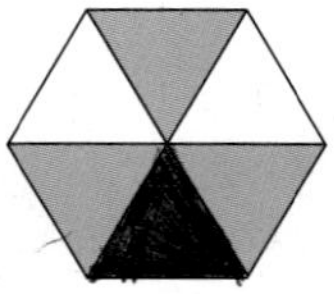

CHAPTER 6 Cumulative Review

For Exercises 1–8, choose the correct letter.

1. Which are rotations of the figure at the right?

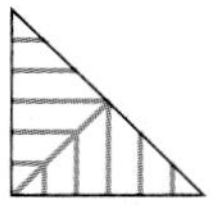

I. 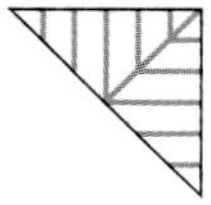

II. 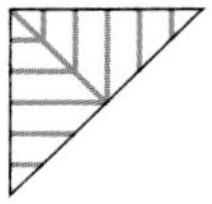

III. 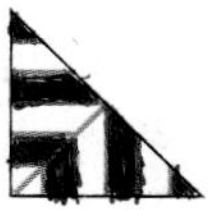

IV. 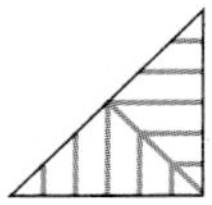

A. I only **B.** IV only **C.** II and IV
D. I, III and IV **E.** I, II, III, and IV

2. What is the volume of the prism?

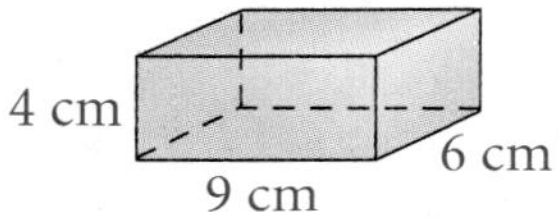

A. 228 cm^3 **B.** 216 cm^3 **C.** 76 cm^3
D. 19 cm^3 **E.** none of the above

3. What is the circumference of a circle with radius 9?
A. 4.5π **B.** 9π **C.** 18π
D. 81π **E.** none of the above

4. Which figure is *not* a convex polygon?

A. 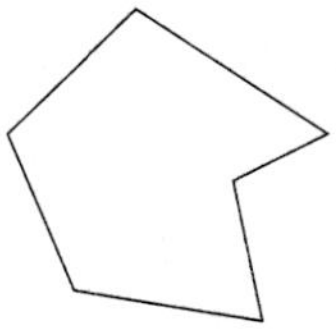

B. 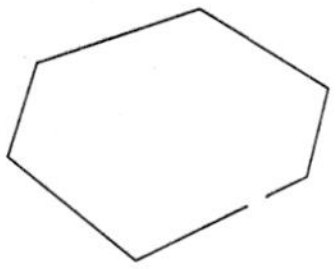

C. 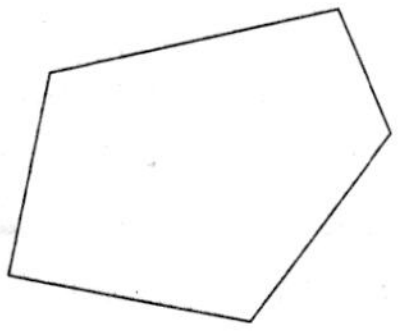

D. 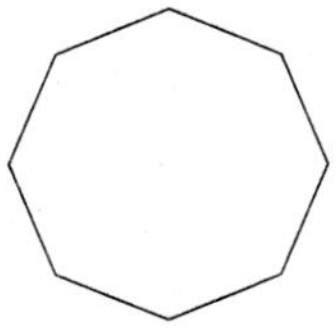

E. none of the above

5. The lengths of the hypotenuse and one leg of a right triangle are 10 and 15. Find the length of the other leg to the nearest whole number.
A. 8 **B.** 9 **C.** 10 **D.** 11 **E.** 12

6. What is the area of the trapezoid?

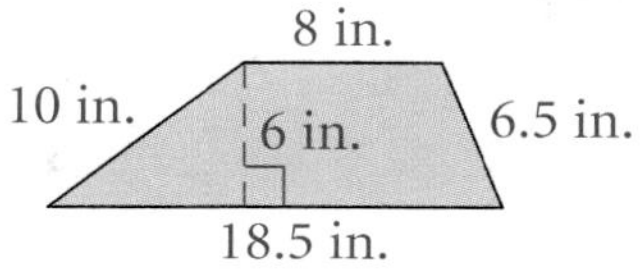

A. 75 $in.^2$ **B.** 43 $in.^2$ **C.** 79.5 $in.^2$
D. 159 $in.^2$ **E.** none of the above

Compare the boxed quantity in Column A with the boxed quantity in Column B. Choose the best answer.

A. The quantity in Column A is greater.
B. The quantity in Column B is greater.
C. The two quantities are equal.
D. The relationship cannot be determined on the basis of the information supplied.

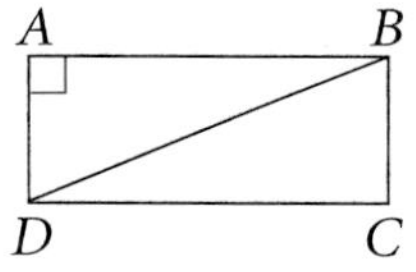

	Column A	Column B
7.	$AB^2 + AD^2$	BD^2
8.	the x-coordinate of the point halfway between (3, 6) and (−5, 2)	the y-coordinate of the point halfway between (9, 1) and (7, −3)

Find each answer.

9. **Open-ended** Sketch a figure with line and point symmetry.

10. What is the measure of the angle formed by the minute hand and the hour hand?

CHAPTER

7 Reasoning and Parallel Lines

Relating to the Real World

Take a look around. Chances are, you can see an example of parallel lines from where you are sitting. But how can you be sure the lines you see are parallel? Architects and builders use the basic geometric concepts in this chapter to ensure that lines are indeed parallel.

Lessons

- 7-1 *Parallel Lines and Related Angles*
- 7-2 *Proving Lines Parallel*
- 7-3 *Constructing Parallel and Perpendicular Lines*
- 7-4 *Parallel Lines and Perspective Drawing*
- 7-5 *Exploring Spherical Geometry*

CHAPTER PROJECT

NETWORK NEWS

Like the strands of an immense spider web, unseen cables join your telephone to millions of others. Similarly, your street is part of a vast system of roads enabling you to travel from your house to just about any other house in the country. We live amid telephone, television, and computer networks, networks of highways and hallways, and networks that join airports, bus stops, and people. Though varied in purpose, each of these networks can be modeled geometrically by a collection of connected points.

In your project for this chapter, you will explore some simple networks to see how economic and social issues can affect their design. You will also design your own network for your school or community. You will see that when it comes to solving complex, real-world problems, practicality and mathematics are constant partners.

To help you complete the project:

- **p. 369** *Find Out by Investigating*
- **p. 382** *Find Out by Designing*
- **p. 391** *Find Out by Investigating*
- **p. 396** *Find Out by Modeling*
- **p. 398** *Finishing the Project*

Math ToolboX

Technology

Exploring Parallel Lines and Related Angles

Before Lesson 7-1

Work in pairs or small groups.

Construct

Use geometry software to construct two parallel lines. Make sure that the lines remain parallel when you manipulate them. Construct a point on each line. Then construct the line through these two points. This line is called a *transversal.*

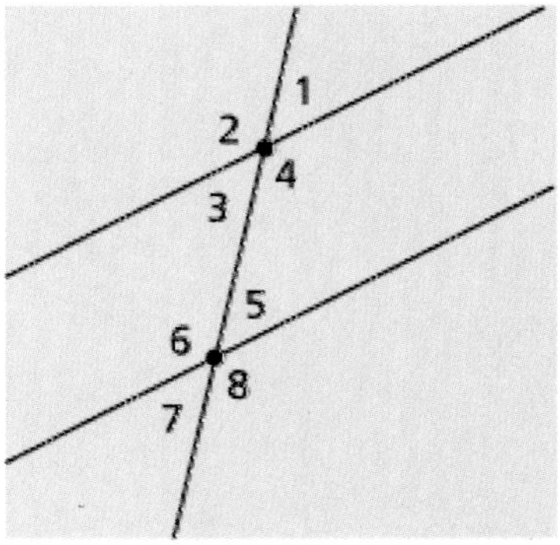

Investigate

Measure each of the eight angles formed by the parallel lines and the transversal and record the measurements. Manipulate the lines and record the new measurements. What relationships do you notice?

Conjecture

When two parallel lines are intersected by a transversal, what are the relationships among the angles formed? Make as many **conjectures** as possible.

Extend

- Use your software to construct three or more parallel lines. Construct a line that intersects one of the lines. Does it intersect the other lines also? If it does, what relationships exist among the angles formed? How many different measures are there?
- Construct two parallel lines and draw a transversal that is perpendicular to one of the parallel lines. What do you discover?

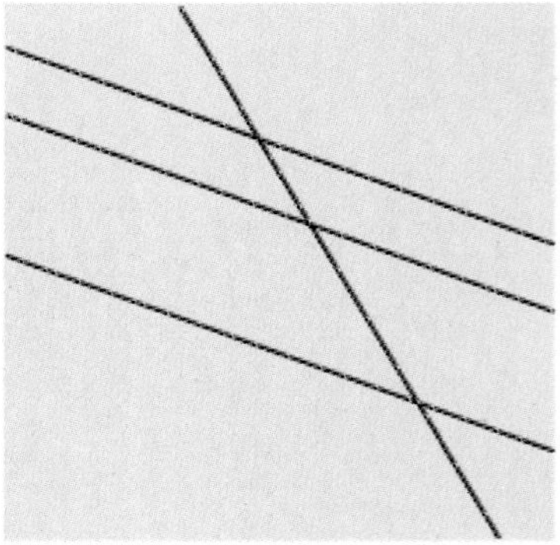

What You'll Learn

- Identifying pairs of angles formed by two lines and a transversal
- Relating the measures of angles formed by parallel lines and a transversal

...And Why

To understand how parallel lines are used in building, city planning, and construction

What You'll Need

- ruler
- protractor

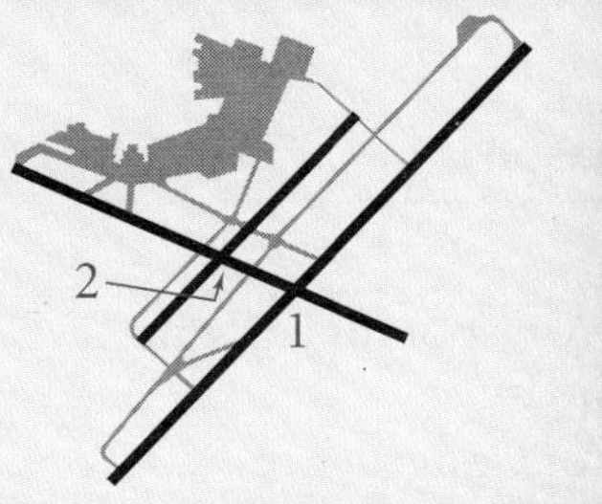

Lafayette Regional Airport
Lafayette, Louisiana

7-1 Parallel Lines and Related Angles

THINK AND DISCUSS

Angles Formed by Intersecting Lines

A **transversal** is a line that intersects two coplanar lines at two distinct points. The diagram shows the eight angles formed by the transversal and the two lines.

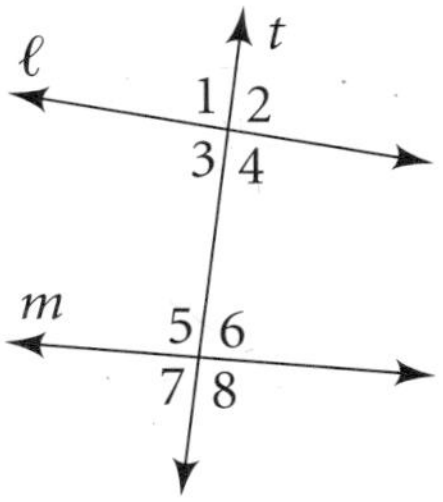

1. **a.** $\angle 3$ and $\angle 6$ are in the *interior* of ℓ and m. Name another pair of angles in the interior of ℓ and m.
   b. Name a pair of angles in the *exterior* of ℓ and m.
2. **a.** $\angle 1$ and $\angle 4$ are on *alternate sides* of the transversal t. Name another pair of angles on alternate sides of t.
   b. Name a pair of angles on the *same side* of t.

You can use the terms in Questions 1 and 2 to describe pairs of angles.

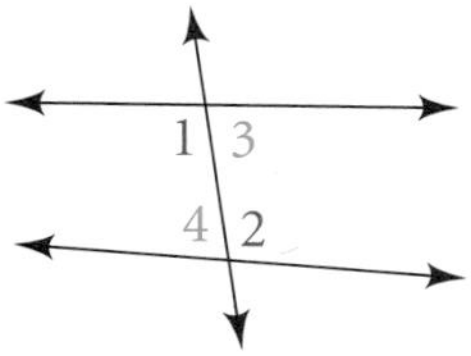

$\angle 1$ and $\angle 2$ are **alternate interior angles**, as are $\angle 3$ and $\angle 4$.

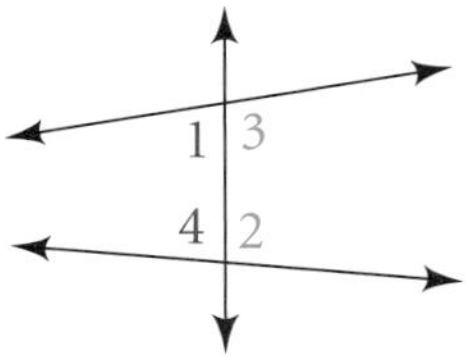

$\angle 1$ and $\angle 4$ are **same-side interior angles**, as are $\angle 2$ and $\angle 3$.

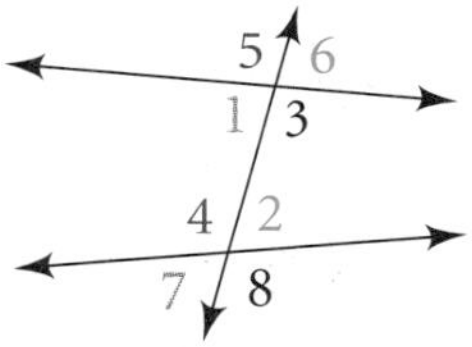

$\angle 5$ and $\angle 4$ are **corresponding angles**, as are $\angle 6$ and $\angle 2$, $\angle 1$ and $\angle 7$, and $\angle 3$ and $\angle 8$.

3. In the diagrams above, $\angle 1$ and $\angle 3$ are not alternate interior angles. What term decribes their positions in relation to each other?

Example 1 **Relating to the Real World**

Aviation In the diagram of Lafayette Regional Airport, the black segments are runways and the grey areas are taxiways and terminal buildings. Classify $\angle 1$ and $\angle 2$ as alternate interior angles, same-side interior angles, or corresponding angles.

$\angle 1$ and $\angle 2$ are corresponding angles.

WORK TOGETHER

Work with a partner.

- Draw two parallel lines using lined paper or the two edges of a ruler. Then draw a transversal that intersects the two parallel lines.
- Use a protractor to measure each of the eight angles formed. Record the measures on your drawing.

4. Make **conjectures** about the measures of corresponding angles, alternate interior angles, and same-side interior angles. Compare your results with those of your classmates.

> ***TECHNOLOGY HINT***
> The Work Together could be done with geometry software.

THINK AND DISCUSS

Angles Formed by Parallel Lines

The results of the Work Together lead to the following postulate and theorems.

Postulate 7-1
Corresponding Angles Postulate

If two parallel lines are cut by a transversal, then corresponding angles are congruent.

If $\ell \parallel m$, then $\angle 1 \cong \angle 5$, $\angle 2 \cong \angle 6$, $\angle 3 \cong \angle 7$, and $\angle 4 \cong \angle 8$.

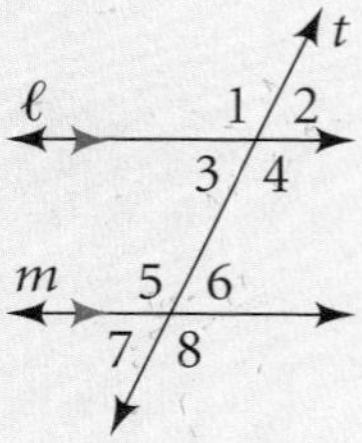

Theorem 7-1
Alternate Interior Angles Theorem

If two parallel lines are cut by a transversal, then alternate interior angles are congruent.

If $\ell \parallel m$, then $\angle 3 \cong \angle 6$ and $\angle 4 \cong \angle 5$.

Theorem 7-2
Same-Side Interior Angles Theorem

If two parallel lines are cut by a transversal, then the pairs of same-side interior angles are supplementary.

If $\ell \parallel m$, then $\angle 3$ and $\angle 5$ are supplementary, and so are $\angle 4$ and $\angle 6$.

Proof of Theorem 7-1

Given: $a \parallel b$

Prove: $\angle 1 \cong \angle 3$

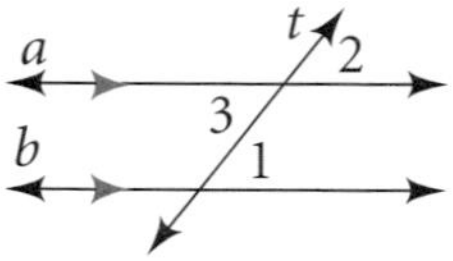

Statements	Reasons
1. $a \parallel b$	**1.** Given
2. $\angle 1 \cong \angle 2$	**2.** If $\parallel$ lines, then corresponding $\angle$s are $\cong$.
3. $\angle 2 \cong \angle 3$	**3.** Vertical angles are $\cong$.
4. $\angle 1 \cong \angle 3$	**4.** Transitive Prop. of Congruence

When writing a proof, it is often helpful to start by writing a plan. The plan should describe how you can reason from the given information to what you want to prove.

Example 2

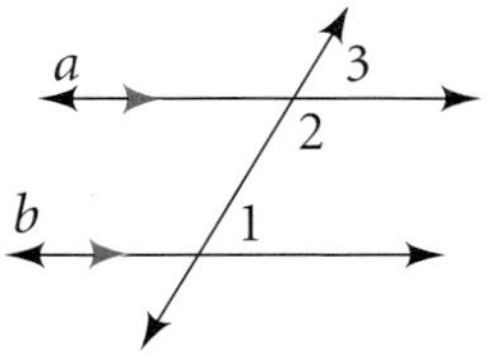

Write a plan for the proof of Theorem 7-2.

Given: $a \parallel b$

Prove: $\angle 1$ and $\angle 2$ are supplementary.

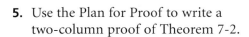

Plan for Proof To show that $m\angle 1 + m\angle 2 = 180$, show that $m\angle 3 + m\angle 2 = 180$ and that $m\angle 1 = m\angle 3$. Then substitute $m\angle 1$ for $m\angle 3$.

5. Use the Plan for Proof to write a two-column proof of Theorem 7-2.
6. **Try This** Find the measure of each angle.

   a. $\angle 1$ b. $\angle 2$ c. $\angle 3$
   d. $\angle 4$ e. $\angle 5$ f. $\angle 6$

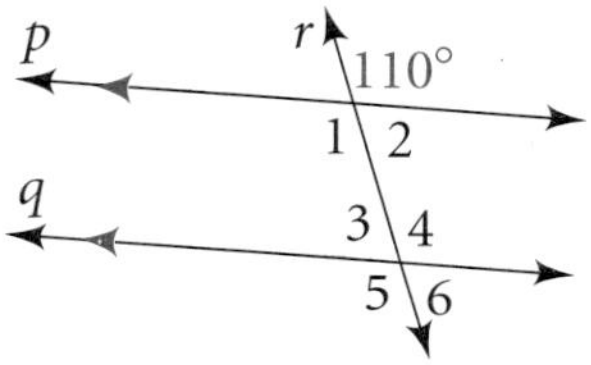

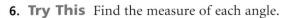

Example 3 Relating to the Real World

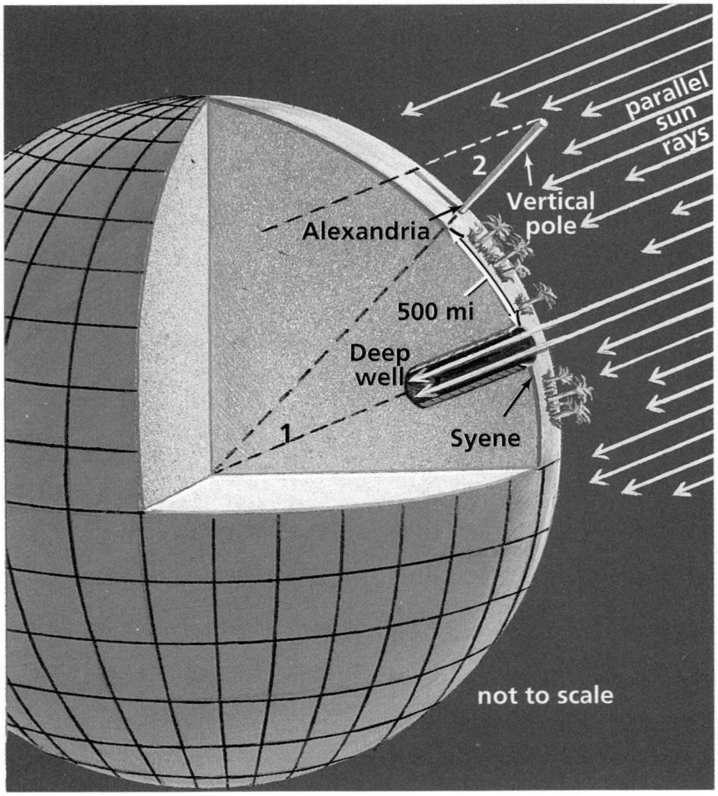

Measuring Earth About 220 B.C., Eratosthenes estimated the circumference of Earth. He achieved this remarkable feat by assuming that Earth is a sphere and that the sun's rays are parallel. He knew that the sun was directly over the town of Syene on the longest day of the year, because sunlight shone directly down a deep well. On that day, he measured the angle of the shadow of a vertical pole in Alexandria, which was 5000 stadia (about 500 miles) north of Syene. How did Eratosthenes know that $\angle 1 \cong \angle 2$? And how could he compute the circumference of Earth knowing $m\angle 1$?

Since $\angle 1$ and $\angle 2$ are alternate interior angles formed by the sun's parallel rays, the angles are congruent. The angle Eratosthenes measured was 7.2°. This is $\frac{1}{50}$ of 360°, so 500 mi is $\frac{1}{50}$ of the circumference of Earth. His estimate of 25,000 mi is very close to the actual value.

Exercises ON YOUR OWN

Classify each pair of angles labeled with the same color as alternate interior angles, same-side interior angles, or corresponding angles.

1.

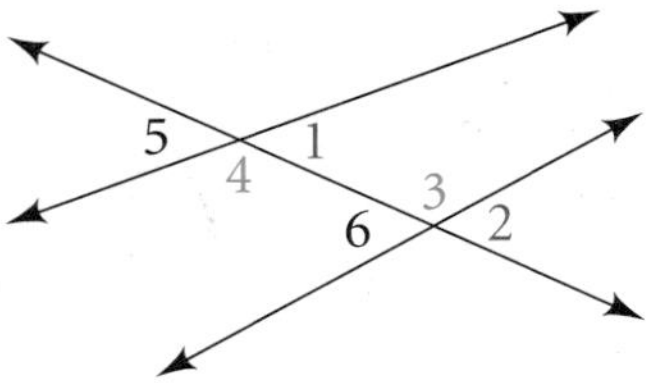

2.

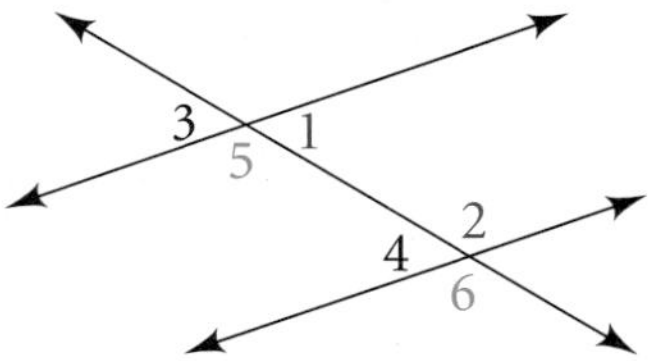

3.

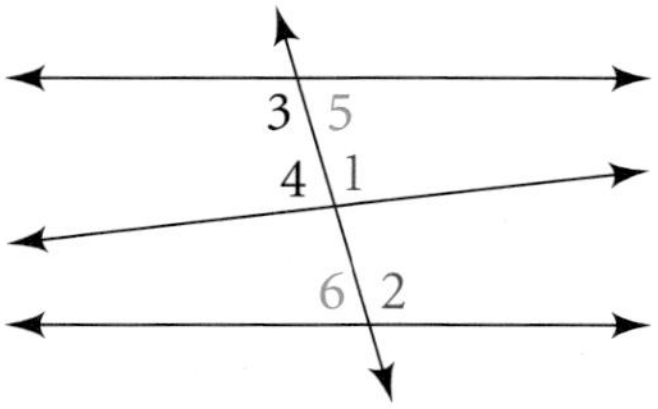

Find $m\angle 1$ and then $m\angle 2$. State the theorems or postulates that justify your answers.

4.

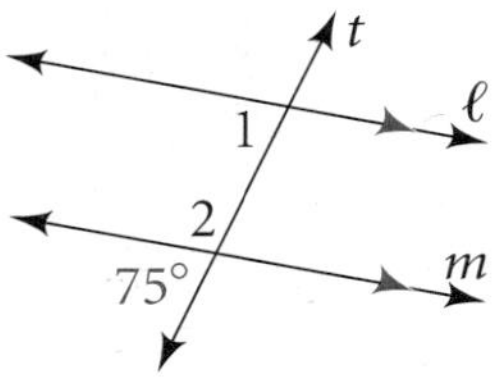

5.

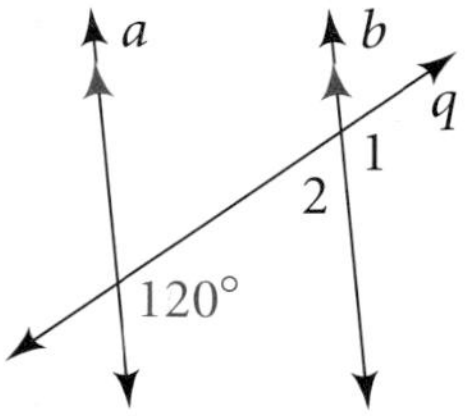

6.

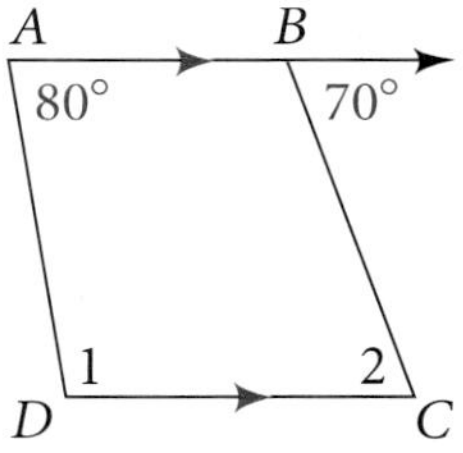

7.

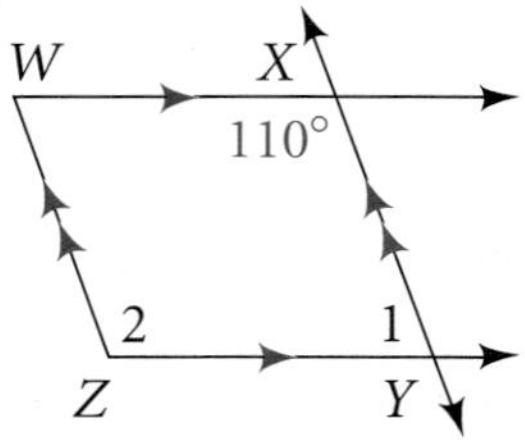

State the theorem or postulate that justifies each statement about the figure at the right.

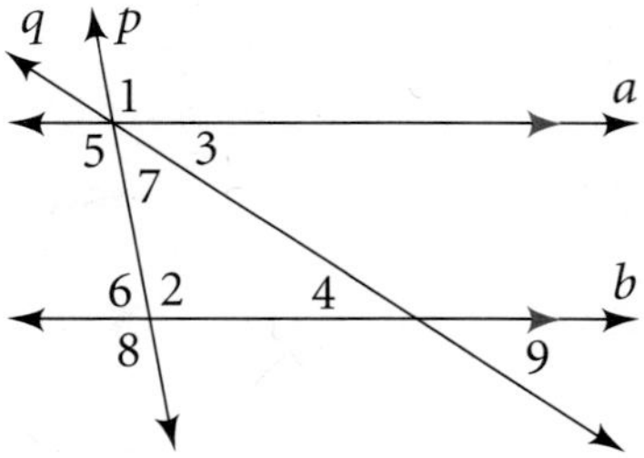

8. $\angle 1 \cong \angle 2$

9. $\angle 3 \cong \angle 4$

10. $m\angle 5 + m\angle 6 = 180$

11. $\angle 5 \cong \angle 2$

12. $\angle 5 \cong \angle 8$

13. $\angle 3 \cong \angle 9$

14. **Open-ended** The letter **Z** illustrates alternate interior angles. Find at least two other letters that illustrate the pairs of angles presented in this lesson. For each letter, show which types of angles are formed.

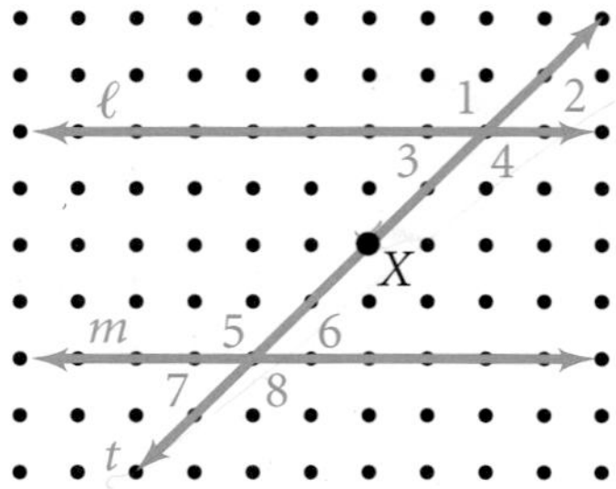

15. a. **Transformational Geometry** Lines ℓ and m are parallel, and line t is a transversal. Under the translation $\langle -4, -4 \rangle$, the image of each of the angles $\angle 1$, $\angle 2$, $\angle 3$, and $\angle 4$ is its __?__ angle.

b. Under a rotation of 180° in point X, the image of each of the angles $\angle 3$, $\angle 4$, $\angle 5$, and $\angle 6$ is its __?__ angle.

16. **Writing** Look up the meaning of the prefix *trans-*. Explain how the meaning of the prefix relates to the word *transversal.*

17. a. **Probability** Suppose that you pick one even-numbered angle and one odd-numbered angle from the diagram. Find the probability that the two angles are congruent.

b. **Open-ended** Write a probability problem of your own based on the diagram. Then solve it.

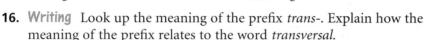

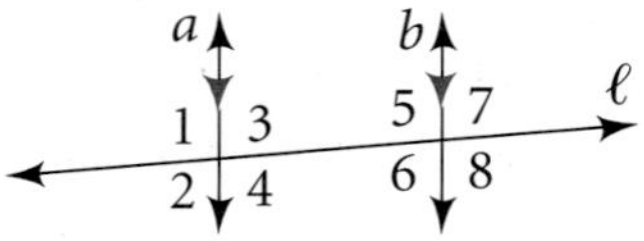

18. **a.** **Architecture** This photograph contains many examples of parallel lines cut by transversals. Given that $a \parallel b$ and $m\angle 1 = 68$, find $m\angle 2$.

b. Given that $c \parallel d$ and $m\angle 3 = 42$, find $m\angle 4$.

For Exercises 19–27, refer to the diagram below right.

19. Name all pairs of corresponding angles formed by the transversal p and lines ℓ and m.

20. Name all pairs of alternate interior angles formed by the transversal ℓ and lines p and q.

Name the relationship between $\angle 2$ and each of the given angles. In each case, state which line is the transversal.

21. $\angle 3$ 22. $\angle 10$ 23. $\angle 7$ 24. $\angle 4$

Find all angles that have the given relationship to $\angle 6$.

25. alternate interior 26. corresponding 27. same-side interior

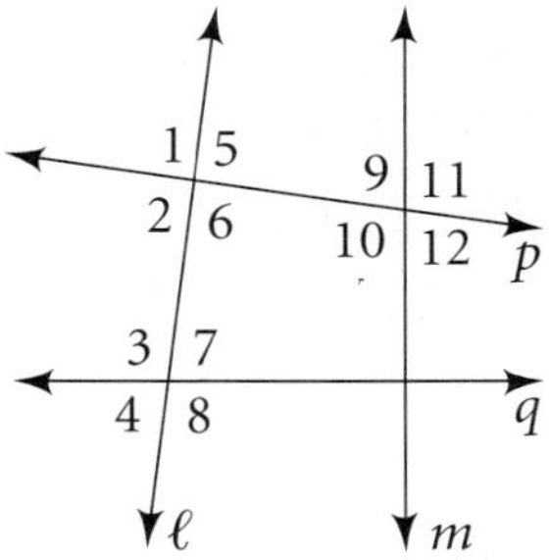

Exs. 19–27

28. **Preparing for Proof** Supply the reasons to complete this proof.

Given: B is the midpoint of $\overline{AC}$.
E is the midpoint of $\overline{AD}$.

Prove: $\angle 1 \cong \angle 2$

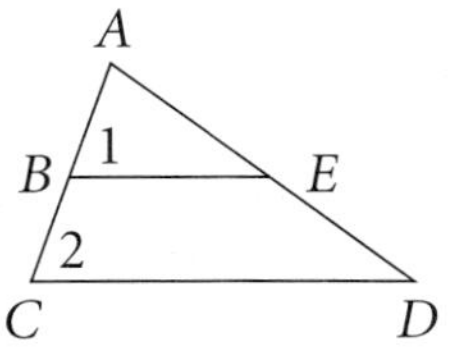

Statements	Reasons
1. B is the midpoint of $\overline{AC}$. E is the midpoint of $\overline{AD}$.	a. ?
2. $\overline{BE} \parallel \overline{CD}$	b. ?
3. $\angle 1 \cong \angle 2$	c. ?

29. **Preparing for Proof** Complete this paragraph proof.

Given: $BERT$ is a parallelogram.

Prove: $\angle 1 \cong \angle 2$

We are given that **a.** __?__. By **b.** __?__, $\overline{ER} \parallel \overline{BT}$. $\angle 1 \cong \angle 2$ because **c.** __?__.

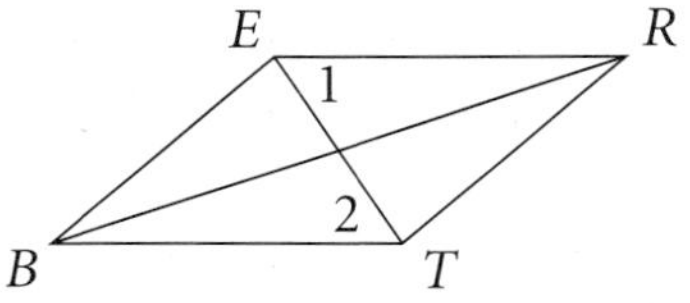

Algebra **Find the values of the variables.**

30.

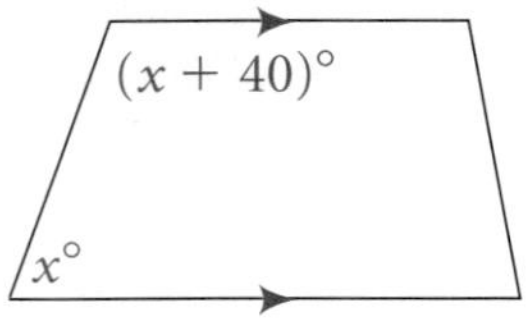

31.

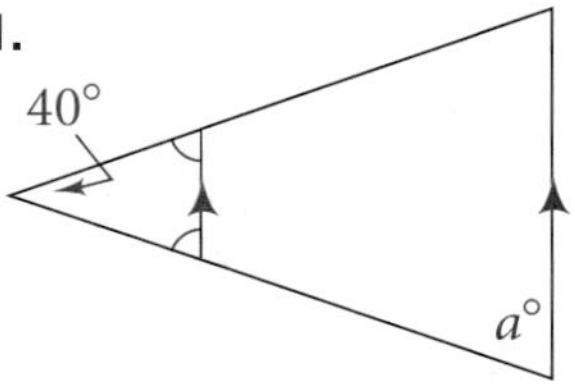

32.

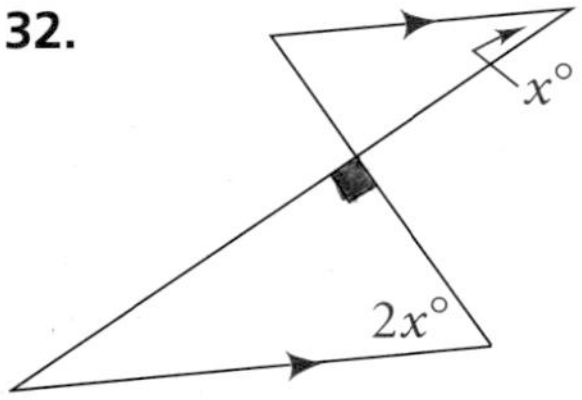

33.

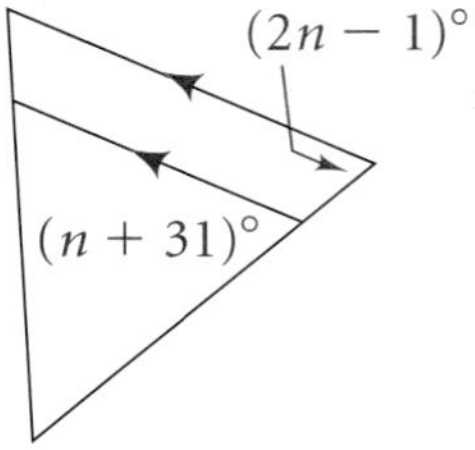

34.

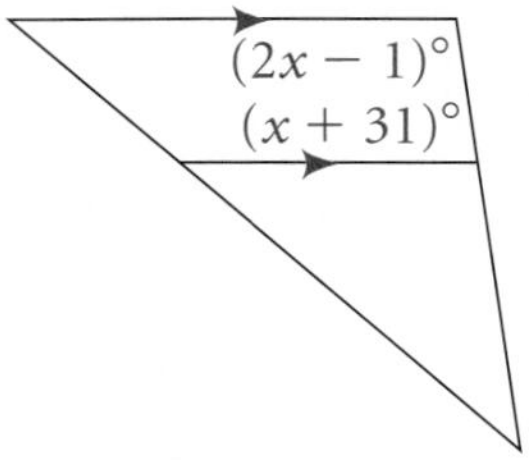

35.

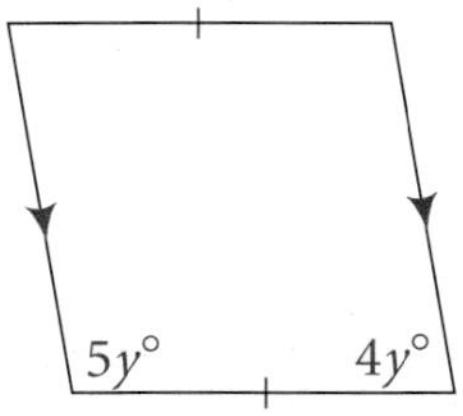

36.

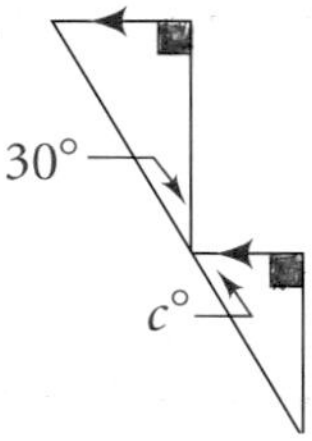

37.

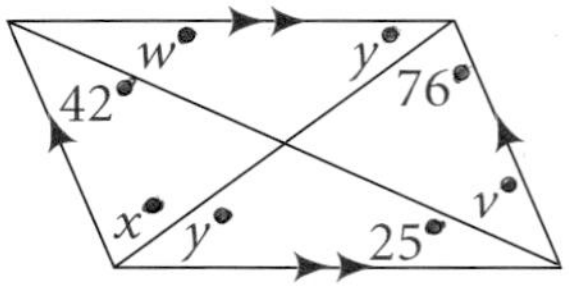

38. $\angle 1$ and $\angle 3$ are *alternate exterior angles.* Write a two-column proof of this statement: If two parallel lines are cut by a transversal, then alternate exterior angles are congruent.

Given: $a \parallel b$

Prove: $\angle 1 \cong \angle 3$

Plan for Proof: Show that $\angle 1 \cong \angle 3$ by showing that both angles are congruent to $\angle 2$.

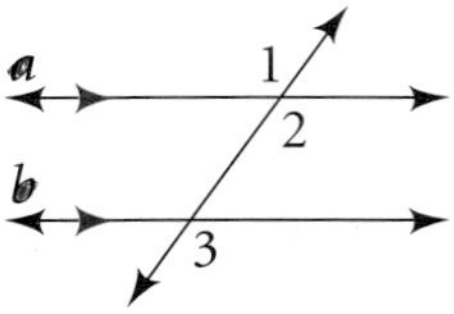

39. **Critical Thinking** $\angle 4$ and $\angle 5$ are *same-side exterior angles.* Make a **conjecture** about the same-side exterior angles formed by two parallel lines and a transversal. Prove your conjecture.

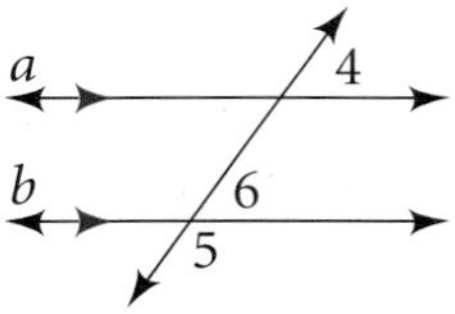

40. Use the diagram to write a paragraph proof of this statement: If a transversal is perpendicular to one of two parallel lines, then it is perpendicular to the other line.

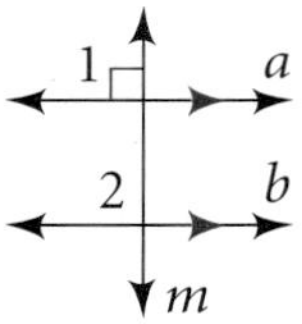

Given: $a \parallel b$, $m \perp a$

Prove: $m \perp b$

41. **Traffic Flow** You are designing the parking lot for a local shopping area, and are considering the two arrangements shown. Give the advantages and disadvantages of each.

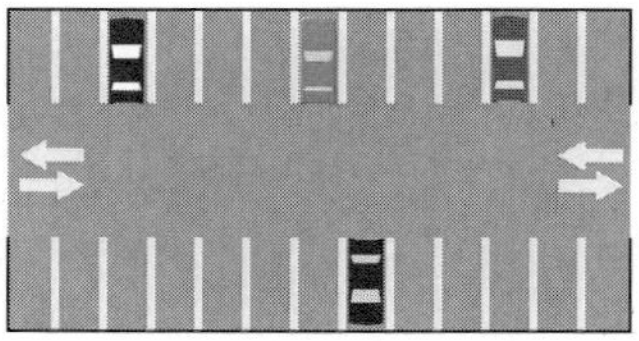

Chapter Project — Find Out by Investigating

Once a month a math coordinator starts at school *A*, drives to the 11 other schools labeled *B* through *L* on the map, and returns to *A*.

- Why must the coordinator backtrack on some of the roads?
- Sketch the shortest route that can be taken. How long is it?
- What additional roads could be built so that the trip could be made without backtracking? Minimize the total length of the new roads. Sketch the shortest route and find its length.

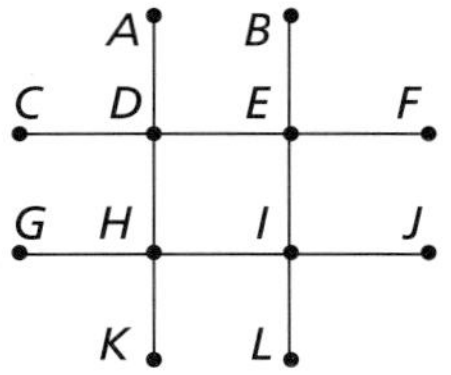

The distance from any school to an adjacent school is 10 mi.

Exercises MIXED REVIEW

Coordinate Geometry **Find the coordinates of the midpoint of $\overline{AB}$.**

42. $A(0, 9), B(1, 5)$ **43.** $A(-3, 8), B(2, -1)$ **44.** $A(10, -1), B(-4, 7)$ **45.** $A(-5, -11), B(2, 6)$

Data Analysis **Use the double bar graph for Exercises 46–48.**

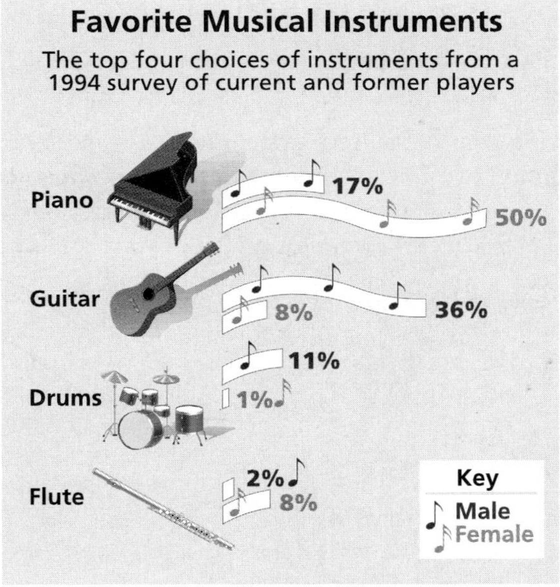

Source: *National Association of Music Merchants*

46. What percent of female players selected the piano as their favorite instrument?

47. Given 50 randomly selected male players, about how many are drummers?

48. Do males or females prefer guitar more?

49. **Sports** The circumference of a regulation basketball is between 75 cm and 78 cm. What are the smallest and largest surface areas a basketball can have? Give your answers to the nearest whole unit.

50. a. **Photography** A regular-sized photo is 3 in. by 5 in. A larger-sized photo is 4 in. by 6 in. What percent more paper is used for a larger-sized photo than for a regular photo?

b. You are making a poster of photos from a class trip. What is the greatest number of regular-sized photos you can fit on a 2 ft-by-3 ft poster? What is the greatest number of larger-sized photos you can fit?

Getting Ready for Lesson 7-2

Write the converse of each conditional statement.

51. If the sky is blue, then there are no clouds in the sky.

52. If ℓ and m are parallel, then corresponding angles $\angle 1$ and $\angle 2$ are congruent.

Math ToolboX

Algebra Review

Systems of Linear Equations

After Lesson 7-1

You can solve a system of equations in two variables by using substitution to create a one-variable equation.

Example 1

Solve the system: $y = 3x + 5$
$y = x + 1$

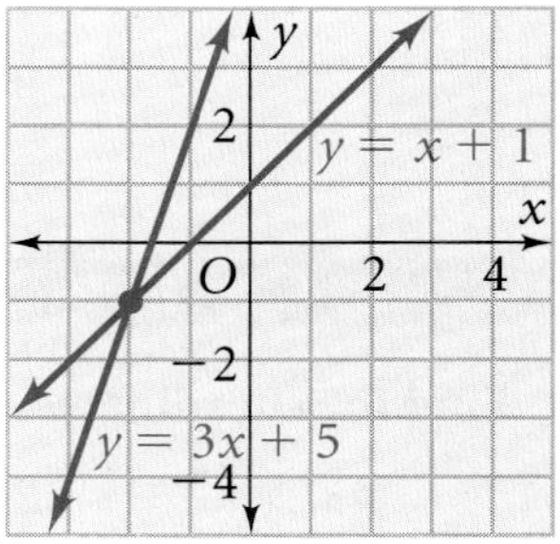

$y = x + 1$ ← Start with one equation.
$3x + 5 = x + 1$ ← Substitute $3x + 5$ for y.
$2x = -4$ ← Solve for x.
$x = -2$

Substitute -2 for x in either equation and solve for y.

$$y = x + 1$$
$$= (-2) + 1 = -1$$

Since $x = -2$ and $y = -1$, the solution is $(-2, -1)$.

The graph of a linear system with *no solution* is two parallel lines, and the graph of a linear system with *infinitely many solutions* is one line.

Example 2

Solve the system: $x + y = 3$
$4x + 4y = 8$

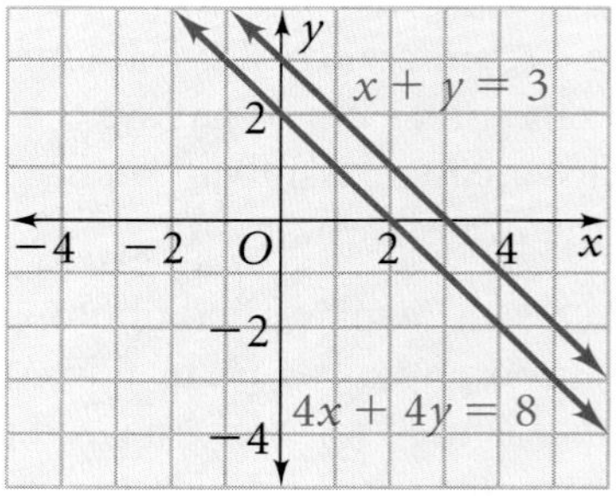

$x + y = 3$
$x = 3 - y$ ← Solve the first equation for x.
$4(3 - y) + 4y = 8$ ← Substitute $3 - y$ for x in the second equation.
$12 - 4y + 4y = 8$ ← Simplify.
$12 = 8$ ← False!

Since $12 = 8$ is a false statement, the system has no solution.

Solve each system of equations.

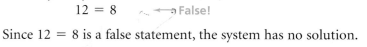

1. $y = x - 4$
$y = 3x + 2$

2. $2x - y = 8$
$x + 2y = 9$

3. $3x + y = 4$
$-6x - 2y = 12$

4. $y = -x + 2$
$2y = 4 - 2x$

5. $y = 2x + 1$
$y = 3x - 7$

6. $x - y = 4$
$3x - 3y = 6$

7-2 Proving Lines Parallel

What You'll Learn

- Recognizing conditions that result in parallel lines
- Writing proofs that involve parallel lines

...And Why

To ensure parallel lines in construction and graphic arts

What You'll Need

- protractor
- ruler
- dot paper

TECHNOLOGY HINT

The Work Together could also be done using geometry software.

WORK TOGETHER

- Draw a segment on dot paper and label it $\overline{AB}$. Draw and label a point C not on $\overleftrightarrow{AB}$. Draw $\overline{BC}$.
- Translate $\angle ABC$ so that the image of B is C. Label the image $\angle A'CC'$.

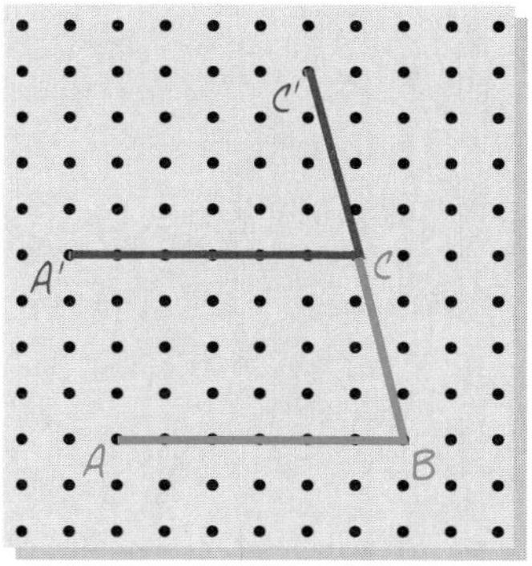

1. What is true of $\angle ABC$ and $\angle A'CC'$? Explain.

2. What appears to be true of $\overleftrightarrow{AB}$ and $\overleftrightarrow{A'C}$?

3. Use your answers to Questions 1 and 2 to complete this statement:
If two lines are cut by a transversal so that a pair of corresponding angles are congruent, then _?_.

THINK AND DISCUSS

In the Work Together, you discovered that the converse of the Corresponding Angles Postulate is true.

Postulate 7-2
Converse of Corresponding Angles Postulate

If two lines are cut by a transversal so that a pair of corresponding angles are congruent, then the lines are parallel.

If $\angle 1 \cong \angle 2$, then $\ell \parallel m$.

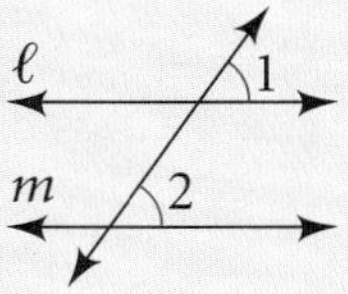

Example 1

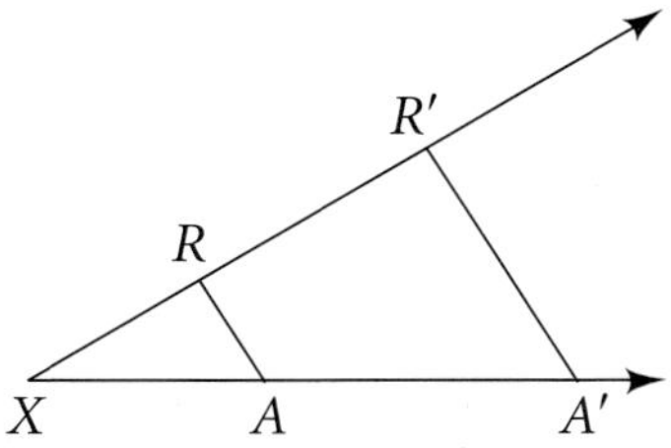

Transformational Geometry $\overline{R'A'}$ is the image of $\overline{RA}$ under a dilation centered at X. Show that $\overline{RA} \parallel \overline{R'A'}$.

Because a dilation is a similarity transformation, $\triangle XRA \sim \triangle XR'A'$. Corresponding angles of similar polygons are congruent, so $\angle RAX \cong \angle R'A'X$. Since $\overrightarrow{XA'}$ is a transversal of $\overline{RA}$ and $\overline{R'A'}$, by the Converse of the Corresponding Angles Postulate, $\overline{RA} \parallel \overline{R'A'}$. ■

In Lesson 7-1, you proved the Alternate Interior Angles Theorem and the Same-Side Interior Angles Theorem by using the Corresponding Angles Postulate. You can prove the converses of these theorems by using the converse of the postulate.

Theorem 7-3
Converse of Alternate Interior Angles Theorem

If two lines are cut by a transversal so that a pair of alternate interior angles are congruent, then the lines are parallel.

If $\angle 1 \cong \angle 2$, then $\ell \parallel m$.

Theorem 7-4
Converse of Same-Side Interior Angles Theorem

If two lines are cut by a transversal so that a pair of same-side interior angles are supplementary, then the lines are parallel.

If $\angle 2$ and $\angle 4$ are supplementary, then $\ell \parallel m$.

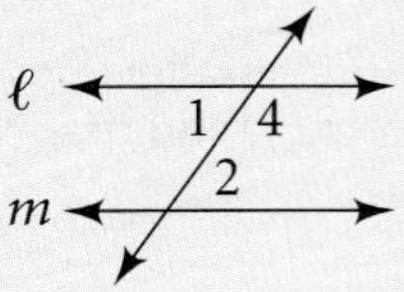

This is a **flow proof** of Theorem 7-3. Arrows show the logical connections between the statements. Reasons are written below the statements.

Flow Proof of Theorem 7-3

Given: $\angle 1 \cong \angle 2$
Prove: $\ell \parallel m$

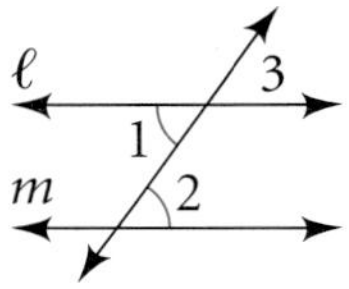

$\angle 1 \cong \angle 2$ — Given → $\angle 3 \cong \angle 2$

$\angle 1 \cong \angle 3$ — Vertical ∠s are ≅. → $\angle 3 \cong \angle 2$

$\angle 3 \cong \angle 2$ — Transitive Prop. of ≅ → $\ell \parallel m$

$\ell \parallel m$ — If ≅ corresponding ∠s, then lines are ∥.

Example 2

Write a paragraph proof.

Given: $\angle BAC$ and $\angle ACD$ are supplementary.
$\angle CDF$ and $\angle DFE$ are supplementary.
Prove: $\overline{AB} \parallel \overline{EF}$

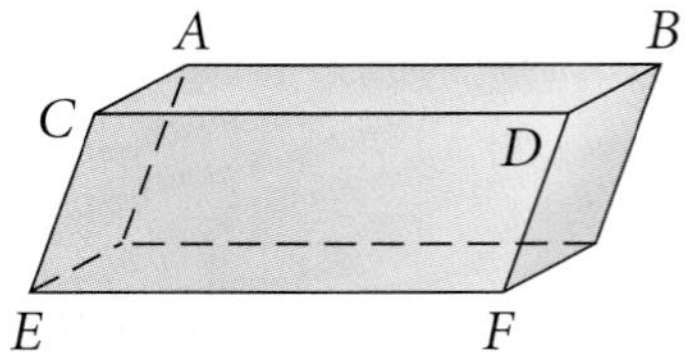

$\angle BAC$ and $\angle ACD$ are supplementary same-side interior angles formed by transversal $\overline{AC}$ with $\overline{AB}$ and $\overline{CD}$. So, $\overline{AB} \parallel \overline{CD}$. Similarly, $\angle CDF$ and $\angle DFE$ are supplementary same-side interior angles formed by transversal $\overline{DF}$ with $\overline{CD}$ and $\overline{EF}$. So, $\overline{CD} \parallel \overline{EF}$. Since $\overline{AB}$ and $\overline{EF}$ are parallel to the same segment, they are parallel to each other (Theorem 2-5). ■

4. **Try This** Write a flow proof for Example 2.

Example 3

What types of quadrilaterals can have angles of 56°, 56°, 124°, and 124°?

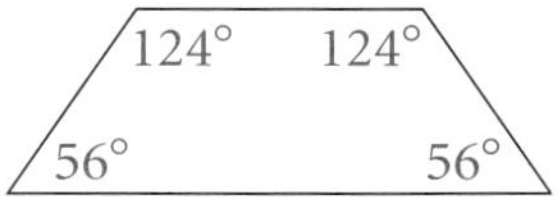

If the angles occur in the given order, then the quadrilateral is a trapezoid. One pair of sides is parallel by the Converse of the Same-Side Interior Angles Theorem.

If the angles occur in the order 124°, 56°, 124°, 56°, then the quadrilateral is a parallelogram. Two pairs of opposite sides are parallel by the Converse of the Same-Side Interior Angles Theorem.

5. **Critical Thinking** What types of quadrilaterals can have angles of 120°, 60°, 150°, and 30°?

Example 4 Relating to the Real World

Woodworking When a frame is made for a painting, a miter box and a backsaw are used to cut framing at 45° angles. Explain why cutting the framing at this angle ensures that opposite sides of the frame will be parallel.

The sum of two 45° angles is a 90° angle. Two 90° angles are supplementary, so by the Converse of the Same-Side Interior Angle Theorem, opposite sides of the frame are parallel.

Exercises ON YOUR OWN

Which lines or segments are parallel? Justify your answer with a theorem or postulate.

1.

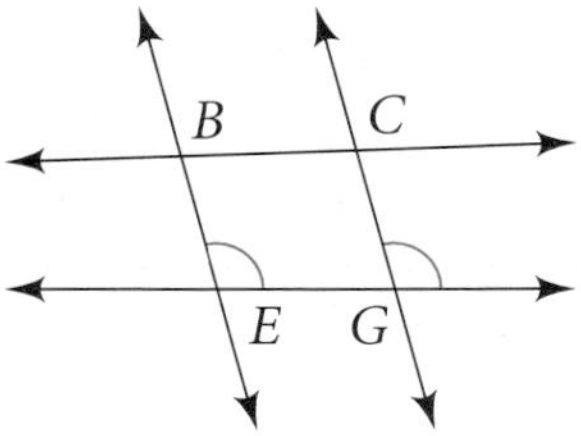

2.

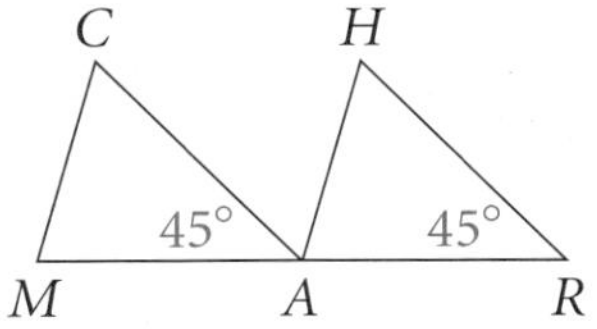

3.

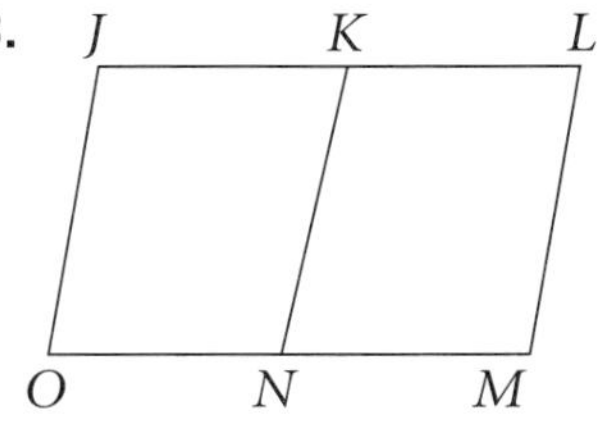

$m\angle J + m\angle L = 180$

4. **Paper cutting** Use the method shown in the diagram to cut out two congruent triangles. Join the two triangles so that they form a parallelogram. Use what you learned in this lesson to **justify** that the figure is indeed a parallelogram.

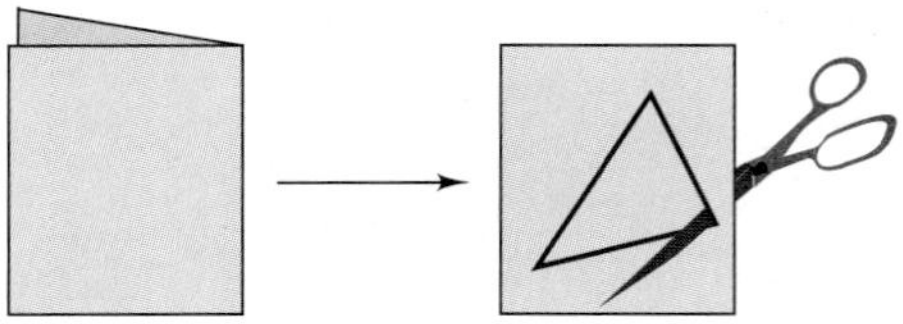

Refer to the diagram. Use the given information to determine which lines, if any, must be parallel. If any lines are parallel, use a theorem or postulate to tell why.

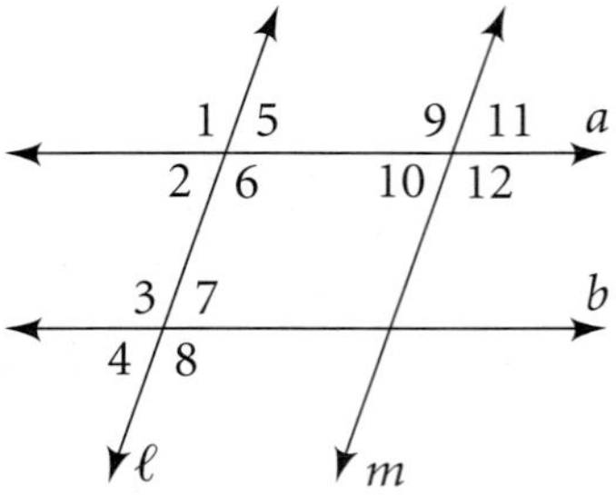

Exs. 5–18

5. $\angle 1 \cong \angle 3$

6. $\angle 1 \cong \angle 6$

7. $\angle 6$ is supplementary to $\angle 7$.

8. $\angle 11 \cong \angle 7$

9. $\angle 5 \cong \angle 10$

10. $m\angle 7 = 70$, $m\angle 9 = 110$

11. $\angle 2$ is supplementary to $\angle 3$.

12. $\angle 8 \cong \angle 6$

13. $\angle 9 \cong \angle 3$

14. $\angle 2 \cong \angle 10$

15. If $\angle 1 \cong \angle 12$, what theorems or postulates can you use to show that $\ell \parallel m$?

16. **Preparing for Proof** Use this Plan for Proof to write a flow proof. Refer to the diagram at the right.

Given: $\ell \parallel m$, $\angle 4 \cong \angle 10$

Prove: $a \parallel b$

Plan for Proof: Show that $a \parallel b$ by showing that the corresponding angles $\angle 2$ and $\angle 4$ are congruent. Show that they are congruent by showing that they are both congruent to $\angle 10$.

Refer to the diagram for Exercises 5–18 to write a flow proof.

17. **Given:** $a \parallel b$, $\angle 7 \cong \angle 10$
Prove: $\ell \parallel m$

18. **Given:** $\ell \parallel m$, $\angle 7$ and $\angle 12$ are supplementary.
Prove: $a \parallel b$

19. **Carpentry** A *T-bevel* is a tool used by carpenters to draw congruent angles. By loosening the locking lever, the carpenter can adjust the angle. Explain how the carpenter knows that the two lines that he has drawn using the T-bevel are parallel.

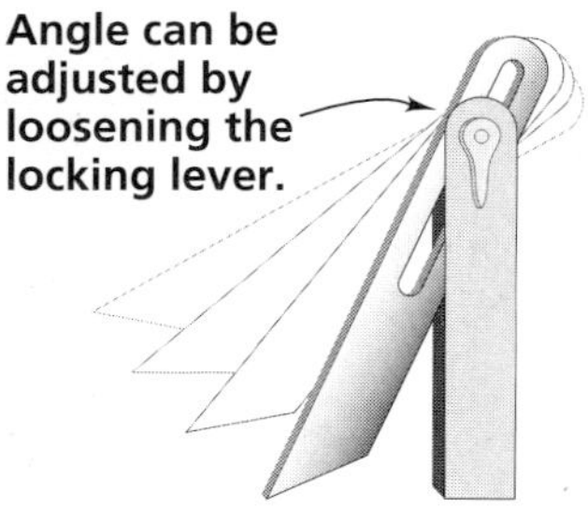

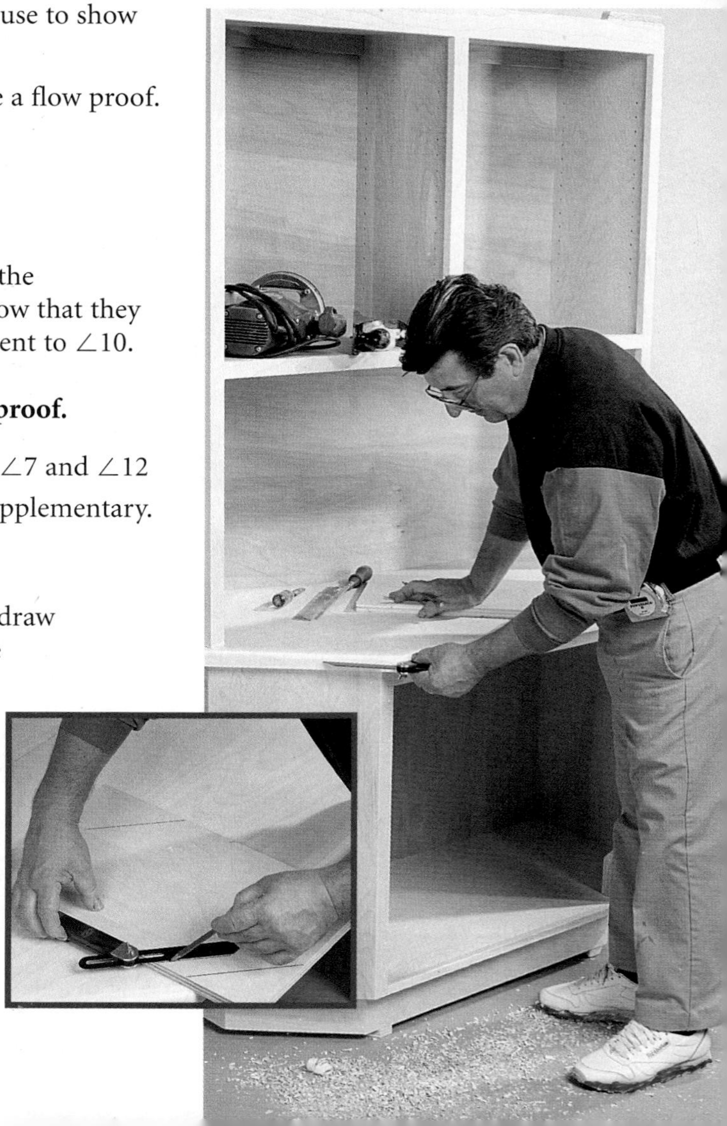

20. a. *Writing* Explain why this statement is true: In a plane, two lines perpendicular to the same line are parallel.

b. *Geometry in 3 Dimensions* Use this rectangular prism to explain why the words "in a plane" are needed in part (a).

c. *Geometry in 3 Dimensions* Use the rectangular prism to find three lines that are parallel to each other, but that are not all in the same plane. Are any two of these lines in the same plane?

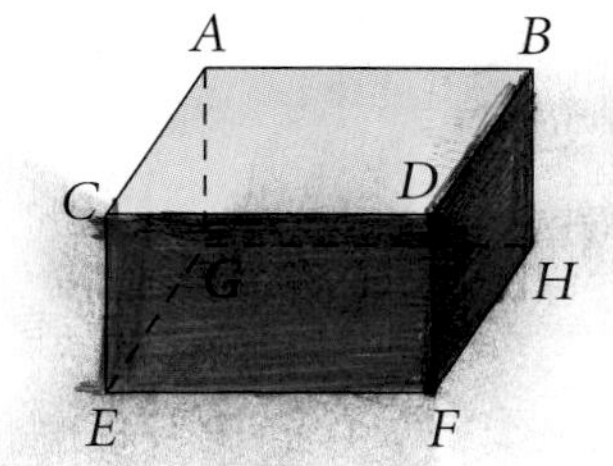

21. *Standardized Test Prep* The measures of the angles of a quadrilateral are $x + 10$, $2x + 20$, $x + 70$, and $2x - 40$. What type or types of quadrilateral could this be?

I. parallelogram **II.** trapezoid **III.** square

A. I only **B.** II only **C.** III only **D.** I and II only **E.** I and III only

Choose **Use paper and pencil, mental math, or a calculator to determine the value of x for which $\ell \parallel m$.**

22.

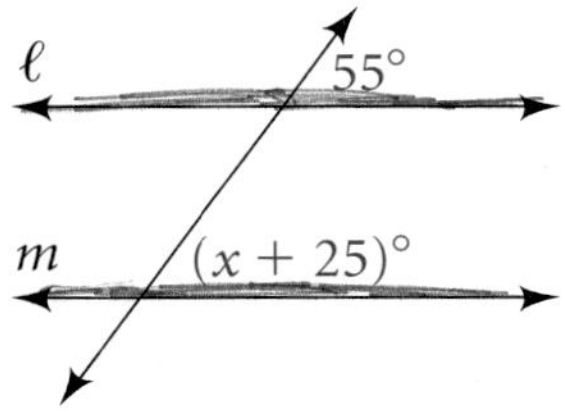

23.

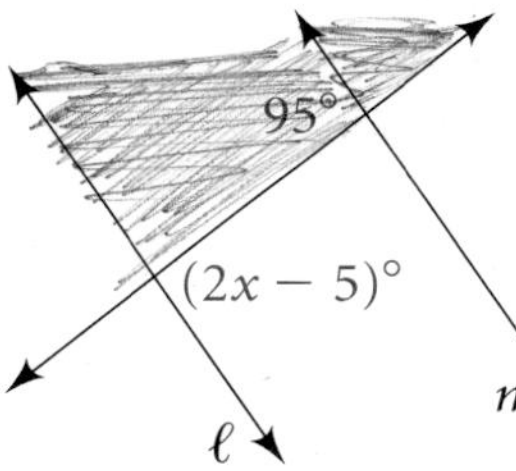

24.

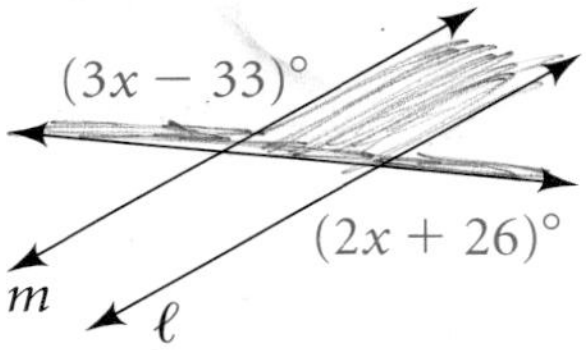

25.

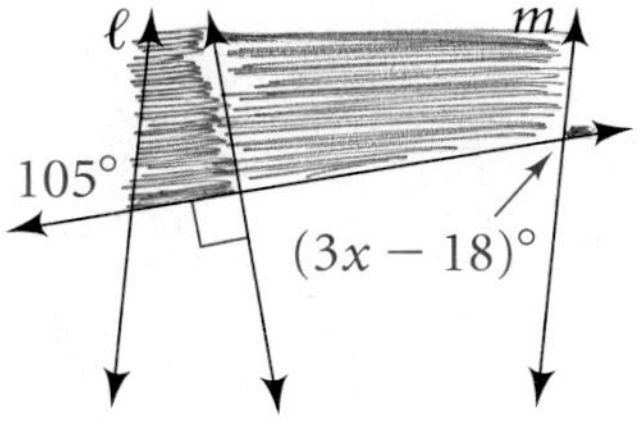

26.

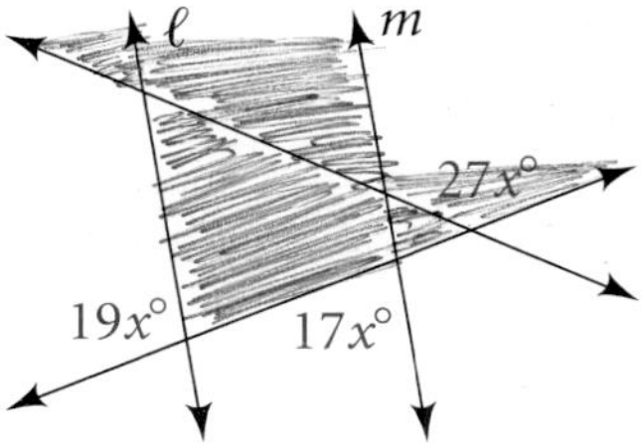

27.

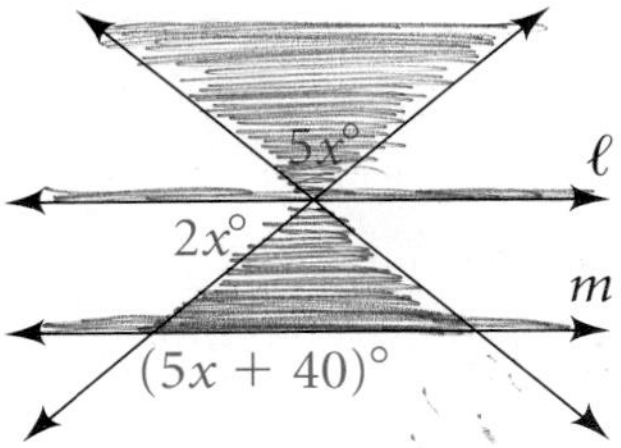

28. Rewrite this paragraph proof of Theorem 7-4 as a flow proof.

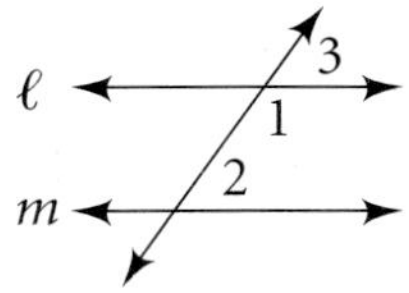

Given: $\angle 1$ and $\angle 2$ are supplementary.
Prove: $\ell \parallel m$

By the Angle Addition Postulate, $m\angle 1 + m\angle 3 = 180$. Therefore, $\angle 1$ and $\angle 3$ are supplementary. It is given that $\angle 1$ and $\angle 2$ are supplementary, so by the Congruent Supplements Theorem, $\angle 2 \cong \angle 3$. $\angle 2$ and $\angle 3$ are corresponding angles, so by the Converse of the Corresponding Angles Postulate, $\ell \parallel m$.

29. a. *Furniture* The legs of the chairs shown form similar isosceles triangles. Explain how you know that the seat of the chair is parallel to the ground.

b. *Open-ended* Give an example of another type of furniture that shows parallel lines. Include a sketch showing the lines.

30. Coordinate Geometry Given the equations of two lines in the coordinate plane, explain how you can prove that they are parallel.

31. Drafting Explain why the lines shown in the diagram are parallel.

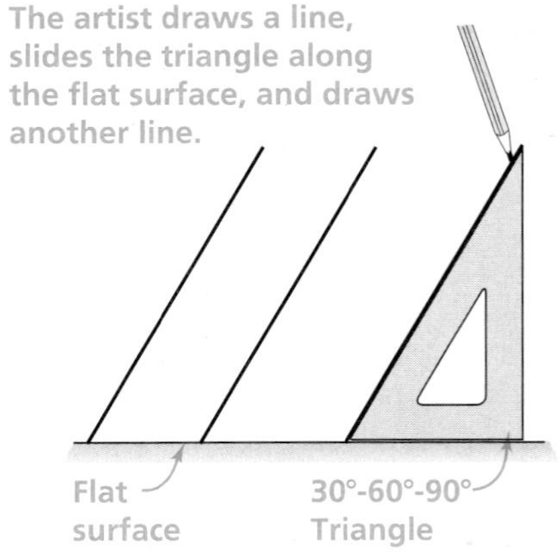

What type of quadrilateral is *PLAN*? Explain your answers.

32. $m\angle P = 72$, $m\angle L = 108$, $m\angle A = 72$, $m\angle N = 108$

33. $m\angle P = 59$, $m\angle L = 37$, $m\angle A = 143$, $m\angle N = 121$

34. $m\angle P = 67$, $m\angle L = 120$, $m\angle A = 73$, $m\angle N = 100$

35. $m\angle P = 90$, $m\angle L = 90$, $m\angle A = 90$, $m\angle N = 90$

36. Transformations Under what circumstances will a line be parallel to its image under a reflection? Draw a diagram to support your explanation.

37. Transformations Write a two-column proof. (*Hint:* Refer to the definition of a reflection in Lesson 3-1.)

Given: $\overline{A'B'}$ is the reflection image of $\overline{AB}$ in line m.
Prove: $\overline{AA'} \parallel \overline{BB'}$

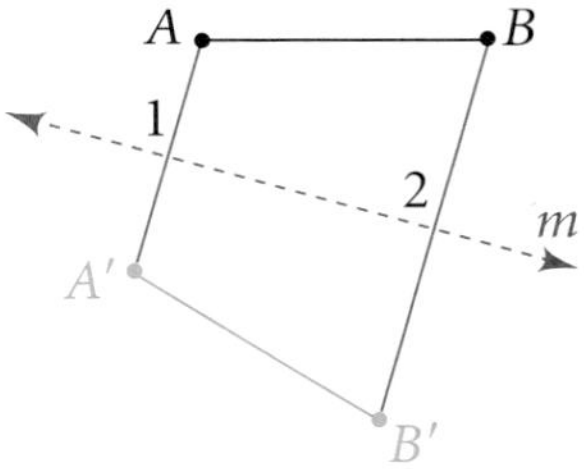

Exercises MIXED REVIEW

Find the area of a semicircular region with the given radius. Round your answer to the nearest tenth.

38. 8 in. 39. 10 cm 40. 4 ft 41. 1.2 m

42. a. Olympics The torch used for the 1996 Olympic Games in Atlanta is approximately cylindrical. Its length is 32 in., and its diameter is 2.25 in. What is its volume? Round your answer to the nearest whole unit.
b. The torch has 22 equally-spaced prongs around the top representing the number of cities that have hosted Olympic Games. How many degrees apart are the prongs? Round your answer to the nearest tenth.

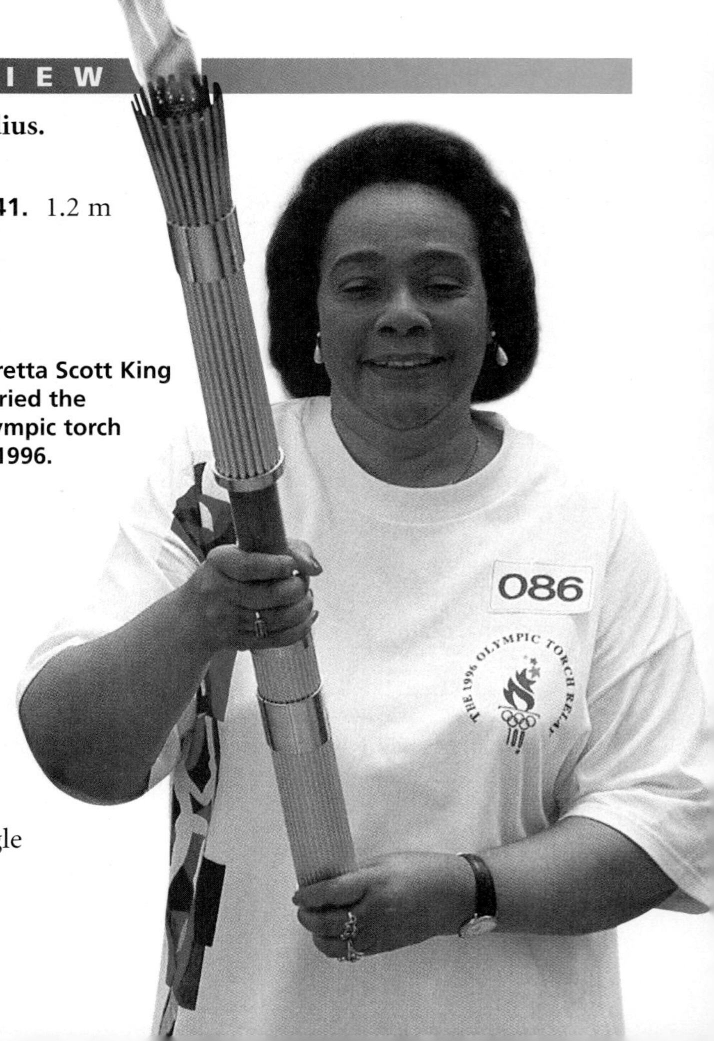

Coretta Scott King carried the Olympic torch in 1996.

Getting Ready for Lesson 7-3

Constructions **Draw each figure and then use a compass and straightedge to construct a figure congruent to the one you drew.**

43. an obtuse angle 44. a segment 45. a triangle

Connections Technology . . . and more

7-3 Constructing Parallel and Perpendicular Lines

What You'll Learn

- Constructing parallel and perpendicular lines

...And Why

To be able to use parallels and perpendiculars in arts and crafts

What You'll Need

straightedge, compass

THINK AND DISCUSS

Constructing Parallel Lines

You can use what you know about corresponding angles and parallel lines to construct parallel lines.

Construction 5
Parallel through a Point Not on a Line

Construct a line parallel to a given line and through a given point not on the line.
Given: Line ℓ and point N not on ℓ

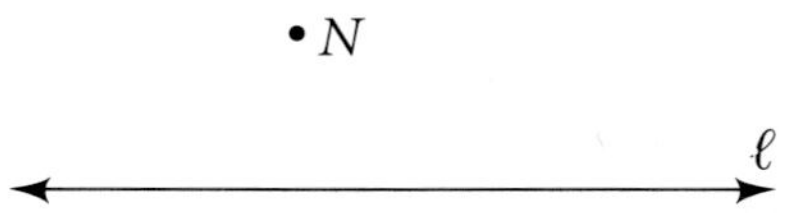

Step 1
Label two points H and J on ℓ.
Draw $\overleftrightarrow{HN}$.

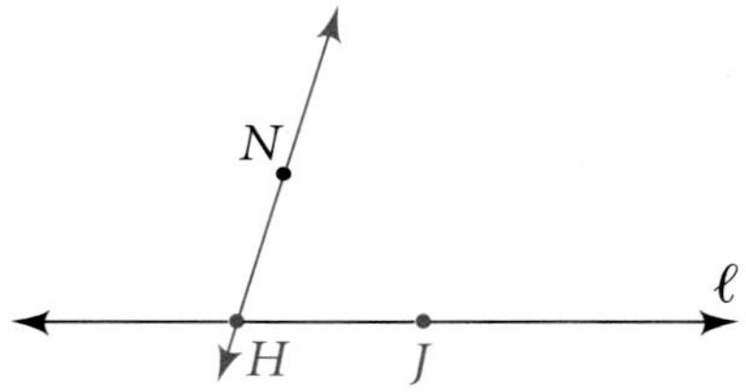

Step 2
Construct $\angle 1$ with vertex at N so that $\angle 1 \cong \angle JHN$ and the two angles are corresponding angles. Label the line you just constructed m.

$\ell \parallel m$

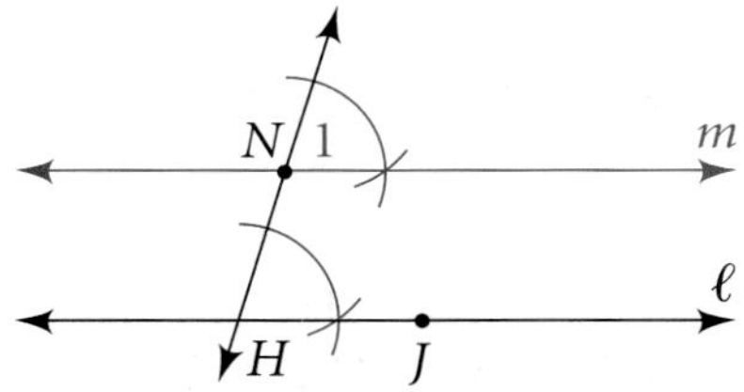

QUICK REVIEW
Instructions for constructing congruent angles are on page 40.

1. Explain why lines ℓ and m are parallel.

Example 1

Use the segments shown at the left to construct a trapezoid with bases of lengths a and b.

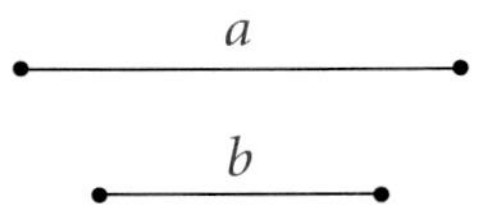

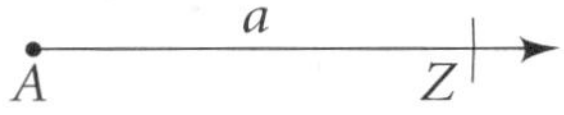

Step 1: Construct $\overline{AZ}$ with length a.

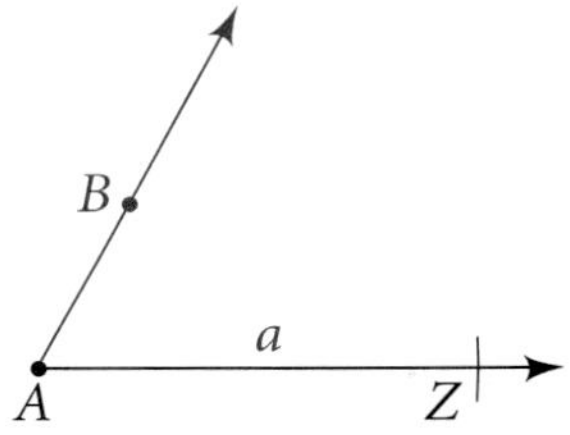

Step 2: Draw a point B not on $\overleftrightarrow{AZ}$. Then draw $\overrightarrow{AB}$.

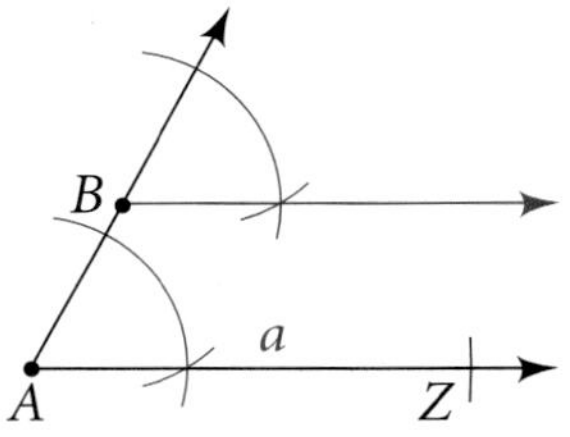

Step 3: Construct a line parallel to $\overleftrightarrow{AZ}$ through B.

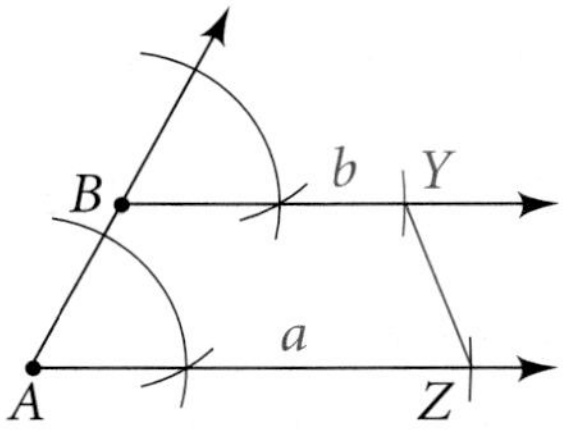

Step 4: Construct Y so that $BY = b$. Then draw $\overline{YZ}$. $ZABY$ is a trapezoid.

Constructing Perpendicular Lines

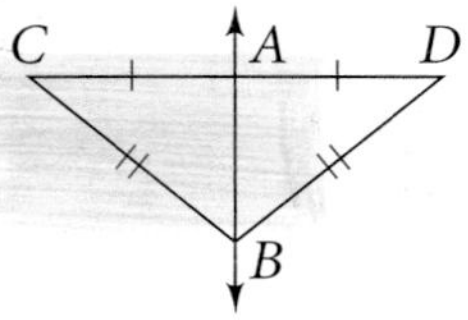

The two constructions that follow are based on Theorem 4-13: If a point is equidistant from the endpoints of a segment, then it is on the perpendicular bisector of the segment.

2. Explain why $\overleftrightarrow{AB}$ must be the perpendicular bisector of $\overline{CD}$.

3. Critical Thinking Turn to page 42 and study the method for constructing a perpendicular bisector of a segment. Explain why that construction works.

Construction 6
Perpendicular through a Point on a Line

Construct the perpendicular to a given line at a given point on the line.
Given: Point P on line ℓ

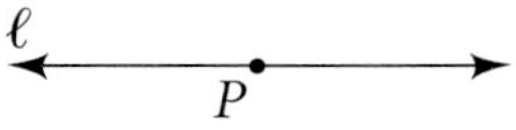

Step 1
Place the compass tip on P. Draw arcs intersecting ℓ in two points. Label the points A and B.

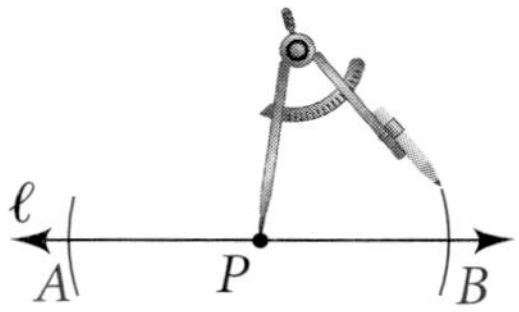

Step 2
Open the compass wider. With the compass tip on A, draw an arc above point P.

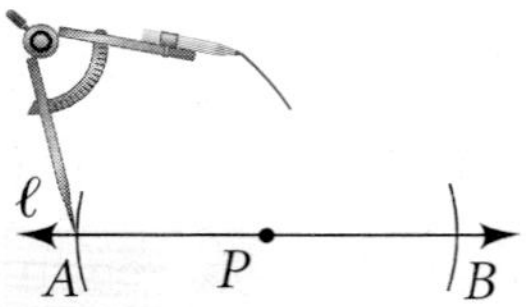

Step 3
Without changing the compass setting, place the compass tip on B. Draw an arc that intersects the arc from Step 2. Label the point of intersection C.

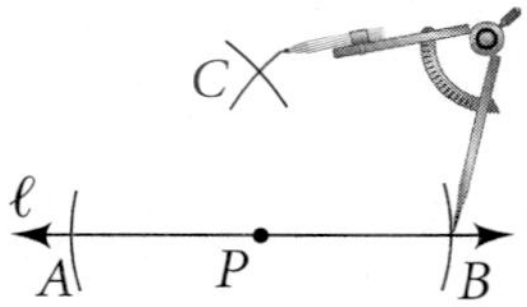

Step 4
Draw $\overleftrightarrow{CP}$.

$\overleftrightarrow{CP} \perp \ell$

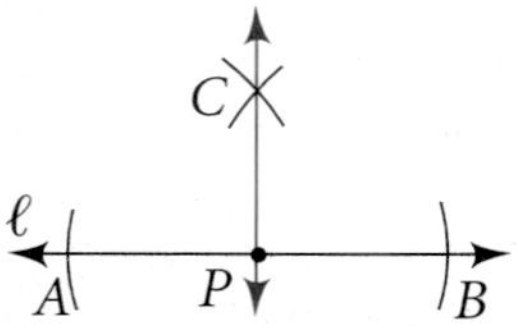

Justification of Construction 6

You constructed A and B so that $AP = BP$. You constructed C so that $AC = BC$. Because P and C are both equidistant from the endpoints of $\overline{AB}$, $\overleftrightarrow{CP}$ is the perpendicular bisector of $\overline{AB}$. So $\overleftrightarrow{CP} \perp \ell$.

4. **Try This** Draw a line $\overleftrightarrow{EF}$. Construct a line $\overleftrightarrow{FG}$ so that $\overleftrightarrow{EF} \perp \overleftrightarrow{FG}$.

American artist Sol LeWitt creates wall-sized drawings that contain geometric figures such as arcs and perpendicular lines. To create arcs, he uses simple, large-scale compasses like the one shown below. He often employs local students to assist in creating his artwork. The two works shown here were on display at the Addison Gallery in Andover, Massachusetts.

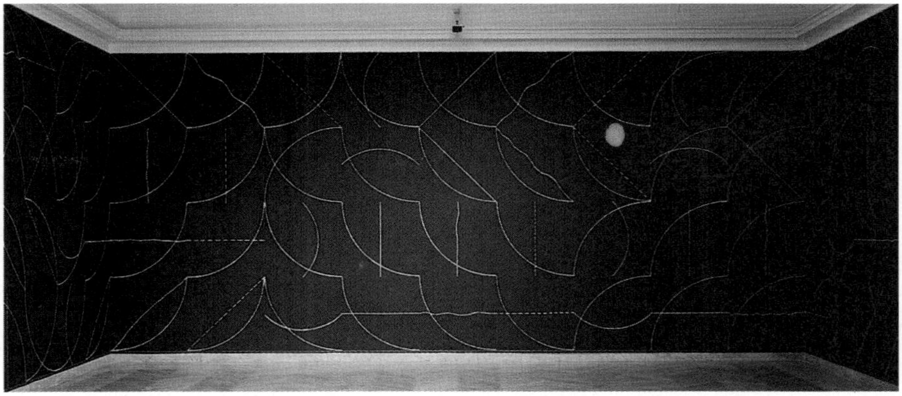

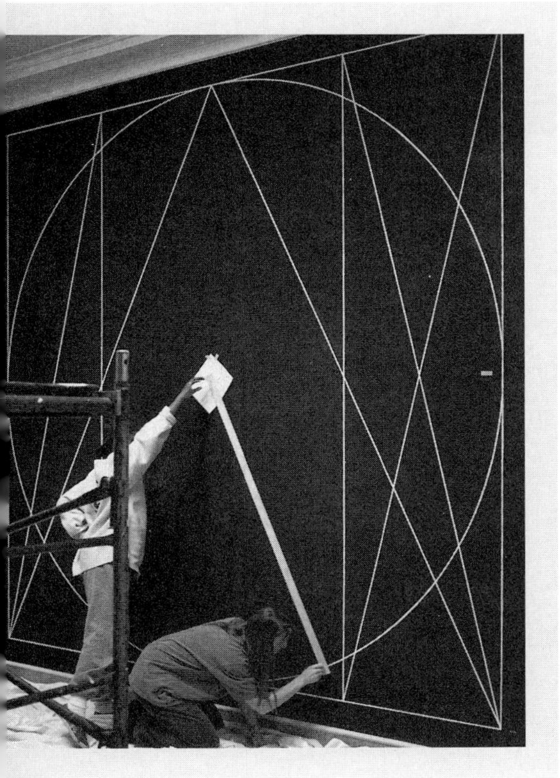

Construction 7

Perpendicular through a Point Not on a Line

Construct the perpendicular to a given line from a given point not on the line.

Given: line ℓ and point R not on ℓ

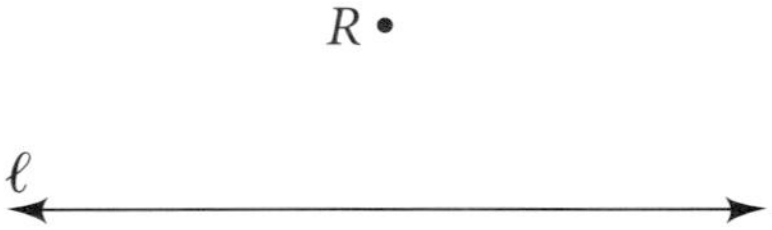

Step 1
Open your compass to a distance greater than the distance from R to ℓ. With the compass tip on R, draw an arc that intersects ℓ at two points. Label the points E and F.

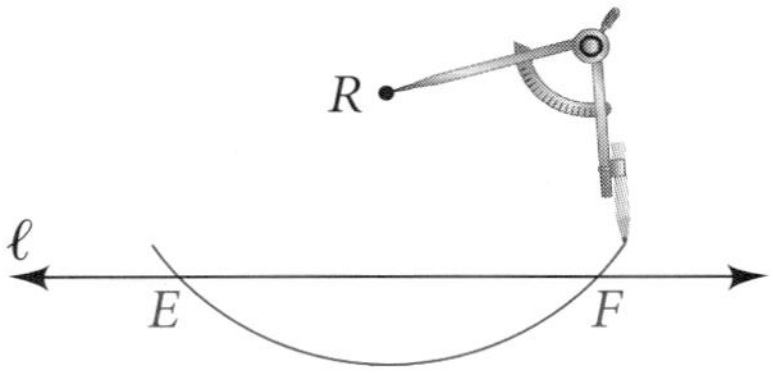

Step 2
Keep the same compass setting. Place the compass point on E and make an arc.

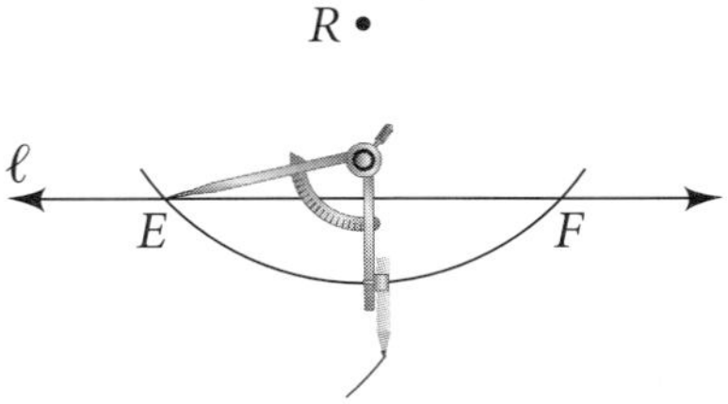

Step 3
Keep the same compass setting. With the compass tip on F, draw an arc that intersects the arc from Step 2. Label the point of intersection G.

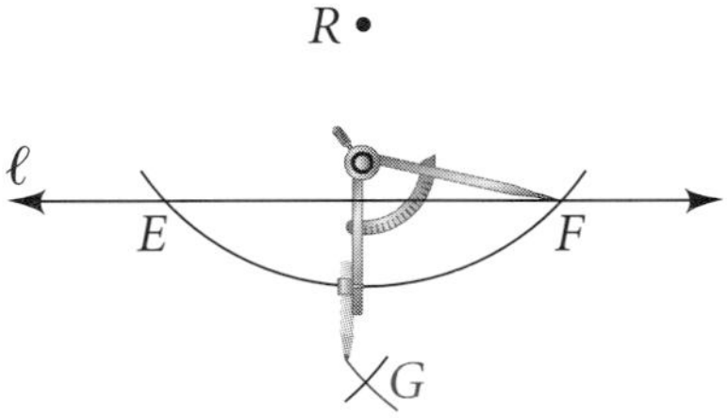

Step 4
Draw $\overleftrightarrow{GR}$.

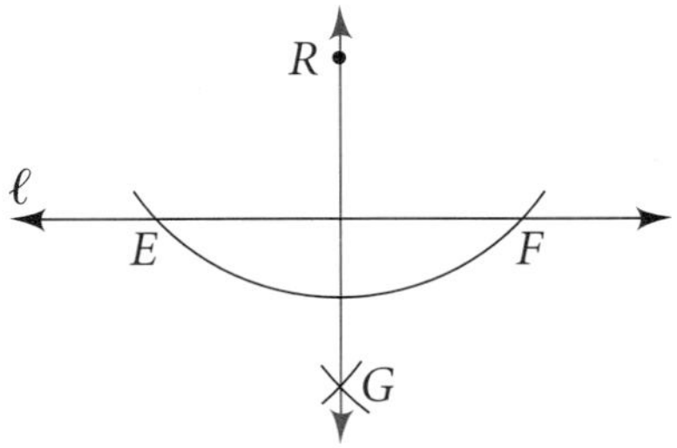

$\overleftrightarrow{GR} \perp \ell$

5. **Try This** Draw a line $\overleftrightarrow{CX}$ and a point Z not on the line. Construct $\overleftrightarrow{ZB}$ so that $\overleftrightarrow{ZB} \perp \overleftrightarrow{CX}$.

6. **Critical Thinking** Explain why Construction 7 works.

Example 2

Technology Use geometry software to construct a right triangle in which one leg is twice as long as the other.

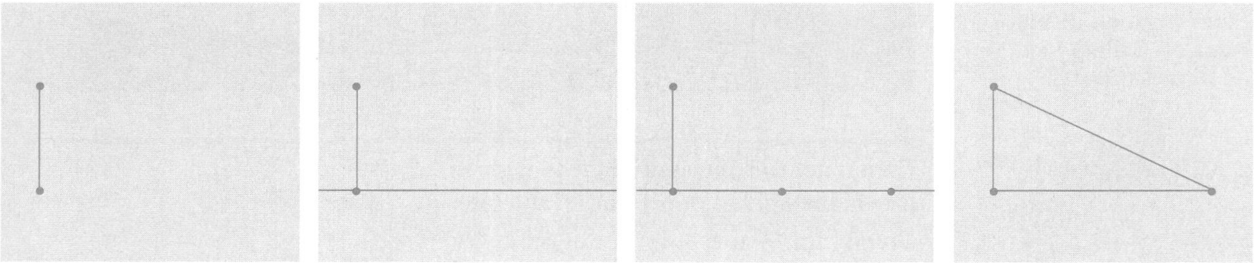

Step 1: Draw a segment of any length.

Step 2: Construct a line perpendicular to the segment through one of its endpoints.

Step 3: On the line, construct two segments congruent to the original segment.

Step 4: Draw the hypotenuse of the triangle. Hide the construction points and lines.

Exercises ON YOUR OWN

1. Draw a line ℓ and a point S not on ℓ. Construct a line m through S so that $m \parallel \ell$.

2. **Transformations** Draw a line ℓ and a segment $\overline{PQ}$ as shown at the right. Construct the reflection image of $\overline{PQ}$ in line ℓ. (*Hint:* Construct a line perpendicular to ℓ through P. Construct P' on the perpendicular so that ℓ bisects $\overline{PP'}$.)

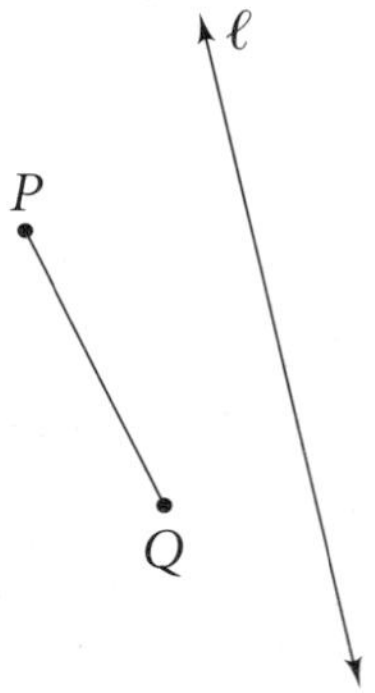

3. **a.** Draw an acute angle on your paper. Construct an angle congruent to the angle you drew so that the two angles are alternate interior angles. (*Hint:* The figure you end up with should look like a **Z**.)
   b. **Writing** Explain how to construct a line parallel to a given line through a point not on the line by using the Converse of the Alternate Interior Angles Theorem.

4. Line m was constructed perpendicular to ℓ through P by this method.
   - Draw two points on ℓ and label them R and S.
   - Construct a circle with center R and radius PR.
   - Construct a circle with center S and radius PS.
   - Draw line m through the intersections of the circles.

   a. **Critical Thinking** Why is line m perpendicular to line ℓ?
   b. Draw a line b and a point C not on b. Construct the perpendicular through C using this two-circle method.

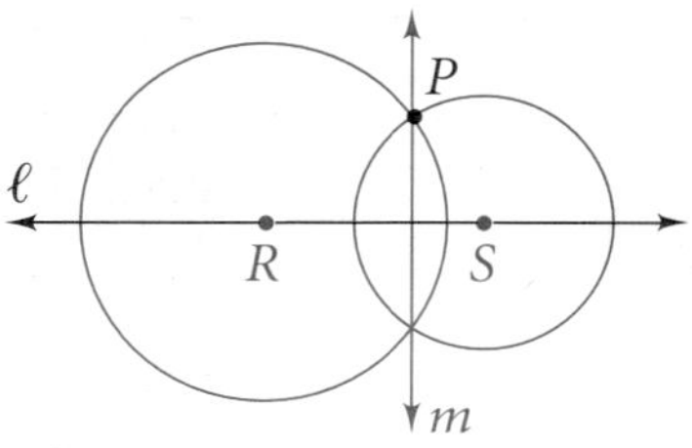

5. **a.** **Technology** In this computer drawing, R and Q are the intersections of $\odot P$ and line n, all three circles are congruent, and line m is determined by the intersection of $\odot Q$ and $\odot R$. Explain why line m must be perpendicular to line n.
   b. **Symmetry** Describe the symmetries of this drawing.

6. **a.** Draw a line t and a point P on t. Construct a line s through P so that $s \perp t$.
   b. Draw a point R that is neither on t nor s. Construct a line q through R so that $q \perp s$.
   c. **Critical Thinking** How are q and t related? Explain.

For Exercises 7–16, use the segments at the right.

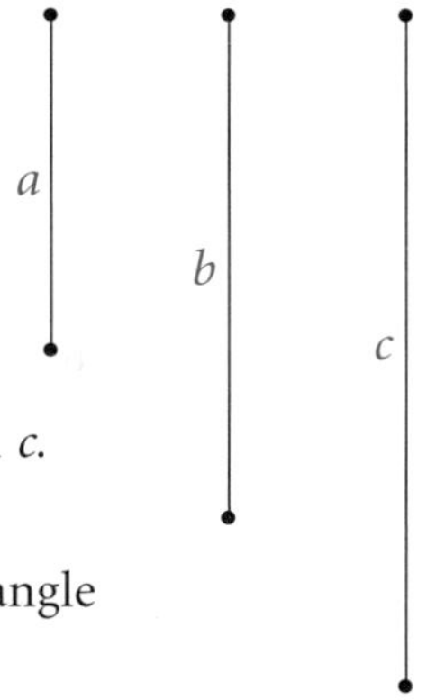

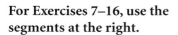

7. Draw a line m. Construct a point C so that the distance from C to m is b. (*Hint:* Construct a perpendicular from a point on the line.)
8. Construct a trapezoid with bases of lengths a and b.
9. Construct an isosceles trapezoid with legs of length a.
10. **a.** Construct a quadrilateral with a pair of parallel sides of length c.
    b. What type of quadrilateral does the figure appear to be?
11. **a.** Construct a triangle with side lengths a, b, and c. Label the triangle $\triangle ABC$, where A is opposite the side of length a, and so on.
    b. Construct the altitude from C.
12. Construct a right triangle in which the length of a leg is a and the length of the hypotenuse is c.
13. Construct a right triangle with legs of lengths b and $\frac{1}{2}b$.
14. **a.** Construct an isosceles right triangle with legs of length b.
    b. What is the length of the hypotenuse of this triangle?
15. Construct a rectangle with base b and height c.
16. **Locus** Draw a segment of any length. Construct the locus of points in a plane a distance a from the segment.

17. Paper Folding You can use paper folding to create a perpendicular to a given line through a given point. Fold the paper so the line overlaps itself and so the fold line contains the point.

  a. Draw a line m and a point W not on the line. Use paper folding to create the perpendicular through W. Label the fold line k.

  b. Use paper folding to create a line perpendicular to k through W. Label this fold line p.

  c. What is true of p and m? **Justify** your answer.

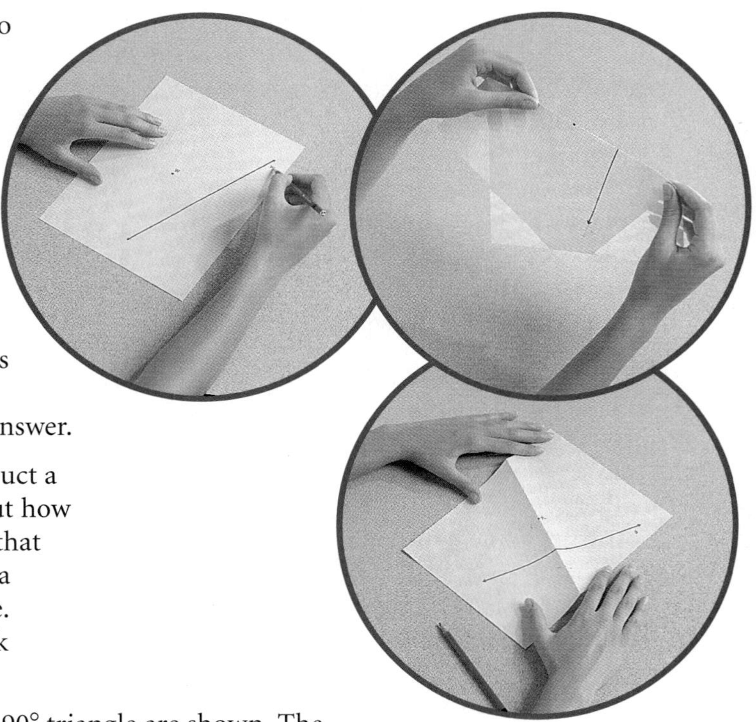

18. Critical Thinking Jane was trying to construct a 45°-45°-90° triangle, but couldn't figure out how to construct a 45° angle. Lenesha told her that she didn't need to know how to construct a 45° angle in order to construct the triangle. What construction procedure do you think Lenesha had in mind?

19. Three methods for constructing a 30°-60°-90° triangle are shown. The numbers indicate the order in which the arcs should be drawn. Explain why each method works.

a.

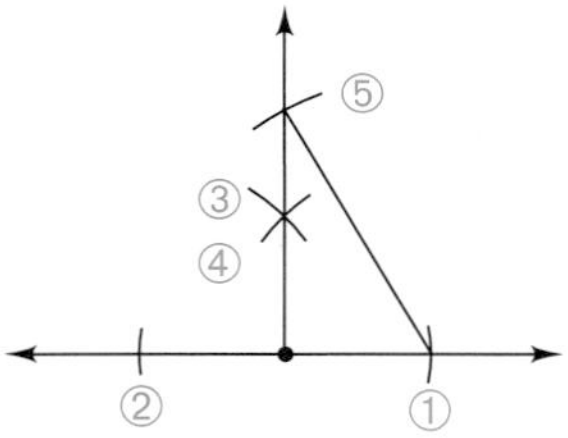

b.

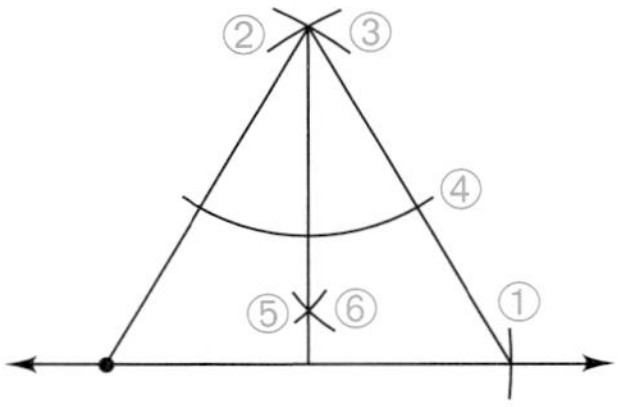

c.

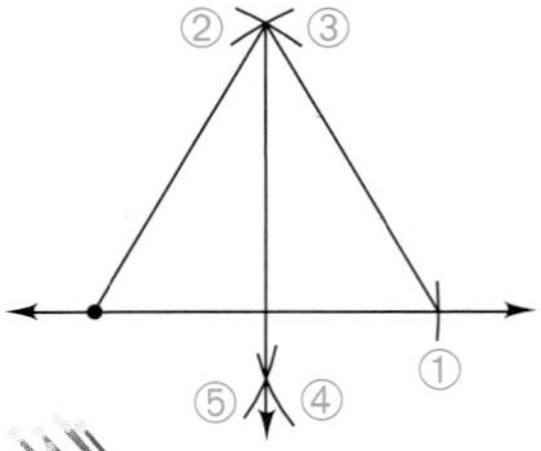

20. **a.** Open-ended Construct a rectangle whose length is twice its width.

  b. Construct a rectangle whose length is four times its width.

  c. Critical Thinking How could you construct a rectangle whose length is 1.5 times its width? (*Hint:* Use a ratio equivalent to 1.5 : 1.)

Chapter Project

Find Out by Designing

Design a network of fiber-optic cables to link the 12 schools labeled *A–L* in the diagram. Assume that a computer at each school can be programmed to route a signal from that school to any other school, even if the signal has to pass through other schools on the way. Assume also that a single cable can handle both incoming and outgoing signals—just as a two-lane road handles traffic in both directions. In your design, try to use the fewest miles of cable possible. The cables do not have to coincide with roads. When you've completed your design, compare it with those of your classmates. Describe the design that you recommend and explain your recommendation.

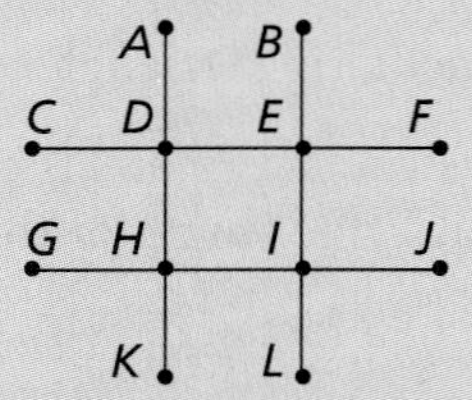

The distance from any school to an adjacent school is 10 mi.

Exercises MIXED REVIEW

If a triangle can be drawn with sides of the given lengths, will the triangle be acute, right, or obtuse? Explain.

21. 2, 3, 5 **22.** 3, 4, 6 **23.** 4, 5, 4

24. Coordinate Geometry Write the equation of a line through (5, −2) and (4, 3).

25. You may leave your answers to parts (a–c) in terms of π.
   a. A circle has radius 8 cm. Two radii of the circle form a 120° angle. What is the arc length of the intercepted arc?
   b. What is the area of the entire circle?
   c. What is the area of the 120° sector?

SELF ASSESSMENT

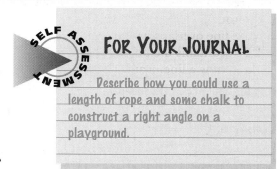

FOR YOUR JOURNAL

Describe how you could use a length of rope and some chalk to construct a right angle on a playground.

Getting Ready for Lesson 7-4

Draw a prism with the given figure as a base.

26.

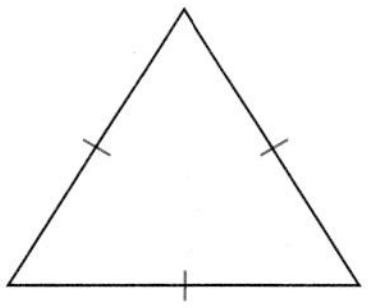

27.

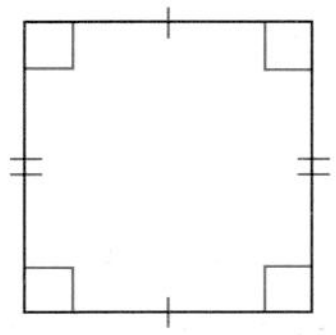

28.

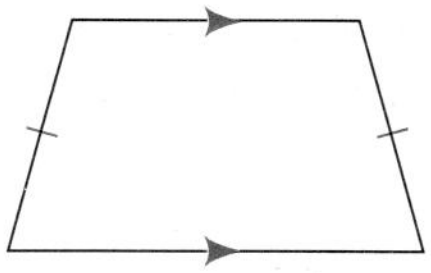

Exercises CHECKPOINT

State the theorem or postulate that justifies each statement.

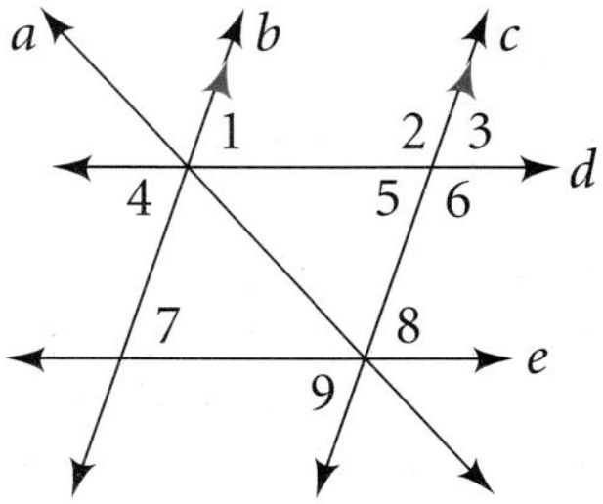

1. $\angle 1 \cong \angle 3$

2. If $\angle 5 \cong \angle 9$, then $d \parallel e$.

3. $m\angle 1 + m\angle 2 = 180$

4. If $\angle 4 \cong \angle 7$, then $d \parallel e$.

5. $\angle 1 \cong \angle 4$

6. $\angle 7 \cong \angle 9$

7. If $\angle 3 \cong \angle 8$, then $d \parallel e$.

8. $\angle 4 \cong \angle 5$

9. If $m\angle 8 + m\angle 6 = 180$, then $d \parallel e$.

10. Draw a line m and a point D on m. Then construct a line n through D so that $n \perp m$.

11. Open-ended Construct a trapezoid with a right angle and with one base twice the length of the other.

12. Standardized Test Prep Which of the following statements is *not* always true if two parallel lines m and n are cut by transversal t?
   A. Corresponding angles are congruent.
   B. Alternate interior angles are congruent.
   C. Exterior angles are congruent.
   D. Four of the eight angles formed are congruent.
   E. Pairs of same-side interior angles are supplementary.

Math ToolboX

Technology

Perspective Drawing

Before Lesson 7-4

Work in pairs or small groups using geometry software to construct a rectangular prism in one-point perspective.

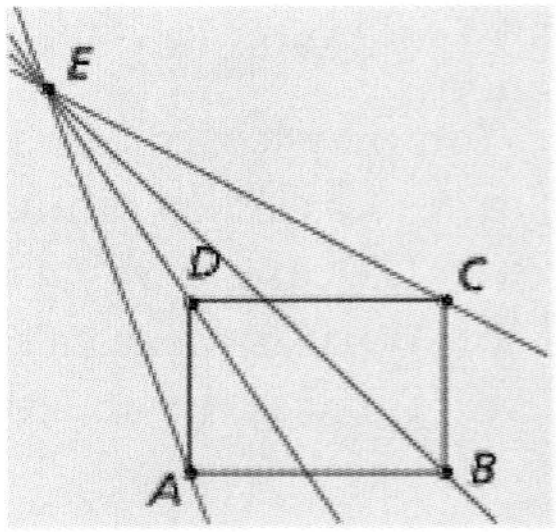

Construct

Construct a rectangle *ABCD* for the front of your prism. Start with $\overline{AB}$ and construct perpendiculars at *A* and *B*. Locate point *C* on the perpendicular through *B*. *D* is the intersection of the perpendicular through *C* with the perpendicular through *A*. Hide the lines and construct the segments for the rectangle.

Draw a point *E* outside the rectangle. Construct $\overleftrightarrow{EA}$, $\overleftrightarrow{EB}$, $\overleftrightarrow{EC}$, and $\overleftrightarrow{ED}$.

Now construct the rectangular back of the prism. Locate a point *G* on $\overline{EA}$ and construct a line through *G* parallel to $\overline{AB}$. Construct the intersection *H* of this line with $\overleftrightarrow{EB}$. Construct the line through *H* parallel to $\overline{BC}$. Continue with similar constructions to locate *J* and *K*. Hide all the lines and draw the sides of rectangle *GHJK*.

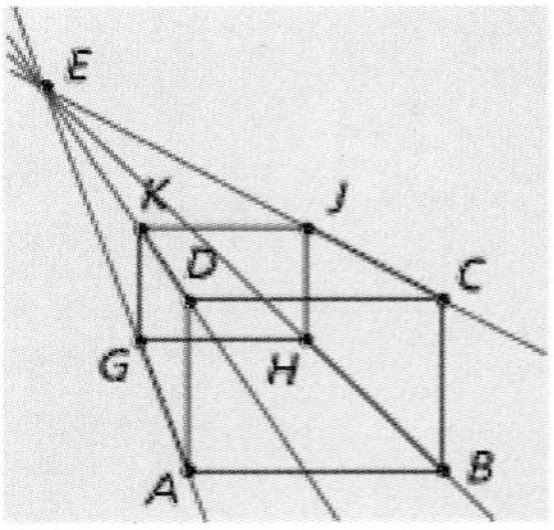

Complete the prism by drawing the remaining edges, $\overline{AG}$, $\overline{BH}$, $\overline{CJ}$, and $\overline{DK}$.

Investigate

Manipulate point *E*. Explore how the position of *E* affects the appearance of the prism. Where is *E* when it seems that you are looking down on the prism and can see its left side? looking directly into the prism? looking up at the prism and can see its right side?

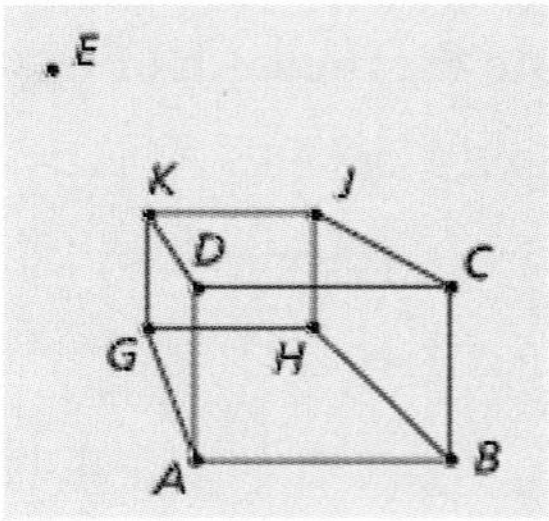

Manipulate points *G*, *C*, *B*, and *A* to change the shape of the prism.

Extend

Draw a house in one-point perspective. Pretend that you are in a helicopter in front of this house and you rise straight up into the sky. Use the software to print several "snapshots" of the house as you would see it from the helicopter.

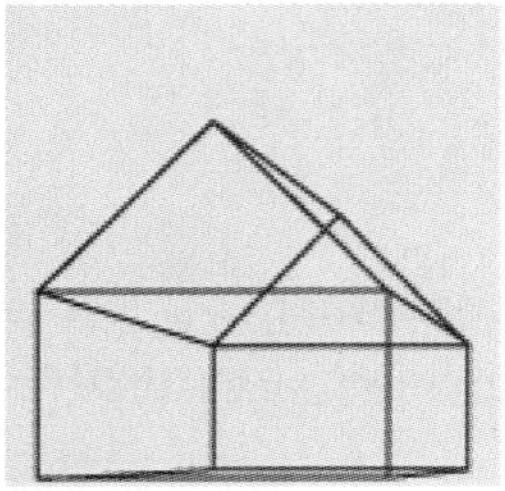

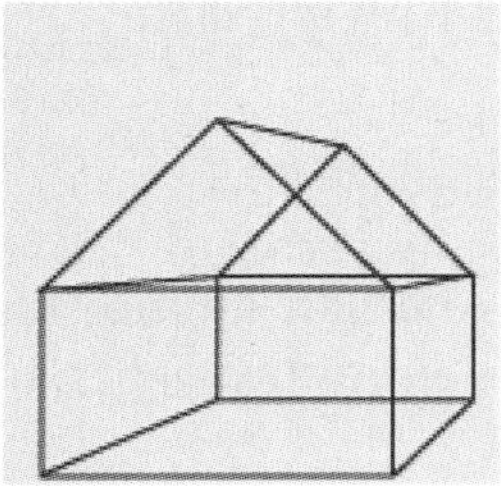

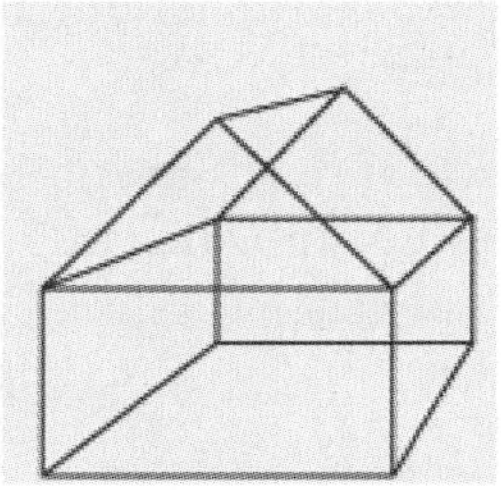

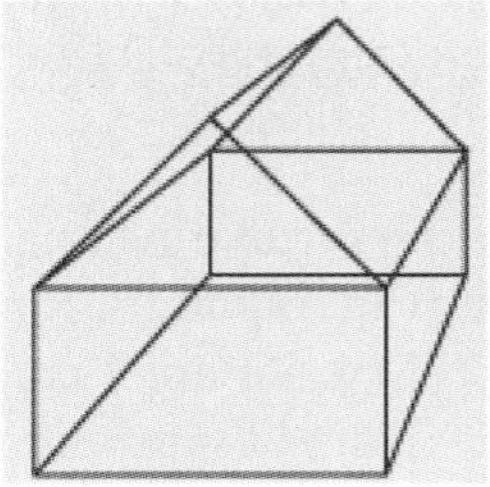

7-4 Parallel Lines and Perspective Drawing

What You'll Learn

- Drawing objects in one- and two-point perspective

...And Why

To create realistic pictures of real-world objects such as buildings

What You'll Need

- ruler

THINK AND DISCUSS

Perspective drawing is a way of drawing objects on a flat surface so that they look the same as they appear to the eye.

Art Techniques for drawing objects in perspective were developed by the ancient Greeks and Romans, but were lost to the Western world until the beginning of the Renaissance in the early 1400s. Renaissance artists used the geometry of perspective to produce works that were far more lifelike and realistic than earlier works.

Before the Renaissance . . .

Artist unknown; English manuscript illumination (c. 1400) depicting Marco Polo setting out from Venice in 1271

The size of objects in pre-Renaissance art had more to do with their importance to the artist than to their positions in space.

After the Renaissance . . .

Jacopo Bellini, *The Palace of Herod,* 15th Century

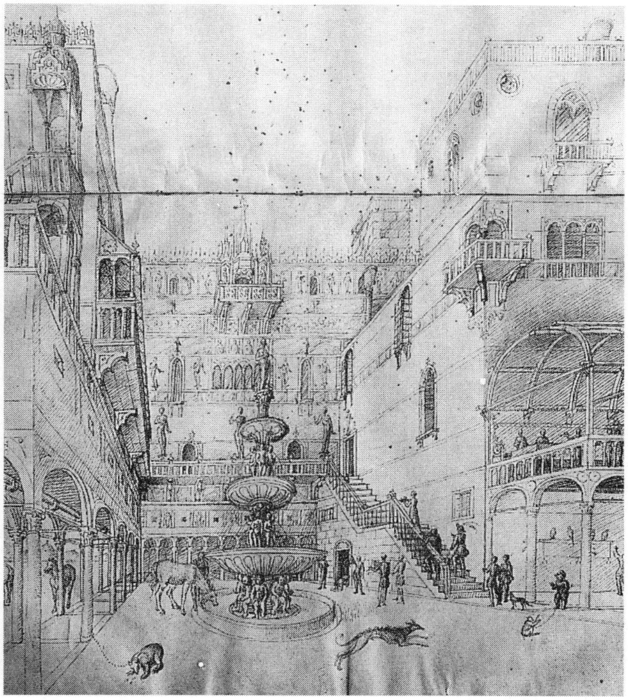

Bellini's sketchbook contains some of the earliest examples of the application of the principles of perspective drawing.

1. Discussion Notice in the painting of Marco Polo's departure from Venice that the ships in the foreground are about the same size as the rowboat behind them. Find other examples of objects that are out of perspective in the painting.

The work of some modern artists broke from a strict rendering of objects in perspective. The French artist Fernand Léger broke the rules of perspective in this painting in order to call attention to specific colors and shapes.

Léger's paintings during the 1930s made strong visual connections to mechanical objects. Even the people in his paintings tended to have a smooth, mechanical look.

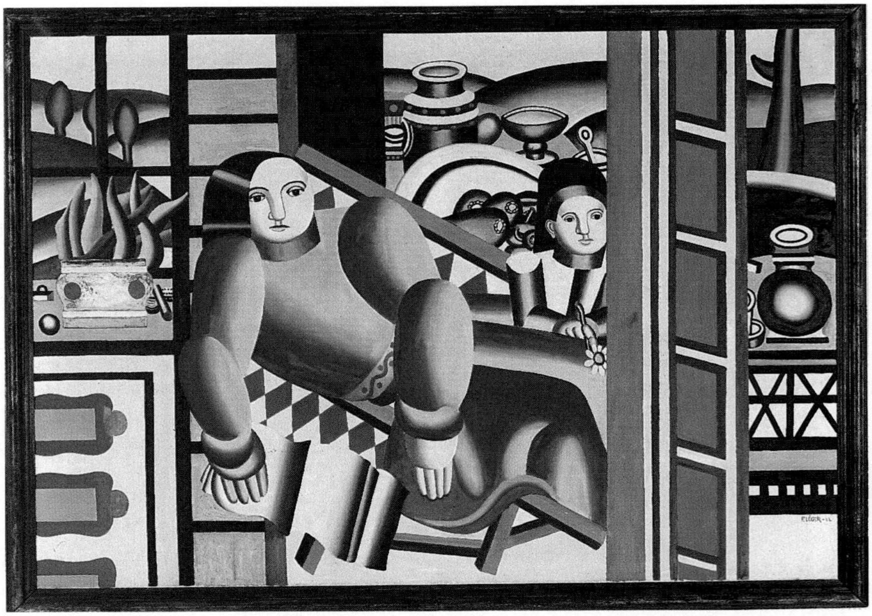

Fernand Léger, *Mother and Child*, 1936

The key to perspective drawing is the use of vanishing points. A **vanishing point** is a point on the "horizon" of a picture where parallel lines "meet." Here is how to draw a cube in **one-point perspective.**

Step 1: Draw a square. Then draw a horizon line and a vanishing point on the line.

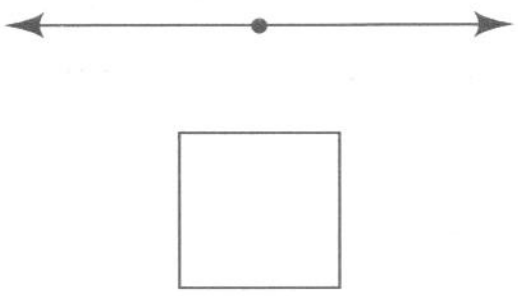

Step 2: Lightly draw segments from the vertices of the square to the vanishing point.

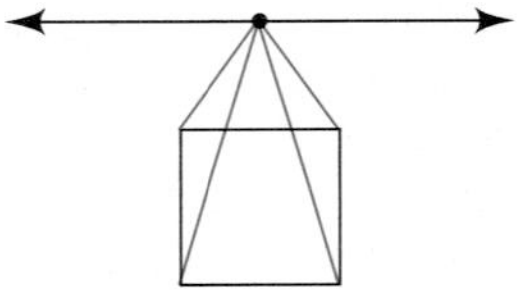

Step 3: Draw a square for the back of the cube. Each vertex should lie on a segment you drew in Step 2.

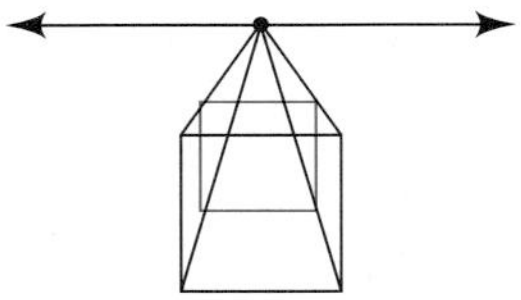

Step 4: Complete the figure by using dashes for hidden sides of the cube. Erase unneeded lines.

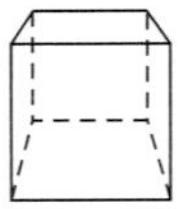

2. Try This Draw a shoe box in one-point perspective.

Two-point perspective involves the use of two vanishing points.

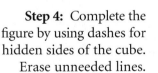

Step 1: Draw a vertical segment. Then draw a horizon line and two vanishing points on the line.

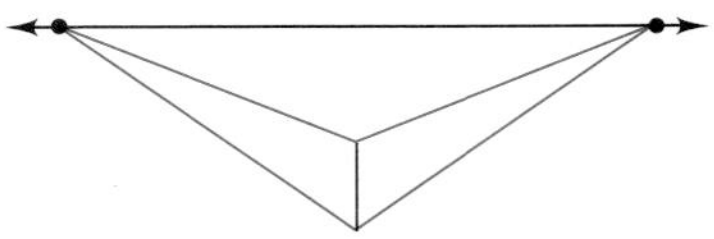

Step 2: Lightly draw segments from the endpoints of the segment to each vanishing point.

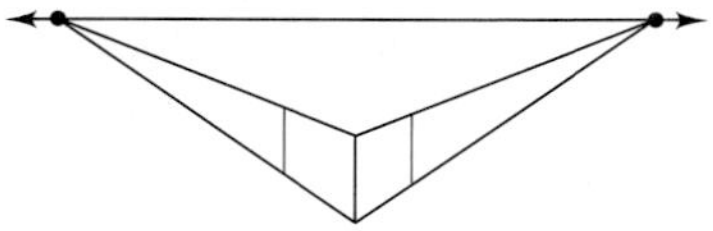

Step 3: Draw two vertical segments between the segments of Step 2.

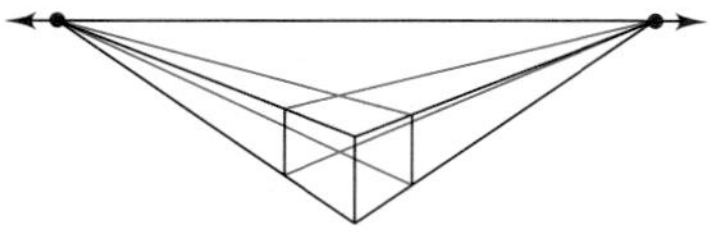

Step 4: Draw segments from the endpoints of the segments you drew in Step 3 to the vanishing points.

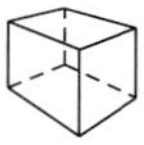

Step 5: Complete the figure by using dashes for hidden sides of the figure. Erase unneeded lines.

3. **Try This** Draw a shoe box in two-point perspective.

Example ***Relating to the Real World***

Technical Art Is the object drawn in one- or two-point perspective?

a.

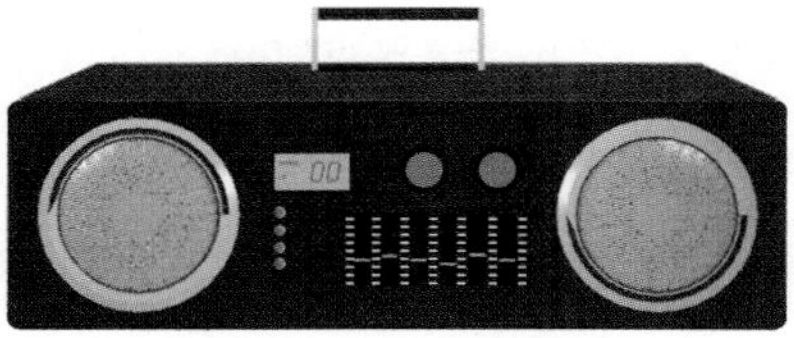

b.

a.

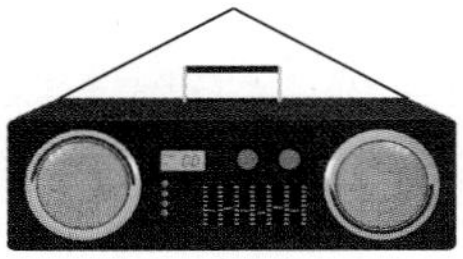

One-point perspective

b.

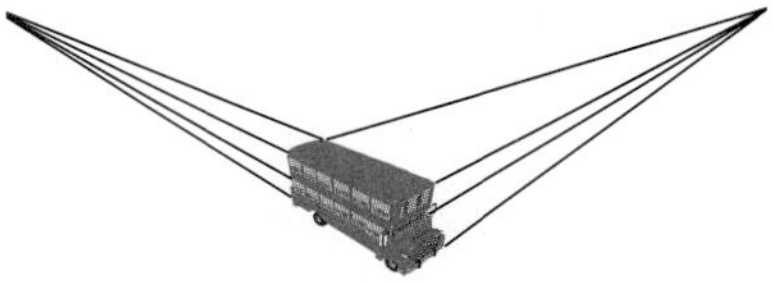

Two-point perspective

Who? Julie Dorsey, an Associate Professor of Architecture at the Massachusetts Institute of Technology, used principles of perspective to develop a computer program for the Metropolitan Opera in New York. The program enables designers to design stage sets and to simulate lighting effects on them.

To decide whether to use two-point perspective or one-point perspective, you need to know the relationship between the objects you are drawing and you.

One-Point Perspective

The faces of the buildings are parallel to you, so use one-point perspective.

Two-Point Perspective

None of the faces of the buildings are parallel to you, so use two-point perspective.

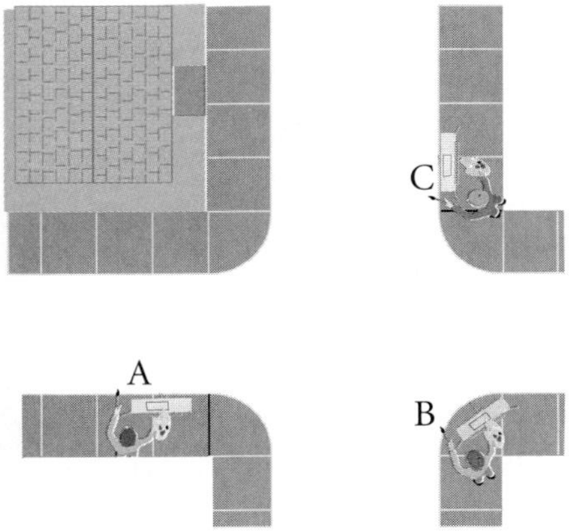

4. *Critical Thinking* Three artists set up easels around a historical building in Paris. Each artist plans to paint a realistic view of the building. Based on the positions of their easels, which type of perspective will each artist's painting show?

WORK TOGETHER

Work in groups to explore how the placement of the vanishing point affects a drawing. Have each member of your group draw a box in one-point perspective. Vary the positions of the horizon lines and vanishing points within your group.

5. Where is the vanishing point when the "viewer" can see three sides of the box? two sides? one side?

Exercises ON YOUR OWN

Is the object drawn in one- or two-point perspective?

1. 

2. 

3. 

4. 

Study each diagram at the left. Match each position in the diagram to what the viewer would see from that point.

5. 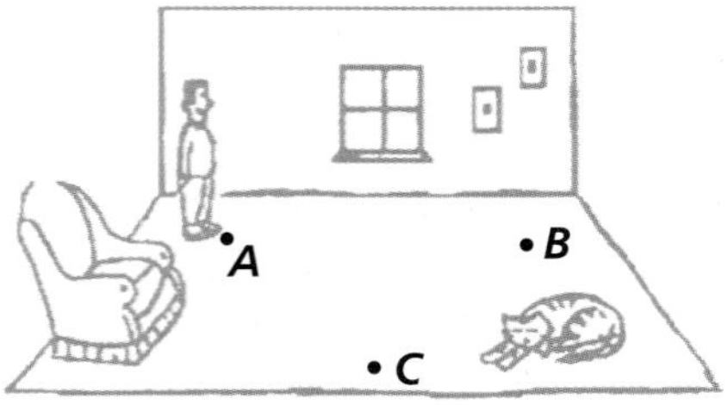

I. 

II. 

III. 

6. 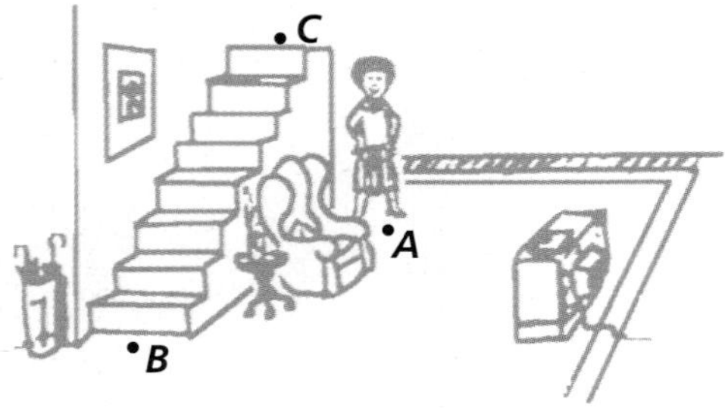

I. 

II. 

III. 

Copy each prism and locate the vanishing point(s).

7. 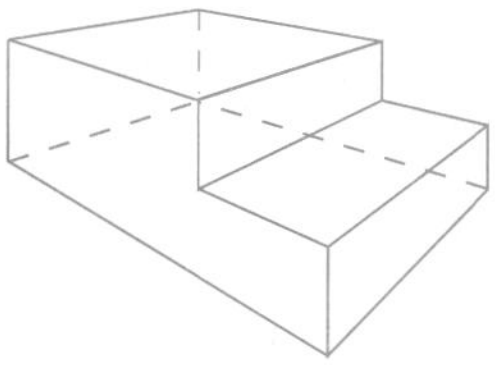

8. 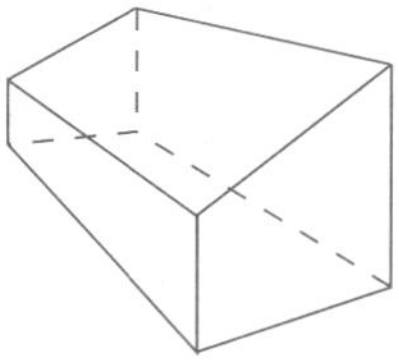

9. 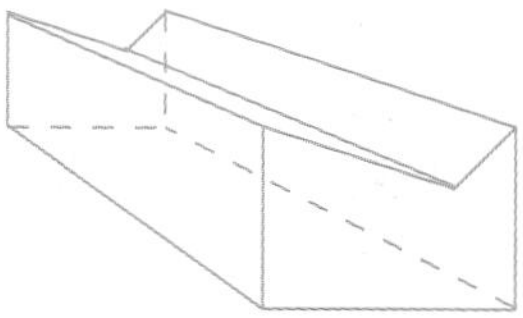

A Computer's Perspective

Ray Dream Designer™ is a software program that allows artists to create 3D objects and scenes. The artist starts by drawing top, front, and side views of an object. The program uses this information to assemble a 3D image of the object.

Each object the artist creates is then placed in a scene such as a room or a street. Finally, the artist positions a "camera" and "lights" within the scene. The computer automatically draws all the objects from the viewpoint of the camera.

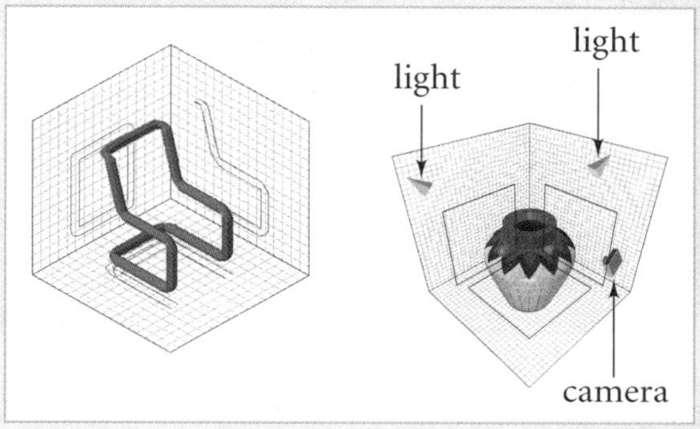

10. Describe the position of the camera for each of these views.

a. 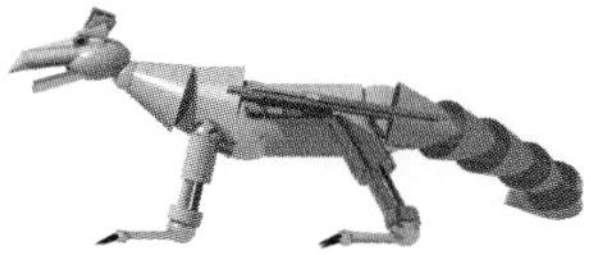

b. 

c. 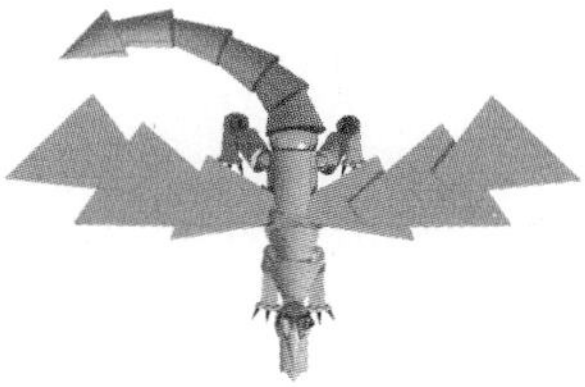

11. a. Refer to the cartoon to explain why an artist might purposefully paint a scene not in proper perspective.

b. Research Find out about the Cubist movement in painting. What techniques did these artists use? What were they trying to communicate?

Fine Art **Explain how each artist used or did not use perspective.**

12. Stuart Davis, *New York/Paris No. 2,* 1931

Stuart Davis once wrote "I never ask the question `Does this picture have depth or is it flat?' I consider such a question irrelevant."

13. Richard Estes, *Holland Hotel,* 1984

Richard Estes' paintings are so realistic that they are easily mistaken for photographs.

Optical Illusions **Explain how the concepts you learned in this lesson relate to each of these optical illusions.**

14.

The students appear to be different heights.

15.

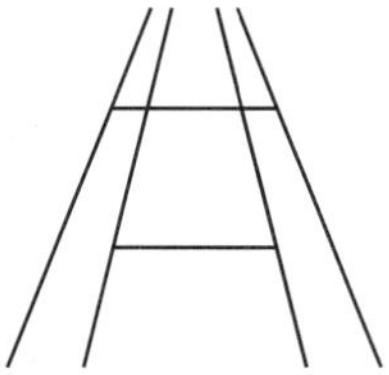

The horizontal segments appear to be different lengths.

16.

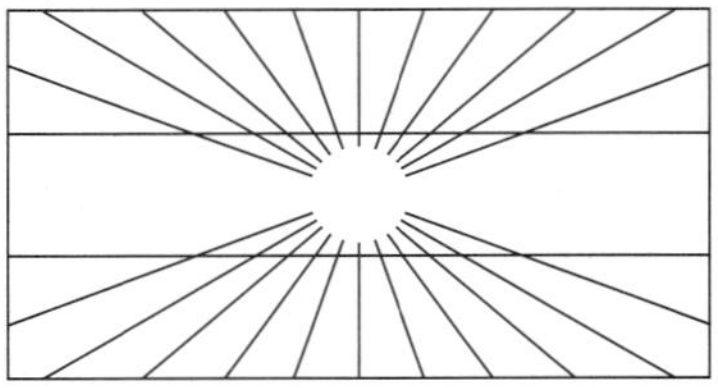

The horizontal lines appear to be curved.

17. **Photography** This photograph was taken with a type of lens called a *fish-eye lens.* Describe how this type of lens affects parallel lines and vanishing points.

18. **a.** Create an isometric drawing of a box.

b. Draw the same box in two-point perspective.

c. **Writing** Compare how parallel lines appear in each drawing.

19. *Open-ended* You can draw block letters in either one-point perspective or two-point perspective. Write your initials in block letters using one-point perspective and two-point perspective.

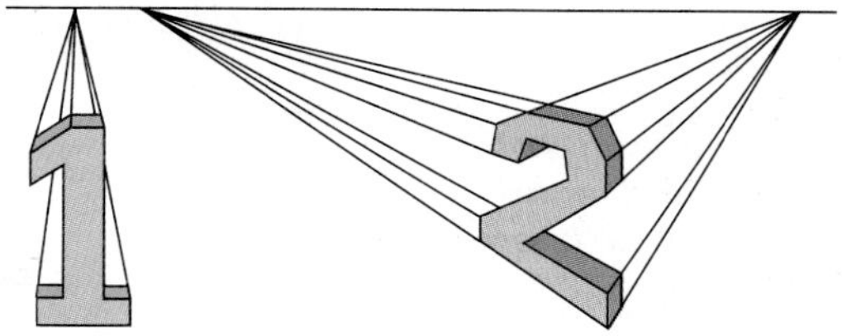

Chapter Project — *Find Out by Investigating*

Each of the points in the two networks at the right is a *vertex* of the network. An *odd vertex* is on an odd number of paths. *O* is an odd vertex. An *even vertex* is on an even number of paths. *E* is an even vertex.

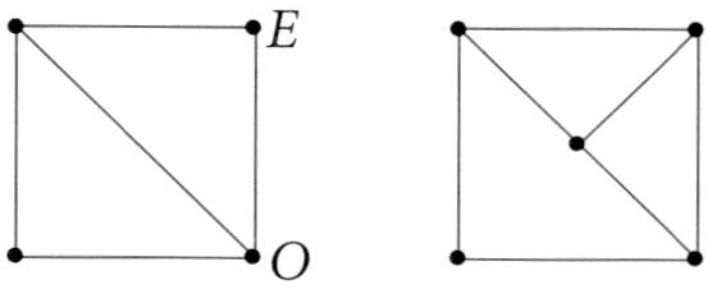

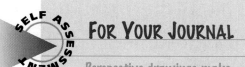

- How many odd and even vertices does each network have?
- Try to trace each network without lifting your pencil from your paper or retracing any path. (You may go over a vertex more than once.)
- Make your own network, and then try to trace it. Record your results in a table like this one. Keep drawing and tracing networks until you see a pattern relating the number of odd and even vertices to whether the network is traceable.

Diagram of Network	Odd Vertices	Even Vertices	Traceable?
	0	5	yes

Exercises MIXED REVIEW

In Exercises 20–22: (a) Write the converse of each conditional. (b) Determine the truth value of the conditional and its converse.

20. If a triangle is a right triangle, then its acute angles are complementary.

21. If it is a national holiday in the United States, then it is July 4.

22. If a figure is a rectangle, then it is a parallelogram.

23. *Transformational Geometry* Describe the image of a circle with center $(4, -2)$ and radius 2 under a dilation with scale factor 3 centered at the origin.

SELF ASSESSMENT — FOR YOUR JOURNAL

Perspective drawings make drawings of three-dimensional objects look more realistic. However, perspective drawings have some disadvantages. Describe some of them.

Getting Ready for Lesson 7-5

Find the sum of the measures of the angles of each spherical triangle.

24.

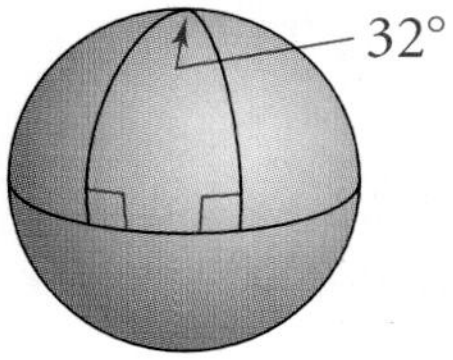

25.

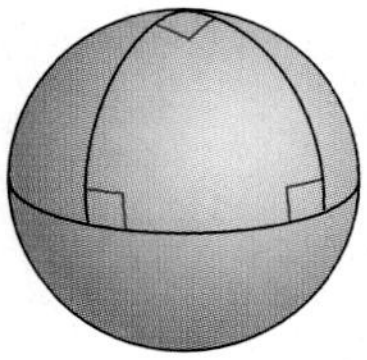

26.

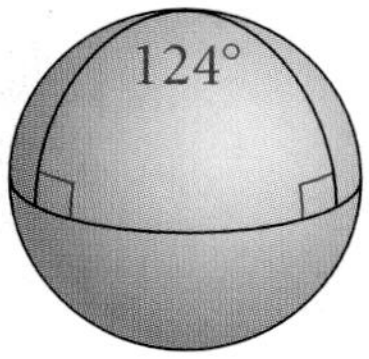

What You'll Learn

- Using some of the basic ideas of spherical geometry

...And Why

To better understand the properties of the sphere on which we live

What You'll Need

- globe, ball, or foam ball
- string
- protractor

7-5 Exploring Spherical Geometry

WORK TOGETHER

1. On a globe or ball, select two points that are opposite each other. That is, select two points that are endpoints of a diameter, such as the North and South poles on a globe. Find the shortest path between the points by holding a string taut between them. Is there one shortest path between the points?
2. Now hold the string taut between two points that are not opposite one another. Is there one shortest path between the points?
3. Form a triangle on a globe or ball by taping string between three points. Use a protractor to estimate the measures of the three angles formed. Is the sum of the measures of the angles of a triangle on a sphere equal to 180?
4. Now form larger triangles on the globe or ball and measure the three angles. Does the sum seem to depend on the size of the triangle? Explain.

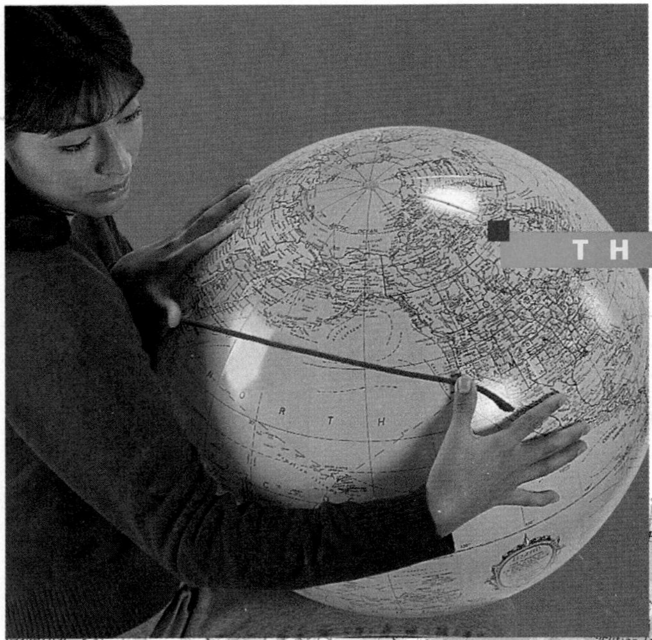

THINK AND DISCUSS

On this map, it looks like the red segment parallel to the lines of latitude is the shortest path from San Francisco to Tokyo. But on the globe, the red string shows that the shortest path is actually quite different.

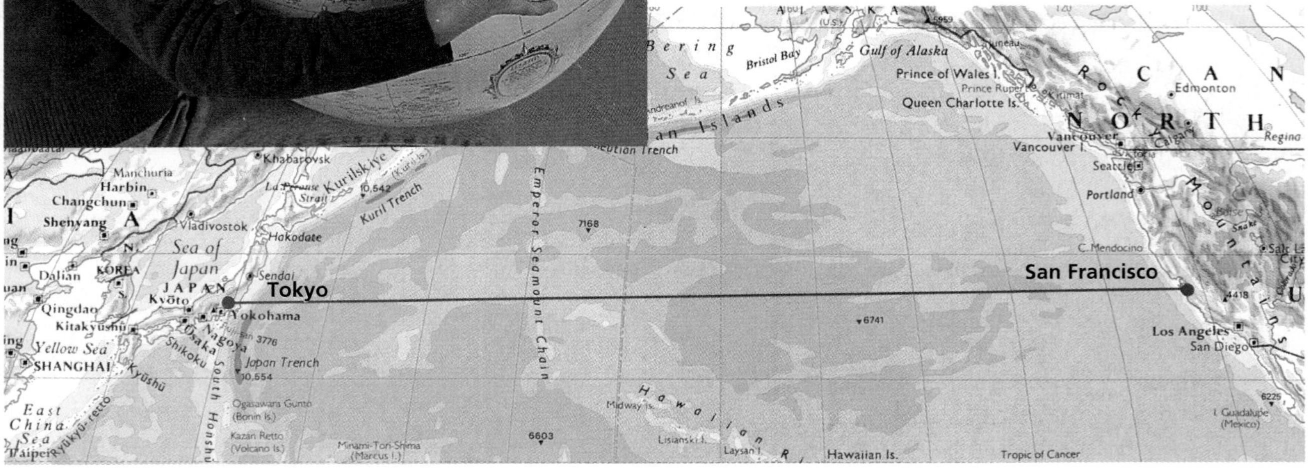

As the travel routes indicate, geometry on a sphere is different from geometry on a plane. The geometry of flat planes, straight lines, and points is called **Euclidean geometry.** In **spherical geometry,** *point* has the same meaning as in Euclidean geometry, but *line* and *plane* do not. In spherical geometry, a "plane" is the surface of a sphere and a "line" is a great circle of a sphere. Recall that a great circle is the intersection of a sphere and a plane that contains the center of the sphere.

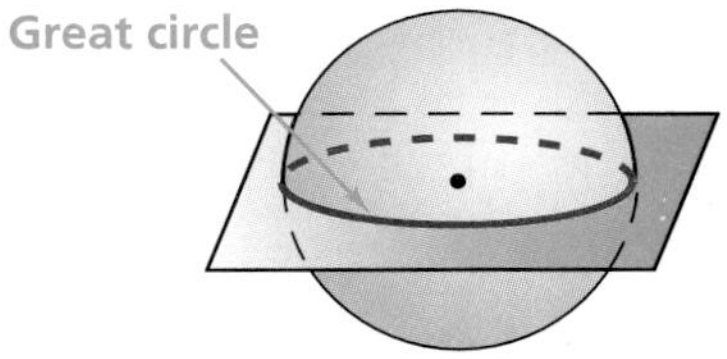

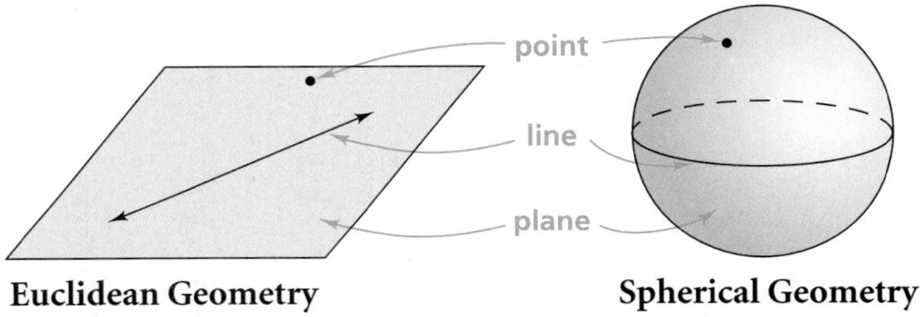

Example ***Relating to the Real World***

Navigation Lines of latitude and longitude are used to identify positions on Earth, much as x- and y-coordinates identify points on the coordinate plane. Which of these lines are great circles?

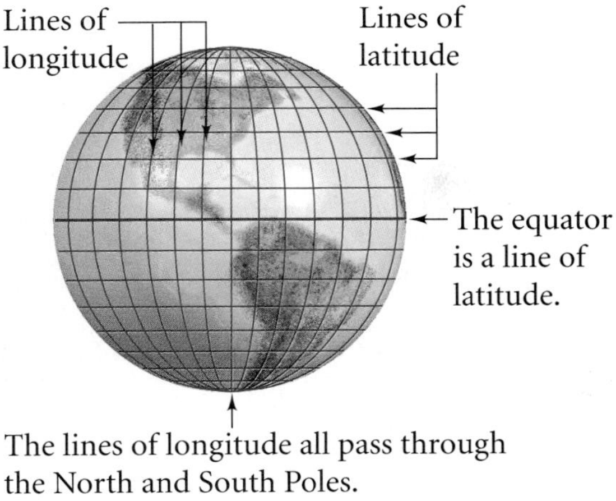

All the lines of longitude are great circles. The equator is the only line of latitude that is a great circle. All the other lines of latitude are circles smaller than a great circle.

Who? The physicist Albert Einstein (1879–1955) found that a non-Euclidean geometry was the best model for the paths of light rays over the huge distances of the universe.

You discovered in the Work Together that there are many shortest paths between any two points that are opposite each other on a sphere. You also discovered that there is only one shortest path between two points that are not opposite each other. Each of these shortest paths is an arc of a great circle.

5. **Try This** How many shortest paths are there between San Francisco and Tokyo? Explain your answer.

6. In Euclidean geometry, if two lines intersect, they intersect in exactly one point. Does this property hold in spherical geometry?

The study of parallel lines is an important part of Euclidean geometry. This study is based on the following postulate.

Postulate 7-3
Euclid's Parallel Postulate

Through a point not on a line, there is one and only one line parallel to the given line.

In spherical geometry, two lines (great circles) always intersect. Therefore, there are no parallel lines. The parallel postulate for spherical geometry is stated below.

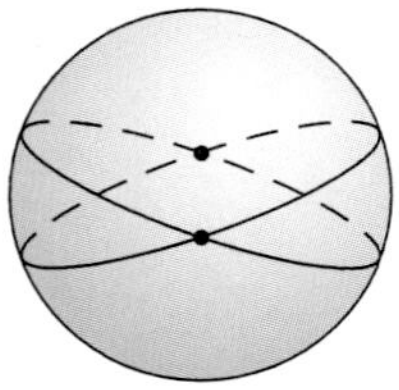

Postulate 7-4
Spherical Geometry Parallel Postulate

Through a point not on a line, there is no line parallel to the given line.

Who? Non–Euclidean geometry has been the focus of mathematician Linda G. Keen's work. She wrote her Ph.D. thesis on Riemann surfaces in 1964. She has returned to the subject in many subsequently published papers.

In the Work Together you found that the sum of the measures of the angles of a spherical triangle is greater than 180. Each of these spherical triangles has two right angles, so the sum of the measures of their angles exceeds 180 by the size of the vertex angle.

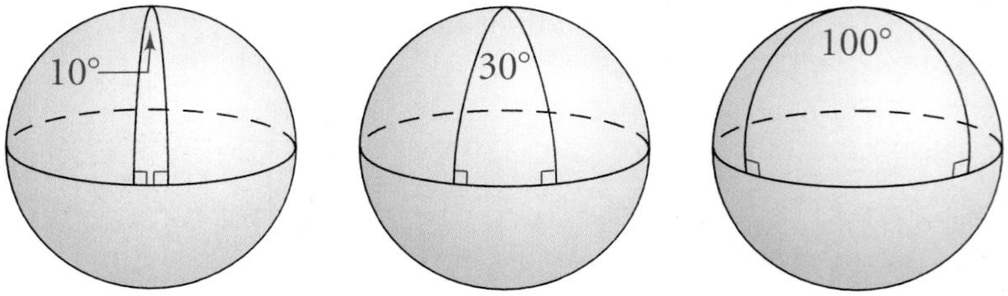

Comparison of Properties of Euclidean and Spherical Geometry

Property	Euclidean Geometry	Spherical Geometry
1. Given a line ℓ and a point P not on ℓ, there exists	one and only one line parallel to ℓ.	no line through P parallel to ℓ.
2. The sum of the measure of the angles of a triangle is	180.	greater than 180.
3. Two distinct lines intersect in	at most one point.	two points.
4. Through two given points, there is	one and only one line.	at least one line.
5. The shortest distance between two points is the length of	a segment.	an arc of a great circle.

Exercises ON YOUR OWN

Draw a sketch to illustrate each property of spherical geometry.

1. There are pairs of points on a sphere through which more than one line can be drawn.
2. A triangle can have more than one right angle.
3. You can draw two equiangular triangles that each have different angle measures.
4. *Geography* Bergen, Norway; Melbourne, Australia; and Montevideo, Uruguay, are on the same great circle. The distance from Bergen to Melbourne is about 9990 miles and the distance from Melbourne to Montevideo is about 7370 miles. Use the fact that the circumference of Earth is about 24,900 miles to find the distance from Montevideo to Bergen.
5. *Open-ended* Find three locations on a globe that are on a great circle and find the distances between them.

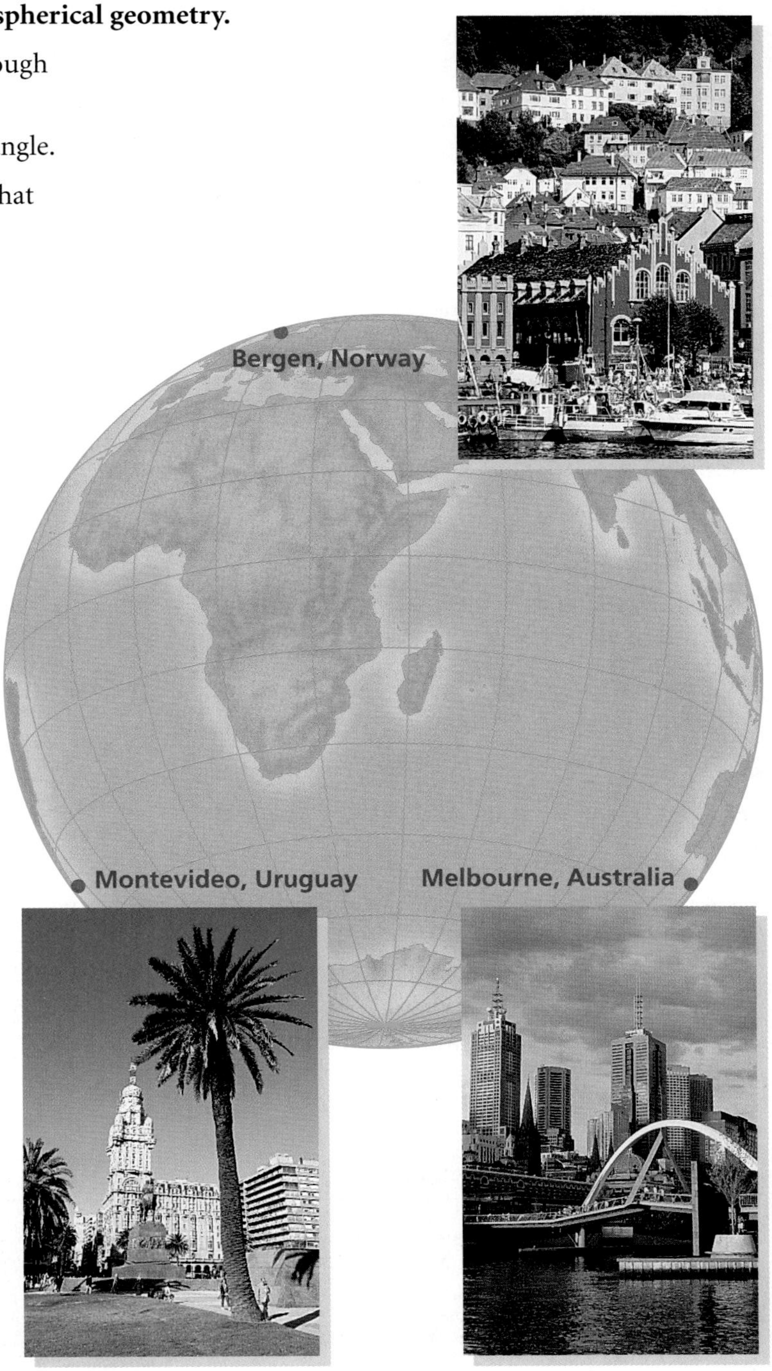

Draw a counterexample to show that the following properties of Euclidean geometry are not true in spherical geometry.

6. Through a point not on line ℓ, there exists one and only one line perpendicular to ℓ.
7. Two coplanar lines that are perpendicular to the same line do not intersect.
8. A triangle can have at most one obtuse angle.
9. If a triangle contains a right angle, then the other two angles in the triangle are complementary.
10. If two angles of one triangle are congruent to two angles of another triangle, then the third angles are congruent.
11. If A, B, and C are three points on a line, then exactly one of the points is between the other two.

12. *Writing* This figure seems to show parallel lines on a sphere. Explain why they are not parallel lines.

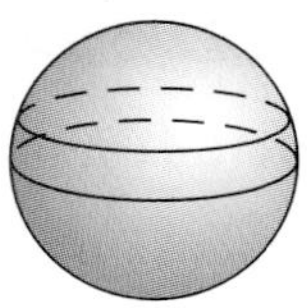

13. *Critical Thinking* Does the concept of a *ray* have any meaning in spherical geometry? Explain.

14. A well-known proof of the Triangle Angle-Sum Theorem is based on the concepts you studied in this chapter. Complete the proof.

Given: $\triangle ABC$
Prove: $m\angle A + m\angle B + m\angle 2 = 180$

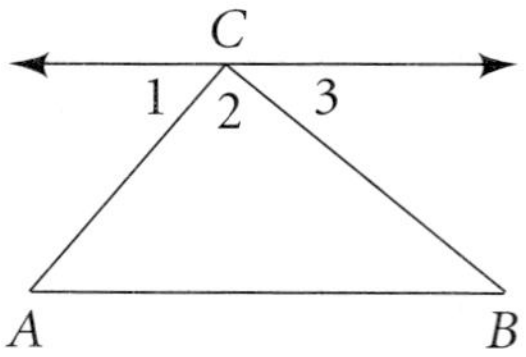

By **a.** __?__, there is exactly one line through C parallel to $\overleftrightarrow{AB}$. If two parallel lines are cut by a transversal, then **b.** __?__, so $m\angle A = m\angle 1$ and $m\angle B = m\angle 3$. By **c.** __?__, $m\angle 1 + m\angle 2 + m\angle 3 = 180$. By Substitution, **d.** __?__ $= 180$.

The following statements are true in Euclidean geometry. Make figures on a globe, ball, or balloon to help you decide which statements are false in spherical geometry.

15. Vertical angles are congruent.

16. An equilateral triangle is equiangular.

17. The base angles of an isosceles triangle are congruent.

18. Through a point on a line ℓ there exists one and only one line perpendicular to ℓ.

19. The measure of an exterior angle of a triangle equals the sum of the measures of its two remote interior angles.

20. In the 19th century, the Russian mathematician Nickolai Lobachevsky (1793–1856) and the Hungarian mathematician Janos Bolyai (1802–1860) independently developed a geometry that kept all of Euclid's postulates except his parallel postulate. They assumed that, in a plane, more than one parallel can be drawn to a given line through a given point. This geometry can be modeled on a saddle-shaped surface. Lines a and b are two lines through X that do not intersect c. So, they are parallel to c. Consider a triangle drawn on such a surface. Is the sum of the measures of the angles of this triangle less than, equal to, or greater than 180?

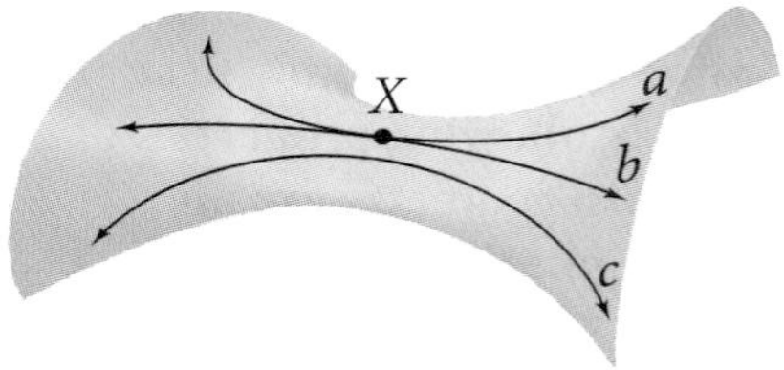

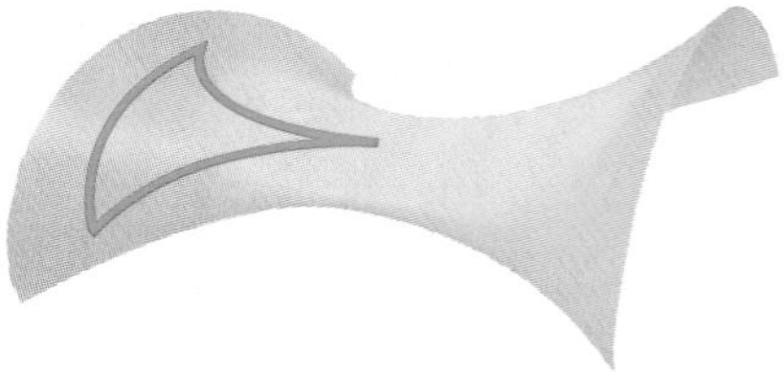

Chapter Project

Find Out by Modeling

A snail lives in a cubical glass terrarium with edges 1 ft long. How far does the snail have to go to visit all eight vertices of the cube? How far does it have to go to travel along all twelve edges of the cube?

Exercises MIXED REVIEW

Does each figure tessellate? If so, create a tessellation with it.

21. 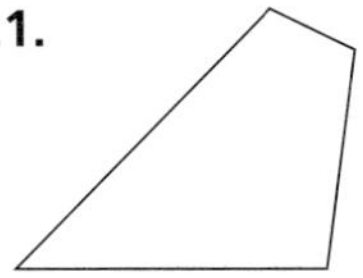

22. 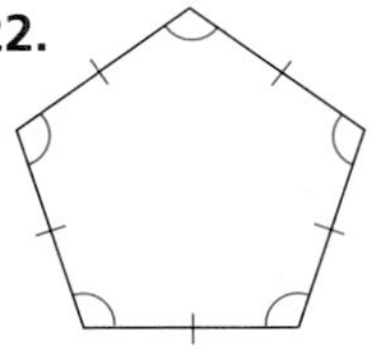

23. 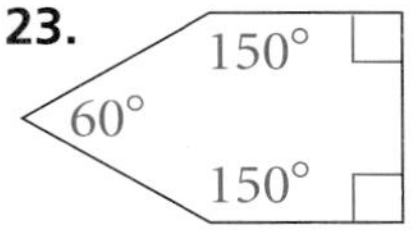

Algebra **Find the *slope of a line* that is perpendicular to the line with the given equation.**

24. $y = -3x - 4$

25. $y = \frac{3}{2}x + 6$

26. $2x - y = 7$

27. $6x + 2y = -3$

28. The altitude to the hypotenuse of an isosceles right triangle has length 5 cm. What is the area of the triangle?

29. The measures of two complementary angles are in the ratio of 2 to 3. Find their measures.

30. Standardized Test Prep A point (x, y) is reflected in a line parallel to the y-axis. Which point could be its reflection image?
A. $(2x, 2y)$ **B.** $(-2x, y)$ **C.** $(x, 2y)$ **D.** $(x, -2y)$ **E.** $(2x, -y)$

SELF ASSESSMENT PORTFOLIO

For your portfolio, select one or two items from your work for this chapter. Here are some possibilities:
- corrected work
- most-improved work
- drawings or models you have made

Explain why you have included each selection.

A Point in Time

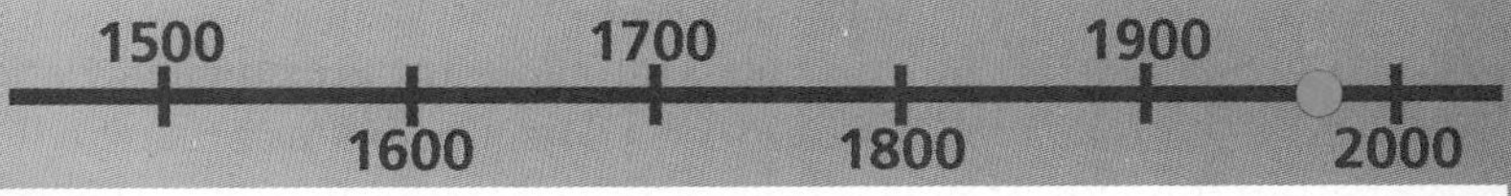

Computer Proofs

Regions that border one another on a map are shaded different colors so that you can tell them apart. Suppose that you wanted to color the regions of a map and, in doing so, use the least number of colors possible. How many would you need? As it turns out, the answer is four. Mathematicians had known this for years, but were unable to *prove* that the minimum number was four.

Then in **1976,** mathematicians Kenneth Appel and Wolfgang Haken found that all maps could be divided into 1936 categories. Using 1000 hours of computer time, they analyzed every category and proved that each could be shaded with four colors. Appel and Haken have shown the value of a computer for proving enormously complex theorems.

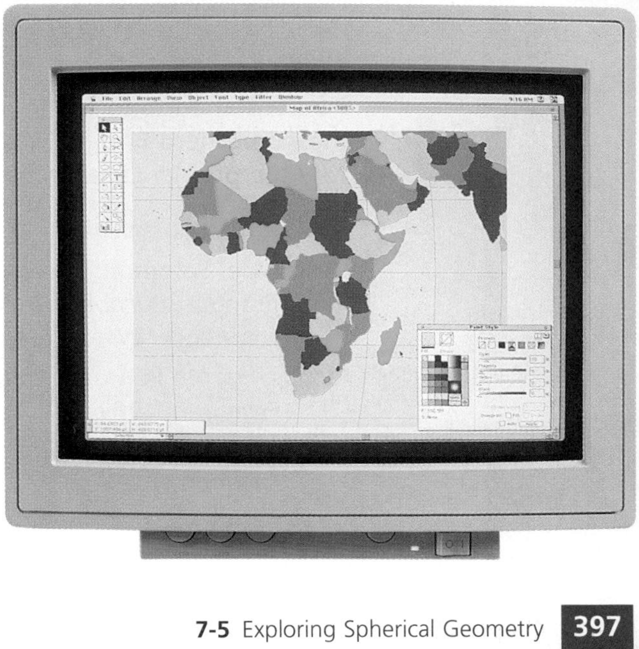

CHAPTER PROJECT

Finishing the Chapter Project

NETWORK NEWS

Find Out activities on pages 369, 382, 391, and 396 should help you complete your project. Design a network for your school, community, or another group. Here are some ideas.

- Computer network linking all the classrooms in the school
- Network of routes for school buses or city buses
- Fiber-optic network linking all cities with over half a million people

State the purpose of the network and any assumptions you made, such as budget limitations. Prepare a proposal to the decision-making body explaining the economic, environmental, and social benefits of your design.

Reflect and Revise

Ask a classmate to review your work with you. Together, check that your diagrams and explanations are clear and your information accurate. Have you taken into account the efficiency and convenience of your network? Could your project be better organized? Revise your work as needed.

Follow Up

Find out about one of the most famous problems involving networks—the Königsberg Bridge Problem. Describe the problem, its solution, and why the problem is so famous.

For More Information

Garfunkle, Solomon; Lynn Arthur Steen; and Joseph Malkevitch. *For All Practical Purposes: Introduction to Contemporary Mathematics.* New York: W. H. Freeman, 1988.

Jacobs, Harold R. *Mathematics, A Human Endeavor*. New York: W. H. Freeman, 1982.

RouteSmart Street Routing Software. Mineola, New York: Bowne Distinct, 1995.

CHAPTER 7

Wrap Up

Key Terms

alternate interior angles (p. 363)
corresponding angles (p. 363)
Euclidean geometry (p. 393)
flow proof (p. 372)
one-point perspective (p. 386)
perspective drawing (p. 385)
same-side interior angles (p. 363)
spherical geometry (p. 393)
transversal (p. 363)
two-point perspective (p. 386)
vanishing point (p. 386)

How am I doing?

SELF ASSESSMENT

- State three ideas from this chapter that you think are important. Explain your choices.
- Describe how you can determine if lines are parallel.

Parallel Lines and Related Angles 7-1

A **transversal** is a line that intersects two coplanar lines at two distinct points. Line t is a transversal of lines ℓ and m. $\angle 1$ and $\angle 4$ are **corresponding angles.** $\angle 3$ and $\angle 4$ are **alternate interior angles.** $\angle 2$ and $\angle 4$ are **same-side interior angles.**

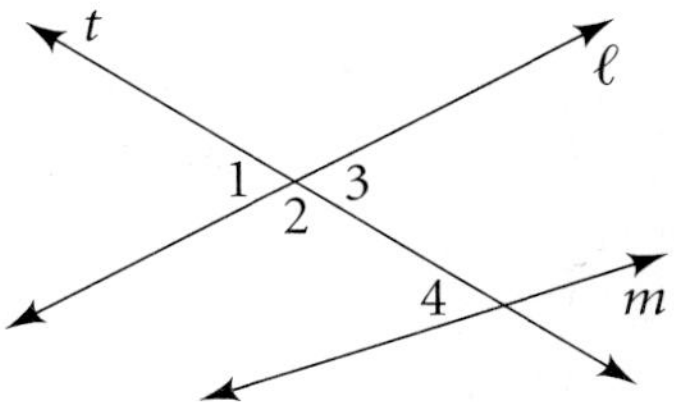

If two parallel lines are cut by a transversal, then the following are true.

- Corresponding angles are congruent.
- Alternate interior angles are congruent.
- Same-side interior angles are supplementary.

Find $m\angle 1$ and then $m\angle 2$. State the theorem or postulate that justifies your answer.

1.

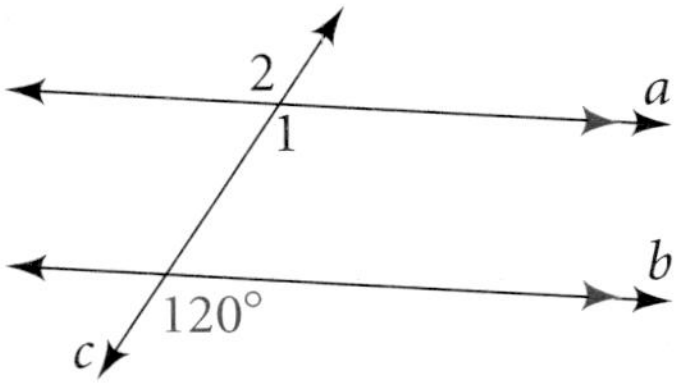

2.

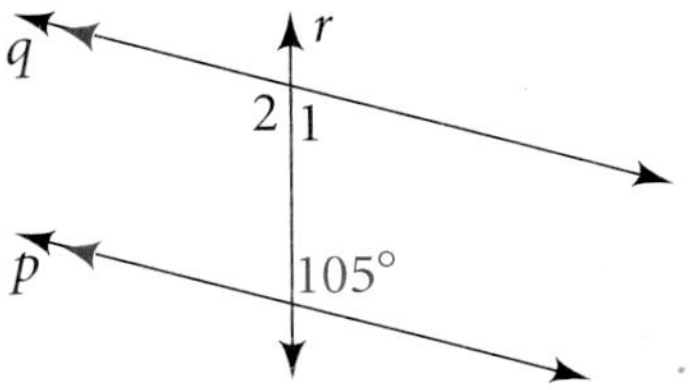

3.

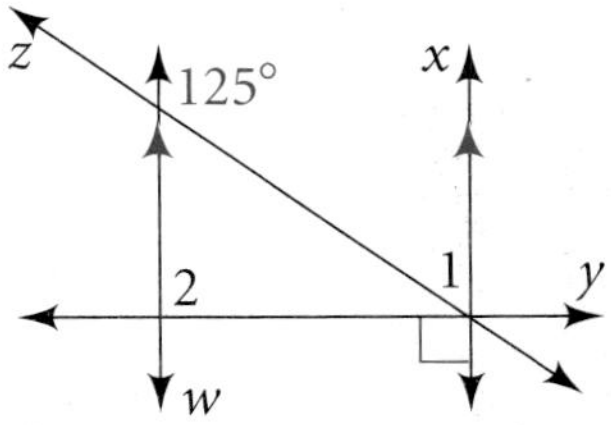

4. *Writing* Which angles of a parallelogram must be supplementary? Explain your answer.

5. *Open-ended* Describe two corresponding angles formed by lines in your classroom.

6. *Standardized Test Prep* If two parallel lines are cut by a transversal, which angles could be complementary?

I. alternate interior angles **II.** same-side interior angles **III.** corresponding angles

A. I only **B.** II only **C.** I and III only **D.** II and III only **E.** I, II, and III

Proving Lines Parallel 7-2

Two lines cut by a transversal are parallel if any of the following are true.

- Corresponding angles are congruent.
- Alternate interior angles are congruent.
- Same-side interior angles are supplementary.

A **flow proof** uses arrows to show the logical connections between the statements. Reasons are written below the statements.

Algebra **Determine the value of x for which $\ell \parallel m$.**

7.

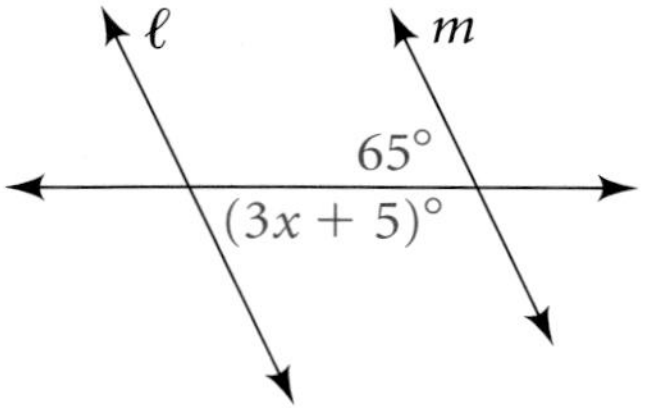

8.

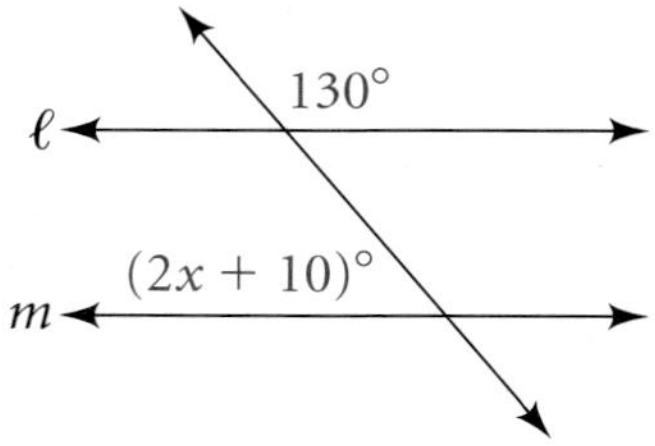

9.

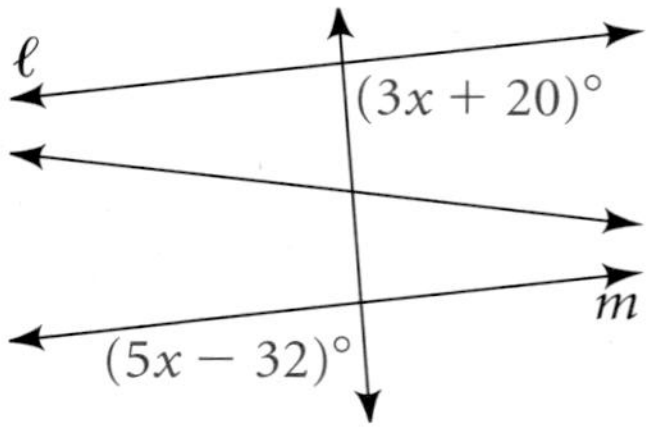

Constructing Parallel and Perpendicular Lines 7-3

You can construct a line parallel to a given line through a given point not on the line. You can also construct the perpendicular to a given line through a given point on the line or through a given point not on the line.

Use the segments at the right.

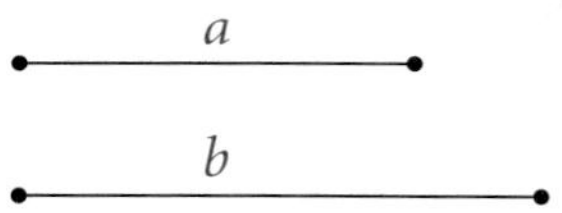

10. Construct a parallelogram with side lengths a and b.

11. a. Construct a right triangle with a leg of length b and a hypotenuse with length $2b$.

b. What is the length of the other leg?

c. Critical Thinking Find the measure of each angle.

Parallel Lines and Perspective Drawing 7-4

You can use **perspective drawing** to draw three-dimensional objects on a flat surface so that they look the same as they appear to the eye. Perspective drawing involves the use of vanishing points. A **vanishing point** is a point on the "horizon" of the drawing where parallel lines "meet." When there is one vanishing point, the drawing is in **one-point perspective.** When the drawing has two vanishing points, it is in **two-point perspective.**

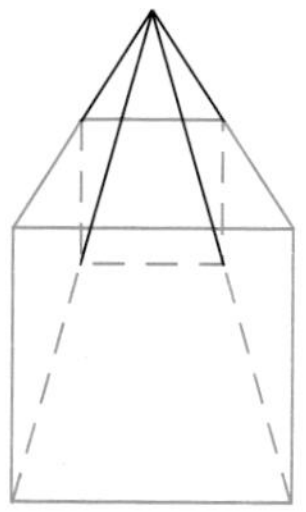

one-point perspective

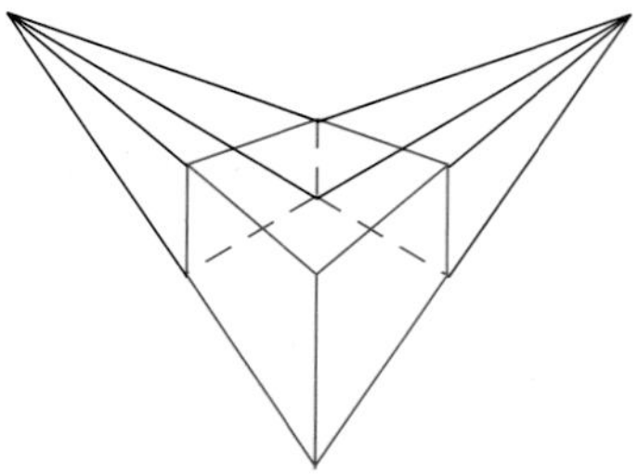

two-point perspective

Copy each figure and locate the vanishing point(s).

12.

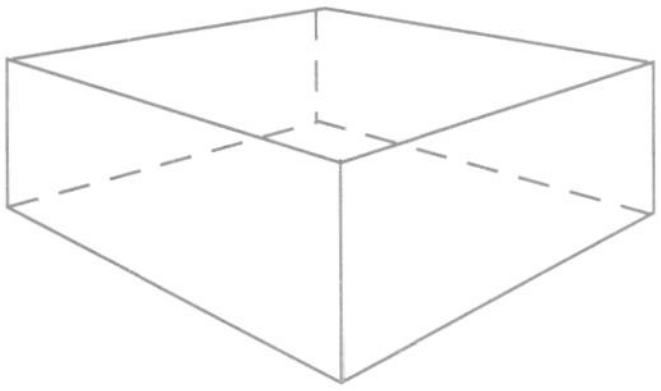

13.

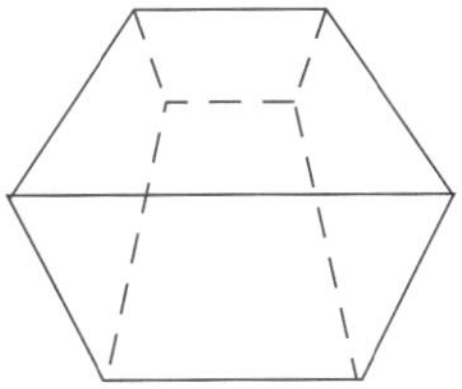

14.

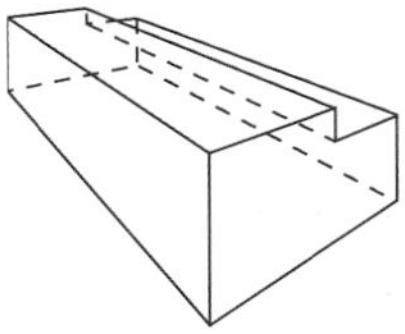

Exploring Spherical Geometry 7-5

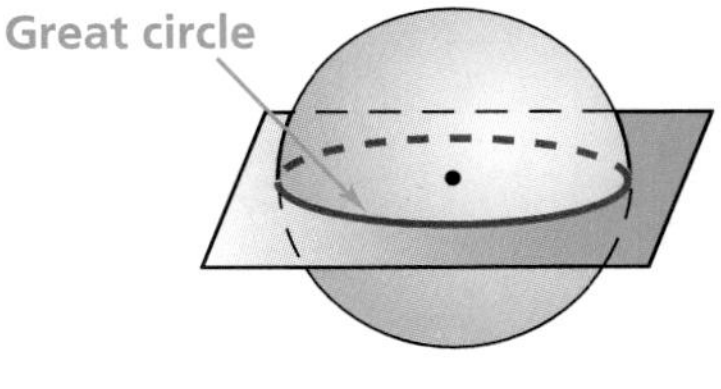

The geometry of points, straight lines, and flat planes is **Euclidean geometry.** Geometry on a sphere is **spherical geometry.** In spherical geometry, "point" has the same meaning as in Euclidean geometry, but a "line" is a great circle of the sphere, and a "plane" is the surface of the sphere.

Some statements that are true in Euclidean geometry are not true in spherical geometry. For example, in Euclidean geometry, there is one and only one line parallel to a given line through a given point not on the line. This is Euclid's Parallel Postulate. In spherical geometry, there is no line parallel to a given line through a point not on the line.

Draw a counterexample to show that the following properties of Euclidean geometry are not true in spherical geometry.

15. An equiangular triangle cannot have a right angle.

16. Perpendicular lines intersect at one point.

17. A right triangle has two acute angles.

18. Two points determine a line.

Getting Ready for CHAPTER 8

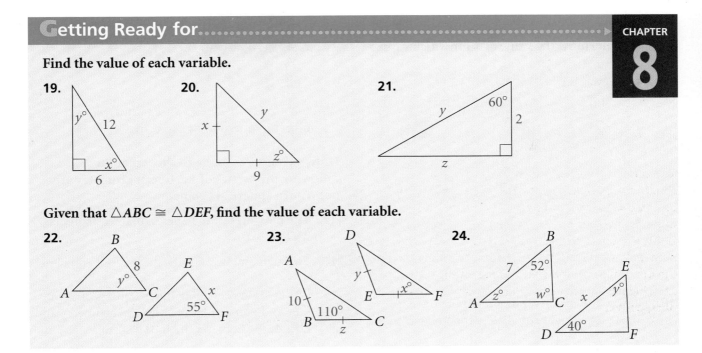

Find the value of each variable.

19.

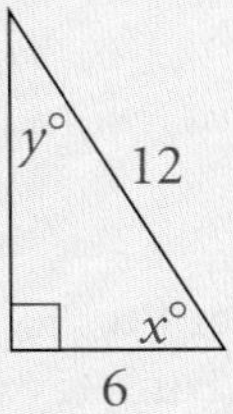

20.

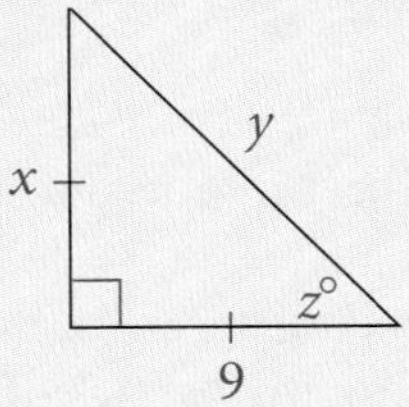

21.

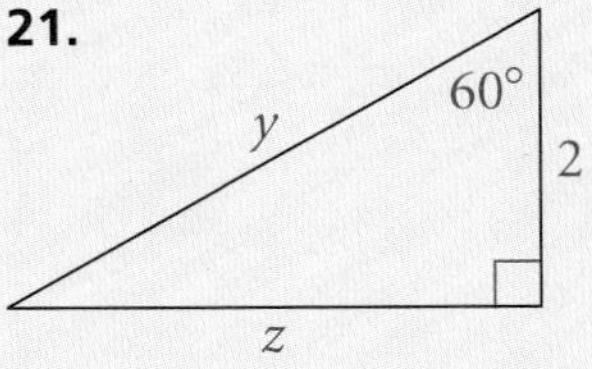

Given that $\triangle ABC \cong \triangle DEF$, find the value of each variable.

22.

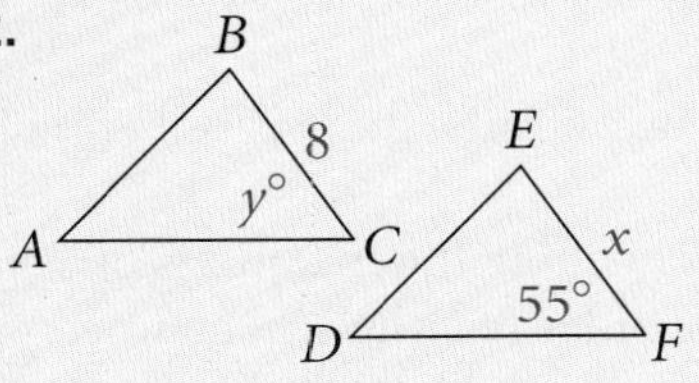

23.

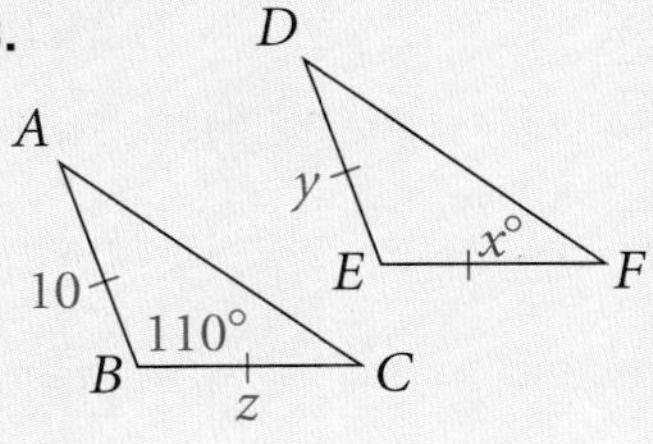

24.

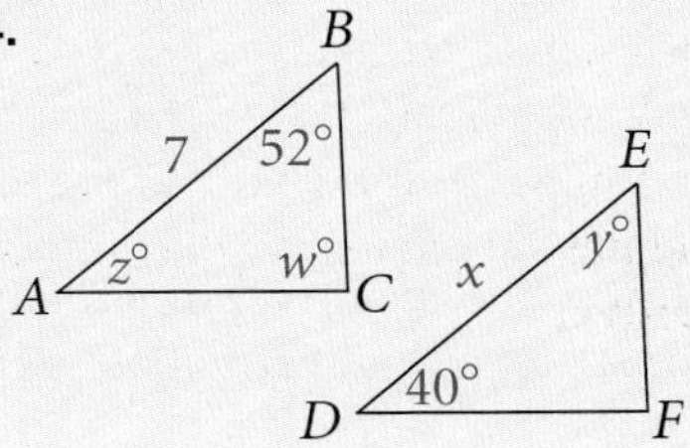

CHAPTER 7 Assessment

Find $m\angle 1$, and then $m\angle 2$. State the theorems or postulates that justify your answers.

1.

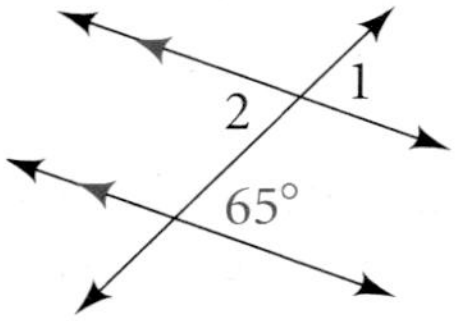

2.

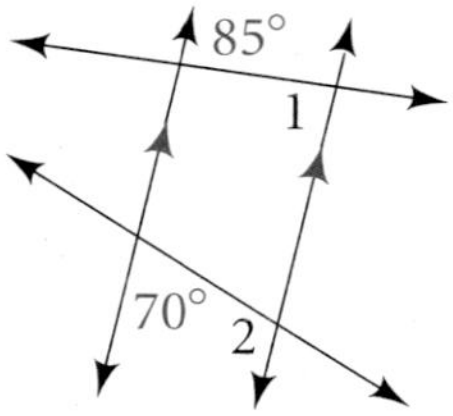

3.

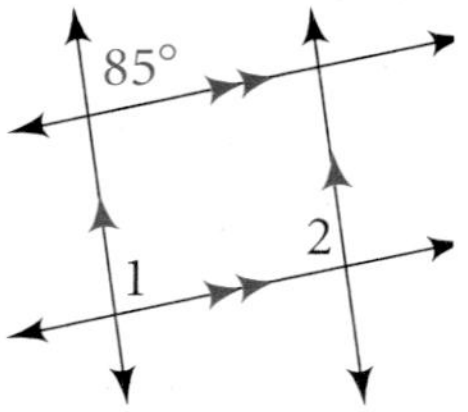

4.

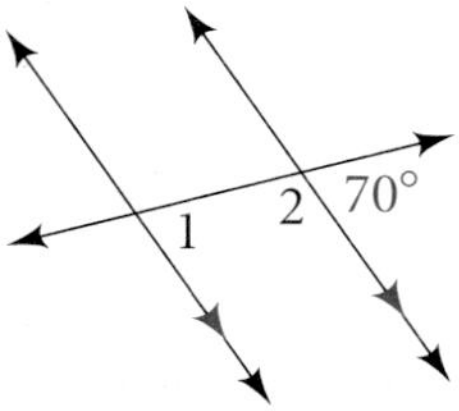

5. *Standardized Test Prep* Two lines are parallel. What could be the measures of two of their same-side interior angles?

I. 40 and 140 **II.** 90 and 90
III. 60 and 60 **IV.** 27 and 27

A. I only **B.** II only **C.** II and III
D. I and II **E.** I, II, and III

Algebra **Determine the value of x for which $\ell \parallel m$.**

6.

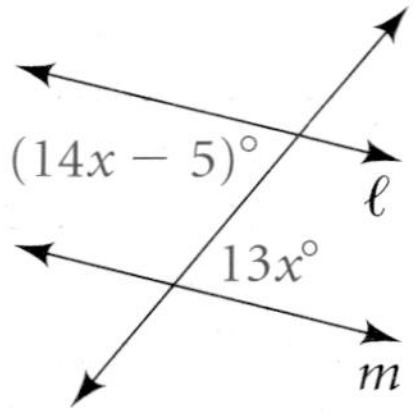

7.

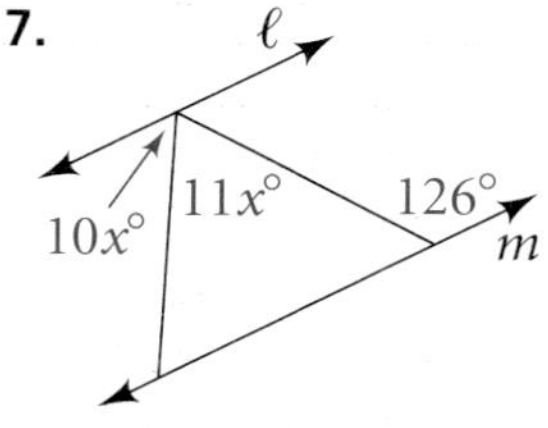

8.

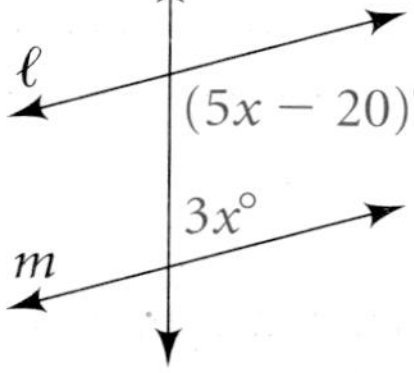

9.

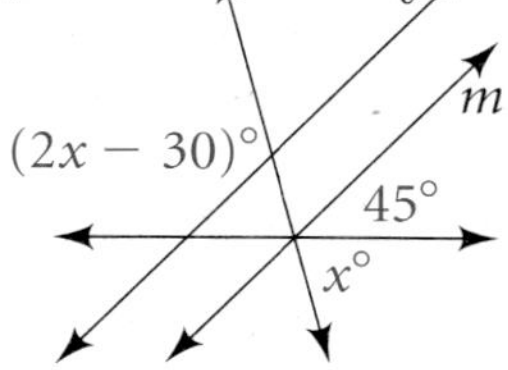

10. Draw a line m and a point T on the line. Construct a line through T perpendicular to m.

11. Draw an angle $\angle ABC$. Then construct line m through C so that $m \parallel \overleftrightarrow{BA}$. Construct D on m so that $ABCD$ is a parallelogram.

12. Construct a rectangle.

Is each object drawn in one- or two-point perspective?

13.

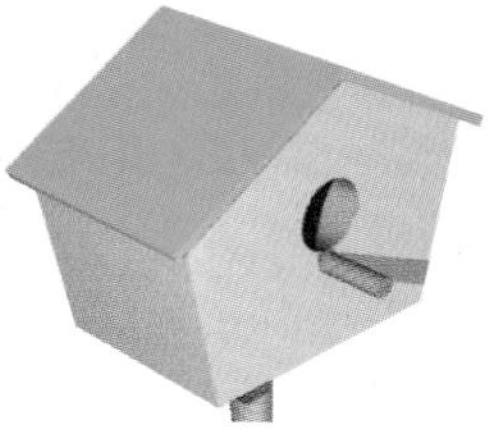

14.

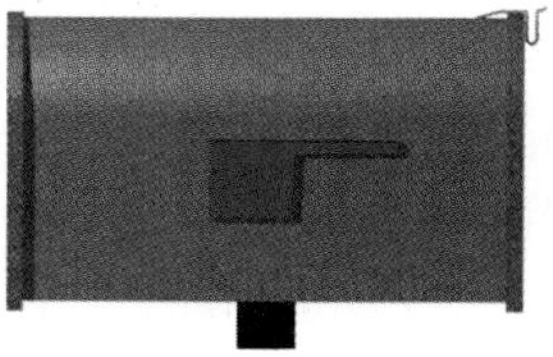

15.

16. *Open-ended* Sketch a three-dimensional object in one-point perspective.

Draw a counterexample to show that the following properties of Euclidean geometry are not true in spherical geometry.

17. The measures of the angles of an equiangular triangle are 60.

18. Through any two points, there is only one line.

19. *Writing* Describe how the meanings of *point, line,* and *plane* differ in spherical geometry and Euclidean geometry. Include a sketch.

CHAPTER 7 Preparing for Standardized Tests

For Exercises 1–9, choose the correct letter.

1. What is the surface area of a sphere with radius 7 cm?

   A. $\frac{196}{3}\pi\,\text{cm}^2$ **B.** $49\pi\,\text{cm}^2$ **C.** $196\pi\,\text{cm}^2$
   D. $14\pi\,\text{cm}^2$ **E.** $\frac{1372}{3}\pi\,\text{cm}^2$

2. Two of the angles of a triangle on a sphere are right angles. What could be the measure of the third angle?

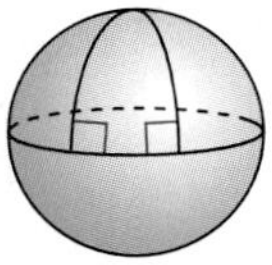

   I. 45 **II.** 90 **III.** 135 **IV.** 180

   A. I and II **B.** II and III **C.** III and IV
   D. I and IV **E.** I, II, and III

3. Which drawing is in two-point perspective?

   A.
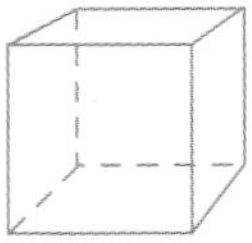

   B.
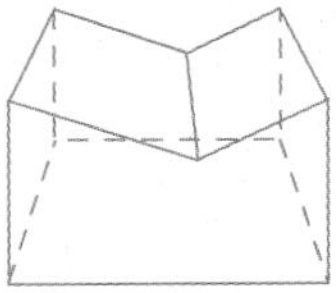

   C.
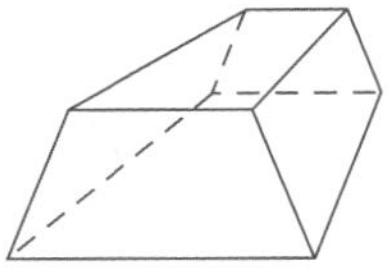

   D.
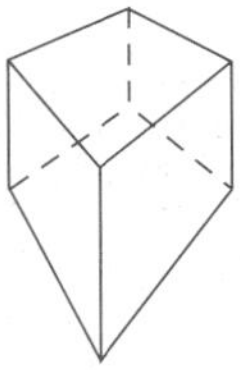

   E. none of the above

4. Which can you use to prove two lines parallel?
   A. supplementary corresponding angles
   B. congruent alternate interior angles
   C. congruent vertical angles
   D. congruent same-side interior angles
   E. none of the above

5. What kind of symmetry does the figure have?
   A. 60° rotational symmetry
   B. 90° rotational symmetry
   C. line symmetry
   D. point symmetry
   E. all of the above

6. One leg of an isosceles right triangle is 3 in. long. What is the length of the hypotenuse?

   A. 3 in. **B.** $3\sqrt{2}$ in. **C.** $3\sqrt{3}$ in.
   D. 6 in. **E.** none of the above

Compare the boxed quantity in Column A with the boxed quantity in Column B. Choose the best answer.

A. The quantity in Column A is greater.
B. The quantity in Column B is greater.
C. The two quantities are equal.
D. The relationship cannot be determined on the basis of the information supplied.

	Column A	Column B
7.	the sum of the measures of the angles of a triangle in Euclidean geometry	the sum of the measures of the angles of a triangle in spherical geometry
8.	the volume of a cone with radius 8 in. and height 9 in.	the volume of a cylinder with radius 3 in. and height 6 in.
9.	the number of pairs of corresponding angles formed by two parallel lines and a transversal	the number of pairs of alternate interior angles formed by three parallel lines and a transversal

Find each answer.

10. A cone has the same radius and height as a cylinder. How are their volumes related?

11. **Writing** Describe two ways in which a triangle on a sphere differs from a triangle in a plane.

CHAPTER 8

Proving Triangles Congruent

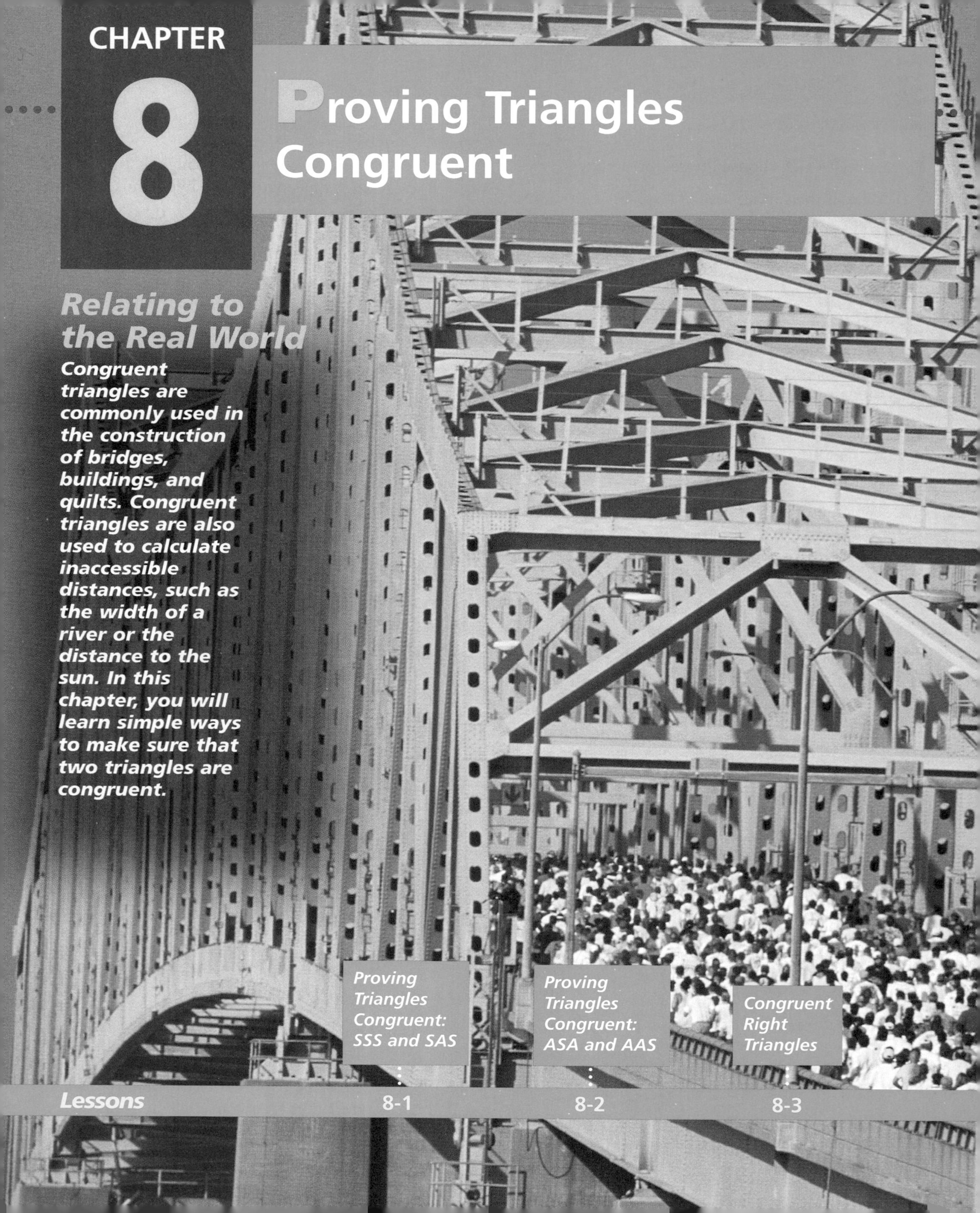

Relating to the Real World

Congruent triangles are commonly used in the construction of bridges, buildings, and quilts. Congruent triangles are also used to calculate inaccessible distances, such as the width of a river or the distance to the sun. In this chapter, you will learn simple ways to make sure that two triangles are congruent.

Lessons

- 8-1 *Proving Triangles Congruent: SSS and SAS*
- 8-2 *Proving Triangles Congruent: ASA and AAS*
- 8-3 *Congruent Right Triangles*

CHAPTER PROJECT

Tri Tri Again

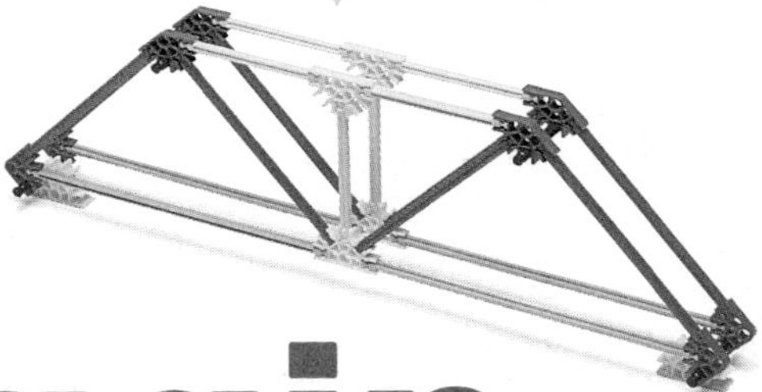

Have you ever wondered how bridges stay up? How do such frail-looking frameworks stretch through the air without falling? How can they withstand the twisting forces of hurricane winds and the rumbling weight of trucks and trains? Part of the answer lies in the natural strength of triangles.

In your project for this chapter, you will explore how engineers use triangles to construct safe, strong, stable structures. You will then get a chance to apply these ideas as you design and build your own bridge with toothpicks or craft sticks. You will see how a simple shape can often be the strongest one.

To help you complete the project:

- **p. 412** *Find Out by Modeling*
- **p. 424** *Find Out by Observing*
- **p. 438** *Find Out by Investigating*
- **p. 440** *Finishing the Project*

What You'll Learn

- Proving two triangles congruent using the SSS and SAS postulates

...And Why

To develop an understanding of congruent triangles, which are important in the design and construction of buildings and bridges

What You'll Need

- protractor
- ruler
- straws
- scissors
- compass
- straightedge

8-1 Proving Triangles Congruent: SSS and SAS

WORK TOGETHER

- Have each person in your group use straws to make a triangle with sides 2 in., 3 in., and 4 in. long. Compare your triangles.

1. Make a **conjecture** about two triangles in which three sides of one triangle are congruent to three sides of the other triangle.
2. **Verify** your conjecture by selecting other lengths for the sides of a triangle. Have each person in your group make the triangle. Compare your triangles.

THINK AND DISCUSS

Using the SSS Postulate

In Lesson 2-6, you learned that if two triangles have three pairs of congruent corresponding angles and three pairs of congruent corresponding sides, then the triangles are congruent.

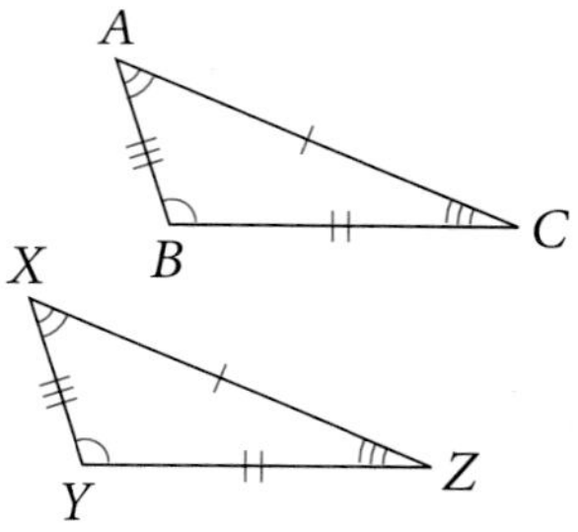

If you know this,	then you know this.
$\angle A \cong \angle X$	$\triangle ABC \cong \triangle XYZ$
$\angle B \cong \angle Y$	
$\angle C \cong \angle Z$	
$\overline{AB} \cong \overline{XY}$	
$\overline{AC} \cong \overline{XZ}$	
$\overline{BC} \cong \overline{YZ}$	

In the Work Together, you discovered that you don't need all six of the congruence statements to ensure that two triangles are congruent. If you know that corresponding sides are congruent, then you know that the triangles are congruent.

Postulate 8-1
Side-Side-Side Postulate (SSS Postulate)

If three sides of one triangle are congruent to three sides of another triangle, then the two triangles are congruent.

$\triangle GHF \cong \triangle PQR$

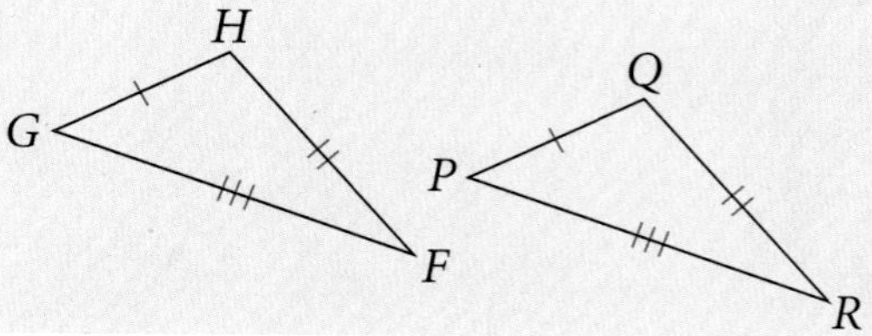

Example 1 Relating to the Real World

Bridges Do you have enough information to prove the two triangles in the bridge congruent? If so, write a two-column proof.

Given: $\overline{AB} \cong \overline{CB}, \overline{AD} \cong \overline{CD}$
Prove: $\triangle ABD \cong \triangle CBD$

Quick Review

The Reflexive Property of Congruence tells you that a segment, such as $\overline{BD}$ in this figure, is congruent to itself. It also tells you that an angle is congruent to itself.

Plan for Proof To prove the triangles are congruent, you can use the SSS Postulate. You are given that two pairs of sides are congruent. You can use the Reflexive Property of Congruence to show that $\overline{BD}$ in $\triangle ABD$ is congruent to $\overline{BD}$ in $\triangle CBD$.

Statements	Reasons
1. $\overline{AB} \cong \overline{CB}$	1. Given
2. $\overline{AD} \cong \overline{CD}$	2. Given
3. $\overline{BD} \cong \overline{BD}$	3. Reflexive Property of $\cong$
4. $\triangle ABD \cong \triangle CBD$	4. SSS Postulate

3. A two-column proof is just one way to write this proof. Rewrite the proof as a paragraph proof or as a flow proof.

4. **Try This** $\overline{RS} \cong \overline{TK}$. What additional information do you need to prove that $\triangle RSK \cong \triangle TKS$ by the SSS Postulate?

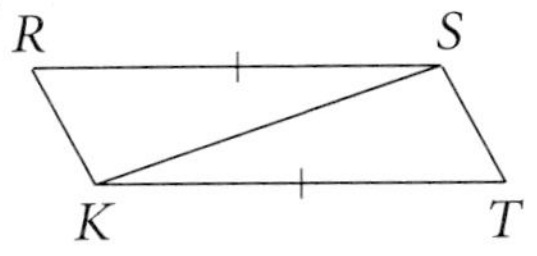

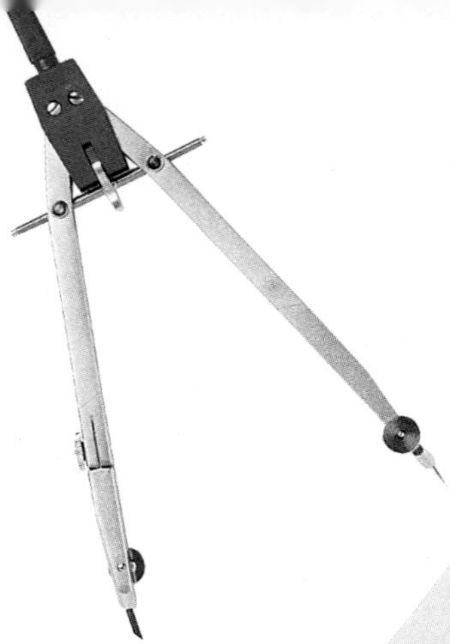

WORK TOGETHER

Work in groups of three.

- Have each person use a straightedge to draw one of the following types of triangles: acute, obtuse, right. Make sure that at least one of the triangles is scalene. Label your triangle $\triangle HAM$.
- Exchange triangles within your group. Use a compass and straightedge to construct a copy of $\angle H$. Label the angle $\angle L$.
- On one side of $\angle L$, construct $\overline{LT}$ congruent to $\overline{HM}$. On the other side of $\angle L$, construct $\overline{LE}$ congruent to $\overline{HA}$. Draw $\overline{ET}$, forming $\triangle LET$.
- Cut out the triangles. Place one triangle over the other so that the corresponding parts match.

5. Make a **conjecture.** What seems to be true when an angle of one triangle is congruent to an angle of another triangle, and the two pairs of sides that form these angles are congruent?

THINK AND DISCUSS

Using the SAS Postulate

The word *included* is used frequently when referring to the angles and the sides of a triangle.

$\overline{AN}$ is included between $\angle N$ and $\angle A$.

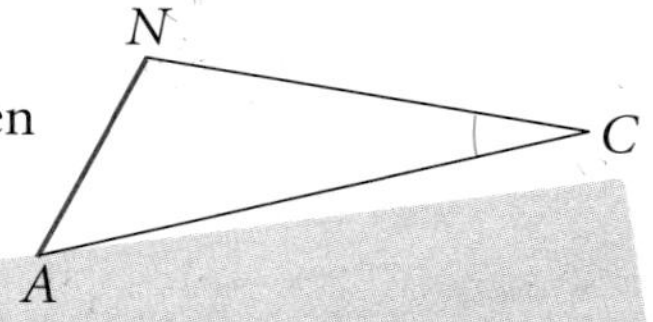

$\angle C$ is included between $\overline{NC}$ and $\overline{AC}$.

6. Which angle is included between $\overline{NC}$ and $\overline{AN}$?

7. Which side is included between $\angle C$ and $\angle N$?

The conjecture you made in the Work Together suggests another important postulate for proving triangles congruent.

Postulate 8-2
Side-Angle-Side Postulate (SAS Postulate)

If two sides and the included angle of one triangle are congruent to two sides and the included angle of another triangle, then the two triangles are congruent.

$\triangle CBA \cong \triangle DFE$

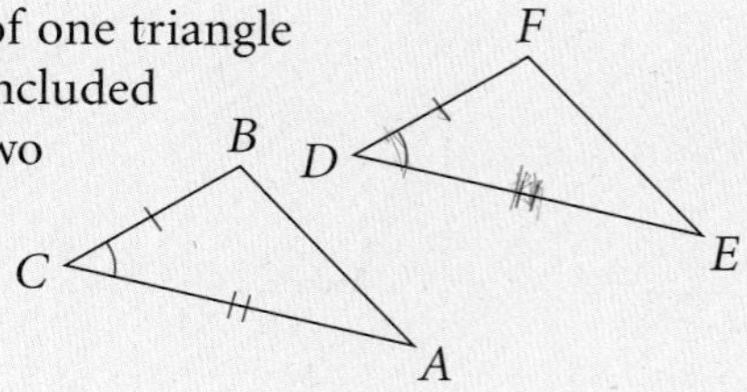

8. **Try This** What additional information do you need to prove the triangles congruent by the SAS Postulate?

a.

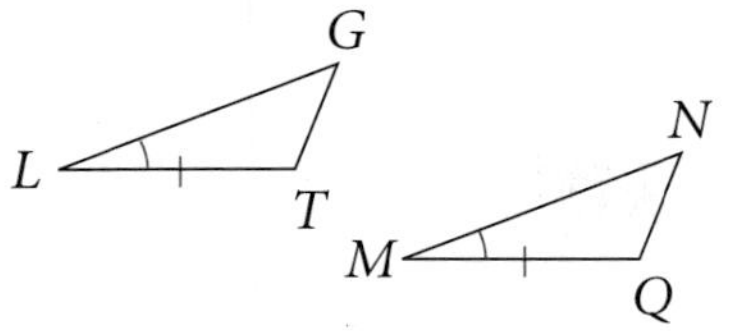

b.

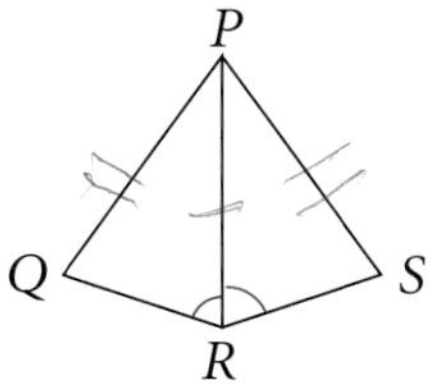

Example 2

Given: $\overline{RE} \cong \overline{CA}$, $\overline{RD} \cong \overline{CT}$,
$\angle R \cong \angle T$

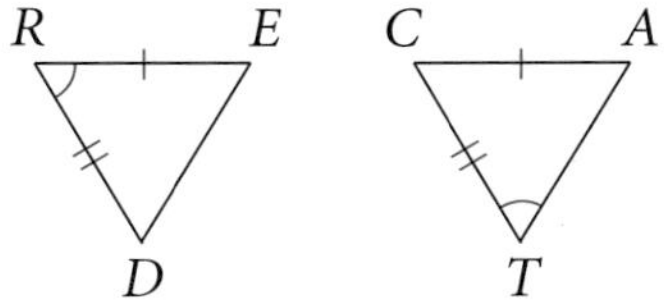

From the information given, can you prove $\triangle RED \cong \triangle CAT$? Explain.

No, there is not enough information to prove $\triangle RED \cong \triangle CAT$. $\angle T$ is not *included* between the congruent sides. $\triangle RED$ may or may not be congruent to $\triangle CAT$.

9. **Try This** In Example 2, suppose you also know that $\triangle CAT$ is equilateral. Can you prove $\triangle RED \cong \triangle CAT$? Explain.

10. **Try This** Suppose that in Example 2, $RE = CA = RD = CT = 7$ and $m\angle R = m\angle T = 58$.
    - **a.** Sketch each figure, labeling angles and sides with their measures.
    - **b.** Can you prove that $\triangle RED \cong \triangle CAT$? Explain.

Exercises ON YOUR OWN

Try to answer these questions without drawing a figure.

1. Which side is included between $\angle X$ and $\angle Z$ in $\triangle XYZ$?
2. Which angle is included between $\overline{XY}$ and $\overline{XZ}$ in $\triangle XYZ$?

Name the triangle congruence postulate you can use to prove each pair of triangles congruent.

3.

4.

5. **Critical Thinking** Suppose a friend insists that since you can prove triangles congruent using the SSS Congruence Postulate, there should be an AAA Congruence Postulate. Draw two triangles that will show that your friend's assumption is incorrect.

Decide whether you can use the SSS or SAS Postulate to prove that the triangles below are congruent. If so, write the congruence and identify the postulate. If not, write *not possible.*

6. 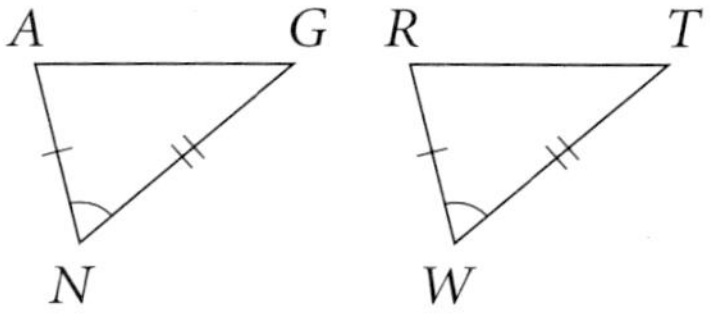

7. 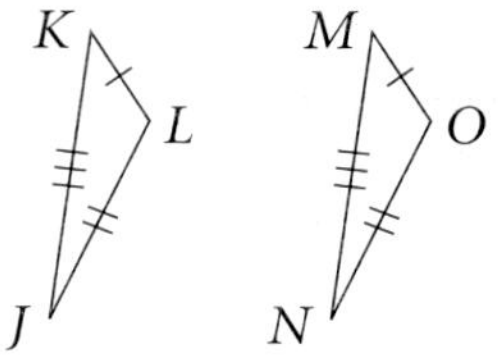

8. 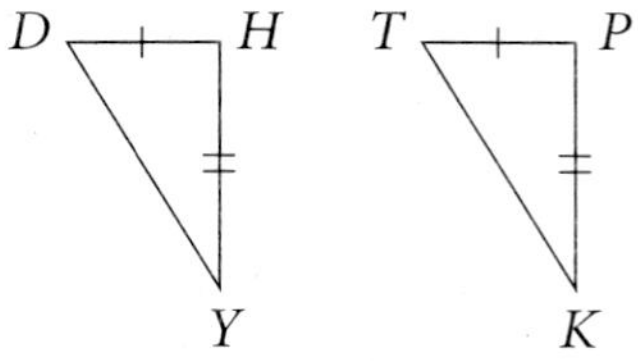

9. 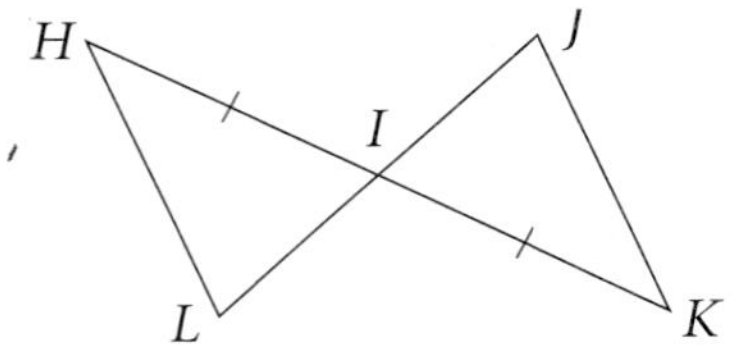

10. 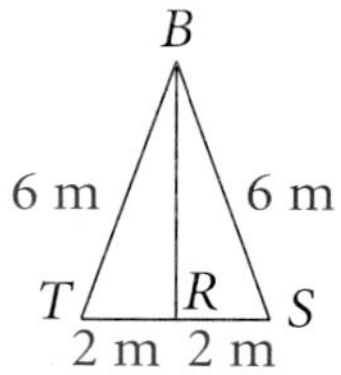

11. 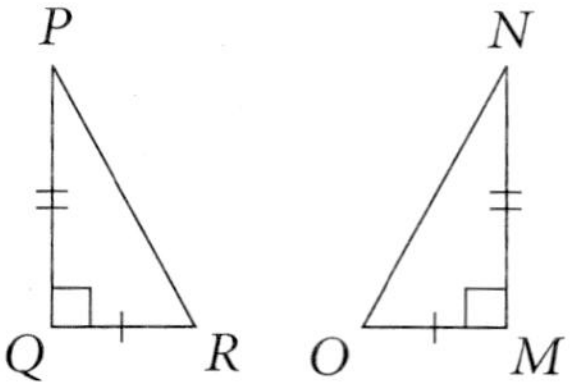

12. 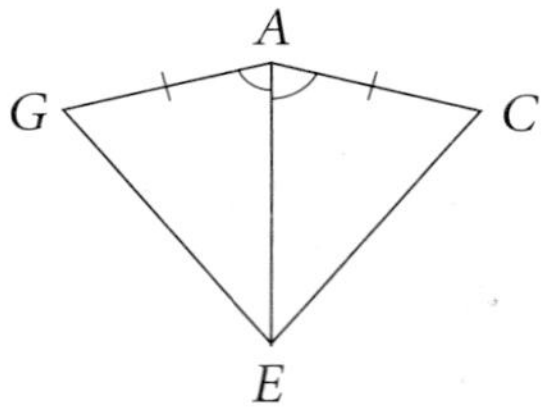

13. 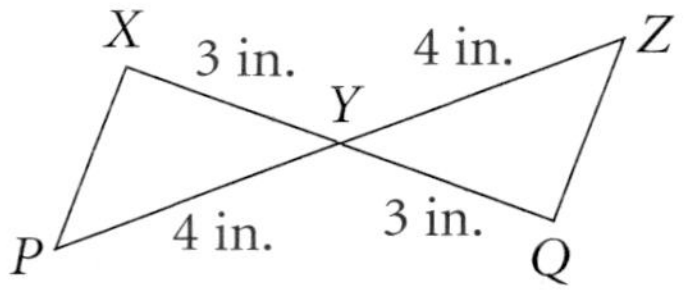

14. 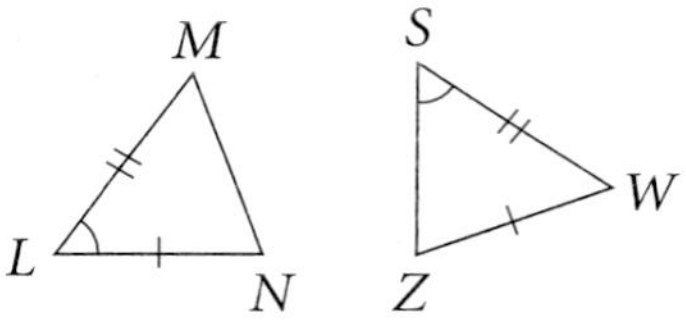

15. **Open-ended** List several places where you have seen congruent triangles used.

16. **Preparing for Proof** Supply the reasons in this proof.

Given: X is the midpoint of $\overline{AG}$ and of $\overline{NR}$.

Prove: $\triangle ANX \cong \triangle GRX$

Statements	Reasons
1. $\angle 1 \cong \angle 2$	a. ?
2. X is the midpoint of $\overline{AG}$.	b. ?
3. $\overline{AX} \cong \overline{GX}$	c. ?
4. X is the midpoint of $\overline{NR}$.	d. ?
5. $\overline{NX} \cong \overline{RX}$	e. ?
6. $\triangle ANX \cong \triangle GRX$	f. ?

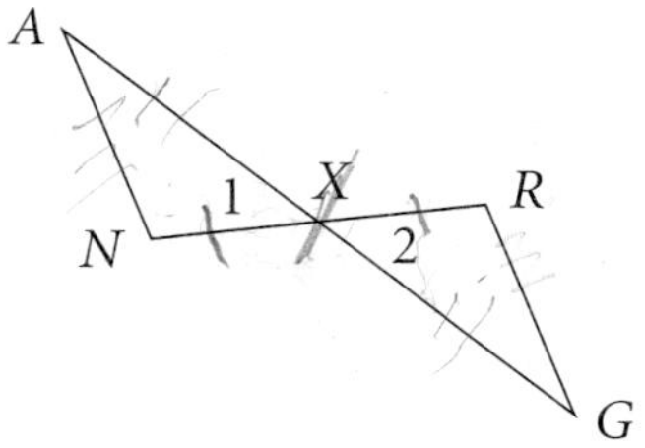

17. **Paper Folding** Draw an isosceles triangle. Fold the bisector of the vertex angle. Are the two triangles formed congruent? How do you know?

18. Complete the flow proof.

Given: $\overline{AE} \cong \overline{DE}$, $\overline{AB} \cong \overline{DC}$

Prove: $\triangle ABE \cong \triangle DCE$

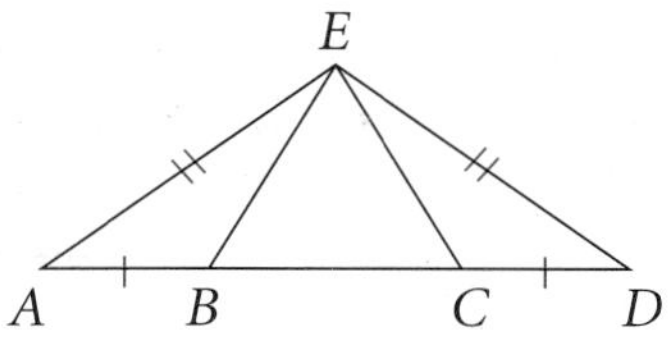

$\overline{AE} \cong \overline{DE}$ → **b.** ?

a. ?

b. ? — Base ∠s of an isosceles △ are ≅.

$\overline{AB} \cong \overline{DC}$ → **d.** ?

c. ?

d. ? — SAS Postulate

Write a two-column proof.

19. Given: $\overline{AE}$ and $\overline{BD}$ bisect each other.

Prove: $\triangle ACB \cong \triangle ECD$

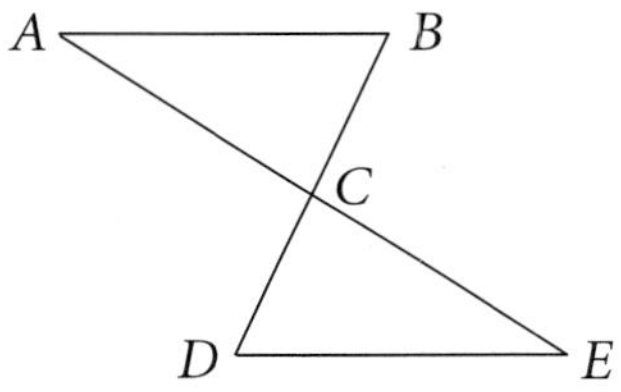

20. Given: $\overline{GK}$ bisects $\angle JGM$, $\overline{GJ} \cong \overline{GM}$

Prove: $\triangle GJK \cong \triangle GMK$

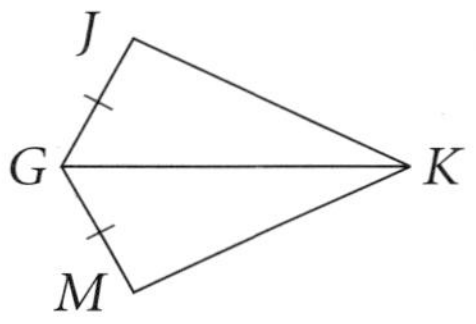

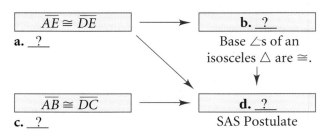

21. a. What two symbols are combined to form the congruence symbol?

b. Writing Explain why the congruence symbol makes sense.

Choose **Write a two-column proof, a paragraph proof, or a flow proof.**

22. Given: ⊙O with A, B, C, D on the circle

Prove: $\triangle AOB \cong \triangle COD$

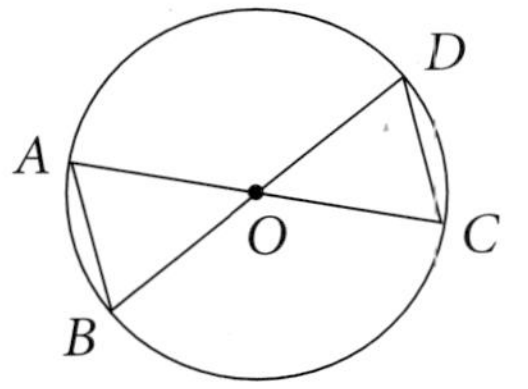

23. Given: $\overline{FG} \parallel \overline{KL}$, $\overline{FG} \cong \overline{KL}$

Prove: $\triangle FGK \cong \triangle KLF$

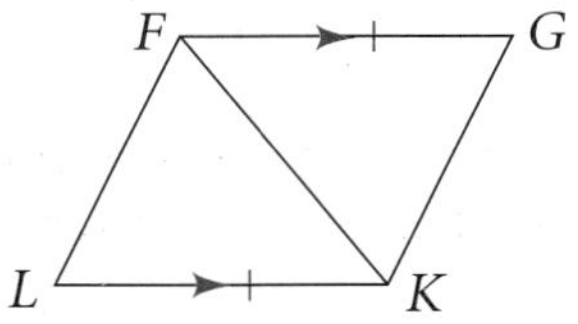

24. **Critical Thinking** Suppose you construct two isosceles triangles so that the four legs are the same length. Can you prove that the triangles are congruent? **Justify** your answer, or give a counterexample.

25. **Critical Thinking** Janet knows that the four sides of quadrilateral *ABCD* are congruent to the four sides of quadrilateral *EFGH*. Are the two quadrilaterals necessarily congruent? Explain.

Chapter Project — *Find Out by Modeling*

Many structures have straight beams that meet at *joints.* You can use models to explore ways to strengthen joints.

- Cut seven cardboard strips about 6 in. by $\frac{1}{2}$ in. Make a square frame and a triangular frame. Staple across the joints, as shown.
- With your fingertips, hold each model flat on a desk or table and try to change its shape. Which shape is more stable?
- Cut another cardboard strip and use it to form a brace for the square frame. Is it more rigid? Why do you think the brace works?

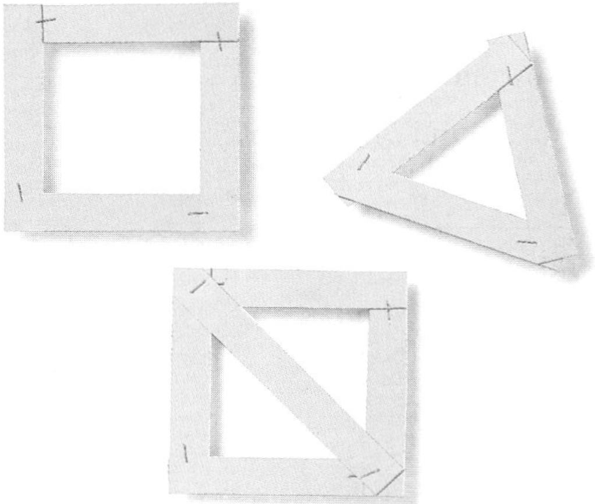

Exercises MIXED REVIEW

Calculator **Find the volume and surface area of each figure to the nearest tenth.**

26.

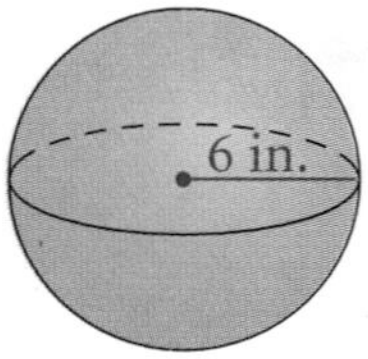

27.

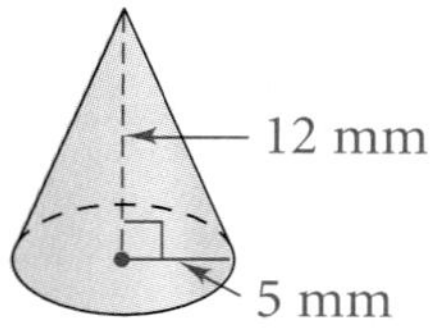

28.

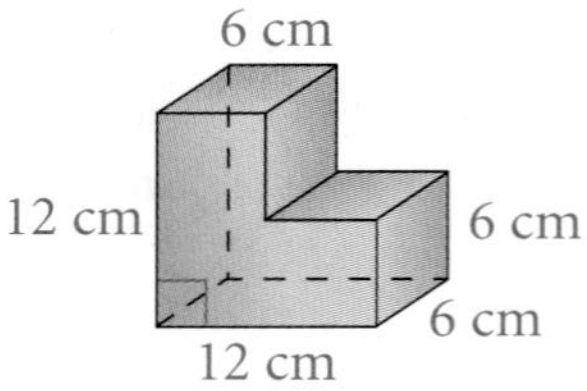

29. One angle of an isosceles triangle is 112°. Find the measures of the other two angles.

Getting Ready for Lesson 8-2

30. For $\triangle STP$, which sides are *not* included between $\angle S$ and $\angle T$?

31. **Constructions** Draw an obtuse angle on your paper. Label it $\angle 1$. Construct $\angle A$ so that $\angle A \cong \angle 1$.

SELF ASSESSMENT — FOR YOUR JOURNAL

Summarize the two ways that you learned to prove triangles congruent in this lesson. Include diagrams.

Technology

Exploring SSA and AAA

After Lesson 8-1

Work in pairs or small groups.

Construct

Use geometry software to construct ray $\overrightarrow{AB}$. Draw a circle with center C that intersects $\overrightarrow{AB}$ in two points. Construct $\overline{AC}$. Construct a point E on the circle and construct $\overline{CE}$.

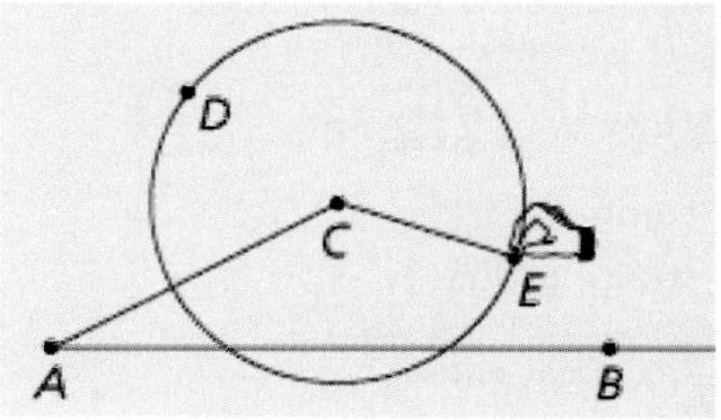

Investigate

Move point E around the circle until E is on $\overrightarrow{AB}$ and forms $\triangle ACE$. Then move E to another point on the circle that is also on $\overrightarrow{AB}$ to form another $\triangle ACE$. In the two triangles, compare the measures of $\angle A$, $\overline{AC}$, and $\overline{CE}$. Are two sides and a nonincluded angle of one triangle congruent to two sides and a nonincluded angle of the other triangle? Are the two triangles congruent? Do you get the same results if you change the size of $\angle A$ and the size of the circle?

Construct

Construct rays $\overrightarrow{AB}$ and $\overrightarrow{AC}$. Construct $\overline{BC}$ to create $\triangle ABC$. Construct a line parallel to $\overline{BC}$ that intersects $\overrightarrow{AB}$ and $\overrightarrow{AC}$ at points D and E to form $\triangle ADE$.

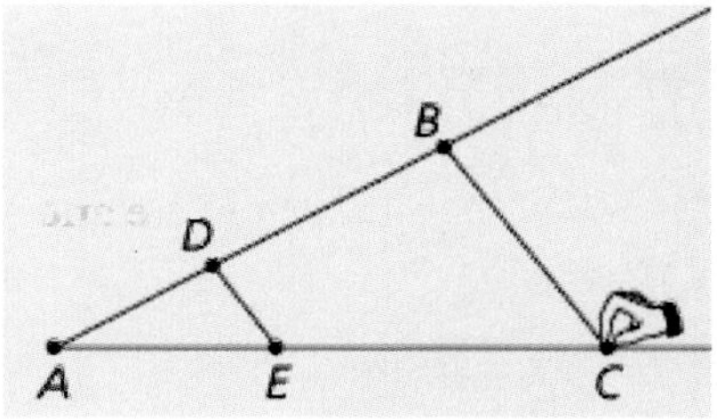

Investigate

Are three angles of $\triangle ABC$ congruent to three angles of $\triangle ADE$? Manipulate the figure to change the positions of sides $\overline{DE}$ and $\overline{BC}$. Do the corresponding angles of the triangles remain congruent? Are the two triangles congruent? Can the two triangles be congruent?

Conjecture

Do you think there is a SSA congruency theorem? Do you think there is an AAA congruency theorem? Explain your answers.

Extend

Manipulate the first figure you drew so that $\angle A$ is obtuse. Now can the circle intersect $\overrightarrow{AB}$ twice? Can two triangles be formed? Could there be a SSA congruency theorem if the congruent angles are obtuse?

What You'll Learn

- Proving two triangles congruent by the ASA Postulate and the AAS Theorem

...And Why

To understand the use of congruent figures in constructed objects such as lacrosse nets

What You'll Need

- compass
- straightedge
- scissors

8-2 Proving Triangles Congruent: ASA and AAS

WORK TOGETHER

Have each person in your group draw a different triangle. Label each triangle $\triangle ABC$.

① Construct $\overline{XZ}$ such that $\overline{XZ} \cong \overline{AC}$.

② At X, construct an angle congruent to $\angle A$.
③ At Z, construct an angle congruent to $\angle C$.
④ Label point Y as shown.

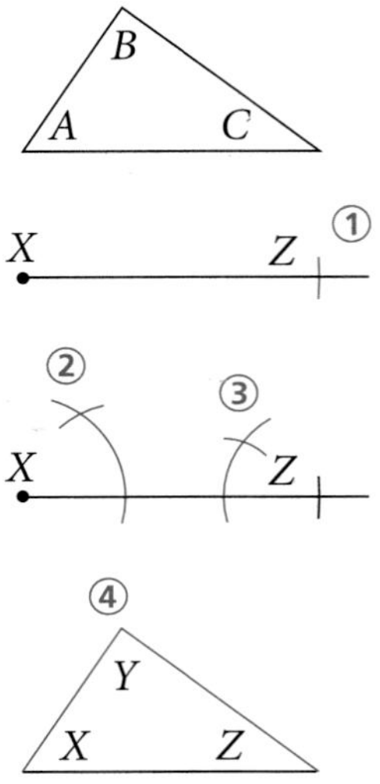

1. Cut out $\triangle ABC$ and $\triangle XYZ$. Place $\triangle ABC$ over $\triangle XYZ$ so that corresponding angles match. Are the triangles congruent? Compare your results with those of others in your group.
2. Make a **conjecture.** What seems to be true when two angles and the included side of one triangle are congruent to two angles and the included side of another triangle?

THINK AND DISCUSS

Using the ASA Postulate

Your investigation in the Work Together involved two angles and an included side.

Postulate 8-3
Angle-Side-Angle Postulate (ASA Postulate)

If two angles and the included side of one triangle are congruent to two angles and the included side of another triangle, then the two triangles are congruent.

$\triangle GBH \cong \triangle KPN$

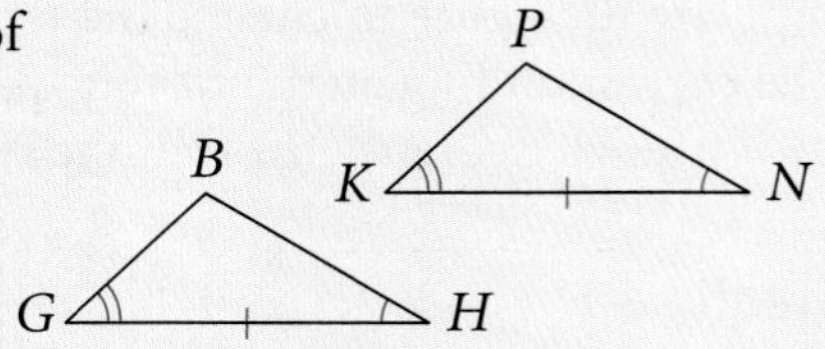

3. **Try This** Write a congruence statement for the two triangles you can prove congruent by the ASA Postulate.

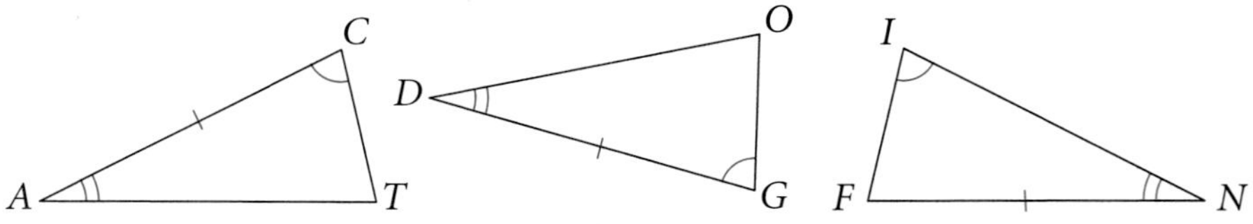

Example 1 — Relating to the Real World

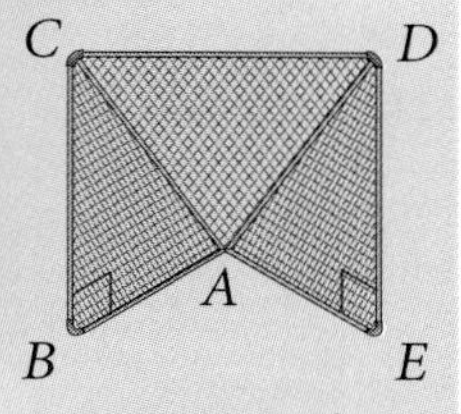

Front view of net

Lacrosse Write a plan for the following proof.

Given: $\angle CAB \cong \angle DAE$, $\overline{AB} \cong \overline{AE}$, $\angle ABC$ and $\angle AED$ are right angles.
Prove: $\triangle ABC \cong \triangle AED$

Plan for Proof To prove $\triangle ABC \cong \triangle AED$, you can use the ASA Postulate. First prove that $\angle ABC \cong \angle AED$.

4. Use the plan in Example 1 to write a two-column proof.

You can use the ASA Postulate to prove that if two angles and a nonincluded side of one triangle are congruent to two angles and the corresponding nonincluded side of another triangle, then the triangles are congruent.

Example 2

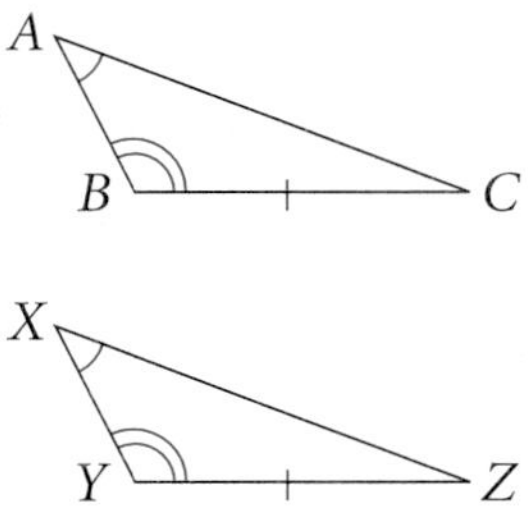

Given: $\angle A \cong \angle X$, $\angle B \cong \angle Y$, $\overline{BC} \cong \overline{YZ}$
Prove: $\triangle ABC \cong \triangle XYZ$

Flow Proof

$\angle A \cong \angle X$ — Given → $\angle C \cong \angle Z$

$\angle B \cong \angle Y$ — Given → $\angle C \cong \angle Z$ and → $\triangle ABC \cong \triangle XYZ$

$\overline{BC} \cong \overline{YZ}$ — Given → $\triangle ABC \cong \triangle XYZ$

$\angle C \cong \angle Z$ — If 2 $\angle$s of one $\triangle$ are $\cong$ to 2 $\angle$s of another $\triangle$, then the 3rd $\angle$s are $\cong$. → $\triangle ABC \cong \triangle XYZ$

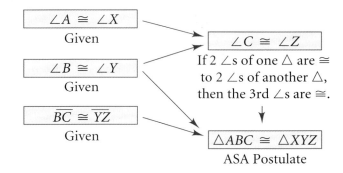

$\triangle ABC \cong \triangle XYZ$ — ASA Postulate

Using the AAS Theorem

In Example 2, the ASA Postulate was used in the flow proof to prove the following theorem.

Theorem 8-1
Angle-Angle-Side Theorem (AAS Theorem)

If two angles and a nonincluded side of one triangle are congruent to two angles and the corresponding nonincluded side of another triangle, then the triangles are congruent.

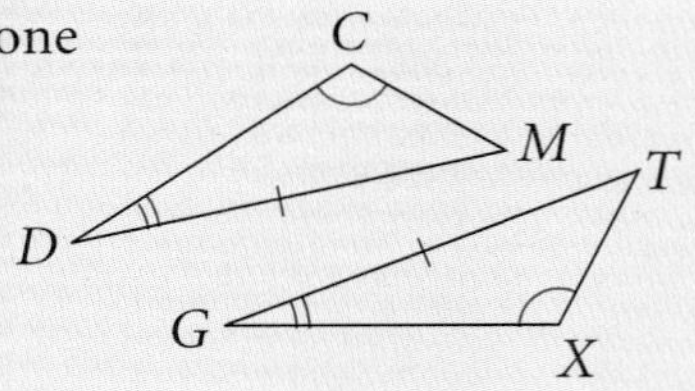

$\triangle DCM \cong \triangle GXT$

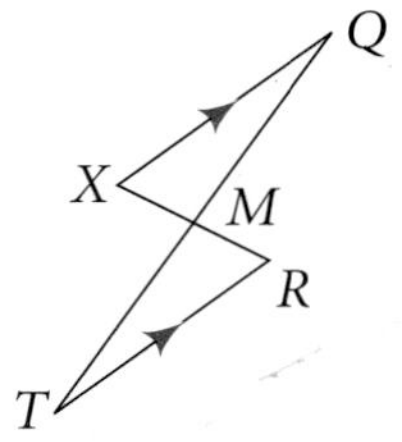

5. **Try This** Supply the missing reasons in the proof.

Given: $\overline{XQ} \parallel \overline{TR}$, $\overline{XR}$ bisects $\overline{QT}$.
Prove: $\triangle XMQ \cong \triangle RMT$

Statements	Reasons
1. $\overline{XQ} \parallel \overline{TR}$	a. __?__
2. $\angle Q \cong \angle T$	b. __?__
3. $\angle X \cong \angle R$	c. __?__
4. $\overline{XR}$ bisects $\overline{QT}$.	d. __?__
5. $\overline{TM} \cong \overline{QM}$	e. __?__
6. $\triangle XMQ \cong \triangle RMT$	f. __?__

6. **Critical Thinking** Explain how you could prove the triangles in Question 5 congruent using the ASA Postulate.

Exercises ON YOUR OWN

Tell whether the ASA Postulate or the AAS Theorem can be applied directly to prove the triangles congruent. If the triangles *cannot* be proven congruent, write *not possible.*

1.

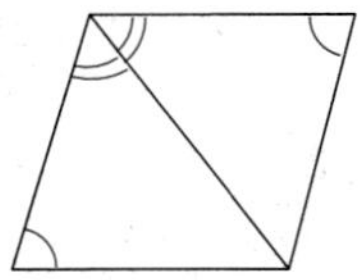

2.

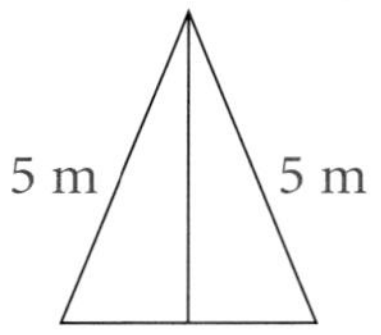

3.

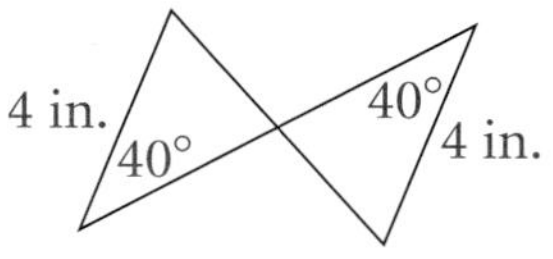

4.

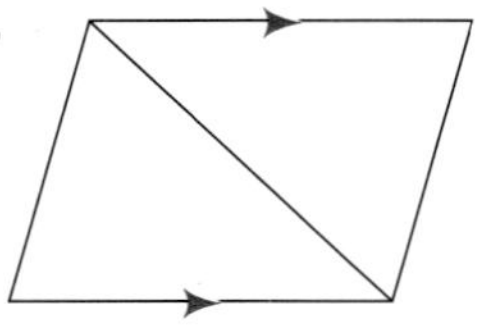

5.

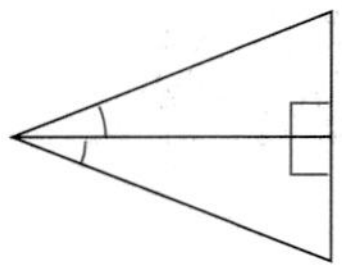

6.

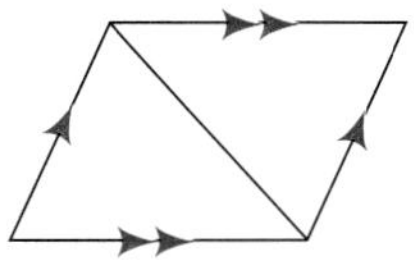

7.

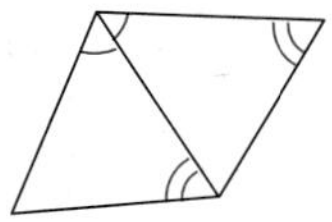

8.

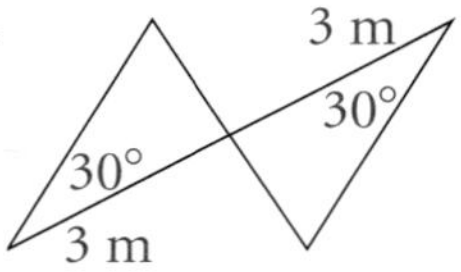

Preparing for Proof **What additional information would you need to prove the triangles congruent by the stated postulate or theorem?**

9. AAS Theorem

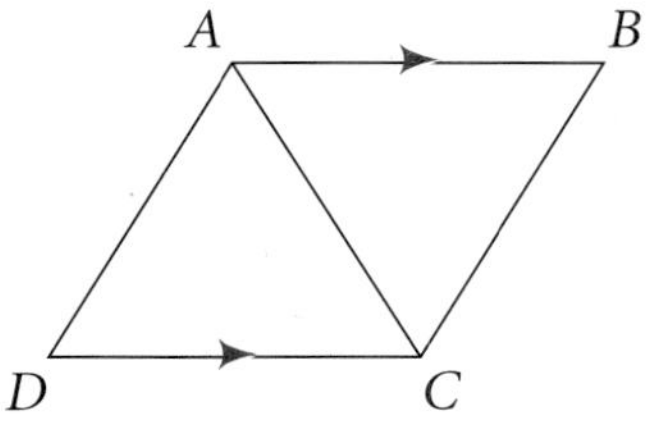

10. ASA Postulate

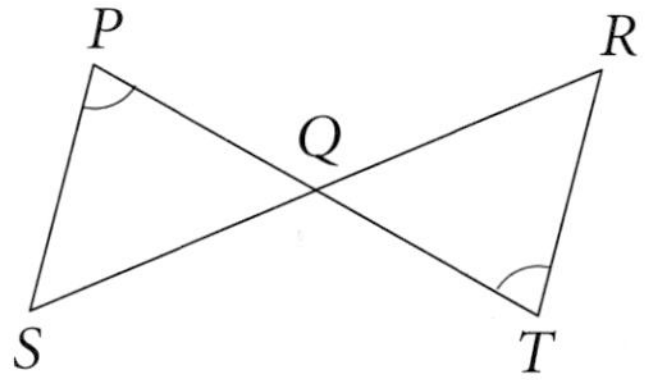

11. AAS Theorem

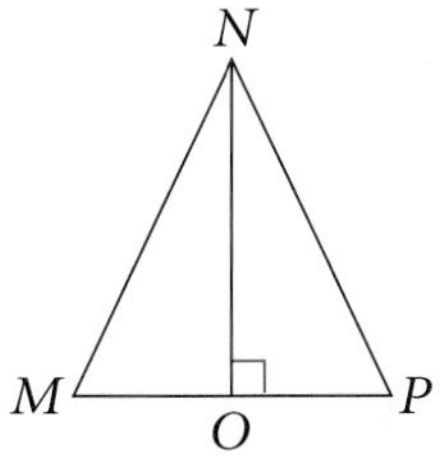

12. SAS Postulate

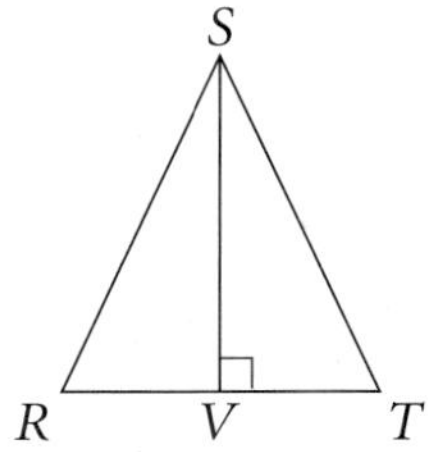

13. a. **Open-ended** Draw a triangle. Draw a second triangle that is congruent to the first one and shares a common side with it.

b. Think about how you drew your second triangle. What congruence postulate or theorem did you use to make the second triangle congruent to the first one?

14. **Preparing for Proof** Supply the missing statements and reasons.

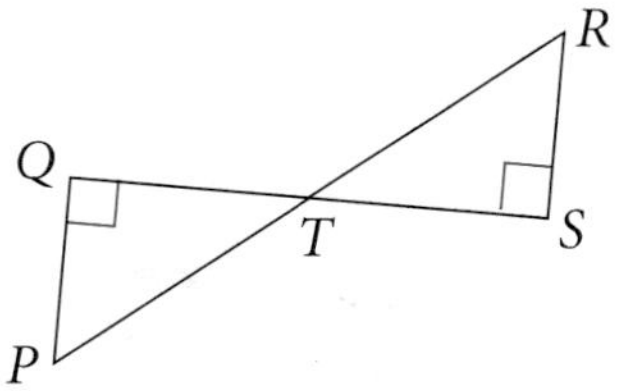

Given: $\overline{PQ} \perp \overline{QS}$, $\overline{RS} \perp \overline{QS}$, T is the midpoint of $\overline{PR}$.
Prove: $\triangle PQT \cong \triangle RST$

Statements	Reasons
1. $\overline{PQ} \perp \overline{QS}$, $\overline{RS} \perp \overline{QS}$	**1.** Given
2. $\angle Q$ and $\angle S$ are right angles.	**a.** ?
3. $\angle Q \cong \angle S$	**b.** ?
c. ?	**4.** Vertical angles are $\cong$.
5. T is the midpoint of $\overline{PR}$.	**5.** Given
6. $\overline{PT} \cong \overline{RT}$	**d.** ?
7. $\triangle PQT \cong \triangle RST$	**e.** ?

15. **Geometry in 3 Dimensions** The top portion of this quartz crystal is a hexagonal pyramid. The lateral edges of the pyramid are congruent and the base edges of the pyramid are congruent. Are the triangles that form the lateral faces of the pyramid congruent? **Justify** your answer.

16. **Probability** Here are six congruence statements about the triangles at the right.

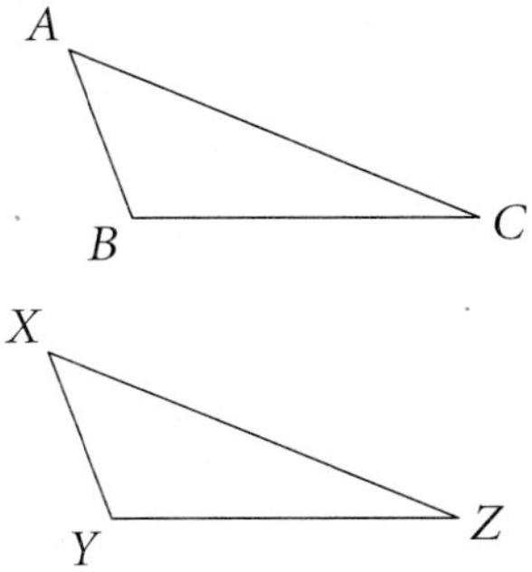

$\angle A \cong \angle X$ $\quad\angle B \cong \angle Y$ $\quad\angle C \cong \angle Z$

$\overline{AB} \cong \overline{XY}$ $\quad\overline{AC} \cong \overline{XZ}$ $\quad\overline{BC} \cong \overline{YZ}$

There are 20 ways to choose groups of three statements from these six statements. What is the probability that a set of three congruence statements chosen at random from the six listed will guarantee that the triangles are congruent?

17. **Writing** Anita claims that it is possible to rewrite any proof that uses the AAS Theorem as a proof that uses the ASA Postulate. Do you agree with Anita? Explain why or why not.

Choose **Write a two-column proof, a paragraph proof, or a flow proof.**

18. **Given:** $\overline{DH}$ bisects $\angle BDF$, $\angle 1 \cong \angle 2$
Prove: $\triangle BDH \cong \triangle FDH$

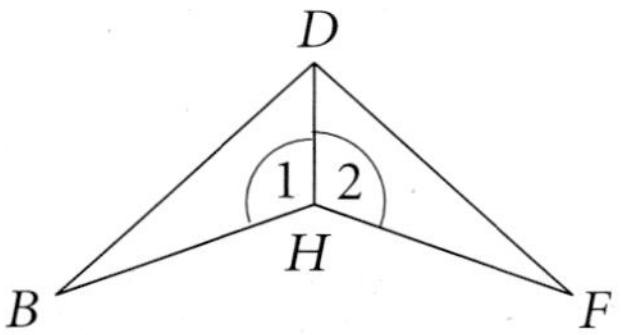

19. **Given:** $\angle F \cong \angle H$, $\overline{FG} \parallel \overline{JH}$
Prove: $\triangle FGJ \cong \triangle HJG$

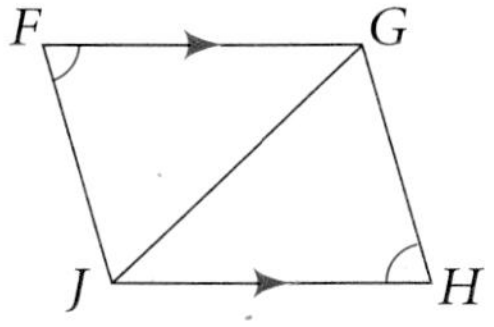

20. **Given:** $\overline{AE} \parallel \overline{BD}$, $\overline{AE} \cong \overline{BD}$, $\angle E \cong \angle D$
Prove: $\triangle AEB \cong \triangle BDC$

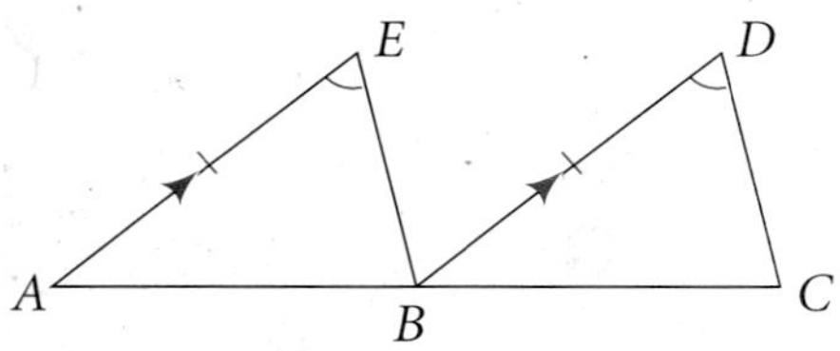

21. **Given:** $\angle N \cong \angle P$, $\overline{MO} \cong \overline{QO}$
Prove: $\triangle MON \cong \triangle QOP$

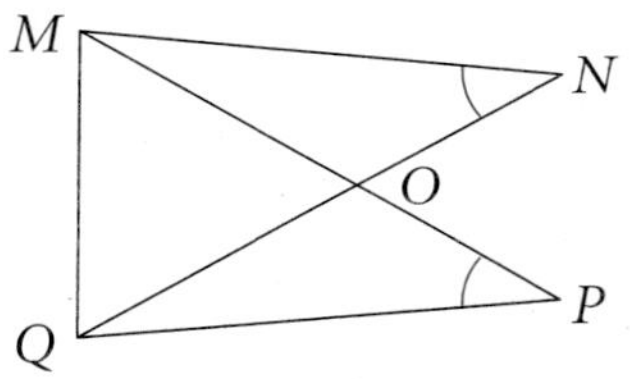

22. **Standardized Test Prep** Which of the following can you conclude based on the figure at the right?

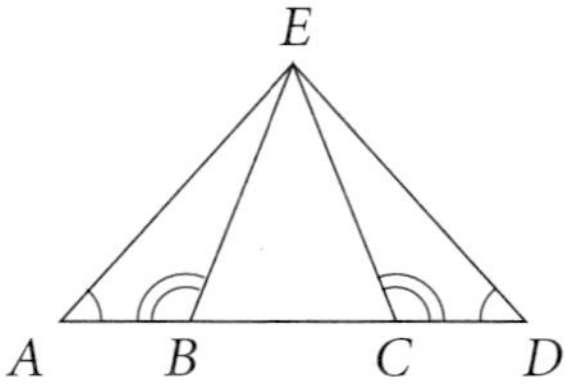

I. $\triangle ADE$ is isosceles. **II.** $\triangle ABE \cong \triangle DCE$
III. $\triangle BDE$ is isosceles. **IV.** $\angle EBC \cong \angle ECB$

A. I, II, and IV only **B.** I, II, and III only **C.** II, III, and IV only
D. I, III, and IV only **E.** III and IV only

23. **Critical Thinking** Explain why $\triangle ABC$ cannot be congruent to $\triangle FDE$.

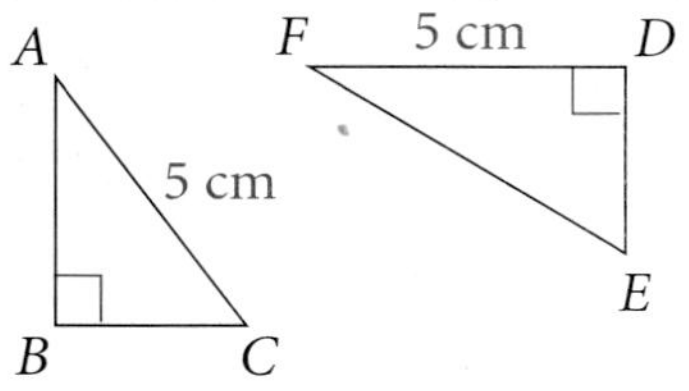

Exercises MIXED REVIEW

Coordinate Geometry **Find the coordinates of the midpoint of the segment with the given endpoints.**

24. (4, 7) and (2, 9)

25. (−1, 8) and (5, −8)

26. Constructions Draw a line ℓ and a point M not on ℓ. Construct a line n through M so that $n \perp \ell$.

SELF ASSESSMENT

FOR YOUR JOURNAL

Explain why "included" and "nonincluded" are necessary terms in the SAS Postulate, the ASA Postulate, and the AAS Theorem.

Getting Ready for Lesson 8-3

Find the value of each variable. If an answer is not a whole number, leave it in simplest radical form.

27.

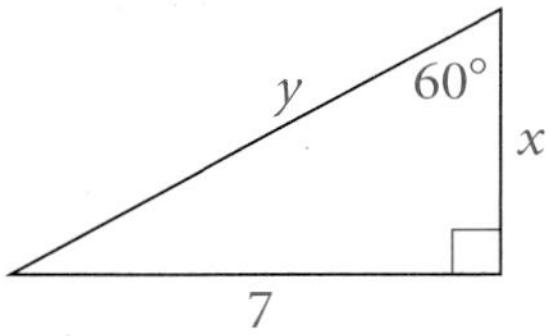

28.

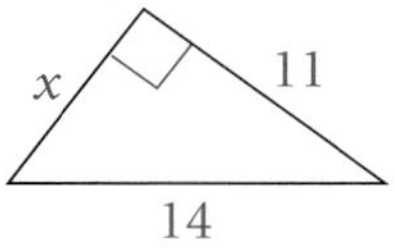

29.

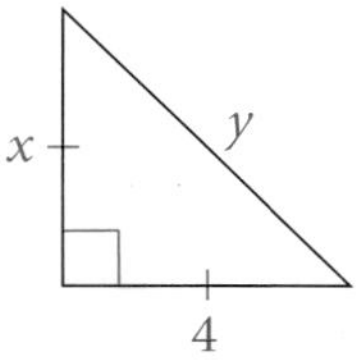

Geometry at Work

Die Maker

Two centuries ago, all manufactured articles were made by hand. Each article produced was slightly different from the others. In 1800, inventor Eli Whitney recognized that manufacturing could be greatly speeded up by using *congruent* parts. Whitney made a *die,* or mold, for each part of a musket he was producing for the U.S. Army. This allowed standard-sized muskets to be put together rapidly, and it ushered in the era of *mass production.*

Today, die makers are highly skilled industrial workers. To create a new product, die makers use metal, plastic, rubber, or other materials to create dies. During manufacture, a part is shaped by its die. This ensures that parts are congruent.

Mini Project: Research Write a paragraph on the use of molds and castings to produce congruent parts.

8-3 Congruent Right Triangles

What You'll Learn

- Proving triangles congruent using the HL Theorem

...And Why

To investigate congruence in applications such as fabric for tents

What You'll Need

- compass
- protractor
- ruler

WORK TOGETHER

- As a group, select two numbers from the set {2, 3, 4, 5, 6, 7}. Let x equal the smaller number and y the larger. Have each group member draw a right triangle at the corner of a piece of paper with a leg x in. long and a hypotenuse y in. long. (An example using 5 and 2 is shown.)

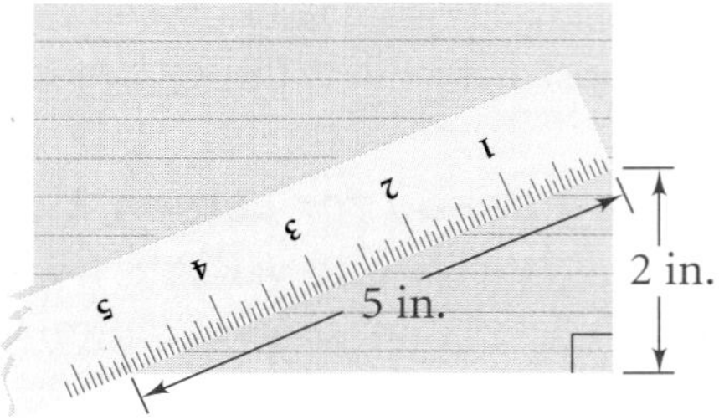

1. Compare triangles within your group and make a **conjecture.** If the hypotenuse and a leg of one right triangle are congruent to the hypotenuse and a leg of another right triangle, then __?__.

- **Verify** your conjecture by selecting two more numbers from the set and creating more right triangles.

QUICK REVIEW

The hypotenuse of a right triangle is the longest side.

THINK AND DISCUSS

In the Work Together, you saw that all triangles formed with a given length for a leg and a given length for a hypotenuse are congruent. This observation leads to the following theorem.

Theorem 8-2
Hypotenuse-Leg Theorem (HL Theorem)

If the hypotenuse and a leg of one right triangle are congruent to the hypotenuse and leg of another right triangle, then the triangles are congruent.

Proof of Theorem 8-2

Given: $\triangle PQR$ and $\triangle XYZ$ are right triangles, $\overline{QR} \cong \overline{YZ}$, and $\overline{PQ} \cong \overline{XY}$

Prove: $\triangle PQR \cong \triangle XYZ$

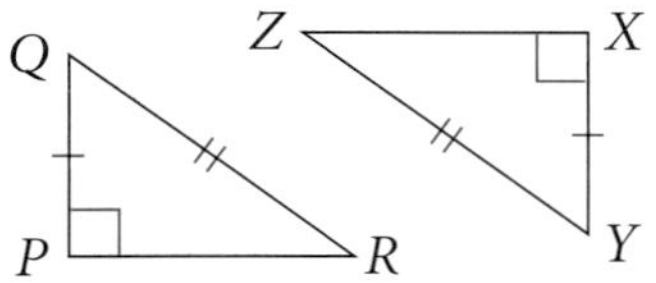

Let $PQ = XY = a$ and $QR = ZY = c$.
By the Pythagorean Theorem, $a^2 + PR^2 = c^2$ and $a^2 + XZ^2 = c^2$. Using the Subtraction Property of Equality, you get $PR^2 = c^2 - a^2$ and $XZ^2 = c^2 - a^2$. So $PR^2 = XZ^2$ and $PR = XZ$. By the SSS Postulate, $\triangle PQR \cong \triangle XYZ$.

Example 1 Relating to the Real World

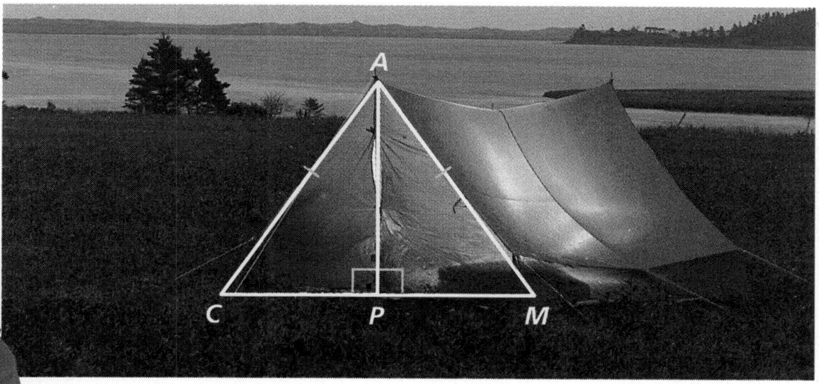

Tents On the tent, $\angle CPA$ and $\angle MPA$ are right angles and $\overline{CA} \cong \overline{MA}$. Can you use the same pattern for both flaps of the tent?

You are given that $\angle CPA$ and $\angle MPA$ are right angles. Therefore, $\triangle CPA$ and $\triangle MPA$ are right triangles. You are also given that $\overline{CA} \cong \overline{MA}$. $\overline{PA}$ is a leg of both $\triangle CPA$ and $\triangle MPA$. $\overline{PA} \cong \overline{PA}$ by the Reflexive Property. The triangles are congruent by the HL Theorem. Therefore, you can use the same pattern for both flaps of the tent.

To use the HL Theorem in proofs, you must show that these three conditions are met.

- There are two right triangles.
- There is one pair of congruent hypotenuses.
- There is one pair of congruent legs.

Example 2

Write a flow proof.

Given: $\overline{CD} \cong \overline{EA}$, $\overline{AD}$ is the perpendicular bisector of $\overline{CE}$.

Prove: $\triangle CBD \cong \triangle EBA$

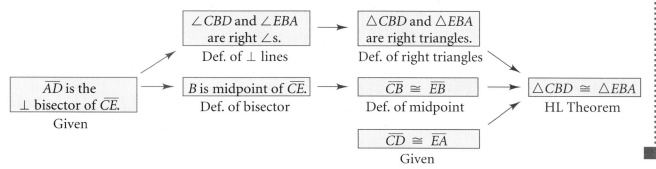

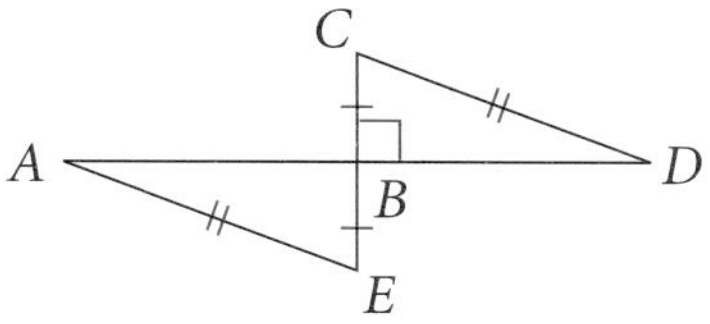

Flow Proof

- $\overline{AD}$ is the $\perp$ bisector of $\overline{CE}$. — Given
  - → $\angle CBD$ and $\angle EBA$ are right $\angle$s. — Def. of $\perp$ lines
    - → $\triangle CBD$ and $\triangle EBA$ are right triangles. — Def. of right triangles
  - → B is midpoint of $\overline{CE}$. — Def. of bisector
    - → $\overline{CB} \cong \overline{EB}$ — Def. of midpoint
- $\overline{CD} \cong \overline{EA}$ — Given
- → $\triangle CBD \cong \triangle EBA$ — HL Theorem

2. **Critical Thinking** Suppose you know that two legs of one right triangle are congruent to two legs of another right triangle. How could you prove the triangles congruent? Explain.

3. **Try This** Complete the proof.

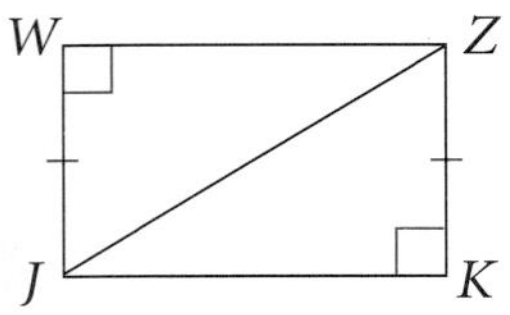

Given: $\overline{WJ} \cong \overline{KZ}$, $\angle JWZ$ and $\angle ZKJ$ right angles.
Prove: $\triangle WJZ \cong \triangle KZJ$

Statements	Reasons
1. $\angle JWZ$ and $\angle ZKJ$ are rt. $\angle$s.	a. __?__
2. $\triangle WJZ$ and $\triangle KZJ$ are rt. $\triangle$s.	b. __?__
c. __?__	3. Reflexive Property of $\cong$
d. __?__	4. Given
5. $\triangle WJZ \cong \triangle KZJ$	e. __?__

Exercises ON YOUR OWN

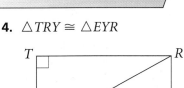

1. **Standardized Test Prep** Which set of conditions does *not* provide enough information to prove that $\triangle ABC \cong \triangle ADC$?
   - **A.** $\angle 1 \cong \angle 2$, $\angle 5 \cong \angle 6$
   - **B.** $\overline{AD} \cong \overline{AB}$, $\angle 3$ and $\angle 4$ are right angles.
   - **C.** $\overline{AD} \cong \overline{AB}$, $\overline{DC} \cong \overline{BC}$
   - **D.** $\angle 5 \cong \angle 6$, $\angle 3$ and $\angle 4$ are right angles.
   - **E.** $\overline{AD} \cong \overline{AB}$, $\angle 5 \cong \angle 6$

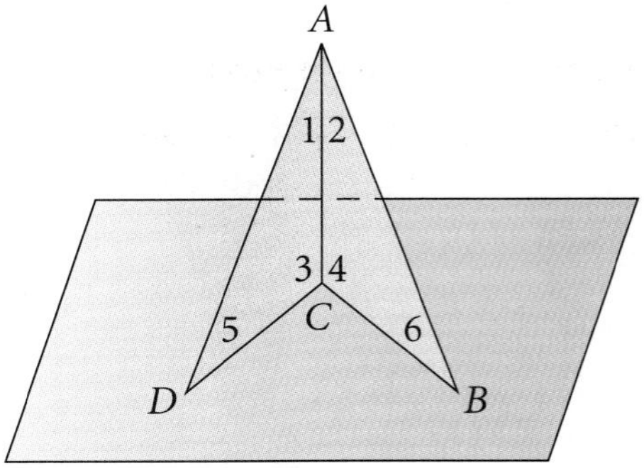

What additional information would you need to prove the triangles congruent by the HL Theorem?

2. $\triangle BLT \cong \triangle RKQ$

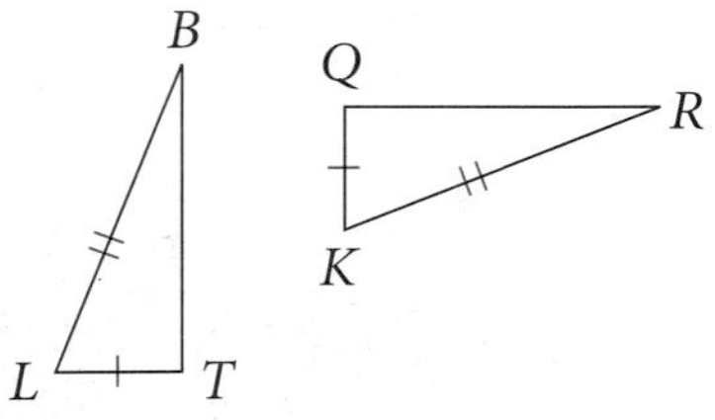

3. $\triangle XRV \cong \triangle TRV$

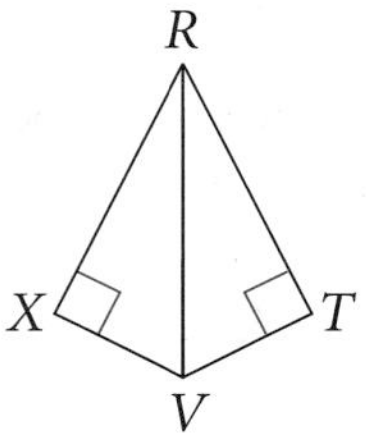

4. $\triangle TRY \cong \triangle EYR$

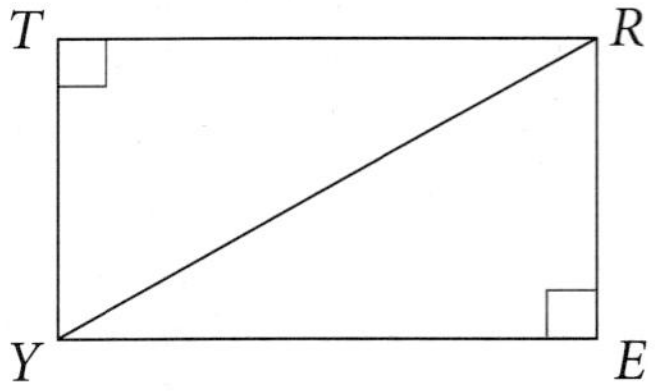

5. $\triangle ACQ \cong \triangle GCJ$

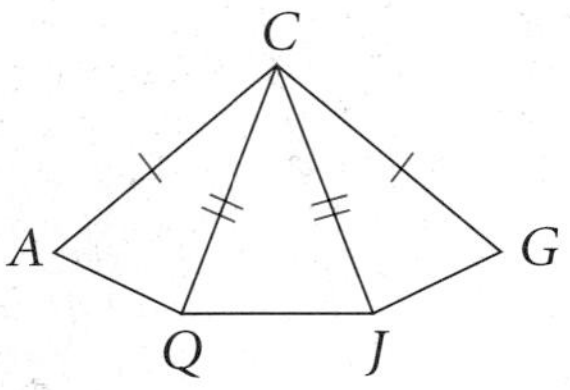

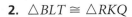

6. $\triangle BDC \cong \triangle FEA$

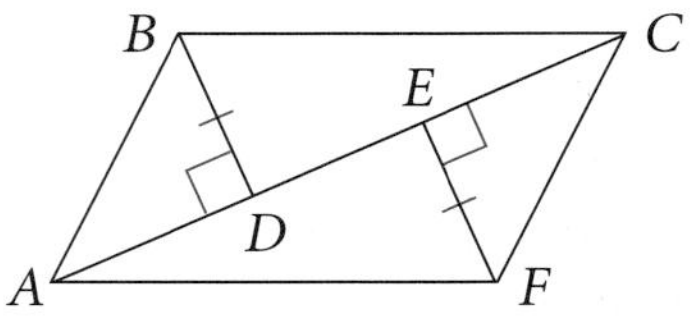

7. $\triangle STR \cong \triangle PQN$

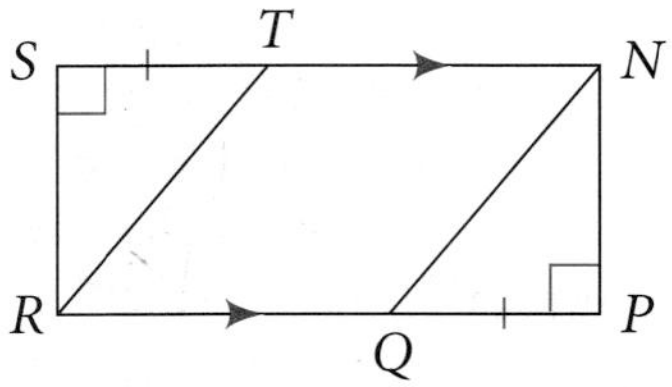

8. **Critical Thinking** Quadrilateral $ABCD$ is a kite with $AB = AD$ and $BC = DC$. $\angle ABC$ and $\angle ADC$ are right angles. Name all the pairs of congruent right triangles in the diagram. For each pair, explain how you know that the triangles are congruent.

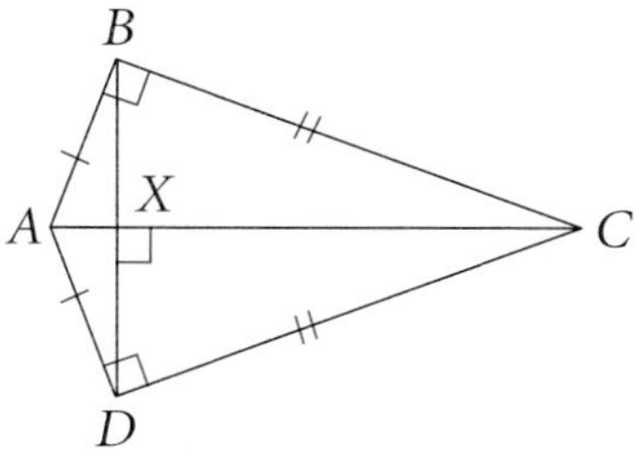

9. **Writing** While working for a landscape architect, you are told to lay out a flower bed in the shape of a right triangle with sides of 3 yd and 7 yd. Explain what else you need to know in order to make the flower bed.

10. **Geometry in 3 Dimensions** Write a paragraph proof.
Given: $\overline{BE} \perp \overline{EA}$, $\overline{BE} \perp \overline{EC}$, $\triangle ABC$ is equilateral.
Prove: $\triangle AEB \cong \triangle CEB$

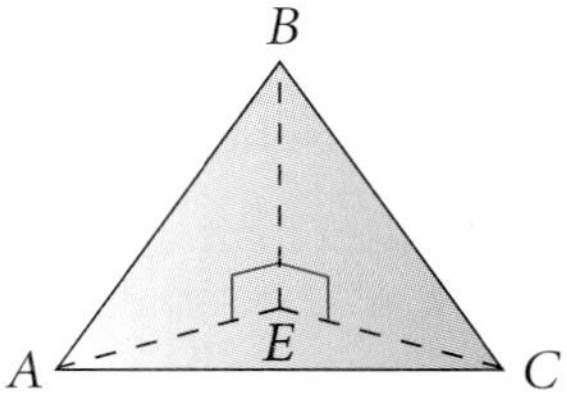

11. a. **Coordinate Geometry** Graph the points $E(-1, -1)$, $F(-2, -6)$, $G(-4, -4)$, and $D(-6, -2)$. Connect the points with segments.
b. Find the slopes of $\overline{DG}$, $\overline{GF}$, and $\overline{GE}$.
c. Use your answer to part (b) to describe $\angle EGD$ and $\angle EGF$.
d. Use the Distance Formula to find DE and FE.
e. Write a paragraph to prove that $\triangle EGD \cong \triangle EGF$.

Write a two-column proof.

12. **Given:** $\overline{EB} \cong \overline{DB}$, $\angle A$ and $\angle C$ are right angles, B is the midpoint of $\overline{AC}$.
Prove: $\triangle BEA \cong \triangle BDC$

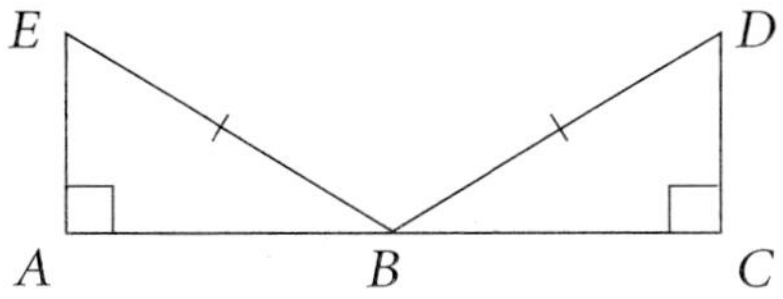

13. **Given:** $\odot O$, $\angle M$ and $\angle N$ are right angles.
Prove: $\triangle LMO \cong \triangle LNO$

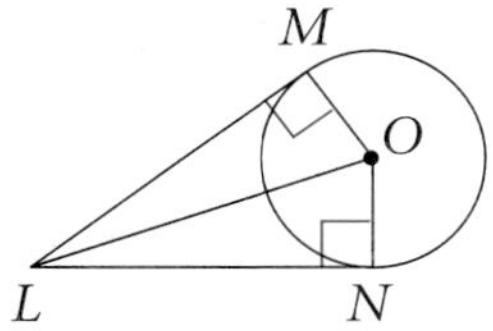

14. **History** About 300 B.C., Aristarchus of Samos devised a method for determining the relative distances to the sun and moon. He reasoned that the angle between the sun, the moon, and Earth is 90° when they are positioned as shown in the diagram.
a. Explain why the two triangles shown must be congruent.
b. **Research** Aristarchus also reasoned that it takes longer for the moon to go from the first quarter to the last quarter than from the last to the first. He attempted the difficult measurement of that time difference and used it to find the size of the angles of the triangles. From this he estimated that the sun is 20 times farther from us than the moon is. How accurate was his estimate?

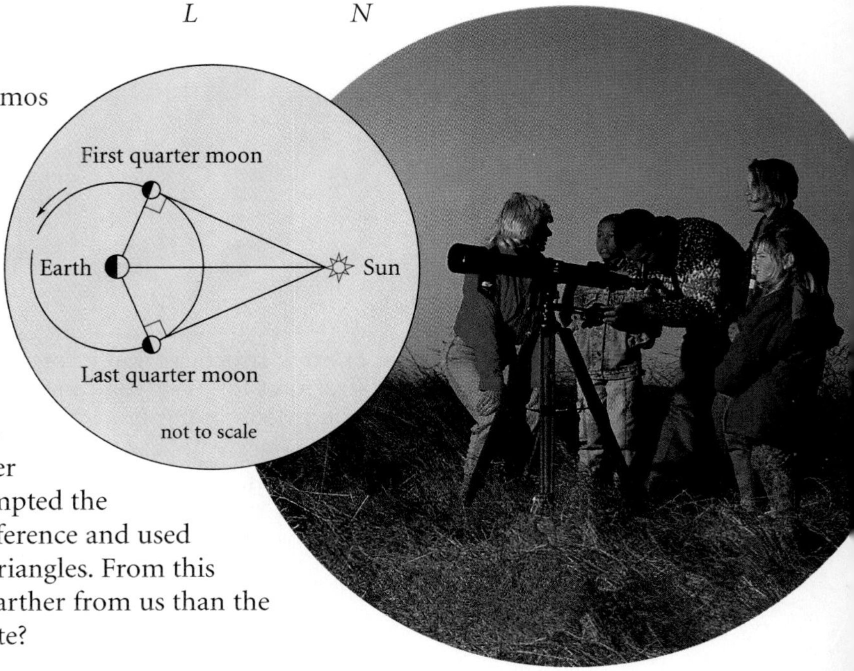

15. Open-ended A table has been placed in the corner of a room. What measurements would you make with a tape measure in order to place a matching table in the other corner at exactly the same angle? Explain why your method works.

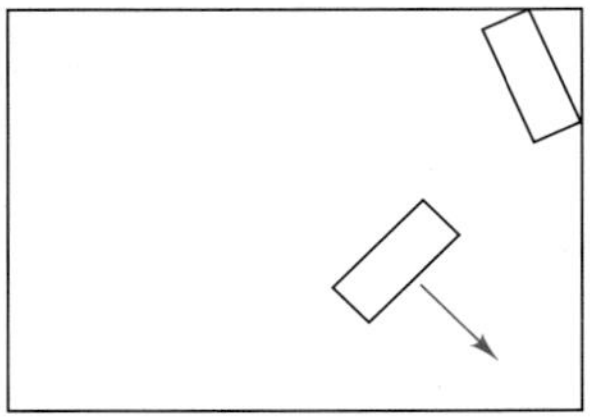

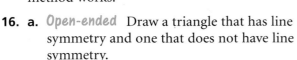

16. a. Open-ended Draw a triangle that has line symmetry and one that does not have line symmetry.

b. Critical Thinking Explain why any triangle with line symmetry must be isosceles.

17. Clock Repair To repair an antique clock, a 12-toothed wheel has to be made by cutting right triangles out of a regular polygon that has twelve 4-cm sides. The hypotenuse of each triangle is a side of the regular polygon, and the shorter leg is 1 cm long. Explain why the twelve triangles must be congruent.

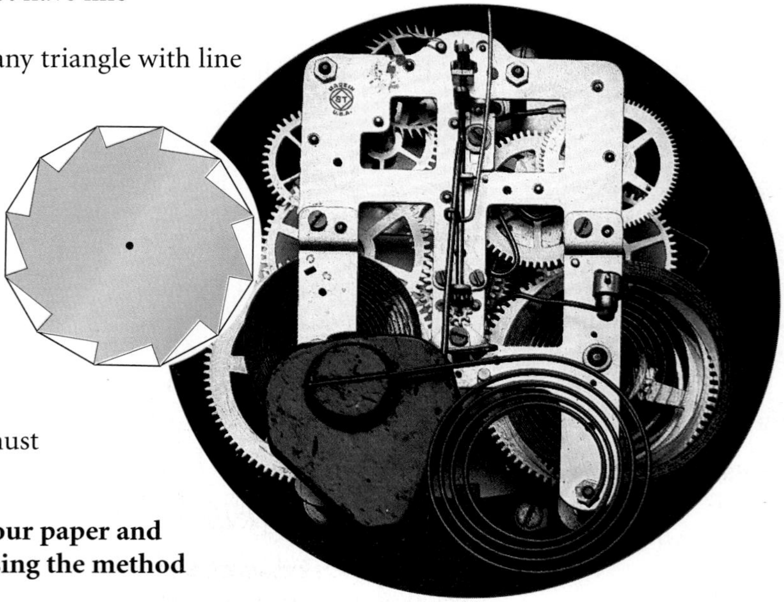

Constructions **Copy the triangle on your paper and construct a triangle congruent to it using the method stated.**

18. by SAS

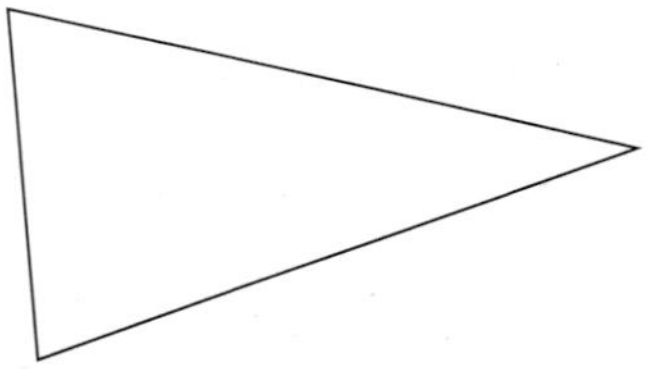

19. by HL

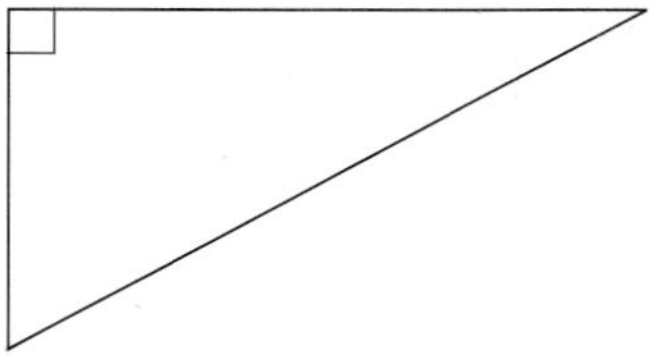

20. by ASA

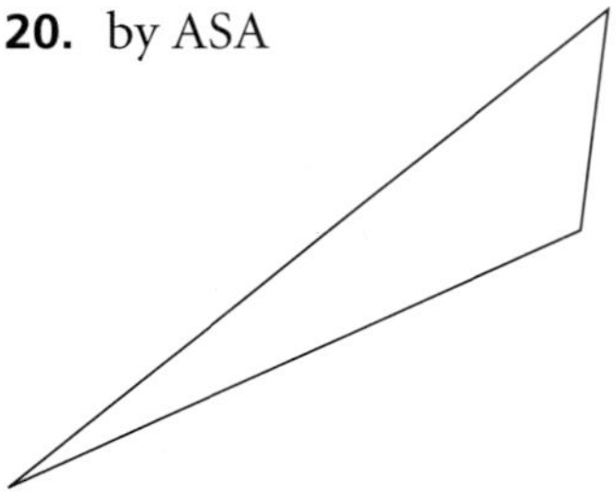

Chapter Project

Find Out by Observing

Visit local bridges, towers, or other structures that have exposed frameworks. Examine these structures for ideas you can use when you design and build a toothpick bridge later in this project. Record your ideas.

Sketch or take pictures of the structures. On the sketches or photos, show where triangles are used for stability.

Exercises MIXED REVIEW

State the postulate or theorem that justifies each statement.

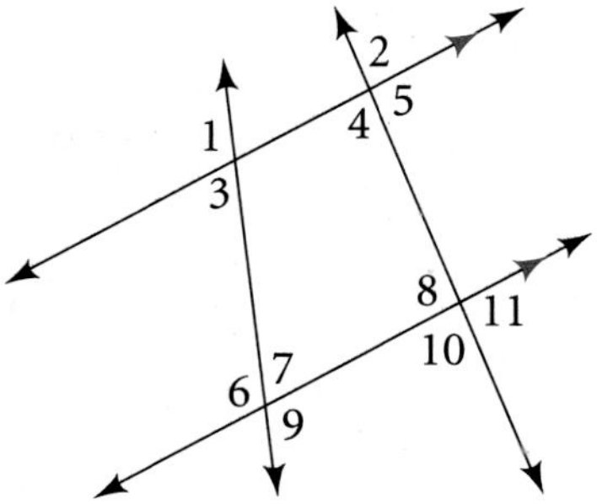

21. $m\angle 1 + m\angle 3 = 180$

22. $\angle 5 \cong \angle 8$

23. $m\angle 4 + m\angle 8 = 180$

24. $\angle 3 \cong \angle 7$

25. $\angle 1 \cong \angle 6$

26. $\angle 5 \cong \angle 11$

27. Calculator A circle has radius 5 m. Two radii form a 35° angle. What is the area of the sector between them to the nearest hundredth?

Getting Ready for Lesson 8-4

28. Given that $\triangle TRC \cong \triangle HGV$, list everything you know about the two triangles.

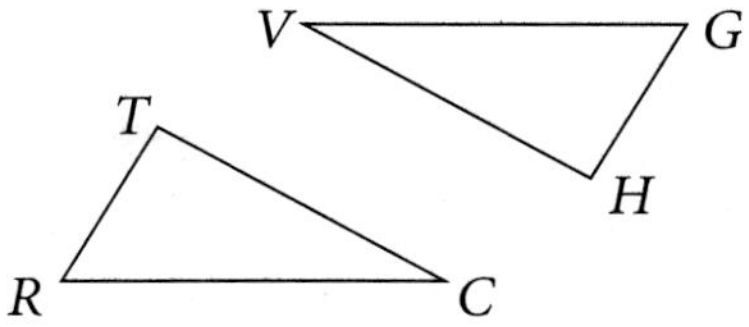

Exercises CHECKPOINT

State the postulate or theorem you can use to prove the triangles congruent. If the triangles *cannot* be proven congruent, write *not possible.*

1.

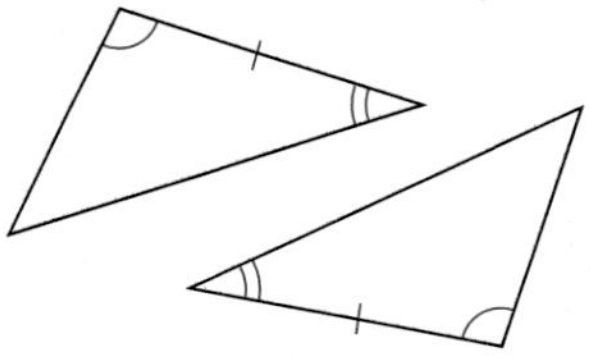

2.

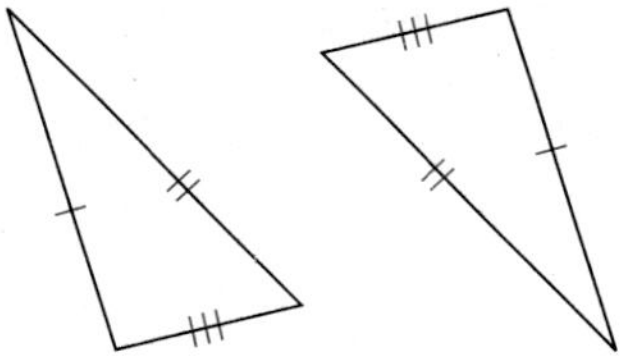

3.

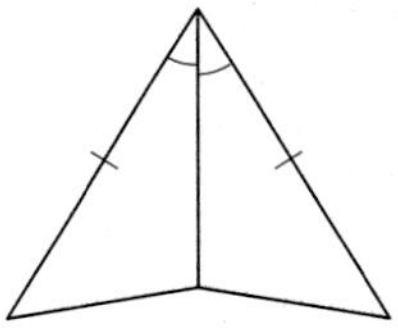

4.

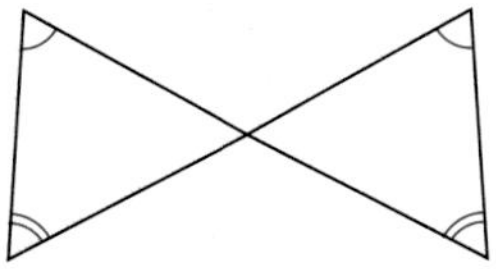

5.

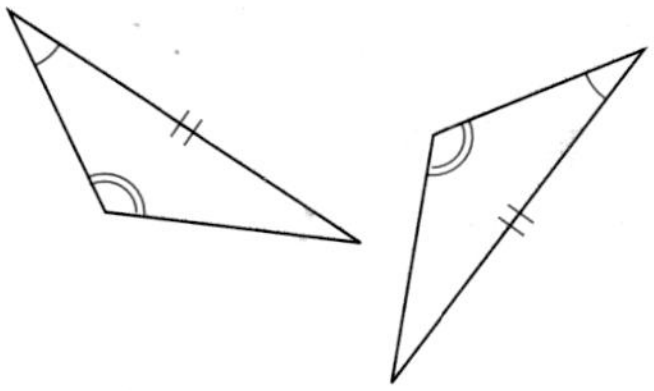

6.

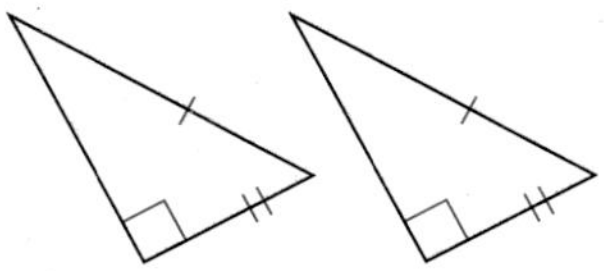

7. a. Open-ended Draw two triangles that have two pairs of congruent sides and one pair of congruent angles, but are not congruent.

b. Can you draw two noncongruent triangles that have two pairs of congruent angles and one pair of congruent sides? Explain.

8. Standardized Test Prep Which congruence statement *cannot* be used to prove two triangles congruent?

A. AAS **B.** SAS **C.** SSS **D.** ASA **E.** AAA

9. Writing Use the HL Theorem as a model to write an HA Theorem. Is the theorem true? **Justify** your answer.

8-4 Using Congruent Triangles in Proofs

What You'll Learn

- Using triangle congruence and CPCTC to prove that parts of two triangles are congruent

...And Why

To see how triangle congruence has been used in the past to measure distances indirectly

THINK AND DISCUSS

Using CPCTC

In the previous lessons you learned to use SSS, SAS, ASA, AAS, and HL to prove that two triangles are congruent. Once you know that triangles are congruent, you can make conclusions about corresponding segments and angles because **C**orresponding **P**arts of **C**ongruent **T**riangles are **C**ongruent. A shorthand way of writing this is **CPCTC.**

1. a. Suppose you know that $\triangle ABD \cong \triangle CDB$ by SAS. Which additional pairs of sides and angles are congruent by CPCTC?

b. *Critical Thinking* What other conclusions can you make about segments in this figure?

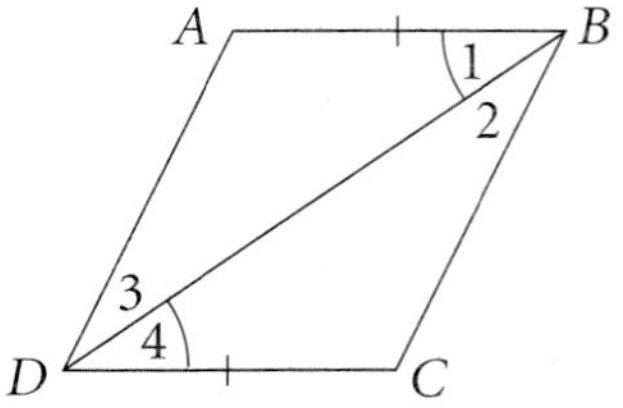

Example 1

Given: $\overline{AC} \cong \overline{EC}$, $\overline{BC} \cong \overline{DC}$
Prove: $\angle A \cong \angle E$

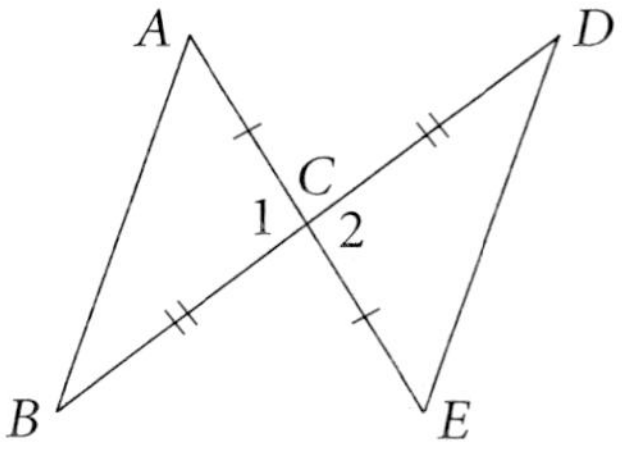

Plan for Proof You can show that $\angle A \cong \angle E$ if you can show that these angles are corresponding parts of congruent triangles. Prove that $\triangle ABC \cong \triangle EDC$ and then use CPCTC.

Statements	Reasons
1. $\overline{AC} \cong \overline{EC}$	1. Given
2. $\overline{BC} \cong \overline{DC}$	2. Given
3. $\angle 1 \cong \angle 2$	3. Vertical angles are $\cong$.
4. $\triangle ABC \cong \triangle EDC$	4. SAS Postulate
5. $\angle A \cong \angle E$	5. CPCTC

2. a. Try This State two relationships that are true for $\overline{AB}$ and $\overline{ED}$ in Example 1.

b. Justify your answers.

According to legend, one of Napoleon's officers used congruent triangles to estimate the width of a river. The officer stood on the riverbank and lowered the visor of his cap until the farthest thing he could see was the edge of the opposite bank of the river. He then turned and noted the spot on his side of the river that was in line with his eye and the tip of his visor. The officer then paced off the distance to this spot and declared that distance to be the width of the river!

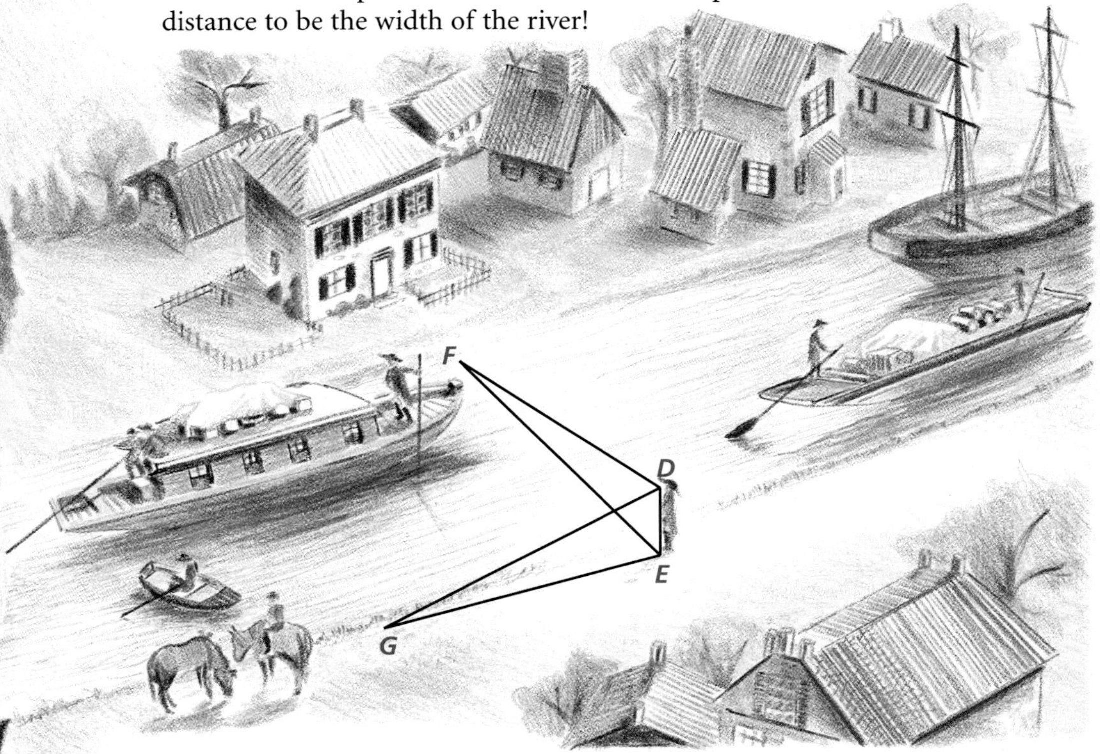

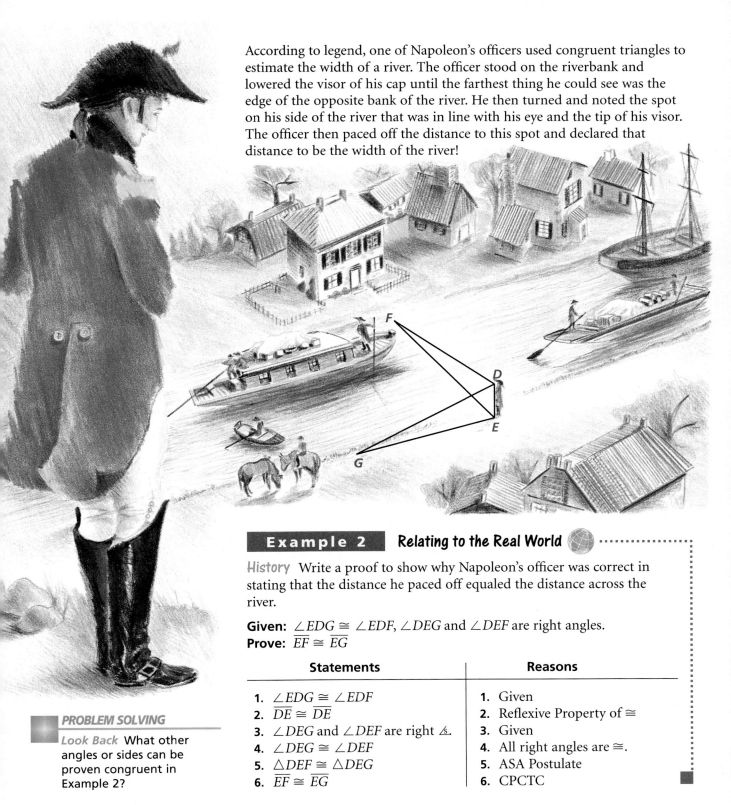

Example 2 **Relating to the Real World**

History Write a proof to show why Napoleon's officer was correct in stating that the distance he paced off equaled the distance across the river.

Given: $\angle EDG \cong \angle EDF$, $\angle DEG$ and $\angle DEF$ are right angles.
Prove: $\overline{EF} \cong \overline{EG}$

Statements	Reasons
1. $\angle EDG \cong \angle EDF$	**1.** Given
2. $\overline{DE} \cong \overline{DE}$	**2.** Reflexive Property of $\cong$
3. $\angle DEG$ and $\angle DEF$ are right $\measuredangle$s.	**3.** Given
4. $\angle DEG \cong \angle DEF$	**4.** All right angles are $\cong$.
5. $\triangle DEF \cong \triangle DEG$	**5.** ASA Postulate
6. $\overline{EF} \cong \overline{EG}$	**6.** CPCTC

PROBLEM SOLVING

Look Back What other angles or sides can be proven congruent in Example 2?

3. About how wide was the river if the officer stepped off 20 paces and each pace was about $2\frac{1}{2}$ ft?

Proving Theorems

You can use congruent triangles to prove many of the theorems you discovered in earlier chapters.

Example 3

Write a paragraph proof of the Isosceles Triangle Theorem.

Begin with isosceles $\triangle XYZ$ with $\overline{XY} \cong \overline{XZ}$. Draw $\overline{XB}$, the bisector of the vertex angle $\angle X$.

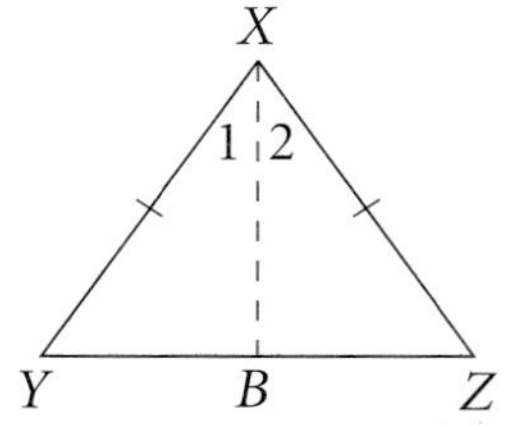

Given: $\overline{XY} \cong \overline{XZ}$, $\overline{XB}$ bisects $\angle YXZ$.
Prove: $\angle Y \cong \angle Z$

Paragraph Proof

By the definition of angle bisector, $\angle 1 \cong \angle 2$. You are given that $\overline{XY} \cong \overline{XZ}$, and by the Reflexive Property of Congruence, $\overline{XB} \cong \overline{XB}$. Thus, $\triangle XYB \cong \triangle XZB$ by the SAS Postulate, and $\angle Y \cong \angle Z$ by CPCTC. ■

4. **Try This** There is more than one way to prove the Isosceles Triangle Theorem. Complete the following steps for a different proof.
   a. Use $\triangle XYZ$ from Example 3 and instead of drawing the bisector of $\angle X$, draw the altitude $\overline{XA}$ from X to $\overline{YZ}$.
   b. Write a new *Given* statement.
   c. What postulate or theorem can you use to prove $\triangle XYA \cong \triangle XZA$?
   d. Write your proof in paragraph form.

QUICK REVIEW

Isosceles Triangle Theorem
If two sides of a triangle are congruent, then the angles opposite those sides are congruent.

Example 4 outlines a proof of the Angle Bisector Theorem.

Example 4

Write a plan for the following proof.

Given: $\overrightarrow{LP}$ bisects $\angle MLN$,
$\overline{PM} \perp \overrightarrow{LM}$,
$\overline{PN} \perp \overrightarrow{LN}$
Prove: $\overline{PM} \cong \overline{PN}$

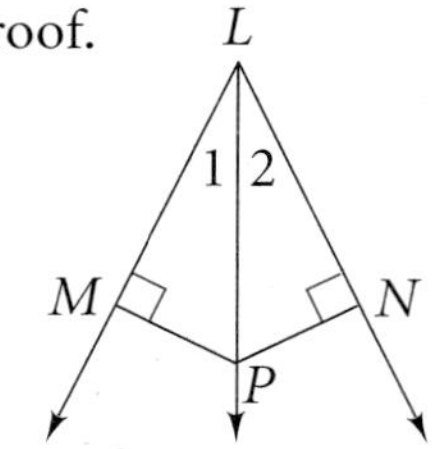

Plan for Proof $\overline{PM}$ and $\overline{PN}$ are corresponding parts of $\triangle LMP$ and $\triangle LNP$. These triangles are congruent by the AAS Theorem because a bisector forms congruent angles, perpendiculars form congruent right angles, and $\overline{LP} \cong \overline{LP}$ by the Reflexive Property of Congruence. ■

5. *Choose* Use the Plan for Proof in Example 4. Write a two-column proof, a paragraph proof, or a flow proof.

QUICK REVIEW

Angle Bisector Theorem
If a point is on the bisector of an angle, then it is equidistant from the sides of the angle.

The **distance from a point to a line** is the length of the perpendicular segment from the point to the line.

Exercises ON YOUR OWN

Explain how you would use SSS, SAS, ASA, AAS, or HL with CPCTC to prove each statement.

1. $\overline{AB} \cong \overline{CB}$

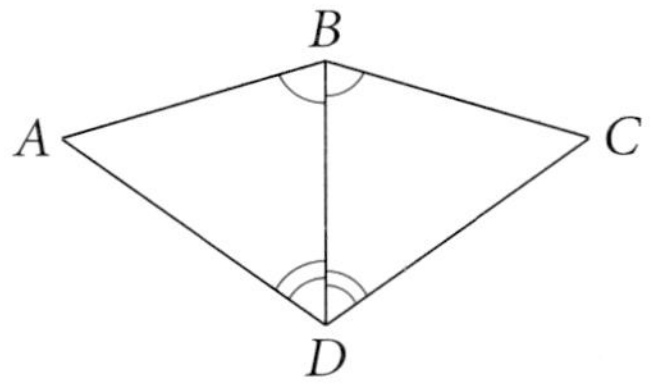

2. $\angle M \cong \angle R$

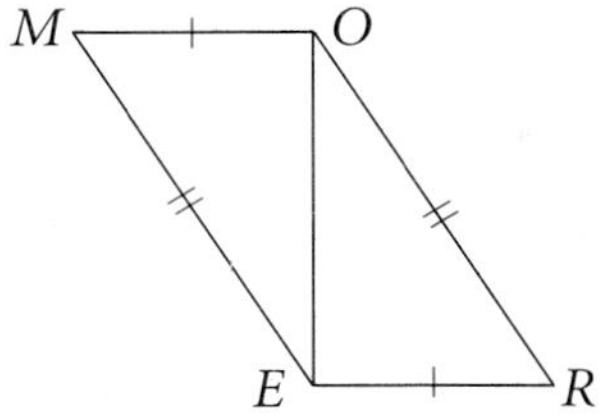

3. $\angle S \cong \angle O$

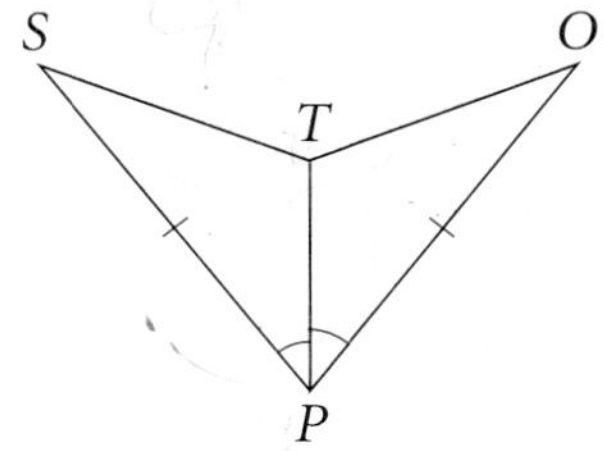

4. $\overline{KP} \cong \overline{LM}$

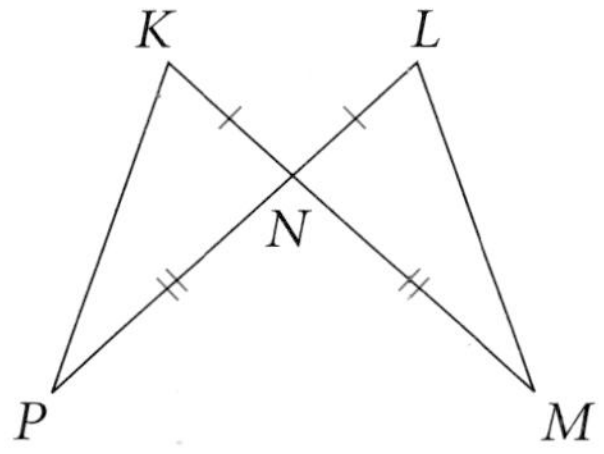

5. $\angle R \cong \angle P$

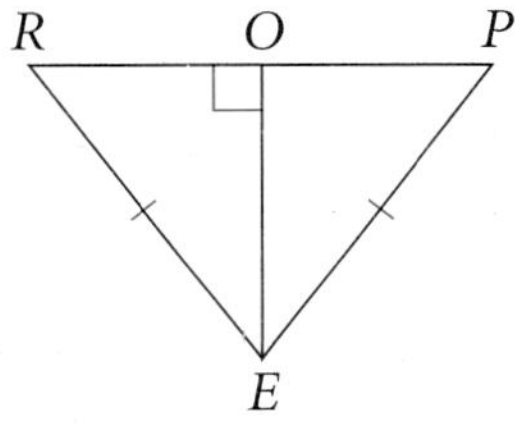

6. $\overline{CT} \cong \overline{RP}$

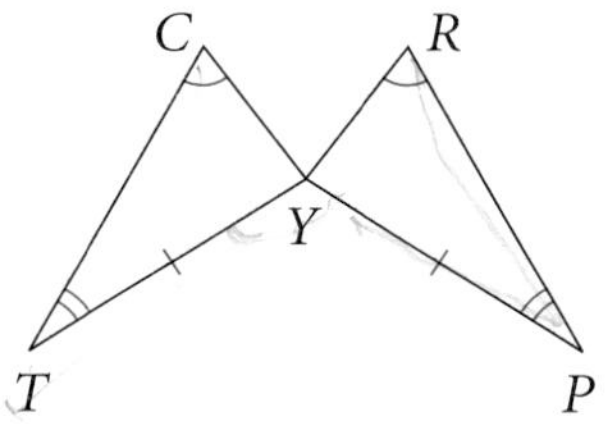

7. Karen cut this pattern for stained glass so that $AB = CB$ and $AD = CD$. Must $\angle A$ be congruent to $\angle C$? How do you know? Explain.

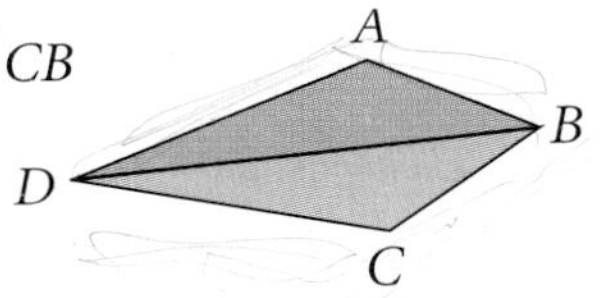

EARTH NEWS

Sinkhole Swallows House

The large sinkhole in this photo occurred suddenly in 1981 in Winter Park, Florida, following a severe drought. Increased water consumption speeds formation of sinkholes by lowering the water table. This leads to the collapse of underground caverns that have formed in underlying limestone.

8. A geometry class indirectly measured the distance across a sinkhole in a nearby field. They measured distances as shown in the diagram. Explain how to use their measurements to find the distance across the sinkhole.

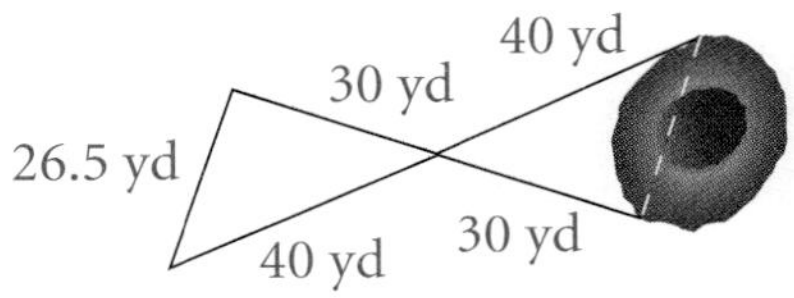

9. Theorem 4-3 states that if two angles of a triangle are congruent, then the sides opposite the angles are congruent.

To prove this theorem, begin with $\triangle ABC$ with $\angle A \cong \angle C$ and draw the bisector of $\angle B$. Supply a reason for each statement in the proof.

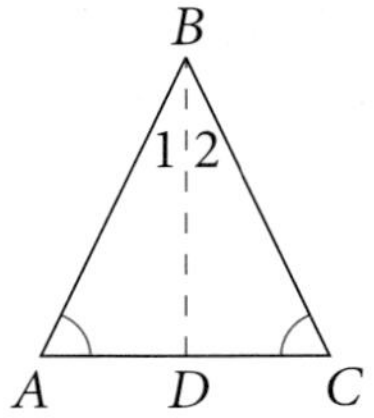

Given: $\angle A \cong \angle C$, $\overline{BD}$ bisects $\angle ABC$.
Prove: $\overline{AB} \cong \overline{CB}$

Statements	Reasons
1. $\angle A \cong \angle C$	a. ?
2. $\overline{BD}$ bisects $\angle ABC$.	b. ?
3. $\angle 1 \cong \angle 2$	c. ?
4. $\overline{BD} \cong \overline{BD}$	d. ?
5. $\triangle ABD \cong \triangle CBD$	e. ?
6. $\overline{AB} \cong \overline{CB}$	f. ?

10. The reasons given in the proof below are correct, but they are in the wrong order. List them in the correct order.

Given: O is the center of the circle.
Prove: $\overline{AB} \cong \overline{CD}$

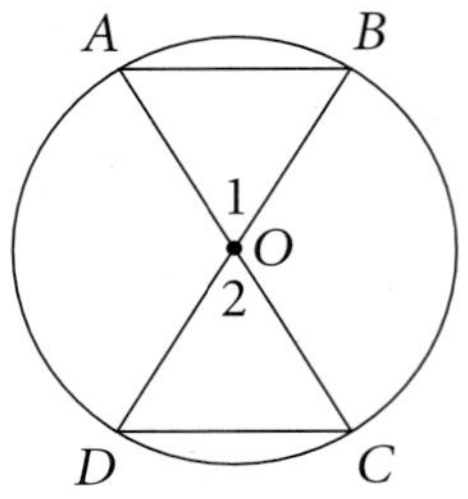

Statements	Reasons
1. O is the center of the circle.	a. CPCTC
2. $\overline{AO} \cong \overline{CO}$, $\overline{BO} \cong \overline{DO}$	b. SAS Postulate
3. $\angle 1 \cong \angle 2$	c. All radii of a circle are $\cong$.
4. $\triangle ABO \cong \triangle CDO$	d. Given
5. $\overline{AB} \cong \overline{CD}$	e. Vertical angles are $\cong$.

11. Theorem 4-12 states that if a point is on the perpendicular bisector of a segment, then it is equidistant from the endpoints of the segment.

Supply the reasons for the following flow proof of the theorem.

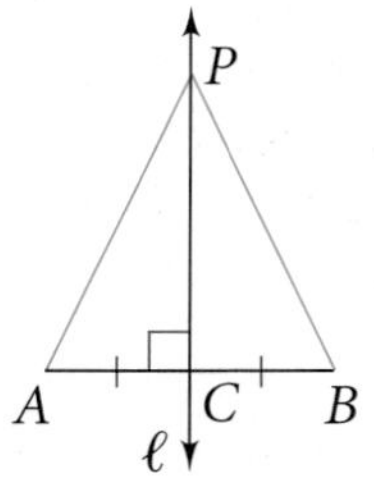

Given: $\ell \perp \overline{AB}$, ℓ bisects $\overline{AB}$ at C, P is on ℓ.
Prove: $PA = PB$

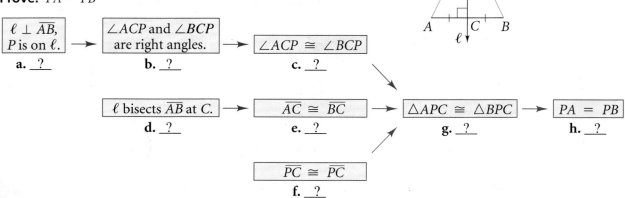

$\ell \perp \overline{AB}$, P is on ℓ. → $\angle ACP$ and $\angle BCP$ are right angles. → $\angle ACP \cong \angle BCP$
a. ? b. ? c. ?

ℓ bisects $\overline{AB}$ at C. → $\overline{AC} \cong \overline{BC}$ → $\triangle APC \cong \triangle BPC$ → $PA = PB$
d. ? e. ? g. ? h. ?

$\overline{PC} \cong \overline{PC}$
f. ?

12. **Choose** Write a two-column proof, a paragraph proof, or a flow proof.

Given: $\overline{FB} \perp \overline{AD}$, $\overline{GC} \perp \overline{AD}$, $\overline{FB} \cong \overline{GC}$, $\overline{AE} \cong \overline{DE}$
Prove: $\overline{AB} \cong \overline{DC}$

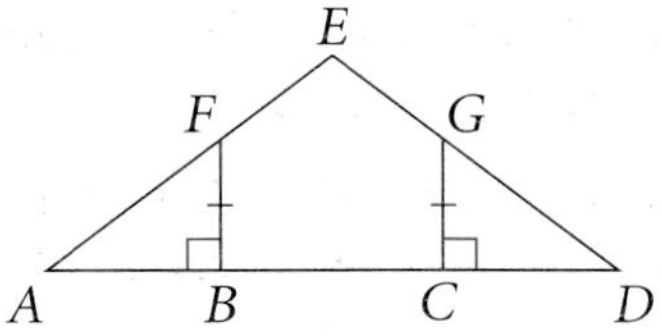

13. Theorem 4-2 states that the bisector of the vertex angle of an isosceles triangle is the perpendicular bisector of the base.

Use the following Plan for Proof to write a paragraph proof for the theorem.

Given: Isosceles $\triangle ABC$ with $\overline{BA} \cong \overline{BC}$, $\overline{BD}$ bisects $\angle ABC$.
Prove: $\overline{BD} \perp \overline{AC}$, $\overline{BD}$ bisects $\overline{AC}$.

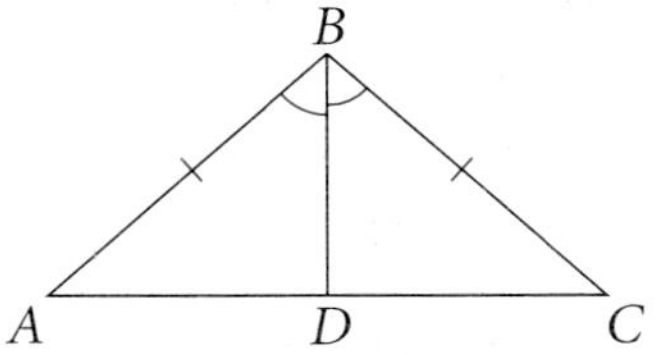

Plan for Proof You can show that $\overline{BD} \perp \overline{AC}$ by showing that $\angle BDA \cong \angle BDC$ and using the fact that congruent supplementary angles are right angles. You can show that $\overline{BD}$ bisects $\overline{AC}$ by showing that $\overline{AD} \cong \overline{CD}$. These angles and segments are corresponding parts of $\triangle ABD$ and $\triangle CBD$. Prove that $\triangle ABD \cong \triangle CBD$.

14. **Choose** Write a two-column proof, a paragraph proof, or a flow proof.

Given: $\overline{ZK} \perp \overline{JQ}$ in $\odot O$
Prove: $\overline{JW} \cong \overline{QW}$

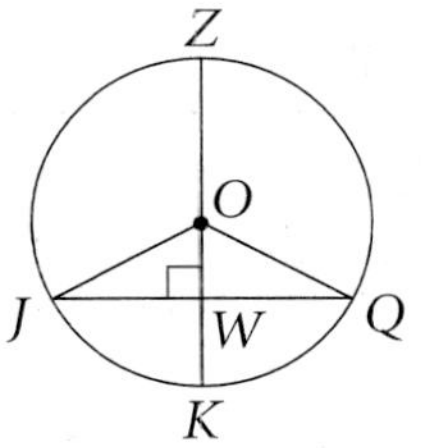

15. **a.** How can you prove $\triangle BGF \cong \triangle BGD$?
b. $BD = 8$ cm. What other segment is 8 cm long? **Justify** your answer.
c. How can you prove $\triangle FAB \cong \triangle BCD$? (*Hint:* Use your results from part (b).)

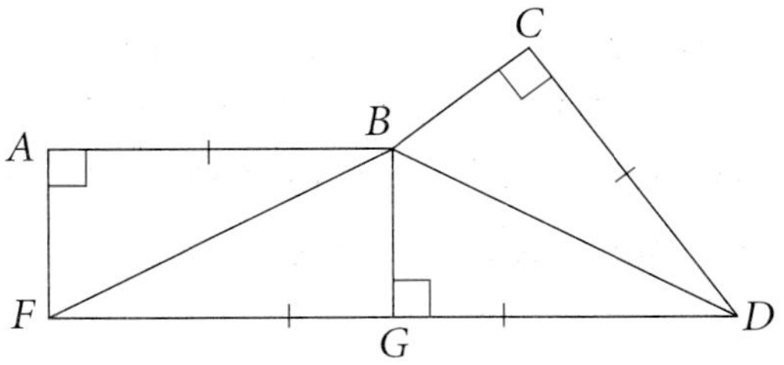

16. **Constructions** The construction of the bisector of $\angle A$ is shown below.

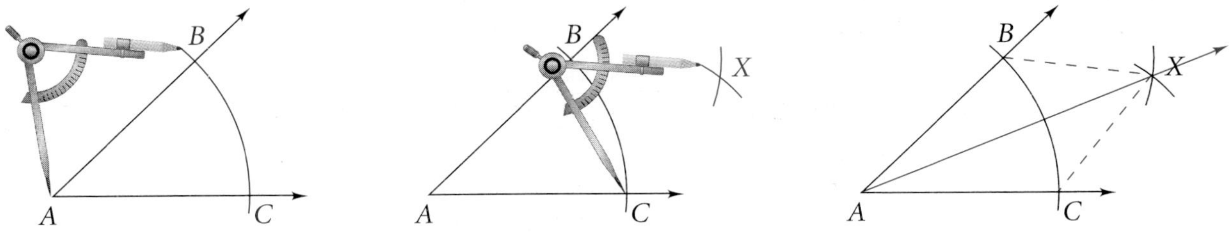

$\overline{AB} \cong \overline{AC}$ because they are radii of the same circle. $\overline{BX} \cong \overline{CX}$ because the same compass setting was used to draw both arcs. Use these facts to prove that $\overrightarrow{AX}$ bisects $\angle BAC$.

17. *Constructions* The construction of a line perpendicular to line ℓ through point P on ℓ is shown here.
   a. Which segments are congruent by construction?
   b. Prove that $\overleftrightarrow{CP}$ is perpendicular to ℓ. (*Hint:* Copy the diagram and draw $\overline{CA}$ and $\overline{CB}$.)

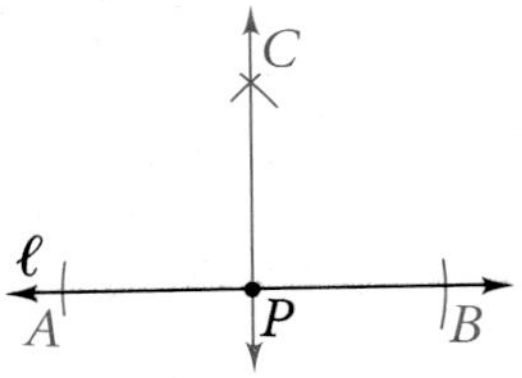

Exercises MIXED REVIEW

Copy each figure and draw its image under the given transformation.

18. reflection in line m

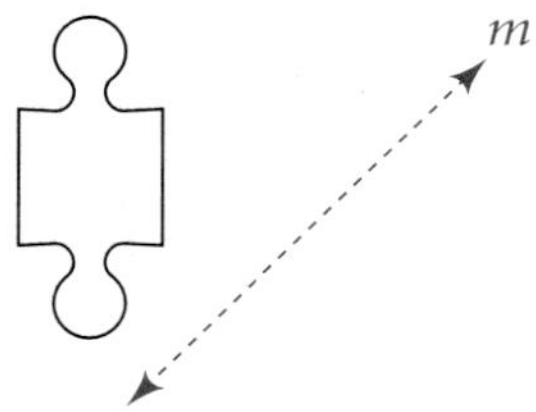

19. 90° counterclockwise rotation about point P

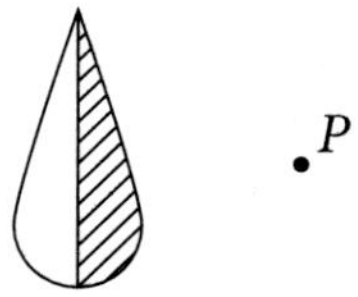

20. a translation of $\langle -2, 4\rangle$

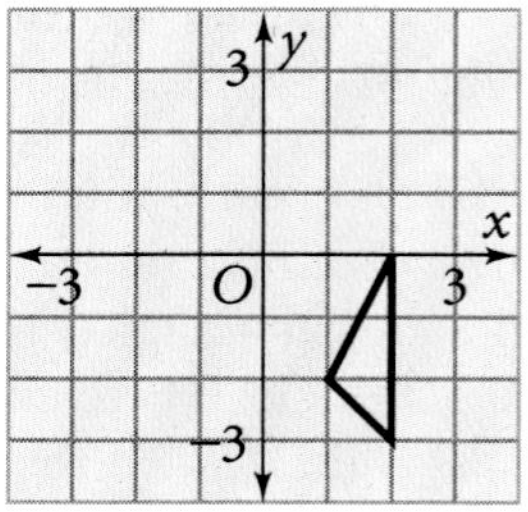

Data Analysis **For Exercises 21–23, use the circle graph.**

21. What percent of people knew the number of days they take off from work because of colds?
22. What percent of the people said they take off four or fewer days annually because of colds?
23. What is the measure of the central angle of the sector for the people who take one or two days off because of colds?

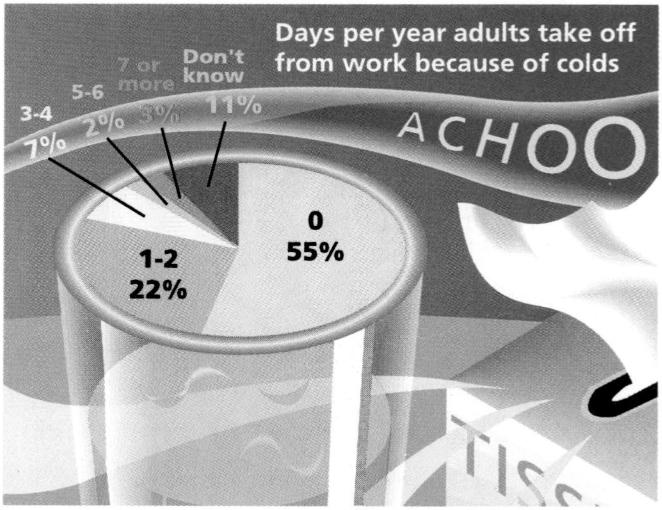

Source: *Ketchum Public Relations for Smith Kline Beecham*

Getting Ready For Lesson 8-5

Copy and label each triangle. Draw its image under a reflection in the red line. State which sides or angles a pair of triangles have in common.

24.

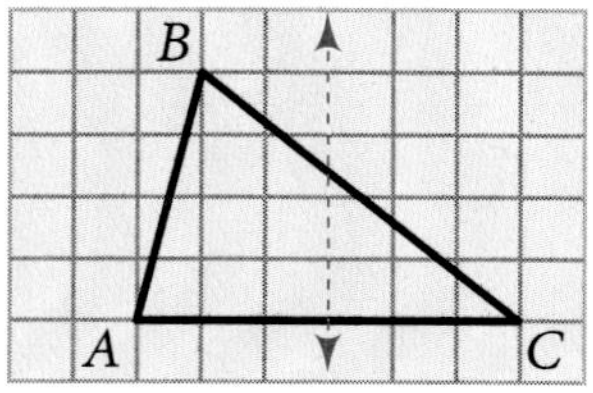

25.

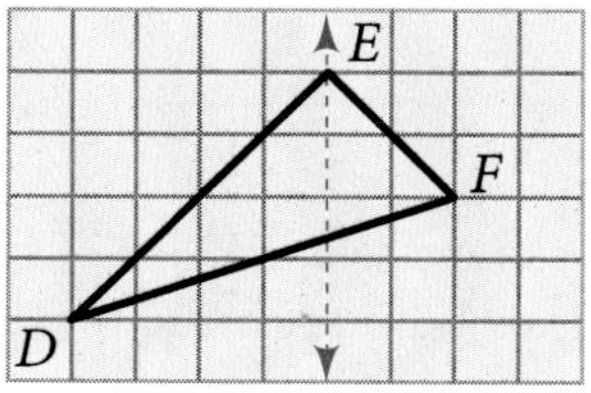

26.

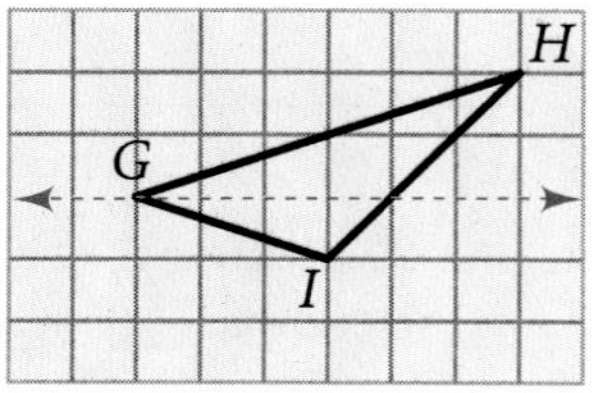

What You'll Learn

- Identifying congruent overlapping triangles
- Proving two triangles congruent by first proving two other triangles congruent

...And Why

To be able to visualize overlapping triangles and use pairs of congruent triangles in real-world situations such as engineering

What You'll Need

- tracing paper
- colored pencils or pens

8-5 Using More than One Pair of Congruent Triangles

WORK TOGETHER

Work with a partner. Trace the diagram below and take turns finding pairs of congruent triangles. Use a different colored pencil to outline each pair you find. Then answer the questions below.

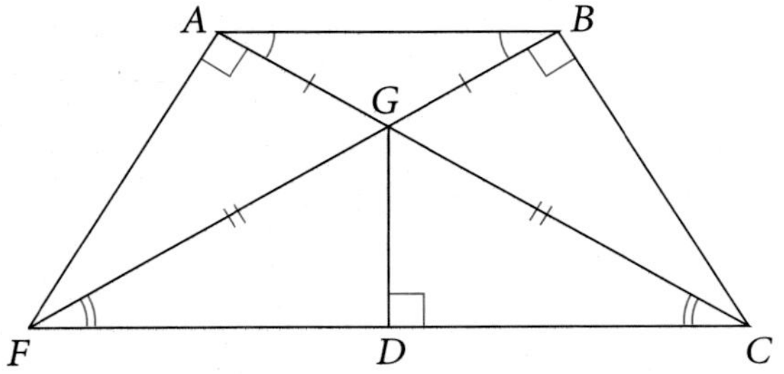

1. Write a congruence statement for each pair of triangles that you colored. State the congruence postulate or theorem you could use to prove the triangles congruent. Compare your results with those of another group.

2. a. Name a pair of congruent triangles that overlap.
b. Identify the common side the triangles share.

THINK AND DISCUSS

Using Overlapping Triangles in Proofs

In the Work Together, you identified pairs of overlapping triangles that share a common side. Some overlapping triangles share a common angle. It is helpful to separate and redraw the overlapping triangles.

Example 1

Separate and redraw $\triangle DFG$ and $\triangle EHG$. Identify the common angle.

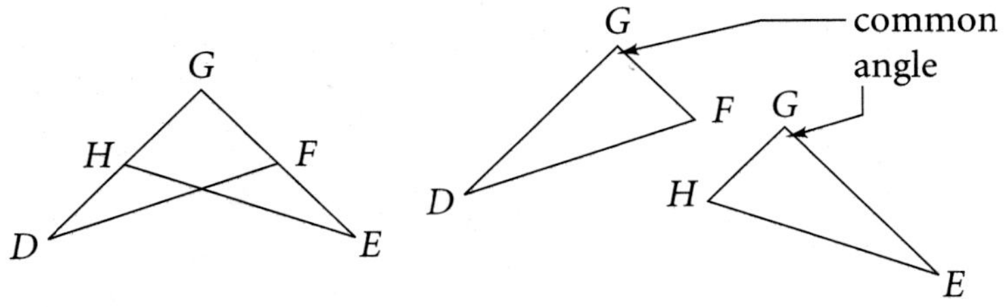

3. *Engineering* The diagram below shows triangles from the scaffolding workers used when they repaired and cleaned the Statue of Liberty.
   a. Identify the common side in $\triangle ADC$ and $\triangle BCD$.
   b. Name another pair of triangles that share a common side.

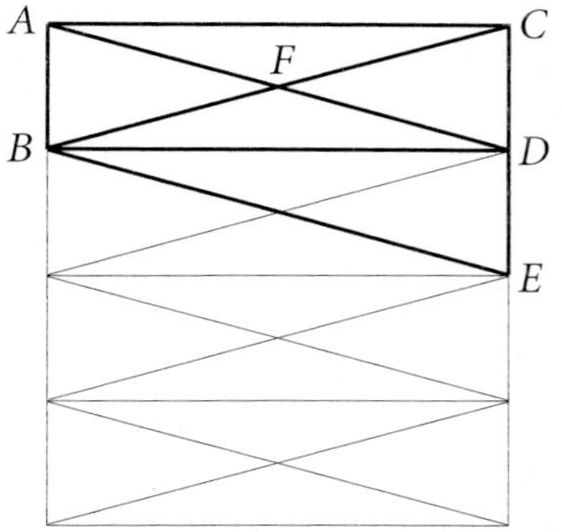

You can use common angles and common sides to prove overlapping triangles congruent.

Example 2

Given: $\angle ZWX \cong \angle YXW$,
$\angle ZXW \cong \angle YWX$
Prove: $\overline{WZ} \cong \overline{XY}$

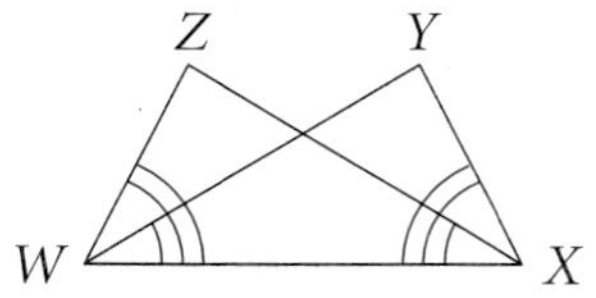

Separate the overlapping triangles. Then write a two-column proof.

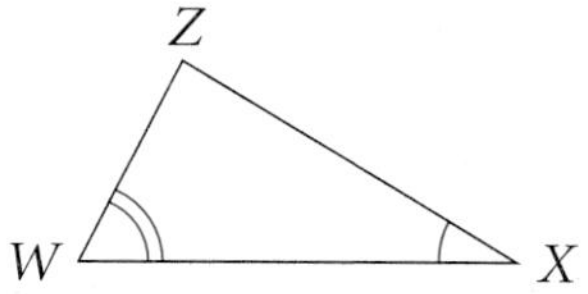

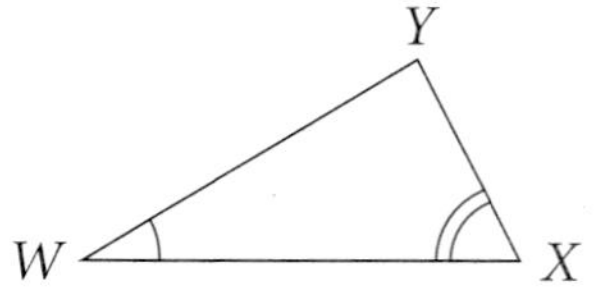

Plan for Proof To use CPCTC to show that $\overline{WZ} \cong \overline{XY}$, you must first prove that $\triangle ZWX \cong \triangle YXW$.

Proof

Statements	Reasons
1. $\angle ZWX \cong \angle YXW$, $\angle ZXW \cong \angle YWX$	1. Given
2. $\overline{WX} \cong \overline{WX}$	2. Reflexive Property of $\cong$
3. $\triangle ZWX \cong \triangle YXW$	3. ASA Postulate
4. $\overline{WZ} \cong \overline{XY}$	4. CPCTC

Using Two Pairs of Congruent Triangles

Sometimes you can prove that one pair of triangles is congruent and then use corresponding congruent parts of those triangles to prove that another pair of triangles is congruent.

Example 3

Given: E is the midpoint of $\overline{AC}$ and $\overline{DB}$.
Prove: $\triangle GED \cong \triangle JEB$

There is not enough information about $\triangle GED$ and $\triangle JEB$ to prove that they are congruent. But you can prove that $\triangle AED \cong \triangle CEB$. Then you can use corresponding congruent parts of these triangles to prove that $\triangle GED \cong \triangle JEB$.

Paragraph Proof

Since E is the midpoint of $\overline{AC}$ and $\overline{DB}$, $\overline{AE} \cong \overline{CE}$ and $\overline{DE} \cong \overline{BE}$. $\angle AED \cong \angle CEB$ because vertical angles are congruent. Therefore, $\triangle AED \cong \triangle CEB$ by SAS. $\angle D \cong \angle B$ by CPCTC, and $\angle GED \cong \angle JEB$ because they are vertical angles. Therefore, $\triangle GED \cong \triangle JEB$ by ASA. ■

WORK TOGETHER

Work with a partner.

4. Copy and complete the two-column proof below using two pairs of congruent triangles. After you prove the first pair of triangles congruent, separate and redraw the second pair of overlapping triangles that you need to use.

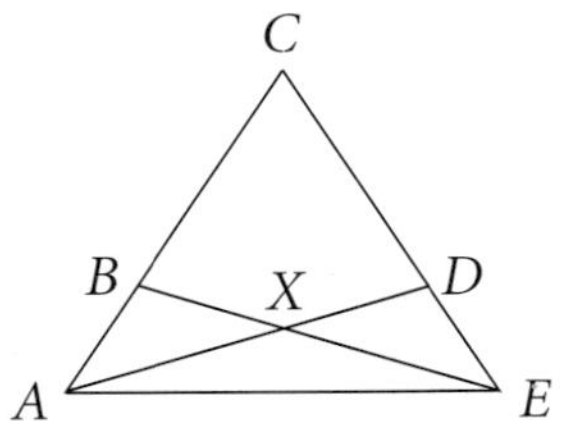

Given: $\overline{CA} \cong \overline{CE}$, $\overline{BA} \cong \overline{DE}$
Prove: $\overline{BX} \cong \overline{DX}$

Statements	Reasons
1. $\overline{CA} \cong \overline{CE}$	**a.** ?
2. $\angle CAE \cong \angle CEA$	**b.** ?
3. $\overline{AE} \cong \overline{AE}$	**c.** ?
4. $\overline{BA} \cong \overline{DE}$	**d.** ?
5. $\triangle BAE \cong \triangle DEA$	**e.** ?
6. $\angle ABE \cong \angle EDA$	**f.** ?
7. $\angle BXA \cong \angle DXE$	**g.** ?
8. $\triangle BXA \cong \triangle DXE$	**h.** ?
9. $\overline{BX} \cong \overline{DX}$	**i.** ?

Exercises ON YOUR OWN

In each diagram, the red and blue triangles are congruent. Identify their common side or angle.

1.

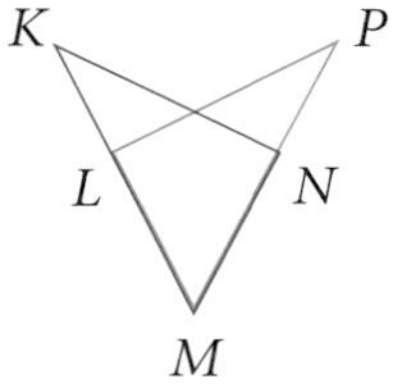

2.

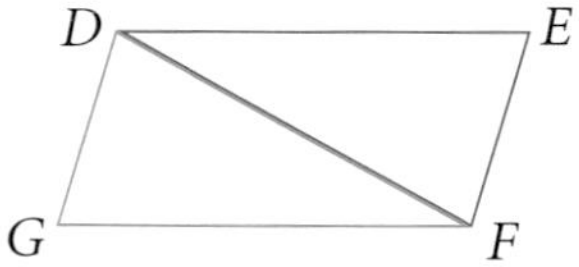

3.

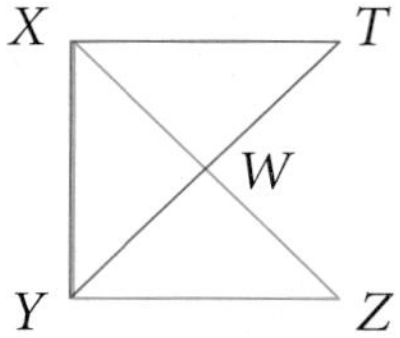

Name a pair of overlapping congruent triangles in each diagram. State whether the triangles are congruent by SSS, SAS, ASA, AAS, or HL.

4. Given: $\overline{MP} \cong \overline{QL}$, $\overline{LP} \perp \overline{LM}$, $\overline{LP} \perp \overline{PQ}$

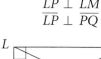

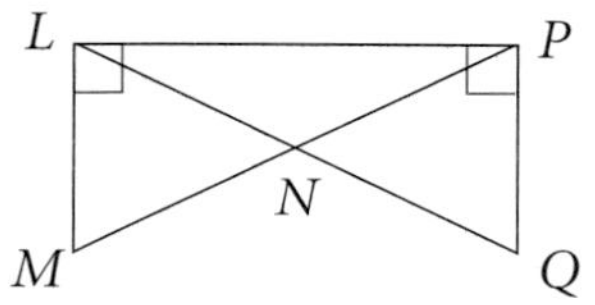

5. Given: $\overline{RS} \cong \overline{UT}$, $\overline{RT} \cong \overline{US}$

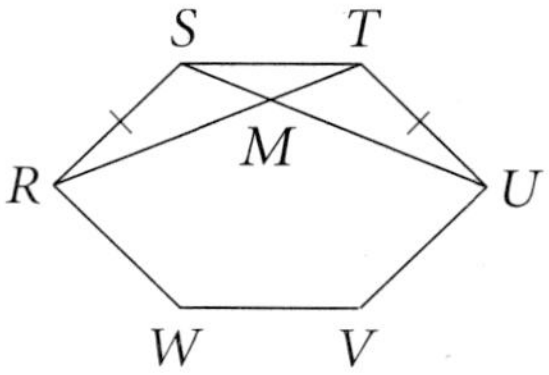

6. Given: $\overline{QD} \cong \overline{UA}$, $\angle QDA \cong \angle UAD$

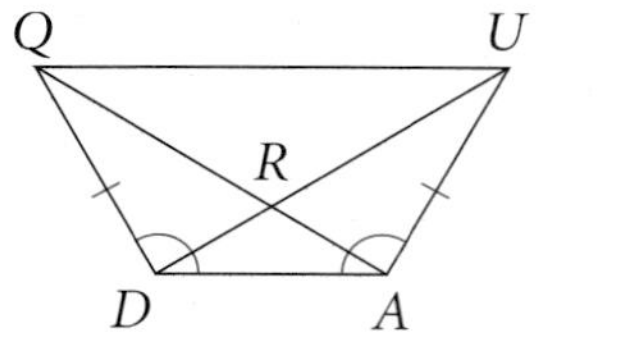

7. Given: $\overline{PQ} \parallel \overline{UR}$, $\overline{QT} \parallel \overline{RS}$, $\overline{QT} \cong \overline{RS}$

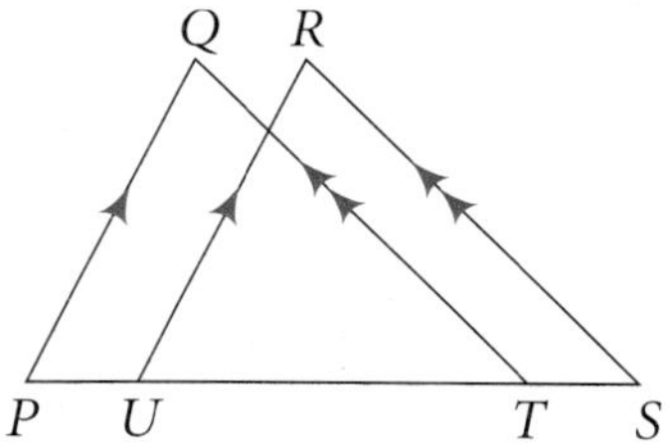

8. Given: $\overline{AC} \cong \overline{BC}$, $\angle A \cong \angle B$

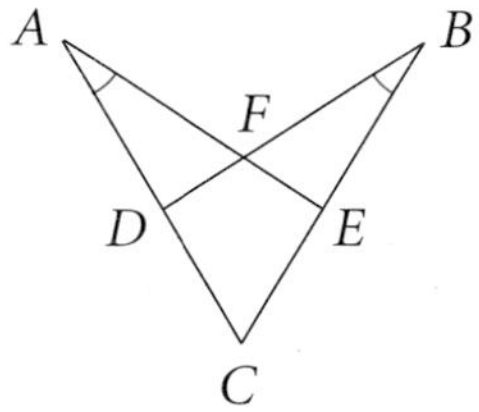

9. Given: $\odot O$, $\overline{WY} \perp \overline{YX}$, $\overline{ZX} \perp \overline{YX}$

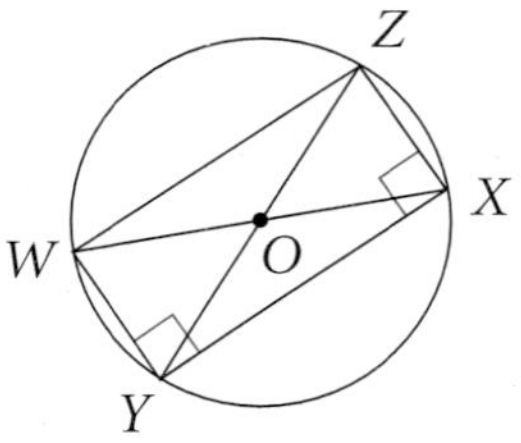

10. a. **Design** The figure below is part of a clothing design pattern. In the figure, $\overline{AB} \parallel \overline{DE} \parallel \overline{FG}$, $\overline{AB} \perp \overline{BC}$, $\overline{GC} \perp \overline{AC}$. $\triangle DEC$ is isosceles with base $\overline{DC}$, and $m\angle A = 56$. Find the measures of all the numbered angles in the figure.

b. $\overline{AB} \cong \overline{FC}$. Name two congruent triangles and tell how you can prove them congruent.

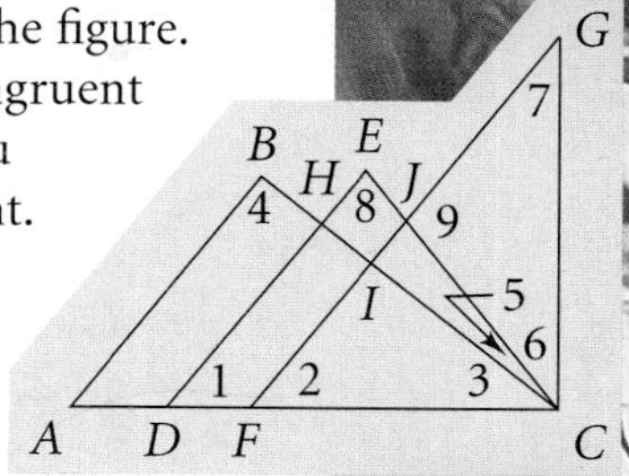

Preparing for Proof **Write a plan for a proof.**

11. Given: $\angle 1 \cong \angle 2$, $\angle 3 \cong \angle 4$
Prove: $\triangle QET \cong \triangle QEU$

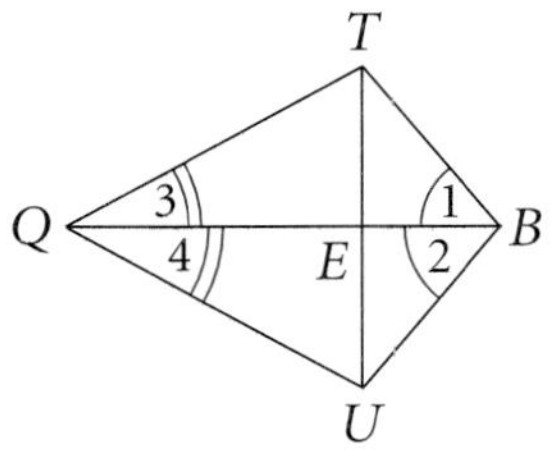

12. Given: $\overline{AD} \cong \overline{ED}$, D is the midpoint of $\overline{BF}$.
Prove: $\triangle ADC \cong \triangle EDG$

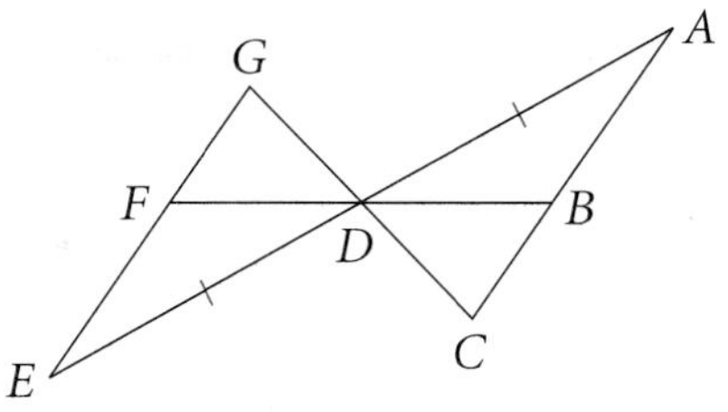

Preparing for Proof **Copy and complete each two-column proof.**

13. Given: $\overline{ER} \cong \overline{IT}$, $\overline{ET} \cong \overline{IR}$, $\angle TDI$ and $\angle ROE$ are right $\angle$s.
Prove: $\overline{TD} \cong \overline{RO}$

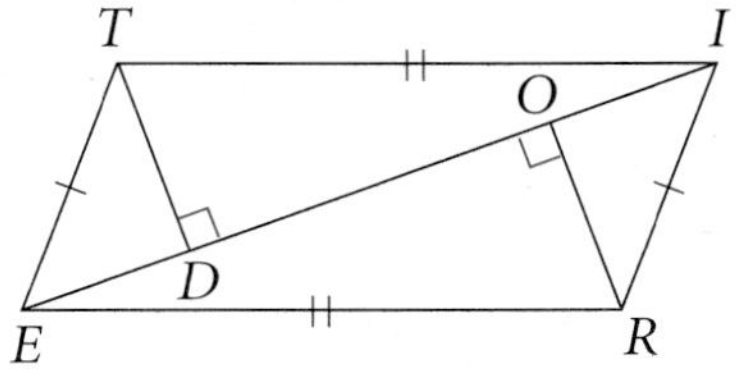

Statements	Reasons
1. $\overline{ER} \cong \overline{IT}$, $\overline{ET} \cong \overline{IR}$	a. ?
2. $\overline{EI} \cong \overline{EI}$	b. ?
c. $\triangle ERI \cong \triangle$?	d. ?
e. $\angle REO \cong \angle$?	4. CPCTC
5. $\angle TDI$ and $\angle ROE$ are right $\angle$s.	f. ?
6. $\angle TDI \cong \angle ROE$	g. ?
7. $\triangle TDI \cong \triangle ROE$	h. ?
i. ? $\cong$?	j. ?

14. Given: $\overline{AB} \perp \overline{BC}$, $\overline{DC} \perp \overline{BC}$, $\overline{AC} \cong \overline{DB}$
Prove: $\overline{AE} \cong \overline{DE}$

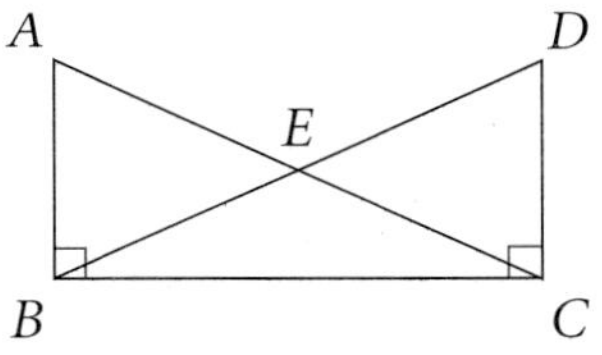

Statements	Reasons
1. $\overline{AB} \perp \overline{BC}$, $\overline{DC} \perp \overline{BC}$	a. ?
2. $\angle ABC$ and $\angle DCB$ are rt. $\angle$s.	b. ?
3. $\triangle ABC$ and $\triangle DCB$ are rt. $\triangle$s.	c. ?
4. $\overline{AC} \cong \overline{DB}$	d. ?
e. ? $\cong$?	5. Reflexive Property of $\cong$
6. $\triangle ABC \cong \triangle DCB$	f. ?
7. $\angle A \cong \angle D$, $\overline{AB} \cong \overline{DC}$	g. ?
h. $\angle AEB \cong \angle$?	i. ?
9. $\triangle ABE \cong \triangle DCE$	j. ?
k. ? $\cong$?	l. ?

15. **Standardized Test Prep** In order to prove $\overline{JW} \cong \overline{QX}$ in the diagram at the right, what would you prove first?

A. $\triangle ZJW \cong \triangle KQX$ **B.** $\triangle KJZ \cong \triangle ZQK$
C. $\triangle JWK \cong \triangle QXZ$ **D.** $\overline{ZW} \cong \overline{KX}$
E. Not enough information is given.

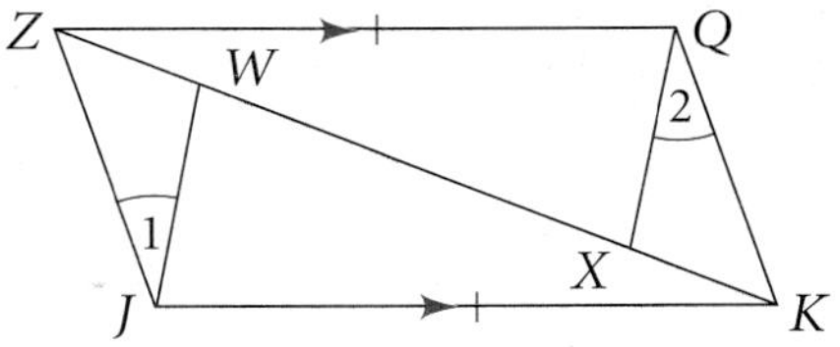

Given: $\overline{ZQ} \parallel \overline{KJ}$, $\overline{ZQ} \cong \overline{KJ}$, $\angle 1 \cong \angle 2$

Choose **Write a two-column proof, a paragraph proof, or a flow proof.**

16. Given: $\overline{AC} \cong \overline{EC}$, $\overline{CB} \cong \overline{CD}$
Prove: $\angle A \cong \angle E$

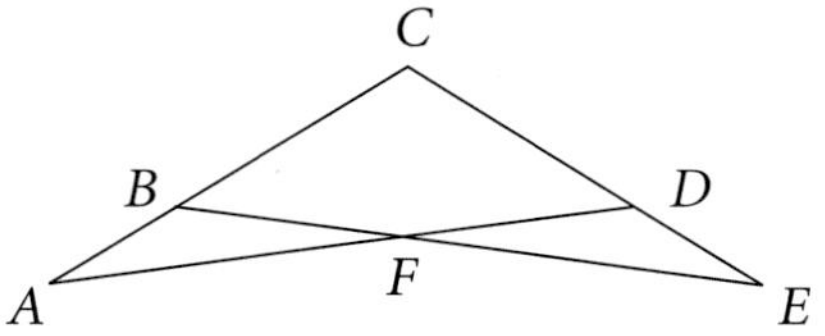

17. Given: $\overline{TQ}$ is the perpendicular bisector of $\overline{PR}$; $\overline{TQ}$ bisects $\angle VQS$.
Prove: $\overline{VQ} \cong \overline{SQ}$

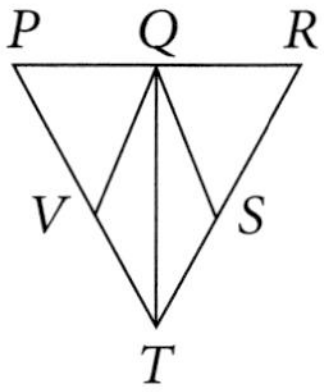

18. **a.** *Open-ended* Draw a parallelogram and its diagonals. Label your diagram and list all the pairs of congruent segments.

b. *Writing* Explain how you can be sure that the segments you listed must be congruent.

Chapter Project — Find Out by Investigating

In the Find Out question on page 412, you tested the strength of two-dimensional models. Now investigate the strength of three-dimensional models.

Use toothpicks or craft sticks and glue to construct a cube and a tetrahedron (a triangular pyramid).

- Which model is stronger?
- Describe how you could strengthen the weaker model.

Use toothpicks or craft sticks and glue to construct a structure that can support the weight of your geometry book.

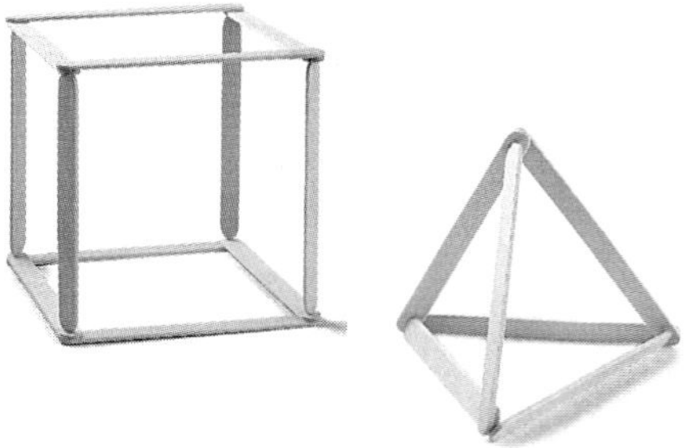

Exercises MIXED REVIEW

Coordinate Geometry **Graph $\triangle ABC$ with vertices $A(5, 5)$, $B(5, 1)$, and $C(1, 1)$.**

19. Classify $\triangle ABC$ by its sides and angles.

20. Find the length of $\overline{AC}$. Leave your answer in simplest radical form.

21. Find the area and perimeter of $\triangle ABC$. Leave your answer in simplest radical form.

22. *Calculator* The interior of a cylindrical mug has diameter 8 cm and is 9 cm deep. How much liquid can it hold? Round your answer to the nearest milliliter. (*Hint:* 1 cm^3 = 1 mL)

SELF ASSESSMENT — PORTFOLIO

For your portfolio, select one or two items from your work for this chapter. Here are some possibilities.
- most improved work
- a journal entry
- a constructed model

Explain why you have included each selection.

Solving Quadratic Equations

After Lesson 8-5

The **standard form of a quadratic equation** is

$$ax^2 + bx + c = 0, a \neq 0.$$

You can solve a quadratic equation by substituting the values for a, b, and c in the **Quadratic Formula.**

$$x = \frac{-b \pm \sqrt{b^2 - 4ac}}{2a}$$

Example 1

Solve $7x^2 + 6x - 1 = 0$. The equation is in standard form.

$$a = 7, b = 6, c = -1$$

$$x = \frac{-6 \pm \sqrt{6^2 - 4(7)(-1)}}{2(7)}$$ Substitute in the Quadratic Formula.

$$x = \frac{-6 \pm \sqrt{36 + 28}}{14}$$ Simplify.

$$x = \frac{-6 \pm \sqrt{64}}{14}$$

$$x = \frac{-6 + 8}{14} \text{ or } x = \frac{-6 - 8}{14}$$

$$x = \frac{1}{7} \text{ or } x = -1$$

Sometimes you may need a calculator to approximate solutions.

Example 2

Calculator Solve $-3x^2 - 5x + 1 = 0$.

$$a = -3, b = -5, c = 1$$

$$x = \frac{-(-5) \pm \sqrt{(-5)^2 - 4(-3)(1)}}{2(-3)}$$ Substitute in the Quadratic Formula.

$$x = \frac{5 \pm \sqrt{25 + 12}}{-6}$$ Simplify.

$$x = \frac{5 + \sqrt{37}}{-6} \text{ or } x = \frac{5 - \sqrt{37}}{-6}$$

$$x \approx -1.85 \text{ or } x \approx 0.18$$ Use a calculator.

Solve. Round answers that are not integers to the nearest hundredth.

1. $x^2 + 5x - 14 = 0$

2. $4x^2 - 13x + 3 = 0$

3. $2x^2 + 7x + 3 = 0$

4. $5x^2 + 2x - 2 = 0$

5. $6x^2 + 20x = -5$

6. $1 = 2x^2 - 6x$

7. $x^2 - 6x = 27$

8. $2x^2 - 10x + 11 = 0$

9. $8x^2 - 2x - 3 = 0$

CHAPTER PROJECT

Finishing the Chapter Project

Tri Tri Again

Find Out questions and activities on pages 412, 424, and 438 should help you complete your project. Design and construct a bridge made entirely of glue and toothpicks or craft sticks. Your bridge must be at least 8 inches long and contain no more than 100 toothpicks or no more than 30 craft sticks. With your classmates, decide how to test the strength of the bridge. Record the dimensions of your bridge, the number of toothpicks or craft sticks used, and the weight it could support. Experiment with as many designs and models as you like—the more the better. Include a summary of your experiments with notes about how each one helped you improve your design.

Reflect and Revise

Ask a classmate to review your project with you. Together, check that your bridge meets all the requirements and that your diagrams and explanations are clear. Have you tried several designs and kept a record of what you learned from each? Can your bridge be stronger, more efficient, or more pleasing to the eye? Revise your work as needed.

Follow Up

Research Buckminster Fuller and geodesic domes. Design and build a geodesic structure, using toothpicks or other materials.

For More Information

Bridge Builder. Baton Rouge, Louisiana: Pre-Engineering Software Corp., 1994. (Software)

Newhouse, Elizabeth L., ed. *The Builders: Marvels of Engineering.* Washington, D.C.: The National Geographic Society, 1992.

Servatius, Bridgitte. *Geometry and Its Applications: Rigidity & Braced Structures.* Lexington, Massachusetts: COMAP, Inc., 1995.

CHAPTER 8 Wrap Up

Key Terms

Angle-Angle-Side Theorem (p. 416)
Angle-Side-Angle Postulate (p. 414)
CPCTC (corresponding parts of congruent triangles are congruent) (p. 426)
Hypotenuse-Leg Theorem (p. 420)
Quadratic Formula (p. 439)
Side-Angle-Side Postulate (p. 408)
Side-Side-Side Postulate (p. 406)
Standard form of a quadratic equation (p. 439)

How am I doing?

SELF ASSESSMENT

- State three ideas from this chapter that you think are important. Explain your choices.
- List the different ways you can prove two triangles congruent.

Proving Triangles Congruent: SSS and SAS — 8-1

If three sides of one triangle are congruent to three sides of another triangle, then the two triangles are congruent by the **Side-Side-Side (SSS) Postulate.**

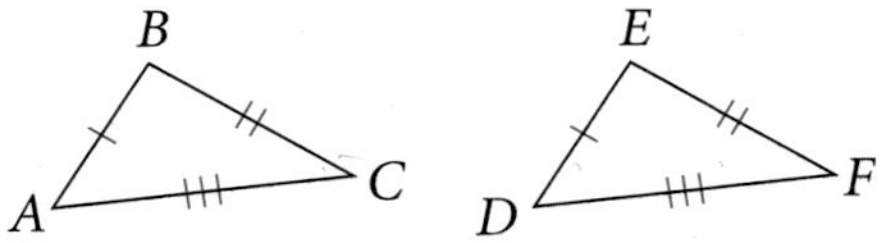

If two sides and the included angle of one triangle are congruent to two sides and the included angle of another triangle, then the two triangles are congruent by the **Side-Angle-Side (SAS) Postulate.**

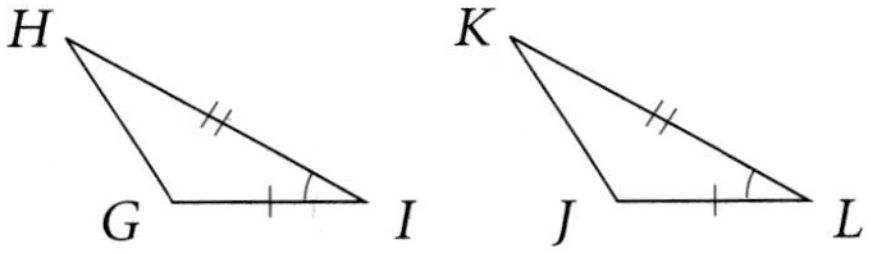

Decide whether you can use SSS or SAS to prove the triangles congruent. Write the congruence. If the triangles *cannot* be proven congruent, write *not possible.*

1.

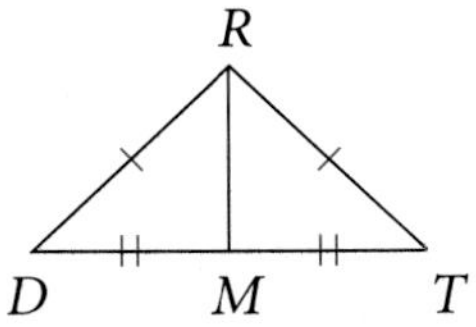

2.

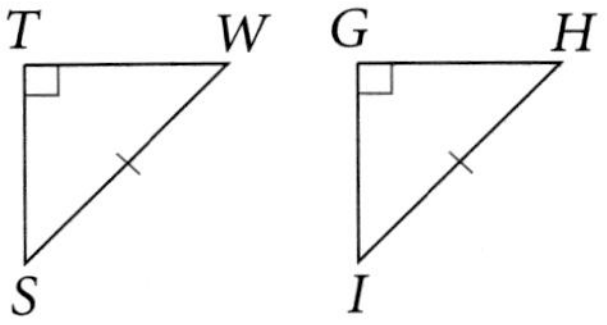

3.

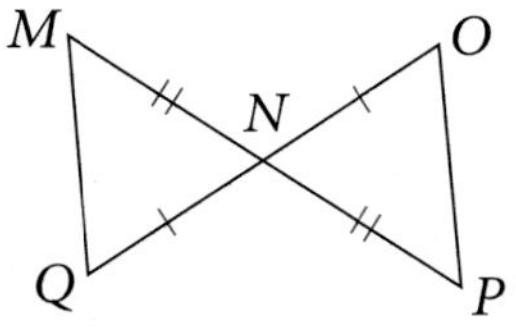

4. Writing Explain why the order of the vertices of a triangle in a congruence statement makes a difference.

5. Standardized Test Prep Which statements can you use to prove the triangles at the right congruent by the SAS Postulate?

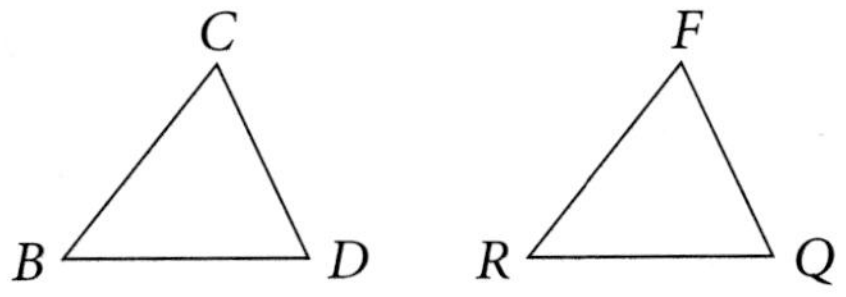

I. $\overline{BC} \cong \overline{RF}$ **II.** $\overline{BD} \cong \overline{RQ}$
III. $\angle D \cong \angle Q$ **IV.** $\angle B \cong \angle R$

A. I, II, and III only **B.** I, III, and IV only **C.** II, III, and IV only
D. I, II, and IV only **E.** I, II, III, and IV

Proving Triangles Congruent: ASA and AAS 8-2

If two angles and the included side of one triangle are congruent to two angles and the included side of another triangle, then the two triangles are congruent by the **Angle-Side-Angle (ASA) Postulate.**

If two angles and a nonincluded side of one triangle are congruent to two angles and the corresponding nonincluded side of another triangle, then the two triangles are congruent by the **Angle-Angle-Side (AAS) Theorem.**

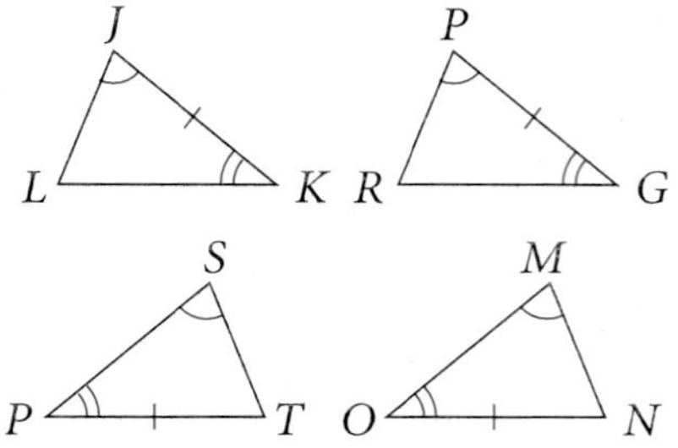

Would it be easier to use the ASA Postulate or the AAS Theorem to prove the triangles congruent? If the triangles *cannot* be proven congruent, write *not possible.*

6.

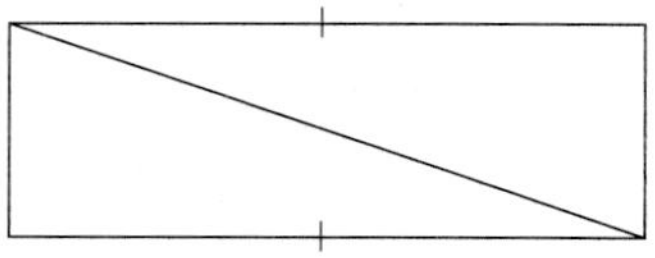

7.

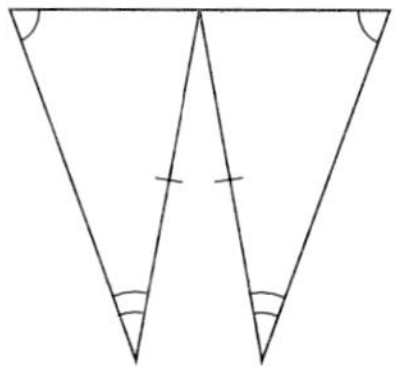

8.

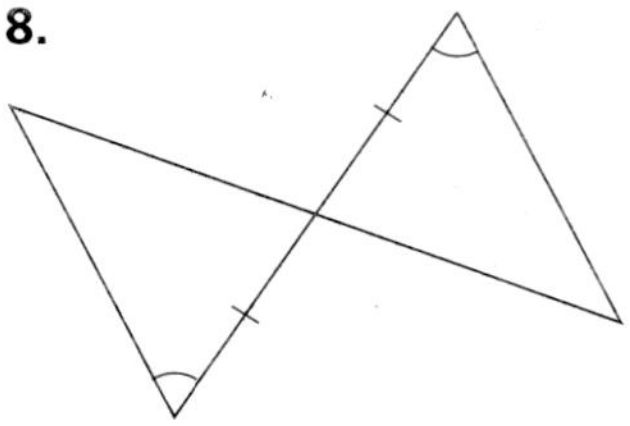

Congruent Right Triangles 8-3

If the hypotenuse and a leg of one right triangle are congruent to the hypotenuse and leg of another right triangle, then the triangles are congruent by the **Hypotenuse-Leg (HL) Theorem.**

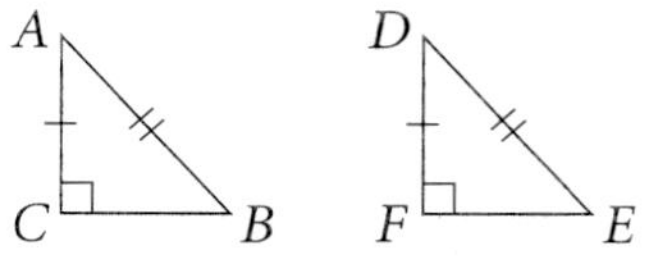

Choose **Write a two-column proof, a paragraph proof, or a flow proof.**

9. Given: $\overline{PS} \perp \overline{SQ}$, $\overline{RQ} \perp \overline{QS}$, $\overline{PQ} \cong \overline{RS}$
Prove: $\triangle PSQ \cong \triangle RQS$

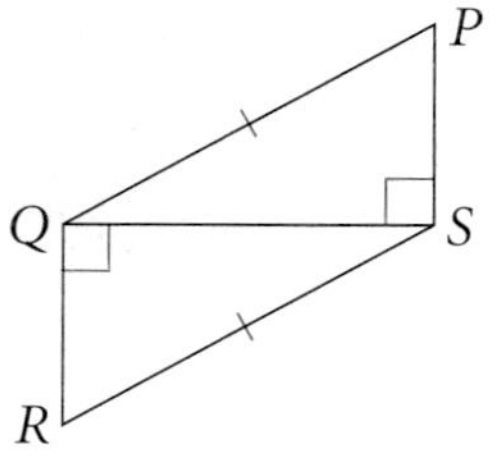

10. Given: $\overline{XU} \perp \overline{TV}$, $\overline{WV} \perp \overline{TV}$, $\overline{TX} \cong \overline{UW}$, U is the midpoint of $\overline{TV}$.
Prove: $\triangle TXU \cong \triangle UWV$

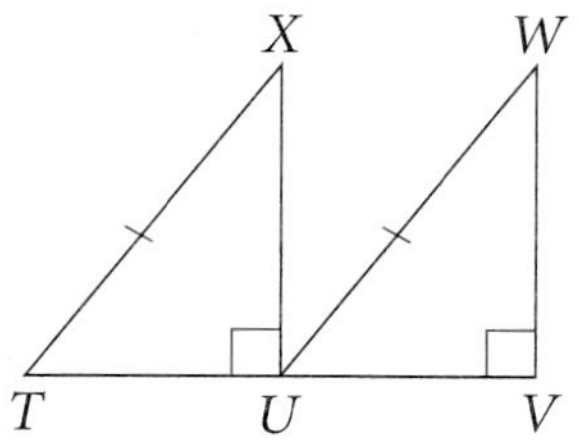

11. Given: $\overline{LN} \perp \overline{KM}$, $\overline{KL} \cong \overline{ML}$
Prove: $\triangle KLN \cong \triangle MLN$

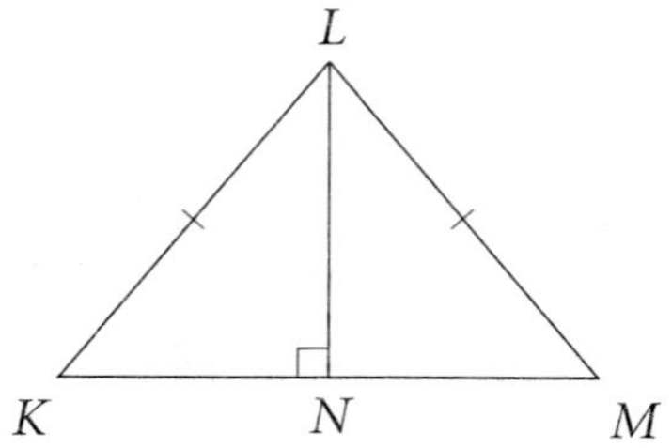

12. Open-ended Find several examples of right triangles in your school, home, or community. Which ones are congruent? What measurements would you make to show that they are congruent?

Using Congruent Triangles in Proofs 8-4

Once you know that triangles are congruent, you can make conclusions about corresponding segments and angles because **corresponding parts of congruent triangles are congruent (CPCTC).** You can use congruent triangles in the proofs of many theorems.

Explain how you would use SSS, SAS, ASA, AAS, or HL with CPCTC to prove each statement.

13. $\overline{TV} \cong \overline{YW}$

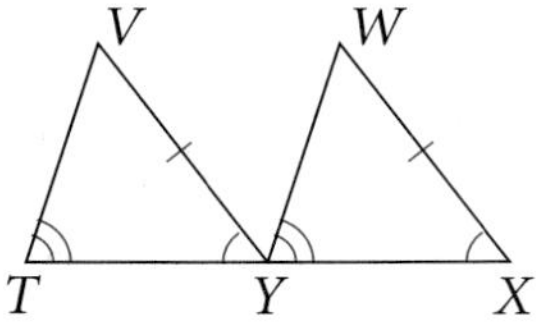

14. $\overline{BE} \cong \overline{DE}$

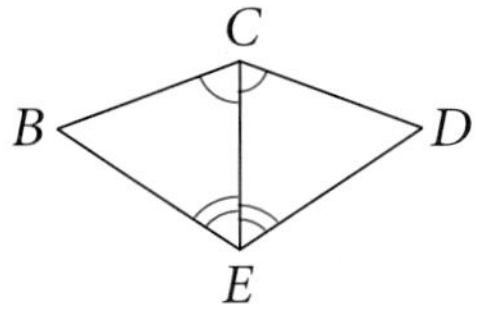

15. $\overline{KN} \cong \overline{ML}$

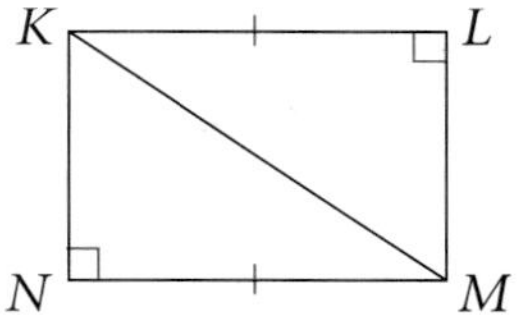

Using More than One Pair of Congruent Triangles 8-5

You can prove overlapping triangles congruent. You can also use the common or shared sides and angles of triangles in congruence proofs.

Name a pair of overlapping congruent triangles in each diagram. State whether the triangles are congruent by SSS, SAS, ASA, AAS, or HL.

16.

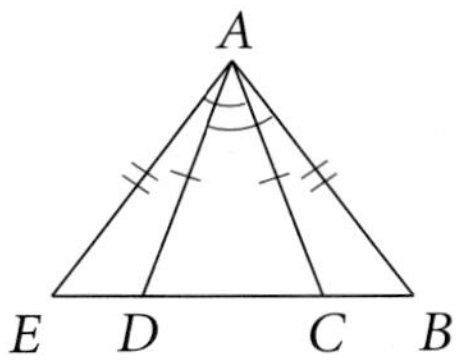

17.

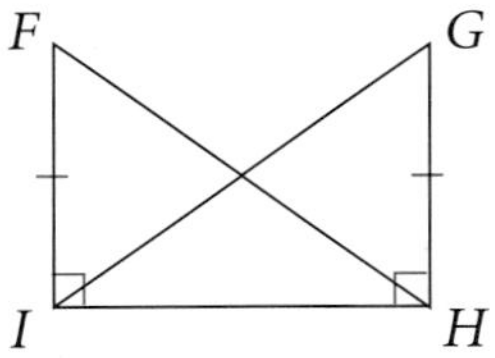

18.

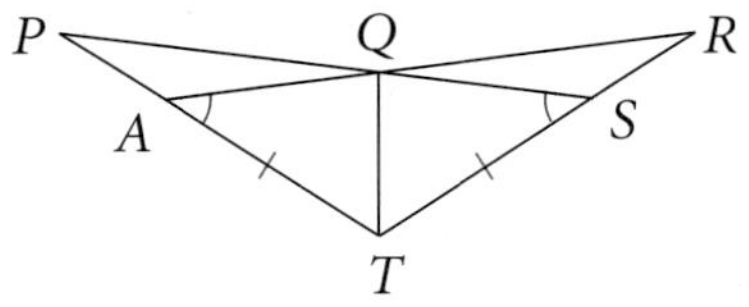

Getting Ready for CHAPTER 9

Coordinate Geometry **Graph and label each quadrilateral with the given vertices. Use slope and/or the Distance Formula to determine the most precise name for each figure.**

19. $A(-2, 3), B(1, 3), C(1, -7), D(-2, -7)$

20. $W(-1, 0), X(2, -2), Y(4, 1), Z(1, 3)$

21. $Q(-1, -2), R(1, 4), S(5, 4), T(3, -2)$

22. $K(-6, 1), L(-3, 5), M(5, 1), N(-3, -3)$

23. The four sides of parallelogram *ABCD* are congruent. Therefore *ABCD* is a rhombus. List as many properties of the diagonals of a rhombus as you can find.

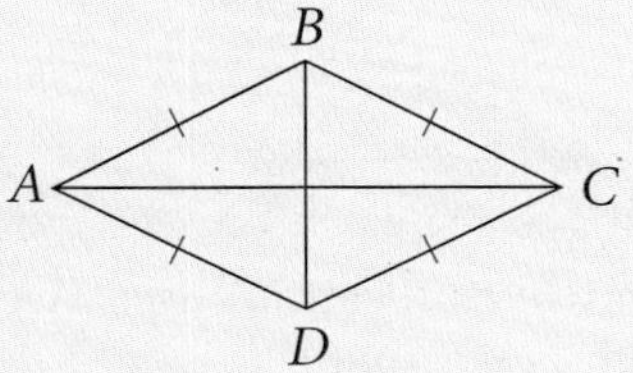

CHAPTER 8

Assessment

State the postulate or theorem you would use to prove each pair of triangles congruent. If the triangles *cannot* be proven congruent, write *not possible.*

1.

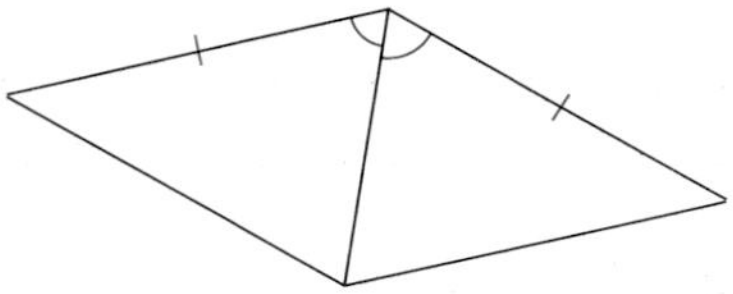

2.

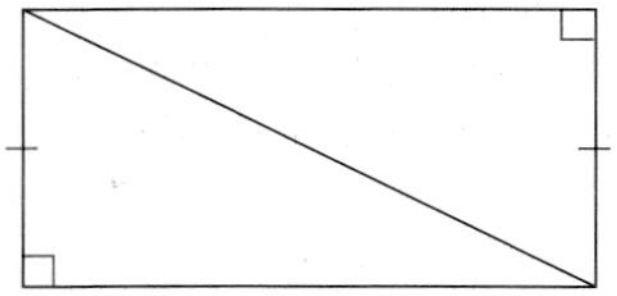

3.

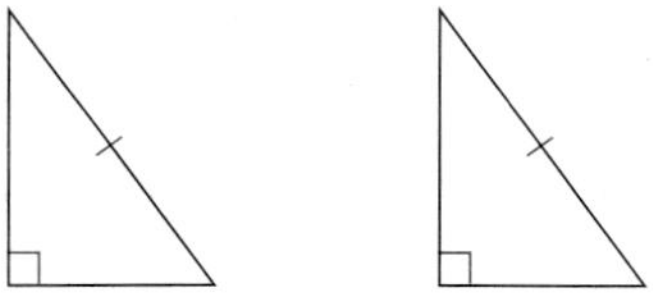

4.

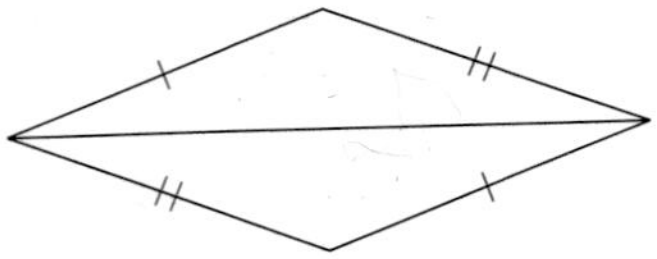

5.

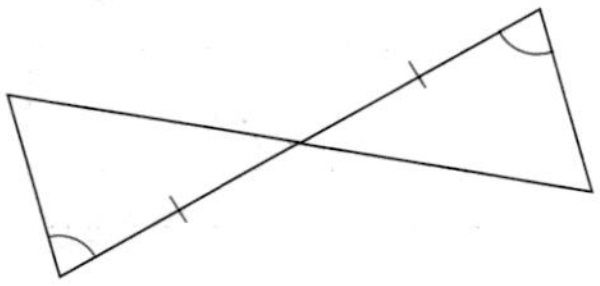

6.

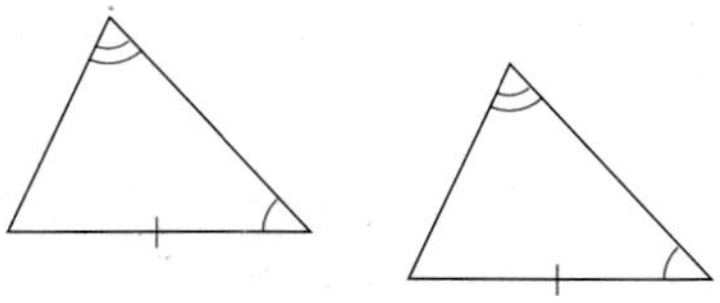

7. **Writing** Explain why you cannot use AAA to prove two triangles congruent.

8. **Standardized Test Prep** $\triangle ABC \cong \triangle DEF$. Which of the following are true?
   - **I.** $\angle A \cong \angle D$ **II.** $\overline{BC} \cong \overline{DF}$
   - **III.** $\angle C \cong \angle E$ **IV.** $\overline{DE} \cong \overline{AB}$

   A. I and II **B.** II and III **C.** I and IV
   D. III and IV **E.** I, II, and IV

Write a two-column proof, a paragraph proof, or a flow proof.

9. **Given:** $\odot A$ and $\odot T$
   Prove: $\triangle GAT \cong \triangle SAT$

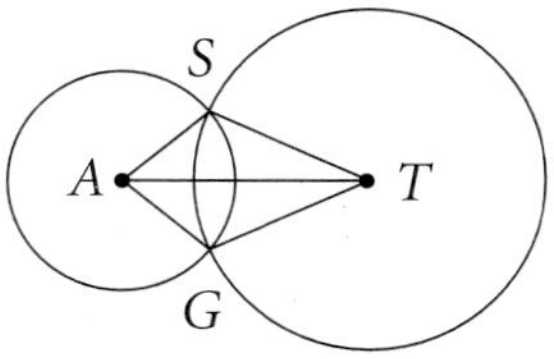

10. **Given:** $\overline{LN}$ bisects $\angle OLM$ and $\angle ONM$.
    Prove: $\triangle OLN \cong \triangle MLN$

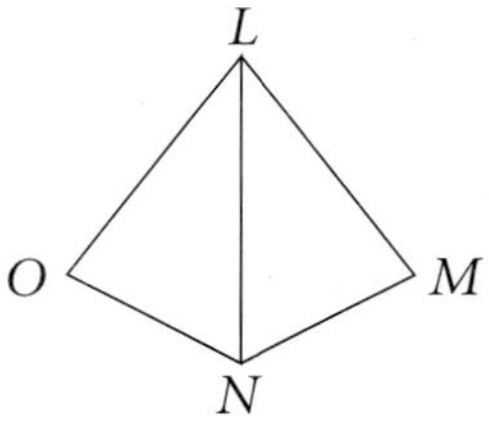

Name a pair of overlapping congruent triangles in each diagram. State whether the triangles are congruent by SSS, SAS, ASA, AAS, or HL.

11. **Given:** $\overline{CE} \cong \overline{DF}$, $\overline{CF} \cong \overline{DE}$

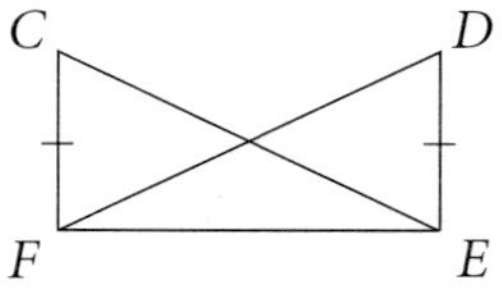

12. **Given:** $\overline{RT} \cong \overline{QT}$, $\overline{AT} \cong \overline{ST}$

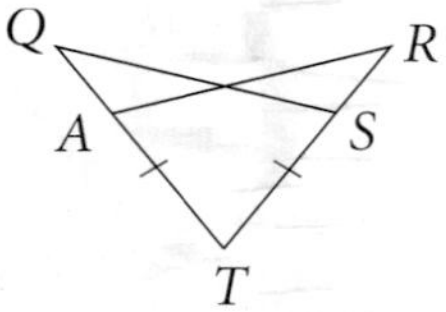

13. **Open-ended** Draw a kite and its diagonals. Which diagonal divides the kite into two congruent triangles? Explain how you know they are congruent.

CHAPTER 8 Cumulative Review

For Exercises 1–8, choose the correct letter.

1. Quadrilateral $ABCD \cong QRST$. Which segment is congruent to $\overline{TS}$?
   A. $\overline{AB}$ **B.** $\overline{BC}$ **C.** $\overline{BA}$ **D.** $\overline{CB}$ **E.** $\overline{DC}$

2. Which condition(s) will allow you to prove that $\ell \parallel m$?

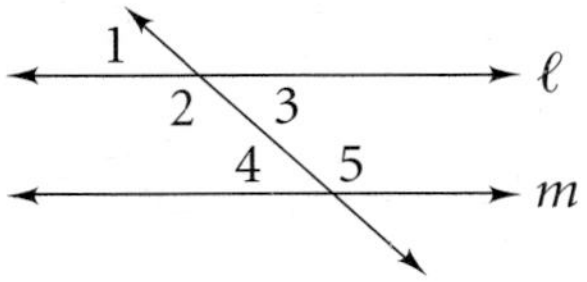

   I. $\angle 1 \cong \angle 4$ **II.** $\angle 2 \cong \angle 5$
   III. $m\angle 2 + m\angle 4 = 180$ **IV.** $\angle 3 \cong \angle 4$

   A. I only **B.** II only
   C. I and II only **D.** III and IV only
   E. I, II, III, and IV

3. By which postulate or theorem are the triangles congruent?

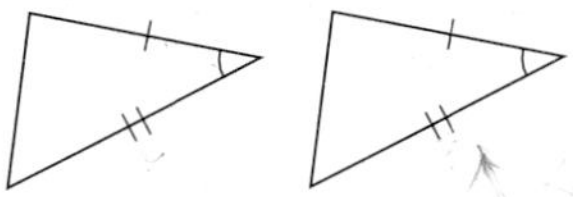

   A. SAS **B.** SSS **C.** ASA
   D. AAS **E.** HL

4. Which solid has the least volume?

   A.

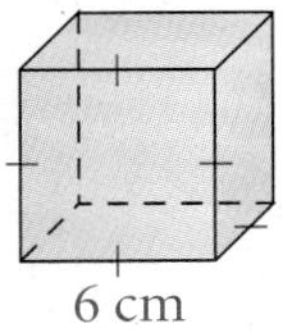

   B.

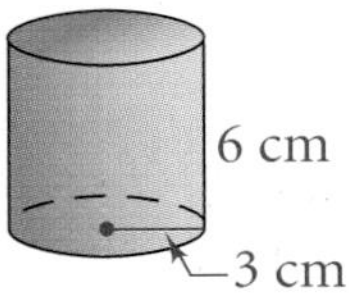

   C.

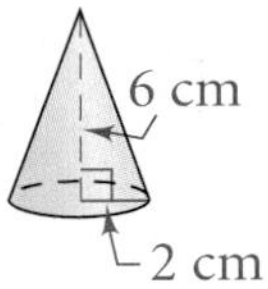

   D.

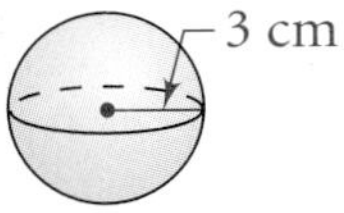

   E. It cannot be determined from the information given.

5. In spherical geometry, how many lines are perpendicular to a line at a given point on the line?
   A. 0 **B.** 1 **C.** 2
   D. 3 **E.** infinitely many

6. $WXYZ$ is a rectangle. Which segment is longest?
   A. $\overline{WX}$ **B.** $\overline{YZ}$ **C.** $\overline{WY}$ **D.** $\overline{ZW}$
   E. It cannot be determined from the information given.

Compare the boxed quantity in Column A with the boxed quantity in Column B. Choose the best answer.

A. The quantity in Column A is greater.
B. The quantity in Column B is greater.
C. The two quantities are equal.
D. The relationship cannot be determined on the basis of the information supplied.

	Column A	**Column B**

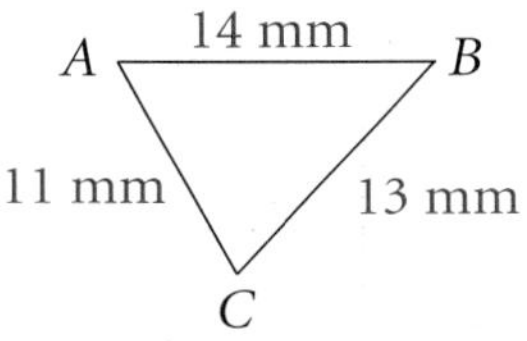

7.	$m\angle A$	$m\angle B$

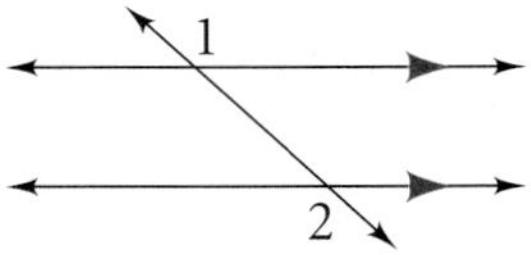

8.	$m\angle 1$	$m\angle 2$

Find each answer.

9. **Open-ended** Sketch an object in two-point perspective.

10. Write the contrapositive of the conditional "If the trees have leaves, it is not winter."

11. **Constructions** Draw a line. Construct a line perpendicular to the first line.

CHAPTER

9

Quadrilaterals

Relating to the Real World

When you were a child you could recognize different shapes. You knew that a door is a rectangle and a sandbox is a square. You may even have thought that a flying kite is a diamond. Now you can define different types of quadrilaterals and deduce some of their properties. In this chapter, you will learn how architects, kite makers, and artists use definitions and properties of special quadrilaterals in their designs and creations.

CHAPTER PROJECT

GO FLY A KITE

If you think kites are mere child's play, think again. From ancient China to modern times, geometric arrangements of fabric and rods have helped people rescue sailors, vanquish enemies, predict the weather, invent the airplane, study wind power, and, of course, entertain with displays of aerodynamic artistry.

In your project for this chapter, you will turn a sheet of paper and couple of staples into a fully functioning kite. You will explore how weight and form determine whether a kite sinks or soars. Finally you will design, build, and fly your own kite. You will see how geometry can make a kite light and strong—spelling the difference between flight and failure.

To help you complete the project:

- **p. 453** *Find Out by Doing*
- **p. 468** *Find Out by Analyzing*
- **p. 476** *Find Out by Researching*
- **p. 488** *Finishing the Project*

What You'll Learn

- Finding relationships among angles, sides, and diagonals of parallelograms

...And Why

To make designs using parallelograms

What You'll Need

- ruler
- protractor
- compass
- scissors

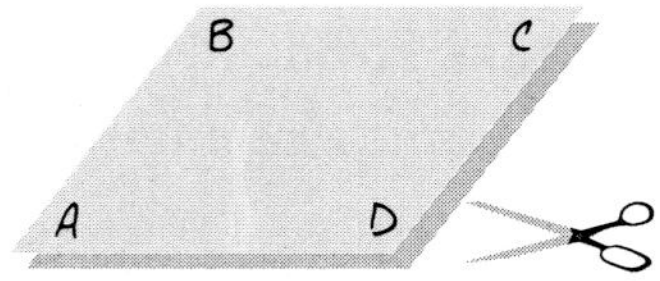

9-1 Properties of Parallelograms

WORK TOGETHER

Work with a partner.

- Draw three noncollinear points to make three vertices of a large parallelogram. Label them *A*, *B*, and *C*. Draw $\overline{AB}$ and $\overline{BC}$.
- Construct a line parallel to $\overline{BC}$ through *A* and a line parallel to $\overline{AB}$ through *C*. Label their point of intersection *D*.
- Cut out ▱*ABCD*. Use a protractor and ruler to compare the opposite sides and opposite angles of *ABCD*.

1. Make **conjectures** about the relationship between opposite sides and the relationship between opposite angles of a parallelogram.

- Draw $\overline{AC}$ and $\overline{BD}$. Label the point of intersection *E*. Use paper folding or a ruler to compare *AE* to *CE* and *BE* to *DE*.

2. Make a conjecture about the relationship between the diagonals of a parallelogram.

THINK AND DISCUSS

The results of the Work Together are the basis of this lesson's theorems.

Theorem 9-1

Opposite sides of a parallelogram are congruent.

If *ABCD* is a parallelogram, then $\overline{AB} \cong \overline{CD}$ and $\overline{BC} \cong \overline{DA}$.

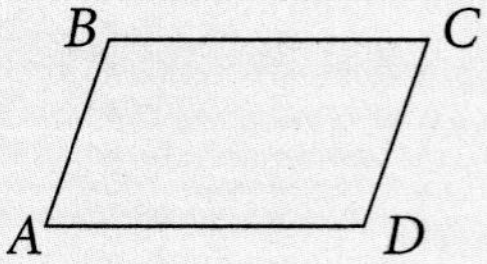

In planning this tessellation, Escher used the fact that opposite sides of a parallelogram are congruent.

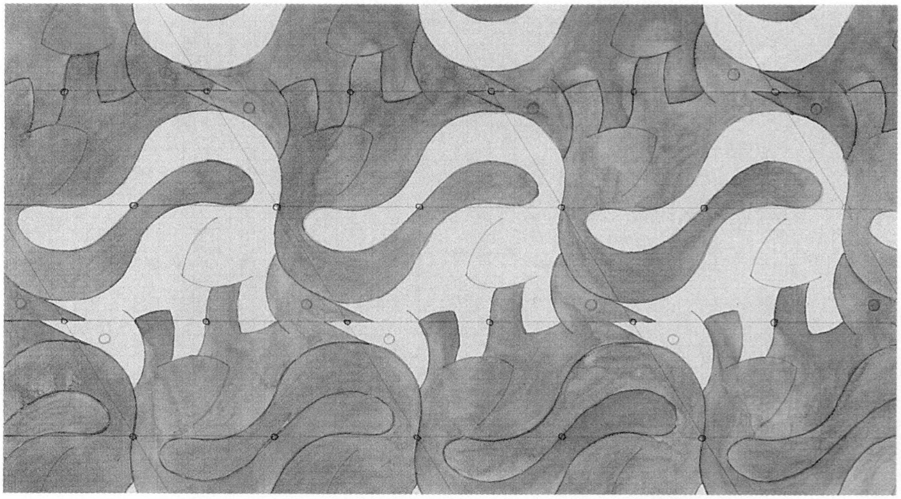

Example 1

Use a flow proof to prove Theorem 9-1: Opposite sides of a parallelogram are congruent.

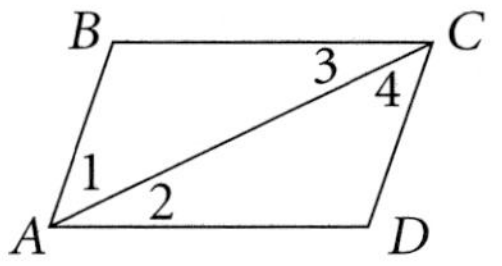

Given: ▱ $ABCD$
Prove: $\overline{AB} \cong \overline{CD}$, $\overline{BC} \cong \overline{DA}$

Flow Proof

- $ABCD$ is a ▱. — Given
  - → $\overline{AB} \parallel \overline{DC}$ — Def. of ▱ → $\angle 1 \cong \angle 4$ — If $\parallel$ lines, then alt. int. $\angle$s are $\cong$.
  - → $\overline{BC} \parallel \overline{AD}$ — Def. of ▱ → $\angle 3 \cong \angle 2$ — If $\parallel$ lines, then alt. int. $\angle$s are $\cong$.
- $\overline{AC} \cong \overline{AC}$ — Reflexive Prop. of $\cong$
- → $\triangle ABC \cong \triangle CDA$ — ASA
- → $\overline{AB} \cong \overline{CD}$, $\overline{BC} \cong \overline{DA}$ — CPCTC

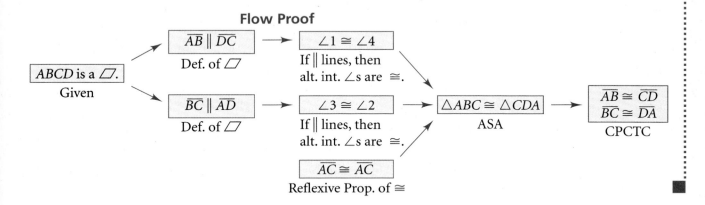

3. a. What is true about the two triangles formed by one diagonal of a parallelogram?
b. Preparing for Proof How would you prove your conclusion in part (a)?

Theorem 9-2

Opposite angles of a parallelogram are congruent.
If $ABCD$ is a parallelogram, then $\angle A \cong \angle C$ and $\angle B \cong \angle D$.

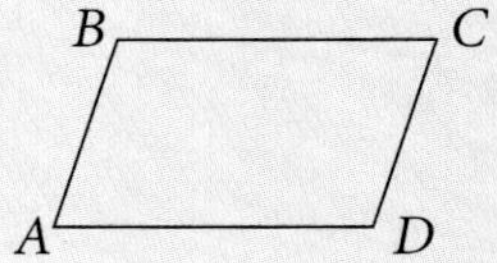

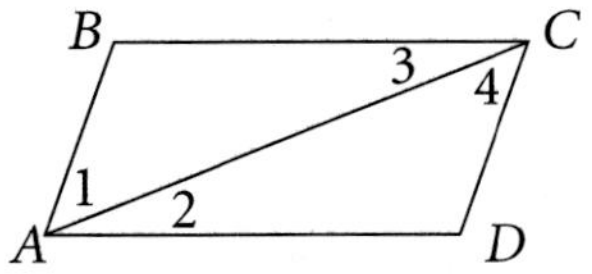

4. Preparing for Proof Give a plan for proof for Theorem 9-2.

5. a. Try This In ▱$ABCD$, $m\angle 1 = 48$ and $m\angle BCD = 70$. Find the measures of $\angle BAD$, $\angle 2$, $\angle 3$, and $\angle 4$.
b. Find the measures of $\angle B$ and $\angle D$.

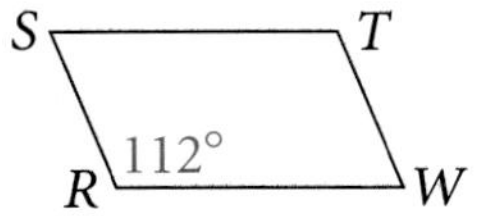

Angles of a polygon that share a common side are **consecutive angles.** In the figure at the left, $\angle S$ and $\angle T$ are consecutive angles. So are $\angle T$ and $\angle W$.

6. a. Suppose the measures of two consecutive angles of a parallelogram are x and y. By Theorem 9-2, the other two angles also have measures of x and y. Use algebra and the fact that the sum of the measures of the angles of a quadrilateral is 360 to show that the consecutive angles are supplementary.
b. Try This In ▱$RSTW$, $m\angle R = 112$. Find the measure of each angle of the parallelogram.

Theorem 9-3 The diagonals of a parallelogram bisect each other.

Example 2

Use a two-column proof to prove Theorem 9-3.

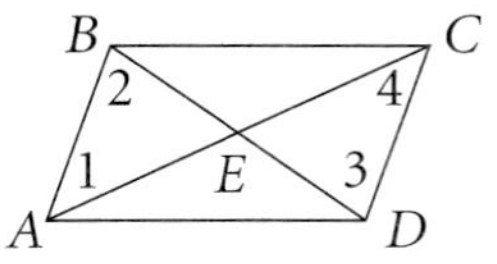

Given: $\square ABCD$
Prove: $\overline{AC}$ and $\overline{BD}$ bisect each other at E.

Statements	Reasons
1. $ABCD$ is a parallelogram.	1. Given
2. $\overline{AB} \parallel \overline{DC}$	2. Definition of parallelogram
3. $\angle 1 \cong \angle 4$; $\angle 2 \cong \angle 3$	3. If $\parallel$ lines, then alt. int. $\angle$s are $\cong$.
4. $\overline{AB} \cong \overline{CD}$	4. Opposite sides of a parallelogram are $\cong$.
5. $\triangle ABE \cong \triangle CDE$	5. ASA
6. $\overline{AE} \cong \overline{CE}$; $\overline{BE} \cong \overline{DE}$	6. CPCTC
7. $\overline{AC}$ and $\overline{BD}$ bisect each other at E.	7. Definition of bisect

7. **Algebra** In the figure for Example 2, suppose that $AE = 4x$, $EC = 3y$, $BE = x + 1$, and $ED = y$. Solve the system of linear equations to find the values of x and y.

In Exercise 28, you will use Theorem 9-1 to prove the following theorem.

Theorem 9-4 If three (or more) parallel lines cut off congruent segments on one transversal, then they cut off congruent segments on every transversal.

If $\overleftrightarrow{AB} \parallel \overleftrightarrow{CD} \parallel \overleftrightarrow{EF}$ and $\overline{AC} \cong \overline{CE}$, then $\overline{BD} \cong \overline{DF}$.

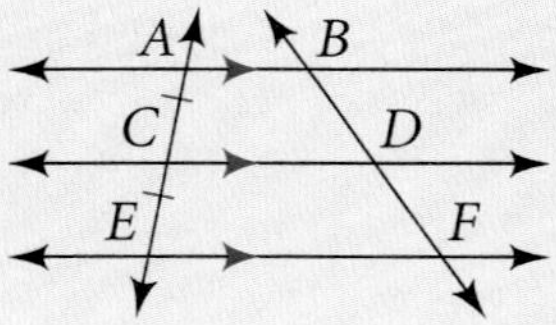

Example 3 **Relating to the Real World**

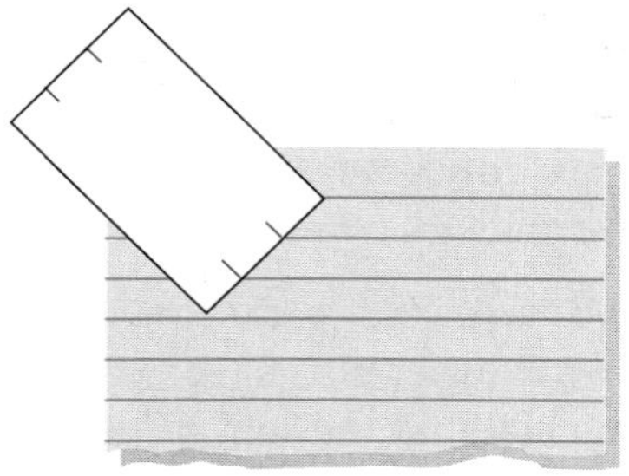

Measuring You want to divide a blank card into three equal rows and you do not have a ruler. Apply Theorem 9-4 to do this using a piece of lined paper and a straightedge.

The lines of the paper are parallel and equally spaced. Place a corner of the top edge of the card on the first line of the paper. Place the corner of the bottom edge on the fourth line. Mark the points where the second and third lines intersect the card. The marks will be equally spaced because the edge of the card is a transversal for the equally spaced parallel lines of the paper. Repeat for the other side of the card. Connect the marks.

Exercises ON YOUR OWN

Find the measures of the numbered angles for each parallelogram.

1.

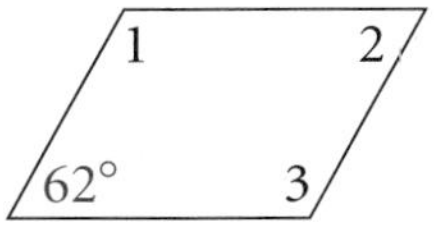

2.

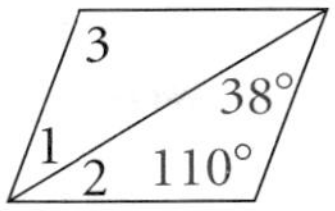

3.

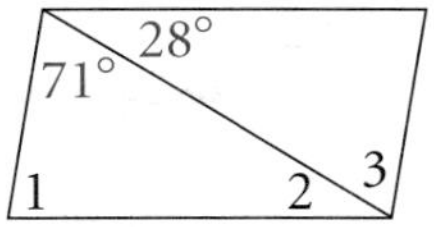

4.

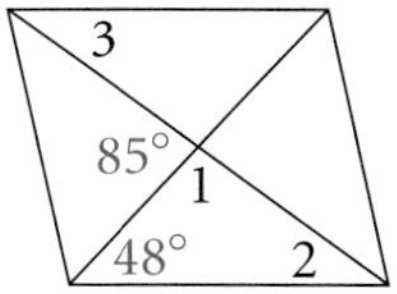

Find the length of $\overline{AP}$ in each parallelogram.

5.

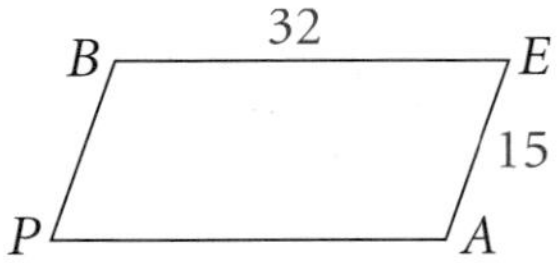

6. $YP = 8$, $TR = 10$

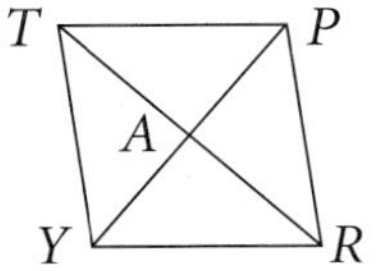

7. $YP = 8$, $TR = 10$

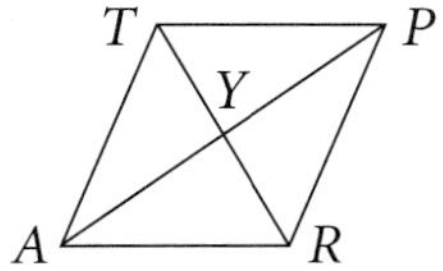

8. $MD = \frac{1}{2} AD$

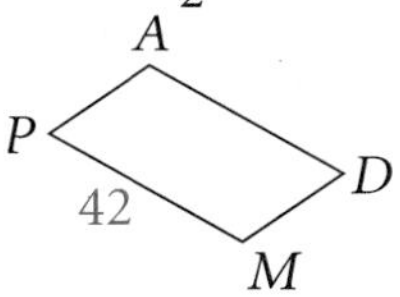

9. The perimeter of parallelogram *RSTW* is 48 cm. If $RS = 17$ cm, find the lengths of the other sides of the parallelogram.

10. The perimeter of parallelogram *EFGH* is 48 in. If *EF* is 5 in. less than *EH*, find the lengths of all four sides.

11. *Sewing* Suppose you don't have a ruler. Explain how to space 6 buttons equally on a baby's sweater if you know where the first and last buttons must be placed and you have a large piece of lined paper.

12. **a.** *Probability* If two angles of a parallelogram that is not a rectangle are randomly selected, what is the probability that they will be supplementary?

b. *Probability* If two angles of a rectangle are randomly selected, what is the probability that they will be supplementary?

13. **a.** *Open-ended* Sketch two parallelograms whose corresponding sides are congruent but whose corresponding angles are not congruent.

b. *Critical Thinking* Is there an SSSS congruence theorem for parallelograms? Explain.

14. Given: ▱ *LENS* and ▱ *NGTH*
Prove: $\angle L \cong \angle T$

15. Given: ▱ *LENS* and ▱ *NGTH*
Prove: $\overline{LS} \parallel \overline{GT}$

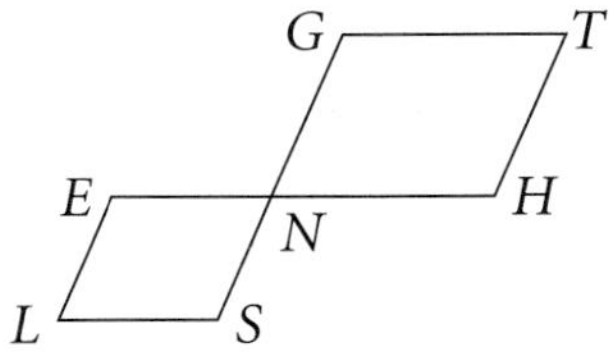

16. Given: ▱ *LENS* and ▱ *NGTH*
Prove: $\angle E$ is supplementary to $\angle T$.

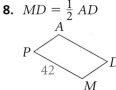

Algebra **Find the value of the variable(s) in each parallelogram.**

17.

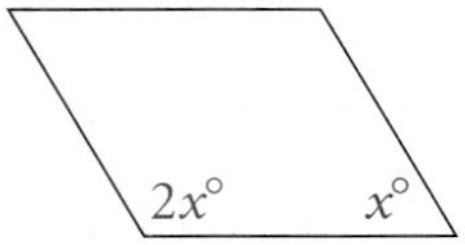

18.

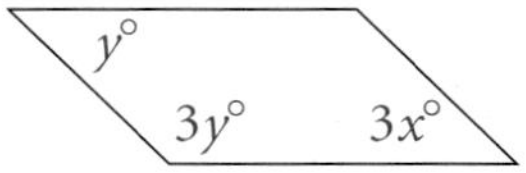

19.

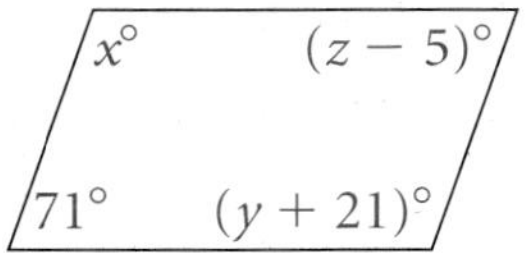

20.

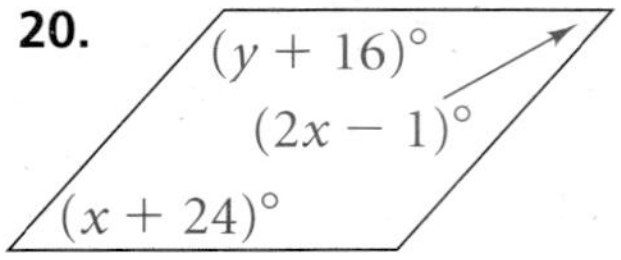

21.

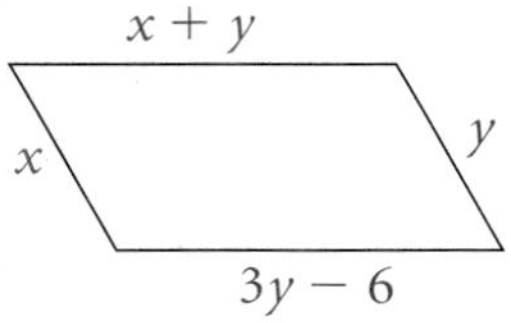

22.

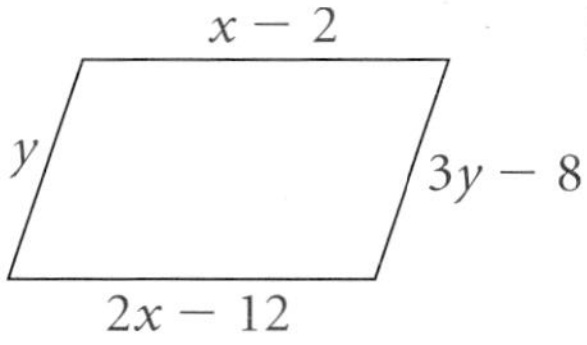

23. $AC = 4x + 10$

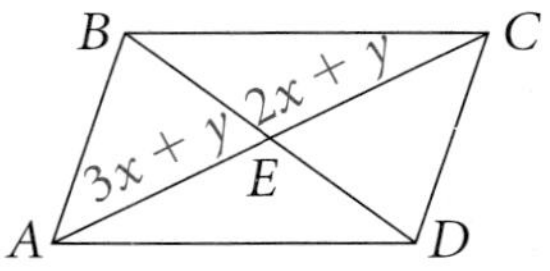

24.

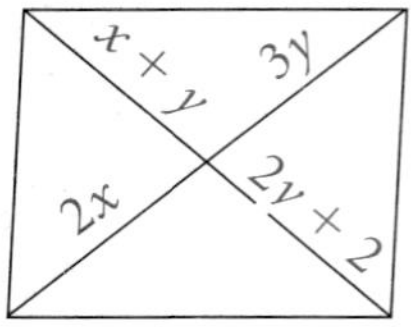

25. **Writing** If you know the measure of one angle of a parallelogram, explain how to find the measures of the remaining three angles.

26. **a.** **Transformations** What types of parallelograms have reflectional symmetry? Explain.

b. Explain why all parallelograms have rotational symmetry.

27. **a.** **Logical Reasoning** Restate Theorem 9-2 (opposite angles of a parallelogram are congruent) in if-then form.

b. Prove Theorem 9-2. Include a figure and state what is given and what is to be proved.

28. The proof of Theorem 9-4 is outlined below. Supply the missing reason for each step.

Given: $\overleftrightarrow{AB} \parallel \overleftrightarrow{CD} \parallel \overleftrightarrow{EF}$ and $\overline{AC} \cong \overline{CE}$

Prove: $\overline{BD} \cong \overline{DF}$

Draw lines through B and D parallel to $\overleftrightarrow{AE}$ and intersecting $\overleftrightarrow{CD}$ and $\overleftrightarrow{EF}$ at G and H.

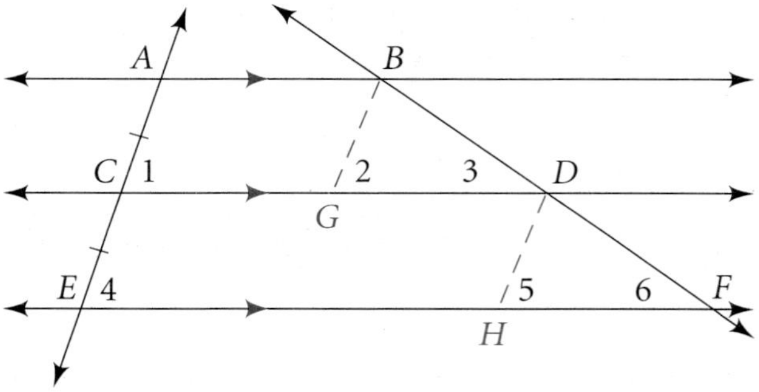

a. $\overleftrightarrow{AB} \parallel \overleftrightarrow{CD} \parallel \overleftrightarrow{EF}$

b. $ABGC$ and $CDHE$ are parallelograms.

c. $\overline{BG} \cong \overline{AC}$, $\overline{DH} \cong \overline{CE}$

d. $\overline{AC} \cong \overline{CE}$

e. $\overline{BG} \cong \overline{DH}$

f. $\angle 2 \cong \angle 1$, $\angle 1 \cong \angle 4$, $\angle 4 \cong \angle 5$, and $\angle 3 \cong \angle 6$

g. $\angle 2 \cong \angle 5$

h. $\triangle BGD \cong \triangle DHF$

i. $\overline{BD} \cong \overline{DF}$

29. **Given:** $\square RSTW$ and $\square XYTZ$

Prove: $\angle R \cong \angle X$

30. **Given:** $\square RSTW$ and $\square XYTZ$

Prove: $\overline{XY} \parallel \overline{RS}$

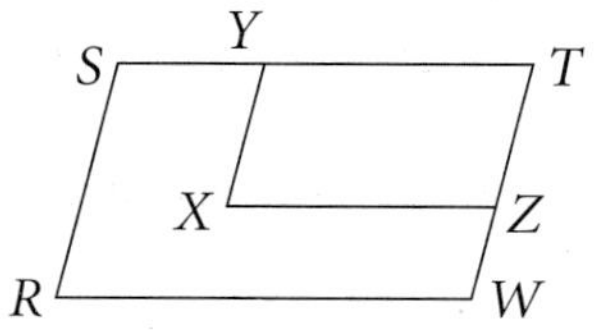

Transformations **Exercises 31–33 refer to $\triangle ABC$ at the right and its image $\triangle A'B'C'$ under each transformation.**

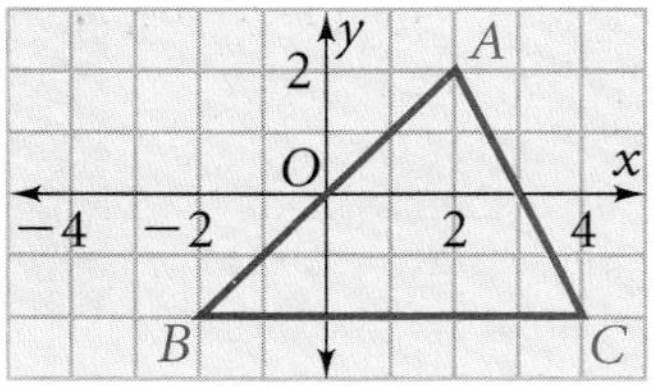

31. Reflect $\triangle ABC$ in $y = -2$. What type of quadrilateral is $ABA'C$?

32. Rotate $\triangle ABC$ 180° about the origin. What type of quadrilateral is $AC'BC$?

33. Translate $\triangle ABC$ by $\langle 4, 4\rangle$. What type of quadrilateral is $A'C'CB$?

Chapter Project *Find Out by Doing*

Building a kite usually takes a lot of time, patience, and care. You can build a simple paper kite by following these directions.

Draw diagonals to find the center of an $8\frac{1}{2}$ by 11 in. sheet of paper. Then fold the paper, creasing only about a half inch past the center, to form the keel.

Bend each of the front corners of the paper out over to the keel. With a single staple, attach both corners to the keel and to each other about 1 in. from the front of the kite.

Attach one staple perpendicular to the keel as shown and tie one end of a spool of thread to it. You are ready to fly!

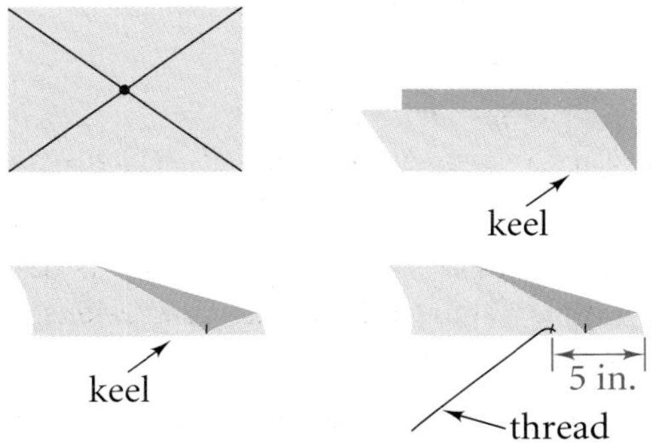

Exercises MIXED REVIEW

Sketch an example of each pair of angles.

34. supplementary angles

35. same-side interior angles

36. alternate interior angles

37. complementary angles

38. corresponding angles

39. vertical angles

40. **Coordinate Geometry** A triangle has coordinates $A(3, 0)$, $B(2, 5)$, and $C(-1, 3)$. Find the image of the triangle after a translation right 3 units and down 2 units followed by a reflection in the x-axis.

Getting Ready for Lesson 9-2

41. **a.** Find the coordinates of the midpoints of $\overline{AC}$ and $\overline{BD}$. What is the relationship between $\overline{AC}$ and $\overline{BD}$?

b. Find the slopes of $\overline{BC}$ and $\overline{AD}$. How do they compare?

c. Are $\overline{AB}$ and $\overline{DC}$ parallel? Explain.

d. What type of figure is $ABCD$?

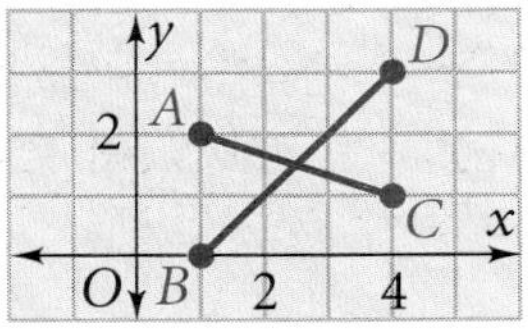

9-2 Proving That a Quadrilateral Is a Parallelogram

What You'll Learn

- Finding characteristics of quadrilaterals that indicate that they are parallelograms

...And Why

To solve problems in navigation and sailing

What You'll Need

- 11-by-11 peg geoboard
- straightedge

WORK TOGETHER

By definition, a quadrilateral is a parallelogram if both pairs of opposite sides are parallel. Work with a partner and a geoboard to explore other ways to determine whether a quadrilateral is a parallelogram.

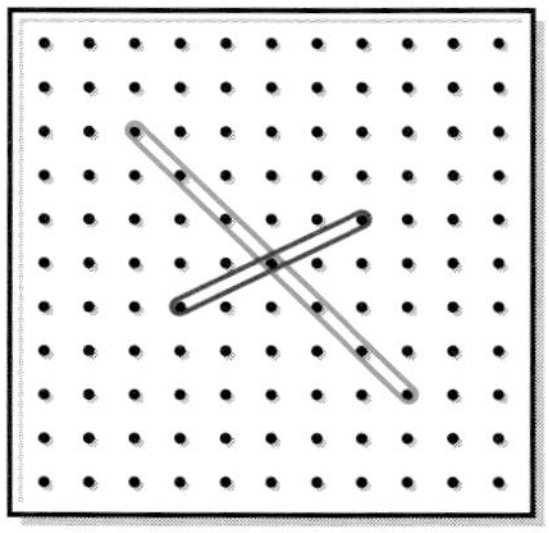

Figure 1

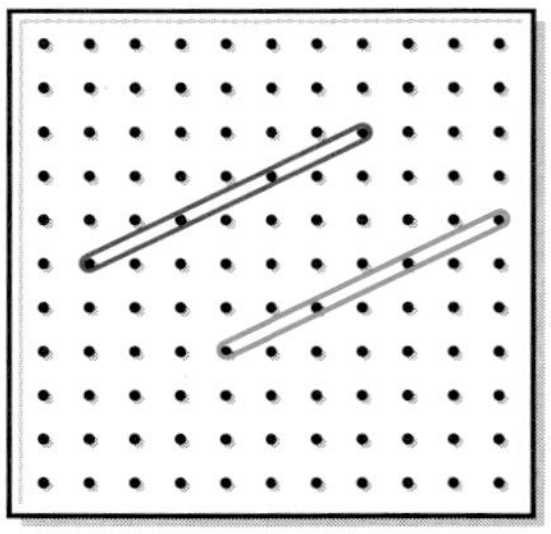

Figure 2

QUICK REVIEW

slope $= \frac{\text{rise}}{\text{run}}$

1. The segments in Figure 1 bisect each other. Connect the endpoints to form a quadrilateral. Find the slopes of opposite pairs of sides and determine what type of quadrilateral it is.

- Make several quadrilaterals by joining the endpoints of pairs of bisecting segments on a geoboard. Classify the quadrilaterals you form.

2. Make a **conjecture** about quadrilaterals whose diagonals bisect each other.

3. The segments in Figure 2 are congruent and parallel. Connect the endpoints to form a quadrilateral. Find the slopes of all four sides and determine what type of quadrilateral it is.

- Make several quadrilaterals by joining the endpoints of pairs of congruent and parallel segments on a geoboard. Classify the quadrilaterals you form.

4. Make a **conjecture** about quadrilaterals that have one pair of congruent and parallel sides.

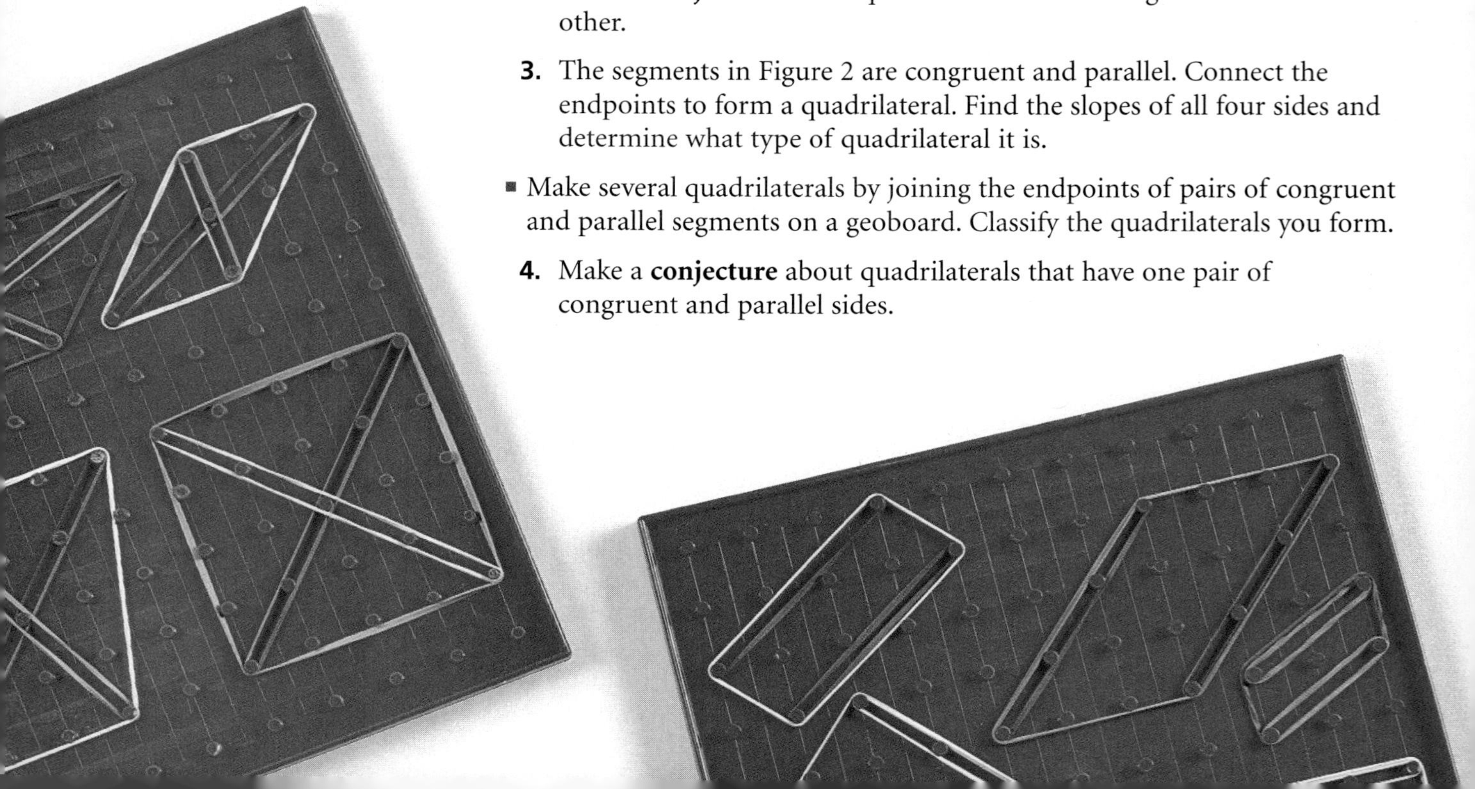

THINK AND DISCUSS

The results you obtained in the Work Together are the basis of Theorem 9-5 and Theorem 9-6.

Theorem 9-5

If the diagonals of a quadrilateral bisect each other, then the quadrilateral is a parallelogram.

Example 1

Use a flow proof to prove Theorem 9-5.

Given: $\overline{AC}$ and $\overline{BD}$ bisect each other at E.
Prove: $ABCD$ is a parallelogram.

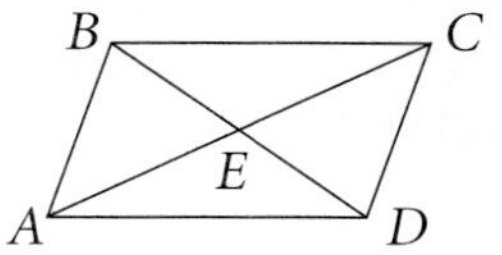

Who? Writing a flow proof is similar to writing a flow chart for a computer program. Ada Byron King, Countess of Lovelace (1815–1852), was the first programmer and the inventor of programming. In 1979, the U.S. Department of Defense named one of its computer languages ADA in her honor.

Flow Proof

$\overline{AC}$ and $\overline{BD}$ bisect each other at E.
Given

↓

$\angle AEB \cong \angle CED$	$\overline{AE} \cong \overline{CE}$, $\overline{BE} \cong \overline{DE}$	$\angle BEC \cong \angle DEA$
Vertical $\angle$s are $\cong$.	Def. of seg. bisector	Vertical $\angle$s are $\cong$.

$\triangle AEB \cong \triangle CED$	$\triangle BEC \cong \triangle DEA$
SAS	SAS
↓	↓
$\angle BAE \cong \angle DCE$	$\angle ECB \cong \angle EAD$
CPCTC	CPCTC
↓	↓
$\overline{AB} \parallel \overline{CD}$	$\overline{BC} \parallel \overline{AD}$
If $\cong$ alt. int. $\angle$s, then lines are $\parallel$.	If $\cong$ alt. int. $\angle$s, then lines are $\parallel$.

↓

$ABCD$ is a parallelogram.
Def. of parallelogram

QUICK REVIEW

A parallelogram is a quadrilateral with both pairs of opposite sides parallel.

5. **Try This** Determine the value of the variables for which $LMNP$ is a parallelogram.

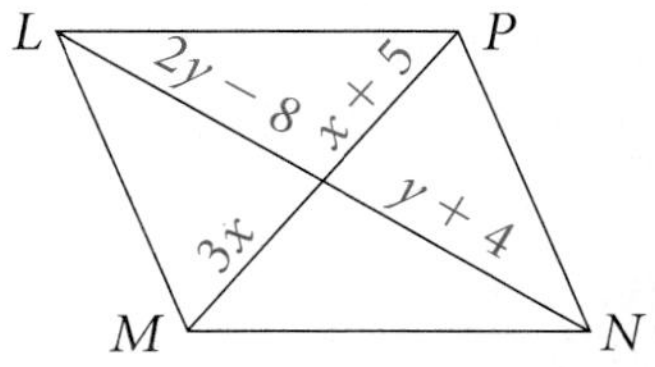

Theorem 9-6

If one pair of opposite sides of a quadrilateral are both congruent and parallel, then the quadrilateral is a parallelogram.

You can use Theorem 9-5 to show that Theorem 9-6 is true.

6. **Given:** $\overline{AB} \cong \overline{DC}$; $\overline{AB} \parallel \overline{DC}$
   a. Which numbered angles must be congruent? Why?
   b. Which triangles must be congruent? Why?
   c. Complete: $\overline{AE} \cong$ ■ and $\overline{BE} \cong$ ■.
   d. What follows from part (c)?
   e. Why is *ABCD* a parallelogram?

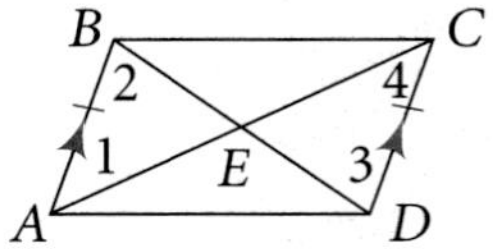

Theorems 9-7 and 9-8 provide additional ways to prove that a quadrilateral is a parallelogram.

Theorem 9-7

If both pairs of opposite sides of a quadrilateral are congruent, then the quadrilateral is a parallelogram.

7. **Logical Reasoning** Write a paragraph proof of Theorem 9-7. Use quadrilateral *ABCD*, with $\overline{AB} \cong \overline{CD}$ and $\overline{BC} \cong \overline{DA}$. (*Hint:* Draw diagonal $\overline{BD}$ and use congruent triangles.)

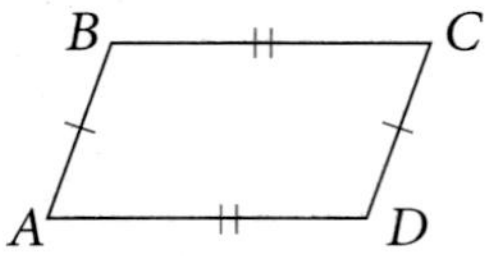

Theorem 9-8

If both pairs of opposite angles of a quadrilateral are congruent, then the quadrilateral is a parallelogram.

You can use the fact that the sum of the measures of the interior angles of a quadrilateral is 360 to show that Theorem 9-8 is true.

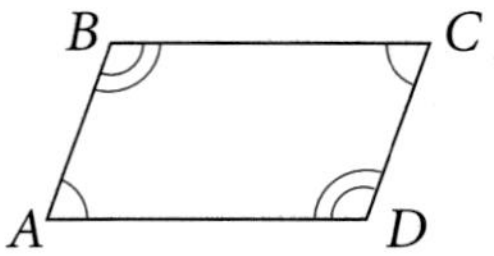

8. **Given:** $\angle A \cong \angle C$ and $\angle B \cong \angle D$
   a. Let $m\angle A = x$ and $m\angle B = y$. Write an equation expressing the sum of the measures of the angles of quadrilateral *ABCD* in terms of *x* and *y*.
   b. Solve for *y* in terms of *x*.
   c. What is the relationship between $\angle A$ and $\angle B$? between $\angle B$ and $\angle C$?
   d. What is the relationship between $\overline{AB}$ and $\overline{DC}$? between $\overline{BC}$ and $\overline{AD}$? Explain.
   e. Classify quadrilateral *ABCD*.

9. **Try This** Based on the markings, decide if each figure must be a parallelogram. Justify your answer.

   a. 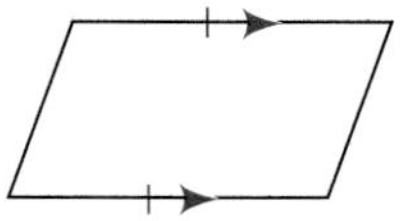

   b. 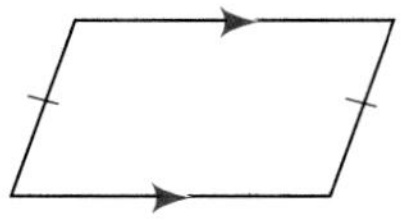

   c. 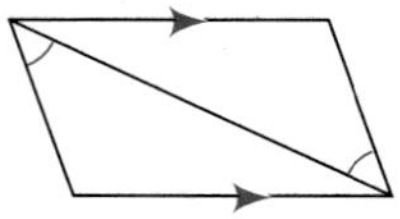

Example 2 Relating to the Real World

Navigation A parallel rule is a navigational tool used to plot ship routes on charts. It is made of two rulers connected with congruent crossbars, such that $AB = DC$ and $AD = BC$. You place one ruler on a line connecting the ship's position to its destination point. Then you move the other ruler onto the chart's compass to find the angle of the route. Explain why this instrument works.

Because both the sections of the rulers and the crossbars are congruent, the rulers and crossbars form parallelogram *ABCD*. Since the figure is a parallelogram, the rulers are parallel. Therefore, the angle shown on the chart's compass is congruent to the angle the ship should travel.

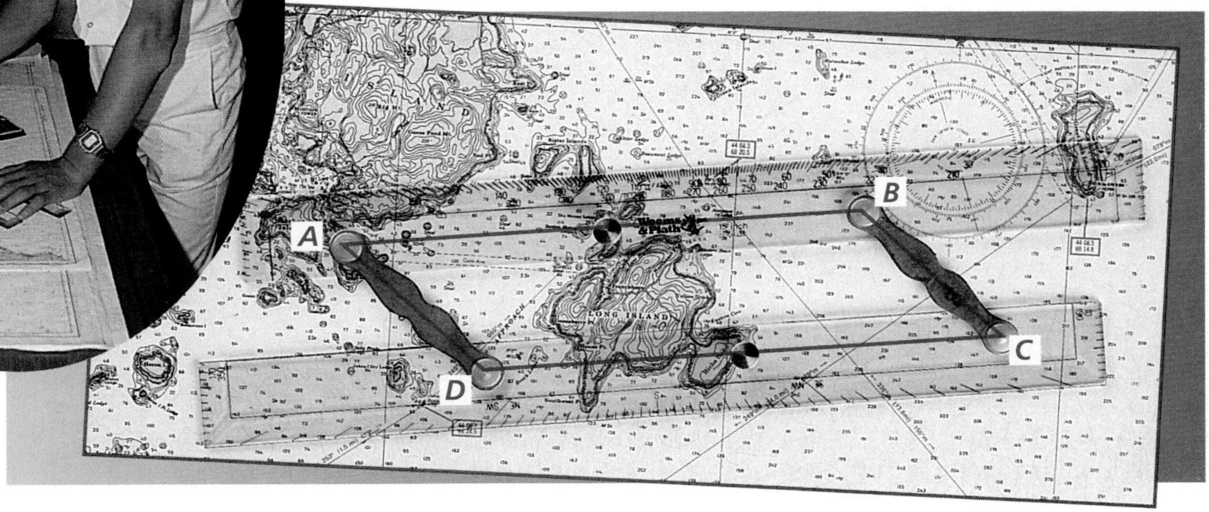

10. Suppose $m\angle B = 60$. Find the measures of the other angles of parallelogram *ABCD* formed by the parallel rule.

Exercises ON YOUR OWN

Based on the markings, decide if each figure must be a parallelogram. Justify your answer.

1. 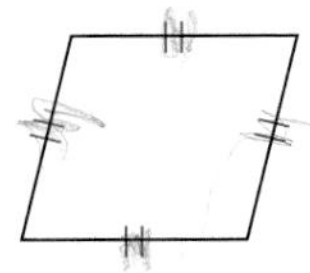

2. 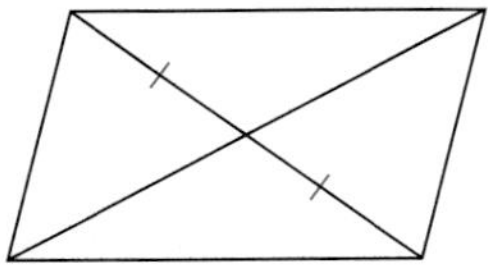

3. 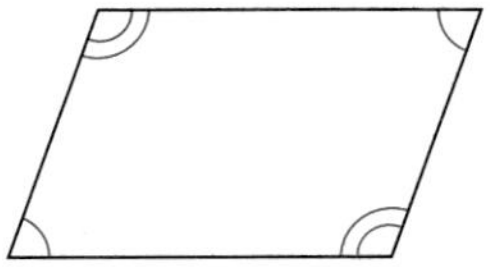

4. 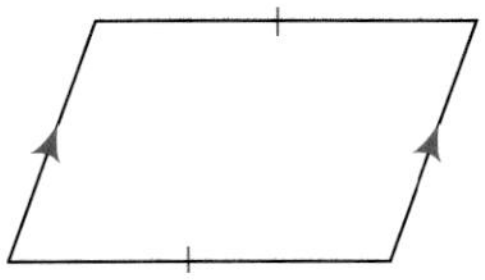

5. 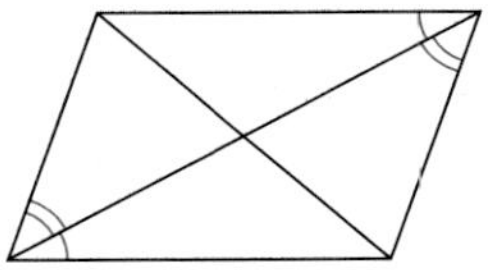

6. 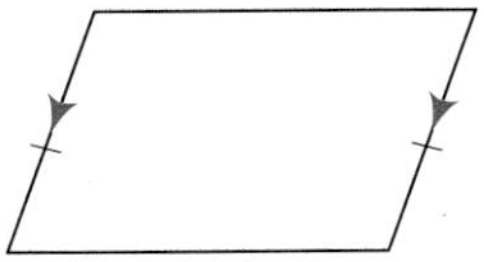

7. 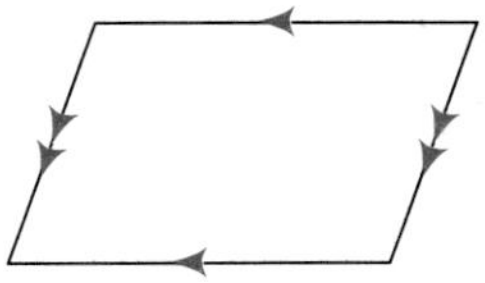

8. 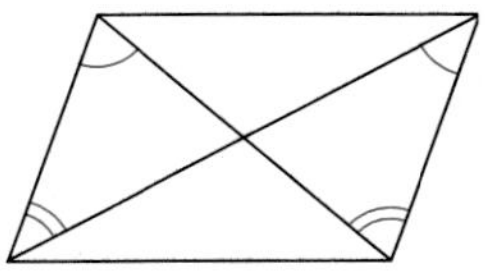

State whether the information given about quadrilateral *RSTW* is sufficient to determine that it is a parallelogram.

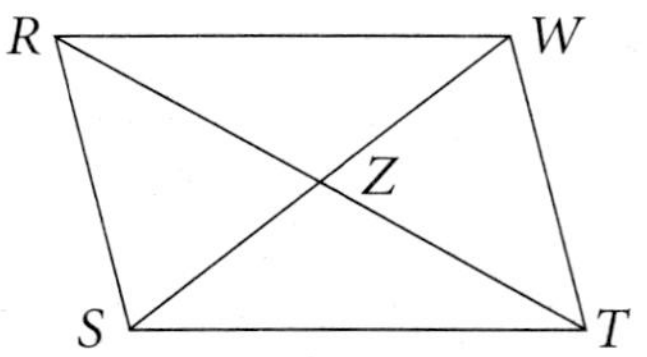

9. $\overline{RS} \parallel \overline{WT}, \overline{RS} \cong \overline{WT}$
10. $\overline{RS} \parallel \overline{WT}, \overline{ST} \cong \overline{RW}$
11. $\overline{RS} \cong \overline{WT}, \overline{ST} \cong \overline{RW}$
12. $\angle SRW \cong \angle WTS, \angle RST \cong \angle TWR$
13. $\overline{RZ} \cong \overline{TZ}, \overline{SZ} \cong \overline{WZ}$
14. $\angle TSZ \cong \angle RSZ, \angle TWZ \cong \angle RWZ$

Algebra **Find the value of *x*. Then tell whether *ABCD* must be a parallelogram. Explain your answer.**

15. 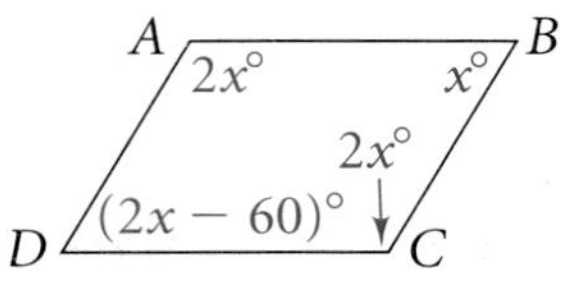

16. 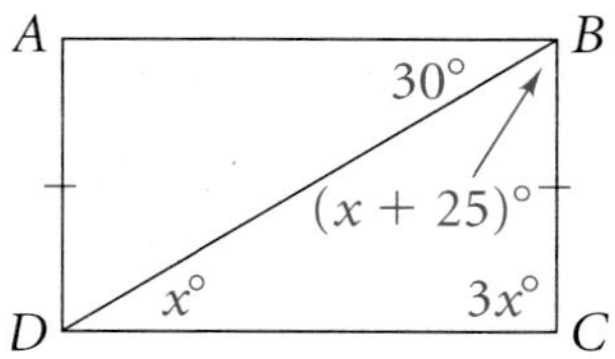

17. 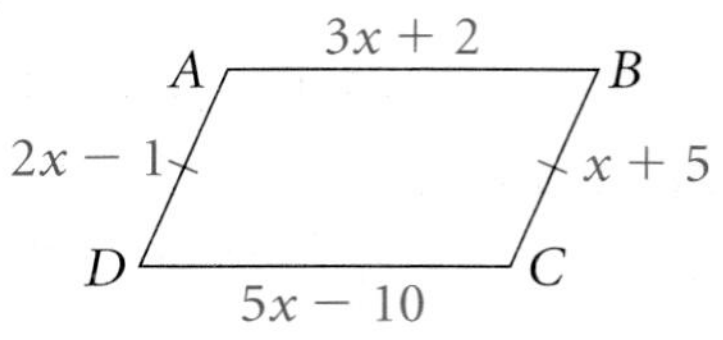

Algebra **Determine the values of the variables for which *ABCD* is a parallelogram.**

18. 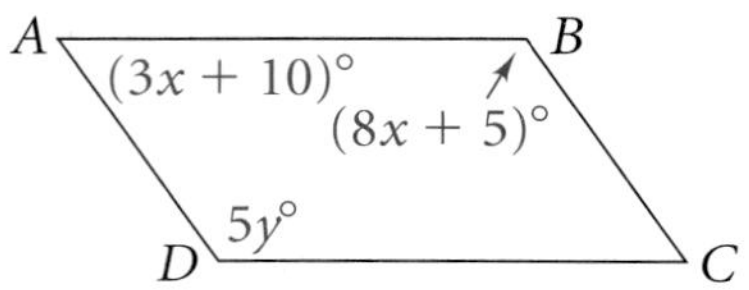

19. 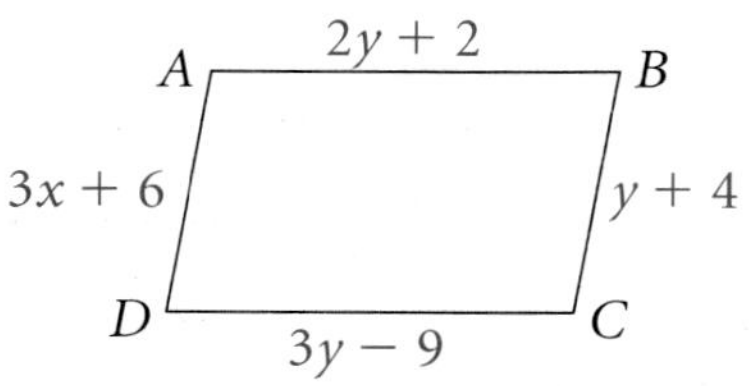

20. 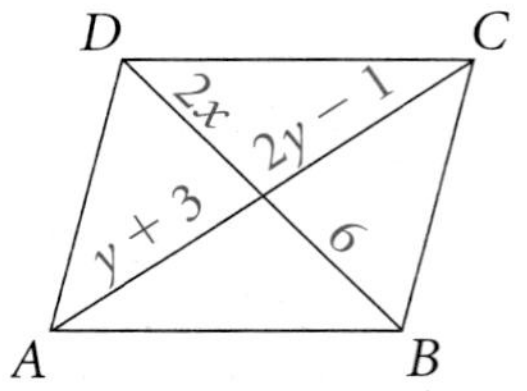

21. **Standardized Test Prep** Which of the following pairs of conditions will *not* be sufficient to prove that quadrilateral *RSTW* is a parallelogram?

I. $\overline{RS} \parallel \overline{WT}$ II. $\overline{RS} \cong \overline{WT}$
III. $\overline{ST} \cong \overline{RW}$ IV. $\angle R$ is supplementary to $\angle S$.

A. I and II B. I and III C. II and III
D. I and IV E. III and IV

22. a. **Transformations** Describe the transformation that maps $\overline{AB}$ to $\overline{A'B'}$.
b. Suppose $\overline{C'D'}$ is the image of $\overline{CD}$ under the transformation in part (a). What kind of quadrilateral is $CDD'C'$? Explain.

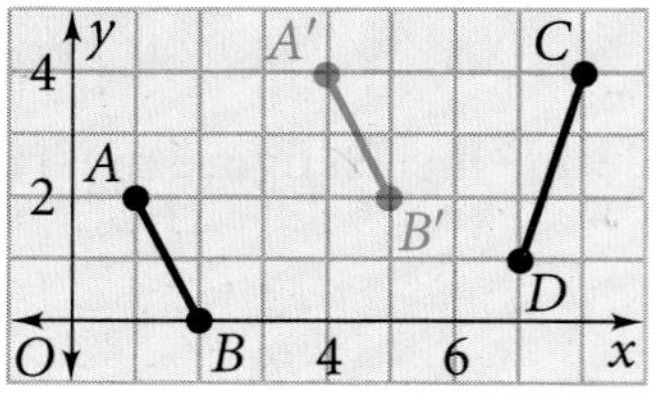

23. **Open-ended** Sketch two noncongruent parallelograms *ABCD* and *EFGH* such that $\overline{AC} \cong \overline{EG}$ and $\overline{BD} \cong \overline{FH}$.

24. **Probability** Two opposite angles of a quadrilateral measure 120. The measures of the other angles are multiples of 10. What is the probability that the quadrilateral is a parallelogram?

25. a. **Coordinate Geometry** Find the coordinates of point D so that $ABCD$ is a parallelogram.

b. Find the coordinates of point E so that $ABEC$ is a parallelogram.

c. Find the coordinates of point F so that $AFBC$ is a parallelogram.

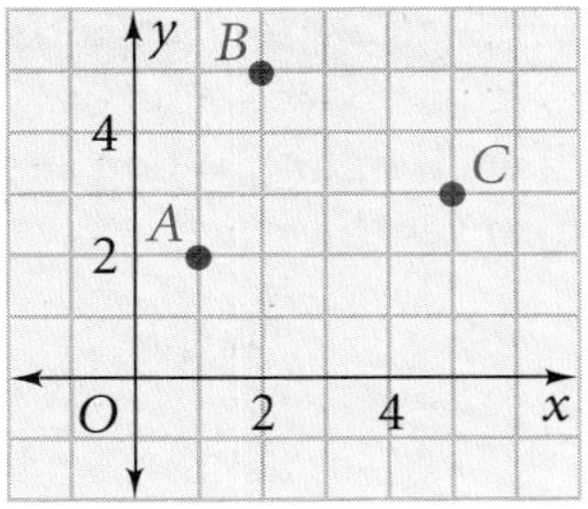

26. **Writing** Which pairs of theorems in Lessons 9-1 and 9-2 can be combined in if-and-only-if statements? Write the statements.

Choose **Write a paragraph proof, a flow proof, or a two-column proof.**

27. Given: $\triangle TRS \cong \triangle RTW$
Prove: $RSTW$ is a parallelogram.

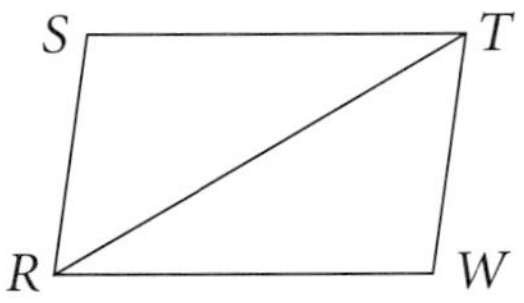

28. Given: $\triangle PKJ \cong \triangle PML$
Prove: $JKLM$ is a parallelogram.

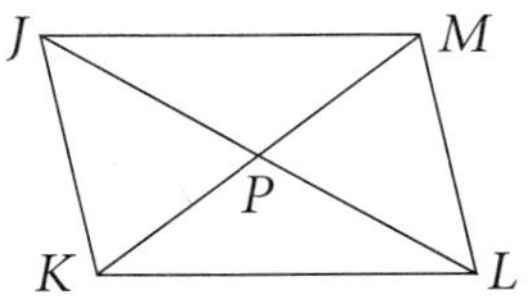

29. Given: Parallelogram $ABCD$;
E is the midpoint of $\overline{BC}$;
F is the midpoint of $\overline{AD}$.

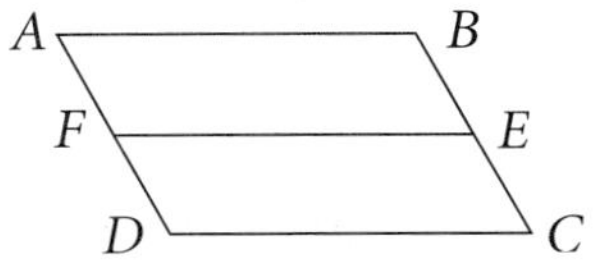

a. Prove: $ABEF$ is a parallelogram.

b. What theorem allows you to conclude that $\angle A \cong \angle BEF$?

c. What theorem allows you to conclude that $\overline{AB} \cong \overline{FE}$?

30. **Coordinate Geometry** The diagonals of quadrilateral $ABCD$ intersect at $E(-1, 4)$. Two vertices of $ABCD$ are $A(2, 7)$ and $B(-3, 5)$. What must the coordinates of C and D be in order to ensure that $ABCD$ is a parallelogram?

31. **Weaving** Fabric is made by weaving threads vertically (the warp) and horizontally (the weft), to form small rectangles. If the fabric is pulled vertically or horizontally, the fabric is rigid; that is, the rectangles keep their shape.

a. If the fabric is pulled along its diagonal (the bias), what happens to the weave of the fabric? What figures appear?

b. How does this affect the shape of the fabric?

c. **Research** Sail makers have learned techniques to cut sails so that the fabric of the sail will not stretch as the force of the wind hits the sail. Research these techniques and prepare a short report summarizing your findings.

Exercises MIXED REVIEW

Calculator **Find the area of each figure. When the answer is not a whole number, round to the nearest tenth.**

32.

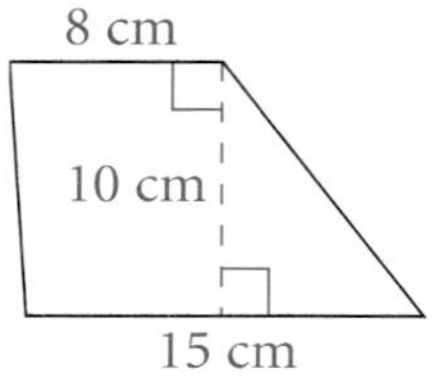

33.

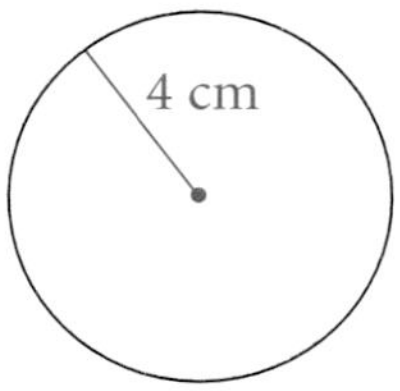

34.

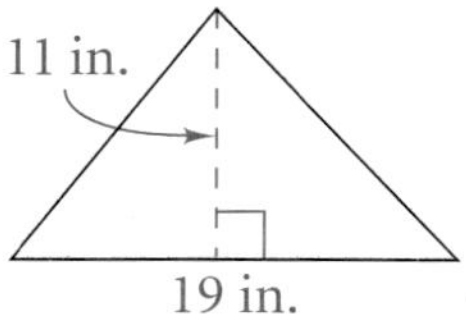

35.

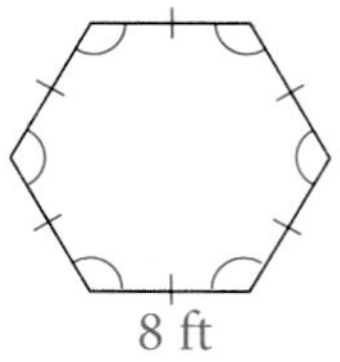

36. **Locus** Sketch the locus of points in space that are equidistant from two points.

Getting Ready for Lesson 9-3

State the definition for each of these special quadrilaterals.

37. a parallelogram

38. a rhombus

39. a rectangle

40. a square

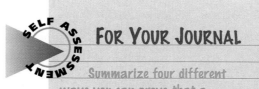

SELF ASSESSMENT

FOR YOUR JOURNAL

Summarize four different ways you can prove that a quadrilateral is a parallelogram.

Exercises CHECKPOINT

Find the measures of the numbered angles for each parallelogram.

1.

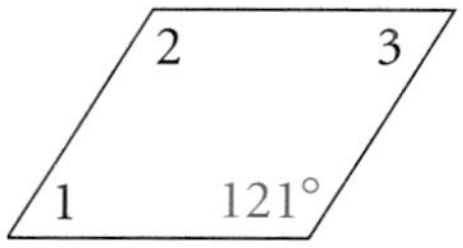

2.

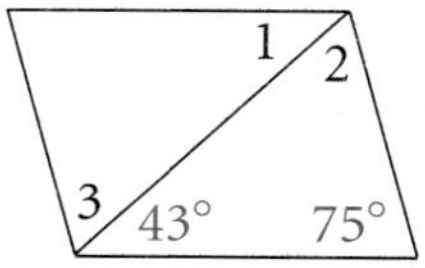

3.

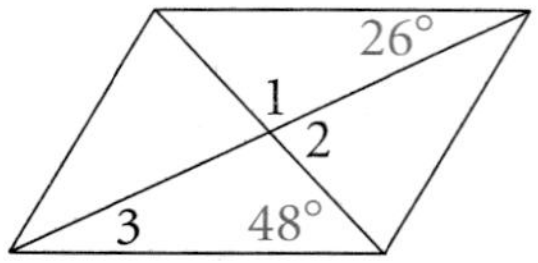

Algebra **Determine the values of the variables for which *ABCD* is a parallelogram.**

4.

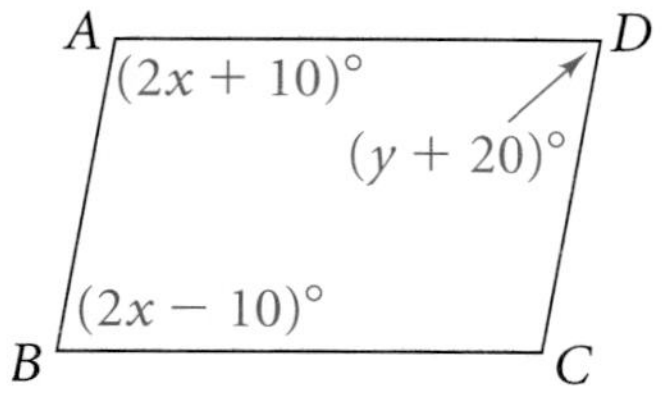

5.

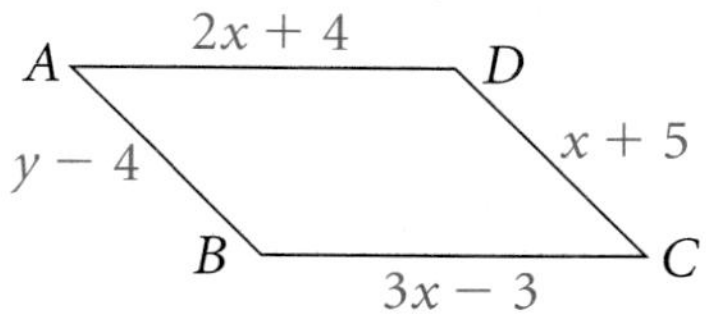

6.

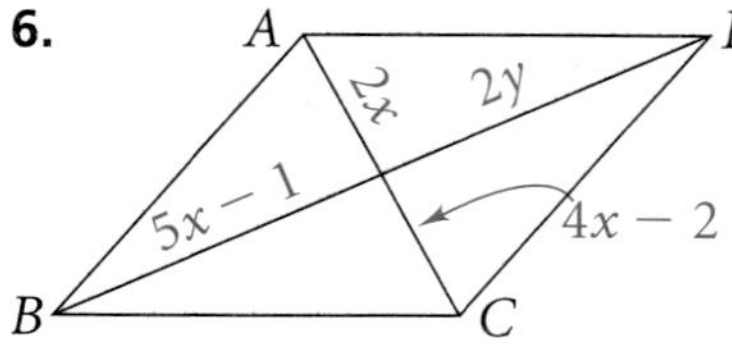

7. **Standardized Test Prep** Which is *not* sufficient to prove that a quadrilateral is a parallelogram?

A. The diagonals bisect each other.

B. Both pairs of opposite sides are congruent.

C. Both pairs of opposite angles are congruent.

D. The diagonals are perpendicular.

E. A pair of opposite sides are congruent and parallel.

Exploring the Diagonals of Parallelograms

Before Lesson 9-3

Work in pairs or small groups.

Construct

Use geometry software to construct a parallelogram. First construct segments $\overline{AB}$ and $\overline{BC}$. Construct a line through C parallel to $\overline{AB}$ and a line through A parallel to $\overline{BC}$. Label the point where the two lines intersect as D. Hide the lines and construct $\overline{AD}$ and $\overline{CD}$. Construct the diagonals of parallelogram $ABCD$ and their point of intersection E.

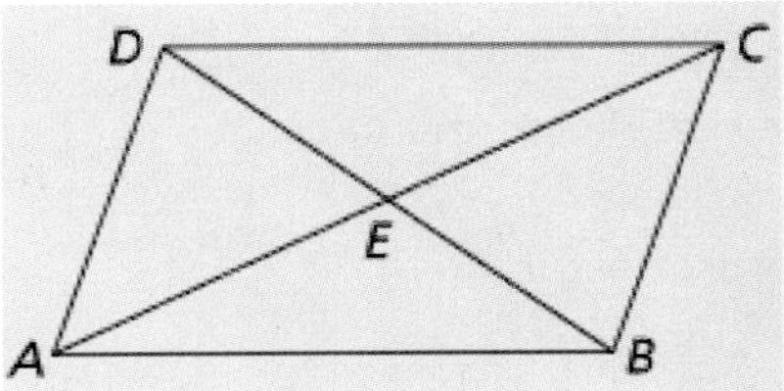

Measure the diagonals of the parallelogram so you can tell when they are congruent. Measure an angle formed by the diagonals so you can tell when the diagonals are perpendicular. Measure the angles formed by the diagonals and the sides of the parallelogram so you can tell when the diagonals bisect the angles of the parallelogram.

Investigate

- Measure an angle of the parallelogram so that when you manipulate the figure you can tell when it becomes a rectangle. Make note of any special properties of the diagonals of a rectangle. Manipulate the rectangle to check whether the properties hold.
- Measure two adjacent sides of the parallelogram so you can tell when it is a rhombus. Make note of any special properties of the diagonals of a rhombus. Manipulate the rhombus to check whether the properties hold.

Conjecture

- Make as many conjectures as you can about the properties of the diagonals of rectangles and rhombuses.
- Use what you know about squares to make **conjectures** about the diagonals of squares.

Extend

- Manipulate the diagonals so they are perpendicular. Make a **conjecture** about the type of parallelogram that is determined by perpendicular diagonals.
- Do the same for congruent diagonals and for diagonals that bisect the angles of the parallelogram.

9-3 Properties of Special Parallelograms

What You'll Learn

- Finding properties of rectangles, rhombuses, and squares

...And Why

To help in designing playgrounds and kites

What You'll Need

- centimeter grid paper
- scissors
- protractor
- ruler

WORK TOGETHER

In Lesson 9-1, you explored the properties of parallelograms. Work in a group of three to explore the properties of rhombuses, rectangles, and squares.

- Each member should choose one of these special quadrilaterals and draw a large figure on grid paper.
- Using paper folding or a protractor and ruler, each member should measure and compare the sides, angles, and diagonals of his or her quadrilateral. Then he or she should share the results with the group.
- Use the results to complete a table like the one below. The properties of a parallelogram have been checked off.

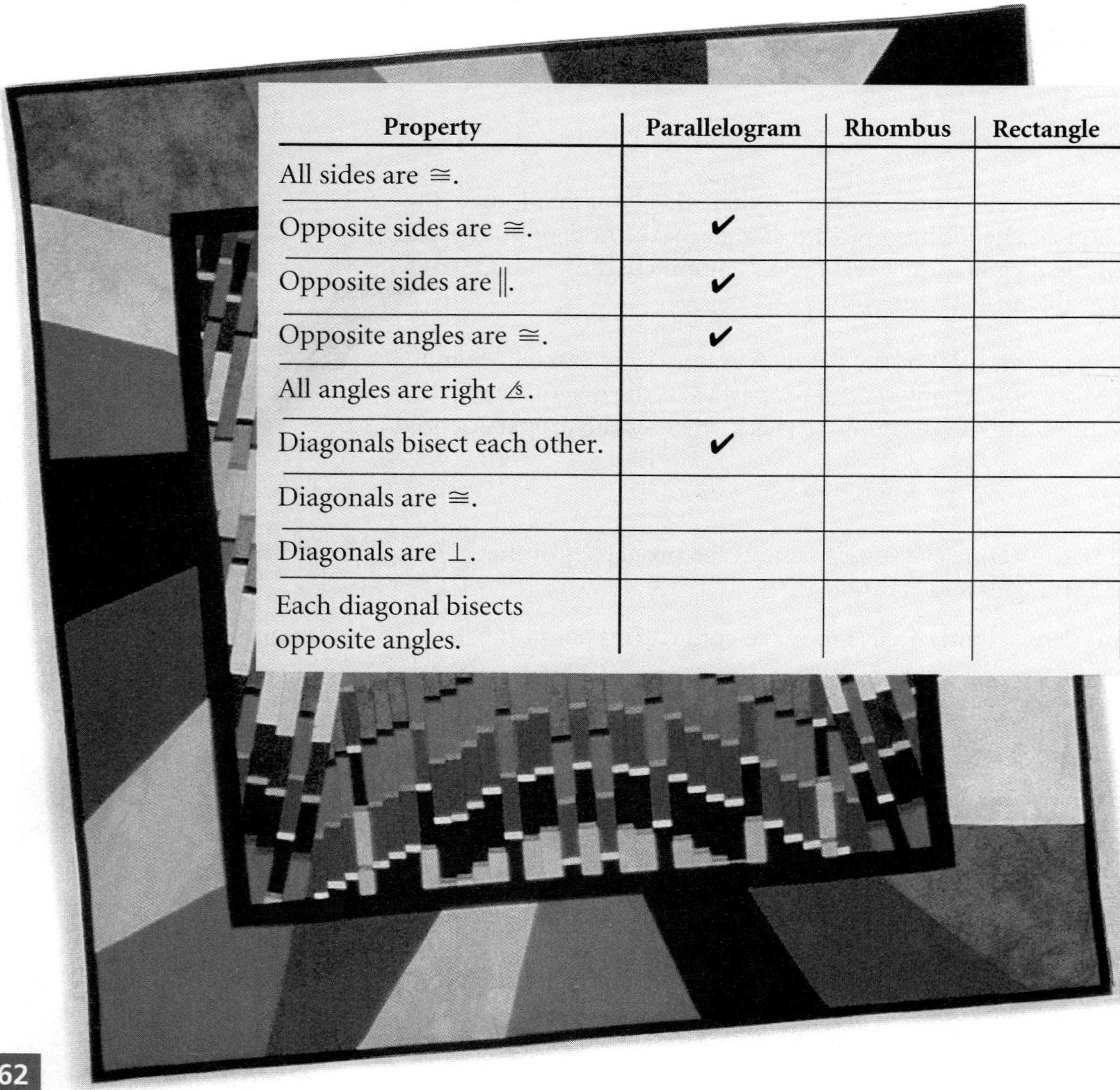

Property	Parallelogram	Rhombus	Rectangle	Square
All sides are ≅.				
Opposite sides are ≅.	✔			
Opposite sides are ∥.	✔			
Opposite angles are ≅.	✔			
All angles are right ∠s.				
Diagonals bisect each other.	✔			
Diagonals are ≅.				
Diagonals are ⊥.				
Each diagonal bisects opposite angles.				

THINK AND DISCUSS

Properties of Rhombuses

The properties you discovered in the Work Together are the basis for many of the theorems in this lesson.

Theorem 9-9 Each diagonal of a rhombus bisects two angles of the rhombus.

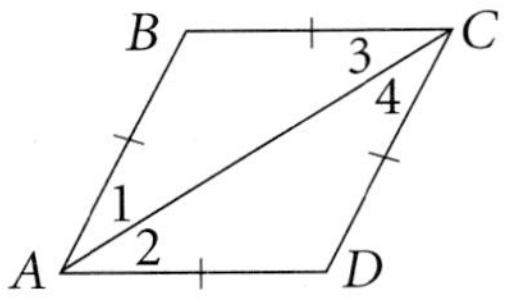

1. Answer the following questions to prove Theorem 9-9.
   a. $ABCD$ is a rhombus. Why is $\triangle ABC \cong \triangle ADC$?
   b. What is the relationship between $\angle 1$ and $\angle 2$? between $\angle 3$ and $\angle 4$? Explain.
2. **Try This** If $m\angle B = 120$, find the measures of the numbered angles.

Theorem 9-10 The diagonals of a rhombus are perpendicular.

Example 1

Write a paragraph proof of Theorem 9-10.

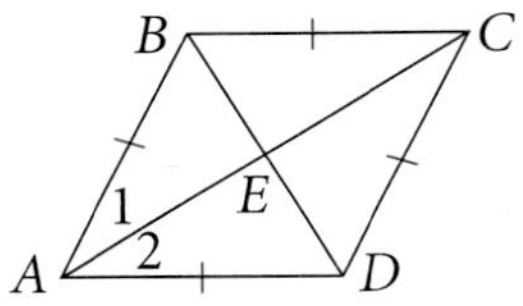

Given: $ABCD$ is a rhombus.
Prove: $\overline{AC} \perp \overline{BD}$

Paragraph Proof

By the definition of rhombus, $\overline{AB} \cong \overline{AD}$. Because $\overline{AC}$ is a diagonal of a rhombus, it bisects $\angle BAD$. Therefore $\angle 1 \cong \angle 2$. $\overline{AE} \cong \overline{AE}$ by the Reflexive Property of Congruence. By the SAS Postulate, $\triangle ABE \cong \triangle ADE$. By CPCTC, $\angle AEB \cong \angle AED$. Because $\angle AEB$ and $\angle AED$ are both congruent and supplementary, they are right angles. By the definition of perpendicular, $\overline{AC} \perp \overline{BD}$.

In Chapter 5, you found the area of a rhombus by using the formula for the area of a parallelogram, $A = bh$. Using Theorems 9-9 and 9-10, you can find another formula for the area of a rhombus. You will derive this formula in Exercise 17.

Theorem 9-11 The area of a rhombus is equal to half the product of the lengths of its diagonals.

$$A = \frac{1}{2}d_1 d_2$$

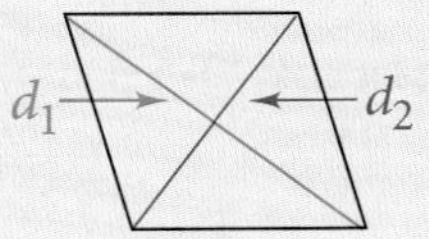

3. **Critical Thinking** Explain why you can use the area formula $A = \frac{1}{2}d_1 d_2$ to find the area of a square.

Example 2

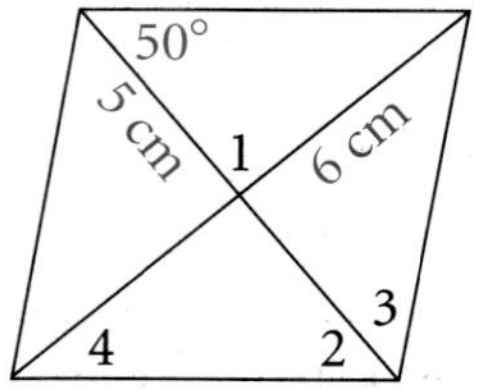

a. Find the measures of the numbered angles in the rhombus.
b. Find the area of the rhombus.

a. $m\angle 1 = 90$ — Diagonals of a rhombus are perpendicular.
$m\angle 2 = 50$ — If $\parallel$ lines, then alt. int. ∠s are $\cong$.
$m\angle 3 = 50$ — Each diagonal of a rhombus bisects two angles.
$m\angle 4 = 40$ — Acute angles of a right △ are complementary.

b. $d_1 = 2(6) = 12$ — Diagonals of a parallelogram bisect each other.
$d_2 = 2(5) = 10$
$A = \frac{1}{2}d_1d_2$ — Formula for the area of a rhombus
$= \frac{1}{2}(12)(10)$ — Substitute the lengths of the diagonals.
$= 60$ — Simplify.

The area of the rhombus is 60 cm^2.

PROBLEM SOLVING

Look Back Explain how you could use triangles to find the area of the rhombus.

Properties of Rectangles and Parallelograms

In the Work Together on page 462, you probably discovered the following properties of rectangles and parallelograms.

Theorem 9-12 The diagonals of a rectangle are congruent.

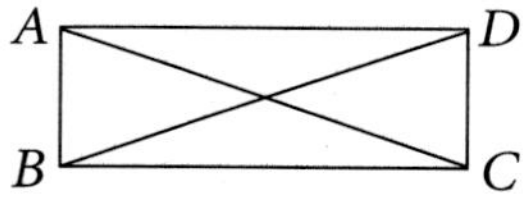

4. Answer the following questions to prove Theorem 9-12.
  a. *ABCD* is a rectangle. What parts of △*ABC* and △*DCB* must be congruent? Explain.
  b. Why is △*ABC* ≅ △*DCB*?
  c. Why is $\overline{AC} \cong \overline{BD}$?

The following theorems are suggested by the converses of Theorems 9-9, 9-10, and 9-12. You will write proofs of these theorems in the exercises.

Theorem 9-13 If one diagonal of a parallelogram bisects two angles of the parallelogram, then the parallelogram is a rhombus.

Theorem 9-14 If the diagonals of a parallelogram are perpendicular, then the parallelogram is a rhombus.

Theorem 9-15 If the diagonals of a parallelogram are congruent, then the parallelogram is a rectangle.

5. *Logical Reasoning* Suppose the diagonals of a parallelogram are both perpendicular and congruent. What type of special quadrilateral is it? Explain your reasoning.

Builders and construction workers use the properties of diagonals to "square off" rectangular foundations and other rectangular constructions.

Example 3 Relating to the Real World

Community Service Volunteers are helping to build a rectangular play area in a city park. Explain how they can use properties of the diagonals of a parallelogram and a rectangle to stake the vertices of the play area.

The volunteers can use these two theorems to design the play area:

- Theorem 9-5: If the diagonals of a quadrilateral bisect each other, then the quadrilateral is a parallelogram.
- Theorem 9-15: If the diagonals of a parallelogram are congruent, then the parallelogram is a rectangle.

First, cut two pieces of rope of equal length for the diagonals. Then, fold each piece to find the midpoint of each diagonal. Next, position the ropes so that the ropes intersect at the midpoints. Any quadrilateral formed by connecting the endpoints of the two ropes will be a rectangle. ■

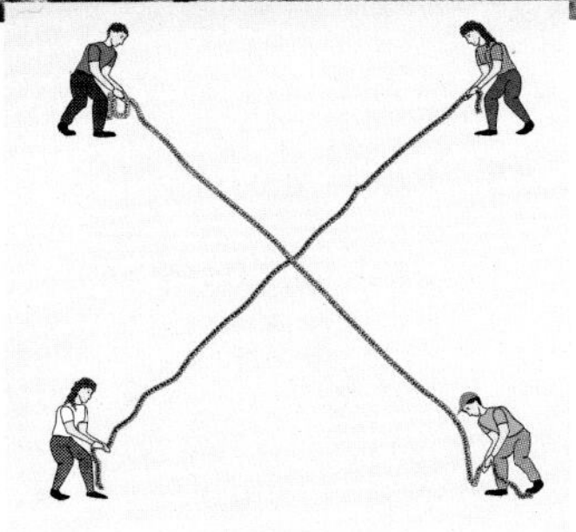

6. **Open-ended** The volunteers use two 20-ft-long pieces of rope as diagonals. Describe possible dimensions of the rectangle they can form.
7. **Critical Thinking** Steven thinks that they can use this same method to stake off a square. Is he right? Explain why or why not.

Exercises ON YOUR OWN

For each parallelogram (a) choose the best name and then (b) find the measures of the numbered angles.

1. 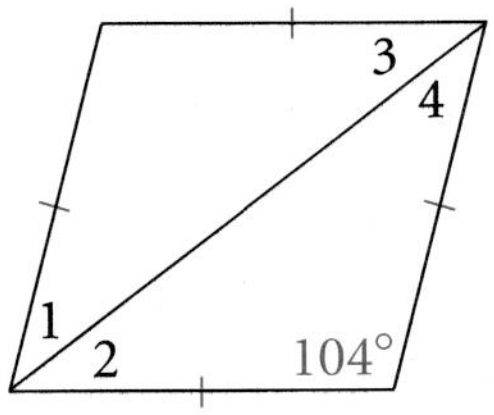

2. 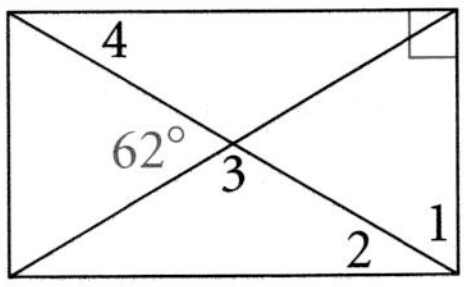

3. 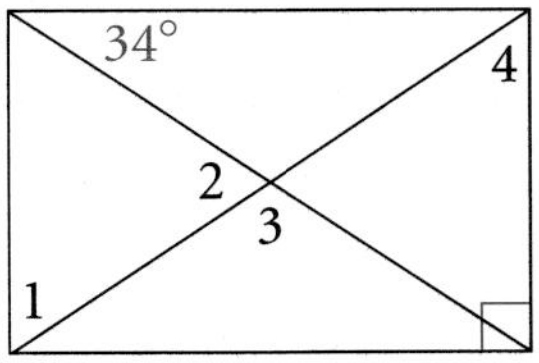

4. 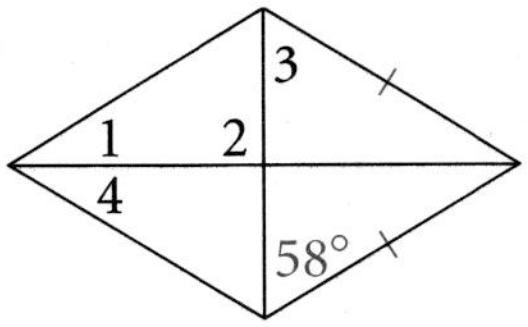

5. **Carpentry** An open-backed bookcase has shelves 24 in. wide and 12 in. high. Diagonal braces help keep the shelves and the sides of the bookcase perpendicular to each other. Find the lengths of the braces.

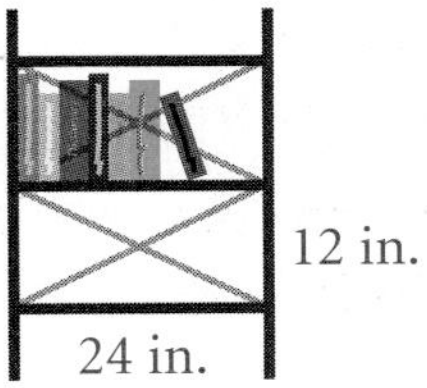

Choose **Use mental math, pencil and paper, or a calculator to find LB, BP, and LM for each parallelogram.**

6. 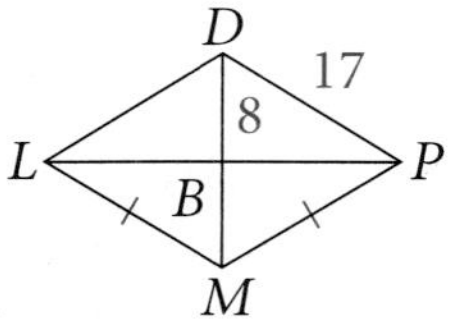

7. 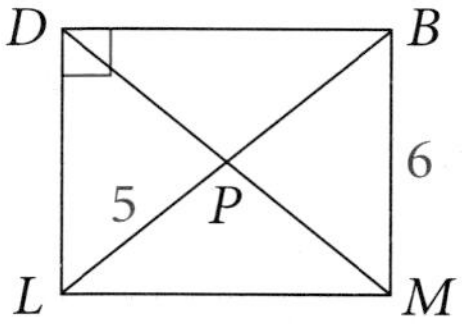

8. 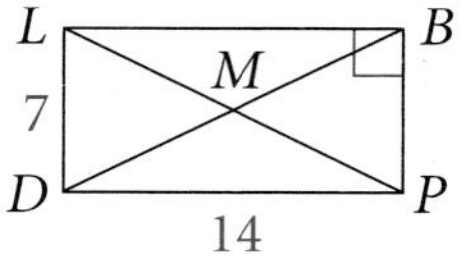

9. 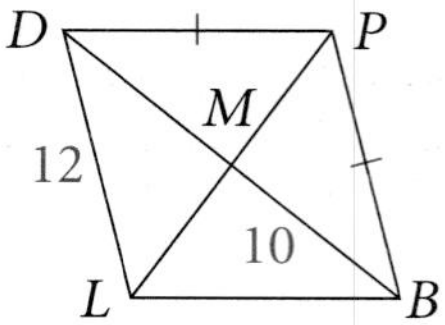

10. **Open-ended** Sketch two noncongruent rectangles with congruent diagonals.

11. **Standardized Test Prep** In quadrilateral $RSTW$, $\overline{RT}$ and $\overline{SW}$ bisect each other at A and $\overline{RA} \cong \overline{WA}$. $RSTW$ must be a

I. parallelogram. **II.** rectangle. **III.** square.

A. I only **B.** II only **C.** I and II **D.** II and III **E.** I, II, and III

Algebra **Find the value of the variable(s) for each parallelogram.**

12. $RZ = 2x + 5$, $SW = 5x - 20$

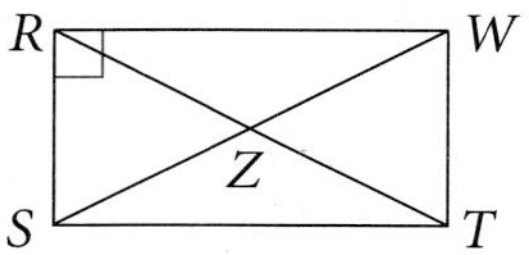

13. 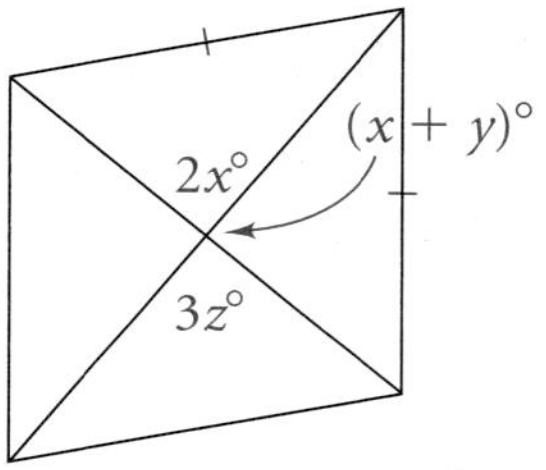

14. $m\angle 1 = 3y - 6$

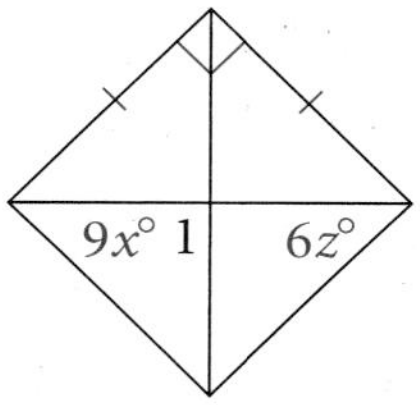

15. $BD = 4x - y + 1$

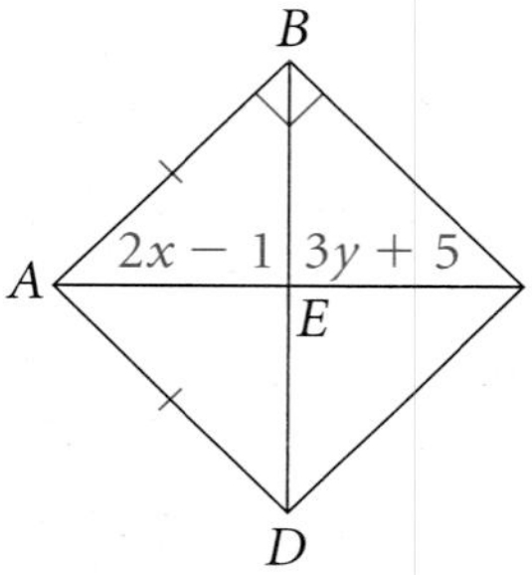

16. **Transformations** Draw $\overline{AB}$ and a line ℓ through B that is not perpendicular to $\overline{AB}$. Reflect $\overline{AB}$ in ℓ. Then reflect $\triangle A'AB$ in $\overline{AA'}$. What type of quadrilateral is $ABA'B'$? Explain.

17. Complete the following steps to derive the formula for the area of a rhombus.

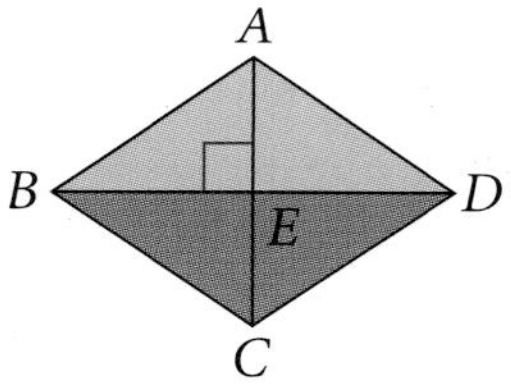

Given: Rhombus $ABCD$

Prove: Area $= \frac{1}{2} \cdot BD \cdot AC$

a. $\overline{AC} \perp$ ■ because the diagonals of a rhombus are perpendicular.

b. Area of $\triangle ABD = \frac{1}{2} \cdot BD \cdot$ ■

c. Area of $\triangle BCD = \frac{1}{2} \cdot BD \cdot$ ■

d. Area of rhombus $ABCD = \frac{1}{2} \cdot BD \cdot$ ■ $+ \frac{1}{2} \cdot BD \cdot$ ■

e. Area of rhombus $ABCD = \frac{1}{2} \cdot BD\,($■ $+$ ■$)$

f. Area of rhombus $ABCD = \frac{1}{2} \cdot BD \cdot$ ■

18. **Writing** Summarize the properties of squares that follow from a square being a parallelogram, a rhombus, and a rectangle.

19. **Calculator** In rhombus $EFGH$, the length of each side is 12 and $EG = 18$. Find FH to the nearest tenth.

PROBLEM SOLVING HINT

Draw a diagram.

20. In $\square RSTW$, $RS = 7$, $ST = 24$, and $RT = 25$. Is $RSTW$ a rectangle? Explain.

21. In $\square JKLM$, $JL = 5$, $KM = 12$, and $JM = 6.5$. Is $JKLM$ a rhombus? Explain.

Find the area of each rhombus.

22.
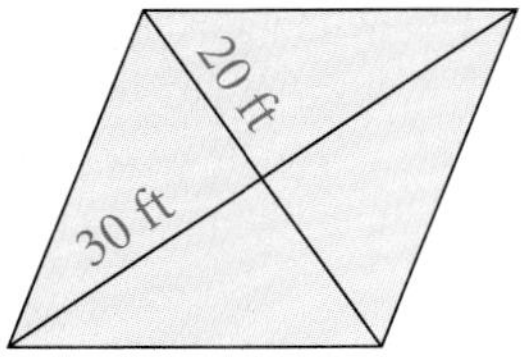

23.
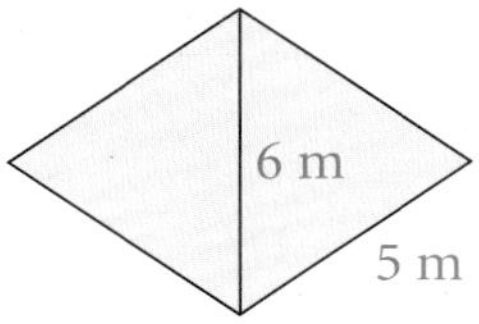

24.
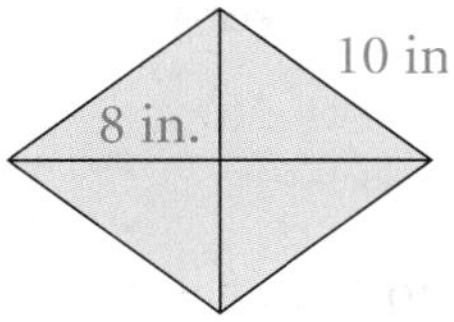

25.
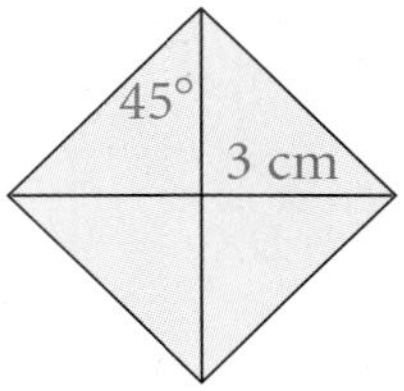

Logical Reasoning **Decide whether each of these is a good definition. Justify your answer.**

26. A rectangle is a quadrilateral with four right angles.

27. A rhombus is a quadrilateral with four congruent sides.

28. A square is a quadrilateral with four right angles and four congruent sides.

Choose **Write a two-column proof, a paragraph proof, or a flow proof.**

29. Theorem 9-13: If one diagonal of a parallelogram bisects two angles of the parallelogram, then the parallelogram is a rhombus.

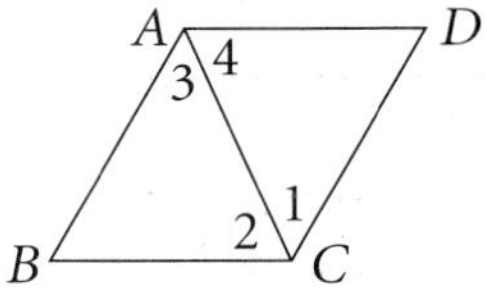

Given: $\square ABCD$; $\overline{AC}$ bisects $\angle BAD$ and $\angle BCD$.
Prove: $ABCD$ is a rhombus.

30. Theorem 9-14: If the diagonals of a parallelogram are perpendicular, then the parallelogram is a rhombus.

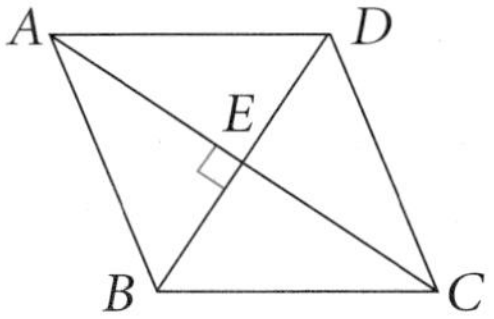

Given: $\square ABCD$; $\overline{AC} \perp \overline{BD}$
Prove: $ABCD$ is a rhombus.

31. Theorem 9-15: If the diagonals of a parallelogram are congruent, then the parallelogram is a rectangle.

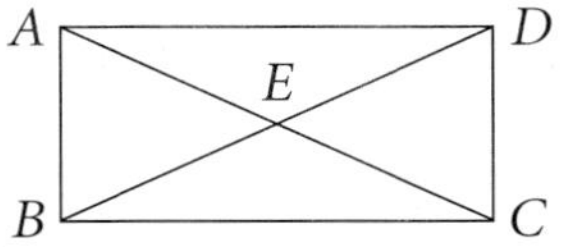

Given: $\square ABCD$; $\overline{AC} \cong \overline{BD}$
Prove: $ABCD$ is a rectangle.

32. **Auto Repair** A car jack is shaped like a rhombus. As two of the opposite vertices get closer together, the other two opposite vertices get farther apart. The sides of a car jack are 17 cm long. When the horizontal distance between the vertices is 30 cm, what is the vertical distance between the other two vertices?

33. a. **Given:** $\square ABCD$; $\angle B$ is a right angle.
Prove: $ABCD$ is a rectangle.
b. **Critical Thinking** State what you proved in part (a) as a theorem in if-then form.

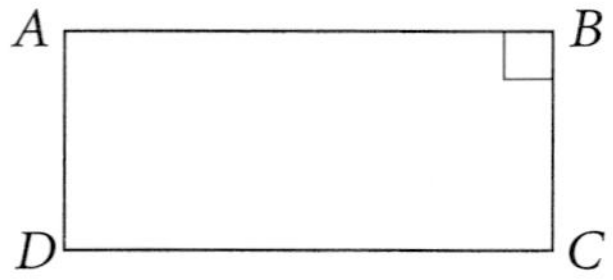

34. a. **Given:** $\square EFGH$; $\overline{EF} \cong \overline{FG}$
Prove: $EFGH$ is a rhombus.
b. **Critical Thinking** State what you proved in part (a) as a theorem in if-then form.

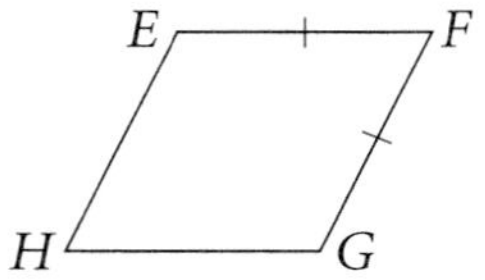

Chapter Project — *Find Out by Analyzing*

Weight and area exposed to the wind are key factors in kite design. The greater the *effective area* facing the wind and the lighter the kite, the less wind you need to get the kite off the ground.

1. In Figure 1, a face of the square box kite is perpendicular to the wind. Describe the *effective area.*
2. In Figure 2, a diagonal brace of the square box kite is perpendicular to the wind. Describe the *effective area.*
3. Where on a box kite would you tie the string to get the greatest *effective area?* Explain.
4. If you use two different lengths of wood for the diagonal braces of a box kite, you can make a *rhomboid* box kite. Explain why changing a square box kite to a rhomboid box kite can increase the *effective area.*

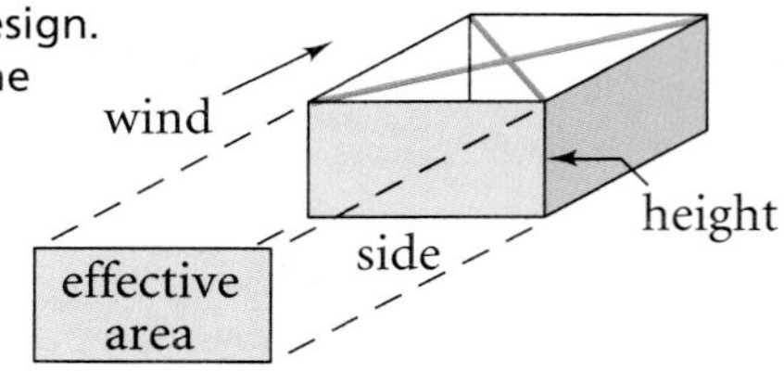

Figure 1

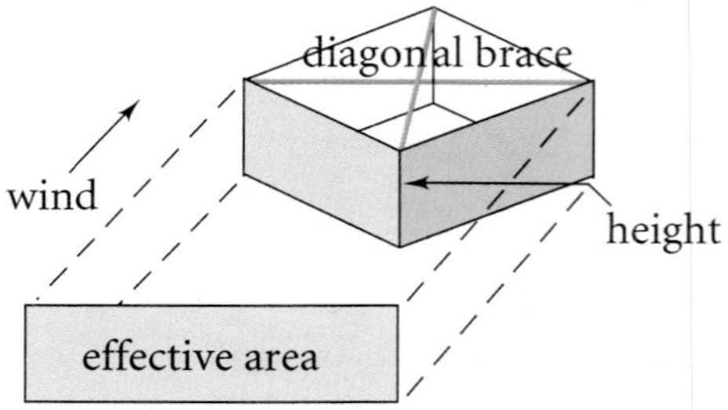

Figure 2

Exercises MIXED REVIEW

Sketch each pair of figures.

35. congruent isosceles triangles
36. similar right triangles
37. congruent regular hexagons

38. **Open-ended** Give the lengths of the sides of an obtuse triangle.

Getting Ready for Lesson 9-4

39. Sketch a quadrilateral with only two consecutive right angles. Classify the quadrilateral.

40. Sketch a quadrilateral in which one diagonal is a perpendicular bisector of the other, but the second diagonal does not bisect the first. Classify the quadrilateral.

Math ToolboX

Technology

Exploring Quadrilaterals within Quadrilaterals

Before Lesson 9-4

Work in pairs or small groups.

Construct

Use geometry software to construct a quadrilateral *ABCD*. Construct the midpoint of each side of *ABCD*. Construct segments joining the adjacent midpoints to form quadrilateral *EFGH*.

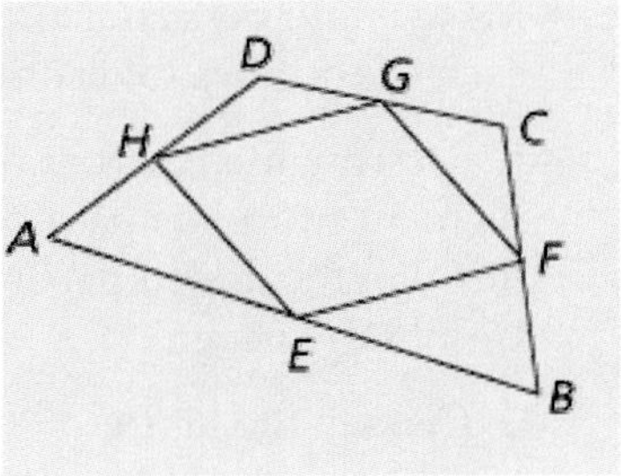

Investigate

Measure the lengths of the sides of *EFGH* and their slopes. Measure the angles of *EFGH*. What kind of quadrilateral does *EFGH* appear to be?

Conjecture

Manipulate quadrilateral *ABCD* and make a conjecture about the quadrilateral whose vertices are the midpoints of a quadrilateral. Does your conjecture hold when *ABCD* is convex? Can you manipulate *ABCD* so that your conjecture doesn't hold?

Extend

- Draw the diagonals of *ABCD*. Describe *EFGH* when the diagonals are perpendicular and when they are congruent.
- Construct the midpoints of *EFGH* and use them to construct quadrilateral *IJKL*. Construct the midpoints of *IJKL* and use them to construct quadrilateral *MNOP*. Compare the ratios of the lengths of the sides, perimeters, and areas of *MNOP* and *EFGH*. How are *MNOP* and *EFGH* related?

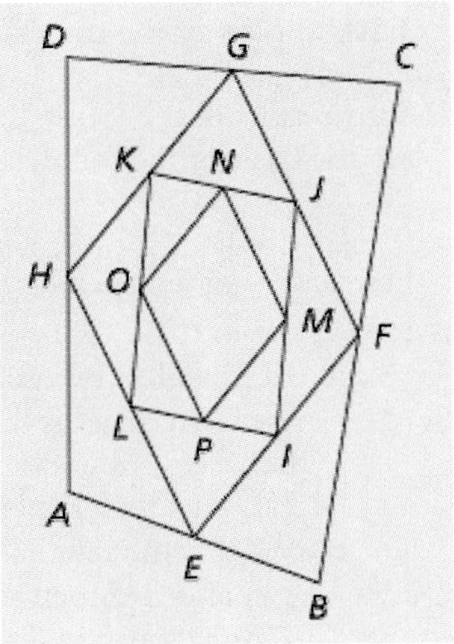

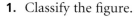

9-4 Trapezoids and Kites

What You'll Learn

- Finding the properties of trapezoids and kites

...And Why

To solve problems in construction and design

What You'll Need

- lined paper
- scissors
- straightedge

WORK TOGETHER

Work with a partner.

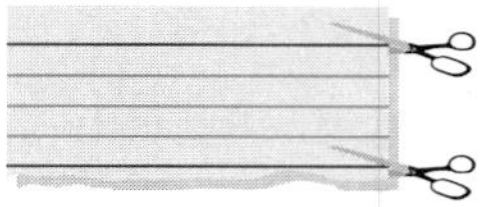

- Choose two lines on a piece of lined paper that are about 2 in. apart. Cut along these lines.

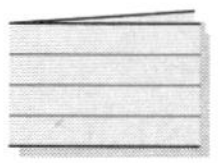

- Make a fold perpendicular to the cut lines so that each of the lines folds onto itself.

- Use a straightedge to draw a nonperpendicular segment from one parallel line to the other. Cut through the folded paper along that segment. Unfold the paper.

1. Classify the figure.
2. a. Describe the relationship between the nonparallel sides of this figure.
   b. Describe the relationship between the acute angles and the relationship between the obtuse angles.

THINK AND DISCUSS

Quick Review

An isosceles trapezoid is a trapezoid whose nonparallel sides are congruent.

In Chapter 5, you learned that the parallel sides of a trapezoid are the bases and the nonparallel sides are the legs. Each pair of angles adjacent to a base of a trapezoid are **base angles** of the trapezoid.

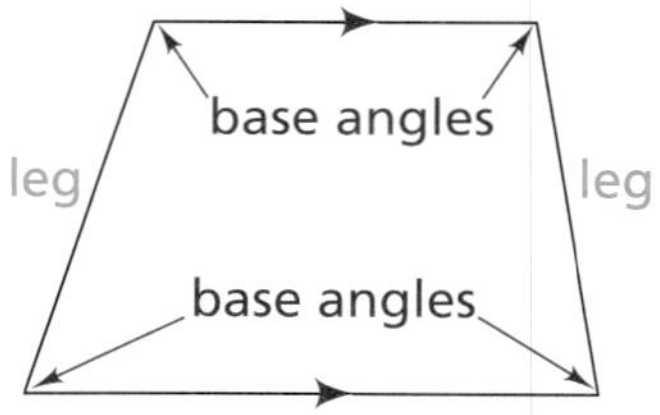

Your results in the Work Together activity are the basis of the following theorem.

Theorem 9-16 Base angles of an isosceles trapezoid are congruent.

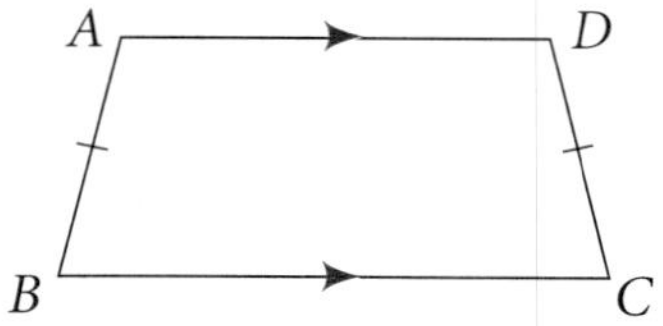

3. a. In isosceles trapezoid *ABCD* at the right, what is the relationship between $\angle A$ and $\angle B$? between $\angle D$ and $\angle C$? Explain.
   b. What is the relationship between $\angle A$ and $\angle C$? between $\angle D$ and $\angle B$? Explain.
   c. If $m\angle A = 102$, find $m\angle B$, $m\angle C$, and $m\angle D$.

Example 1 Relating to the Real World

Architecture Part of the window of the World Financial Center in New York City is made from eight congruent isosceles trapezoids that create the illusion of a semicircle. What is the measure of the base angles of these trapezoids?

Each trapezoid is part of an isosceles triangle whose vertex angle is at the center of the circle and whose base angles are the acute base angles of the trapezoid.

The measure of $\angle 1$ is $\frac{180}{8}$, or 22.5.

The measure of each acute base angle is $\frac{180 - 22.5}{2}$, or 78.75.

The measure of each obtuse base angle of the trapezoid is $180 - 78.75$, or 101.25.

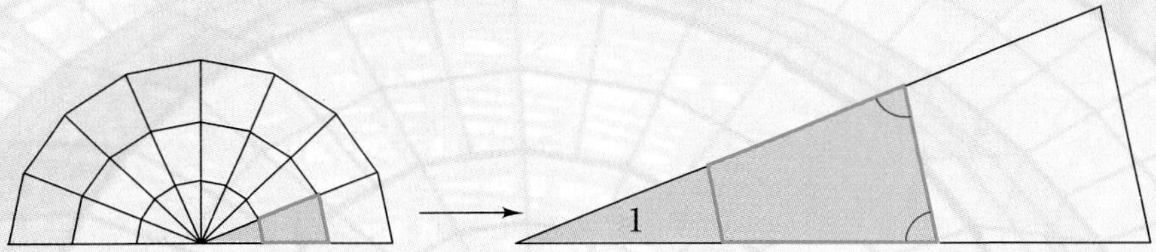

PROBLEM SOLVING HINT

Draw $\triangle ABC$ and $\triangle DCB$ as separate triangles and look for congruent parts.

4. a. *ABCD* is an isosceles trapezoid. How do $\triangle ABC$ and $\triangle DCB$ compare? Explain.

b. What must be true of the diagonals of isosceles trapezoid *ABCD*? Explain.

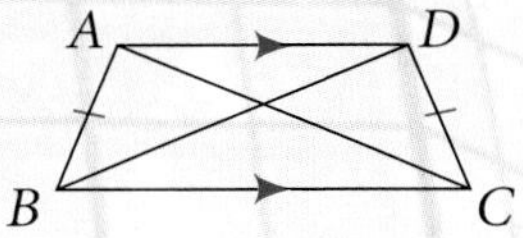

Theorem 9-17 The diagonals of an isosceles trapezoid are congruent.

Example 2

Write a paragraph proof of Theorem 9-17.

Given: Isosceles trapezoid *ABCD* with $\overline{AB} \cong \overline{DC}$

Prove: $\overline{AC} \cong \overline{DB}$

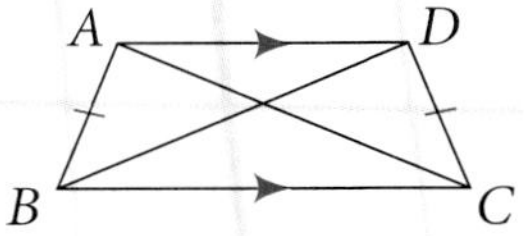

Paragraph Proof

It is given that $\overline{AB} \cong \overline{DC}$. Because the base angles of an isosceles trapezoid are congruent, $\angle ABC \cong \angle DCB$. By the Reflexive Property of Congruence, $\overline{BC} \cong \overline{BC}$. Then by the SAS Postulate, $\triangle ABC \cong \triangle DCB$. Therefore $\overline{AC} \cong \overline{DB}$ by CPCTC.

You can use the lengths of the sides of an isosceles trapezoid to find the height of the trapezoid.

Example 3

Find the height of isosceles trapezoid *FLAG.*

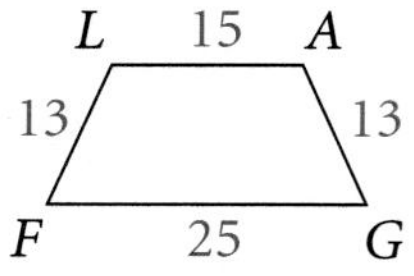

Draw altitudes $\overline{LM}$ and $\overline{AB}$, creating rectangle *MLAB,* with $MB = 15$.
$\triangle FLM \cong \triangle GAB$ by AAS.

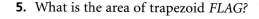

By CPCTC, $\overline{FM} \cong \overline{GB}$.
$FM = (25 - 15) \div 2 = 5$.

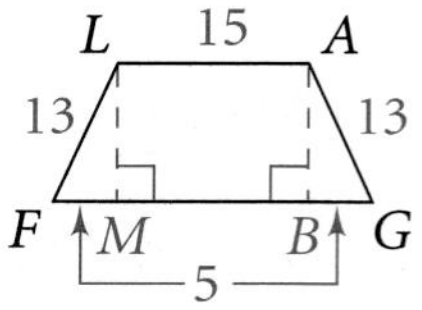

Use the Pythagorean Theorem to solve for *LM.*

$$\begin{aligned} 5^2 + (LM)^2 &= 13^2 \\ 25 + (LM)^2 &= 169 \\ (LM)^2 &= 144 \\ LM &= 12 \end{aligned}$$

The height of isosceles trapezoid *FLAG* is 12 cm.

5. What is the area of trapezoid *FLAG?*

Another special quadrilateral that is not a parallelogram is a kite.

6. a. *Coordinate Geometry* Find the lengths of the sides of *OBCD.*
b. Classify *OBCD.*
c. Find the slopes of $\overline{OC}$ and $\overline{BD}$.
d. What is the relationship between $\overline{OC}$ and $\overline{BD}$?

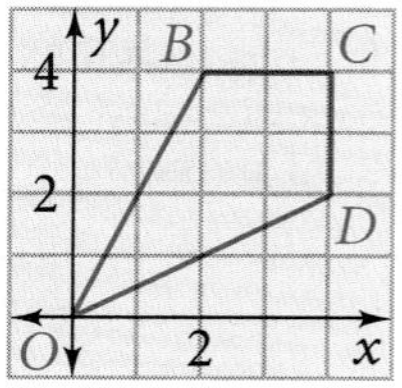

Theorem 9-18 The diagonals of a kite are perpendicular.

7. Answer the following questions to prove Theorem 9-18.

Given: Kite *RSTW* with $\overline{RS} \cong \overline{RW}$ and $\overline{ST} \cong \overline{WT}$
Prove: $\overline{RT} \perp \overline{SW}$

a. What triangle is congruent to $\triangle RST$? Justify your answer.
b. Why is $\angle 1 \cong \angle 2$?
c. Why is $\triangle STZ \cong \triangle WTZ$?
d. Why is $\angle 3 \cong \angle 4$?
e. Why are $\angle 3$ and $\angle 4$ right angles?
f. Why is $\overline{RT} \perp \overline{SW}$?

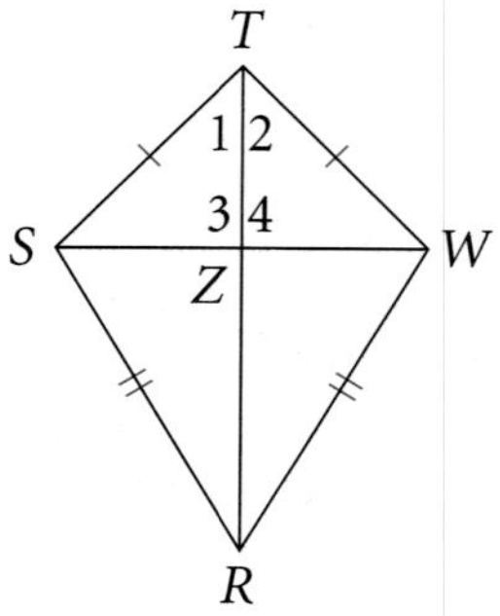

8. Try This If $RS = 15$, $ST = 13$, and $SW = 24$, find *RT.* (*Hint:* Use Pythagorean triples.)

PROBLEM SOLVING

Look Back Explain why Theorem 9-18 could easily be a corollary of Theorem 4-13: Converse of Perpendicular Bisector Theorem.

The following statements are suggested by the converses of the theorems in this lesson. Draw figures to show that the statements are false.

9. A quadrilateral with perpendicular diagonals is a kite or a rhombus.

10. A quadrilateral with congruent diagonals is an isosceles trapezoid or a rectangle.

11. A quadrilateral with diagonals that are congruent and perpendicular is a square.

Exercises ON YOUR OWN

Find the measures of ∠1 and ∠2.

1. 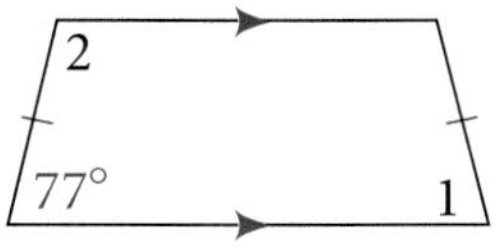

2. 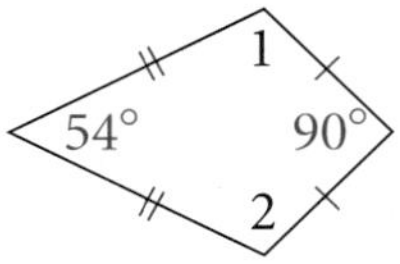

3. 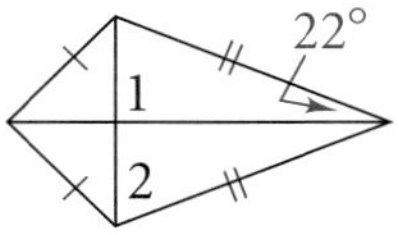

4. 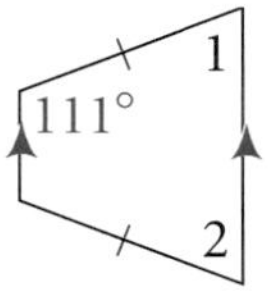

5. Find UI, IT, and UT, if $AS = 16$.

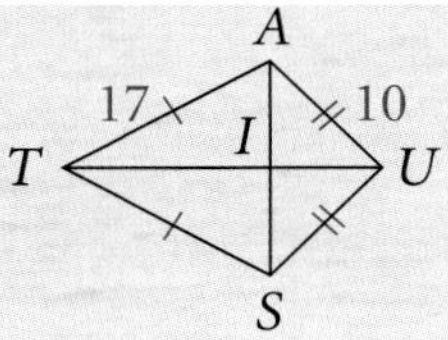

6. **a.** Find the height of an isosceles trapezoid with bases 30 cm and 70 cm long and legs 29 cm long.
   b. Find the area of the trapezoid.

7. The perimeter of a kite is 66 cm. The length of one of its sides is 3 cm less than twice the length of another. Find the length of each side of the kite.

8. **Design** To make a beach umbrella, Taheisha cut eight panels from material with parallel stripes. The panels were congruent isosceles triangles with a vertex angle of 40° as shown at the right.
   a. Classify each yellow quadrilateral.
   b. Find the measures of the interior angles of each yellow quadrilateral.

9. a. *Construction* At the right, a quadrilateral formed by the beams of the bridge is outlined. Classify the quadrilateral. Explain your reasoning.

b. Find the measures of the other interior angles of the figure.

10. a. *Transformations* Sketch a kite and describe its symmetries.

b. *Critical Thinking* A classmate claims that the longer diagonal of a kite is a line of symmetry for the kite. Is he correct? Explain.

11. a. *Critical Thinking* Can two consecutive angles of a kite be supplementary? Explain.

b. Can two opposite angles of a kite be supplementary? Explain.

12. *Standardized Test Prep* In quadrilateral $DEAL$, $\angle D$ is supplementary to $\angle E$. Quadrilateral $DEAL$ could be a __?__.

I. trapezoid **II.** kite **III.** rhombus

A. I only **B.** III only **C.** I or III
D. II or III **E.** I, II, or III

Algebra **Find the value of the variable(s).**

13.

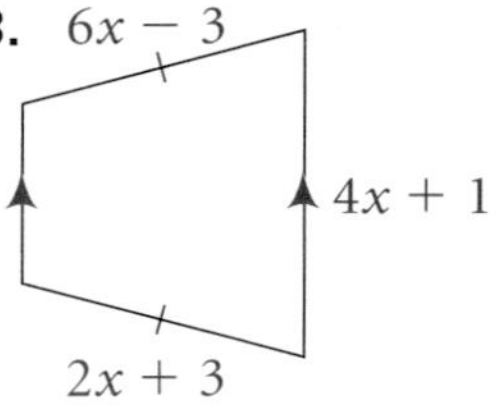

14.

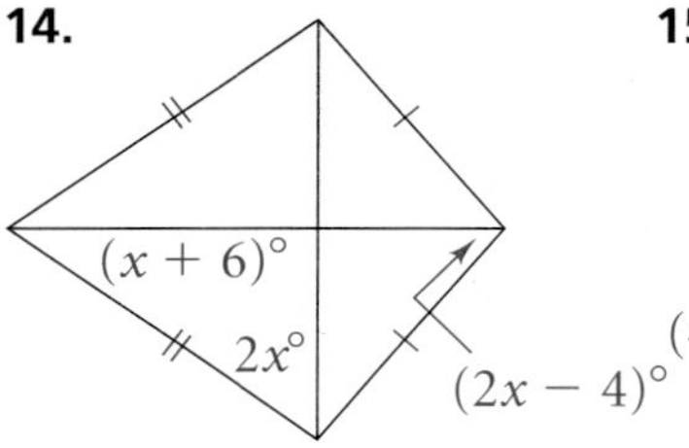

15.

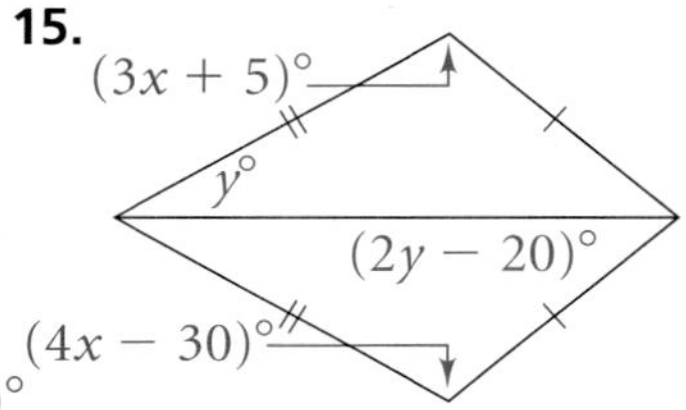

16. $AC = 4x$, $BD = 3y$

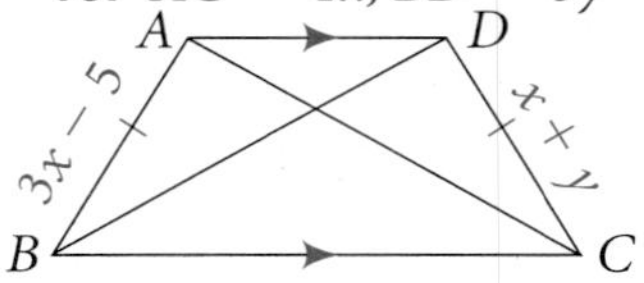

17. *Open-ended* Sketch two kites that are not congruent, but such that the diagonals of one are congruent to the diagonals of the other.

18. a. *Calculator* If $SW = 8$, find RT to the nearest tenth.

b. If $RT = 12$, find SW to the nearest tenth. (*Hint:* Let $TZ = x$, $RZ = 12 - x$, and $SZ = y$. Write two equations and then solve.)

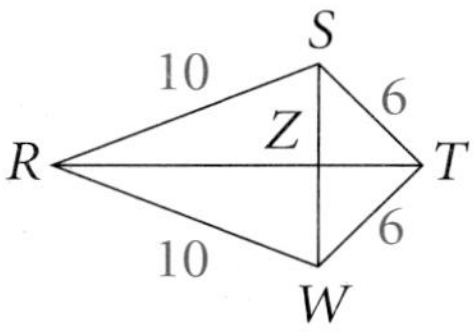

19. *Transformations* Draw $\overline{AB}$. Rotate $\overline{AB}$ $x°$ about B, where $0 < x < 90$. Reflect $\triangle ABA'$ in $\overline{AB}$, labeling the image of A' as A''. What type of quadrilateral is $AA'BA''$? Explain.

20. a. *Geometry in 3 Dimensions* If a plane parallel to the base intersects a square pyramid, what shape is the cross section?

b. If the plane is not parallel to the base, what shapes can the cross section have?

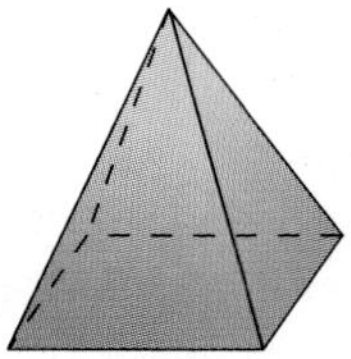

21. **Open-ended** Charlie Brown is talking to Lucy about a "three-foot flat kite with a sail area of four and one-half square feet." Draw a kite with two 3-ft diagonals that has an area of $4\frac{1}{2}$ ft^2. Show the lengths of the segments of both diagonals.

22. **Critical Thinking** If *KLMN* is an isosceles trapezoid, is it possible for $\overline{KM}$ to bisect $\angle LMN$ and $\angle LKN$? Explain.

23. **Writing** A *kite* is sometimes defined as a quadrilateral with two pairs of congruent adjacent sides. Compare this to the definition on page 91. Are parallelograms, trapezoids, rhombuses, rectangles, or squares special kinds of kites according to the changed definition? Explain.

24. Complete the following two-column proof of Theorem 9-16: The base angles of an isosceles trapezoid are congruent.

Given: *ABCD* is an isosceles trapezoid with $\overline{AB} \cong \overline{DC}$.
Prove: $\angle B \cong \angle C$ and $\angle BAD \cong \angle D$
Begin by drawing $\overline{AE} \parallel \overline{DC}$.

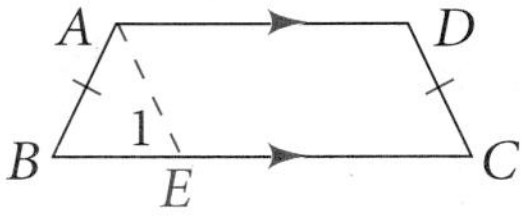

Statements	Reasons
1. *ABCD* is an isosceles trapezoid; $\overline{AB} \cong \overline{DC}$.	1. Given
a. ■ $\parallel \overline{EC}$	2. Definition of trapezoid
3. *AECD* is a ▱.	b. ?
4. $\angle C \cong \angle 1$	c. ?
d. $\overline{DC} \cong$ ■	5. Opposite sides of a ▱ are ≅.
6. $\overline{AE} \cong \overline{AB}$	e. ? Property of ≅
7. $\angle B \cong \angle 1$	f. ?
8. $\angle B \cong \angle C$	g. ?
9. $\angle B$ and $\angle BAD$ are supplements; $\angle C$ and $\angle ADC$ are supplements.	h. ?
10. $\angle BAD \cong \angle CDA$	i. Supplements of ≅ ∠s are ? .

25. **a.** **Picture Framing** Describe the shape of the four pieces of wood that are joined to form the picture frame at the right.
b. Find the angle measures of each piece of wood.

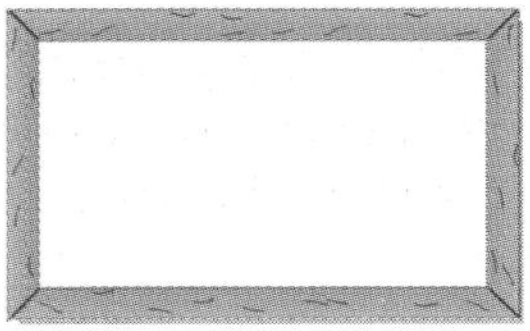

Choose **Write a two-column proof, a paragraph proof, or a flow proof.**

26. Given: Kite $ABCD$ with $\overline{AB} \cong \overline{AD}$ and $\overline{BC} \cong \overline{DC}$
Prove: Area of kite $ABCD = \frac{1}{2} \cdot AC \cdot BD$

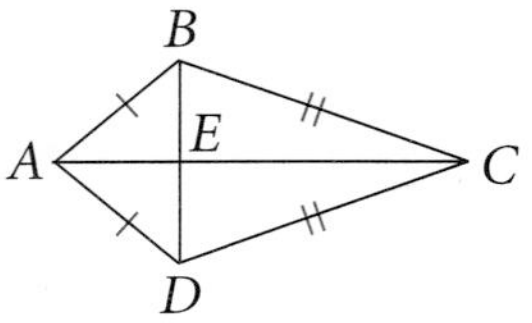

27. Given: Isosceles trapezoid $TRAP$ with $\overline{TR} \cong \overline{PA}$
Prove: $\angle 1 \cong \angle 2$

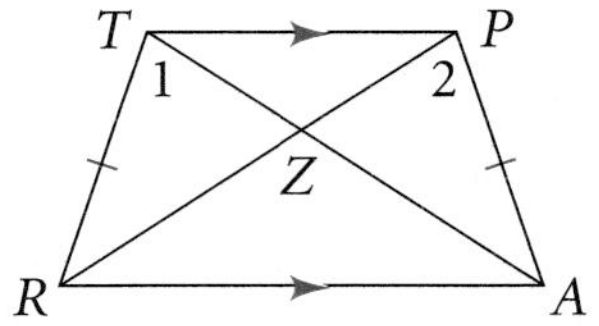

Chapter Project *Find Out by Researching*

To help decide what kind of kite you will design and build for this project, read about three different kinds of kites. Use the books listed on page 488 or other books or magazines. Summarize your research. Include the following in your report:

- a description of each kite
- a list of materials needed to build each kite
- the wind conditions each kite is suited for

Exercises MIXED REVIEW

Probability **You randomly toss a dart at each dartboard. Find the probability the dart lands in a red area.**

28.

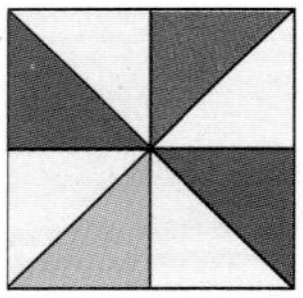

29.

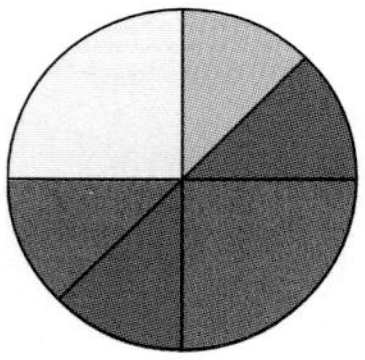

30.

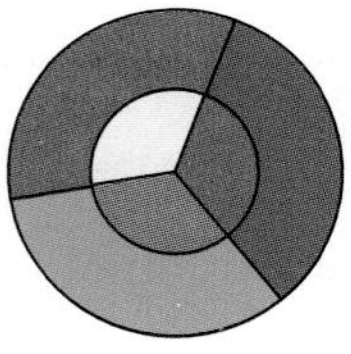

SELF ASSESSMENT **FOR YOUR JOURNAL**

Describe the properties of isosceles trapezoids.

31. a. **Algebra** Write an algebraic expression for the area of a rectangle with width x and length $x + 7$.

b. If the perimeter of this rectangle is 38 cm, find its area.

Getting Ready for Lesson 9-5

Coordinate Geometry **Draw quadrilateral *QUAD* with the given vertices. Then classify the quadrilateral.**

32. $Q(-5, 0)$, $U(-3, 2)$, $A(3, 2)$, $D(5, 0)$

33. $Q(0, 0)$, $U(4, 0)$, $A(3, 2)$, $D(-1, 2)$

34. $Q(0, 0)$, $U(5, 5)$, $A(8, 4)$, $D(7, 1)$

35. $Q(-3, 0)$, $U(0, 3)$, $A(3, 0)$, $D(0, -3)$

Math ToolboX

Algebra Review

Rational Expressions

After Lesson 9-4

A **rational expression** is an expression that can be written in the form $\frac{\text{polynomial}}{\text{polynomial}}$, where a variable is in the denominator. A rational expression is in simplest form if the numerator and denominator have no common factors except 1.

To multiply rational expressions $\frac{a}{b}$ and $\frac{c}{d}$, where b and $d \neq 0$, multiply the numerators and multiply the denominators. Then write the product in simplest form.

$$\frac{a}{b} \cdot \frac{c}{d} = \frac{ac}{bd}$$

To divide $\frac{a}{b}$ by $\frac{c}{d}$, where b, c, and $d \neq 0$, multiply by the reciprocal of $\frac{c}{d}$.

$$\frac{a}{b} \div \frac{c}{d} = \frac{a}{b} \cdot \frac{c}{d}$$

Example 1

Simplify $\frac{3x-6}{x-2}$.

$$\frac{3x-6}{x-2} = \frac{3(\cancel{x-2})^{1}}{\cancel{x-2}_{1}}$$

$$= 3,\ x \neq 2$$

Example 2

Multiply $\frac{3x}{x-1} \cdot \frac{2x-2}{9x^2}$.

$$\frac{3x}{x-1} \cdot \frac{2x-2}{9x^2} = \frac{{}^{1}\cancel{3x}(2)(\cancel{x-1})^{1}}{(\cancel{x-1})_{1}(\cancel{3x})_{1}(3x)}$$

$$= \frac{2}{3x},\ x \neq 0 \text{ or } 1$$

Example 3

Divide $\frac{4x^5}{3}$ by $16x^7$.

$$\frac{4x^5}{3} \div 16x^7 = \frac{\overset{1}{\cancel{4x^5}}}{3} \cdot \frac{1}{\cancel{16x^7}_{4x^2}}$$

$$= \frac{1}{12x^2},\ x \neq 0$$

Simplify each expression.

1. $\frac{4x+12}{x+3}$
2. $\frac{6c^3}{3c^4}$
3. $\frac{3x-9}{3x+9}$
4. $\frac{2a^6}{16a}$
5. $\frac{28w+12}{4}$
6. $\frac{m+7}{m^2-49}$
7. $\frac{5t^2}{25t^3}$
8. $\frac{6x^2+6x}{4x^2+4x}$
9. $\frac{4x^3-12x^2}{x-3}$
10. $\frac{3s^2+s}{s^3}$
11. $\frac{v^2+2v+1}{v+1}$
12. $\frac{8r^7}{56r^2}$

Find each product or quotient.

13. $\frac{3r^2}{4} \cdot \frac{20}{5r^3}$
14. $\frac{x-1}{x+2} \div (2x-2)$
15. $\frac{3c+6}{5} \cdot \frac{25c}{c+2}$
16. $\frac{4}{w^2} \div \frac{16}{w}$
17. $20x^2 \div \frac{4x}{3}$
18. $\frac{3x^2+x}{2x} \cdot \frac{4x^2}{3x+1}$
19. $\frac{9a^5}{2a} \div \frac{12a^3}{4}$
20. $\frac{x+2}{x-7} \div \frac{x}{x-7}$
21. $\frac{t+3}{t-2} \cdot \frac{t^2-4}{5t+15}$
22. $\frac{x^4-x^3}{x-1} \cdot \frac{2}{x^2}$
23. $\frac{a-5}{a+5} \div (2a-10)$
24. $\frac{4r^2}{r^5} \cdot \frac{r^2}{2}$
25. $9y^4 \div \frac{3y}{y+1}$
26. $\frac{9w^3-w}{2w-1} \cdot \frac{1-2w}{w}$
27. $\frac{6c+2}{c^2+1} \div \frac{2}{c-1}$
28. $\frac{4m+2}{m-3} \cdot \frac{3-m}{4m^2-1}$

What You'll Learn

- Choosing convenient placement of coordinate axes on figures

...And Why

To use coordinate proofs to solve design problems

What You'll Need

- graph paper
- straightedge

9-5 Organizing Coordinate Proofs

WORK TOGETHER

Work with a small group.

- Use graph paper to draw a number of coordinate planes with x- and y-coordinates from -12 to 12. On each, draw one or more squares with sides 10 units long. Draw the squares in different positions, with sides on one or both axes, parallel to the axes, or not parallel to either axis. Use more than one quadrant. Label each square *ABCD*.
- On each square, record the slope of each side, the slope of each diagonal, and the coordinates of the midpoints of the diagonals.
- Discuss which type of squares made your calculations easiest. **Generalize** your conclusions and explain your reasons.

THINK AND DISCUSS

In the Work Together, you saw that the position of a figure has an effect on how easy the figure is to work with. In coordinate proofs, it is generally easiest to use the origin as the center for the figure or to place a vertex at the origin and at least one side of the figure on an axis.

Example 1

Use the properties of each figure to find the missing coordinates.

a. rectangle *KLMN*

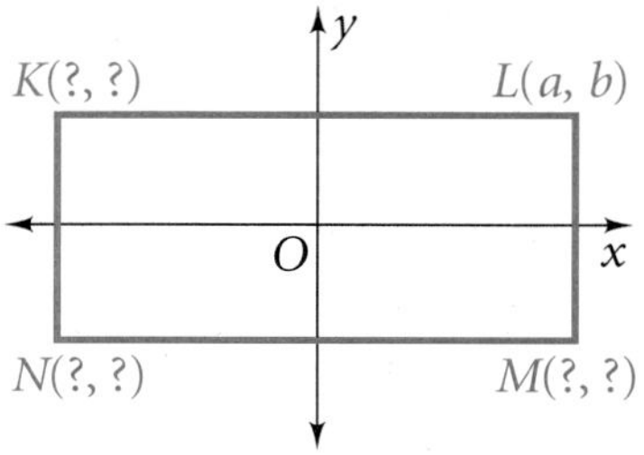

The x- and y-axes are horizontal and vertical lines of symmetry. The coordinates are $K(-a, b)$, $M(a, -b)$, and $N(-a, -b)$.

b. ▱*OPQR*

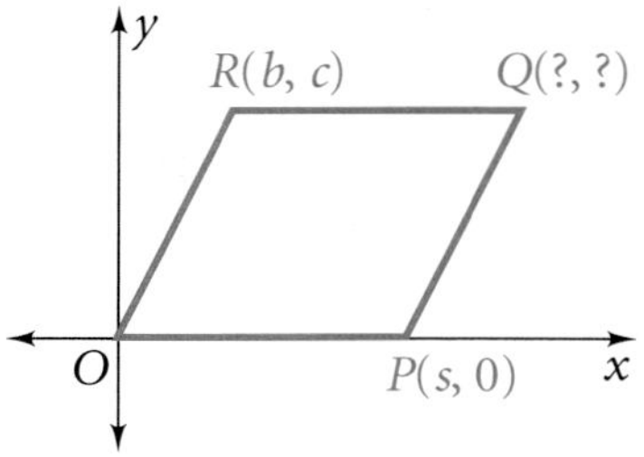

To go from point P to point Q keeping $OP = RQ$, go right b units and up c units. Point Q has coordinates $Q(s + b, c)$.

1. **Try This** When you place two sides of a figure on the coordinate axes, what are you assuming about the figure?

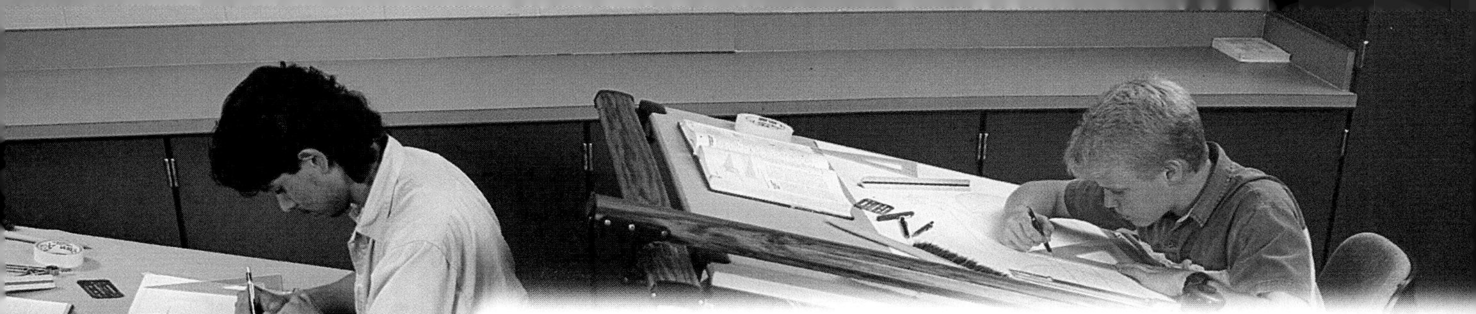

It is often convenient to use coordinates that are multiples of 2 so that when you find midpoints you will not need to use fractions.

Example 2 **Relating to the Real World**

Design The assignment in art class is to create a T-shirt design by drawing any quadrilateral, connecting its midpoints to form another quadrilateral, and coloring the regions. Elena claims that everyone's inner quadrilateral will be a parallelogram. Is she correct? Explain.

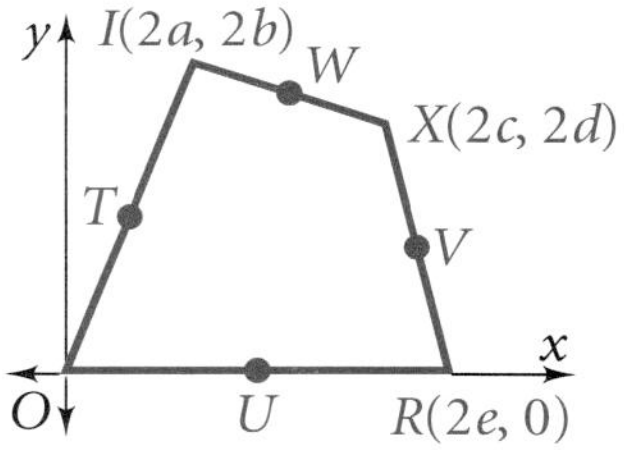

Draw a quadrilateral on a coordinate plane. Locate one vertex at the origin and one side on the x-axis. Since you are finding midpoints, use coordinates that are multiples of 2. Find the coordinates of the midpoints T, W, V, and U, and the slopes of the sides of $TWVU$.

T = midpoint of $\overline{OI} = \left(\frac{2a+0}{2}, \frac{2b+0}{2}\right) = (a, b)$

W = midpoint of $\overline{IX} = \left(\frac{2a+2c}{2}, \frac{2b+2d}{2}\right) = (a+c, b+d)$

V = midpoint of $\overline{XR} = \left(\frac{2c+2e}{2}, \frac{2d+0}{2}\right) = (c+e, d)$

U = midpoint of $\overline{OR} = \left(\frac{0+2e}{2}, \frac{0+0}{2}\right) = (e, 0)$

slope of $\overline{TW} = \frac{b-(b+d)}{a-(a+c)} = \frac{d}{c}$

slope of $\overline{VU} = \frac{d-0}{(c+e)-e} = \frac{d}{c}$ — The slopes are equal, so the lines are parallel.

slope of $\overline{WV} = \frac{(b+d)-d}{(a+c)-(c+e)} = \frac{b}{a-e}$

slope of $\overline{TU} = \frac{b-0}{a-e} = \frac{b}{a-e}$ — The slopes are equal, so the lines are parallel.

Since both pairs of opposite sides of $TWVU$ are parallel, $TWVU$ is a parallelogram and Elena is correct!

PROBLEM SOLVING

Look Back Explain how you could prove that Elena is correct without using coordinate geometry.

2. **Try This** Use a different method to show that $TWVU$ is a parallelogram by finding the midpoints of the diagonals.

Exercises ON YOUR OWN

Give coordinates for points *W* and *Z* without using any new variables.

1. rectangle

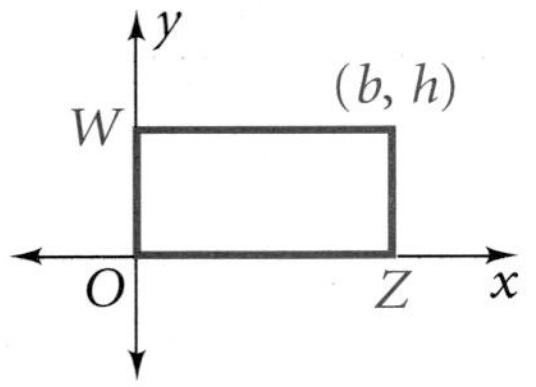

2. square

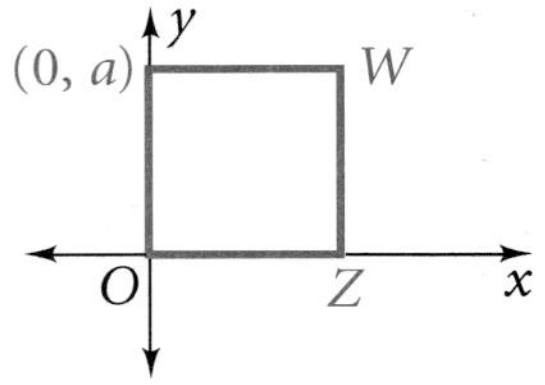

3. square

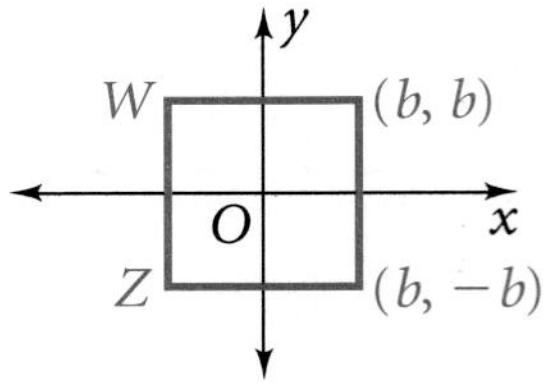

4. parallelogram

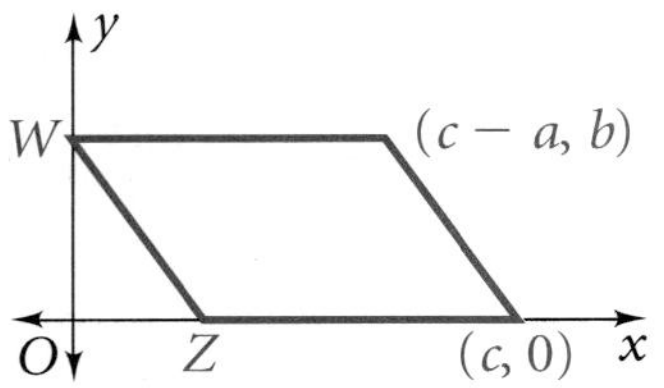

5. rhombus

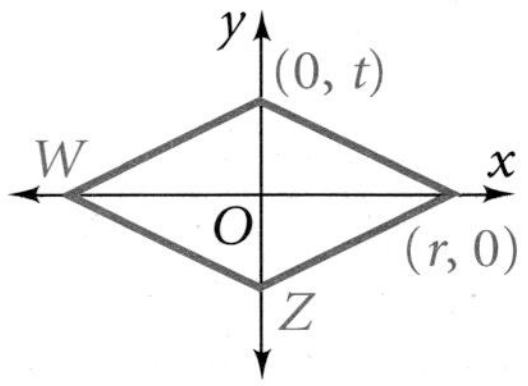

6. isosceles trapezoid

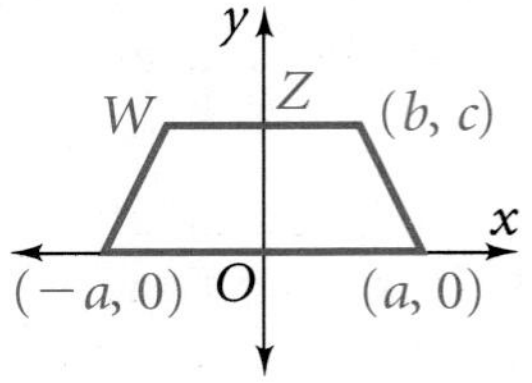

Find the coordinates of the midpoint of $\overline{WZ}$ and find *WZ* in the exercise named.

7. Exercise 1

8. Exercise 2

9. Exercise 3

10. Exercise 4

11. Exercise 5

12. Exercise 6

13. **a.** *Open-ended* Choose values for r and t in Exercise 5. Find the slope and length of each side.

b. Does the figure satisfy the definition of a rhombus? Explain.

14. Draw a square with sides $2a$ units long on the coordinate plane. Give coordinates for each vertex.

15. **a.** Draw a square centered at the origin whose diagonals of length $2b$ units lie on the x- and y-axes.

b. Give the coordinates of the vertices of the square.

c. Compute the length of a side of the square.

d. Find the slopes of two adjacent sides of the square.

e. Do the slopes show that the sides are perpendicular? Explain.

16. **a.** Draw an isosceles triangle with base length $2b$ units and height $2c$, placing one vertex on the origin.

b. Draw an isosceles triangle with base length $2b$ units and height $2c$, placing the base on the x-axis and the line of symmetry on the y-axis.

c. Find the lengths of the legs of the triangle in part (a).

d. Find the lengths of the legs of the triangle in part (b).

e. How do the results of parts (c) and (d) compare? Explain.

Give the coordinates for point *P* without using any new variables.

17. isosceles trapezoid

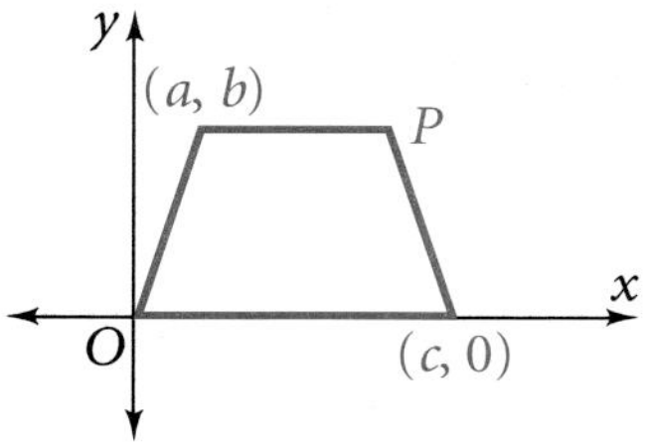

18. trapezoid with a right angle

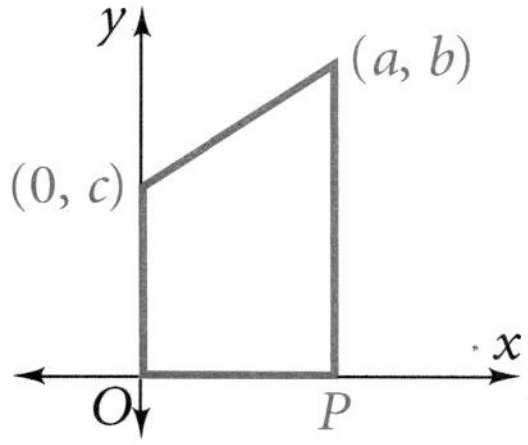

19. kite

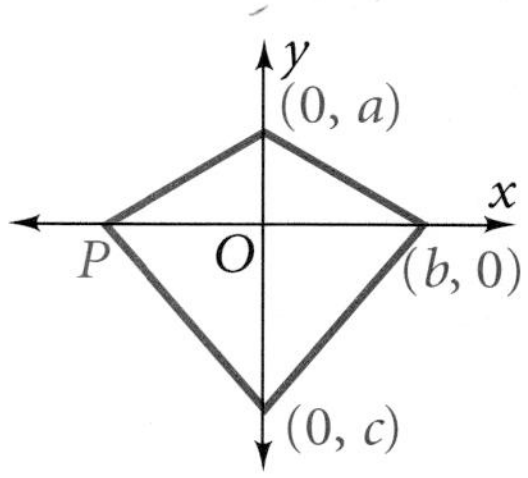

20. Geometry in 3 Dimensions Choose coordinates (x, y, z) for the vertices of a rectangular prism whose dimensions are a, b, and c. Use the origin as one of the vertices.

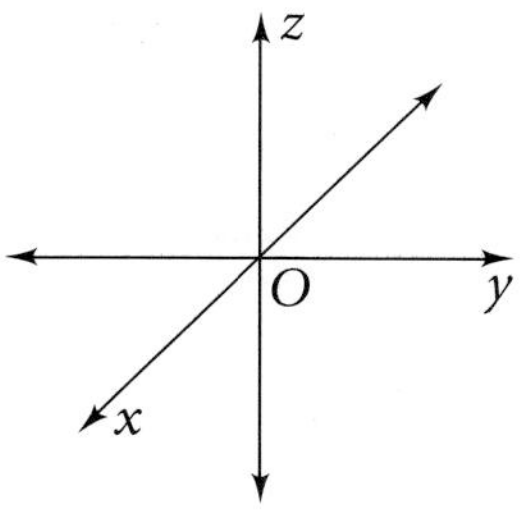

21. Geometry in 3 Dimensions A cube has edges $2a$ units long.

a. Choose coordinates for the vertices, using the origin as one vertex.

b. Choose coordinates for the vertices, using the origin as the center of the cube and making each of the cube's faces parallel to two of the three axes.

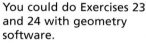

22. Draw a model for an equilateral triangle with sides $2a$ units long. Give coordinates of the vertices without introducing new variables.

23. Transformations In Quadrant I draw $\overline{AB}$ that is not parallel to either axis. Let $\overline{A'B'}$ be its reflection over the y-axis. What type of quadrilateral is $BAA'B'$? Explain.

24. Transformations In Quadrant I draw $\overline{AB}$ that is parallel to the y-axis. Let $\overline{A'B'}$ be its reflection over the y-axis. What type of quadrilateral is $BAA'B'$? Explain.

TECHNOLOGY HINT

You could do Exercises 23 and 24 with geometry software.

25. a. What property of a rhombus makes it convenient to place its diagonals on the x- and y-axes?

b. Writing Suppose a parallelogram is not a rhombus. Explain why it is inconvenient to place opposite vertices on the y-axis.

26. Archaeology Divers searching a shipwreck sometimes use gridded-plastic sheets to establish a coordinate system on the ocean floor. They record the coordinates of points where artifacts are found. Assume that divers search a square area and can go no farther than b units from their starting points. Draw a model for the area one diver can search. Assign coordinates to the vertices without using any new variables.

Exercises MIXED REVIEW

Find the area of each figure.

27. an equilateral triangle with altitude $6\sqrt{3}$ in.

28. a square with diagonals $5\sqrt{2}$ ft

29. How much cardboard do you need to make a box 4 ft by 5 ft by 2 ft?

Getting Ready for Lesson 9-6

30. a. Coordinate Geometry Graph the rhombus with vertices $A(2, 2)$, $B(7, 2)$, $C(4, -2)$, $D(-1, -2)$.

b. Connect the midpoints of consecutive sides to form a quadrilateral. What do you notice about the figure formed?

Exercises CHECKPOINT

Find the value of the variable(s).

1.

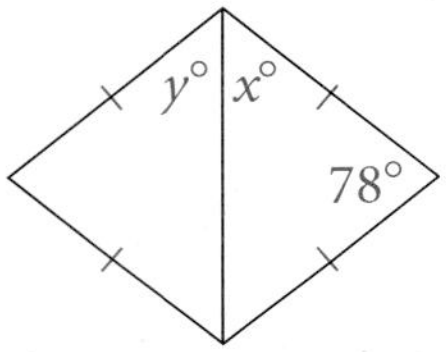

2.

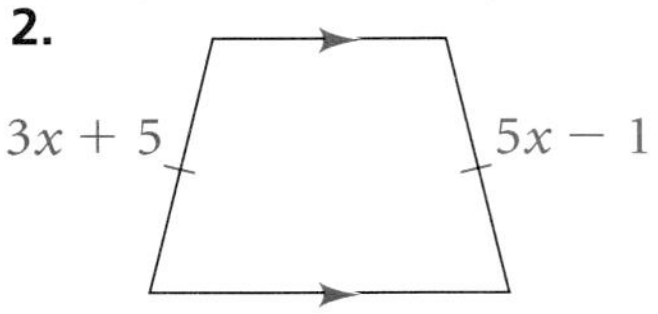

3.

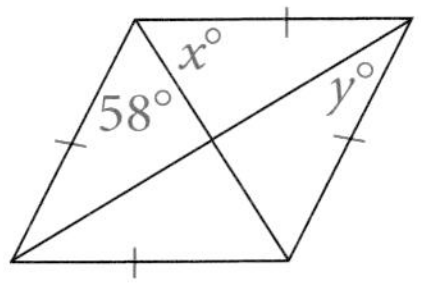

4.

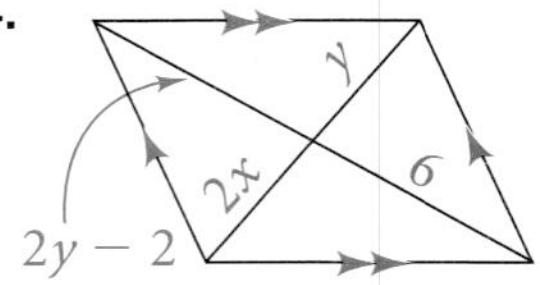

5. Writing Summarize the characteristics of the diagonals of a rhombus that are not true for the diagonals of every parallelogram.

A Point in Time

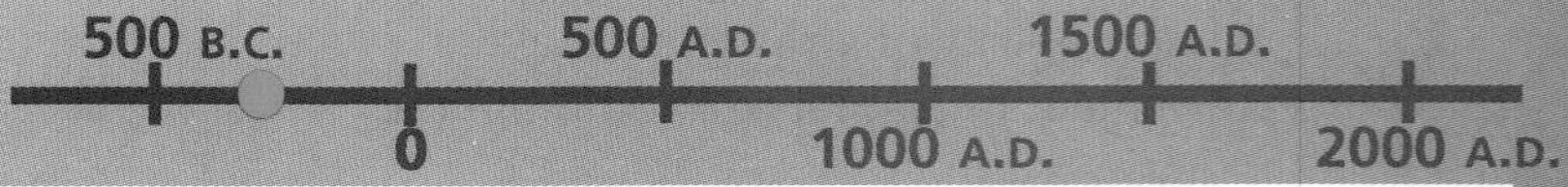

Egyptian Relief Sculpture

Many walls in ancient Egypt were decorated with relief sculptures. The relief sculpture in the photo was created in the year **255 B.C.** First the artist sketched the scene on papyrus overlaid with a grid. Next the wall was marked with a grid the size of the intended sculpture. To draw each line, a tightly stretched string that had been dipped in red ochre was plucked, like a guitar string.

Using the grid squares as guides, the artist transferred the drawing to the wall. Then a relief sculptor cut the background away, leaving the scene slightly raised. Finally an artist painted the scene.

9-6 Using Coordinate Geometry in Proofs

What You'll Learn
- Proving theorems using figures in the coordinate plane

...And Why
To use coordinate geometry to design flags

What You'll Need
- graph paper
- straightedge

WORK TOGETHER

Work with a partner. Each of you should make your own figures.

- On graph paper, draw a trapezoid with one vertex at the origin and one base on the positive *x*-axis. Find the coordinates of each vertex.
- Find the coordinates of the midpoints *M* and *N* of the two nonparallel sides. Draw $\overline{MN}$.
- Find the slopes of $\overline{MN}$ and both bases of the trapezoid. Compare the slopes to find a relationship between $\overline{MN}$ and the bases.
- Compare the length of $\overline{MN}$ and the sum of the lengths of the bases.
- Compare your results with your partner. Make **conjectures** about the relationships between $\overline{MN}$ and the bases of a trapezoid.

THINK AND DISCUSS

The segment that joins the midpoints of the nonparallel sides of a trapezoid is the **midsegment of the trapezoid.** Your conjectures in the Work Together lead to the following theorem about the midsegment of a trapezoid.

Theorem 9-19 The midsegment of a trapezoid is (1) parallel to the bases and (2) half as long as the sum of the lengths of the bases.

Your main tools in coordinate geometry proofs are the formulas for slope, midpoint, and distance. Use these tools to prove Theorem 9-19 by answering the following questions.

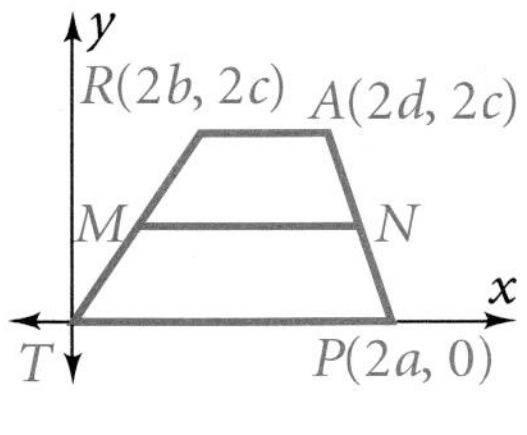

Given: $\overline{MN}$ is the midsegment of trapezoid *TRAP.*

Prove: $\overline{MN} \parallel \overline{TP}$, $\overline{MN} \parallel \overline{RA}$, and $MN = \frac{1}{2}(TP + RA)$.

1. Find the coordinates of midpoints *M* and *N*.
2. **a.** Find the slopes of the bases of the trapezoid.
   b. Find the slope of $\overline{MN}$.
   c. Compare the slopes.
3. **a.** Find the sum of the lengths of the bases.
   b. Find the length of $\overline{MN}$.
   c. Compare the lengths.

Example — Relating to the Real World

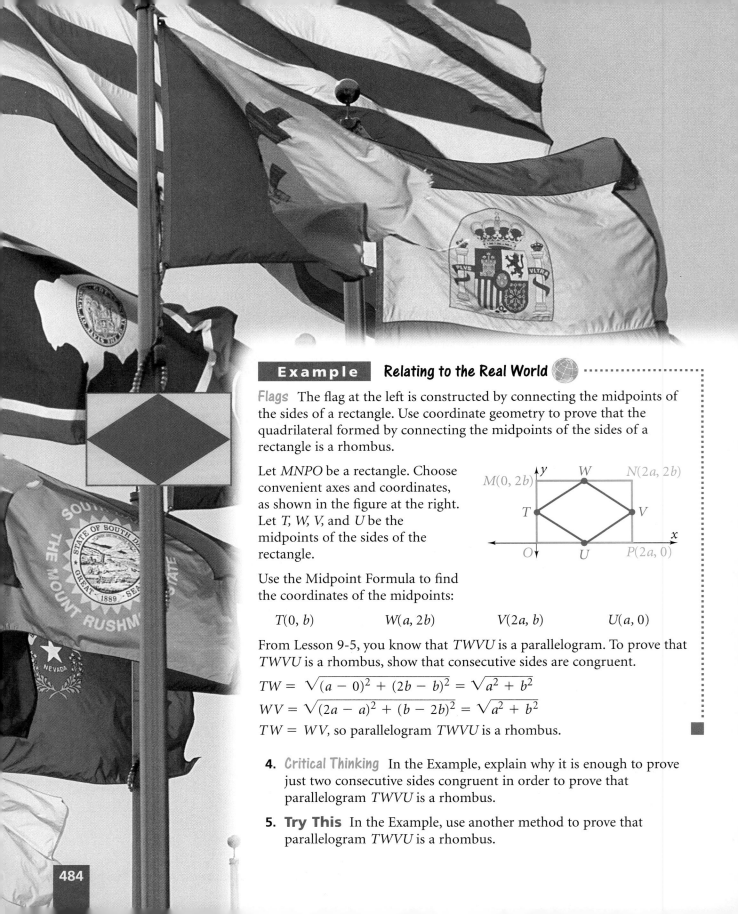

Flags The flag at the left is constructed by connecting the midpoints of the sides of a rectangle. Use coordinate geometry to prove that the quadrilateral formed by connecting the midpoints of the sides of a rectangle is a rhombus.

Let *MNPO* be a rectangle. Choose convenient axes and coordinates, as shown in the figure at the right. Let *T, W, V,* and *U* be the midpoints of the sides of the rectangle.

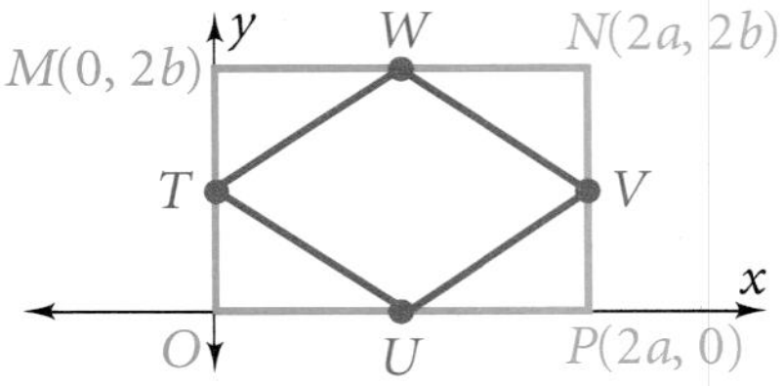

Use the Midpoint Formula to find the coordinates of the midpoints:

$T(0, b)$ $\quad W(a, 2b)$ $\quad V(2a, b)$ $\quad U(a, 0)$

From Lesson 9-5, you know that *TWVU* is a parallelogram. To prove that *TWVU* is a rhombus, show that consecutive sides are congruent.

$TW = \sqrt{(a - 0)^2 + (2b - b)^2} = \sqrt{a^2 + b^2}$

$WV = \sqrt{(2a - a)^2 + (b - 2b)^2} = \sqrt{a^2 + b^2}$

$TW = WV$, so parallelogram *TWVU* is a rhombus.

4. **Critical Thinking** In the Example, explain why it is enough to prove just two consecutive sides congruent in order to prove that parallelogram *TWVU* is a rhombus.

5. **Try This** In the Example, use another method to prove that parallelogram *TWVU* is a rhombus.

1. Writing The midpoints of $\overline{OR}$ and $\overline{ST}$ are W and Z, respectively. Which of the following figures would you prefer to use to find the coordinates of the midpoint of $\overline{WZ}$? Explain your choice.

A.
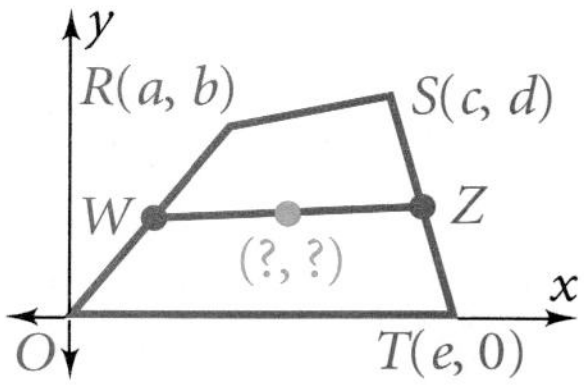

B.
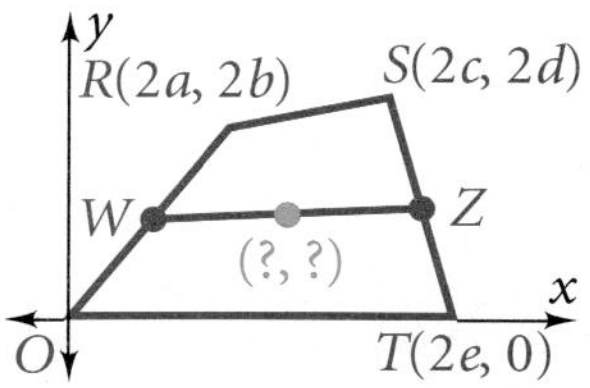

C.
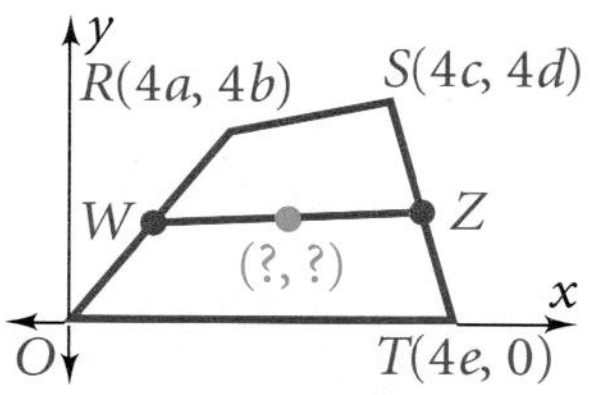

Complete the coordinates for each figure. Then use coordinate geometry to prove the statement.

2. The diagonals of a parallelogram bisect each other (Theorem 9-3).

Given: Parallelogram $ABCD$
Prove: $\overline{AC}$ bisects $\overline{BD}$, and $\overline{BD}$ bisects $\overline{AC}$.

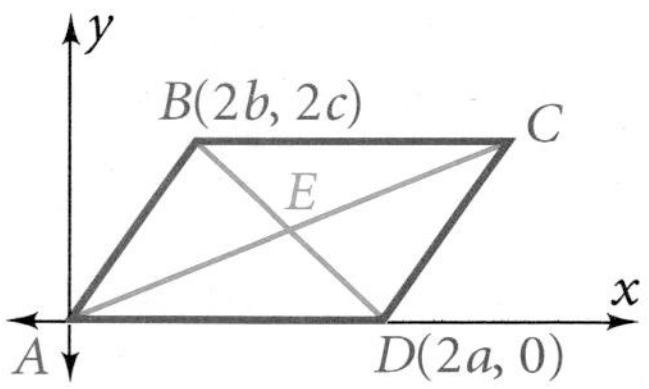

3. The diagonals of an isosceles trapezoid are congruent (Theorem 9-17).

Given: Trapezoid $EFGH$, $\overline{FE} \cong \overline{GH}$
Prove: $\overline{EG} \cong \overline{HF}$

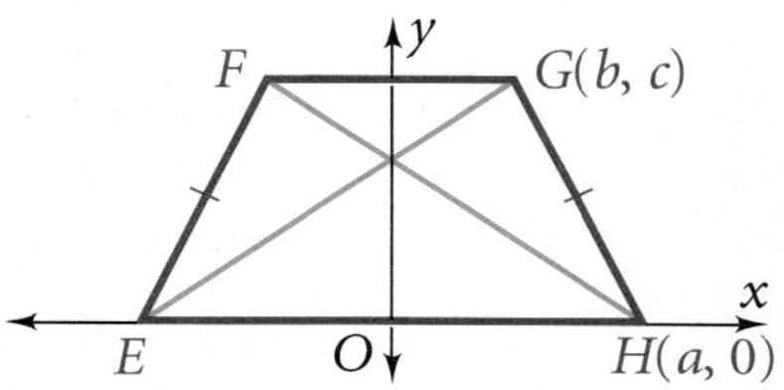

4. The midpoint of the hypotenuse of a right triangle is equidistant from the vertices.

Given: $\angle MON$ is a right angle, P is the midpoint of $\overline{MN}$.
Prove: $MP = PN = OP$

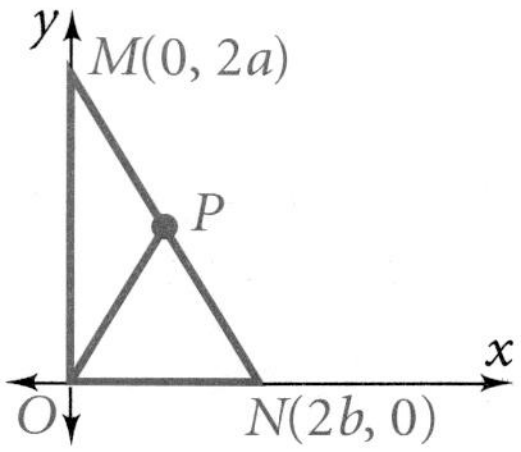

5. The segments joining the midpoints of consecutive sides of an isosceles trapezoid form a rhombus.

Given: Trapezoid $TRAP$ with $\overline{TR} \cong \overline{AP}$, D, E, F, and G are midpoints of the indicated sides.
Prove: $DEFG$ is a rhombus.

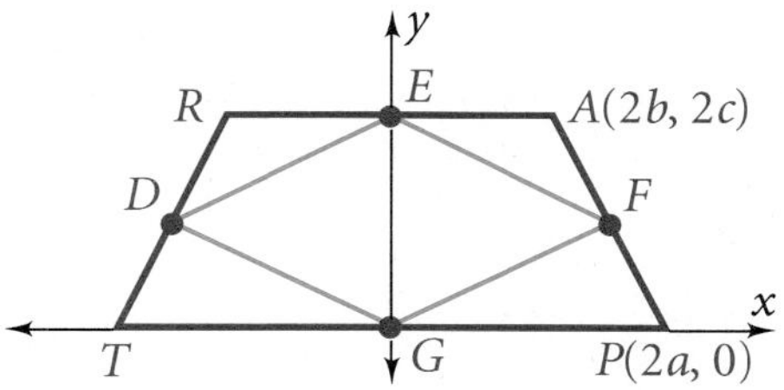

6. Compare your coordinate proof of Theorem 9-3 in Exercise 2 with the two-column proof on page 450. Which proof do you think is easier to understand? Explain.

7. Use coordinate geometry to prove that the midpoints of the sides of a kite determine a rectangle.

Given: Kite *DEFG*, $DE = EF$, $DG = GF$, *K*, *L*, *M*, and *N* are midpoints of the indicated sides.
Prove: *KLMN* is a rectangle.

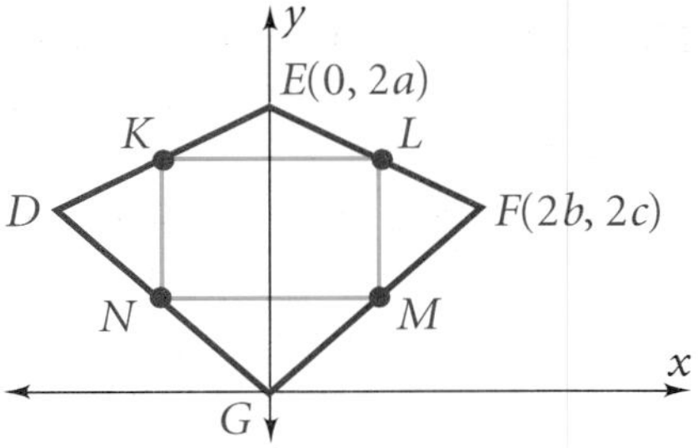

8. Open-ended Give an example of a statement that you think is easier to prove with coordinate geometry than with a two-column proof. Explain your choice.

Alexander Calder

Alexander Calder (1898–1976) was an American sculptor who is considered the founder of modern kinetic art. Kinetic art, which is art that involves motion, is perhaps best seen in Calder's mobiles. His mobiles consist of a series of carefully balanced metal shapes hanging from wires. Calder's early mobiles are motorized. In his later ones, however, the shapes float gracefully and move in different directions with currents of air.

Calder was inspired to make mobiles by watching circus performers. His initial attempts at kinetic art were motorized circus figures made of wire and wood.

9. Research For a mobile to be in balance, the artist must suspend the mobile from its center of gravity. The center of gravity, or *centroid*, is the point around which the weight of an object is evenly distributed. Choose an object from Calder's mobile. Research how you would find the center of gravity of the object.

10. The centroid of a triangle is the point where the medians meet. The centroid is $\frac{2}{3}$ of the distance from each vertex to the midpoint of the opposite side. Complete the following steps to find the centroid of the triangle at the right.

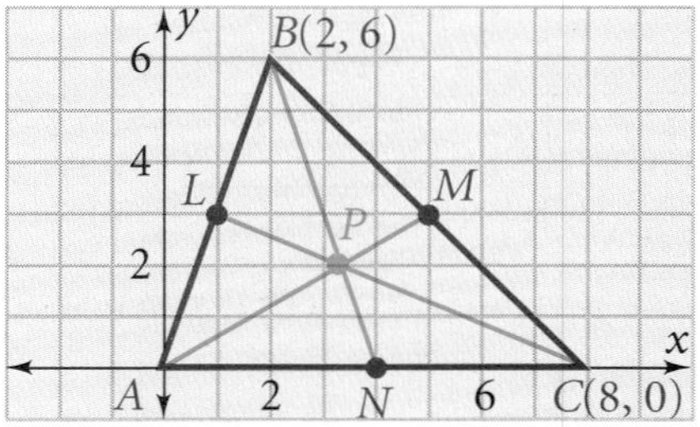

a. Find the coordinates of points *L*, *M*, and *N*, the midpoints of the sides of the triangle.
b. Find the equations of lines $\overleftrightarrow{AM}$, $\overleftrightarrow{BN}$, and $\overleftrightarrow{CL}$.
c. Find the coordinates of point *P*, the of intersection of lines $\overleftrightarrow{AM}$ and $\overleftrightarrow{BN}$.
d. **Verify** that point *P* is the intersection of lines $\overleftrightarrow{AM}$ and $\overleftrightarrow{CL}$.
e. Use the distance formula to **verify** that point *P* is $\frac{2}{3}$ of the distance from each vertex to the midpoint of the opposite side.

11. Given $\triangle ABC$ with altitudes p, q, and r, complete the following to show that p, q, and r intersect in a point. (This point is called the *orthocenter* of the triangle.)

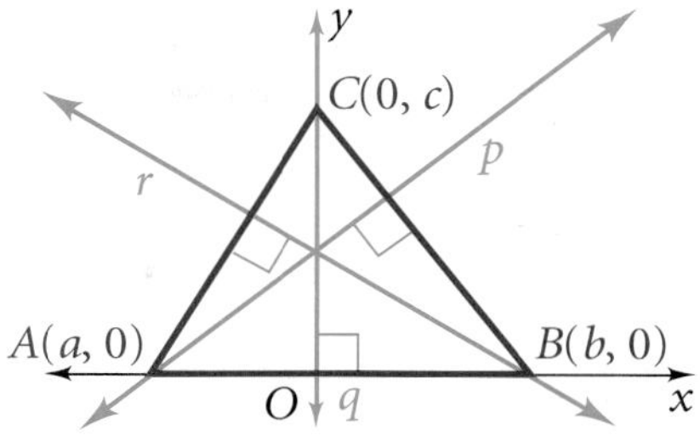

a. The slope of $\overline{BC}$ is $\frac{c}{-b}$. What is the slope of line p?
b. Show that the equation of line p is $y = \frac{b}{c}(x - a)$.
c. What is the equation of line q?
d. Show that lines p and q intersect at $(0, \frac{-ab}{c})$.
e. The slope of $\overline{AC}$ is $\frac{c}{-a}$. What is the slope of line r?
f. Show that the equation of line r is $y = \frac{a}{c}(x - b)$.
g. Show that lines r and q intersect at $(0, \frac{-ab}{c})$.
h. Give the coordinates of the orthocenter of $\triangle ABC$.

12. *RSTW* is a rhombus. Give the coordinates of *S* and *T*. In part (c) give the coordinates in terms of *a* and *c*, without introducing any new variables.

a.

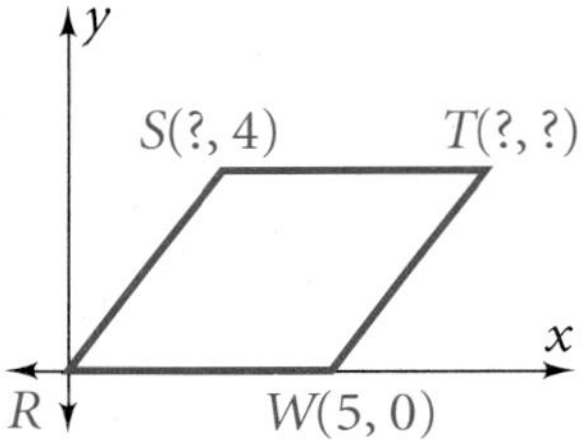

b.

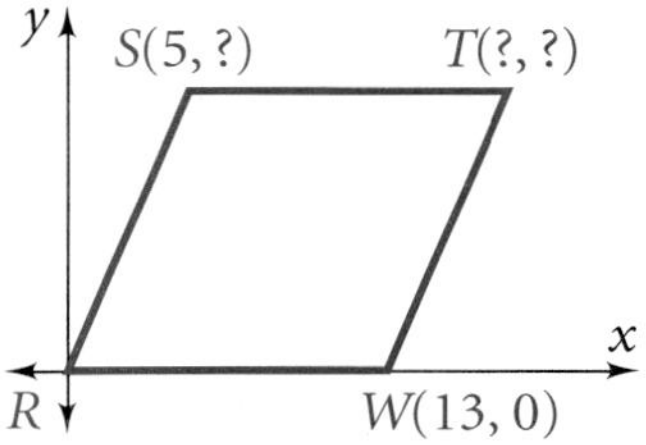

c.

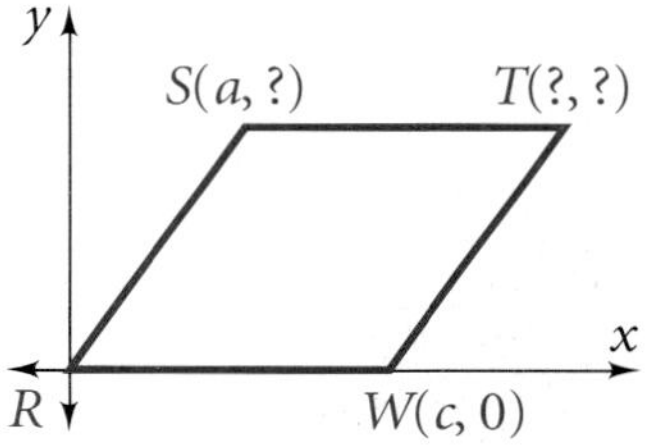

d. Use your figure in part (c) for a coordinate proof that the diagonals of a rhombus are perpendicular.

Exercises MIXED REVIEW

1995 Cellular Phone Buyers

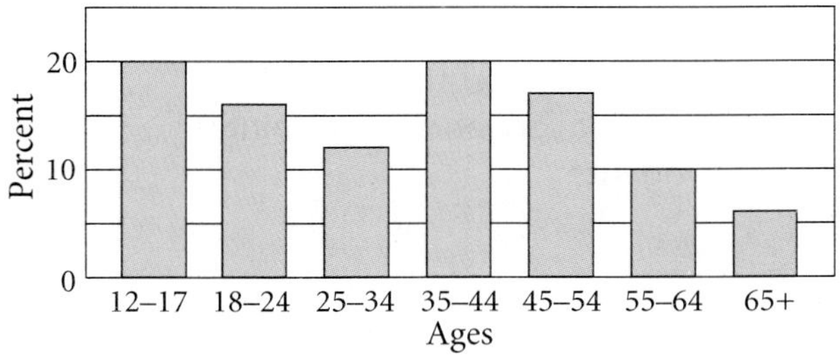

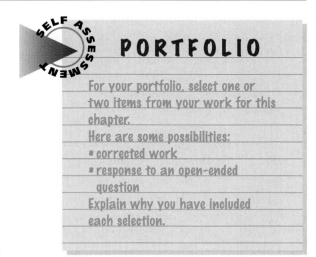

SELF ASSESSMENT

PORTFOLIO

For your portfolio, select one or two items from your work for this chapter.
Here are some possibilities:
- corrected work
- response to an open-ended question

Explain why you have included each selection.

13. **Data Analysis** What percent of cellular phone buyers were younger than 25?

14. **Writing** Describe two ways to find what percent of cellular phone buyers were younger than 65. Use both methods. Are the answers the same? Explain.

15. **Critical Thinking** Explain why the total of all cellular phone buyers is greater than 100%.

CHAPTER PROJECT

Finishing the Chapter Project

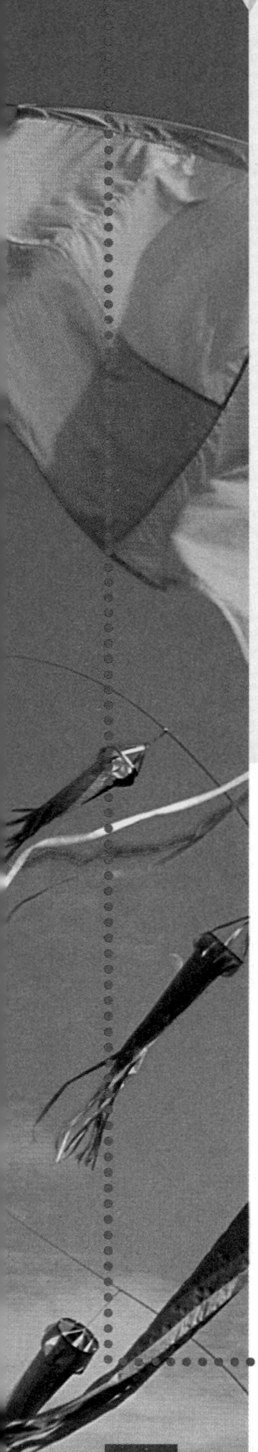

GO FLY A KITE

Find Out activities on pages 453, 468, and 476 should help you complete your project. Design and construct a kite and then fly it. Include complete plans for building the kite, specifying the size, shape, and material for each piece. Use geometric terms in your plans. Experiment with different designs and models. Describe how your model performs and what you would do to improve it.

Reflect and Revise

Ask a classmate to review your project with you. Together, check that the diagrams and explanations are clear, complete, and accurate. Have you tried several designs and kept a record of what you learned from each? Could your kite be stronger, more efficient, or more pleasing to the eye? Revise your work as needed.

Follow Up

Research the history of kites. Try to find examples of each use mentioned on page 447. You might make an illustrated time line to show special events like Benjamin Franklin's discovery about lightning and electricity and Samuel Cody's man-lifting kite.

For More Information

Baker, Rhoda, and Miles Denyer. *Flying Kites*. Edison, New Jersey: Chartwell Books, 1995.

Eden, Maxwell. *Kiteworks: Explorations in Kite Building & Flying*. New York: Sterling Publishing Co., 1989.

Kremer, Ron. *From Crystals to Kites*. Palo Alto, California: Dale Seymour Publications, 1995.

Morgan, Paul and Helene. *The Ultimate Kite Book*. New York: Simon & Schuster, 1992.

CHAPTER 9 Wrap Up

Key Terms

base angles of a trapezoid (p. 470)
consecutive angles (p. 449)
midsegment of a trapezoid (p. 483)

How am I doing?

SELF ASSESSMENT

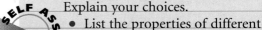

- State three ideas from this chapter that you think are important. Explain your choices.
- List the properties of different quadrilaterals.

Properties of Parallelograms — 9-1

Opposite sides and opposite angles of a parallelogram are congruent. The diagonals of a parallelogram bisect each other.

If three or more parallel lines cut off congruent segments on one transversal, then they cut off congruent segments on every transversal.

Find the measures of the numbered angles for each parallelogram.

1.

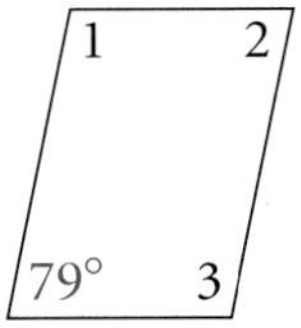

2.

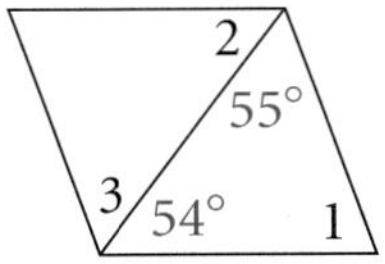

3.

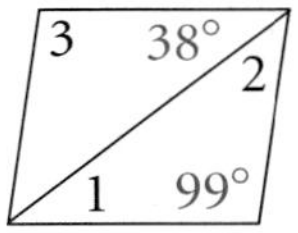

4.

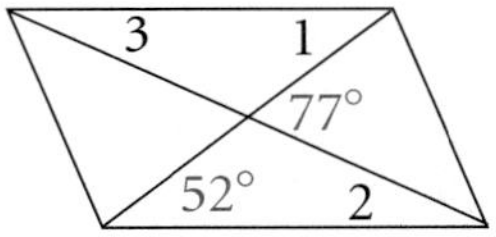

Proving That a Quadrilateral Is a Parallelogram — 9-2

A quadrilateral is a parallelogram if any of the following are true.

- The diagonals of the quadrilateral bisect each other.
- One pair of opposite sides of the quadrilateral are both congruent and parallel.
- Both pairs of opposite sides of the quadrilateral are congruent.
- Both pairs of opposite angles of the quadrilateral are congruent.

5. *Standardized Test Prep* Which quadrilaterals must be parallelograms?

I.

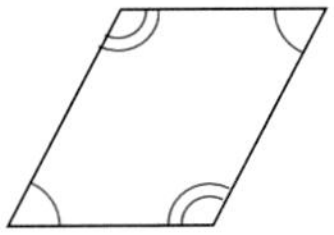

II.

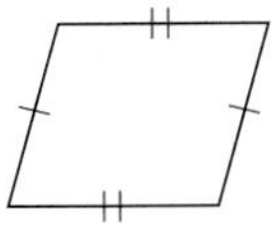

III.

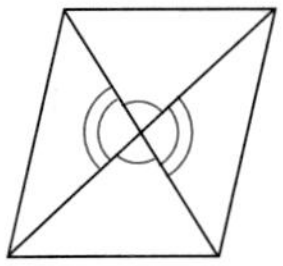

IV.

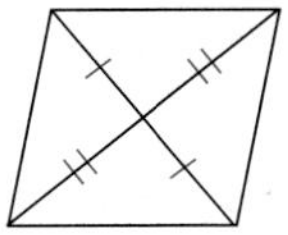

A. I only **B.** III only **C.** I and II **D.** II and IV **E.** I, II, and IV

Algebra **Determine the values of the variables for which *ABCD* is a parallelogram.**

6. 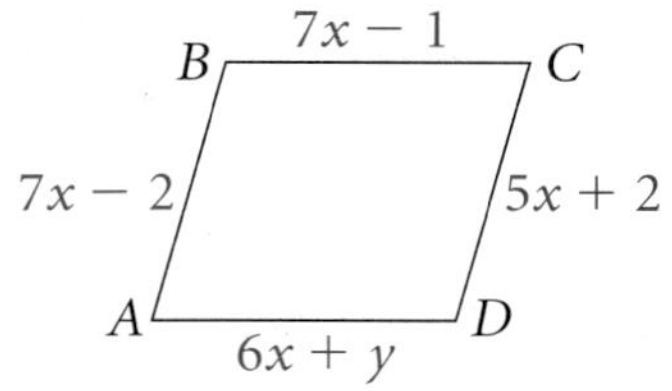

7. 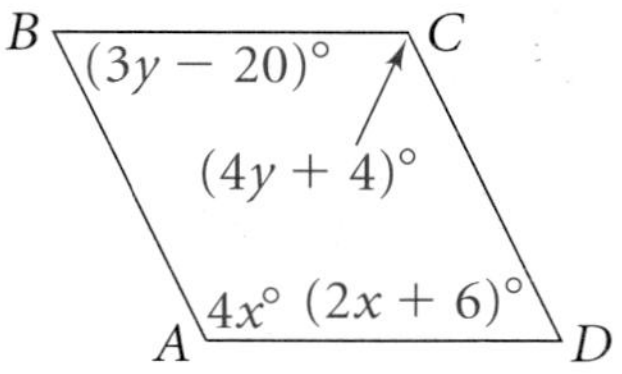

8. 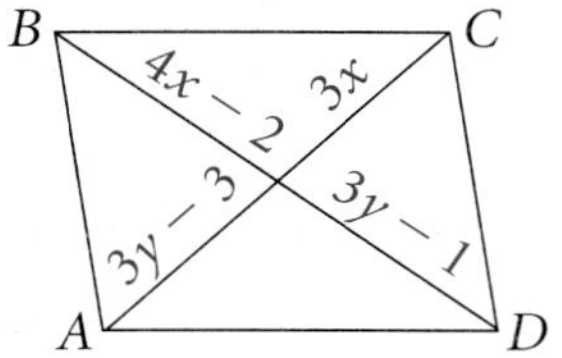

Properties of Special Parallelograms, Trapezoids, and Kites 9-3, 9-4

The diagonals of a rhombus are perpendicular. Each diagonal of a rhombus bisects two angles of the rhombus. The area of a rhombus is half the product of the lengths of its diagonals, $A = \frac{1}{2}d_1d_2$.

The diagonals of a rectangle are congruent.

If one diagonal of a parallelogram bisects two angles of the parallelogram, it is a rhombus. If the diagonals of a parallelogram are perpendicular, it is a rhombus. If the diagonals of a parallelogram are congruent, it is a rectangle.

The nonparallel sides of a trapezoid are the legs. Each pair of angles adjacent to a base of a trapezoid are **base angles** of the trapezoid.

Base angles of an isosceles trapezoid are congruent. The diagonals of an isosceles trapezoid are congruent.

The diagonals of a kite are perpendicular.

Find the measures of the numbered angles for each quadrilateral.

9. 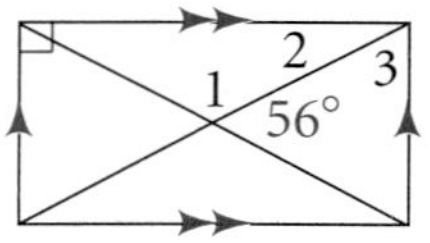

10. 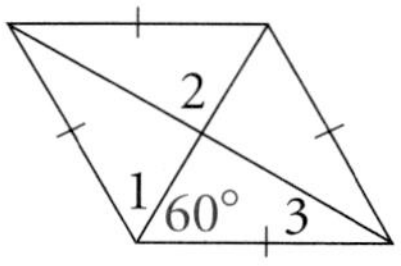

11. 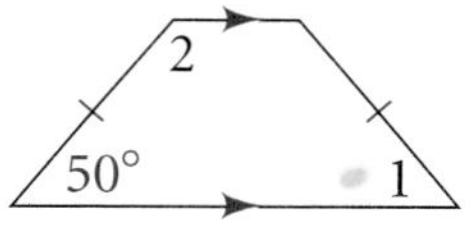

12. 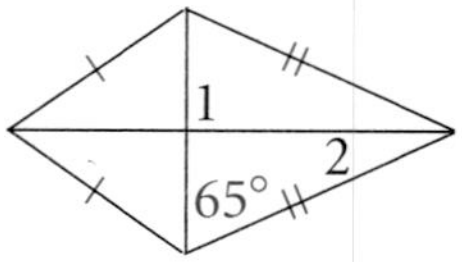

Find *AC* for each quadrilateral.

13. 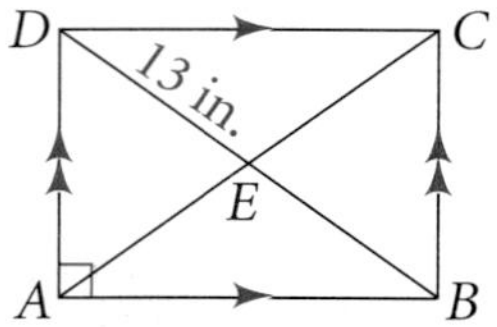

14. 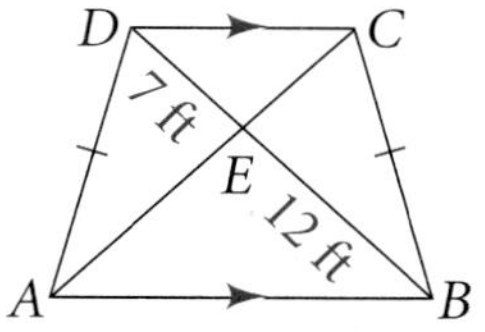

15. 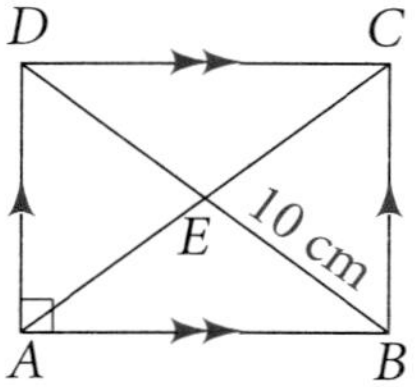

16. A rhombus has sides that are 10 ft long and a diagonal that is 16 ft long. Find its area.

17. **Writing** Explain how you could use a piece of lined paper to draw four lines on a blank index card that divide it into five equal sections.

Organizing Coordinate Proofs 9-5

In coordinate proofs, it is usually easiest to use the origin as the center for the figure or to place a vertex at the origin and at least one side of the figure on an axis.

Give the coordinates of point *P* without using any new variables.

18. rectangle

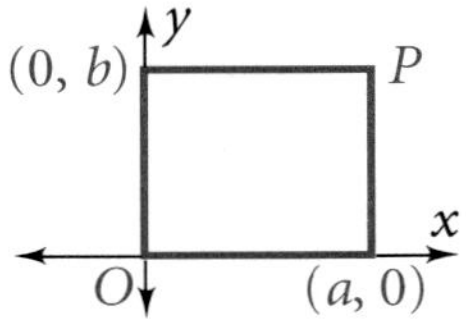

19. rhombus

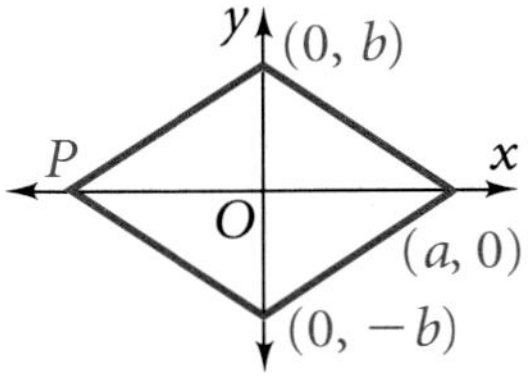

20. square

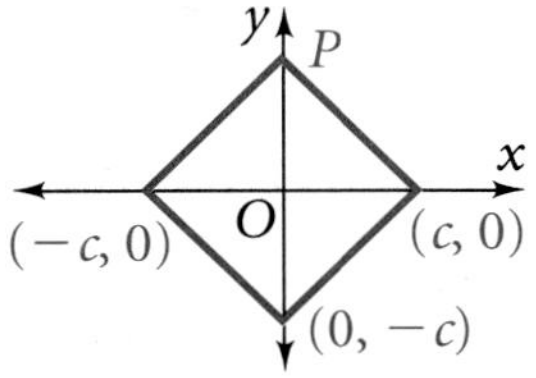

21. parallelogram

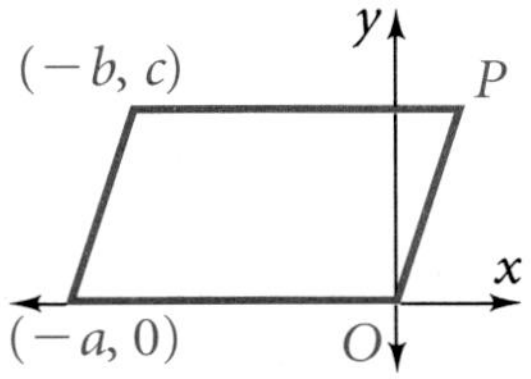

Using Coordinate Geometry in Proofs 9-6

The segment that joins the midpoints of the nonparallel sides of a trapezoid is the **midsegment of the trapezoid.** It is parallel to the bases and half as long as the sum of the lengths of the bases.

The formulas for slope, midpoint, and distance are used in coordinate proofs.

Choose coordinates for each figure. Then use coordinate geometry to prove the statement.

22. The diagonals of a rectangle are congruent.
Given: Rectangle *ABCD*
Prove: $\overline{AC} \cong \overline{BD}$

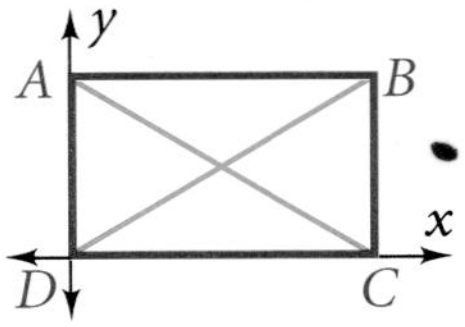

23. The diagonals of a square are perpendicular.
Given: Square *FGHI*
Prove: $\overline{FH} \perp \overline{GI}$

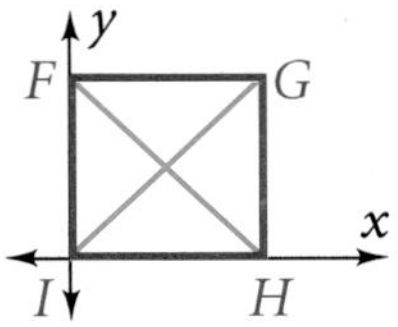

24. *Open-ended* Show two different ways to draw an isosceles triangle for a coordinate proof.

Getting Ready for CHAPTER 10

Write a fraction equivalent to the given fraction.

25. $\frac{4}{5}$ **26.** $\frac{9}{4}$ **27.** $\frac{6}{8}$ **28.** $\frac{1}{10}$ **29.** $\frac{4}{7}$ **30.** $\frac{3}{8}$

Solve each proportion.

31. $\frac{3}{4} = \frac{x}{8}$ **32.** $\frac{2}{x} = \frac{8}{24}$ **33.** $\frac{x}{9} = \frac{1}{3}$ **34.** $\frac{10}{25} = \frac{2}{x}$

Assessment

Find *AN* in each parallelogram.

1.

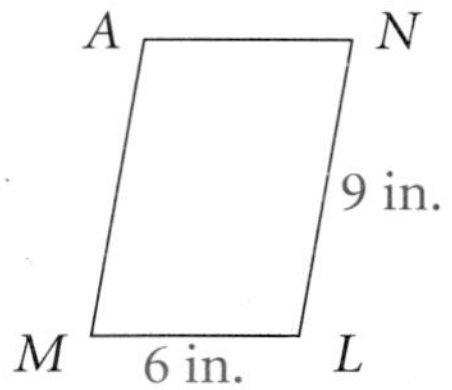

2.

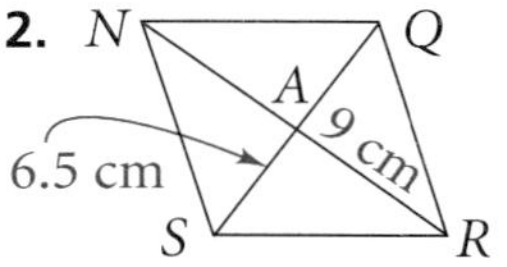

3. *Open-ended* Sketch two noncongruent parallelograms *ABCD* and *EFGH* such that $\overline{AC} \cong \overline{BD} \cong \overline{EG} \cong \overline{FH}$.

Algebra **Find the values of the variables for each parallelogram.**

4.

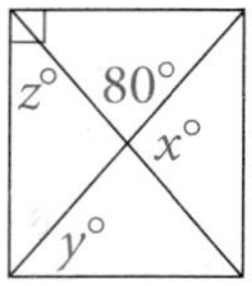

5.

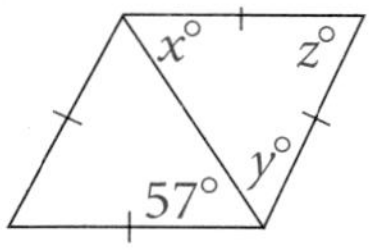

6.

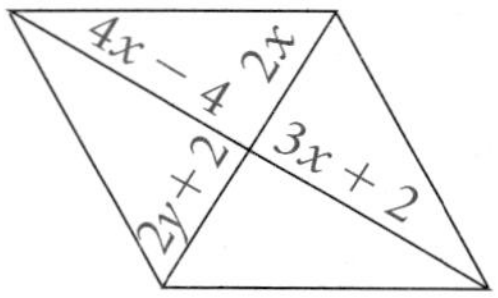

7.

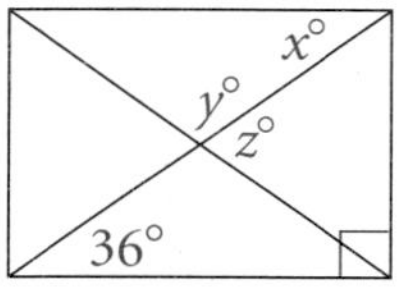

8. *Standardized Test Prep* What is sufficient to prove that *ABCD* is a parallelogram?

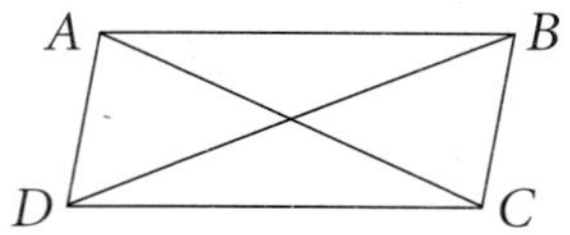

I. $\overline{AC}$ bisects $\overline{BD}$.
II. $\overline{AB} \cong \overline{DC}$; $\overline{AB} \parallel \overline{DC}$
III. $\overline{AB} \cong \overline{DC}$; $\overline{BC} \cong \overline{AD}$
IV. $\angle DAB \cong \angle BCD$ and $\angle ABC \cong \angle CDA$

A. I and III　**B.** II and IV　**C.** II and III
D. I and IV　**E.** II, III, and IV

9. *Writing* Explain why a square cannot be a kite.

Find the measures of ∠1 and ∠2.

10.

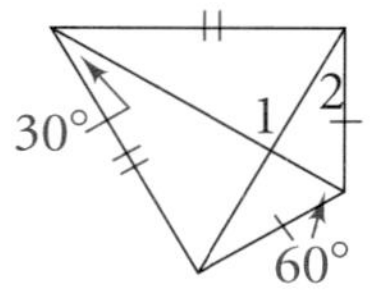

11.

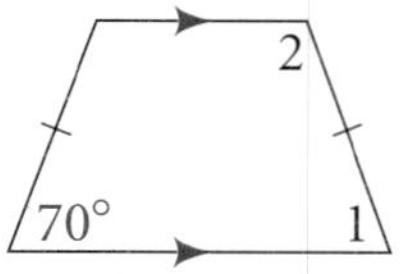

12. *Transformations* Draw $\overline{AB}$ with midpoint *M*. Rotate $\overline{AB}$ $x°$ about *M*, where $0 < x < 180$. What type of quadrilateral is $AA'BB'$? Explain.

Find the area of each rhombus. If the answer is not an integer, round to the nearest tenth.

13.

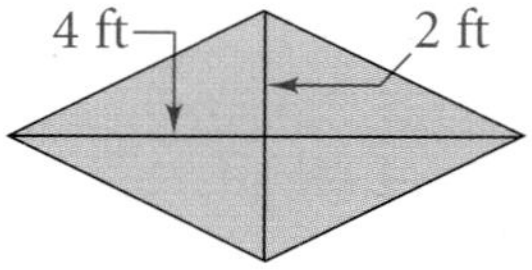

14.

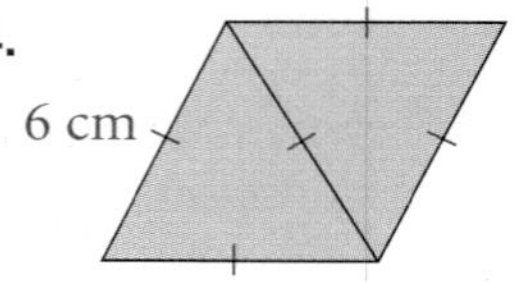

Give the coordinates for points *S* and *T* without using any new variables. Then find the coordinates of the midpoint of $\overline{ST}$.

15.

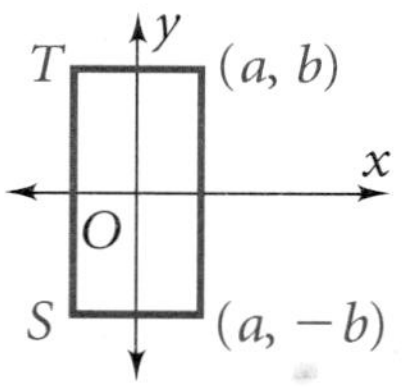

16.

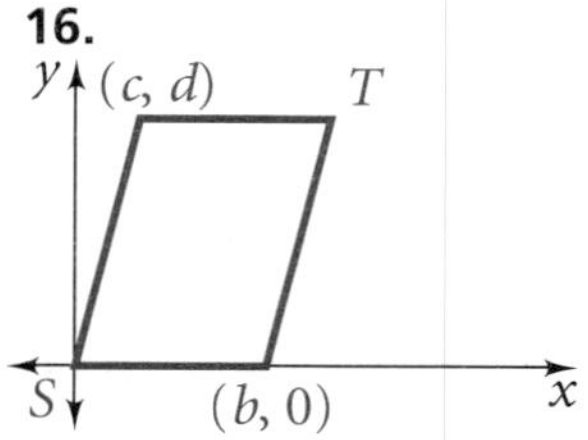

17. Use coordinate geometry to prove that the diagonals of a square are congruent.

Given: *ABCD* is a square with vertices $A(0, 0)$, $B(a, 0)$, $C(a, a)$, and $D(0, a)$.
Prove: $AC = BD$

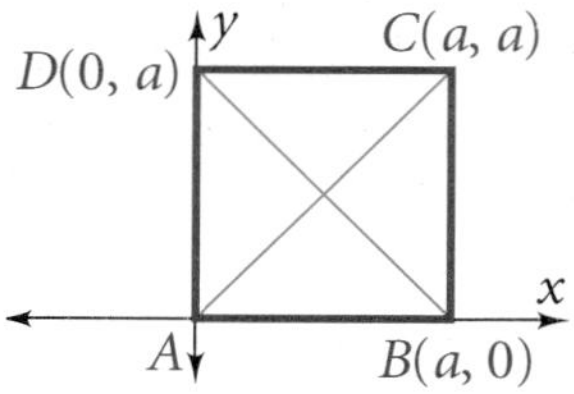

CHAPTER 9 Preparing for Standardized Tests

For Exercises 1–8, choose the correct letter.

1. The diagonals of which quadrilateral are congruent?
   A. trapezoid **B.** rhombus
   C. parallelogram **D.** kite
   E. rectangle

2. For which value of x are lines g and h parallel?

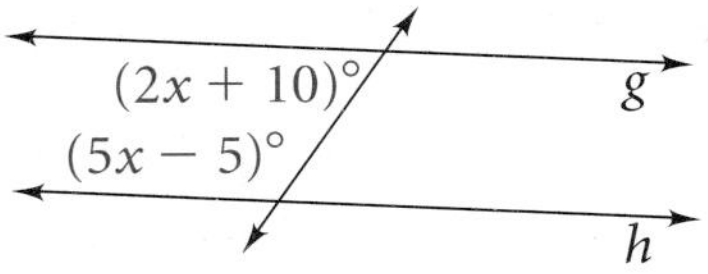

   A. 5 **B.** 12 **C.** 18 **D.** 25
   E. none of the above

3. What is the surface area of a sphere with radius 6 in.?
   A. 12π in.^{2} **B.** 36π in.^{2} **C.** 144π in.^{2}
   D. 216π in.^{2} **E.** 288π in.^{2}

4. Which trapezoid has an 8-in. midsegment?

   A.
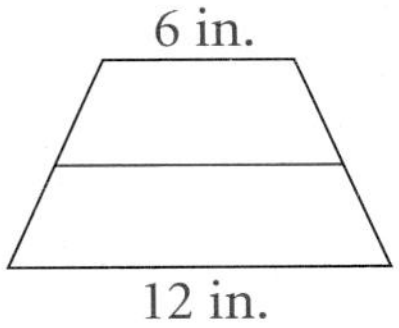

   B.
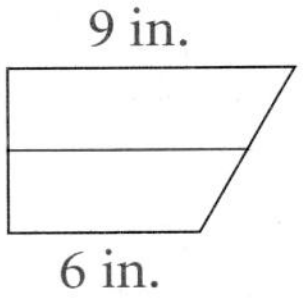

   C.
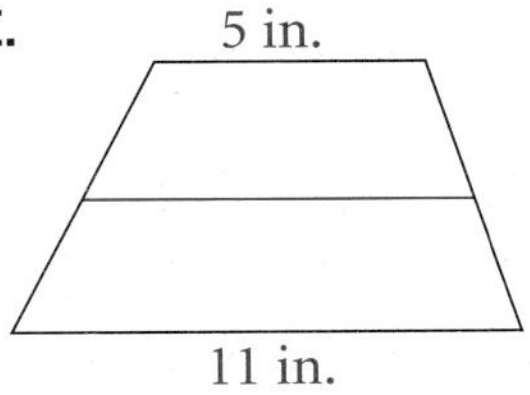

   D.
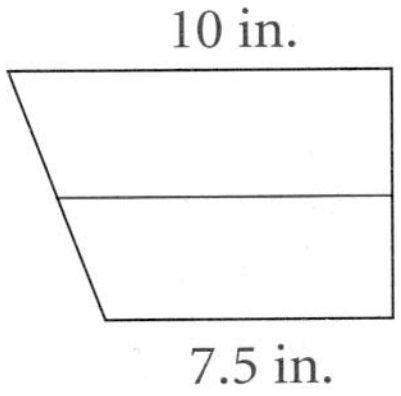

   E. none of the above

5. *Transformations* $\triangle ABC$ has vertices $A(3, 1)$, $B(4, -2)$, and $C(4, 7)$. Find the vertices of the image of $\triangle ABC$ under a translation 4 units left and 1 unit down.
   A. $A'(7, 2)$, $B'(8, -1)$, $C'(8, 8)$
   B. $A'(-1, 0)$, $B'(0, -3)$, $C'(0, 6)$
   C. $A'(7, 0)$, $B'(8, -3)$, $C'(8, 6)$
   D. $A'(-1, 2)$, $B'(0, -1)$, $C'(0, 8)$
   E. $A'(7, 0)$, $B'(8, -1)$, $C'(0, 6)$

6. How can you prove the triangles congruent?

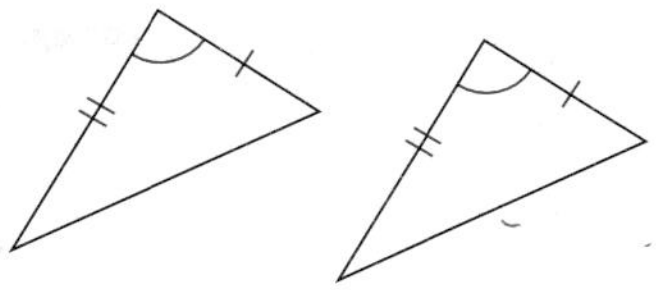

   A. ASA **B.** SSS **C.** SAS **D.** HL
   E. The triangles are not congruent.

Compare the boxed quantity in Column A with the boxed quantity in Column B. Choose the best answer.

A. The quantity in Column A is greater.
B. The quantity in Column B is greater.
C. The two quantities are equal.
D. The relationship cannot be determined on the basis of the information supplied.

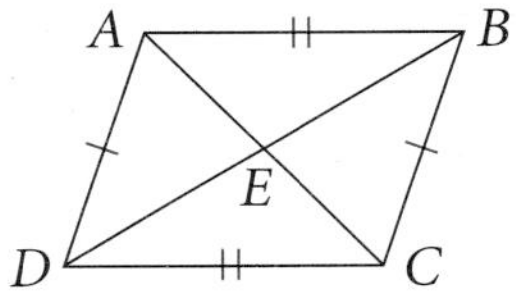

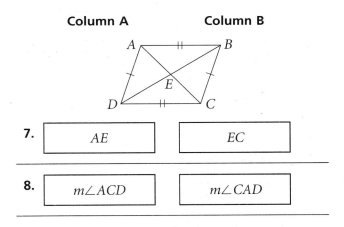

	Column A	Column B
7.	AE	EC
8.	$m\angle ACD$	$m\angle CAD$

Find each answer.

9. What is the area of the rhombus?

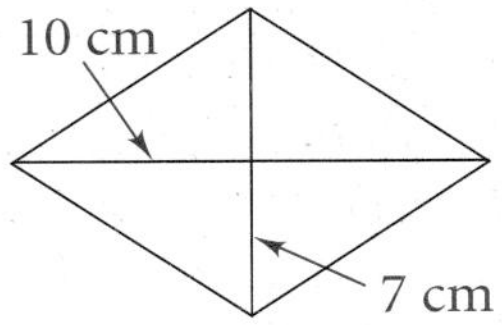

10. *Constructions* Draw a segment with length a. Construct a rectangle with width a and length $2a$.

11. *Open-ended* Give three numbers that could be the lengths of the sides of an acute triangle.

CHAPTER

10 Similarity

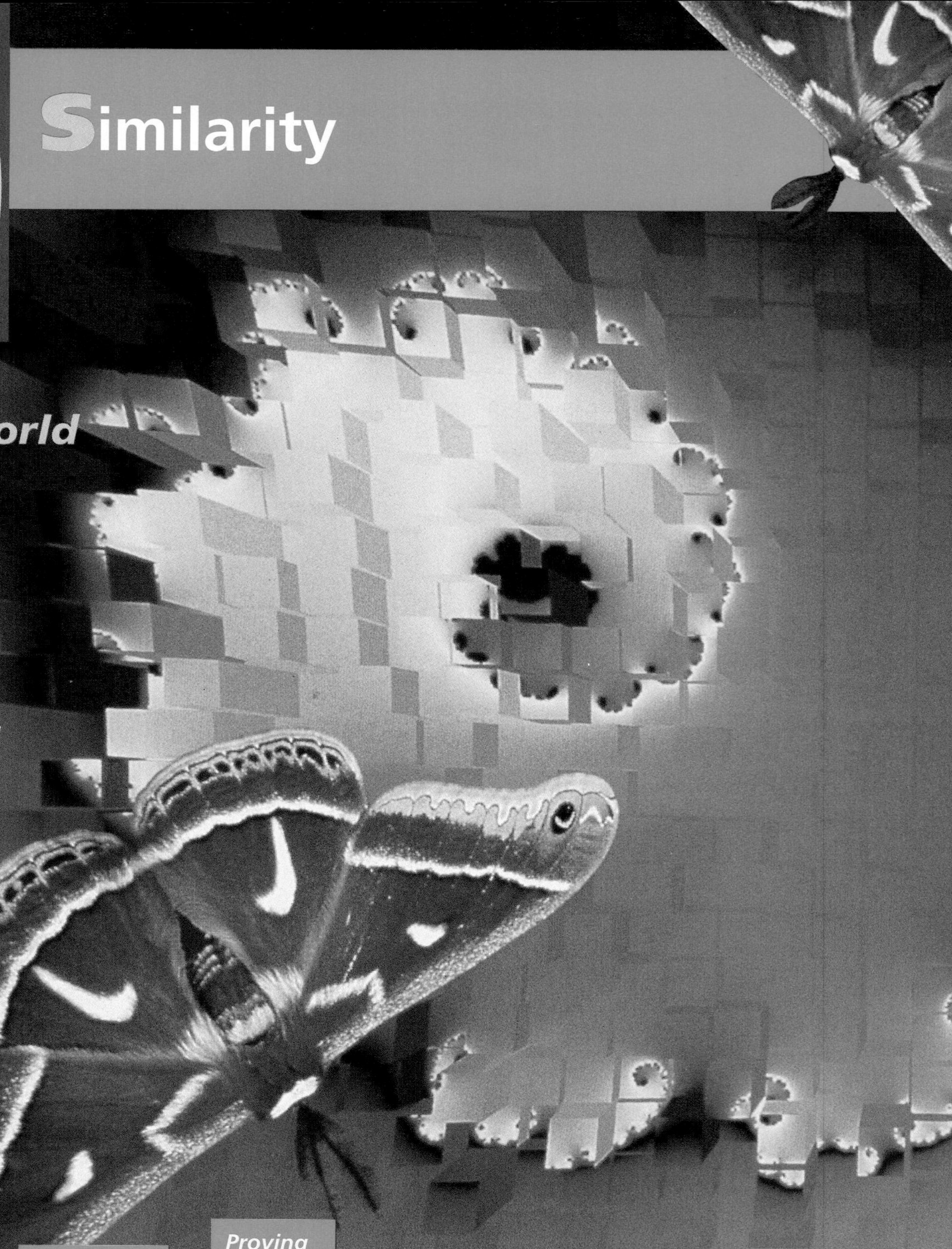

Relating to the Real World

This chapter involves the geometry of size changes. You often see size changes in everyday life, including those in photographs, computer drawing programs, scale models, and photocopy machines. You also see them in the swirling designs of fractals—an exciting and relatively new area of mathematics with many practical applications.

Lessons

- 10-1 *Ratio, Proportion, and Similarity*
- 10-2 *Proving Triangles Similar: AA, SAS, and SSS*
- 10-3 *Similarity in Right Triangles*
- 10-4 *Proportions and Similar Triangles*
- 10-5 *Perimeters and Areas of Similar Figures*

CHAPTER PROJECT

Fractals FOREVER

Nature is full of shapes that are not straight lines, smooth curves, or flat surfaces. Just look at a cloud bank or the bark on a tree. Many shapes contain patterns that repeat themselves on different scales—a head of cauliflower, a fern frond, and details of a coastline, to name but a few. Fractal geometry, developed in the last 20 years, is the study of these irregular, *self-similar* shapes, called *fractals.*

In your project for this chapter, you will create fractals. You will learn how to do an iterative process—one in which steps are repeated in a regular cycle. Finally, you will investigate properties of fractals, including some surprising facts about length and area.

To help you complete the project:

- **p. 502** *Find Out by Doing*
- **p. 516** *Find Out by Analyzing*
- **p. 523** *Find Out by Doing*
- **p. 529** *Find Out by Thinking*
- **p. 536** *Finishing the Project*

Areas and Volumes of Similar Solids

10-6

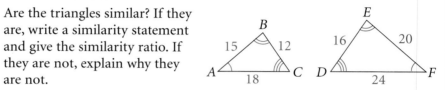

What You'll Learn

- Finding how to use ratio and proportion with similar polygons

...And Why

To use similarity in scale drawings, art, and architecture

What You'll Need

- ruler
- calculator
- graph paper
- compass
- straightedge

10-1 Ratio, Proportion, and Similarity

THINK AND DISCUSS

Ratio and Proportion in Similar Figures

In Chapter 2 you learned that two polygons are *similar* if (1) corresponding angles are congruent **and** (2) corresponding sides are proportional. The ratio of the lengths of corresponding sides is the *similarity ratio.* You can write the ratio of a to b as $a : b$ or as $\frac{a}{b}$.

Example 1

Are the triangles similar? If they are, write a similarity statement and give the similarity ratio. If they are not, explain why they are not.

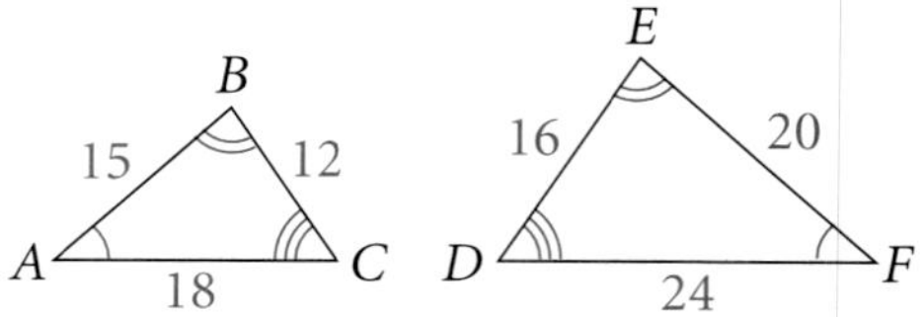

Three pairs of angles are congruent. To compare the ratios of corresponding side lengths, express the ratios in simplest form.

$$\frac{AC}{FD} = \frac{18}{24} = \frac{3}{4} \qquad \frac{AB}{FE} = \frac{15}{20} = \frac{3}{4} \qquad \frac{BC}{ED} = \frac{12}{16} = \frac{3}{4}$$

$\triangle ABC \sim \triangle FED$ with a similarity ratio $\frac{3}{4}$, or 3 : 4.

1. What is the similarity ratio for the similarity $\triangle FED \sim \triangle ABC$?
2. **Try This** If the triangles at the left are similar, write a similarity statement and give the similarity ratio. If they are not, explain why they are not.

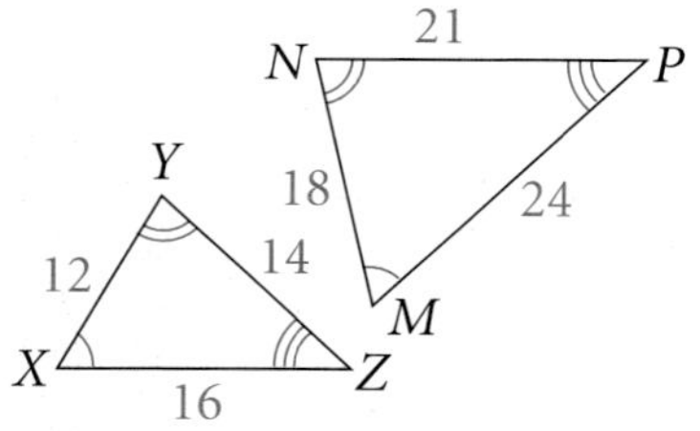

A **proportion** is a statement that two ratios are equal. You can read both

$$\frac{a}{b} = \frac{c}{d} \quad \text{and} \quad a : b = c : d$$

as "a is to b as c is to d." When three or more ratios are equal, you can write an **extended proportion.** For Example 1 you could write the following:

$$\frac{18}{24} = \frac{15}{20} = \frac{12}{16}$$

The following properties are helpful in solving proportions.

Properties of Proportions

$\frac{a}{b} = \frac{c}{d}$ is equivalent to

(1) $ad = bc$ (2) $\frac{b}{a} = \frac{d}{c}$

(3) $\frac{a}{c} = \frac{b}{d}$ (4) $\frac{a+b}{b} = \frac{c+d}{d}$

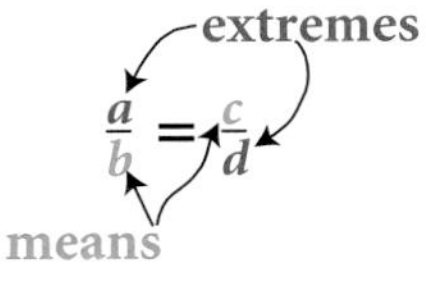

$ad = bc$

The first property is called the **Cross-Product Property.** It can be stated as "The product of the extremes is equal to the product of the means."

3. To eliminate the denominators in $\frac{a}{b} = \frac{c}{d}$, by what would you multiply both sides of the equation? What would be the result? Use your answers to **justify** the Cross-Product Property.

Who? As a 21-year-old architecture student, Maya Ying Lin won the design competition for the Vietnam War Memorial. She also designed the Civil Rights Memorial in Montgomery, Alabama.

Using Proportions

A similarity ratio always compares dimensions in the same unit, such as meters. In a **scale drawing,** the **scale** compares each length in the drawing to the actual length being represented. The scale in a scale drawing can be in different units. A scale might be written as 1 in. to 100 mi, 1 in. = 12 ft, or 1 mm : 1 m.

Example 2 **Relating to the Real World**

Architecture What are the actual dimensions of the bedroom?

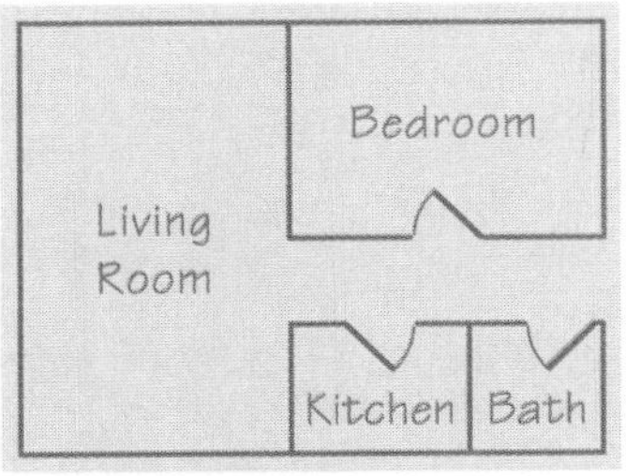

Scale 1 in. = 16 ft

Use a ruler to find that the bedroom is $\frac{7}{8}$ in. by $\frac{5}{8}$ in. on the scale drawing.

$$\frac{1}{16} = \frac{\frac{7}{8}}{x} \qquad \frac{1}{16} = \frac{\frac{5}{8}}{y} \qquad \begin{array}{l}\leftarrow \text{drawing length (in.)} \\ \leftarrow \text{actual length (ft)}\end{array}$$

$$x = 16\left(\tfrac{7}{8}\right) \qquad y = 16\left(\tfrac{5}{8}\right) \qquad \text{Cross-Product Property}$$

$$x = 14 \qquad y = 10$$

The actual bedroom is 14 ft by 10 ft.

4. a. Try This What are the dimensions of the actual living room?

b. What is the area of the actual living room?

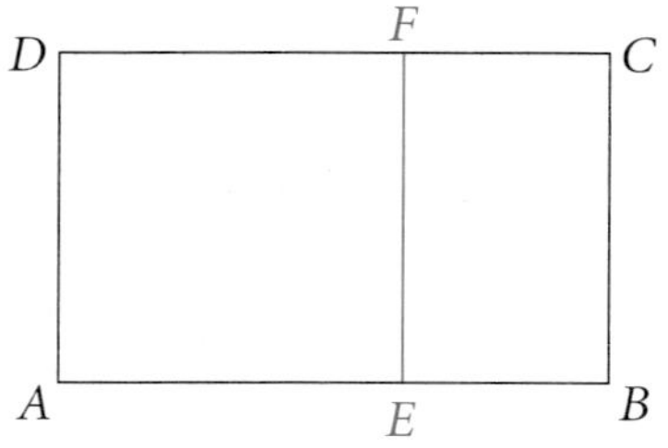

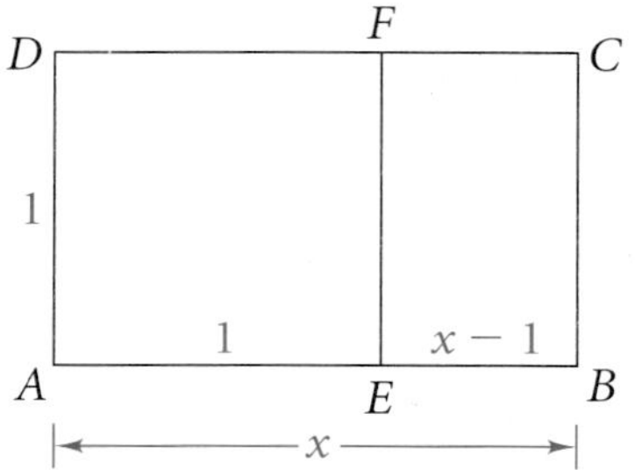

A **Golden Rectangle** is a rectangle that can be divided into a square and a rectangle that is similar to the original rectangle. The ratio of the length of a Golden Rectangle to the width is called the **Golden Ratio.** To find the Golden Ratio, you can find the length x of a Golden Rectangle whose width is 1.

$ABCD \sim BCFE$	Definition of Golden Rectangle
$\frac{AB}{BC} = \frac{BC}{CF}$	Corresponding sides of $\sim$ polygons are proportional.
$\frac{x}{1} = \frac{1}{x - 1}$	Substitution
$x^2 - x = 1$	Cross-Product Property
$x^2 - x - 1 = 0$	Subtract 1 from each side.

Using the Quadratic Formula to solve this equation, you get $x = \frac{1 \pm \sqrt{5}}{2}$.

5. Critical Thinking Why does the answer $x = \frac{1 - \sqrt{5}}{2}$ not apply to this problem?

6. a. Calculator Find the value of x to the nearest thousandth.
b. What is the length of a Golden Rectangle whose width is 3 cm?
c. What is the width of a Golden Rectangle whose length is 5 cm?

The Golden Rectangle is considered pleasing to the human eye and has appeared in architecture and art since ancient times. It has intrigued artists including Leonardo da Vinci (1452–1519), who illustrated *The Divine Proportion,* a book about the Golden Rectangle. The Golden Rectangle also occurs frequently in nature and in the proportions of the human body.

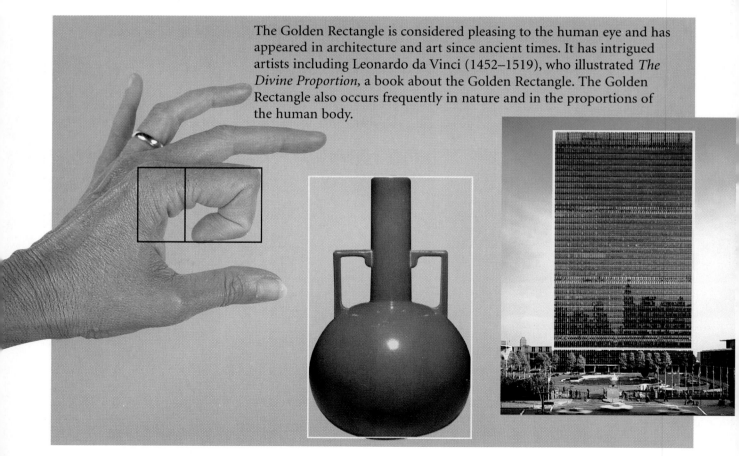

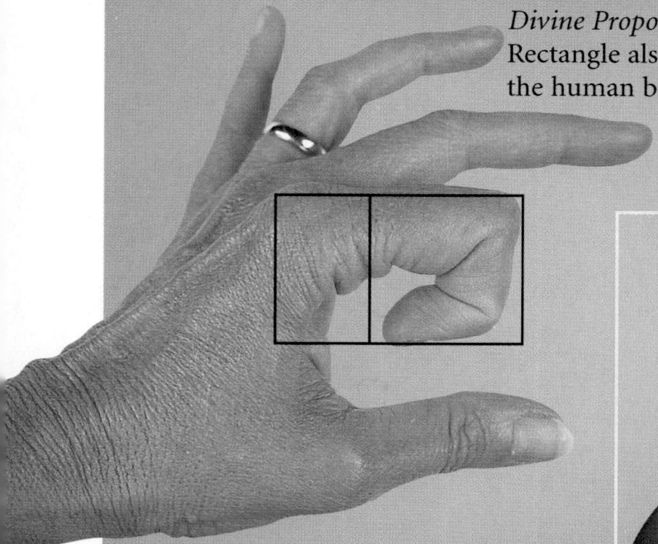

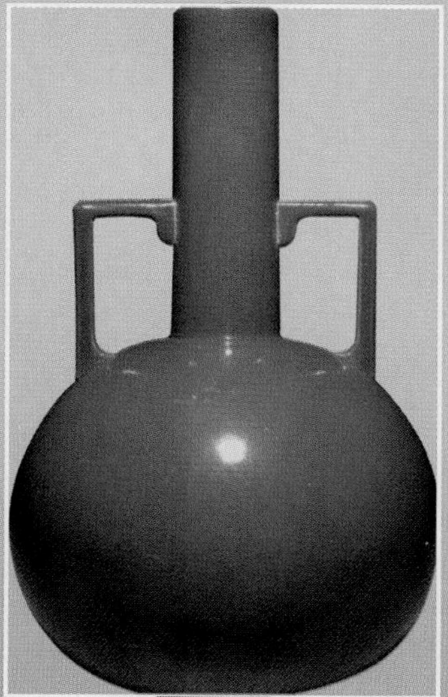

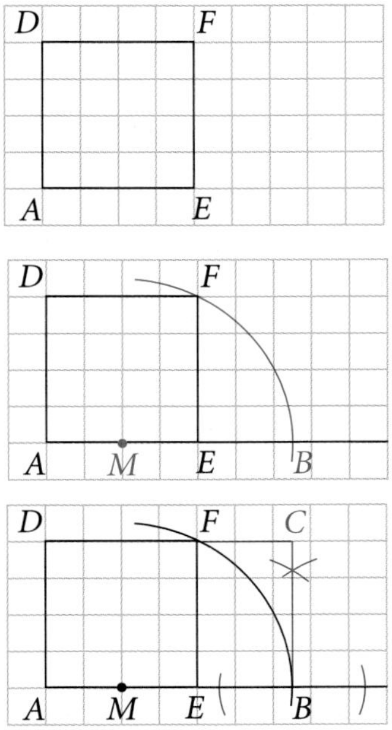

WORK TOGETHER

Work with a partner to draw a Golden Rectangle using graph paper, a compass, and a straightedge.

- On graph paper draw a square *AEFD* with an even number of units on each side.
- Find the midpoint *M* of $\overline{AE}$.
- With radius *MF* draw an arc that intersects $\overrightarrow{AE}$ at point *B*.
- Construct the perpendicular to $\overleftrightarrow{AB}$ at point *B*.
- Let point *C* be the intersection of this perpendicular and $\overrightarrow{DF}$.
- Draw the rectangle *ABCD*.

7. a. Use *ME*, *EF*, and the Pythagorean Theorem to find *MF*.
b. Use *MF* to find *EB*.
c. What are the dimensions of rectangle *ABCD?*
d. **Verify** that *ABCD* is a Golden Rectangle.

Exercises ON YOUR OWN

Algebra **If $\frac{a}{b} = \frac{3}{4}$, which of the following must be true?**

1. $4a = 3b$

2. $3a = 4b$

3. $\frac{a}{3} = \frac{b}{4}$

4. $\frac{4}{3} = \frac{b}{a}$

5. $\frac{4}{b} = \frac{3}{a}$

6. $ab = 3(4)$

7. $\frac{a+b}{b} = \frac{3+4}{4}$

8. $\frac{a+b}{a} = \frac{7}{3}$

9. $\frac{a+3}{3} = \frac{b+4}{4}$

10. $\frac{a}{b} = \frac{6}{8} = \frac{0.75}{1}$

11. Models The Leaning Tower of Pisa in Italy is about 180 ft tall. A souvenir paperweight of the Leaning Tower is 6 in. tall. What is the similarity ratio of the paperweight to the real tower?

12. Suppose the designer of the pyramid at the right wanted the base to be a 675-m square. The sides of the base of the pyramid that was built measure 0.675 m. What is the similarity ratio of the pyramid that was built to the one the designer had planned?

"We had a little problem with the decimal point."

Are the polygons similar? If they are, write a similarity statement and give the similarity ratio. If they are not, explain.

13.

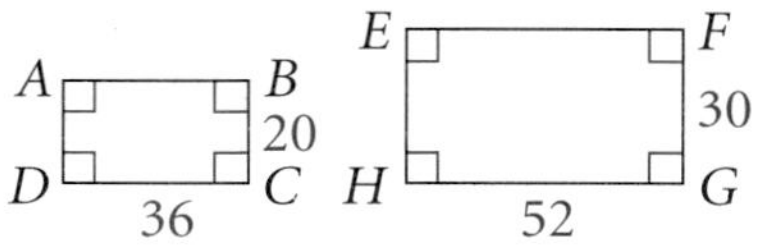

14.

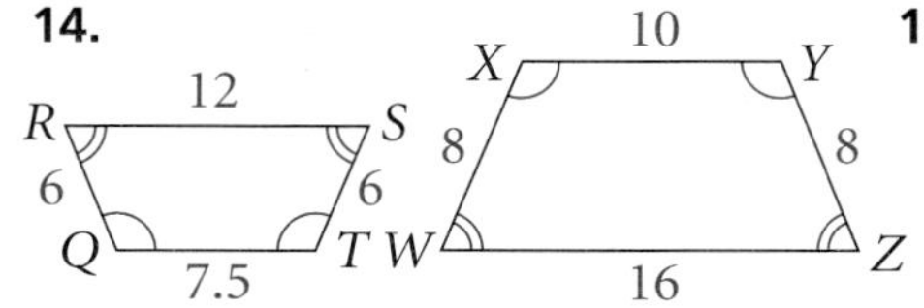

15.

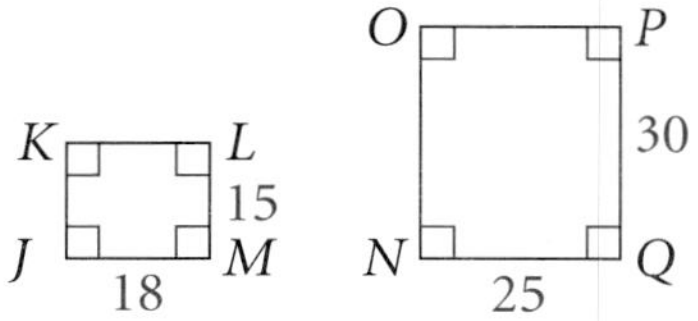

16.

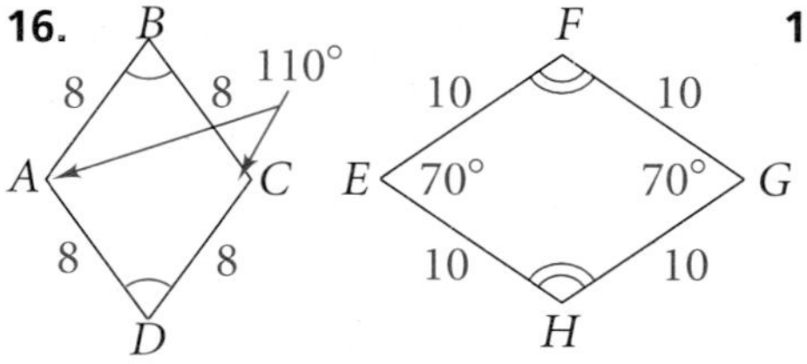

17.

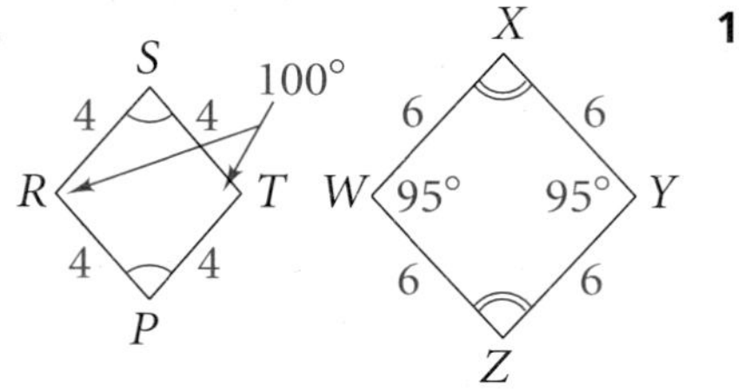

18.

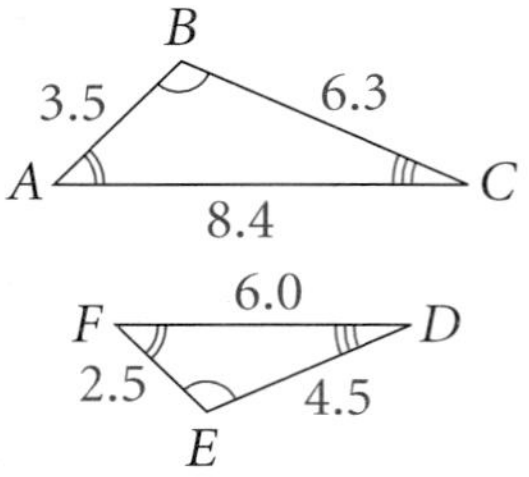

Algebra **Solve for *x*.**

19. $\frac{x}{10} = \frac{15}{25}$

20. $\frac{9}{24} = \frac{12}{x}$

21. $\frac{11}{14} = \frac{x}{21}$

22. $\frac{5}{x} = \frac{8}{11}$

23. $\frac{x + 3}{3} = \frac{10 + 4}{4}$

24. $\frac{x + 7}{7} = \frac{15}{5}$

25. $\frac{4}{x} = \frac{x}{9}$

26. $\frac{x}{2} = \frac{50}{x}$

27. **Mental Math** A postcard is 6 in. by 4 in. A commercial printing shop will enlarge it to any size up to 3 ft on the longer dimension. What are the dimensions of the largest possible enlargement for this postcard?

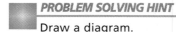

PROBLEM SOLVING HINT

Draw a diagram.

28. **Open-ended** Measure the dimensions of the fronts of several boxes in your home, such as cereal boxes and laundry detergent boxes. Which box is closest to the shape of a Golden Rectangle?

29. **Coordinate Geometry** $\triangle ABC$ with vertices $A(2, 3)$, $B(2, 6)$, and $C(4, 6)$ is similar to $\triangle QRS$ with vertices $Q(6, 9)$, $R(6, 24)$, and S. Give two possibilities for the coordinates of vertex S.

30. **Geography** A map of Louisiana is drawn to the scale 1 in. = 40 mi. On the map, the distance from Lake Charles to Baton Rouge is about $3\frac{1}{4}$ in. About how far apart are the two cities?

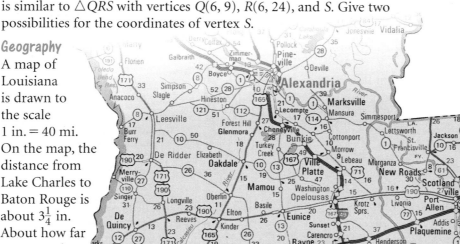

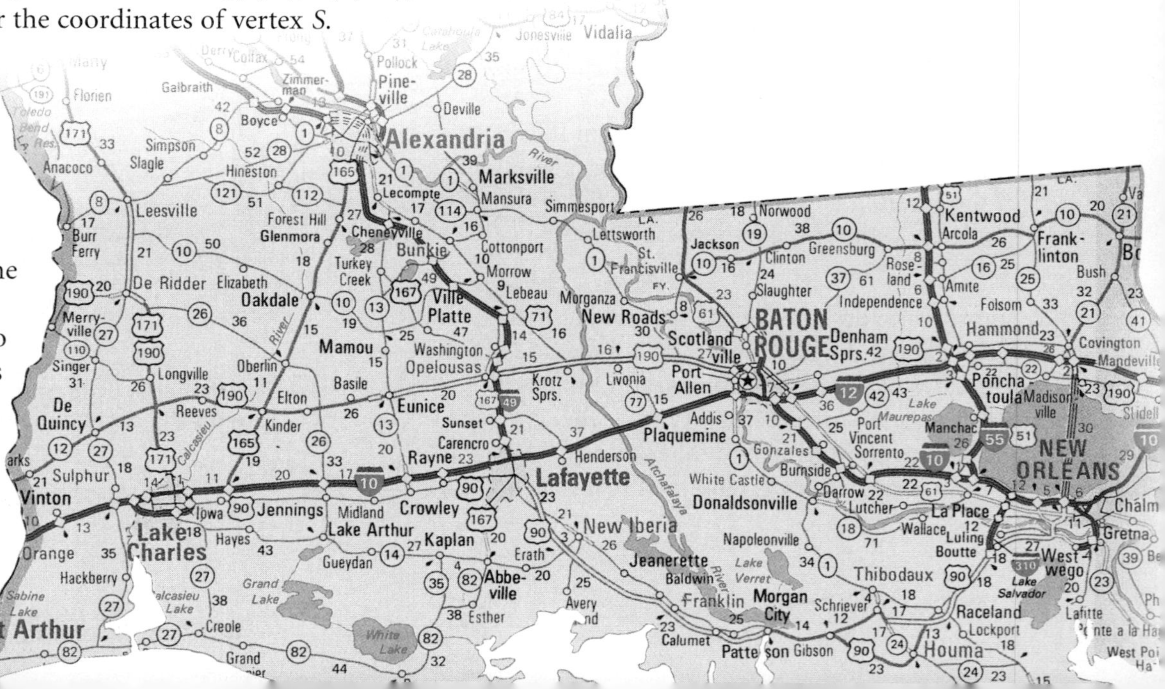

The polygons are similar. Find the values of the variables.

31.

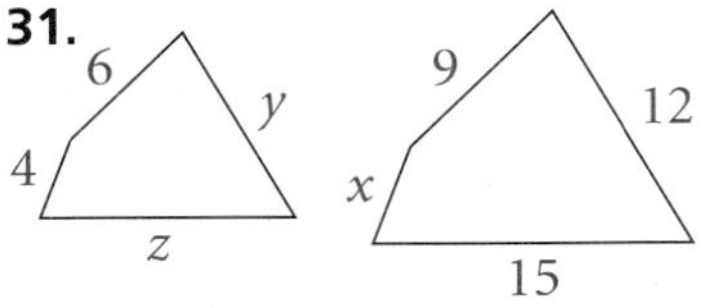

32.

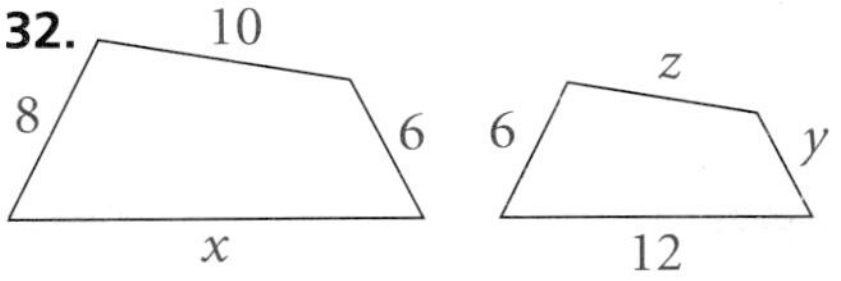

33.

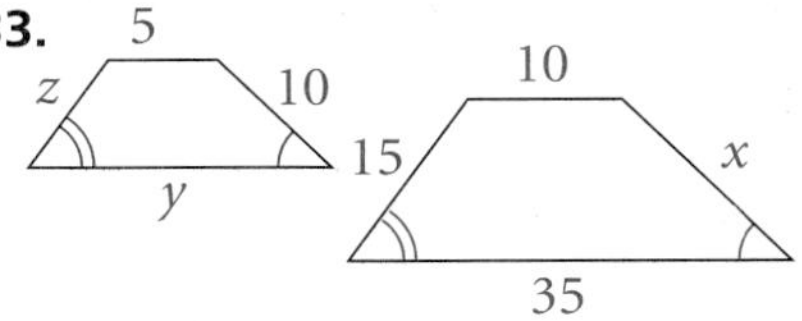

34. **Calculator** The switch plate for a standard electric light switch is in the shape of a Golden Rectangle. The longer side of a standard switch plate is about 114 mm. What is the length of the shorter side? Round your answer to the nearest millimeter.

35. **Money** From 1861 to 1928 the dollar bill measured $7\frac{7}{16}$ in. by $3\frac{1}{8}$ in. The dimensions of the current dollar bill are shown at the right. Are the two dollar bills similar rectangles? Explain.

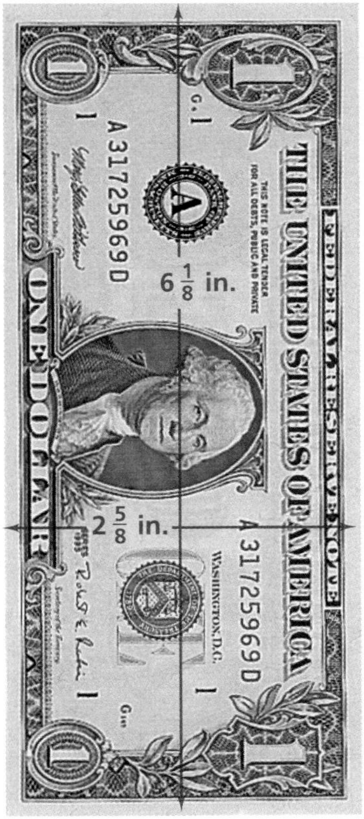

36. **Geography** Students on the campus of the University of Minnesota in Minneapolis built a model globe 42 ft in diameter on a scale of 1 : 1,000,000. About how tall would Mount Everest be on the model? (Mount Everest is about 29,000 ft tall.)

37. **Data Collection** On a single sheet of white paper, draw rectangles with the following dimensions: 3 in. by 3 in., 3 in. by 4 in., 3 in. by 5 in., and 3 in. by 6 in. Show the sheet to ten different people and ask them to select their first and second choices for most pleasing rectangles. Combine your findings with those of your classmates. Describe how your results do or do not confirm that the Golden Rectangle is the most pleasing rectangle.

38. a. **Sports** A basketball court is 84 ft by 50 ft. Choose a scale and draw a scale drawing of a basketball court.

b. **Writing** Explain how you chose the scale for your drawing.

39. a. **Transformations** Draw $JKLM$ with vertices $J(-2, -2)$, $K(6, -2)$, $L(6, 4)$, and $M(-2, 4)$. Then draw its image $J'K'L'M'$ under the dilation with center $O(0, 0)$ and scale factor $\frac{1}{2}$.

b. How are $JKLM$ and $J'K'L'M'$ related? Explain.

c. Is $J'K'L'M'$ an enlargement or reduction of $JKLM$?

40. **Standardized Test Prep** Which rectangles are similar to rectangle $ABCD$?

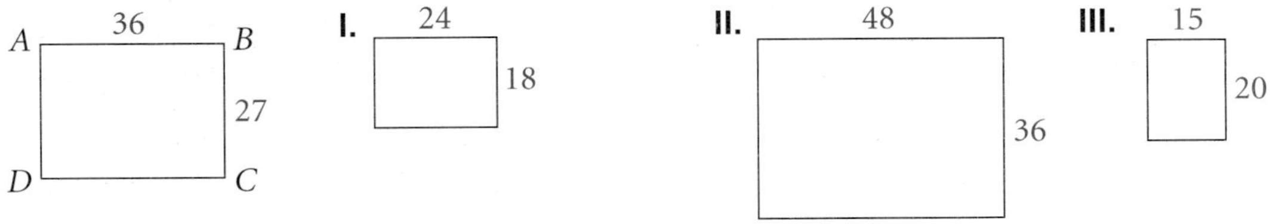

A. I only **B.** II only **C.** I and II **D.** I and III **E.** I, II, and III

#32

Chapter Project — Find Out by Doing

In 1904, Swedish mathematician Helge von Koch created a fractal "snowflake." Draw one by starting with an equilateral triangle (Stage 0).

- Divide each side into three congruent segments.
- Draw an equilateral triangle on the middle segment of each side.
- Erase the middle segments on which you drew the smaller triangles.

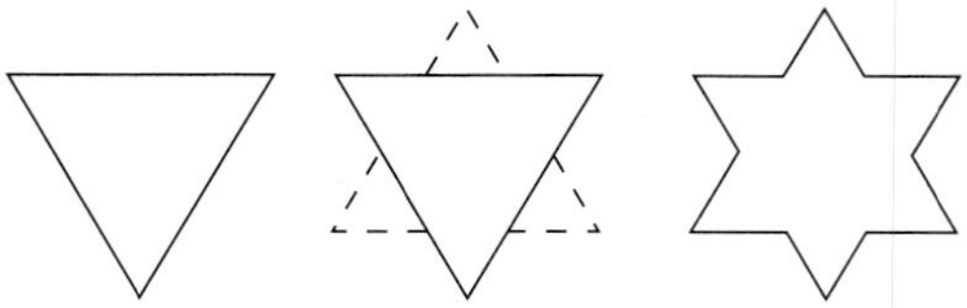

Stage 0 Stage 1

You have now drawn Stage 1. Repeat the steps to create Stage 2. (Divide all twelve sides of Stage 1 into three congruent segments.)

Exercises MIXED REVIEW

Sketch a quadrilateral with the given property.

41. congruent diagonals

42. perpendicular diagonals

43. only two parallel sides

44. An isosceles right triangle has a leg 9 in. long. How long is the hypotenuse?

Getting Ready for Lesson 10-2

What postulate or theorem can you use to prove the triangles congruent?

45.

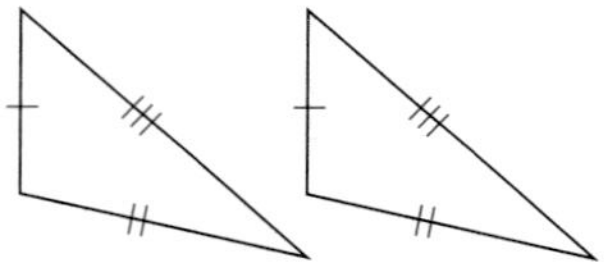

46.

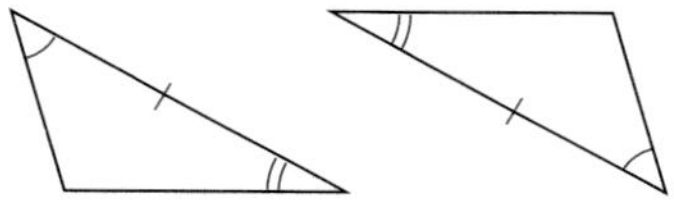

47.

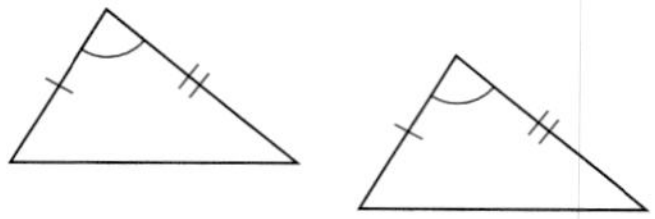

A Point in Time

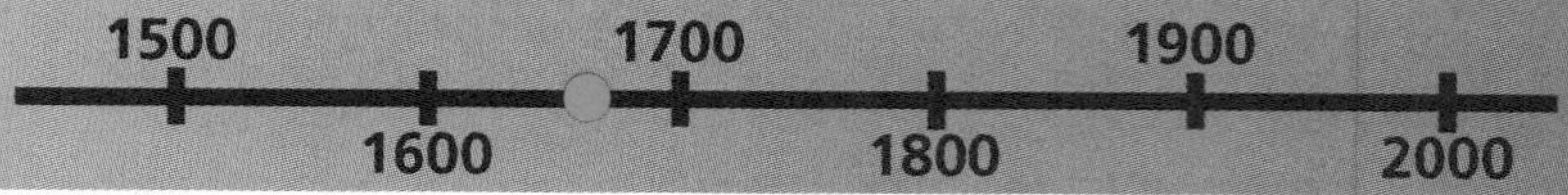

The Speed of Light

In **1675,** Danish astronomer Ole Römer used proportions to estimate the speed of light. He carefully measured the rotations of Jupiter's moons. With Earth at point *B,* a moon emerged from behind Jupiter 16.6 minutes later than when Earth was at point *A.* He reasoned that it must have taken 16.6 minutes for the light to travel from *A* to *B.* Using proportions, Römer estimated the speed of light is 150,000 mi/s, an estimate that compares favorably with today's accepted value of 186,282 mi/s.

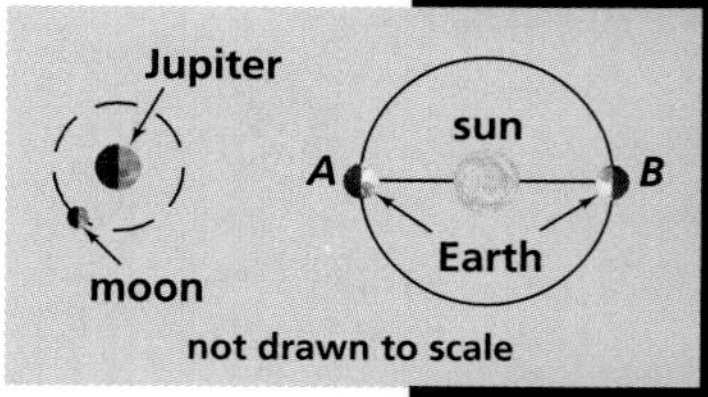

Algebra Review

Direct Variation

Before Lesson 10-2

The perimeter p of an equilateral triangle depends on its side length s. The table shows perimeters for various side lengths. An equation that describes this relationship is $p = 3s$. This type of relationship is called a direct variation.

s	1	2	5	10
p	3	6	15	30

A **direct variation** is a relationship that can be described by an equation in the form $y = kx$, where $k \neq 0$. In the equation, k is called the **constant of variation.** You say that "y varies directly as x."

Example

Retail Sales Juan worked 16 hours last week at a department store and earned \$92.00.

a. Assuming that his earnings e vary directly as the number of hours w he works, write an equation that relates these two variables.
b. How much will Juan earn if he works 12.5 hours in a week?

a.
$e = kw$ — Write a direct variation equation.
$92 = k(16)$ — Substitute for e and w.
$k = 5.75$ — Solve for k.

The equation $e = 5.75w$ describes this direct variation.

b.
$e = 5.75w$ — Use the equation from part (a).
$= 5.75(12.5)$ — Substitute 12.5 for w.
$= 71.875$ — Simplify.

Juan will earn \$71.88 for working 12.5 hours.

Solve each problem.

1. **Taxes** The sales tax t on a purchase varies directly as the cost c of the items purchased. Write an equation relating these two variables if the sales tax rate is 6.5%.
2. The circumference C of a circle varies directly as its diameter d. Write an equation that relates these two variables.
3. **Consumer Issues** Tandrell spends \$16.48 for 12.3 gallons of gasoline.
   a. Write an equation that relates the total amount t spent on gasoline to the number of gallons purchased n.
   b. How many gallons can Tandrell buy for \$20.00?
4. **Astronomy** An object that weighs 6 lb on Earth weighs 1 lb on the moon. If the weight w of an object on the moon varies directly as its weight on Earth e, write an equation relating these variables.

10-2 Proving Triangles Similar: AA, SAS, and SSS

What You'll Learn

- Proving two triangles similar using the AA ~ Postulate and the SAS ~ and SSS ~ Theorems
- Using similarity in indirect measurement to find distances

...And Why

To find a missing length

What You'll Need

metric ruler, protractor, calculator

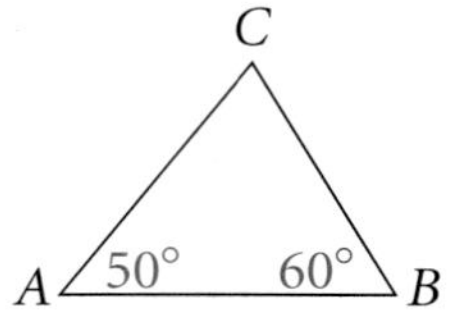

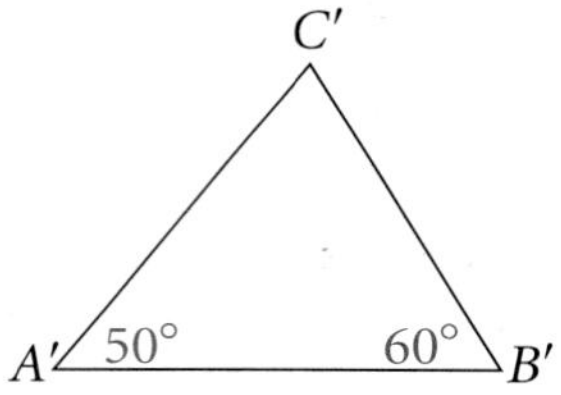

WORK TOGETHER

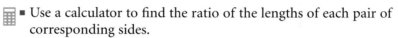

Work in groups.

- Draw two large triangles of different sizes, each with a 50° and a 60° angle. Label them as shown at the left below. Use a protractor to make sure that the angles are 50° and 60°.
- Measure the sides of each triangle to the nearest millimeter.
- Use a calculator to find the ratio of the lengths of each pair of corresponding sides.

1. What conclusion can you make about your triangles?

2. Compare your results with those of other groups. Complete this **conjecture:**

If two angles of one triangle are congruent to two angles of another triangle, then the triangles are __?__.

THINK AND DISCUSS

Using the AA Similarity Postulate

In Chapter 8, you learned ways to show that two triangles are congruent. In this lesson you will learn three ways to show that two triangles are similar without relying on the definition of similarity. In the Work Together you discovered the following theorem.

Postulate 10-1
Angle-Angle Similarity Postulate (AA ~ Postulate)

If two angles of one triangle are congruent to two angles of another triangle, then the triangles are similar.

$\triangle RST \sim \triangle LMP$

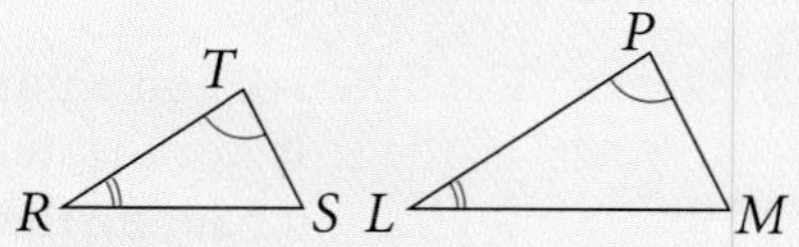

Example 1

Explain why the triangles are similar. Write a similarity statement.

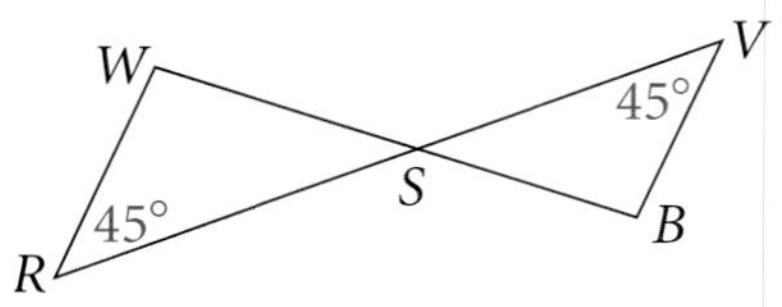

$\angle RSW \cong \angle VSB$ because vertical angles are congruent. $\angle R \cong \angle V$ because their measures are equal. $\triangle RSW \sim \triangle VSB$ by the Angle-Angle Similarity Postulate.

You can use similar triangles and measurements to compute distances that are difficult to measure directly. These methods are called **indirect measurement.** One method uses the fact that light reflects off a mirror at the same angle at which it hits the mirror.

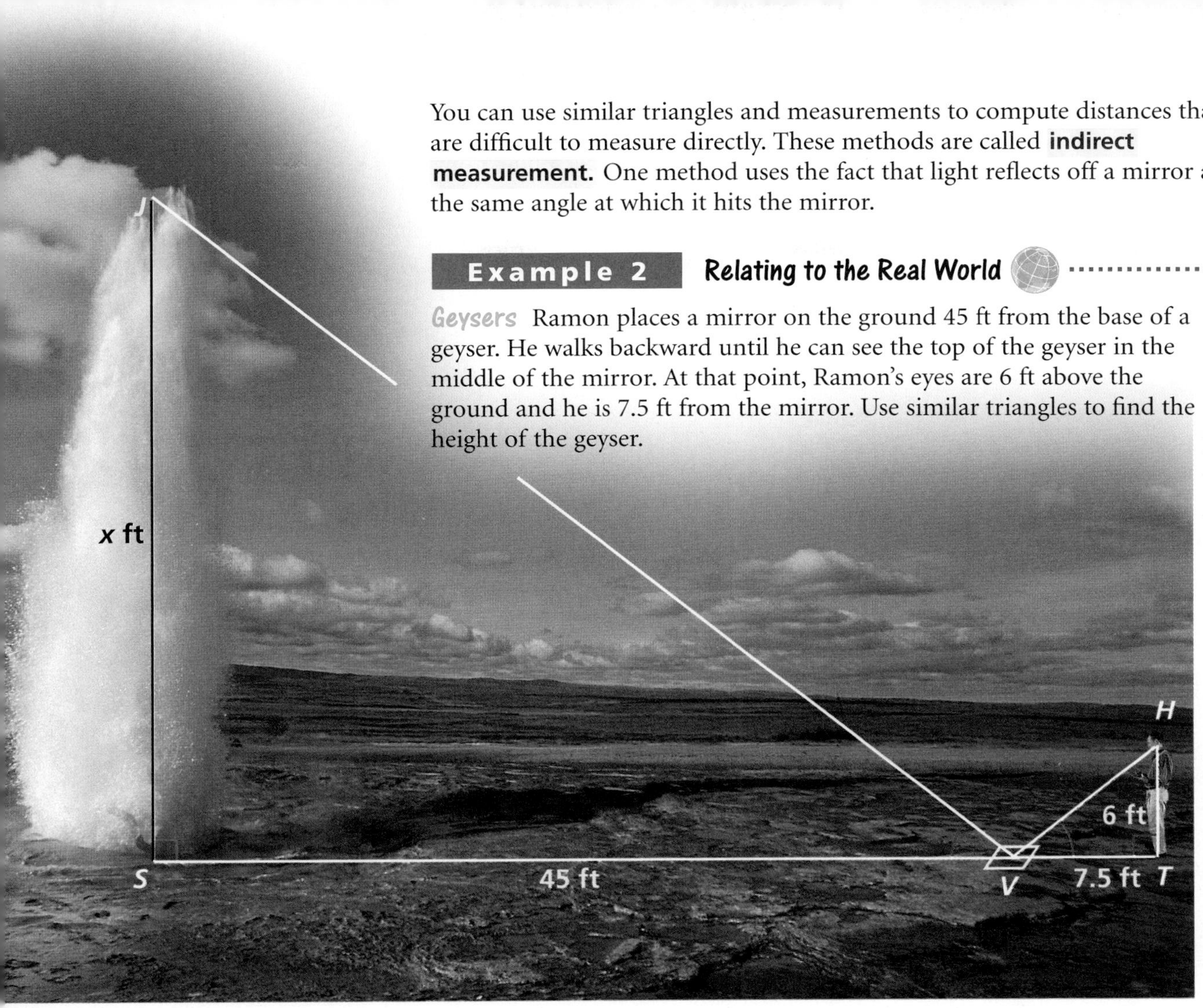

Example 2 *Relating to the Real World*

Geysers Ramon places a mirror on the ground 45 ft from the base of a geyser. He walks backward until he can see the top of the geyser in the middle of the mirror. At that point, Ramon's eyes are 6 ft above the ground and he is 7.5 ft from the mirror. Use similar triangles to find the height of the geyser.

$\triangle HTV \sim \triangle JSV$	AA ~ Postulate
$\frac{HT}{JS} = \frac{TV}{SV}$	Corr. sides of ~ triangles are proportional.
$\frac{6}{x} = \frac{7.5}{45}$	Substitute.
$270 = 7.5x$	Cross-Product Property
$36 = x$	Divide each side by 7.5.

The geyser is 36 ft high.

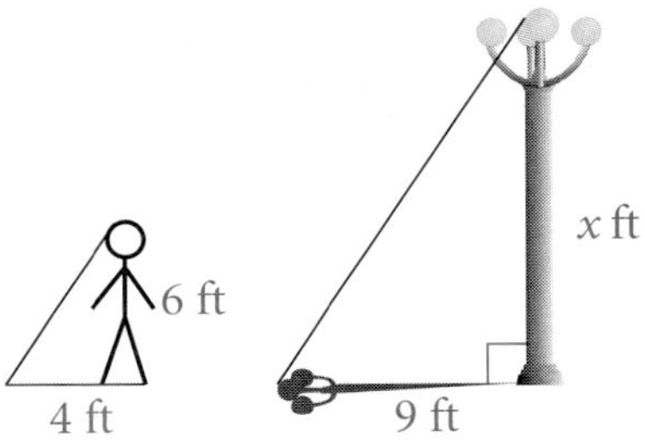

3. Try This To find the height of a fire tower, Latisha places a mirror on the ground 40 ft from the base of the tower. Latisha's eyes are $5\frac{1}{2}$ ft above the ground. When Latisha stands 4 ft from the mirror, she can see the top of the tower. How tall is the fire tower?

4. A lamppost casts a 9-ft shadow at the same time a person 6 ft tall casts a 4-ft shadow. Use similar triangles to find the height of the lamppost.

Using the SAS and SSS Similarity Theorems

The following two theorems follow from the AA Similarity Postulate.

Theorem 10-1
Side-Angle-Side Similarity Theorem (SAS ~ Theorem)

If an angle of one triangle is congruent to an angle of a second triangle, and the sides including the two angles are proportional, then the triangles are similar.

Given: $\angle A \cong \angle Q$, $\frac{AB}{QR} = \frac{AC}{QS}$
Prove: $\triangle ABC \sim \triangle QRS$

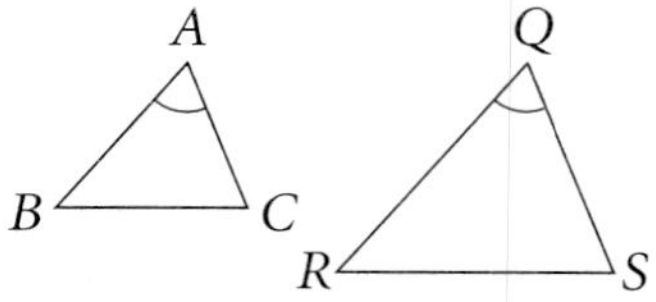

Outline of Proof of Theorem 10-1:

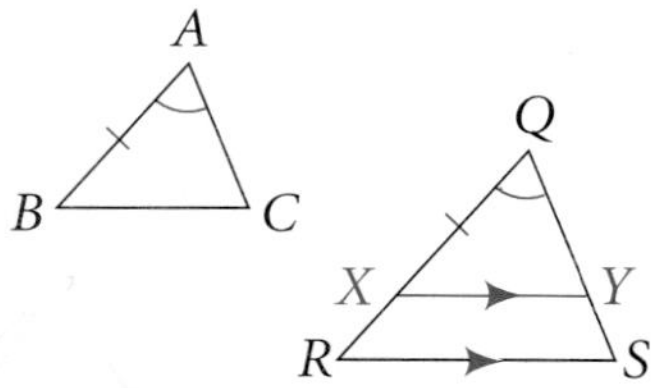

- Choose X on $\overline{QR}$ so that $AB = QX$. Draw $\overline{XY} \parallel \overline{RS}$. Use the AA ~ Postulate to show that $\triangle QXY \sim \triangle QRS$ and thus $\frac{QX}{QR} = \frac{QY}{QS}$.
- Combine this proportion with the given proportion and the fact that $AB = QX$ to show that $AC = QY$.
- Prove that $\triangle ABC \cong \triangle QXY$ and $\triangle ABC \sim \triangle QRS$.

Theorem 10-2
Side-Side-Side Similarity Theorem (SSS ~ Theorem)

If the corresponding sides of two triangles are proportional, then the triangles are similar.

Given: $\frac{AB}{QR} = \frac{BC}{RS} = \frac{AC}{QS}$
Prove: $\triangle ABC \sim \triangle QRS$

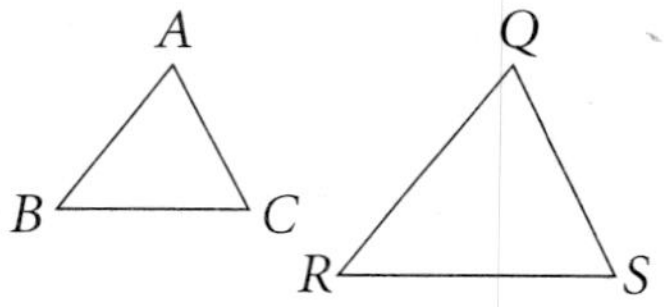

Outline of Proof of Theorem 10-2:

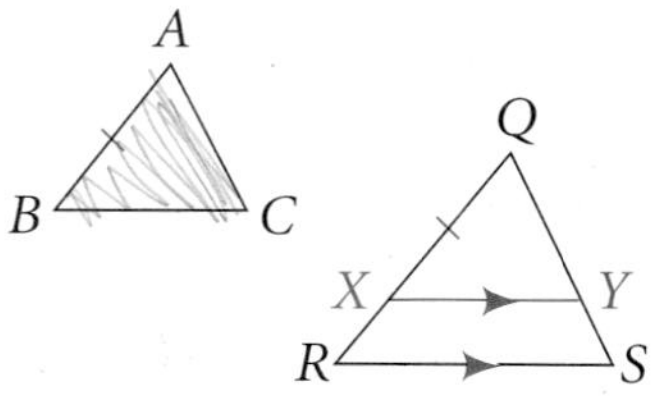

- Draw $\overline{XY}$ as in the proof for Theorem 10-1. Show that $\triangle QXY \sim \triangle QRS$ and thus $\frac{QX}{QR} = \frac{XY}{RS} = \frac{QY}{QS}$.
- Combine this proportion with the given proportion and the fact that $AB = QX$ to show that $BC = XY$ and $AC = QY$. Show that $\triangle ABC \cong \triangle QXY$ and $\triangle ABC \sim \triangle QRS$.

5. What postulate or theorem proves the pairs of triangles similar?

a.

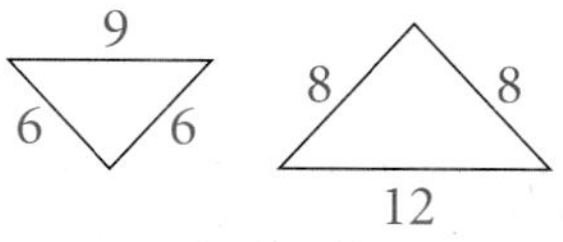

b.

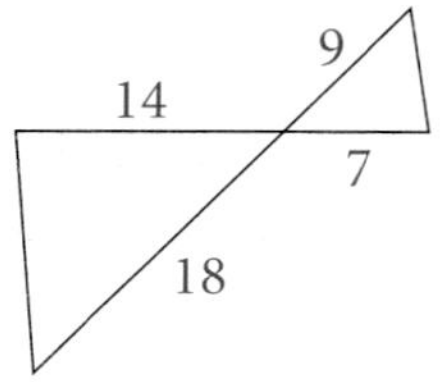

Example 3

Write a similarity statement and explain why the triangles are similar. Then find the length of $\overline{DE}$.

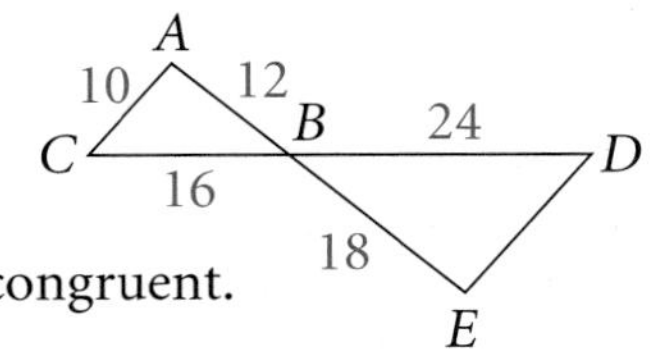

$\angle ABC \cong \angle EBD$ because vertical angles are congruent.

$\frac{AB}{EB} = \frac{12}{18} = \frac{2}{3}$ and $\frac{CB}{DB} = \frac{16}{24} = \frac{2}{3}$.

Therefore, $\triangle ABC \sim \triangle EBD$ by the SAS $\sim$ Postulate.

$\frac{AC}{DE} = \frac{2}{3}$	Corr. sides of similar triangles are proportional.
$\frac{10}{DE} = \frac{2}{3}$	Substitution
$30 = 2 \cdot DE$	Cross-Product Property
$15 = DE$	Divide each side by 2.

The length of $\overline{DE}$ is 15.

6. Find $m\angle E$ if $m\angle C = 49$ and $m\angle ABC = 39$.

Exercises ON YOUR OWN

Can you prove the triangles similar? If so, write the similarity statement and name the postulate or theorem you used.

1.

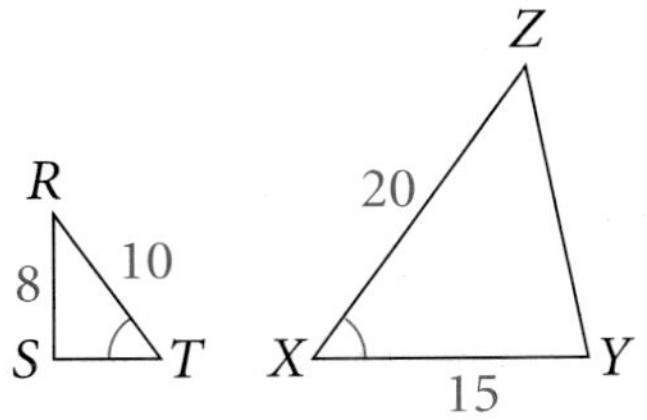

2.

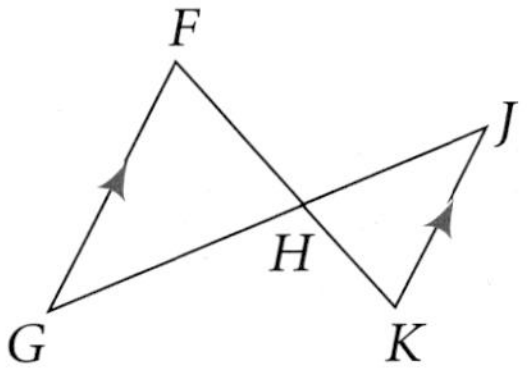

3.

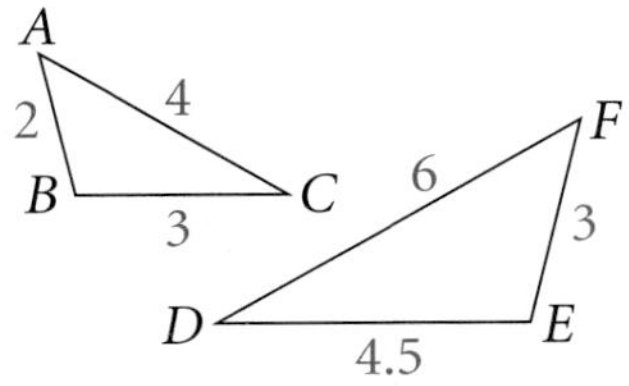

4.

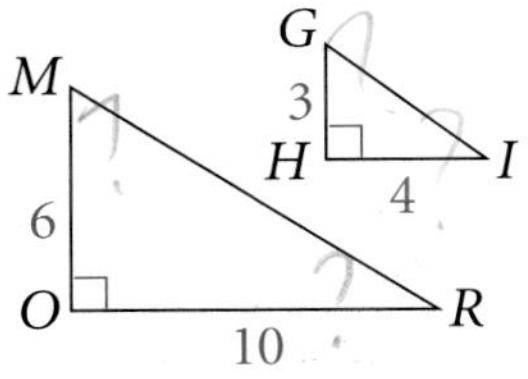

5.

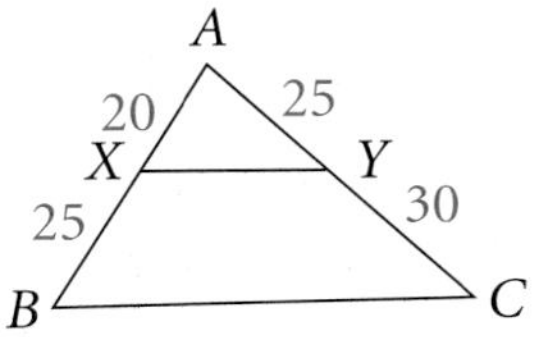

6.

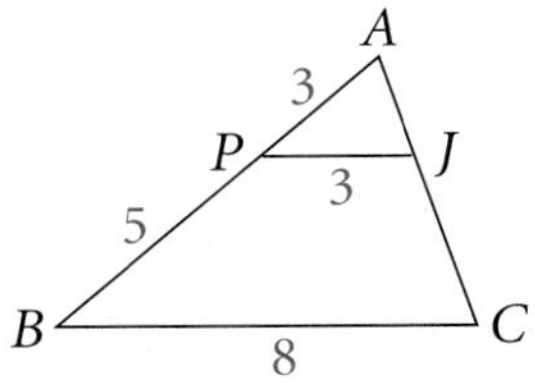

7.

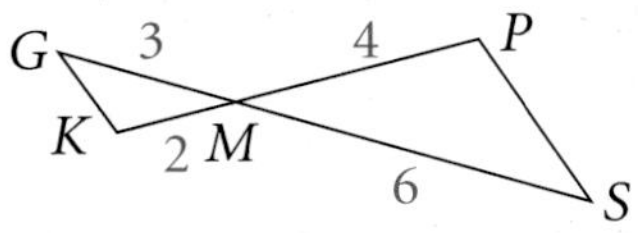

8.

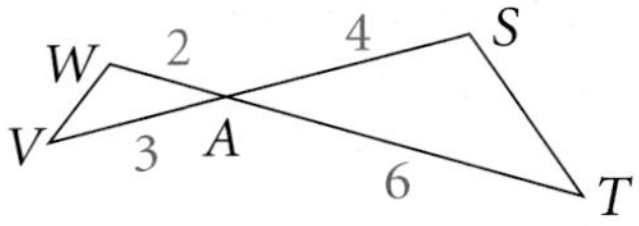

9.

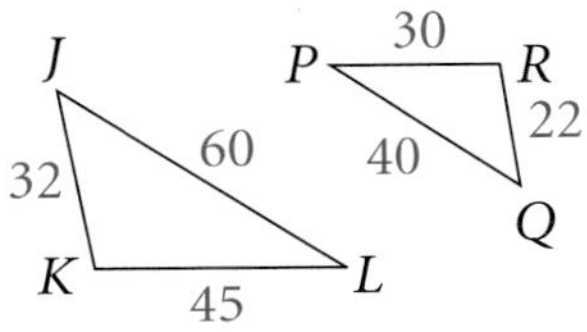

10. *Constructions* Draw any $\triangle ABC$. Use a straightedge and a compass to construct $\triangle RST$ so that $\triangle ABC \sim \triangle RST$ with similarity ratio 1 : 3.

11. **Standardized Test Prep** What is the value of x?
A. $5\frac{1}{3}$ **B.** 6 **C.** 6.75 **D.** 7 **E.** 8

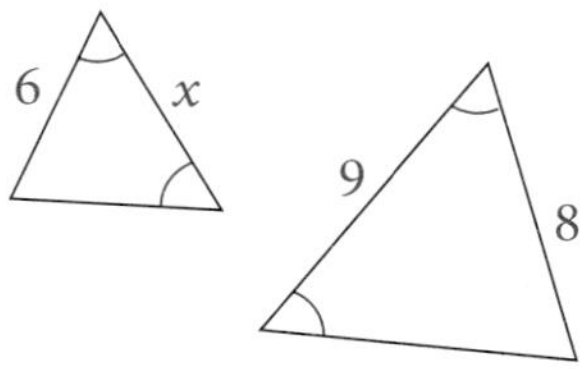

12. a. **Critical Thinking** Are two isosceles triangles always similar? Explain.
b. Are two isosceles right triangles always similar? Explain.

Algebra **Solve for x.**

13.

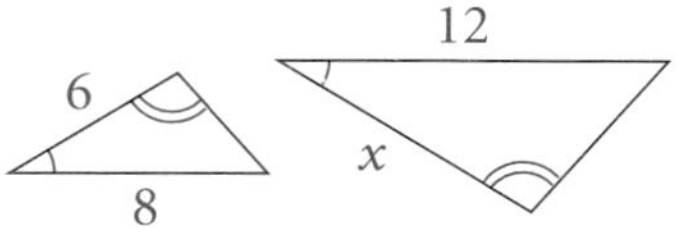

14.

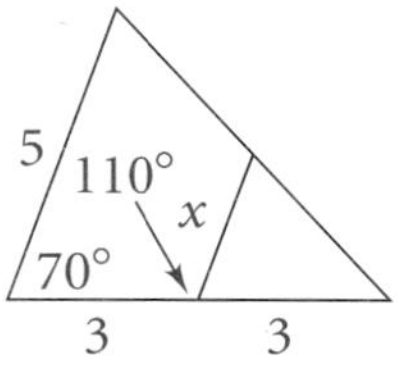

15.

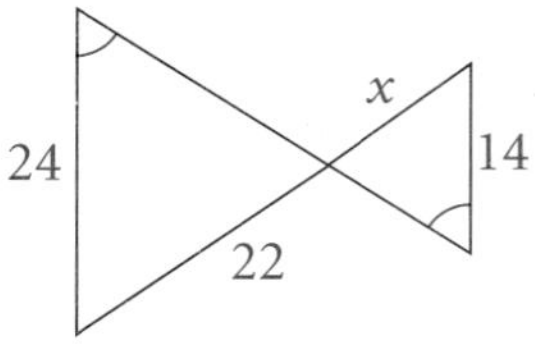

16.

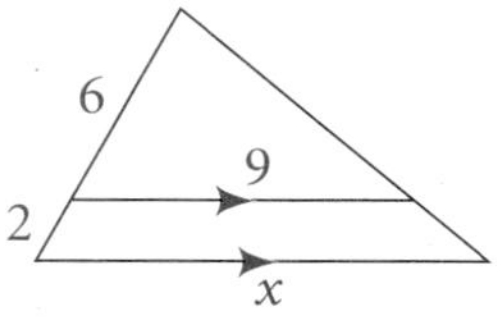

17.

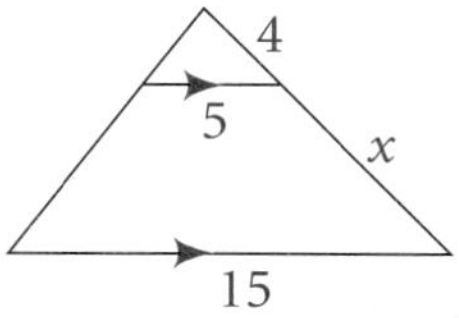

18.

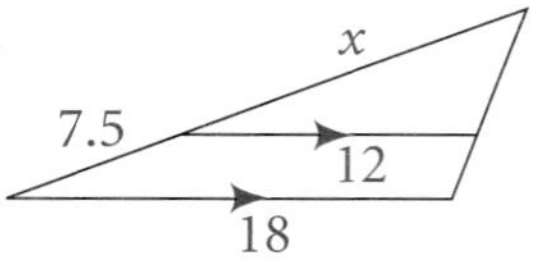

19. **Open-ended** Name something that would be difficult to measure directly, and describe how you could measure it indirectly.

Indirect Measurement **Find the distance represented by x.**

20.

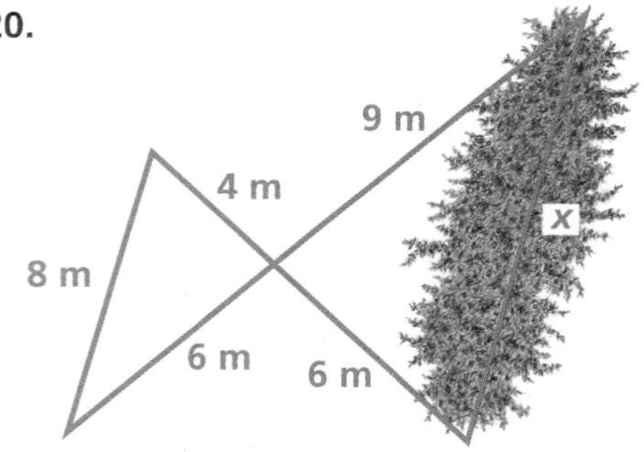

21.

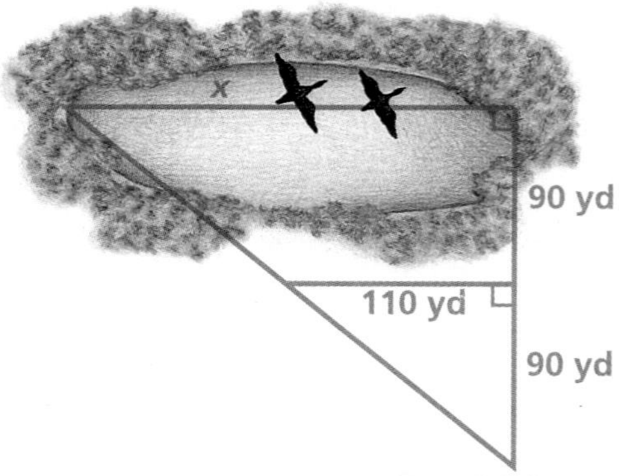

22.

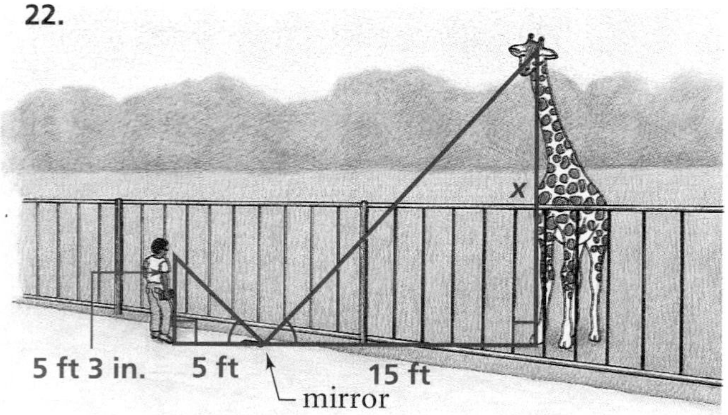

23.

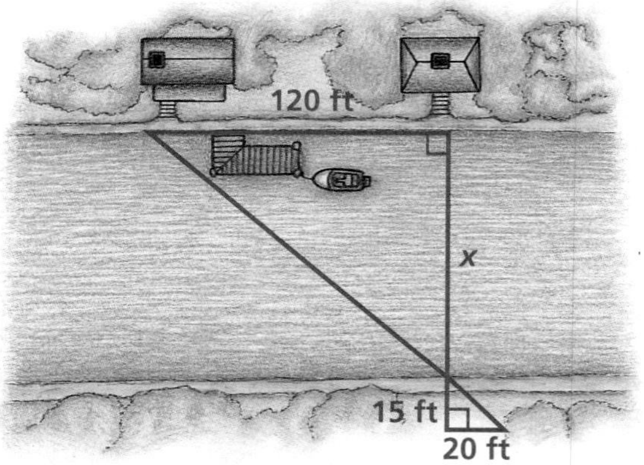

24. **a.** Classify *RSTW*.
b. Must any of the triangles shown be similar? Explain.

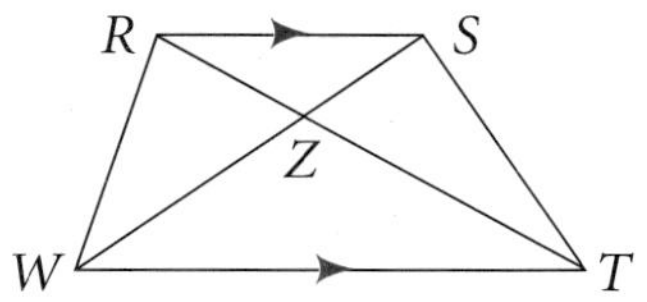

25. **Mental Math** A yardstick casts a 1-ft shadow at the same time that a nearby tree casts a 15-ft shadow. How tall is the tree?

26. **a.** Write similarity statements involving two different pairs of similar triangles for the figure shown at the right.
b. **Algebra** Find the missing lengths.

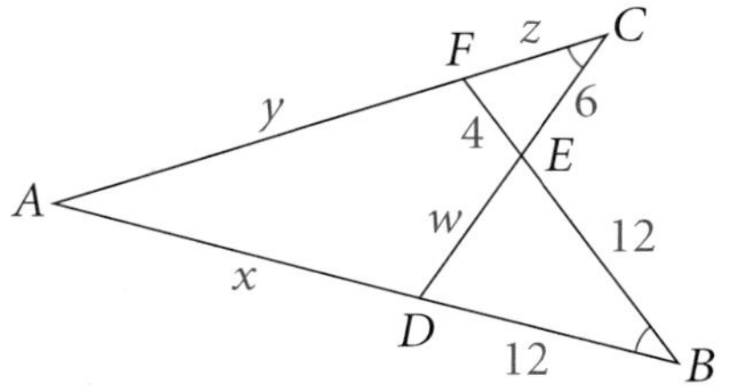

Skyscraper? *Nein!*

Hannalore Krause of Frankfurt, Germany, stopped the development of what was to be the tallest skyscraper in Europe because she wanted her apartment to get its fair amount of sunlight. To halt the construction she used a German law that specifies that every homeowner is entitled to sunlight.

Krause was offered 1.6 million dollars to drop her lawsuit, but she refused. The skyscraper, which would have cost about 400 million dollars, was to be built on a site that was only 60 m from Krause's apartment. Scheduled to be 265 m tall, it would have been slightly less than three-fourths the height of the Empire State Building.

Use the magazine article to answer Exercises 27 and 28.

27. **Writing** Explain how Hannalore Krause can use indirect measurement to estimate the length of the shadow of the building at a particular time of day.

28. Suppose Hannalore is 1.75 m tall. When her shadow is 1 m long, about how long would the shadow of the proposed building be?

Choose **Write a paragraph proof, a two-column proof, or a flow proof.**

29. **Given:** $\overline{BC} \parallel \overline{DF}$
Prove: $\triangle BYC \sim \triangle DYF$

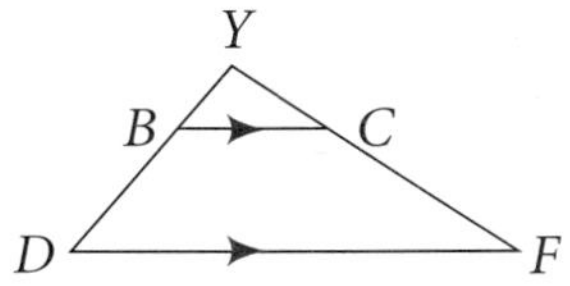

30. **Given:** $RT \cdot TQ = MT \cdot TS$
Prove: $\triangle RTM \sim \triangle STQ$

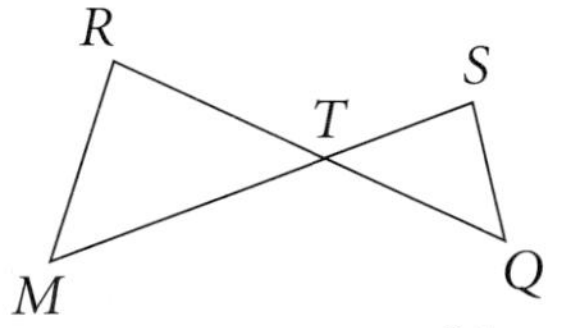

31. **a.** Find the perimeters of the triangles at the right.
b. Find the areas of the triangles at the right.
c. **Critical Thinking** Can you conclude that two triangles with equal perimeters and equal areas are similar? Explain.

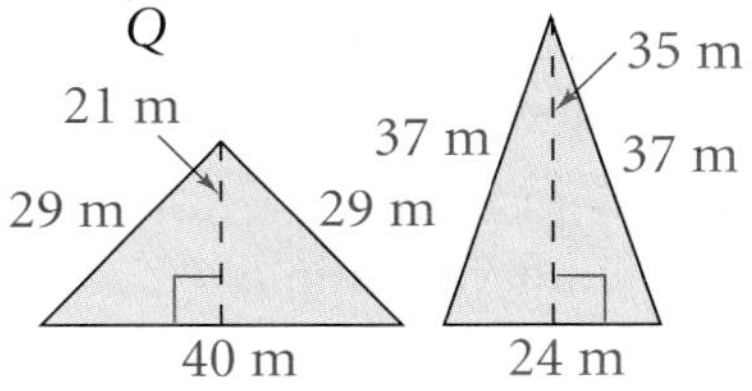

Exercises MIXED REVIEW

Sketch each figure.

32. a regular hexagon

33. opposite rays $\overrightarrow{RS}$ and $\overrightarrow{RB}$

34. line ℓ intersecting plane P at point E

Draw a net for each space figure.

35. a cube **36.** a pyramid with a triangular base

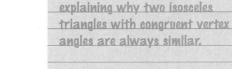

SELF ASSESSMENT — FOR YOUR JOURNAL

Write a brief paragraph explaining why two isosceles triangles with congruent vertex angles are always similar.

Getting Ready for Lesson 10-3

Solve each proportion. Leave your answers in simplest radical form.

37. $\frac{4}{x} = \frac{x}{5}$ **38.** $\frac{3}{m} = \frac{m}{8}$ **39.** $\frac{w}{2} = \frac{20}{w}$ **40.** $\frac{a}{6} = \frac{27}{a}$

Exercises CHECKPOINT

Are the triangles similar? If so, write the similarity statement and name the postulate or theorem that proves they are similar.

1.

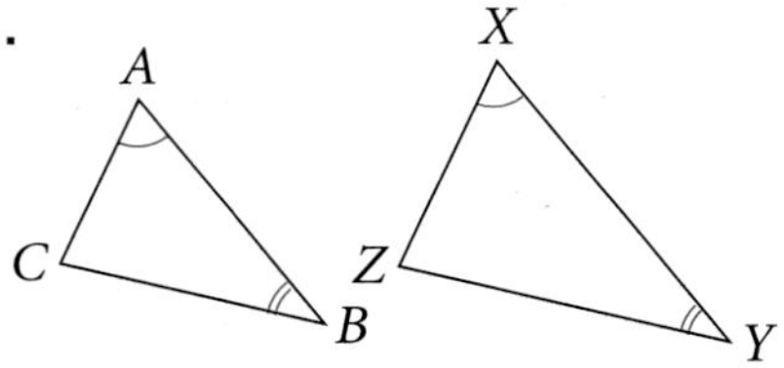

2.

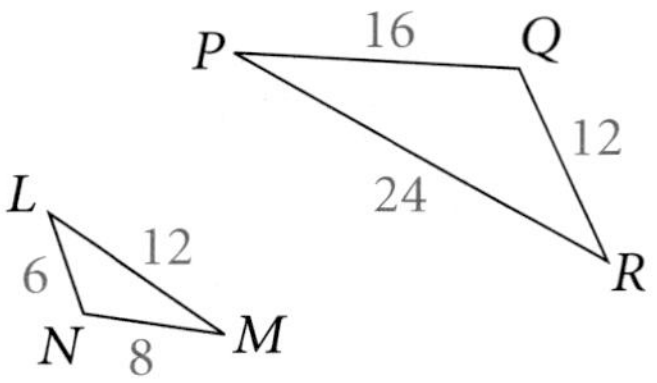

3.

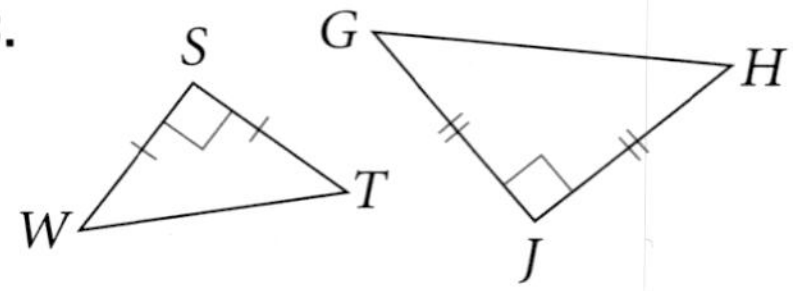

The polygons are similar. Find the values of the variables.

4.

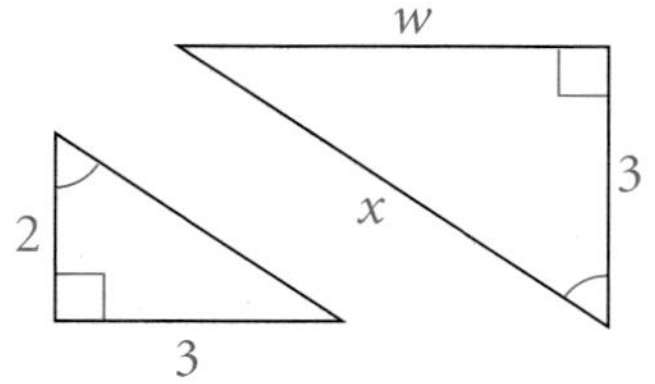

5.

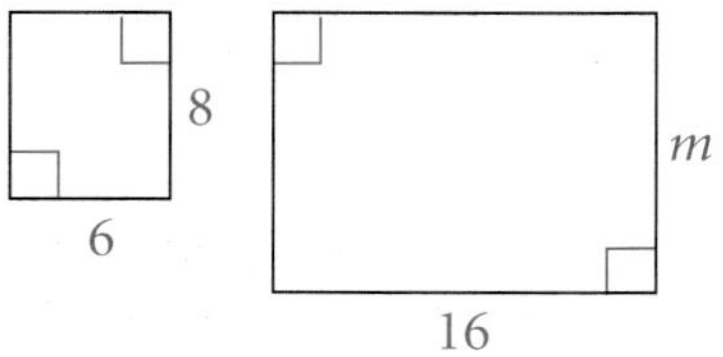

6. In the movie *Ghostbusters,* a giant made of marshmallow roams the streets of New York City. Suppose that a man 6 ft 3 in. tall stands so that the tip of his shadow coincides with the tip of the giant's shadow. The man is 4 ft from the tip of the shadows and 24 ft from the marshmallow giant. How tall is the giant?

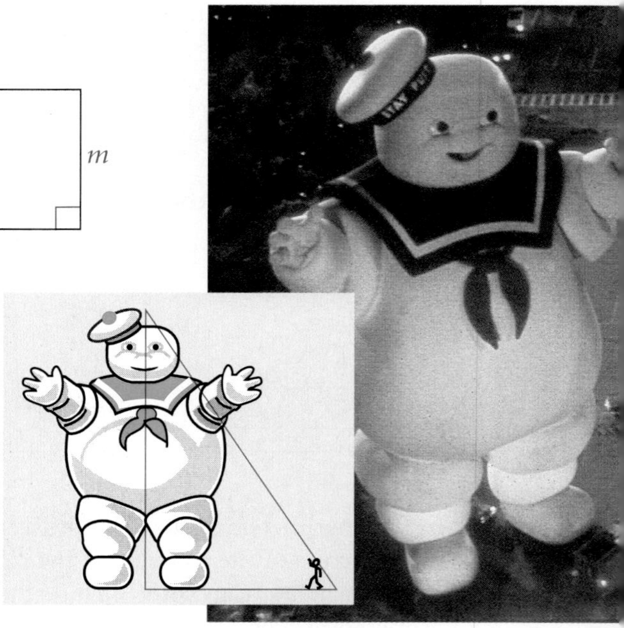

7. *Open-ended* Sketch two figures that are similar and have a similarity ratio of 2 : 3.

10-3 Similarity in Right Triangles

What You'll Learn

- Finding relationships among the lengths of the sides of a right triangle and the altitude to the hypotenuse

...And Why

To find distances by indirect measurement

What You'll Need

- straightedge
- scissors

WORK TOGETHER

Work with a partner.

- Draw one diagonal of a rectangular sheet of paper. Cut the paper on the diagonal to make two congruent right triangles.
- In one of the triangles, use paper folding to locate the altitude to the hypotenuse. Cut the triangle along the altitude to make two smaller right triangles.
- Label the angles of the three triangles as shown.

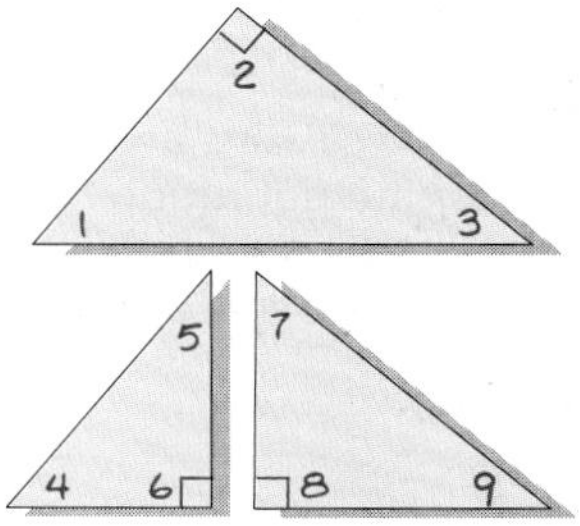

1. Compare the angles of the three triangles by placing the triangles on top of one another. Which angles have the same measure as $\angle 1$?
2. Which angles have the same measure as $\angle 2$?
3. Which angles have the same measure as $\angle 3$?
4. Based on your results, what is true about the three triangles?
5. Use the diagram at the left to complete the similarity statement. $\triangle RST \sim \triangle ■ \sim \triangle ■$

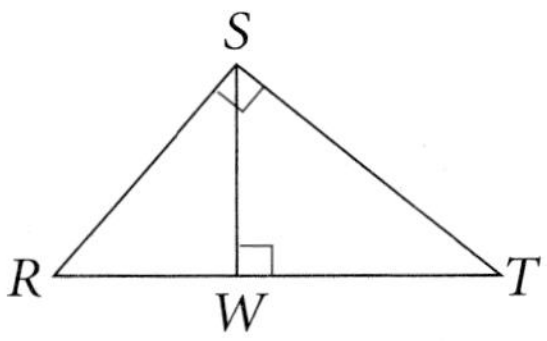

THINK AND DISCUSS

Your findings in the Work Together suggest the following theorem.

Theorem 10-3

The altitude to the hypotenuse of a right triangle divides the triangle into two triangles that are similar to the original triangle and to each other.

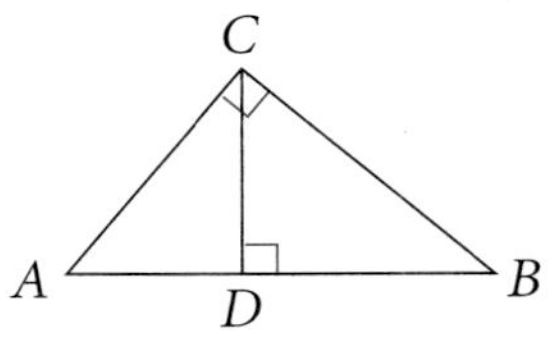

Given: Right $\triangle ABC$, $\overline{CD}$ is the altitude to the hypotenuse.
Prove: $\triangle ABC \sim \triangle ACD \sim \triangle CBD$

Proof of Theorem 10-3

Both smaller triangles are similar to $\triangle ABC$ by the AA Similarity Postulate because they each share an acute angle with $\triangle ABC$ and all three triangles are right triangles. Since both smaller triangles are similar to $\triangle ABC$, their corresponding angles are congruent. Thus they are similar to each other.

QUICK REVIEW

In the proportion

$\frac{a}{b} = \frac{c}{d}$,

b and c are the means.

Proportions in which the means are equal occur frequently in geometry. For any two positive numbers a and b, the **geometric mean** of a and b is the positive number x such that $\frac{a}{x} = \frac{x}{b}$.

Example 1

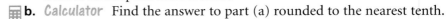

Algebra Find the geometric mean of 4 and 18.

$\frac{4}{x} = \frac{x}{18}$ Write a proportion.

$x^2 = 72$ Use the Cross-Product Property

$x = \sqrt{72}$ Find the positive square root.

$x = 6\sqrt{2}$ Write in simplest radical form.

PROBLEM SOLVING

Look Back Give an alternate definition of geometric mean that does not involve a proportion.

6. a. **Try This** Find the geometric mean of 15 and 20. Leave your answer in simplest radical form.

b. **Calculator** Find the answer to part (a) rounded to the nearest tenth.

Two important corollaries of Theorem 10-3 involve the geometric mean.

Corollary 1

The length of the altitude to the hypotenuse of a right triangle is the geometric mean of the lengths of the segments of the hypotenuse.

Proof of Corollary 1

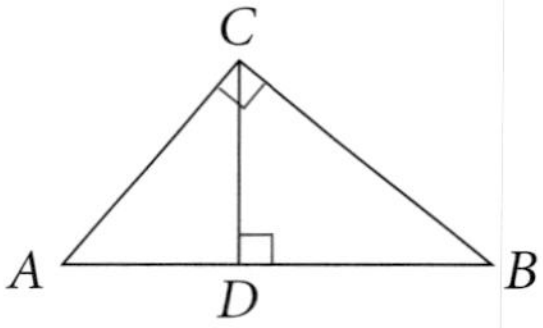

Given: Right $\triangle ABC$,
$\overline{CD}$ is the altitude to the hypotenuse.

Prove: $\frac{AD}{CD} = \frac{CD}{DB}$

By Theorem 10-3, $\triangle CDB \sim \triangle ADC$. Since corresponding sides of similar triangles are proportional, $\frac{AD}{CD} = \frac{CD}{DB}$.

Corollary 2

The altitude to the hypotenuse of a right triangle intersects it so that the length of each leg is the geometric mean of the length of its adjacent segment of the hypotenuse and the length of the entire hypotenuse.

Proof of Corollary 2

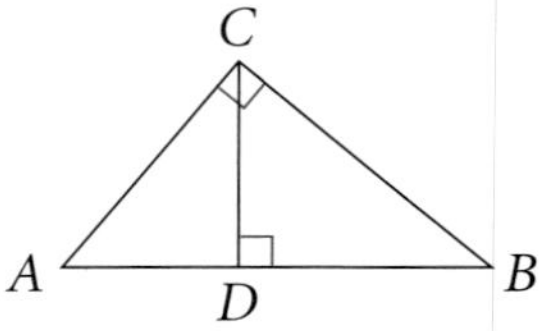

Given: Right $\triangle ABC$,
$\overline{CD}$ is the altitude to the hypotenuse.

Prove: $\frac{AB}{AC} = \frac{AC}{AD}$, $\frac{AB}{CB} = \frac{CB}{DB}$

By Theorem 10-3, $\triangle ABC \sim \triangle ACD$. Their corresponding sides are proportional, so $\frac{AB}{AC} = \frac{AC}{AD}$. Similarly, $\triangle ABC \sim \triangle CBD$ and $\frac{AB}{CB} = \frac{CB}{DB}$.

Example 2

Solve for x and y.

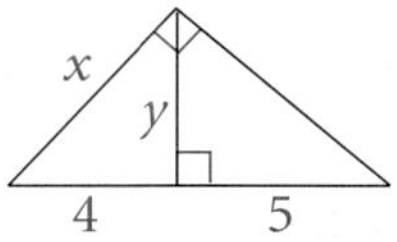

Use Corollary 2 to solve for x.

$$\frac{4+5}{x} = \frac{x}{4}$$
$$x^2 = 36$$
$$x = 6$$

Use Corollary 1 to solve for y.

$$\frac{4}{y} = \frac{y}{5}$$
$$y^2 = 20$$
$$y = 2\sqrt{5}$$

7. **Try This** Solve for x and y. If the answers are not whole numbers, leave them in simplest radical form.

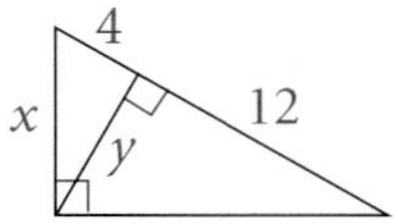

You can use the Corollaries to Theorem 10-3 to find distances.

Example 3 Relating to the Real World

Recreation At the parking lot of a State Park, the 300-m path to the snack bar and the 400-m path to the boat rental shop meet at a right angle. Marla walks straight from the parking lot to the ocean. How far is Marla from the snack bar?

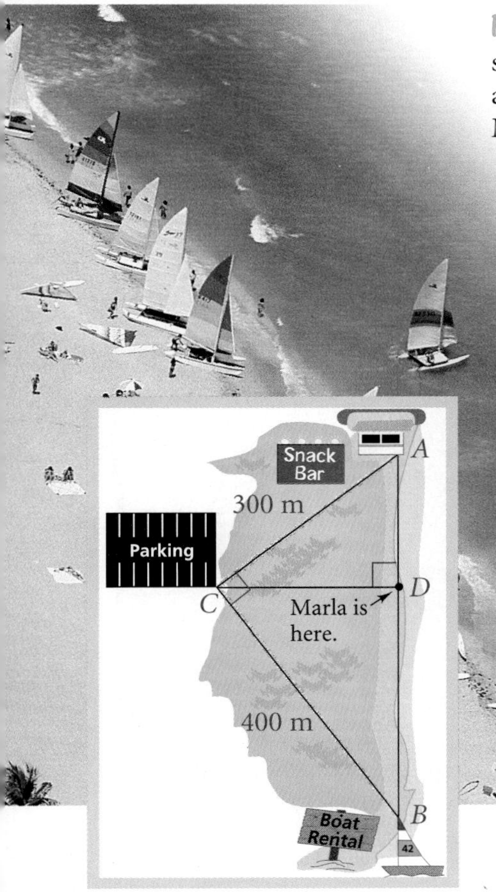

$\overline{CD}$, the perpendicular segment from point C to $\overline{AB}$, is the shortest path to the ocean. $\triangle ABC$ is a right triangle. Using Pythagorean triples, $AB = 500$. To find AD, apply Corollary 2.

$\frac{AD}{AC} = \frac{AC}{AB}$	Corollary 2
$\frac{AD}{300} = \frac{300}{500}$	Substitute.
$500(AD) = 90{,}000$	Cross-Product Property
$AD = 180$	Divide each side by 500.

Marla is 180 m from the snack bar.

8. How far did Marla walk from the parking lot to the ocean?

Exercises ON YOUR OWN

1. Complete: $\triangle JKL \sim \triangle$■ $\sim \triangle$■

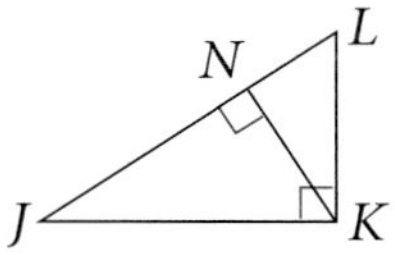

Algebra **Find the geometric mean of each pair of numbers. If an answer is not a whole number, leave it in simplest radical form.**

2. 4 and 9 **3.** 4 and 10 **4.** 4 and 12 **5.** 3 and 48 **6.** 5 and 125 **7.** 11 and 1331

Algebra **Refer to the figure to complete each proportion.**

8. $\frac{a}{c} = \frac{r}{\blacksquare}$ **9.** $\frac{a}{c} = \frac{h}{\blacksquare}$ **10.** $\frac{r}{h} = \frac{a}{\blacksquare}$ **11.** $\frac{\blacksquare}{h} = \frac{h}{\blacksquare}$

12. $\frac{\blacksquare}{a} = \frac{a}{\blacksquare}$ **13.** $\frac{\blacksquare}{b} = \frac{b}{\blacksquare}$ **14.** $\frac{h}{a} = \frac{\blacksquare}{c}$ **15.** $\frac{h}{a} = \frac{\blacksquare}{b}$

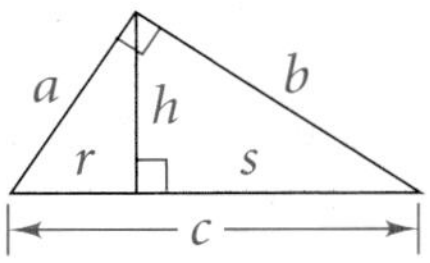

Algebra **Solve for x. If an answer is not a whole number, leave it in simplest radical form.**

16.

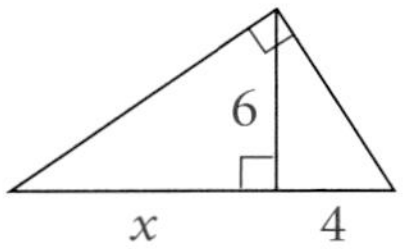

17.

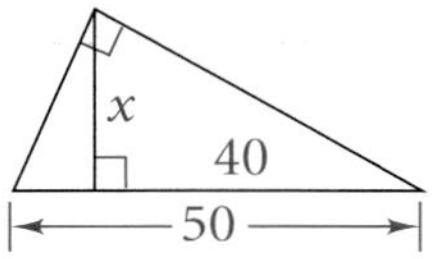

18.

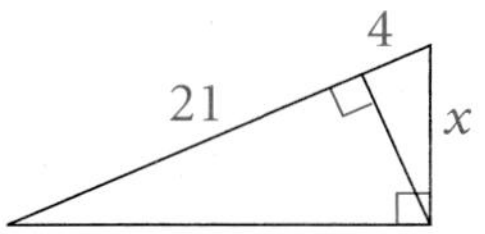

19.

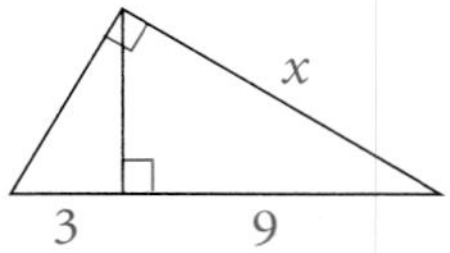

20. a. The altitude to the hypotenuse of a right triangle divides the hypotenuse into segments 2 cm and 8 cm long. Find the length h of the altitude.

b. **Drawing** Use the value you found for h in part (a), along with the lengths 2 cm and 8 cm, to draw the right triangle accurately.

c. **Writing** Explain how you drew the triangle in part (b).

21. **Coordinate Geometry** $\overline{CD}$ is the altitude to the hypotenuse of right $\triangle ABC$. The coordinates of A, D, and B are (4, 2), (4, 6), and (4, 15), respectively. Find all possible coordinates of point C.

22. **Algebra** The altitude to the hypotenuse of a right triangle divides the hypotenuse into segments whose lengths are in the ratio 1 : 2. The length of the altitude is 8. How long is the hypotenuse?

23. **Civil Engineering** Study the plan at the right. A service station will be built on the interstate highway and a road will connect it with Cray. How far from Blare should the service station be located so that the proposed road will be perpendicular to the interstate? How long will the new road be?

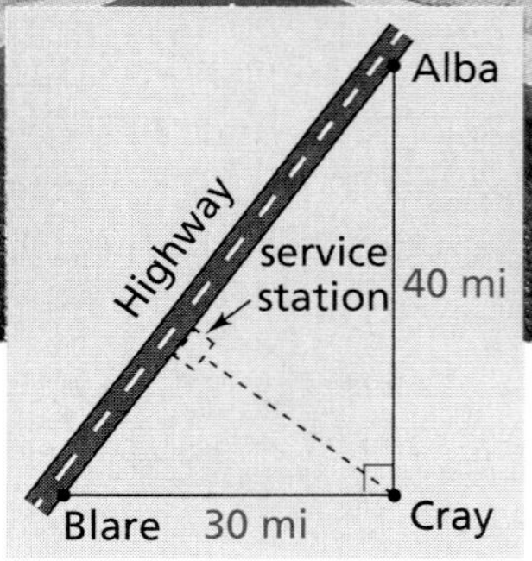

Algebra **Find the values of the variables. If an answer is not a whole number, leave it in simplest radical form.**

24.

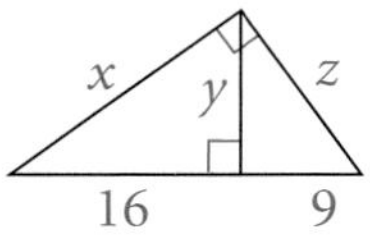

25.

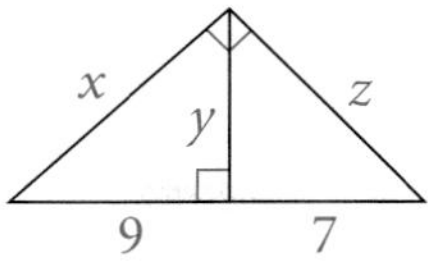

26.

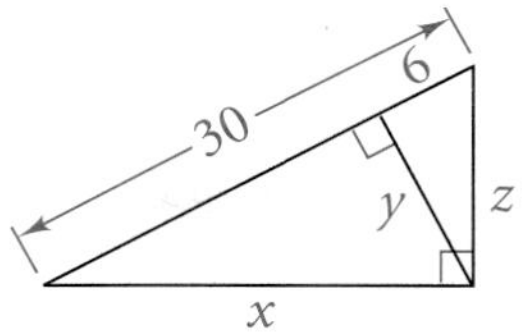

27.

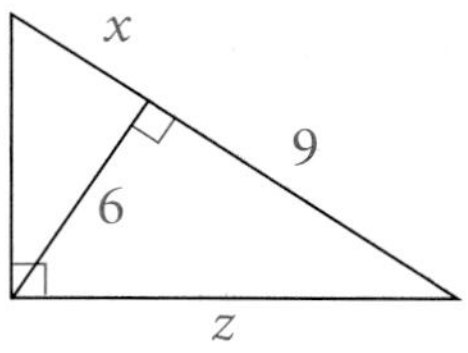

28. **Standardized Test Prep** The length of the altitude to the hypotenuse of a right triangle is 8. The length of the hypotenuse is 16. What is the length of one of the segments into which the altitude divides the hypotenuse?

A. 8 **B.** $8\sqrt{2}$ **C.** 4 **D.** $4\sqrt{2}$ **E.** cannot be determined

Algebra **Find the value of *x*. If an answer is not a whole number, leave it in simplest radical form.**

PROBLEM SOLVING HINT

Use the Quadratic Formula. (See page 439.)

29.

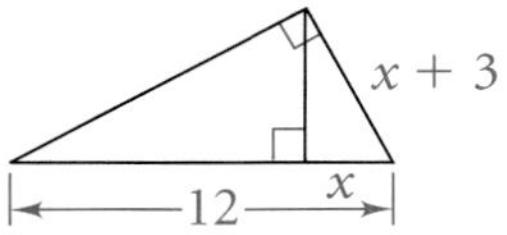

30.

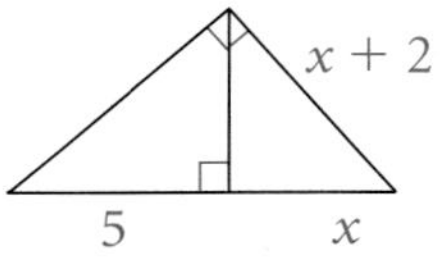

31.

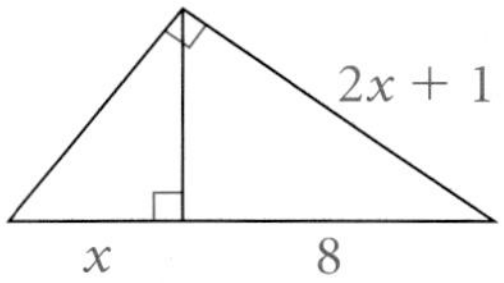

32. **Pythagorean Theorem** You can use Corollary 2 of Theorem 10-3 to prove the Pythagorean Theorem. Complete the following proof.

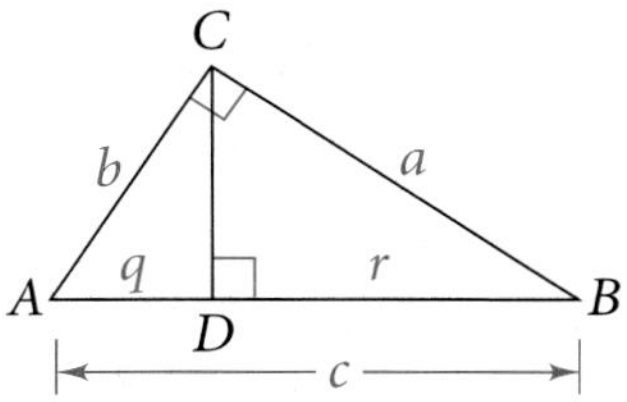

Given: Right $\triangle ABC$ with altitude $\overline{CD}$
Prove: $c^2 = a^2 + b^2$

Statements	Reasons
1. Right $\triangle ABC$ with altitude $\overline{CD}$	a. ?
2. $\frac{c}{a} = \frac{a}{r}, \frac{c}{b} = \frac{b}{q}$	b. ?
c. $■ = a^2, ■ = b^2$	d. ?
3. $cr + cq = a^2 + b^2$	e. ?
4. $c(r + q) = a^2 + b^2$	f. ?
5. $r + q = c$	g. ?
6. $c^2 = a^2 + b^2$	h. ?

33. The length of the shorter leg of a 30°-60°-90° triangle is 10 cm. What is the length of the altitude to the hypotenuse?

34. **a.** Lauren thinks she has found a new corollary: The product of the lengths of the two legs of a right triangle is equal to the product of the lengths of the hypotenuse and the altitude from the right angle. Draw a figure for this corollary and write the *Given* and *To Prove.*

b. **Critical Thinking** Is Lauren's corollary true? Explain.

35. **Open-ended** Draw a right triangle so that the altitude from the right angle to the hypotenuse bisects the hypotenuse.

Chapter Project

Find Out by Analyzing

Use the Stage 3 Koch snowflake shown here and the earlier stages you made in the Find Out question on page 502.

- At each stage, is the snowflake equilateral?
- Suppose each side of the original triangle is one unit. Complete the table to find the perimeter of each stage.

Stage	Number of sides	Length of a side	Perimeter
0	3	1	3
1	■	$\frac{1}{3}$	■
2	48	■	■
3	■	■	■

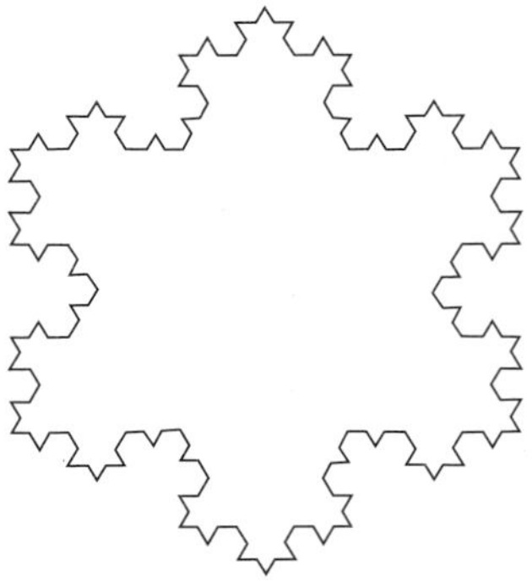

Stage 3

- Patterns Can you **predict** the perimeter at Stage 4?
- Inductive Reasoning Will there be a stage with a perimeter greater than 100 units? Explain.

Exercises MIXED REVIEW

Data Analysis **Use the graph showing the percent of people in the United States who received dental services.**

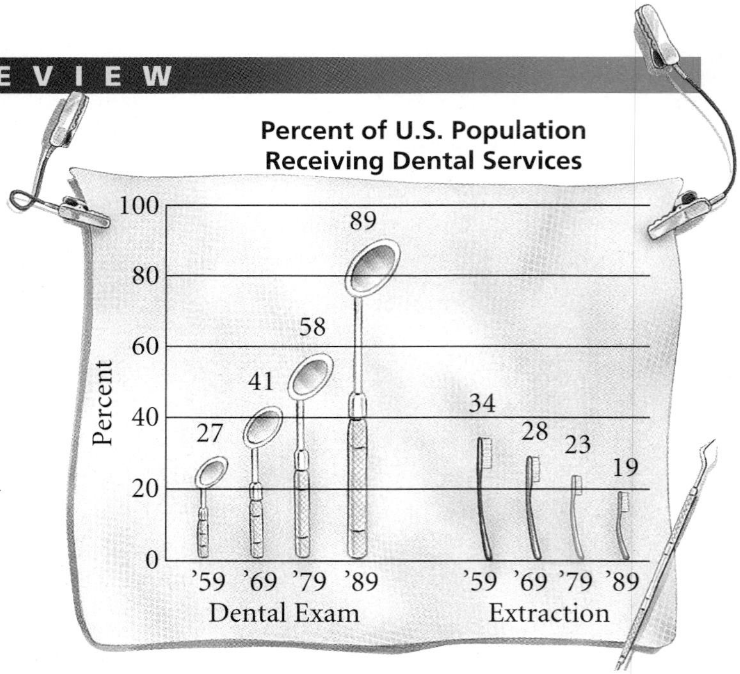

36. In which year did more people have an extraction than a dental exam?

37. Between which years did the percent of the population having a dental exam increase the most?

38. Critical Thinking Compare the ratio of dental exams to extractions for each year. What trend do you see in dental services?

39. Draw a rectangular prism in two-point perspective.

Getting Ready for Lesson 10-4

Find the length of the midsegment of each trapezoid or triangle.

40.

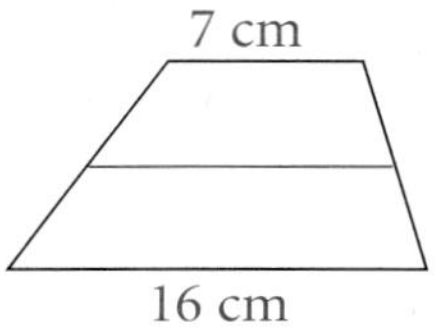

41.

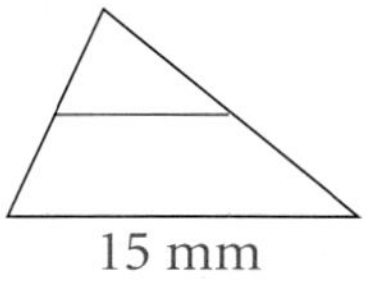

42.

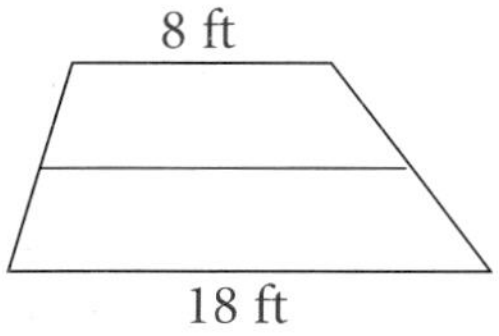

43.

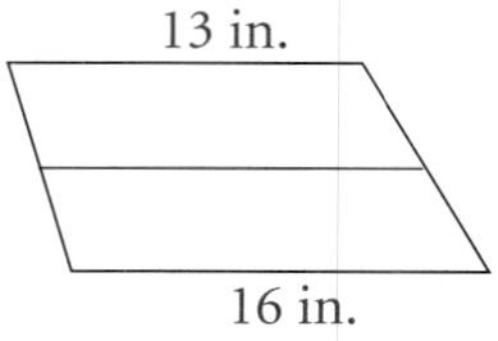

What You'll Learn

- Investigating proportional relationships in triangles

...And Why

To find lengths of material needed to make a sail and to determine distances on a city map

What You'll Need

- geometry software

10–4 Proportions and Similar Triangles

WORK TOGETHER

Work with a partner. Use geometry software.

- Construct $\triangle ABC$ and a point D on $\overline{AB}$.
- Construct a line through D parallel to $\overline{AC}$.
- Construct the intersection E of the parallel line with $\overline{BC}$.
- Measure $\overline{BD}$, $\overline{DA}$, $\overline{BE}$, and $\overline{EC}$.
- Calculate the ratios $\frac{BD}{DA}$ and $\frac{BE}{EC}$.

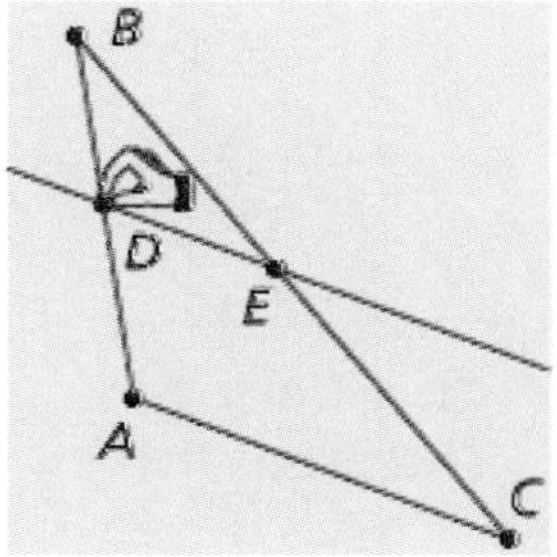

1. **Patterns** Compare the ratios $\frac{BD}{DA}$ and $\frac{BE}{EC}$ as you move the parallel line $\overleftrightarrow{DE}$ and as you change the shape of the triangle.

THINK AND DISCUSS

Using the Side-Splitter Theorem

In the Work Together you discovered the following relationship.

Theorem 10-4
Side-Splitter Theorem

If a line is parallel to one side of a triangle and intersects the other two sides, then it divides those sides proportionally.

Proof of Theorem 10-4

Given: $\triangle QXY$ with $\overleftrightarrow{RS} \parallel \overleftrightarrow{XY}$
Prove: $\frac{XR}{RQ} = \frac{YS}{SQ}$

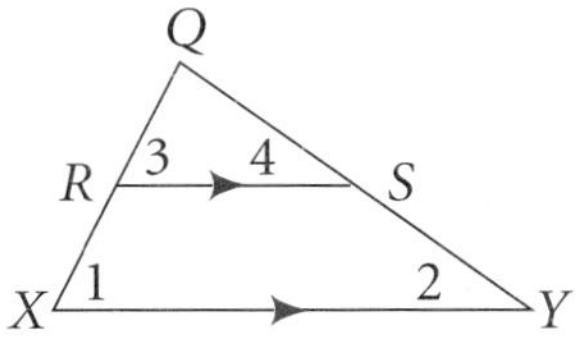

Statements	Reasons
1. $\overleftrightarrow{RS} \parallel \overleftrightarrow{XY}$	1. Given
2. $\angle 1 \cong \angle 3$, $\angle 2 \cong \angle 4$	2. If $\parallel$ lines, then corr. $\angle$s are $\cong$.
3. $\triangle QXY \sim \triangle QRS$	3. AA $\sim$ Postulate
4. $\frac{XQ}{RQ} = \frac{YQ}{SQ}$	4. Corresponding sides of $\sim$ triangles are proportional.
5. $XQ = XR + RQ$, $YQ = YS + SQ$	5. Segment Addition Postulate
6. $\frac{XR + RQ}{RQ} = \frac{YS + SQ}{SQ}$	6. Substitution
7. $\frac{XR}{RQ} = \frac{YS}{SQ}$	7. A Property of Proportions

2. Try This Use the Side-Splitter Theorem to find the value of x in each diagram.

a.

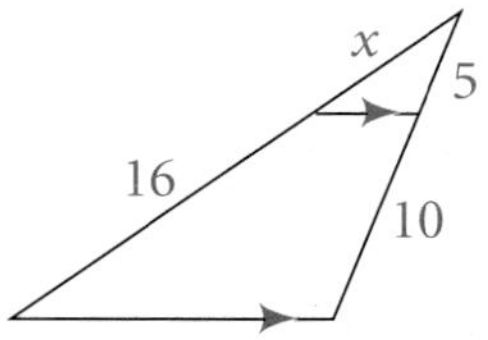

b.

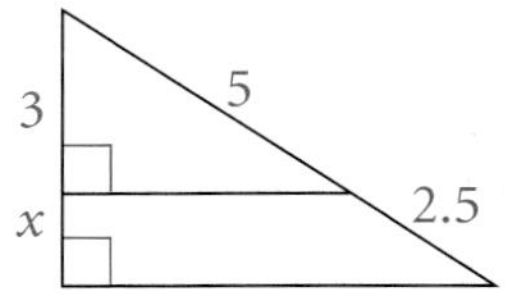

The following corollary can be easily derived from the Side-Splitter Theorem. You will prove this corollary in Exercise 24.

Corollary to Theorem 10-4

If three parallel lines intersect two transversals, then the segments intercepted on the transversals are proportional.

$$\frac{a}{b} = \frac{c}{d}$$

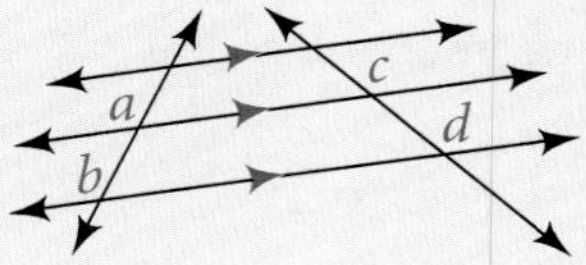

Example 1 Relating to the Real World

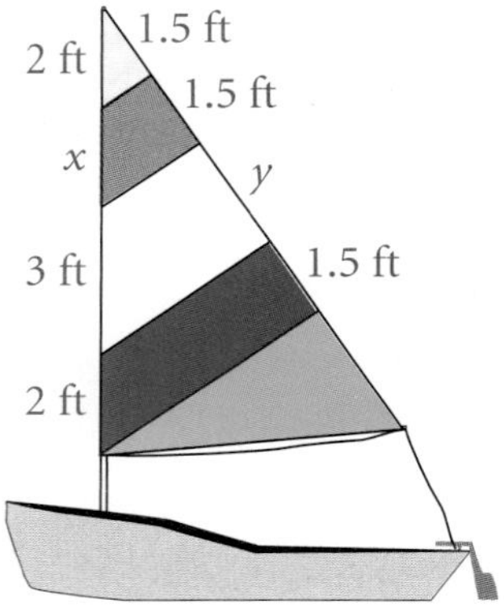

Sail Making Sail makers use computers to create a pattern for every sail they make. Then they draw a chalk outline on the cutting floor. After they cut out the panels of the sail, they sew them together to form the sail. The panel seams are parallel. Find the lengths x and y.

$\frac{2}{x} = \frac{1.5}{1.5}$ Use the Side-Splitter Theorem.

$x = 2$

$\frac{3}{2} = \frac{y}{1.5}$ Use the Corollary to Theorem 10-4.

$3(1.5) = 2y$ Use the Cross-Product Property.

$\frac{3(1.5)}{2} = y$ Divide each side by 2.

$2.25 = y$ Simplify.

Length x is 2 ft and length y is 2.25 ft.

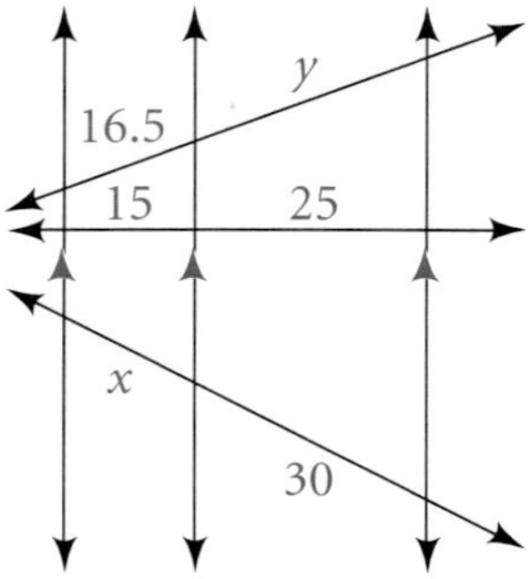

3. Try This Solve for x and y.

WORK TOGETHER

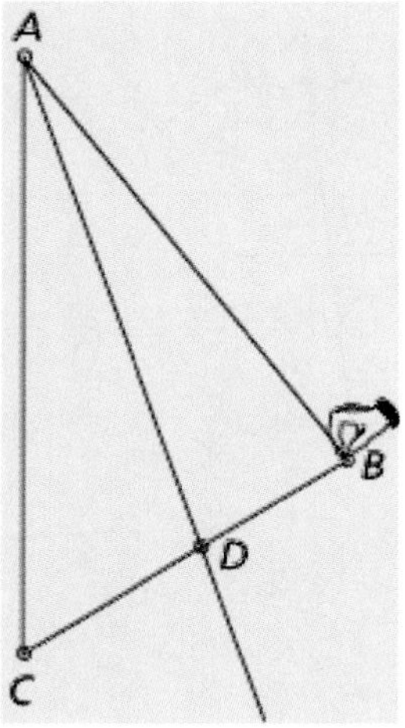

Work with a partner. Use geometry software.

- Construct $\triangle ABC$ and the bisector of $\angle A$. Label point D, the intersection of the bisector and $\overline{CB}$.
- Measure $\overline{BD}$, $\overline{BA}$, $\overline{CD}$, and $\overline{CA}$.
- Calculate the ratios $\frac{CD}{DB}$ and $\frac{CA}{BA}$.

4. *Patterns* Compare the ratios as you change the triangle.

THINK AND DISCUSS

Using the Triangle-Angle-Bisector Theorem

In the Work Together you discovered the following relationship.

Theorem 10-5
Triangle-Angle-Bisector Theorem

If a ray bisects an angle of a triangle, then it divides the opposite side into two segments that are proportional to the other two sides of the triangle.

Proof of Theorem 10-5

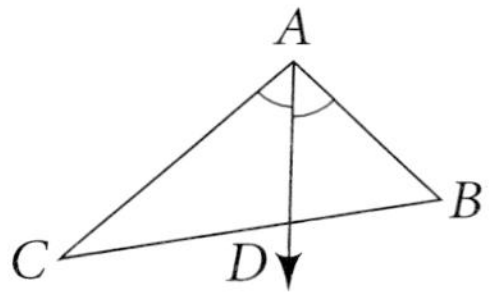

Given: $\triangle ABC$, $\overrightarrow{AD}$ bisects $\angle CAB$.

Prove: $\frac{CD}{DB} = \frac{CA}{BA}$

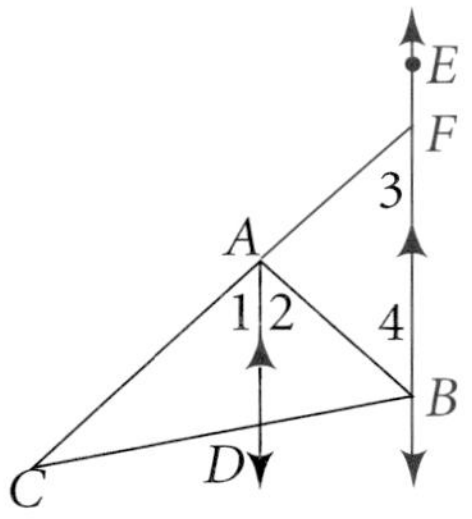

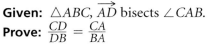

Draw $\overleftrightarrow{BE} \parallel \overline{DA}$.

Extend $\overline{CA}$ to meet $\overleftrightarrow{BE}$ at point F.
By the Side-Splitter Theorem, $\frac{CD}{DB} = \frac{CA}{AF}$. By the Corresponding Angles Theorem, $\angle 3 \cong \angle 1$. Since $\overrightarrow{AD}$ bisects $\angle CAB$, $\angle 1 \cong \angle 2$. By the Alternate Interior Angles Theorem, $\angle 2 \cong \angle 4$. Using the Transitive Property of Congruence, you know that $\angle 3 \cong \angle 4$. By the Converse of the Isosceles Triangle Theorem, $AF = AB$. Substituting BA for AF, $\frac{CD}{DB} = \frac{CA}{BA}$.

Example 2

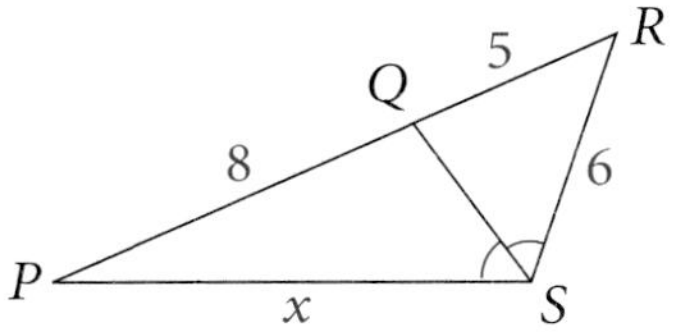

Find the value of x.

$\frac{PS}{SR} = \frac{PQ}{RQ}$ Use the Triangle-Angle-Bisector Theorem.

$\frac{x}{6} = \frac{8}{5}$ Substitute.

$x = \frac{6 \cdot 8}{5} = 9.6$

5. Use the diagram in Example 2 and the properties of proportions to write four equivalent proportions.

Exercises ON YOUR OWN

Use the figure at the right to complete each proportion.

1. $\frac{RS}{\blacksquare} = \frac{JR}{KJ}$

2. $\frac{KJ}{JP} = \frac{KS}{\blacksquare}$

3. $\frac{QL}{PM} = \frac{SQ}{\blacksquare}$

4. $\frac{PT}{\blacksquare} = \frac{TQ}{KQ}$

5. $\frac{KL}{LW} = \frac{\blacksquare}{MW}$

6. $\frac{\blacksquare}{KP} = \frac{LQ}{KQ}$

7. $\frac{\blacksquare}{SQ} = \frac{JK}{KS}$

8. $\frac{KL}{KM} = \frac{\blacksquare}{MW}$

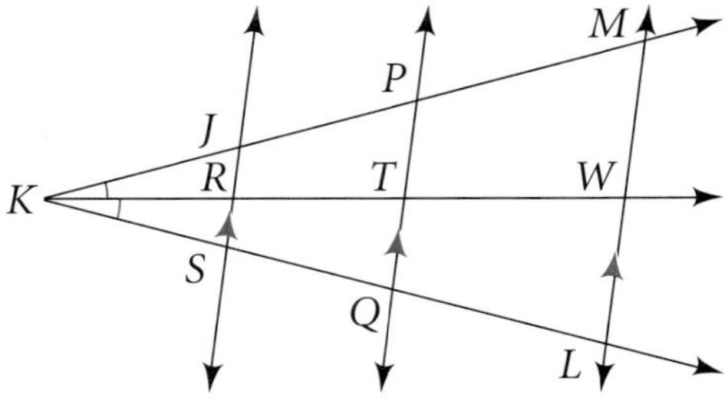

Algebra **Solve for *x*.**

9.

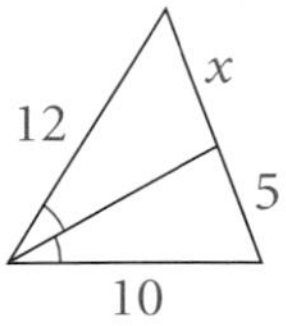

10.

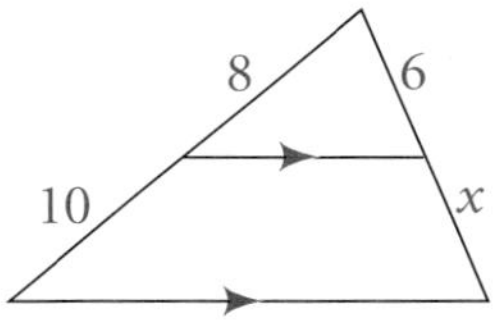

11.

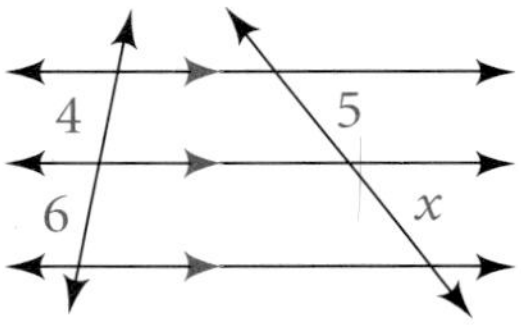

12.

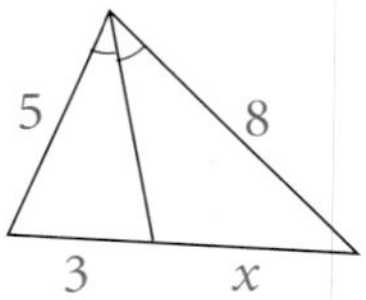

13.

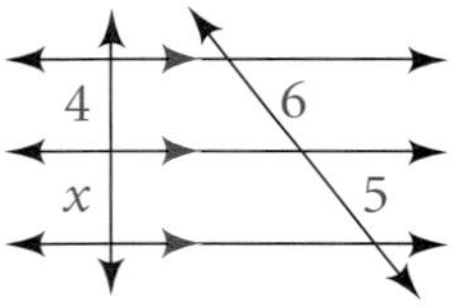

14.

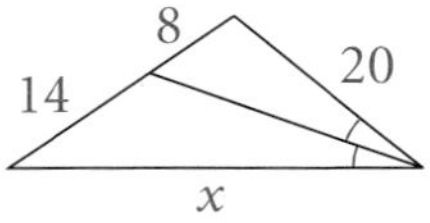

15.

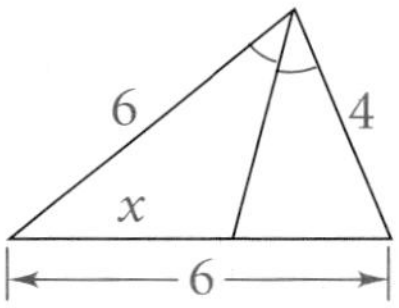

16.

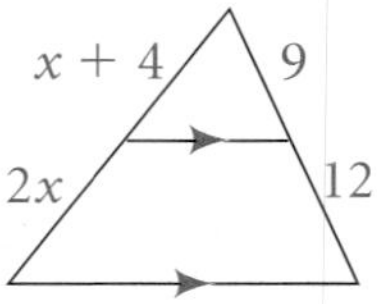

17. **Geography** In Washington, D.C., 17th, 18th, 19th, and 20th Streets are parallel streets that intersect Pennsylvania Avenue and I Street.

a. How long (to the nearest foot) is Pennsylvania Avenue between 19th Street and 18th Street?

b. How long (to the nearest foot) is Pennsylvania Avenue between 18th Street and 17th Street?

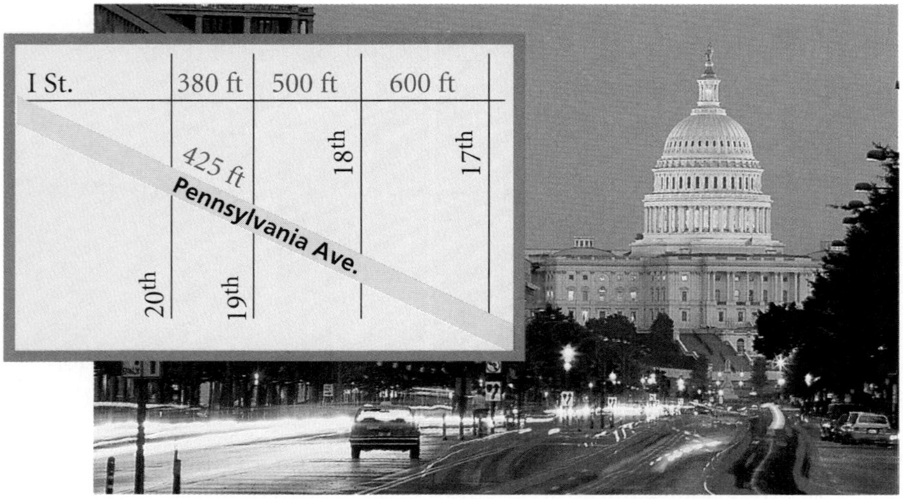

The United States Capitol viewed from Pennsylvania Avenue

18. The legs of a right triangle are 5 cm and 12 cm long. Find the lengths, to the nearest tenth, of the segments into which the bisector of the right angle divides the hypotenuse.

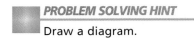

PROBLEM SOLVING HINT
Draw a diagram.

19. *Standardized Test Prep* In a triangle, the bisector of an angle divides the opposite side into two segments with lengths 6 cm and 9 cm. Which of the following can be the lengths of the other two sides of the triangle?

I. 4 cm and 6 cm **II.** 20 cm and 30 cm **III.** 12 cm and 18 cm

A. I only **B.** II only **C.** I and II **D.** II and III **E.** I, II, and III

Algebra **Solve for x.**

20.

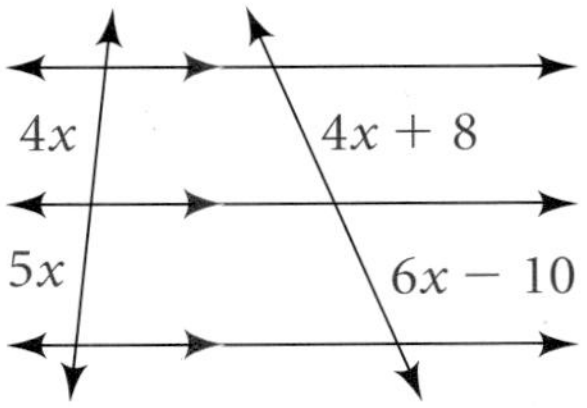

21.

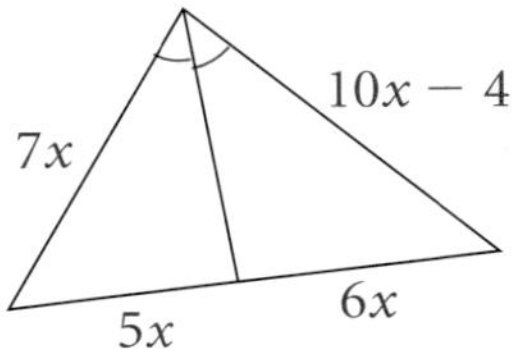

22.

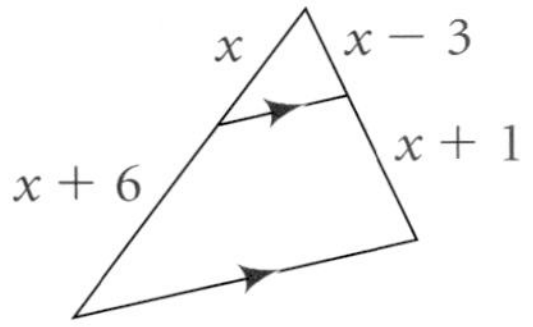

23. *Critical Thinking* Sharell draws $\triangle ABC$. She finds that the bisector of $\angle C$ bisects the opposite side.

a. Sketch $\triangle ABC$ and the bisector.

b. What type of triangle is $\triangle ABC$? Explain your reasoning.

24. Answer the following questions to prove the Corollary to the Side-Splitter Theorem.

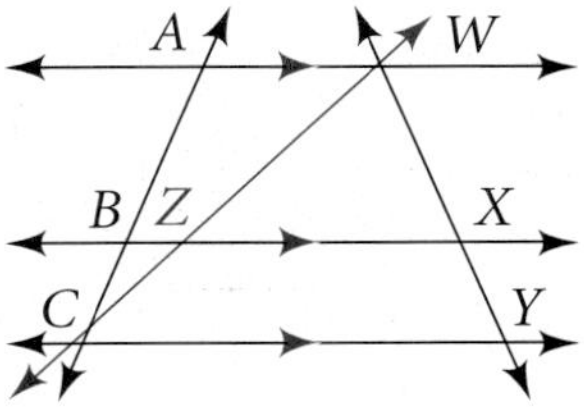

Given: $\overleftrightarrow{AW} \parallel \overleftrightarrow{BX} \parallel \overleftrightarrow{CY}$

Prove: $\frac{AB}{BC} = \frac{WX}{XY}$

Begin by drawing $\overleftrightarrow{WC}$, intersecting $\overline{BX}$ at point Z.

a. Apply the Side-Splitter Theorem to $\triangle ACW$: $\frac{■}{■} = \frac{WZ}{ZC}$.

b. Apply the Side-Splitter Theorem to $\triangle CWY$: $\frac{WZ}{ZC} = \frac{■}{■}$.

c. Substitute to prove the corollary.

25. An angle bisector of a triangle divides the opposite side of the triangle into segments 5 cm and 3 cm long. One side of the triangle is 7.5 cm long. Find all possible lengths for the third side of the triangle.

26. *Surveying* The perimeter of the triangular lot at the right is 50 m. The surveyor's tape bisects an angle. Find lengths x and y. (*Hint:* You can use a system of linear equations. See Skills Handbook page 667.)

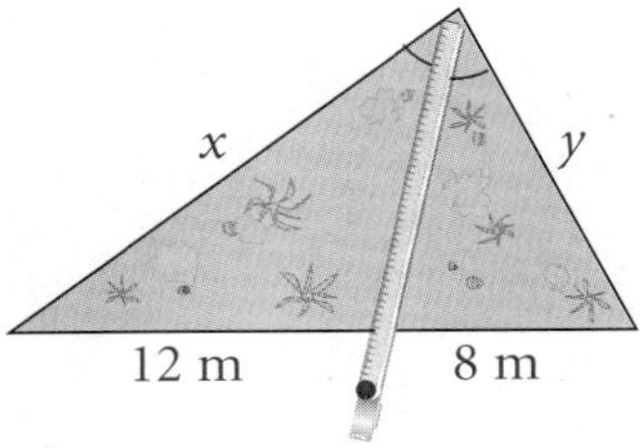

27. a. Logical Reasoning Write the converse of the Side-Splitter Theorem.

b. Draw a diagram, state the *Given* and *To Prove* for the converse, and write a Plan for Proof.

Determine whether the red segments are parallel. Explain each answer. You can use the result of Exercise 27.

28.
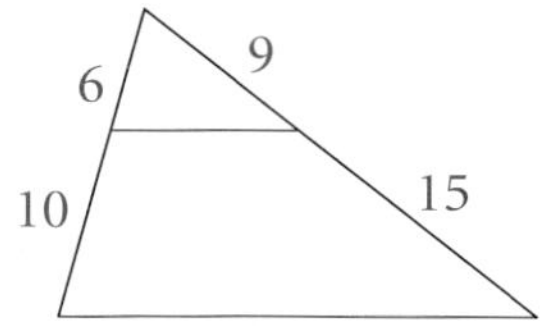

29.
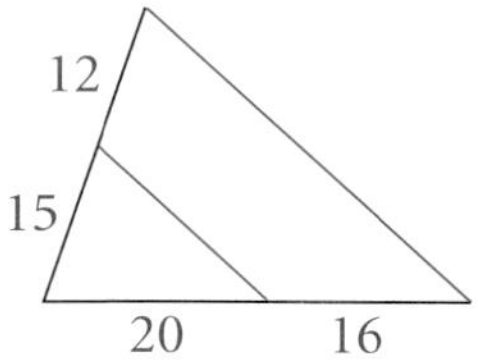

30.
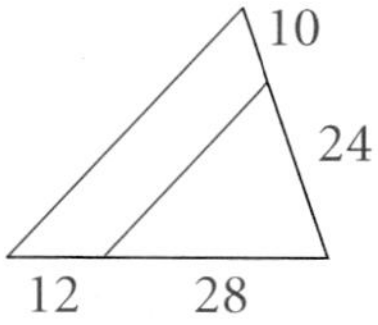

31.
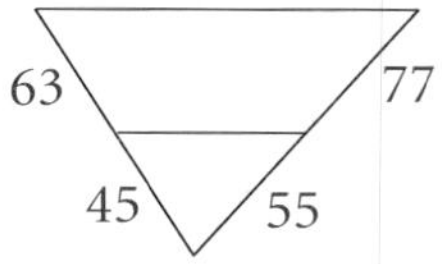

32. Technology Claire used geometry software to draw the red triangle. She dilated it with center (0, 0) and scale factor $\frac{3}{2}$ to get the blue triangle. Then she dilated the blue triangle with center (0, 0) and scale factor $\frac{1}{3}$ to get the green triangle.

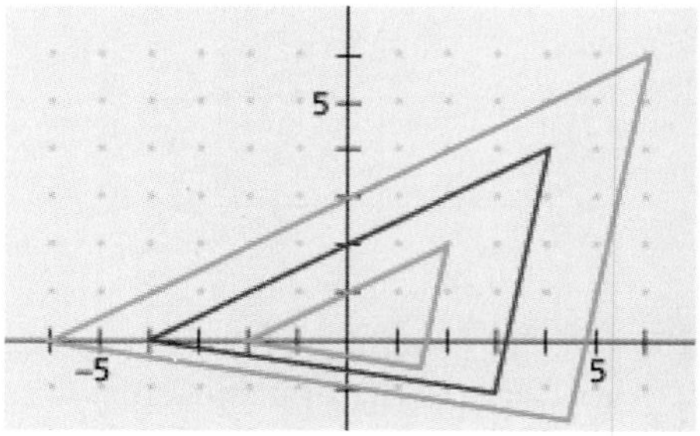

a. What is the scale factor of the dilation of the red triangle to the green one?

b. The scale factor of the dilation of the green triangle to the red one is 2. What is the scale factor of the dilation of the green triangle to the blue one?

33. Geometry in 3 Dimensions In the pyramid at the right, $\overleftrightarrow{FG} \parallel \overleftrightarrow{AB}$ and $\overleftrightarrow{GH} \parallel \overleftrightarrow{BC}$, $AF = 2$, $FE = 4$, and $BG = 3$.

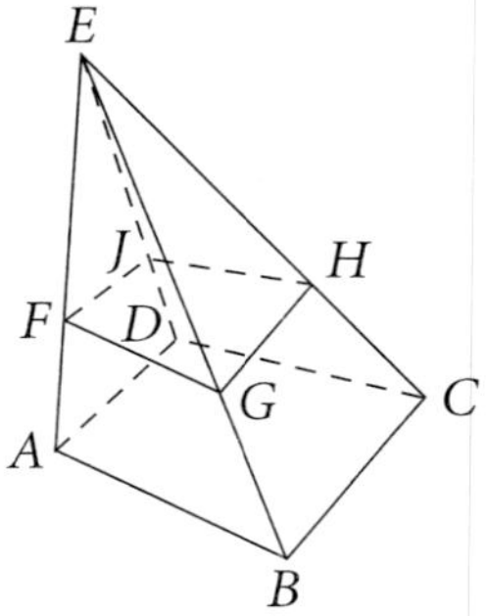

a. Find GE.

b. If $EH = 5$, find HC.

c. If $FG = 3$, find the perimeter of $\triangle ABE$.

34. The legs of a right triangle are 6 cm and 8 cm long. The bisector of the right angle divides the hypotenuse into two segments.

a. Draw a sketch of the right triangle with its angle bisector.

b. Find the lengths of these segments to the nearest tenth.

35. Writing Describe how you could use the figure at the right to find the length of the oil spill indirectly. What measurements and calculations would you use?

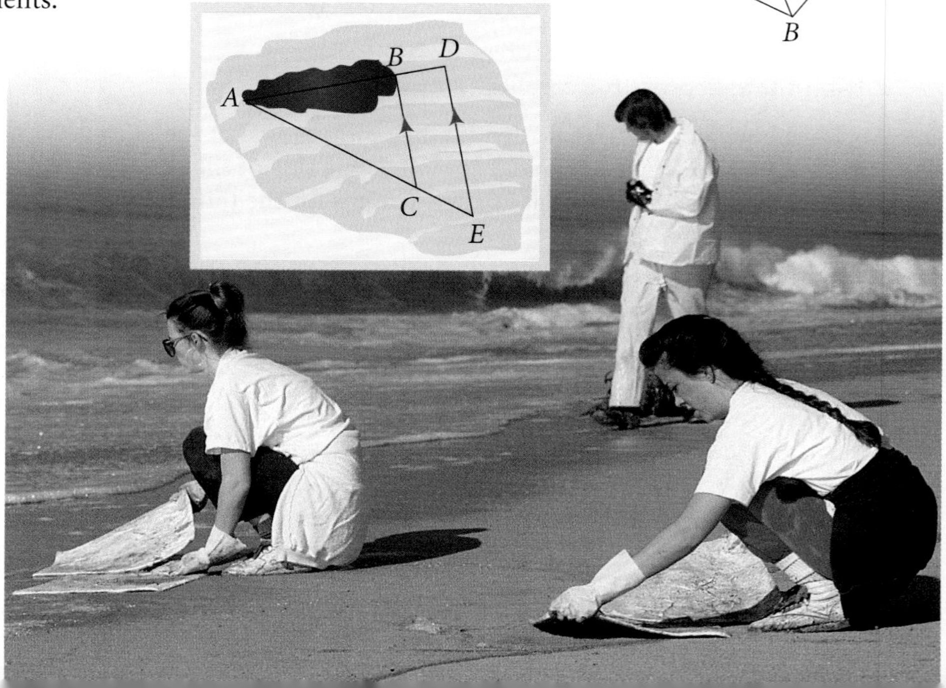

Exercises ON YOUR OWN

For each pair of similar figures, give the ratio of the perimeters and the ratio of the areas of the first figure to the second one.

1.

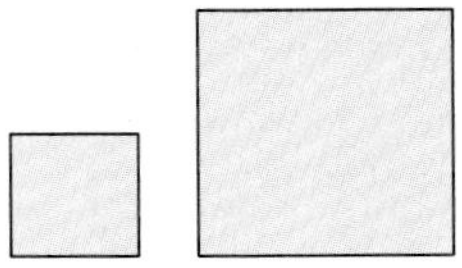

2.

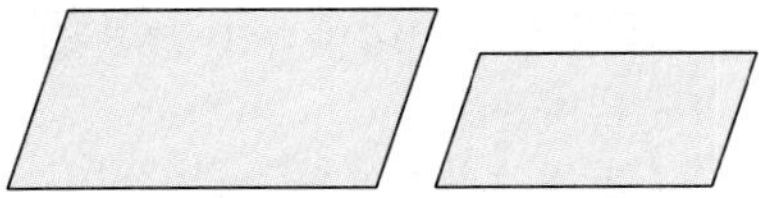

3.

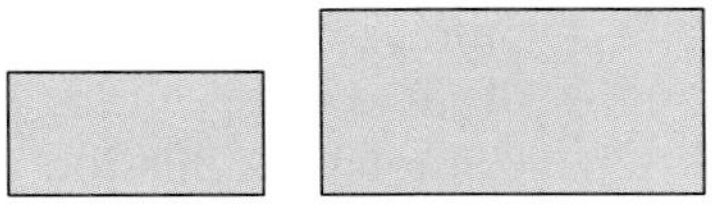

What is the similarity ratio of each pair of similar figures?

4. two circles with areas 2π cm^2 and 200π cm^2

5. two regular octagons with areas 4 ft^2 and 16 ft^2

6. two triangles with areas 80 m^2 and 20 m^2

7. two trapezoids with areas 49 cm^2 and 9 cm^2

8. a. Data Analysis A reporter used the pictograph at the right to show that the number of houses with more than two televisions has doubled in the past few years. Why is such a pictograph misleading?

b. Research Find examples in magazines or newspapers in which areas of similar figures give misleading information.

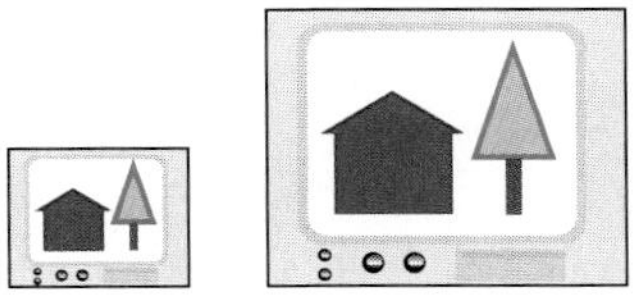

9. Remodeling It costs the Johnsons \$216 to have a 9 ft-by-12 ft wooden floor refinished. At that rate, how much would it cost them to have a 12 ft-by-16 ft wooden floor refinished?

10. Transformations $R'S'T'W'$ is the image of $RSTW$ under a dilation with center $(1, 1)$. How do the areas of $RSTW$ and $R'S'T'W'$ compare? Explain.

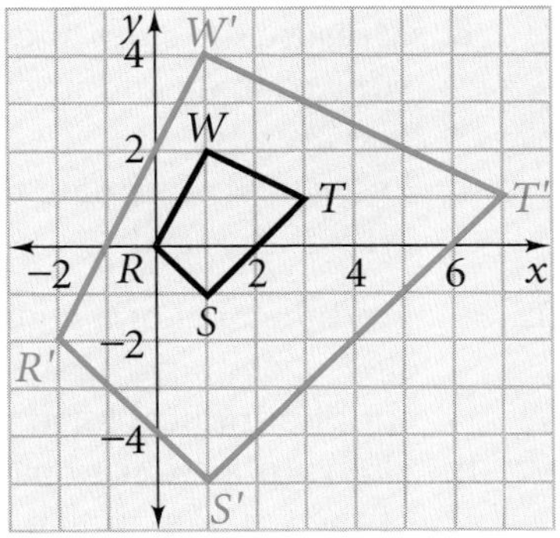

11. Drawing Draw a square with an area of 4 in.^{2}. Draw a second square with an area that is four times as large. What is the ratio of their perimeters?

12. The area of a regular decagon is 50 cm^2. What is the area of a regular decagon with sides four times the length of the smaller decagon?

13. In $\triangle RST$, $RS = 20$ m, $ST = 25$ m, and $RT = 40$ m.

a. Open-ended Choose a convenient scale. Then use a compass and a metric ruler to draw $\triangle R'S'T' \sim \triangle RST$.

b. Constructions Construct an altitude of $\triangle R'S'T'$ and measure its length. Find the area of $\triangle R'S'T'$.

c. Estimation Estimate the area of $\triangle RST$.

14. The area of the smaller triangle shown at the right is 24 cm^2. Solve for x and y.

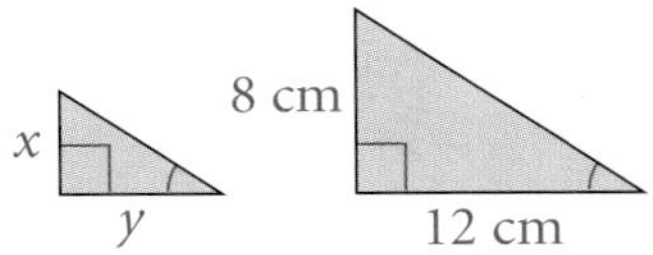

15. The areas of two similar rectangles are 27 in.^{2} and 48 in.^{2}. The longer side of the larger rectangle is 16 in. long. What is the length of the longer side of the smaller rectangle?

16. a. *Transformations* If a dilation with scale factor k maps a polygon onto a second polygon, what is the relationship between the lengths of the corresponding sides of the polygons?

b. Complete: A dilation with scale factor k maps any polygon to a similar polygon whose area is ■ times as large.

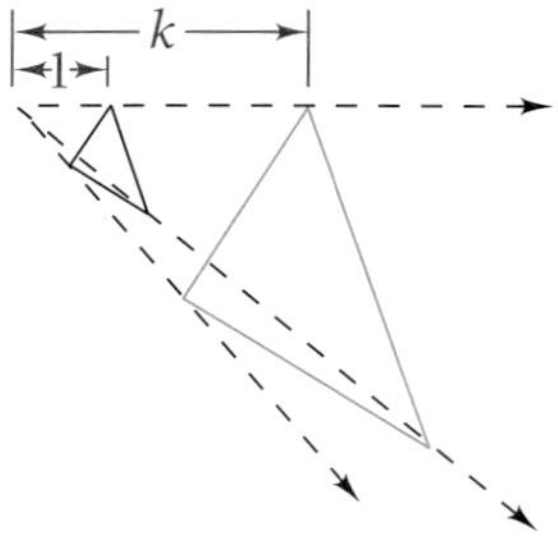

17. a. *Transformations* Graph $ABCD$ with vertices $A(4, 0)$, $B(-2, 4)$, $C(-4, 2)$, and $D(0, -2)$. Then graph its image $A'B'C'D'$ under the dilation with center $(0, 0)$ and scale factor $\frac{3}{2}$.

b. The area of $ABCD$ is 22 square units. What is the area of $A'B'C'D'$?

Find the ratio of the perimeters and the ratio of the areas of the blue figure to the red one.

18.

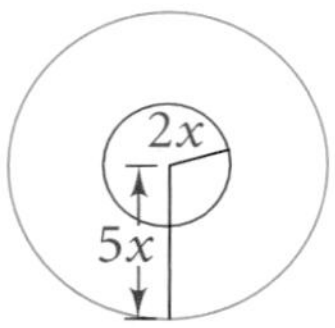

19.

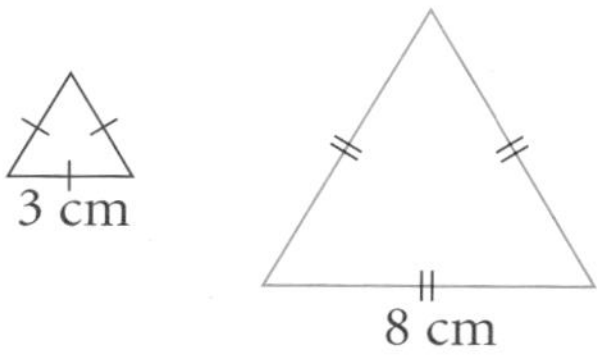

20.

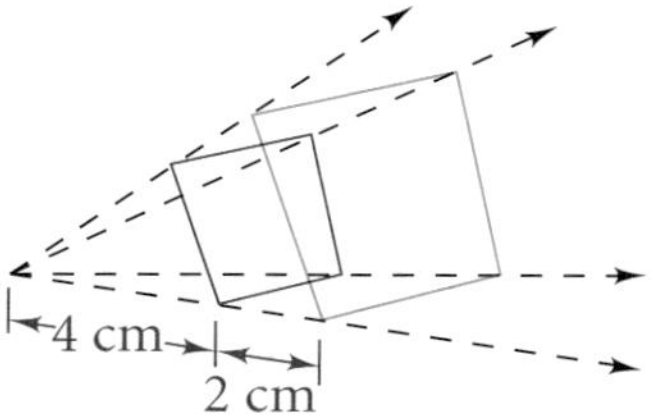

21. a. *Surveying* A surveyor measured one side and two angles of a field as shown in the diagram. Use a ruler and a protractor to draw a similar triangle.

b. Measure the sides and an altitude of your triangle and find its perimeter and area.

c. *Estimation* Estimate the perimeter and area of the field.

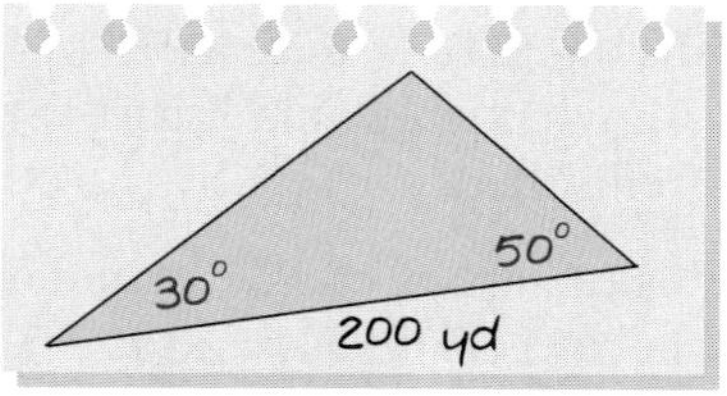

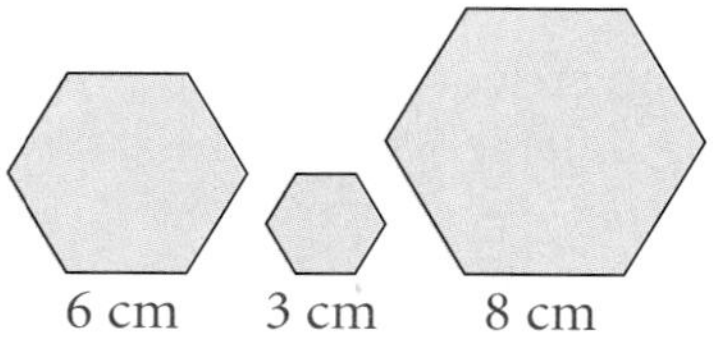

22. a. Find the area of a regular hexagon with sides 2 cm long. Leave your answer in simplest radical form.

b. Use your answer to part (a) and the ratio of the areas of similar polygons to find the areas of these regular hexagons.

23. *Writing* The enrollment at an elementary school is going to increase from 200 students to 395 students. The local parents' group is planning to increase the playground area from 100 ft by 200 ft to a larger area that is 200 ft by 400 ft. What would you tell the parents' group when they ask your opinion about whether the new playground area will be large enough?

Chapter Project

Find Out by Thinking

What is the area of the Koch snowflake? At each stage you increase the area by adding more and more equilateral triangles. Suppose the area of Stage 0 is 1 square unit. Copy the diagrams and explain why the area of the Koch snowflake will never be greater than 2 square units.

Stage 0
1 square
unit of area

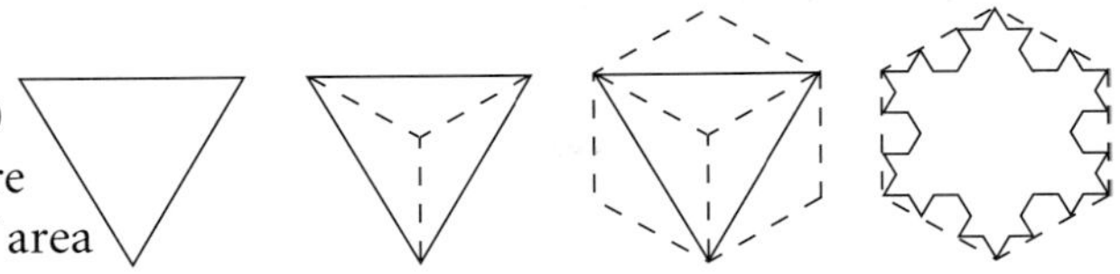

Exercises MIXED REVIEW

Find the length of the hypotenuse of a right triangle with the given legs. Leave your answers in simplest radical form.

24. 8 cm, 9 cm **25.** 3 in., 5 in. **26.** 10 mm, 5mm

27. Drawing Use a protractor to draw a regular octagon.

Getting Ready for Lesson 10-6

Find the volume and surface area of each space figure.

28. cube with a 3-in. side

29. 3 m-by-5 m-by-9 m rectangular prism

SELF ASSESSMENT

FOR YOUR JOURNAL

In your own words, explain the relationship among the similarity ratio of two similar polygons, the ratio of their perimeters, and the ratio of their areas. Include an example.

Exercises CHECKPOINT

Algebra **Find the values of the variables.**

1.

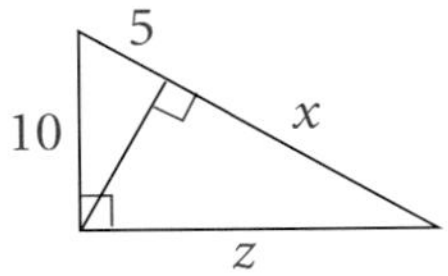

2.

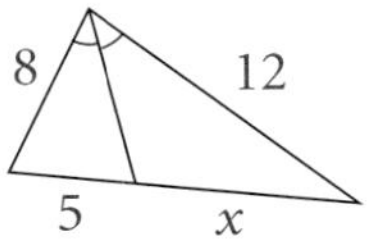

3.

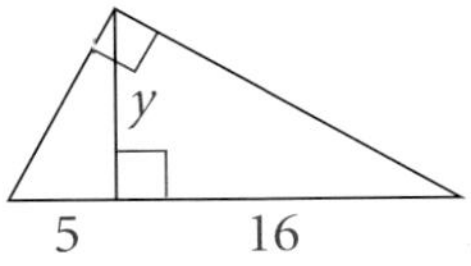

4.

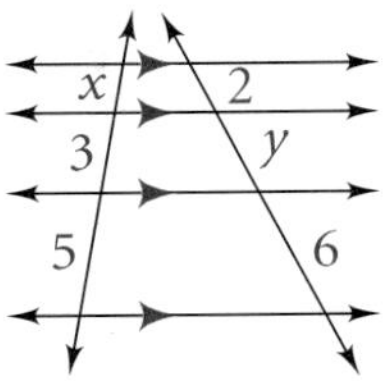

Find the ratios of the perimeters of these similar figures.

5. two triangles with areas 4 ft^2 and 16 ft^2

6. two kites with areas 100 cm^2 and 25 cm^2

7. two squares with areas 63 m^2 and 7 m^2

8. two hexagons with areas 18 $in.^2$ and 128 $in.^2$

9. Standardized Test Prep In $\triangle ABC$, $\overrightarrow{BD}$ bisects $\angle B$. Which of the following is true?

A. $\frac{AD}{CD} = \frac{CD}{DB}$ **B.** $\frac{AB}{CB} = \frac{CB}{DB}$ **C.** $\frac{AB}{AC} = \frac{AC}{AD}$

D. $\frac{AB}{BC} = \frac{AD}{DC}$ **E.** all of the above

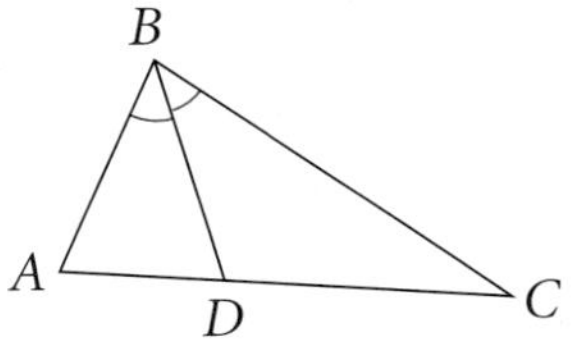

Technology

Exploring Similar Solids

With Lesson 10-6

Work in pairs or small groups.

Input

To explore surface areas and volumes of similar rectangular prisms, set up a spreadsheet like the one below. You choose numbers for the length, width, height, and similarity ratio. All other numbers will be calculated by formulas.

	A	B	C	D	E	F	G	H	I
1					Surface		Similarity		
2		Length	Width	Height	Area	Volume	Ratio (II : I)	Ratio of	Ratio of
3	Rectangular Prism I	6	4	23	508	552	2	Surface	Volumes
4								Areas (II : I)	(II : I)
5	Similar Prism II	12	8	46	2032	4416		4	8

In cell E3 enter the formula $= 2 * (B3 * C3 + B3 * D3 + C3 * D3)$, which will calculate the sum of the areas of the six faces of Prism I. In cell F3 enter the formula $= B3 * C3 * D3$, which will calculate the volume of Prism I.

In cells B5, C5, and D5 enter the formulas $= G3 * B3$, $= G3 * C3$, and $= G3 * D3$, respectively, which will calculate the dimensions of similar Prism II. Copy the formulas from E3 and F3 into E5 and F5 to calculate the surface area and volume of Prism II.

In cell H5 enter the formula $= E5/E3$ and in cell I5 enter the formula $= F5/F3$ to calculate the ratios of the surface areas and volumes.

Investigate

In row 3, enter numbers for the length, width, height, and similarity ratio. Change those numbers to investigate how the ratios of the surface areas and volumes are related to the similarity ratio.

Conjecture

Make conjectures about the relationships you have discovered.

Extend

- Set up a spreadsheet to investigate the ratios of the surface areas and volumes of similar right cylinders. Do the same for similar square pyramids.

10-6 Areas and Volumes of Similar Solids

What You'll Learn

- Finding the relationships between the similarity ratio and the ratios of the areas and volumes of similar solids

...And Why

To determine the area or volume of a solid when you know the area or volume of a similar solid

What You'll Need

- isometric dot paper
- calculator

WORK TOGETHER

Work with a group.

- Draw a 2-by-3-by-2 rectangular prism on isometric dot paper.
- Choose a value of k from the set {2, 3, 4, 5} and draw a $2k$-by-$3k$-by-$2k$ rectangular prism and call it Prism I. Choose two other values of k and draw Prisms II and III.
- Find the surface area and volume of each prism. Compare the original prism to each of the three prisms. Organize your data in a table like this.

	Value of k	Ratio of Corresponding Edges	Ratio of Surface Areas	Ratio of Volumes
Original to I	■	■ : ■	■ : ■	■ : ■
Original to II	■	■ : ■	■ : ■	■ : ■
Original to III	■	■ : ■	■ : ■	■ : ■

1. a. If the corresponding dimensions of two rectangular prisms are in the ratio 1 : k, what will be the ratio of their surface areas?
   b. What will be the ratio of their volumes?
2. a. If the corresponding dimensions of two rectangular prisms are in the ratio $a : b$, what will be the ratio of their surface areas?
   b. What will be the ratio of their volumes?

THINK AND DISCUSS

These nested Russian dolls are similar solids. **Similar solids** have the same shape and all their corresponding dimensions are proportional.

3. **Critical Thinking** Which of the following *must* be similar solids?
   a. two spheres b. two cones c. two cubes

4. Can a right triangular prism be similar to an oblique triangular prism? Explain why or why not.

5. Can a triangular pyramid be similar to a square pyramid? Explain why or why not.

The ratio of corresponding dimensions of two similar solids is the **similarity ratio.** In the Work Together, you discovered that the ratio of the surface areas of similar solids equals the square of their similarity ratio, and the ratio of their volumes equals the cube of their similarity ratio. The ratios you have investigated for similar rectangular prisms apply to *all* similar solids.

Theorem 10-7
Areas and Volumes of Similar Solids

If the similarity ratio of two similar solids is $a : b$, then
(1) the ratio of their corresponding areas is $a^2 : b^2$, and
(2) the ratio of their volumes is $a^3 : b^3$.

Example 1

The lateral areas of two similar cylinders are 196π in.^{2} and 324π in.^{2}. The volume of the smaller cylinder is 686π in.^{3}. Find the volume of the larger cylinder.

First find the similarity ratio $a : b$.

$$\frac{a^2}{b^2} = \frac{196\pi}{324\pi}$$ The ratio of the surface areas is $a^2 : b^2$.

$$\frac{a^2}{b^2} = \frac{49}{81}$$ Divide the numerator and denominator by the common factor 4π.

$$\frac{a}{b} = \frac{7}{9}$$ Find the square root of each side.

$$\frac{V_1}{V_2} = \frac{7^3}{9^3}$$ The ratio of the volumes is $a^3 : b^3$.

$$\frac{686\pi}{V_2} = \frac{343}{729}$$ Substitute 686π for V_1.

$$343V_2 = 686\pi \cdot 729$$ Cross-Product Property

$$V_2 = 1458\pi$$ Divide both sides by 343.

The volume of the larger cylinder is 1458π in.^{3}.

6. **Try This** The surface areas of two similar solids are 160 m^2 and 250 m^2. The volume of the larger one is 250 m^3. What is the volume of the smaller one?

The weights of objects made of the same material are proportional to their volumes.

Example 2 *Relating to the Real World*

Paperweights A marble paperweight shaped like a pyramid weighs 0.15 lb. How much does a similarly-shaped paperweight weigh if each dimension is three times as large?

The similarity ratio is 1 : 3, so the ratio of the volumes is $1^3 : 3^3$, or 1 : 27.

$$\frac{1}{27} = \frac{0.15}{x} \qquad \text{Let } x = \text{weight of the larger paperweight.}$$

$$x = 27\,(0.15) \qquad \text{Cross-Product Property}$$

$$x = 4.05$$

The larger marble paperweight weighs about 4 lb.

7. **Try This** There are 750 toothpicks in a regular-size box. If a jumbo box is made by doubling all the dimensions of the regular box, how many toothpicks will the jumbo box hold?

8. **Try This** A regular pentagonal prism with base edges 9 cm long is enlarged to a similar prism with base edges 36 cm long. By what factor is its volume increased?

Exercises ON YOUR OWN

Are the solids similar? Explain.

1.

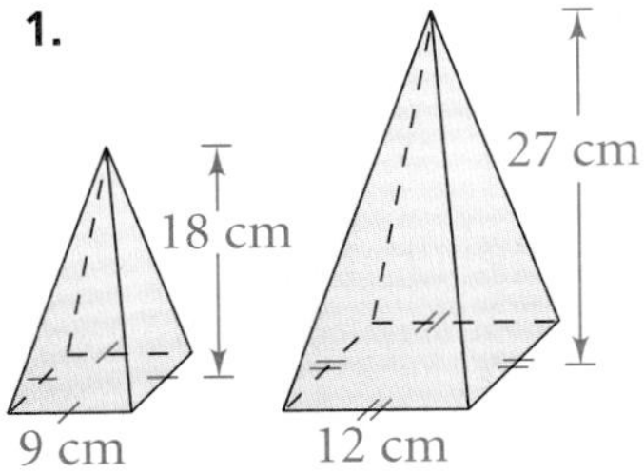

2.

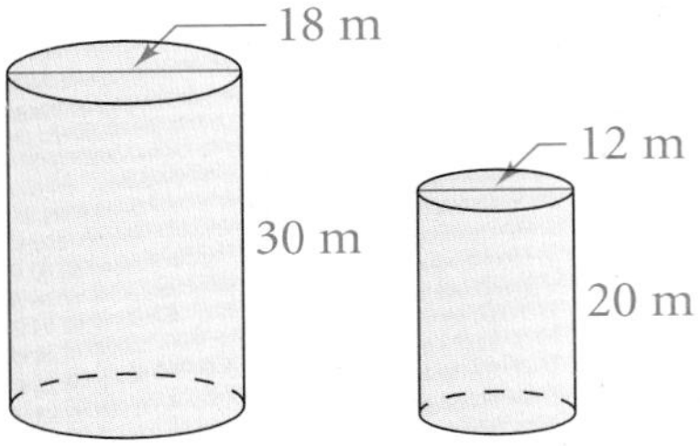

3.

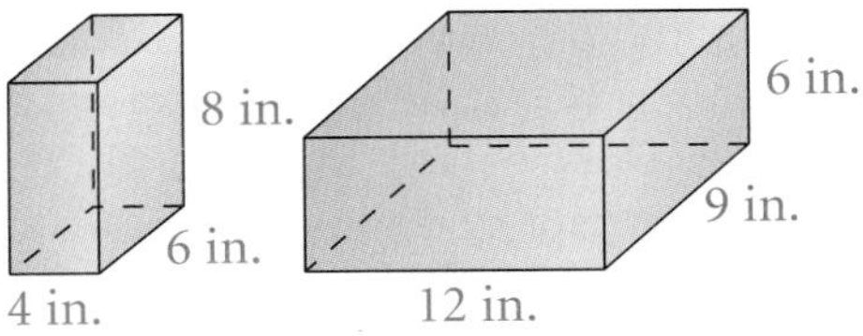

4. Two similar prisms have heights 4 cm and 10 cm.
   a. What is the similarity ratio?
   b. What is the ratio of the surface areas?
   c. What is the ratio of the volumes?

5. *Standardized Test Prep* The ratio of the surface areas of two similar solids is 16 : 49. What is the volume of the smaller one?
   A. 4 **B.** 7 **C.** 64 **D.** 343 **E.** cannot be determined

6. Is there a value of x for which the rectangular solids at the right are similar? Explain.

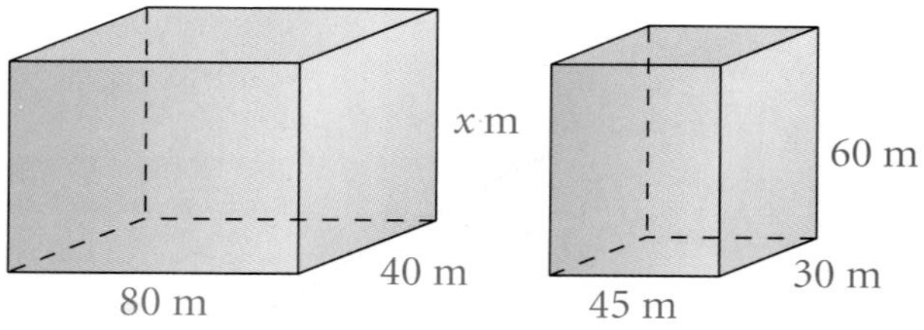

7. The volumes of two spheres are 729 in.^{3} and 81 in.^{3}. Find the ratio of their radii.

8. A carpenter is making a copy of an antique blanket chest that has the shape of a rectangular solid. The length, width, and height of the copy will be 4 in. greater than the original dimensions. Will the chests be similar? Explain.

9. **Estimation** The volume of a spherical balloon with radius 3.1 cm is about 125 cm^3. Estimate the volume of a similar balloon with radius 6 cm.

Copy and complete the table for two similar solids.

	Similarity Ratio	Ratio of Surface Areas	Ratio of Volumes
10.	1 : 2	■ : ■	■ : ■
11.	3 : 5	■ : ■	■ : ■
12.	■ : ■	49 : 81	■ : ■
13.	■ : ■	■ : ■	125 : 512

14. A clown's face on a balloon is 4 in. high when the balloon holds 108 in.^{3} of air. How much air must the balloon hold for the face to be 8 in. high?

15. **Critical Thinking** A company recently announced that it had developed the technology to reduce the size of *atomic clocks,* which are used in electronic devices that transmit data. The company claims that the smaller clock will be $\frac{1}{10}$ the size of existing atomic clocks and $\frac{1}{100}$ the weight. Do these ratios make sense? Explain.

16. **Packaging** A cylinder 4 in. in diameter and 6 in. high holds 1 lb of oatmeal. To the nearest ounce, how much oatmeal will a similar 10-in. high cylinder hold?

17. **Estimation** A teapot 4 in. tall and 4 in. in diameter holds about 1 quart of tea. About how many quarts would this teapot-shaped building hold?

18. Literature In *Gulliver's Travels* by Jonathan Swift, Gulliver first traveled to Lilliput. The average height of a Lilliputian was one twelfth Gulliver's height.

a. How many Lilliputian coats could be made from the material in Gulliver's coat? (*Hint:* Use the ratio of surface areas.)
b. How many Lilliputian meals would be needed to make a meal for Gulliver? (*Hint:* Use the ratio of volumes.)
c. Research Visit your local library and read about Gulliver's voyage to Brobdingnag in *Gulliver's Travels.* Describe the size of the people and objects in this land. Use ratios of areas and volumes to describe Gulliver's life there.

19. Giants such as King Kong, Godzilla, and Paul Bunyan cannot exist because their bone structure could not support them. The legendary Paul Bunyan is ten times as tall as the average human.
a. The strength of bones is proportional to the area of their cross-section. How many times stronger than the average person's bones would Paul Bunyan's bones be?
b. Weights of objects made of the same material are proportional to their volumes. How many times the average person's weight would Paul Bunyan's weight be?
c. Human leg bones can support about 6 times the average person's weight. How many times the average person's weight could Paul Bunyan's legs support?
d. Use your answers to parts (b) and (c) to explain why Paul Bunyan's legs could not support his weight.
e. Writing Explain why massive animals such as elephants have such thick legs.

Exercises MIXED REVIEW

Coordinate Geometry **Draw the image of each figure for the given transformation.**

20. $90°$ rotation clockwise about the origin

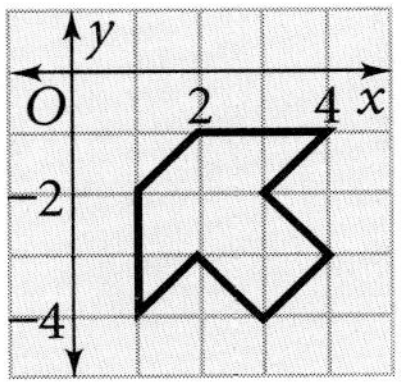

21. reflection in the line $y = x$

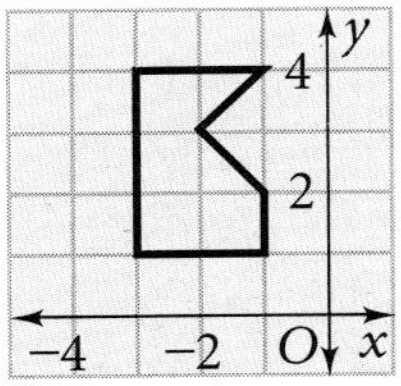

22. What is the area of a $30°$-$60°$-$90°$ triangle with a shorter leg that is 4 cm long? Leave your answer in simplest radical form.

SELF ASSESSMENT

PORTFOLIO

For your portfolio, select two or three items from your work for this chapter. Some possibilities:
- corrected work
- work that you think best shows what you learned or did in this chapter

Explain why you included each selection.

CHAPTER PROJECT

Finishing the Chapter Project

Fractals FOREVER

Find Out questions on pages 502, 516, 523, and 529 should help you complete your project. Prepare a report on fractals. Include the three basic properties of fractals and find nature photographs that illustrate these properties. Include a discussion of the perimeter and area of the Koch snowflake. Create a fractal tree or a fractal of your own design. Write directions for one iteration of your fractal by listing the steps used to create Stage 1.

Reflect and Revise

Ask a classmate to review your report. Ask this reviewer to test your directions for creating a fractal. Also, check that you have used geometric terms correctly and that your report is attractive as well as informative.

Follow Up

In studying the Koch snowflake, you noticed that the total perimeter increases beyond any fixed number (it is infinite), while the area remains less than some fixed number (it is finite). Find or create another fractal with this property and include it in your report.

For More Information

Coes, III, Loring. "Building Fractal Models with Manipulatives." *Mathematics Teacher* (November 1993): 646–651.

Peitgen, Heinz-Otto, et al. *Fractals for the Classroom: Strategic Activities Volume One.* New York: Springer-Verlag, 1991.

Peterson, Ivers. *The Mathematical Tourist.* New York: W.H. Freeman, 1988.

CHAPTER 10

Wrap Up

How am I doing?

- State three ideas from this chapter that you think are important. Explain your choices.
- Describe the properties of similar figures.

SELF ASSESSMENT

Key Terms

Angle-Angle Similarity Postulate (AA~) (p. 504)
constant of variation (p. 503)
Cross-Product Property (p. 497)
direct variation (p. 503)
extended proportion (p. 496)
geometric mean (p. 512)
Golden Ratio (p. 498)
Golden Rectangle (p. 498)
indirect measurement (p. 505)
proportion (p. 496)
scale (p. 497)
scale drawing (p. 497)
Side-Angle-Side Similarity Theorem (SAS~) (p. 506)
Side-Side-Side Similarity Theorem (SSS~) (p. 506)
similar solids (p. 531)
similarity ratio (p. 532)

Ratio, Proportion, and Similarity 10-1

A **proportion** is a statement that two ratios are equal.

$\frac{a}{b} = \frac{c}{d}$ is equivalent to:

(1) $ad = bc$ (2) $\frac{b}{a} = \frac{d}{c}$ (3) $\frac{a}{c} = \frac{b}{d}$ (4) $\frac{a+b}{b} = \frac{c+d}{d}$

In a **scale drawing,** the **scale** compares each length in the drawing to the actual length being represented.

A **Golden Rectangle** is a rectangle that can be divided into a square and a rectangle that is similar to the original rectangle. The lengths of the adjacent sides of a Golden Rectangle are in a ratio called the **Golden Ratio,** which is $1 : \frac{1+\sqrt{5}}{2}$.

The polygons below are similar. Find the values of the variables.

1.

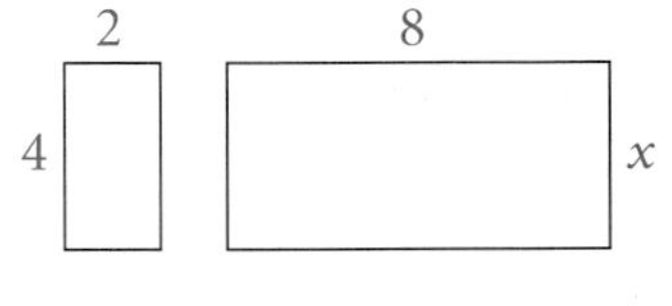

2.

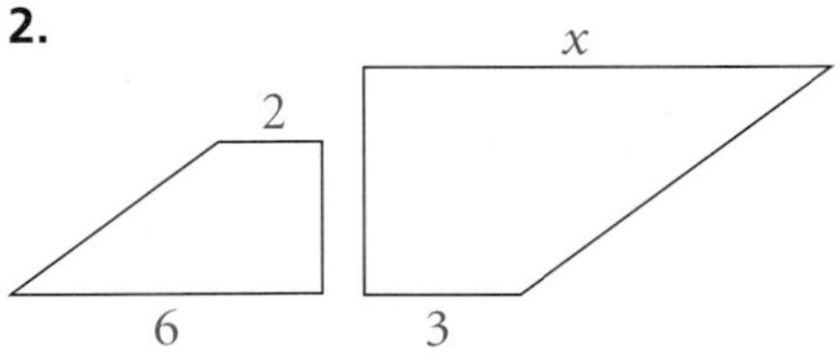

3.

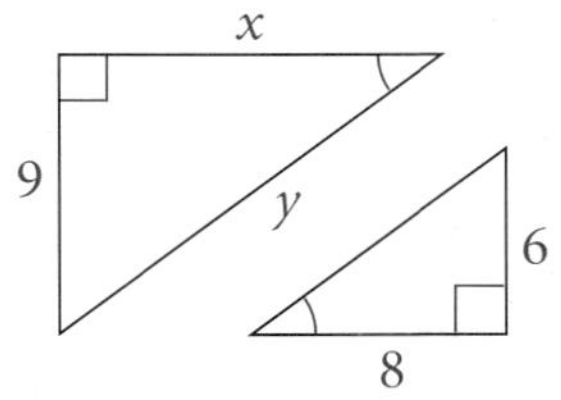

Proving Triangles Similar: AA, SAS, and SSS 10-2

If two angles of one triangle are congruent to two angles of another triangle, then the two triangles are similar by the **Angle-Angle Similarity Postulate (AA~).** If an angle of one triangle is congruent to an angle of a second triangle, and the sides including the two angles are proportional, then the triangles are similar by the **Side-Angle-Side Similarity Theorem (SAS~).**

If the corresponding sides of two triangles are proportional, then the triangles are similar by the **Side-Side-Side Similarity Theorem (SSS~).**

Are the triangles similar? If so, write the similarity statement and name the postulate or theorem that proves they are similar.

4.

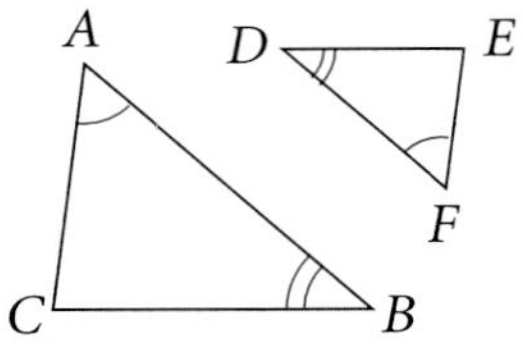

5.

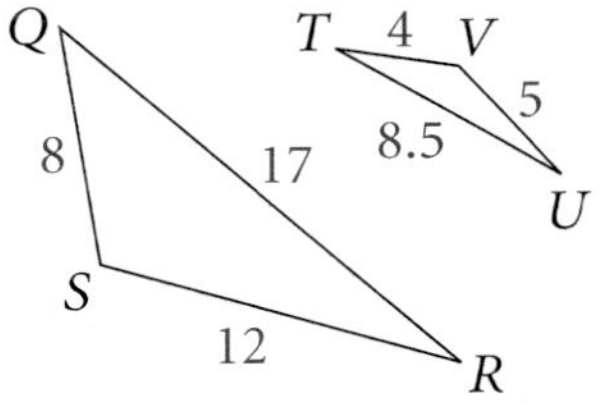

6.

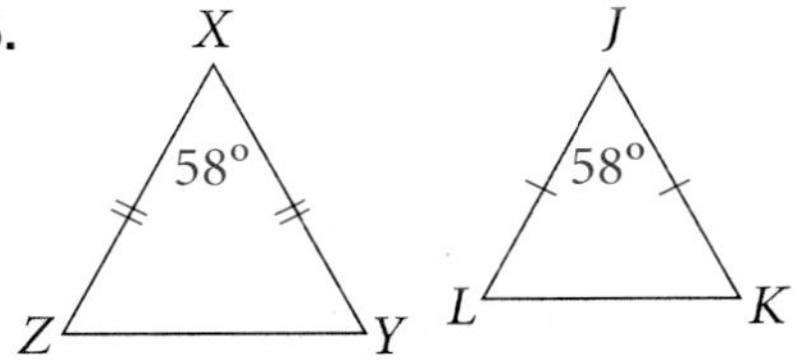

7. *Standardized Test Prep* Two right triangles have an acute angle with the same measure. What is the most direct way to prove that the triangles are similar?

A. SSS~ **B.** SAS~ **C.** AA~ **D.** SSA~ **E.** cannot be proven

Similarity in Right Triangles 10-3

The **geometric mean** of two positive numbers a and b is the positive number x such that $\frac{a}{x} = \frac{x}{b}$.

When the altitude is drawn to the hypotenuse of a right triangle:

- the two triangles formed are similar to the original triangle and to each other;
- the length of the altitude is the geometric mean of the lengths of the segments of the hypotenuse; and
- the length of each leg is the geometric mean of the lengths of the hypotenuse and the segment of the hypotenuse that is adjacent to that leg.

Find the values of the variables. When an answer is not a whole number, leave it in simplest radical form.

8.

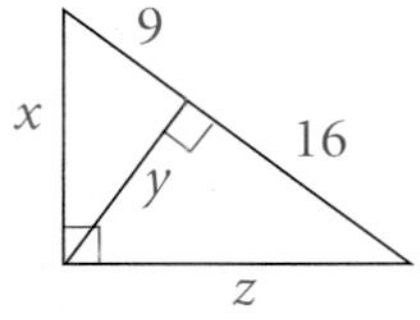

9.

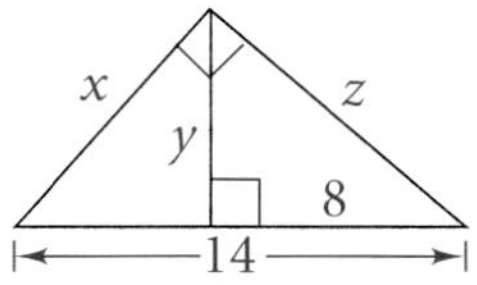

10.

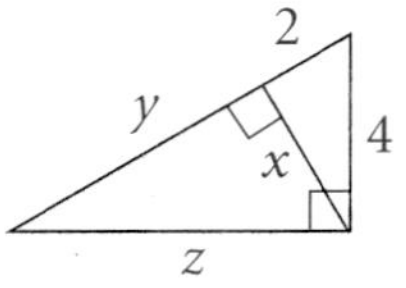

Proportions and Similar Triangles 10-4

If a line is parallel to one side of a triangle and intersects the other two sides, then it divides those sides proportionally. If three parallel lines intersect two transversals, then the segments intercepted on the transversals are proportional. The bisector of an angle of a triangle divides the opposite side into two segments that are proportional to the sides of the triangle.

11. *Writing* Explain why a line that intersects the legs of a trapezoid and is parallel to the bases divides the legs proportionally.

Find the value of x.

12.

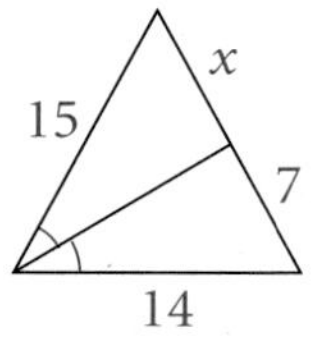

13.

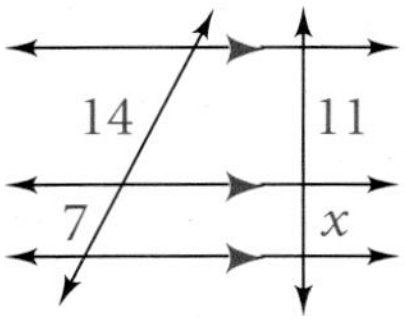

14.

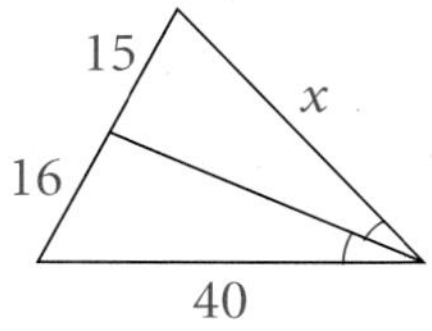

Perimeters, Areas, and Volumes of Similar Figures and Solids 10-5, 10-6

If the similarity ratio of two similar plane figures is $a : b$, then the ratio of their perimeters is $a : b$, and the ratio of their areas is $a^2 : b^2$.

Similar solids have the same shape and all their corresponding dimensions are proportional. If the similarity ratio of two similar solids is $a : b$, then the ratio of their surface areas is $a^2 : b^2$, and the ratio of their volumes is $a^3 : b^3$.

For each pair of similar figures, find the ratio of the area of the first figure to the area of the second.

15.

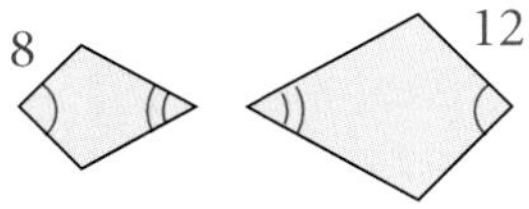

16.

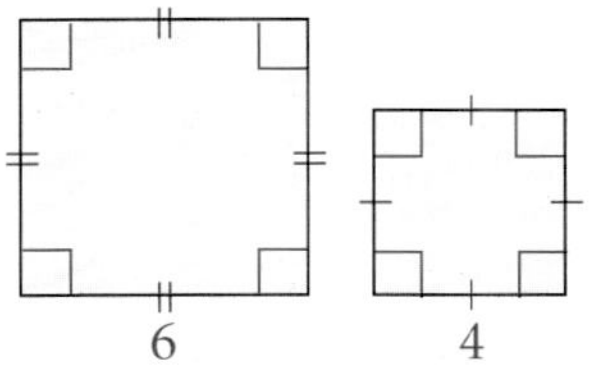

17.

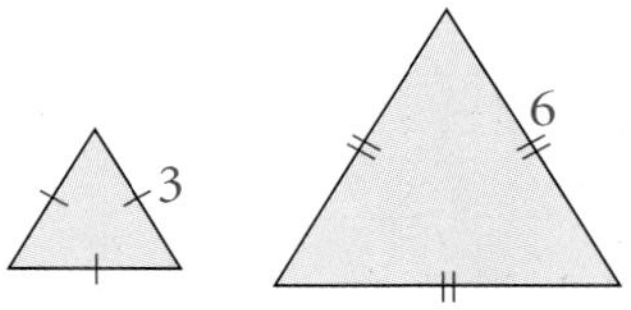

For each pair of similar solids, find the ratio of the volume of the first figure to the volume of the second.

18.

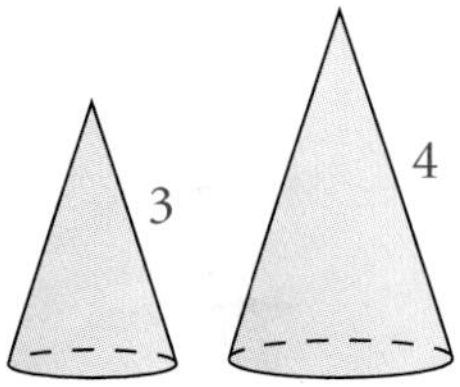

19.

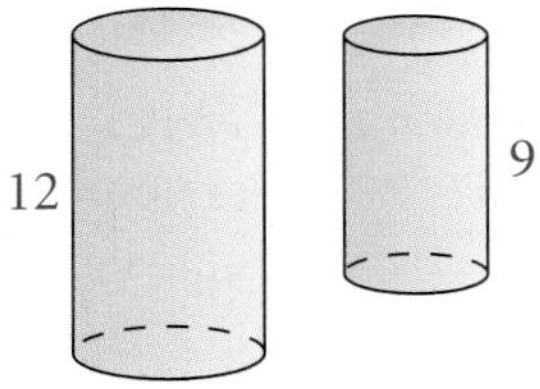

20.

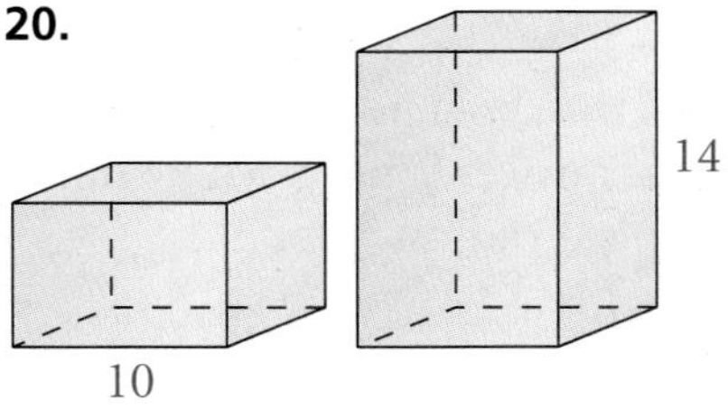

21. *Open-ended* Sketch two similar figures whose areas are in the ratio 16 : 25. Include dimensions.

Getting Ready for CHAPTER 11

Find the ratios $\frac{BC}{AB}$, $\frac{AC}{AB}$, and $\frac{BC}{AC}$. Leave your answers in simplest radical form.

22.

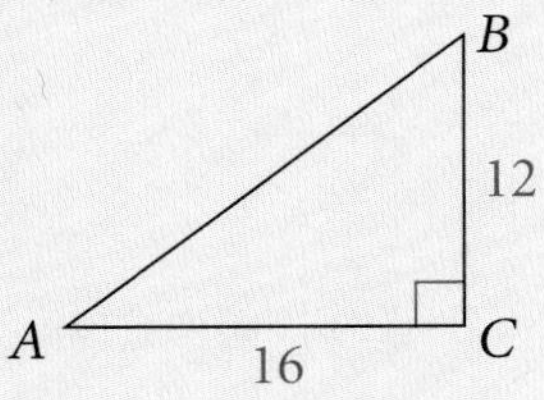

23.

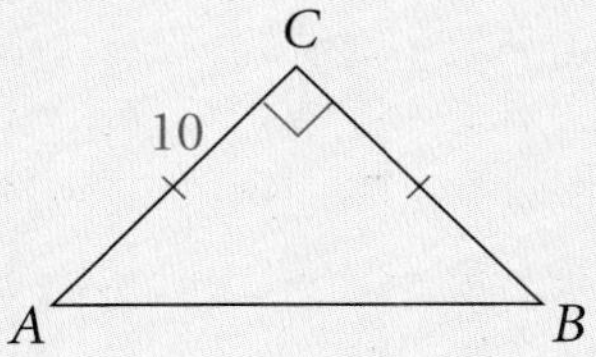

24.

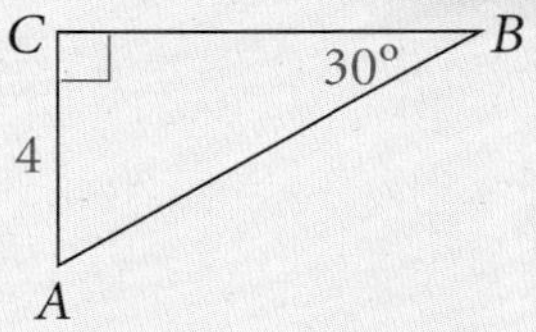

CHAPTER 10

Assessment

Solve each proportion.

1. $\frac{4}{5} = \frac{x}{20}$ **2.** $\frac{6}{x} = \frac{10}{7}$ **3.** $\frac{x}{3} = \frac{8}{12}$

Are the triangles similar? If so, write the similarity statement and name the postulate or theorem that you can use to prove they are similar.

4.

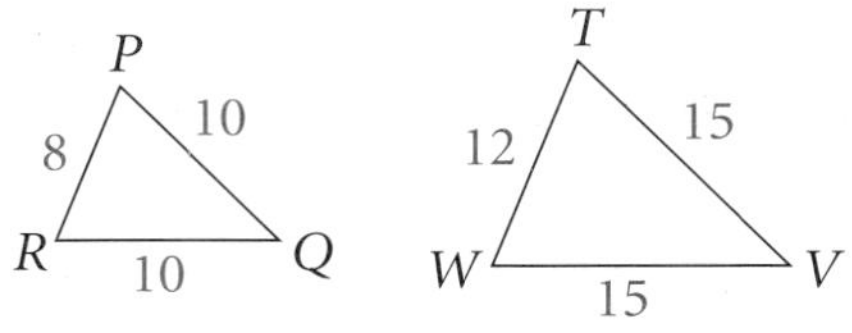

5.

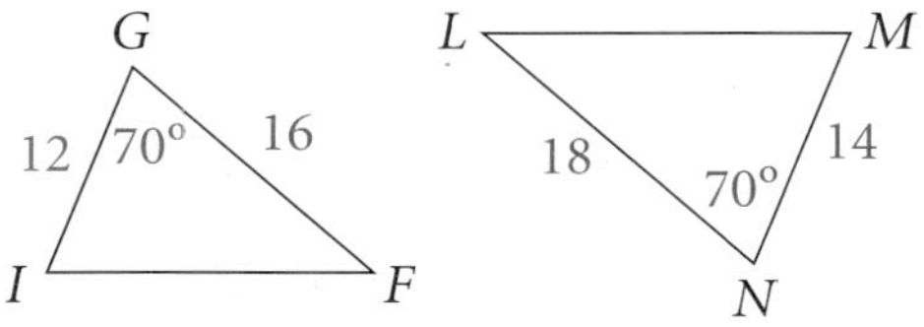

6. A meter stick is held perpendicular to the ground. It casts a shadow 1.5 m long. At the same time, a telephone pole casts a shadow that is 9 m long. How tall is the telephone pole?

Find the geometric mean of each pair of numbers. If the answer is not a whole number, leave it in simplest radical form.

7. 10, 15 **8.** 4, 9 **9.** 6, 12

10. *Standardized Test Prep* Which proportion is true for the triangle below?

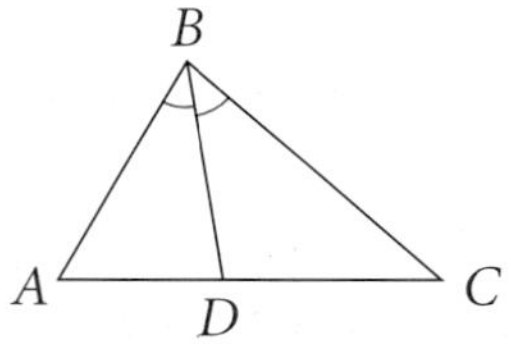

A. $\frac{AB}{BC} = \frac{AD}{DC}$ **B.** $\frac{AD}{BD} = \frac{BD}{CD}$ **C.** $\frac{AB}{AC} = \frac{AC}{AD}$

D. $\frac{AB}{BC} = \frac{AD}{BD}$ **E.** $\frac{AC}{BC} = \frac{CB}{DC}$

11. *Open-ended* Draw an isosceles triangle $\triangle ABC$. Then draw a triangle $\triangle DEF$ so that $\triangle ABC \sim \triangle DEF$. State the similarity ratio of $\triangle ABC$ to $\triangle DEF$.

Find the value of *x*.

12.

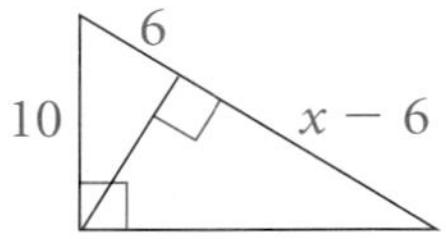

13.

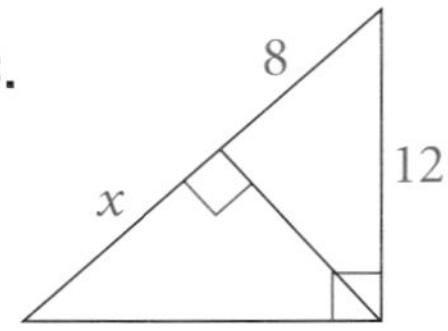

14.

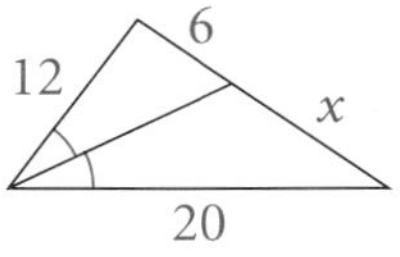

15.

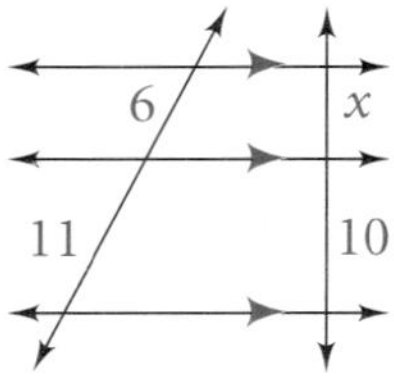

16. *Writing* Describe an object whose height or length would be difficult to measure directly. Then describe a method for measuring the object that involves using similar triangles.

For each pair of similar figures, find the ratio of the area of the first figure to the area of the second.

17.

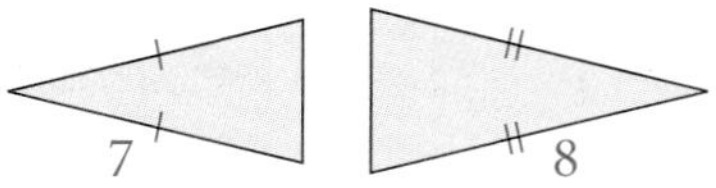

18.

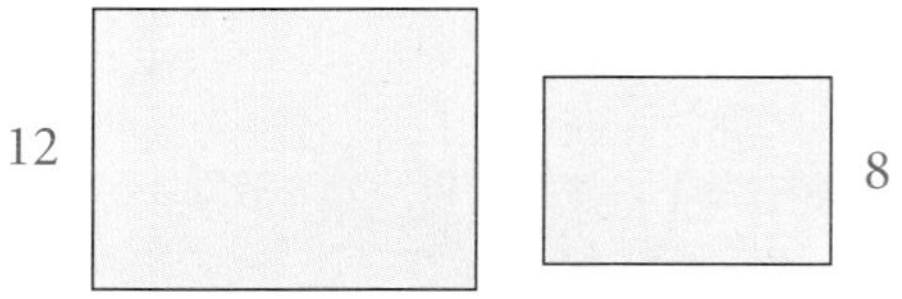

19. The two solids are similar. Find the ratio of the volume of the first figure to the volume of the second.

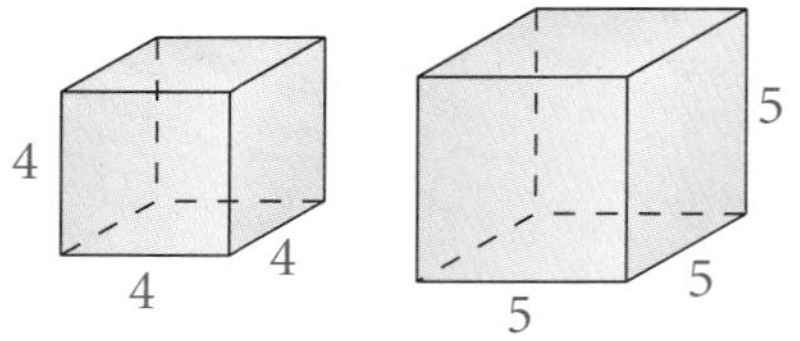

Cumulative Review

For Exercises 1–7, choose the correct letter.

1. Which figure has exactly one line of symmetry?

A. 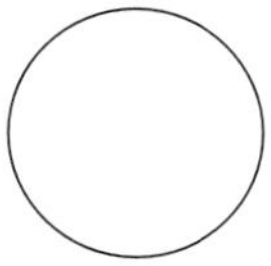

B. 

C. 

D. 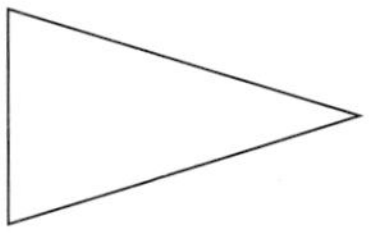

E. none of them

2. What is the ratio of the volumes of similar solids whose similarity ratio is 4 : 9?

A. $\sqrt{2} : \sqrt{3}$ B. 2 : 3 C. 4 : 9
D. 16 : 81 E. 64 : 729

3. What is the surface area of the cylinder?

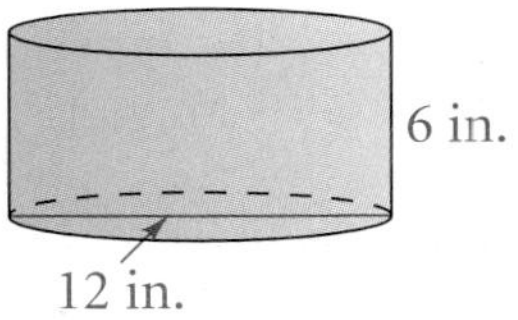

A. 96π in.^{2} B. 144π in.^{2} C. 192π in.^{2}
D. 216π in.^{2} E. 432π in.^{2}

4. Which triangle is *not* necessarily congruent to the given triangle?

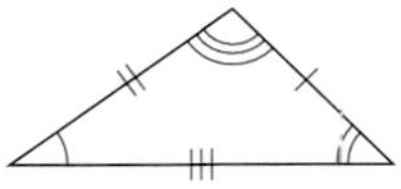

A. 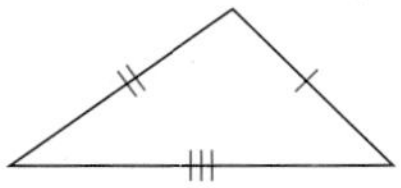

B. 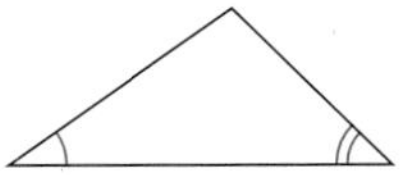

C. 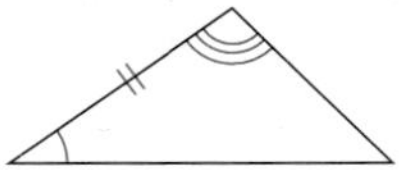

D. 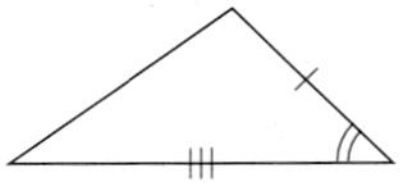

E. All of the triangles are congruent.

5. What is the most precise name of the figure?

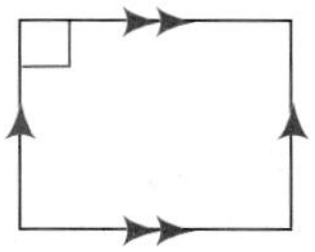

A. quadrilateral
B. parallelogram
C. rectangle
D. rhombus
E. square

Compare the boxed quantity in Column A with the boxed quantity in Column B. Choose the best answer.

A. The quantity in Column A is greater.
B. The quantity in Column B is greater.
C. The two quantities are equal.
D. The relationship cannot be determined on the basis of the information supplied.

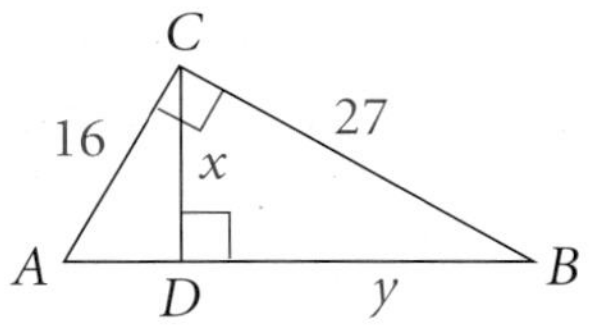

	Column A	Column B
6.	x	y
7.	$m\angle ACD$	$m\angle B$

Find each answer.

8. Which special quadrilaterals have perpendicular diagonals?

9. **Writing** Explain the difference between a median and an altitude of a triangle.

10. How many nonoverlapping circles with 1-in. radius can you fit on a 4-in. square? Find the area of the square not covered by circles.

11. **Constructions** Draw an angle of about 150°. Construct its angle bisector.

CHAPTER 11

Right Triangle Trigonometry

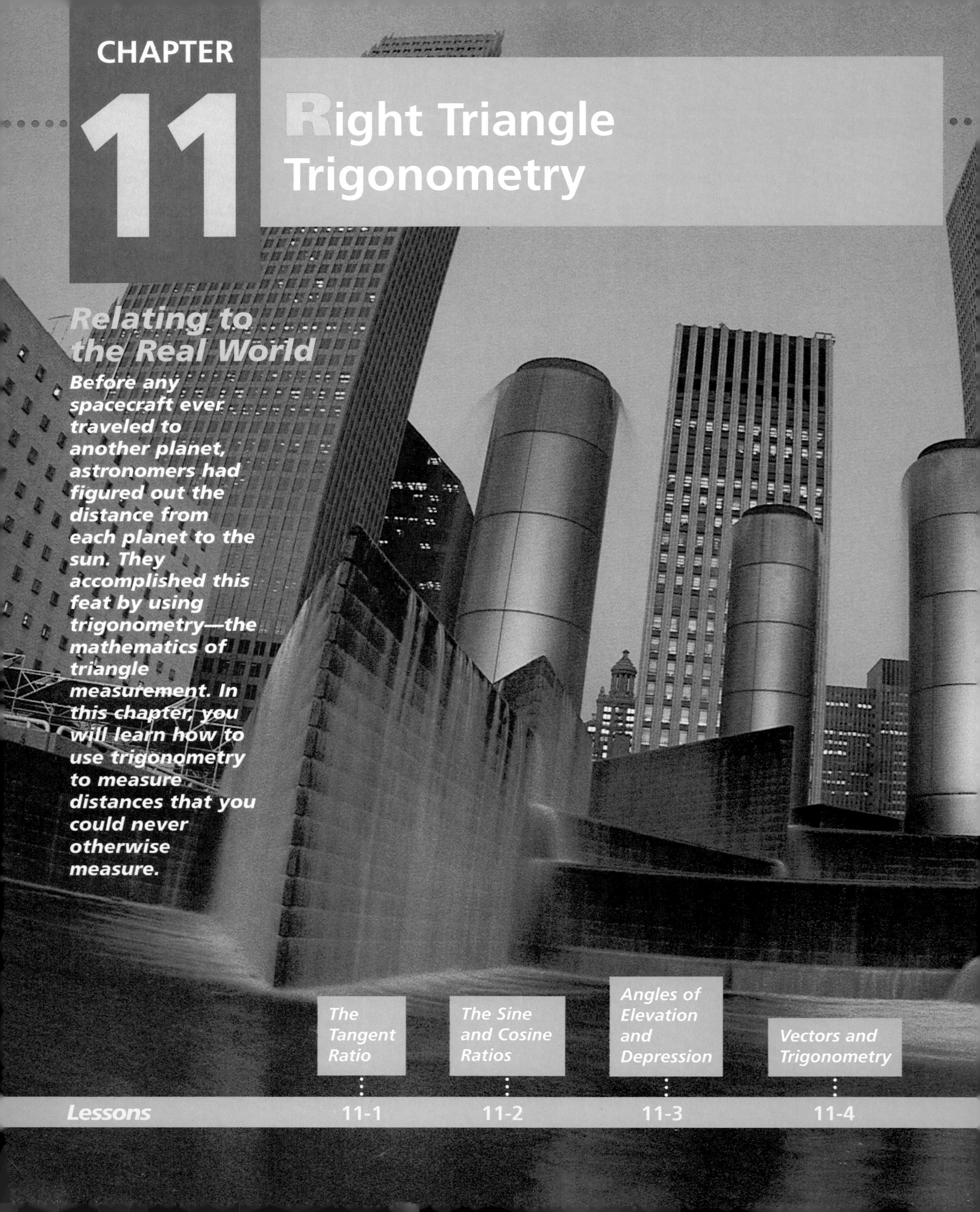

Relating to the Real World

Before any spacecraft ever traveled to another planet, astronomers had figured out the distance from each planet to the sun. They accomplished this feat by using trigonometry—the mathematics of triangle measurement. In this chapter, you will learn how to use trigonometry to measure distances that you could never otherwise measure.

Lessons	11-1	11-2	11-3	11-4
	The Tangent Ratio	*The Sine and Cosine Ratios*	*Angles of Elevation and Depression*	*Vectors and Trigonometry*

CHAPTER PROJECT

MEASURE FOR MEASURE

How do we know how far away the sun is without laying a tape measure across millions of miles? How do we know the mass of an electron when we can't even see one? Since ancient times, people have found ways to measure indirectly what they could not measure directly.

In your project for this chapter, you will construct an instrument similar to ones used by ancient astronomers and travelers. You will use your device to measure heights you cannot easily reach. You will see how mathematics can extend your power to measure things far beyond your physical grasp.

To help you complete the project:

- **p. 549** *Find Out by Building*
- **p. 560** *Find Out by Measuring*
- **p. 567** *Find Out by Comparing*
- **p. 577** *Find Out by Researching*
- **p. 578** *Finishing the Project*

11-1 The Tangent Ratio

What You'll Learn

- Calculating tangents of acute angles in right triangles
- Using tangents to determine side lengths in triangles

...And Why

To calculate distances that cannot be measured directly

What You'll Need

- protractor
- centimeter ruler
- calculator

WORK TOGETHER

- Work in small groups. Your teacher will assign each group a different angle measure from the set {10°, 20°, . . . , 80°}.
- Have each member of your group draw a right triangle containing the assigned angle. Make the triangles different sizes. Label your triangle $\triangle PAW$, where $\angle P$ is the assigned angle.
- Use a ruler to measure and label the lengths of the legs of $\triangle PAW$ to the nearest millimeter.

1. *Calculator* Use a calculator to compute the ratio $\frac{\text{leg opposite } \angle P}{\text{leg adjacent to } \angle P}$. Round your answer to two decimal places.
2. Compare the results of Question 1 within your group. Make a **conjecture** based on your comparison.
3. Compare your results with those of other groups. What do you notice?

THINK AND DISCUSS

Using the Tangent Ratio

When? The word *trigonometry* was first used in a publication in 1595 by German clergyman and mathematician Bartholomaus Pitiscus (1561–1613).

The word *trigonometry* comes from the Greek words meaning "triangle measurement." In the Work Together, you made a discovery about a ratio of the lengths of sides of a right triangle. This ratio is called the tangent ratio.

tangent of $\angle A = \dfrac{\text{leg opposite } \angle A}{\text{leg adjacent to } \angle A}$

This equation can be abbreviated:

$$\tan A = \frac{\text{opposite}}{\text{adjacent}}$$

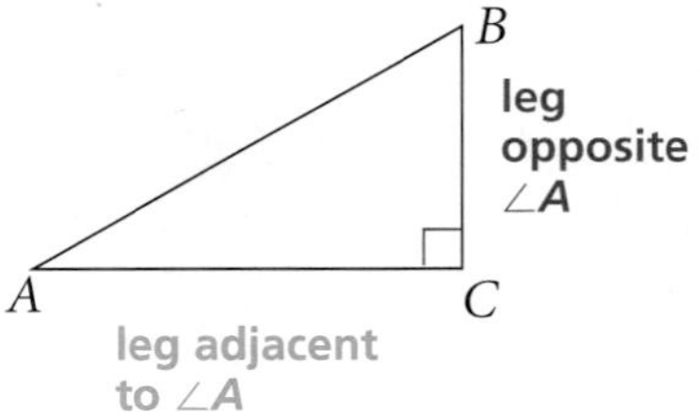

Example 1

Write the tangent ratios for $\angle U$ and $\angle T$.

$$\tan U = \frac{\text{opposite}}{\text{adjacent}} = \frac{TV}{UV} = \frac{4}{3}$$

$$\tan T = \frac{\text{opposite}}{\text{adjacent}} = \frac{UV}{TV} = \frac{3}{4}$$

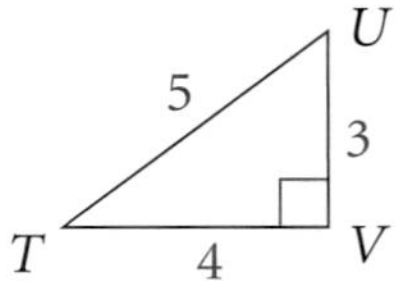

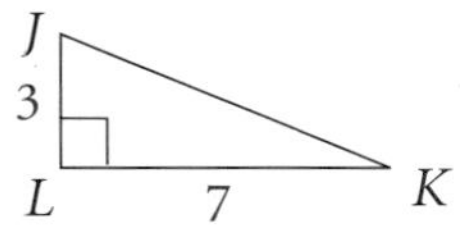

4. a. **Try This** Write the tangent ratios for $\angle K$ and $\angle J$.
   b. How is $\tan K$ related to $\tan J$?

You discovered in the Work Together that the tangent ratio for a given acute angle does not depend on the size of the triangle. Why is this true? Consider the two right triangles shown below.

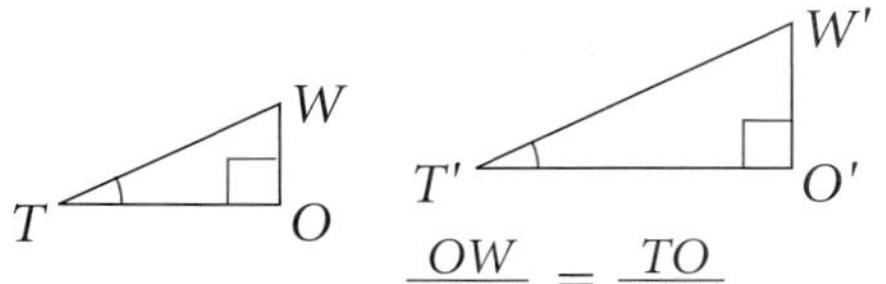

$$\frac{OW}{O'W'} = \frac{TO}{T'O'}$$ By the AA Similarity Postulate, the two triangles are similar.

Because the triangles are similar, these corresponding sides are in proportion.

$$\frac{OW}{TO} = \frac{O'W'}{T'O'}$$ Applying a property of proportions yields the tangent ratios for $\angle T$ and $\angle T'$.

You can use the tangent ratio to measure distances that would be difficult to measure directly.

Example 2 *Relating to the Real World*

Hiking You are hiking in the Rocky Mountains. You come to a canyon and follow these steps to estimate its width.

Step 1: Point your compass at an object on the opposite edge of the canyon and note the reading.

Step 2: Turn 90° and walk off a distance in a straight line along the edge of the canyon.

Step 3: Turn and point the compass at the object again, taking another reading.

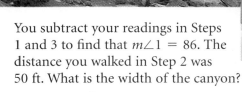

You subtract your readings in Steps 1 and 3 to find that $m\angle 1 = 86$. The distance you walked in Step 2 was 50 ft. What is the width of the canyon?

$\tan 86° = \frac{x}{50}$ Use the tangent ratio.

$x = 50(\tan 86°)$ Solve for x.

50 [×] 86 [TAN] [=] 715.03331

The canyon is about 700 ft wide.

GRAPHING CALCULATOR HINT

The keystrokes for a scientific calculator are shown at the right. On a graphing calculator, enter 50 [×] [TAN] 86 [ENTER].

5. Try This Find the value of w to the nearest tenth.

a.
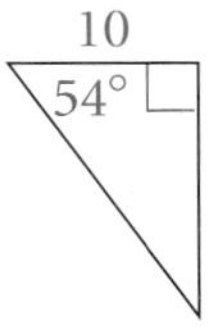

b.
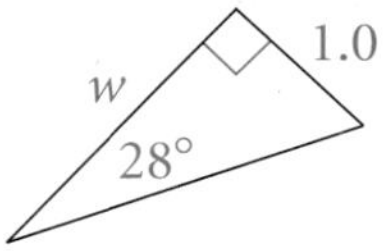

c.
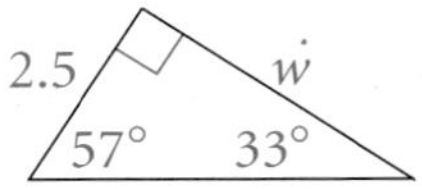

Using the Inverse Tangent Function

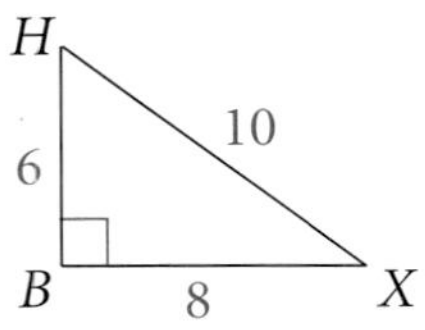

Consider $\triangle BHX$ at the left. Suppose you want to find $m\angle X$. You can use the tangent ratio to do so. You know that $\tan X = \frac{6}{8} = 0.75$. To find $m\angle X$, use the *inverse tangent* function on your calculator.

$\tan X = 0.75$

$m\angle X = \tan^{-1}(0.75)$ "$\tan^{-1}$" means "inverse tangent."

0.75 TAN-1 36.869898 Use the inverse tangent function on a calculator.

So $m\angle X \approx 37$.

GRAPHING CALCULATOR HINT

On most calculators, you find the inverse tangent of 0.75 by entering .75 2nd TAN or .75 TAN-1. On a graphing calculator, enter 2nd TAN .75 ENTER.

6. Find $m\angle J$ to the nearest degree.

a. $\tan J = 0.5$ **b.** $\tan J = 0.34$ **c.** $\tan J = 100$

7. **Critical Thinking** Given that $\tan C = 1$, explain how to find $m\angle C$ without using a calculator.

8. Find $m\angle Y$ to the nearest degree.

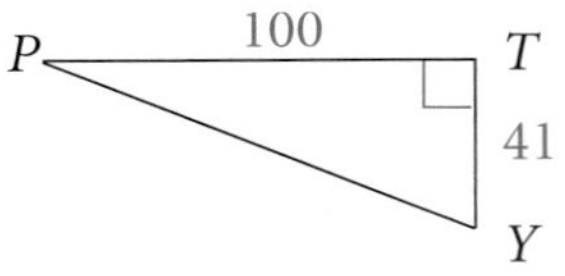

Example 3 shows how the tangent ratio is related to slope.

Example 3

Algebra Find the measure of the acute angle that the line $y = 3x + 2$ makes with a horizontal line.

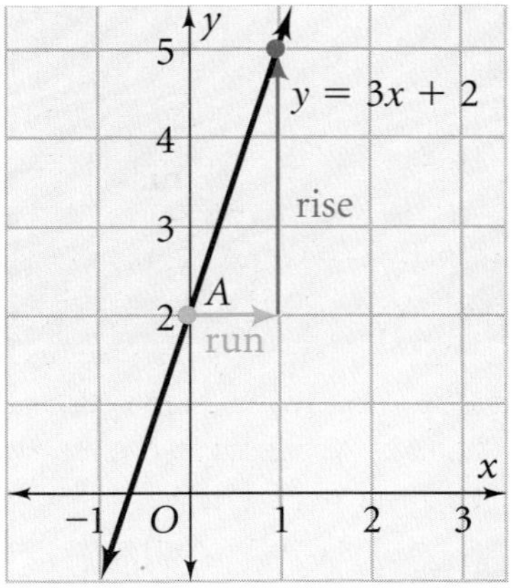

The line $y = 3x + 2$ has a slope of 3. Notice that the ratio $\frac{\text{rise}}{\text{run}}$ is the same as $\frac{\text{leg opposite } \angle A}{\text{leg adjacent to } \angle A}$. So $\tan A = 3$, and $m\angle A = \tan^{-1}(3)$.

3 TAN-1 71.565051

$m\angle A \approx 71.6$

9. Try This Find the measure of the acute angle that each line makes with a horizontal line. Round your answer to the nearest tenth.

a. $y = \frac{1}{2}x + 6$ **b.** $y = 6x - 1$

Exercises ON YOUR OWN

Express tan *A* and tan *B* as ratios.

1. 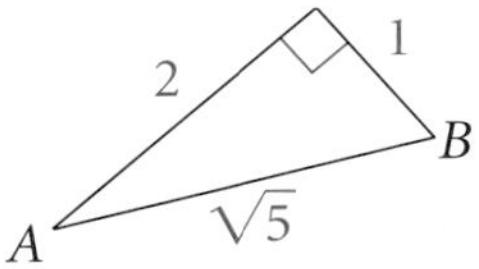

2. 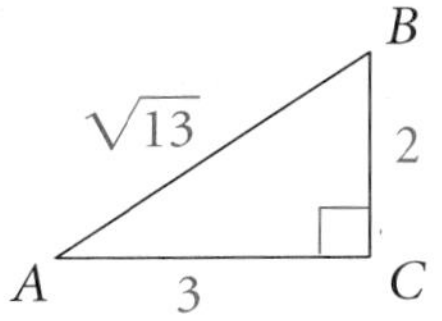

3. 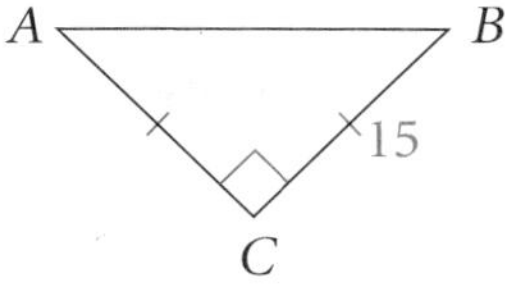

4. 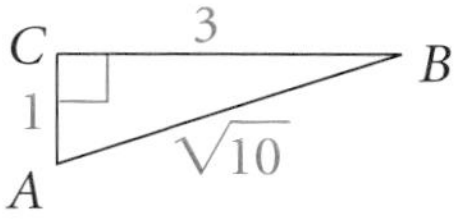

Find each missing value. Round your answer to the nearest tenth.

5. $\tan \blacksquare° = 3.5$ 6. $\tan 34° = \frac{\blacksquare}{20}$ 7. $\tan 2° = \frac{4}{\blacksquare}$ 8. $\tan \blacksquare° = 90$

Find the value of *x*. Round lengths of segments to the nearest tenth and angle measures to the nearest degree.

9. 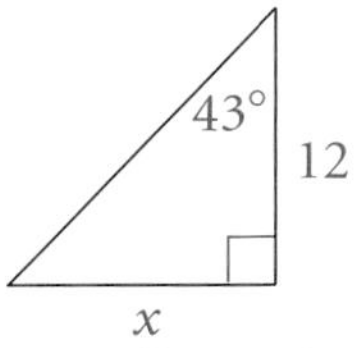

10. 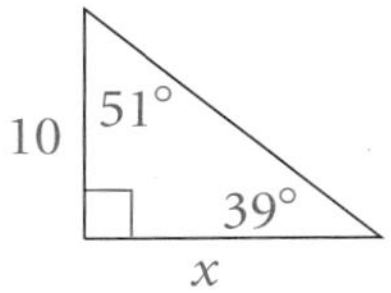

11. 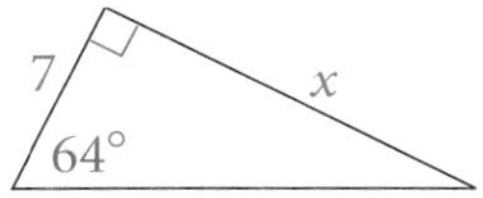

12. 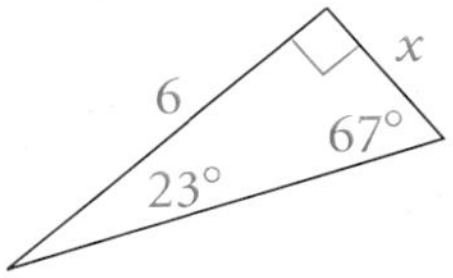

13. 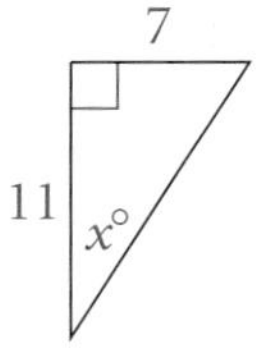

14. 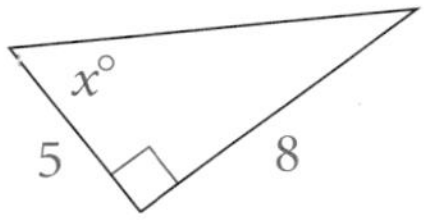

15. 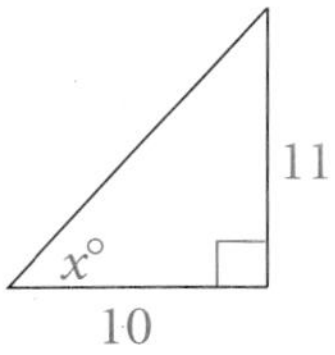

16. 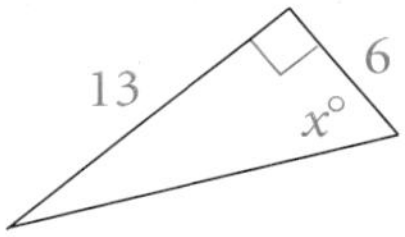

17. 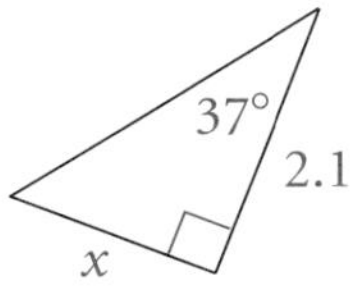

18. 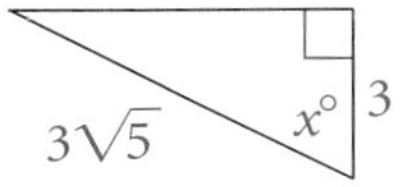

19. 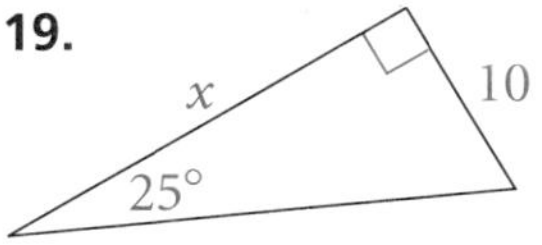

20. 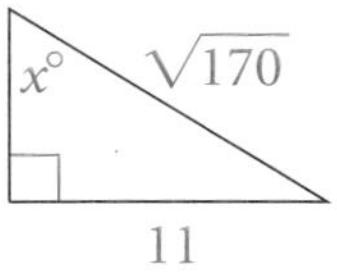

21. **Engineering** The *grade* of a road or railway is the ratio $\frac{\text{rise}}{\text{run}}$, usually expressed as a percent. For example, a railway with a grade of 5% rises 5 ft for every 100 ft of horizontal distance. The world's steepest railway is the Katoomba Scenic Railway in the Blue Mountains of Australia. It has a grade of 122%. At what angle does this railway go up?

22. The lengths of the diagonals of a rhombus are 2 in. and 5 in. Find the measures of the angles of the rhombus to the nearest degree.

23. **Pyramids** All but two of the pyramids built by the ancient Egyptians have faces inclined at 52° angles. (The remaining two have faces inclined at $43\frac{1}{2}$°.) Suppose an archaeologist discovers the ruins of a pyramid. Most of the pyramid has eroded, but she is able to determine that the length of a side of the square base is 82 m. How tall was the pyramid, assuming its faces were inclined at 52°? Round your answer to the nearest meter.

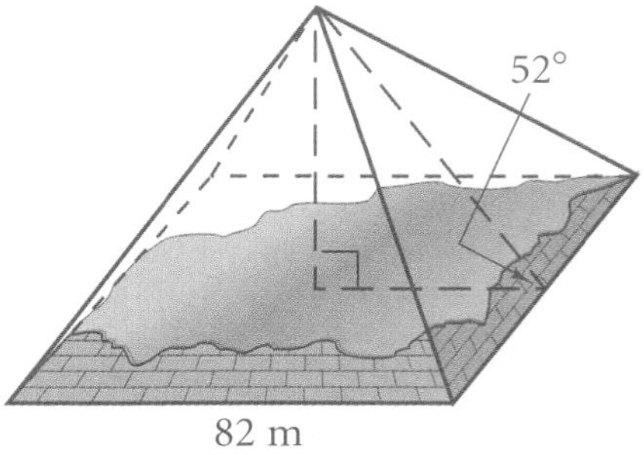

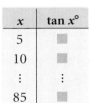

24. **Writing** Explain why $\tan 60° = \sqrt{3}$. Include a diagram with your explanation.

25. a. **Coordinate Geometry** Use a calculator to complete the table of values at the right. Give your answers to the nearest tenth.
    b. Plot the points $(x, \tan x°)$ on the coordinate plane. Connect them with a smooth curve.
    c. What happens to the tangent ratio as the measure of the angle approaches 0? as it approaches 90?
    d. Use the graph to estimate each value.
    $\tan ■° = 7$ $\quad$ $\tan 68° = ■$ $\quad$ $\tan ■° = 3.5$

x	$\tan x°$
5	■
10	■
⋮	⋮
85	■

Find w, then x. Round lengths of segments to the nearest tenth and angle measures to the nearest degree.

26.
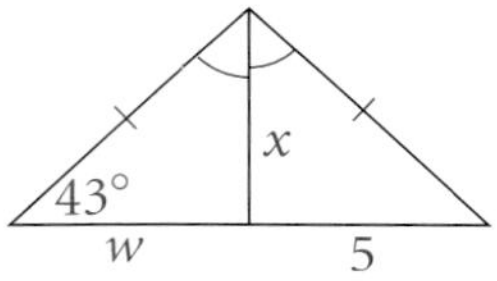

27.
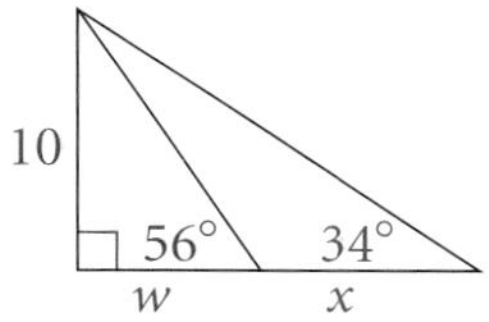

28.
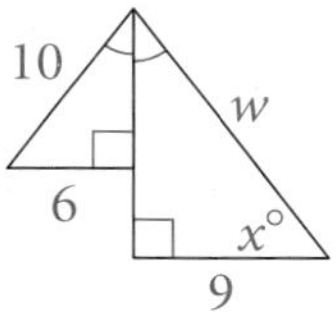

29.
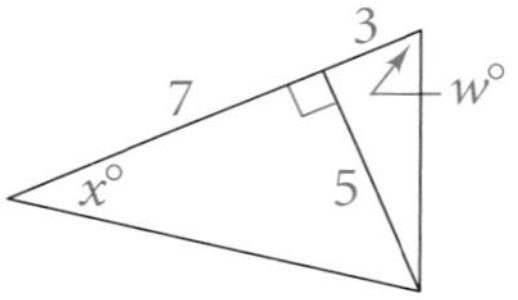

Algebra **Find the measure of the acute angle that each line makes with a horizontal line. Round your answer to the nearest tenth.**

30. $y = 5x - 7$

31. $y = \frac{4}{3}x - 1$

32. $3x - 4y = 8$

33. $-2x + 3y = 6$

34. **Construction** The roadway of a suspension bridge is supported by cables, as shown. The cables extend from the roadway to a point on the tower 50 ft above the roadway. The cables are set at 10-ft intervals along the roadway. Find the angles from the roadway to the top of the tower for the six cables nearest the tower. Round your answers to the nearest tenth.

35. **Open-ended** Select a Pythagorean triple other than a multiple of 3, 4, 5. Find the measures of the acute angles of the right triangle associated with your Pythagorean triple. Round each measure to the nearest tenth.

Chapter Project — *Find Out by Building*

Use the diagram at the right to build a *clinometer*, an angle-measuring device similar to instruments used by navigators for hundreds of years. You will need a protractor, string, tape, a piece of heavy cardboard, a small weight, and a straw.

To use the clinometer, look through the straw at the object you want to measure (the top of a building, for example). Have someone else read the angle marked by the hanging string.

- Why is the angle measure indicated by the string on the protractor the same as $m\angle 1$? (*Hint:* The string is always perpendicular to the horizon.)

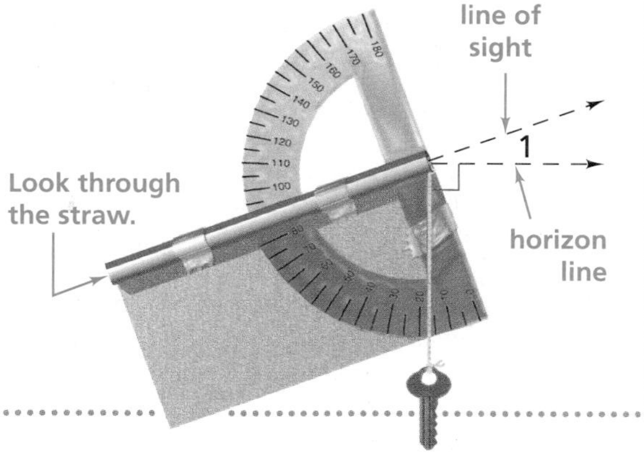

Exercises — MIXED REVIEW

In Exercises 36–39: (a) Write the converse of each conditional. (b) Determine the truth value of the statement and its converse.

36. If a quadrilateral is a parallelogram, then the quadrilateral has congruent diagonals.

37. If a flag contains the colors red, white, and blue, then the flag is a United States flag.

38. If a quadrilateral is a rhombus, then its diagonals are perpendicular bisectors of one another.

39. If you live in Hawaii, then you live on an island.

40. *Locus* Sketch the locus of points in a plane that are equidistant from the diagonals of a kite.

41. a. *Constructions* Draw a triangle. Then construct the perpendicular bisectors of two of its sides.

b. Draw the circle that circumscribes the triangle. (*Hint:* The intersection of the perpendicular bisectors is its center.)

SELF ASSESSMENT

FOR YOUR JOURNAL

Two different right triangles each have an angle that measures 60. Explain why tan 60° is the same for both of these triangles.

Getting Ready for Lesson 11-2

For each triangle, find the ratios $\frac{\text{leg opposite } \angle B}{\text{hypotenuse}}$ **and** $\frac{\text{leg adjacent to } \angle B}{\text{hypotenuse}}$.

42.

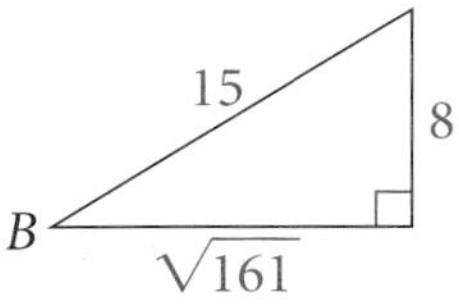

43.

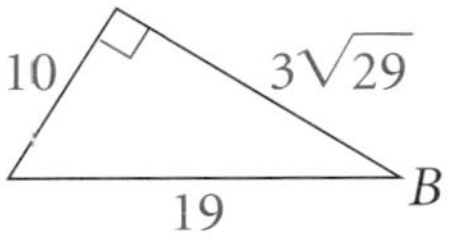

44.

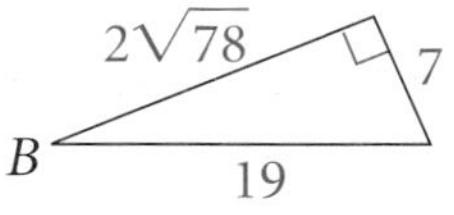

45.

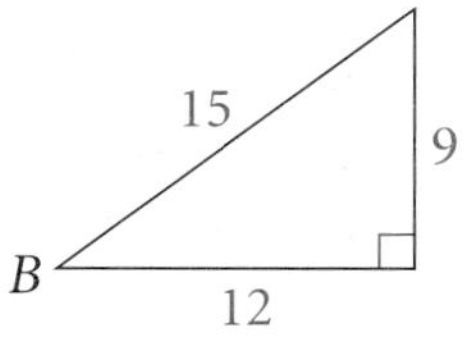

Exploring Trigonometric Ratios

Before Lesson 11-2

Work in pairs or small groups.

Construct

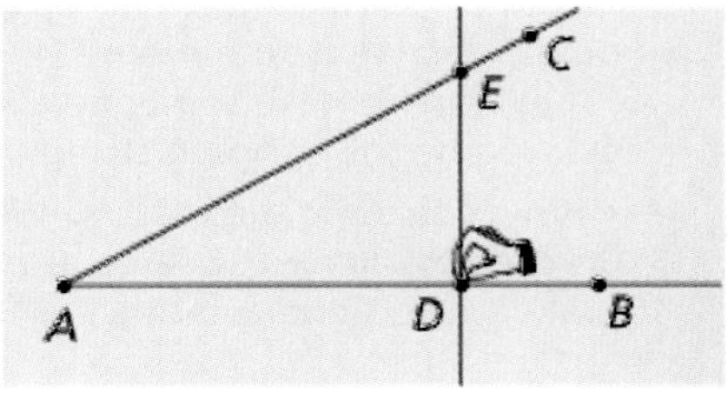

Use geometry software to construct $\overrightarrow{AB}$ and $\overrightarrow{AC}$ so that $\angle A$ is acute. Through a point D on $\overrightarrow{AB}$ construct a line perpendicular to $\overrightarrow{AB}$ that intersects $\overrightarrow{AC}$ in point E. Moving point D enlarges or reduces $\triangle ADE$. Moving point C changes the size of $\angle A$.

Investigate

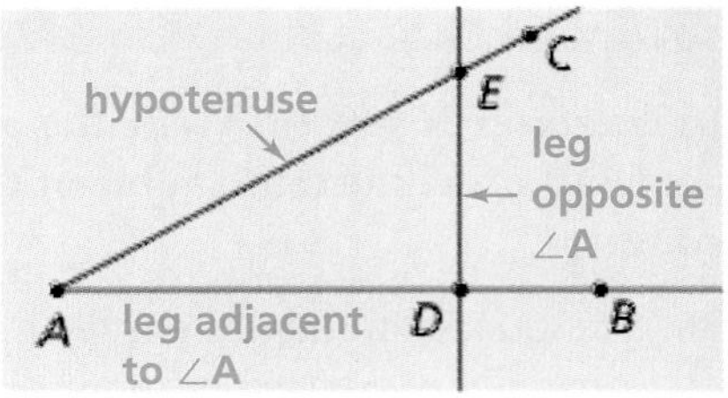

- Measure $\angle A$ and then find the lengths of the sides of the triangle. Calculate the ratio $\frac{\text{length of leg opposite } \angle A}{\text{length of hypotenuse}}$, which is $\frac{DE}{AE}$.
- Move point D to change the size of the right triangle without changing the size of $\angle A$. Does the ratio change as the size of the triangle changes?
- Move point C to change the size of $\angle A$. How does the ratio change as the size of $\angle A$ changes? What value does the ratio approach as $m\angle A$ approaches 0? as $m\angle A$ approaches 90?
- Make a table that shows $m\angle A$ and the ratio $\frac{\text{length of leg opposite } \angle A}{\text{length of hypotenuse}}$. In your table, include values of 10, 20, 30, . . . , 80 for $m\angle A$.

Conjecture

Compare your table with the table of trigonometric ratios on page 674. Do your values for $\frac{\text{length of leg opposite } \angle A}{\text{length of hypotenuse}}$ match the values in one of the columns of the table? What is the name of this ratio in the table?

Extend

- Repeat the investigation for the ratio $\frac{\text{length of leg adjacent to } \angle A}{\text{length of hypotenuse}}$, which is $\frac{DA}{EA}$.
- Repeat the investigation for the ratio $\frac{\text{length of leg opposite } \angle A}{\text{length of leg adjacent to } \angle A}$, which is $\frac{ED}{DA}$.

What You'll Learn

- Calculating sines and cosines of acute angles in right triangles
- Using sine and cosine to determine unknown measures in right triangles

...And Why

To understand how the ratios were used to calculate the sizes of orbits of planets

What You'll Need

- centimeter ruler
- graph paper
- calculator

11-2 The Sine and Cosine Ratios

THINK AND DISCUSS

The tangent ratio, as you've seen, involves both legs of a right triangle. The sine and cosine ratios involve one leg and the hypotenuse.

sine of $\angle A = \dfrac{\text{leg opposite } \angle A}{\text{hypotenuse}}$

cosine of $\angle A = \dfrac{\text{leg adjacent to } \angle A}{\text{hypotenuse}}$

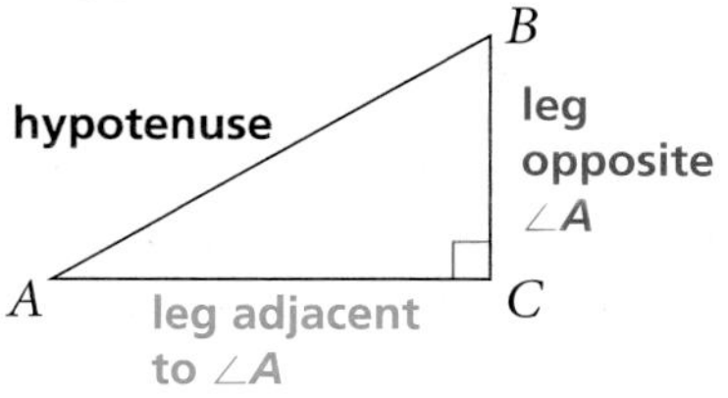

These equations can be abbreviated:

$$\sin A = \frac{\text{opposite}}{\text{hypotenuse}}$$

$$\cos A = \frac{\text{adjacent}}{\text{hypotenuse}}$$

Example 1

Refer to $\triangle GRT$ at the left. Find each ratio.

a. $\sin T$ **b.** $\cos T$
c. $\sin G$ **d.** $\cos G$

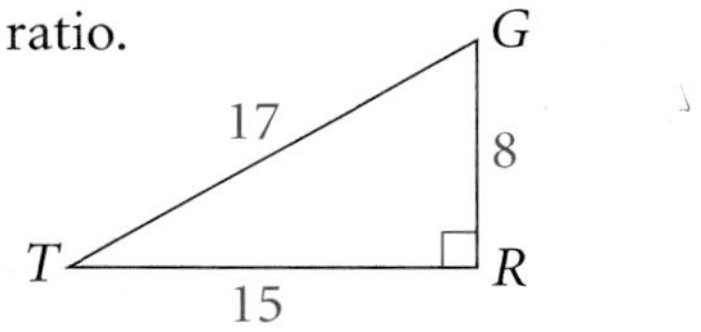

a. $\sin T = \dfrac{\text{opposite}}{\text{hypotenuse}} = \dfrac{8}{17}$ **b.** $\cos T = \dfrac{\text{adjacent}}{\text{hypotenuse}} = \dfrac{15}{17}$

c. $\sin G = \dfrac{\text{opposite}}{\text{hypotenuse}} = \dfrac{15}{17}$ **d.** $\cos G = \dfrac{\text{adjacent}}{\text{hypotenuse}} = \dfrac{8}{17}$

1. a. Explain why $\sin T = \cos G = \frac{8}{17}$.
b. The word *cosine* is derived from the words *co*mplement's *sine.* Which angle in $\triangle GRT$ is the complement of $\angle T$? of $\angle G$?
c. Explain why the derivation of the word *cosine* makes sense.

Another way of describing the pattern in Question 1 is to say that $\sin x° = \cos (90 - x)°$ for values of x between 0 and 90. This type of equation is called an **identity** because it is always true for the allowed values of the variable. You will discover other identities in the exercises.

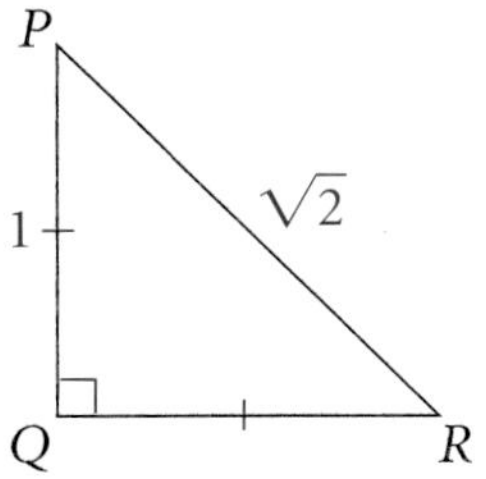

2. Refer to $\triangle PQR$ at the left. Find each ratio.
a. $\sin P$ **b.** $\cos P$
c. $\sin R$ **d.** $\cos R$

3. Refer to $\triangle TSN$ at the right. Find each ratio.
a. $\sin S$ **b.** $\cos S$
c. $\sin T$ **d.** $\cos T$

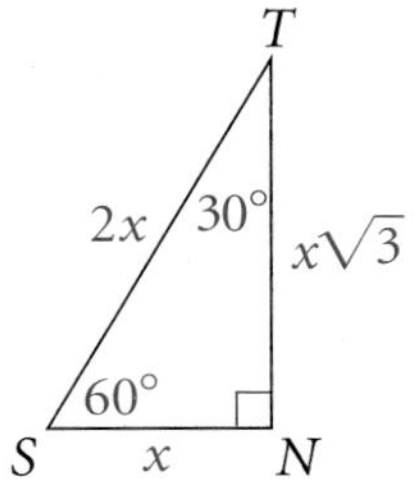

Who? Vera Rubin (1928–) used techniques of indirect measurement to determine the speeds at which stars in far-off galaxies travel.

What? 1 AU (astronomical unit) is defined as the average distance from Earth to the sun (about 93 million miles).

GRAPHING CALCULATOR HINT

On a scientific calculator, enter 22.3 SIN. On a graphing calculator, enter SIN 22.3 ENTER.

The trigonometric ratios have been known for centuries. People in many cultures have used them to solve problems—especially problems involving distances that cannot be measured directly.

Example 2 Relating to the Real World

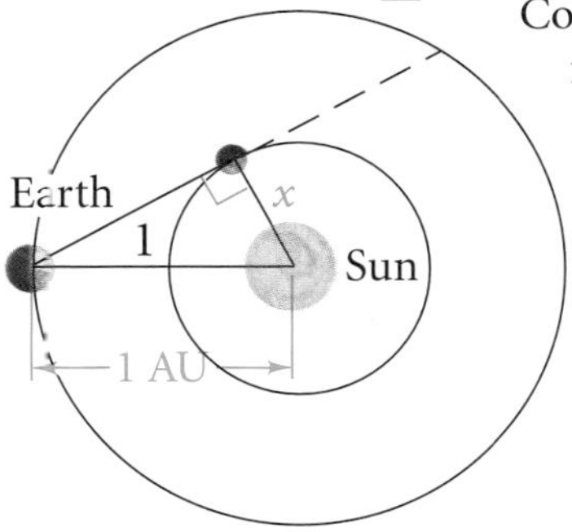

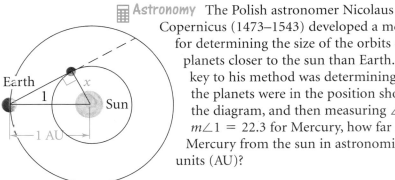

Astronomy The Polish astronomer Nicolaus Copernicus (1473–1543) developed a method for determining the size of the orbits of planets closer to the sun than Earth. The key to his method was determining when the planets were in the position shown in the diagram, and then measuring $\angle 1$. If $m\angle 1 = 22.3$ for Mercury, how far is Mercury from the sun in astronomical units (AU)?

$\sin 22.3° = \frac{x}{1}$ Use the sine ratio.

$0.38 \approx x$ Use a calculator to find sin 22.3°.

So Mercury is about 0.38 AU from the sun.

4. Try This If $m\angle 1 = 46.1$ for Venus, how far is Venus from the sun?

5. Find the value of x to the nearest tenth.

a.
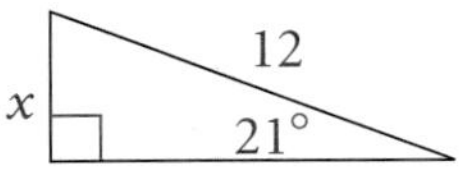

b.
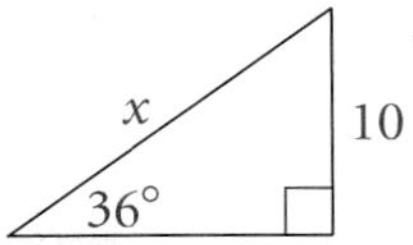

c.
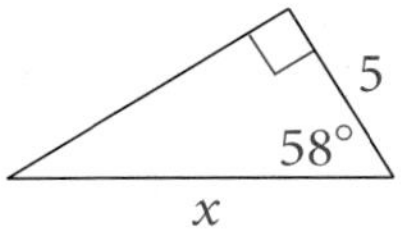

When you know the lengths of a leg and the hypotenuse of a right triangle, you can use inverse sine or inverse cosine functions on a calculator to find the measures of the acute angles.

Example 3

Find $m\angle L$ to the nearest degree.

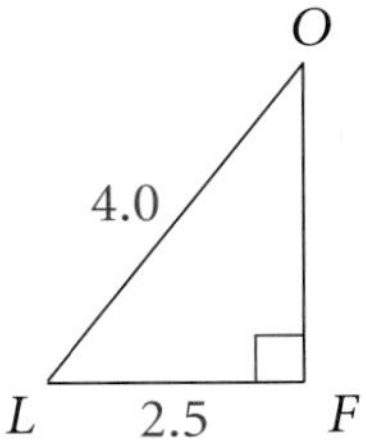

$\cos L = \frac{2.5}{4.0}$ Use the cosine ratio.

$= \frac{5}{8}$ Simplify the right side.

$m\angle L = \cos^{-1}\left(\frac{5}{8}\right)$ Use inverse cosine.

(5 ÷ 8) COS⁻¹ 51.317813

So $m\angle L \approx 51$.

6. **Try This** Find the value of x. Round your answer to the nearest degree.

a.
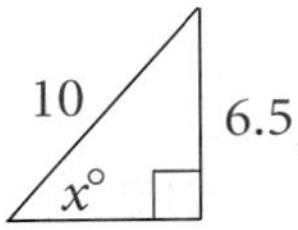

b.
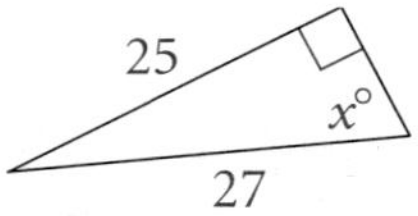

c.
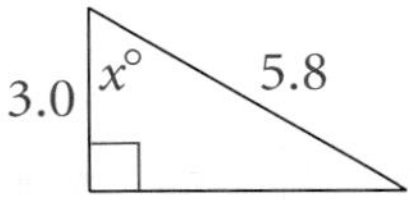

WORK TOGETHER

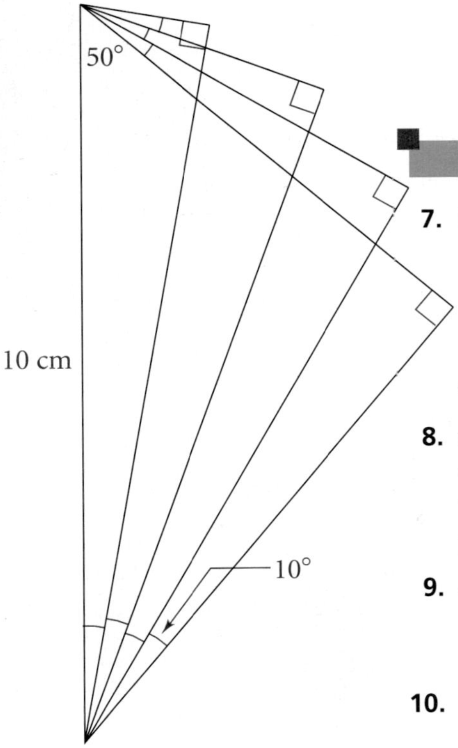

7. Copy the table at the right. Complete it by using a centimeter ruler to measure the triangles at the left to the nearest millimeter. Because the hypotenuse of each triangle is 10 cm, you shouldn't need a calculator to compute the ratios.

x	$\sin x°$	$\cos x°$
10	■	■
20	■	■
30	■	■
40	■	■
50	■	■
60	■	■
70	■	■
80	■	■

8. **Algebra** Graph the ordered pairs $(x, \sin x°)$ on the coordinate plane for values of x in the domain $0 < x < 90$. Connect the points with a smooth curve.

9. **Algebra** Graph the ordered pairs $(x, \cos x°)$ on the same coordinate plane. Again, connect the points with a smooth curve.

10. **Transformations** What transformation maps one of the curves onto the other?

Exercises ON YOUR OWN

Express sin M and cos M as ratios.

1. 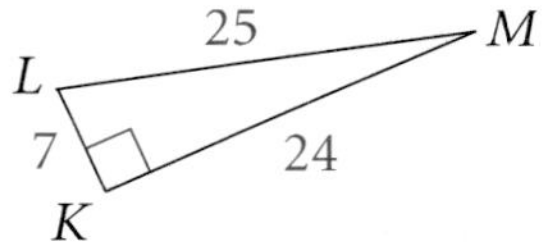

2. 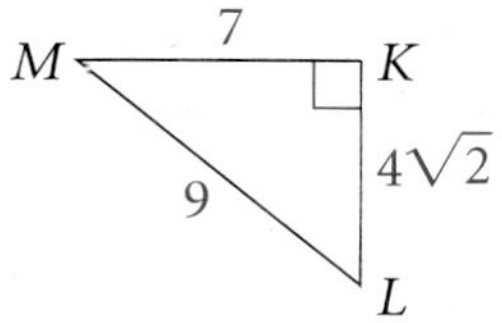

3. 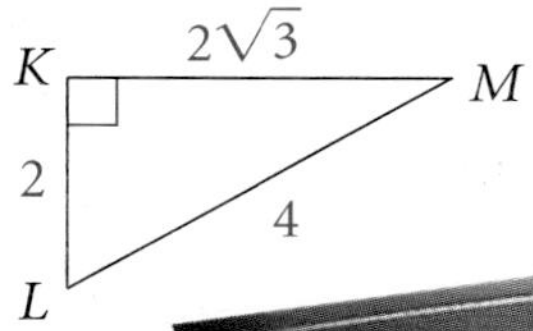

4. **Escalators** The world's longest escalator is in the subway system of St. Petersburg, Russia. The escalator has a vertical rise of 195 ft 9.5 in. and rises at an angle of 10.4°. How long is the escalator? Round your answer to the nearest foot.

5. **Writing** Leona Halfmoon said that if she had a diagram that showed the measure of one acute angle and the length of one side of a right triangle, she could find the measure of the other acute angle and the lengths of the other sides. Is she right? Explain.

Find the value of x. Round lengths of segments to the nearest tenth and angle measures to the nearest degree.

6. 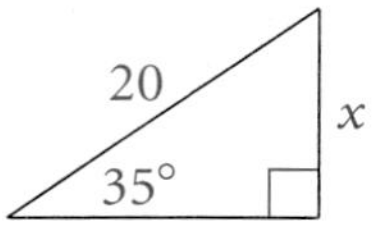

7. 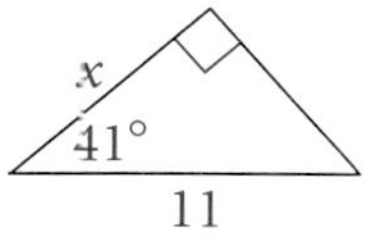

8. 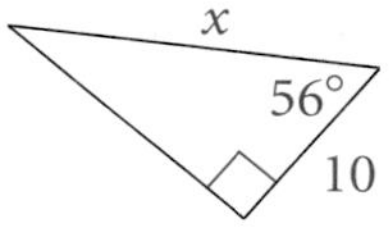

9. 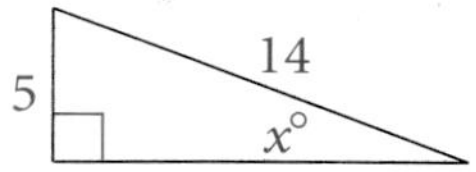

10. 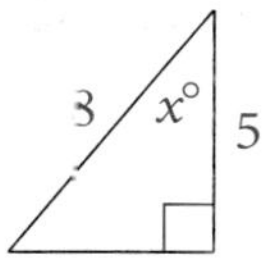

11. 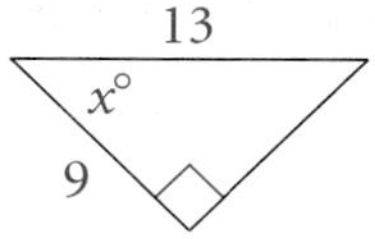

12. 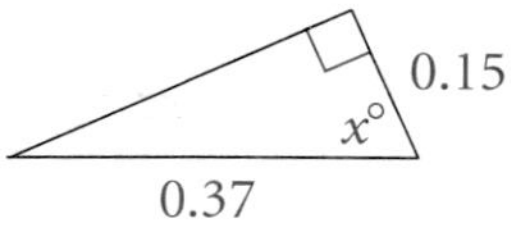

13. 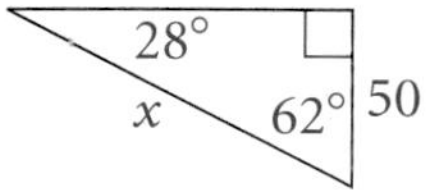

14. 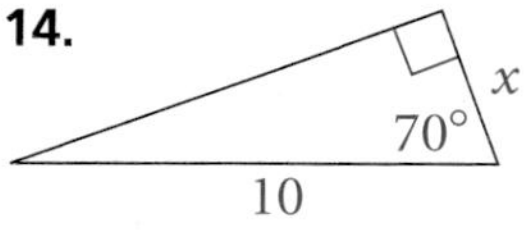

15. **Standardized Test Prep** In $\triangle ABC$, $m\angle B = 90$. Find sin C.

A. $\frac{BC}{AC}$ **B.** $\frac{BC}{AB}$ **C.** $\frac{AB}{BC}$ **D.** $\frac{AB}{AC}$ **E.** $\frac{AC}{AB}$

16. **Agriculture** Jane is planning to build a new grain silo with a radius of 15 ft. She reads that the recommended slope of the roof is 22°. She wants the roof to overhang the edge of the silo by 1 ft. What should the slant height of the roof be? Give your answer in feet and inches.

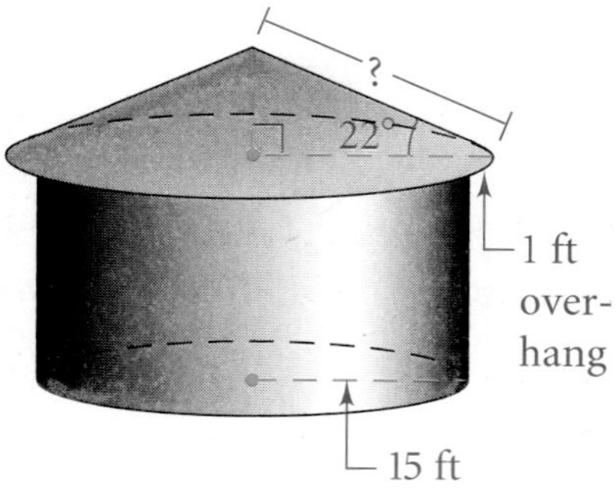

17. **Critical Thinking** Use what you know about trigonometric ratios to show that the following equation is an identity.

$$\tan A = \sin A \div \cos A$$

18. a. **Open-ended** Pick three values of x between 0 and 90. For each value, evaluate the expression $(\sin x°)^2 + (\cos x°)^2$.

b. **Patterns** Use your results from part (a) to make a **conjecture.**

19. **Astronomy** Copernicus had to devise a method different from the one in Example 2 in order to find the size of the orbits of planets farther from the sun than Earth. His method involved noting the number of days between the times that the planets were in the positions labeled A and B in the diagram. Using this time and the number of days in each planet's year, he calculated the measures of the angles labeled 1 and 2.

a. For Mars, $m\angle 1 = 55.2$ and $m\angle 2 = 103.8$. How far is Mars from the sun in astronomical units (AU)?

b. For Jupiter, $m\angle 1 = 21.9$ and $m\angle 2 = 100.8$. How far is Jupiter from the sun in astronomical units (AU)?

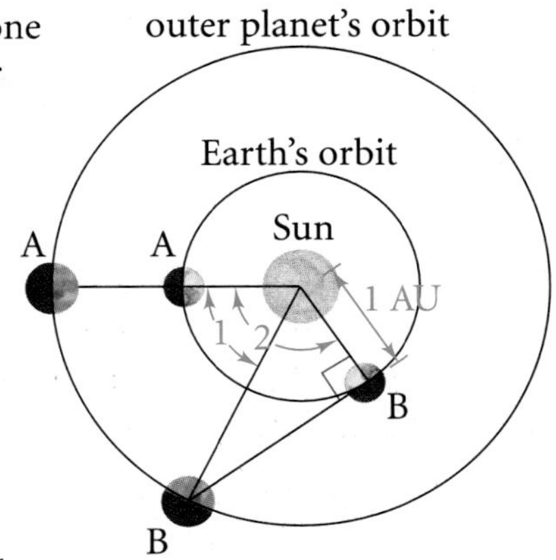

Find w and then x. Round lengths of segments to the nearest tenth and angle measures to the nearest degree.

20.

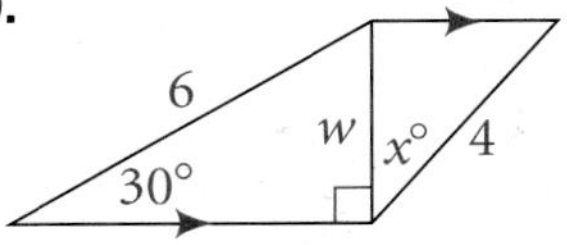

21.

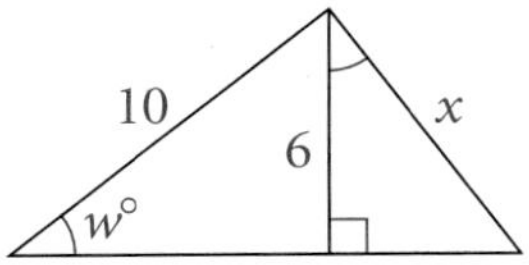

22.

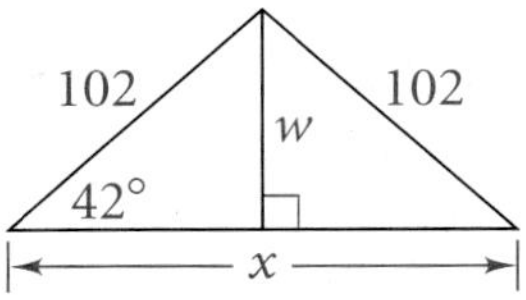

23.

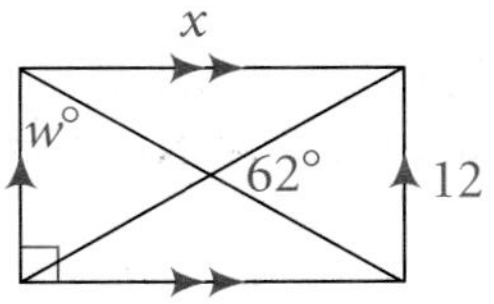

Exercises MIXED REVIEW

Find the value of each variable. When an answer is not a whole number, round to the nearest tenth.

24. ▱*ABCD*

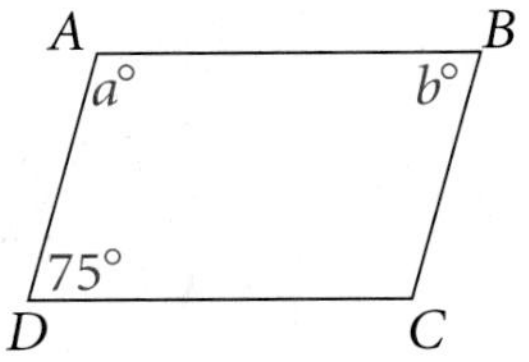

25. rhombus *QRST*

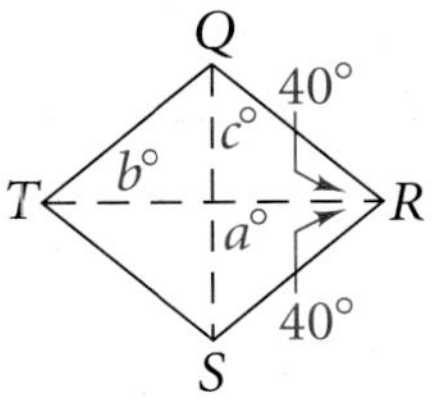

26. rectangle *JKLM*

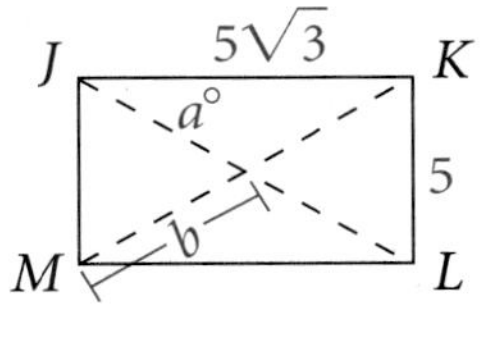

27. kite *WXYZ*

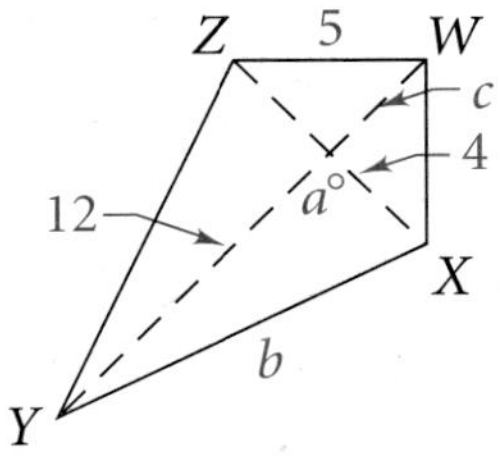

28. **Sports** The circumference of a softball is 12 in. The circumference of a field hockey ball is 9 in. How many times larger than the volume of the field hockey ball is the volume of the softball?

Getting Ready for Lesson 11-3

Refer to rectangle *ABCD* to complete the statements.

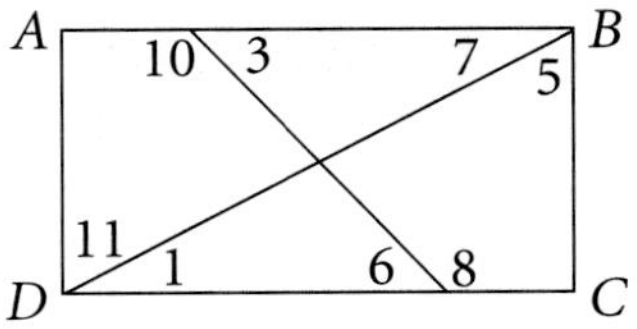

29. $\angle 1 \cong$ ■

30. $\angle 5 \cong$ ■

31. $\angle 3 \cong$ ■

32. $m\angle 1 + m\angle 5 =$ ■

33. $m\angle 10 + m\angle 3 =$ ■

34. $\angle 10 \cong$ ■

What You'll Learn

- Identifying angles of elevation and depression
- Using angles of elevation and depression and trigonometric ratios to solve problems

...And Why

To calculate distances indirectly in a variety of real-world settings

What You'll Need

- calculator

11-3 Angles of Elevation and Depression

THINK AND DISCUSS

Suppose a person in a hot-air balloon sees a person at an angle 38° *below* a horizontal line. This angle is an **angle of depression.** At the same time, the person on the ground sees the hot-air balloon at an angle 38° *above* a horizontal line. This angle is an **angle of elevation.**

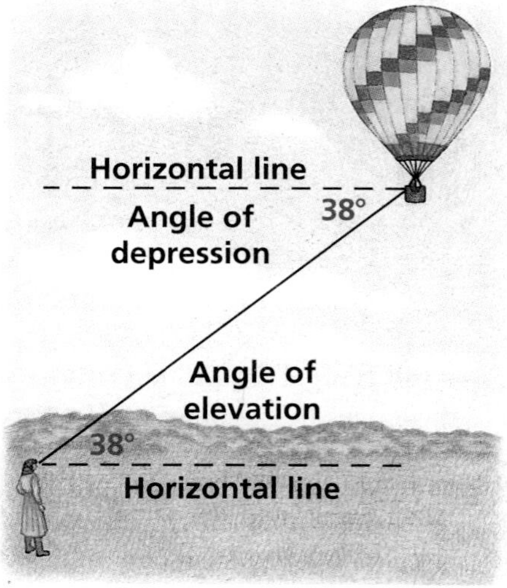

1. Refer to the diagram. What property of parallel lines ensures that the angle of elevation is congruent to the angle of depression?

Example 1

Describe each angle as it relates to the objects in the diagram.

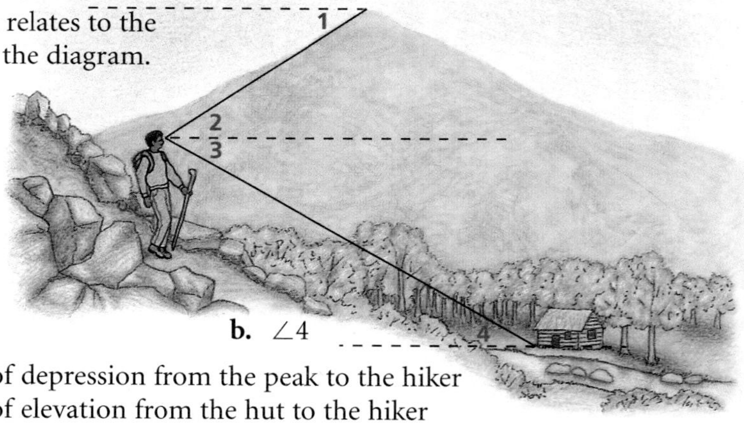

a. ∠1 **b.** ∠4

a. angle of depression from the peak to the hiker
b. angle of elevation from the hut to the hiker

2. **Try This** Describe each angle as it relates to the diagram in Example 1.
   a. ∠2 **b.** ∠3

Surveyors use two main instruments to measure angles of elevation and depression. They are the *transit* and the *theodolite*. On both instruments, the surveyor sets the horizon line perpendicular to the direction of gravity. Using gravity to find the horizon line ensures accurate measures even on sloping surfaces.

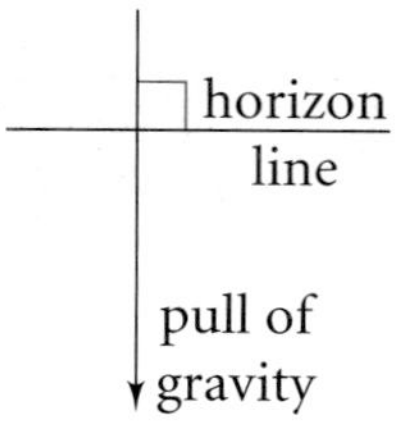

Example 2 Relating to the Real World

Surveying To find the height of Delicate Arch in Arches National Park in Utah, a surveyor places a theodolite at the same level as the bottom of the arch. From there, she measures the angle of elevation to the top of the arch. She then uses a steel tape to measure the distance from her location to the spot directly under the arch. The results of her survey are shown in the diagram. How tall is Delicate Arch?

$\tan 48° = \frac{x}{36}$ — Use the tangent ratio.

$x = 36(\tan 48°)$ — Solve for *x*.

36 [×] 48 [TAN] [=] 39.982051 — Use a calculator.

So $x \approx 40$. To find the height of the arch, add the height of the theodolite. $40 + 5 = 45$, so Delicate Arch is about 45 feet tall.

3. Try This A surveyor uses a theodolite to measure the angle of elevation to the top of a cliff. He then uses an electronic distance measurement device to measure the distance from the theodolite to the top of the cliff. His measurements are shown in the diagram. How tall is the cliff? Round your answer to the nearest foot.

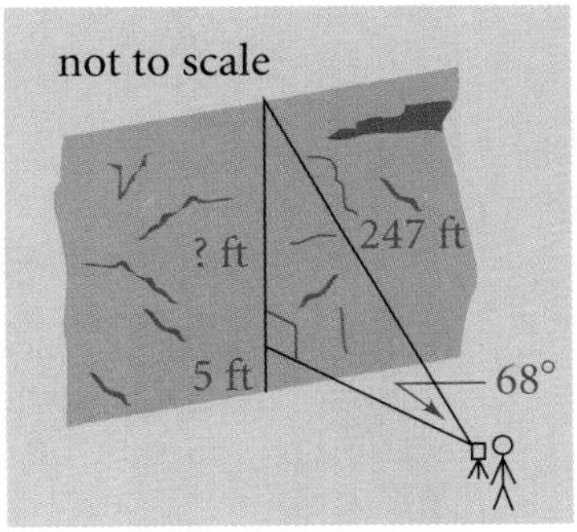

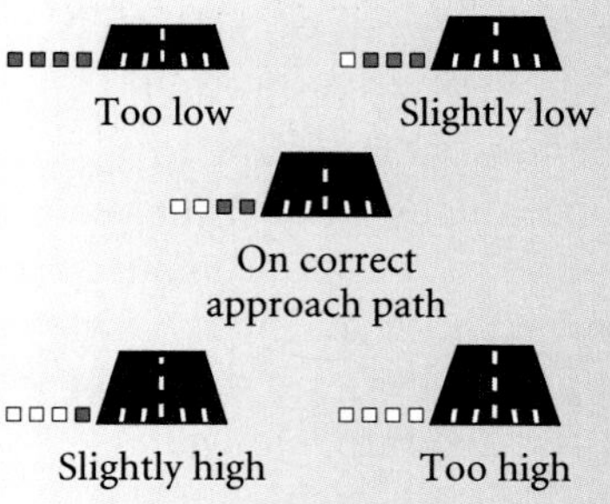

How? Most airports place lights near the runway that signal the pilot if the angle of descent is too great or too small. The lights shown here are red when viewed from above and white when viewed from below. Each of the four lights is tilted at a different angle.

Example 3 — Relating to the Real World

Aviation If you live near an airport, you have probably noticed that airplanes follow certain paths as they prepare for landing. These paths are called *approaches.* The approach to runway 17 of the Ponca City Municipal Airport in Oklahoma calls for the pilot to begin a 3° descent starting from an altitude of 2714 ft. How many miles from the runway is an airplane at the beginning of this approach?

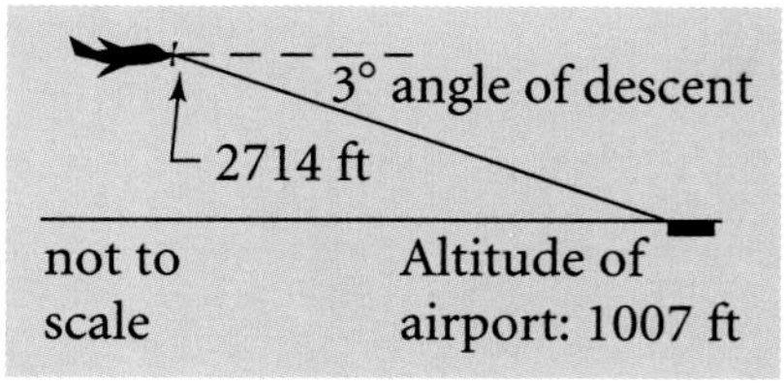

The airplane is 2714 − 1007, or 1707 ft above the level of the airport. Use trigonometry to find the desired distance.

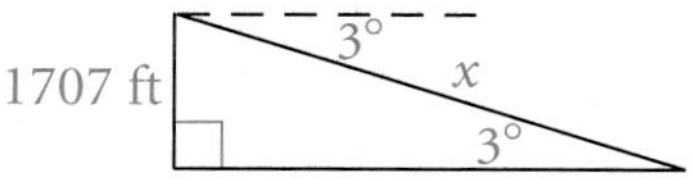

$\sin 3^\circ = \dfrac{1707}{x}$ — Use the sine ratio.

$x = \dfrac{1707}{\sin 3^\circ}$ — Solve for x.

1707 ÷ 3 SIN = 32616.2 — Use a calculator to find x.

÷ 5280 = 6.1773105 — Convert feet to miles by dividing by 5280.

An airplane is about 6.2 miles from the runway at the beginning of the approach to runway 17 at Ponca City.

Exercises ON YOUR OWN

Describe each angle as it relates to the objects in the diagram.

1. **a.** ∠1
b. ∠2
c. ∠3
d. ∠4

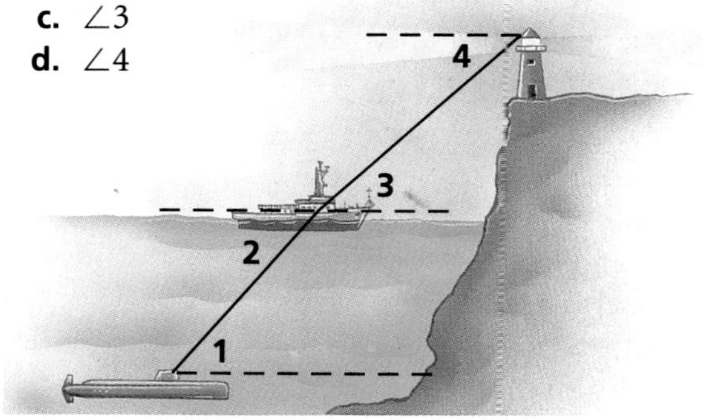

2. **a.** ∠1
b. ∠2
c. ∠3
d. ∠4

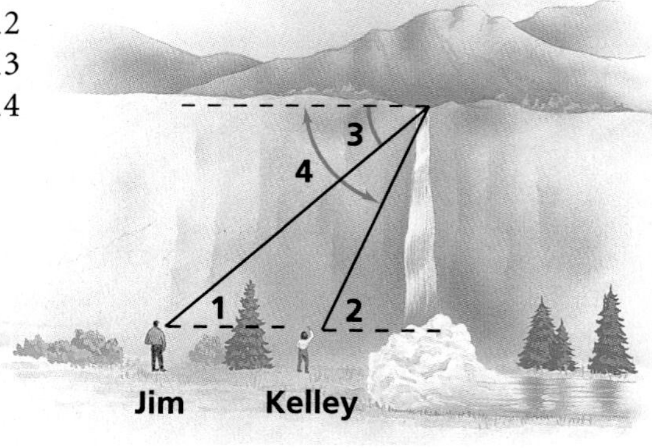

3. **Engineering** The Americans with Disabilities Act states that wheelchair ramps can have a slope no greater than $\frac{1}{12}$. Find the maximum angle of elevation of a ramp with this slope. Round your answer to the nearest tenth.

Solve each problem. Round your answer to the nearest unit unless instructed otherwise.

4. Two office buildings are 51 m apart. The height of the taller building is 207 m. The angle of depression from the top of the taller building to the top of the shorter building is 15°. Find the height of the shorter building.

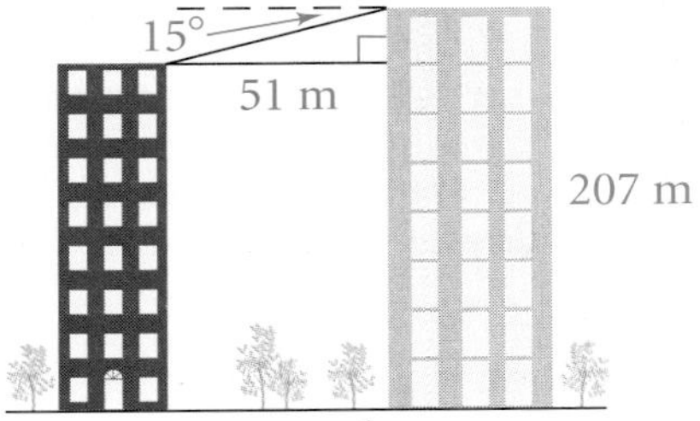

5. A surveyor is 980 ft from the base of the world's tallest fountain at Fountain Hills, Arizona. The angle of elevation to the top of the column of water is 29.7°. His angle measuring device is at the same level as the base of the fountain. Find the height of the column of water to the nearest 10 ft.

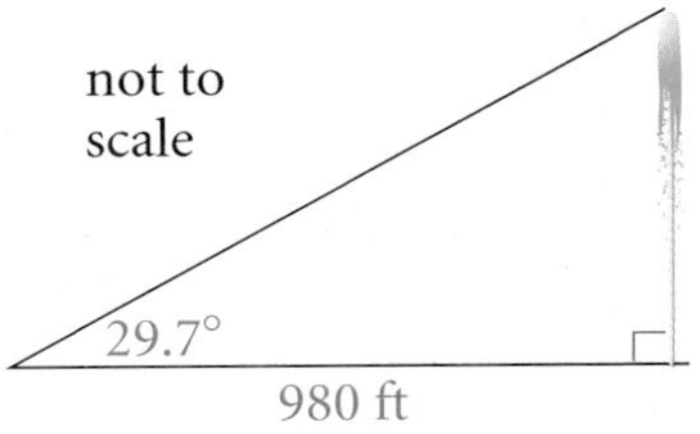

6. On the observation platform in the crown of the Statue of Liberty, Miguel is approximately 250 ft above ground. He sights a ship in New York harbor and measures the angle of depression as 18°. Find the distance from the ship to the base of the statue.

7. A meteorologist measures the angle of elevation of a weather balloon as 41°. A radio signal from the balloon indicates that it is 1503 m from her location. How high is the weather balloon above the ground?

8. The world's tallest unsupported flagpole is a 282-ft-tall steel pole in Surrey, British Columbia. The shortest shadow cast by the pole during the year is 137 ft long. What is the angle of elevation of the sun when the shortest shadow is cast?

9. A blimp is flying to cover a football game. The pilot sights the stadium at a 7° angle of depression. The blimp is flying at an altitude of 400 m. How many kilometers is the blimp from the point 400 m above the stadium? Round your answer to the nearest tenth.

PROBLEM SOLVING HINT

Draw a diagram using the given information. Then decide which trigonometric ratio you can use to solve the problem.

10. a. **Navigation** A simple method for finding a north-south line is shown in the diagram. How could you use this method to also find the angle of elevation of the sun at noon?

b. **Research** For locations in the continental United States, the relationship between the latitude ℓ and angle of elevation a of the sun at noon on the first day of summer is $a = 90° - \ell + 23\frac{1}{2}°$. Find the latitude of your town and determine the angle of elevation of the sun on the first day of summer.

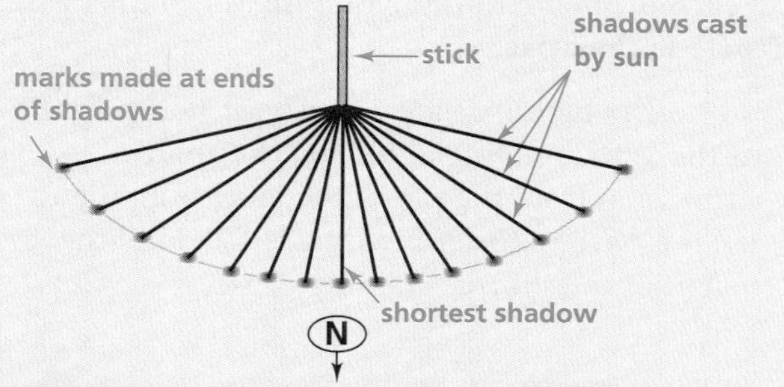

Put a stick in the ground before noon and mark the end of its shadow every 15 minutes. When the shadows begin to lengthen, stop marking the shadows. The mark closest to the stick is directly north of the stick.

11. **Critical Thinking** A television tower is located on a flat plot of land. The tower is supported by several guy wires. Explain how you could find the length of any of these wires. Assume that you are able to measure distances along the ground as well as angles formed by wires and the ground.

12. **Meteorology** One method that meteorologists use to find the height of a layer of clouds above the ground is to shine a bright spotlight directly up onto the cloud layer and measure the angle of elevation from a known distance away. Find the height of the cloud layer in the diagram to the nearest 10 m.

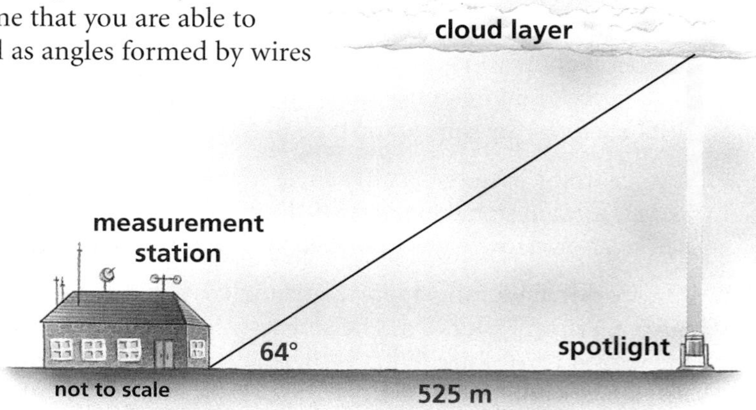

13. a. **Open-ended** Draw and label a diagram that shows a real-world example of an angle of elevation and an angle of depression.

b. **Writing** Write a word problem that uses the angle of depression from your diagram. Include a detailed solution to your problem.

14. **Standardized Test Prep** In the diagram, $m\angle Y > m\angle Z$. Which statement is *not* true?

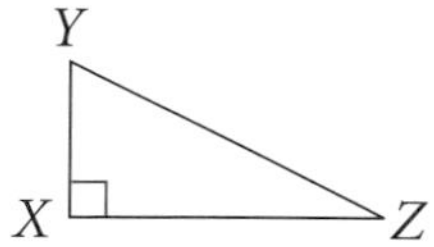

A. $\sin Y > \cos Y$ **B.** $\cos Y < \cos Z$ **C.** $\tan Y > \sin Y$
D. $\tan Z > \cos Y$ **E.** $\cos Z > \sin Y$

Chapter Project *Find Out by Measuring*

Use your clinometer to measure the angle of elevation to an object such as the roof of your school or the top of a flagpole. Then determine how far away you are from the object that you measured. Use what you've learned about trigonometry to find the height of the object. Be sure to add your own height to the height you come up with!

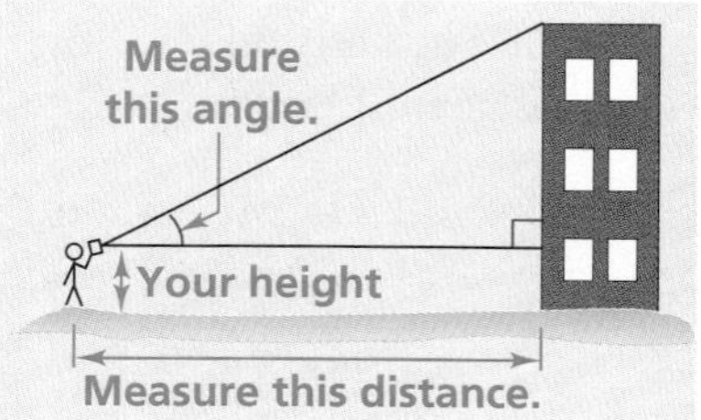

Exercises MIXED REVIEW

Data Analysis **Refer to the graph to answer the questions.**

Source: *International Bluegrass Music Association, Owensboro, KY*

15. What percent of adults are 18–24 years old?

16. What percent of bluegrass purchases did adults 18–24 make?

17. Which age group contains the least number of people?

18. Create two circle graphs using the data from the double bar graph.

19. a. Which age group buys more bluegrass: 35–44-year-olds or 55–64-year-olds?

b. Which of these two age groups likes bluegrass more? Explain.

20. **Coordinate Geometry** Find the equation of the line that passes through (3, 7) and is parallel to the line $y = \frac{3}{2}x - 5$.

Getting Ready for Lesson 11-4

Transformations **Find the image of each point under a translation with the given translation vector.**

21. $(2, 7); \langle 1, -9 \rangle$ **22.** $(-3, 0); \langle 6, -1 \rangle$ **23.** $(4, -7); \langle 5, 0 \rangle$ **24.** $(-5, 12); \langle -8, -11 \rangle$

Exercises CHECKPOINT

1. **Standardized Test Prep** Which value is greatest?

A. sin 30° **B.** cos 45° **C.** tan 60°
D. sin 10° **E.** cos 70°

2. **Architecture** The Leaning Tower of Pisa leans about 5.5° from vertical. How far from the base of the tower will an object dropped from the tower land?

3. A captain of a sailboat sights the top of a lighthouse at a 17° angle of elevation. A navigation chart shows the height of the lighthouse to be 120 m. How far is the sailboat from the lighthouse?

Find the value of *x*. Round lengths of segments to the nearest tenth and angle measures to the nearest degree.

4.

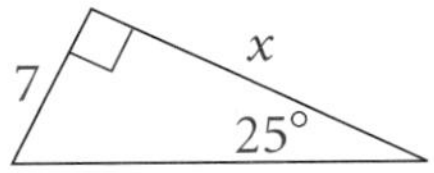

5.

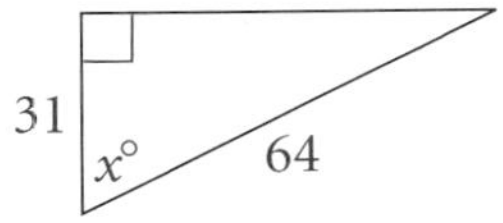

6.

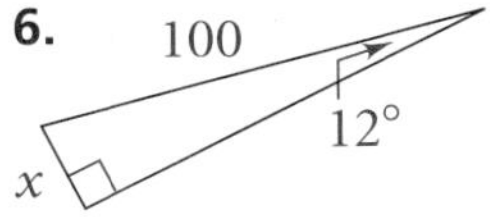

Math ToolboX

Algebra Review

Literal Equations

Before Lesson 11-4

A **literal equation** is an equation involving two or more variables. Formulas are special types of literal equations. You solve for one variable in terms of the others when you transform a literal equation.

Example 1

The formula for the volume of a cylinder is $V = \pi r^2 h$. Find a formula for the height in terms of the radius and volume.

$$V = \pi r^2 h$$

$$\frac{V}{\pi r^2} = \frac{\pi r^2 h}{\pi r^2} \quad \leftarrow \text{Divide each side by } \pi r^2, r \neq 0.$$

$$\frac{V}{\pi r^2} = h \quad \leftarrow \text{Simplify.}$$

The formula for the height is $h = \frac{V}{\pi r^2}$.

Example 2

The formula for the surface area of a cylinder is $A = 2\pi rh + 2\pi r^2$. Solve for h.

$$A = 2\pi rh + 2\pi r^2$$

$$A - 2\pi r^2 = 2\pi rh + 2\pi r^2 - 2\pi r^2 \quad \leftarrow \text{Subtract } 2\pi r^2 \text{ from each side.}$$

$$A - 2\pi r^2 = 2\pi rh \quad \leftarrow \text{Simplify.}$$

$$\frac{A - 2\pi r^2}{2\pi r} = \frac{2\pi rh}{2\pi r} \quad \leftarrow \text{Divide each side by } 2\pi r, r \neq 0.$$

$$\frac{A - 2\pi r^2}{2\pi r} = h \quad \leftarrow \text{Simplify.}$$

The formula for the height is $h = \frac{A - 2\pi r^2}{2\pi r}$.

Solve each equation for the variable in red.

1. $P = 2w + 2\ell$
2. $\tan A = \frac{y}{x}$
3. $A = \frac{1}{2}bh$
4. $C = 2\pi r$
5. $A = \frac{1}{2}(b_1 + b_2)h$
6. $V = \ell wh$
7. $A = \frac{1}{2}ap$
8. $m\angle C + m\angle D = 180$
9. $A = \pi r^2$
10. $S = 180(n - 2)$
11. $V = \frac{1}{3}\ell wh$
12. $a^2 + b^2 = c^2$
13. $\cos A = \frac{b}{c}$
14. $V = \frac{1}{3}\pi r^2 h$
15. $S = \pi r^2 + \pi r\ell$

Connections Air Travel . . . and more

11-4 Vectors and Trigonometry

What You'll Learn

- Describing vectors using ordered pair notation
- Describing the magnitude and direction of vectors

...And Why

To model real-world situations such as those involving velocity

What You'll Need

- calculator

THINK AND DISCUSS

A **vector** is any quantity that has *magnitude* (size) and *direction.* In Chapter 3, you used vectors to describe translations. In this lesson you will explore some other applications of vectors. Here are a few.

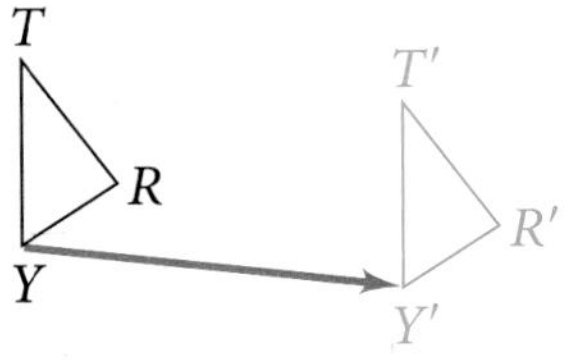

This vector describes the translation $\triangle TRY \rightarrow \triangle T'R'Y'$.

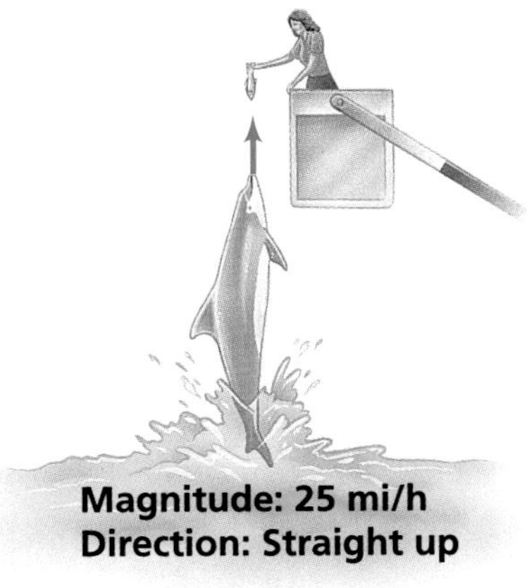
Magnitude: 25 mi/h
Direction: Straight up

Magnitude: 1.62 m/s²
Direction: Toward center of the moon

Magnitude: 34 yd
Direction: Straight through the uprights

1. *Critical Thinking* Does the sentence contain enough information to describe a vector? Explain.
   a. A car heads northwest.
   b. A hiker walks south at 5 miles per hour.
   c. The wind blows at 45 kilometers per hour.

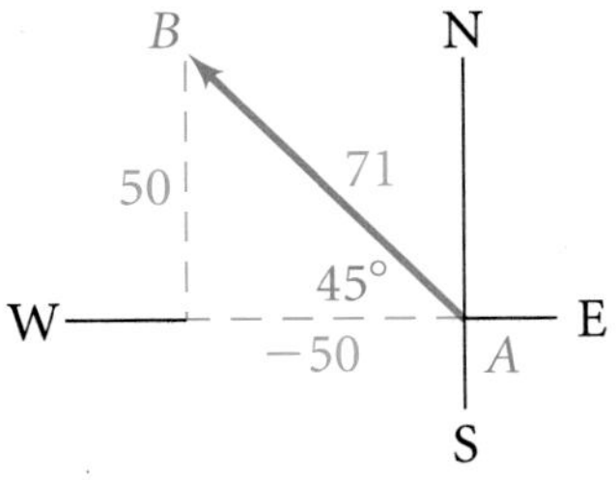

2. Two sisters, Altheia and Chenice, drive from point A to point B. In describing their trip, Altheia says, "We ended up 71 miles northwest of where we started." Chenice says, "We ended up 50 miles west and 50 miles north of where we started." Explain why both sisters' descriptions of their trip are accurate.

Question 2 suggests two different ways that you can describe a vector.

- *Use its size and direction.* The magnitude of $\overrightarrow{AB}$ is 71 (about $50\sqrt{2}$), and its direction is northwest (or 45° north of west).
- *Use its horizontal and vertical change.* The ordered pair notation $\langle -50, 50\rangle$ describes $\overrightarrow{AB}$, where -50 represents the horizontal change from A to B, and 50 represents the vertical change.

Example 1

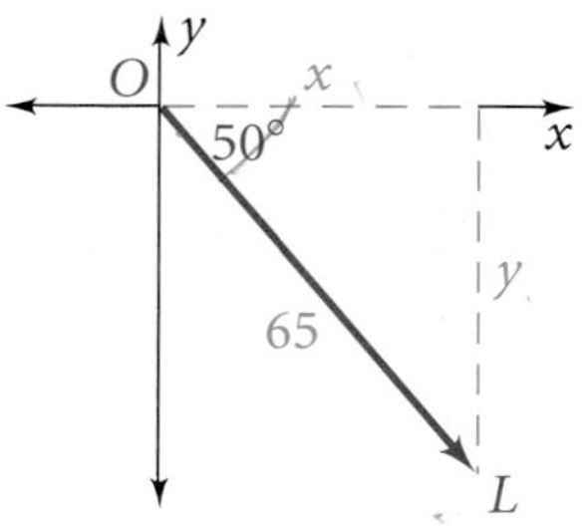

Coordinate Geometry Describe $\overrightarrow{OL}$ by using ordered pair notation. Give the coordinates to the nearest tenth.

Use the sine and cosine ratios to find the values of x and y.

$\cos 50^\circ = \frac{x}{65}$	$\sin 50^\circ = \frac{y}{65}$	Use sine and cosine.
$x = 65(\cos 50^\circ)$	$y = 65(\sin 50^\circ)$	Solve for the variable.
≈ 41.8	≈ 49.8	Use a calculator.

The y-coordinate of the ordered pair is negative since L is in the fourth quadrant. So $\overrightarrow{OL} = \langle 41.8, -49.8 \rangle$.

QUICK REVIEW

In the diagram, I is the *initial point* of $\overrightarrow{IT}$ and T is its *terminal point.*

In many applications of vectors, the direction of the vector is described in relation to north, south, east, and west. Here are a few examples.

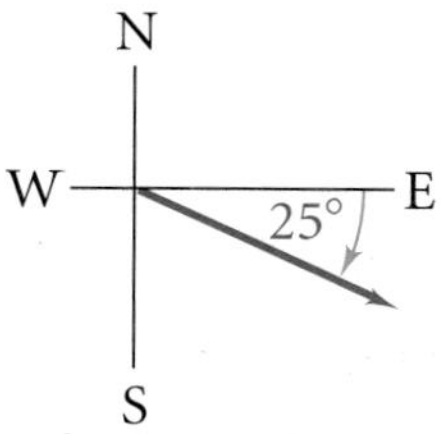

25° south of east

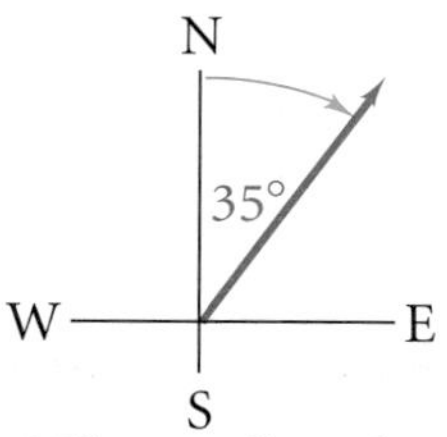

35° east of north

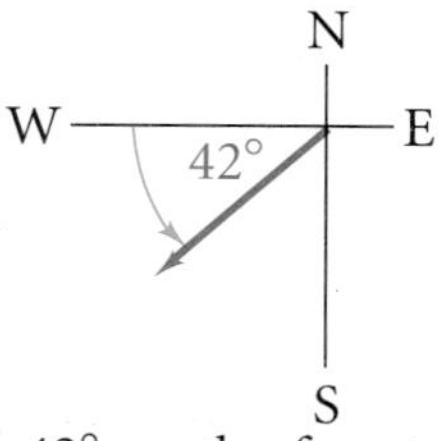

42° south of west

3. Try This Sketch a vector that has a direction of 30° west of north.

Example 2 Relating to the Real World

Air Travel A helicopter lands at a point 40 km west and 25 km south of the point at which it took off. Find the distance that the helicopter flew and the direction of its flight.

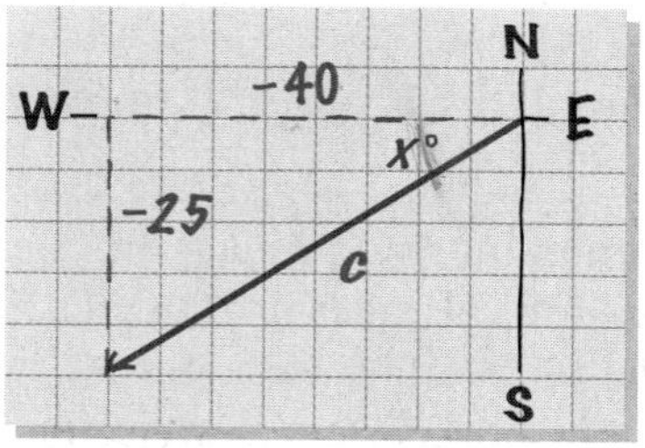

Start by drawing a diagram of the vector. To find the distance flown c, use the Pythagorean Theorem.

$40^2 + 25^2 = c^2$	Substitute 40 and 25 for the lengths of the legs.
$2225 = c^2$	Simplify the left side.
$c \approx 47$	Take the square root.

To find the direction of the helicopter's flight, find the angle that the vector makes with west.

$\tan x^\circ = \frac{25}{40}$	Use the tangent ratio.
(25 ÷ 40) TAN⁻¹ 32.005383	Use a calculator.

The helicopter flew about 47 km at 32° south of west.

4. Try This A small plane lands at a point 246 mi east and 76 mi north of the point at which it took off. Find the distance that the plane flew and the direction of its flight.

Exercises ON YOUR OWN

Coordinate Geometry **Describe each vector by using ordered pair notation. Give the coordinates to the nearest unit.**

1.

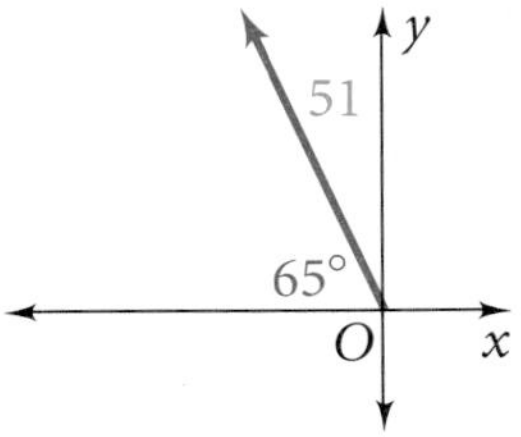

2.

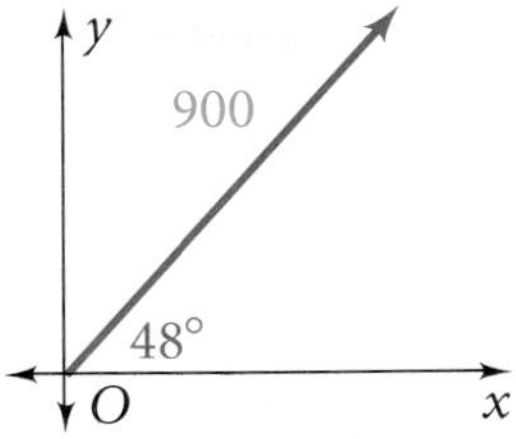

3.

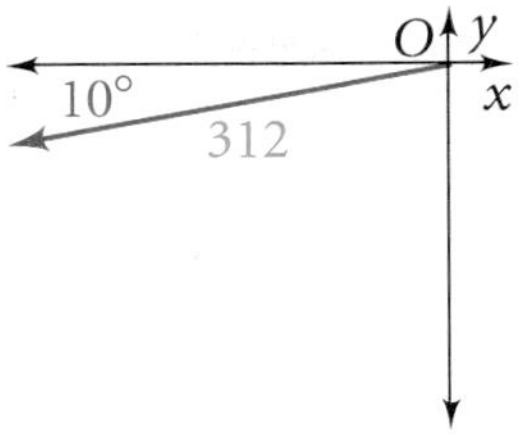

4.

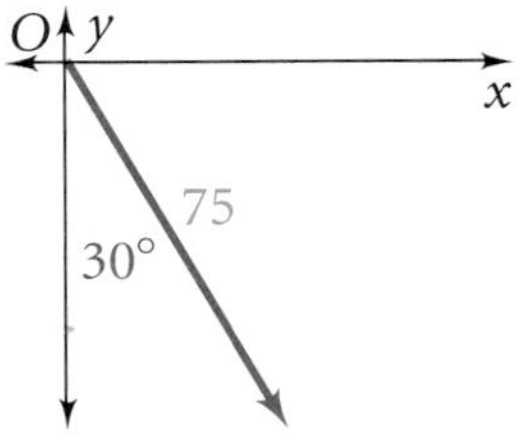

Describe the direction of each vector.

5.

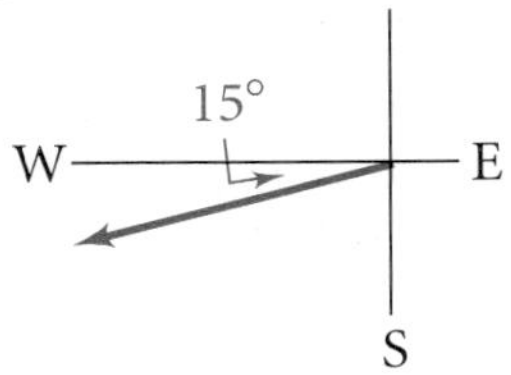

6.

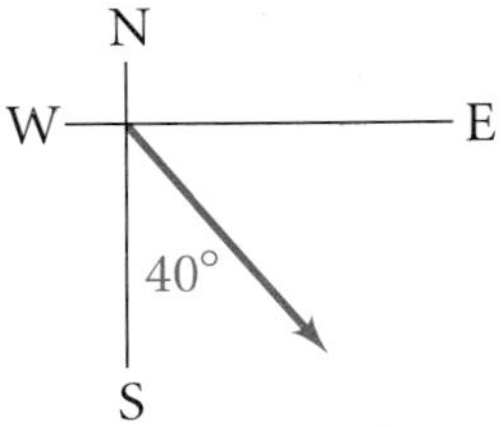

7.

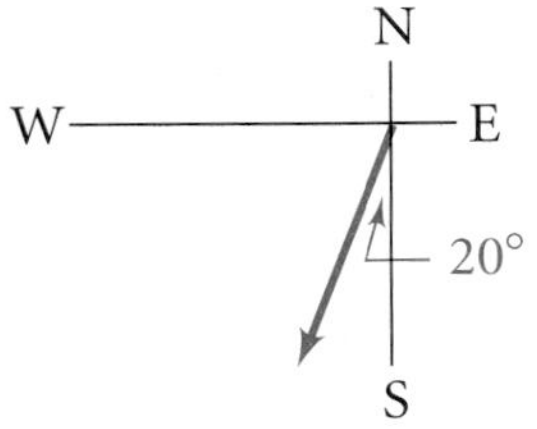

8.

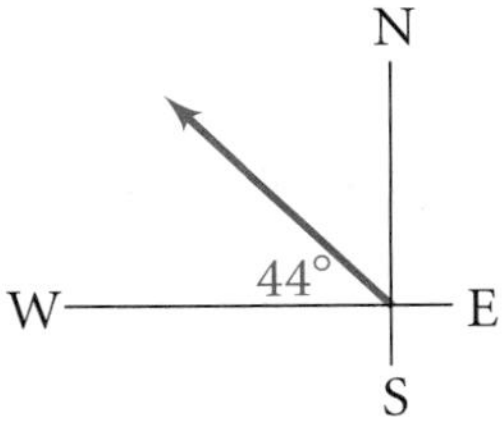

Biology **Solve each problem. Round your answers to the nearest unit, unless otherwise directed.**

9. Homing pigeons are capable of returning to their home upon release. Homing pigeons carried news of Olympic victories to various cities in ancient Greece. Suppose one such pigeon took off from Athens and landed in Sparta, which is 73 mi west and 64 mi south of Athens. Find the distance and direction of its flight.

10. Honeybees describe vectors through movement. A bee that discovers a new source of pollen flies back to the hive and does a "dance" for the other bees. The movements of this dance tell other bees how to get to the pollen. Suppose a bee describes a source of pollen 1000 ft away and 60° counterclockwise from the direction of the sun. Draw a diagram of this vector. Include labels for the sun, the hive, and the pollen.

11. During its adult life, a European eel travels from the Belgian coast to breeding grounds in the Sargasso Sea. This part of the Atlantic Ocean is about 2300 mi south and 4800 mi west of Belgium. Find the distance and direction of the eel's journey. Round the distance to the nearest 100 mi.

12. The owners of a golden retriever named Ginger took her from her home to a location 17 mi north and 29 miles east. There, they somehow lost her. Ginger, however, managed to find her way home after 3 days. Find the distance and direction of her journey home (assuming she took the shortest route).

13. **Writing** How are the vectors $\overrightarrow{AB}$ and $\overrightarrow{BA}$ alike? How are they different?

14. **Critical Thinking** Valerie described the direction of a vector as 35° south of east. Pablo described it as 55° east of south. Could the two be describing the same vector? Explain.

15. **Transformational Geometry** Point A' is the image of A under a dilation with scale factor k and center O. How are the direction and magnitude of $\overrightarrow{OA}$ related to the direction and magnitude of $\overrightarrow{OA'}$?

16. **Open-ended** Name four other vectors with the same magnitude as $\langle -7, -24 \rangle$.

"Well, lemme think. ... You've stumped me, son. Most folks only wanna know how to go the other way."

Critical Thinking **Does each sentence contain enough information to describe a vector? Explain your answer.**

17. The angle of elevation of a baseball hit by Frank Thomas is 35°.

18. A rocket is launched upward with an initial thrust of 400,000 lb.

19. A cheetah runs at 71 miles per hour.

Find the magnitude and direction of each vector.

20.
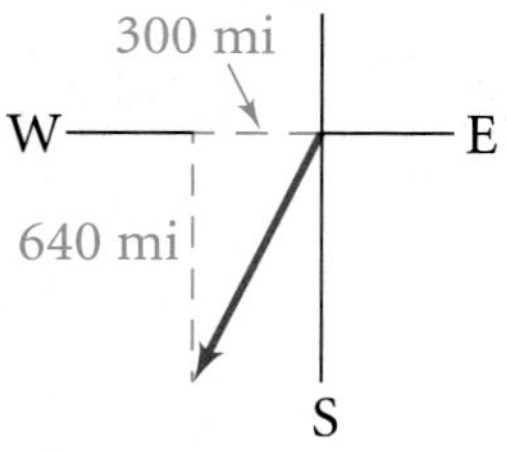

21.
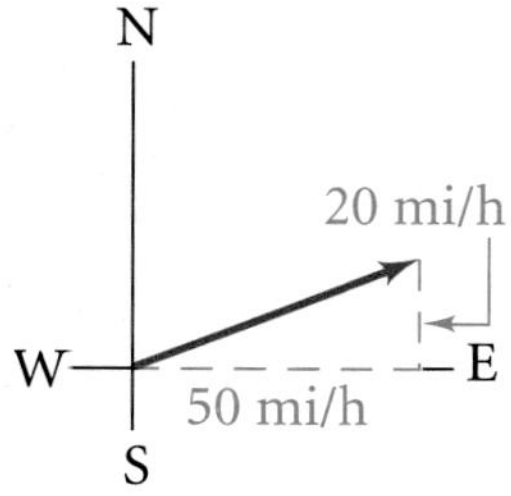

22.
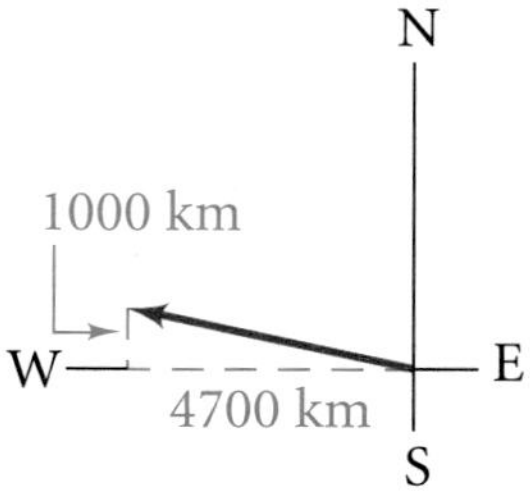

23.
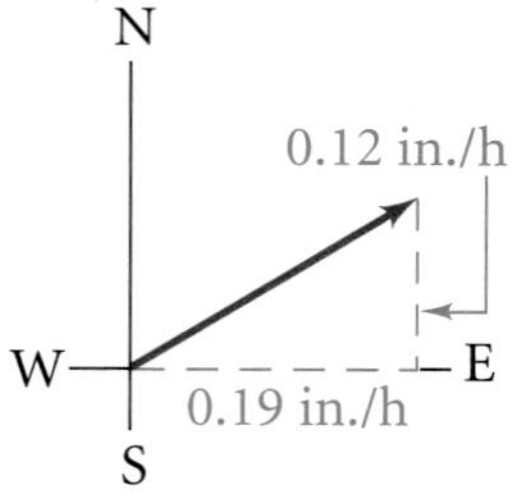

24. **Patterns** Use the diagrams below to write a definition of *equal vectors.*

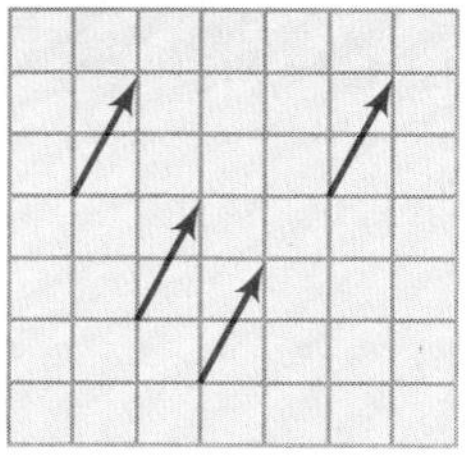
All of these vectors are equal.

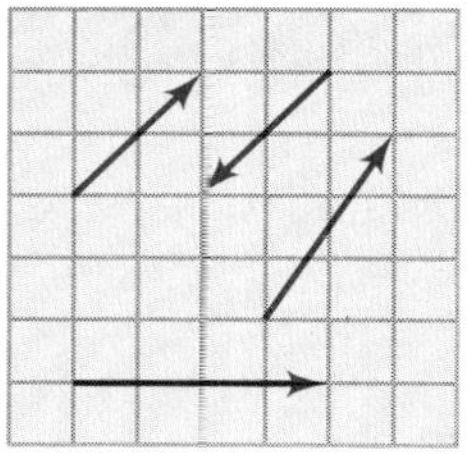
None of these vectors are equal.

25. **Patterns** Use the diagrams below to write a definition of *parallel vectors.*

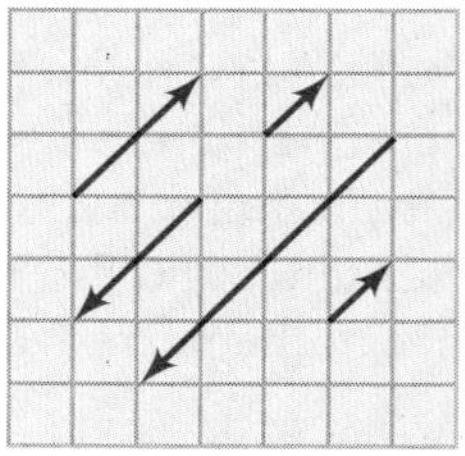
All of these vectors are parallel.

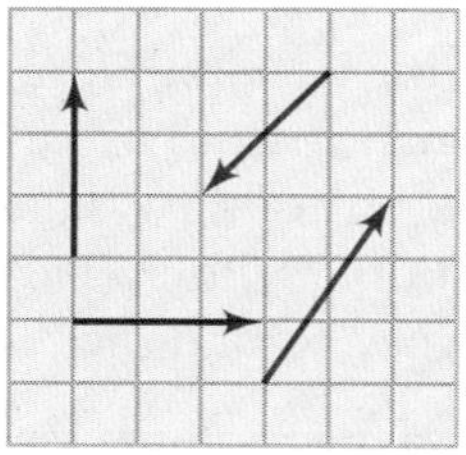
None of these vectors are parallel.

26. Air Travel A plane takes off on a runway in the direction 10° east of south. When it reaches 5000 ft, it turns right 45°. It cruises at this altitude for 60 mi. It then turns left 160°, descends, and lands. Match each velocity vector with the appropriate portion of the flight.

A. The plane ascends.
B. The plane cruises.
C. The plane descends.

I.

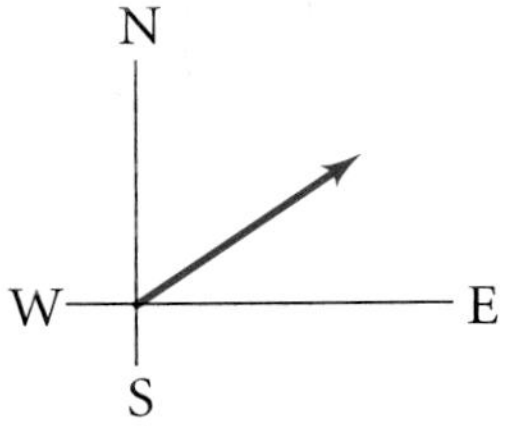

II.

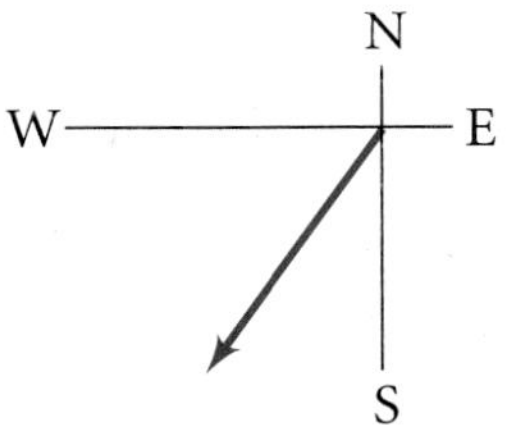

III.

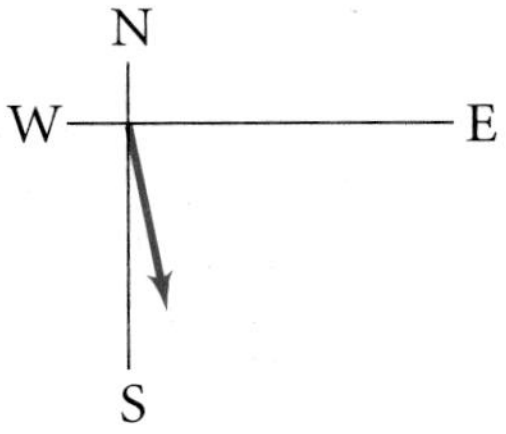

Chapter Project *Find Out by Comparing*

Repeat the activity on page 560, but this time work in groups to estimate the height of a single object on your school grounds. Have each person in your group measure the angle of elevation from a different distance. Then compare results within the group. Write a brief report explaining the variations in your estimates of the object's height. Decide as a group on a single estimate of the object's height.

Exercises MIXED REVIEW

Is each pair of triangles *congruent, similar,* or *neither?* Justify your answers.

27.

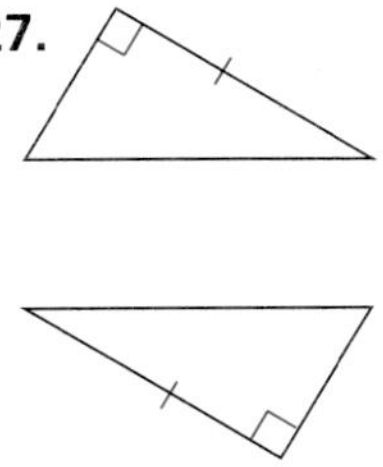

28.

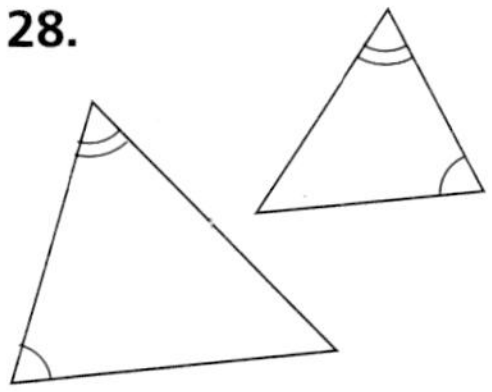

29.

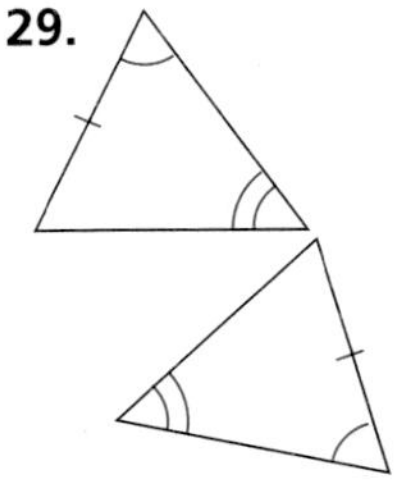

SELF ASSESSMENT

FOR YOUR JOURNAL

Create a diagram or table that summarizes what you learned in this lesson.

30. Sports A cylindrical hockey puck is 1 in. high and 3 in. in diameter. What is its volume? Round your answer to the nearest tenth.

31. A circle has radius 7 m. Two radii form a 100° angle. What is the length of the minor arc? Round your answer to the nearest tenth.

Getting Ready for Lesson 11-5

32. a. Find A', the image of $A(4, 3)$ under the translation $\langle 2, 5\rangle$.
b. Find A'', the image of A' under the translation $\langle -5, 6\rangle$.
c. What single translation maps A to A''?

What You'll Learn

- Solving problems that involve vector addition

...And Why

To find the speed and direction of an object that has two forces acting on it

What You'll Need

- graph paper
- a marble
- three straws
- calculator

11-5 Adding Vectors

THINK AND DISCUSS

Until now, you've named vectors by using the initial point and terminal point ($\overrightarrow{AB}$, for example). You can also name vectors by using a single lowercase letter, such as $\vec{\mathbf{u}}$.

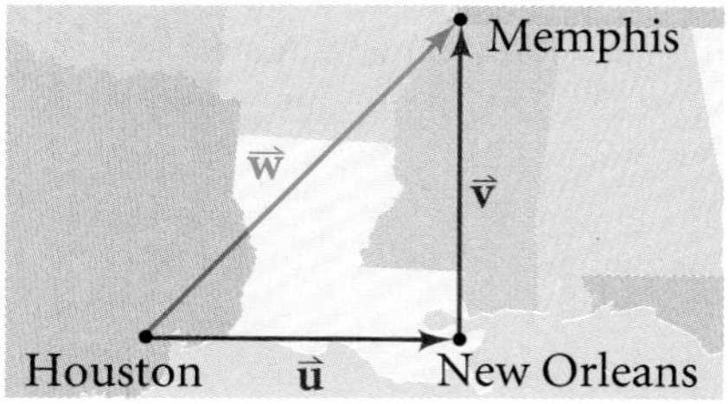

The map shows vectors representing a flight from Houston to Memphis with a stopover in New Orleans. The vector from Houston to Memphis is called the *sum* of the other two vectors. You write this as $\vec{\mathbf{u}} + \vec{\mathbf{v}} = \vec{\mathbf{w}}$.

1. **Critical Thinking** The sum of two vectors is called a **resultant.** Use the example of the trip from Houston to Memphis to explain why this name makes sense.

Example 1

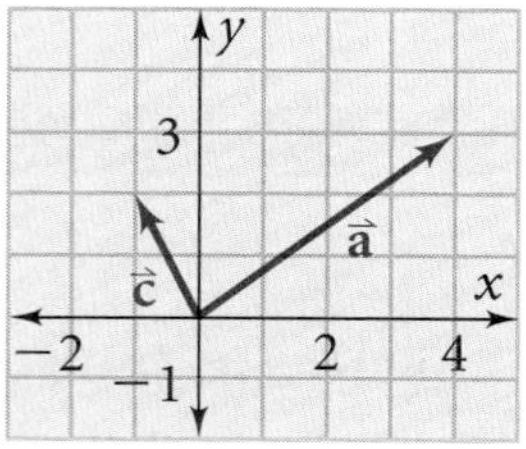

Coordinate Geometry Refer to the diagram at the left. Find $\vec{\mathbf{e}}$, the sum of $\vec{\mathbf{a}}$ and $\vec{\mathbf{c}}$.

Draw $\vec{\mathbf{a}}$ with its initial point at the origin. Then draw $\vec{\mathbf{c}}$ so that its initial point is the terminal point of $\vec{\mathbf{a}}$. Finally, draw a vector from the initial point of $\vec{\mathbf{a}}$ to the terminal point of $\vec{\mathbf{c}}$. This vector is $\vec{\mathbf{e}}$, the resultant.

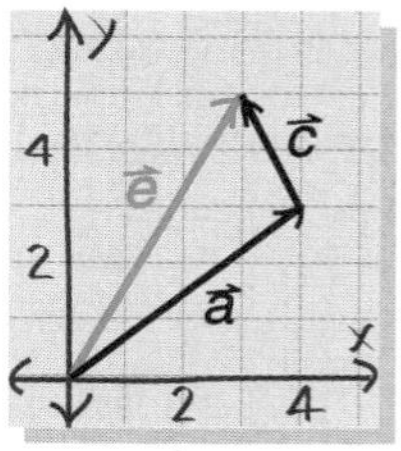

The method of adding vectors shown in Example 1 is called the *head-to-tail method.* (Do you see why?) You can describe the head-to-tail method mathematically by using ordered pair notation.

$\vec{\mathbf{a}} + \vec{\mathbf{c}} = \langle 4, 3\rangle + \langle -1, 2\rangle$ — Express $\vec{a}$ and $\vec{c}$ in ordered pair notation.

$= \langle 4 + (-1), 3 + 2\rangle$ — Add the x- and y-coordinates.

$= \langle 3, 5\rangle$ — Simplify.

A look back at the diagram confirms that $\langle 3, 5\rangle$ is indeed the resultant.

Adding Vectors

For $\vec{\mathbf{a}} = \langle x_1, y_1\rangle$ and $\vec{\mathbf{c}} = \langle x_2, y_2\rangle$, $\vec{\mathbf{a}} + \vec{\mathbf{c}} = \langle x_1 + x_2, y_1 + y_2\rangle$.

Example: For $\vec{\mathbf{a}} = \langle -3, 2\rangle$ and $\vec{\mathbf{c}} = \langle 1, 5\rangle$, $\vec{\mathbf{a}} + \vec{\mathbf{c}} = \langle -2, 7\rangle$.

The diagram at the right shows that you can add vectors in any order. That is, $\vec{u} + \vec{v} = \vec{v} + \vec{u}$. Notice also that the four vectors shown in red form a quadrilateral. This quadrilateral is a parallelogram because its opposite sides are congruent. The result $\vec{w}$ is a diagonal of this parallelogram. This characteristic of vector addition is called the *parallelogram rule.*

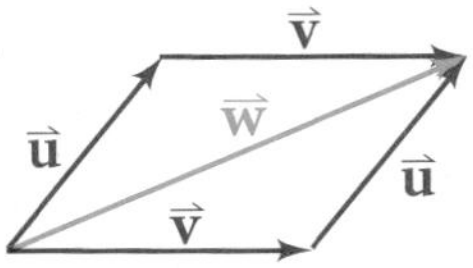

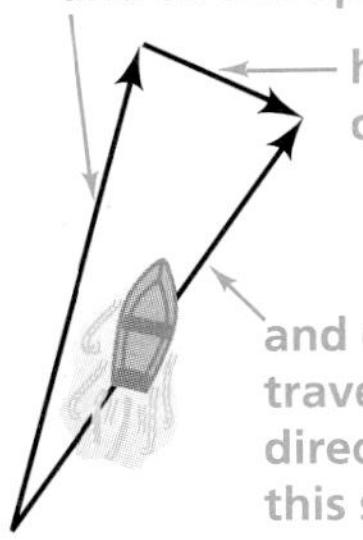

Vector sums can show the result of actions that take place one after the other, such as a trip from Houston to Memphis by way of New Orleans. Vector sums can also show the result of two forces that act at the same time on an object. An example of this is the two forces that act on a boat—the forces of its engine and the current. As shown in the diagram, the force of the current combines with the force of the engine to give the boat its velocity.

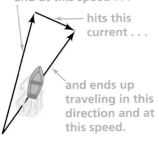

Example 2 Relating to the Real World

Navigation A ferry shuttles people from one side of a river to the other. The speed of the ferry in still water is 25 mi/h, and the river flows directly south at 7 mi/h. Suppose the ferry heads directly west. What are the ferry's resulting speed and direction?

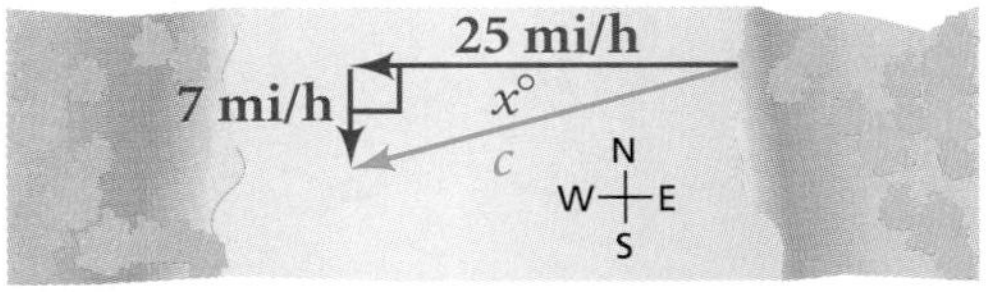

The diagram shows the sum of the two vectors. To find the ferry's speed, use the Pythagorean Theorem to find the length of the resultant.

$c^2 = 25^2 + 7^2$ The lengths of the legs are 25 and 7.

$c^2 = 674$ Simplify.

$c \approx 25.96151$ Use a calculator to take the square root.

To find the ferry's direction, use trigonometry.

$\tan x° = \frac{7}{25}$ Use the tangent ratio.

$x = \tan^{-1}\left(\frac{7}{25}\right)$

$x \approx 15.642246$ Use a calculator.

The ferry's speed is about 26 mi/h, and its direction is about 16° south of west.

2. **Try This** Use the diagram to find the angle at which the ferry must head upriver in order to travel directly across the river.

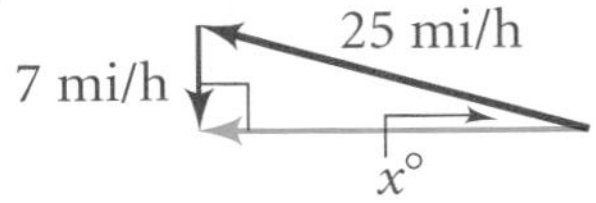

WORK TOGETHER

Work in groups of at least three. Draw a set of axes on a piece of graph paper, as shown in the photo. Place a marble at the origin. Have two members of your group aim straws along the x- and y-axes. When the third member of the group gives a signal, blow through the straws at the marble. Stop the marble before it rolls off the paper, and mark its stopping point.

3. Draw a vector from the origin to the stopping point. Then draw the vectors that represent the two forces on the marble. Which of the forces has the greater magnitude?

4. Come up with a way to tell who in your group can exert the most force on the marble. Explain why your method works.

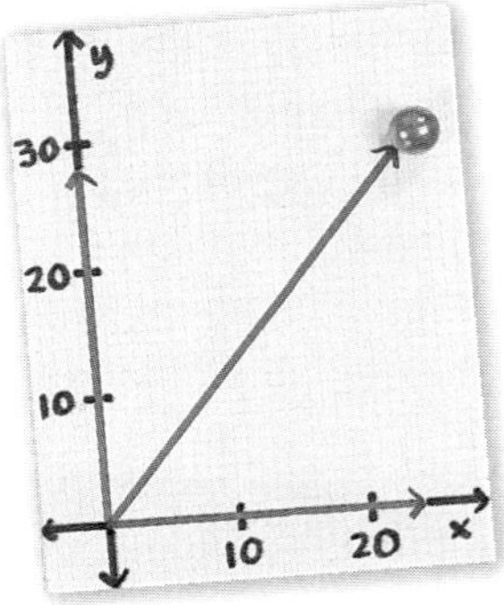

Exercises ON YOUR OWN

Coordinate Geometry **Find the sum of each pair of vectors. Give your answers in ordered pair notation.**

1.
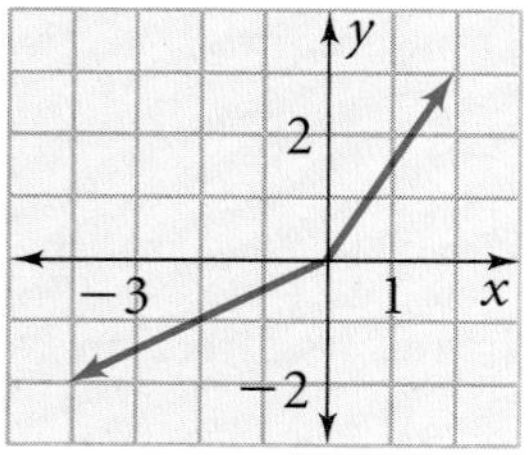

2.
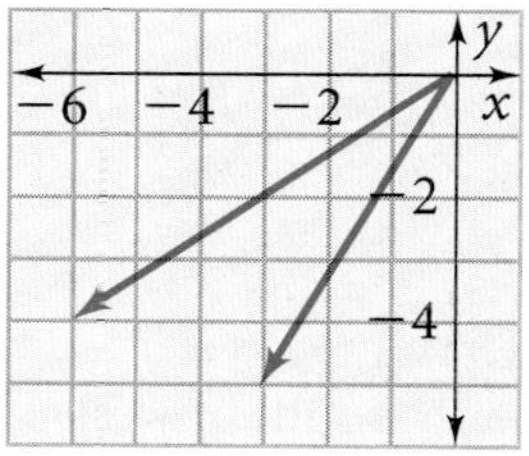

3.
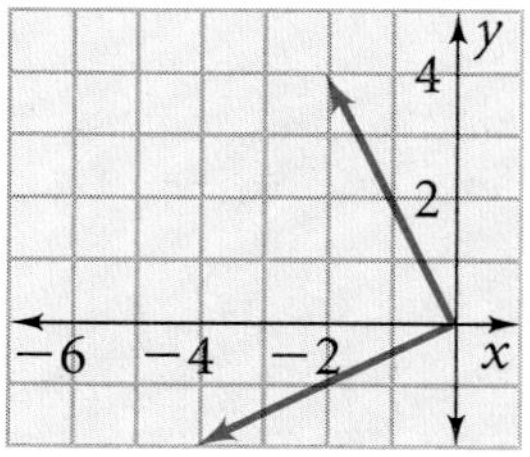

4.
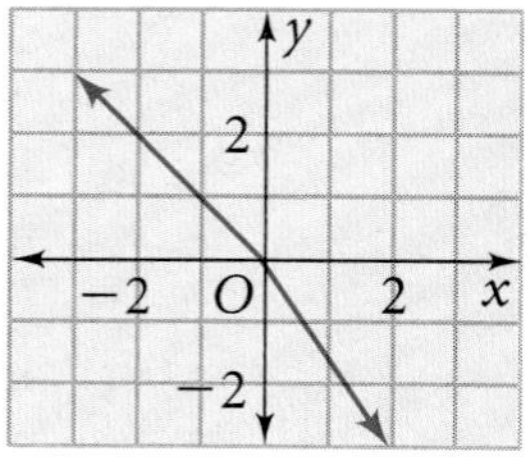

5.
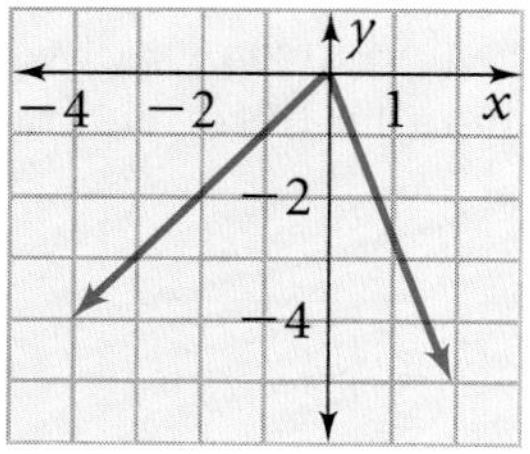

6.
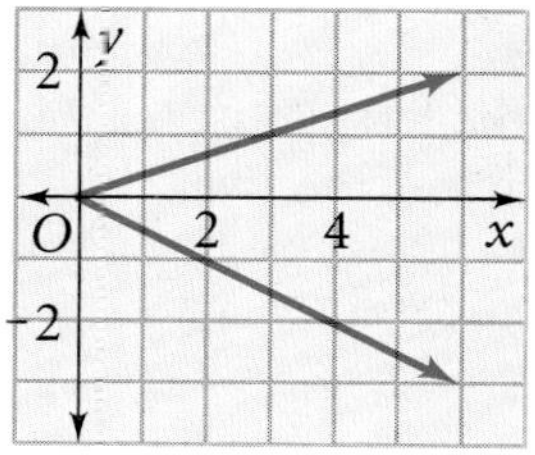

7.
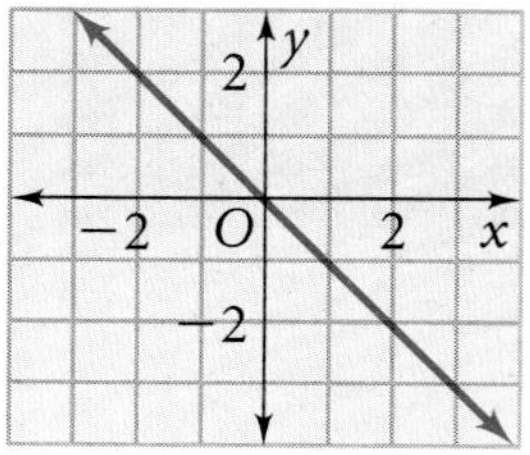

8.
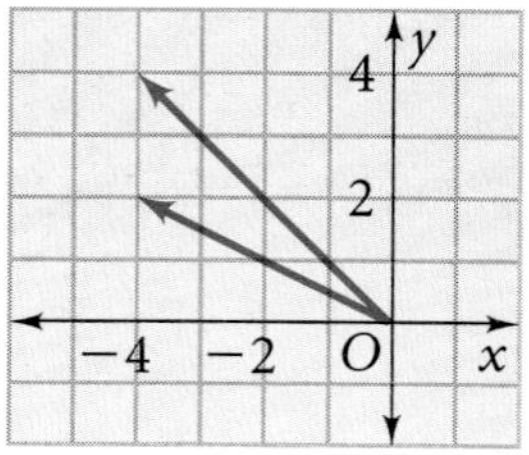

9. a. **Probability** Find the probability that the sum of any two of the vectors at the right has a greater magnitude than the third.
   b. **Open-ended** Draw three vectors of your own. Then use your diagram to find the answer to part (a).

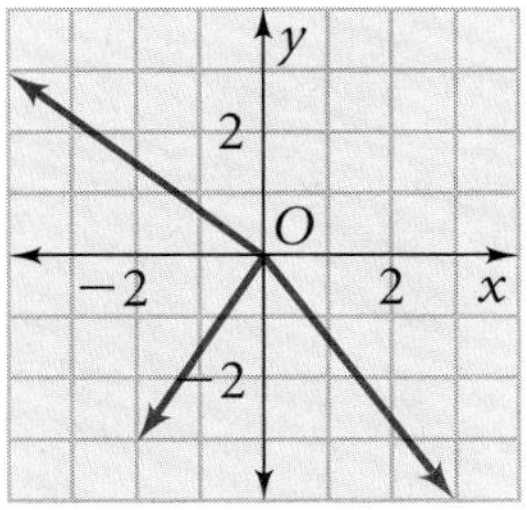

10. **Boat Travel** The speed of a powerboat in still water is 35 mi/h. This powerboat heads directly west across the Trinity River at a point at which it flows directly south at 8 mi/h. What are the resulting speed and direction of the boat? Round your answers to the nearest tenth.

11. **Navigation** A fishing trawler leaves its home port and travels 150 mi directly east. It then changes course and travels 40 mi due north.
   a. In what direction should the trawler head to return to its home port?
   b. How long will the return trip take if the trawler averages 23 mi/h?

12. a. Find the sum $\vec{a} + \vec{c}$, where $\vec{a} = \langle 45, -60 \rangle$ and $\vec{c} = \langle -45, 60 \rangle$.
   b. **Writing** What does your answer to part (a) tell you about $\vec{a}$ and $\vec{c}$?

In Exercises 13–16: (a) Copy the diagram and then draw a parallelogram that has the given vectors as adjacent sides. (b) Find the magnitude and direction of the resultant.

13. 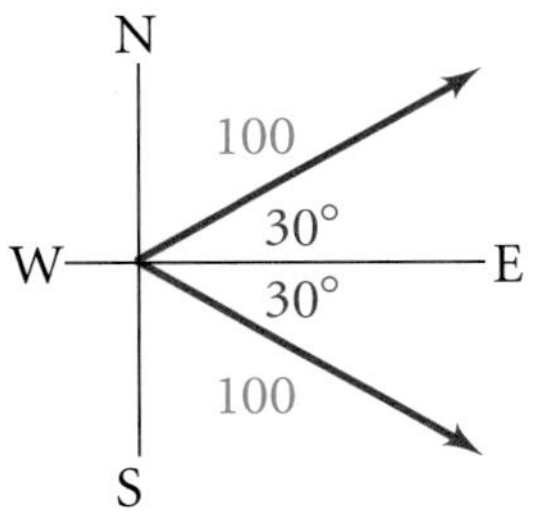

14. 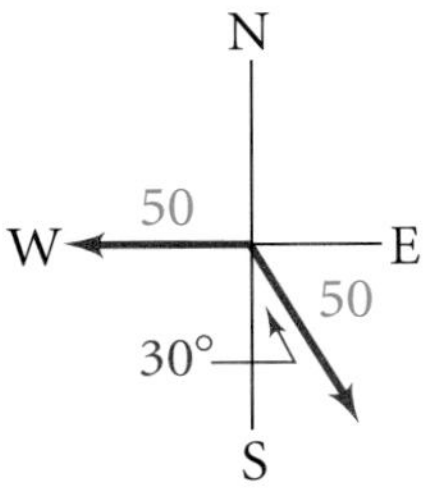

15. 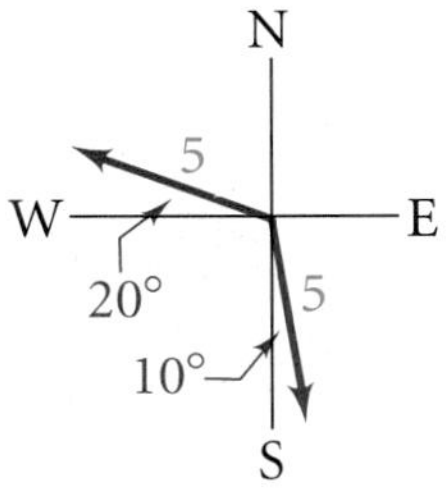

16. 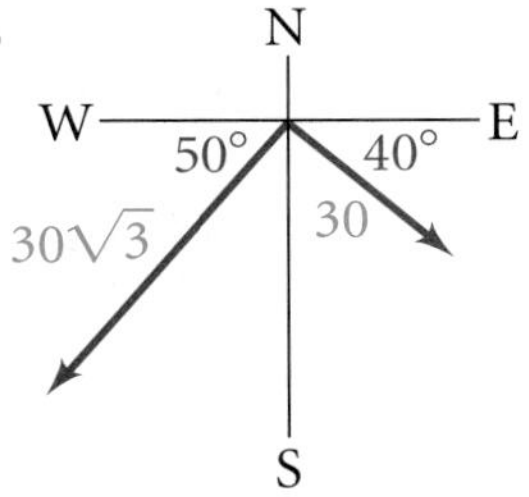

17. A Red Cross helicopter takes off and flies 75 km at 20° south of west. There, it drops off some relief supplies. It then flies 125 km at 10° west of north to pick up three medics.
   a. Make an accurate head-to-tail drawing of the two vectors described.
   b. Draw the resultant and measure it to find the helicopter's distance from its point of origin and the direction it should head to get back.

18. **Air Travel** The cruising speed of a Boeing 767 in still air is 530 mi/h. Suppose that a 767 cruising directly east encounters a 80 mi/h wind blowing 40° south of west.
   a. Sketch the vectors for the velocity of the plane and the wind.
   b. Express both vectors from part (a) in ordered pair notation.
   c. Find the sum of the vectors from part (b).
   d. Find the magnitude and direction of the vector from part (c).

19. **Geometry in 3 Dimensions** A bear leaves home and ambles 10 km straight south. She then turns and walks another 10 km straight west. After a short rest, she turns and wanders straight north another 10 km, ending up at the point at which she started. What color is the bear? (*Hint:* Figure out where on Earth she is.)

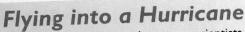

Flying into a Hurricane

When most pilots hear a forecast for gale force winds, they don't think, "Time to fly." Then again, most pilots don't work for the National Oceanic and Atmospheric Administration. NOAA operates two WP-3D airplanes that fly directly into hurricanes. These four-engine turboprops, which carry eight crew members and up to ten scientists, are loaded with data-collection equipment. Some of this equipment is in the WP-3D's long "snout," which also serves as a lightning rod. In a routine flight, the WP-3D is struck by lightning 3 or 4 times. Surprisingly, small burn holes are the only damage from these strikes. To help overcome temporary blindness caused by lightning flashes, the pilots set the cockpit lights at the brightest level.

TIMELESS MAGAZINE **72**

20. In still air, the WP-3D flies at 374 mi/h. Suppose that a WP-3D flies due west into a hurricane and encounters a 95-mi/h wind blowing due south. What are the resulting speed and direction of the plane?

Exercises MIXED REVIEW

Find the area of each figure to the nearest tenth.

21.

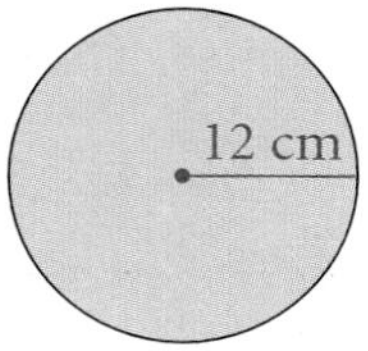

22.

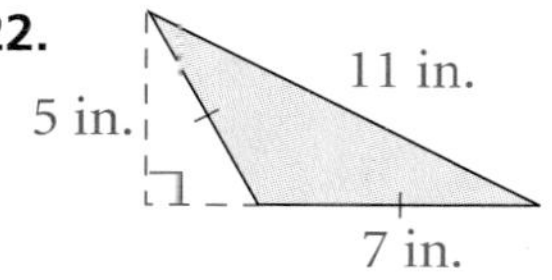

23.

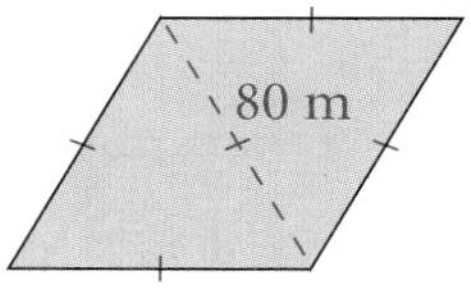

24.

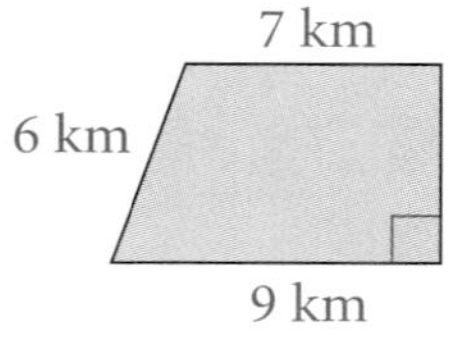

25. *Open-ended* Draw a figure in one-point perspective.

Getting Ready for Lesson 11-6

The hexagon at the right is regular. Name each point or segment.

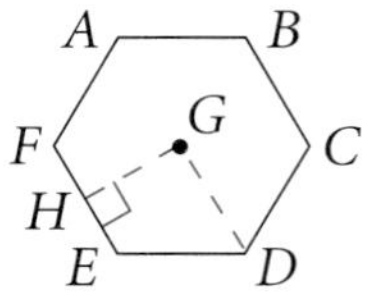

26. the center **27.** a radius **28.** an apothem

Exercises CHECKPOINT

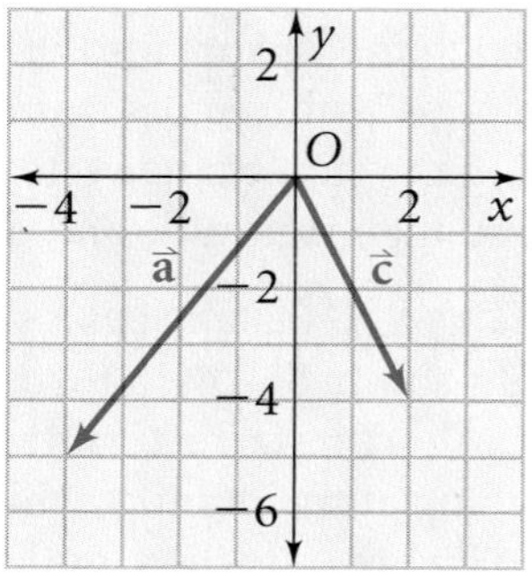

1. Find the magnitude of vectors $\vec{a}$ and $\vec{c}$. Round your answers to the nearest tenth.

2. Find the sum $\vec{a} + \vec{c}$. Give your answer in ordered pair notation.

3. *Air Travel* A twin-engine plane has a speed of 300 mi/h in still air. Suppose this plane travels directly south and encounters a 50 mi/h wind blowing due east. Find the resulting speed and direction of the plane. Round your answers to the nearest unit.

4. *Open-ended* Sketch two vectors $\vec{u}$ and $\vec{v}$ and their sum $\vec{w}$ on graph paper. Give the ordered pair notation for each vector.

11-6 Trigonometry and Area

What You'll Learn

- Using trigonometry to find the areas of regular polygons
- Using the sine ratio to find the areas of acute triangles

...And Why

To solve problems involving angle measures and areas

What You'll Need

- calculator

THINK AND DISCUSS

Finding Areas of Regular Polygons

In Chapter 5, you learned to find the area of a regular polygon by using the formula $A = \frac{1}{2}ap$, where a is the apothem and p is the perimeter. By using the trigonometric ratios together with this formula, you can solve other types of problems. Before you do so, however, take a minute to review the parts of a regular polygon in the diagram below.

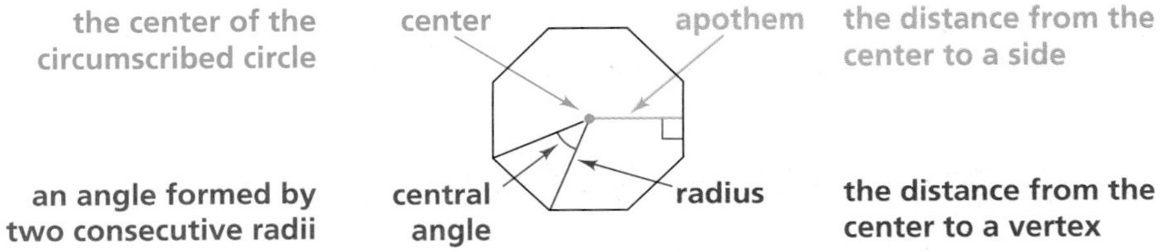

Example 1

Find the area of a regular pentagon with side 8 cm.

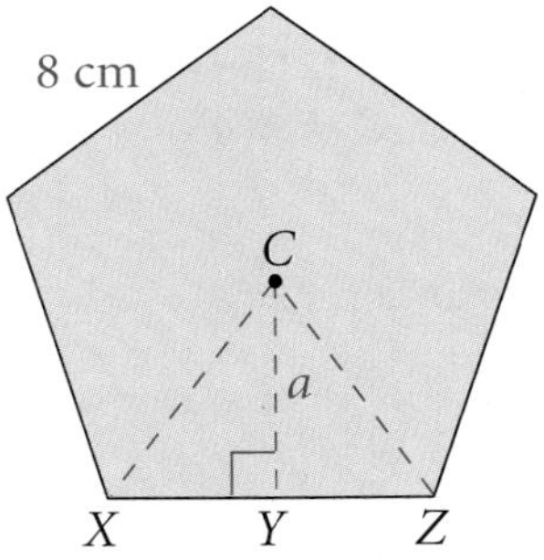

To use the formula $A = \frac{1}{2}ap$, you need the perimeter and apothem. The perimeter is $5 \cdot 8$, or 40. To find the apothem, use trigonometry.

The measure of central angle $\angle XCZ$ is $\frac{360}{5}$, or 72.
$m\angle XCY = \frac{1}{2}m\angle XCZ$, so $m\angle XCY = 36$.
$XY = \frac{1}{2}XZ$, so $XY = 4$.

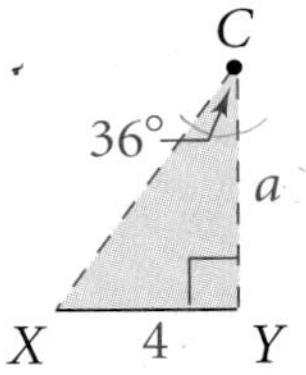

$\tan 36° = \frac{4}{a}$ Use the tangent ratio.

$a = \frac{4}{\tan 36°}$ Solve for a.

Now substitute into the area formula.

$A = \frac{1}{2}ap$

$= \frac{1}{2} \cdot \frac{4}{\tan 36°} \cdot 40$ Substitute for a and p.

$= \frac{80}{\tan 36°}$ Simplify.

80 [÷] 36 [TAN] [=] 110.11055 Use a calculator.

The area of a regular pentagon with side 8 cm is about 110 cm^2.

CALCULATOR HINT

Do all the rounding at the end of your solution. This method gives a more accurate answer than you would get by rounding several times before the final computation.

1. **Try This** Find the area of a regular octagon with a perimeter of 80 in. Give your answer to the nearest tenth.

The Castel del Monte, situated on a hill in southern Italy, contains many regular octagons. It was built in the 1200s as a hunting residence for Frederick II, the king of Sicily.

Example 2 *Relating to the Real World*

Architecture The radius of the castle's inner courtyard is 16 m. Find the area of the courtyard.

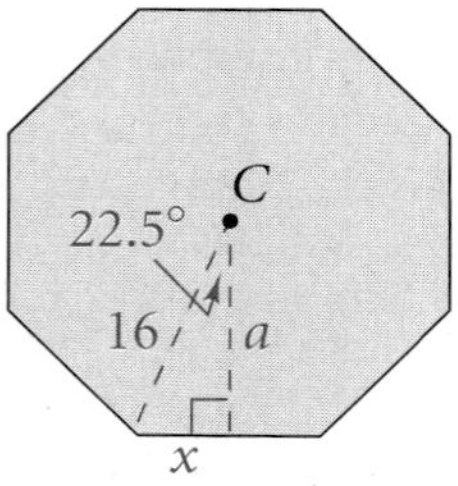

Use trigonometry to find the apothem and the perimeter. The measure of the central angle of an octagon is $\frac{360}{8}$, or 45. So $m\angle C = \frac{1}{2}(45) = 22.5$.

$\cos 22.5° = \frac{a}{16}$ $\qquad$ $\sin 22.5° = \frac{x}{16}$

$a = 16(\cos 22.5°)$ $\qquad$ $x = 16(\sin 22.5°)$

Now use x to find the perimeter p.

$p = 8 \cdot \text{length of side}$

$= 8 \cdot 2x$ — The length of each side is $2x$.

$= 16 \cdot 16(\sin 22.5°)$ — Simplify and substitute for x.

$= 256(\sin 22.5°)$ — Simplify.

Finally, substitute into the area formula, $A = \frac{1}{2}ap$.

$A = \frac{1}{2} \cdot 16(\cos 22.5°) \cdot 256(\sin 22.5°)$ — Substitute for a and p.

$= 2048(\cos 22.5°)(\sin 22.5°)$ — Simplify.

$= 724.07734$ — Use a calculator.

The area of the courtyard of the Castel del Monte is about 724 m^2.

2. **Try This** All of the eight small towers around the castle are also regular octagons. The radius of each is 7.3 m. Find the area of a horizontal cross section of each tower to the nearest square meter.

Finding the Area of a Triangle Given SAS

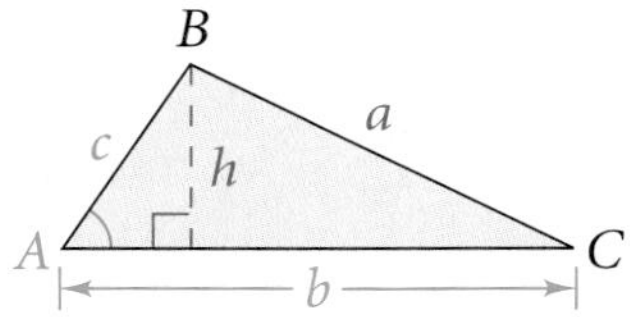

Suppose you want to find the area of $\triangle ABC$, but you know only the lengths b and c, and $m\angle A$. To use the formula Area $= \frac{1}{2}bh$, you need to know the height. You can find the height by using trigonometry.

$\sin A = \frac{h}{c}$ Use the sine ratio.

$h = c(\sin A)$ Solve for *h*.

Now substitute for h in the formula Area $= \frac{1}{2}bh$.

Area $= \frac{1}{2}bc(\sin A)$

Theorem 11-1
Area of a Triangle Given SAS

The area of a triangle is one half the product of the lengths of two sides and the sine of the included angle.

Area of triangle $= \frac{1}{2} \cdot$ side length $\cdot$ side length $\cdot$ sine of included angle

The proof of this theorem for the case in which the given angle is obtuse requires a more complete definition of sine. You will study this definition in later math courses.

Example 3 Relating to the Real World

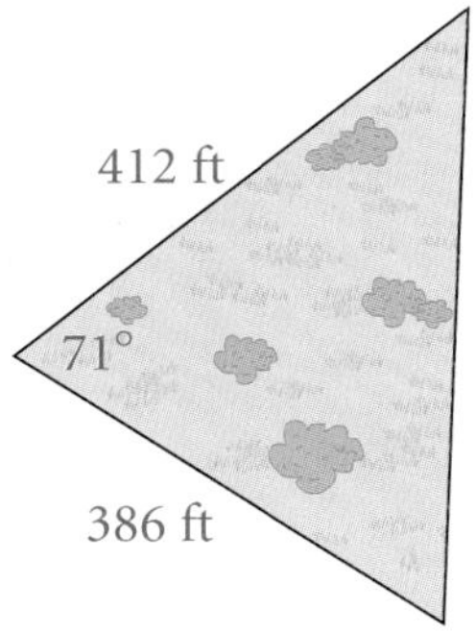

Surveying When surveyed, two adjacent sides of a triangular plot of land measured 412 ft and 386 ft. The angle between the sides was 71°. Find the area of the plot.

Area $= \frac{1}{2} \cdot$ side length $\cdot$ side length $\cdot$ sine of included angle

$= \frac{1}{2} \cdot 412 \cdot 386 \cdot \sin 71^\circ$ Substitute the given information.

$= 75183.855$ Use a calculator.

$\approx 75{,}200\ \text{ft}^2$

3. Try This Two adjacent sides of a triangular building plot measure 120 ft and 85 ft. The angle between them measures 85°. Find the area of the land. Round your answer to the nearest square foot.

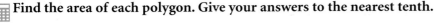

Exercises ON YOUR OWN

Find the area of each polygon. Give your answers to the nearest tenth.

1. equilateral triangle with apothem 4 in.
2. regular octagon with apothem 6 cm
3. regular hexagon with radius 10 ft
4. regular dodecagon with apothem 10 ft
5. regular decagon with radius 4 in.
6. regular pentagon with apothem 7 cm
7. regular 15-gon with side length 12 yd
8. regular decagon with perimeter 80 cm

Find the area of each triangle. Give your answers to the nearest tenth.

9. 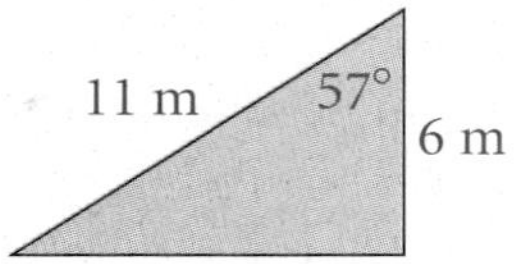

10. 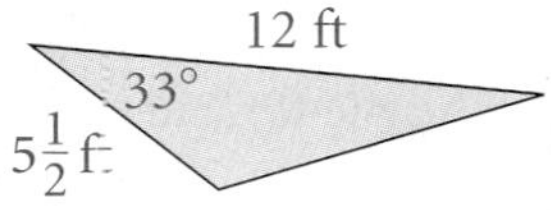

11. 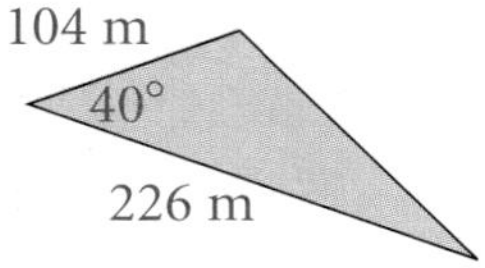

12. 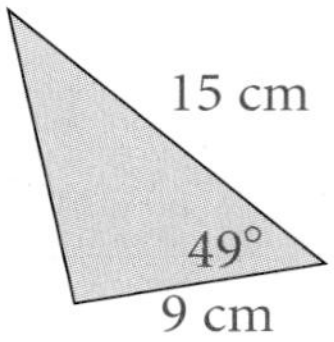

13. 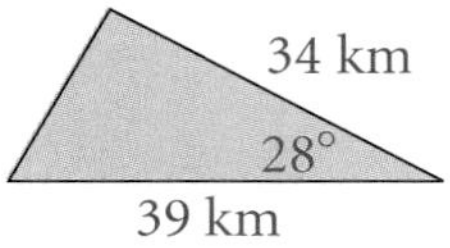

14. 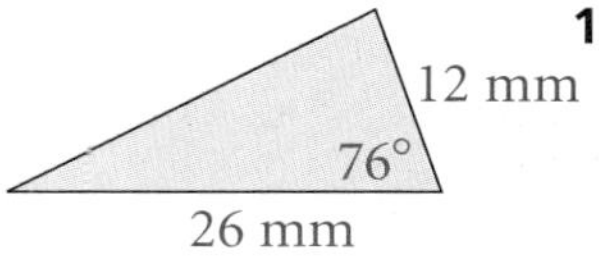

15. 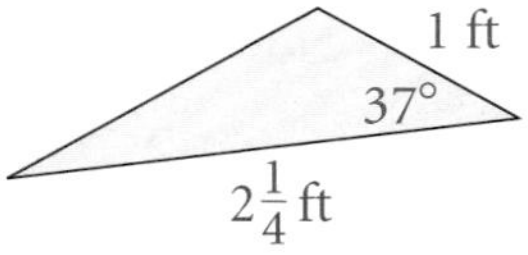

16. 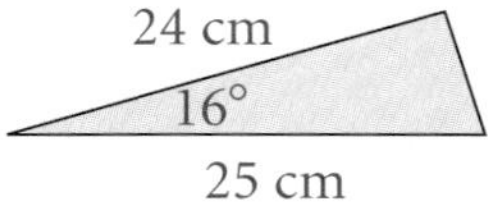

17. **Industrial Design** Refer to the diagram of the hexagonal nut. Round each answer to the nearest unit.
   a. Find the area of the circular space in the nut.
   b. Find the area of the hexagon minus the area of the circle.
   c. **Geometry in 3 Dimensions** Find the volume of the nut.

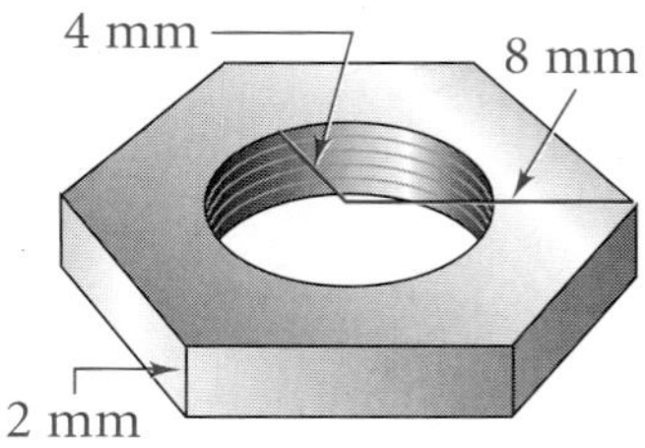

18. Suppose a circle is inscribed in a regular n-gon with center C and apothem 1, as shown in the diagram.
   a. Explain why $m\angle C = \frac{1}{2}\left(\frac{360}{n}\right) = \frac{180}{n}$.
   b. Explain why $s = \tan C$.
   c. Explain why the perimeter of the n-gon is $2n(\tan C)$.
   d. Explain why the area of the n-gon is $n\left(\tan \frac{180}{n}\right)$.
   e. **Open-ended** Pick a value for n between 0 and 10. Use a calculator to find the area of the n-gon. Then find the area of regular polygons with $10n$ sides and $100n$ sides.
   f. What do you notice about your answers to part (e)? Explain.

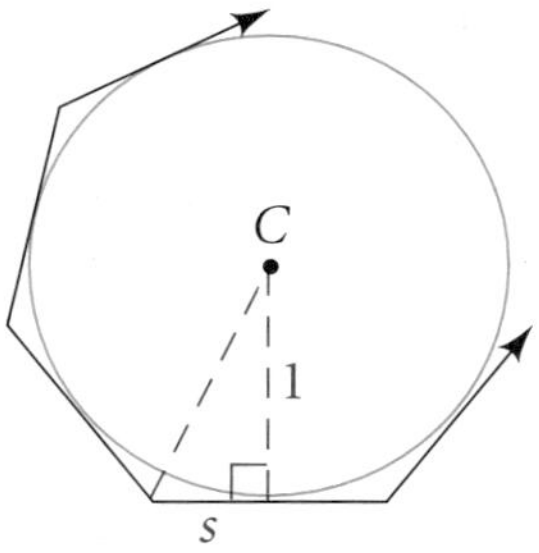

19. **Architecture** The Pentagon, in Arlington, Virginia, is one of the world's largest office buildings. It is a regular pentagon, and the length of each of its sides is 921 ft. Find the amount of space covered by the Pentagon. Round your answer to the nearest thousand square feet.

20. The standard length of a side of a stop sign is 1 ft $\frac{1}{4}$ in. Find the area of this stop sign to the nearest tenth of a square foot.

21. **Surveying** A surveyor wants to mark off a triangular parcel with an area of 1 acre. One boundary of the triangle extends 300 ft along a straight road. A second boundary extends at an angle of 65° from one end of this boundary. Draw a triangle to represent the piece of land and determine the length of the second boundary line to the nearest foot. (1 acre = 43,560 ft^2)

22. **Writing** Explain why each apothem of a regular polygon bisects a central angle and a side of the polygon.

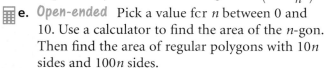

Chapter Project — Find Out by Researching

Find the star Polaris in the evening sky. Use your clinometer to measure the angle of elevation to Polaris. The angle that you measure should be approximately equal to your latitude. Use a map to find your latitude. How accurate was your estimate?

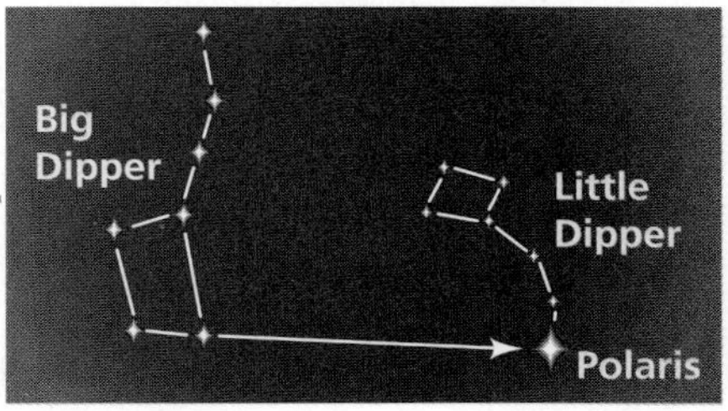

Exercises MIXED REVIEW

Find the length of each midsegment.

23. 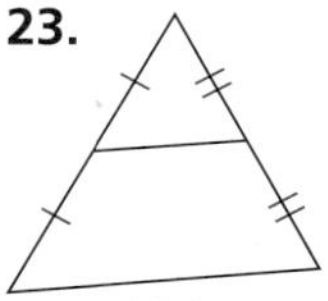

24. 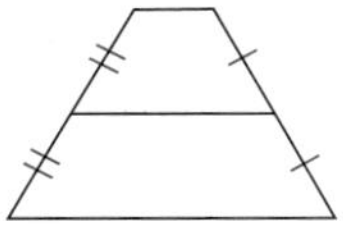

25. 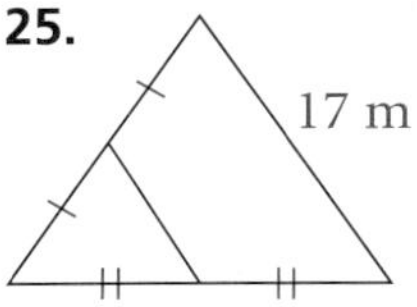

26. *Constructions* Draw a line and a point not on the line. Construct a line parallel to the first line through the point.

27. *Geometry in 3 Dimensions* Find the volume and lateral surface area of a cylindrical can with radius 4.0 cm and height 5.6 cm. Give your answers to the nearest tenth.

SELF ASSESSMENT — PORTFOLIO

For your portfolio, select one or two items from your work for this chapter. Here are some possibilities.
- an interesting example of indirect measurement
- work that shows originality

Explain why you have included each selection.

Geometry at Work

Surveyor

Surveyors calculate the locations, shapes, and areas of plots of land. A survey begins with a benchmark—a reference point whose latitude, longitude, and elevation are known. The surveyor uses a device called a *transit* to measure the angles of the plot of land and the distances of key points from the benchmark. Using trigonometry, the surveyor finds the latitude, longitude, and elevation of each key point in the survey. These points are located on a map and an accurate sketch of the plot is drawn. Finally, the area is calculated. One method involves dividing the plot into triangles and measuring the lengths of two sides and the included angle of each. The formula $A = \frac{1}{2}ab(\sin C)$ gives the area of each triangle. The area of the entire plot is the sum of the areas of the triangles.

Mini Project: On a flat area outside your school, place four objects to mark the vertices of an irregular quadrilateral. Calculate the area of the quadrilateral. Explain your method, and tell how you measured the sides and angles of the figure.

CHAPTER PROJECT

Finishing the Chapter Project

MEASURE FOR MEASURE

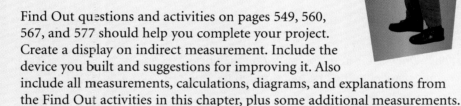

Find Out questions and activities on pages 549, 560, 567, and 577 should help you complete your project. Create a display on indirect measurement. Include the device you built and suggestions for improving it. Also include all measurements, calculations, diagrams, and explanations from the Find Out activities in this chapter, plus some additional measurements.

Reflect and Revise

Ask a classmate to review your project with you. Together, check that the diagrams and explanations are clear, complete, and accurate. Have you explained why your suggestions for changing the device you built will improve it? Have you made measurements in addition to those called for in the Find Out activities? Revise your work as needed.

Follow Up

Research various navigation devices, including ones invented long ago, such as the sextant or the astrolabe, and modern ones, such as GPS (Global Positioning Satellites). Explain how each navigation device works.

For More Information

Bishop, Owen. *Yardsticks of the Universe.* New York: Peter Bedrick Books, 1982.

Boyer, Carl B., and Uta C. Merzbach, rev. editor. *A History of Mathematics*. New York: John Wiley & Sons, 1991.

Murdoch, John E. *Album of Science: Antiquity and the Middle Ages.* New York: Charles Scribner's Sons, 1984.

Wrap Up

Key Terms

angle of depression (p. 556)
angle of elevation (p. 556)
cosine (p. 551)
identity (p. 551)
literal equation (p. 562)
resultant (p. 568)
sine (p. 551)
tangent (p. 544)
vector (p. 563)

How am I doing?

SELF ASSESSMENT

- List the three trigonometric ratios described in this chapter. Include a diagram.
- State two ways you can describe a vector and explain how to convert from one description to the other.

The Sine, Cosine, and Tangent Ratios 11-1, 11-2

In right triangle $\triangle ABC$, **sine** of $\angle A = \sin A = \frac{\text{leg opposite } \angle A}{\text{hypotenuse}}$.

cosine of $\angle A = \cos A = \frac{\text{leg adjacent to } \angle A}{\text{hypotenuse}}$,

and **tangent** of $\angle A = \tan A = \frac{\text{leg opposite } \angle A}{\text{leg adjacent to } \angle A}$.

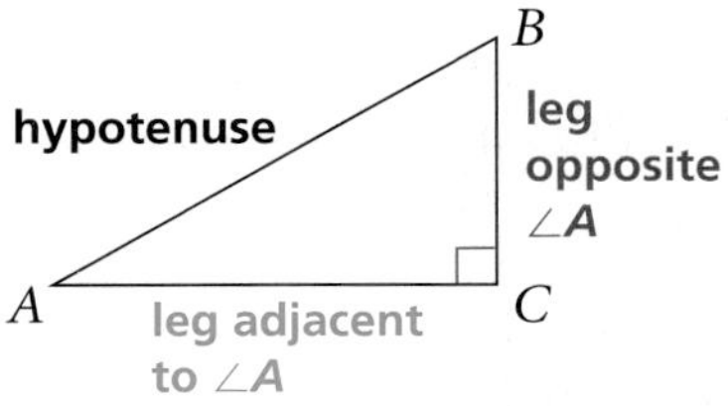

You can use the inverses of the trigonometric ratios to find the measure of an angle of a right triangle.

There are many trigonometric **identities,** including $\sin x° = \cos (90 - x)°$.

Express sin *A*, cos *A*, and tan *A* as ratios.

1.

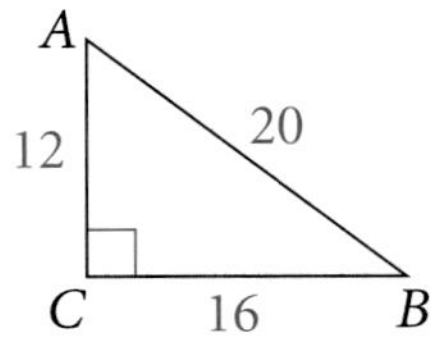

2.

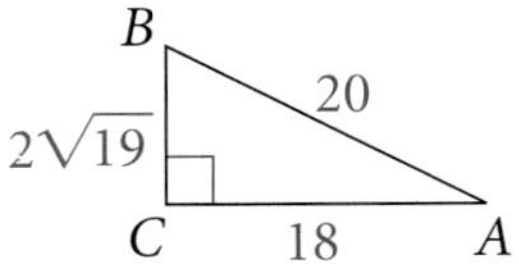

3.

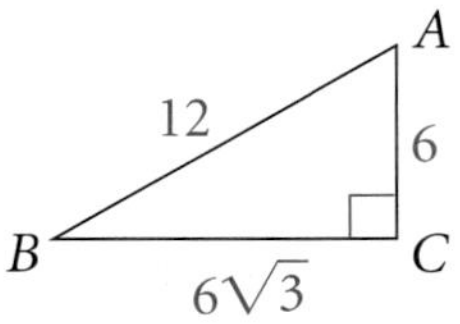

4.

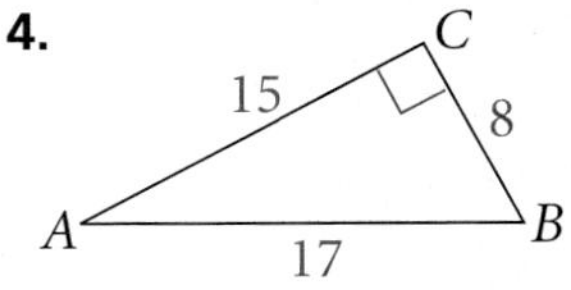

Find the value of *x*. Round lengths of segments to the nearest tenth and angle measures to the nearest degree.

5.

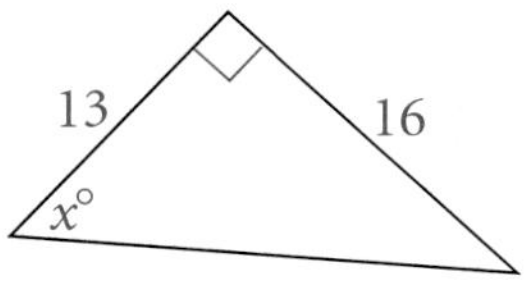

6.

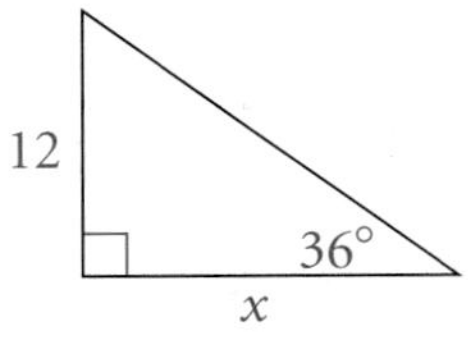

7.

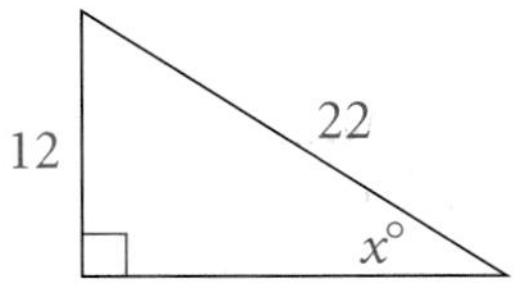

8.

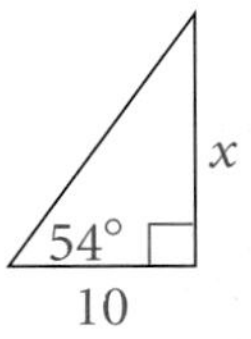

9. *Standardized Test Prep* In which triangle is $m\angle X$ the greatest?

A.

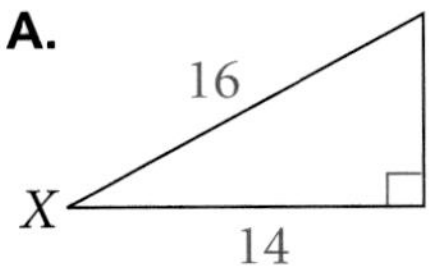

B.

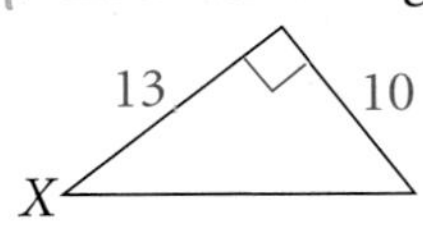

C.

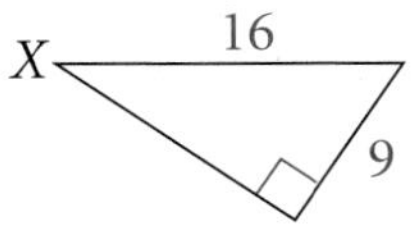

D.

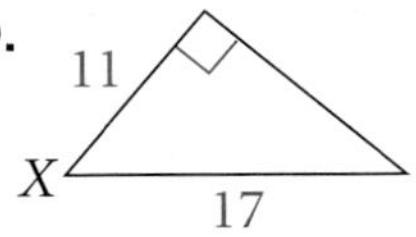

E.

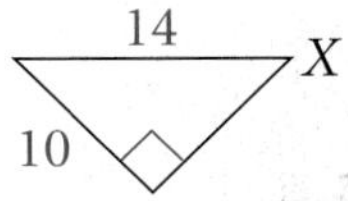

Angles of Elevation and Depression 11-3

An angle below a horizontal line is an **angle of depression**. An angle above a horizontal line is an **angle of elevation.**

Solve each problem.

10. Two hills are 2 mi apart. The elevation of the taller hill is 2707 ft. The angle of depression from the top of the taller hill to the top of the shorter hill is 7°. Find the elevation of the shorter hill to the nearest foot. (1 mi = 5280 ft)

not to scale

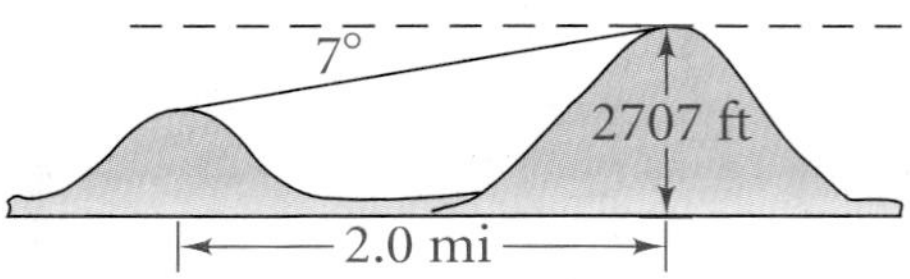

11. A surveyor is 305 ft from the base of the new courthouse. Her angle measuring device is 5 ft above the ground. The angle of elevation to the top of the building is 42°. Find the height of the courthouse to the nearest foot.

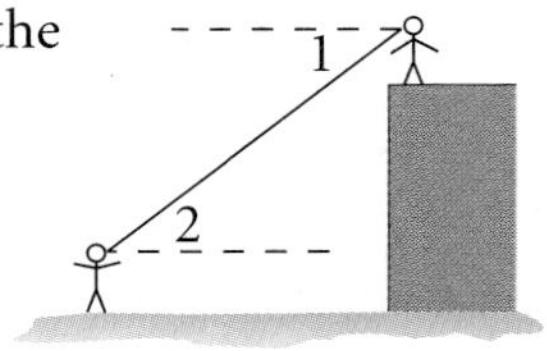

12. *Writing* Explain why the angle of depression ∠1 and the angle of elevation ∠2 in the diagram are congruent.

Vectors and Trigonometry 11-4

A **vector** is any quantity that has magnitude and direction.

Describe each vector by using ordered pair notation. Give the coordinates to the nearest unit.

13.

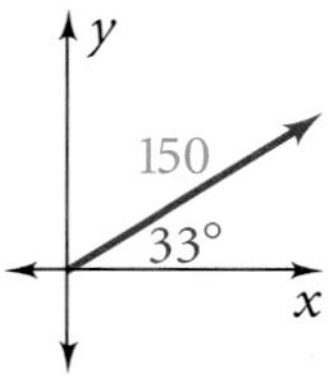

14.

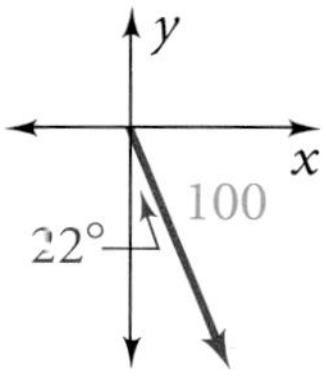

15.

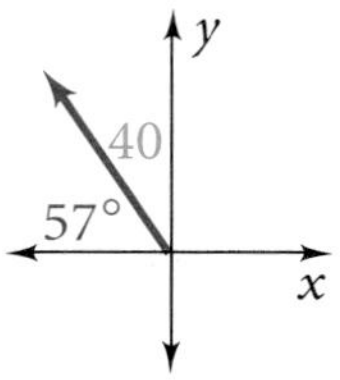

16.

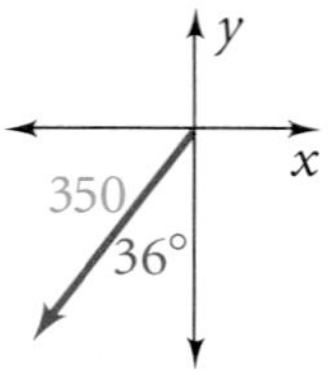

Find the magnitude and direction of each vector.

17.

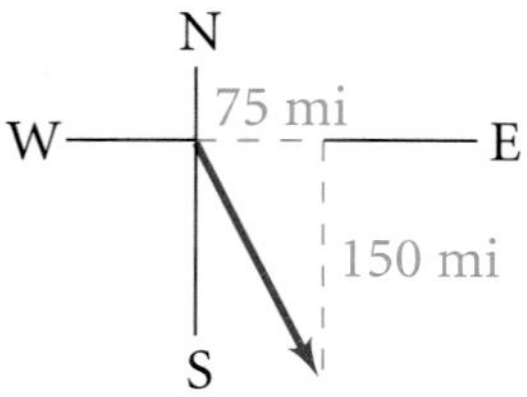

18.

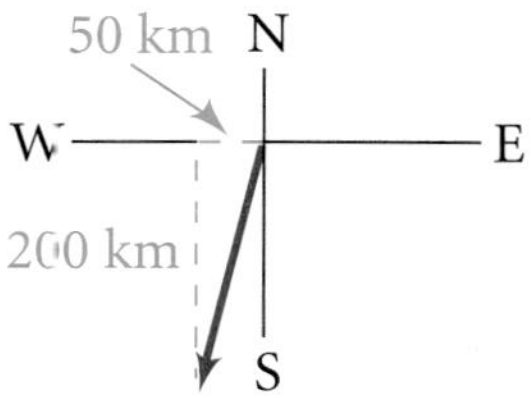

19.

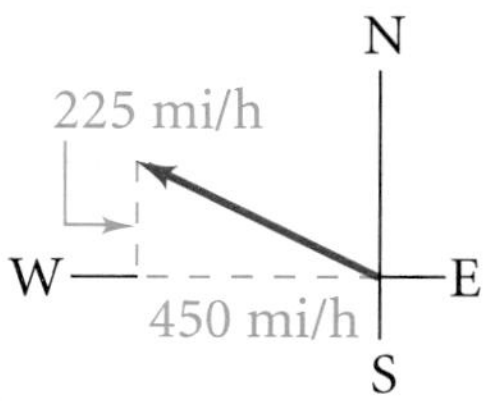

20.

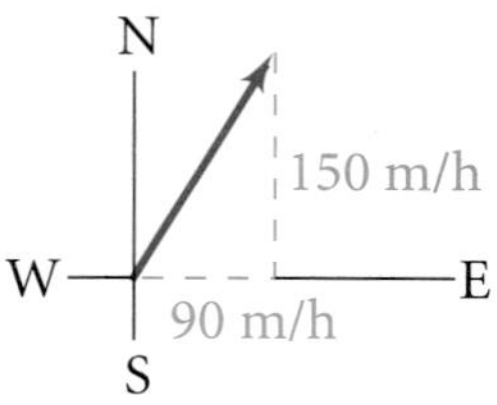

21. *Open-ended* Write three vectors with the same direction as ⟨3, 4⟩. Explain how you found your answers.

Critical Thinking **Does each sentence contain enough information to describe a vector? Explain.**

22. A person bicycles at 15 miles per hour.

23. A butterfly flies 60 km northeast.

24. A manatee migrates 100 km to the Atlantic coast.

Adding Vectors 11-5

The sum of two vectors is a **resultant.** You can add vectors by using the head-to-tail method or by using ordered pair notation. Vector sums can show the result of actions that take place one after the other or the result of two forces that act at the same time on an object.

Coordinate Geometry **Find the sum of each pair of vectors. Give your answers in ordered pair notation.**

25.

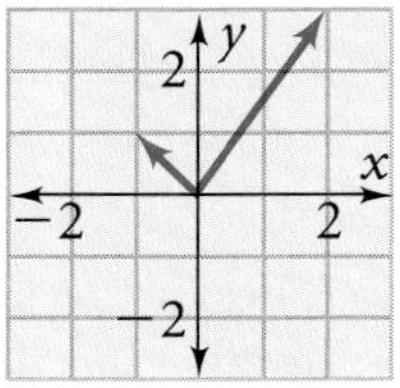

26.

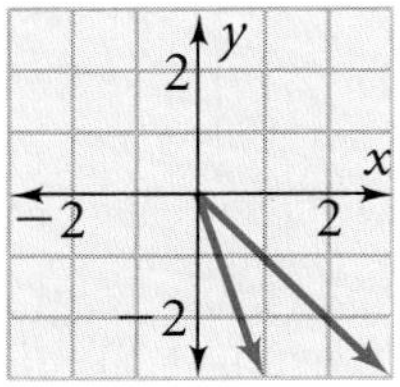

27.

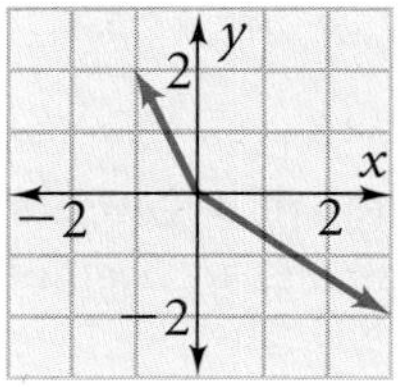

28.

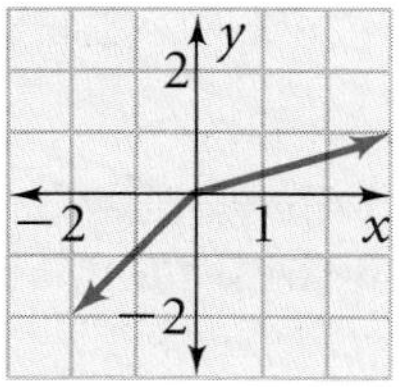

29. A whale-watching tour leaves port and travels 60 mi directly north. The tour then travels 5 mi due east.

a. In what direction should the boat head to return to port?

b. How long will the return trip take if the boat averages 20 mi/h?

Trigonometry and Area 11-6

You can use trigonometry to find the area of regular polygons. You can also use trigonometry to find the area of a triangle when you know the lengths of two sides and the measure of the included angle.

Area of triangle $= \frac{1}{2} \cdot$ side length $\cdot$ side length $\cdot$ sine of included angle

Find the area of each polygon. Round your answers to the nearest tenth.

30. regular decagon with radius 5 ft

31. regular pentagon with apothem 8 cm

32.

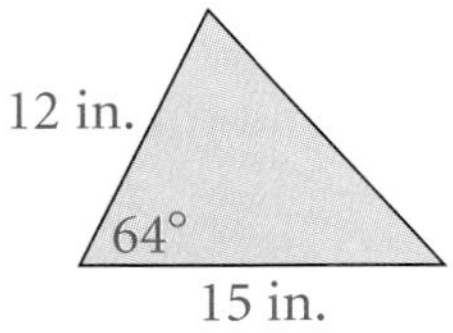

33.

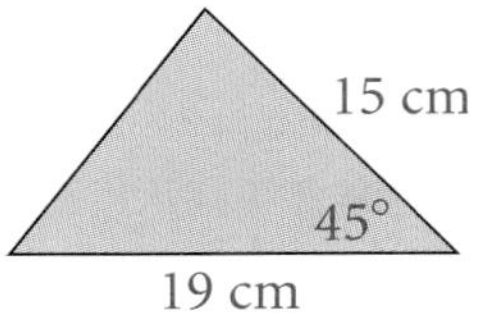

34.

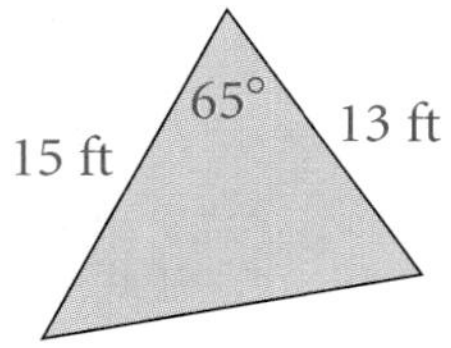

35.

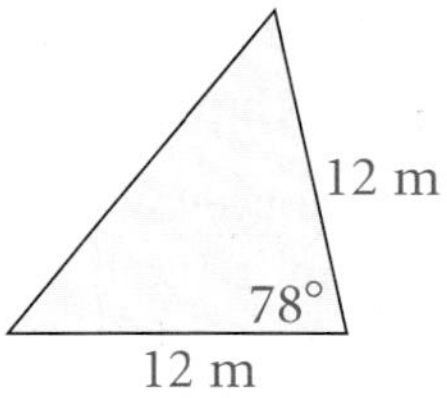

Getting Ready for CHAPTER 12

36. *Open-ended* Draw a circle. Then draw two central angles, one twice the measure of the other.

Find the area and circumference of a circle with the given radius or diameter. Leave your answer in terms of π.

37. $r = 5$ in.

38. $d = 8$ cm

39. $r = 11$ m

40. $d = 4$ ft

Assessment

Express sin *B*, cos *B*, and tan *B* as ratios.

1.

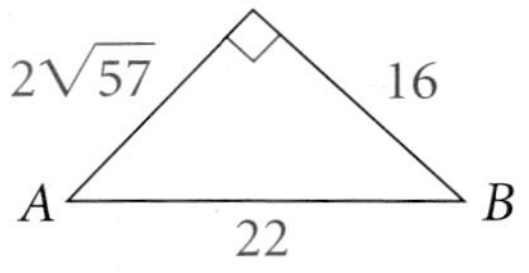

2.

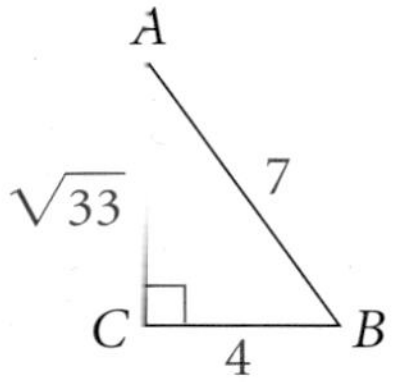

Find the value of *x*. Round lengths of segments to the nearest tenth and angle measures to the nearest degree.

3.

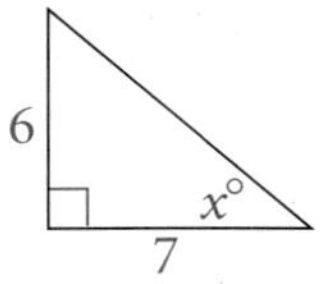

4.

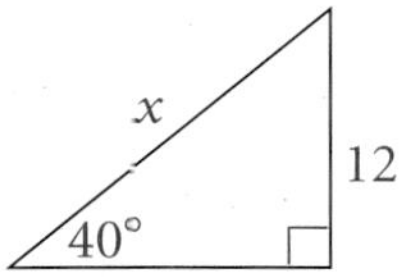

5.

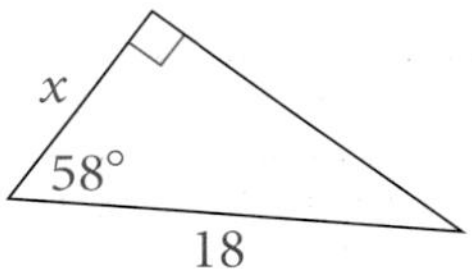

6.

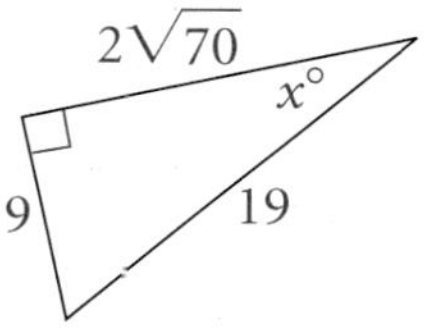

Algebra Find the measure of the acute angle that each line makes with a horizontal line. Round your answer to the nearest tenth.

7. $y = 3x - 2$

8. $y = x + 1$

9. $y = \frac{1}{2}x + 5$

10. $x - 2y = 6$

Solve each problem.

11. A surveyor measuring the tallest tree in a park stands 100 ft from it. His angle-measuring device is 5 ft above the ground. The angle of elevation to the top of the tree is 48°. How tall is the tree?

12. A hot-air balloon is competing in a race. After 20 min, the balloon is at an altitude of 300 m. The pilot can still see the starting point at a 25° angle of depression. How many meters is the balloon from the starting point on the ground?

13. **Writing** Explain why $\sin x° = \cos (90 - x)°$. Include a diagram with your explanation.

Describe each vector using ordered pair notation. Round the coordinates to the nearest unit.

14.

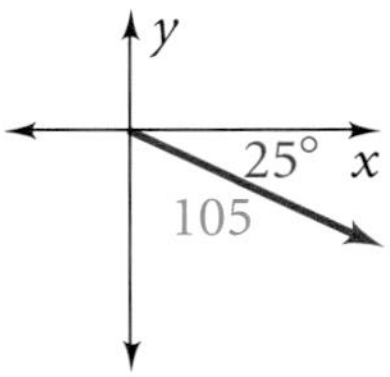

15.

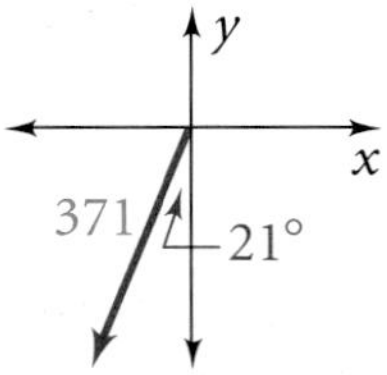

16. A family went on vacation to a beach 120 mi east and 30 mi south of their home. Find the distance and the direction the beach is from their home.

Find the sum of each pair of vectors. Give your answers in ordered pair notation.

17.

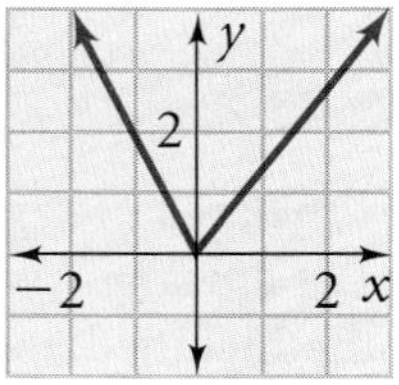

18.

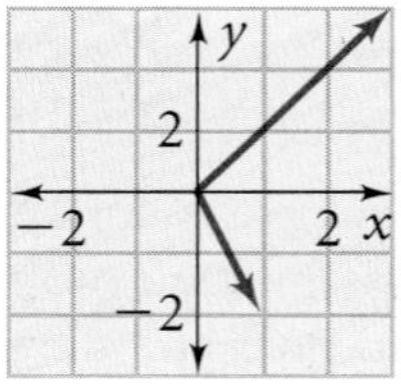

19. **Open-ended** Draw two vectors with different directions on a coordinate grid. Then draw their resultant and describe the sum using ordered pair notation.

Find the area of each polygon. Round your answer to the nearest tenth.

20.

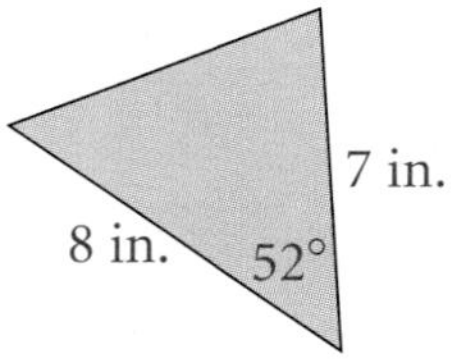

21.

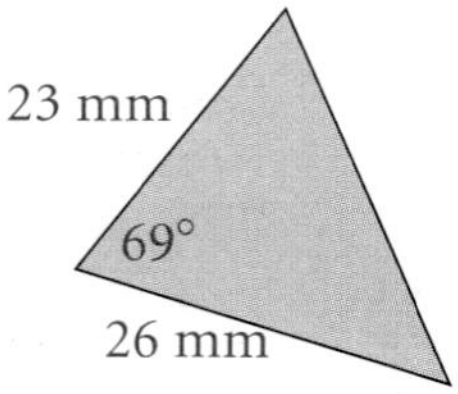

22. a regular hexagon with apothem 5 ft

23. a regular pentagon with radius 3 cm

24. a regular decagon with perimeter 90 in.

Preparing for Standardized Tests

For Exercises 1–9, choose the correct letter.

1. In which transformation is the image *not* necessarily congruent to the preimage?
A. rotation **B.** reflection
C. translation **D.** dilation
E. glide translation

2. Which quadrilaterals have congruent diagonals?
I. square **II.** trapezoid
III. parallelogram **IV.** kite

A. I only **B.** III only
C. I and II only **D.** III and IV only
E. I, II, and IV only

3. Which is greatest in the triangle below?

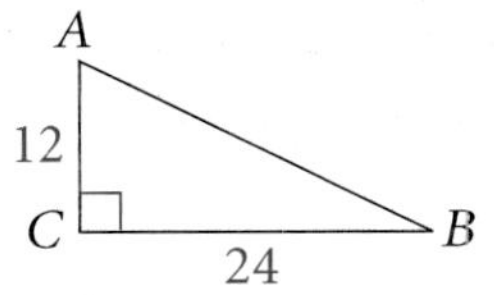

A. sin *A* **B.** cos *A* **C.** tan *A*
D. sin *B* **E.** cos *B*

4. What is the surface area of a rectangular prism with length 9 cm, width 8 cm, and height 10 cm?
A. 27 cm^2 **B.** 240 cm^2 **C.** 242 cm^2
D. 484 cm^2 **E.** 720 cm^2

5. For what value of *x* will the two triangles be similar?

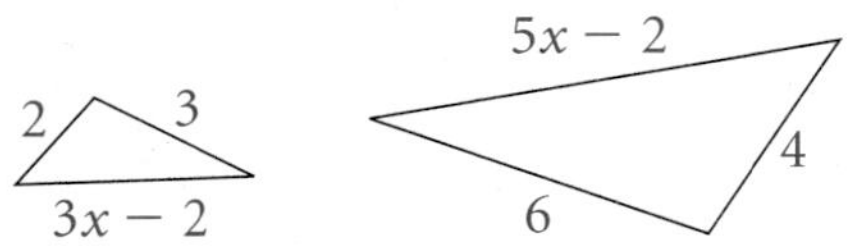

A. 2 **B.** 4 **C.** 8
D. $10\frac{2}{3}$ **E.** 12

6. Which information *cannot* be used to prove two triangles congruent?
A. SSS **B.** SAS **C.** ASA
D. AAS **E.** AAA

7. What is the locus of points in a plane a given distance from a given line?
A. a line **B.** a circle **C.** a cylinder
D. a ray **E.** two parallel lines

Compare the boxed quantity in Column A with the boxed quantity in Column B. Choose the best answer.

A. The quantity in Column A is greater.
B. The quantity in Column B is greater.
C. The two quantities are equal.
D. The relationship cannot be determined on the basis of the information supplied.

	Column A	Column B
8.	sin 30°	cos 30°
9.	the magnitude of $\vec{c} + \vec{a}$	the magnitude of $\vec{a} + \vec{c}$

Find each answer.

10. Aviation Pilots often refer to the angle of an object outside the plane in terms of a clock face. Thus, an object at 12 o'clock is straight ahead, an object at 3 o'clock is 90° to the right, and so on.
a. Suppose that two pilots flying in the same direction spot the same object. One reports it at 1 o'clock, the other at 2 o'clock. Draw a diagram showing the possible locations of the two planes and the object.
b. Suppose that at the same time one pilot sights the object, she sights the other plane at 9 o'clock. Draw a diagram of the positions of the planes and the object.

11. Open-ended Draw a triangle. Measure two sides and the angle between them. Then find the area of the triangle.

12. Constructions Construct a square with side length *a*.

CHAPTER 12

Chords, Secants, and Tangents

Relating to the Real World

Geometric figures drawn in circles and drawn about circles have many interesting characteristics. You will rely on your knowledge of congruent triangles and similar triangles as you discover some of these characteristics. The principles you will study in this chapter have applications in photography, architecture, and communications.

Lessons	12-1	12-2	12-3	12-4
	Circles in the Coordinate Plane	*Properties of Tangents*	*Properties of Chords and Arcs*	*Inscribed Angles*

CHAPTER PROJECT

GO FOR A SPIN

For centuries, artists have used the simple elegance of the circle in their designs. Some have crafted intertwining patterns that like the circle itself have no beginning and no end. Some have disturbed the symmetry of the circle to create optical illusions.

In your chapter project, you will explore some of the techniques used through the ages to produce circular art such as the painting at the left by Alma Thomas. You will then apply your discoveries to create a dizzying design. You will see why some artists find that "going in circles" may be the best way to reach their objective.

To help you complete the project:

- **p. 591** *Find Out by Doing*
- **p. 605** *Find Out by Exploring*
- **p. 618** *Find Out by Constructing*
- **p. 627** *Finishing the Project*

What You'll Learn

- Writing the equation of a circle
- Using the equation of a circle to solve real-world problems

...And Why

To investigate situations in which equations of circles can be used, such as expressing the range of cellular telephone towers

What You'll Need

- graph paper

12-1 Circles in the Coordinate Plane

WORK TOGETHER

Work with a small group. Consider the equation $x^2 + y^2 = 25$.

1. What are the values of y when $x = 0$?
2. What are the values of x when $y = 4$?
3. Make a table of at least eight ordered pairs (x, y) that satisfy the equation $x^2 + y^2 = 25$. Then graph the ordered pairs.
4. **a.** The points you graphed lie on a circle. What is the center of the circle? What is the radius?
   b. What is the square of the radius?
   c. What connection do you see between the square of the radius and the equation of the circle?
5. Find the center and radius of each circle.
   a. $x^2 + y^2 = 9$ **b.** $x^2 + y^2 = 36$ **c.** $x^2 + y^2 = 121$

THINK AND DISCUSS

Writing the Equation of a Circle

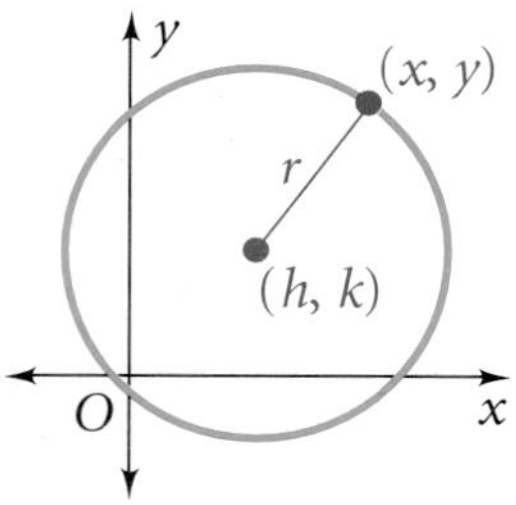

You can use the Distance Formula to find an equation of a circle with center (h, k) and radius r. Let (x, y) be any point on the circle. The radius is the distance from (h, k) to (x, y).

$r = \sqrt{(x - h)^2 + (y - k)^2}$ Distance Formula

$r^2 = (x - h)^2 + (y - k)^2$ Square both sides.

This proves the following theorem.

Theorem 12-1
Equation of a Circle

The **standard form of an equation of a circle** with center (h, k) and radius r is $(x - h)^2 + (y - k)^2 = r^2$.

Example 1

Write the standard equation of the circle with center $(5, -2)$ and radius 7.

Use the equation of a circle with $(h, k) = (5, -2)$ and $r = 7$.

$(x - 5)^2 + [(y - (-2))]^2 = 7^2$ Substitute.

$(x - 5)^2 + (y + 2)^2 = 49$ Simplify.

6. **Try This** Write the equation of each circle.
   a. center (3, 5); $r = 6$ **b.** center (−2, −1); $r = \sqrt{2}$

7. **Try This** Find the center and radius of each circle.
   a. $(x - 7)^2 + (y + 2)^2 = 64$ **b.** $(x - 2)^2 + (y - 3)^2 = 100$

8. What is the standard form of the equation of a circle with center at the origin and radius *r*?

Using the Equation of a Circle

You can use equations of circles to model real-world situations.

Example 2 **Relating to the Real World**

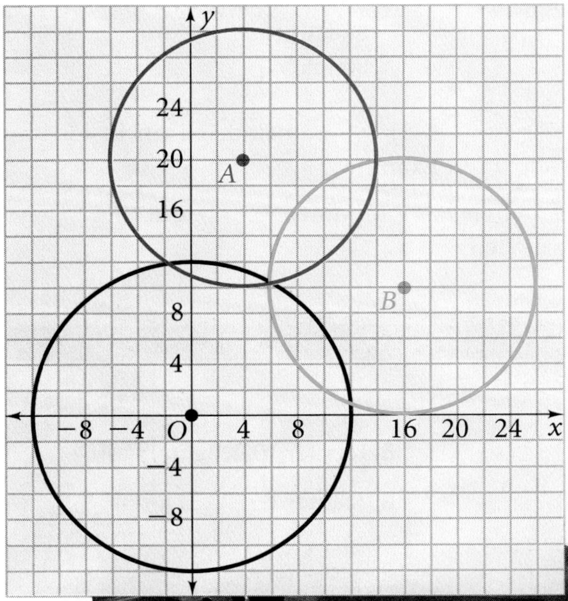

Communications When you make a call from a cellular phone, a tower receives the call. In the graph, the center of each circle is the location of a cellular telephone tower. The equation of a circle can describe the receiving and transmitting range of a tower. Find the equation that describes the position and the range of Tower *A*.

The center of ⊙*A* is (4, 20), and the radius is 10. Use the equation of a circle with $(h, k) = (4, 20)$ and $r = 10$.

$(x - 4)^2 + (y - 20)^2 = 10^2$ Substitute.
$(x - 4)^2 + (y - 20)^2 = 100$ Simplify.

9. Find the equation of the positions and ranges of the other two towers.

10. **Transformations** ⊙*B* is a translation of ⊙*A*. Use ordered pair notation to describe the translation vector.

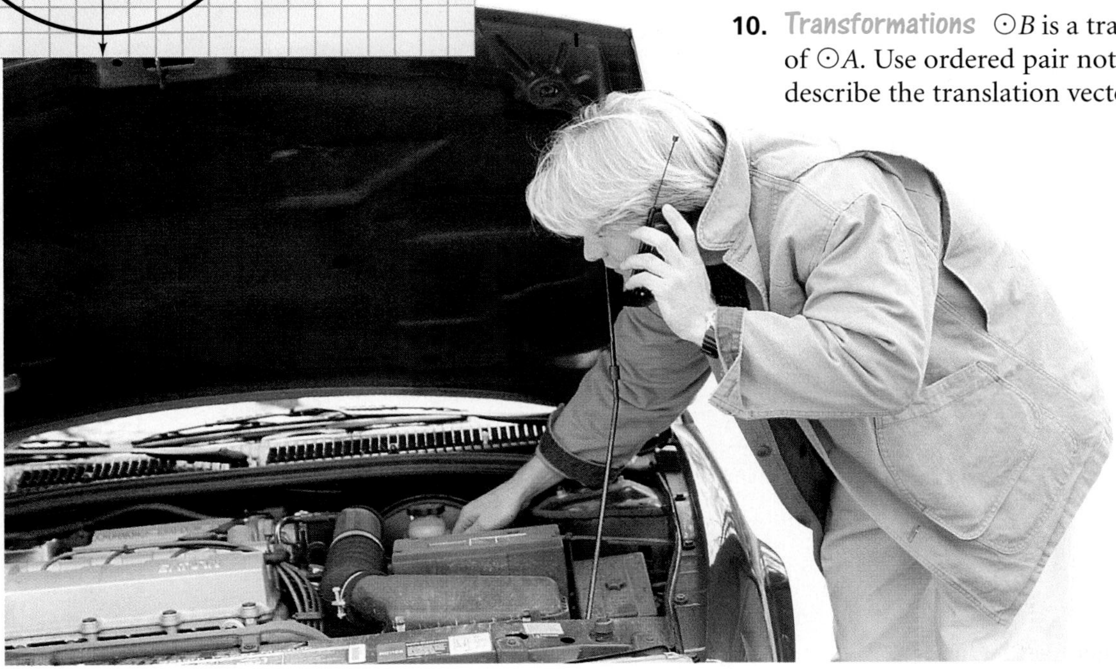

If you know the center of a circle and a point on the circle, you can write an equation of the circle.

Example 3

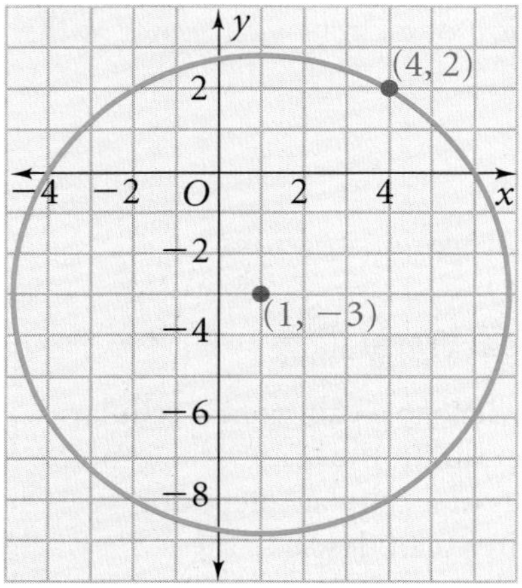

Write an equation of the circle with center $(1, -3)$ that passes through the point $(4, 2)$.

The radius is the distance from the center to any point on the circle.

$$\begin{aligned} r &= \sqrt{(1-4)^2 + (-3-2)^2} && \text{Use the Distance Formula.} \\ &= \sqrt{(-3)^2 + (-5)^2} && \text{Simplify.} \\ &= \sqrt{9 + 25} = \sqrt{34} \end{aligned}$$

Use the equation of a circle with $(h, k) = (1, -3)$ and $r = \sqrt{34}$.

$$\begin{aligned} (x-1)^2 + (y-(-3)^2) &= (\sqrt{34})^2 && \text{Substitute.} \\ (x-1)^2 + (y+3)^2 &= 34 && \text{Simplify.} \end{aligned}$$

11. Does the graph of $(x-1)^2 + (y+3)^2 \leq 34$ include points inside or points outside the circle shown in Example 3?

12. **Try This** A diameter of a circle has endpoints $(-3, 7)$ and $(5, 5)$. Write an equation of the circle.

Exercises ON YOUR OWN

Write an equation of each circle.

1. center $(2, -8)$; $r = 9$

2. center $(0, 3)$; $r = 7$

3. center $(0.2, 1.1)$; $r = 0.4$

4. center $(5, -1)$; $r = 12$

5. center $(-6, 3)$; $r = 8$

6. center $(-9, -4)$; $r = \sqrt{5}$

Match each equation or inequality with a graph.

7. $(x-1)^2 + (y-3)^2 = 9$

8. $(x+1)^2 + (y-1)^2 > 4$

9. $x^2 + y^2 = 4$

10. $(x-2)^2 + y^2 < 9$

A.

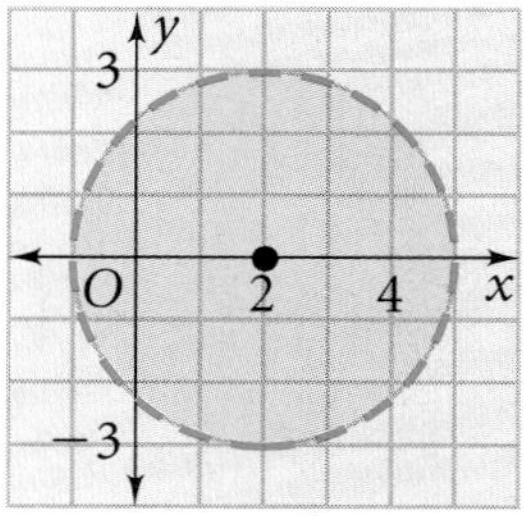

B.

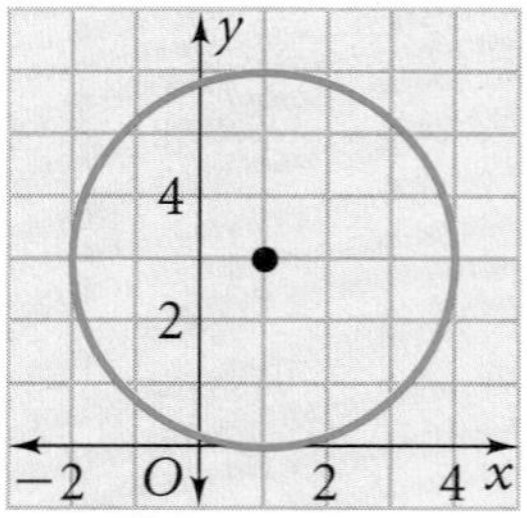

C.

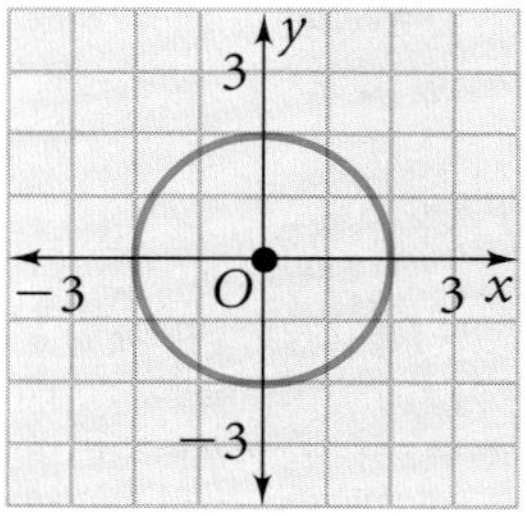

D.

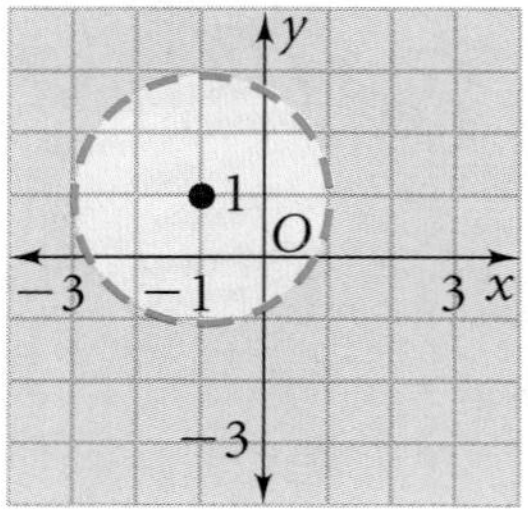

Find the center and the radius of each circle.

11. $(x + 7)^2 + (y - 5)^2 = 16$

12. $(x - 3)^2 + (y + 8)^2 = 100$

13. $(x - 0.3)^2 + y^2 = 0.04$

14. $(x + 5)^2 + (y + 2)^2 = 48$

Graph each circle. Label its center and state its radius.

15. $x^2 + y^2 = 36$

16. $x^2 + (y - 4)^2 = 16$

17. $(x + 2)^2 + y^2 = 9$

18. $(x + 4)^2 + (y - 1)^2 = 25$

19. a. **Transformations** The circle with equation $(x + 5)^2 + (y - 3)^2 = 64$ is a translation image of $x^2 + y^2 = 64$. Use ordered pair notation to describe the translation vector.

b. **Transformations** The circle $x^2 + y^2 = 64$ is the dilation image of $x^2 + y^2 = 1$. What is the center and scale factor of the dilation?

c. **Writing** Write *true* or *false*. Every circle in the coordinate plane is a transformation image of $x^2 + y^2 = 1$. Explain.

20. a. **Weather** Consider a point on the edge of a cup of an anemometer. Describe the locus of points that this point passes through as the cup spins in the wind.

b. **Locus** Suppose the distance from the center of the anemometer to a point on the edge of a cup is 6 in. Write an equation for the locus of points you described in part (a). Use the center of the anemometer as the origin.

The speed at which the cups of an anemometer spin indicates the speed of the wind.

Write an equation of each circle.

21.

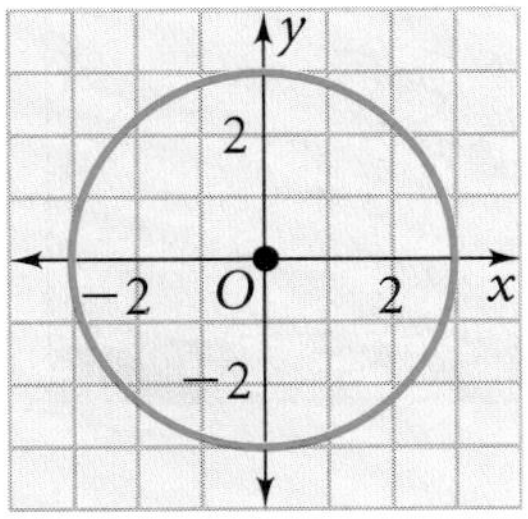

22.

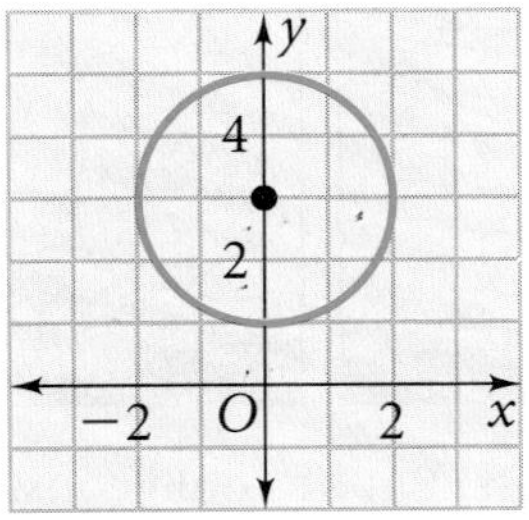

23.

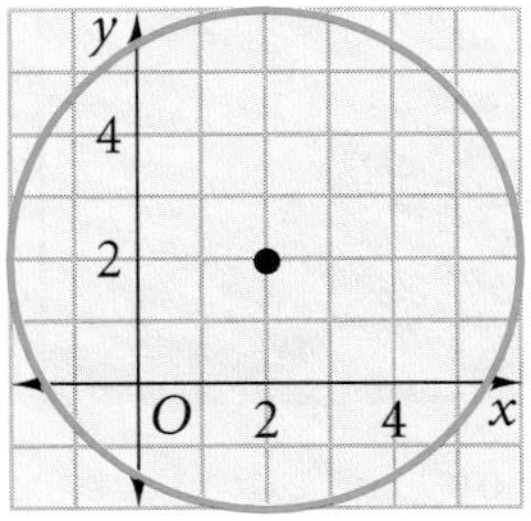

24.

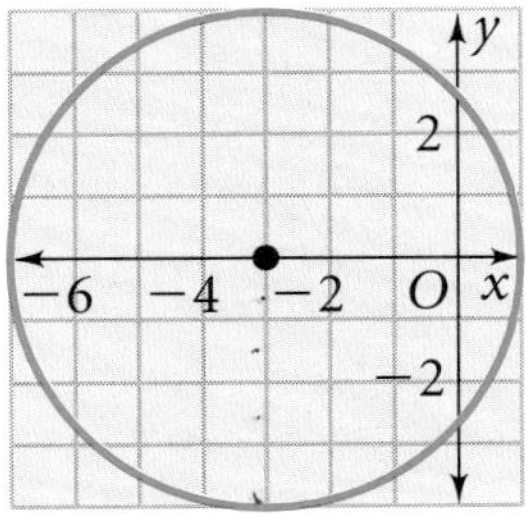

25.

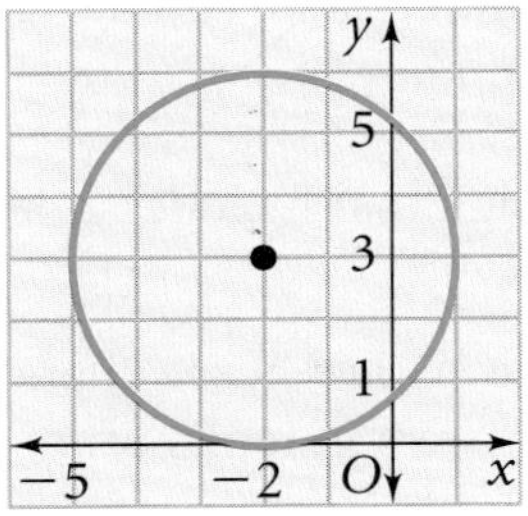

26.

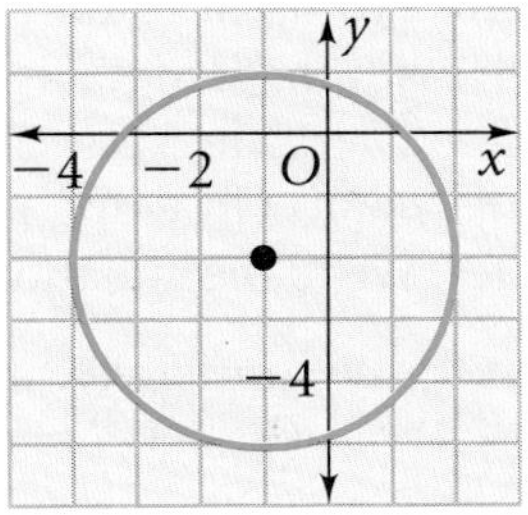

Write an equation of the circle with the given center passing through the given point.

27. center $(-2, 6)$; through $(-2, 10)$ **28.** center $(1, 2)$; through $(0, 6)$

29. center $(7, -2)$; through $(1, -6)$ **30.** center $(-10, -5)$; through $(-5, 5)$

31. Transformations The graph of $x^2 + y^2 = 100$ is translated so that its center is $(3, -5)$. What is the equation of the image?

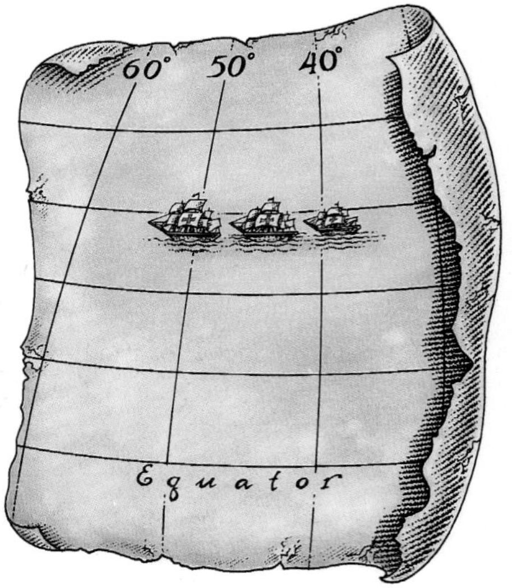

32. Critical Thinking Determine whether each equation is an equation of a circle. If it is not, explain why.

a. $(x - 1)^2 + (y + 2)^2 = 9$ **b.** $x + y = 9$

c. $x + (y - 3)^2 = 9$ **d.** $x^2 + (y - 1)^2 = 9$

33. a. Write the equation of the equator with the center of Earth as the origin. The equator's radius is about 3960 mi long.

b. Find the length of a 1° arc on the equator to the nearest tenth.

c. At the equator, a 1° arc in latitude is 60 nautical miles long. How many miles are in a nautical mile? Round to the nearest tenth.

d. History Columbus planned his trip to the east by going west. He thought each 1° arc was 45 miles long. He estimated that the trip would take 21 days. Use your answer to part (b) to find a better estimate.

Write an equation of a circle with diameter $\overline{AB}$.

34. $A(0, 0)$, and $B(8, 6)$ **35.** $A(3, 0)$, and $B(7, 6)$ **36.** $A(1, 1)$, and $B(5, 5)$

37. $A(-1, 0)$, and $B(-5, -3)$ **38.** $A(-3, 1)$, and $B(0, 9)$ **39.** $A(-2, 3)$, and $B(6, -7)$

PEANUTS® by Charles M. Schulz

40. Snoopy can't draw a square with a compass. But he could construct a square and use it to find the length $5\sqrt{2}$ in. Then he could draw a circle with that length as the radius. Explain how.

41. Open-ended On graph paper, make a design that includes at least three circles. Write the equations of your circles.

42. Find the circumference and the area of the circle whose equation is $(x - 9)^2 + (y - 3)^2 = 64$. Leave your answers in terms of π.

43. *Geometry in 3 Dimensions* The equation of a sphere is similar to the equation of a circle. The equation of a sphere with center (h, j, k) and radius r is $(x - h)^2 + (y - j)^2 + (z - k)^2 = r^2$.

a. $M(-1, 3, 2)$ is the center of a sphere passing through $T(0, 5, 1)$. What is the radius of the sphere?

b. Write an equation of the sphere.

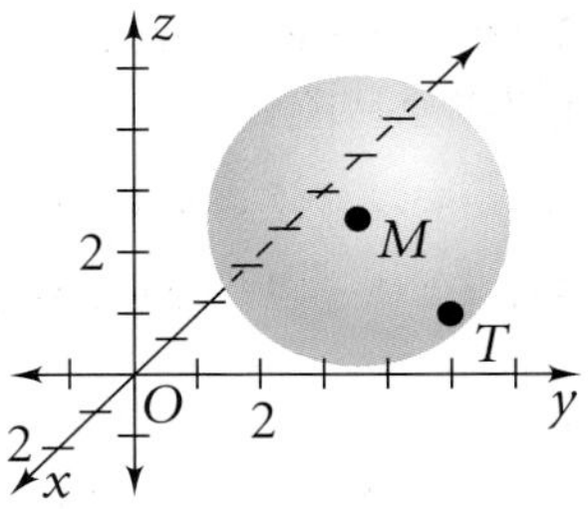

Chapter Project — Find Out by Doing

For many centuries, artists throughout the world have used ropelike patterns called *knots* on jewelry, clothing, stone carvings, and other items. You can create a knot design using graph paper and a compass. Use a pencil because you will need to erase portions of your drawing.

- Mark the origin at the center of a sheet of graph paper, but do not draw any axes. Draw four circles with centers (0, 5), (5, 0), (0, −5), and (−5, 0) and with radius $5\sqrt{2}$. Using the same centers draw four circles with radius $4\sqrt{2}$.
- Connect the four centers to form a square. (shown in red)
- Draw segments through the intersections of the smaller and larger circles. (shown in green)
- Erase arcs to make bands that appear to weave in and out. Color your design.

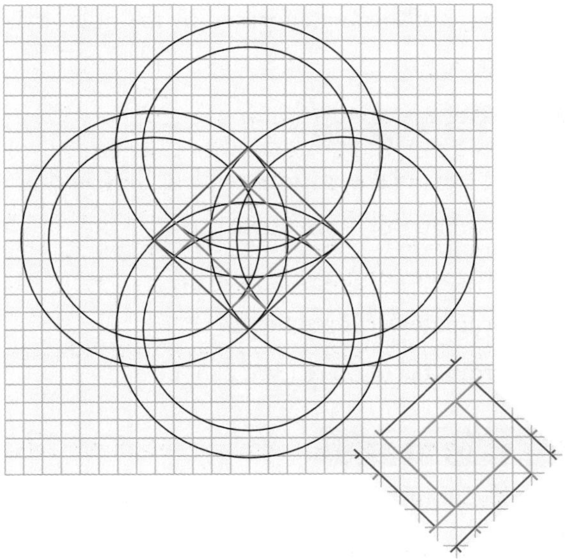

Exercises MIXED REVIEW

Use the given information. Explain how to prove $\triangle ABC \cong \triangle DCB$.

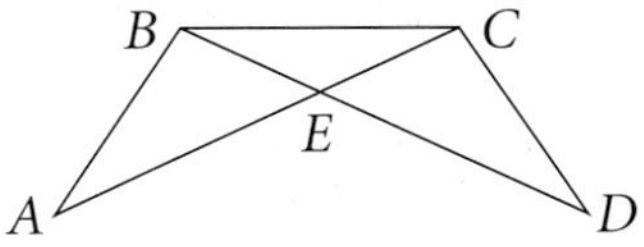

44. Given: $\angle BCA \cong \angle CBD$, $\overline{BD} \cong \overline{CA}$

45. Given: $\angle ABE \cong \angle DCE$, $\overline{AB} \cong \overline{DC}$

46. A cube has surface area 96 in.^{2}. What is the surface area of a cube with sides twice as long?

47. The volume of a cube is 64 in.^{3}. What is the volume of a cube with sides twice as long?

Getting Ready for Lesson 12-2

Find the complement and supplement of each angle.

48. 48° 49. 73° 50. 21° 51. 87°

Polar Coordinates

Extends Lesson 12-1

The polar-coordinate system locates a point using its distance from the origin, or **pole,** and the measure of a central angle formed by the polar axis and a ray from the pole through the point. The coordinates of A are $(2, 60°)$ because A is 2 units from the pole O and the measure of the angle formed by the polar axis and $\overrightarrow{OA}$ is 60.

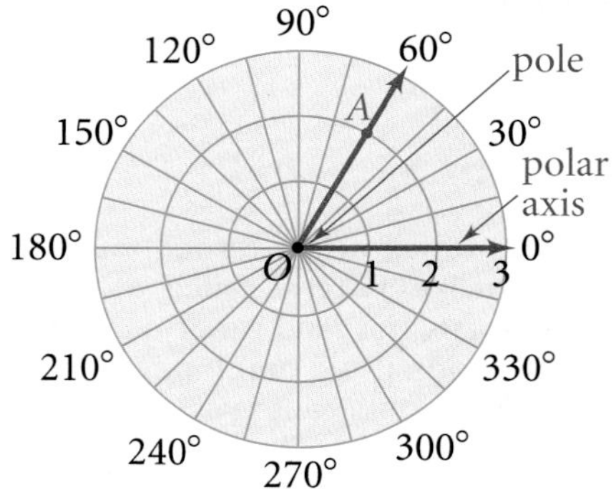

Example 1

What are the polar coordinates of P, Q, R, and S?

The coordinates of P are $(3, 90°)$.

The coordinates of Q are $(2, 210°)$.

The coordinates of R are $(4, 45°)$.

The coordinates of S are $(3.5, 285°)$.

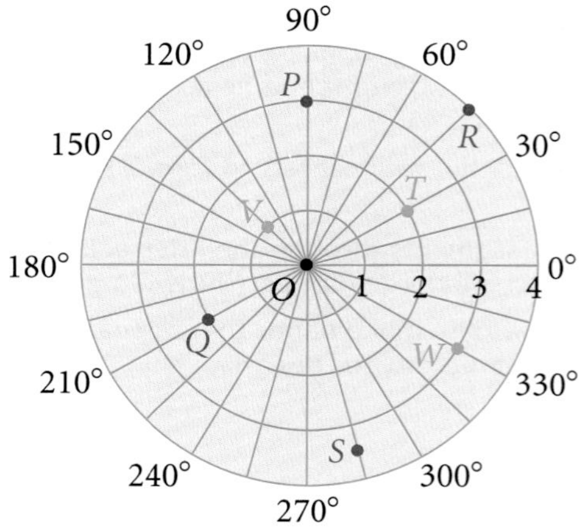

Example 2

Graph the points $T(2, 30°)$, $V(1, 135°)$, and $W(3, 330°)$.

Find the polar coordinates of each point.

1. A **2.** B **3.** C

4. D **5.** E **6.** F

Graph each point.

7. $G(5, 180°)$ **8.** $H(4, 60°)$

9. $I(3, 45°)$ **10.** $J(1, 225°)$

11. $K(3.5, 150°)$ **12.** $L(5, 270°)$

13. Explain how you can use the polar coordinates of X and Y to find the measure of $\angle XOY$.

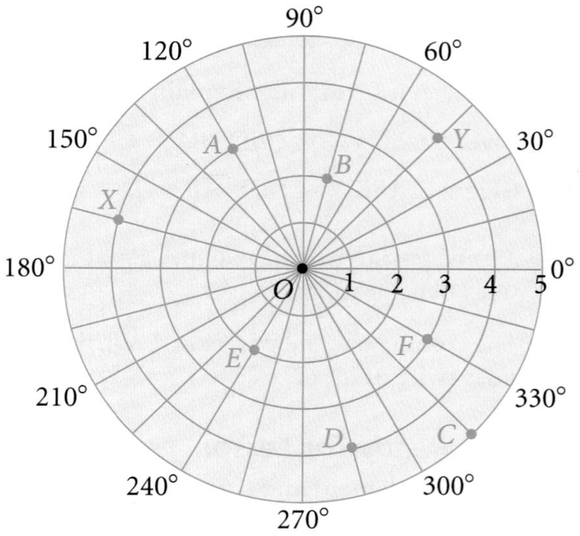

12-2 Properties of Tangents

What You'll Learn

- Finding the relationship between a radius and a tangent, and between two tangents drawn from the same point
- Circumscribing a circle

...And Why

To use tangents to circles in real-world situations, such as working in a machine shop

What You'll Need

- compass
- straightedge
- centimeter ruler

WORK TOGETHER

Work in a group.

- Use a compass to draw a circle. Label the center *O*.
- Use a straightedge to draw a line that intersects the circle in only one point. Label the point *B*.
- Draw $\overline{OB}$.

1. What seems to be true about the angle formed by the radius and the line that you drew? Compare results within your group.

THINK AND DISCUSS

Tangents to Circles

In Chapter 11, you studied *the tangent of an angle in a right triangle.* Now you will study properties of *a tangent to a circle.*

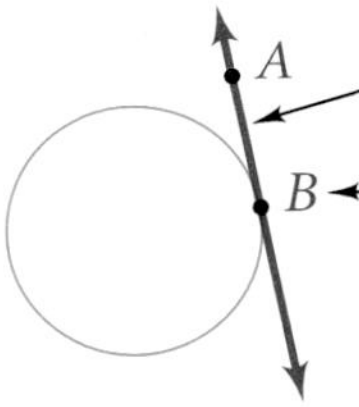

A **tangent to a circle** is a line in the plane of the circle that intersects the circle in exactly one point.

The point where a circle and a tangent intersect is the **point of tangency**.

$\overrightarrow{BA}$ is a tangent ray and $\overline{BA}$ is a tangent segment. The word *tangent* may refer to a tangent line, a tangent ray, or a tangent segment.

Your observation in the Work Together suggests the following theorem.

Theorem 12-2

If a line is tangent to a circle, then it is perpendicular to the radius drawn to the point of tangency.

If $\overleftrightarrow{AB}$ is tangent to $\odot O$ at *P*, then $\overleftrightarrow{AB} \perp \overline{OP}$.

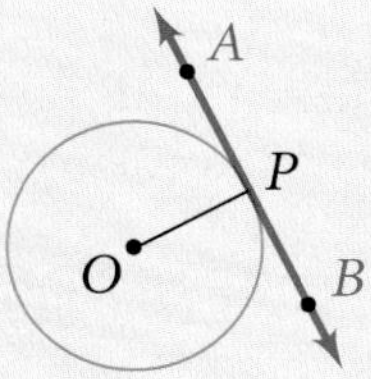

2. How many lines can you draw that are tangent to a circle at a given point on the circle?

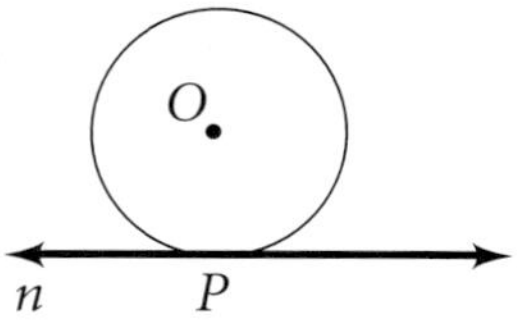

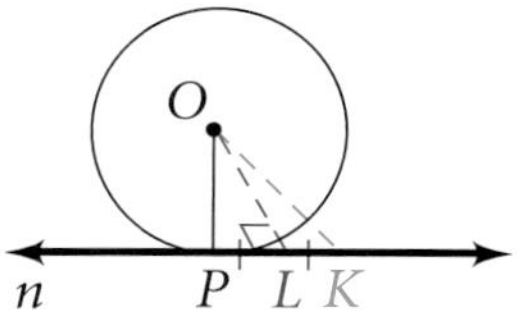

Indirect Proof of Theorem 12-2

Given: n is tangent to $\odot O$ at P.

Prove: $n \perp \overline{OP}$

Step 1: Assume that n is not perpendicular to $\overline{OP}$.

Step 2: If n is not perpendicular to $\overline{OP}$, some other segment $\overline{OL}$ must be perpendicular to n. Also there is a point K on n such that $\overline{LK} \cong \overline{LP}$. $\angle OLK \cong \angle OLP$ because perpendicular lines form congruent adjacent angles. $\overline{OL} \cong \overline{OL}$. So, $\triangle KLO \cong \triangle PLO$ by SAS. By CPCTC, $\overline{OK} \cong \overline{OP}$, which means K and P are both on $\odot O$ by the definition of a circle. For two points on n to also be on $\odot O$ contradicts the given fact that n is tangent to $\odot O$ at P. So the assumption that n is not perpendicular to $\overline{OP}$ must be false.

Step 3: Therefore, $n \perp \overline{OP}$ must be true.

Who? Alma Thomas (1891–1978) used geometric designs in many of her paintings. In *A Joyful Scene of Spring* (p. 585), Thomas used wedges of color to represent the many circular gardens planted throughout Washington, D.C.

You can use the Pythagorean Theorem to solve problems involving tangents.

Example 1

$\overline{PQ}$ is tangent to $\odot S$ at P. Find the length of a radius of $\odot S$.

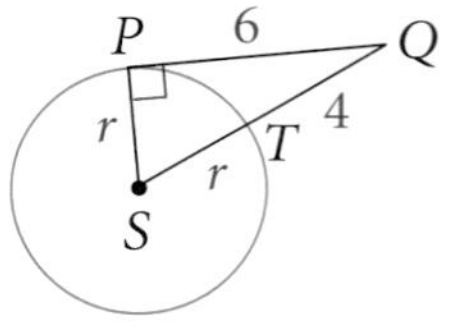

Since $\overline{PQ}$ is tangent to $\odot S$ at P, $\triangle PQS$ is a right triangle with hypotenuse $\overline{SQ}$. $\overline{PS}$ and $\overline{ST}$ are radii of $\odot S$.

$PS^2 + PQ^2 = SQ^2$	Pythagorean Theorem
$r^2 + 6^2 = (r + 4)^2$	$PS = ST = r$; $ST + TQ = SQ$
$r^2 + 36 = r^2 + 8r + 16$	Simplify.
$20 = 8r$	Subtract r^2 and 16 from each side.
$r = 2.5$	Divide each side by 8.

Tangents are formed by belts and ropes that pass over pulleys and wheels.

Example 2

Relating to the Real World

Machine Shop A belt fits tightly around two circular pulleys. Find the distance between the centers of the pulleys.

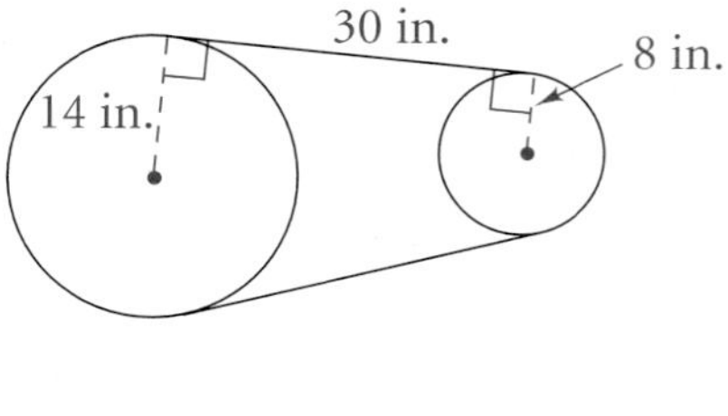

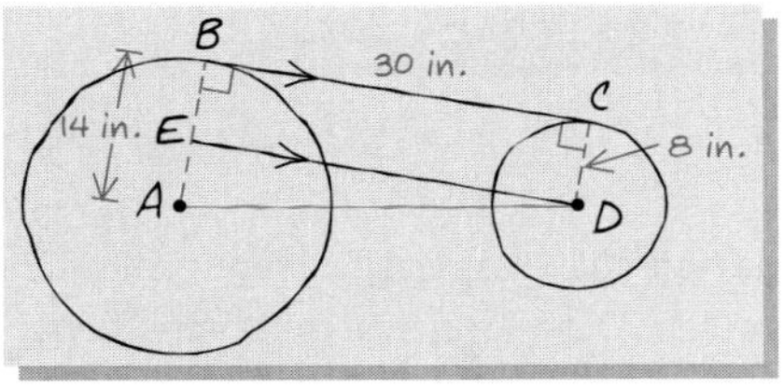

Label the diagram. Draw a segment parallel to $\overline{CB}$ from D to $\overline{AB}$. Label the intersection E.

$EBCD$ is a rectangle. $\triangle AED$ is a right triangle with $ED = 30$ in. and $AE = 14 - 8 = 6$ in.

Use the Pythagorean Theorem to find the distance between the centers, AD.

$AE^2 + ED^2 = AD^2$ Pythagorean Theorem

$6^2 + 30^2 = AD^2$ Substitute.

$936 = AD^2$ Simplify.

$30.594117 = AD$ Use a calculator.

The distance between the centers is about 30.6 in.

PROBLEM SOLVING

Look Back Explain how you could solve the problem in Example 2 by drawing a line from *C* rather than from *D*.

3. Explain why *EBCD* in Example 2 is a rectangle.

4. Explain why $\triangle AED$ in Example 2 is a right triangle.

5. **Try This** In the diagram at the right, $\overline{PQ}$ is tangent to $\odot O$ at *P*. Find the radius of $\odot O$.

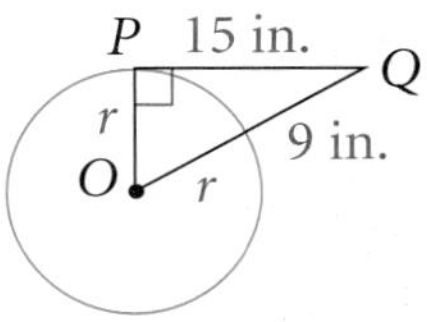

The converse of Theorem 12-2 is also true. You can use the Converse of Theorem 12-2 to determine if a segment is tangent to a circle. You can also use the converse to construct a tangent to a circle.

Theorem 12-3
Converse of Theorem 12-2

If a line in the same plane as a circle is perpendicular to a radius at its endpoint on the circle, then the line is tangent to the circle.

If $\overleftrightarrow{AB} \perp \overline{OP}$ at *P*, then $\overleftrightarrow{AB}$ is tangent to $\odot O$.

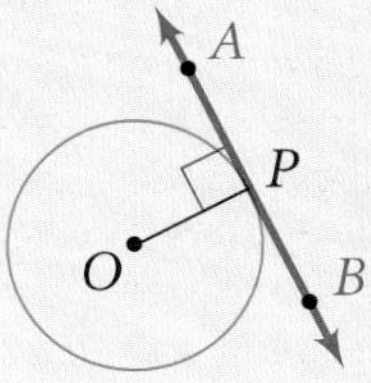

6. In the diagram at the right, is $\overline{LM}$ tangent to $\odot N$ at *L*? Explain.

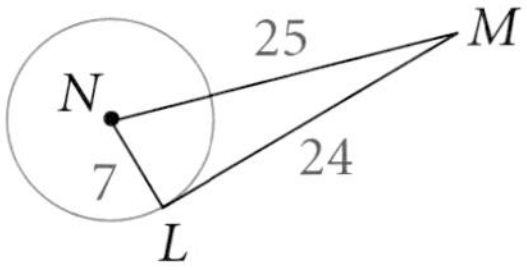

WORK TOGETHER

TECHNOLOGY HINT

The Work Together could be done using geometry software.

Have each member of your group construct a circle and select a point outside the circle. From the point, draw two tangent segments to the circle.

7. a. Measure the tangent segments. What do you notice? Compare results within your group.

b. Write a **conjecture** based on your results.

THINK AND DISCUSS

Circumscribing Circles

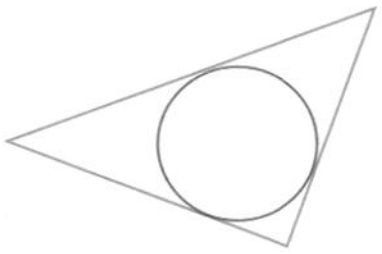

In the figure at the left, the sides of the triangle are tangent to the circle. The triangle is **circumscribed about** the circle. The circle is **inscribed in** the triangle.

Theorem 12-4 allows you to investigate figures that circumscribe a circle. You will prove this theorem in Exercise 10.

Theorem 12-4

Two segments tangent to a circle from a point outside the circle are congruent.

If $\overline{AB}$ and $\overline{CB}$ are tangent to $\odot O$ at A and C, respectively, then $\overline{AB} \cong \overline{CB}$.

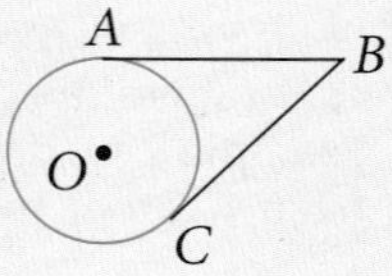

Example 3

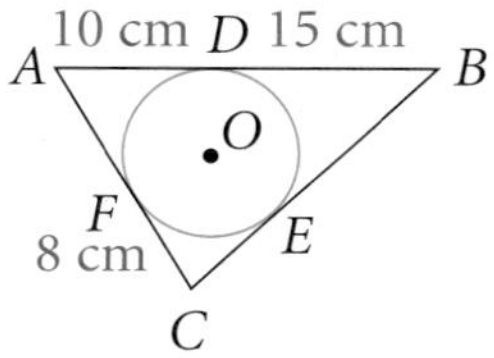

$\odot O$ is inscribed in $\triangle ABC$. Find the perimeter of $\triangle ABC$.

$AD = AF = 10$ cm
$BD = BE = 15$ cm
$CF = CE = 8$ cm

Two segments tangent to a circle from a point outside the circle are congruent.

$AD + AF + BD + BE + CF + CE =$ the perimeter

$10 + 10 + 15 + 15 + 8 + 8 = 66$

The perimeter is 66 cm.

8. a. **Try This** $\odot O$ is inscribed in $\triangle PQR$. Copy the diagram and label each length you know based on Theorem 12-4.
   b. $\triangle PQR$ has a perimeter 88 cm. Find QY.

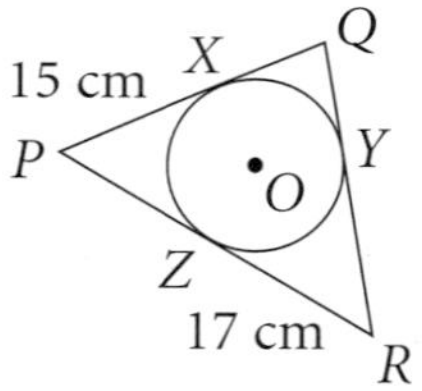

Exercises ON YOUR OWN

Assume that lines that appear to be tangent are tangent. *O* is the center of each circle. Find the value of *x*.

1.

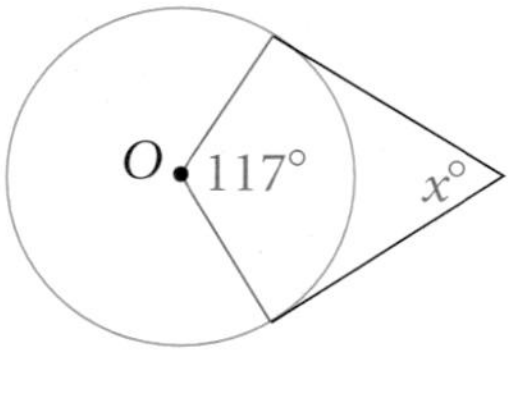

2.

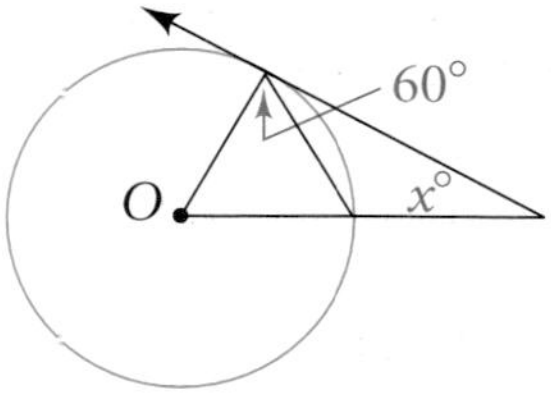

3.

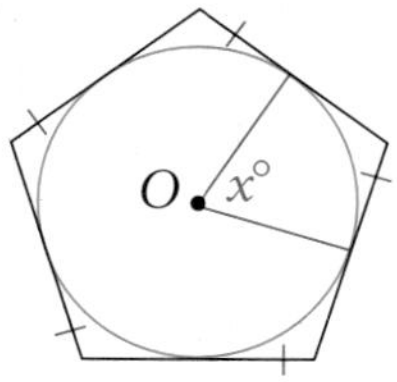

4.

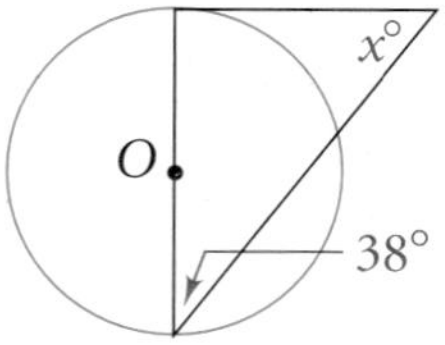

In each diagram, $\overline{PQ}$ is tangent to $\odot O$ at P. Find the value of x.

5.

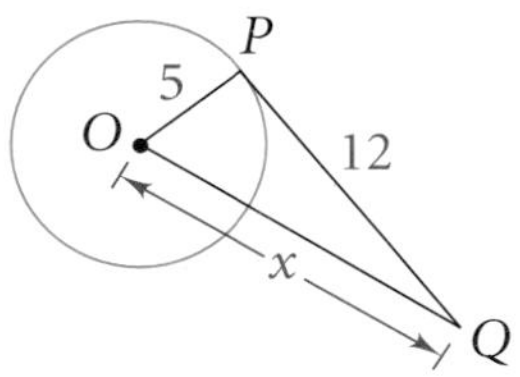

6.

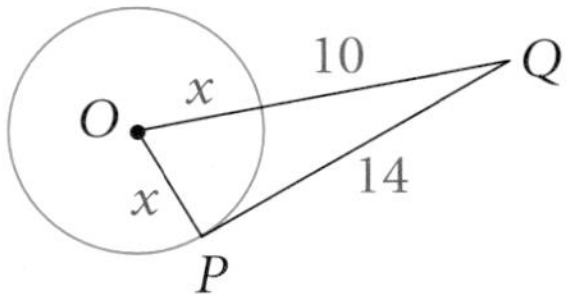

7.

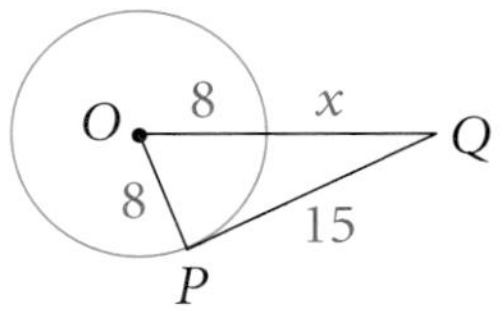

8. **Air Conditioning** Mr. Gonzales is replacing a cylindrical air-conditioning duct in a building. He estimates the radius of the duct by folding a ruler to form two 6-in. tangents to the duct. The tangents form an angle. Mr. Gonzales measures the angle bisector from the vertex to the duct. It is about $2\frac{3}{4}$ in. long. What is the radius of the air-conditioning duct?

9. a. **Coordinate Geometry** Draw several circles tangent to both coordinate axes.

b. **Writing** Describe the locus of the centers of all such circles.

c. **Algebra** Write the equation(s) of the locus of points you described in part (b).

10. **Choose** Write a two-column proof, paragraph proof, or flow proof of Theorem 12-4.

Given: $\overline{BA}$ and $\overline{BC}$ are tangent to $\odot O$ at A and C, respectively.

Prove: $\overline{BA} \cong \overline{BC}$

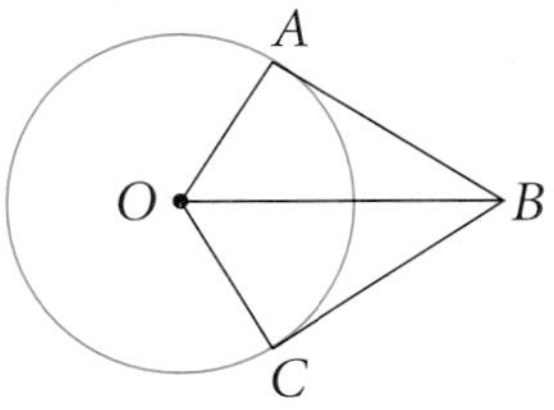

11. **Standardized Test Prep** $\overline{BD}$ and $\overline{CK}$ are diameters of $\odot A$. $\overline{BP}$ and $\overline{QP}$ are tangents to $\odot A$. Which of the following is *not* true?

A. $m\angle QPA = 25$ **B.** $m\angle BAP = 65$ **C.** $m\angle CDA = 57.5$
D. $m\angle DAQ = 25$ **E.** $m\angle CAB = 115$

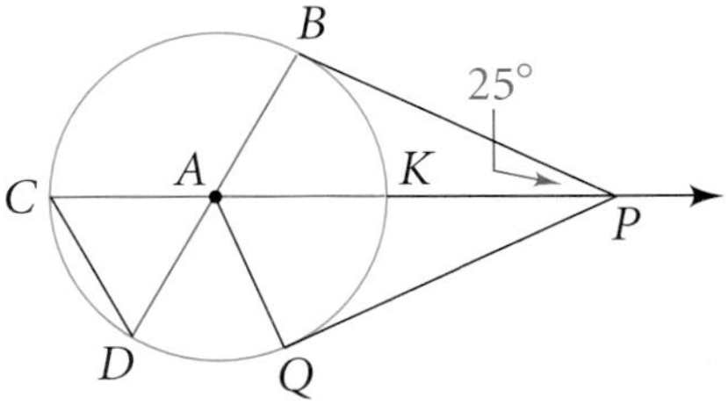

12. a. **Constructions** Use a compass to draw a circle. Draw a diameter of the circle. Use the construction of a perpendicular through a point on a line to construct tangents at the endpoints of the diameter.

b. What seems to be true of the tangents? **Justify** your answer.

13. A plumber installs a pipe with diameter 6 in. in a corner. The board securing the pipe in place is 1 in. thick. The shortest length of board she can use forms an isosceles right triangle with the two walls.

a. What is the length (to the nearest tenth) of the side of the board tangent to the pipe?

b. What is the outside length of the board to the nearest tenth?

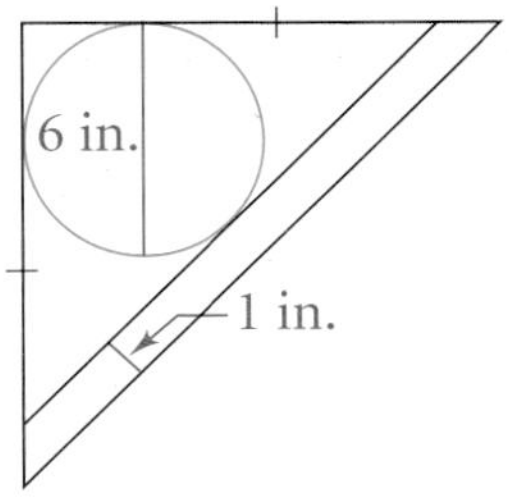

In each diagram, a polygon circumscribes a circle. Find the perimeter of each polygon.

14.

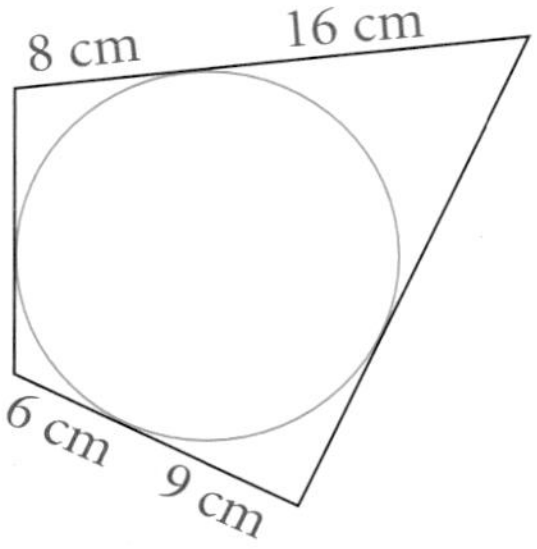

15.

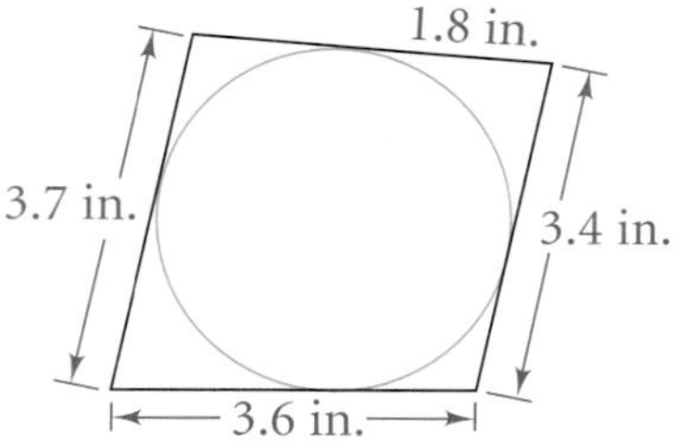

16.

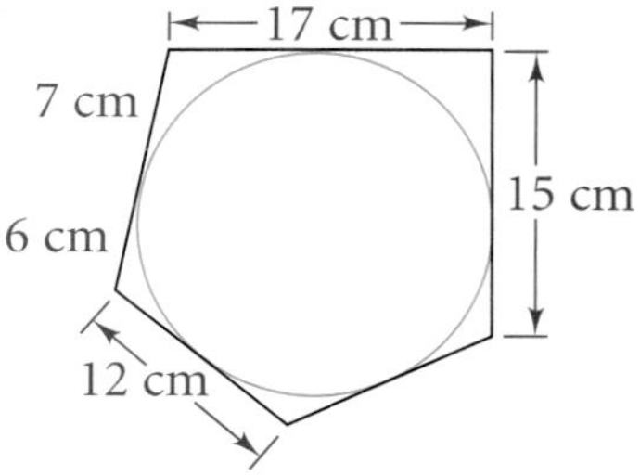

Choose **Write a two-column proof, paragraph proof, or flow proof.**

17. Given: $\overline{BC}$ is tangent to $\odot A$ at D.
$\overline{DB} \cong \overline{DC}$
Prove: $\overline{AB} \cong \overline{AC}$

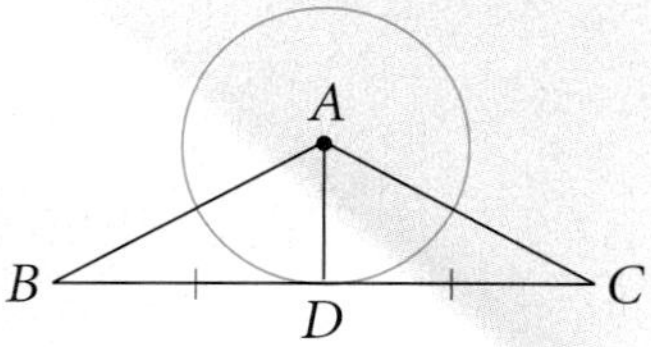

18. Given: $\odot A$ and $\odot B$ with common tangents $\overline{DF}$ and $\overline{CE}$
Prove: $\triangle GDC \sim \triangle GFE$

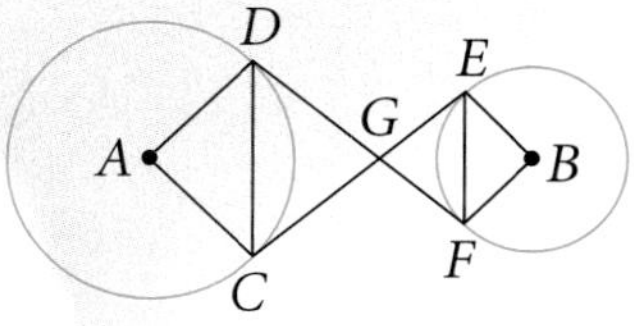

19. **Critical Thinking** $\overline{EF}$ is tangent to both $\odot A$ and $\odot B$ at F. $\overline{CD}$ is tangent to $\odot A$ at C and to $\odot B$ at D. What can you conclude about $\overline{CE}$, $\overline{DE}$, and $\overline{FE}$? **Justify** your answer.

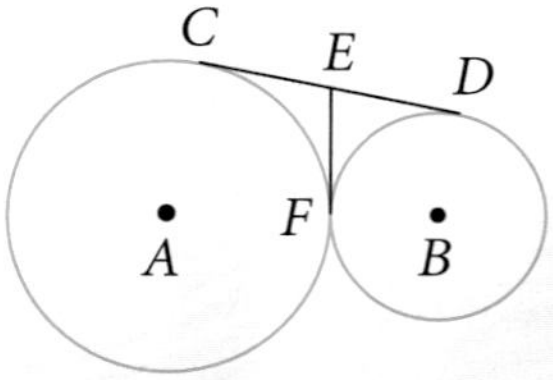

20. Leonardo da Vinci wrote, "When each of two squares touch the same circle at four points, one is double the other."

a. Sketch a figure to illustrate this statement.

b. **Justify** the statement.

21. **Solar Eclipse** Common tangents to two circles may be internal or external. If a line were drawn through the centers of the two circles, a *common internal tangent* would intersect the line and a *common external tangent* would *not* intersect the line. In the diagram of a solar eclipse, which are common tangents to the sun and the moon? to the sun and Earth? State whether the tangents are internal or external.

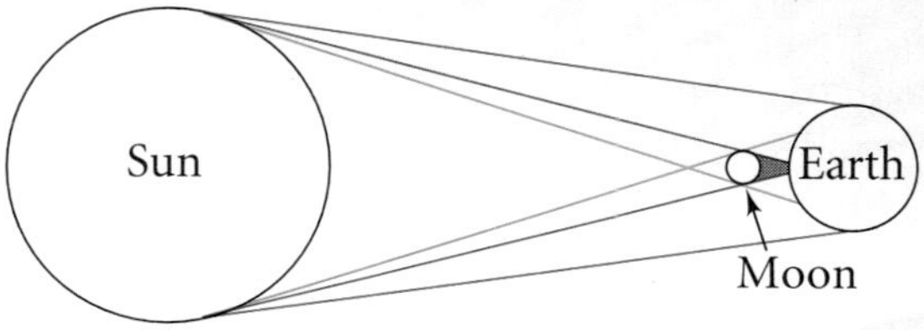

22. **a.** $\overline{AC}$ is tangent to $\odot O$ at A and $m\angle 1 = 70$. Find $m\angle 4$.
b. Let $m\angle 1 = x$. Find $m\angle 4$ in terms of x. What is the relationship between $\angle 1$ and $\angle 4$? Explain.

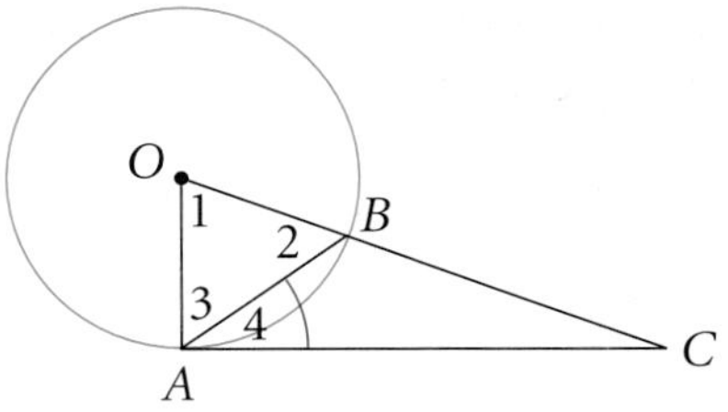

23. **Coordinate Geometry** Graph the equation $x^2 + y^2 = 9$. Then draw a segment from (0, 5) tangent to the circle. Find the length of the segment.

24. **a.** **Open-ended** Draw a quadrilateral circumscribed about a circle.
b. **Patterns** Compare the sums of the lengths of the opposite sides. Repeat the experiment with another circle and quadrilateral. Make a **conjecture** based on your observations.

25. **Clock Making** Mr. Franklin is building a case for a clock. The case will be in the shape of a regular hexagon. The diameter of the circular face is 10 in. Find the inside perimeter of the clock case.

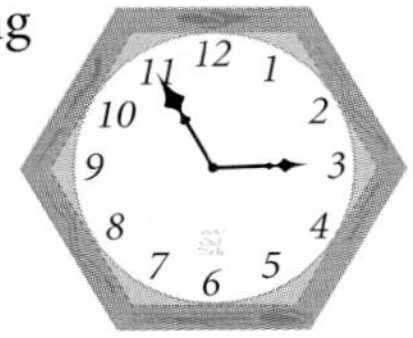

Exercises MIXED REVIEW

Find sin *A* and cos *A*. Leave your answers in simplest radical form.

26.
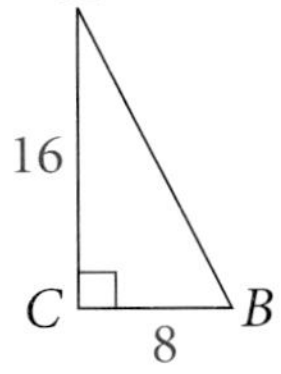

27.
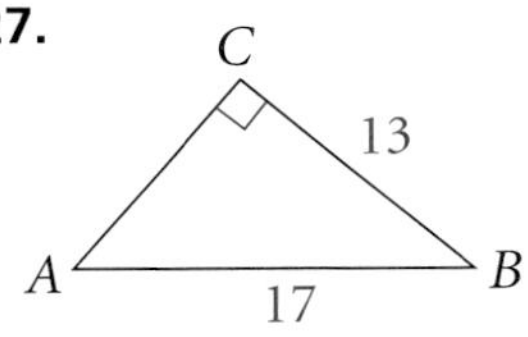

28.
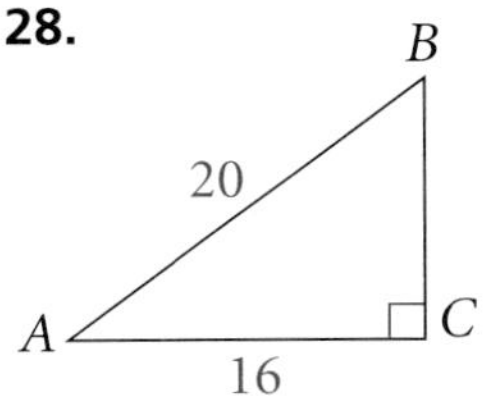

29. **a.** Sketch a quadrilateral whose diagonals are congruent and bisect each other. What kind of quadrilateral did you sketch?
b. **Coordinate Geometry** Decide where you would place coordinate axes on the quadrilateral if you were using it in a coordinate proof. Sketch your axes.

SELF ASSESSMENT

FOR YOUR JOURNAL

Describe places at school or at home where you see circles and their tangents.

Getting Ready for Lesson 12-3

Choose **Use mental math, paper and pencil, or a calculator. Find the value of *x*. If your answer is not an integer, round to the nearest tenth.**

30.
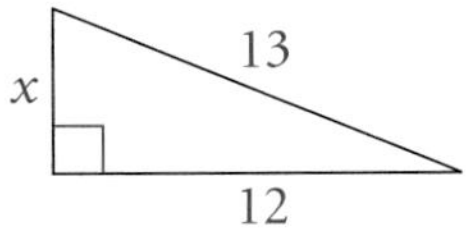

31.
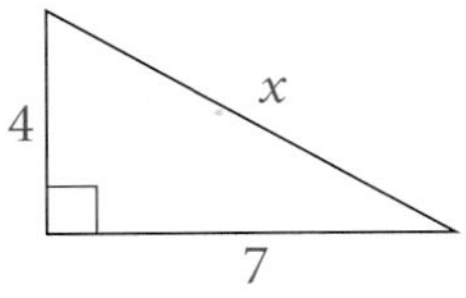

32.
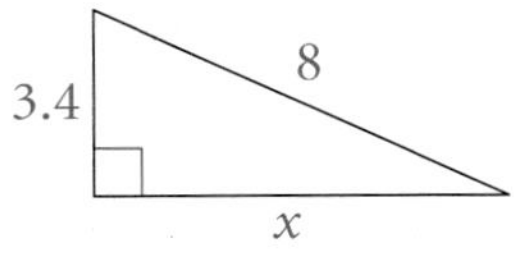

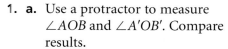

12-3 Properties of Chords and Arcs

What You'll Learn

- Finding the lengths of chords and measures of arcs of a circle
- Locating the center of a circle using chords

...And Why

To find the radius of a circle in real-life situations such as archaeology

What You'll Need

compass, MIRA™, protractor, centimeter ruler

WORK TOGETHER

Have each member of your group use a compass to draw a large circle. Label its center O. Create $\widehat{AB}$ by labeling two points A and B on your circle.

- Construct an arc congruent to $\widehat{AB}$ by placing a MIRA on $\odot O$ and moving it until the circle maps on itself and you see the images of A and B. Label the images A' and B'. Draw radii $\overline{OA}$, $\overline{OB}$, $\overline{OA'}$, and $\overline{OB'}$.

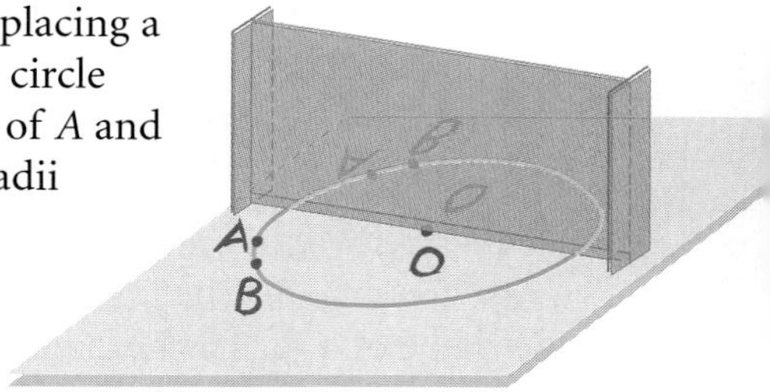

1. a. Use a protractor to measure $\angle AOB$ and $\angle A'OB'$. Compare results.
   b. Write a **conjecture** about the central angles of congruent arcs.

TECHNOLOGY HINT

The Work Together could be done using geometry software.

- Draw $\overline{AB}$ and $\overline{A'B'}$.

2. a. Measure $\overline{AB}$ and $\overline{A'B'}$. Compare results within your group.
   b. Write a **conjecture** about the segments joining the endpoints of congruent arcs.

QUICK REVIEW

The measure of a minor arc is the measure of its corresponding central angle. The measure of a major arc is 360 minus the measure of its related minor arc.

- Use a compass to draw another circle. Label its center M. Label two points G and H on your circle. Draw $\overline{GH}$. While keeping an edge of the MIRA on point M, construct a diameter perpendicular to $\overline{GH}$ by moving the MIRA until the image of G maps onto H. Draw the MIRA line. Label its intersection with $\overline{GH}$ as F and label its intersection with the circle as J so that F is between J and M.

3. Measure $\overline{GF}$ and $\overline{HF}$. What do you notice? Compare results.
4. Find $m\widehat{GJ}$ and $m\widehat{HJ}$. What do you notice? Compare results.

THINK AND DISCUSS

Congruent Arcs and Chords

QUICK REVIEW

Congruent arcs have the same measure and are in the same circle or in congruent circles.

In the Work Together, you created segments with endpoints on a circle. These segments are **chords** of the circle. The diagram shows arc $\widehat{PQ}$ and its related chord $\overline{PQ}$.

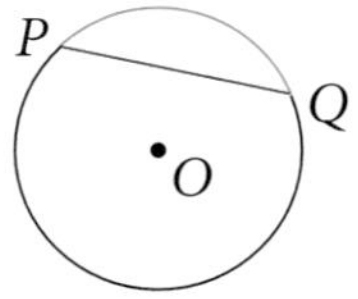

5. Is a diameter a chord? Explain why or why not.
6. Is a radius a chord? Explain why or why not.

Your observations in the Work Together suggest the following theorems. You will prove Theorem 12-7 in Exercise 18.

Theorem 12-5 In the same circle or in congruent circles, (1) congruent central angles have congruent arcs; and (2) congruent arcs have congruent central angles.

Theorem 12-6 In the same circle or in congruent circles, (1) congruent chords have congruent arcs; and (2) congruent arcs have congruent chords.

Theorem 12-7 A diameter that is perpendicular to a chord bisects the chord and its arc.

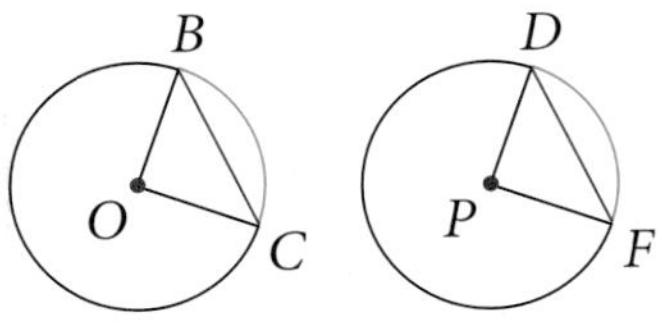

7. *Logical Reasoning* In the diagram at the left, $\odot O \cong \odot P$.

a. Given that $\overline{BC} \cong \overline{DF}$, **justify** Theorem 12-6, part (1), by explaining why $\overset{\frown}{BC} \cong \overset{\frown}{DF}$.

b. Given that $\overset{\frown}{BC} \cong \overset{\frown}{DF}$, **justify** Theorem 12-6, part (2), by explaining why $\overline{BC} \cong \overline{DF}$.

A segment that contains the center of a circle is part of a diameter. So, you may use Theorem 12-7 for a segment that is part of a diameter and is perpendicular to a chord.

QUICK REVIEW

The distance from a point to a line is the length of the perpendicular segment from the point to the line.

Example 1

Chord $\overline{CE}$ is 24 in. long and 5 in. from the center of $\odot A$.

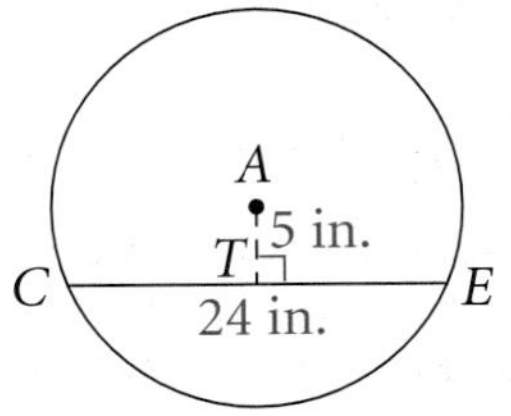

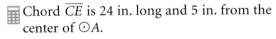

a. Find the radius of $\odot A$.
b. Find $m\overset{\frown}{CE}$.

a. Copy the diagram and draw radius $\overline{AC}$ as shown at the left.

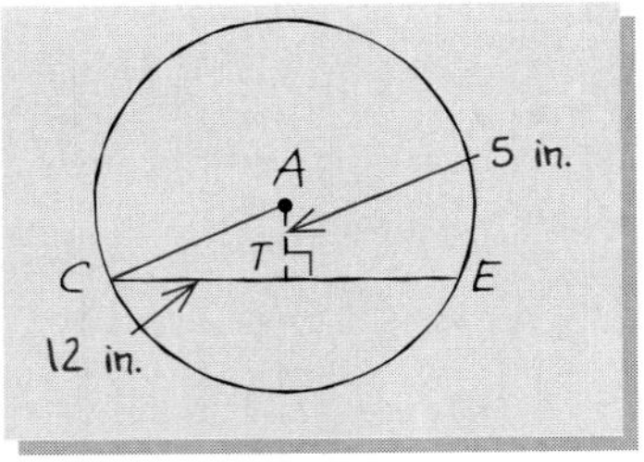

$TC = \frac{1}{2}(24) = 12$ — The diameter ⊥ to a chord bisects the chord.

$AC^2 = 5^2 + 12^2 = 169$ — Use the Pythagorean Theorem.

$AC = 13$ in.

b. Extend $\overline{AT}$ to intersect the circle at D as shown below.

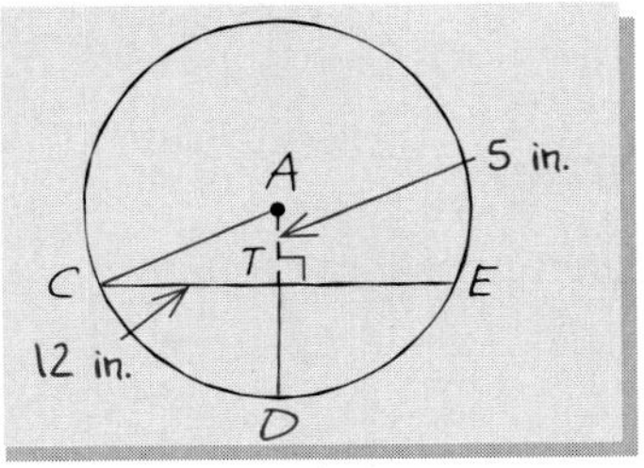

$\frac{1}{2}m\overset{\frown}{CE} = m\overset{\frown}{CD}$ — A diameter ⊥ to a chord bisects the chord and its arc.

$m\overset{\frown}{CE} = 2m\overset{\frown}{CD}$

$m\overset{\frown}{CD} = m\angle CAT$ — The measure of a minor arc = the measure of its central angle.

$m\overset{\frown}{CE} = 2m\angle CAT$ — Substitution

$\tan \angle CAT = \frac{12}{5} = 2.4$ — $\tan \angle A = \frac{\text{opp.}}{\text{adj.}} = \frac{CT}{AT}$

$m\angle CAT \approx 67.380135$ — Use a calculator to find $\tan^{-1}(2.4)$.

$m\overset{\frown}{CE} \approx 2$ ✖ 67.380135

$m\overset{\frown}{CE} \approx 134.8$

8. a. **Try This** Find the radius of $\odot C$ to the nearest tenth.
   b. Find $m\overset{\frown}{AB}$ to the nearest tenth.

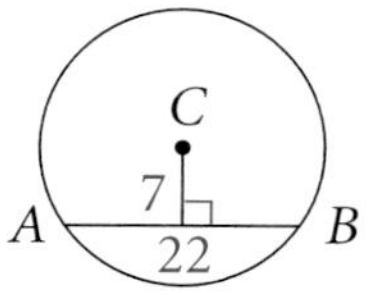

You can use the following theorem to reconstruct a circle from an arc.

Theorem 12-8 The perpendicular bisector of a chord contains the center of the circle.

9. **Logical Reasoning** Use the Converse of the Perpendicular Bisector Theorem (Theorem 4-13 on page 219) to **justify** Theorem 12-8.

Example 2 **Relating to the Real World**

Archaeology An archaeologist found pieces of a jar. She wants to find the radius of the rim of the jar to help guide her as she reassembles the pieces. How can she find the center and radius of the rim?

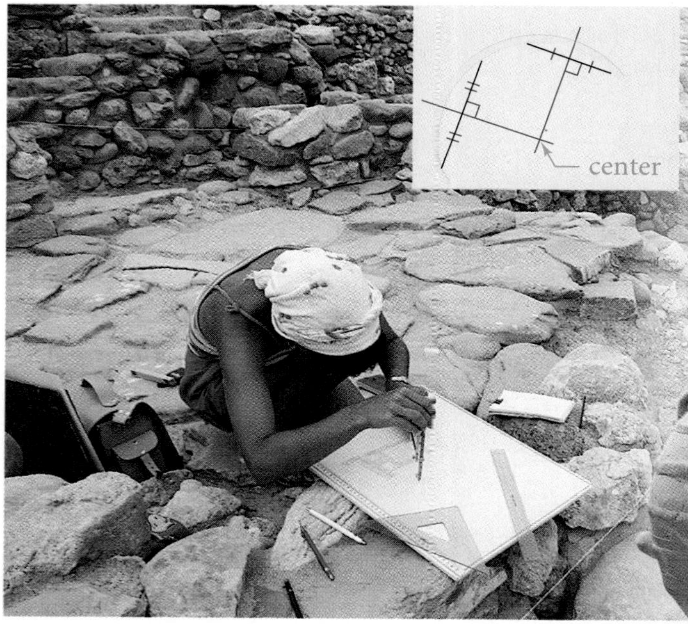

First, she should carefully trace a piece of the rim. Then she should draw two chords and construct the perpendicular bisector of each chord. Since each perpendicular bisector contains the center of the circle, the point where they intersect must be the center. (Two lines intersect in exactly one point.) After finding the center, she can measure the radius.

Chords Equidistant from the Center of a Circle

The following theorem shows the relationship between two congruent chords and their distances from the center of a circle.

Theorem 12-9 In the same circle or in congruent circles, (1) chords equidistant from the center are congruent; and (2) congruent chords are equidistant from the center.

Example 3

Prove Theorem 12-9, part (1).

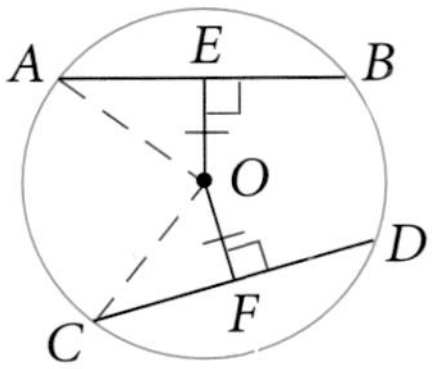

Given: $\odot O$, $\overline{OE} \perp \overline{AB}$, $\overline{OF} \perp \overline{CD}$, $\overline{OE} \cong \overline{OF}$
Prove: $\overline{AB} \cong \overline{CD}$

Begin by drawing $\overline{OA}$ and $\overline{OC}$.

Statements	Reasons
1. $\overline{OA} \cong \overline{OC}$	**1.** All radii of a circle are $\cong$.
2. $\overline{OE} \cong \overline{OF}$	**2.** Given
3. $\overline{OE} \perp \overline{AB}$, $\overline{OF} \perp \overline{CD}$	**3.** Given
4. $\angle AEO$ and $\angle CFO$ are right angles.	**4.** Definition of right angles
5. $\triangle AEO \cong \triangle CFO$	**5.** HL Theorem
6. $AE = CF$	**6.** CPCTC
7. $AE = \frac{1}{2}AB$, $CF = \frac{1}{2}CD$	**7.** A diameter $\perp$ to a chord bisects the chord.
8. $\frac{1}{2}AB = \frac{1}{2}CD$	**8.** Substitution
9. $AB = CD$, or $\overline{AB} \cong \overline{CD}$	**9.** Multiplication Property of Equality

Exercises ON YOUR OWN

Calculator **Find the value of x. If your answer is not an integer, round it to the nearest tenth.**

1.

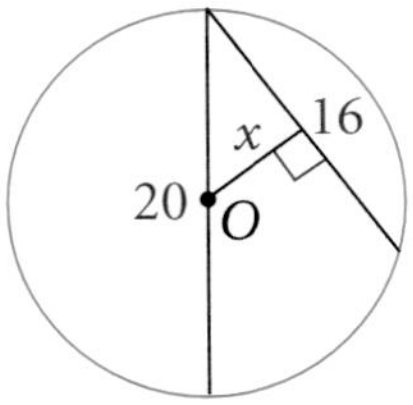

2.

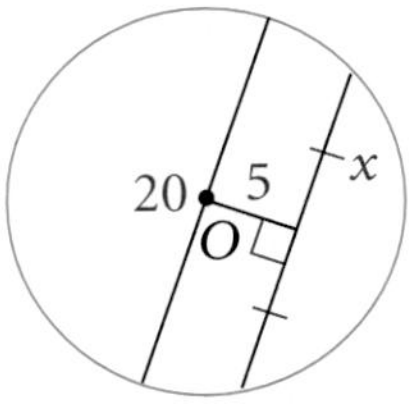

3.

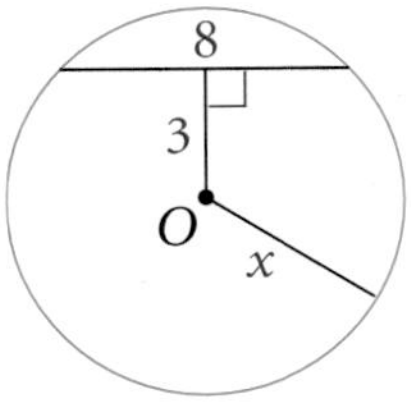

4.

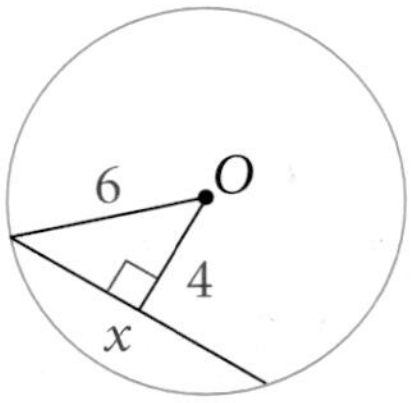

5.

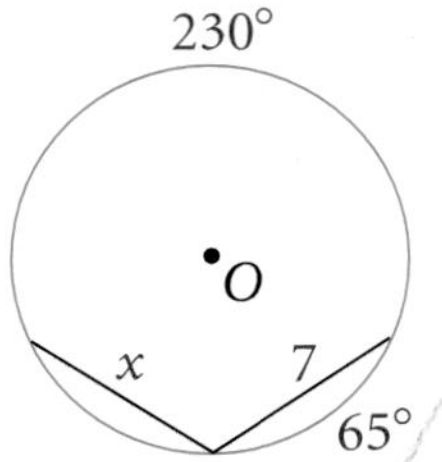

6.

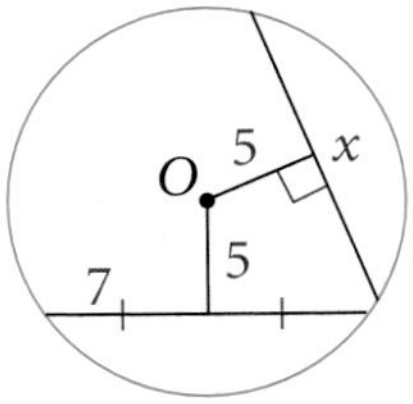

7.

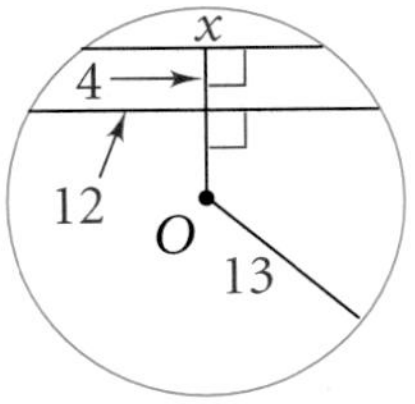

8.

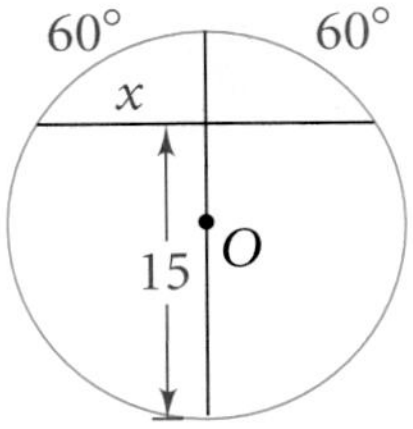

9. **a.** The photo at the right shows part of a compact disc. Trace the disc. Find its radius in centimeters.
b. **Writing** Explain how you found the radius of the compact disc.

Find $m\widehat{AB}$.

10.

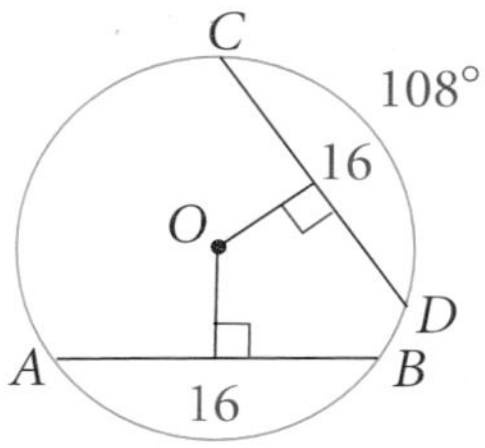

11.

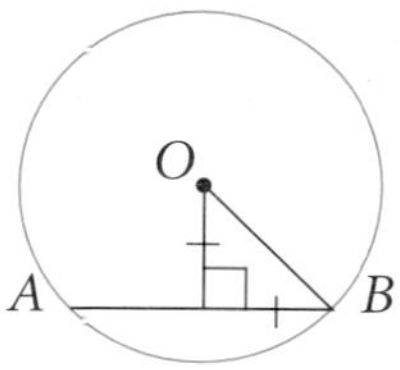

12.

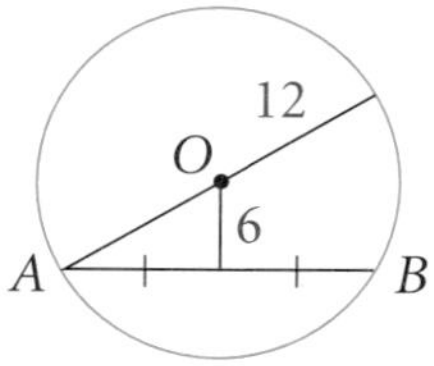

13.

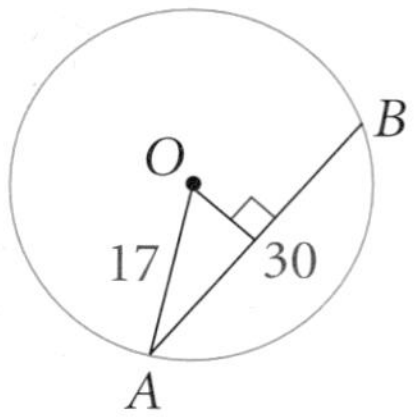

14. Geometry in 3 Dimensions Sphere O with radius 13 cm is intersected by a plane 5 cm from its center forming $\odot A$. Find the radius of $\odot A$.

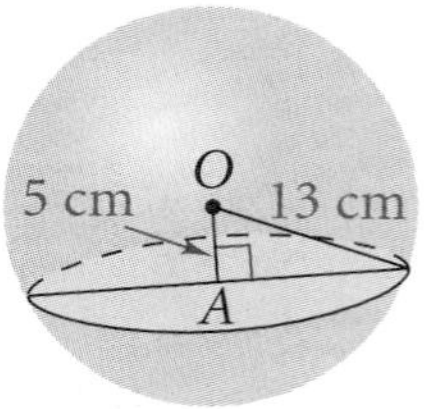

15. A plane intersects a sphere that has radius 10 in. forming $\odot B$ with radius 8 in. How far is the plane from the center of the sphere?

16. A plane intersects a sphere 12 in. from the sphere's center forming $\odot C$ with radius 18 in. What is the radius of the sphere? Round to the nearest tenth.

17. Two concentric circles have radii of 4 cm and 8 cm. A segment tangent to the smaller circle is a chord of the larger circle. What is the length of the segment?

PROBLEM SOLVING HINT
Draw a diagram.

18. Supply the reasons for each statement in this proof of Theorem 12-7.

Given: $\odot O$ with diameter $\overline{ED} \perp \overline{AB}$ at C
Prove: $\overline{AC} \cong \overline{BC}$ and $\widehat{AD} \cong \widehat{BD}$

Begin by drawing $\overline{OA}$ and $\overline{OB}$.

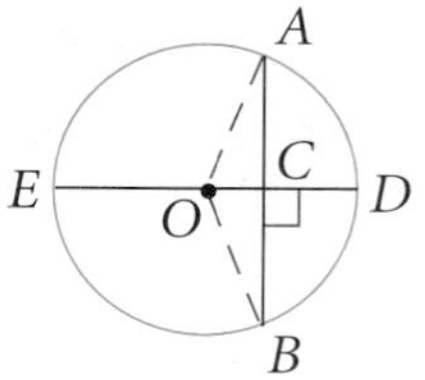

Statements	Reasons
1. $\overline{OA} \cong \overline{OB}$	a. ?
2. $\overline{ED} \perp \overline{AB}$	b. ?
3. $\angle ACO$ and $\angle BCO$ are right angles.	c. ?
4. $\overline{OC} \cong \overline{OC}$	d. ?
5. $\triangle AOC \cong \triangle BOC$	e. ?
6. $\overline{AC} \cong \overline{BC}$	f. ?
7. $\angle AOC \cong \angle BOC$	g. ?
8. $\widehat{AD} \cong \widehat{BD}$	h. ?

19. Constructions Use a circular object such as a can or a saucer to draw a circle. Construct the center of the circle.

20. Transporting Milk The diameter of the base of a cylindrical milk tank is 59 in. The height of the tank is 470 in. A delivery person estimates that the depth of the milk in the tank is 20 in. Find the number of gallons of milk in the tank to the nearest gallon. ($231 \text{ in.}^3 = 1$ gal)

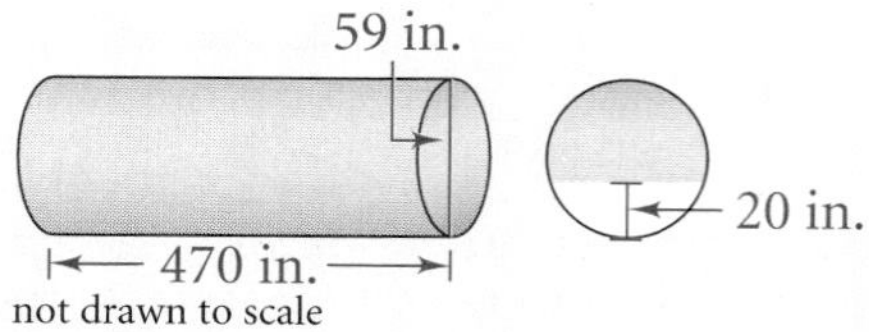

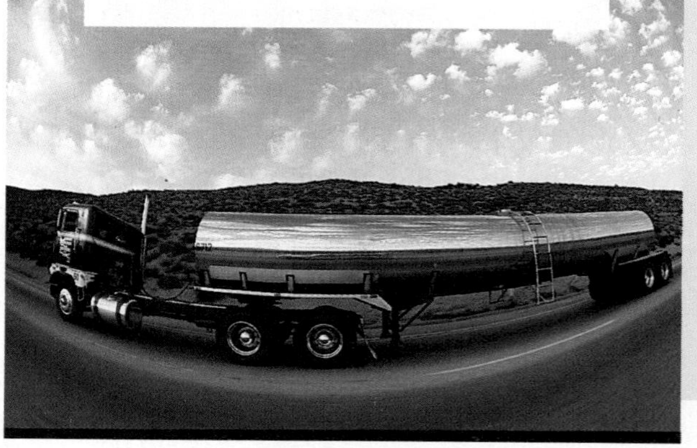

Choose **Write a two-column proof, paragraph proof, or flow proof.**

21. **Given:** $\odot A$ with $\overline{CE} \perp \overline{BD}$
Prove: $\widehat{BC} \cong \widehat{DC}$

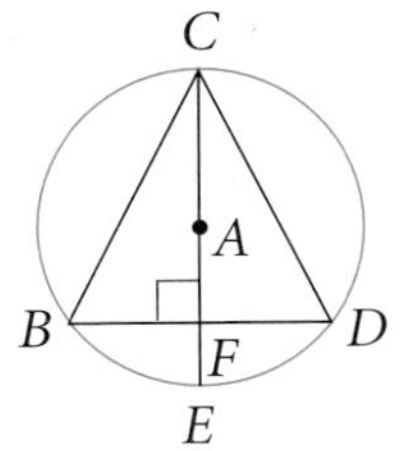

22. **Given:** $\odot A$ with $\widehat{BC} \cong \widehat{DE}$, $\overline{AF} \perp \overline{BC}$, $\overline{AG} \perp \overline{DE}$
Prove: $\angle AFG \cong \angle AGF$

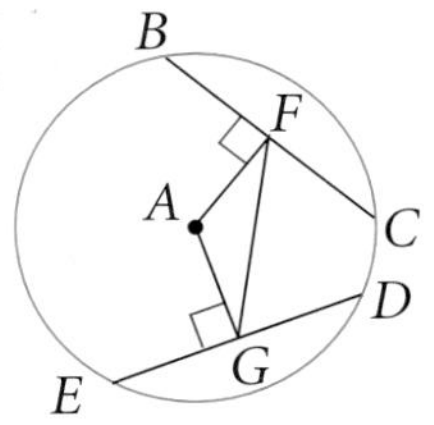

23. **a.** Use the diagram at the right. Complete each statement. Given that $\overline{AB}$ is a diameter of the circle and $\overline{AB} \perp \overline{CD}$, then ■ ≅ ■, and ■ ≅ ■.

b. Given that $\overline{AB}$ is the perpendicular bisector of $\overline{CD}$, then $\overline{AB}$ contains the __?__ .

c. Which theorems are illustrated in parts (a) and (b)?

d. **Open-ended** Draw a figure or figures to illustrate the other theorems presented in this lesson.

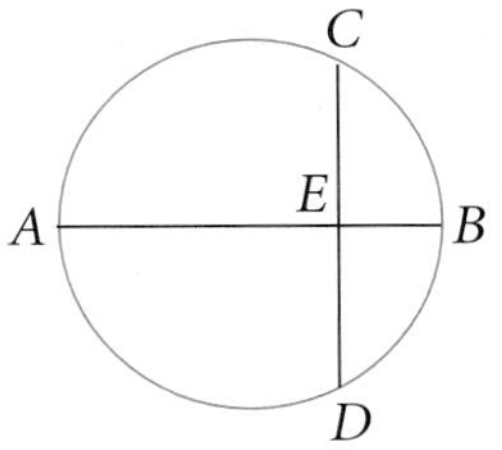

24. $\odot A$ and $\odot B$ are congruent. $AB = 8$ in. $\overline{CD}$ is a chord of both circles. $CD = 6$ in. How long is a radius?

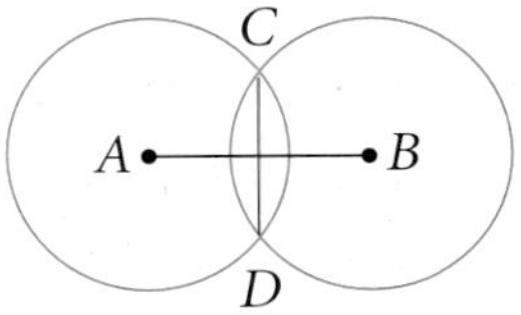

25. **Writing** Explain why the phrase "in the same circle or in congruent circles" is essential to Theorems 12-5, 12-6, and 12-9.

26. **Coordinate Geometry** Find the length of the chord of the circle $x^2 + y^2 = 25$ that is determined by the line $x = 3$.

27. **a.** **Critical Thinking** The diameter of a circle is 20 cm. Two chords in the circle are 6 cm and 16 cm. In how many ways can you draw the chords so that they are parallel to the same diameter? (*Hint:* Draw a diagram.)

b. What are the possible distances between the chords to the nearest tenth of a centimeter?

Chapter Project **Find Out by Exploring**

Stare at these circular patterns for a few seconds. Notice how they seem to pulsate. In contrast to the ancient art form you explored in the Find Out activity on page 591, op art is a twentieth-century phenomenon.

- Use geometric terms to describe how Figures A and B are related.
- Create your own op art by transforming a *target* design like Figure A. Use geometric terms to describe your design.

Figure A

Figure B

Exercises MIXED REVIEW

Determine whether a triangle with the given side lengths is acute, right, or obtuse.

28. 4, 4, 4 **29.** 3, 5, 6 **30.** 6, 8, 10 **31.** 2, 3, 4 **32.** 7, $\sqrt{63}$, 10 **33.** 5, 8, $\sqrt{70}$

34. You want to find out how tall a tree in a park is. Your shadow length is $\frac{3}{4}$ of your height. The tree's shadow is 57 ft long. How tall is the tree?

Getting Ready for Lesson 12-4

Find the measure of each arc in ⊙*O*.

35. $\widehat{AB}$ **36.** $\widehat{BC}$ **37.** $\widehat{AD}$

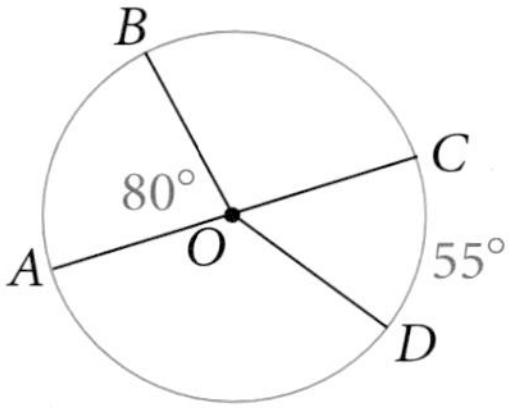

Exercises CHECKPOINT

Coordinate Geometry **The points are endpoints of the diameter of a circle. Write the equation of the circle.**

1. (3, 1) and (0, 0) **2.** (−2, 5) and (9, −3) **3.** (−4, −8) and (1, 0)

In each diagram, the polygon circumscribes a circle. Find the perimeter of the polygon.

4.

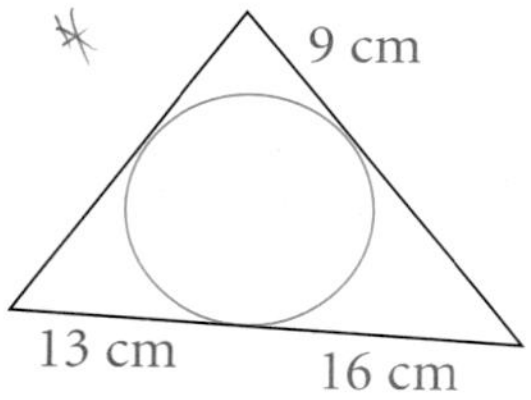

5.

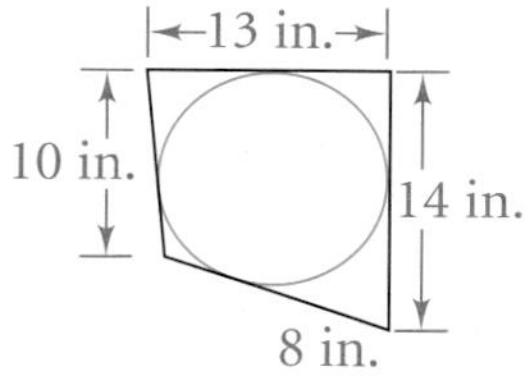

6.

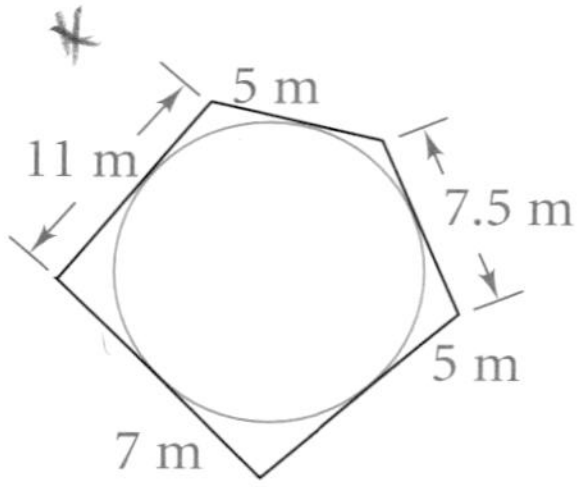

Find the value of *x*.

7.

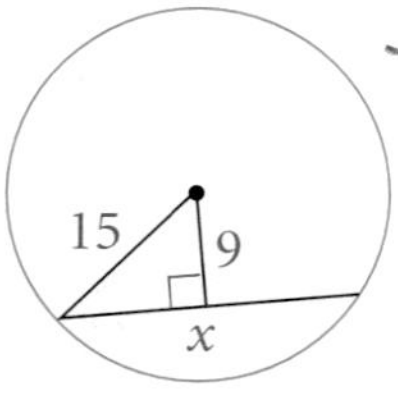

8.

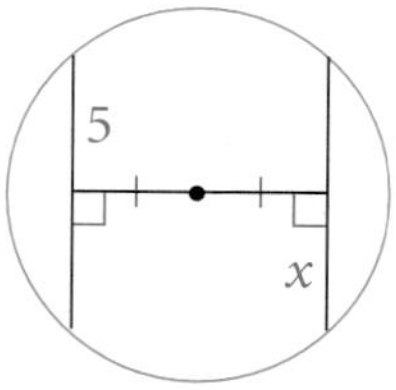

9.

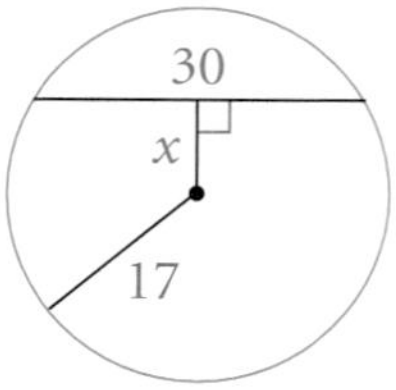

10. a. *Open-ended* Draw a triangle circumscribed about a circle. Then draw the radii to each tangent.

b. How many convex quadrilaterals are in the figure you drew in part (a)?

c. What is the name of these special quadrilaterals? **Justify** your answer.

TECHNOLOGY HINT

This exercise could be done using geometry software.

12-4 Inscribed Angles

What You'll Learn

- Finding the measure of inscribed angles and the arcs they intercept

...And Why

To use the relationships between inscribed angles and arcs in real-world situations, such as motion pictures

What You'll Need

compass, protractor, ruler

WORK TOGETHER

Have each member of your group draw two large circles. Label the centers X and Y. Draw the following diagrams on the circles.

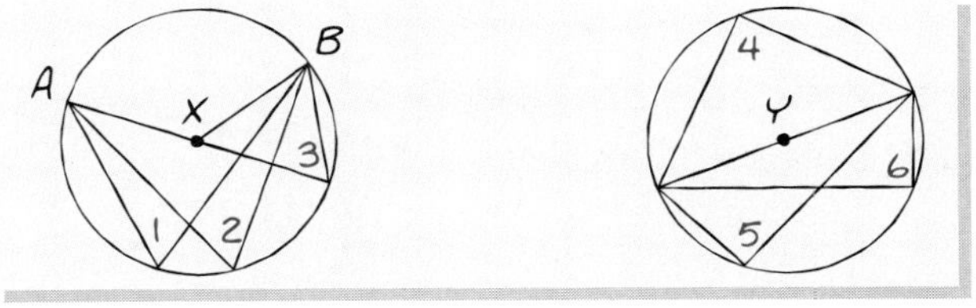

1. **a.** **Patterns** In $\odot X$, use a protractor to measure $\angle AXB$ and each numbered angle. Determine $m\widehat{AB}$. Record your results and look for patterns. Compare results within your group.
   b. Write a **conjecture** about the relationship between $m\angle 1$ and $m\widehat{AB}$.
   c. Write a **conjecture** about the measures of $\angle 1$, $\angle 2$, and $\angle 3$.
2. **a.** **Patterns** Use a protractor to measure the numbered angles in $\odot Y$. Record your results and look for patterns. Compare your results.
   b. Write a **conjecture** about an angle whose vertex is on a circle and whose sides intersect the endpoints of a diameter of the circle.

TECHNOLOGY HINT

The Work Together could be done using geometry software.

THINK AND DISCUSS

Measuring Inscribed Angles

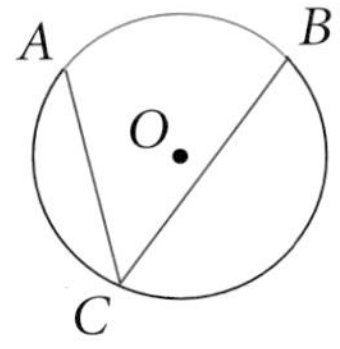

At the left, the vertex of $\angle C$ is on $\odot O$, and the sides of $\angle C$ are chords of the circle. $\angle C$ is an **inscribed angle.** $\widehat{AB}$ is the **intercepted arc** of $\angle C$.

A polygon is **inscribed in** a circle if all its vertices lie on the circle. $\triangle DEF$ is inscribed in $\odot Q$. $\odot Q$ is **circumscribed about** $\triangle DEF$.

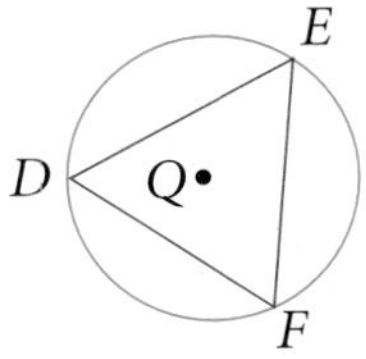

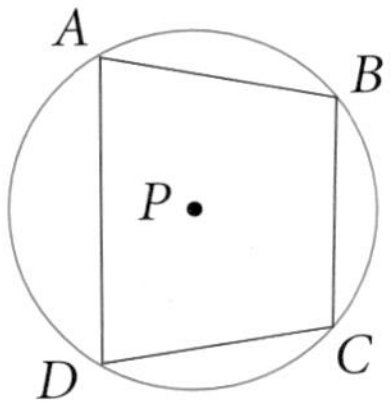

3. In the diagram at the left, which arc does $\angle A$ intercept?
4. Which angle intercepts $\widehat{DAB}$?
5. Is quadrilateral $ABCD$ inscribed in the circle or is the circle inscribed in $ABCD$?
6. **a.** Which angles appear to intercept major arcs?
   b. What kind of angles do these appear to be?

Your observations in the Work Together suggest the following theorem.

Theorem 12-10
Inscribed Angle Theorem

The measure of an inscribed angle is half the measure of its intercepted arc.

$m\angle B = \frac{1}{2}m\widehat{AC}$

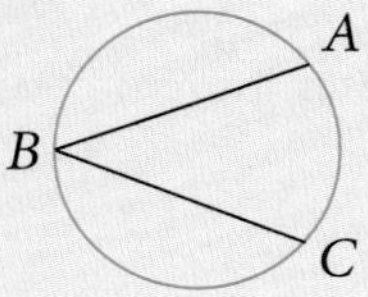

There are three cases of this theorem to consider.

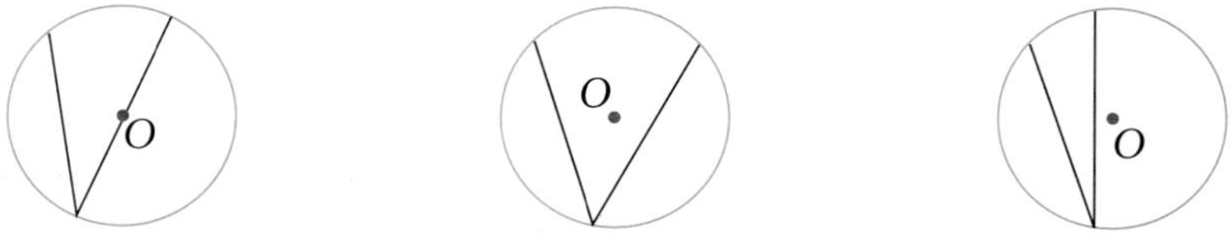

Case I: The center is on a side of the angle.

Case II: The center is inside the angle.

Case III: The center is outside the angle.

Case I is proved below. You will prove Case II in Exercise 13.

Proof of Theorem 12-10, Case I

Given: $\odot O$ with inscribed $\angle ABC$ and diameter $\overline{BC}$

Prove: $m\angle ABC = \frac{1}{2}m\widehat{AC}$

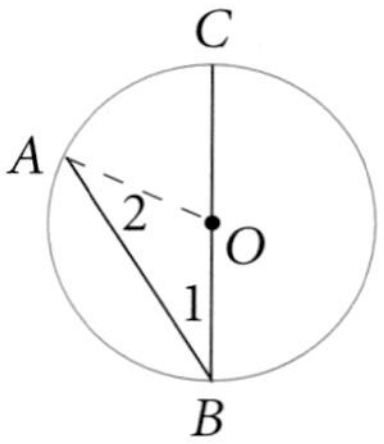

Draw radius $\overline{OA}$. By the Exterior Angle Theorem, $m\angle 1 + m\angle 2 = m\angle AOC$. Since $OA = OB$, $\triangle AOB$ is isosceles and $m\angle 1 = m\angle 2$. Therefore, by substitution, $2(m\angle 1) = m\angle AOC$, or $m\angle 1 = \frac{1}{2}m\angle AOC$. Since $m\angle AOC = m\widehat{AC}$, $m\angle 1 = \frac{1}{2}m\widehat{AC}$ by substitution.

Example 1

Find the values of a and b in the diagram at the right.

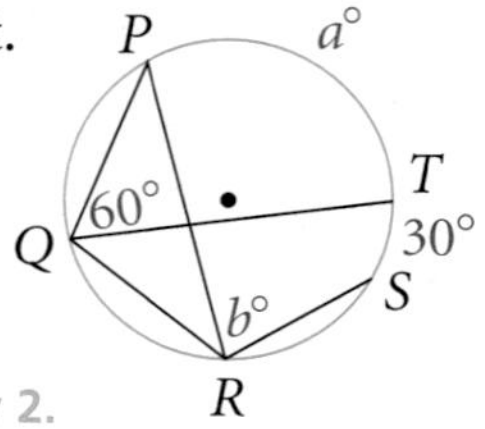

Use the Inscribed Angle Theorem.

$m\angle PQT = \frac{1}{2}m\widehat{PT}$	
$60 = \frac{1}{2}a$	Substitution
$a = 120$	Multiply each side by 2.
$m\angle PRS = \frac{1}{2}m\widehat{PS}$	Inscribed Angle Theorem
$m\angle PRS = \frac{1}{2}(m\widehat{PT} + m\widehat{TS})$	Arc Addition Postulate
$b = \frac{1}{2}(120 + 30)$	Substitution
$b = \frac{1}{2}(150) = 75$	Simplify.

7. **Try This** Find $m\angle PQR$ if $m\widehat{PTR} = 230$.

In the Work Together, you investigated the first two of the following three corollaries that follow from the Inscribed Angle Theorem. You will justify the corollaries in Exercises 14, 20, and 21.

Corollary 1 Two inscribed angles that intercept the same arc are congruent.

Corollary 2 An angle inscribed in a semicircle is a right angle.

Corollary 3 The opposite angles of a quadrilateral inscribed in a circle are supplementary.

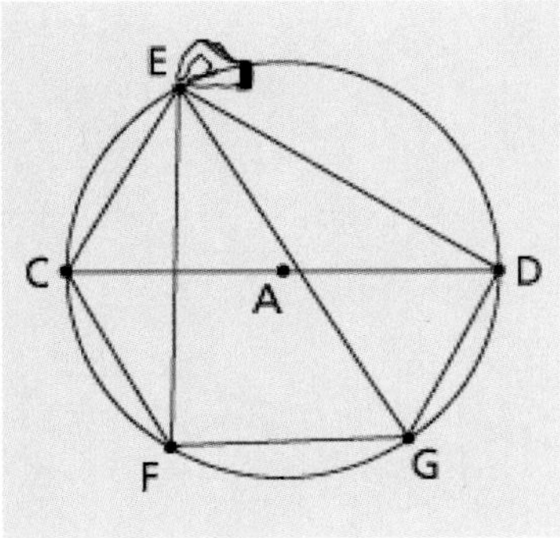

Example 2

Technology Dolores constructed ⊙A and the chords shown with geometry software.

a. As Dolores moves E on $\overset{\frown}{CED}$ between C and D, which pairs of inscribed angles will remain congruent to each other?

b. Which inscribed angle will remain a right angle?

c. Which pairs of inscribed angles will remain supplementary in quadrilateral $EFGD$?

a. $\angle CFE \cong \angle CDE$ — Corollary 1
$\angle ECD \cong \angle EGD$

b. $\angle CED$ is a right angle. — Corollary 2

c. $\angle DEF$ and $\angle FGD$ are supplementary. — Corollary 3
$\angle EFG$ and $\angle GDE$ are supplementary.

8. Try This Find the value of each variable.

a.

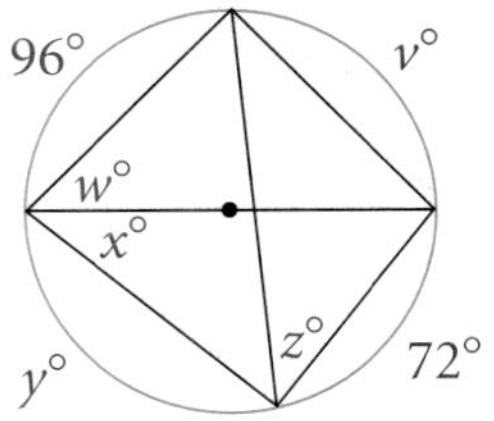

b.

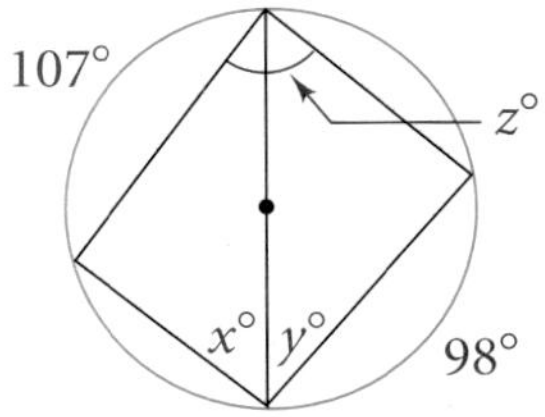

Angles Formed by Tangents and Chords

In the diagram, B and C are fixed points, and point A moves along the circle. From the Inscribed Angle Theorem, you know that as A moves, $m\angle A$ remains the same, and that $m\angle A = \frac{1}{2}m\widehat{BC}$. As the last diagram suggests, this is also true when A and C coincide.

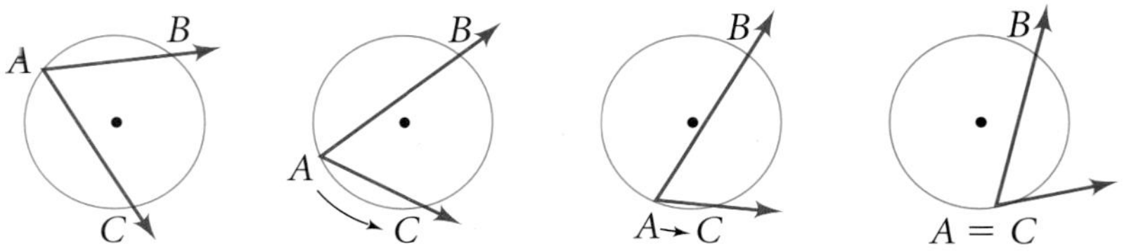

Theorem 12-11

The measure of an angle formed by a chord and a tangent that intersect on a circle is half the measure of the intercepted arc.

$m\angle C = \frac{1}{2}m\widehat{BDC}$

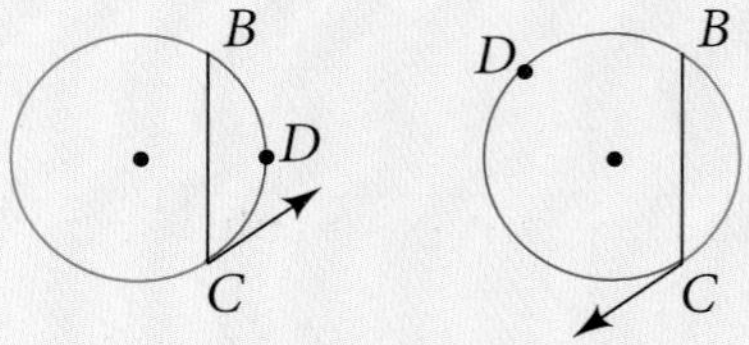

Example 3

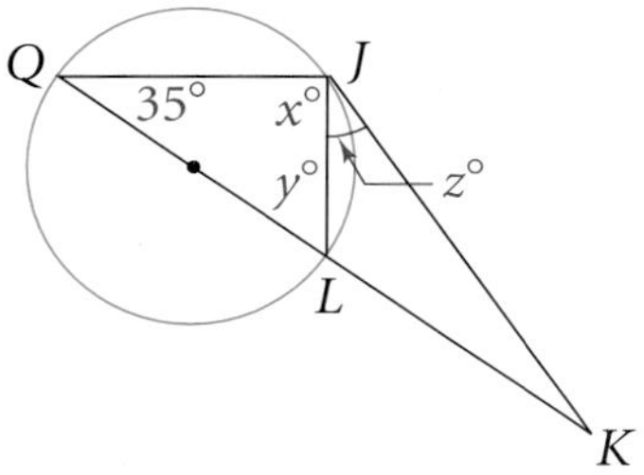

Find the values of x, y, and z.

$x = 90$	An angle inscribed in a semicircle is a rt. ∠.
$y = 90 - 35 = 55$	The acute ∠s of a rt. △ are complementary.
$z = \frac{1}{2}m\widehat{JL}$	Theorem 12-11
$\frac{1}{2}m\widehat{JL} = m\angle Q = 35$	Inscribed Angle Theorem
$z = 35$	Substitution

9. Try This $\overrightarrow{AB}$ is a tangent. $m\angle BAC = 38$. Find $m\angle D$.

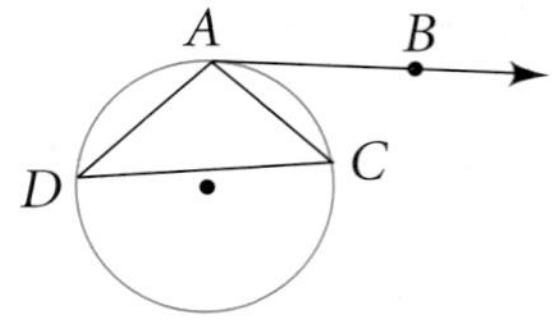

Exercises ON YOUR OWN

Tell whether each polygon is *inscribed in* or *circumscribed about* the circle. If neither, explain why.

1.

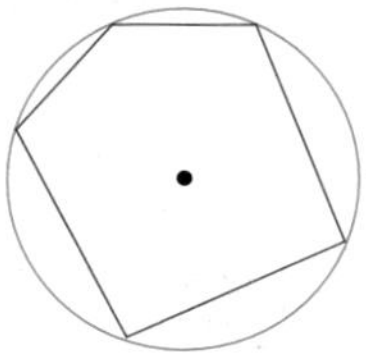

2.

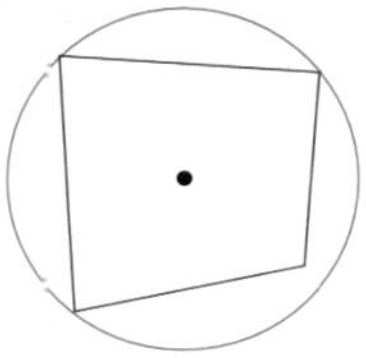

3.

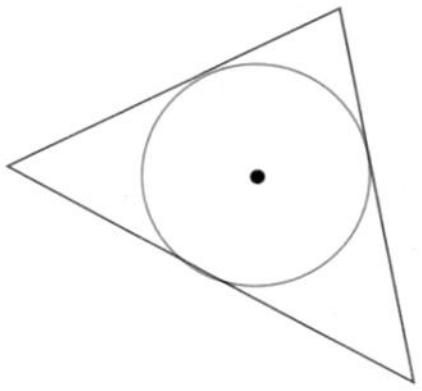

4.

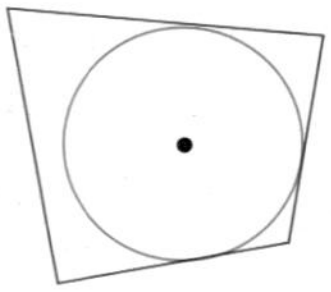

Find the value of each variable.

5.

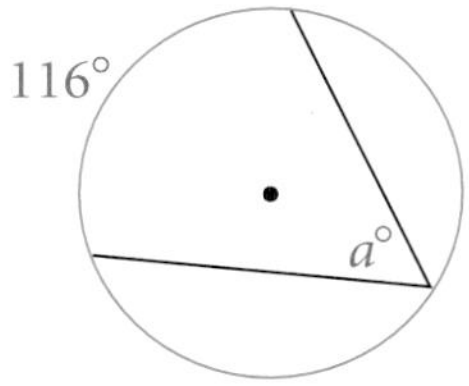

6.

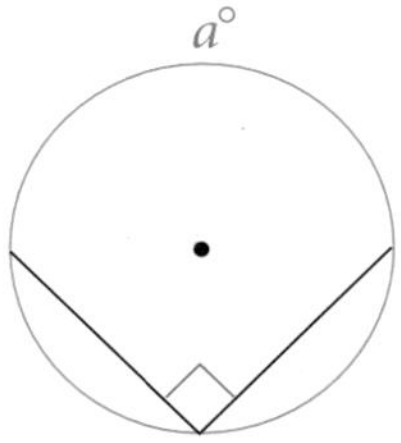

7.

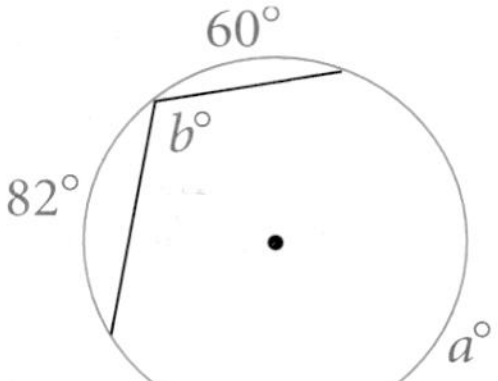

8.

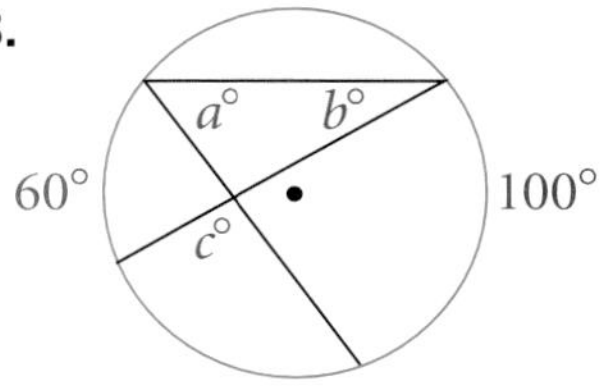

9.

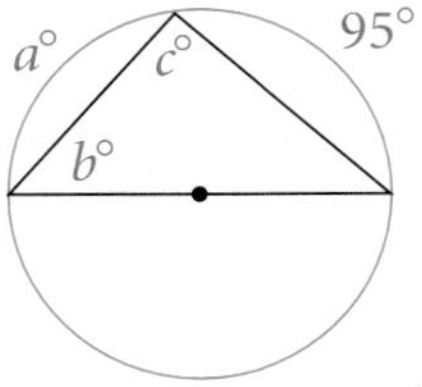

10.

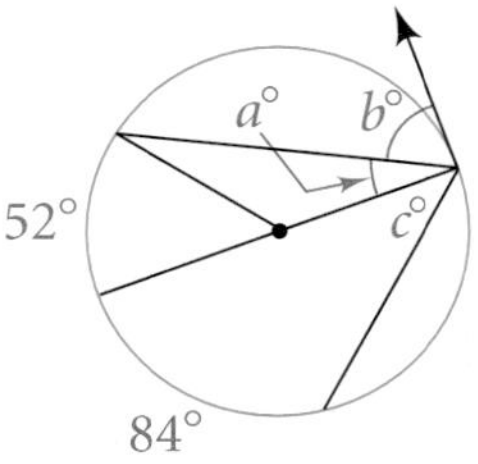

11.

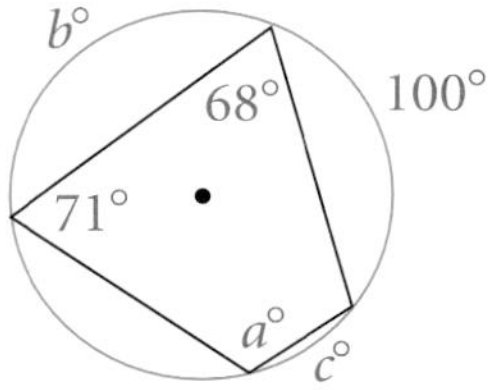

12.

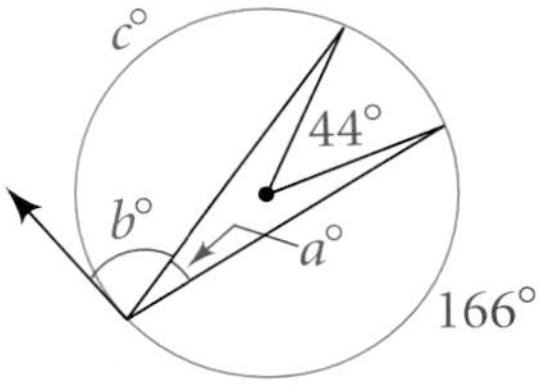

Choose **Write a two-column proof, paragraph proof, or flow proof.**

13. Prove the Inscribed Angle Theorem, Case II.

Given: $\odot O$ with inscribed $\angle ABC$

Prove: $m\angle ABC = \frac{1}{2}m\widehat{AC}$

Plan for Proof: Use the Inscribed Angle Theorem, Case I, to show that $m\angle ABP = \frac{1}{2}m\widehat{AP}$ and $m\angle PBC = \frac{1}{2}m\widehat{PC}$.

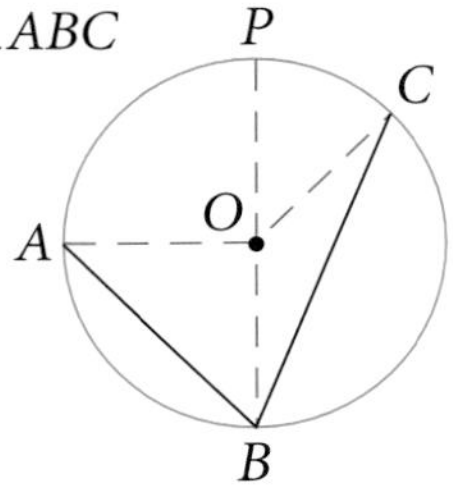

14. Prove Corollary 1 to the Inscribed Angle Theorem.

Given: $\odot O$, $\angle A$ intercepts $\widehat{BC}$, and $\angle D$ intercepts $\widehat{BC}$.

Prove: $\angle A \cong \angle D$

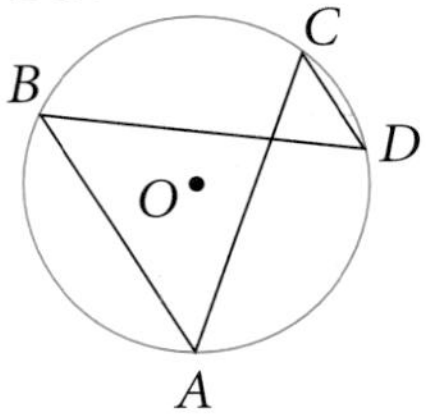

15. **Given:** In $\odot O$, $m\widehat{AD} = m\widehat{BC}$.

Prove: $\triangle ABD \cong \triangle BAC$

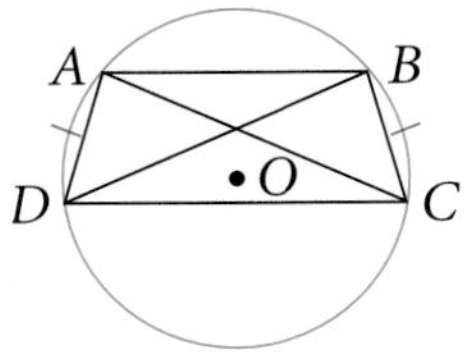

16. **Given:** In $\odot O$, $m\widehat{BC} = m\widehat{ED}$.

Prove: $\triangle EFB \sim \triangle CFD$

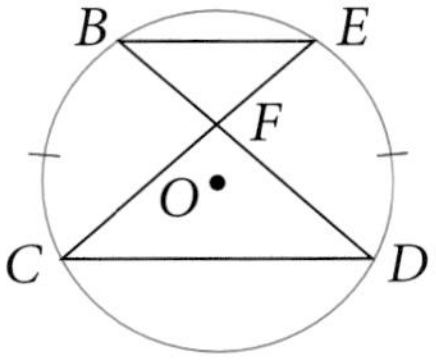

17. a. *Critical Thinking* A parallelogram is inscribed in a circle. What kind of parallelogram must it be?

b. Explain why your statement in part (a) is true.

18. Copy the diagram at the right on your paper. Draw chord $\overline{RQ}$. Explain why $m\widehat{PR} = m\widehat{QS}$.

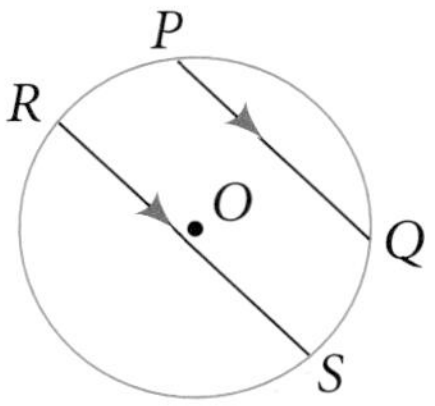

19. **Constructions** Explain how you would construct the tangent to a circle at any point on the circle.

20. **Logical Reasoning** In the diagram at the right, $\angle A$ is inscribed in a semicircle. Write a justification of Corollary 2 of the Inscribed Angle Theorem.

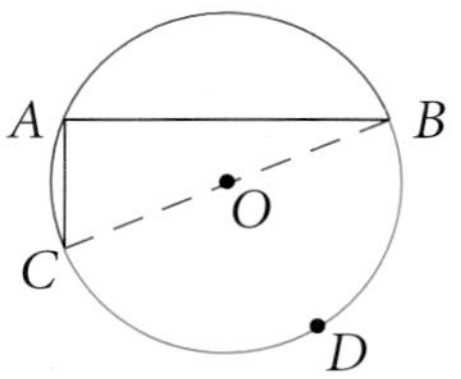

21. In the diagram at the right, quadrilateral $ABCD$ is inscribed in $\odot O$.

a. Find the sum of the measures of the arcs intercepted by $\angle A$ and $\angle C$.

b. **Logical Reasoning** Explain why Corollary 3 to the Inscribed Angle Theorem must be true.

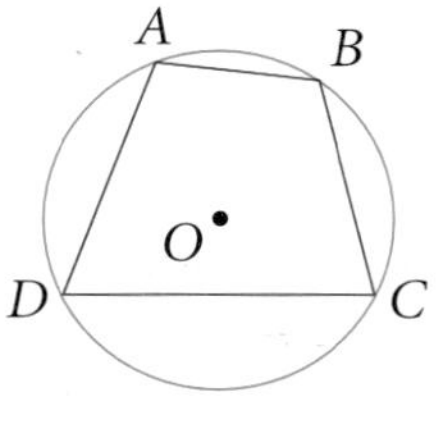

22. **Critical Thinking** Decide whether the following statements are true or false. Give a counterexample for each false statement.

a. If two angles inscribed in a circle are congruent, then they intercept the same arc.

b. If a right angle is an inscribed angle, then it is inscribed in a semicircle.

c. A circle can always be circumscribed about a quadrilateral whose opposite angles are supplementary.

23. **Constructions** The diagrams below show the construction of a tangent to a circle from a point outside the circle. Explain why $\overleftrightarrow{BC}$ must be tangent to $\odot A$. (*Hint:* Copy the third diagram and draw $\overline{AC}$.)

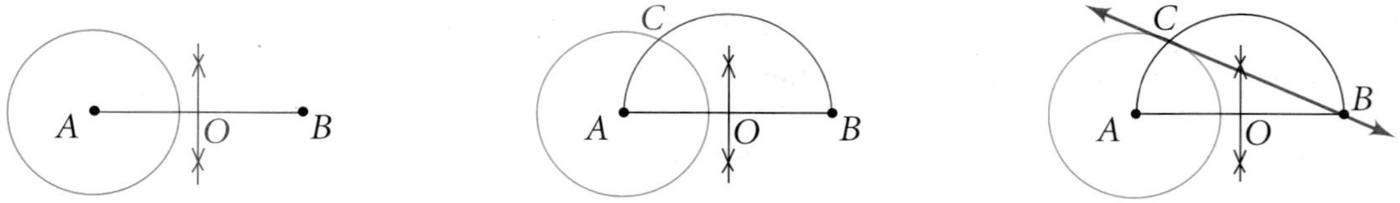

Given: $\odot A$ and point B
Construct the midpoint of $\overline{AB}$.
Label the point O.

Construct a semicircle with radius OA and center O. Label its intersection with $\odot A$ as C.

Draw $\overleftrightarrow{BC}$.

Find each indicated measure for $\odot O$.

24. **a.** $m\overset{\frown}{BC}$
b. $m\angle B$
c. $m\angle C$
d. $m\overset{\frown}{AB}$

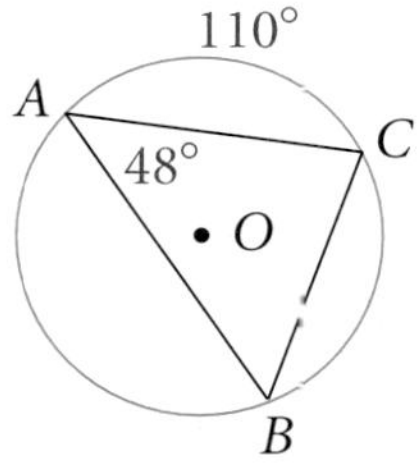

25. **a.** $m\overset{\frown}{EA}$
b. $m\overset{\frown}{BC}$
c. $m\angle A$
d. $m\angle B$
e. $m\angle BCD$
f. $m\angle D$

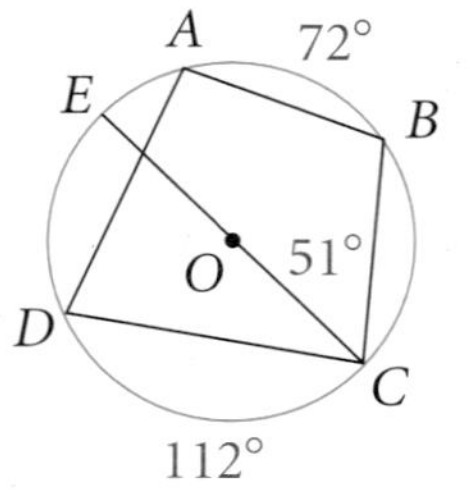

26. **a.** $m\angle A$
b. $m\overset{\frown}{CE}$
c. $m\angle C$
d. $m\angle D$
e. $m\angle ABE$

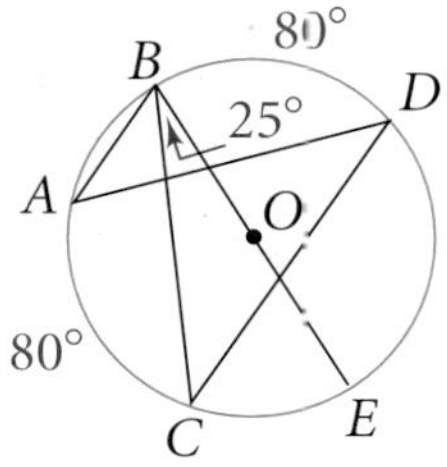

27. **a.** $m\overset{\frown}{DC}$
b. $m\overset{\frown}{BD}$
c. $m\angle BCD$
d. $m\angle BDA$
e. $m\angle ABD$
f. $m\angle A$
g. $m\angle CDE$

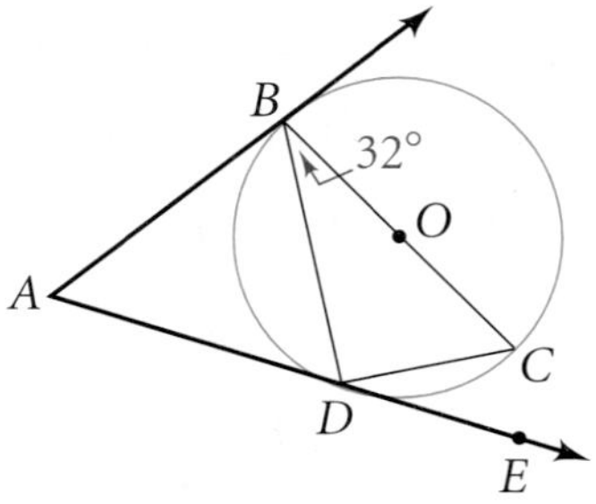

28. a. **Motion Pictures** A director wants to capture the same scene on film from three different viewpoints. Explain why the cameras in the positions shown will record the same scene.

b. **Critical Thinking** Will the scenes look the same when they are shown? Explain.

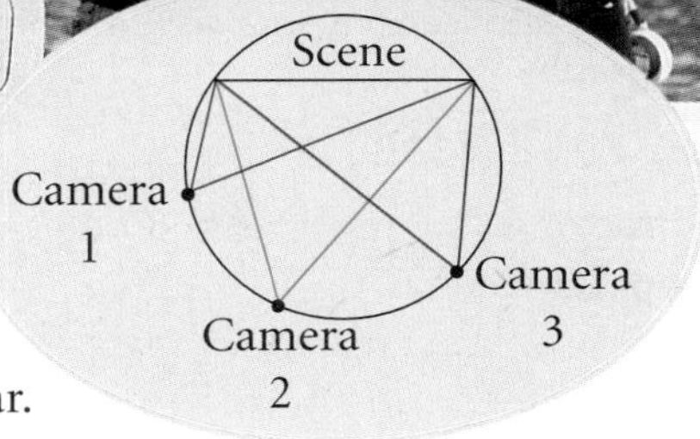

29. a. **Patterns** Sketch a trapezoid inscribed in a circle. Repeat several times using different circles.

b. Make a **conjecture** about the kind of trapezoid that can be inscribed in a circle.

30. **Writing** Explain how to construct a circle circumscribed about a right triangle.

31. **Mental Math** Draw a circle and inscribe an irregular five-pointed star. Find the sum of the measures of the five inscribed angles.

32. a. **Constructions** Draw two segments. Label their lengths x and y. Construct the geometric mean of x and y by following these steps.

Step 1: On a line, construct $\overline{RS}$ with length x and $\overline{ST}$ with length y such that S is between R and T.

Step 2: Bisect $\overline{RT}$. Label the midpoint O. Use the length RO to construct a semicircle with diameter $\overline{RT}$.

Step 3: Construct the perpendicular to $\overline{RT}$ at S. Label the point of intersection of the perpendicular with the semicircle as Q.

b. What kind of triangle is $\triangle RQT$?

c. **Critical Thinking** Why is QS the geometric mean of x and y?

Exercises MIXED REVIEW

Find the length of the midsegment of a trapezoid with the given bases.

33. $b_1 = 3$ cm, $b_2 = 7$ cm **34.** $b_1 = 8$ in., $b_2 = 13$ in.

35. What is the measure of an interior angle of a regular pentagon?

SELF ASSESSMENT

FOR YOUR JOURNAL

Describe the relationships between arcs and angles on a circle.

Getting Ready for Lesson 12-5

O is the center of each circle. Find the measure of $\overset{\frown}{AB}$.

36.

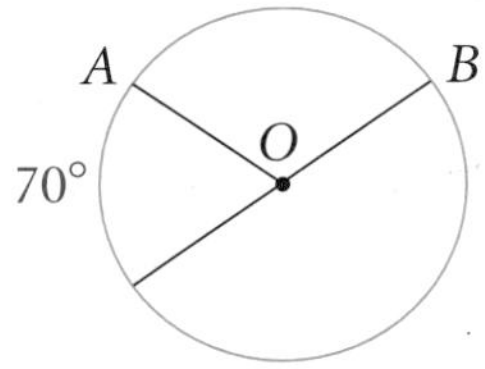

37.

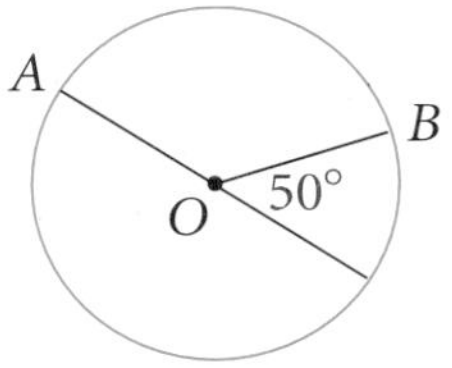

38.

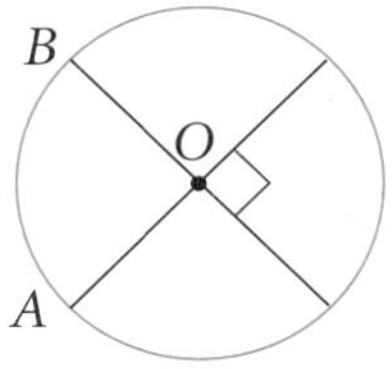

What You'll Learn

- Finding the measures of angles formed by chords, secants, and tangents

...And Why

To use measures of angles in real-world applications such as photography

What You'll Need

- protractor
- compass

12-5 Angles Formed by Chords, Secants, and Tangents

WORK TOGETHER

- Have each member of your group draw four circles. Label the centers M, N, P, and Q. Draw segments intersecting inside and outside the circles as shown below.

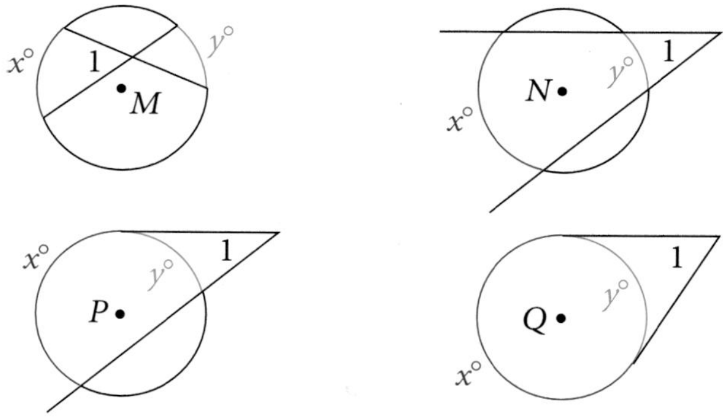

TECHNOLOGY HINT

The Work Together could be done using geometry software.

1. Copy and complete the chart below. In each diagram, use a protractor to measure $\angle 1$, and find $x°$ and $y°$ by measuring central angles.

Circle	$m\angle 1$	x	y	$x + y$	$x - y$
$\odot M$					
$\odot N$					
$\odot P$					
$\odot Q$					

2. In $\odot M$, what do you notice about $m\angle 1$ and the sum or the difference of the measures of the arcs? Compare results within your group.

3. In $\odot N$, $\odot P$, and $\odot Q$, what do you notice about $m\angle 1$ and the sum or the difference of the measures of the arcs? Compare results within your group.

THINK AND DISCUSS

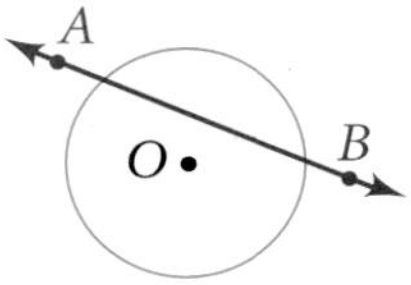

A **secant** is a line that intersects a circle at two points. $\overrightarrow{AB}$ is a secant ray, and $\overline{AB}$ is a secant segment. The word *secant* may refer to a secant line, secant ray, or secant segment.

4. Does a secant always contain a chord of a circle? Explain.

The diagrams below show arcs intercepted by chords, secants, and tangents.

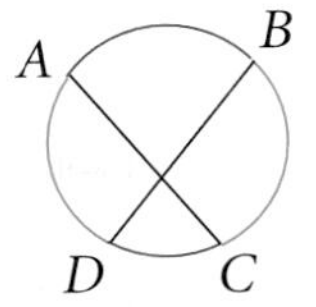

Chords $\overline{AC}$ and $\overline{BD}$ intercept two pairs of opposite arcs: $\overset{\frown}{AB}$ and $\overset{\frown}{DC}$, $\overset{\frown}{AD}$ and $\overset{\frown}{BC}$.

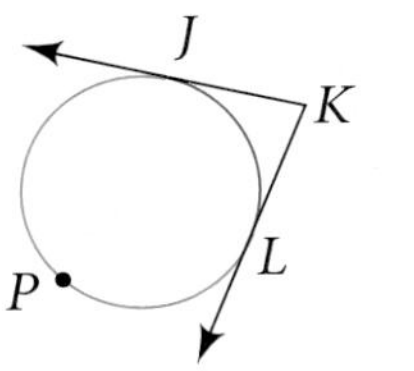

Tangents $\overrightarrow{KJ}$ and $\overrightarrow{KL}$ intercept $\overset{\frown}{JL}$ and $\overset{\frown}{JPL}$.

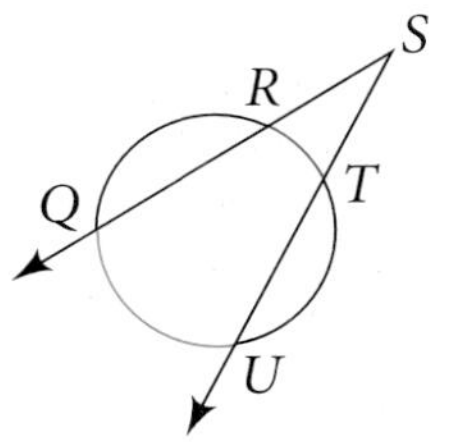

Secants $\overrightarrow{SQ}$ and $\overrightarrow{SU}$ intercept $\overset{\frown}{RT}$ and $\overset{\frown}{QU}$.

Your observations in the Work Together suggest the following theorems.

Theorem 12-12

The measure of an angle formed by two chords that intersect inside a circle is half the sum of the measures of the intercepted arcs.

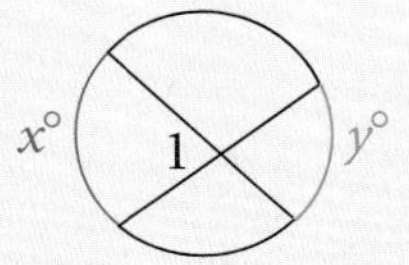

$m\angle 1 = \frac{1}{2}(x + y)$

Theorem 12-13

The measure of an angle formed by two secants, two tangents, or a secant and a tangent drawn from a point outside the circle is half the difference of the measures of the intercepted arcs.

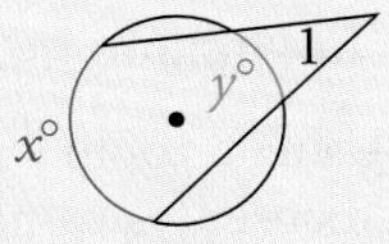

$m\angle 1 = \frac{1}{2}(x - y)$

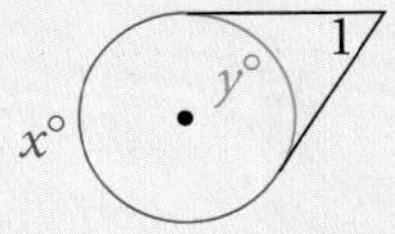

$m\angle 1 = \frac{1}{2}(x - y)$

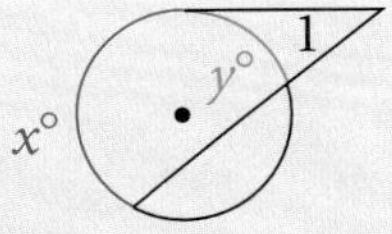

$m\angle 1 = \frac{1}{2}(x - y)$

Proof of Theorem 12-12

Given: $\odot O$ with intersecting chords $\overline{AC}$ and $\overline{BD}$

Prove: $m\angle 1 = \frac{1}{2}(m\overset{\frown}{AB} + m\overset{\frown}{CD})$

Begin by drawing $\overline{AD}$.

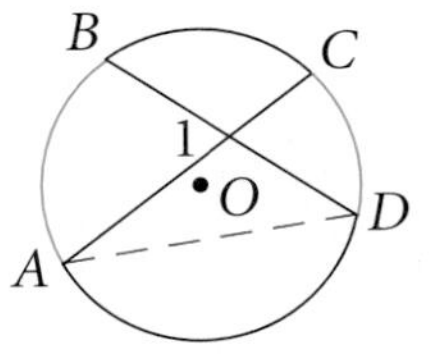

Statements	Reasons
1. $m\angle D = \frac{1}{2}m\overset{\frown}{AB}$, $m\angle A = \frac{1}{2}m\overset{\frown}{CD}$	**1.** Inscribed Angle Theorem
2. $m\angle 1 = m\angle D + m\angle A$	**2.** Exterior Angle Theorem
3. $m\angle 1 = \frac{1}{2}m\overset{\frown}{AB} + \frac{1}{2}m\overset{\frown}{CD}$	**3.** Substitution
4. $m\angle 1 = \frac{1}{2}(m\overset{\frown}{AB} + m\overset{\frown}{CD})$	**4.** Distributive Property

QUICK REVIEW

The Exterior Angle Theorem says that the measure of each exterior angle of a triangle equals the sum of the measures of the two remote interior angles.

Example 1

Find the value of each variable.

a.

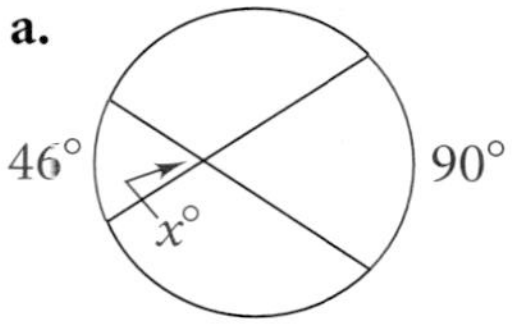

b.

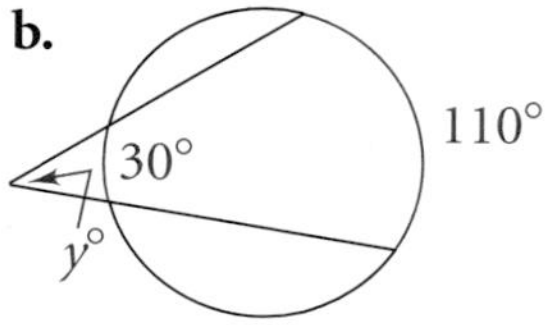

c.

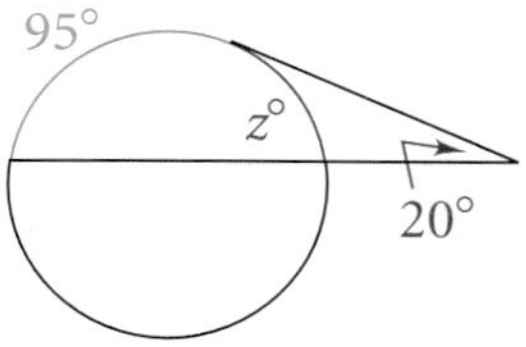

a. $x = \frac{1}{2}(46 + 90)$

$x = 68$

b. $y = \frac{1}{2}(110 - 30)$

$y = 40$

c. $20 = \frac{1}{2}(95 - z)$

$40 = 95 - z$

$z = 55$

5. Try This Find the value of each variable.

a.

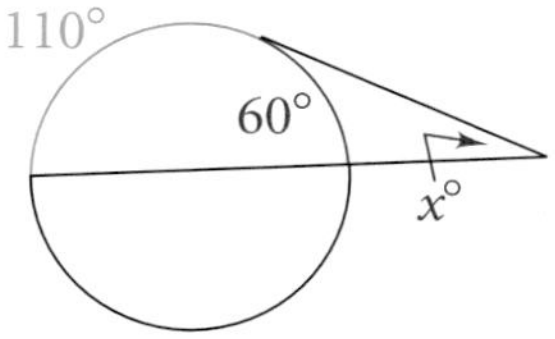

b.

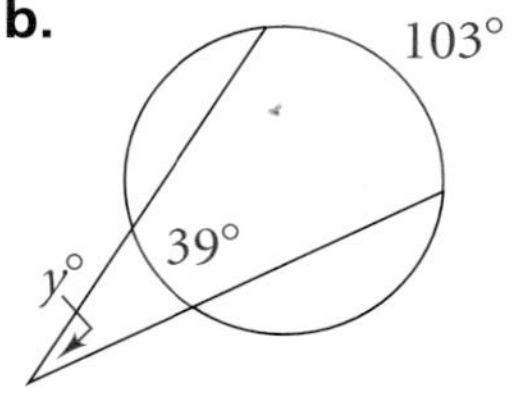

c.

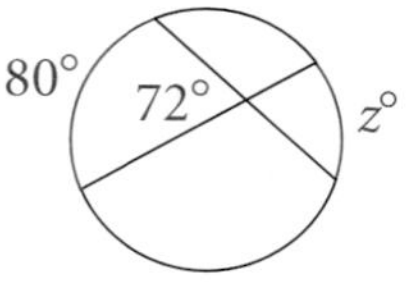

You can find the measures of arcs of a circle intercepted by tangents if you know the measure of the angle formed by the tangents.

Example 2 — Relating to the Real World

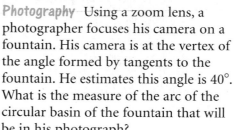

Photography Using a zoom lens, a photographer focuses his camera on a fountain. His camera is at the vertex of the angle formed by tangents to the fountain. He estimates this angle is 40°. What is the measure of the arc of the circular basin of the fountain that will be in his photograph?

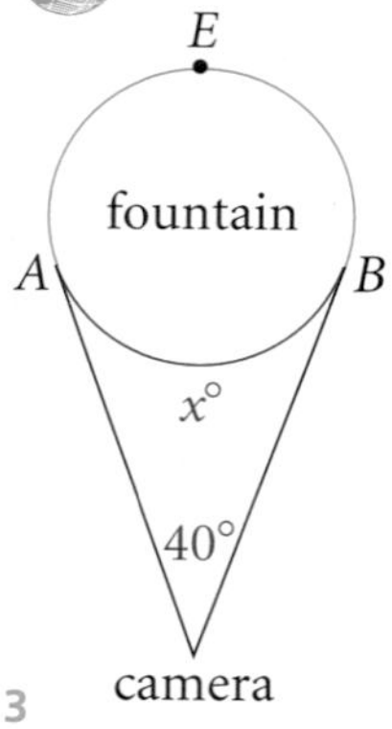

Let $m\widehat{AB} = x$. Then $m\widehat{AEB} = 360 - x$.

$40 = \frac{1}{2}(m\widehat{AEB} - m\widehat{AB})$	Theorem 12-13
$40 = \frac{1}{2}[(360 - x) - x]$	Substitution
$40 = \frac{1}{2}(360 - 2x)$	Simplify.
$40 = 180 - x$	Distributive Property
$x = 140$	Solve for x.

An arc of 140° will be in the photograph.

6. In Example 2, what is $m\widehat{AEB}$?

7. Critical Thinking To capture a greater arc of the circular basin in the photo, should the photographer step closer to or farther away from the fountain? Explain.

8. What should the angle at the camera be in order to photograph a 70° arc of the basin?

Exercises ON YOUR OWN

Assume that lines that appear tangent are tangent. Find the value of each variable.

1.
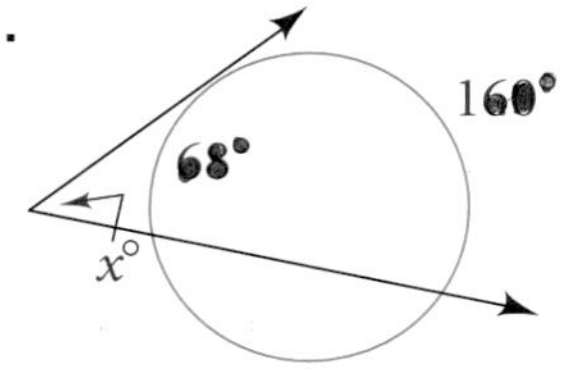

2.
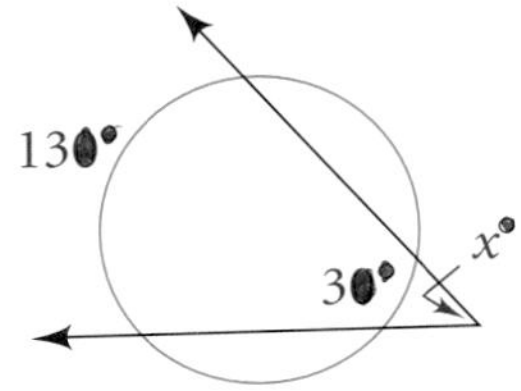

3.
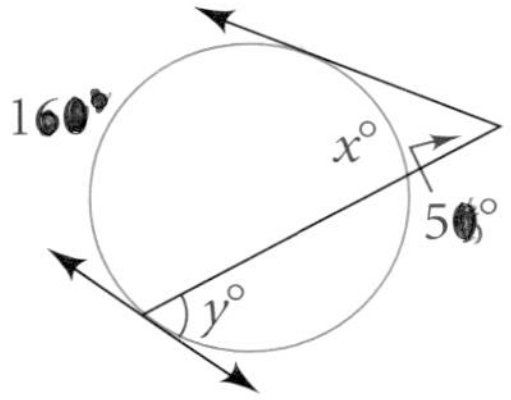

4.
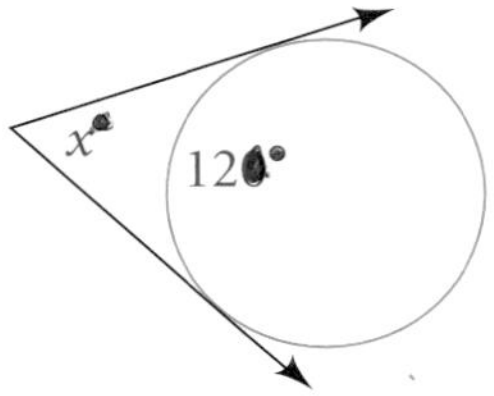

5.
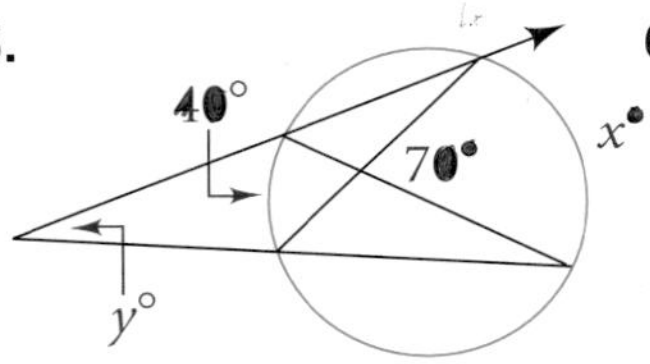

6.
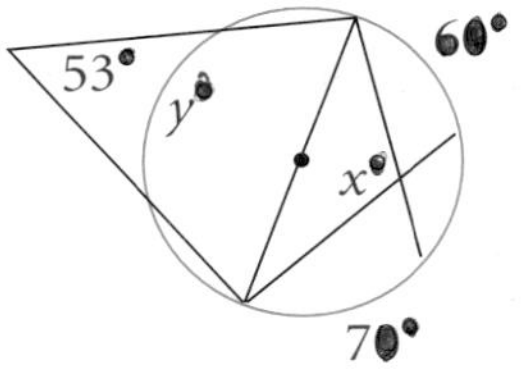

7.
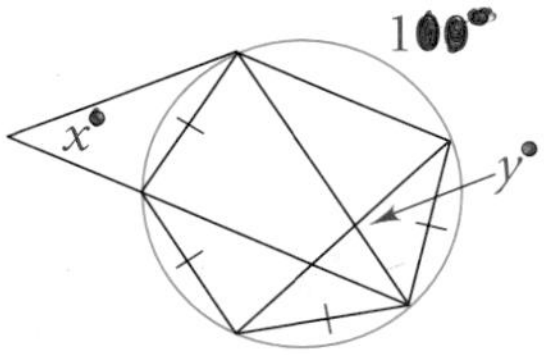

8.
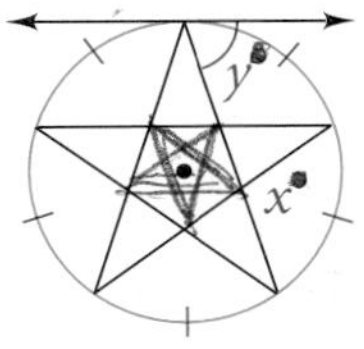

9. a. Space Travel An astronaut directly over the equator estimates that the angle formed by the two tangents to the equator is 20°. What arc of the equator is visible to the astronaut?

b. The radius of Earth is about 3960 mi. About how far is the spaceship from Earth?

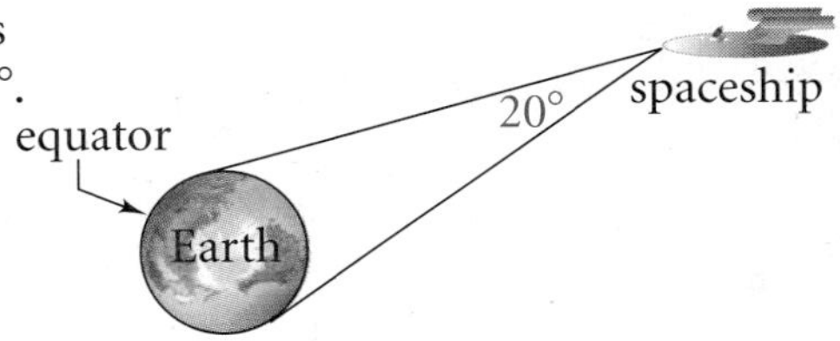

10. Open-ended Create a figure using tangents and secants. Find the measures of the angles and arcs in your figure.

11. Standardized Test Prep $\overline{AC}$ is tangent to $\odot O$, $m\widehat{AD} = 90$, and the center of $\odot O$ is inside $\angle ABC$. What is $m\angle C$?

A. less than 45
B. 45
C. between 45 and 60
D. between 60 and 90
E. none of the above

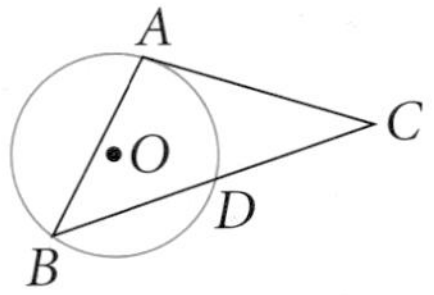

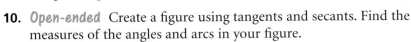

12. Writing Two chords intersect in a circle forming right angles. What is the sum of the measures of a pair of opposite arcs? **Justify** your answer.

13. a. Algebra $\overline{CA}$ and $\overline{CB}$ are tangents to $\odot O$. Write an expression for $m\widehat{ADB}$ in terms of x.

b. Algebra Write an expression for $m\angle C$ in terms of x.

c. Algebra Write an expression for $m\widehat{AB}$ in terms of y.

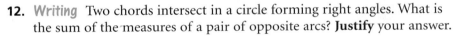

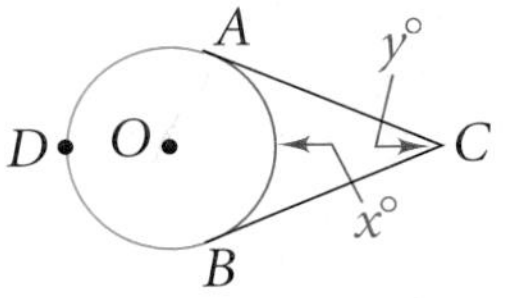

14. Write a Plan for Proof for Theorem 12-13 as it applies to two secants drawn from a point outside a circle.

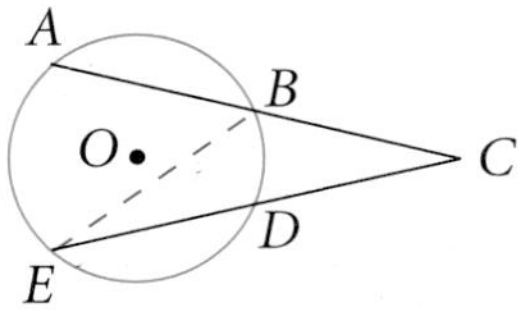

Given: $\odot O$ with secants $\overline{CA}$ and $\overline{CE}$ intersecting at C.
Prove: $m\angle ACE = \frac{1}{2}(m\widehat{AE} - m\widehat{BD})$

15. **Critical Thinking** A navigational map shows that the waters near lighthouses A and B are dangerous. The unsafe area is within $\widehat{AXB}$, a 300° arc.

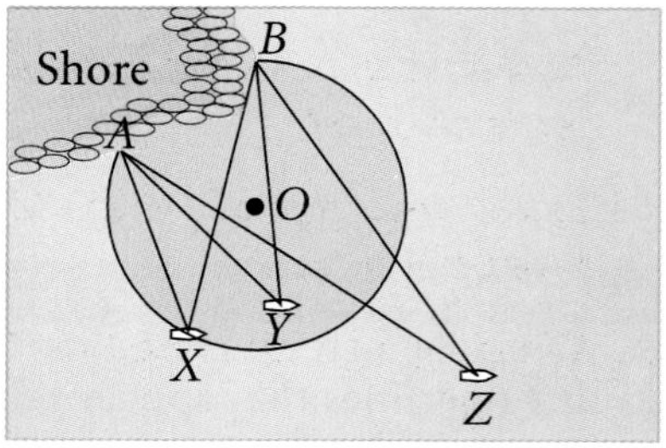

a. X represents locations of a ship on $\odot O$. Y represents locations of a ship inside $\odot O$. Z represents locations of a ship outside $\odot O$. What is the measure of $\angle X$? What is the possible range of measures of $\angle Y$? of $\angle Z$?

b. Using the angle a ship makes with the lighthouses, how can a ship's navigator be sure the ship is in safe waters?

Find the value of each variable.

16.

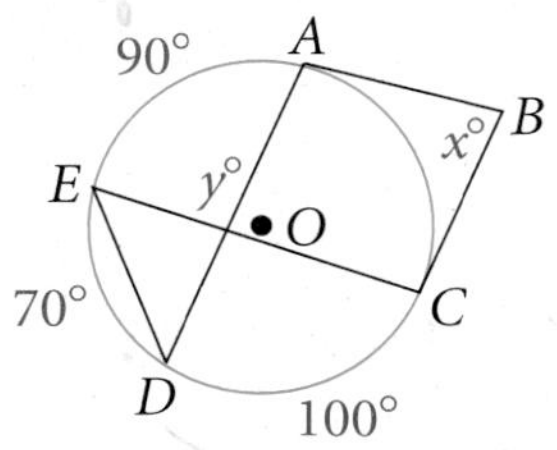

17.

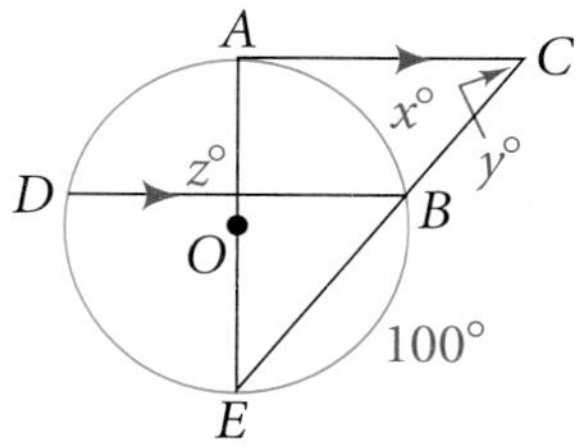

18. The angles of a quadrilateral measure 85, 76, 94, and 105. A circle is inscribed in the quadrilateral. What are the measures of the arcs between each two consecutive points of tangency?

Chapter Project — Find Out by Constructing

Follow the steps below to create a pattern commonly found in 14th-century Islamic art and furniture.

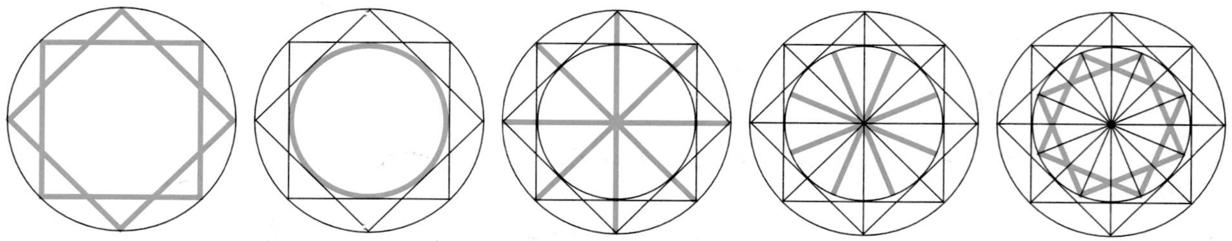

Step 1: In a circle, inscribe two squares rotated 45° from each other.

Step 2: Inscribe a circle in the squares.

Step 3: In each square, draw the diagonals.

Step 4: Bisect the central angles of the smaller circle.

Step 5: Inscribe two squares as shown. Then color to make the desired design.

Exercises MIXED REVIEW

Data Analysis **Use the graph showing the number of living siblings, including step-siblings and half-siblings, adults of different ages have.**

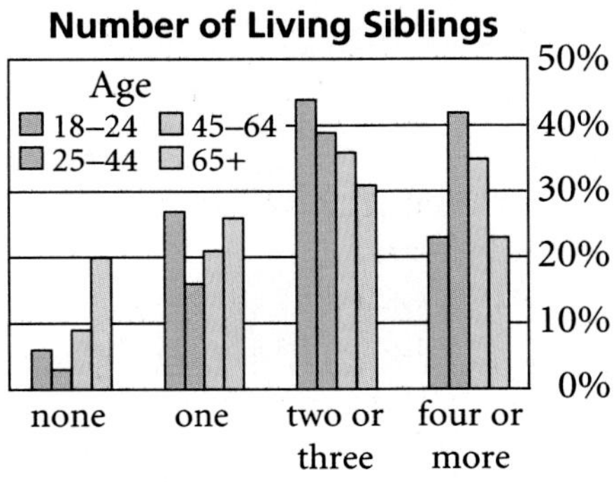

Source: *American Demographics*

19. About what percent of 45–64-year-olds have four or more siblings?

20. About what percent of 25–44-year-olds have two or more siblings?

21. Which age group is least likely to have only one sibling?

Getting Ready for Lesson 12-6

Each pair of figures is similar. Complete each statement.

22.

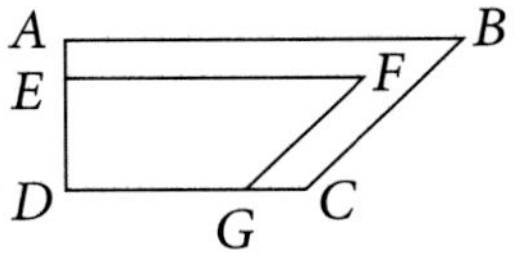

$\frac{AB}{EF} = \frac{\blacksquare}{FG}$

23.

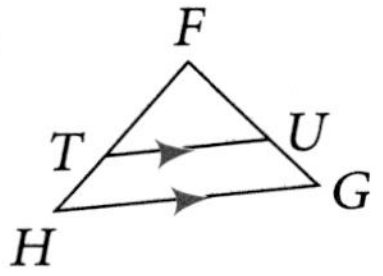

$\frac{TF}{\blacksquare} = \frac{\blacksquare}{GH}$

SELF ASSESSMENT

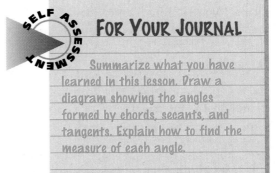

FOR YOUR JOURNAL

Summarize what you have learned in this lesson. Draw a diagram showing the angles formed by chords, secants, and tangents. Explain how to find the measure of each angle.

Exercises CHECKPOINT

Assume that lines that appear tangent are tangent. Find the value of each variable.

1.

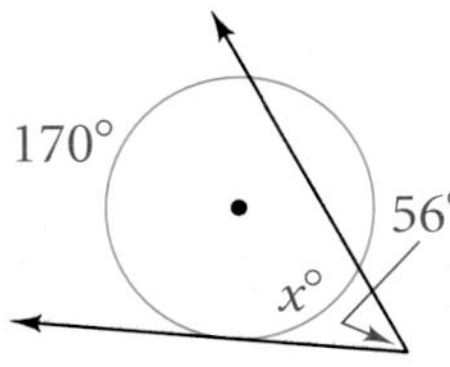

2.

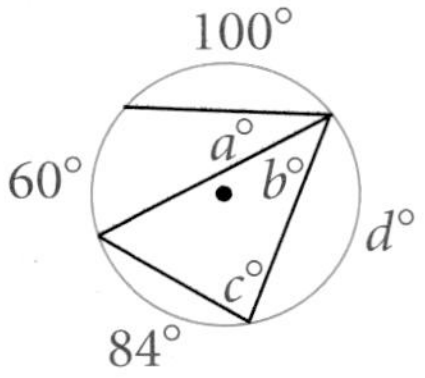

3.

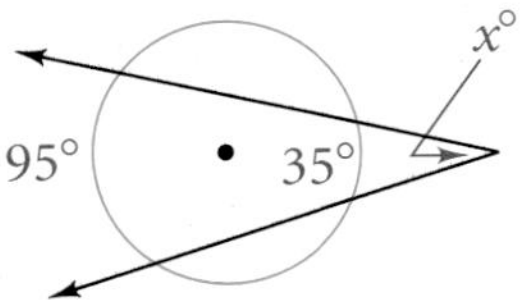

4.

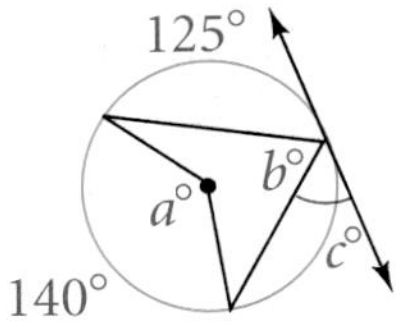

5. **Writing** Explain the difference between a chord and a secant. Include a diagram.

6. **Standardized Test Prep** In the circle at the right, which measure is greatest?

A. $m\angle C$ **B.** $m\angle EBD$ **C.** $m\angle ABE$
D. $m\widehat{BF}$ **E.** $m\angle BDF$

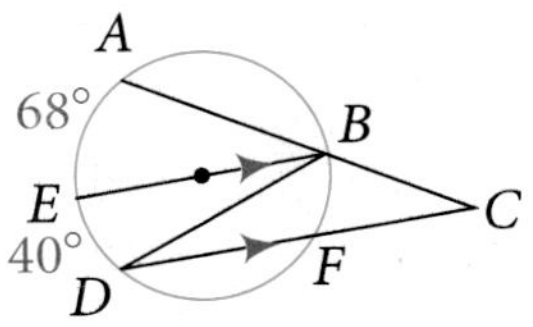

Math ToolboX

Technology

Exploring Chords and Secants

Before Lesson 12-6

Work in pairs or small groups.

Construct

Construct a circle with center A and two chords $\overline{BC}$ and $\overline{DE}$ that intersect each other at F.

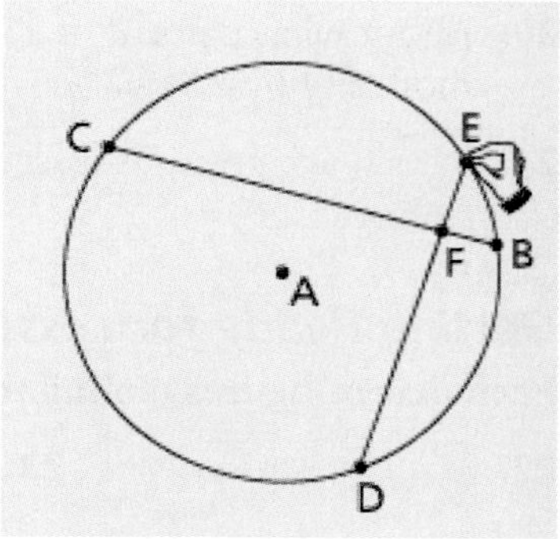

Investigate

Measure the segments $\overline{BF}$, $\overline{FC}$, $\overline{EF}$, and $\overline{FD}$. Use the calculator program of your software to find the products $BF \cdot FC$ and $EF \cdot FD$. Manipulate your construction and observe the products. What do you discover?

Conjecture

Make a conjecture about the products of the lengths of the segments formed by the intersection of two chords in a circle.

Construct

Construct another circle and two secants that intersect in a point outside the circle. Label your construction as shown in the diagram.

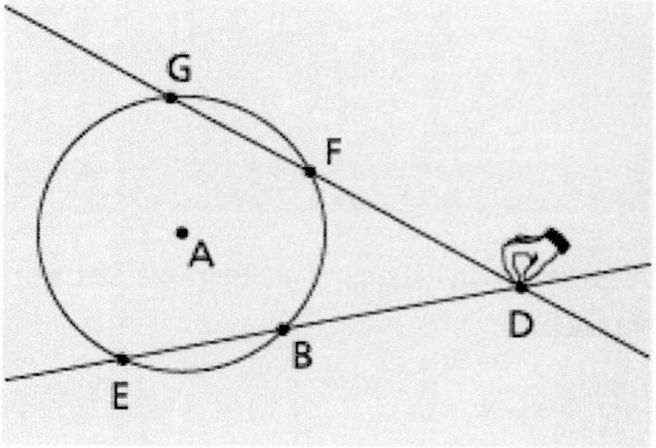

Investigate

Measure the segments $\overline{DG}$, $\overline{DF}$, $\overline{DE}$, and $\overline{DB}$. Calculate the products $DG \cdot DF$ and $DE \cdot DB$. Manipulate your construction and observe the products. What do you discover?

Conjecture

Make a **conjecture** about the products of the lengths of the segments formed by the intersection of two secants.

Extend

Manipulate your construction so that secant $\overleftrightarrow{DG}$ becomes a tangent. What happens to the products you calculated? Can you explain how this special case is related to the case of two secants?

Connections **Architecture . . . and more**

12-6 Circles and Lengths of Segments

What You'll Learn

- Finding the lengths of segments associated with circles

...And Why

To use the lengths of segments associated with circles in real-world applications such as architecture

What You'll Need

- ruler
- compass

WORK TOGETHER

Have members of your group draw two large circles. Draw and label segments as shown. Note that segment t is tangent to $\odot Q$.

1. In $\odot P$, measure lengths a, b, c, and d. Then find $a \cdot b$ and $c \cdot d$. What do you notice? Compare results within your group.

2. **a.** In $\odot Q$, measure lengths w, x, y, and z. Then find $(w + x)w$ and $(y + z)y$. What do you notice? Compare results within your group.
   b. Measure length t. Find t^2. What do you notice about t^2 and $(w + x)w$ and $(y + z)y$? Compare results within your group.

TECHNOLOGY HINT

The Work Together could be done using geometry software.

THINK AND DISCUSS

In the Work Together, you measured segments associated with circles. The diagram below will give you the vocabulary you need to refer to these segments.

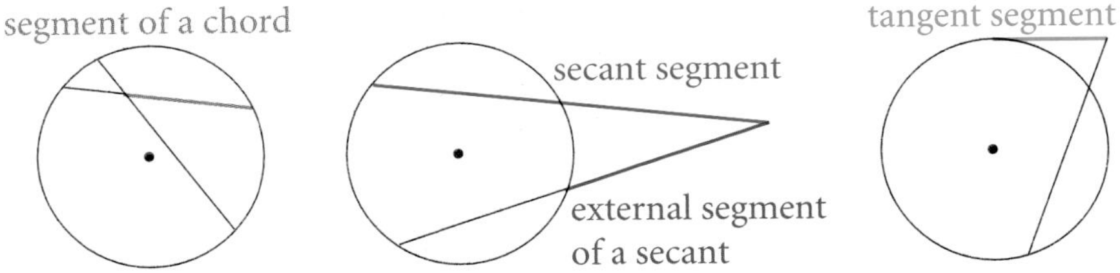

Your observations in the Work Together suggest the following theorems.

Theorem 12-14

If two chords intersect inside a circle, then the product of the lengths of the segments of one chord equals the product of the lengths of the segments of the other chord.

$$a \cdot b = c \cdot d$$

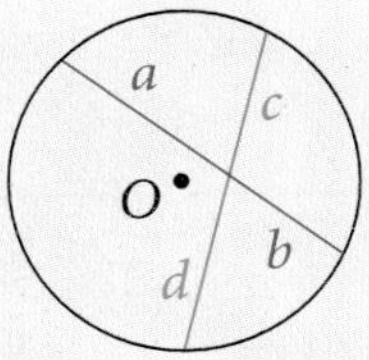

Theorem 12-15 If two secant segments are drawn from a point outside a circle, the product of the lengths of one secant segment and its external segment equals the product of the lengths of the other secant segment and its external segment.

$(w + x)w = (y + z)y$

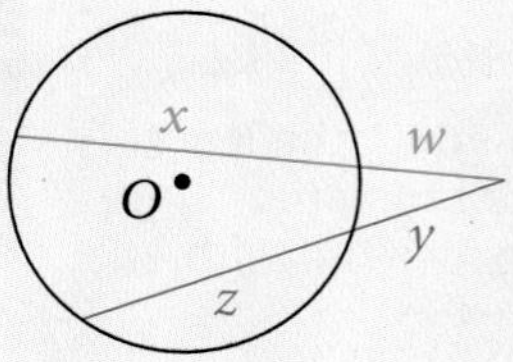

Theorem 12-16 If a tangent and a secant are drawn from a point outside a circle, then the product of the lengths of the secant segment and its external segment equals the square of the length of the tangent segment.

$(y + z)y = t^2$

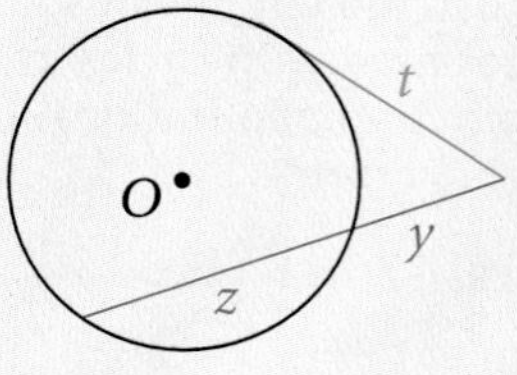

Following is a proof of Theorem 12-14. You will prove Theorems 12-15 and 12-16 in Exercises 12 and 13.

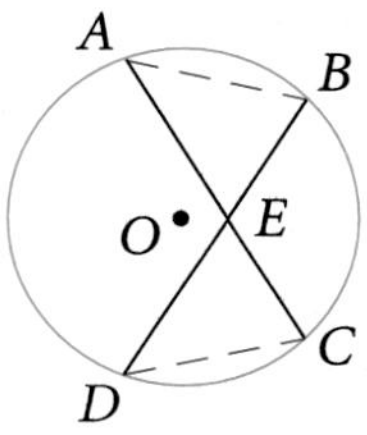

Proof of Theorem 12-14

Given: $\odot O$ with chords $\overline{AC}$ and $\overline{BD}$ intersecting at E

Prove: $AE \cdot EC = DE \cdot EB$

Draw $\overline{AB}$ and $\overline{CD}$. $\angle AEB \cong \angle CED$ because vertical angles are congruent. $\angle C \cong \angle B$ because they are inscribed angles that intercept the same arc. So, $\triangle AEB \sim \triangle DEC$ by the Angle-Angle Similarity Postulate. Because the lengths of corresponding sides of similar figures are proportional, $\frac{AE}{DE} = \frac{EB}{EC}$. Therefore, $AE \cdot EC = DE \cdot EB$.

Example 1

Find the value of each variable.

a.

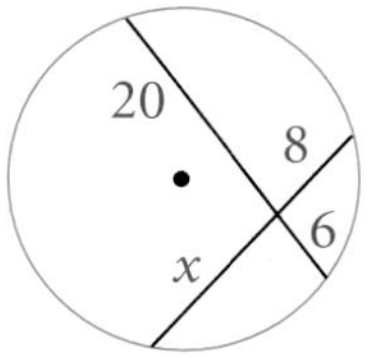

b.

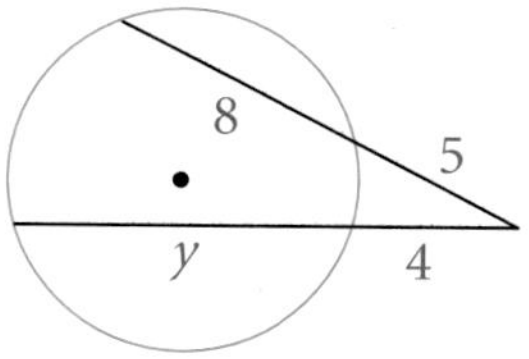

c.

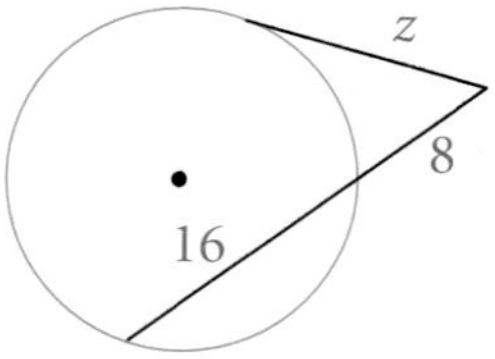

a. $8x = 20 \cdot 6$
$8x = 120$
$x = 15$

b. $(8 + 5)5 = (y + 4)4$
$65 = 4y + 16$
$49 = 4y$
$y \approx 12.3$

c. $(16 + 8)8 = z^2$
$192 = z^2$
$13.9 \approx z$

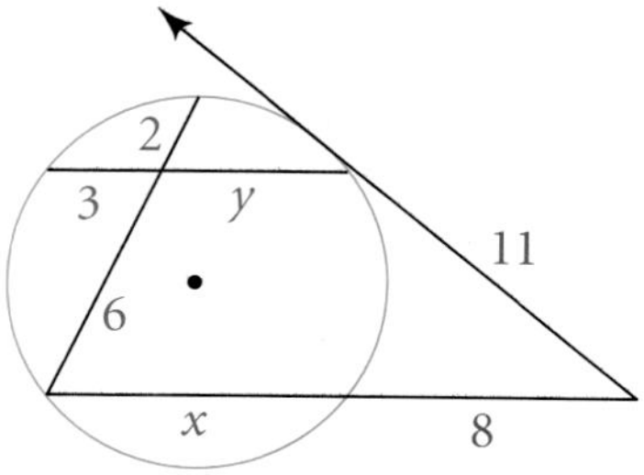

3. Try This Find the value of x and y in the diagram at the left. If your answers are not integers, round them to the nearest tenth.

You can use the theorems in the lesson to solve real-world problems.

Example 2 Relating to the Real World

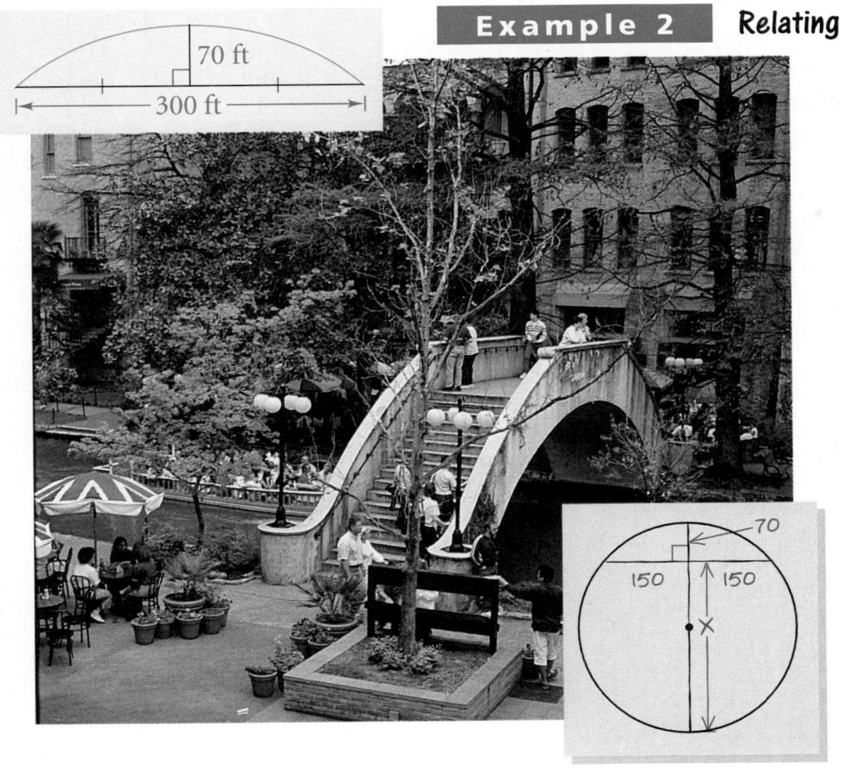

Architecture The arch of a pedestrian bridge that crosses the San Antonio River is an arc of a circle. Find the radius of the circle.

The 300-ft chord is bisected by the perpendicular from the midpoint of the arch. Draw a sketch showing the chords of a circle. Then find the value of the unknown segment of the diameter.

$$70x = 150 \cdot 150$$
$$70x = 22{,}500$$
$$x \approx 321.4$$
$$\text{diameter} \approx 70 + 321.4 = 391.4 \text{ ft}$$
$$\text{radius} \approx 195.7 \text{ ft}$$

The radius is about 196 ft.

4. **Critical Thinking** Explain why the chord that contains the 70-ft segment is a diameter of the circle.

Exercises ON YOUR OWN

Choose **Use mental math, paper and pencil, or a calculator. Find the value of each variable using the given chords, secants, and tangents. If your answer is not an integer, round it to the nearest tenth.**

1.

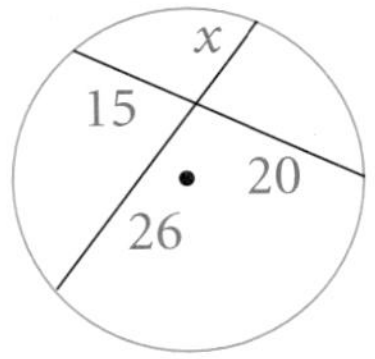

2.

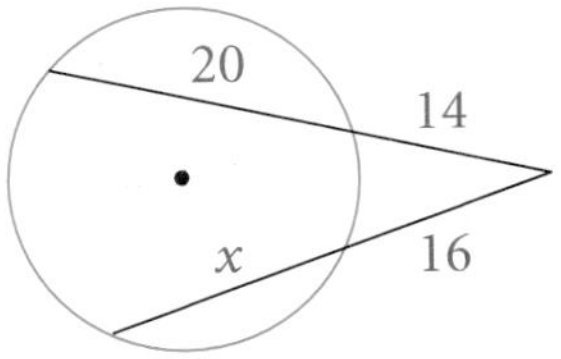

3.

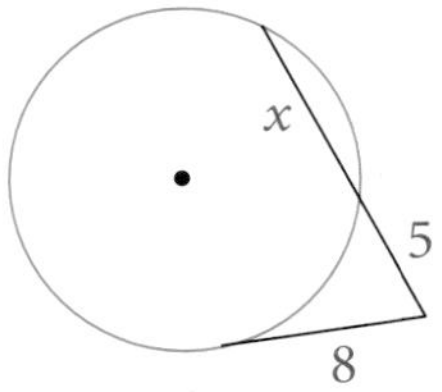

4.

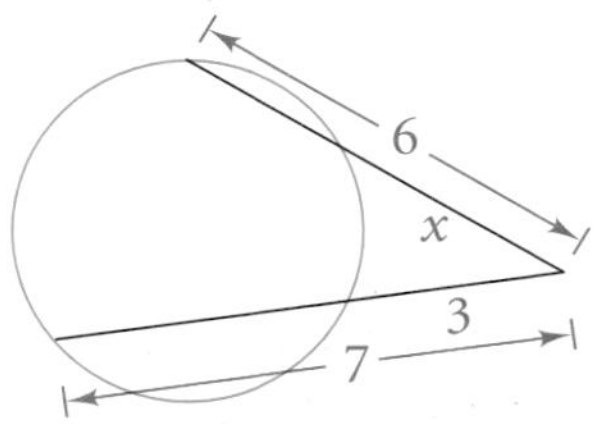

5.

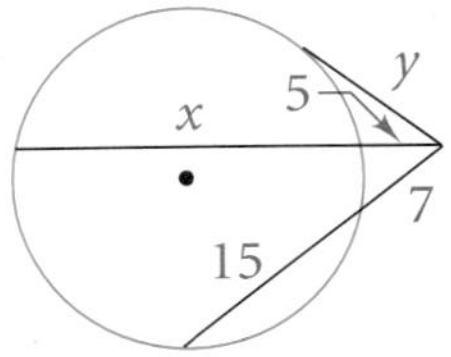

6.

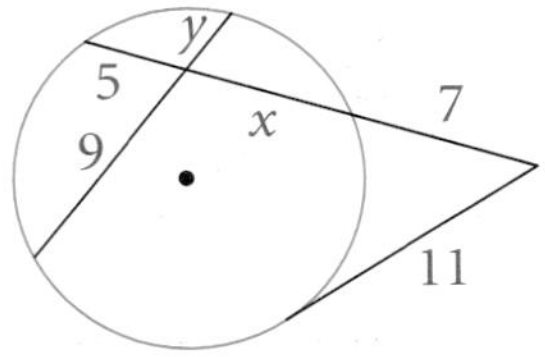

7. *Geology* This natural arch, in Arches National Park, Utah, is an arc of a circle. Find the diameter of the circle.

8. Two chords of a circle intersect. The segments of the first chord are 16 cm and 18 cm long. The lengths of the segments of the second chord are in the ratio 2 : 1. Find the lengths of the segments of the second chord.

Choose **Write a two-column proof, paragraph proof, or flow proof.**

9. Prove Theorem 12-15.

Given: $\odot O$ with secants $\overline{AC}$ and $\overline{BC}$ (Chords $\overline{AE}$ and $\overline{BD}$ are drawn for you.)
Prove: $BC \cdot EC = AC \cdot DC$

Plan for Proof: In order to use the properties of proportions, prove $\triangle AEC \sim \triangle BDC$.

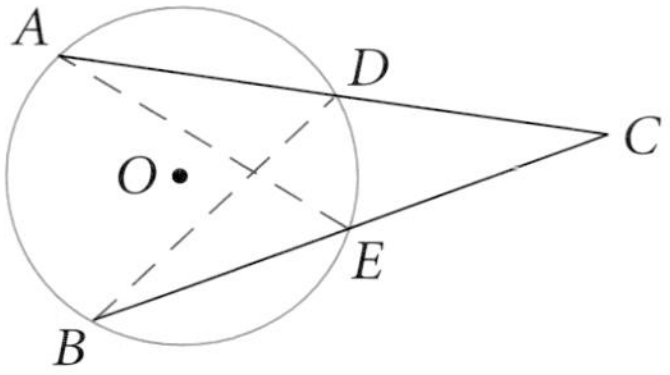

10. Prove Theorem 12-16.

Given: $\odot O$ with tangent $\overline{AC}$ and secant $\overline{BC}$ (Chords $\overline{AB}$ and $\overline{AD}$ are drawn for you.)
Prove: $BC \cdot DC = AC^2$

Plan for Proof: In order to use the properties of proportions, prove $\triangle BAC \sim \triangle ADC$.

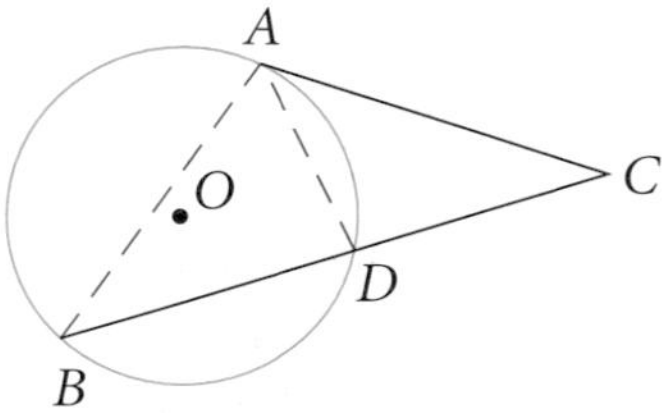

Find the diameter of $\odot O$ given the chords, secants, and tangents. Round your answer to the nearest tenth.

11.

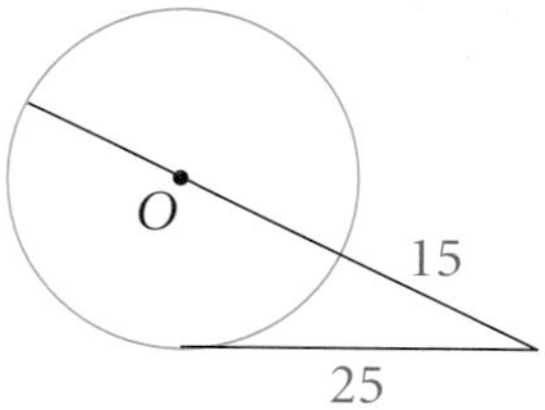

12.

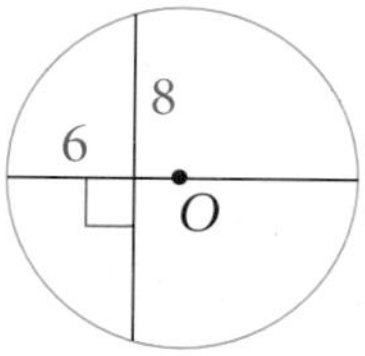

13.

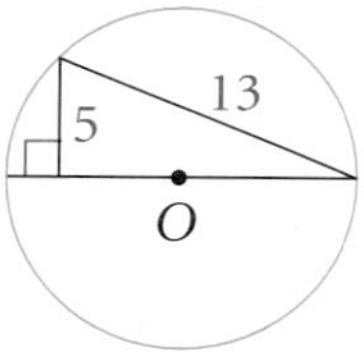

14. *Standardized Test Prep* $\overline{AB}$ is tangent to $\odot O$ at A. $\overline{AB} \perp \overline{EB}$. $AB = BF$. $BC = 1$ and $CE = 7$. Find AF.

A. 2 **B.** $3\frac{1}{2}$ **C.** 4
D. $4\frac{1}{3}$ **E.** none of the above

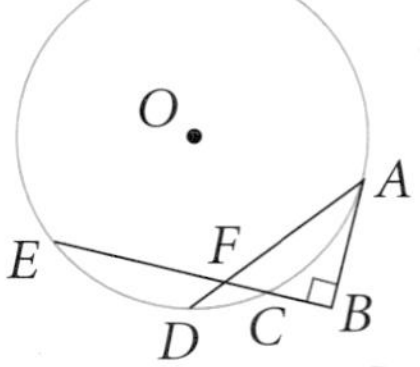

15. *Space Exploration* The line of sight from an object above a planet to the horizon of the planet is a tangent line. The *Galileo* orbiter is a spacecraft that circles 133,300 mi above Jupiter. Jupiter's diameter is 88,700 mi. What is the distance from the *Galileo* orbiter to Jupiter's horizon? Round your answer to the nearest hundred miles.

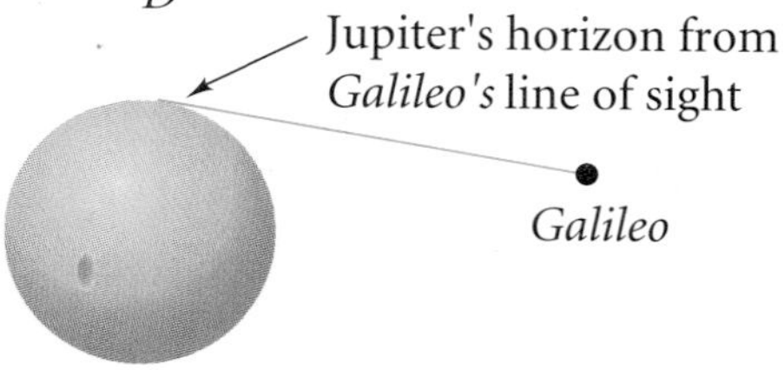

16. **Writing** To find the value of x, a student wrote the equation $(7.5)6 = x^2$. What error did the student make?

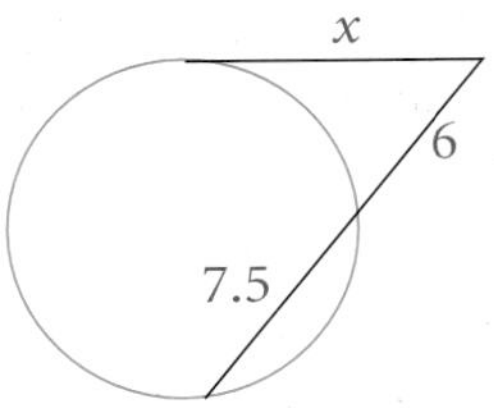

17. a. **Open-ended** Draw a circle. Place a point inside the circle. Draw a chord through the point. Measure the segments of the chord that you drew. Find the product of the lengths.
    b. Choose three other points. Find the product of the lengths of the segments of a chord through each of these points.
    c. **Patterns** Use your answers to parts (a) and (b) to make a **conjecture.** What point inside a circle has the greatest product of the lengths of segments of chords?

Find the value of x and y using the given chords, secants, and tangents. If your answer is not an integer, round to the nearest tenth.

18.

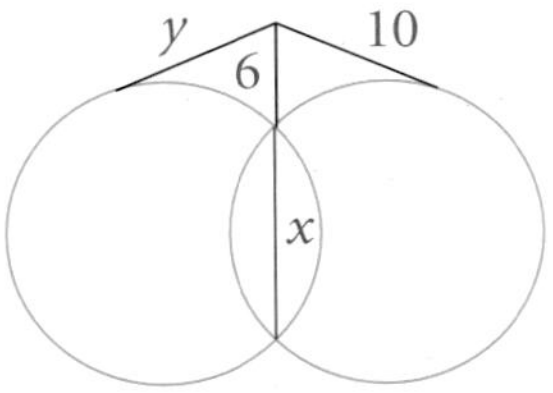

19.

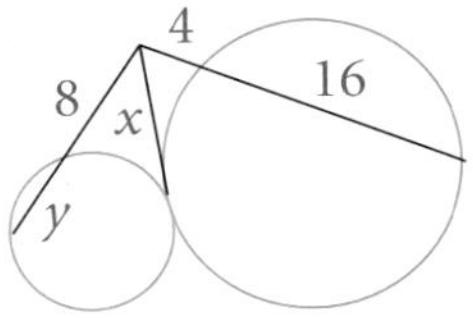

20.

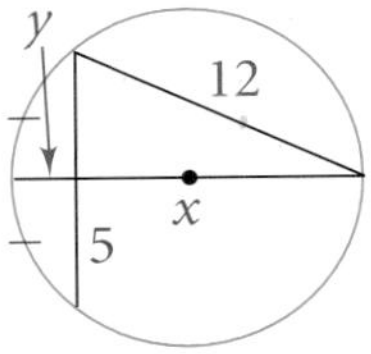

21. a. **Engineering** The basis of a design of a rotor for a Wankel engine is an equilateral triangle. Each side of the triangle is a chord of an arc of a circle. The center of each circle is a vertex of the triangle. In the diagram at the right, each side of the equilateral triangle is 8 in. long. Find the value of x to the nearest tenth.
    b. **Research** Rotary engines are used in some snowmobiles. Find other uses of rotary engines.
    c. **Constructions** Construct an equilateral triangle. Then construct the arcs to draw a rotor.

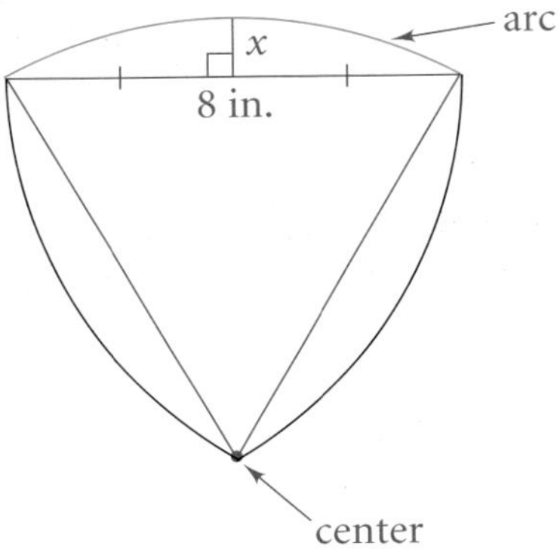

22. **Calculator** The diameter of a circle is 24 cm. A chord perpendicular to the diameter is 5 cm from the center of the circle. What is the length of the chord? Round your answer to the nearest tenth.

23. Following is a proof of the Pythagorean Theorem that uses a theorem presented in this lesson. Supply the missing reasons.

Given: $\odot O$ with tangent $\overline{PQ}$

Prove: $a^2 + b^2 = c^2$

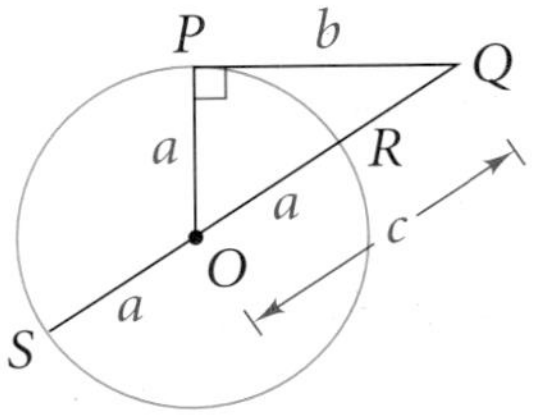

Statements	Reasons
1. $PQ^2 = (QR)(QS)$	a. ?
2. $b^2 = (c - a)(c + a)$	b. ?
3. $b^2 = c^2 - a^2$	c. ?
4. $a^2 + b^2 = c^2$	d. ?

Exercises MIXED REVIEW

Find the value of x to the nearest tenth.

24.

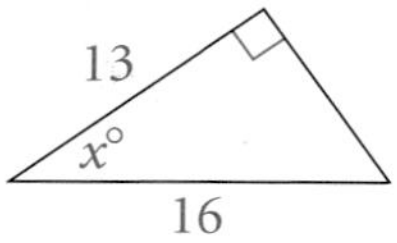

25.

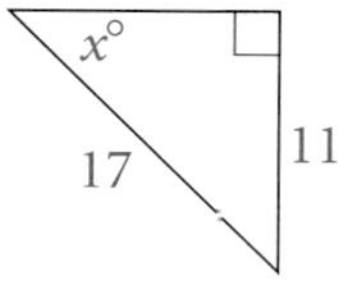

26. a. Calculator A ball has circumference 60 cm. What is its radius? Round your answer to the nearest millimeter.

b. What is the length of an edge of the smallest box that could contain this ball?

SELF ASSESSMENT

PORTFOLIO

For your portfolio, select one or two items from your work for this chapter. Here are some possibilities:

- corrected work
- a journal entry
- results from a Work Together activity

Explain why you have included each selection.

Geometry at Work

Aerospace Engineer

Aerospace engineers design and build all types of spacecraft, from the low-orbit space shuttle to interplanetary probes. Much of today's aerospace work involves communications satellites that relay TV, telephone, computer, and other signals to receivers around the world. The portion of Earth's surface that can communicate with a satellite increases as the height of the orbit increases.

The figure shows a satellite 12,000 miles above Earth, which has a radius of about 3960 miles. $\widehat{AB}$ is the arc of Earth that is in the range of the satellite. You can find $m\widehat{AB}$ by finding $m\angle AEB$, which is twice $m\angle AES$.

cosine of $\angle AES = \frac{AE}{SE} = \frac{3960}{3960 + 12{,}000} \approx 0.2481$

$m\angle AES \approx \cos^{-1}(0.2481) \approx 75.63$

$m\widehat{AB} = m\angle AEB \approx 2 \cdot 75.63 \approx 151.3$

The arc of Earth in the range of the satellite is about 151.3°.

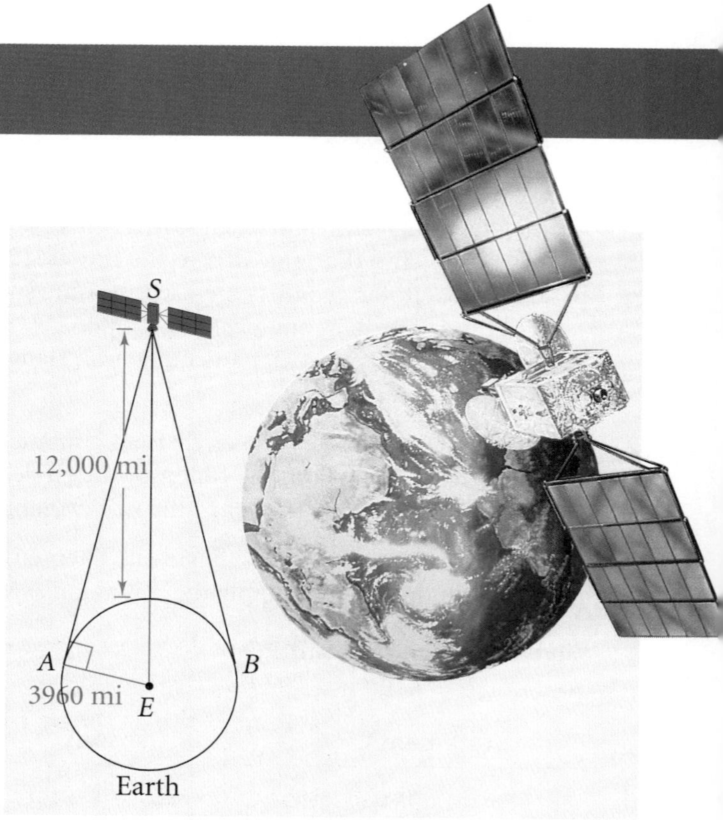

Mini Project: A geosynchronous orbit is used by communications satellites. Research geosynchronous orbits. Find the height of a geosynchronous orbit and find the arc of Earth that is in the range of a satellite in a geosynchronous orbit.

Finishing the Chapter Project

GO FOR A SPIN

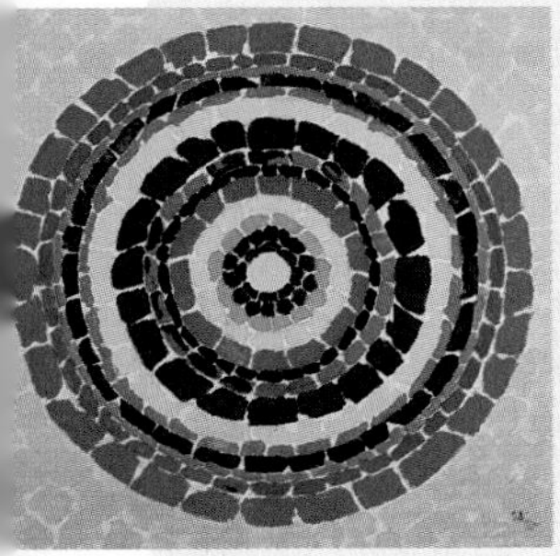

Find Out questions and activities on pages 591, 605, and 618 will help you complete your project. Using one or more of the techniques you explored, create your own design. First decide how the design will be used. Some possibilities are a poster, a school logo, and a tile pattern for a floor. State the purpose of your design, give instructions for drawing it, and explain the geometric concepts incorporated in it.

Reflect and Revise

Ask a classmate to review your project. Together, check that the diagrams and explanations are clear and accurate. Can you improve your design? Can someone follow your instructions? Revise your work as needed.

Follow Up

Leonardo da Vinci explored regions bounded by arcs. He showed that the first region at the right could be cut apart and reassembled into a rectangle. Read about da Vinci's efforts and try the other figures yourself.

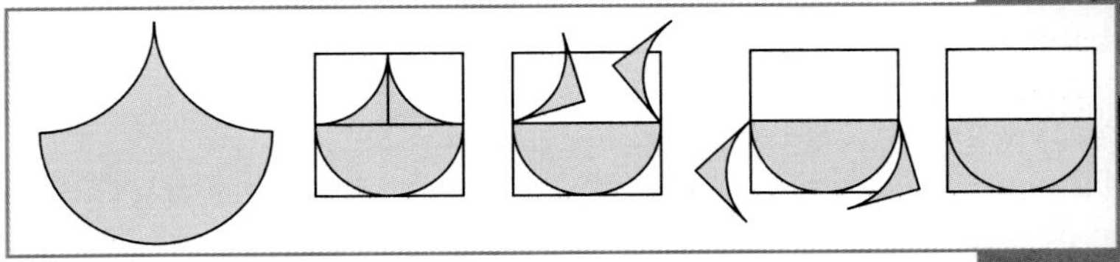

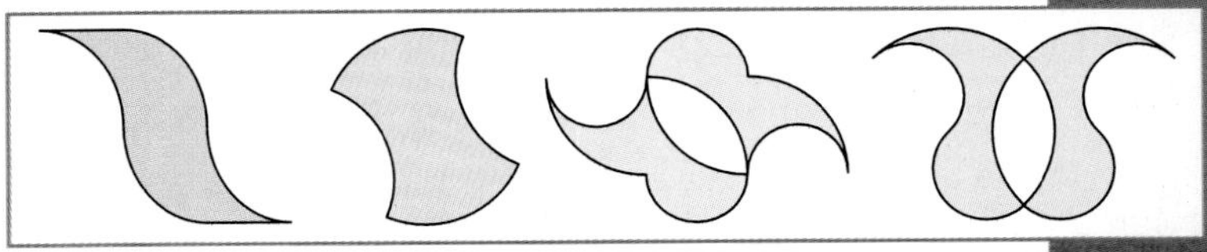

For More Information

Bain, George. *Celtic Art: The Methods of Construction.* New York: Dover, 1973.

Horemis, Spyros. *Optical and Geometrical Patterns and Designs.* New York: Dover, 1970.

Wills, Herbert. *Leonardo's Dessert.* Reston, Virginia: National Council of Teachers of Mathematics, 1985.

CHAPTER 12

Wrap Up

Key Terms

chord (p. 600)
circumscribed about (pp. 596, 607)
inscribed in (pp. 596, 607)
inscribed angle (p. 607)
intercepted arc (p. 607)
point of tangency (p. 593)
secant (p. 614)
standard form of an equation of a circle (p. 586)
tangent to a circle (p. 593)

How am I doing?

SELF ASSESSMENT

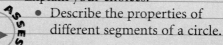

- State three ideas from this chapter that you think are important. Explain your choices.
- Describe the properties of different segments of a circle.

Circles in the Coordinate Plane — 12-1

The **standard form of an equation of a circle** with center (h, k) and radius r is $(x - h)^2 + (y - k)^2 = r^2$.

If you know the center and a point on a circle, you can find an equation of the circle. Find the radius by using the Distance Formula to find the distance from the center to the given point. Then substitute the coordinates of the center for (h, k) and the radius for r in the standard form of the equation of a circle.

Find an equation of a circle with the given center and radius.

1. center $= (2, 5)$; $r = 3$

2. center $= (-3, 1)$; $r = \sqrt{5}$

3. center $= (9, -4)$; $r = 3.5$

Find an equation of the circle with the given center passing through the given point.

4. center $(0, 1)$, through $(4, 9)$

5. center $(-2, 3)$, through $(4, -4)$

6. center $(10, 7)$, through $(-8, -5)$

7. *Standardized Test Prep* Which circle has the least area?

A. $(x - 1)^2 + (y - 3)^2 = 4$
B. $(x + 2)^2 + y^2 = 7$
C. $x^2 + (y - 5)^2 = 9$
D. $x^2 + y^2 = 10$
E. $(x + 3)^2 + (y + 2)^2 = 3$

Properties of Tangents — 12-2

A **tangent to a circle** is a line, ray, or segment in the plane of the circle that intersects the circle in exactly one point, the **point of tangency**.

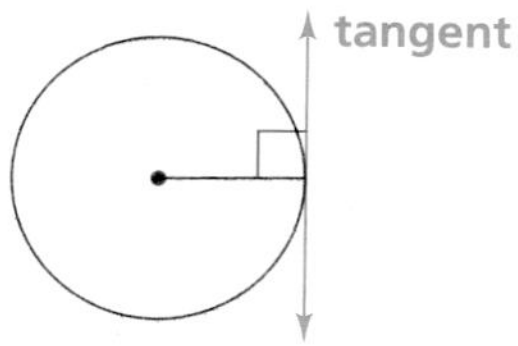

If a line is tangent to a circle, then the line is perpendicular to the radius drawn to the point of tangency. If a line is perpendicular to a radius at its endpoint on the circle, then the line is tangent to the circle.

Two segments tangent to a circle from a point outside the circle are congruent.

Each polygon circumscribes a circle. Find the perimeter of the polygon.

8.

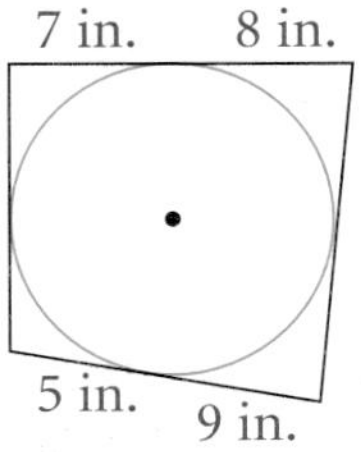

9.

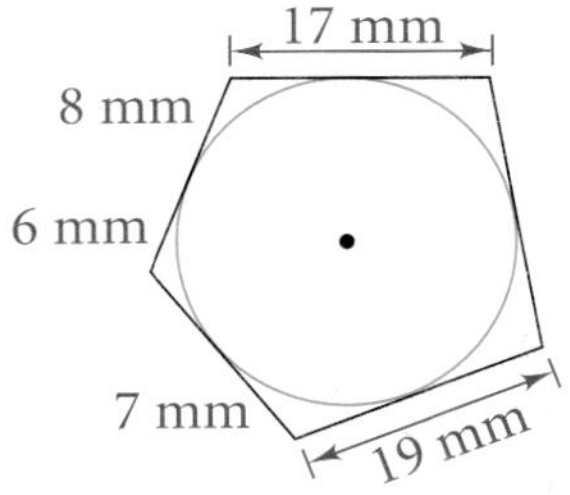

10.

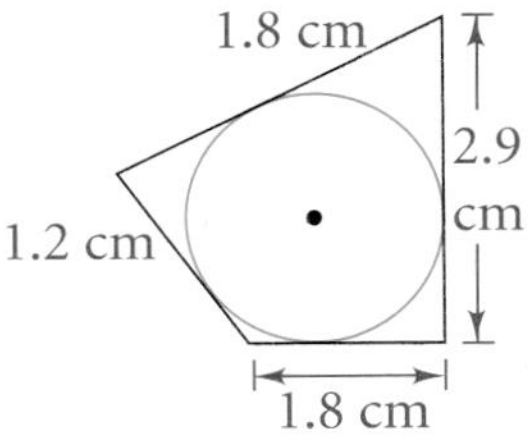

11. Writing The word *tangent* is used in different ways in Chapters 11 and 12. Explain each use. Include diagrams in your explanation.

Properties of Chords and Arcs 12-3

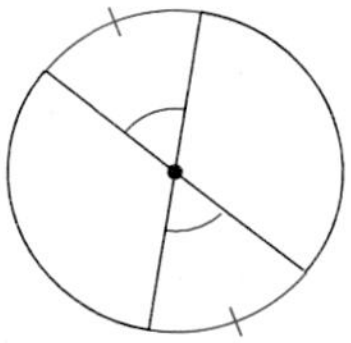

In the same circle or in congruent circles,

- congruent central angles intercept congruent arcs;
- congruent arcs have congruent central angles;
- congruent chords have congruent arcs;
- congruent arcs have congruent chords;
- chords equidistant from the center are congruent; and
- congruent chords are equidistant from the center.

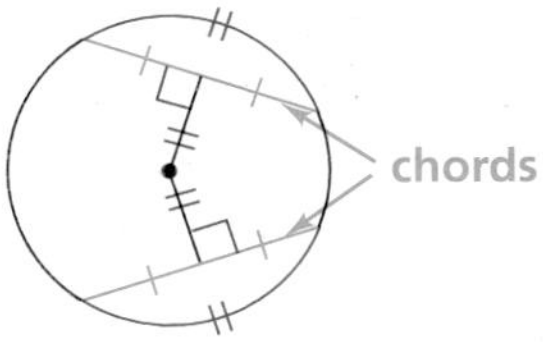

A diameter that is perpendicular to a chord bisects the chord and its arc. The perpendicular bisector of a chord contains the center of the circle.

Calculator **Find the value of x. Round to the nearest tenth.**

12.

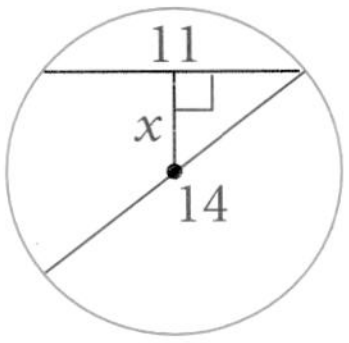

13.

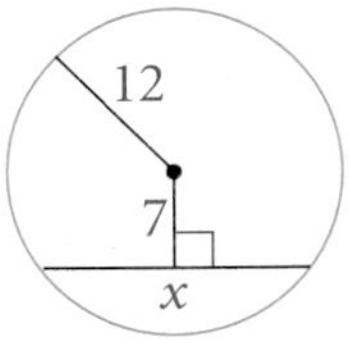

14.

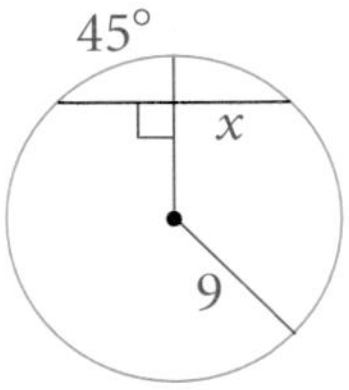

15.

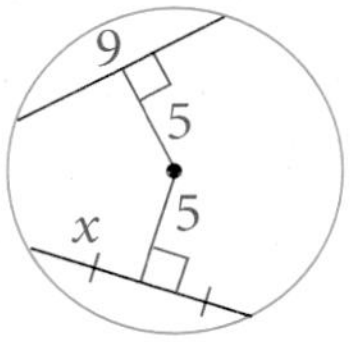

16. Open-ended Draw two circles with different radii. Draw a chord in the smaller circle and a congruent chord in the larger circle. Are the arcs of these chords congruent? Explain.

Inscribed Angles 12-4

The vertex of an **inscribed angle** lies on a circle. Its sides **intercept** an arc of the circle. All the vertices of an **inscribed polygon** lie on a circle.

The measure of an inscribed angle is half the measure of its intercepted arc. The measure of an angle formed by a chord and a tangent that intersect on a circle is half the measure of the intercepted arc.

Two inscribed angles that intercept the same arc are congruent. The opposite angles of a quadrilateral inscribed in a circle are supplementary.

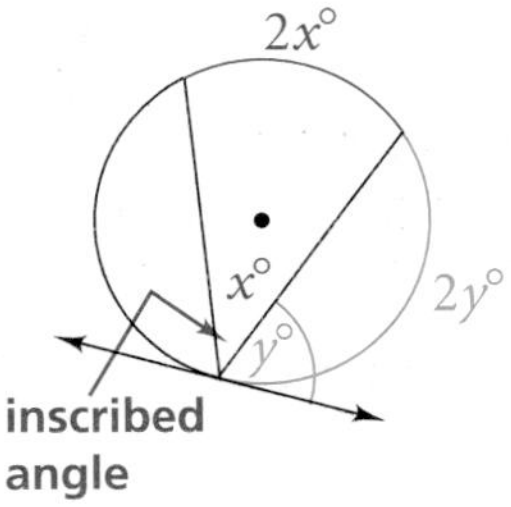

Assume that lines that appear tangent are tangent. Find the value of each variable.

17.

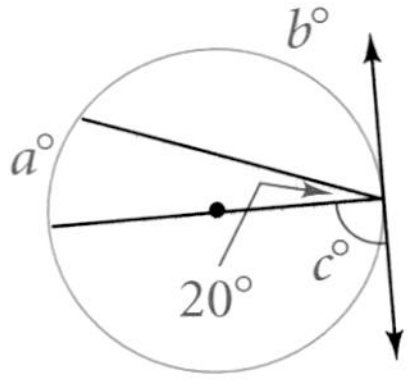

18.

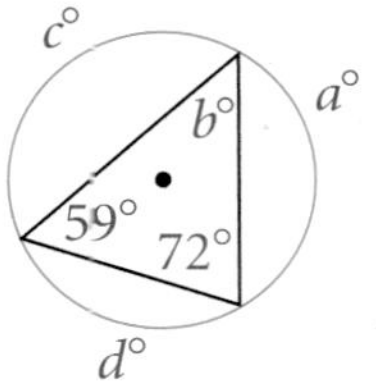

19.

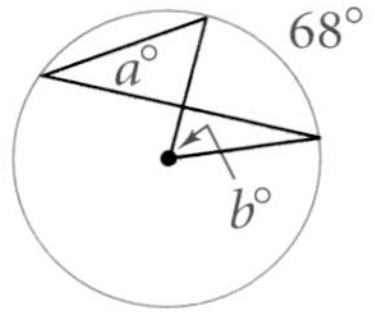

20.

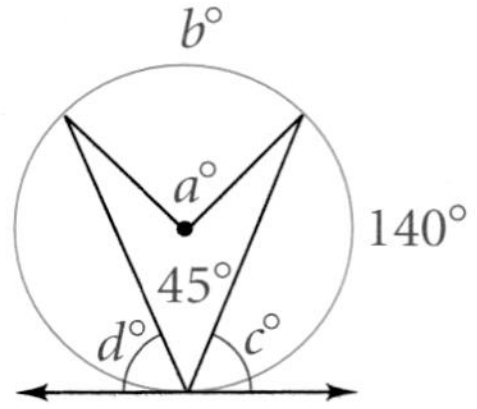

Angles Formed by Chords, Secants, and Tangents 12-5

The measure of an angle formed by two intersecting chords in a circle is half the sum of the measures of the intercepted arcs.

A **secant** is a line, ray, or segment that intersects a circle at two points.

The measure of an angle formed by two secants, two tangents, or a secant and a tangent drawn from a point outside a circle is half the difference of the measures of the intercepted arcs.

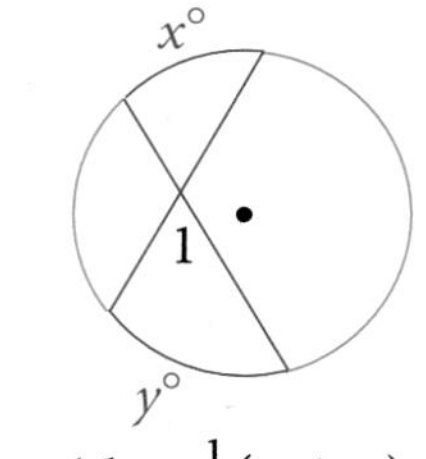

$m\angle 1 = \frac{1}{2}(x + y)$

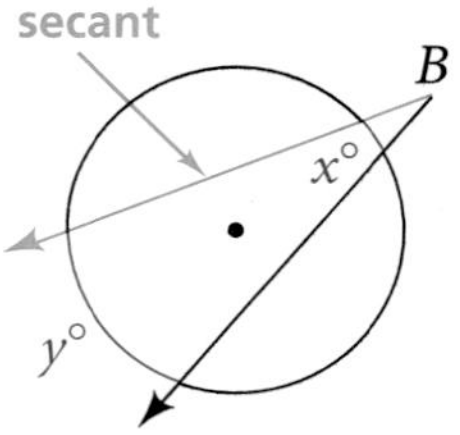

$m\angle B = \frac{1}{2}(y - x)$

Assume that lines that appear tangent are tangent. Find the value of each variable.

21.

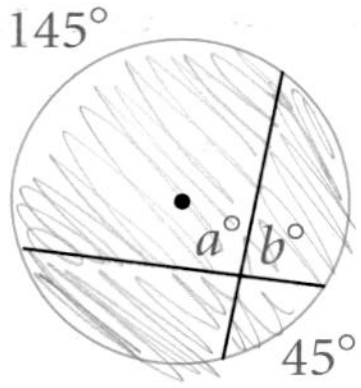

22.

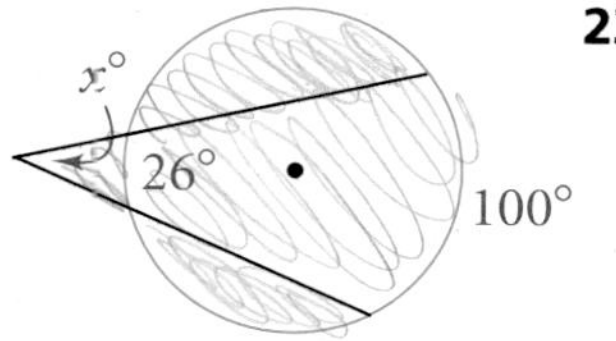

23.

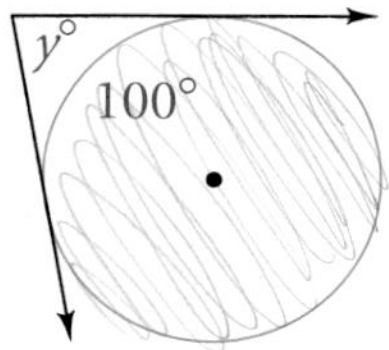

24.

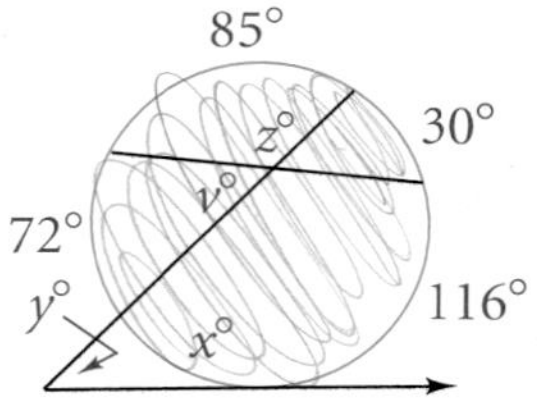

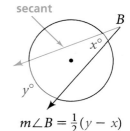

Circles and Lengths of Segments 12-6

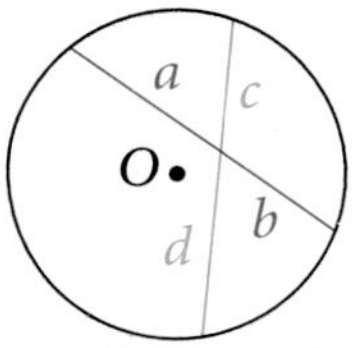

$a \cdot b = c \cdot d$

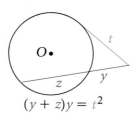

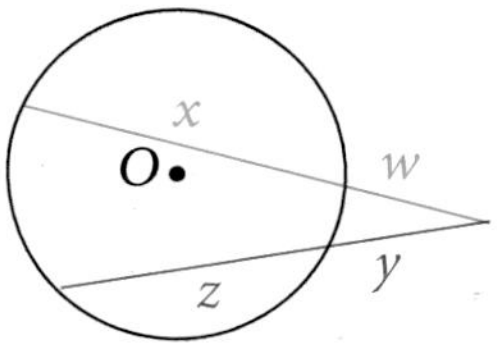

$(w + x)w = (y + z)y$

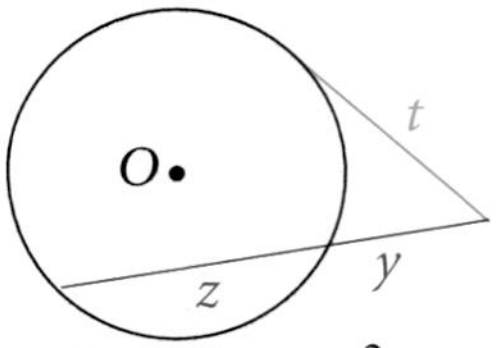

$(y + z)y = t^2$

Find the value of each variable using the given chords, secants, and tangents. If your answer is not an integer, round to the nearest tenth.

25.

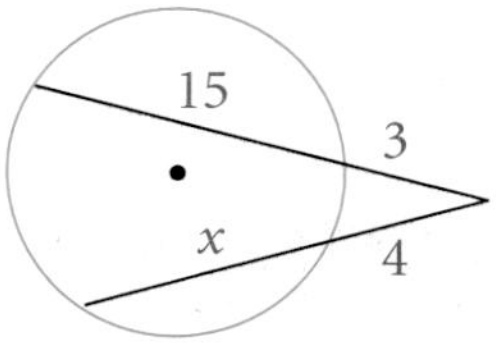

26.

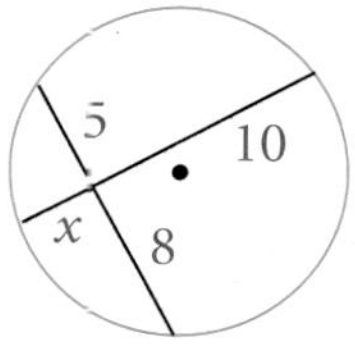

27.

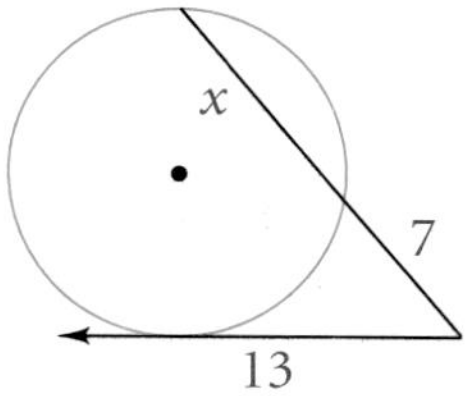

28.

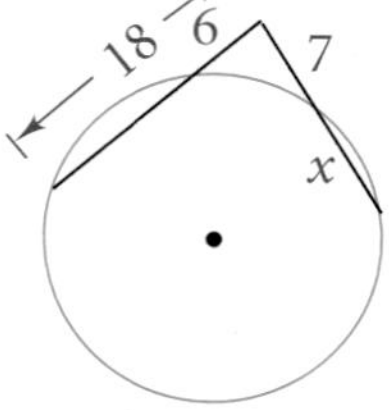

CHAPTER 12 Assessment

Find the center and radius of each circle.

1. $(x+3)^2 + (y-2)^2 = 9$
2. $(x-5)^2 + (y-9)^2 = 225$

Write the equation of each circle.

3. 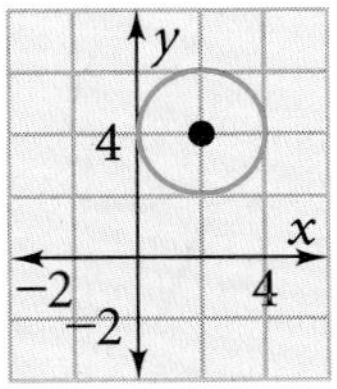

4. 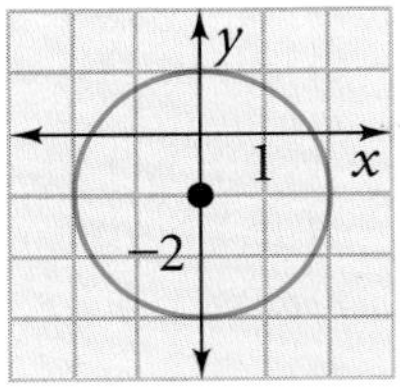

5. Find the circumference and area of the circle whose equation is $(x-2)^2 + (y-7)^2 = 81$. Round to the nearest tenth.

Assume that lines that appear tangent are tangent. Find the value of x.

6. 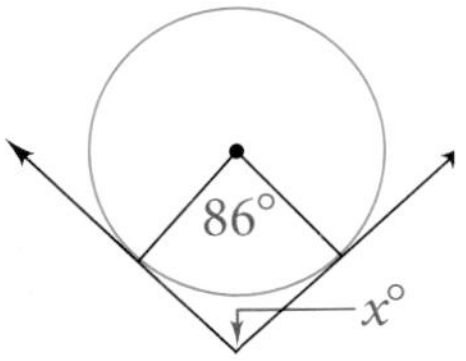

7. 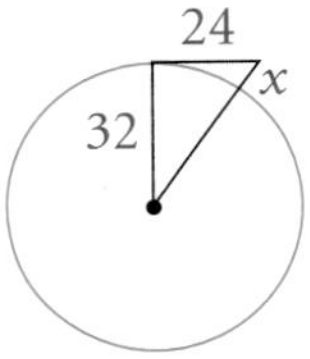

8. a. **Open-ended** Draw a circle with two congruent chords that form an inscribed angle.
   b. **Constructions** Construct the bisector of the inscribed angle. What do you notice?

Find the value of x. If your answer is not an integer, round to the nearest tenth.

9. 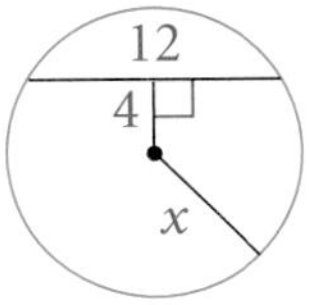

10. 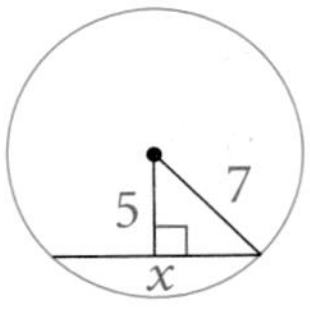

11. **Coordinate Geometry** Graph the circle $(x-4)^2 + (y-3)^2 = 25$. State the center and the radius of the circle.

Find $m\widehat{AB}$.

12. 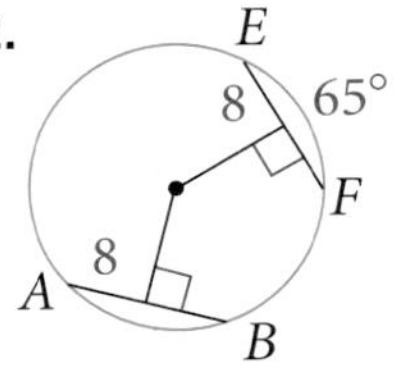

13. 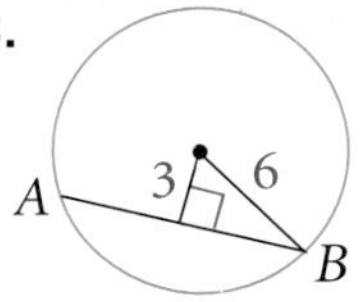

14. **Standardized Test Prep** In the figure at the right, a square is circumscribed about $\odot A$. Find the area of the square.

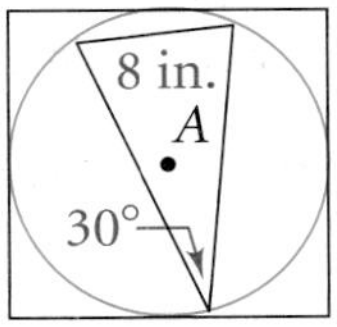

A. $192\ \text{cm}^2$
B. $64\ \text{cm}^2$
C. $256 + 16\sqrt{3}\ \text{cm}^2$
D. $256\ \text{cm}^2$
E. It cannot be determined from the information given.

Assume that lines that appear tangent are tangent. Find the value of each variable. If your answer is not an integer, round to the nearest tenth.

15. 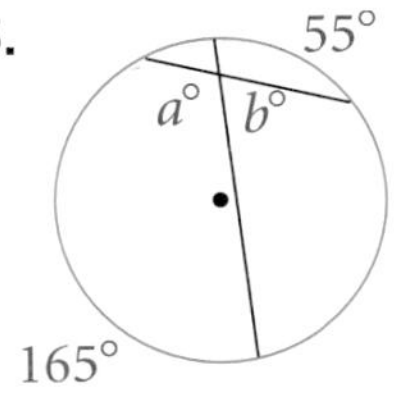

16. 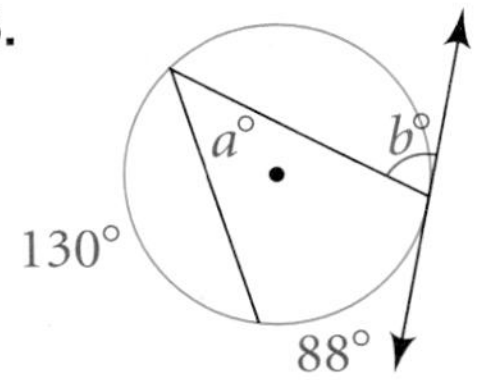

17. 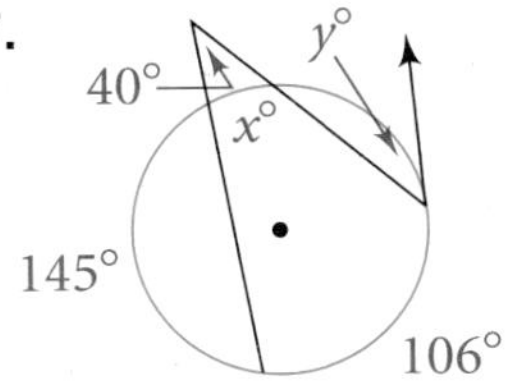

18. 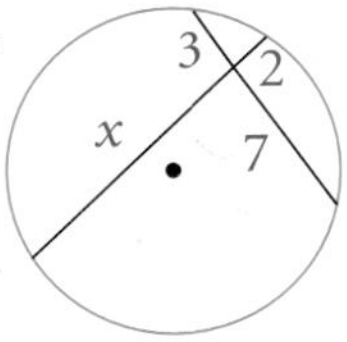

19. 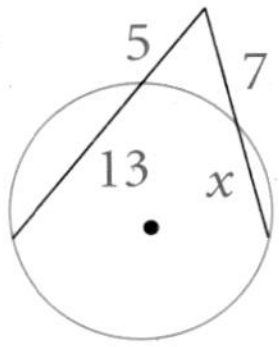

20. 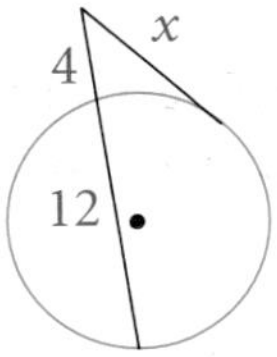

21. **Writing** What is special about a rhombus inscribed in a circle? **Justify** your answer.

CHAPTER 12 Cumulative Review

For Exercises 1–15, choose the correct letter.

1. What is the best description of the triangle below?

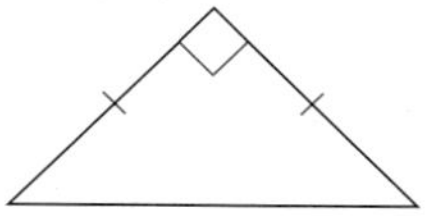

A. acute scalene
B. right
C. equilateral
D. obtuse scalene
E. isosceles right

2. What are the coordinates of the midpoint of $\overline{QS}$ with endpoints $Q(-2, -5)$ and $S(3, -8)$?
A. $(-2.5, 6.5)$ B. $(0.5, -6.5)$ C. $(0.5, 1.5)$
D. $(-2.5, 1.5)$ E. $(-3.5, -5.5)$

3. Which theorem or postulate can you *not* use to prove $\triangle ABC \cong \triangle CDA$?

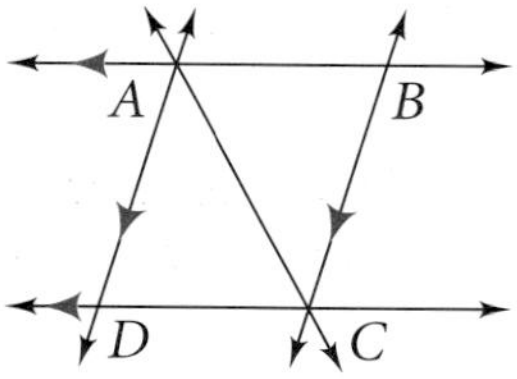

A. HL B. ASA C. AAS
D. SAS E. SSS

4. What is the area of an isosceles right triangle with hypotenuse length $5\sqrt{2}$ in.?
A. 5 in.^{2} B. 12.5 in.^{2} C. 25 in.^{2}
D. $25\sqrt{2}$ in.^{2} E. $(10 + 5\sqrt{2})$ in.^{2}

5. What is the volume of the figure?

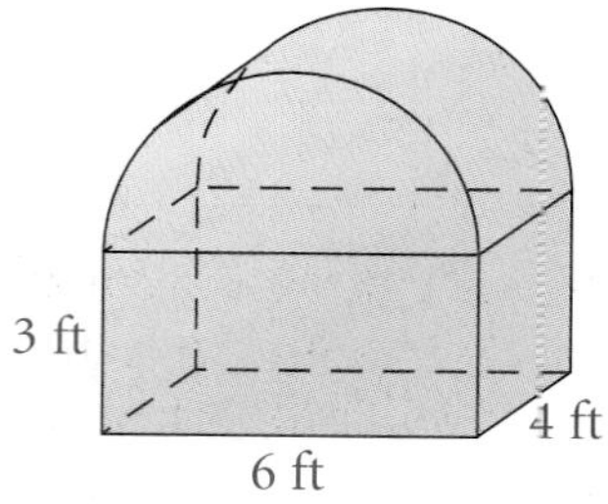

A. 288 ft^3 B. 72 ft^3 C. 72π ft^3
D. $(13 + 72\pi)$ ft^3 E. $(72 + 18\pi)$ ft^3

6. Which of the following statements can be derived from the biconditional statement "The day is long if and only if it is summer"?
A. If the day is long, then it is summer.
B. If it is summer, then the day is long.
C. If the day is not long, then it is not summer.
D. If it is not summer, then the day is not long.
E. all of the above

7. Which is a reflection of the figure in the x-axis?

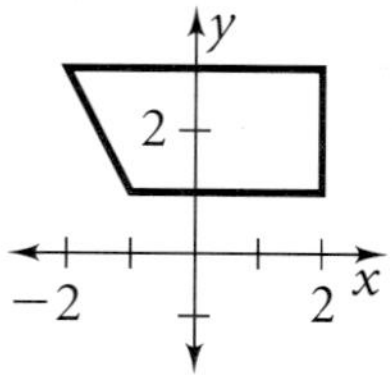

A.
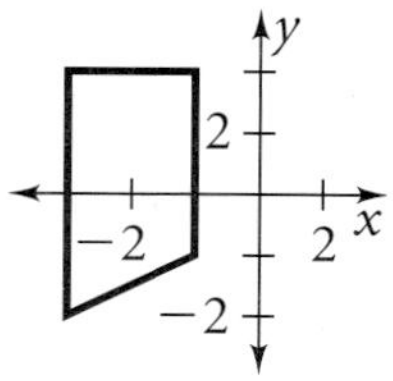

B.
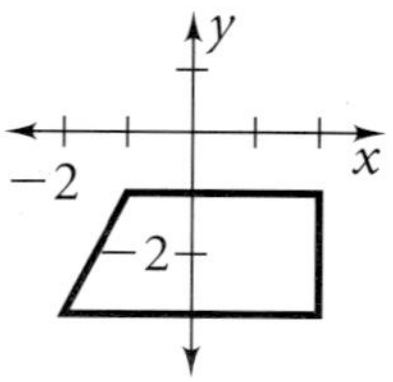

C.
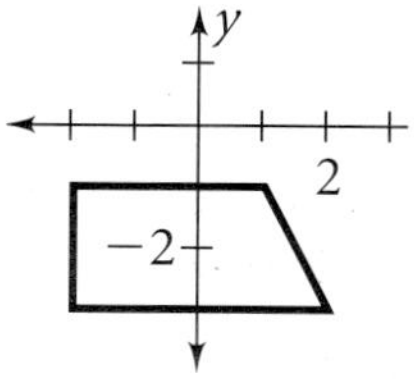

D.
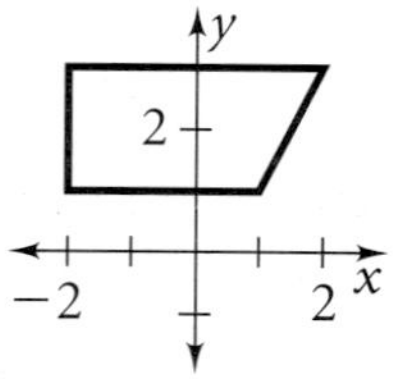

E. none of the above

8. Which line(s) is(are) perpendicular to the line $y = 4x - 1$?
I. $y = 4x + 7$ II. $y = \frac{1}{4}x + 3$
III. $y = -\frac{1}{4}x - 5$ IV. $x + 4y = 16$
A. I only B. II only C. I and II
D. III and IV E. I, II, III, and IV

9. Which statement is true for both a rhombus and a kite?
A. Opposite angles are congruent.
B. The diagonals are congruent.
C. Opposite sides are congruent.
D. The diagonals are perpendicular.
E. Opposite sides are parallel.

10. What is the value of x to the nearest tenth?

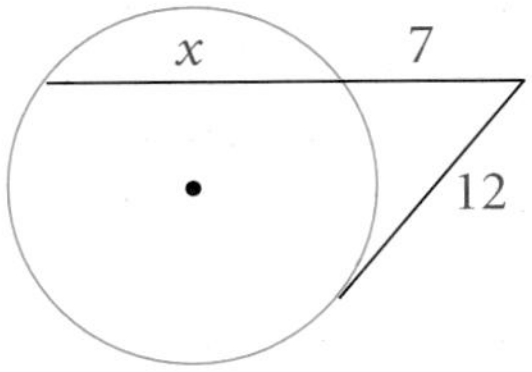

A. 5 **B.** 9 **C.** 13.6 **D.** 15 **E.** 20.6

11. Which triangle is drawn with its medians?

A.

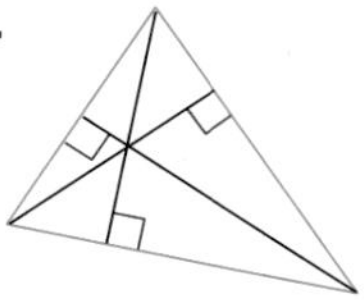

B.

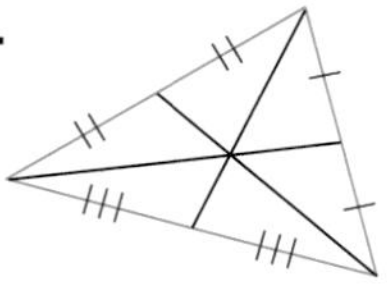

C.

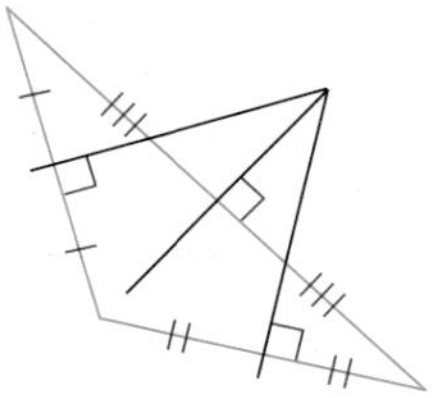

D.

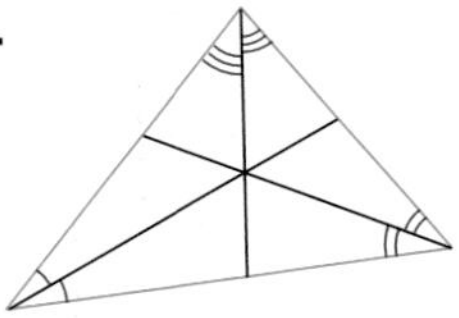

E. none of the above

Compare the boxed quantity in Column A with the one in Column B. Choose the best answer.

A. The quantity in Column A is greater.
B. The quantity in Column B is greater.
C. The two quantities are equal.
D. The relationship cannot be determined on the basis of the information supplied.

	Column A	Column B
12.	the magnitude of $\langle 5, 1 \rangle$	the magnitude of $\langle -4, 2 \rangle$
13.	the measure of an inscribed angle of a circle that intercepts a 78° arc	the measure of a central angle of a circle that intercepts a 78° arc

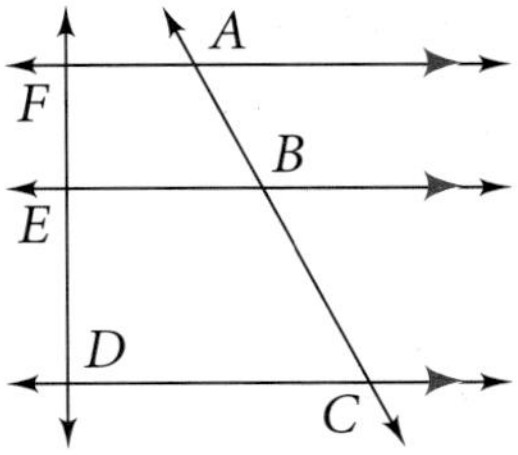

14.	$\frac{AB}{FE}$	$\frac{BC}{ED}$
15.	the distance between (7, 3) and (10, −2)	the distance between (3, −1) and (0, 4)

Find each answer.

16. Open-ended Sketch a figure with two lines of symmetry.

17. Constructions Draw a segment. Construct a kite whose diagonals are congruent to the segment you drew.

18. Writing What information is provided by a flow proof that is not in a two-column proof?

19. Constructions Copy the trapezoid. Construct a trapezoid similar to it whose perimeter is twice that of the original figure. What is the ratio of the area of the smaller figure to the area of the larger figure?

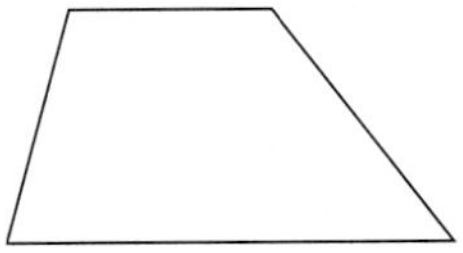

20. You are 5 ft 6 in. tall. When your shadow is 6 ft long, the shadow of a sculpture is 30 ft long. How tall is the sculpture?

21. A diagonal of a rectangular field makes a 70° angle with the side of the field that is 100 ft long. What is the area of the rectangle?

Extra Practice

Find the next two terms in each sequence. — Lesson 1-1

1. 12, 17, 22, 27, 32, . . .

2. 1, 1.1, 1.11, 1.111, 1.1111, . . .

3. 5000, 1000, 200, 40, . . .

4. 1, 12, 123, 1234, . . .

5. 3, 0.3, 0.03, 0.003, . . .

6. 1, 4, 9, 16, 25, . . .

Write *true* or *false*. — Lessons 1-2 and 1-3

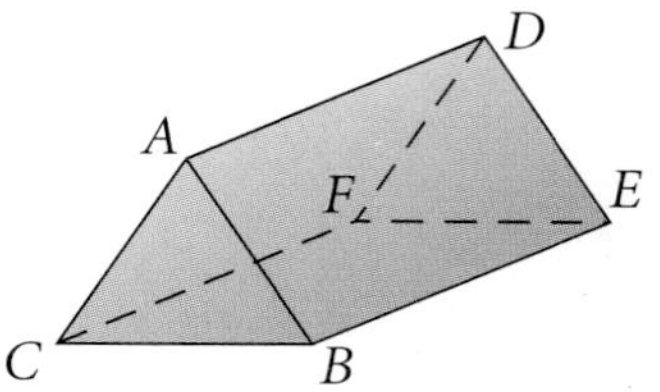

7. *A*, *D*, *F* are coplanar.

8. $\overleftrightarrow{AC}$ and $\overleftrightarrow{FE}$ are coplanar.

9. *A*, *B*, *E* are collinear.

10. *D*, *A*, *B*, *E* are coplanar.

11. $\overleftrightarrow{FC} \parallel \overleftrightarrow{EF}$

12. plane *ABC* $\parallel$ plane *FDE*

13. $\overleftrightarrow{BC}$ and $\overleftrightarrow{DF}$ are skew lines.

14. $\overleftrightarrow{AD}$ and $\overleftrightarrow{EB}$ are skew lines.

Use the figure at the right for Exercises 15–19. — Lessons 1-4 and 1-5

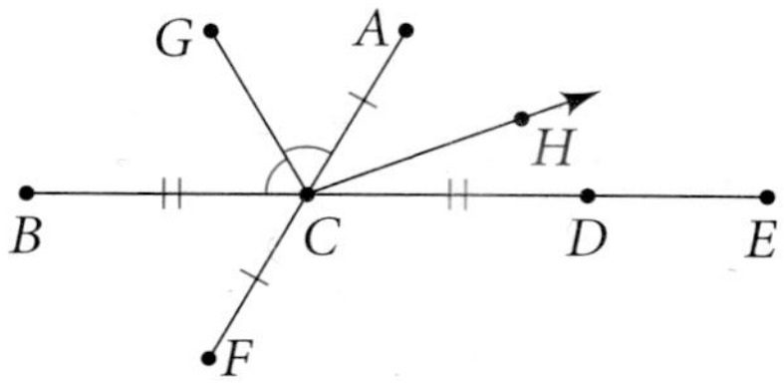

15. If $BC = 12$ and $CE = 15$, then $BE =$ ■.

16. ■ is the angle bisector of ■.

17. $BC = 3x + 2$ and $CD = 5x - 10$. Solve for *x*.

18. $m\angle BCG = 60$, $m\angle GCA =$ ■, $m\angle BCA =$ ■

19. $m\angle ACD = 60$ and $m\angle DCH = 20$. Find $m\angle HCA$.

Draw a diagram larger than the given one. Then do the construction. — Lesson 1-6

20. Construct the perpendicular bisector of $\overline{AB}$.

21. Construct $\angle A$ so that $m\angle A = m\angle 1 + m\angle 2$.

22. Construct the angle bisector of $\angle 1$.

23. Construct $\overline{FG}$ so that $FG = AB + CD$.

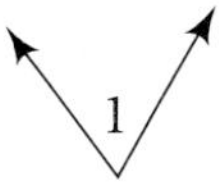

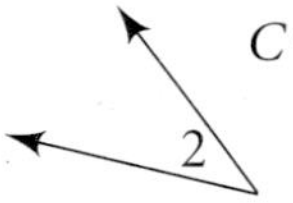

Algebra **Find the value of *x*.** — Lesson 1-7

24.

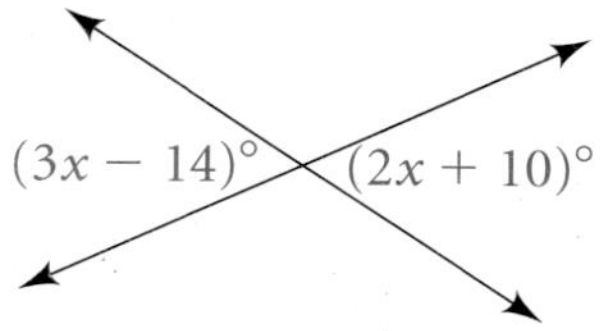

25.

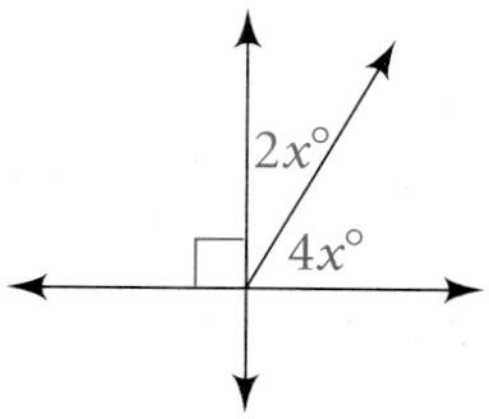

26.

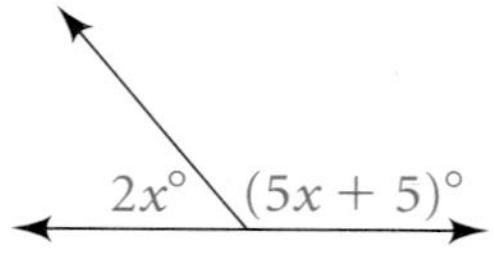

In Exercises 27–32: (a) Find the distance between the points to the nearest tenth. (b) Find the coordinates of the midpoint of the segment with the given endpoints. — Lesson 1-8

27. *A*(2, 1), *B*(3, 0)

28. *R*(5, 2), *S*(−2, 4)

29. *Q*(−7, −4), *T*(6, 10)

30. *C*(−8, −1), *D*(−5, −11)

31. *J*(0, −5), *N*(3, 4)

32. *Y*(−2, 8), *Z*(3, −5)

Extra Practice

Classify each triangle by its sides and angles. To do so, use a ruler to measure lengths of sides and a protractor to measure angles.

■ Lessons 2-1 and 2-2

1. 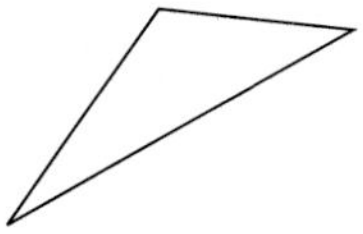

2. 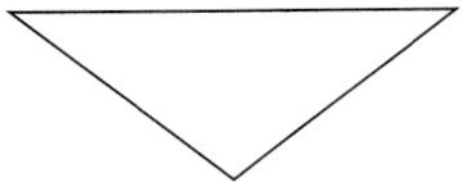

3. 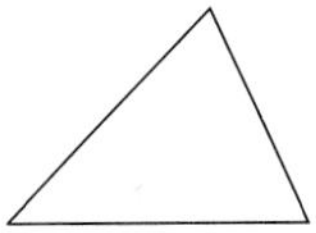

4. 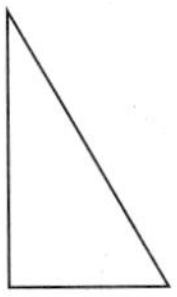

Algebra **Find the value of each variable.**

5. 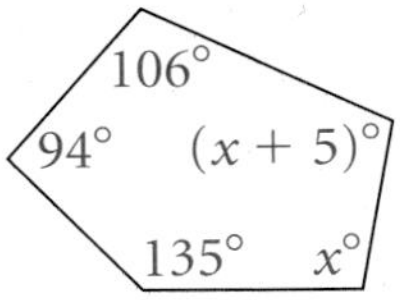

6. 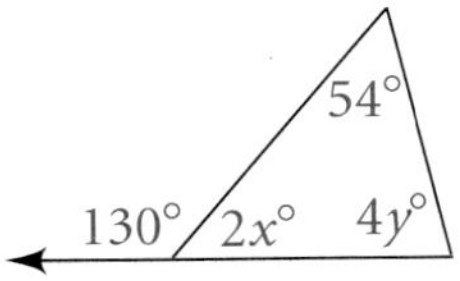

7. 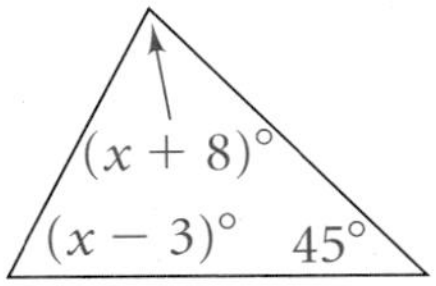

8. 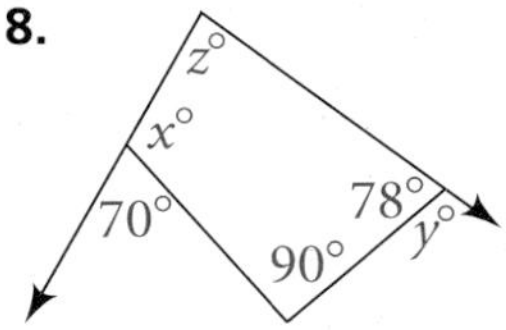

Graph the given points. Use the slope formula and/or the distance formula to determine the most precise name for quadrilateral *ABCD*.

■ Lessons 2-3 and 2-4

9. $A(3, 5), B(6, 5), C(2, 1), D(1, 3)$

10. $A(-1, 1), B(3, -1), C(-1, -3), D(-5, -1)$

11. $A(2, 1), B(5, -1), C(4, -4), D(1, -2)$

12. $A(-4, 5), B(-1, 3), C(-3, 0), D(-6, 2)$

Identify the following in $\odot P$.

■ Lesson 2-5

13. three minor arcs

14. two major arcs

15. two adjacent arcs

16. two radii

17. an acute central angle

18. two diameters

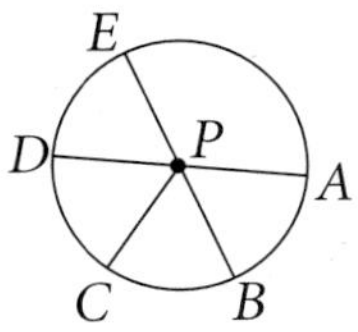

$\triangle SAT \cong \triangle GRE$. Complete the congruence statements.

■ Lesson 2-6

19. $\angle S \cong$ ■

20. $\overline{GR} \cong$ ■

21. $\angle E \cong$ ■

22. $\overline{AT} \cong$ ■

23. $\triangle ERG \cong$ ■

24. $\overline{EG} \cong$ ■

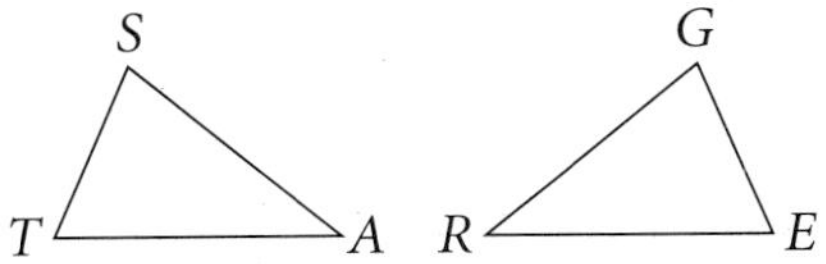

Create an isometric drawing and an orthographic drawing for each foundation plan.

■ Lesson 2-7

25. 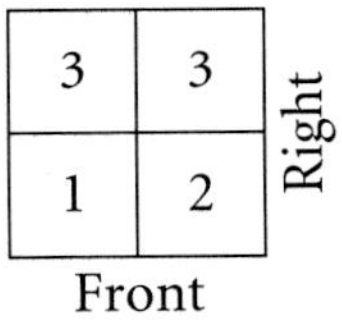

26. 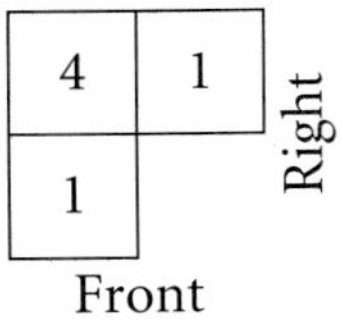

27. 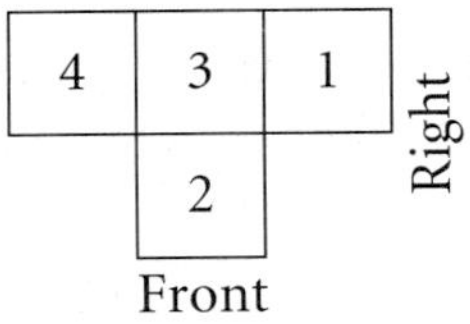

28. 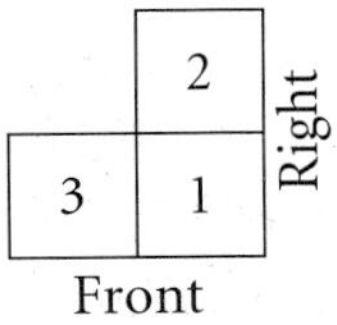

Extra Practice

Lesson 3-1

Coordinate Geometry **Given points $S(6, 1)$, $U(2, 5)$, and $B(-1, 2)$, draw $\triangle SUB$ and its reflection image in the given line.**

1. $y = 5$
2. $x = 7$
3. $y = -1$
4. the x-axis
5. $y = x$
6. $x = -1$
7. $y = 3$
8. the y-axis

Lesson 3-2

In Exercises 9–14, refer to the figure at the right.

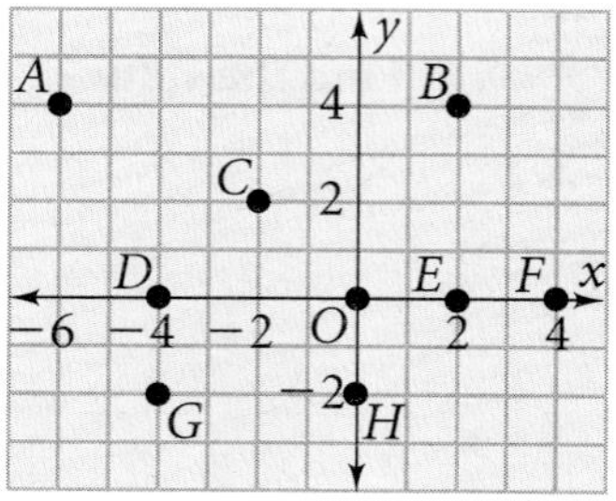

9. What is the image of C under the translation $\langle 4, -2 \rangle$?
10. What vector describes the translation $F \longrightarrow B$?
11. What is the image of H under the translation $\langle -2, 4 \rangle$?
12. What vector describes the translation $D \longrightarrow H$?
13. What is the image of C under the translation $\langle -2, -4 \rangle$?
14. What vector describes the translation $B \longrightarrow A$?

Lesson 3-3

Copy each figure and point P. Rotate the figure the given number of degrees about P. Label the vertices of the image.

15. 60°

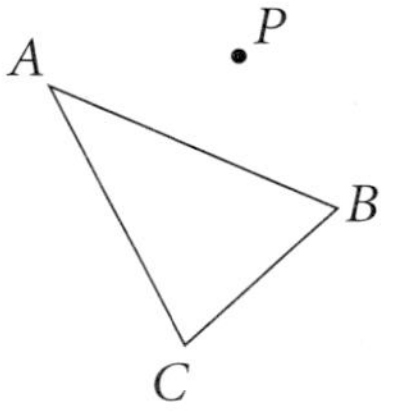

16. 90°

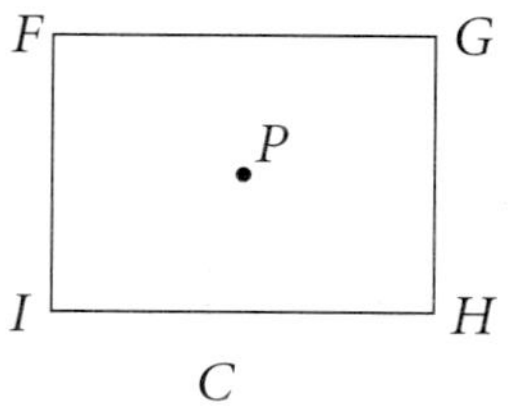

17. 180°

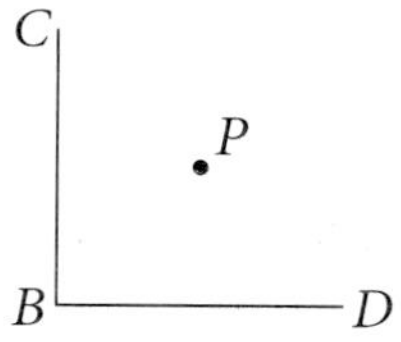

18. 45°

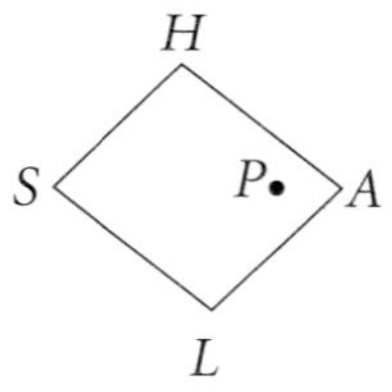

Lessons 3-4 and 3-7

The blue figure is the image of the black figure. State whether the mapping is a reflection, rotation, translation, glide reflection, or dilation.

19.

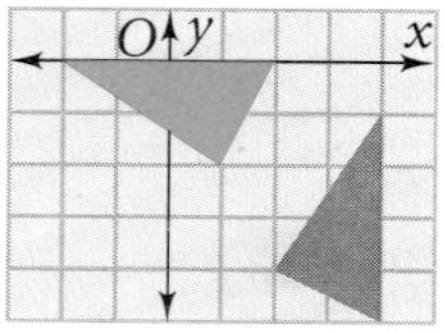

20.

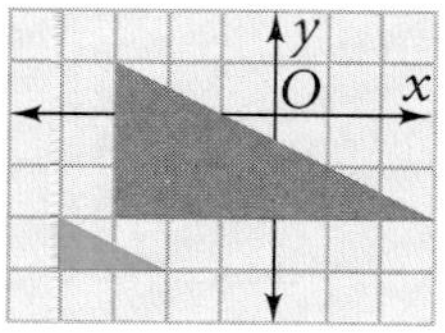

21.

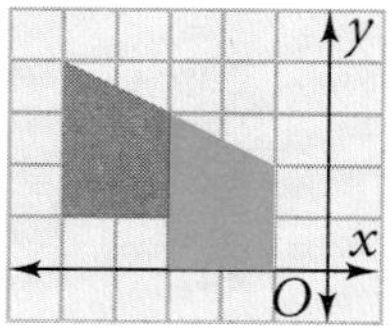

22.

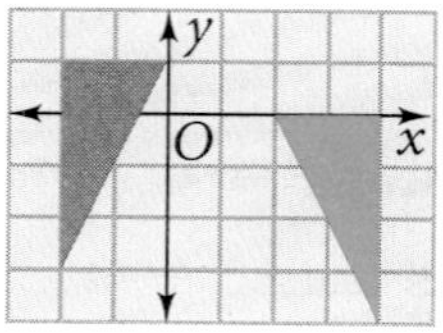

Lessons 3-5 and 3-6

In Exercises 23–26: (a) State what kind of symmetry each figure has. (b) State whether each figure tessellates.

23.

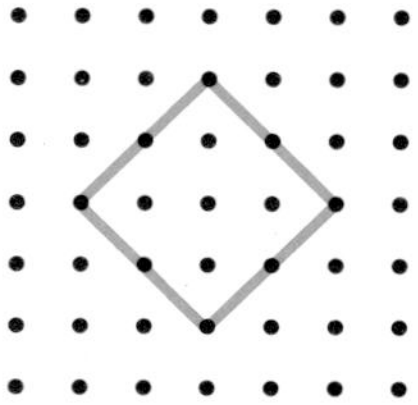

24.

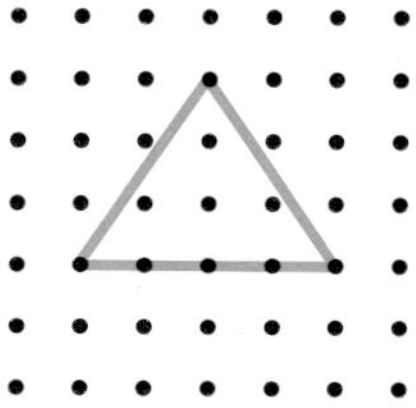

25.

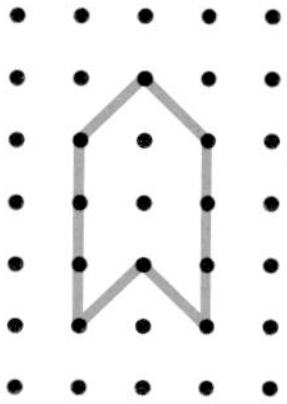

26.

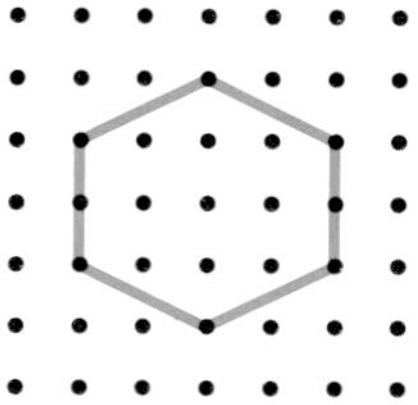

Extra Practice

For each statement, write the converse, the inverse, and the contrapositive. ■ Lesson 4-1

1. If two angles are vertical angles, then they are congruent.
2. If figures are similar, then their side lengths are proportional.
3. If a car is blue, then it has no doors.

Find the value of each variable. ■ Lessons 4-2 and 4-4

4. 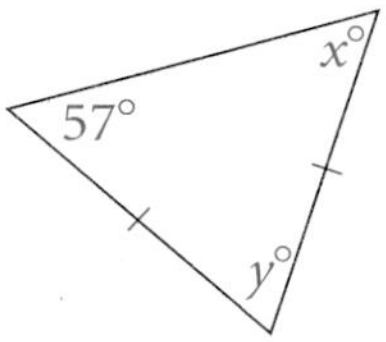

5. 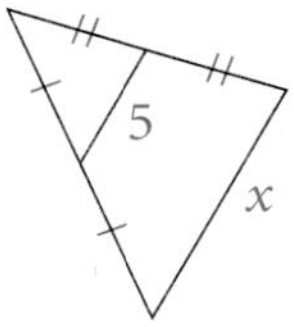

6. 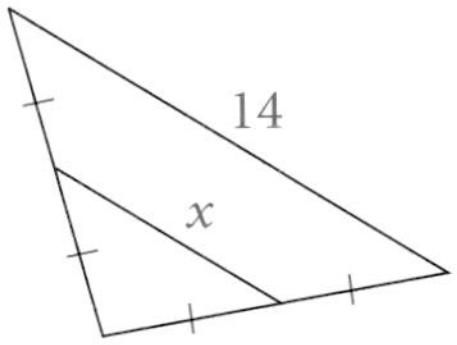

7. 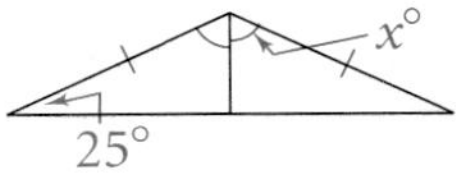

■ Lesson 4-3

8. Rewrite this paragraph proof as a two-column proof.
Given: $\square BGKM$, $m\angle B = m\angle G = m\angle K = m\angle M$
Prove: $\square BGKM$ is a rectangle.

By the Polygon Interior Angle-Sum Thm., $m\angle B + m\angle G + m\angle K + m\angle M = 360$. We are given that $m\angle B = m\angle G = m\angle K = m\angle M$, so by Substitution, $4(m\angle B) = 360$. Dividing each side by 4 yields $m\angle B = 90$. By Substitution, $m\angle G = m\angle K = m\angle M = 90$. So $\angle B$, $\angle G$, $\angle K$, and $\angle M$ are rt. angles. $\square BGKM$ is a rectangle by definition.

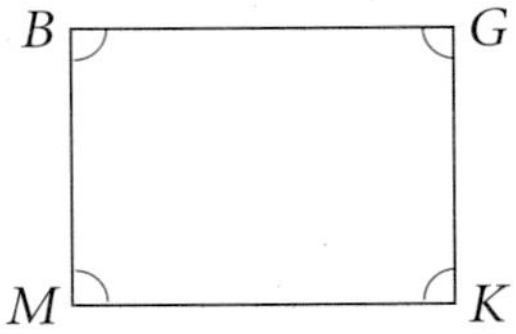

Write the first step of an indirect proof of each statement. ■ Lesson 4-5

9. $\triangle ABC$ is a right triangle.
10. Points *J*, *K*, and *L* are collinear.
11. Lines ℓ and *m* are parallel.

List the sides of each triangle in order from shortest to longest. ■ Lesson 4-6

12. 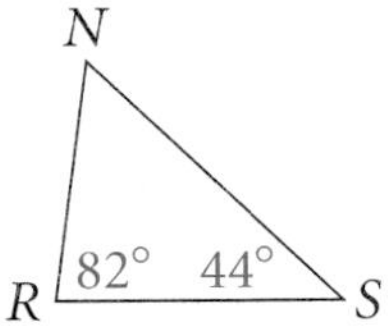

13. 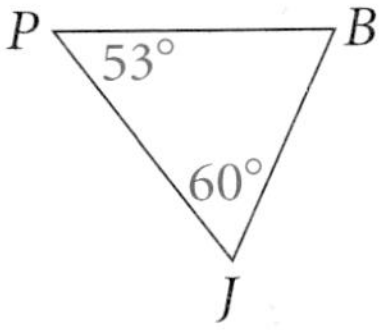

14. 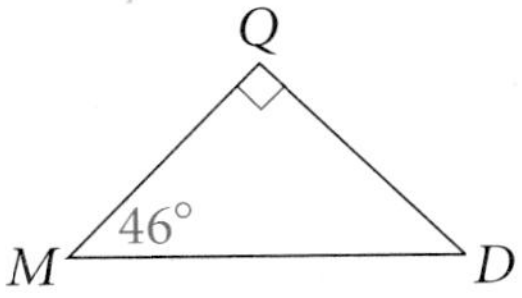

15. 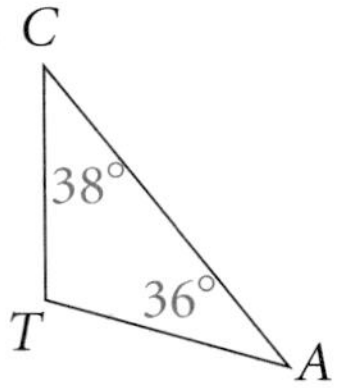

Sketch and label each locus. ■ Lesson 4-7

16. all points in a plane 2 cm from $\overrightarrow{AB}$
17. all points in space 1.5 in. from point *Q*
18. all points in a plane 3 cm from a circle with radius 2 cm

Is $\overline{AB}$ an angle bisector, altitude, median, or perpendicular bisector? ■ Lesson 4-8

19. 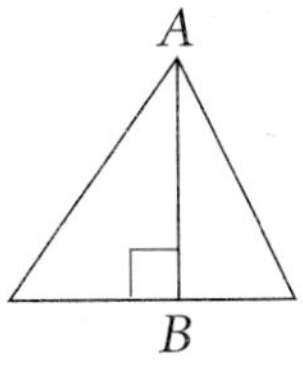

20. 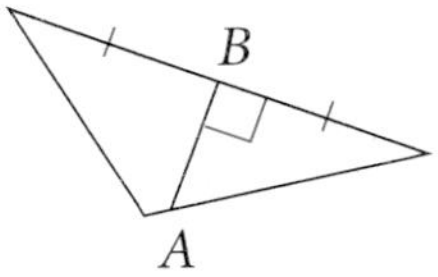

21. 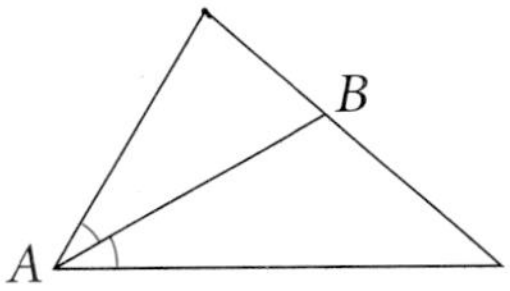

22. 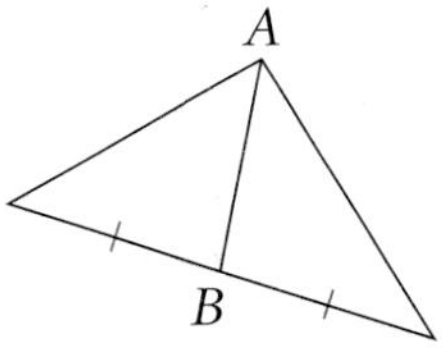

Extra Practice

Find the perimeter and area of each figure. ■ Lessons 5-1 and 5-2

1.
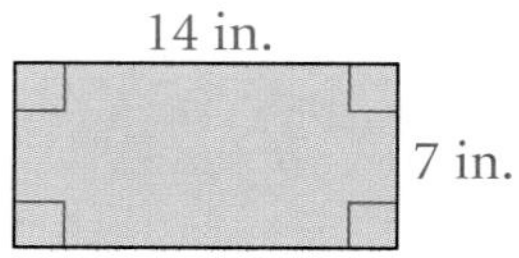

2.
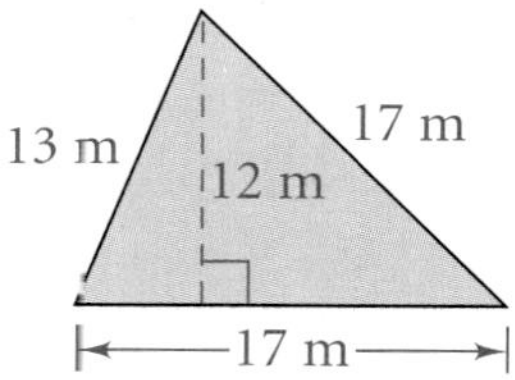

3.
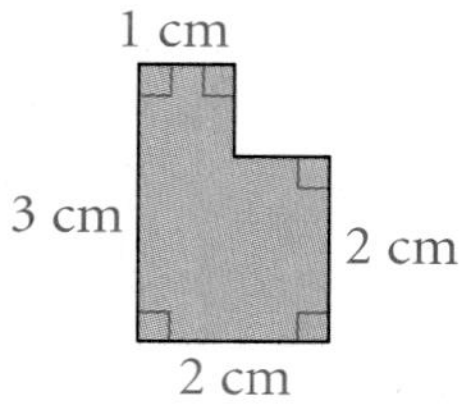

4.
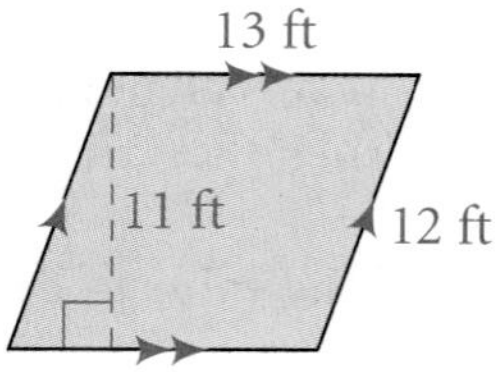

Algebra **Find the value of x. If your answer is not a whole number, leave it in simplest radical form.** ■ Lessons 5-3 and 5-4

5.
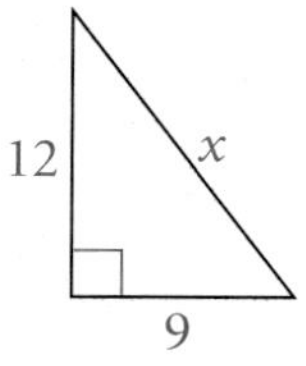

6.
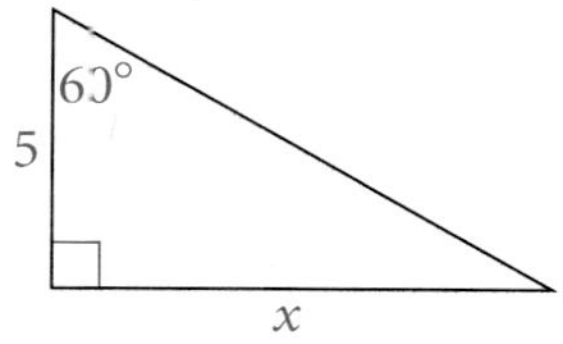

7.
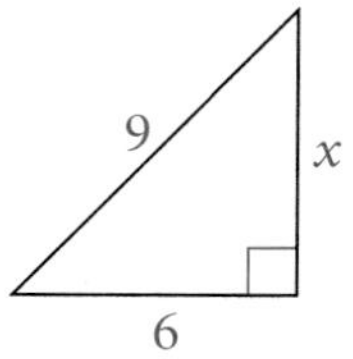

8.
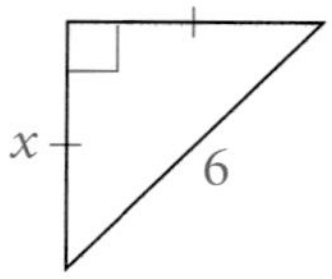

Find the area of each trapezoid or regular polygon. You may leave your answer in simplest radical form. ■ Lessons 5-5 and 5-6

9.
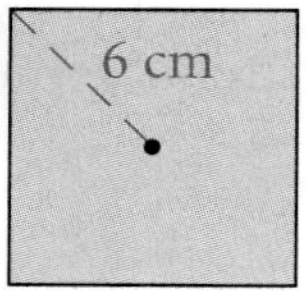

10.
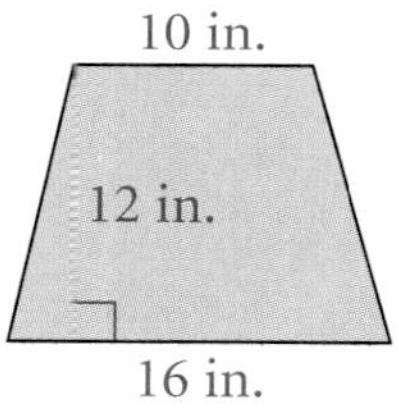

11.
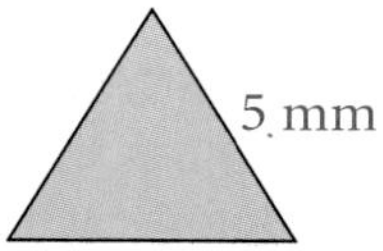

12.
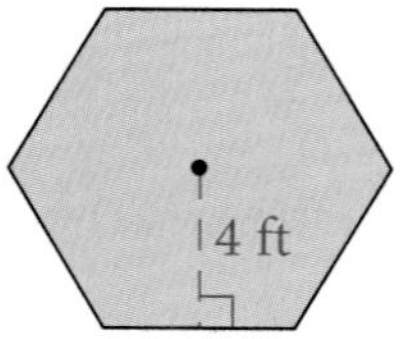

In Exercises 13–16: (a) Find the circumference of each circle. (b) Find the length of the arc shown in red. Leave your answers in terms of π. ■ Lesson 5-7

13.
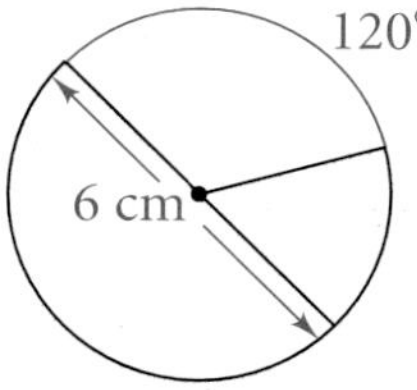

14.
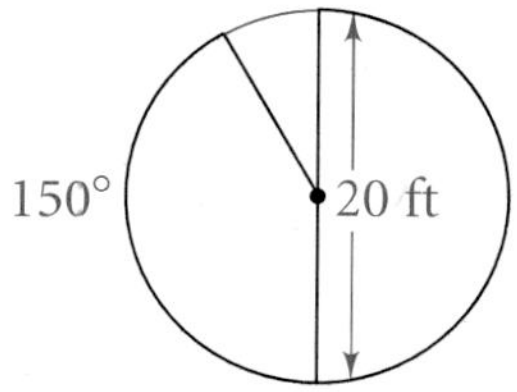

15.
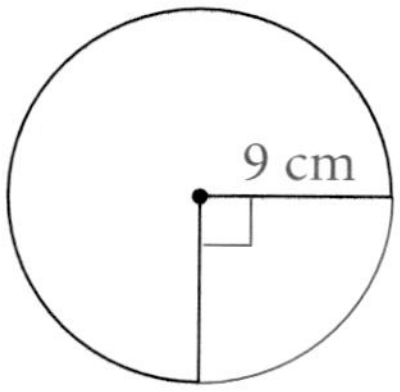

16.
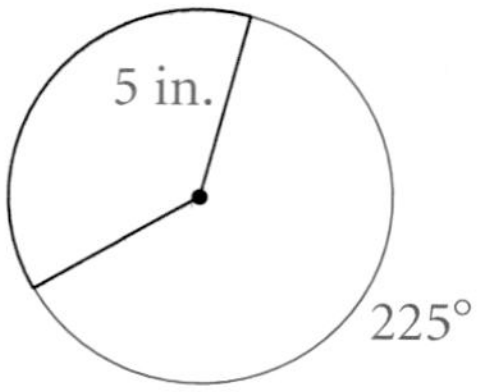

Find the area of each shaded sector or segment. Leave your answers in terms of π. ■ Lesson 5-8

17.
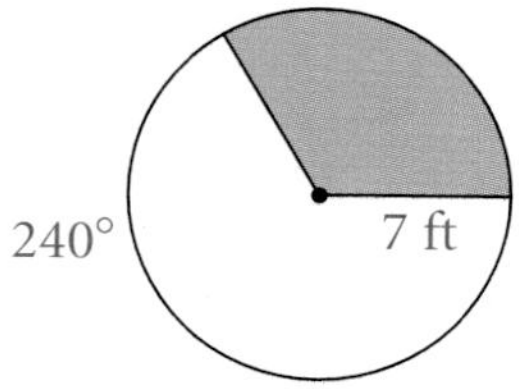

18.
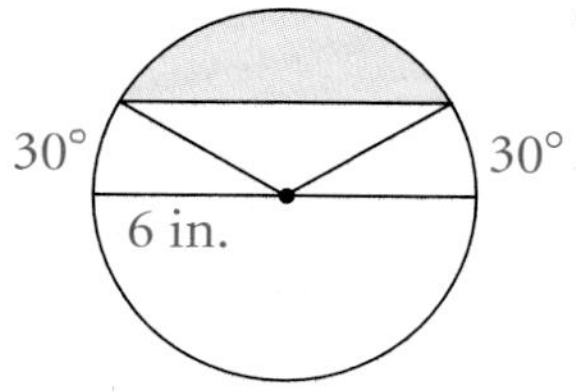

19.
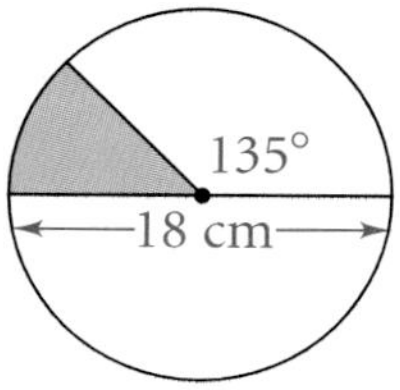

20.
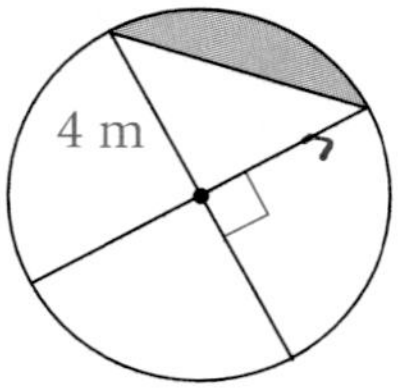

Lesson 6-1

Name the space figure that can be formed by folding each net.

1.
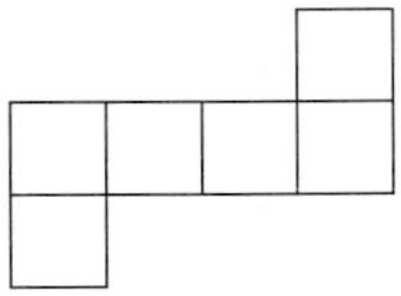

2.
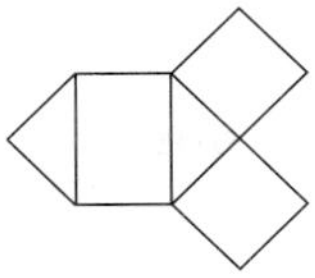

3.
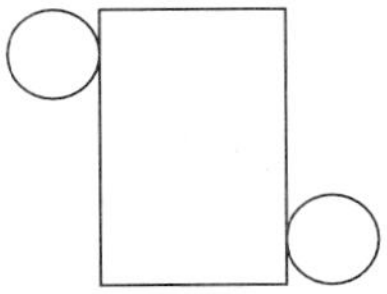

4.
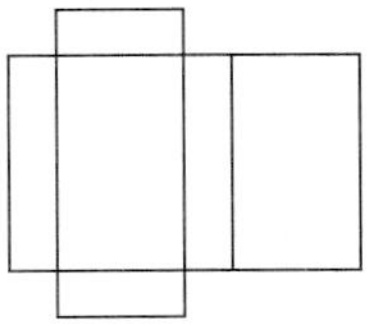

Lessons 6-2 and 6-3

In Exercises 5–25, you may leave your answers in terms of π.
Find the lateral area and surface area of each figure.

5.
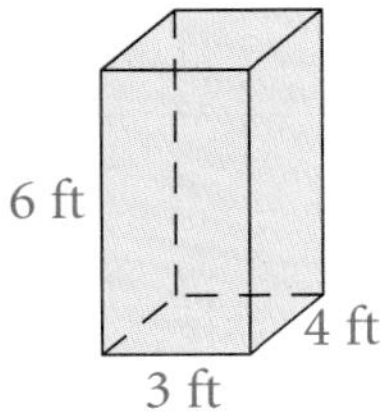

6.
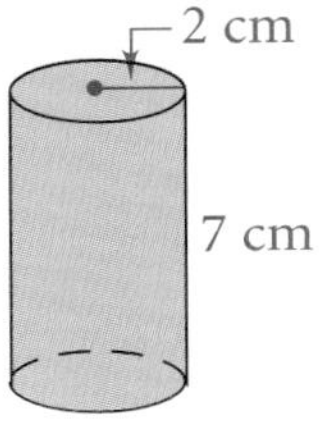

7.
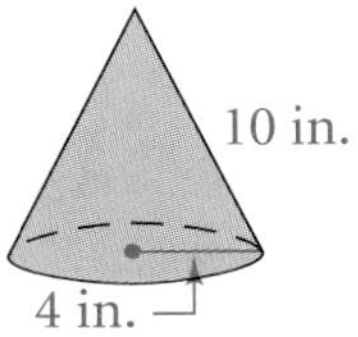

8.
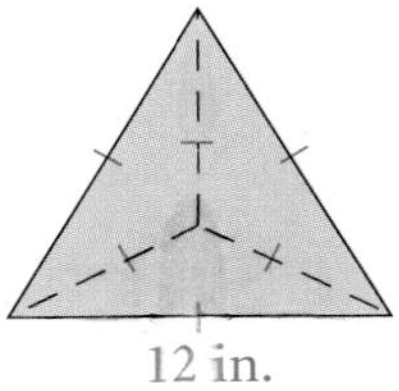

Lessons 6-4 and 6-5

Find the volume of each figure.

9.
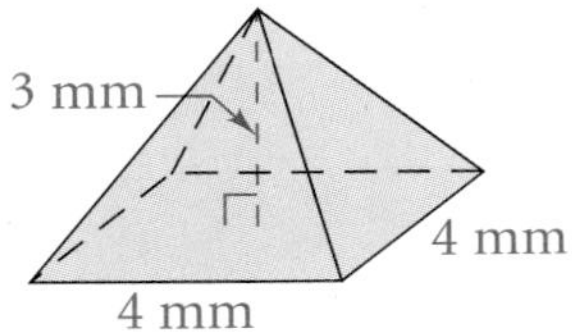

10.
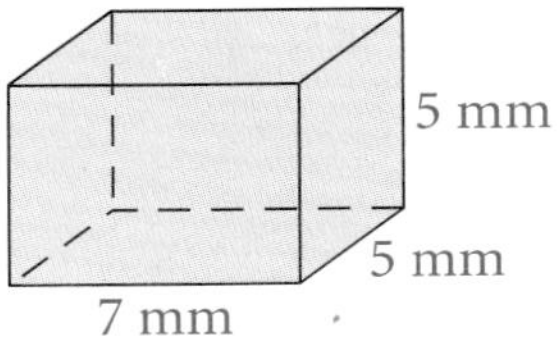

11.
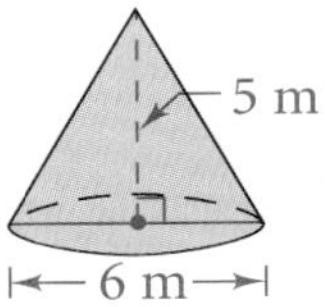

12.
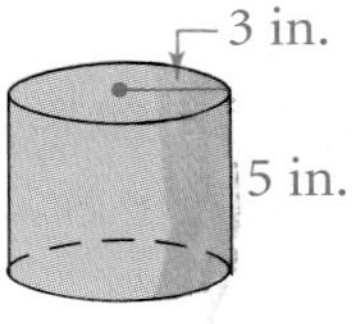

Lesson 6-6

Find the volume and surface area of a sphere with the given radius or diameter.

13. $r = 5$ cm **14.** $d = 8$ in. **15.** $d = 2$ ft **16.** $r = 0.5$ in. **17.** $d = 9$ m

Lesson 6-7

Find the volume of each composite space figure.

18.
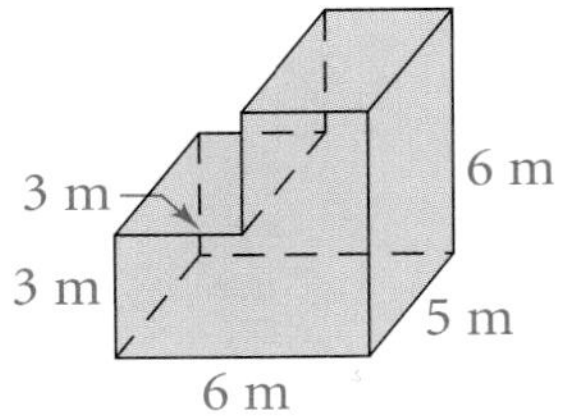

19.
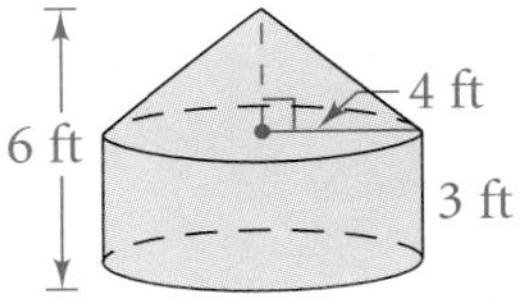

20.
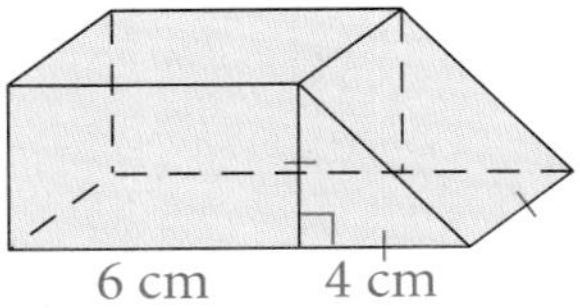

21.
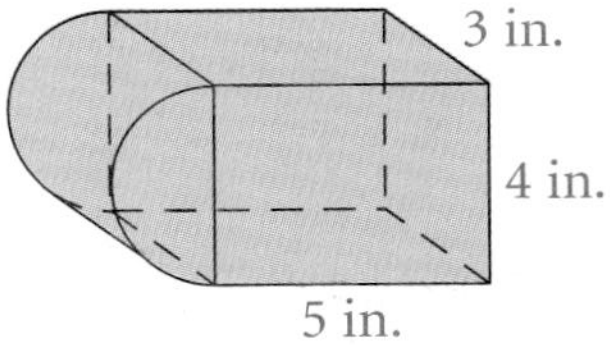

Lesson 6-8

Darts are thrown at random at each of the boards shown. If a dart hits the board, find the probability that it will land in the shaded area.

22.
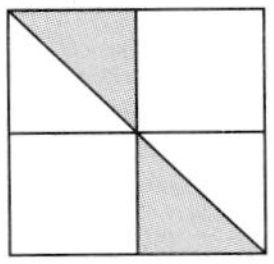

23.
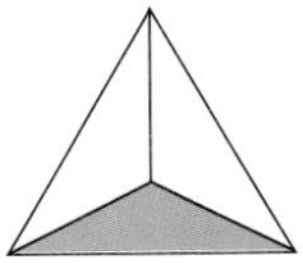

24.
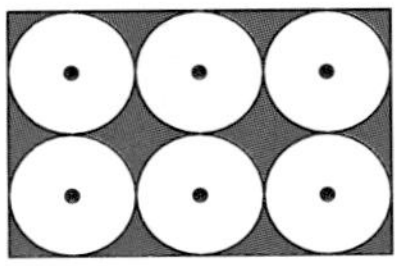

25.
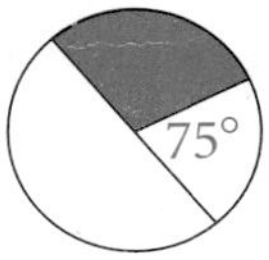

Extra Practice

Extra Practice

Find $m\angle 1$ and then $m\angle 2$. State the theorems or postulates that justify your answers.

Lesson 7-1

1.

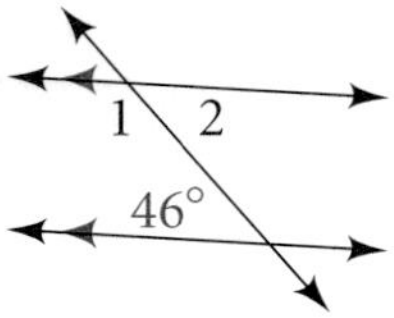

2.

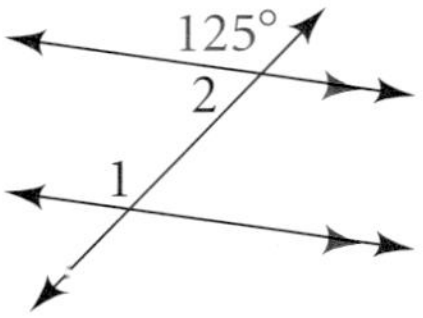

3.

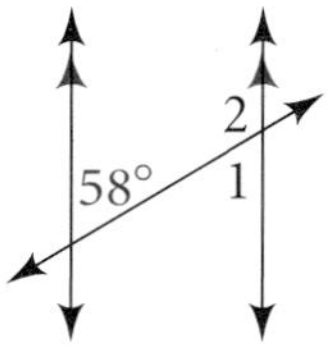

4.

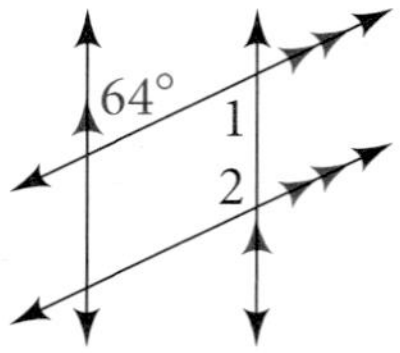

Refer to the diagram. Use the given information to determine which lines, if any, must be parallel. If any lines are parallel, use a theorem or postulate to tell why.

Lesson 7-2

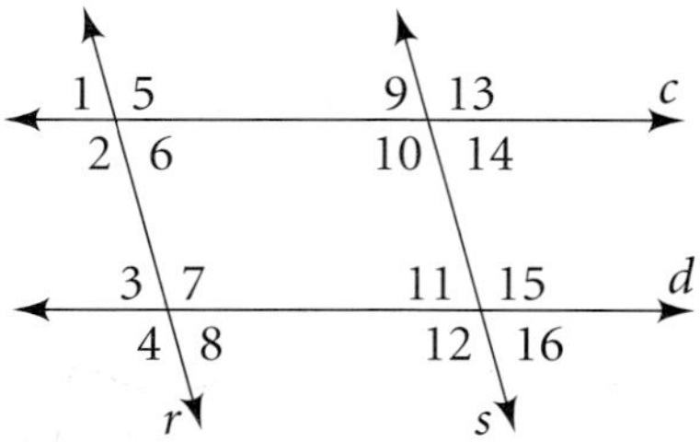

5. $\angle 9 \cong \angle 14$

6. $\angle 1 \cong \angle 9$

7. $\angle 2$ is supplementary to $\angle 3$.

8. $\angle 7 \cong \angle 14$

9. $m\angle 6 = 60$, $m\angle 13 = 120$

10. $\angle 4 \cong \angle 13$

11. $\angle 3$ is supplementary to $\angle 10$.

12. $\angle 10 \cong \angle 15$

Use the segments at the right for each construction.

Lesson 7-3

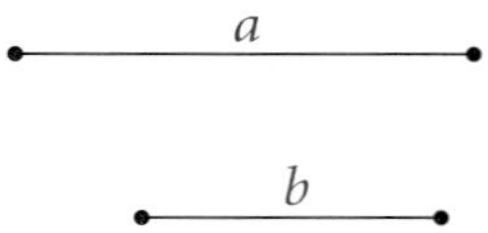

13. Construct a square with side length a.

14. Construct an isosceles right triangle with legs of length b.

15. Construct a trapezoid with bases of lengths a and b.

16. Construct a right triangle in which the length of a leg is b and the length of the hypotenuse is a.

Is each object drawn in one- or two-point perspective?

Lesson 7-4

17.

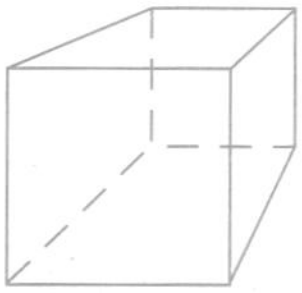

18.

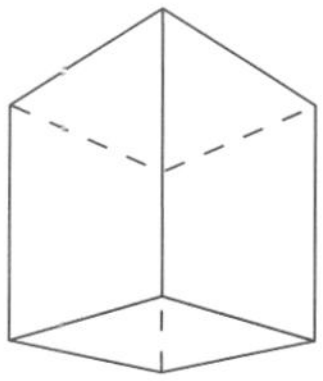

19.

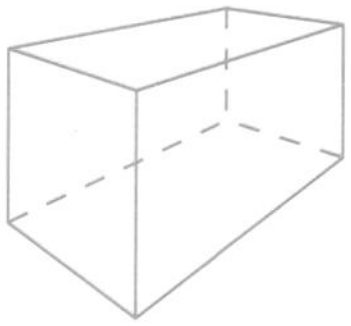

20.

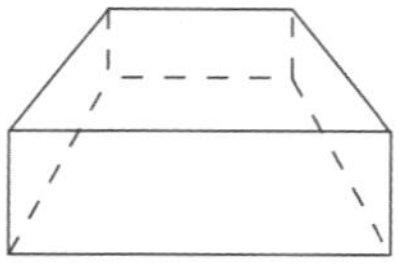

Draw a sketch to illustrate each property of spherical geometry.

Lesson 7-5

21. Two distinct lines intersect in two points.

22. A triangle can have three right angles.

23. The measure of an exterior angle of a triangle is less than the sum of the measures of the remote interior angles.

Extra Practice

Lessons 8-1 and 8-2

Can you prove that the triangles are congruent? If so, write the congruence and tell whether you would use SSS, SAS, ASA, or AAS. If not, write *not possible.*

1.

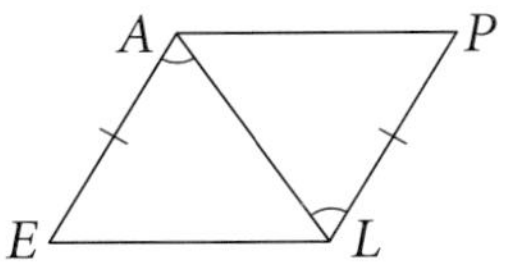

2.

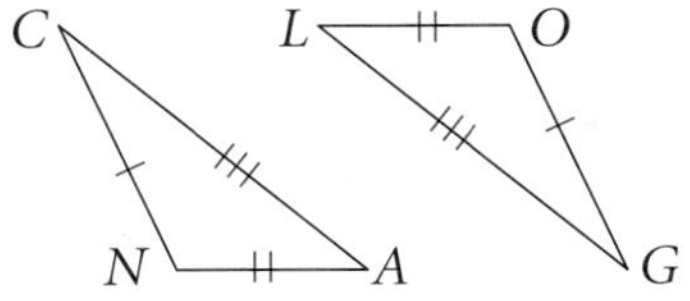

3.

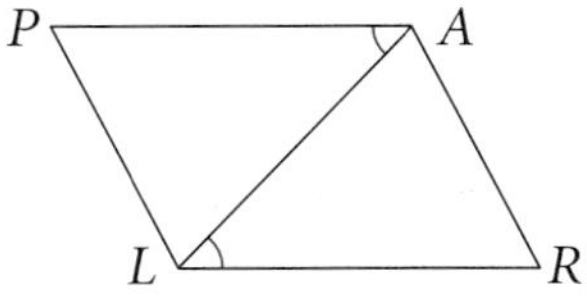

4.

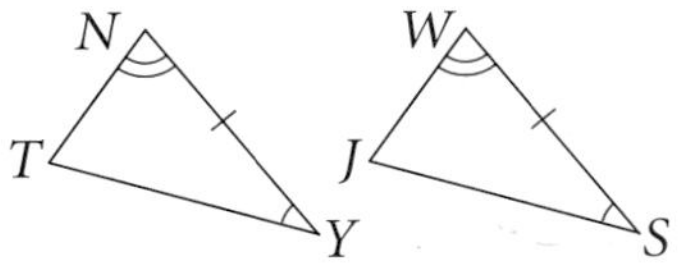

5.

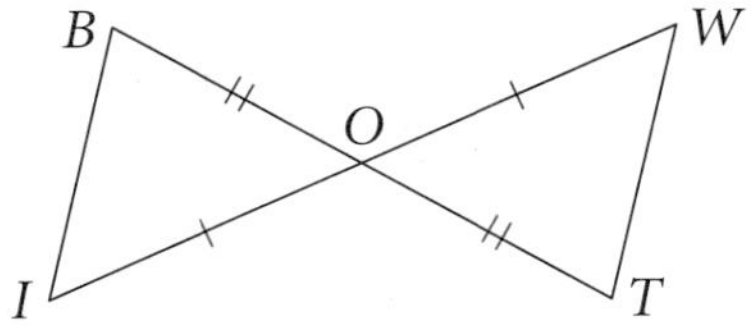

6.

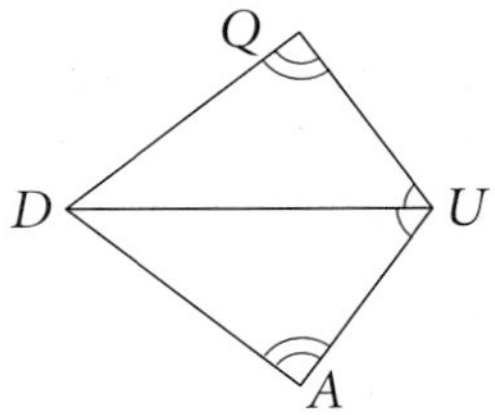

Lesson 8-3

What additional information would you need to prove the triangles congruent by the HL Theorem?

7.

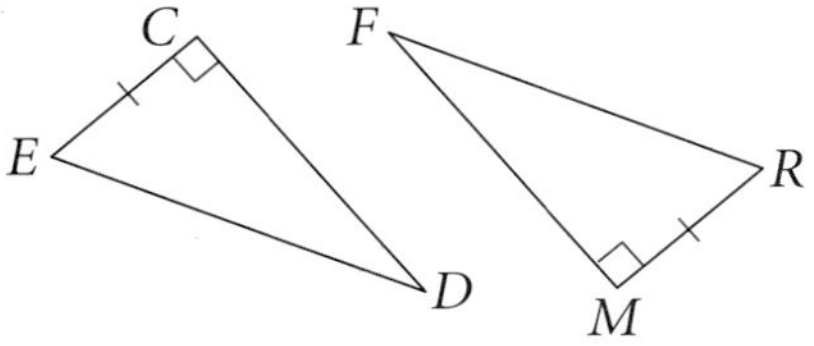

8.

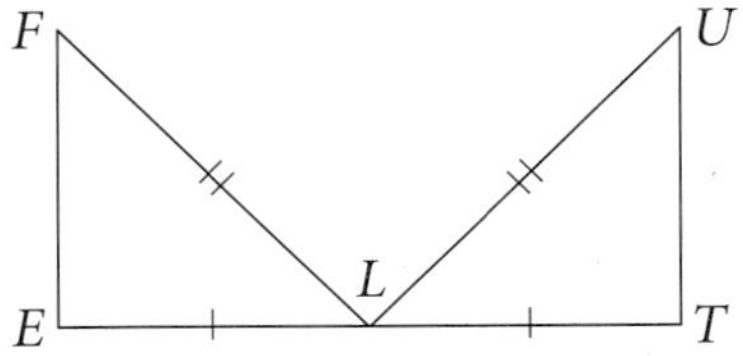

9.

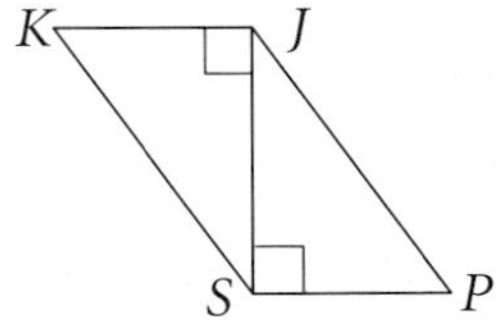

Lesson 8-4

Explain how you would use SSS, SAS, ASA, AAS, or HL with CPCTC to prove each statement.

10. $\angle MLN \cong \angle ONL$

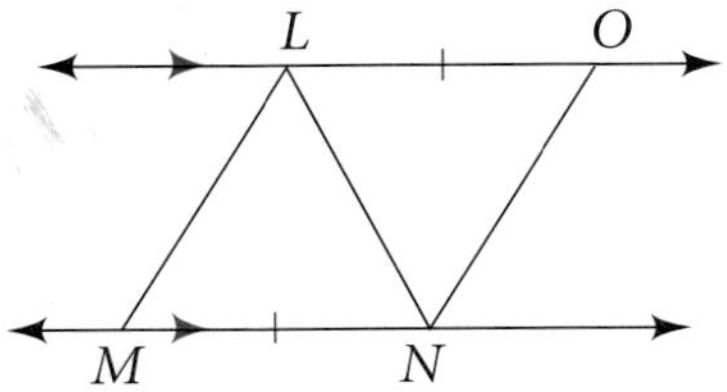

11. $\overline{TO} \cong \overline{ES}$

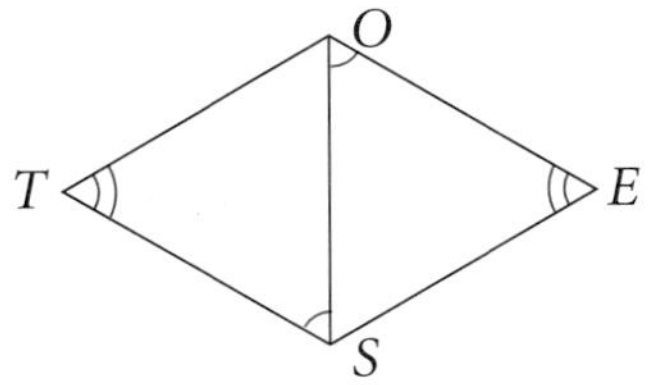

12. $\overline{MB} \cong \overline{RI}$

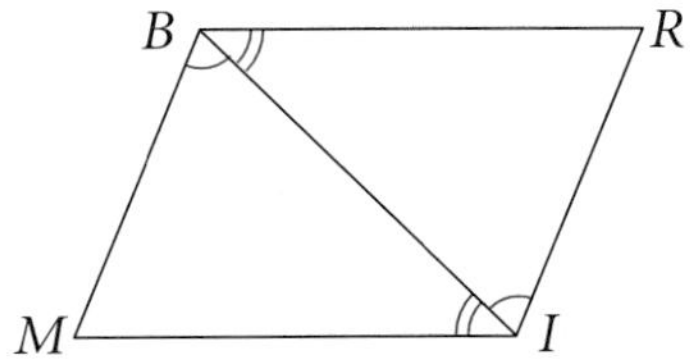

Lesson 8-5

Name a pair of overlapping congruent triangles in each diagram. State whether the triangles are congruent by SSS, SAS, ASA, AAS, or HL.

13.

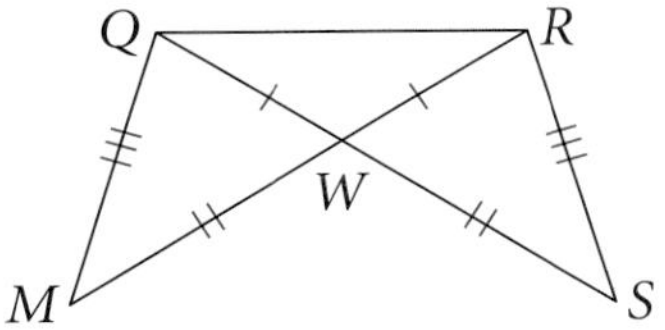

14.

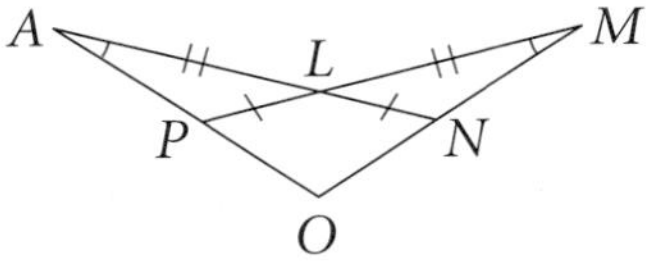

15. $\overline{AF} \cong \overline{BE}$

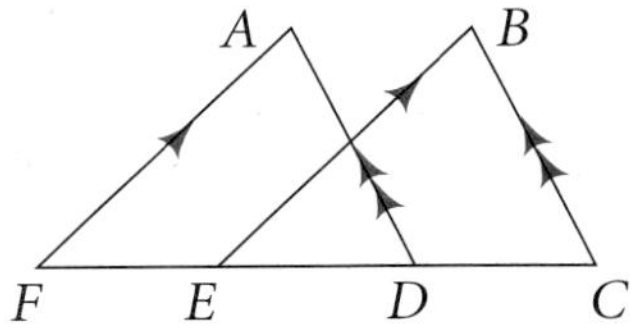

Extra Practice

Algebra **Find the values of the variables in each parallelogram.** ■ **Lesson 9-1**

1.
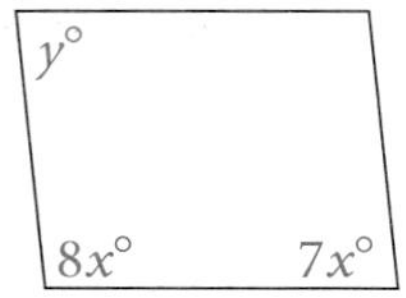

2.
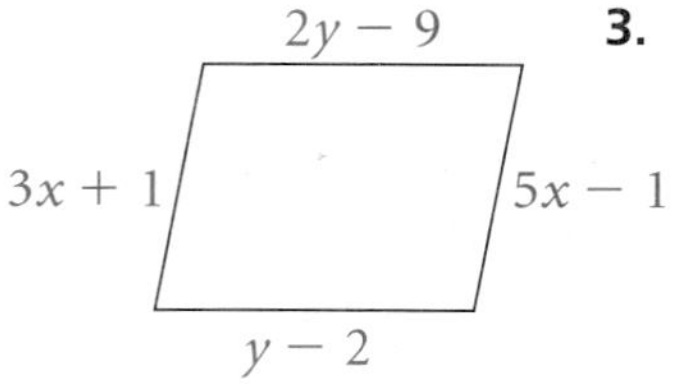

3.
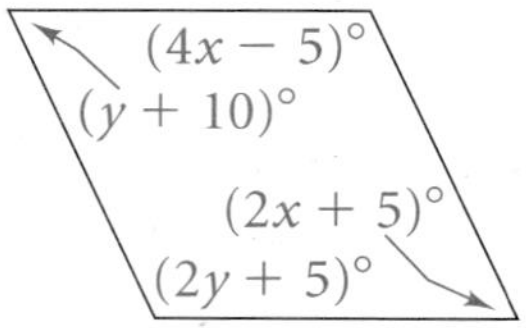

4.
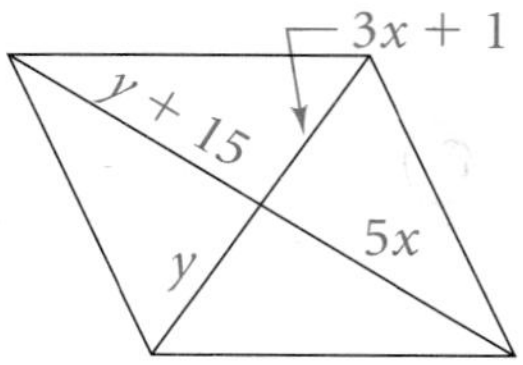

Based on the markings, decide whether each figure is a parallelogram. Justify your answers. ■ **Lesson 9-2**

5.
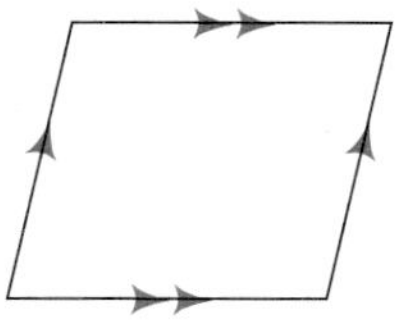

6.
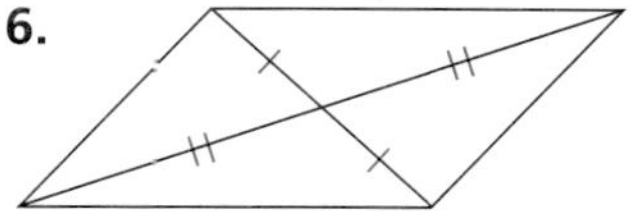

7.
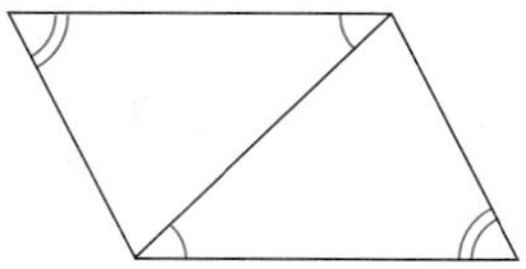

8.
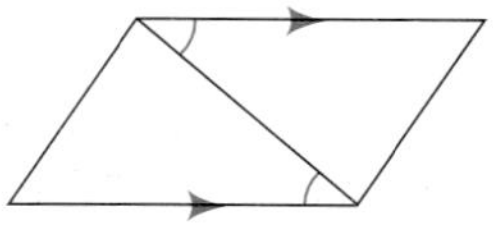

For each parallelogram, determine the most precise name and find the measures of the numbered angles. ■ **Lesson 9-3**

9.
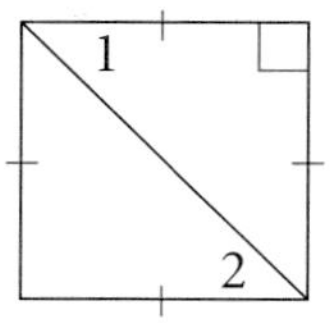

10.
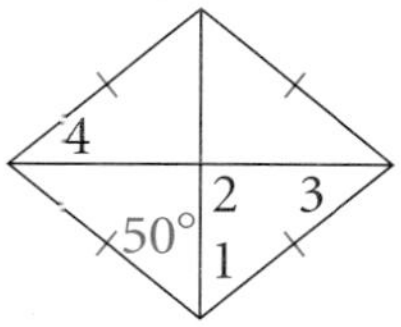

11.
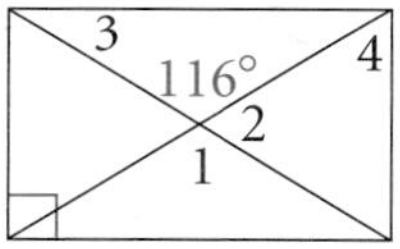

12.
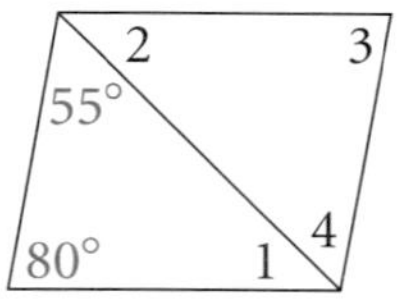

Find $m\angle 1$ and $m\angle 2$. ■ **Lesson 9-4**

13.
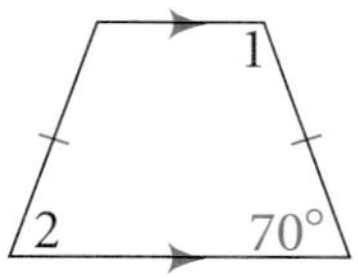

14.
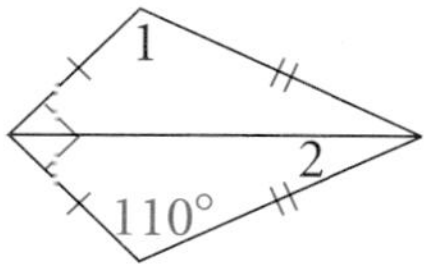

15.
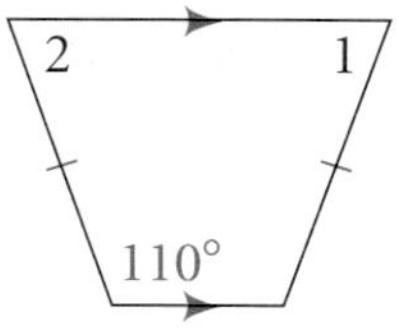

16.
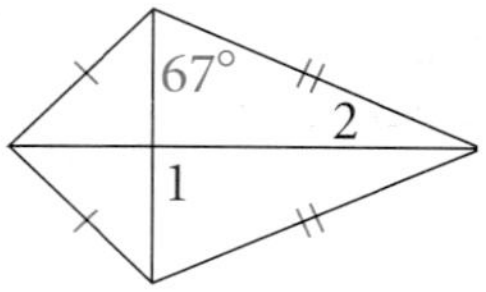

Give coordinates for points D and S without using any new variables. ■ **Lesson 9-5**

17. rectangle
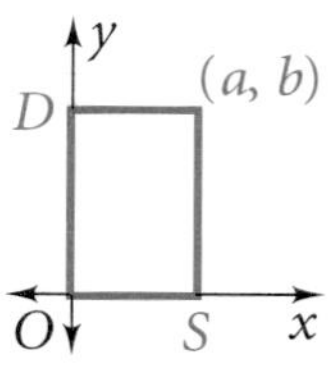

18. parallelogram
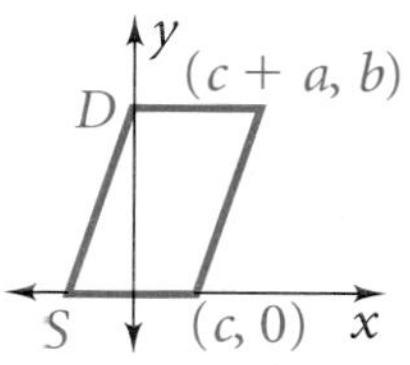

19. rhombus
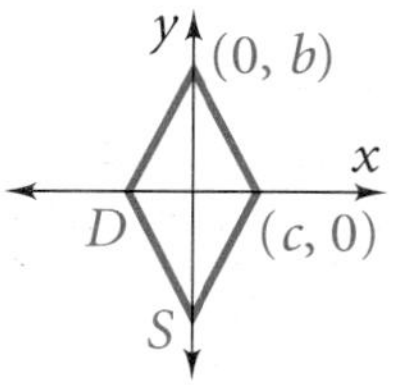

20. square
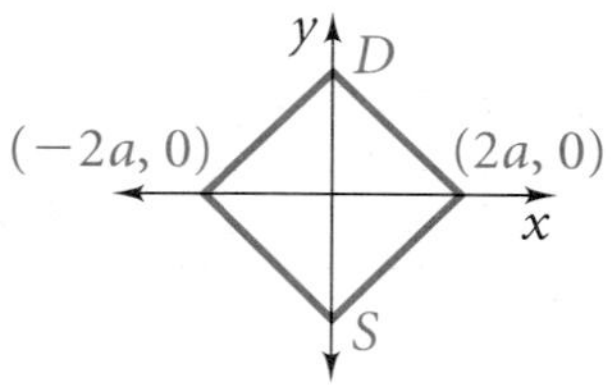

21. For the figure in Exercise 20, use coordinate geometry to prove that the midpoints of the sides of a square determine a square. ■ **Lesson 9-6**

Extra Practice

Algebra **Solve for x.** ■ Lesson 10-1

1. $\frac{2}{3} = \frac{x}{15}$ **2.** $\frac{4}{9} = \frac{16}{x}$ **3.** $\frac{x}{4} = \frac{6}{12}$ **4.** $\frac{2}{x} = \frac{3}{9}$ **5.** $\frac{3}{4} = \frac{x}{6}$ **6.** $\frac{3}{7} = \frac{9}{x}$

Can you prove that the triangles are similar? If so, write the similarity statement and tell whether you would use AA~, SAS~, or SSS~.

7.

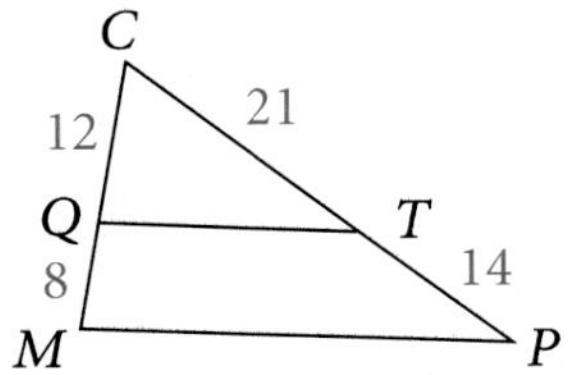

8.

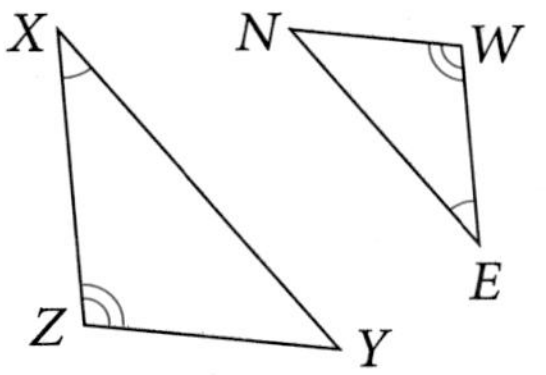

9.

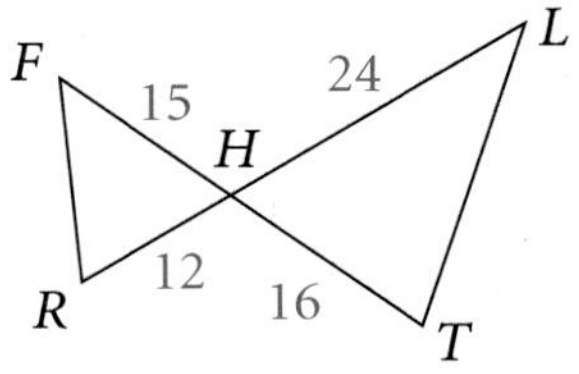

Algebra **Find the value of each variable. If an answer is not a whole number, leave it in simplest radical form.** ■ Lessons 10-3 and 10-4

10.

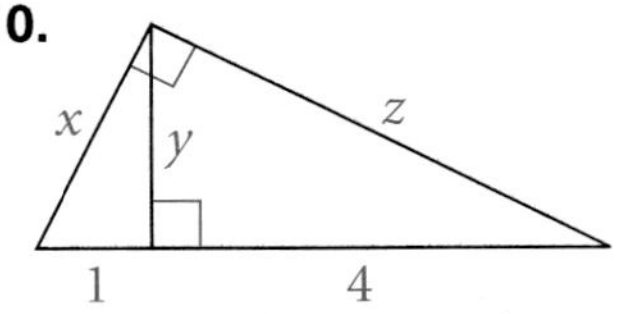

11.

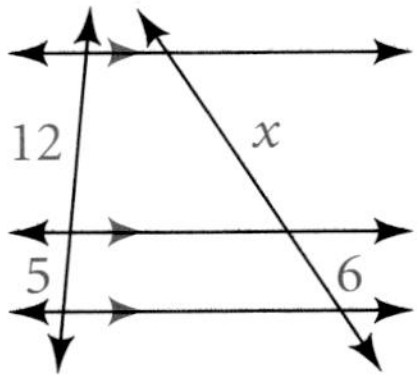

12.

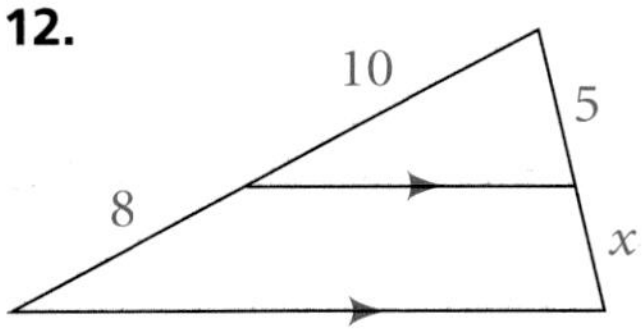

13.

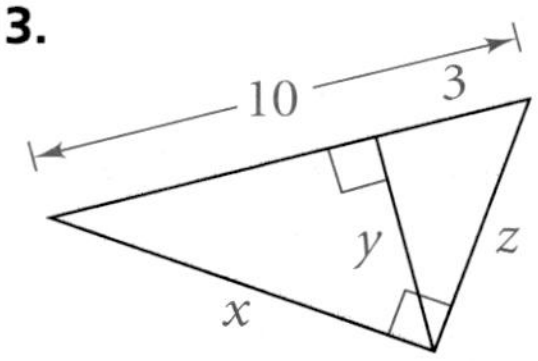

14.

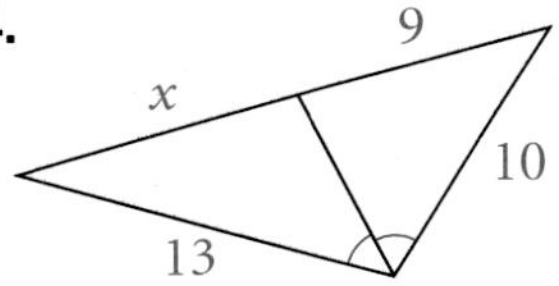

15.

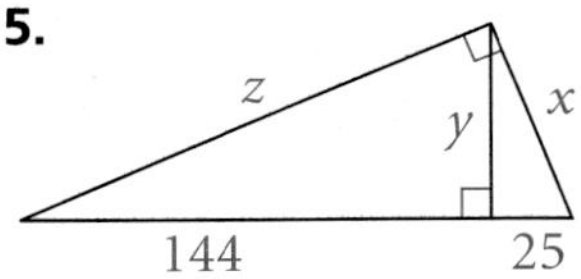

16.

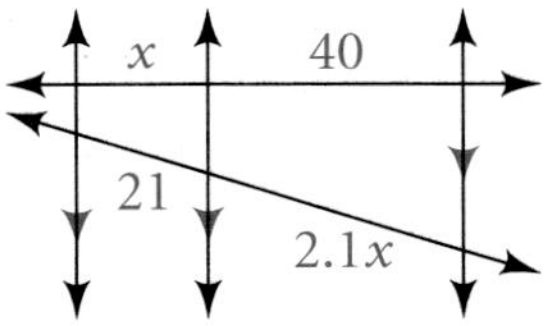

17.

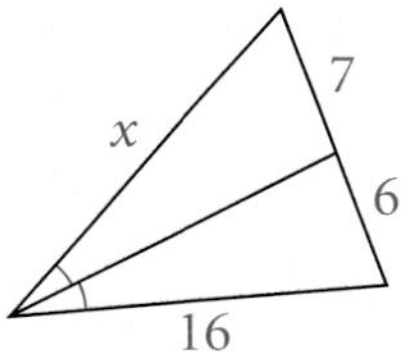

Find the ratio of the perimeters and the ratio of the areas of the blue figure to the red one. ■ Lesson 10-5

18.

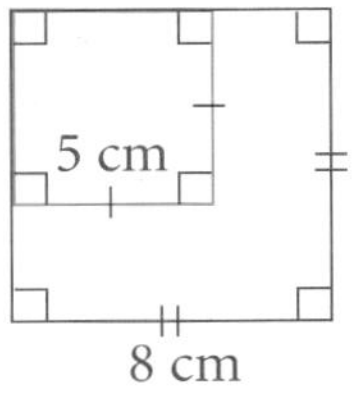

19.

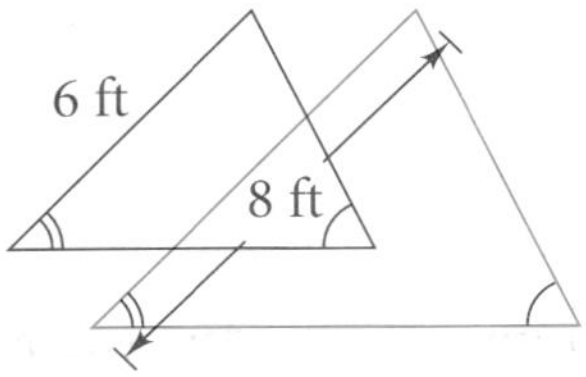

20.

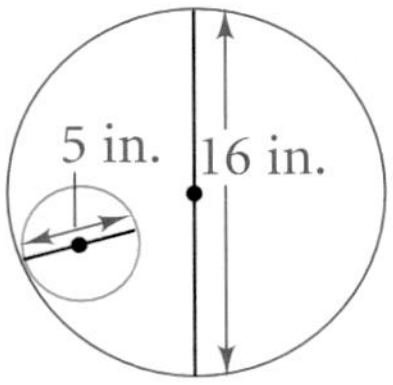

Copy and complete for two similar solids. ■ Lesson 10-6

	Similarity Ratio	Ratio of Surface Areas	Ratio of Volumes
21.	2 : 3	■ : ■	■ : ■
22.	■ : ■	25 : 64	■ : ■
23.	■ : ■	■ : ■	27 : 64

Extra Practice

Find the value of x. Round lengths of segments to the nearest tenth and angle measures to the nearest degree. ■ Lessons 11-1 and 11-2

1.
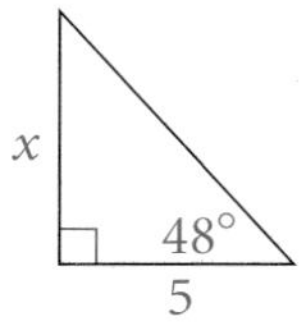

2.
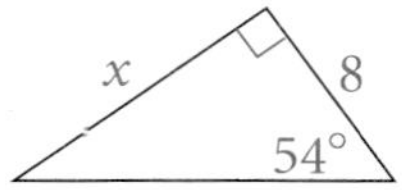

3.
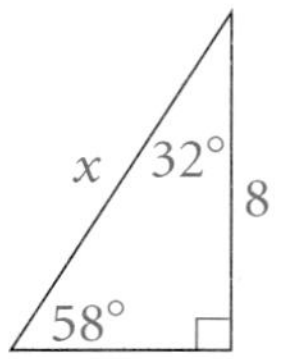

4.
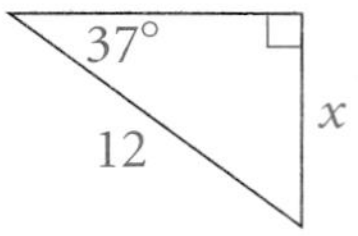

5.
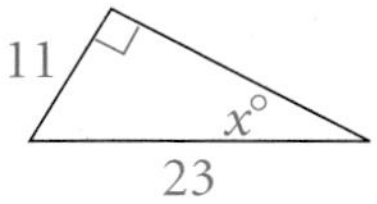

6.
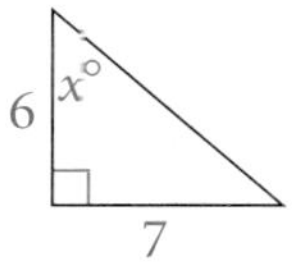

7.
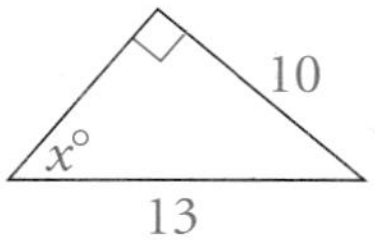

8.
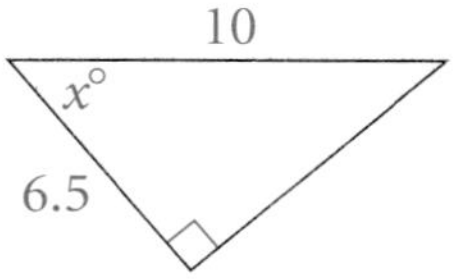

Solve each problem. Round your answers to the nearest foot. ■ Lesson 11-3

9. A couple is taking a balloon ride. After 25 minutes aloft, they measure the angle of depression from the balloon to its launch place as 16°. They are 180 ft above ground. Find the distance from the balloon to its launch place.

10. A surveyor is 300 ft from the base of an apartment building. The angle of elevation to the top of the building is 24°, and her angle-measuring device is 5 ft above the ground. Find the height of the building.

Coordinate Geometry **In Exercises 11–14: (a) Describe each vector by using ordered pair notation. Give the coordinates to the nearest unit. (b) Find the ordered pair notation for the sum of each pair of vectors.** ■ Lessons 11-4 and 11-5

11.
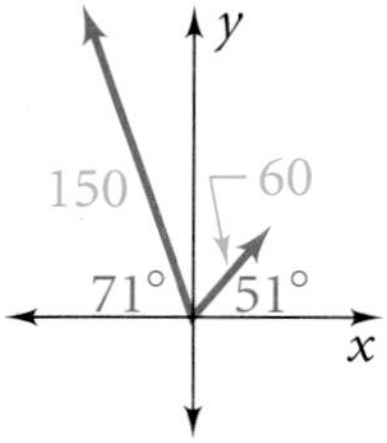

12.
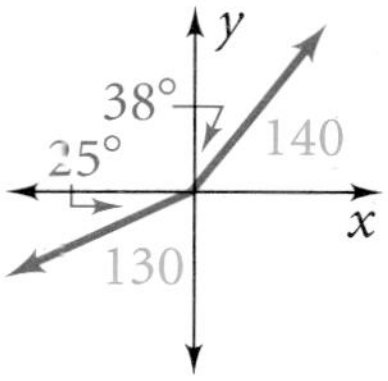

13.
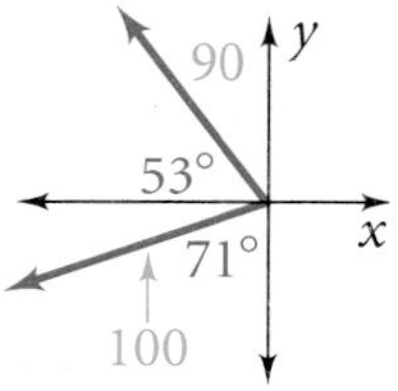

14.
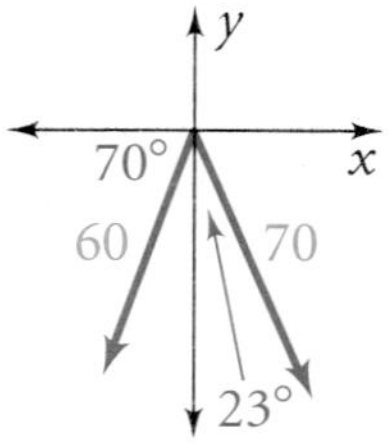

Find the area of each polygon. Give your answers to the nearest tenth. ■ Lesson 11-6

15.
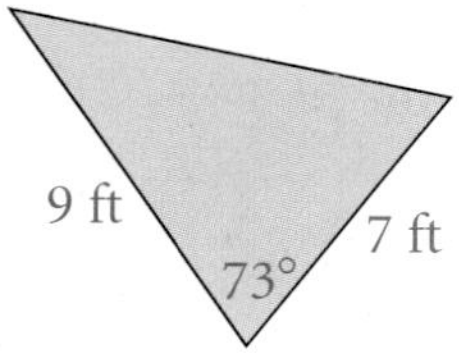

16.
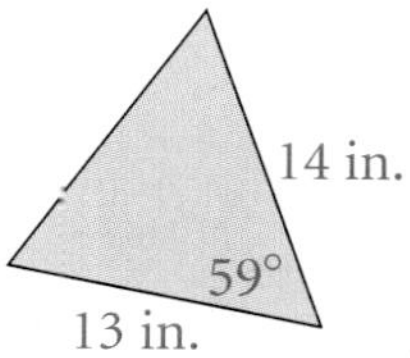

17.
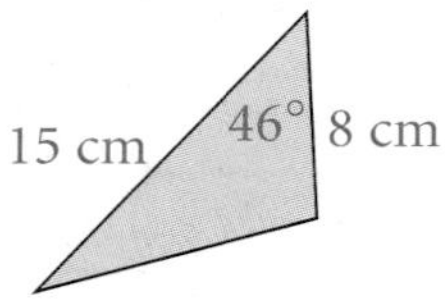

18.
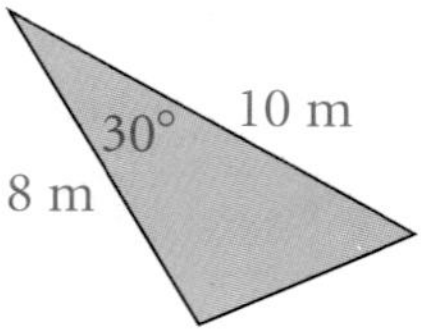

19. regular hexagon with apothem 3 ft

20. regular octagon with radius 5 ft

Extra Practice

Write an equation of each circle. ■ Lesson 12-1

1. center $(0, 0)$; $r = 4$
2. center $(-1, 4)$; through $(-1, 9)$
3. center $(9, -3)$; $r = 7$
4. center $(-4, 0)$; through $(2, 1)$
5. center $(-6, -2)$; through $(-8, 1)$
6. center $(0, 5)$; $r = 3$

Assume that lines that appear to be tangent are tangent. *P* is the center of each circle. Find the value of *x*. ■ Lesson 12-2

7. 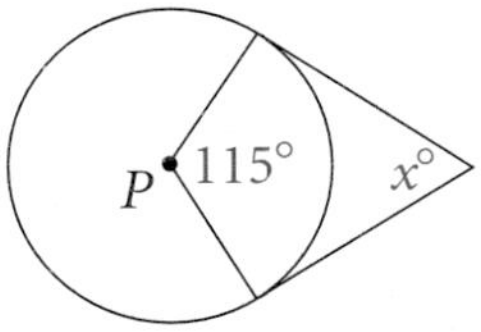

8. 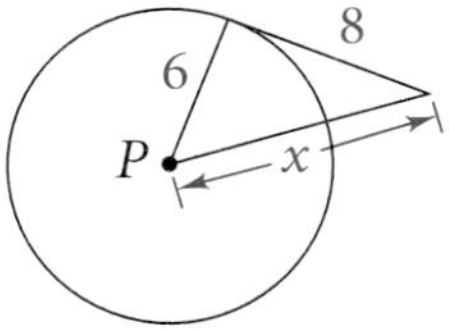

9. 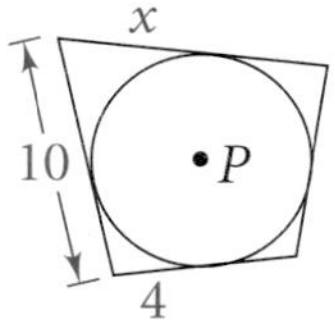

10. 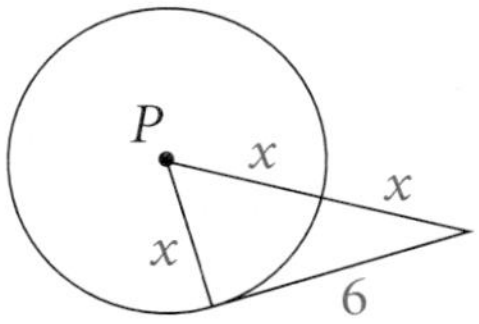

Find the value of each variable. If your answer is not an integer, round it to the nearest tenth. ■ Lessons 12-3 and 12-4

11. 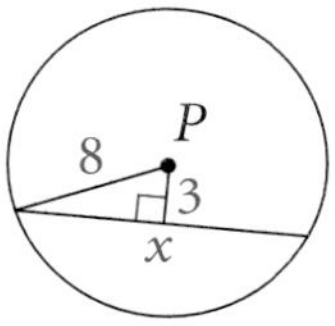

12. 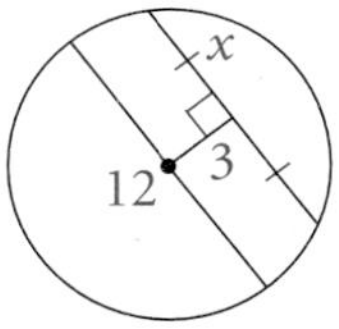

13. 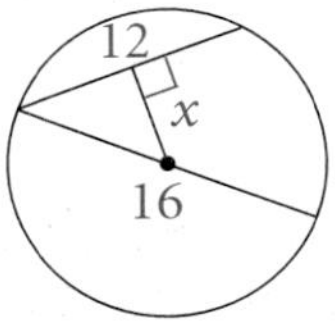

14. 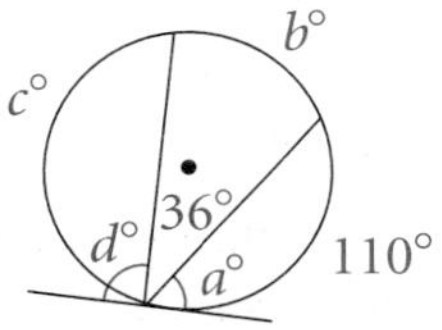

15. 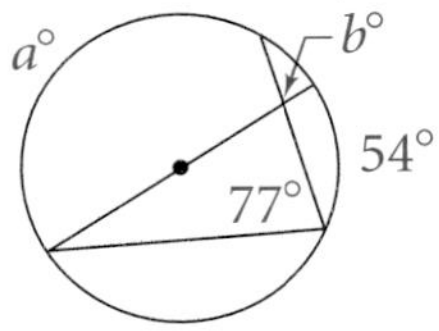

16. 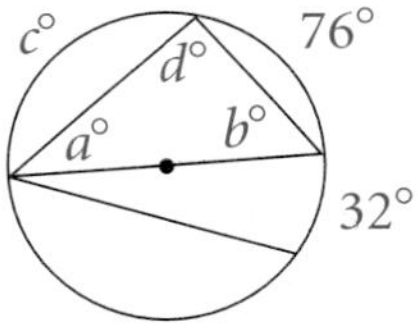

17. 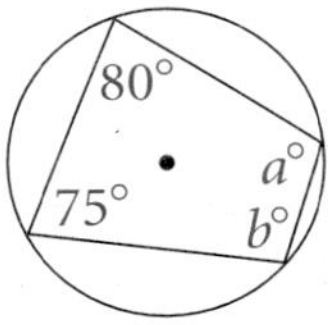

18. 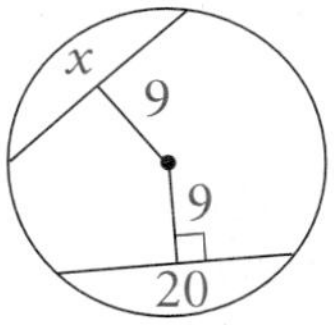

Assume that lines that appear to be tangent are tangent. Find the value of each variable. If your answer is not an integer, round it to the nearest tenth. ■ Lessons 12-5 and 12-6

19. 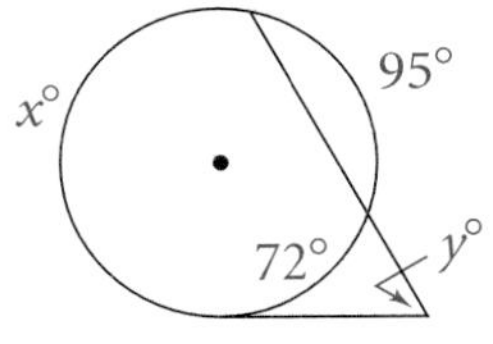

20. 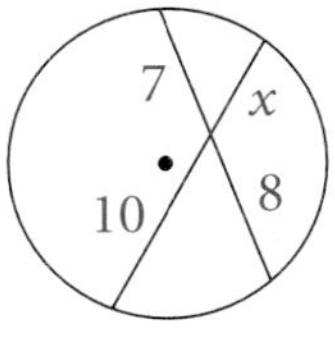

21. 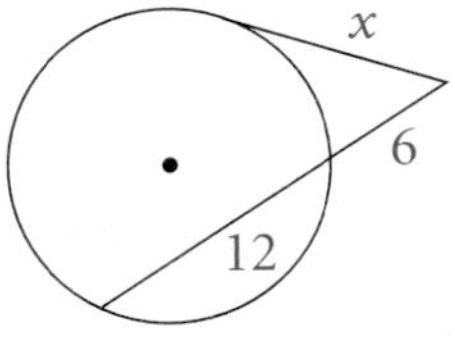

22. 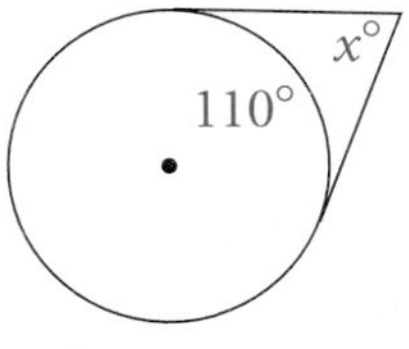

23. 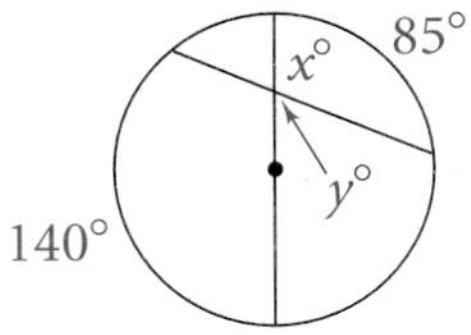

24. 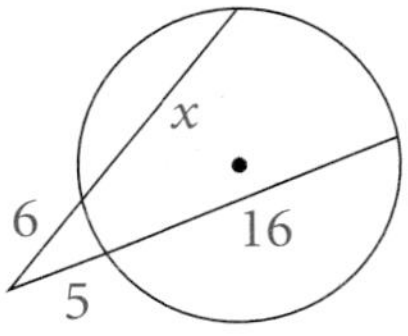

25. 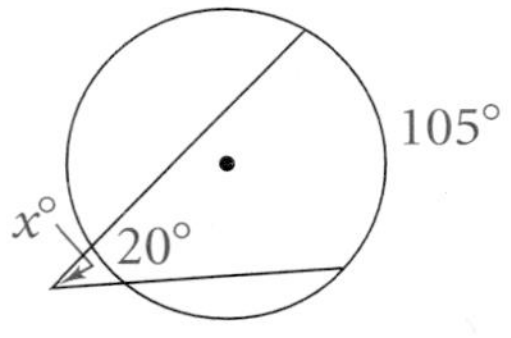

26. 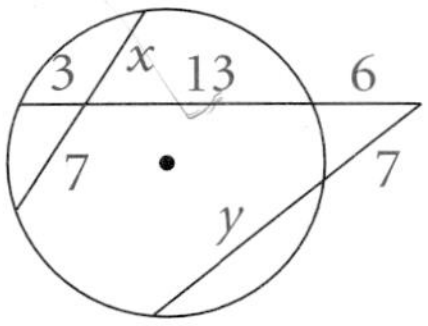

Problem Solving Strategies

You may find one or more of these strategies helpful in solving a word problem.

STRATEGY	WHEN TO USE IT
Draw a Diagram	You need help in visualizing the problem.
Guess and Test	Solving the problem directly is too complicated.
Look for a Pattern	The problem describes a relationship.
Make a Table	The problem has data that need organizing.
Solve a Simpler Problem	The problem is complex or has numbers that are too unmanageable to use at first.
Use Logical Reasoning	You need to reach a conclusion from some given information.
Work Backward	The answer can be arrived at by undoing various operations.

Problem Solving: Draw a Diagram

Example **Antoine is 1.91 m tall. He measures his shadow and finds that it is 2.34 m long. He then measures the length of the shadow of a flagpole and finds that it is 13.2 m long. How tall is the flagpole?**

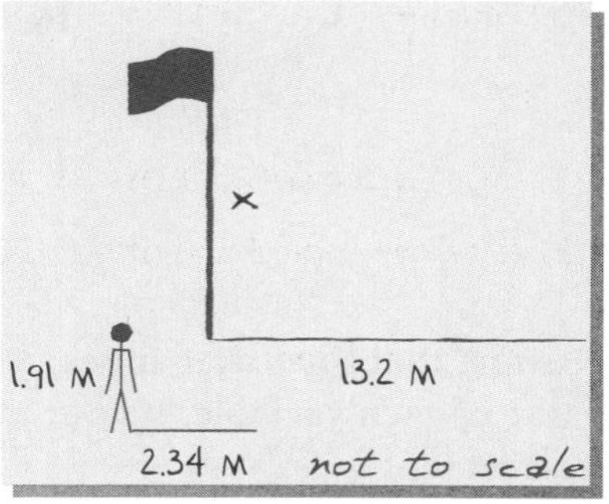

Start by drawing a diagram showing the given information. The diagram shows that the problem can be solved by using a proportion.

$\frac{1.91}{2.34} = \frac{x}{13.2}$ ← Write a proportion.

$x \approx 10.77$ ← Solve for x.

The flagpole is about 10.8 m tall.

EXERCISES

1. Five people meet and shake hands with one another. How many handshakes are there in all?
2. Three tennis balls fit snugly in a standard cylindrical container. Which is greater, the circumference of a ball or the height of the container?
3. Three lines that each intersect a circle can determine at most 7 regions within the circle, as shown in the diagram. What is the greatest number of regions that can be determined by 5 lines?
4. A triangle has vertices (1, 3), (2, 3), and (7, 5). Find its area.

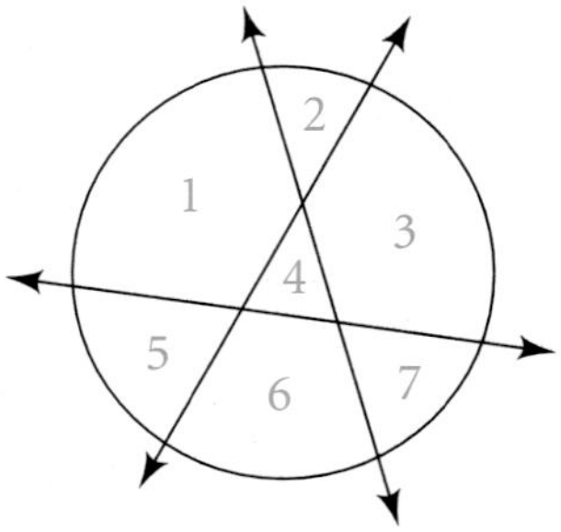

Problem Solving: Guess and Test

Have you ever weighed yourself on a balance scale at a doctor's office? You start by guessing your weight, then you see if the scale balances. If it doesn't, you slide the weights back and forth until the scale does balance. This is an example of the *Guess and Test* strategy, a strategy that is helpful for solving many types of problems.

Example **You are offered two payment plans for a CD club. Plan 1 involves paying a \$20 membership fee and \$7 per CD. Plan 2 involves paying no membership fee and \$10 per CD. What is the least number of CDs you would have to buy to make Plan 1 the less expensive plan?**

	Plan 1	Plan 2	
Guess 4 CDs.	$20 + 7(4)$	$10(4)$	
	$48 >$	40	← Too low. Guess higher.
Guess 7 CDs.	$20 + 7(7)$	$10(7)$	
	$69 <$	70	← Plan 1 is less expensive!

You need to check 6 CDs, however, to be sure that 7 CDs is the *least* number you would have to buy to make Plan 1 less expensive.

$20 + 7(6)$	$10(6)$	
$62 >$	60	← Plan 1 is more expensive.

You need to buy at least 7 CDs to make Plan 1 the less expensive plan.

EXERCISES

1. Find three consecutive even integers whose product is 480.
2. The combined age of a father and his twin daughters is 54 years. The father was 24 years old when the twins were born. How old is each of the three people?
3. What numbers can x represent in the rectangle?

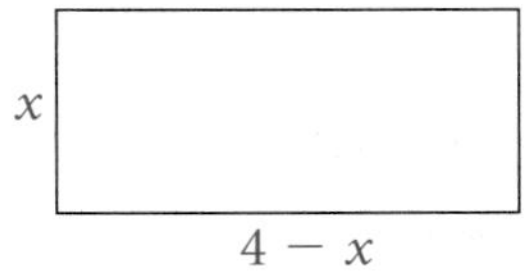

4. You are offered two payment plans for a video rental store. Plan 1 involves paying a \$30 membership fee and \$2 per rental. Plan 2 involves paying \$3.50 per rental. What is the least number of videos you would have to rent to make Plan 1 the less expensive plan?
5. Ruisa bought 7 rolls of film to take 192 pictures on a field trip. Some rolls had 36 exposures and the rest had 24 exposures. How many rolls of each type did Ruisa buy?
6. The sum of five consecutive integers is 5. Find the integers.
7. Paul buys a coupon for \$20 that allows him to see movies for half price at a local theater over the course of one year. The cost of seeing a movie is normally \$7.50. What is the least number of movies Paul would have to see to pay less than the normal price?

Problem Solving: Make a Table and Look for a Pattern

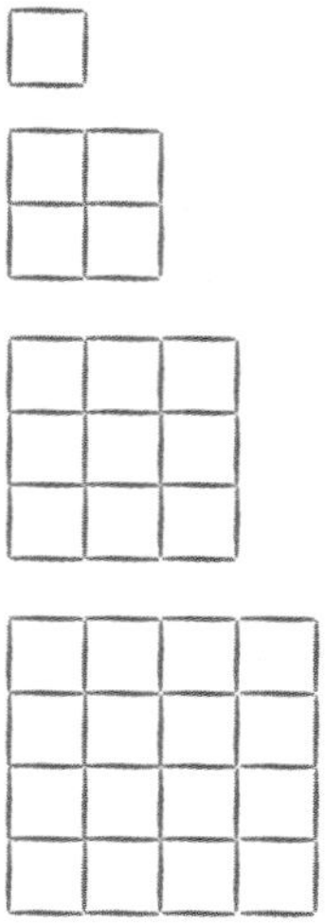

There are two important ways that making a table can help you solve a problem. First, a table is a handy method of organizing information. Second, once the information is in a table, it is easier to find patterns.

Example **The squares at the right are made of toothpicks. How many toothpicks are in the square with 7 toothpicks on a side?**

Make a table to organize the information.

No. of toothpicks on a side	1	2	3	4
Total no. of toothpicks in square	4	12	24	40

+8 +12 +16

Notice the pattern in the increases in the total number of toothpicks in each figure. The total number in the 5th square is 40 + 20, or 60. The number in the 6th is 60 + 24, or 84, and the number in the 7th is 84 + 28, or 112.

EXERCISES

1. The triangles are made up of toothpicks. How many toothpicks are in Figure 10?

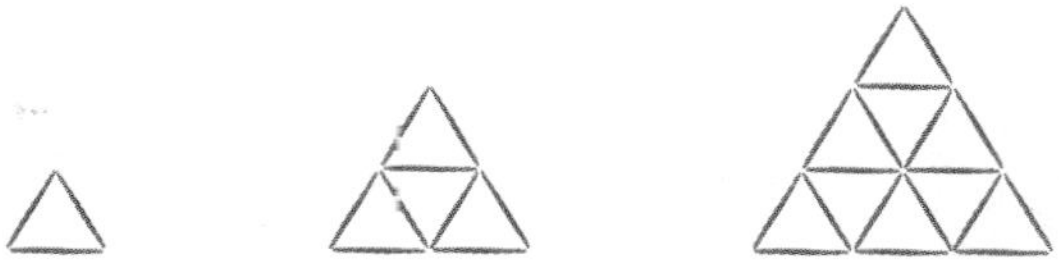

Figure 1 Figure 2 Figure 3

2. In each figure, the vertices of the smallest square are midpoints of the sides of the next larger square. Find the area of the ninth shaded square.

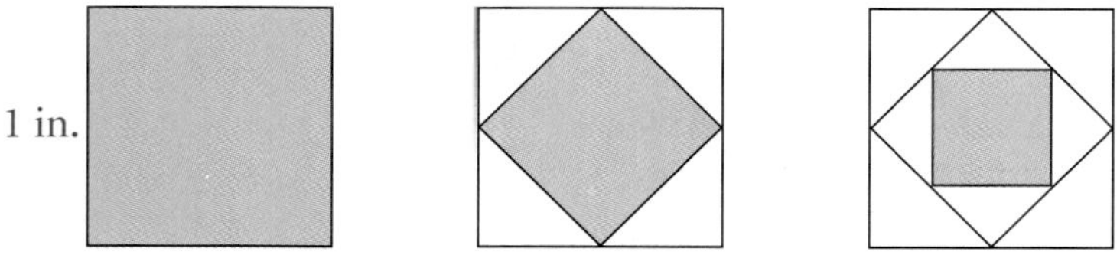

3. This pattern is known as the Sierpinski triangle. Find the total number of shaded triangles in Figure 8.

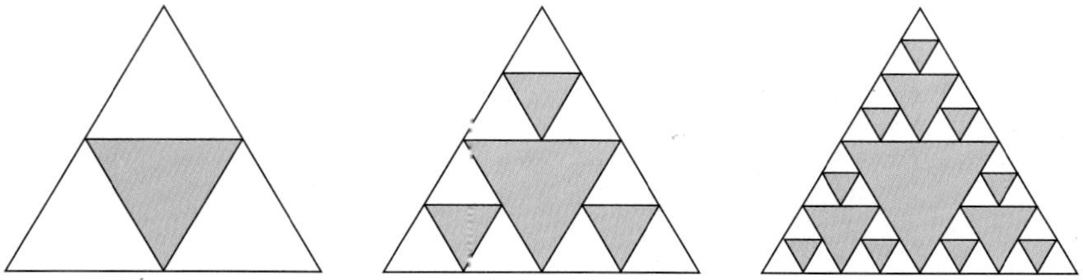

Figure 1 Figure 2 Figure 3

Problem Solving: Solve a Simpler Problem

Looking at a simpler version of a problem can be helpful in suggesting a problem-solving approach.

Example **A fence along the highway is 570 meters long. There is a fence post every 10 meters. How many fence posts are there?**

You may be tempted to divide 570 by 10, getting 57 as an answer, but looking at a simpler problem suggests that this answer isn't right. Suppose there are just 10 or 20 meters of fencing.

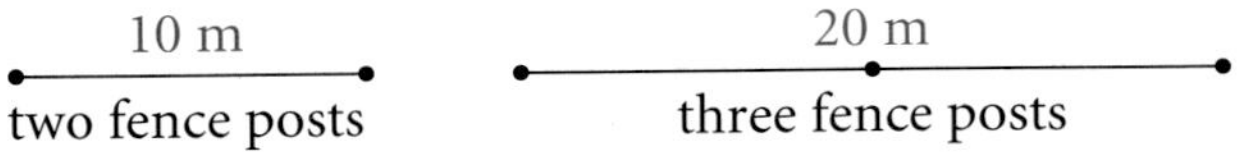

These easier problems suggest that there is always *one more* fence post than one tenth the length. So for a 570 meter fence, there are $\frac{570}{10} + 1$, or 58 fence posts.

EXERCISES

1. A farmer wishes to fence in a square lot with dimensions 70 yards by 70 yards. He will install a fence post every 10 yards. How many fence posts will he need?
2. Janette is planning to walk from her house to her friend Barbara's house. How many different paths can she take to get there? Assume that she walks only east and south.

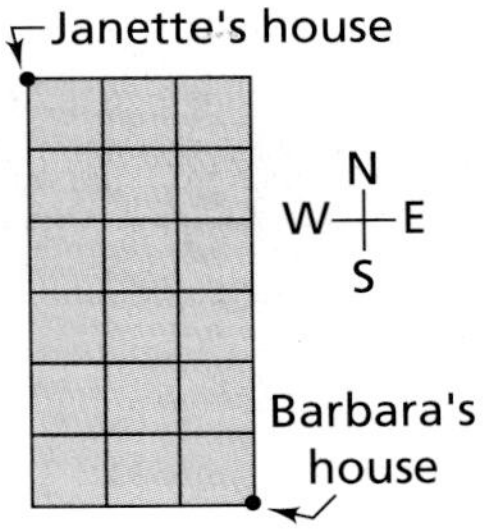

3. A square table can seat four people. For a banquet, a long rectangular table is formed by placing 14 such tables edge-to-edge in a straight line. How many people can sit at the long table?
4. Find the sum of the whole numbers from 1 to 999.
5. How many trapezoids are in the figure below? (*Hint:* Solve several simpler problems, then look for a pattern.)

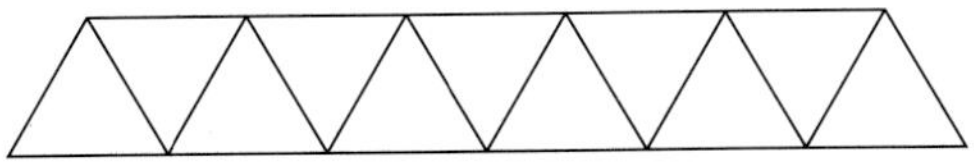

6. At a business luncheon, 424 handshakes took place. No two people shook hands with each other more than once. What is the least number of people in attendance at the luncheon?
7. On his fiftieth birthday, the President was honored with a 21-gun salute. The sound of each gunshot lasted 1 second, and 4 seconds elapsed between shots. How long did the salute last?

Problem Solving: Use Logical Reasoning

Some problems can be solved without the use of numbers. They can be solved by the use of logical reasoning, given some information.

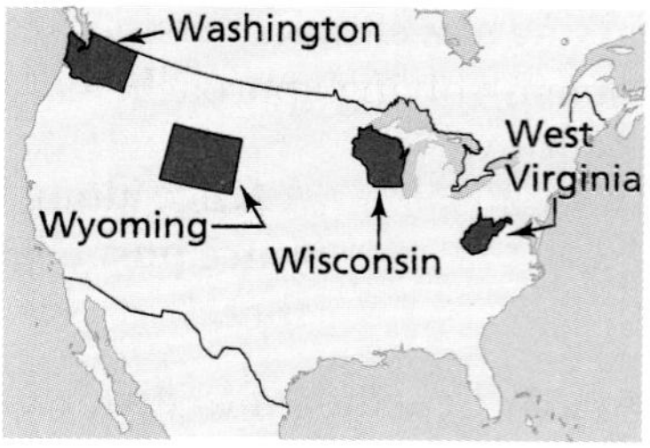

Example **Anna, Bill, Carla, and Doug are siblings. Each lives in a different state beginning with W. Use these clues to determine where each sibling lives:**
(1) Neither sister lives in a state containing two words.
(2) Bill lives west of his sisters.
(3) Anna doesn't cross the Mississippi River when she visits Doug.

Make a table to organize what you know. Use an initial for each name.

State	A	B	C	D
West Virginia	✗	✗	✗	
Wisconsin		✗		
Wyoming		✗		
Washington	✗	✓	✗	✗

← From Clue 1, you know that neither Anna nor Carla lives in West Virginia.

← Using Clues 1 and 2, you know that Bill must live in Washington if he lives west of his sisters.

Use logical reasoning to complete the table.

State	A	B	C	D
West Virginia	✗	✗	✗	✓
Wisconsin	✓	✗	✗	✗
Wyoming	✗	✗	✓	✗
Washington	✗	✓	✗	✗

← Doug lives in West Virginia because none of his siblings do.

← From Clue 3, you know that Anna must live in Wisconsin.

← Carla, therefore, lives in Wyoming.

EXERCISES

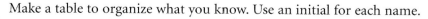

1. Harold has a dog, a canary, a goldfish, and a hamster. Their names are J.T., Izzy, Arf, and Blinky. Izzy has neither feathers nor fins. Arf can't bark. J.T. weighs less than the four-legged pets. Neither the goldfish nor the dog has the longest name. Arf and Blinky don't get along well with the canary. What is each pet's name?

2. At the state basketball championship tournament, 42 basketball games are played to determine the winner of the tournament. After each game, the loser is eliminated from the tournament. How many teams are in the tournament?

3. The sophomore class has 124 students. Of these students, 47 are involved in muscial activities: 25 in band and 36 in choir. How many students are involved in both band and choir?

4. Tina's height is between Kimiko's and Ignacio's. Ignacio's height is between Jerome's and Kimiko's. Tina is taller than Jerome. List the people in order from shortest to tallest.

Problem Solving: Work Backward

In some situations it is easier to start with the end result and work backward to find the solution. You work backward in order to solve linear equations. The equation $2x + 3 = 11$ means "double x and add 3 to get 11." To find x, you "undo" those steps in reverse order.

$2x + 3 = 11$

$2x = 8$ ← Subtract 3 from each side.

$x = 4$ ← Divide each side by 2.

Another time it is convenient to work backward is when you want to "reverse" a set of directions.

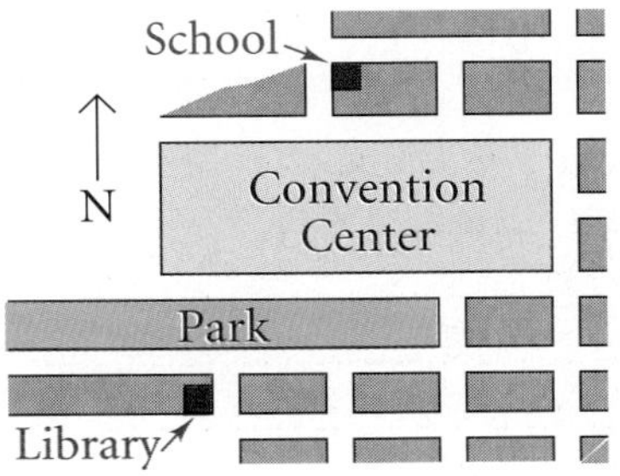

Example **To get from the library to the school, go 3 blocks east, then 5 blocks north, then 2 blocks west. How do you go from the school to the library?**

To reverse the directions, start at the school and work backward. Go 2 blocks east, 5 blocks south, and 3 blocks west.

EXERCISES

1. To go from Bedford to Worcester, take Route 4 south, then Route 128 south, then Route 90 west. How do you get from Worcester to Bedford?

2. Sandy spent $\frac{1}{10}$ of the money in her purse for lunch. She then spent \$23.50 for a gift for her brother, then half of what she had left on a new CD. If Sandy has \$13 left in her purse, how much money did she have in it before lunch?

3. Algae are growing on a pond's surface. The area covered doubles each day. It takes 24 days to cover the pond completely. After how many days will the pond be half covered with algae?

4. Don sold $\frac{1}{5}$ as many raffle tickets as Carlita. Carlita sold 3 times as many as Ranesha. Ranesha sold 7 fewer than Russell. If Russell sold 12 tickets, how many did Don sell?

5. At 6% interest compounded annually, the balance in a bank account will double about every 12 years. If such an account has a balance of \$16,000 now, how much was deposited when the account was opened 36 years ago?

6. Solve the puzzle that Yuan gave to Inez: I am thinking of a number. If I triple the number and then halve the result, I get 12. What number am I thinking of?

7. Carlos paid \$14.60 for a taxi fare from a hotel to the airport, including a \$2.00 tip. Green Cab Co. charges \$1.20 per passenger plus \$0.20 for each additional $\frac{1}{5}$ mile. How many miles is the hotel from the airport?

Using a Ruler and Protractor

Knowing how to use a ruler and protractor is crucial for success in Geometry.

Example **Draw a triangle that has sides of length 5.2 cm and 3.0 cm and a 68° angle between these two sides.**

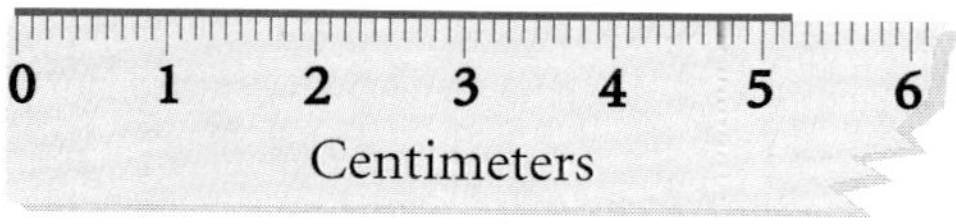

Step 1: Use a ruler to draw a segment 5.2 cm long.

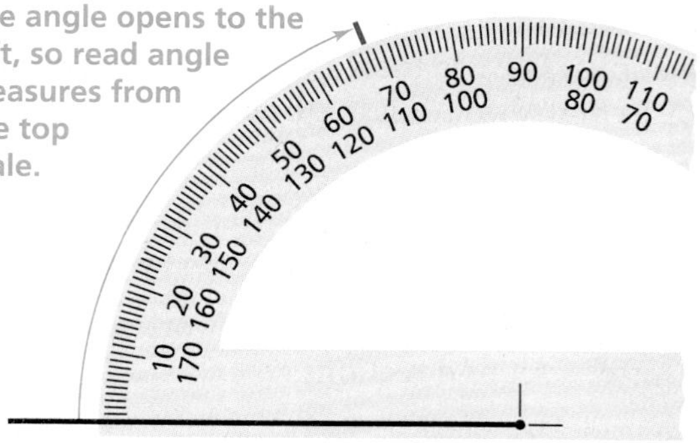

The angle opens to the left, so read angle measures from the top scale.

Step 2: Place the crosshairs of a protractor at one endpoint of the segment. Make a small mark at the 68° position along the protractor.

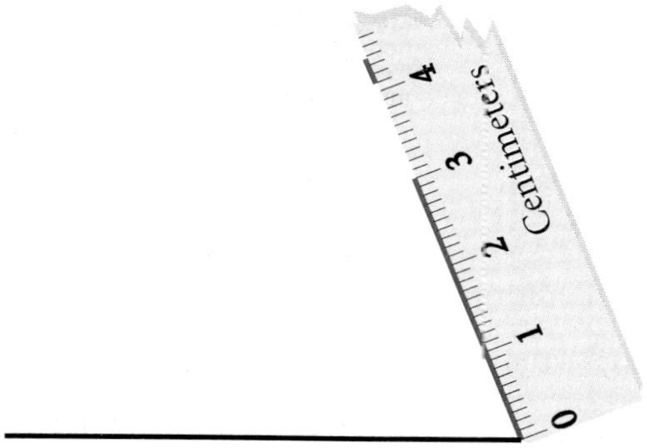

Step 3: Align the ruler along the small mark and the endpoint you used in Step 2. Place the zero point of the ruler at the endpoint. Draw a segment 3.0 cm long.

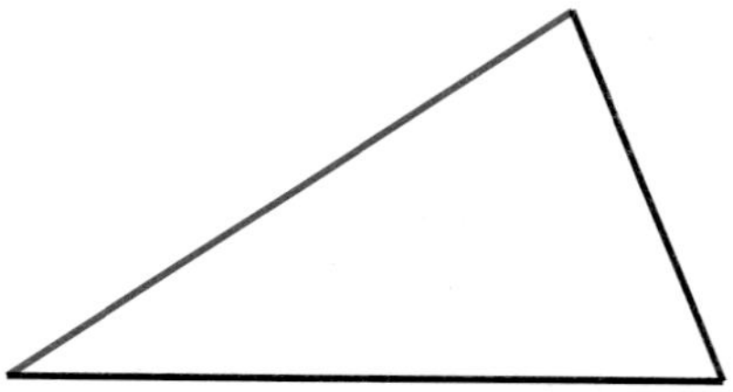

Step 4: Complete the triangle by connecting the endpoints of the first segment and the second.

EXERCISES

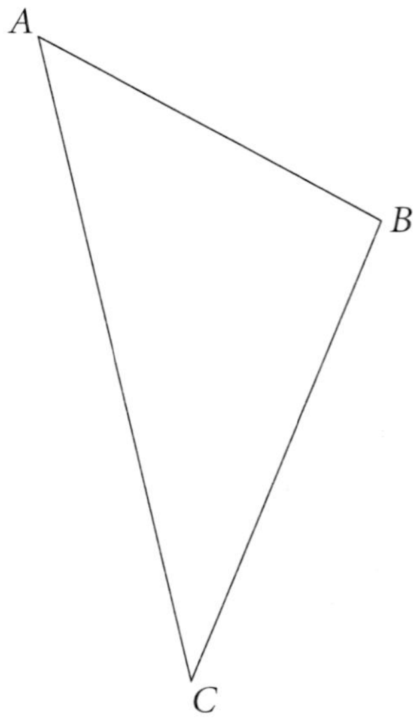

1. Measure sides $\overline{AB}$ and $\overline{BC}$ to the nearest millimeter.
2. Measure each angle of $\triangle ABC$ to the nearest degree.
3. Draw a triangle that has sides of length 4.8 cm and 3.7 cm and a 34° angle between these two sides.
4. Draw a triangle that has 43° and 102° angles and a side of length 5.4 cm between these two angles.
5. Draw a rhombus that has sides of length $2\frac{1}{4}$ in. and 68° and 112° angles.
6. Draw an isosceles trapezoid that has a pair of 48° base angles and a base of length 2 in. between these two base angles.
7. Draw an isosceles triangle that has two congruent sides $3\frac{1}{2}$ in. long and a 134° vertex angle.

Measurement Conversions

To convert from one unit of measure to another, you multiply by a conversion factor. A *conversion factor* is a fraction equal to 1 that has different units in the numerator and the denominator. An example of a conversion factor is $\frac{1 \text{ ft}}{12 \text{ in.}}$. Other conversion factors are given in the table on page 668.

Example 1 **Complete.**

a. **88 in. = ■ ft** **b.** **5.3 m = ■ cm**

c. **3700 mm = ■ cm** **d.** **90 in. = ■ yd**

a. $88 \text{ in.} \cdot \frac{1 \text{ ft}}{12 \text{ in.}} = \frac{88}{12} \text{ ft} = 7\frac{1}{3} \text{ ft}$

b. $5.3 \text{ m} \cdot \frac{100 \text{ cm}}{1 \text{ m}} = 5.3(100) \text{ cm} = 530 \text{ cm}$

c. $3700 \text{ mm} \cdot \frac{1 \text{ cm}}{10 \text{ mm}} = 370 \text{ cm}$

d. $90 \text{ in.} \cdot \frac{1 \text{ ft}}{12 \text{ in.}} \cdot \frac{1 \text{ yd}}{3 \text{ ft}} = \frac{90}{36} \text{ yd} = 2\frac{1}{2} \text{ yd}$

Area is always in square units, and volume is always in cubic units.

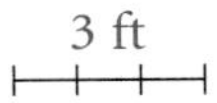

1 yd = 3 ft

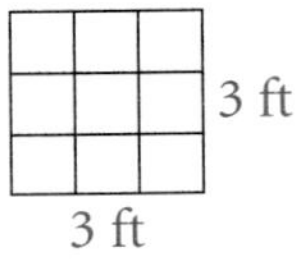

$1 \text{ yd}^2 = 9 \text{ ft}^2$

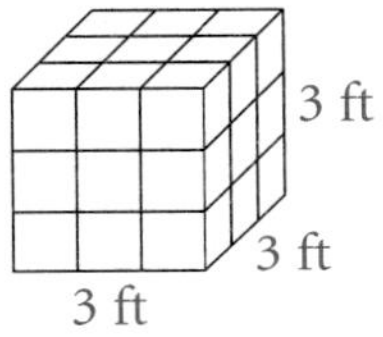

$1 \text{ yd}^3 = 27 \text{ ft}^3$

Example 2 **Complete.**

a. $\mathbf{300 \text{ in.}^2 = ■ \text{ ft}^2}$ **b.** $\mathbf{200{,}000 \text{ cm}^3 = ■ \text{ m}^3}$

a. $1 \text{ ft} = 12 \text{ in.}$, so $1 \text{ ft}^2 = (12 \text{ in.})^2$ or 144 in.^2

$300 \text{ in.}^2 \cdot \frac{1 \text{ ft}^2}{144 \text{ in.}^2} = 2\frac{1}{12} \text{ ft}^2$

b. $1 \text{ m} = 100 \text{ cm}$, so $1 \text{ m}^3 = (100 \text{ cm})^3$ or $1{,}000{,}000 \text{ cm}^3$

$200{,}000 \text{ cm}^3 \cdot \frac{1 \text{ m}^3}{1{,}000{,}000 \text{ cm}^3} = 0.2 \text{ m}^3$

EXERCISES

Complete.

1. 40 cm = ■ m
2. 1.5 kg = ■ g
3. 60 cm = ■ mm
4. 200 in. = ■ ft
5. 28 yd = ■ in.
6. 1.5 mi = ■ ft
7. 42 fl oz = ■ qt
8. 430 mg = ■ g
9. 34 L = ■ mL
10. 1.2 m = ■ cm
11. 43 mm = ■ cm
12. 3600 s = ■ min
13. 15 g = ■ mg
14. 12 qt = ■ c
15. 0.03 kg = ■ mg
16. 14 gal = ■ qt
17. 4500 lb = ■ t
18. 234 min = ■ h
19. 12 mL = ■ L
20. 2 pt = ■ fl oz
21. 20 m/s = ■ km/h
22. $3 \text{ ft}^2 = ■ \text{ in.}^2$
23. $108 \text{ m}^2 = ■ \text{ cm}^2$
24. $2100 \text{ mm}^2 = ■ \text{ cm}^2$
25. $1.4 \text{ yd}^2 = ■ \text{ ft}^2$
26. $0.45 \text{ km}^2 = ■ \text{ m}^2$
27. $1300 \text{ ft}^2 = ■ \text{ yd}^2$
28. $1030 \text{ in.}^2 = ■ \text{ ft}^2$
29. $20{,}000{,}000 \text{ ft}^2 = ■ \text{ mi}^2$
30. $1000 \text{ cm}^3 = ■ \text{ m}^3$
31. $1.4 \text{ ft}^3 = ■ \text{ in.}^3$
32. $3.56 \text{ cm}^3 = ■ \text{ mm}^3$
33. $0.013 \text{ km}^3 = ■ \text{ m}^3$

Measurement, Rounding Error, and Reasonableness

There is no such thing as an *exact* measurement. Measurements are always approximate. No matter how precise it is, a measurement actually represents a range of values.

The possible difference between a measurement and the actual value is called the error. The *error* is equal to half the unit of greatest precision.

Example 1 **Chris's height, to the nearest inch, is 5 ft 8 in. Find the range of values this measurement represents.**

The height is given to the nearest inch, so the error is $\frac{1}{2}$ in. Chris's height, then, is between 5 ft $7\frac{1}{2}$ in. and 5 ft $8\frac{1}{2}$ in., or it is 5 ft 8 in. $\pm\ \frac{1}{2}$ in. Within this range are all measures that, when rounded to the nearest inch, equal 5 ft 8 in.

As you calculate with measurements, error can accumulate.

Example 2 **Jean drives 18 km to work each day. The distance is given to the nearest kilometer.**
a. Find the range of values this measurement represents.
b. Find the error in the round-trip distance.

a. The driving distance is between 17.5 and 18.5 km, or 18 ± 0.5 km.

b. Double the lower limit, 17.5, and the upper limit, 18.5. Thus, the round trip can be anywhere between 35 and 37 km, or 36 ± 1 km. The error for the round trip is twice the error of a single leg of the trip.

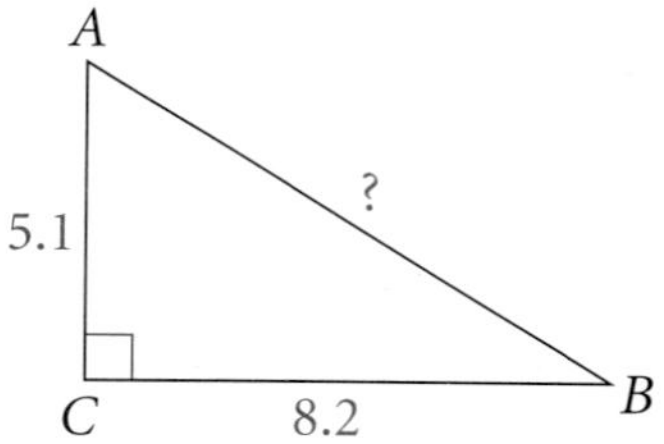

So that your answers will be reasonable, keep precision and error in mind as you calculate. For example, in finding *AB*, the length of the hypotenuse of $\triangle ABC$, it would be inappropriate to give the answer as 9.6566 if the sides are given to the nearest tenth. Round your answer to 9.7.

EXERCISES

Each measurement is followed by its unit of greatest precision. Find the range of values that each measurement represents.

1. 24 ft (ft)
2. 124 cm (cm)
3. 340 mL (mL)
4. $5\frac{1}{2}$ mi ($\frac{1}{2}$ mi)
5. 73.2 cm (mm)
6. 34 yd^2 (yd^2)
7. 5.4 mi (0.1 mi)
8. 6 ft 5 in. (0.5 in.)

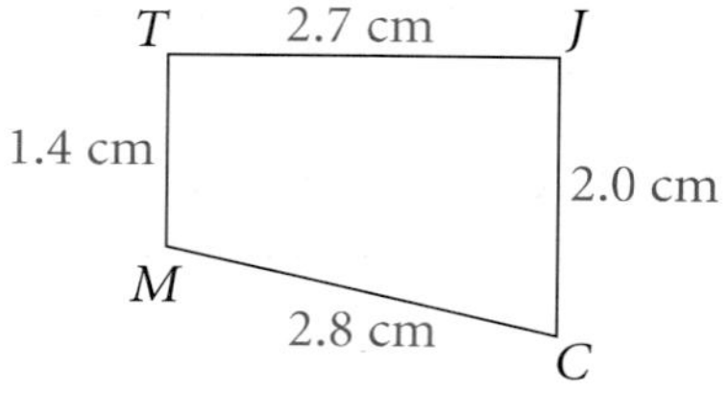

9. The lengths of the sides of *TJCM* are given to the nearest millimeter. Find the range of values for the figure's perimeter.

10. To the nearest degree, two angles of a triangle are 49° and 73°. What is the range of values for the measure of the third angle?

11. The lengths of the legs of a right triangle are given as 131 m and 162 m. You use a calculator to find the length of the hypotenuse. The calculator display reads 208.33867. What should your answer be?

Mean, Median, and Mode

Measures of central tendency, such as mean, median, and mode, are numbers that describe a set of data.

Example **Eighteen students were asked to measure the angle formed by the three objects in the diagram. Their answers, in order from least to greatest, are as follows:**

65, 66, 66, 66, 66, 66, 66, 67, 67, 67, 67, 67, 68, 68, 69, 70, 74, 113

Find the mean, median, and mode of the data.

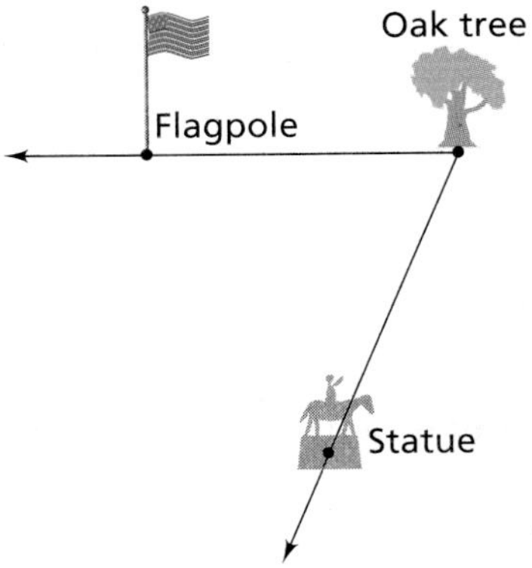

To find the *mean* (sometimes called the average), add the numbers and divide by the number of items.

$$\text{Mean} = \frac{\text{sum of the 18 measures}}{18} = \frac{1258}{18} = 69\tfrac{8}{9}$$

The *median* is the middle number when the items are placed in order. (When there is an even number of items, take the mean of the two middle numbers.) Here, the two middle numbers—the ninth and the tenth numbers on the list—are both 67. So the median is 67.

The *mode* is the number that appears most frequently. A set of data may have more than one mode or no modes. There are more 66's than any other number, so the mode is 66.

EXERCISES

Find the mean, median, and mode of each set of data.

1. Numbers of students per school in Newtown: 234, 341, 253, 313, 273, 301, 760
2. Amounts spent for lunch: \$4.50, \$3.26, \$5.02, \$3.58, \$1.25, \$3.05, \$4.24, \$3.56, \$3.31
3. Salaries at D. B. Widget & Co.: \$15,000; \$18,000; \$18,000; \$21,700; \$26,500; \$27,000; \$29,300; \$31,100; \$43,000; \$47,800; \$69,000; \$140,000
4. Population of towns in Brower County: 567, 632, 781, 902, 1034, 1100, 1598, 2164, 2193, 3062, 3074, 3108, 3800, 3721, 4104
5. In Exercise 3, which measure or measures of central tendency do you think best represent the data? Explain.
6. Find the mean, median, and mode of the exam scores at the right.
7. In the example, the student who reported the angle measure as 113 most likely made an error. If this measure is dropped from the list, what are the mean, median, and mode of the remaining 17 scores?
8. In the example, if the smallest measurement were 51 instead of 65, would the mean decrease? Would the median? Would the mode?
9. In the example, if the two students who measured the angle at 68 both reduced their measurements to 67, would the mode be affected? How?

Final Exam Scores

34, 47, 53, 56, 57, 62, 62, 64, 67, 70, 74, 74, 74, 78, 82, 85, 85, 85, 85, 86, 88, 92, 93, 93, 94, 95, 97

Bar Graphs and Line Graphs

Data displayed in a table can be very useful, but a table is not as easy to interpret as a graph. A bar graph and a line graph can show the same data, but sometimes one type of graph will have advantages over the other.

Example **Make a bar graph and a line graph showing the data in the table at the left.**

Revenue of HJL Co.

Year	Revenue
1993	$39,780
1994	$40,019
1995	$51,772
1996	$63,444
1997	$79,855

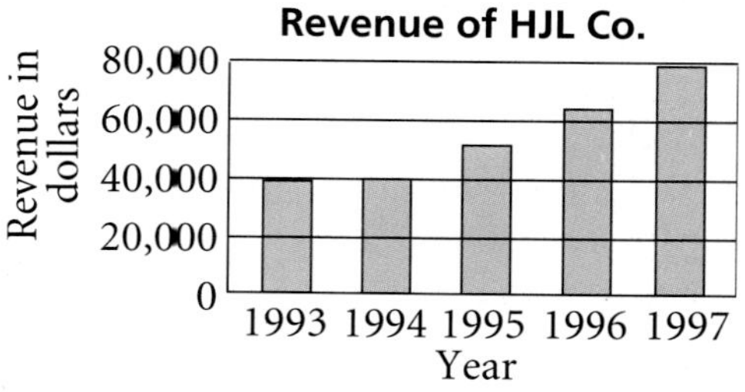

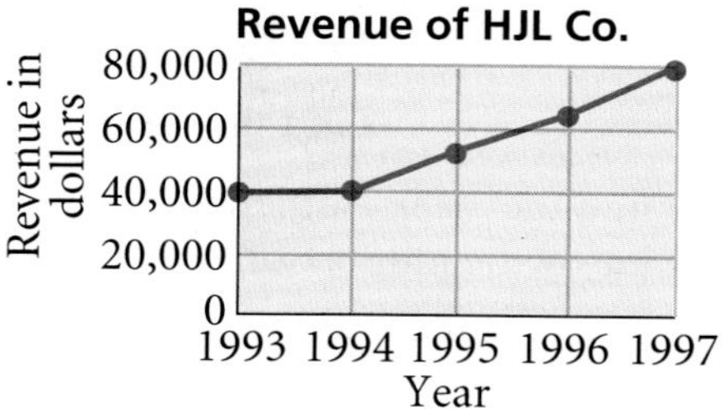

Bar graphs are useful when you wish to compare amounts. In the example, the tallest bar is clearly twice the height of the shortest bar. At a glance, it is evident that in four years the revenue approximately doubled.

Line graphs allow you to see how a set of data changes over time. In the example, the slope of the line shows that revenue has increased at a steady rate since 1994.

Did revenue increase from 1993 to 1994? You can't tell by looking at either graph; for that information, you would have to look back at the table.

EXERCISES

1. Create a bar graph and a line graph to display the data in the table. Show years along the horizontal axis and sales along the vertical axis.

Sales of Rock Music (in millions of dollars)

Year	1989	1990	1991	1992	1993	1994
Sales	$2833	$2722	$2726	$2852	$3034	$4236

Source: *Recording Industry Association of America*

For Exercises 2–6, refer to the line graph at the right.

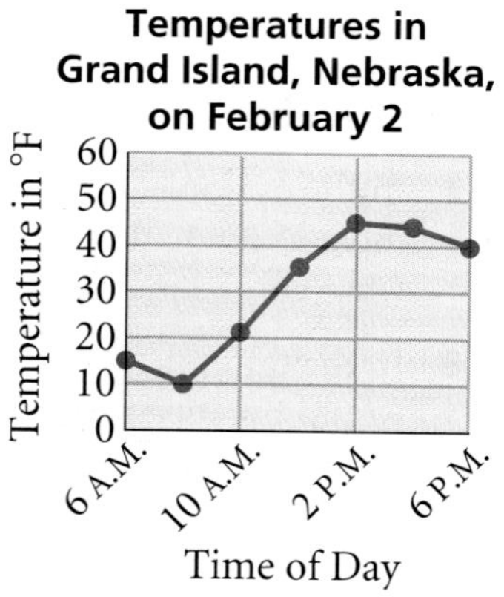

2. What was the lowest temperature recorded between 6 A.M. and 6 P.M.?
3. During which time periods did the temperature appear to increase?
4. Estimate the temperature at 11 A.M. and at 5 P.M.
5. Can you tell from the graph what the actual maximum and minimum temperatures were between 6 A.M. and 6 P.M.? Explain.
6. The same data could be presented in a bar graph. Which presentation is better for these data, a line graph or a bar graph? Why?

Circle Graphs

A circle graph is used for presenting data as a fraction of a total.

Example **Create a circle graph to display the data in the table.**

Plans of Cranston High School Graduates

Attend 4-yr college	59%
Attend 2-yr college	21%
Attend technical college	8%
Enter work force	8%
Undecided/Other	4%

To find the angle for each wedge, or sector, recall that there are 360° in the whole circle. So to find, for example, the angle for 21%, calculate as follows:

$$21\% \text{ of } 360^\circ = 0.21 \cdot 360^\circ$$
$$\approx 76^\circ$$

The number of degrees in the wedge for 59% is greater than 180°. To create this wedge, draw a semicircle (to represent 50%) and add to the semicircle a wedge representing 9%.

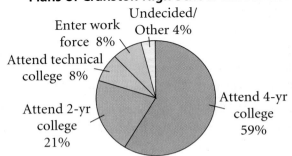

Plans of Cranston High School Graduates

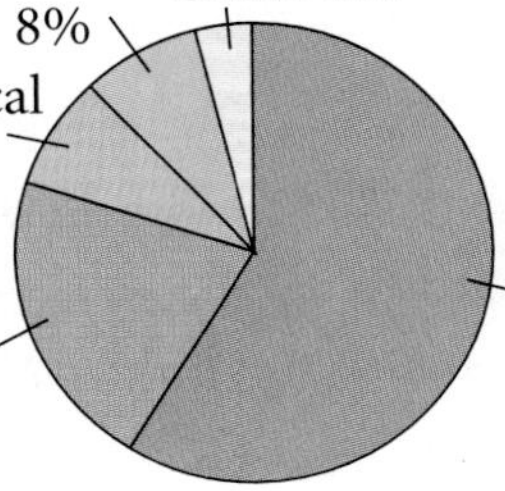

EXERCISES

In Exercises 1–3, refer to the Example.

1. Find the measure of the angle for each of the five categories in the circle graph.
2. What is the ratio of graduates who plan to attend some type of college to those who don't or who are undecided?
3. Suppose there are 358 graduates of Cranston High School. Make a table showing the number of students in each category.

Create a circle graph to display each set of data.

4. **Age of Cars on the Road**

Less than 1 yr	6%
1–5 yr	34%
5–10 yr	30%
10–15 yr	20%
Over 15 yr	10%

Source: *The Unofficial U.S. Census*

5. **Favorite Type of Pet**

Dog	41%
Cat	39%
Fish	8%
Bird	6%
Rabbit	2%
Other	4%

6. **Number of People in Family**

2	7%
3	17%
4	30%
5	24%
6	13%
7	6%
>7	3%

Skills Handbook

Box-and-Whisker Plots

A *box-and-whisker plot* is a way to display data on a number line. It provides a picture of how tightly the data cluster around the median and how wide a range the data have. The diagram below shows the various points associated with a box-and-whisker plot.

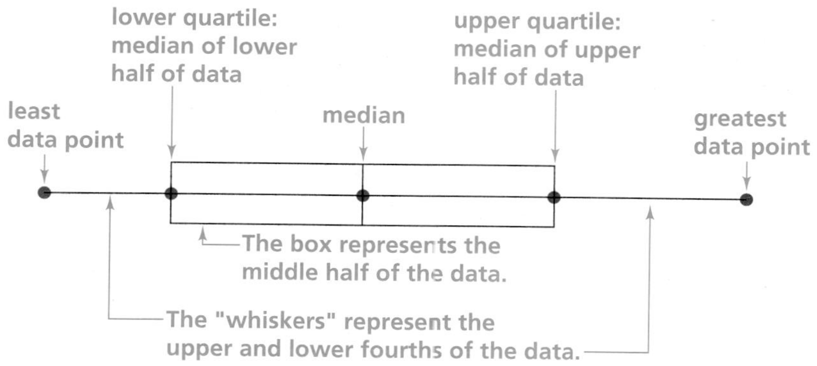
lower quartile: median of lower half of data
upper quartile: median of upper half of data
least data point
median
greatest data point
The box represents the middle half of the data.
The "whiskers" represent the upper and lower fourths of the data.

Example **The heights, in inches, of 23 students are as follows:**

58, 61, 63, 63, 63, 64, 64, 65, 65, 65, 67, 68, 68, 68, 69, 70, 70, 70, 72, 72, 72, 74, 75

Draw a box-and-whisker plot.

The heights vary from 58 in. to 75 in. The median is 68. The lower quartile (the median of the lower eleven heights) is 64. The upper quartile (the median of the upper eleven heights) is 70.

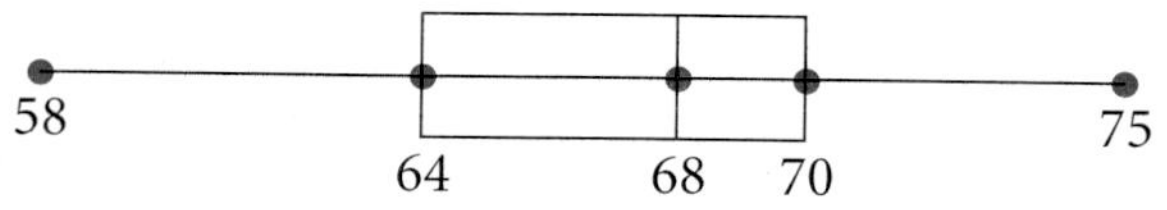
58
64
68
70
75

EXERCISES

1. All of the physical education students at Martin Luther King, Jr., High School were timed sprinting the 100-meter dash. The box-and-whisker plot below summarizes the data. Use it to find the following:
   a. Median **b.** Lower quartile **c.** Upper quartile

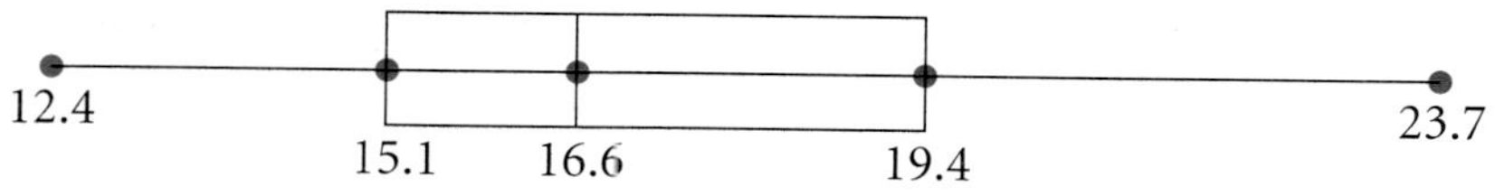
12.4
15.1
16.6
19.4
23.7

2. Make a box-and-whisker plot for the following set of data, which lists the weights, in pounds, of the students trying out for the wrestling team at South Side High School.

   104, 121, 122, 130, 130, 131, 140, 144, 147, 147, 148, 155, 160, 163, 171

Evaluating and Simplifying Expressions

You *evaluate* an expression with variables by substituting a number for each variable. Then simplify the expression using the order of operations.

Be especially careful with exponents and negative signs. For example, the expression $-x^2$ always yields a negative or zero value, and $(-x)^2$ is always positive or zero.

Order of Operations

1. Perform any operation(s) inside grouping symbols.
2. Simplify any term with exponents.
3. Multiply and divide in order from left to right.
4. Add and subtract in order from left to right.

Example 1 Evaluate each expression for $r = 4$.

a. $-r^2$ b. $-3r^2$ c. $(-3r)^2$

a. $-r^2 = -(4)^2 = -16$

b. $-3r^2 = -3(4)^2 = -3(16) = -48$

c. $(-3r)^2 = (-3 \cdot 4)^2 = (-12)^2 = 144$

To simplify an expression, you combine like terms and eliminate any parentheses.

Example 2 Simplify each expression.

a. $5r - 2r + 1$ b. $\pi(3r - 1)$ c. $(r + \pi)(r - \pi)$

a. Combine like terms: $5r - 2r + 1 = 3r + 1$

b. Use the distributive property: $\pi(3r - 1) = 3\pi r - \pi$

c. Multiply polynomials: $(r + \pi)(r - \pi) = r^2 - \pi^2$

EXERCISES

Evaluate each expression for $x = 5$ and $y = -3$.

1. $-2x^2$
2. $-y + x$
3. $-xy$
4. $(x + 5y) \div x$
5. $x + 5y \div x$
6. $(-2y)^2$
7. $(2y)^2$
8. $(x - y)^2$
9. $\dfrac{x + 1}{y}$
10. $y - (x - y)$
11. $-y^x$
12. $\dfrac{2(1 - x)}{y - x}$
13. $x \cdot y - x$
14. $x - y \cdot x$
15. $\dfrac{y^3 - x}{x - y}$
16. $-y(x - 3)^2$
17. Which expression gives the area of the shaded region of the figure at the right?
    A. $\pi(R - S)^2$ B. $\pi(R^2 - S^2)$
    C. $\pi(S^2 - R^2)$ D. $\pi R^2 - 2\pi S$

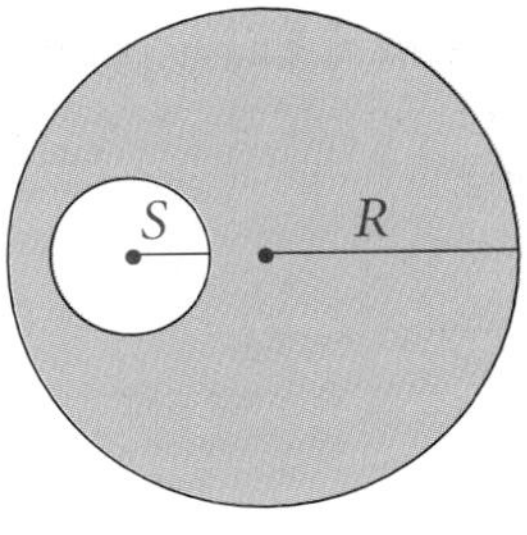

Simplify.

18. $6x - 4x + 8 - 5$
19. $2(\ell + w)$
20. $3x - 5 + 2x$
21. $-(4x + 7)$
22. $-4x(x - 2)$
23. $3x - (5 + 2x)$
24. $2t^2 + 4t - 5t^2$
25. $(r - 1)^2$
26. $(1 - r)^2$
27. $(y + 1)(y - 3)$
28. $4h + 3h - 4 + 3$
29. $\pi r - (1 + \pi r)$
30. $(x + 4)(2x - 1)$
31. $2\pi h(1 - r)^2$
32. $3y^2 - (y^2 + 3y)$
33. $-(x + 4)^2$

Simplifying Radicals

A radical expression is in its *simplest form* when all three of the following statements are true.

1. The expression under the radical sign contains no perfect square factors (other than 1).
2. The expression under the radical sign does not contain a fraction.
3. The denominator does not contain a radical expression.

Example 1 **Simplify.**

a. $\sqrt{\frac{4}{9}}$ b. $\sqrt{12}$

a. $\sqrt{\frac{4}{9}} = \frac{\sqrt{4}}{\sqrt{9}} = \frac{2}{3}$ b. $\sqrt{12} = \sqrt{4} \cdot \sqrt{3} = 2\sqrt{3}$

Example 2 **Find the length of the diagonal of rectangle *HJKL*.**

$c^2 = 7^2 + 1^2$ ← Use the Pythagorean Theorem.
$c^2 = 50$ ← Simplify the right side.
$c = \sqrt{50}$ ← Find the square root of each side.
$= \sqrt{25 \cdot 2}$
$= 5\sqrt{2}$ ← Simplify the radical.

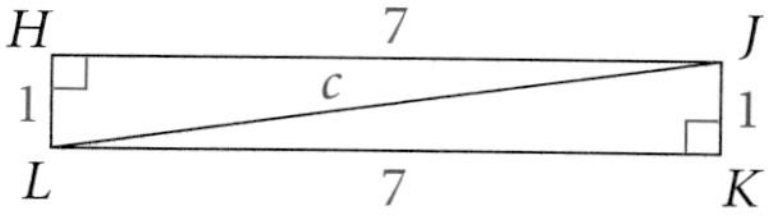

Example 3 **Simplify $\frac{1}{\sqrt{3}}$.**

$\frac{1}{\sqrt{3}} \cdot \frac{\sqrt{3}}{\sqrt{3}} = \frac{\sqrt{3}}{3}$ ← Multiply by $\frac{\sqrt{3}}{\sqrt{3}}$, or 1, to eliminate the radical in the denominator.

EXERCISES

Simplify each radical expression.

1. $\sqrt{27}$
2. $\sqrt{24}$
3. $\sqrt{150}$
4. $\sqrt{\frac{1}{9}}$
5. $\sqrt{\frac{72}{9}}$
6. $\frac{\sqrt{228}}{\sqrt{16}}$
7. $\sqrt{\frac{2}{5}}$
8. $\sqrt{\frac{27}{75}}$
9. $\frac{3}{\sqrt{8}}$
10. $\frac{6\sqrt{18}}{\sqrt{48}}$

Find the value of *x*. Leave your answer in simplest radical form.

11.
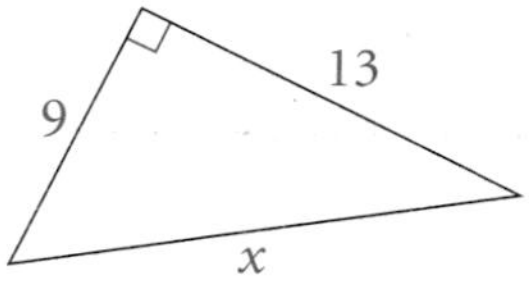

12.
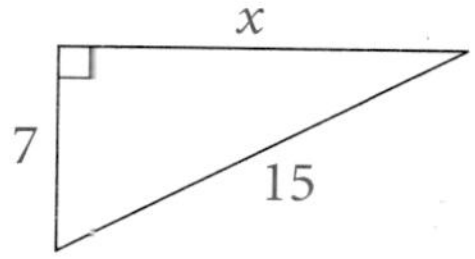

13.
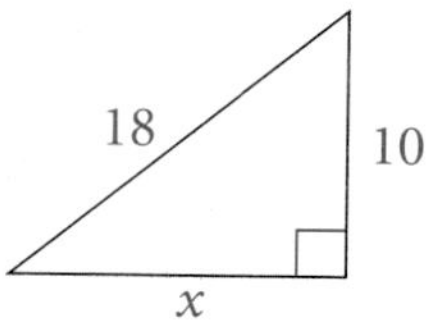

14.
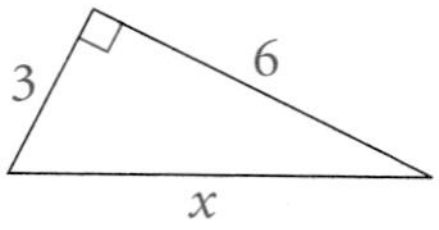

15.
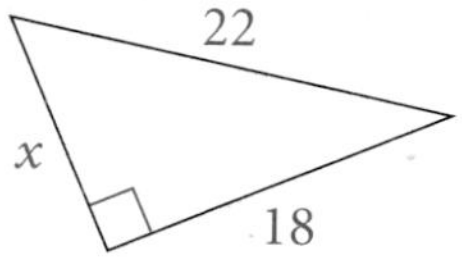

16.
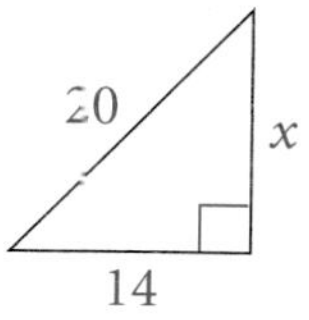

17.
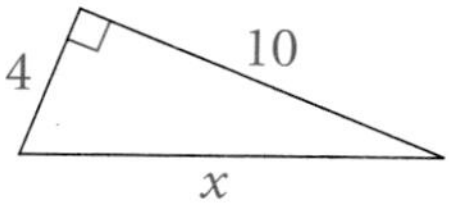

18.
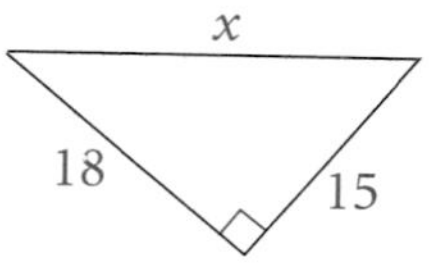

Simplifying Ratios

The ratio of the shorter leg to the longer leg of this right triangle is 4 to 6. This ratio can be written in several ways:

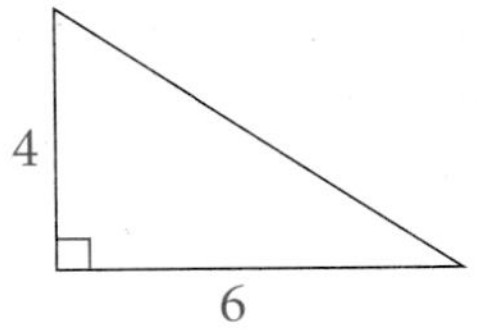

4 to 6 $\quad \frac{4}{6} \quad$ 4 : 6

You simplify ratios the same way you simplify fractions, by dividing out common factors from the numerator and denominator.

Example **Simplify each ratio.**

a. 4 to 6 $\quad$ **b.** $3ab : 27ab$ $\quad$ **c.** $\frac{4a + 4b}{8a + 8b}$

a. $4 \text{ to } 6 = \frac{4}{6}$

$= \frac{2 \cdot \cancel{2}}{3 \cdot \cancel{2}}$ ← Divide out the common factor 2.

$= \frac{2}{3}$

b. $3ab : 27ab = \frac{\cancel{3ab}^{\,1}}{\cancel{27ab}_{\,9}}$ ← Divide out the common factor $3ab$.

$= \frac{1}{9}$

c. $\frac{4a + 4b}{a + b} = \frac{4\cancel{(a + b)}}{\cancel{a + b}}$ ← Factor the numerator. The denominator cannot be factored. Divide out the common factor $(a + b)$.

$= 4$

EXERCISES

Simplify each ratio.

1. 25 to 15

2. 6 : 9

3. $\frac{7}{14x}$

4. 0.8 to 2.4

5. $\frac{12c}{14c}$

6. $22x^2$ to $35x$

7. $0.5ab : 8ab$

8. $\frac{4xy}{0.25x}$

9. $1\frac{1}{2}x$ to $5x$

10. $\frac{x^2 + x}{2x}$

11. $\frac{1}{4}r^2$ to $6r$

12. $0.72t : 7.2t^2$

13. $(2x - 6) : (6x - 4)$

14. $12xy : 8xy$

15. $(9x - 9y)$ to $(x - y)$

16. $\frac{\pi r}{r^2 + \pi r}$

Express each ratio in simplest form.

17. shorter leg : longer leg

18. hypotenuse to shorter leg

19. $\frac{\text{shorter leg}}{\text{hypotenuse}}$

20. hypotenuse : longer leg

21. longer leg to shorter leg

22. $\frac{\text{longer leg}}{\text{hypotenuse}}$

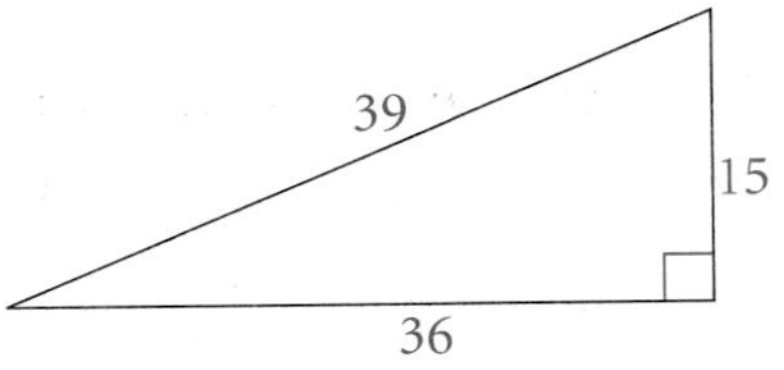

Write an expression in simplest form for $\frac{\text{area of shaded figure}}{\text{area of blue figure}}$.

23.

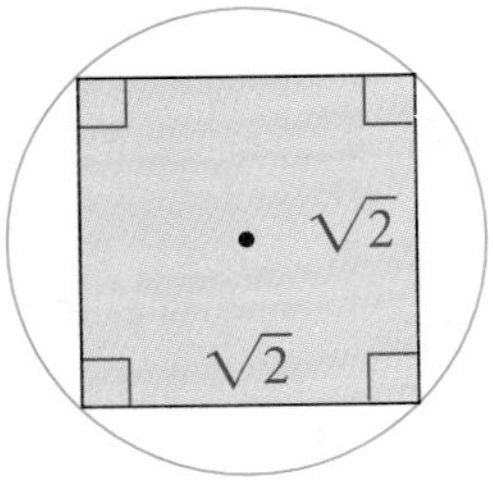

24.

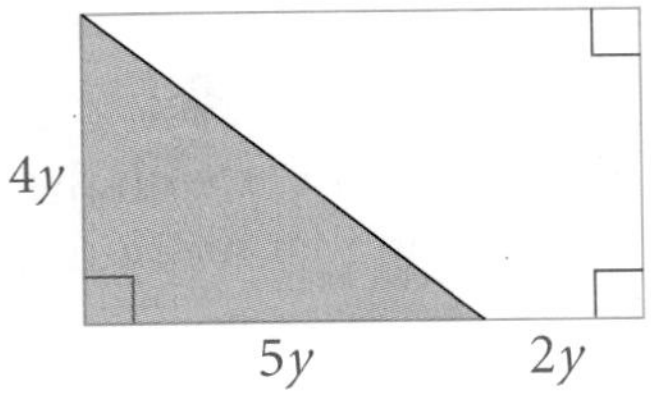

25.

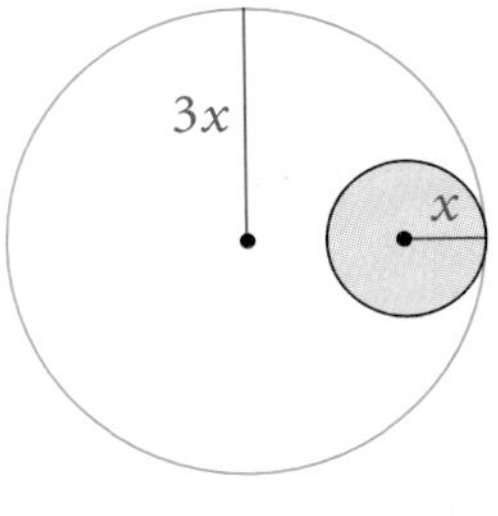

Solving Proportions

An equation in which both sides are ratios is called a *proportion.* A proportion can be written in three equivalent ways:

$\frac{a}{b} = \frac{c}{d}$ a is to b as c is to d

Because the product of the *means* (in this case, b and c) is equal to the product of the *extremes* (a and d), you can solve proportions by cross-multiplying.

Example 1   **Solve for n.**

a. $\frac{n}{8} = \frac{5}{2}$ **b.** $\frac{(n+1)}{4} = \frac{5}{9}$

a.

$$\frac{n}{8} = \frac{5}{2}$$
$$2n = 8 \cdot 5$$
$$2n = 40$$
$$n = 20$$

← Cross-multiply. →

← Simplify and solve for n. →

b.

$$\frac{(n+1)}{4} = \frac{5}{9}$$
$$9(n+1) = 4 \cdot 5$$
$$9n + 9 = 20$$
$$9n = 11$$
$$n = \frac{11}{9}$$

Example 2   **A map has the scale 1 in. = 4 mi. What actual distance does $3\frac{1}{4}$ in. represent on the map?**

$\frac{1\text{ in.}}{4\text{ mi}} = \frac{3.25\text{ in.}}{x\text{ mi}}$ ← Write a proportion.

$x \cdot 1 = 4(3.25)$ ← Cross-multiply.

$x = 13$ ← Solve for x.

$3\frac{1}{4}$ inches represents 13 miles.

EXERCISES

Solve each proportion.

1. $\frac{5}{2} = \frac{x}{7}$
2. $\frac{x+2}{3} = \frac{8}{15}$
3. $\frac{8}{x} = \frac{x}{2}$
4. $\frac{9}{w} = \frac{16}{144}$
5. $\frac{18}{x} = 6$
6. $\frac{13t}{26} = \frac{40}{16}$
7. $\frac{8}{11} = \frac{12}{x}$
8. $\frac{3}{4} = \frac{x}{48}$
9. $\frac{52}{p} = \frac{2}{3}$
10. $12 = \frac{36}{p}$
11. $\frac{7}{3} = \frac{28}{t+9}$
12. $\frac{5}{7} = \frac{10}{y+2}$
13. $\frac{a+5}{9} = \frac{14}{4}$
14. $\frac{x+10}{6} = \frac{3}{2}$
15. $\frac{2}{c+1} = \frac{9}{8c+1}$
16. $\frac{m+4}{7} = \frac{15-m}{12}$
17. A map has the scale 1 cm = 1200 km. What actual distance does 10.4 cm represent on the map?
18. A model airplane has the scale 1 : 72. The wingspan of the model is 11.5 in. What is the wingspan, in feet, of the actual plane?
19. An architect builds a model of an apartment complex with the scale 1 in. = 5 ft. Find the area of a rectangular patio that measures 5 in.-by-7.25 in. on the model.

Solving Linear Equations and Inequalities

An equation or inequality is *linear* if its variables are raised only to the power of 1. So $5x - 3 = 2$ is linear, but $x^2 - x + 1 = 0$ is not. To solve an equation, use the properties of equality and properties of real numbers (see page 672) to find all the values of the variable that satisfy the equation.

Example 1 **Solve each equation.**

a. $\mathbf{5x - 3 = 2}$ **b.** $\mathbf{1 - 2(x + 1) = x}$

a.
$5x - 3 = 2$
$5x = 5$ ← Add 3 to each side.
$x = 1$ ← Divide each side by 5.

b.
$1 - 2(x + 1) = x$
$1 - 2x - 2 = x$ ← Use the Distributive Property.
$-1 - 2x = x$ ← Simplify the left side.
$-1 = 3x$ ← Add $2x$ to each side.
$-\frac{1}{3} = x$ ← Divide each side by 3.

To solve a linear inequality, use properties of inequality (see page 672). Remember that when you multiply or divide both sides by a negative number, you reverse the order of the inequality.

Example 2 **Solve. Graph the solution on a number line.**

a. $\mathbf{2x - 1 \geq 5}$ **b.** $\mathbf{1 - x > -1}$

a.
$2x - 1 \geq 5$
$2x \geq 6$ ← Add 1 to each side.
$x \geq 3$ ← Divide each side by 2.

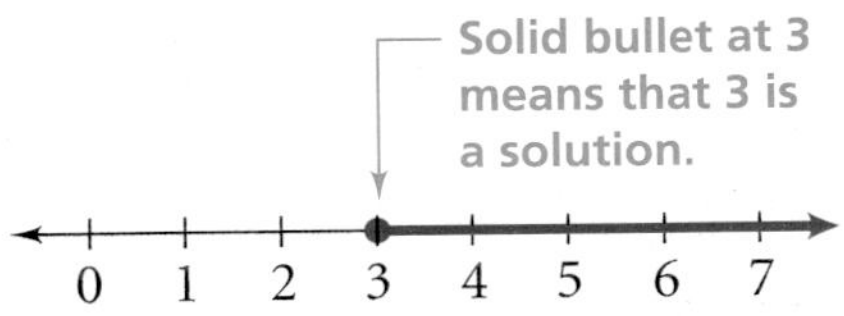

Solid bullet at 3 means that 3 is a solution.
0 1 2 3 4 5 6 7

b.
$1 - x > -1$
$-x > -2$ ← Subtract 1 from each side.
$x < 2$ ← Divide each side by -1. Reverse the order of the inequality.

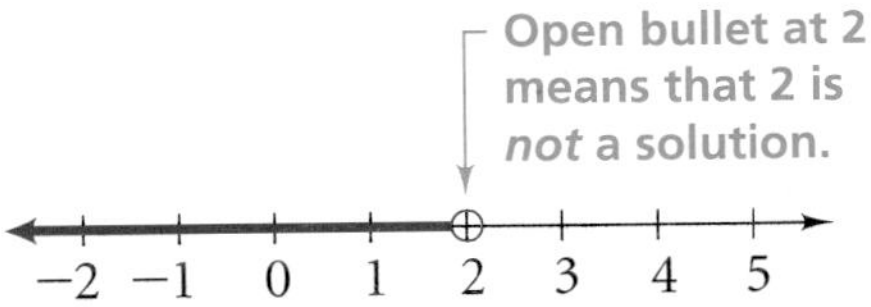

Open bullet at 2 means that 2 is *not* a solution.
−2 −1 0 1 2 3 4 5

EXERCISES

Solve each equation.

1. $3n + 2 = 17$
2. $3n - 4 = -6$
3. $2x + 4 = 10$
4. $3(n - 4) = 15$
5. $5a - 2 = -12$
6. $4 - 2y = 8$
7. $-6z + 1 = 13$
8. $\frac{m}{-2} - 3 = 1$
9. $\frac{2}{3}(n + 1) = -\frac{1}{4}$
10. $7 = -2(4n - 4.5)$
11. $2(1 - 3n) = 2n + 4$
12. $5k + 2(k + 1) = 23$
13. $\frac{5}{7}p - 10 = 30$
14. $6 - (3t + 4) = -17$
15. $(w + 5) - (2w + 5) = 5$

Solve each inequality. Graph the solution on a number line.

16. $2n < 8$
17. $6 - x \leq 4$
18. $\frac{1}{2}n \leq -\frac{3}{8}$
19. $5t + 3 \geq 23$
20. $3 - z \geq 7$
21. $y - 4 > 2 + 3y$
22. $-6m - 4 < 32$
23. $2(k - 4) \leq -12$
24. $5 - (n - 3) \geq -4n$

Slope and Intercepts

The slope of a line is a number that describes how steep it is. A line with a positive slope goes upward from left to right, and a line with a negative slope goes downward from left to right. The *slope* of a line is defined as $\frac{\text{vertical change (rise)}}{\text{horizontal change (run)}}$ or, more formally, as follows:

The slope m of a line through the points (x_1, y_1) and (x_2, y_2) is $m = \frac{y_2 - y_1}{x_2 - x_1}$.

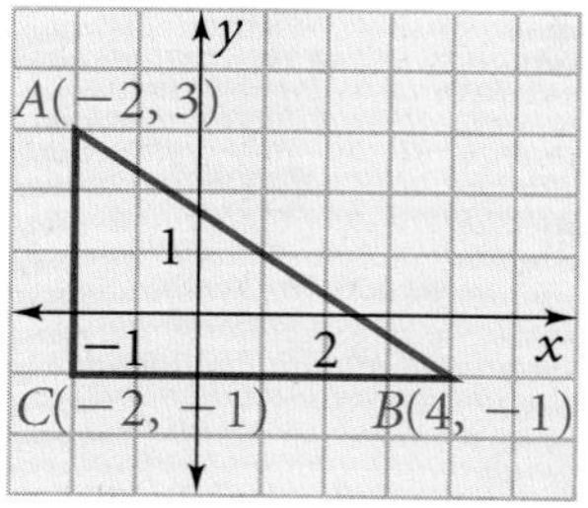

Example 1 **Find the slope of the hypotenuse of $\triangle ABC$.**

Let $(x_1, y_1) = (-2, 3)$ and let $(x_2, y_2) = (4, -1)$.

Then $m = \frac{y_2 - y_1}{x_2 - x_1} = \frac{-1 - 3}{4 - (-2)} = \frac{-4}{6} = -\frac{2}{3}$

An intercept is the x- or y-value where the line crosses an axis.

Example 2 **For the line $2y - x = 4$, find the following:**

a. the x-intercept **b. the y-intercept**

a. To find the x-intercept, let $y = 0$.

$$\begin{aligned} 2y - x &= 4 \\ 2(0) - x &= 4 \\ x &= -4 \end{aligned}$$

The x-intercept is -4.

b. To find the y-intercept, let $x = 0$.

$$\begin{aligned} 2y - x &= 4 \\ 2y - 0 &= 4 \\ y &= 2 \end{aligned}$$

The y-intercept is 2.

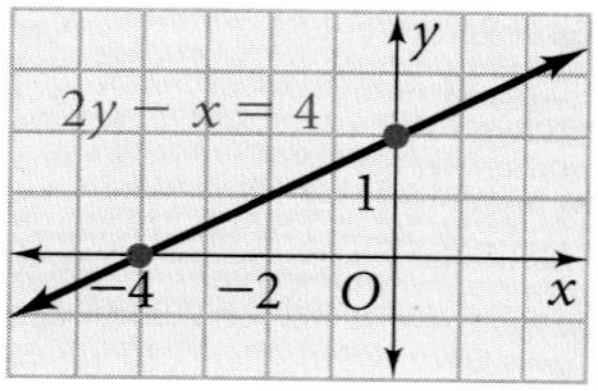

Thus, the graph of $2y - x = 4$ passes through $(-4, 0)$ and $(0, 2)$.

Lines with positive or negative slopes have both x- and y- intercepts. However, horizontal and vertical lines have only one intercept. The slope of a horizontal line is zero, and the slope of a vertical line is undefined.

EXERCISES

Find the slope of the line through each pair of points.

1. $(5, 1), (2, 7)$ **2.** $(-2, 3), (4, -7)$ **3.** $(0, 5), (5, 0)$ **4.** $(-4, -2), (-5, 8)$

5. $(5, 7), (-2, 7)$ **6.** $(0, 0), (4, -3)$ **7.** $(3, -2), (3, 9)$ **8.** $(\frac{1}{2}, -2), (-\frac{3}{2}, 1)$

Find the x- and y-intercepts for each linear equation.

9. $3x - y = 1$ **10.** $2y + x = 2$ **11.** $y = -1$ **12.** $-4x - y = 8$

13. $5x + 5y = 5$ **14.** $y = x$ **15.** $x = -4$ **16.** $\frac{1}{2}x - \frac{1}{4}y = \frac{3}{4}$

17. $1 = x - 2y$ **18.** $y = 4x - 5$ **19.** $0.9y - 0.5x = 2$ **20.** $y = -2x - 31$

21. The vertices of a triangle are $(-3, -1)$, $(-3, 5)$, and $(6, -1)$. Find the slope of each side of the triangle.

22. Graph the line with y-intercept 3 and no x-intercept. Find its slope.

Graphing Linear Equations and Inequalities

Examples 1 and 2 give two methods for graphing a linear equation.

Example 1 **Graph the equation $3x - y = 4$.**

Make a table by selecting several x-values and substituting each in the equation to find the corresponding y-value. Only two points are needed to graph a line, but finding a third is a good check for accuracy. Plot the points and draw the line.

x	0	1	2
y	−4	−1	2

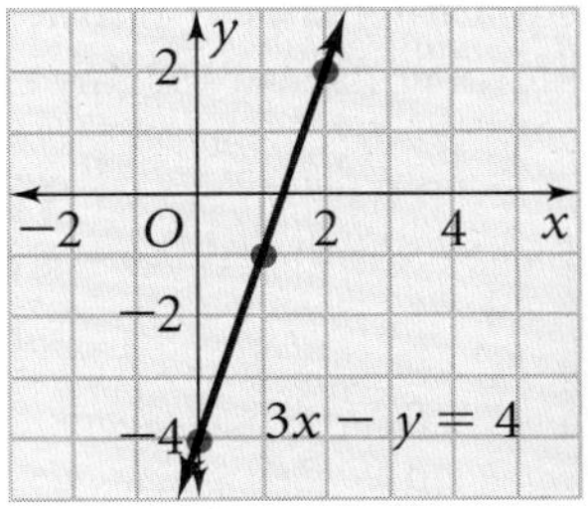

Example 2 **Graph the equation $2y - 3x = 2$.**

You can use a table as in Example 1, or you can use the slope-intercept form $y = mx + b$. In this form, m is the slope and b is the y-intercept. To write the equation in slope-intercept form, solve for y.

$2y - 3x = 2$

$2y = 3x + 2$ ← Subtract $3x$ from each side.

$y = \frac{3}{2}x + 1$ ← Multiply each side by $\frac{1}{2}$.

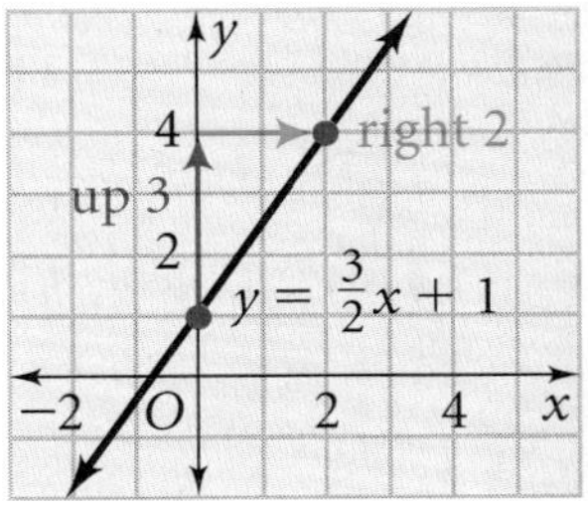

So the slope m is $\frac{3}{2}$, and the y-intercept b is 1. Plot (0, 1). From there, use the fact that $\frac{\text{rise}}{\text{run}} = \frac{3}{2}$ to move up 3 units and **right 2** units. Plot a second point there. Draw a line connecting the two points.

Example 3 **Graph the inequality $y > x - 1$.**

Start by graphing the equation $y = x - 1$. Points on this line are *not* solutions of the inequality. You show this by using a dashed line. (In general, use a dashed line for $<$ or $>$ and a solid line for $\leq$ or $\geq$.) To find which side of the line contains solutions of the inequality, test a point not on the line.

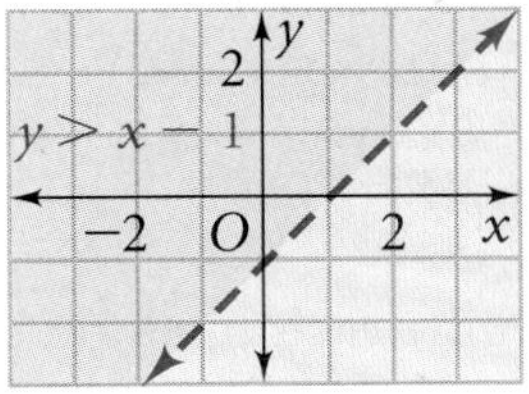

$(0) > (0) - 1$ ← Test (0, 0) in the inequality $y > x - 1$.

$0 > -1$ ← The point makes the inequality true, so shade the region containing (0, 0).

EXERCISES

Graph each equation.

1. $2x + y = 3$
2. $3y - x = 6$
3. $2x + 3y = 12$
4. $y = \frac{2}{3}x + 5$
5. $y = -\frac{1}{3}x$
6. $x - 2y = 4$
7. $x = 4$
8. $y = -2x - 1$
9. $y = -3$
10. $y = \frac{3}{4}x - 3$
11. $3x + y = 9$
12. $4x - 6y = 6$

Graph each inequality.

13. $x + y \leq 4$
14. $y > x + 2$
15. $y < x - 3$
16. $y \geq x + 9$
17. $2x + y \leq 5$
18. $y \leq 5$
19. $y - 3 < 2x + 2$
20. $x < 0$

Solving Literal Equations

An equation with two or more variables is called a *literal equation.* These equations appear frequently in geometry, often as formulas. You can solve a literal equation for any of its variables.

Example 1 **The formula $p = 2(\ell + w)$ gives the perimeter p of a rectangle with length ℓ and width w. Solve the equation for ℓ.**

$p = 2(\ell + w)$

$p = 2\ell + 2w$ ← Use the Distributive Property.

$p - 2w = 2\ell$ ← Subtract $2w$ from each side.

$\frac{p - 2w}{2} = \ell$ ← Divide each side by 2.

Example 2 **The formula $A = \frac{1}{2}(b_1 + b_2)h$ gives the area A of a trapezoid with bases b_1 and b_2 and height h. Solve for h.**

$A = \frac{1}{2}(b_1 + b_2)h$

$2A = h(b_1 + b_2)$ ← Multiply each side by 2.

$\frac{2A}{b_1 + b_2} = h$ ← Divide each side by $(b_1 + b_2)$.

Example 3 **The formula for converting from degrees Celsius C to degrees Fahrenheit F is $F = \frac{9}{5}C + 32$. Solve for C.**

$F = \frac{9}{5}C + 32$

$F - 32 = \frac{9}{5}C$ ← Subtract 32 from each side.

$\frac{5}{9}(F - 32) = C$ ← Multiply each side by $\frac{5}{9}$.

EXERCISES

Solve each equation for the variable in red.

1. Perimeter of rectangle: $p = 2w + 2\ell$
2. Volume of prism: $V = \ell wh$
3. Surface area of sphere: $S = 4\pi r^2$
4. Lateral area of cylinder: $A = 2\pi rh$
5. Area of kite or rhombus: $A = \frac{1}{2}d_1d_2$
6. Area of circle: $A = \pi r^2$
7. Area of regular polygon: $A = \frac{1}{2}ap$
8. Volume of cylinder: $V = \pi r^2h$
9. Area of triangle: $A = \frac{1}{2}bh$
10. Tangent of $\angle A$: $\tan A = \frac{y}{x}$
11. Volume of rectangular pyramid: $V = \frac{1}{3}\ell wh$
12. Circumference of circle: $C = 2\pi r$
13. Cosine of $\angle A$: $\cos A = \frac{b}{c}$
14. Volume of cone: $V = \frac{1}{3}\pi r^2h$
15. Surface area of right cone: $S = \pi r^2 + \pi r\ell$
16. Area of trapezoid: $A = \frac{1}{2}(b_1 + b_2)h$
17. Volume of pyramid: $V = \frac{1}{3}Bh$
18. Pythagorean Theorem: $a^2 + b^2 = c^2$
19. Surface area of regular pyramid: $S = B + \frac{1}{2}p\ell$
20. Surface area of right cylinder: $S = 2\pi r^2 + 2\pi rh$

Systems of Linear Equations

Normally, there are many ordered pairs that satisfy a given equation. For example, (3, 4), (4, 5), (5, 6), and infinitely many other pairs all satisfy the equation $y = x + 1$. To solve a system of two linear equations, however, you need to find ordered pairs that satisfy both equations at once. Ordinarily, there is just one such ordered pair; it is the point where the equations of the two lines intersect.

One method you can always use to solve a system of linear equations is the substitution method.

Example **Solve the system.** $\mathbf{2x - y = -10}$
$\mathbf{-3x - 2y = 1}$

Solve one of the equations for a variable. Looking at the two equations, it seems easiest to solve the first equation for *y*.

$2x - y = -10$
$-y = -2x - 10$ ← Subtract $2x$ from each side.
$y = 2x + 10$ ← Multiply each side by -1.

Now substitute $2x + 10$ for *y* in the other equation.

$-3x - 2y = 1$ ← Write the other equation.
$-3x - 2(2x + 10) = 1$ ← Substitute $(2x + 10)$ for *y*.
$-3x - 4x - 20 = 1$ ← Use the distributive property.
$-7x = 21$ ← Simplify the left side and add 20 to each side.
$x = -3$ ← Divide each side by -7.

So $x = -3$. To find *y*, substitute -3 for *x* in either equation.

$2x - y = -10$ ← Write one of the equations.
$2(-3) - y = -10$ ← Substitute -3 for *x*.
$-6 - y = -10$ ← Simplify.
$-y = -4$ ← Add 6 to each side.
$y = 4$ ← Multiply each side by -1.

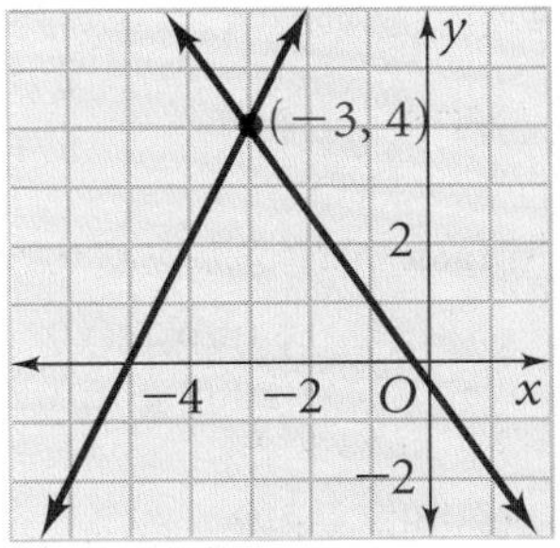

So the solution is $x = -3$ and $y = 4$, or $(-3, 4)$. If you graph $2x - y = -10$ and $-3x - 2y = 1$, you will find that the lines intersect at $(-3, 4)$.

EXERCISES

Solve each system.

1. $x + y = 3$
$x - y = 5$

2. $y - x = 4$
$x + 3 = y$

3. $y = 1$
$5x - 2y = 18$

4. $4y - x = -3$
$2x - 6 = 8y$

5. $8x - 1 = 4y$
$3x = y + 1$

6. $2x + 2y = -4$
$-x + 3y = 6$

7. $12y - 3x = 11$
$x - 2y = -2$

8. $5x + 7y = 1$
$4x - 2y = 16$

9. Give an example of a system of linear equations with no solution. What do you know about the slopes of the lines of such a system?

Tables

Measures

United States Customary	Metric
Length	
12 inches (in.) = 1 foot (ft)	10 millimeters (mm) = 1 centimeter (cm)
36 in. = 1 yard (yd)	100 cm = 1 meter (m)
3 ft = 1 yard	1000 mm = 1 meter
5280 ft = 1 mile (mi)	1000 m = 1 kilometer (km)
1760 yd = 1 mile	
Area	
144 square inches ($in.^2$) = 1 square foot (ft^2)	100 square millimeters (mm^2) = 1 square centimeter (cm^2)
9 ft^2 = 1 square yard (yd^2)	10,000 cm^2 = 1 square meter (m^2)
43,560 ft^2 = 1 acre (a)	10,000 m^2 = 1 hectare (ha)
4840 yd^2 = 1 acre	
Volume	
1728 cubic inches ($in.^3$) = 1 cubic foot (ft^3)	1000 cubic millimeters (mm^3) = 1 cubic centimeter (cm^3)
27 ft^3 = 1 cubic yard (yd^3)	1,000,000 cm^3 = 1 cubic meter (m^3)
Liquid Capacity	
8 fluid ounces (fl oz) = 1 cup (c)	1000 milliliters (mL) = 1 liter (L)
2 c = 1 pint (pt)	1000 L = 1 kiloliter (kL)
2 pt = 1 quart (qt)	
4 qt = 1 gallon (gal)	
Mass	
16 ounces (oz) = 1 pound (lb)	1000 milligrams (mg) = 1 gram (g)
2000 pounds = 1 ton (t)	1000 g = 1 kilogram (kg)
	1000 kg = 1 metric ton (t)
Temperature	
32°F = freezing point of water	0°C = freezing point of water
98.6°F = normal body temperature	37°C = normal body temperature
212°F = boiling point of water	100°C = boiling point of water
Time	
60 seconds (s) = 1 minute (min)	365 days = 1 year (yr)
60 minutes = 1 hour (h)	52 weeks (approx.) = 1 year
24 hours = 1 day (da)	12 months = 1 year
7 days = 1 week (wk)	10 years = 1 decade
4 weeks (approx.) = 1 month (mo)	100 years = 1 century

Symbols

Symbol	Meaning	Page
...	and so on	p. 5
+	plus (addition)	p. 6
=	is equal to, equality	p. 6
n^2	square of n	p. 6
( )	parentheses for grouping	p. 8
−	minus (subtraction)	p. 8
×, ·	times (multiplication)	p. 8
$-a$	opposite of a	p. 9
a^n	nth power of a	p. 10
P(event)	probability of the event	p. 11
$\overleftrightarrow{AB}$	line through points A and B	p. 13
$\overline{AB}$	segment with endpoints A and B	p. 18
$\overrightarrow{AB}$	ray with endpoint A and through point B	p. 18
$\parallel$	is parallel to	p. 18
$>$	is greater than	p. 22
$<$	is less than	p. 22
$\geq$	is greater than or equal to	p. 22
$\leq$	is less than or equal to	p. 22
AB	length of $\overline{AB}$	p. 25
$\lvert a \rvert$	absolute value of a	p. 25
$\cong$	is congruent to	p. 25
$\angle A$	angle with vertex A	p. 26
$\angle ABC$	angle with sides $\overrightarrow{BA}$ and $\overrightarrow{BC}$	p. 26
$m\angle A$	measure of angle A	p. 26
$^\circ$	degree(s)	p. 26
⌝	right angle symbol	p. 27
$\perp$	is perpendicular to	p. 33
m	slope of a linear function	p. 54
b	y-intercept of a linear function	p. 54
x_1, x_2	specific values of the variable x	p. 54
d	distance	p. 54
$\sqrt{x}$	nonnegative square root of x	p. 54
(a, b)	ordered pair with x-coordinate a and y-coordinate b	p. 54
$\triangle ABC$	triangle with vertices A, B, and C	p. 69
$a : b, \frac{a}{b}$	ratio of a to b	p. 72
n-gon	polygon with n sides	p. 80
▱$ABCD$	parallelogram with vertices A, B, C, and D	p. 91
$\odot A$	circle with center A	p. 96
d	diameter	p. 96
r	radius	p. 96
%	percent	p. 97
$\overset{\frown}{AB}$	arc with endpoints A and B	p. 98
$\overset{\frown}{ABC}$	arc with endpoints A and C and containing B	p. 98
$m\overset{\frown}{AB}$	measure of $\overset{\frown}{AB}$	p. 98
$\sim$	is similar to	p. 103
A'	image of A, A prime	p. 125
$\longrightarrow$	maps to	p. 125
$\begin{bmatrix} 1 & 2 \\ 3 & 4 \end{bmatrix}$	matrix	p. 131
$\overrightarrow{AB}$	vector with initial point A and terminal point B	p. 133
$\langle x, y \rangle$	ordered pair notation for a vector	p. 133
{ }	set brackets	p. 166
$\measuredangle$s	angles	p. 197
$\neq$	is not equal to	p. 210
$\ncong$	is not congruent to	p. 210
$\ngtr$	is not greater than	p. 214
$\nless$	is not less than	p. 214
A	area	p. 243
s	length of a side	p. 243
b	base length	p. 244
h	height	p. 244
$\approx$	is approximately equal to	p. 250
b_1, b_2	bases of a trapezoid	p. 269
a	apothem	p. 274
p	perimeter	p. 274
π	pi, ratio of the circumference of a circle to its diameter	p. 279
C	circumference	p. 279
B	area of a base	p. 309
h	length of an altitude	p. 309
L.A.	lateral area	p. 309
S.A.	surface area	p. 310
V	volume	p. 315
ℓ	slant height	p. 316
$\triangle$s	triangles	p. 422
$\pm$	plus or minus	p. 439
d_1, d_2	lengths of diagonals	p. 463
tan A	tangent of $\angle A$	p. 544
sin A	sine of $\angle A$	p. 551
cos A	cosine of $\angle A$	p. 551
$\vec{\mathbf{v}}$	vector **v**	p. 568
[ ]	brackets for grouping	p. 586

Formulas

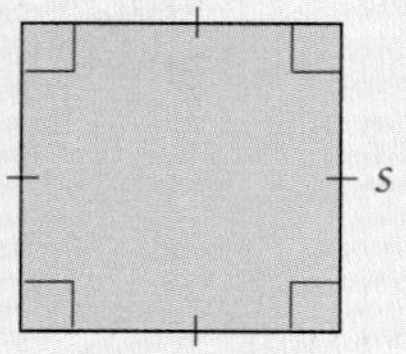

$A = s^2$

Square

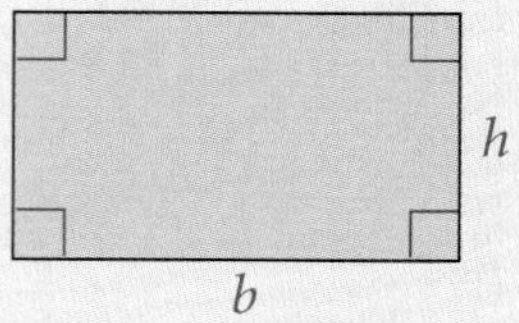

$A = bh$

Rectangle

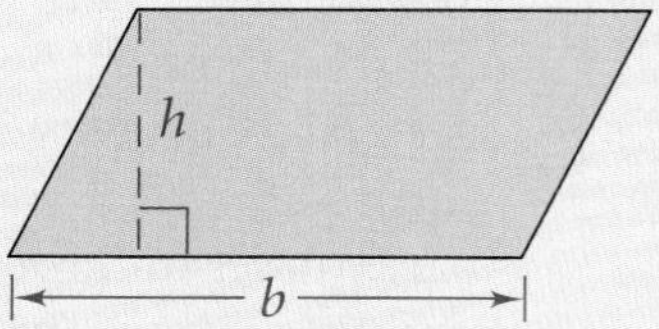

$A = bh$

Parallelogram

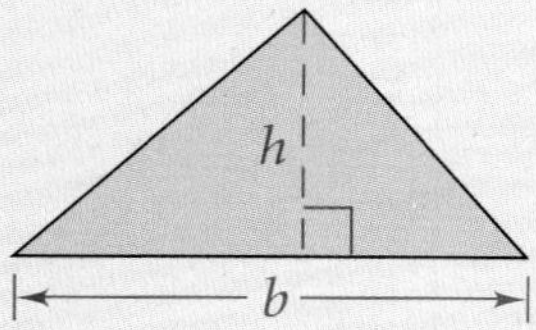

$A = \frac{1}{2}bh$

Triangle

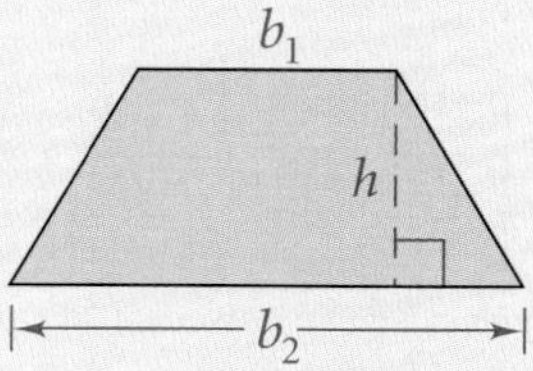

$A = \frac{1}{2}h(b_1 + b_2)$

Trapezoid

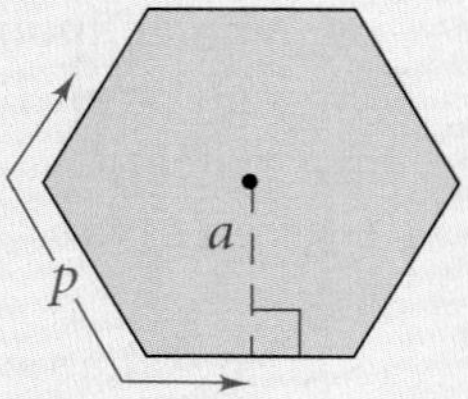

$A = \frac{1}{2}ap$

Regular Polygon

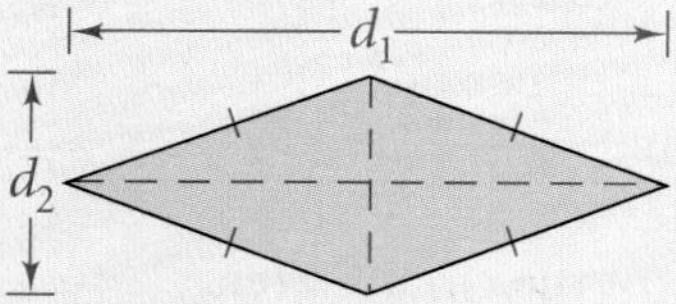

$A = \frac{1}{2}d_1d_2$

Rhombus

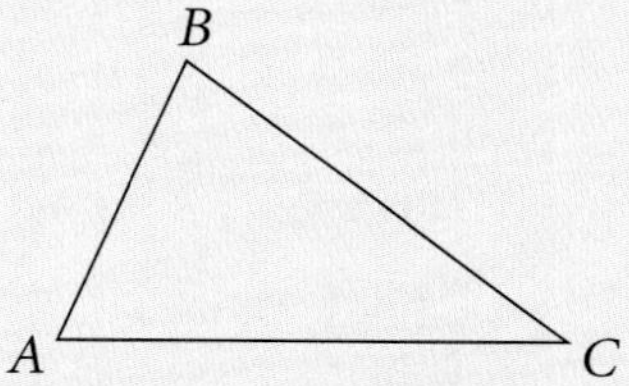

$m\angle A + m\angle B + m\angle C = 180$

Triangle Angle Sum

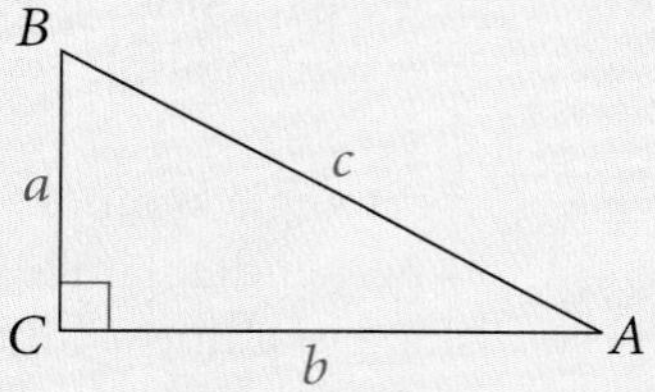

$a^2 + b^2 = c^2$

Pythagorean Theorem

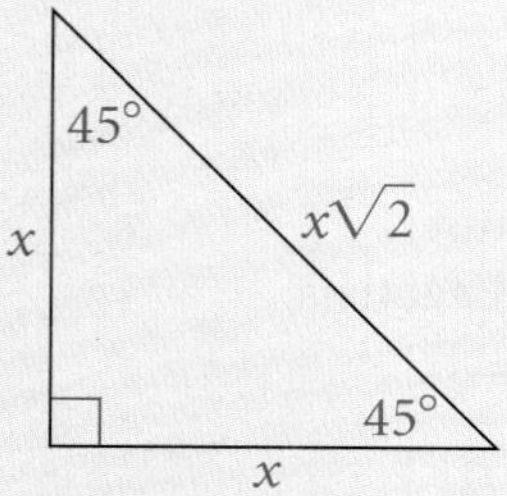

Ratio of sides $= 1 : 1 : \sqrt{2}$

45°-45°-90° Triangle

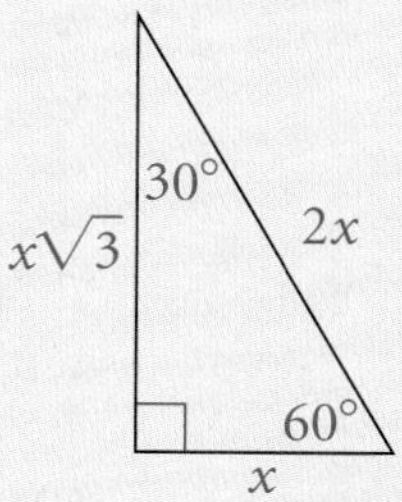

Ratio of sides $= 1 : \sqrt{3} : 2$

30°-60°-90° Triangle

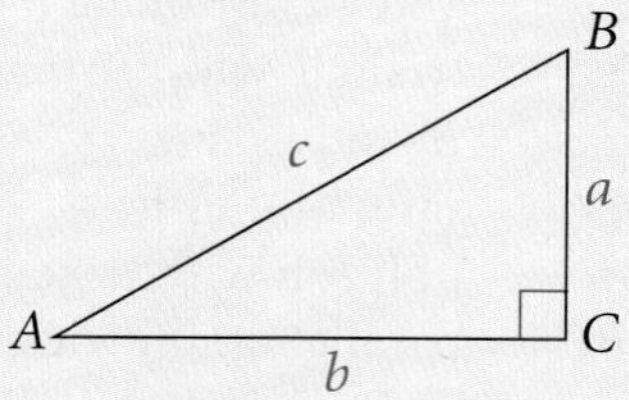

$\tan A = \frac{a}{b}$

$\sin A = \frac{a}{c} \quad \cos A = \frac{b}{c}$

Trigonometric Ratios

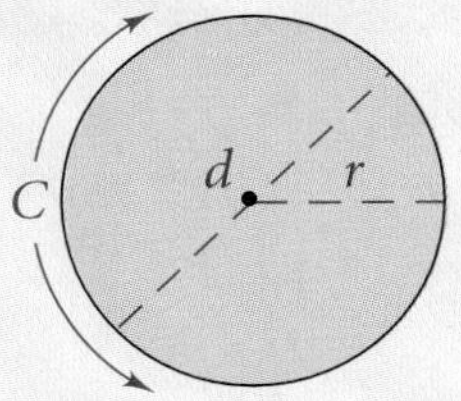

$C = \pi d$ or $C = 2\pi r$

$A = \pi r^2$

Circle

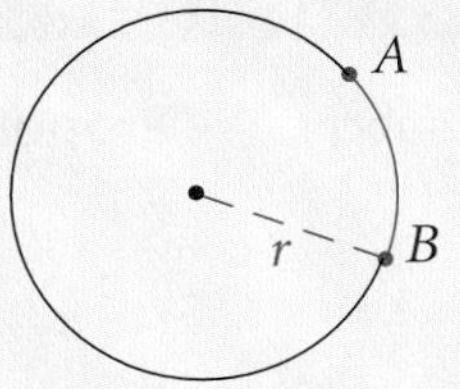

Length of $\widehat{AB} = \dfrac{m\widehat{AB}}{360} \cdot 2\pi r$

Arc

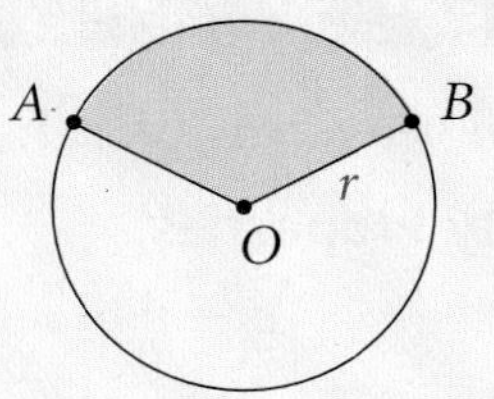

Area sector $AOB = \dfrac{m\widehat{AB}}{360} \cdot \pi r^2$

Sector of a Circle

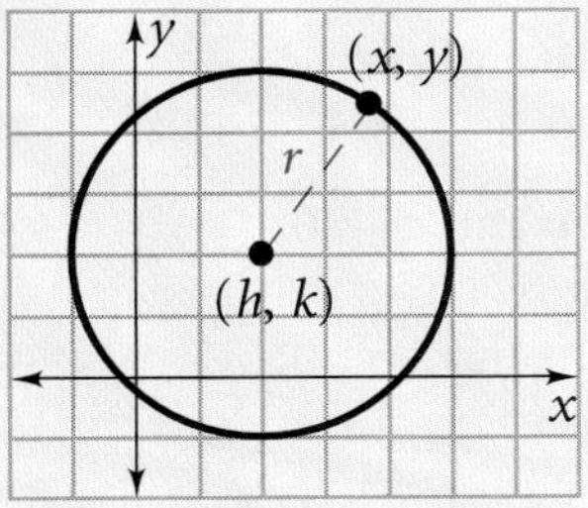

$r^2 = (x - h)^2 + (y - k)^2$

Equation of Circle

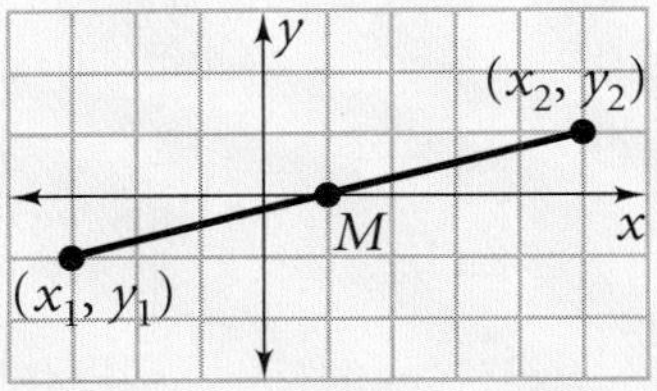

$d = \sqrt{(x_2 - x_1)^2 + (y_2 - y_1)^2}$

$M = \left(\dfrac{x_1 + x_2}{2}, \dfrac{y_1 + y_2}{2}\right)$

Distance and Midpoint

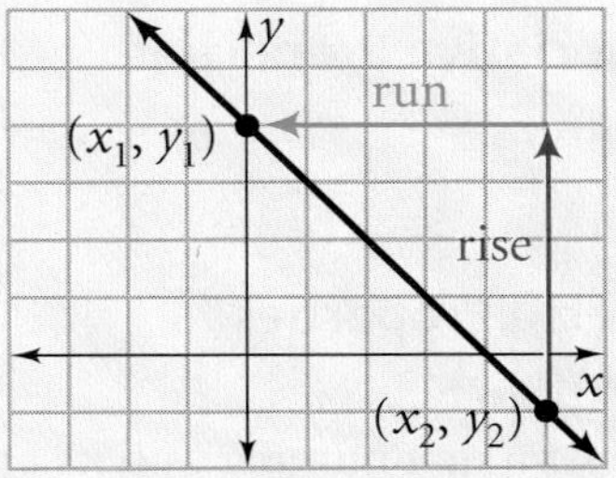

$m = \dfrac{\text{rise}}{\text{run}} = \dfrac{y_2 - y_1}{x_2 - x_1}$

Slope

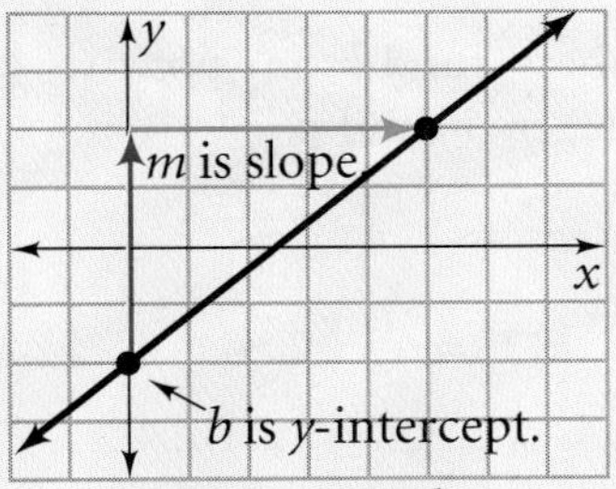

$y = mx + b$

Slope-intercept Form of Linear Equation

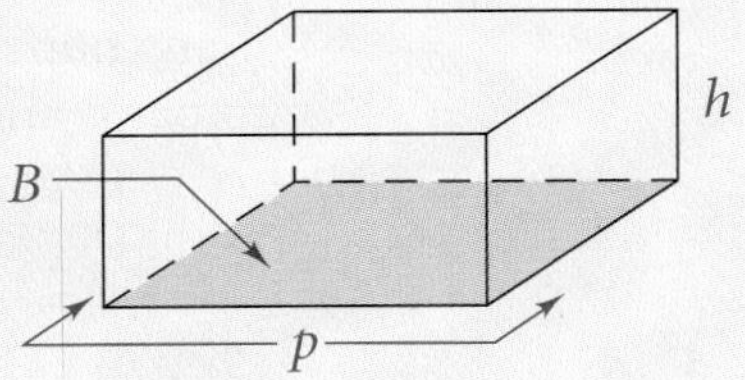

L.A. $= ph$

S.A. $=$ L.A. $+ 2B$

$V = Bh$

Right Prism

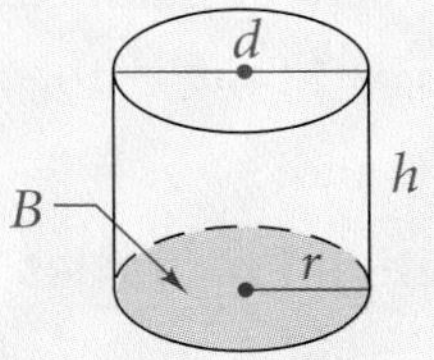

L.A. $= 2\pi rh$ or L.A. $= \pi dh$

S.A. $=$ L.A. $+ 2B$

$V = Bh$ or $V = \pi r^2 h$

Right Cylinder

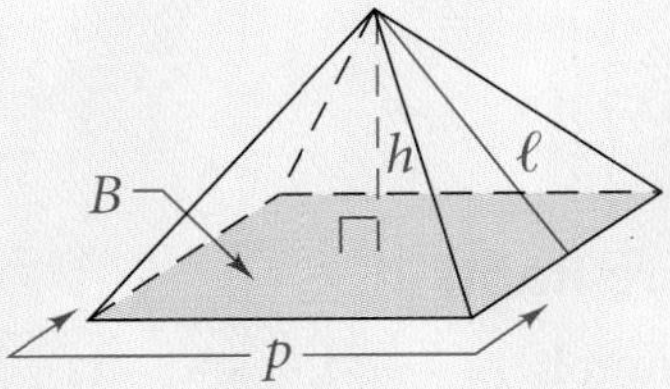

L.A. $= \frac{1}{2}p\ell$

S.A. $=$ L.A. $+ B$

$V = \frac{1}{3}Bh$

Regular Pyramid

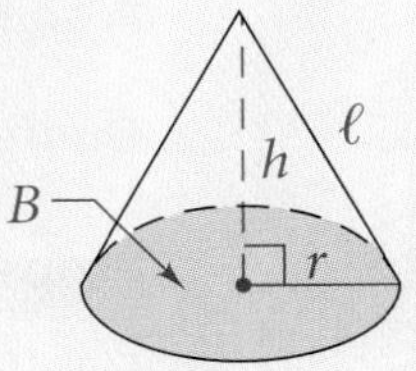

L.A. $= \pi r\ell$

S.A. $=$ L.A. $+ B$

$V = \frac{1}{3}Bh$ or $V = \frac{1}{3}\pi r^2 h$

Right Cone

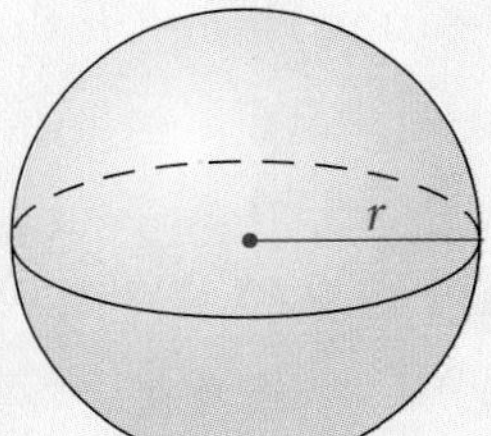

S.A. $= 4\pi r^2$

$V = \frac{4}{3}\pi r^3$

Sphere

Properties of Real Numbers

Unless otherwise stated, *a*, *b*, *c*, and *d* are real numbers.

Identity Properties

Addition $a + 0 = a$ and $0 + a = a$

Multiplication $a \cdot 1 = a$ and $1 \cdot a = a$

Commutative Properties

Addition $a + b = b + a$

Multiplication $a \cdot b = b \cdot a$

Associative Properties

Addition $(a + b) + c = a + (b + c)$

Multiplication $(a \cdot b) \cdot c = a \cdot (b \cdot c)$

Inverse Properties

Addition
The sum of a number and its *opposite,* or *additive inverse,* is zero.

$a + (-a) = 0$ and $-a + a = 0$

Multiplication
The *reciprocal,* or *multiplicative inverse,* of a rational number $\frac{a}{b}$ is $\frac{b}{a}$ $(a, b \neq 0)$.

$a \cdot \frac{1}{a} = 1$ and $\frac{1}{a} \cdot a = 1$ $(a \neq 0)$

Distributive Properties

$a(b + c) = ab + ac$ $\quad (b + c)a = ba + ca$
$a(b - c) = ab - ac$ $\quad (b - c)a = ba - ca$

Properties of Equality

Addition If $a = b$, then $a + c = b + c$.

Subtraction If $a = b$, then $a - c = b - c$.

Multiplication If $a = b$, then $a \cdot c = b \cdot c$.

Division If $a = b$ and $c \neq 0$, then $\frac{a}{c} = \frac{b}{c}$.

Substitution If $a = b$, then b can replace a in any expression.

Reflexive $a = a$

Symmetric If $a = b$, then $b = a$.

Transitive If $a = b$ and $b = c$, then $a = c$.

Properties of Proportions

$\frac{a}{b} = \frac{c}{d}$ is equivalent to

(1) $ad = bc$ $\quad$ (2) $\frac{b}{a} = \frac{d}{c}$

(3) $\frac{a}{c} = \frac{b}{d}$ $\quad$ (4) $\frac{a + b}{b} = \frac{c + d}{d}$

Zero-Product Property

If $ab = 0$, then $a = 0$ or $b = 0$.

Properties of Inequality

Addition If $a > b$ and $c \geq d$, then $a + c > b + d$.

Multiplication If $a > b$ and $c > 0$, then $ac > bc$.
If $a > b$ and $c < 0$, then $ac < bc$.

Transitive If $a > b$ and $b > c$, then $a > c$.

Comparison If $a = b + c$ and $c > 0$, then $a > b$.

Properties of Exponents

For any nonzero numbers *a* and *b*, any positive number *c*, and any integers *m* and *n*,

Zero Exponent $a^0 = 1$

Negative Exponent $a^{-n} = \frac{1}{a^n}$

Product of Powers $a^m \cdot a^n = a^{m+n}$

Quotient of Powers $\frac{a^m}{a^n} = a^{m-n}$

Power to a Power $(c^m)^n = c^{mn}$

Product to a Power $(ab)^n = a^n b^n$

Quotient to a Power $\left(\frac{a}{b}\right)^n = \frac{a^n}{b^n}$

Properties of Square Roots

For any nonnegative numbers *a* and *b*, and any positive number *c*,

Product of Square Roots $\sqrt{a} \cdot \sqrt{b} = \sqrt{ab}$

Quotient of Square Roots $\frac{\sqrt{a}}{\sqrt{c}} = \sqrt{\frac{a}{c}}$

Squares and Square Roots

Number n	Square n^2	Positive Square Root $\sqrt{n}$	Number n	Square n^2	Positive Square Root $\sqrt{n}$	Number n	Square n^2	Positive Square Root $\sqrt{n}$
1	1	1.000	**51**	2601	7.141	**101**	10,201	10.050
2	4	1.414	**52**	2704	7.211	**102**	10,404	10.100
3	9	1.732	**53**	2809	7.280	**103**	10,609	10.149
4	16	2.000	**54**	2916	7.348	**104**	10,816	10.198
5	25	2.236	**55**	3025	7.416	**105**	11,025	10.247
6	36	2.449	**56**	3136	7.483	**106**	11,236	10.296
7	49	2.646	**57**	3249	7.550	**107**	11,449	10.344
8	64	2.828	**58**	3364	7.616	**108**	11,664	10.392
9	81	3.000	**59**	3481	7.681	**109**	11,881	10.440
10	100	3.162	**60**	3600	7.746	**110**	12,100	10.488
11	121	3.317	**61**	3721	7.810	**111**	12,321	10.536
12	144	3.464	**62**	3844	7.874	**112**	12,544	10.583
13	169	3.606	**63**	3969	7.937	**113**	12,769	10.630
14	196	3.742	**64**	4096	8.000	**114**	12,996	10.677
15	225	3.873	**65**	4225	8.062	**115**	13,225	10.724
16	256	4.000	**66**	4356	8.124	**116**	13,456	10.770
17	289	4.123	**67**	4489	8.185	**117**	13,689	10.817
18	324	4.243	**68**	4624	8.246	**118**	13,924	10.863
19	361	4.359	**69**	4761	8.307	**119**	14,161	10.909
20	400	4.472	**70**	4900	8.367	**120**	14,400	10.954
21	441	4.583	**71**	5041	8.426	**121**	14,641	11.000
22	484	4.690	**72**	5184	8.485	**122**	14,884	11.045
23	529	4.796	**73**	5329	8.544	**123**	15,129	11.091
24	576	4.899	**74**	5476	8.602	**124**	15,376	11.136
25	625	5.000	**75**	5625	8.660	**125**	15,625	11.180
26	676	5.099	**76**	5776	8.718	**126**	15,876	11.225
27	729	5.196	**77**	5929	8.775	**127**	16,129	11.269
28	784	5.292	**78**	6084	8.832	**128**	16,384	11.314
29	841	5.385	**79**	6241	8.888	**129**	16,641	11.358
30	900	5.477	**80**	6400	8.944	**130**	16,900	11.402
31	961	5.568	**81**	6561	9.000	**131**	17,161	11.446
32	1024	5.657	**82**	6724	9.055	**132**	17,424	11.489
33	1089	5.745	**83**	6889	9.110	**133**	17,689	11.533
34	1156	5.831	**84**	7056	9.165	**134**	17,956	11.576
35	1225	5.916	**85**	7225	9.220	**135**	18,225	11.619
36	1296	6.000	**86**	7396	9.274	**136**	18,496	11.662
37	1369	6.083	**87**	7569	9.327	**137**	18,769	11.705
38	1444	6.164	**88**	7744	9.381	**138**	19,044	11.747
39	1521	6.245	**89**	7921	9.434	**139**	19,321	11.790
40	1600	6.325	**90**	8100	9.487	**140**	19,600	11.832
41	1681	6.403	**91**	8281	9.539	**141**	19,881	11.874
42	1764	6.481	**92**	8464	9.592	**142**	20,164	11.916
43	1849	6.557	**93**	8649	9.644	**143**	20,449	11.958
44	1936	6.633	**94**	8836	9.695	**144**	20,736	12.000
45	2025	6.708	**95**	9025	9.747	**145**	21,025	12.042
46	2116	6.782	**96**	9216	9.798	**146**	21,316	12.083
47	2209	6.856	**97**	9409	9.849	**147**	21,609	12.124
48	2304	6.928	**98**	9604	9.899	**148**	21,904	12.166
49	2401	7.000	**99**	9801	9.950	**149**	22,201	12.207
50	2500	7.071	**100**	10,000	10.000	**150**	22,500	12.247

Trigonometric Ratios

Angle	Sine	Cosine	Tangent
1°	0.0175	0.9998	0.0175
2°	0.0349	0.9994	0.0349
3°	0.0523	0.9986	0.0524
4°	0.0698	0.9976	0.0699
5°	0.0872	0.9962	0.0875
6°	0.1045	0.9945	0.1051
7°	0.1219	0.9925	0.1228
8°	0.1392	0.9903	0.1405
9°	0.1564	0.9877	0.1584
10°	0.1736	0.9848	0.1763
11°	0.1908	0.9816	0.1944
12°	0.2079	0.9781	0.2126
13°	0.2250	0.9744	0.2309
14°	0.2419	0.9703	0.2493
15°	0.2588	0.9659	0.2679
16°	0.2756	0.9613	0.2867
17°	0.2924	0.9563	0.3057
18°	0.3090	0.9511	0.3249
19°	0.3256	0.9455	0.3443
20°	0.3420	0.9397	0.3640
21°	0.3584	0.9336	0.3839
22°	0.3746	0.9272	0.4040
23°	0.3907	0.9205	0.4245
24°	0.4067	0.9135	0.4452
25°	0.4226	0.9063	0.4663
26°	0.4384	0.8988	0.4877
27°	0.4540	0.8910	0.5095
28°	0.4695	0.8829	0.5317
29°	0.4848	0.8746	0.5543
30°	0.5000	0.8660	0.5774
31°	0.5150	0.8572	0.6009
32°	0.5299	0.8480	0.6249
33°	0.5446	0.8387	0.6494
34°	0.5592	0.8290	0.6745
35°	0.5736	0.8192	0.7002
36°	0.5878	0.8090	0.7265
37°	0.6018	0.7986	0.7536
38°	0.6157	0.7880	0.7813
39°	0.6293	0.7771	0.8098
40°	0.6428	0.7660	0.8391
41°	0.6561	0.7547	0.8693
42°	0.6691	0.7431	0.9004
43°	0.6820	0.7314	0.9325
44°	0.6947	0.7193	0.9657
45°	0.7071	0.7071	1.0000
46°	0.7193	0.6947	1.0355
47°	0.7314	0.6820	1.0724
48°	0.7431	0.6691	1.1106
49°	0.7547	0.6561	1.1504
50°	0.7660	0.6428	1.1918
51°	0.7771	0.6293	1.2349
52°	0.7880	0.6157	1.2799
53°	0.7986	0.6018	1.3270
54°	0.8090	0.5878	1.3764
55°	0.8192	0.5736	1.4281
56°	0.8290	0.5592	1.4826
57°	0.8387	0.5446	1.5399
58°	0.8480	0.5299	1.6003
59°	0.8572	0.5150	1.6643
60°	0.8660	0.5000	1.7321
61°	0.8746	0.4848	1.8040
62°	0.8829	0.4695	1.8807
63°	0.8910	0.4540	1.9626
64°	0.8988	0.4384	2.0503
65°	0.9063	0.4226	2.1445
66°	0.9135	0.4067	2.2460
67°	0.9205	0.3907	2.3559
68°	0.9272	0.3746	2.4751
69°	0.9336	0.3584	2.6051
70°	0.9397	0.3420	2.7475
71°	0.9455	0.3256	2.9042
72°	0.9511	0.3090	3.0777
73°	0.9563	0.2924	3.2709
74°	0.9613	0.2756	3.4874
75°	0.9659	0.2588	3.7321
76°	0.9703	0.2419	4.0108
77°	0.9744	0.2250	4.3315
78°	0.9781	0.2079	4.7046
79°	0.9816	0.1908	5.1446
80°	0.9848	0.1736	5.6713
81°	0.9877	0.1564	6.3138
82°	0.9903	0.1392	7.1154
83°	0.9925	0.1219	8.1443
84°	0.9945	0.1045	9.5144
85°	0.9962	0.0872	11.4301
86°	0.9976	0.0698	14.3007
87°	0.9986	0.0523	19.0811
88°	0.9994	0.0349	28.6363
89°	0.9998	0.0175	57.2900
90°	1.0000	0.0000	

Chapter 1: Tools of Geometry

Postulate 1-1
Through any two points there is exactly one line. (p. 14)

Postulate 1-2
If two lines intersect, then they intersect in exactly one point. (p. 14)

Postulate 1-3
If two planes intersect, then they intersect in a line. (p. 14)

Postulate 1-4
Through any three noncollinear points there is exactly one plane. (p. 14)

Postulate 1-5
Ruler Postulate
The points of a line can be put into a one-to-one correspondence with the real numbers so that the distance between any two points is the absolute value of the difference of the corresponding numbers. (p. 25)

Postulate 1-6
Segment Addition Postulate
If three points A, B, and C are collinear and B is between A and C, then $AB + BC = AC$. (p. 26)

Postulate 1-7
Protractor Postulate
Let $\overrightarrow{OA}$ and $\overrightarrow{OB}$ be opposite rays in a plane. $\overrightarrow{OA}$, $\overrightarrow{OB}$, and all the rays with endpoint O that can be drawn on one side of $\overleftrightarrow{AB}$ can be paired with the real numbers from 0 to 180 in such a way that:

a. $\overrightarrow{OA}$ is paired with 0 and $\overrightarrow{OB}$ is paired with 180.
b. If $\overrightarrow{OC}$ is paired with x and $\overrightarrow{OD}$ is paired with y, then $m\angle COD = |x - y|$. (p. 27)

Postulate 1-8
Angle Addition Postulate
If point B is in the interior of $\angle AOC$, then $m\angle AOB + m\angle BOC = m\angle AOC$.
If $\angle AOC$ is a straight angle, then $m\angle AOB + m\angle BOC = 180$. (p. 28)

Properties of Congruence
Reflexive Property
$\overline{AB} \cong \overline{AB}$ and $\angle A \cong \angle A$
Symmetric Property
If $\overline{AB} \cong \overline{CD}$, then $\overline{CD} \cong \overline{AB}$.
If $\angle A \cong \angle B$, then $\angle B \cong \angle A$.
Transitive Property
If $\overline{AB} \cong \overline{CD}$ and $\overline{CD} \cong \overline{EF}$, then $\overline{AB} \cong \overline{EF}$.
If $\angle A \cong \angle B$ and $\angle B \cong \angle C$, then $\angle A \cong \angle C$. (p. 47)

Theorem 1-1
Vertical Angles Theorem
Vertical angles are congruent. (p. 48)

Theorem 1-2
Congruent Supplements Theorem
If two angles are supplements of congruent angles (or of the same angle), then the two angles are congruent. (p. 49)

Theorem 1-3
Congruent Complements Theorem
If two angles are complements of congruent angles (or of the same angle), then the two angles are congruent. (p. 49)

The Distance Formula
The distance d between two points $A(x_1, y_1)$ and $B(x_2, y_2)$ is $d = \sqrt{(x_2 - x_1)^2 + (y_2 - y_1)^2}$. (p. 54)

The Midpoint Formula
The coordinates of the midpoint M of $\overline{AB}$ with endpoints $A(x_1, y_1)$ and $B(x_2, y_2)$ are the following:
$M = \left(\frac{x_1 + x_2}{2}, \frac{y_1 + y_2}{2}\right)$ (p. 55)

Chapter 2: Investigating Geometric Figures

Theorem 2-1
Triangle Angle-Sum Theorem
The sum of the measures of the angles of a triangle is 180. (p. 68)

Theorem 2-2
Exterior Angle Theorem
The measure of each exterior angle of a triangle equals the sum of the measures of its two remote interior angles. (p. 70)

Corollary
The measure of an exterior angle of a triangle is greater than the measure of either of its remote interior angles. (p. 70)

Theorem 2-3
Polygon Interior Angle-Sum Theorem
The sum of the measures of the interior angles of an n-gon is $(n - 2)180$. (p. 77)

Theorem 2-4
Polygon Exterior Angle-Sum Theorem
The sum of the measures of the exterior angles of a polygon, one at each vertex, is 360. (p. 78)

Slopes of Parallel Lines
The slopes of two nonvertical parallel lines are equal. Two lines with the same slope are parallel. Vertical lines are parallel. (p. 85)

Slopes of Perpendicular Lines
The product of the slopes of two perpendicular lines, neither of which is vertical, is -1. If the product of the slopes of two lines is -1, then the lines are perpendicular. A horizontal and a vertical line are perpendicular. (p. 85)

Theorem 2-5
Two lines parallel to a third are parallel to each other. (p. 85)

Theorem 2-6
In a plane, two lines perpendicular to a third line are parallel to each other. (p. 85)

Postulate 2-1
Arc Addition Postulate
The measure of the arc formed by two adjacent arcs is the sum of the measure of two arcs. (p. 98)

Chapter 3: Transformations: Shapes in Motion

Properties of a Reflection
A reflection reverses orientation. A reflection is an isometry. (p. 126)

Properties of a Translation
A translation is an isometry. A translation does not change orientation. (p. 132)

Properties of a Rotation
A rotation is an isometry. A rotation does not change orientation. (p. 139)

Theorem 3-1
A composition of reflections in two parallel lines is a translation. (p. 144)

Theorem 3-2
A composition of reflections in two intersecting lines is a rotation. (p. 144)

Theorem 3-3
In a plane, two congruent figures can be mapped onto one another by a composition of at most three reflections. (p. 146)

Theorem 3-4
Isometry Classification Theorem
There are only four isometries. They are reflection, translation, rotation, and glide reflection. (p. 147)

Chapter 4: Triangle Relationships

Theorem 4-1
Isosceles Triangle Theorem
If two sides of a triangle are congruent, then the angles opposite those sides are also congruent. (p. 189)

Corollary
If a triangle is equilateral, then it is equiangular. (p. 190)

Theorem 4-2
The bisector of the vertex angle of an isosceles triangle is the perpendicular bisector of the base. (p. 189)

Theorem 4-3
Converse of the Isosceles Triangle Theorem
If two angles of a triangle are congruent, then the sides opposite the angles are congruent. (p. 189)

Corollary
If a triangle is equiangular, then it is equilateral. (p. 190)

Theorem 4-4
If a triangle is a right triangle, then the acute angles are complementary. (p. 195)

Theorem 4-5
If two angles of one triangle are congruent to two angles of another triangle, then the third angles are congruent. (p. 195)

Theorem 4-6
All right angles are congruent. (p. 195)

Theorem 4-7
If two angles are congruent and supplementary, then each is a right angle. (p. 195)

Theorem 4-8
Triangle Midsegment Theorem
If a segment joins the midpoints of two sides of a triangle, then the segment is parallel to the third side and half its length. (p. 201)

Theorem 4-9
Triangle Inequality Theorem
The sum of the lengths of any two sides of a triangle is greater than the length of the third side. (p. 214)

Theorem 4-10
If two sides of a triangle are not congruent, then the larger angle lies opposite the larger side. (p. 215)

Theorem 4-11
If two angles of a triangle are not congruent, then the longer side lies opposite the larger angle. (p. 215)

Theorem 4-12
Perpendicular Bisector Theorem
If a point is on the perpendicular bisector of a segment, then it is equidistant from the endpoints of the segment. (p. 219)

Theorem 4-13
Converse of Perpendicular Bisector Theorem
If a point is equidistant from the endpoints of a segment, then it is on the perpendicular bisector of the segment. (p. 219)

Theorem 4-14
Angle Bisector Theorem
If a point is on the bisector of an angle, then it is equidistant from the sides of the angle. (p. 222)

Theorem 4-15
Converse of Angle Bisector Theorem
If a point in the interior of an angle is equidistant from the sides of the angle, then it is on the angle bisector. (p. 222)

Theorem 4-16
The perpendicular bisectors of the sides of a triangle are concurrent at a point equidistant from the vertices. (p. 227)

Theorem 4-17
The bisectors of the angles of a triangle are concurrent at a point equidistant from the sides. (p. 227)

Theorem 4-18
The lines that contain the altitudes of a triangle are concurrent. (p. 230)

Theorem 4-19
The medians of a triangle are concurrent. (p. 230)

Chapter 5: Measuring in the Plane

Postulate 5-1
The area of a square is the square of the length of a side.
$A = s^2$ (p. 243)

Postulate 5-2
If two figures are congruent, their areas are equal. (p. 243)

Postulate 5-3
The area of a region is the sum of the areas of its nonoverlapping parts. (p. 243)

Theorem 5-1
Area of a Rectangle
The area of a rectangle is the product of its base and height.
$A = bh$ (p. 244)

Postulates & Theorems

Theorem 5-2
Area of a Parallelogram
The area of a parallelogram is the product of any base and the corresponding height.
$A = bh$ (p. 249)

Theorem 5-3
Area of a Triangle
The area of a triangle is half the product of any base and the corresponding height.
$A = \frac{1}{2}bh$ (p. 251)

Theorem 5-4
Pythagorean Theorem
In a right triangle, the sum of the squares of the lengths of the legs is equal to the square of the length of the hypotenuse.
$a^2 + b^2 = c^2$ (p. 257)

Theorem 5-5
Converse of the Pythagorean Theorem
If the square of the length of one side of a triangle is equal to the sum of the squares of the lengths of the other two sides, then the triangle is a right triangle. (p. 258)

Theorem 5-6
45°-45°-90° Triangle Theorem
In a 45°-45°-90° triangle, both legs are congruent and the length of the hypotenuse is $\sqrt{2}$ times the length of a leg.
hypotenuse $= \sqrt{2} \cdot$ leg (p. 264)

Theorem 5-7
30°-60°-90° Triangle Theorem
In a 30°-60°-90° triangle, the length of the hypotenuse is twice the length of the shorter leg. The length of the longer leg is $\sqrt{3}$ times the length of the shorter leg.
hypotenuse $= 2 \cdot$ shorter leg
longer leg $= \sqrt{3} \cdot$ shorter leg (p. 265)

Theorem 5-8
Area of a Trapezoid
The area of a trapezoid is half the product of the height and the sum of the lengths of the bases.
$A = \frac{1}{2}h(b_1 + b_2)$ (p. 269)

Theorem 5-9
Area of a Regular Polygon
The area of a regular polygon is half the product of the apothem and the perimeter.
$A = \frac{1}{2}ap$ (p. 274)

Theorem 5-10
Circumference of a Circle
The circumference of a circle is π times the diameter.
$C = \pi d$ or $C = 2\pi r$ (p. 279)

Theorem 5-11
Arc Length
The length of an arc of a circle is the product of the ratio $\frac{\text{measure of the arc}}{360}$ and the circumference of the circle.
Length of $\widehat{AB} = \frac{m\widehat{AB}}{360} \cdot 2\pi r$ (p. 281)

Theorem 5-12
Area of a Circle
The area of a circle is the product of π and the square of the radius.
$A = \pi r^2$ (p. 285)

Theorem 5-13
Area of a Sector of a Circle
The area of a sector of a circle is the product of the ratio $\frac{\text{measure of the arc}}{360}$ and the area of the circle.
Area of sector $AOB = \frac{m\widehat{AB}}{360} \cdot \pi r^2$ (p. 286)

Chapter 6: Measuring in Space

Theorem 6-1
Lateral and Surface Areas of a Right Prism
The lateral area of a right prism is the product of the perimeter of the base and the height.
L.A. $= ph$
The surface area of a right prism is the sum of the lateral area and the areas of the two bases.
S.A. $=$ L.A. $+ 2B$ (p. 310)

Theorem 6-2
Lateral and Surface Areas of a Right Cylinder
The lateral area of a right cylinder is the product of the circumference of the base and the height of the cylinder.
$\text{L.A.} = 2\pi rh$ or $\text{L.A.} = \pi dh$
The surface area of a right cylinder is the sum of the lateral area and the areas of the two bases.
$\text{S.A.} = \text{L.A.} + 2B$ (p. 311)

Theorem 6-3
Lateral and Surface Areas of a Regular Pyramid
The lateral area of a regular pyramid is half the product of the perimeter of the base and the slant height.
$\text{L.A.} = \frac{1}{2}p\ell$
The surface area of a regular pyramid is the sum of the lateral area and the area of the base.
$\text{S.A.} = \text{L.A.} + B$ (p. 317)

Theorem 6-4
Lateral and Surface Areas of a Right Cone
The lateral area of a right cone is half the product of the circumference of the base and the slant height.
$\text{L.A.} = \frac{1}{2} \cdot 2\pi r\ell$ or $\text{L.A.} = \pi r\ell$
The surface area of a right cone is the sum of the lateral area and the area of the base.
$\text{S.A.} = \text{L.A.} + B$ (p. 318)

Theorem 6-5
Cavalieri's Principle
If two space figures have the same height and the same cross-sectional area at every level, then they have the same volume. (p. 324)

Theorem 6-6
Volume of a Prism
The volume of a prism is the product of the area of a base and the height of the prism.
$V = Bh$ (p. 324)

Theorem 6-7
Volume of a Cylinder
The volume of a cylinder is the product of the area of a base and the height of the cylinder.
$V = Bh$ or $V = \pi r^2h$ (p. 325)

Theorem 6-8
Volume of a Pyramid
The volume of a pyramid is one third the product of the area of the base and the height of the pyramid.
$V = \frac{1}{3}Bh$ (p. 331)

Theorem 6-9
Volume of a Cone
The volume of a cone is one third the product of the area of the base and the height of the cone.
$V = \frac{1}{3}Bh$ or $V = \frac{1}{3}\pi r^2h$ (p. 332)

Theorem 6-10
Surface Area of a Sphere
The surface area of a sphere is four times the product of π and the square of the radius of the sphere.
$\text{S.A.} = 4\pi r^2$ (p. 338)

Theorem 6-11
Volume of a Sphere
The volume of a sphere is $\frac{4}{3}$ the product of π and the cube of the radius of the sphere.
$V = \frac{4}{3}\pi r^3$ (p. 339)

Chapter 7: Reasoning and Parallel Lines

Postulate 7-1
Corresponding Angles Postulate
If two parallel lines are cut by a transversal, then corresponding angles are congruent. (p. 364)

Theorem 7-1
Alternate Interior Angles Theorem
If two parallel lines are cut by a transversal, then alternate interior angles are congruent. (p. 364)

Theorem 7-2
Same-Side Interior Angles Theorem
If two parallel lines are cut by a transversal, then the pairs of same-side interior angles are supplementary. (p. 364)

Postulate 7-2
Converse of Corresponding Angles Postulate
If two lines are cut by a transversal so that a pair of corresponding angles are congruent, then the lines are parallel. (p. 371)

Theorem 7-3
Converse of Alternate Interior Angles Theorem
If two lines are cut by a transversal so that a pair of alternate interior angles are congruent, then the lines are parallel. (p. 372)

Theorem 7-4
Converse of Same-Side Interior Angles Theorem
If two lines are cut by a transversal so that a pair of same-side interior angles are supplementary, then the lines are parallel. (p. 372)

Postulate 7-3
Euclid's Parallel Postulate
Through a point not on a line, there is one and only one line parallel to the given line. (p. 394)

Postulate 7-4
Spherical Geometry Parallel Postulate
Through a point not on a line, there is no line parallel to the given line. (p. 394)

Chapter 8: Proving Triangles Congruent

Postulate 8-1
Side-Side-Side Postulate (SSS Postulate)
If three sides of one triangle are congruent to three sides of another triangle, then the two triangles are congruent. (p. 406)

Postulate 8-2
Side-Angle-Side Postulate (SAS Postulate)
If two sides and the included angle of one triangle are congruent to two sides and the included angle of another triangle, then the two triangles are congruent. (p. 408)

Postulate 8-3
Angle-Side-Angle Postulate (ASA Postulate)
If two angles and the included side of one triangle are congruent to two angles and the included side of another triangle, then the two triangles are congruent. (p. 414)

Theorem 8-1
Angle-Angle-Side Theorem (AAS Theorem)
If two angles and a nonincluded side of one triangle are congruent to two angles and the corresponding nonincluded side of another triangle, then the triangles are congruent. (p. 416)

Theorem 8-2
Hypotenuse-Leg Theorem (HL Theorem)
If the hypotenuse and a leg of one right triangle are congruent to the hypotenuse and a leg of another right triangle, then the triangles are congruent. (p. 420)

Chapter 9: Quadrilaterals

Theorem 9-1
Opposite sides of a parallelogram are congruent. (p. 448)

Theorem 9-2
Opposite angles of a parallelogram are congruent. (p. 449)

Theorem 9-3
The diagonals of a parallelogram bisect each other. (p. 450)

Theorem 9-4
If three (or more) parallel lines cut off congruent segments on one transversal, then they cut off congruent segments on every transversal. (p. 450)

Theorem 9-5
If the diagonals of a quadrilateral bisect each other, then the quadrilateral is a parallelogram. (p. 455)

Theorem 9-6
If one pair of opposite sides of a quadrilateral are both congruent and parallel, then the quadrilateral is a parallelogram. (p. 456)

Theorem 9-7
If both pairs of opposite sides of a quadrilateral are congruent, then the quadrilateral is a parallelogram. (p. 456)

Theorem 9-8
If both pairs of opposite angles of a quadrilateral are congruent, then the quadrilateral is a parallelogram. (p. 456)

Theorem 9-9
Each diagonal of a rhombus bisects two angles of the rhombus. (p. 463)

Examples

Indirect proof (p. 208) See *indirect reasoning* and *proof.*

Indirect reasoning (p. 207) In indirect reasoning, all possibilities are considered and then all but one are proved false. The remaining possibility must be true.

Eduardo spent more than $60 on two books at a store. Prove that at least one book costs more than $30.

Proof: Suppose neither costs more than $30. Then he spent no more than $60 at the store. Since this contradicts the given information, at least one book costs $30 or more.

Inductive reasoning (p. 5) Inductive reasoning is a type of reasoning that reaches conclusions based on a pattern of specific examples or past events.

You see four people walk into a building. Each person emerges with a small bag containing hot food. You use inductive reasoning to conclude that this building contains a restaurant.

Initial point (p. 133) See *vector.*

Inscribe (pp. 228, 596, 607) A circle is inscribed in a polygon if the sides of the polygon are tangent to the circle. A polygon is inscribed in a circle if the vertices of the polygon are on the circle.

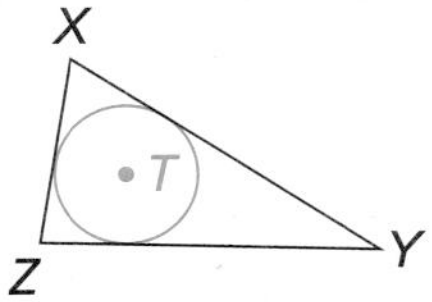

$\odot T$ is inscribed in $\triangle XYZ$.

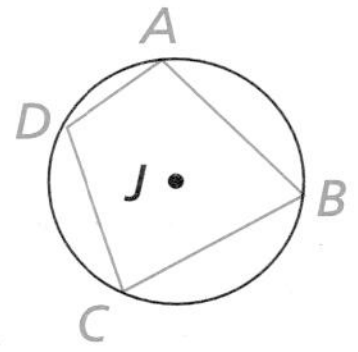

ABCD is inscribed in $\odot J$.

Inscribed angle (p. 607) An angle is inscribed in a circle if the vertex of the angle is on the circle and the sides of the angle are chords of the circle.

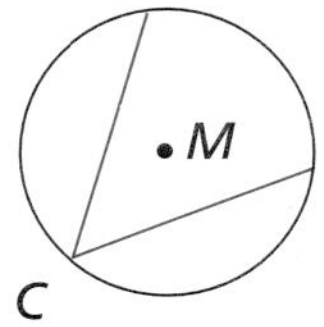

$\angle C$ is inscribed in $\odot M$.

Intercepted arc (p. 607) An intercepted arc of an inscribed angle is an arc whose endpoints are on the sides of the angle and whose remaining points lie in the interior of the angle.

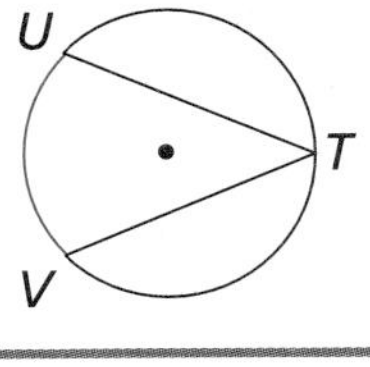

$\widehat{UV}$ is the intercepted arc of inscribed angle $\angle T$.

Inverse (p. 184) The inverse of the conditional "if *p*, then *q*" is the conditional "if not *p*, then not *q*."

Conditional: If a figure is a square, then it is a parallelogram.

Inverse: If a figure is not a square, then it is not a parallelogram.

Examples

Isometric drawing (p. 109) An isometric drawing of a three-dimensional object shows a corner view of the figure.

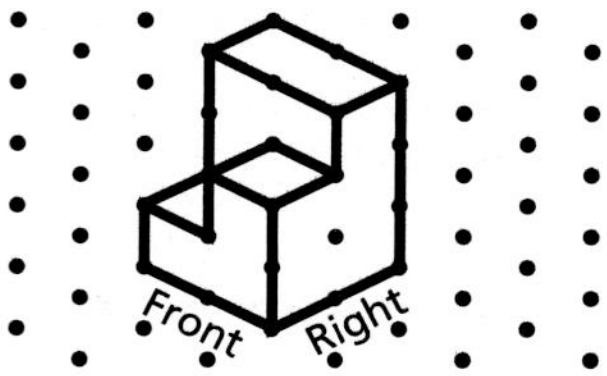

Isometry (p. 124) An isometry, also known as a *congruence transformation,* is a transformation in which the original figure and its image are congruent.

The four isometries are reflections, rotations, translations, and glide reflections.

Isosceles trapezoid (p. 91, 470) An isosceles trapezoid is a trapezoid whose legs are congruent.

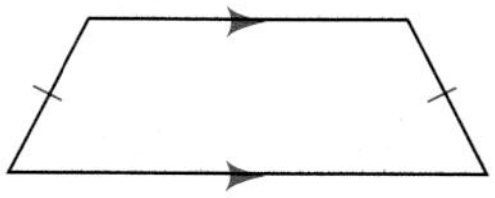

Isosceles triangle (pp. 71, 188) An isosceles triangle is a triangle that has at least two congruent sides. In an isosceles triangle that is not equilateral, the two congruent sides are called *legs* and the third side is called the *base.* The two angles with the base as a side are *base angles* and the third angle is the *vertex angle.*

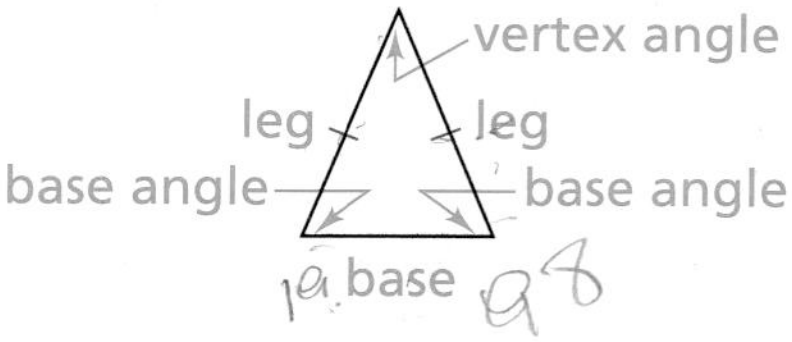

K

Kite (p. 91) A kite is a quadrilateral with two pairs of congruent adjacent sides and no opposite sides congruent.

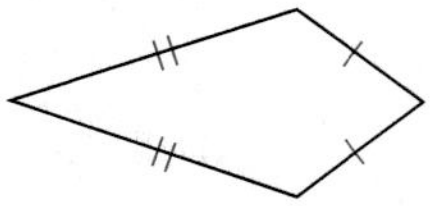

L

Lateral area (pp. 309, 311, 316, 318) The lateral area of a prism or pyramid is the sum of the areas of the lateral faces. The lateral area of a cylinder or cone is the area of the curved surface. A list of lateral area formulas is on pages 670–671.

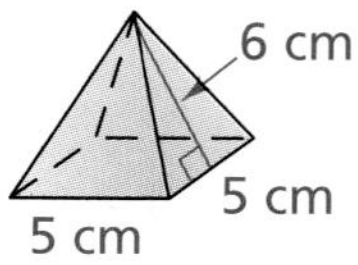

L.A. of pyramid $= \frac{1}{2}p\ell$
$= \frac{1}{2}(20)(6)$
$= 60 \text{ cm}^2$

Lateral face See *prism* and *pyramid.*

Leg See *isosceles triangle, right triangle,* and *trapezoid.*

Examples

Line (pp. 13, 393) In Euclidean geometry, you can think of a line as a series of points that extends in two opposite directions without end. In spherical geometry, you can think of a line as a great circle of a sphere.

Locus (p. 220) A locus is the set of points that meet a stated condition.

Example: The blue figure is the locus of points in a plane 1 cm from $\overline{DC}$.

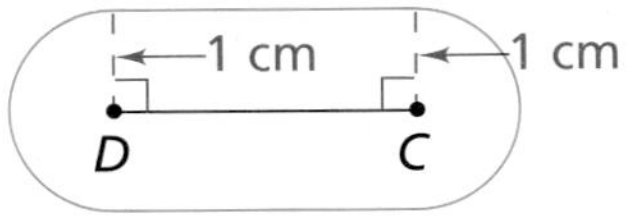

M

Major arc (p. 98) A major arc of a circle is any arc longer than a semicircle.

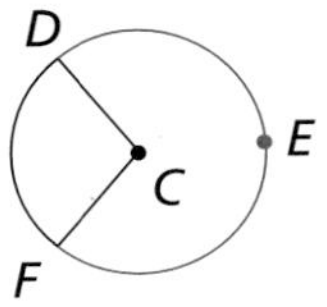

$\overset{\frown}{DEF}$ is a major arc of $\odot C$.

Map (p. 124) See *transformation*.

Matrix (p. 131) A matrix is a rectangular array of numbers. Each item in a matrix is called an *entry*.

The matrix $\begin{bmatrix} 1 & -2 \\ 0 & 13 \end{bmatrix}$ has dimensions 2×2. The number 1 is the entry in the first row and first column.

Measure of an angle (p. 26) Angles are measured in degrees. An angle can be measured with a *protractor*.

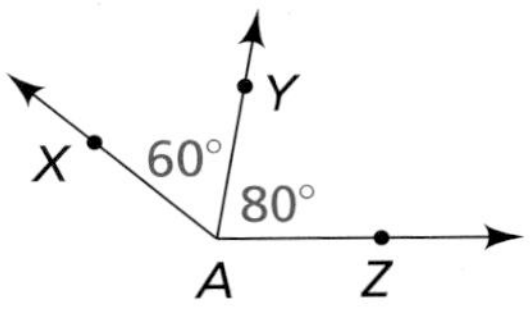

$m\angle ZAY = 80$, $m\angle YAX = 60$, and $m\angle ZAX = 140$.

Measure of an arc (p. 98) The measure of a minor arc is the measure of its central angle. The measure of a major arc is 360 minus the measure of its related minor arc.

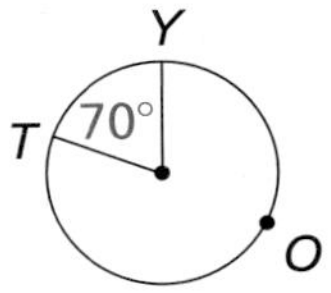

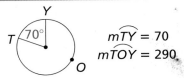

$m\overset{\frown}{TY} = 70$
$m\overset{\frown}{TOY} = 290$

Median of a triangle (p. 229) A median of a triangle is a segment that joins a vertex of the triangle and the midpoint of the side opposite that vertex.

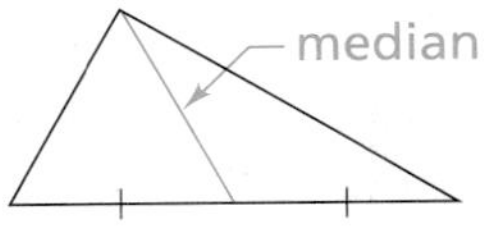

Examples

Midpoint of a segment (p. 33) A midpoint of a segment is the point that divides the segment into two congruent segments.

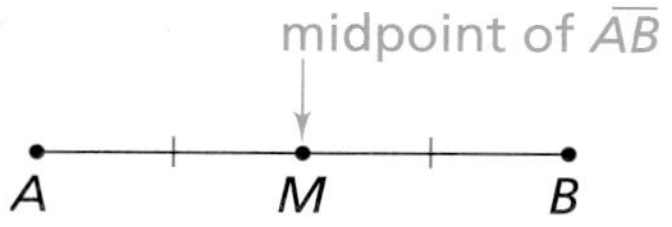

Midsegment of a trapezoid (p. 483) The segment that joins the midpoints of the legs of a trapezoid is the midsegment of the trapezoid.

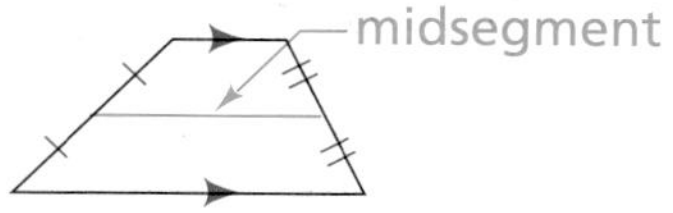

Midsegment of a triangle (pp. 201, 235) A midsegment of a triangle is a segment that joins the midpoints of two sides of the triangle.

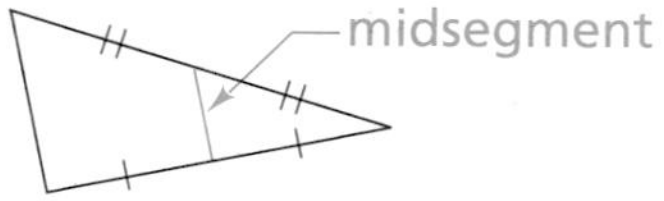

Minor arc (p. 98) A minor arc of a circle is any arc shorter than a semicircle.

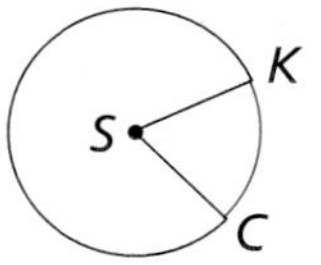

$\overset{\frown}{KC}$ is a minor arc of $\odot S$.

N

Negation (p. 184) A negation of a statement has the opposite meaning of the original statement.

Statement: The angle is obtuse.

Negation: The angle is not obtuse.

Net (p. 302) A net is a two-dimensional pattern that you can fold to form a three-dimensional figure.

Example: The net shown can be folded into a prism with pentagonal bases.

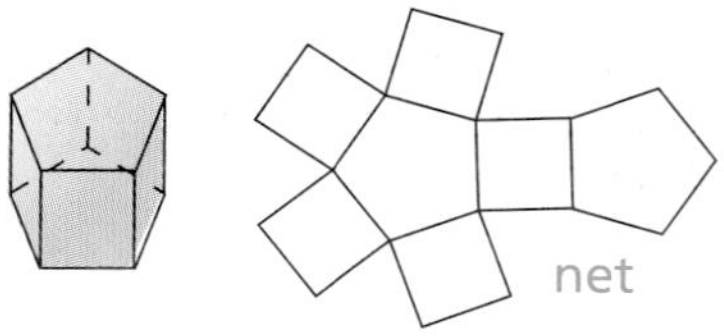

***n*-gon (p. 76)** An *n*-gon is a polygon with *n* sides.

Nonagon (p. 76) A nonagon is a polygon with nine sides.

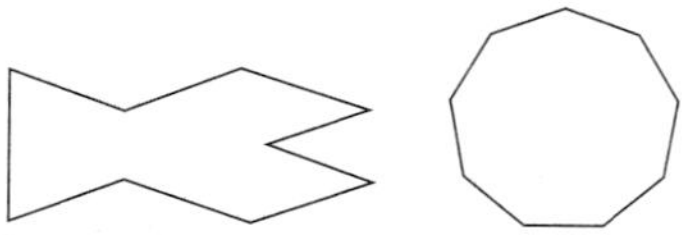

O

Oblique cylinder or prism See *cylinder* and *prism*.

Obtuse angle (p. 27) An obtuse angle is an angle whose measure is between 90 and 180.

Examples

Obtuse triangle (p. 71) An obtuse triangle has one obtuse angle.

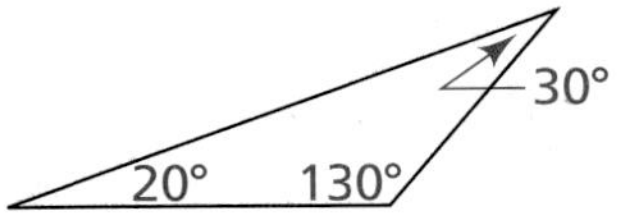

Octagon (p. 76) An octagon is a polygon with eight sides.

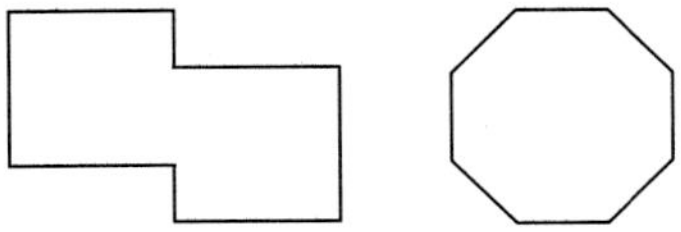

Opposite rays (p. 18) Opposite rays are collinear rays with the same endpoint. They form a line.

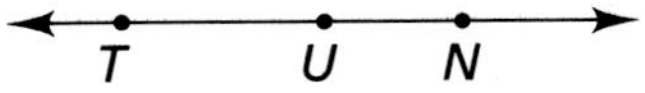

$\overrightarrow{UT}$ and $\overrightarrow{UN}$ are opposite rays.

Orientation (p. 126) Two figures have *opposite* orientation if a reflection is needed to map one onto the other. If a reflection is not needed to map one figure onto the other, the figures have the *same* orientation.

The two R's have opposite orientation.

Origin (p. 54) See *coordinate plane.*

Orthographic drawing (p. 111) An orthographic drawing shows the top view, front view, and right-side view of a three-dimensional figure.

Example: The diagram shows an isometric drawing (upper right) and the three views that make up an orthographic drawing.

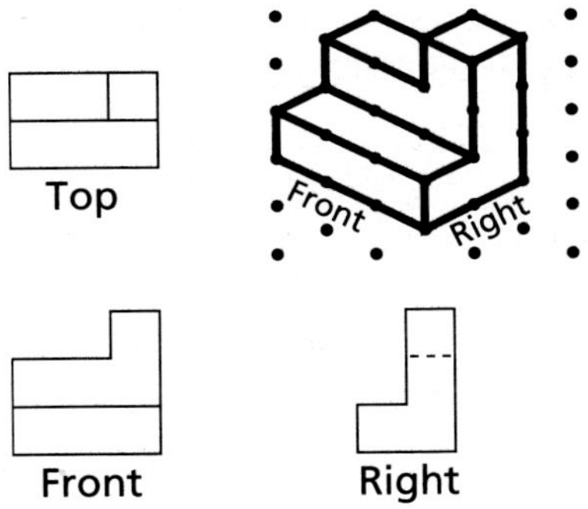

P

Paragraph proof (p. 195) See *proof.*

Parallel lines (pp. 18, 363) Two lines are parallel if they lie in the same plane and do not intersect. The symbol $\parallel$ means "is parallel to."

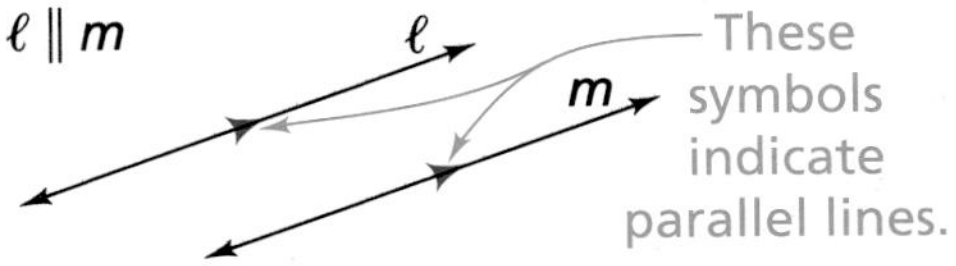

Parallelogram (pp. 91, 249–250) A parallelogram is a quadrilateral with two pairs of parallel sides. You can choose any side to be the *base.* An *altitude* is any segment perpendicular to the line containing the base drawn from the side opposite the base. The *height* is the length of an altitude.

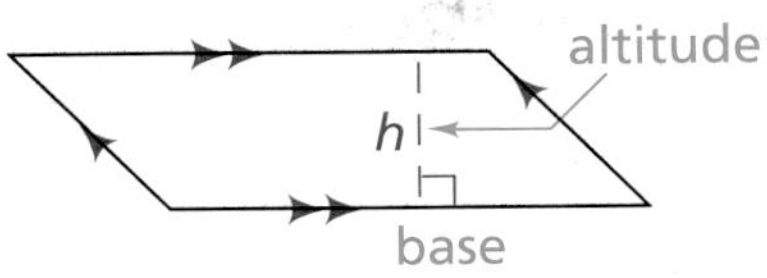

Examples

Parallel planes (p. 19) Parallel planes are planes that do not intersect.

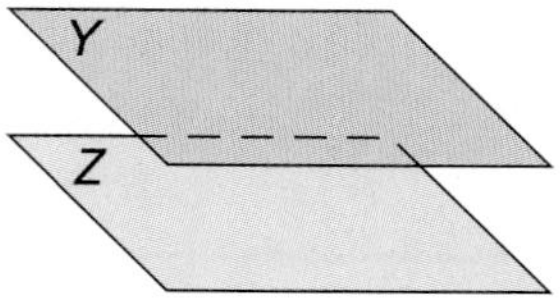

Planes Y and Z are parallel.

Pentagon (p. 76) A pentagon is a polygon with five sides.

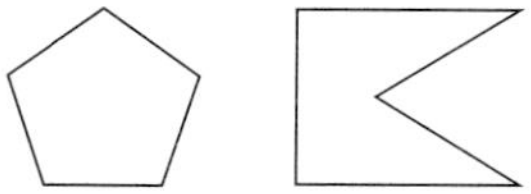

Perimeter of a polygon (p. 242) The perimeter of a polygon is the sum of the lengths of its sides.

$p = 4 + 4 + 5 + 3$
$= 16$ in.

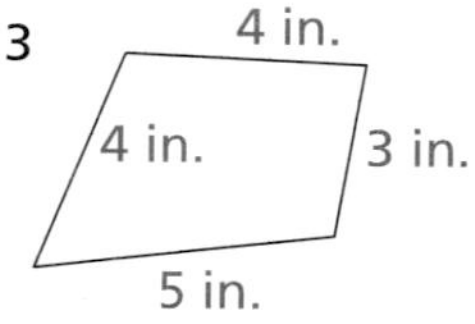

Perpendicular bisector (p. 34) The perpendicular bisector of a segment is a segment, ray, line, or plane that is perpendicular to the segment at its midpoint.

Example: $\overleftrightarrow{YZ}$ is the perpendicular bisector of $\overline{AB}$. It is perpendicular to $\overline{AB}$ and intersects it at its midpoint M.

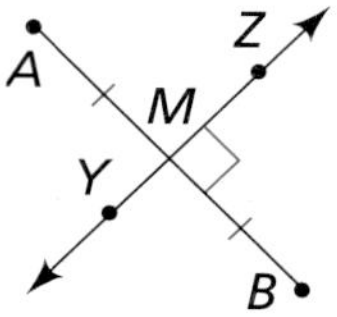

Perpendicular lines (p. 33) Two lines are perpendicular if they intersect and form right angles. The symbol $\perp$ means "is perpendicular to."

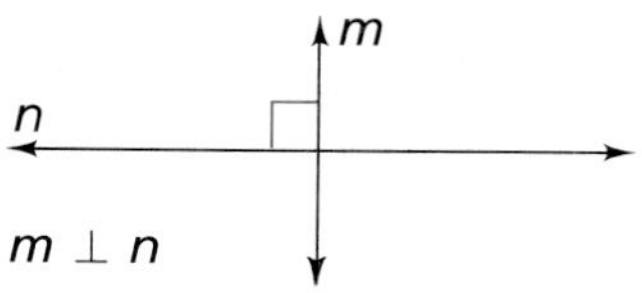

Perspective drawing (p. 385) Perspective drawing is a way of drawing objects on a flat surface so that they look the same way as they appear to the eye. In *one-point perspective,* there is one *vanishing point.* In *two-point perspective,* there are two vanishing points.

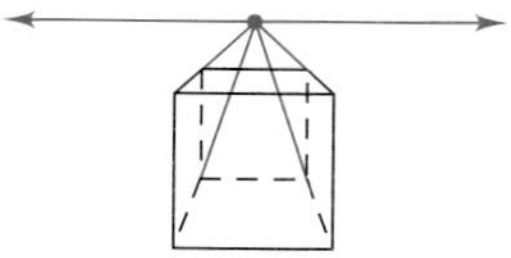

one–point perspective

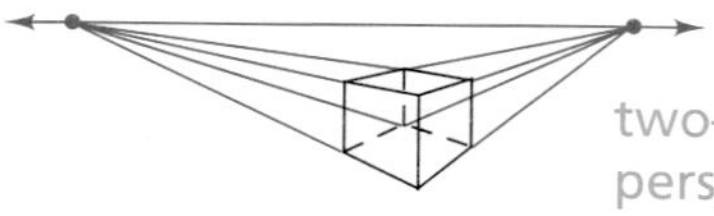

two–point perspective

Pi (p. 279) Pi (π) is the ratio of the circumference of any circle to its diameter. The number π is irrational and can be approximated by $\pi \approx 3.14159$.

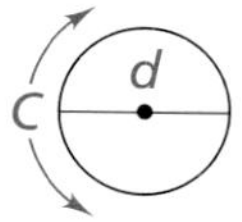

$\pi = \frac{C}{d}$

Examples

Plane (pp. 13, 393) In Euclidean geometry, you can think of a plane as a flat surface that extends in all directions without end. It has no thickness. In spherical geometry, you can think of a plane as the surface of a sphere.

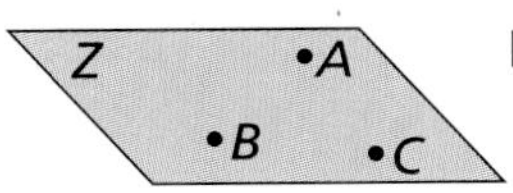

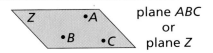

plane *ABC*
or
plane *Z*

Point (p. 12) You can think of a point as a location. A point has no size.

• *P*

Point of concurrency (p. 227) See *concurrent.*

Point of tangency (p. 593) See *tangent to a circle.*

Point symmetry (p. 154) A figure with rotational symmetry of 180° has point symmetry.

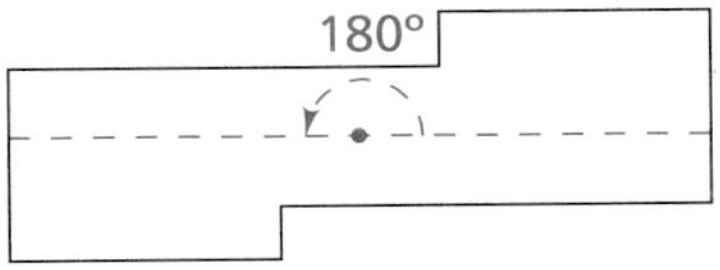

Polygon (p. 76) A polygon is a closed plane figure with at least three *sides.* The sides are segments and intersect only at their endpoints and no adjacent sides are collinear. The *vertices* of the polygon are the endpoints of the sides. A *diagonal* is a segment that connects two nonconsecutive vertices. A polygon is *convex* if no diagonal contains points outside the polygon. A polygon is *concave* if you can draw a diagonal that contains points outside the polygon.

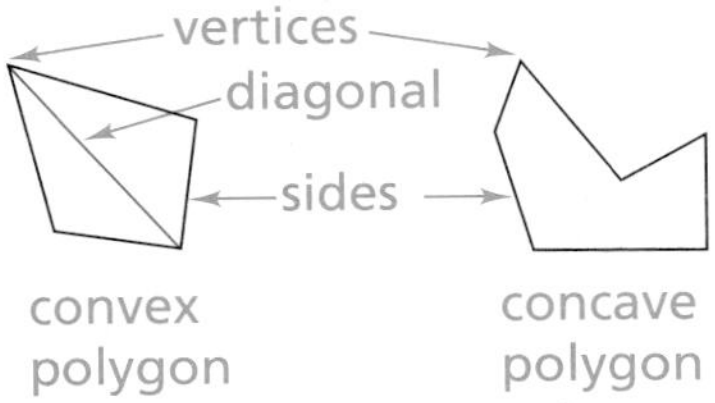

Polyhedron (p. 302) A polyhedron is a three-dimensional figure whose surfaces, or *faces,* are polygons. The vertices of the polygons are the *vertices* of the polyhedron. The intersections of the faces are the *edges* of the polyhedron.

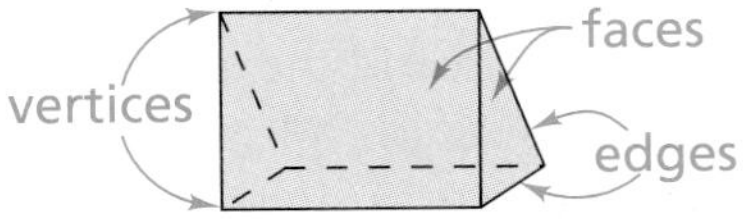

Postulate (p. 14) A postulate is an accepted statement of fact.

Postulate: Through any two points there is exactly one line.

Preimage (p. 124) See *transformation.*

Prime notation (p. 124) See *transformation.*

Examples

Prism (p. 309) A prism is a polyhedron with two congruent, parallel faces, called the *bases.* The other faces are parallelograms and are called the *lateral faces.* An *altitude* of a prism is a perpendicular segment that joins the planes of the bases. Its length is the *height* of the prism. In a *right prism,* the lateral faces are rectangular and a lateral edge is an altitude. In an *oblique prism,* some or all of the lateral faces are nonrectangular.

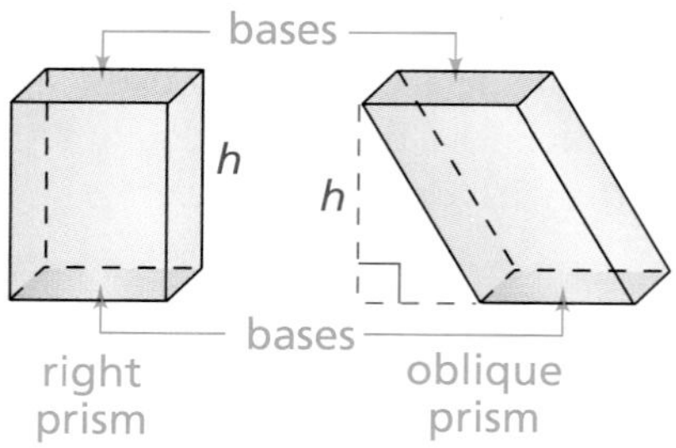

Proof (pp. 49, 194, 208, 372, 483) A proof is a convincing argument that uses deductive reasoning. A proof can be written in many forms. In a *two-column proof,* the statements and reasons are aligned in columns. In a *paragraph proof,* the statements and reasons are connected in sentences. In a *flow proof,* arrows show the logical connections between the statements. In a *coordinate proof,* a figure is drawn on a coordinate plane and the formulas for slope, midpoint, and distance are used to prove properties of the figure. An *indirect proof* involves the use of indirect reasoning.

Given: $\triangle EFG$, with right angle $\angle F$
Prove: $\angle E$ and $\angle G$ are complementary.

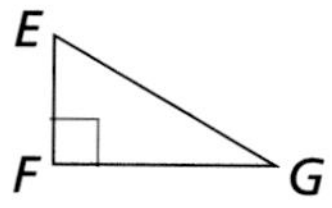

Paragraph Proof: Because $\angle F$ is a right angle, $m\angle F = 90$. By the Triangle Angle-Sum Thm., $m\angle E + m\angle F + m\angle G = 180$. By Substitution, $m\angle E + 90 + m\angle G = 180$. Subtracting 90 from each side yields $m\angle E + m\angle G = 90$. $\angle E$ and $\angle G$ are complementary by definition.

Proportion (p. 496) A proportion is a statement that two ratios are equal. The *extremes* of the proportion $\frac{x}{5} = \frac{3}{4}$ are the first and last terms, x and 4. The *means* are the middle two terms, 5 and 3. An *extended proportion* is a statement that three or more ratios are equal.

$\frac{x}{5} = \frac{3}{4}$ is a proportion.

Pyramid (p. 316) A pyramid is a polyhedron in which one face, the *base,* is a polygon and the other faces, the *lateral faces,* are triangles with a common vertex, called the *vertex* of the pyramid. An *altitude* of a pyramid is the perpendicular segment from the *vertex* to the plane of the base. Its length is the *height* of the pyramid. A *regular pyramid* is a pyramid whose base is a regular polygon and whose lateral faces are congruent isosceles triangles. The *slant height* of a regular pyramid is the length of an altitude of a lateral face.

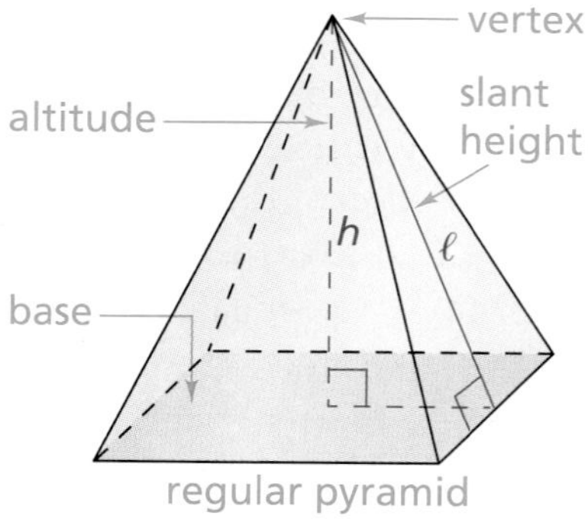

Pythagorean triple (p. 258) A Pythagorean triple is a set of three positive integers that satisfy the Pythagorean Theorem.

The numbers 5, 12, and 13 form a Pythagorean triple because $5^2 + 12^2 = 13^2$.

Q

Quadrant (p. 54) See *coordinate plane.*

Examples

Quadrilateral (p. 76) A quadrilateral is a polygon with four sides.

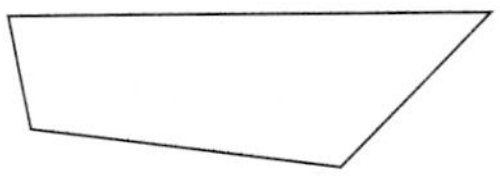

R

Radius of a circle (p. 96) A radius of a circle is any segment with one endpoint on the circle and the other endpoint at the center of the circle. *Radius* can also mean the length of this segment.

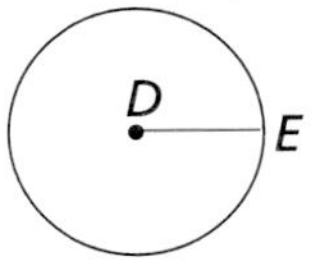

$\overline{DE}$ is a radius of $\odot D$.

Radius of a regular polygon (p. 274) The radius of a regular polygon is the distance from the center to a vertex.

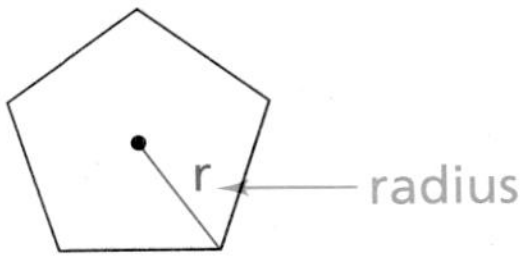

Ray (p. 18) A ray is a part of a line consisting of one *endpoint* and all the points of the line on one side of the endpoint.

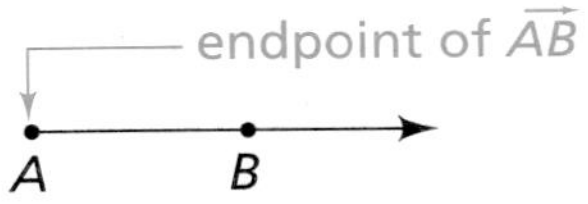

Rectangle (p. 91) A rectangle is a parallelogram with four right angles.

Reduction (p. 167) See *dilation.*

Reflection (p. 126) A reflection in line r is a transformation such that if a point A is on line r, then the image of A is itself, and if a point B is not on line r, then its image B' is the point such that r is the perpendicular bisector of $\overline{BB'}$.

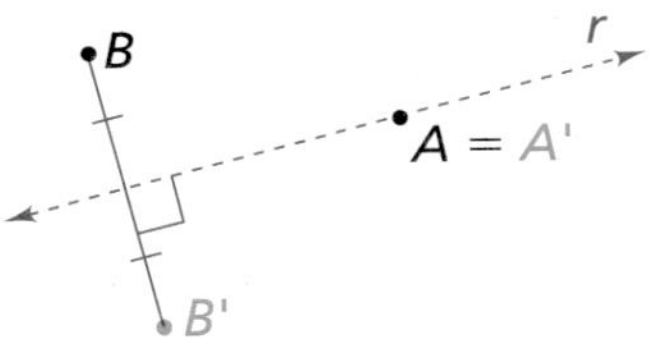

Reflectional symmetry (p. 152) A figure has reflectional symmetry, or *line symmetry,* if there is a reflection that maps the figure onto itself.

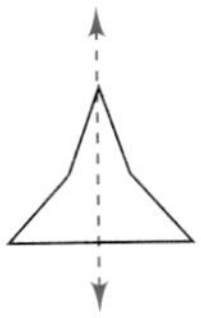

A reflection in the given line maps the figure onto itself.

Regular polygon (pp. 78, 274) A regular polygon is a polygon that is both equilateral and equiangular. Its *center* is the center of the circumscribed circle.

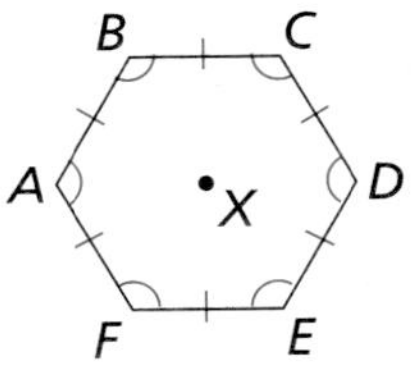

ABCDEF is a regular hexagon. Point *X* is its center.

Regular pyramid (p. 316) See *pyramid.*

Examples

Remote interior angles (p. 69) For each exterior angle of a triangle, the two nonadjacent interior angles are called its remote interior angles.

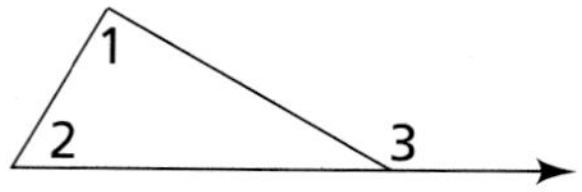

∠1 and ∠2 are remote interior angles of ∠3.

Resultant (p. 568) The sum of two vectors is called a resultant.

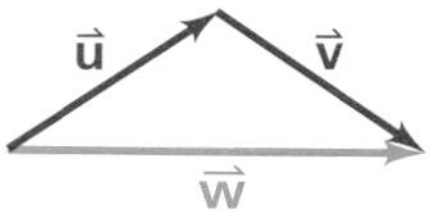

$\vec{w}$ is the resultant of $\vec{u} + \vec{v}$.

Rhombus (p. 91) A rhombus is a parallelogram with four congruent sides.

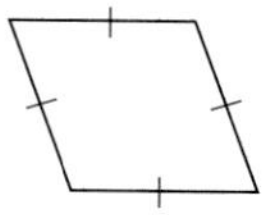

Right angle (p. 27) A right angle is an angle whose measure is 90.

Right cone (p. 318) See *cone.*

Right cylinder (p. 310) See *cylinder.*

Right prism (p. 309) See *prism.*

Right triangle (pp. 71, 256) A right triangle contains one right angle. The side opposite the right angle is the *hypotenuse* and the other two sides are the *legs.*

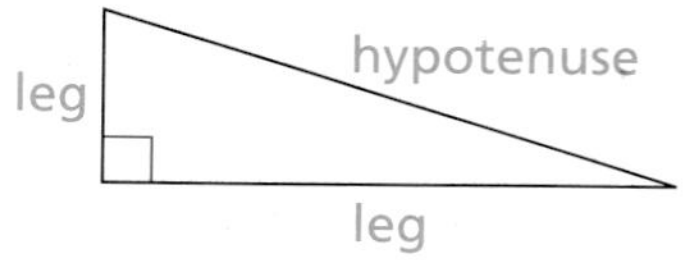

Rotation (p. 139) A rotation of $x°$ about a point R is a transformation such that for any point V, its image is the point V' where $RV = RV'$ and $m\angle VRV' = x$. The image of R is itself. All rotations in this text are *counterclockwise* rotations.

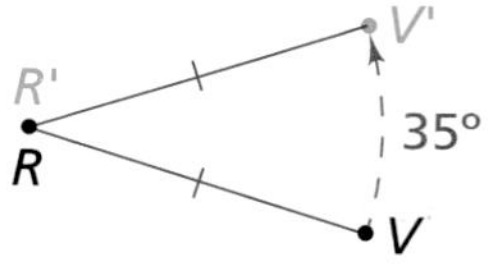

Rotational symmetry (p. 154) A figure has rotational symmetry if there is a rotation of 180° or less that maps the figure onto itself.

The figure has 120° rotational symmetry.

S

Same-side interior angles (p. 363) Same-side interior angles lie on the same side of the transversal t and between ℓ and m.

Example: ∠1 and ∠2 are same-side interior angles, as are ∠3 and ∠4.

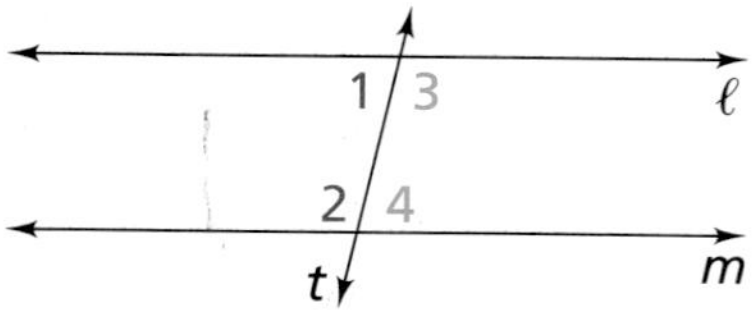

Examples

Scalar multiplication (p. 168) In scalar multiplication, each entry in a matrix is multiplied by the same number, the *scalar.*

$$2 \cdot \begin{bmatrix} 1 & 0 \\ -2 & 3 \end{bmatrix} = \begin{bmatrix} 2(1) & 2(0) \\ 2(-2) & 2(3) \end{bmatrix} = \begin{bmatrix} 2 & 0 \\ -4 & 6 \end{bmatrix}$$

Scale drawing (p. 497) In a scale drawing the *scale* compares each length in the drawing to the actual length being represented.

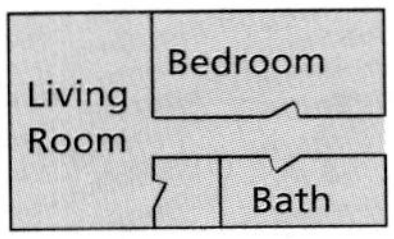

Scale:
1 in. = 30 ft

Scale factor (p. 166) The scale factor of a dilation is the number that describes the size change from an original figure to its image. See also *dilation.*

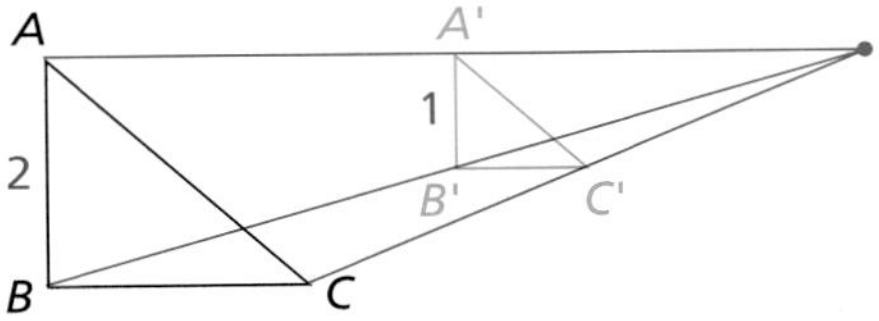

Example: The scale factor of the dilation that maps $\triangle ABC$ to $\triangle A'B'C'$ is $\frac{1}{2}$.

Scalene triangle (p. 71) A scalene triangle has no sides congruent.

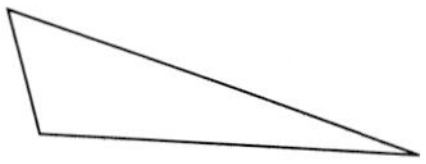

Secant (p. 614) A secant is a line, ray, or segment that intersects a circle at two points.

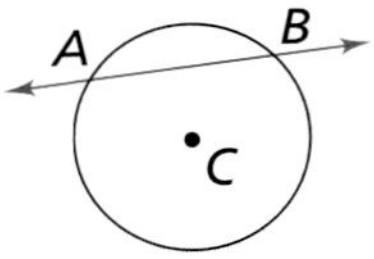

$\overleftrightarrow{AB}$ is a secant of $\odot C$.

Sector of a circle (p. 286) A sector of a circle is the region bounded by two radii and their intercepted arc.

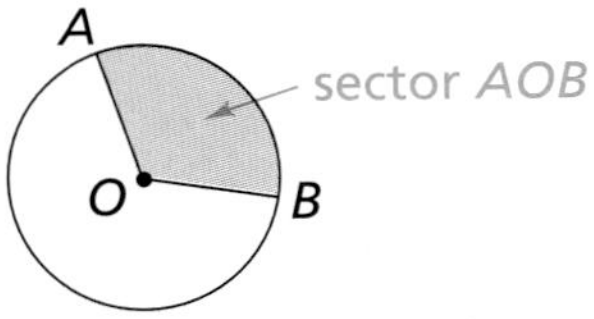

Segment (p. 18) A segment is a part of a line consisting of two points, called *endpoints,* and all points between them.

Segment bisector (p. 33) A segment bisector is a line, segment, ray, or plane that intersects a segment at its midpoint.

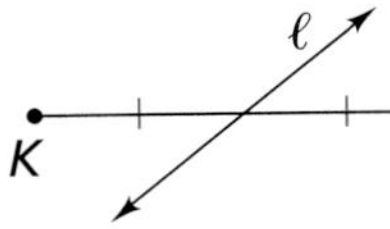

ℓ bisects $\overline{KJ}$.

Segment of a circle (p. 287) The part of a circle bounded by an arc and the segment joining its endpoints is a segment of a circle.

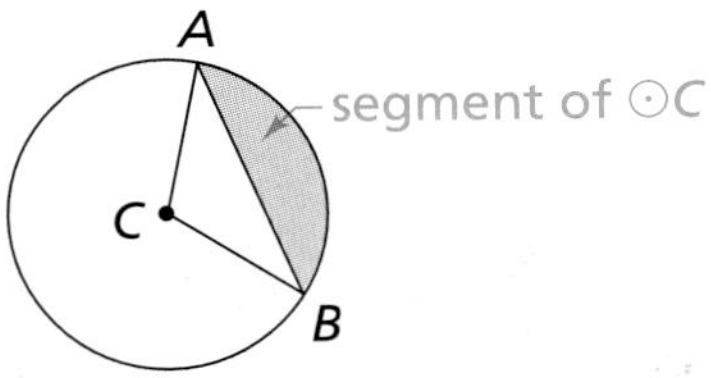

Example: The portion of the interior of the circle bounded by $\angle ACB$ and outside $\triangle ACB$ is a segment of $\odot C$.

Examples

Semicircle (p. 98) A semicircle is half a circle.

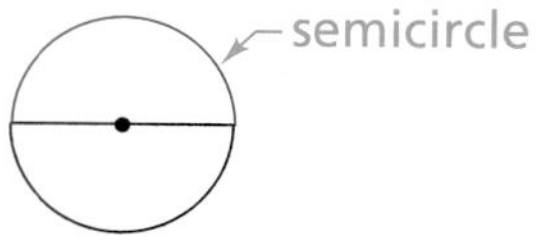

Side See *angle* and *polygon.*

Similarity ratio (p. 103) The ratio of the lengths of corresponding sides of similar polygons or solids is the similarity ratio.

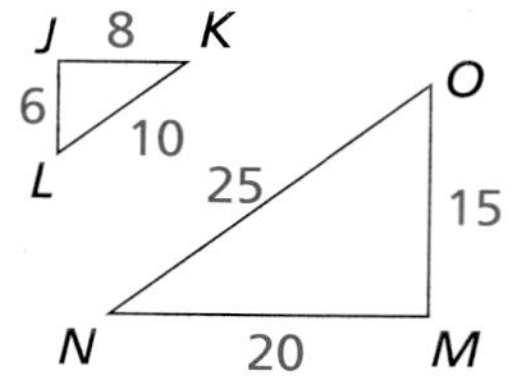

$\triangle JKL \sim \triangle MNO$

Similarity ratio $= \frac{2}{5}$

Similarity transformation (p. 167) See *dilation.*

Similar polygons (p. 103) Two polygons are similar if corresponding angles are congruent and corresponding sides are proportional. The symbol $\sim$ means "is similar to."

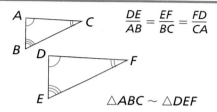

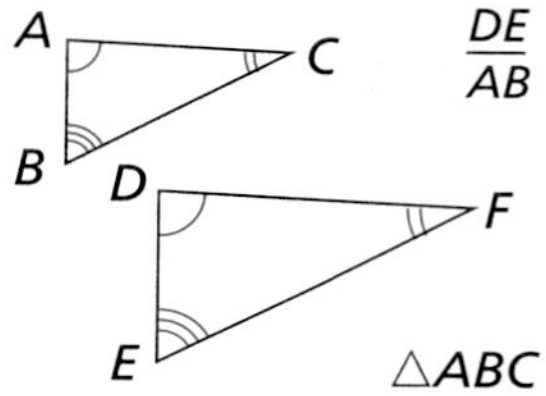

$\frac{DE}{AB} = \frac{EF}{BC} = \frac{FD}{CA}$

$\triangle ABC \sim \triangle DEF$

Similar solids (p. 531) Similar solids have the same shape and all their corresponding dimensions are proportional.

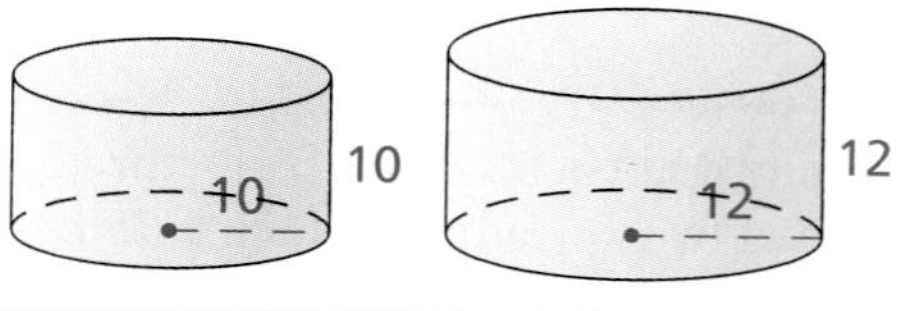

Sine ratio (p. 551) See *trigonometric ratios.*

Skew (p. 19) Two lines are skew if they do not lie in the same plane.

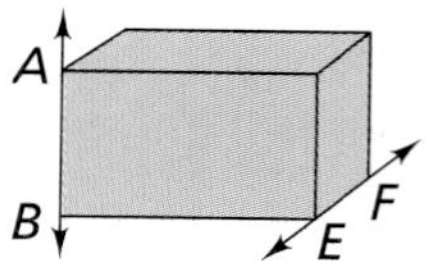

$\overleftrightarrow{AB}$ and $\overleftrightarrow{EF}$ are skew.

Slant height See *cone* and *pyramid.*

Slope of a line (p. 83) The slope of a line in the coordinate plane is the ratio of vertical change to the corresponding horizontal change. If (x_1, y_1) and (x_2, y_2) are points on a nonvertical line, then the slope is $\frac{y_2 - y_1}{x_2 - x_1}$. The slope of a horizontal line is 0, and the slope of a vertical line is undefined.

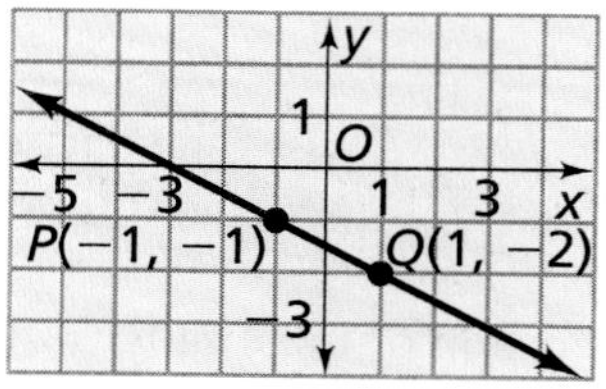

Example: The line containing $P(-1, -1)$ and $Q(1, -2)$ has slope $\frac{-2-(-1)}{1-(-1)} = \frac{-1}{2} = -\frac{1}{2}$.

Examples

Space (p. 12) Space is the set of all points.

Sphere (pp. 337, 340) A sphere is the set of all points in space a given distance *r*, the *radius*, from a given point *C*, the *center.* A *great circle* is the intersection of a sphere and a plane containing the center of the sphere. The *circumference* of a sphere is the circumference of any great circle of the sphere.

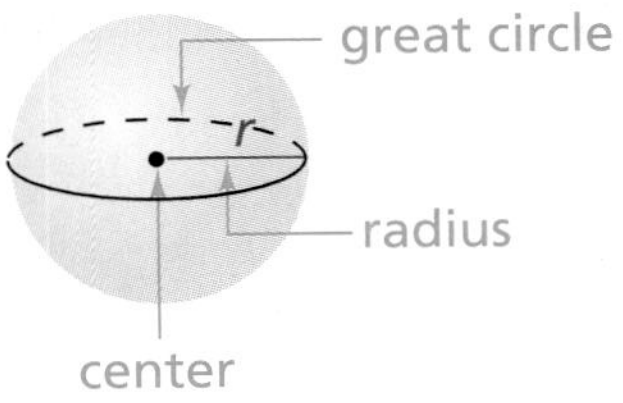

Spherical geometry (p. 393) In spherical geometry, a plane is considered to be the surface of a sphere and a line is considered to be a great circle of the sphere. In spherical geometry, through a point not on a given line, there is no line parallel to the given line.

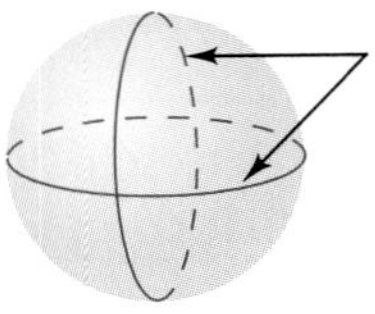

In spherical geometry, lines are represented by great circles of a sphere.

Square (p. 91) A square is a parallelogram with four congruent sides and four right angles.

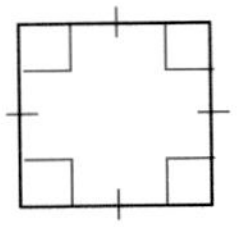

Straight angle (p. 27) A straight angle is an angle whose measure is 180.

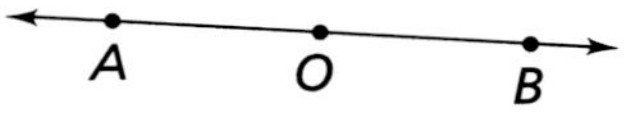

$m\angle AOB = 180$

Supplementary angles (p. 48) Two angles are supplementary if the sum of their measures is 180.

Example: $\angle MNP$ and $\angle ONP$ are supplementary, as are $\angle MNP$ and $\angle QRS$.

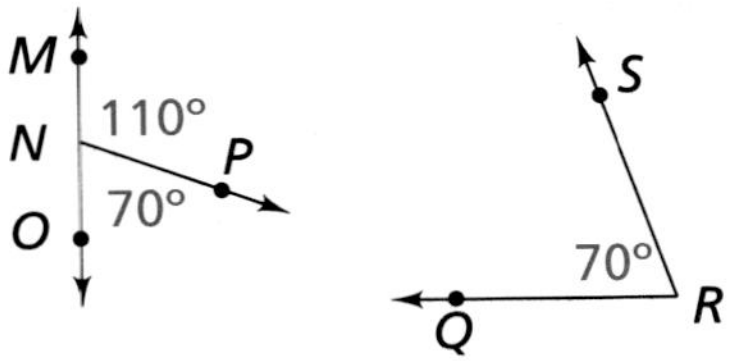

Surface area (pp. 309, 311, 316, 318, 338) The surface area of a prism, pyramid, cylinder, or cone is the sum of the lateral area and the areas of the bases. A list of surface-area formulas is on pages 670–671.

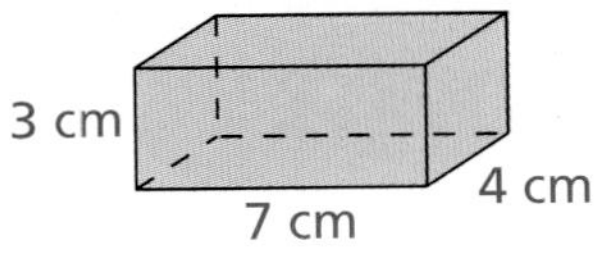

S.A. of prism = L.A. + 2*B*
= 66 + 2(28)
= 122 cm^2

Symmetry (pp. 152–154) A figure has symmetry if there is an isometry that maps the figure onto itself. See *glide reflectional symmetry, point symmetry, reflectional symmetry, rotational symmetry,* and *translational symmetry.*

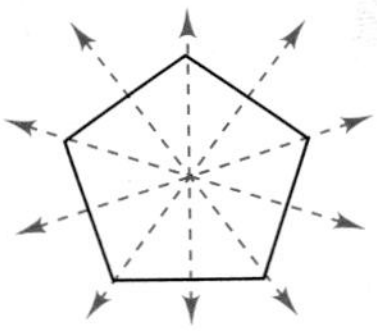

A regular pentagon has reflectional symmetry and 72° rotational symmetry.

T

Examples

Tangent ratio (p. 544) See *trigonometric ratios.*

Tangent to a circle (p. 593) A tangent to a circle is a line, segment, or ray in the plane of the circle that intersects the circle in exactly one point. That point is the *point of tangency.*

Example: Line ℓ is tangent to $\odot C$ at D.

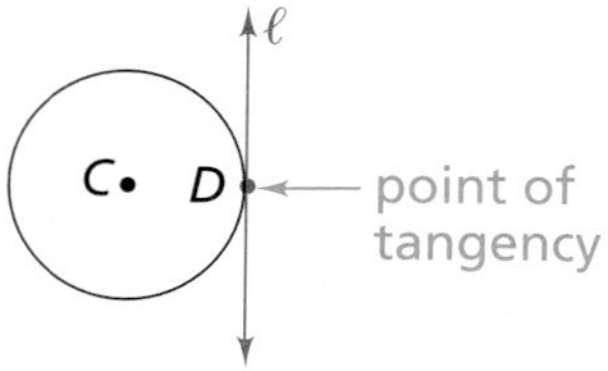

Terminal point (p. 133) See *vector.*

Tessellation (pp. 159–160) A tessellation, or *tiling,* is a repeating pattern of figures that completely covers a plane without gaps or overlap. A *pure tessellation* is a tessellation that consists of congruent copies of one figure.

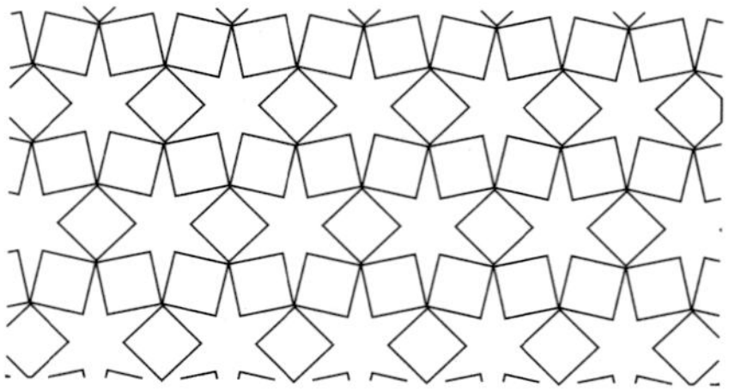

Theorem (p. 49) A conjecture that is proven is a theorem.

The theorem "Vertical angles are congruent" can be proven by using postulates, definitions, properties, and previously stated theorems.

Transformation (p. 124) A transformation is a change in the position, size, or shape of a figure. The given figure is called the *preimage* and the resulting figure is called the *image.* A transformation *maps* a figure onto its image. *Prime notation* is sometimes used to identify image points. In the diagram, X' (read "X prime") is the image of X.

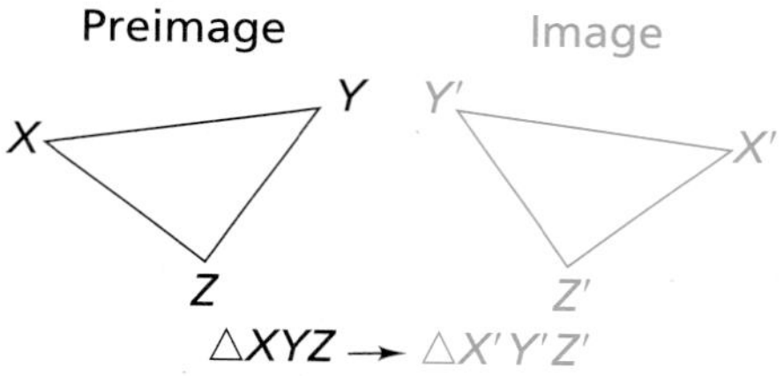

Translation (p. 132) A translation is a transformation that moves points the same distance and in the same direction. A transformation can be described by a vector.

Example: The blue triangle in the diagram is the image of the black triangle under the translation $\langle -5, -2 \rangle$.

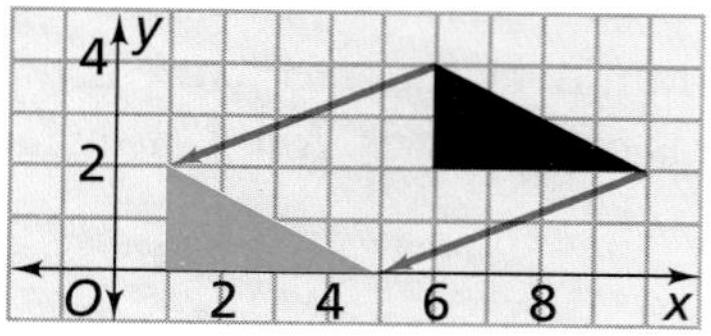

Translational symmetry (p. 161) A repeating pattern has translational symmetry if there is a translation that maps the pattern onto itself.

Example: The tessellation shown can be mapped onto itself by the given translation.

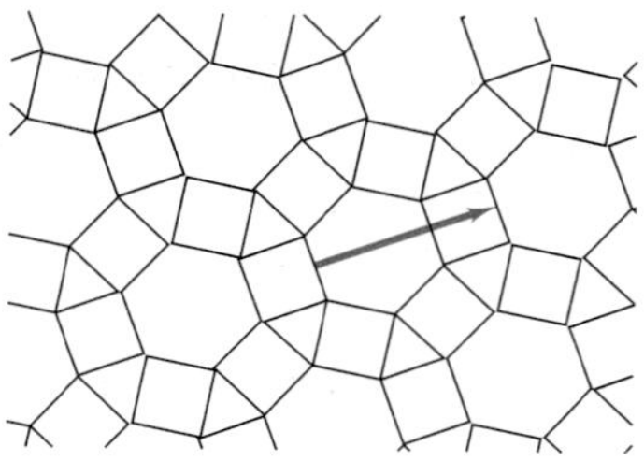

Examples

Transversal (p. 363) A transversal is a line that intersects two coplanar lines in two points.

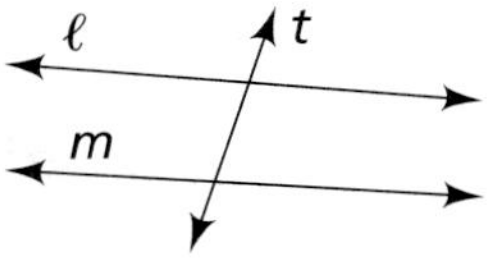

t is a transversal of ℓ and m.

Trapezoid (pp. 91, 269, 470) A trapezoid is a quadrilateral with exactly one pair of parallel sides, the *bases.* The nonparallel sides are called the *legs* of the trapezoid. Each pair of angles adjacent to a base are *base angles* of the trapezoid. An *altitude* of a trapezoid is a perpendicular segment from one base to the line containing the other base. Its length is called the *height* of the trapezoid.

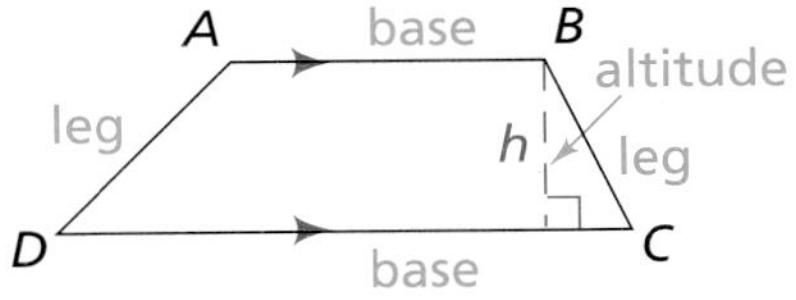

Example: In trapezoid $ABCD$, $\angle ADC$ and $\angle BCD$ are one pair of base angles, and $\angle DAB$ and $\angle ABC$ are the other.

Triangle (pp. 71, 251) A triangle is a polygon with three sides. You can choose any side to be the *base.* Then the *height* is the length of the altitude drawn to the line containing that base.

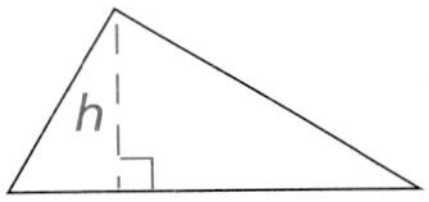

Trigonometric ratios (pp. 544, 551) In right triangle $\triangle ABC$ with acute angle $\angle A$:

$$\text{sine of } \angle A = \sin A = \frac{\text{leg opposite } \angle A}{\text{hypotenuse}}$$

$$\text{cosine of } \angle A = \cos A = \frac{\text{leg adjacent to } \angle A}{\text{hypotenuse}}$$

$$\text{tangent of } \angle A = \tan A = \frac{\text{leg opposite } \angle A}{\text{leg adjacent to } \angle A}$$

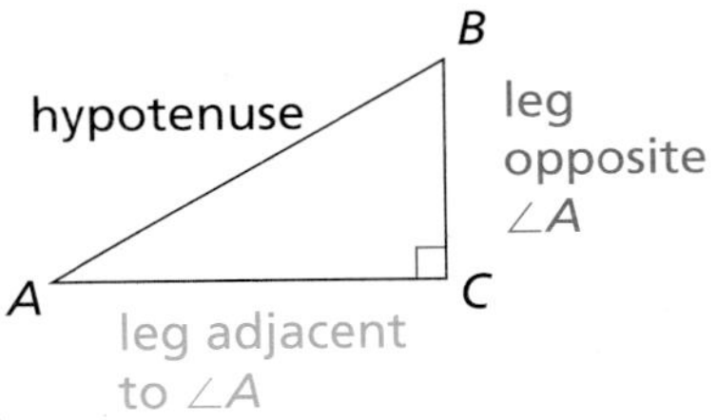

Truth value (p. 182) When you determine whether a conditional statement is true or false, you determine its truth value.

The truth value of the statement "If a figure is a triangle, then it has four sides." is **false.**

Two-column proof (p. 194) See *proof.*

V

Vector (pp. 133, 563) A vector is any quantity that has magnitude (size) and direction. You can represent a vector as an arrow that starts at one point, the *initial point,* and goes to a second point, the *terminal point.* A vector can be described by *ordered pair notation* $\langle x, y\rangle$, where x represents horizontal change from the initial point to the terminal point, and y represents vertical change from the initial point to the terminal point.

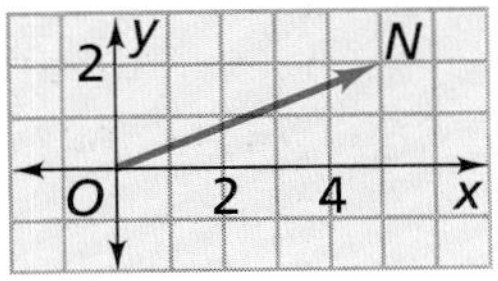

Vector $\overrightarrow{ON}$ has initial point O and terminal point N. The ordered pair notation for the vector is $\langle 5, 2\rangle$.

Examples

Vertex See *angle, cone, polygon, polyhedron,* and *pyramid.* The plural form of vertex is *vertices.*

Vertex angle (p. 188) See *isosceles triangle.*

Vertical angles (p. 48) Two angles are vertical angles if their sides are opposite rays.

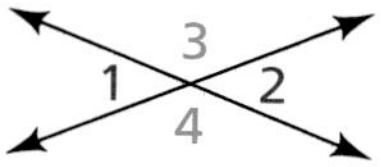

$\angle 1$ and $\angle 2$ are vertical angles, as are $\angle 3$ and $\angle 4$.

Volume (p. 324) Volume is a measure of the space a figure occupies. A list of volume formulas is on pages 670–671.

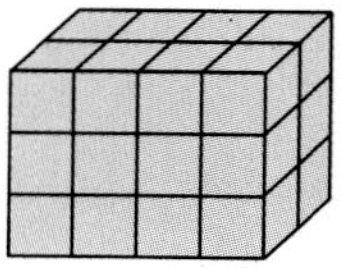

The volume of this prism is 24 cubic units, or 24 unit^3.

CHAPTER 1

Lesson 1-1 pages 7–10

ON YOUR OWN **1.** 80, 160 **3.** −3, 4 **5.** 3, 0 **7.** N, T **9.** 720, 5040 **11.** $\frac{1}{36}, \frac{1}{49}$ **15.** The trip takes about 25 min.

17.  **19.** 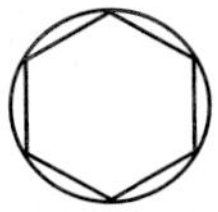 **21.** 

23. a line parallel to the first two and midway between them **25.** It's possible but not likely. As he grows older, his growth will slow down and eventually stop.

27a. 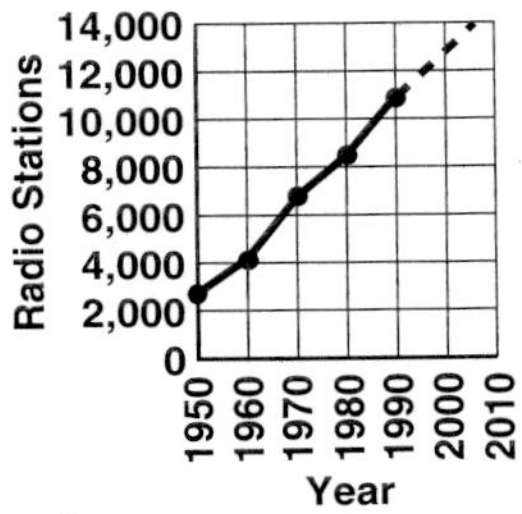

There will be about 15,000 radio stations. **29.** 123454321 **31.** 75° **33a.** 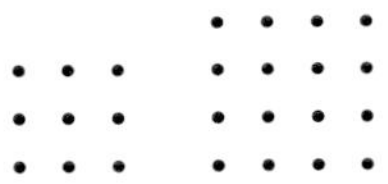

33b. 20^2, or 400; the sequence is the squares of successive counting numbers. **33c.** n^2 **35a.** Women may soon outrun men in running competitions. **35b.** The conclusion was based on continuing the trend shown in past records. **35c.** The conclusions are based on fairly recent records for women, and those rates of improvement may not continue. The conclusion about the marathon is most suspect because records date only from 1955. **37a.** Answers may vary. Sample: Leap years are divisible by 4. **37b.** Answers may vary. Sample: 2020, 2100, 2400 **37c.** Leap years are divisible by 4 except years ending in 00, which are leap years only if they are divisible by 400. **39.** 2

MIXED REVIEW **53a.** *B* and *W* **53b.** *N* and *T*

Toolbox page 11

1. $\frac{3}{10}$ **5.** $\frac{9}{10}$ **9.** $\frac{1}{10}$ **13.** $\frac{1}{3}$ **15.** $\frac{2}{3}$ **17.** $\frac{2}{3}$

Lesson 1-2 pages 15–17

ON YOUR OWN **1.** no **3.** no **5.** no **7.** no **9.** yes **11.** *C* **13.** yes **15.** no **17.** yes **19.** yes **21.** no **23.** *U* **25.** Answers may vary. Sample: plane *XWST* and plane *UVST* **29.** An infinite number; infinitely many planes can intersect in one line. **31.** C **33.** 1; points *A*, *B*, and *C* are points on the 2 lines and these 3 points are noncollinear, so exactly 1 plane contains them. **35.** never **37.** always **39.** never **41a.** $\frac{1}{4}$ **41b.** 1 **43.** collinear **45.** noncollinear

MIXED REVIEW **49.** 34 **51.** 20 **53.** 3 **55.** yes **57.** no

Lesson 1-3 pages 20–22

ON YOUR OWN **1.** $\overline{DF}$ **3.** $\overline{CF}, \overline{BE}$ **5.** plane *ABC* and plane *DEF* **7.** $\overline{RS}, \overline{RT}, \overline{RW}, \overline{ST}, \overline{SW}, \overline{TW}$

9. 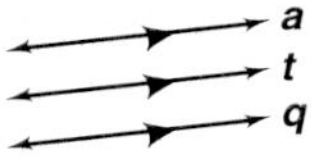

11. 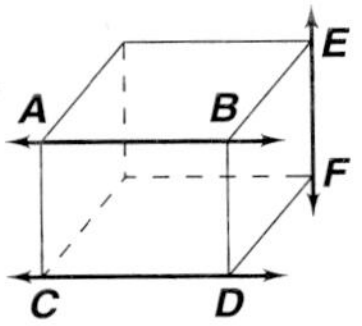

13. false **15.** true **17.** true **19.** true **21.** never **23.** always **25.** always **27.** always **29.** always **35.** The lines of intersection are parallel. Answers will vary. Sample: the ceiling and floor intersect a wall in two parallel lines. **39.** E

MIXED REVIEW **41.** −22, −29 **43.** by 2 points on the line or with a single lower-case letter **45.** with the word *plane* followed either by a single capital letter or the names of at least 3 noncollinear points in the plane **47.** 6 **49.** 3 **51.** 3 **53.** 9 **55.** 9 **57.** 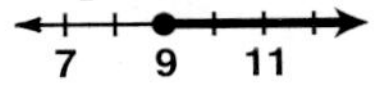

61. 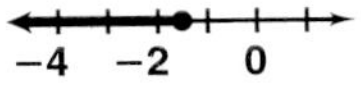

Toolbox page 23

1. −9 **5.** 16 **9.** 6 **13.** $-\frac{7}{3}$ **17.** 4

Lesson 1-4 pages 28–31

ON YOUR OWN **1.** 9 **3.** 11 **5.** false **7.** false **9.** $\overline{AB} \cong \overline{CD}, \overline{AC} \cong \overline{BD}$ **11.** 24 **13.** 13; 40; 24 **15.** 125 **17.** Answers may vary. Samples: **17a.** $\angle QVM$, $\angle PVN$ **17b.** $\angle QVP$, $\angle MVN$ **17c.** $\angle MQV$, $\angle QNP$ **21–23.** Estimates may vary slightly. **21.** 60; acute **23.** 135; obtuse **27.** 15 **33.** 8 **35.** 7

MIXED REVIEW **37.** 25 **39.** 30 **41.** coplanar **43.** collinear **45.** 

47. 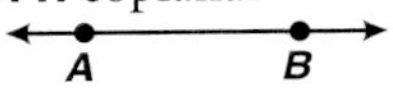

CHECKPOINT **1.** 29, 31.5 **2.** 3.45678, 3.456789 **3.** −162, 486 **5.** yes **6.** no **7.** yes **8.** yes **9.** yes **10.** *H*

11. $\overline{DC}, \overline{EF}, \overline{AB}$ **12.** Sample: $\overleftrightarrow{AB}, \overleftrightarrow{EH}$ **13.** Sample: plane *ABFE* ∥ plane *DCGH* or plane *HEFG* ∥ plane *ABCD* **14.** Sample: $\angle EAB$; $\angle AEF$; $\angle EHG$ **15.** 17

Lesson 1-5 pages 35–38

ON YOUR OWN **1.** b **3.** 6 **5.** 5 **7.** CM (or DM) **9.** 20; 40 **11.** AOC **13.** $\overrightarrow{OB}$; $\angle AOC$ **15.** true **17.** true **19.** false **21.** false

23.

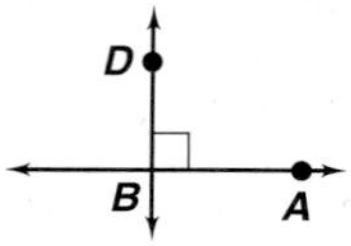

25.

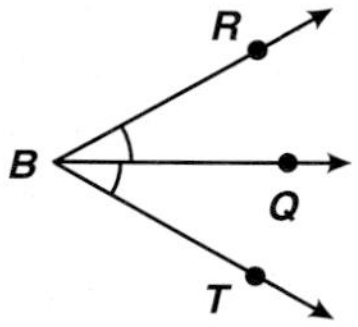

27.

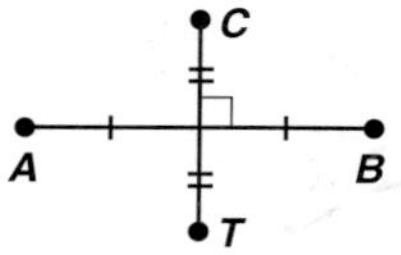

29. Q **31.** -4 **33.** 12 **35.** 4 **37.** 6 **39.** 20 **41a.** one; infinitely many **41b.** one **41c.** infinitely many **47.** b **49.** 15 **51.** 48 **53.** D **55.** perpendicular; it intersects

MIXED REVIEW **57.** Answers may vary. Sample: $\angle AOD$ **59.** $\angle BOE$ **61.** 3 **63.** $\angle APT \cong \angle TPR$

Lesson 1-6 pages 43–44

ON YOUR OWN **1.**

3.

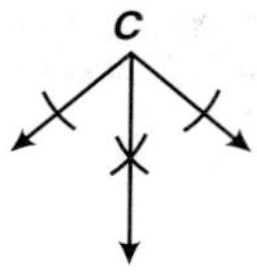

5.

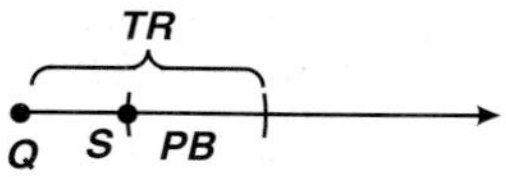

7.

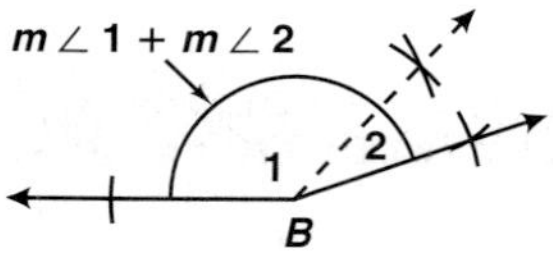

9.

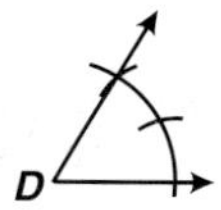

11.

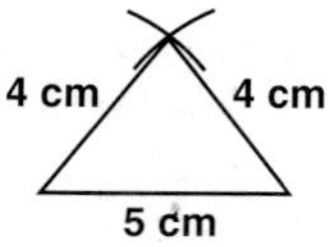

15. The angle bisectors of the 3 angles of any triangle intersect in a single point.

17a. Sample:

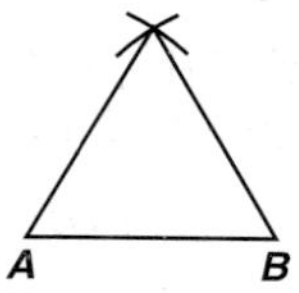

17b. 60

19a–b. Answers may vary. Sample:

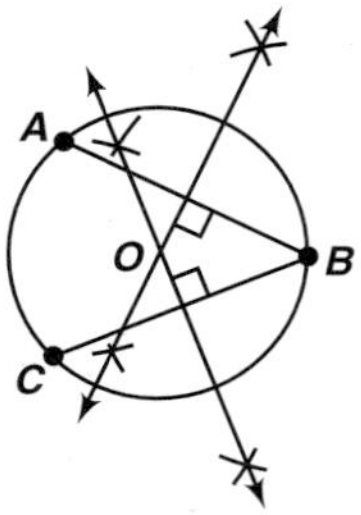

19c. Point O is the center of the circle.

MIXED REVIEW **21.** 16 **23.** 8 **25.** 50 **27.** 45

Toolbox page 45

INVESTIGATE yes; $\overleftrightarrow{HG}$ intersects $\overline{EF}$, but it is not the $\perp$ bisector of $\overline{EF}$.

SUMMARIZE A figure created by *draw* has no constraints. A figure created by *construct* is dependent upon an existing object.

Since $\overleftrightarrow{DC}$ was constructed as the $\perp$ bisector of $\overline{AB}$, it remains the $\perp$ bisector through any manipulation. Since point G was constructed on $\overline{EF}$, the only restriction on $\overleftrightarrow{HG}$ during any manipulation is that it must contain point G which has to be on $\overline{EF}$.

EXTEND $\overrightarrow{KM}$ is always the angle bisector of $\angle JKL$. $\overrightarrow{OQ}$ is not always the bisector of $\angle NOP$.

Lesson 1-7 pages 50–52

ON YOUR OWN **1.** Reflexive Prop. of $\cong$ **3.** Symmetric Prop. of $\cong$ **5.** Substitution Prop. **7.** Mult. Prop. of $=$ **9.** Trans. Prop. of $\cong$ **13.** 9 **15.** 18 **17.** 10 **19a.** B can be any point on the positive y-axis, for example, (0, 5). **21.** $x = 14$; $y = 15$ **23a.** 90 **23b.** 45 **23c.** Not possible; all vert. angles are $\cong$. **25.** 30 and 60 **27.** $\angle EIG$ and $\angle FIH$ are right angles by the markings. $\angle EIF \cong \angle GIH$ because they are complements of the same angle. **29.** Add. Prop. of $=$; Div. Prop. of $=$ **31.** Mult. Prop. of $=$; Distr. Prop.; Add. Prop. of $=$ **33.** Because $\angle 1$ and $\angle 2$ are supplementary, $m\angle 1 + m\angle 2 = 180$. Because $\angle 3$ and $\angle 4$ are supplementary, $m\angle 3 + m\angle 4 = 180$. So, $m\angle 1 + m\angle 2 = m\angle 3 + m\angle 4$. Because $\angle 2 \cong \angle 4$, $m\angle 2 = m\angle 4$. Thus, by Subtraction Prop. of $=$, $m\angle 1 = m\angle 3$ and $\angle 1 \cong \angle 3$. **35.** No; guys with beards may not park on Mon. **37.** No; parking is not allowed from 6:49 A.M. to 9:11 A.M. on Tues.

MIXED REVIEW **39.**

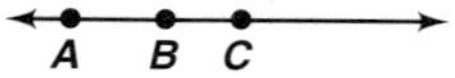

41.

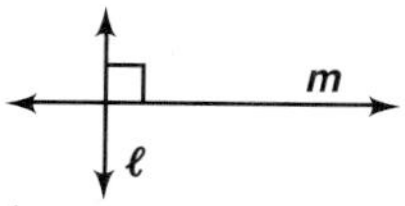

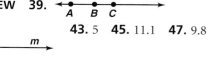

43. 5 **45.** 11.1 **47.** 9.8

CHECKPOINT **1.** A good definition states precisely what a term is, using commonly understood or previously defined terms. **2a.**

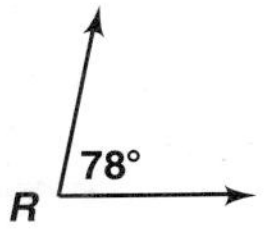

2b.

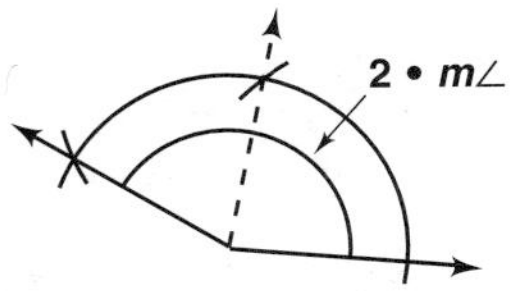

2c.

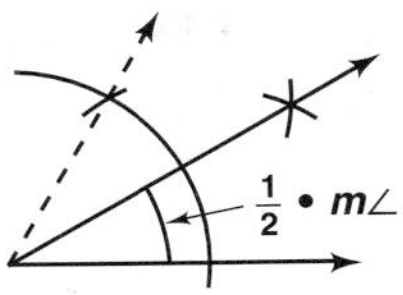

3. $m\angle 1 + m\angle 2 = 90$

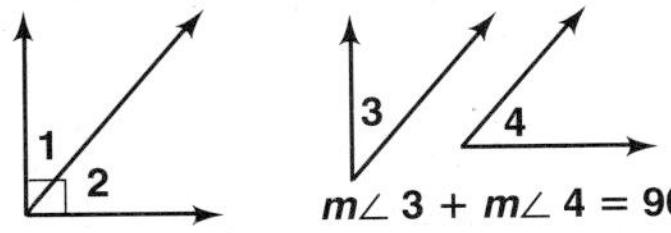

$m\angle 3 + m\angle 4 = 90$

4. C

Lesson 1-8 pages 56–57

ON YOUR OWN **7.** 6 **9.** 8 **11.** 23.3 **13.** 25 **15.** 12.0 **17.** (3, 1) **19.** (6, 1) **21.** $\left(3\frac{7}{8}, -3\right)$ **23.** (8, 18) **25.** No; $AD = AB \approx 4.2$, $DC = CB \approx 3.2$. **29a.** 19.2 **29b.** (−1.5, 0) **31a.** 5.4 **31b.** (−1, 0.5) **33a.** $A(0, 0, 0)$, $B(3, 0, 0)$, $C(3, -3, 0)$, $D(0, -3, 0)$ $E(0, 0, 5)$, $F(3, 0, 5)$, $G(0, -3, 5)$

33b.

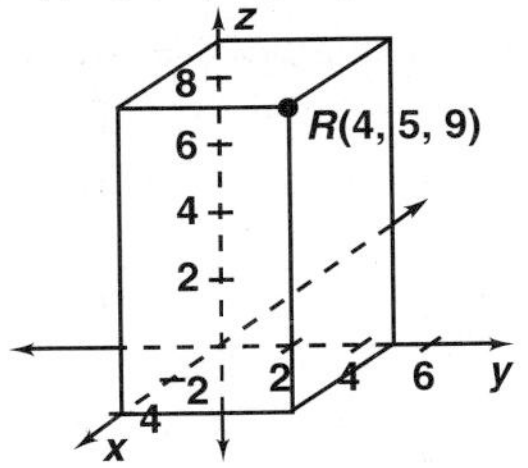

MIXED REVIEW **35.** 72; 162 **37.** 66.5; 156.5 **39.** 12; 102

Wrap Up pages 59–62

1. 17, 21; add 4 to the previous term to get the next term. **2.** 63, 127; add consecutively increasing powers of 2 to the previous term to get the next term. **3.** $\frac{5}{6}, \frac{6}{7}$; add 1 to the numerator and denominator of the preceding term to get the next term. **4.** 5, −6; write the sequence of whole numbers and then change the signs of the even whole numbers.

5.

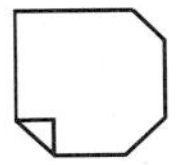

6a. 76 **6b.** The last two digits will always be 76. **7.** If the points were collinear, an infinite number of planes would pass through them. **8–13.** Answers may vary. Samples are given. **8.** $\overleftrightarrow{QR}$ and $\overleftrightarrow{RS}$ **9.** $\overleftrightarrow{QR}$ and $\overleftrightarrow{SC}$ **10.** Q, R, and S **11.** Q, R, S, C **12.** plane $QRST$ and plane $ABCD$ **13.** $\overleftrightarrow{AD}$, $\overleftrightarrow{CD}$, $\overleftrightarrow{TD}$ **14.** always **15.** sometimes **16.** never **17.** always **18.** always **19.** never **21.** 3 or −7 **22.** 18 **23.** 31 **24.** 20 **25.** $m\angle KJD + m\angle DJH = m\angle KJH$ by the Angle Add. Post.; $m\angle KJD = m\angle DJH$ by the markings; $\overrightarrow{JD}$ bisects $\angle KJH$ by the definition of angle bisector. **26.** $AB = CD$ by the markings; $AC = BD$ by the Seg. Add. Post. **27.** $\angle 1 \cong \angle 4$ by the markings; $\angle 1 \cong \angle 2$ and $\angle 3 \cong \angle 4$ because vert. angles are ≅; $\angle 2 \cong \angle 3$ by the Trans. Prop. of ≅. **29.** No; $\overleftrightarrow{BK}$ may not bisect $\overline{LJ}$. **30.** 3 **31.** 1 **32.** D **33.**

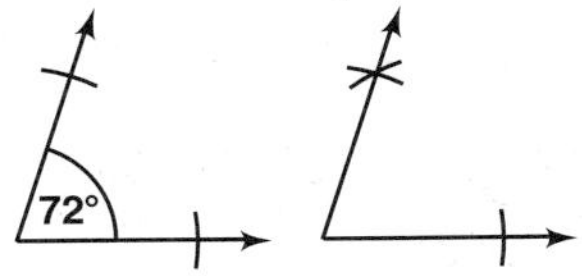

34a–b.

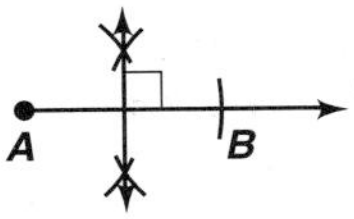

42. (0, 0) **43.** 3.2

44a.

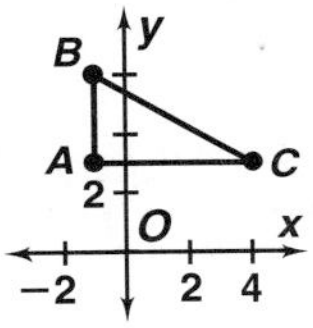

44b. $AB = 3$, $AC = 5$, $BC \approx 5.8$ **44c.** $\overline{BC}$, $\overline{AC}$, $\overline{AB}$

Preparing for Standardized Tests page 65

1. D **3.** E **5.** E **7.** B **9.** D **11.** D **13.** $\left(\frac{15}{2}, 2\right)$

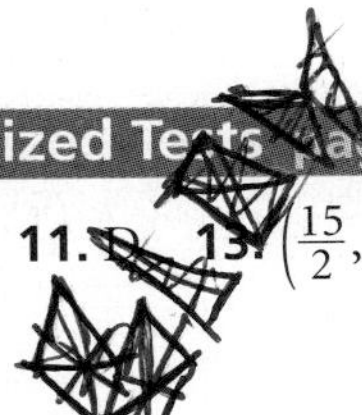

CHAPTER 2

Lesson 2-1 pages 71–74

ON YOUR OWN **1.** acute isosceles **3.** right scalene **9a.** 60; the sum of measures is 180 and the 3 measures are =. **9b.** 90; the sum of measures is 180 and the measure of the 3rd angle is 90. **11.** 115.5 **13.** $t = 60$; $w = 60$ **15.** 83.1 **17.** $a = 67$; $b = 58$; $c = 125$; $d = 23$; $e = 90$ **19.** 103 **21.** C **25.** 33° **27.** 37, 78, 65; acute **29.** > 180; measures of both angles at the equator = 90 and the angle at the pole has pos. measure. **31a.** $\frac{1}{3}$ **31b.** $\frac{1}{7}$ **33a.** 900; 30, 60, 90 **33b.** right

MIXED REVIEW **35.**

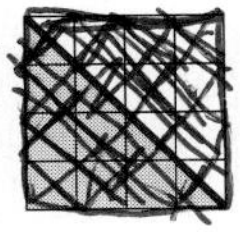

37. 10.0 **39.** 7 **41a.** 60, 120, 60, 120 **41b.** 360

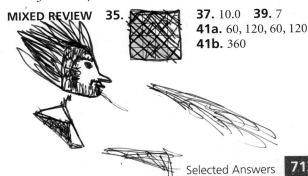

Toolbox page 75

INVESTIGATE 360

CONJECTURE The sum of measures of the exterior angles of a polygon is always 360.

EXTEND When the polygon "disappears," the angles become adjacent. The sum of their measures is 360.

Lesson 2-2 pages 79–81

ON YOUR OWN **1.** convex dodecagon **3.** convex octagon **5.**

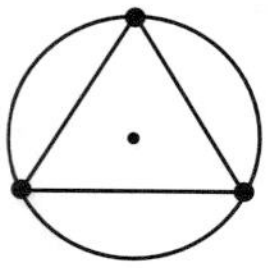

7.

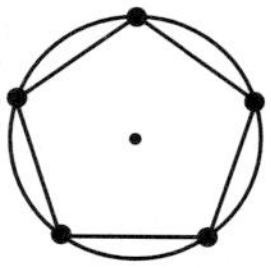

9. octagon; $m\angle 1 = 135$; $m\angle 2 = 45$ **11.** 140; 40 **13.** $\frac{180y - 360}{y}$; $\frac{360}{y}$ **15.** 10 **17.** $\frac{360}{x}$ **19.** 102 **21.** $y = 103$; $z = 70$ **23.** $x = 69$; $w = 111$ **25.** 113 **27a.** (20, 162), (40, 171), (60, 174), (80, 175.5), (100, 176.4), (120, 177), (140, 177.4), (160, 177.8), (180, 178), (200, 178.2)

27b.

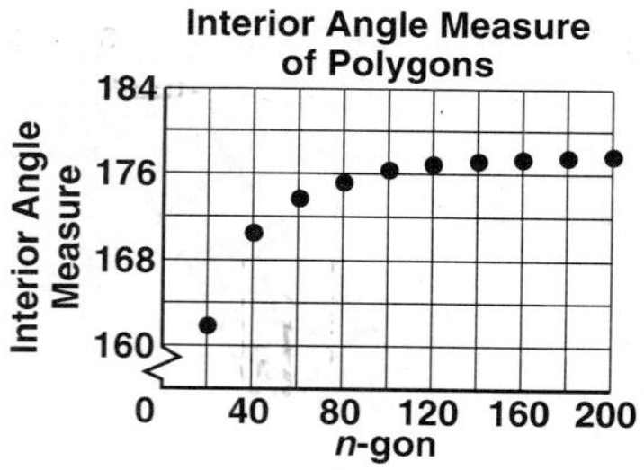

27c. very close to 180 **27d.** No; a regular polygon with all straight angles would have all its vertices on a straight line.

MIXED REVIEW **31.** $\overrightarrow{RT}$ and $\overrightarrow{RK}$ **33.** Answers may vary. Sample: $\angle BRT$ and $\angle BRK$ **35.** 40.25 **37.** $x = 104$; $y = 35$ **39.** 72 and 18 **41.** −1 **43.** 0

Toolbox page 82

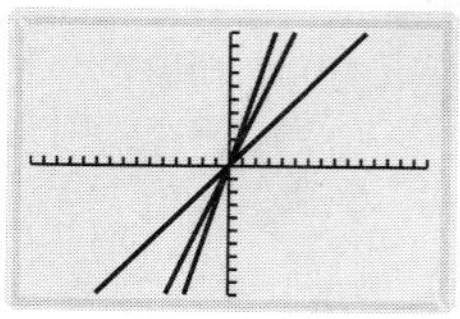

INVESTIGATE The value of m affects the steepness of the line.

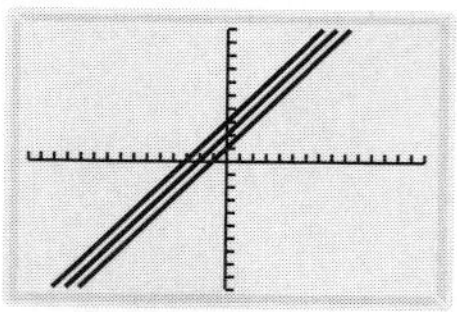

Changing the value of b shifts the line vertically.

CONJECTURE Answers may vary. Samples: A line with pos. value of m goes from the lower left to the upper right. The greater the abs. value of m the steeper the line. For neg. values of b, the line shifts down by the number of units = to the abs. value of b. The line passes through the origin if $b = 0$.

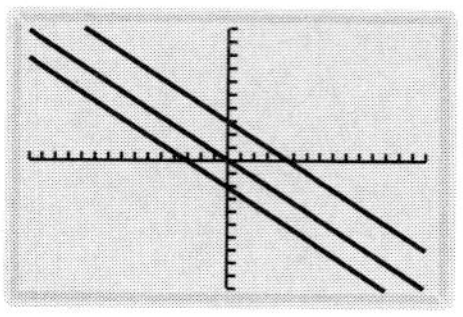

EXTEND The lines are $\parallel$; the values of m are =; lines with equations that have = values of m are $\parallel$.

a.

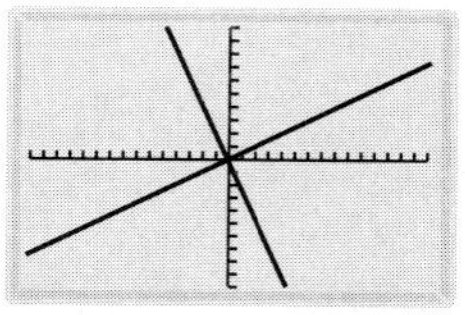

b.

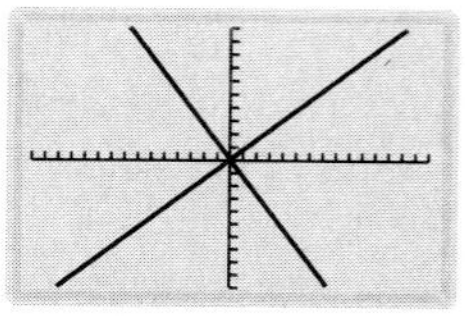

c.

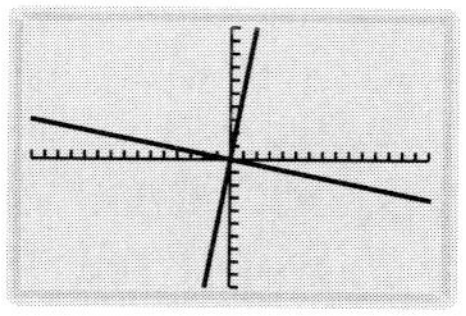

The lines are $\perp$; the product of the values of m is −1; 2 lines with equations in which the product of values of m is −1 are $\perp$.

Lesson 2-3 pages 86–88

ON YOUR OWN **1.** k: pos.; ℓ: neg.; s: 0; t: undef. **3.** No; lines with no slope are vert. Lines with slope 0 are horizontal. **5.** undef.; 0; perpendicular **7.** $-\frac{1}{8}$; 8; perpendicular **9.** 0; 0; parallel **11.** No; the slopes of the sides are $\frac{3}{5}$, $-\frac{5}{8}$, and $-\frac{8}{3}$. No 2 sides are $\perp$.

13.

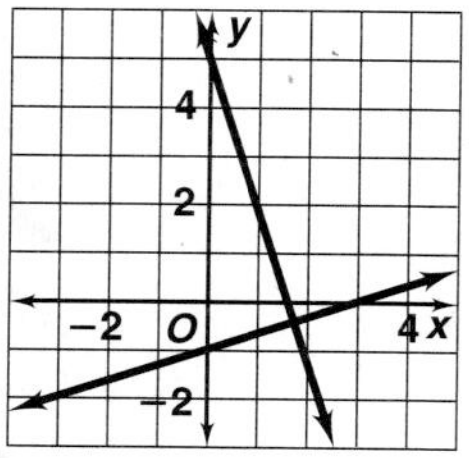

perpendicular

15.

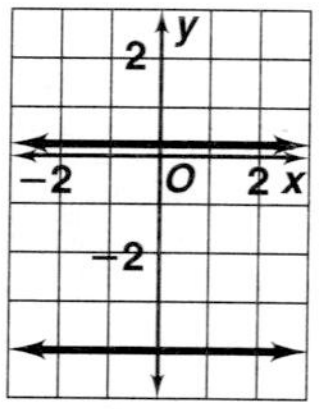

parallel

17.

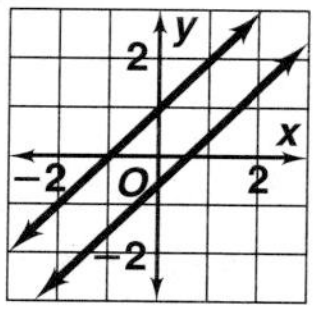

parallel

19.

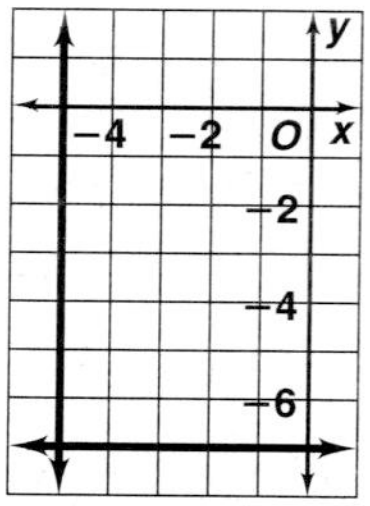

perpendicular

21a.

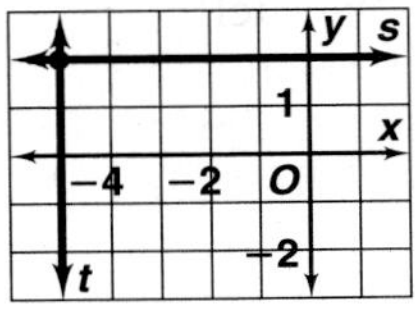

21b. $x = -5$ **21d.** $y = 2$ **21e.** The lines are $\perp$; a horizontal line is always $\perp$ to a vert. line. **23.** yes

25.

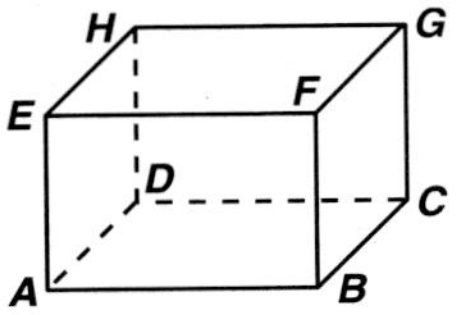

25a. $\overleftrightarrow{AB} \perp \overleftrightarrow{FB}, \overleftrightarrow{BC} \perp \overleftrightarrow{FB}, \overleftrightarrow{AB} \perp \overleftrightarrow{BC}$
25b. $\overleftrightarrow{AB} \perp \overleftrightarrow{BC}, \overleftrightarrow{BC} \perp \overleftrightarrow{CG}, \overleftrightarrow{AB}$ and $\overleftrightarrow{CG}$ are skew lines.
27. Answers may vary. Sample: No; the "lines" intersect twice. **29a.** (1, 10), (2, 20), (3, 30), (4, 40), (5, 50), (6, 60), (7, 70), (8, 80), (9, 90), (10, 100)

29b.

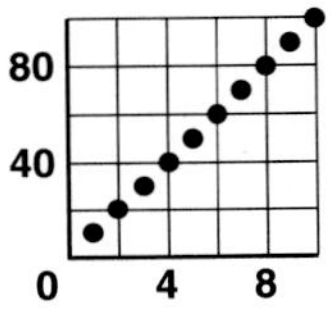

29c. The slope is 1000.
29d. Sample: Yes; you can make the slope as steep as you want, but it will never be vertical.

MIXED REVIEW **31.** 25.5 **33.** 22.5 and 67.5
35. $\overline{FG}$ and $\overline{EH}$, $\overline{EF}$ and $\overline{GH}$

CHECKPOINT **1.**

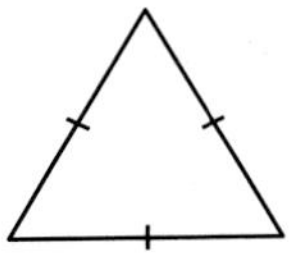

2.

3.

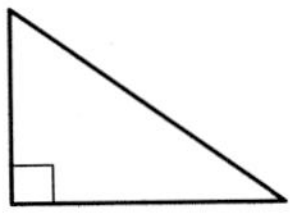

4.

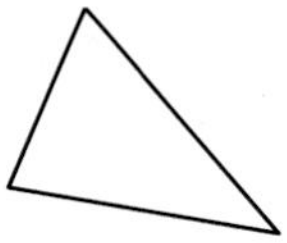

5.

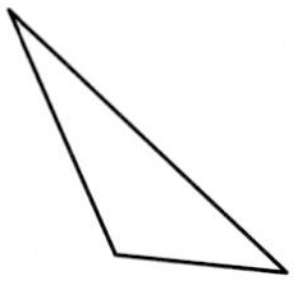

6.

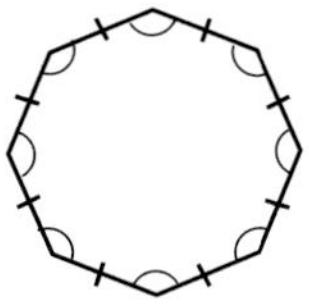

7.

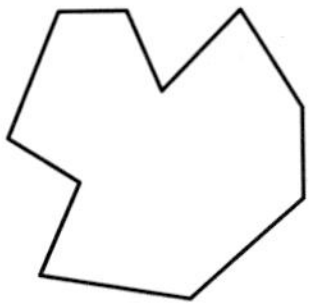

8.

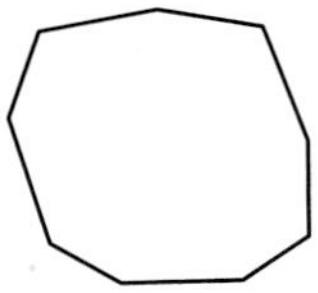

9. (1) Divide 360 by n to find the exterior angle measure. Subtract the result from 180.
(2) Multiply 180 by $n - 2$ and divide the result by n.
10. $-\frac{2}{5}; \frac{5}{2}$; perpendicular **11.** 1; −1; perpendicular
12. $-\frac{5}{4}; \frac{4}{5}$ perpendicular **13.** $\frac{2}{9}; \frac{2}{9}$; parallel

Toolbox page 89

1. $y = 3x + 5$ **5.** $y = -\frac{5}{4}x + 8$ **9.** $y = -x + 7$

13. $y = 5x - 10$

Lesson 2-4 pages 93–95

ON YOUR OWN **1.** parallelogram, rhombus, rectangle, square **3.** trapezoid **5.** true **7.** false **9.** false

17.

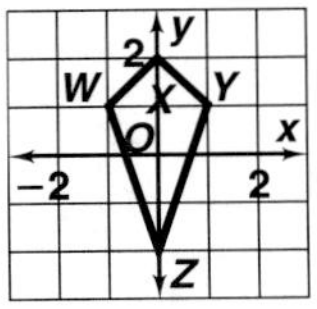

kite

19.

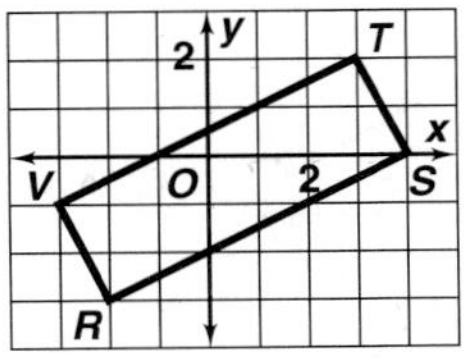

rectangle

21. some isosceles trapezoids some trapezoids

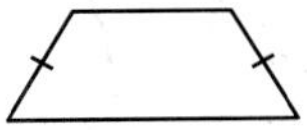

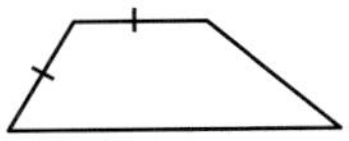

23. rectangle square

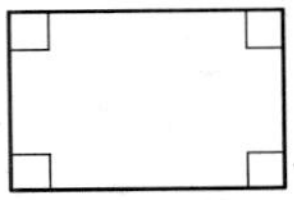

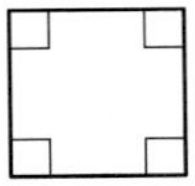

25. $x = 11$; $y = 21$; 13, 13, 15, 15 **27.** $b = 9$; $r = 5$; 6, 6, 6, 6 **29.** parallelogram, kite, rhombus, trapezoid, isos. trapezoid **31.** parallelogram, rectangle, square, kite, trapezoid

MIXED REVIEW **33.** 8.2

35a.

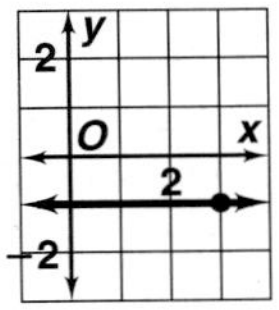

35b.

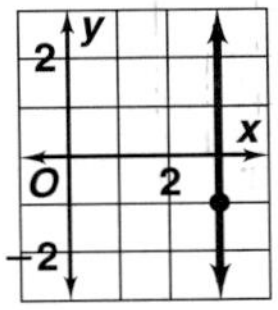

37. 33%
39. 75%

Lesson 2-5 pages 99–101

ON YOUR OWN **1, 3, 7.** Answers may vary. **1.** Sample: $\widehat{BC}, \widehat{CD}$ **3.** Sample: $\widehat{BCE}, \widehat{BFE}$ **5.** $\widehat{BE}, \widehat{CF}$ **7.** Sample: $\angle BOC$ **9.** $\angle BOC, \angle EOF$ **11.** 10 cm **13.** $12\sqrt{2}$ in.
15. 6.5 cm **17.** $\frac{5\sqrt{3}}{2}$ in. **19.** $\frac{d}{2}$ km **23.** (−2, 5); $\sqrt{5}$
25. (−1, 4); $\sqrt{10}$ **27.** (3, −4.5); 8.5 **29.** 180 **31.** 52 **33.** 180 **35.** 90 **39a.** 6°; 30°; 120° **39b.** 2.5°; 5°; 10° **39c.** 102.5 **41a.** 90 **41b.** 30 **41c.** 145 **41d.** 125 **41e.** 235 **41f.** 215 **43a.** 80 **43b.** 100 **43c.** 150 **43d.** 210 **43e.** 280 **45.** 160 **47.** Stay in the circle for a 220° arc before exiting.

MIXED REVIEW **51.** 95 **53.** $t = 120$; $y = 60$ **55.** 37

Lesson 2-6 pages 105–108

ON YOUR OWN **1.** A and H, B and G, C and E, F and D **3.** $\overline{CM}$ **5.** $\angle B$ **7.** $\angle J$ **9.** $\triangle CLM$ **11.** *HY* **13.** *HY* **15.** $\angle R$ **17.** $\frac{2}{3}$ **19.** 50 **21.** 70 **23.** 7.5 cm **25.** $\angle P \cong \angle S; \angle O \cong \angle I; \angle L \cong \angle D; \angle Y \cong \angle E$ **27a.** *IDES* **27b.** *LYPO* **31.** Yes; the ratios of radii, diameters, and circumferences of 2 circles are =. **33.** $\triangle BEC \cong \triangle AED$ **35.** $t = 2$ in.; $x = 15$ **37.** 2.3 cm **39a.** 7.2 cm; 9.6 cm; 12 cm **39b.** 53; 90; 37 **39c.** 9 cm; 12 cm; 15 cm **41.** $\frac{3}{4}$ **43a.** yes; $\frac{8}{16} = \frac{10}{20}$ **43b.** no; $\frac{4}{5} \neq \frac{5}{7}$

MIXED REVIEW

47. 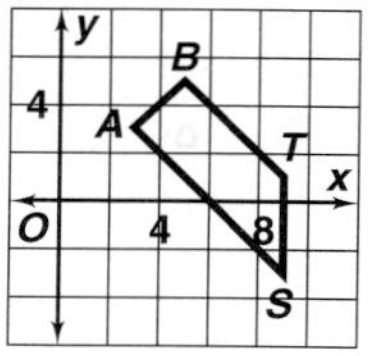

Trapezoid; slope of $\overline{BT}$ = slope of $\overline{AS}$.
Trapezoid; slope of $\overline{AB} \neq$ slope of $\overline{ST}$.
49. 160 **51.** a rectangle

CHECKPOINT **1.** (5.5, 3); 2.5 **2.** (1.5, 7); $\frac{\sqrt{13}}{2}$ **3.** (0.5, 3.5); $\frac{\sqrt{170}}{2}$ **4.** (−5, −0.5); $\frac{\sqrt{117}}{2}$ **5.** 65 **6.** 90 **7.** 25 **8.** 245 **9.** 115 **10.** 90 **11.** 25 **12.** 180 **13.** 180

14. 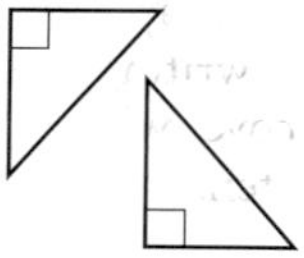

15. 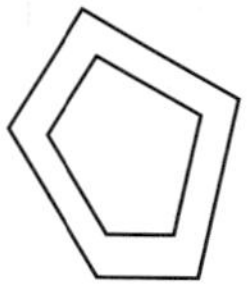

16. 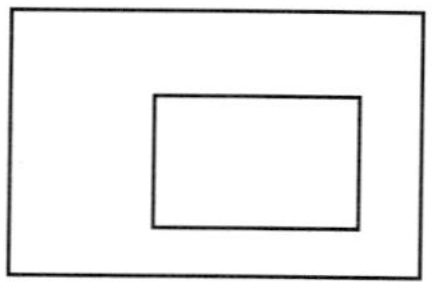

17. A

Lesson 2-7 pages 112–115

ON YOUR OWN

1a. 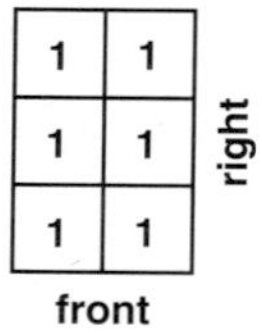

1b. 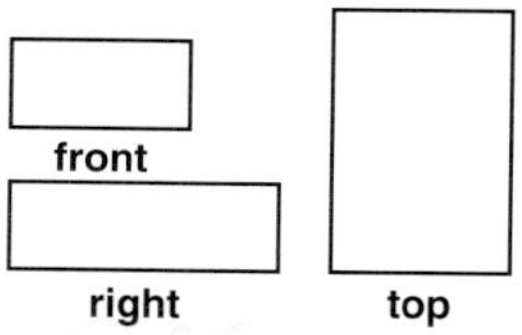

3a. 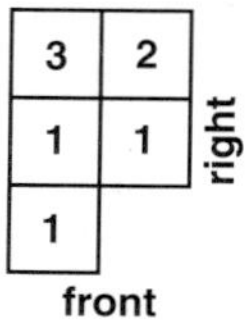

3b. 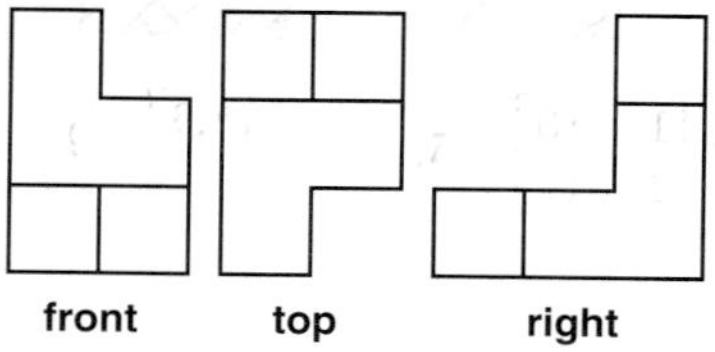

5a. 

5b. 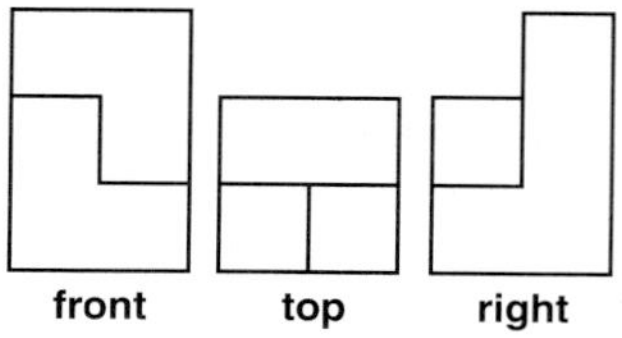

7a. 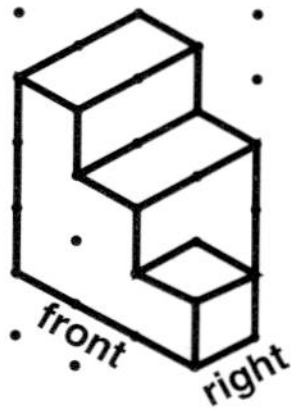

7b. 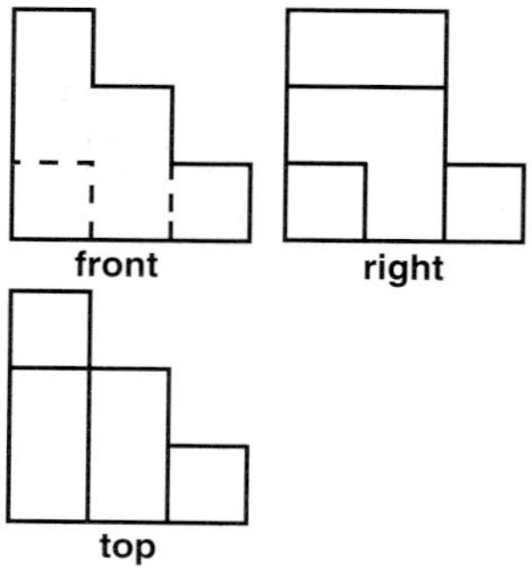

13. triangle **15.** isosceles triangle **17.** B **19.** D

21. 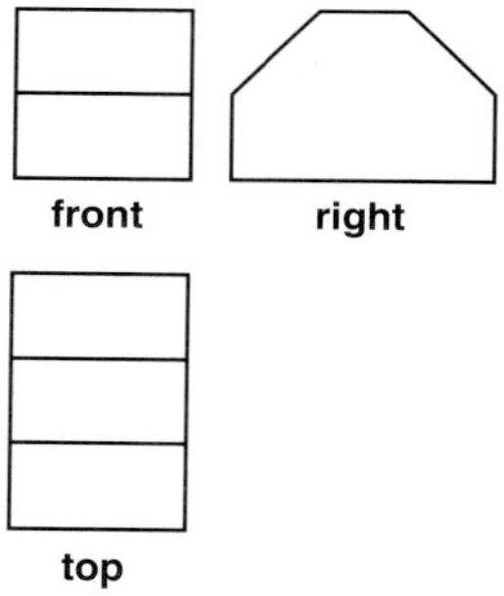

23. 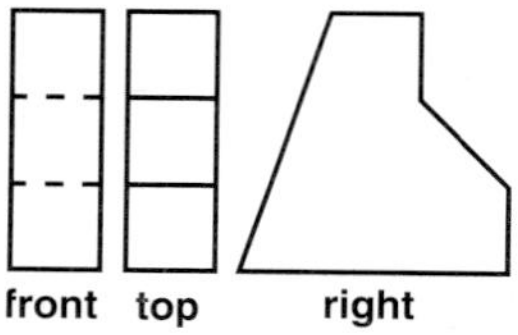

25. 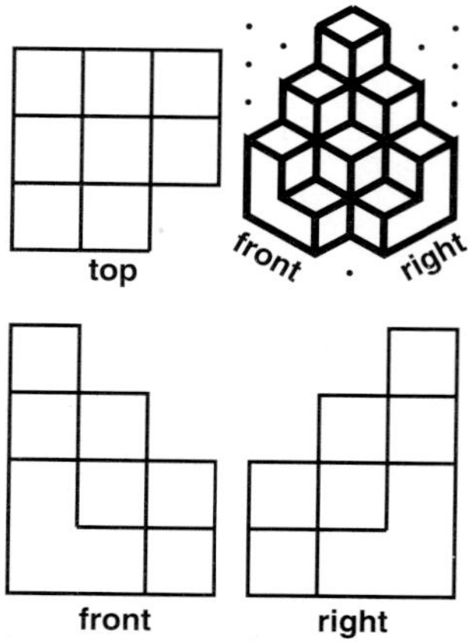

MIXED REVIEW **27.** 5; (−1, 0.5)

Wrap Up pages 117–119

1. 61; scalene acute **2.** 35; isosceles obtuse **3.** $x = 60$; $y = 60$; equilateral; equiangular **4.** $x = 45$; $y = 45$; isosceles; right **5.** D **6.** 120; 60 **7.** 135; 45 **8.** 144; 36 **9.** 165; 15 **10.** 8; 14, 9, 7, 9 **11.** $m = 4$; $t = 5$; 7, 14, 14, 7 **12.** $a = 1$; $b = 2$; 6, 6, 6, 6 **13.** 5; 3; neither **14.** 4; 4; parallel **15.** $-\frac{1}{3}$; 3; perpendicular **16.** 1; 1; parallel **17.** (4, 3); 3 **18.** (1, −1); $\sqrt{5}$ **19.** (−1, −3); $\sqrt{13}$ **20.** (5.5, 5); $\frac{3\sqrt{5}}{2}$ **21.** 30 **22.** 120 **23.** 330 **24.** 120 **25.** $\overline{ML}$ **26.** $\angle U$ **27.** $\overline{ST}$ **28.** *ONMLK*

30a.

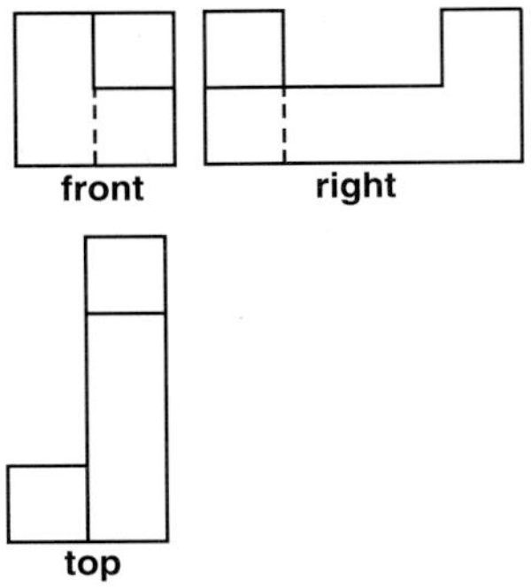

30b.

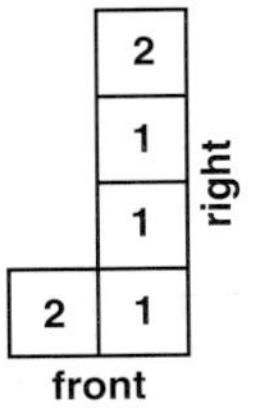

32.

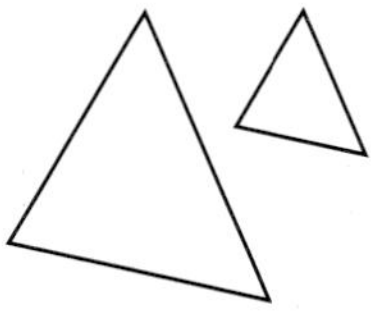

33.

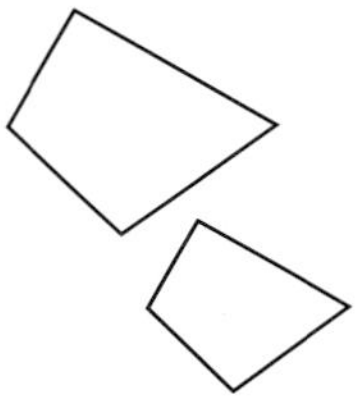

34.

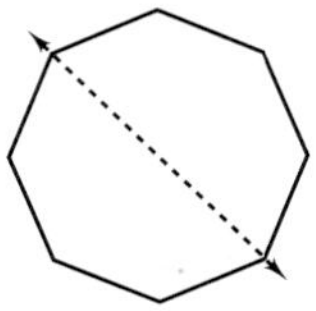

35.

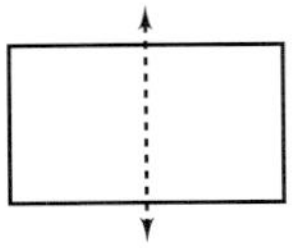

36.

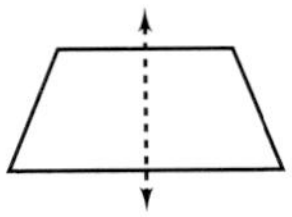

37.

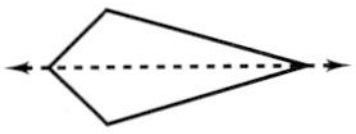

Cumulative Review page 121

1. D **3.** E **5.** C **7.** B **9.** D **11.** (2.5, −2)

13.

CHAPTER 3

Lesson 3-1 pages 128–130

ON YOUR OWN

1.

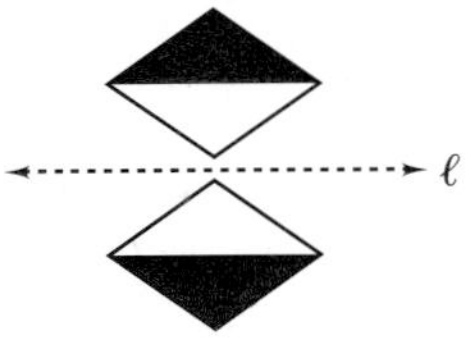

3.

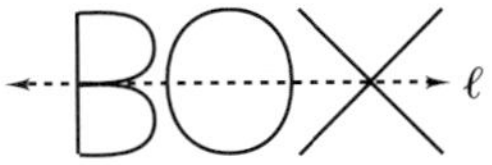

5a. $\overline{PQ}$ and $\overline{P'Q'}$, $\overline{QR}$ and $\overline{Q'R'}$, $\overline{RS}$ and $\overline{R'S'}$, $\overline{SP}$ and $\overline{S'P'}$ **5b.** isometry **5c.** opposite **7a.** $\overline{AR}$ and $\overline{A'R'}$, $\overline{RT}$ and $\overline{R'T'}$, $\overline{TA}$ and $\overline{T'A'}$ **7b.** not isometry **7c.** opposite **9a.** $\overline{RI}$ and $\overline{R'I'}$, $\overline{IT}$ and $\overline{I'T'}$, $\overline{TR}$ and $\overline{T'R'}$ **9b.** not isometry **9c.** opposite

11.

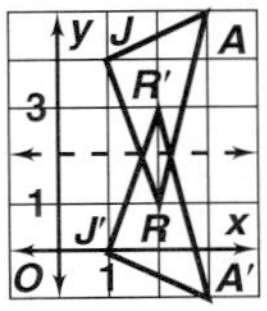

15.

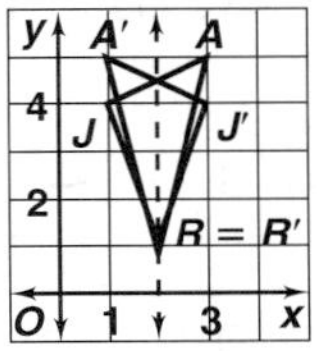

17.

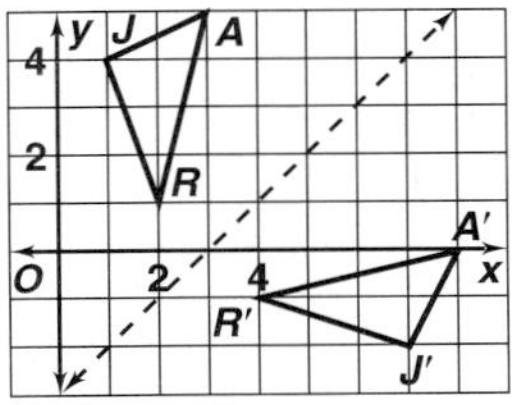

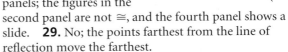

21a. Answers may vary. Sample: The writing hand would not cover what was already written.

21b.

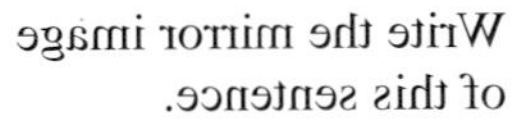

23.

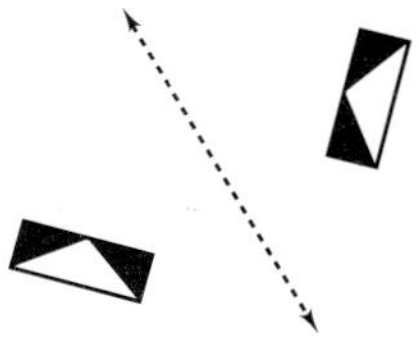

25a. S-Isomer
25b. Samples: gloves, shoes, scissors **27.** First and third panels; the figures in the second panel are not ≅, and the fourth panel shows a slide. **29.** No; the points farthest from the line of reflection move the farthest.

MIXED REVIEW **31.** rectangle **33a.** (−1, −1) **33b.** 2 **33c.** $-\frac{1}{2}$ **33d.** $y = -\frac{1}{2}x - \frac{3}{2}$

Toolbox page 131

1. $\begin{bmatrix} 11 & 10 \\ 1 & 12 \end{bmatrix}$ **5.** $\begin{bmatrix} 8 & 11.3 \\ 15 & 11.1 \end{bmatrix}$ **9.** $\begin{bmatrix} 447 & 18 & 20 \\ 546 & 23 & 10 \\ 450 & 22 & 18 \\ 396 & 30 & 22 \end{bmatrix}$

Selected Answers

Lesson 3-2 pages 135–137

ON YOUR OWN **1.** ⟨1, −3⟩ **3.** ⟨1, −1⟩ **5.** ⟨4, −2⟩ **7.** *C* **9.** *I* **11.** *H* **13.**

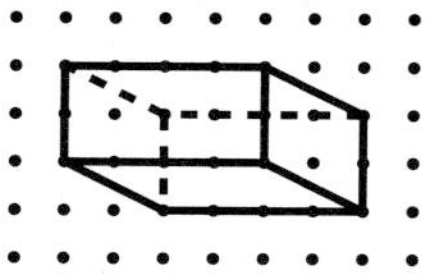

15.

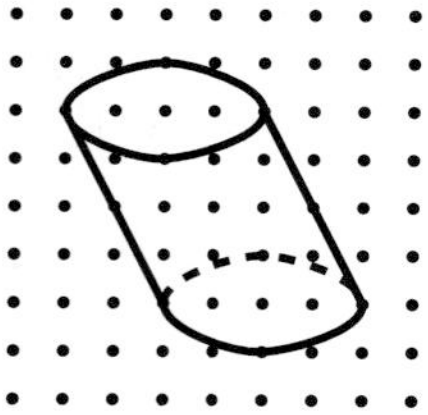

17.
$$\begin{array}{ccc} A' & C' & E' \end{array}$$
$$\begin{bmatrix} -2 & -17 & -9 \\ 6 & 9 & -2 \end{bmatrix}$$

19.
$$\begin{array}{cccc} N' & I' & L' & E' \end{array}$$
$$\begin{bmatrix} -1 & -1 & -6 & -6 \\ -9 & -2 & 0 & -7 \end{bmatrix}$$

21. $U'(1, 16)$; $G'(2, 12)$ **25.** ⟨0, 0⟩ **27a.** ⟨−4, −2⟩

27b.

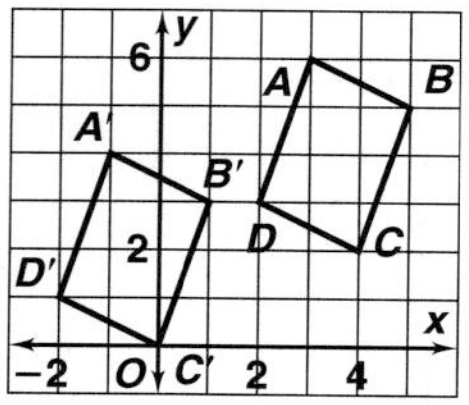

MIXED REVIEW **29.** bisector **31a.**

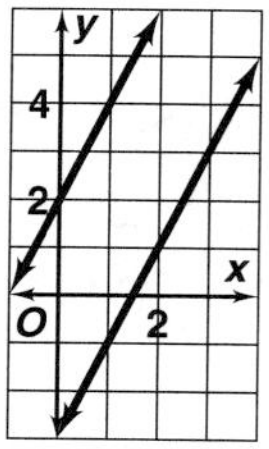

31b. 2, −3; 2, 2
31c. The lines are ∥.
33. 300°
35. 450°

Lesson 3-3 pages 141–143

1.

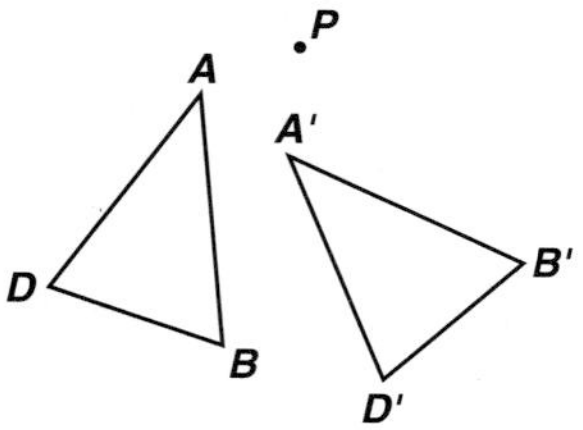

3.

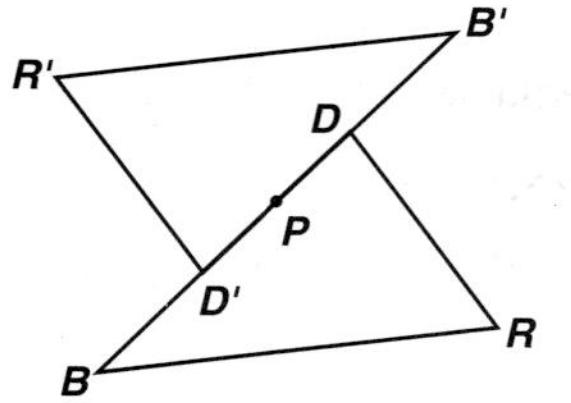

5.

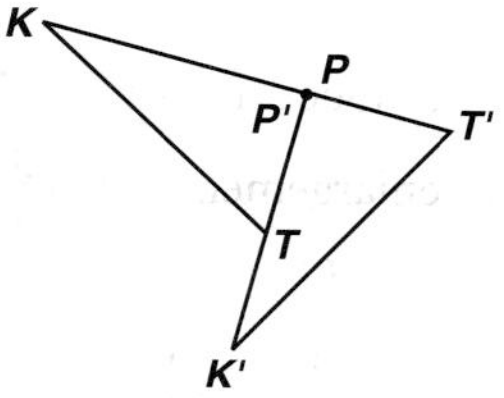

7. 110 **9.** 180 **11.** *M* **13.** $\overline{BC}$ **15.** *I* **17.** *J*
21. $\overline{MN} \cong \overline{M'N'}$, $\overline{ME} \cong \overline{M'E}$, $\overline{EN} \cong \overline{EN'}$, $\angle MEN \cong \angle M'EN'$, $\angle MNE \cong \angle M'N'E$, $\angle EMN \cong \angle EM'N'$, $\angle MEM' \cong \angle NEN'$

23.

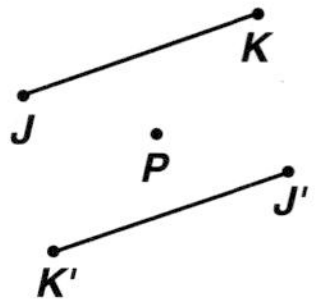

25.

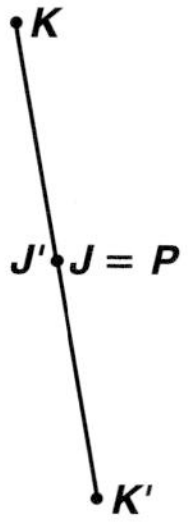

29. 108

MIXED REVIEW **31.** Sample:

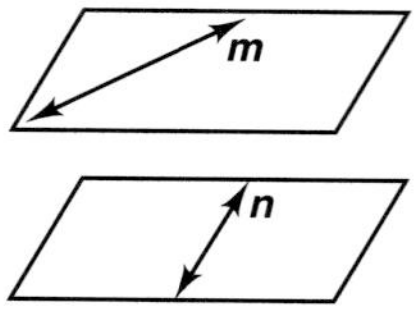

35. Sample:

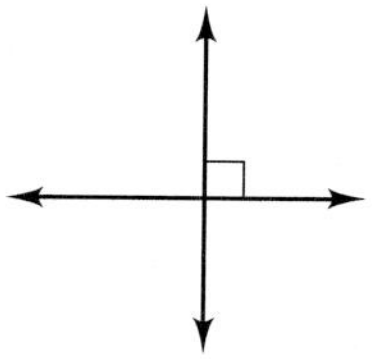

CHECKPOINT **1.** (3, −4) **2.** (−4, 3) **3.** (1, 11) **4.** (4, −3) **5.** (4, −1) **6.** (4, 3) **7.** Images and preimages under translations, reflections, and rotations are congruent to each other. Translations and rotations do not affect orientation. Reflection reverses orientation. **8a.** $\overline{AD}$ and $\overline{A'D'}$, $\overline{AF}$ and $\overline{A'F'}$, $\overline{DF}$ and $\overline{D'F'}$ **8b.** No; the image is not ≅ to the preimage.

Lesson 3-4 pages 147–150

ON YOUR OWN **1.** 60° **3.** 30° **5a.** III **5b.** IV **5c.** II **5d.** I

7.

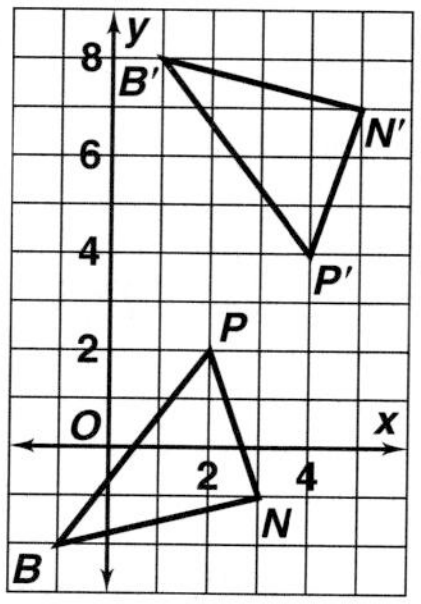

11.

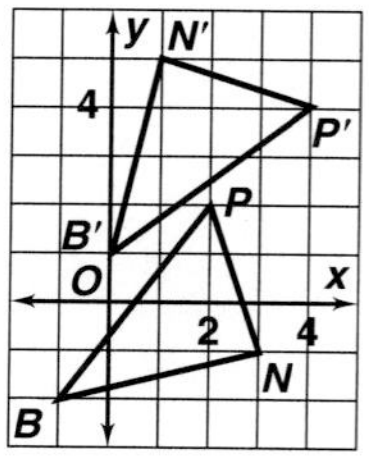

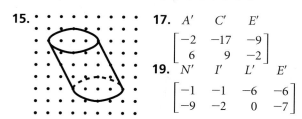

13. translation **15.** rotation **17.** glide reflection **19.** glide reflection **21.** rotation **23.** translation **25.** reflection **27.** rotation **31.** Rotations and glide reflections are equally likely to occur. They are more likely

to occur than translations and reflections.

MIXED REVIEW **33.** 79, 130 and 151 **35a.** true **35b.** false **35c.** true **37.** Each figure maps onto itself.

Toolbox page 151

INVESTIGATE All 6 polygons change in the same way; yes.

Lesson 3-5 pages 155–158

ON YOUR OWN

1.
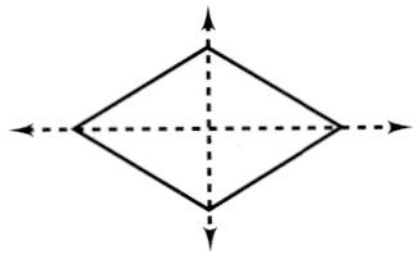

3. rotational: 90°, point **5.** point **7.** no symmetry **9.** reflectional **11.** rotational **13.** any isosceles but not equilateral △

15.
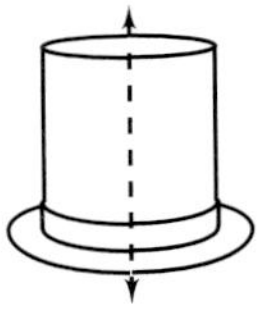

17.

19. reflectional, rotational **21.** point **23.** reflectional, rotational **25.** reflectional, rotational **27.** reflectional, point

31.
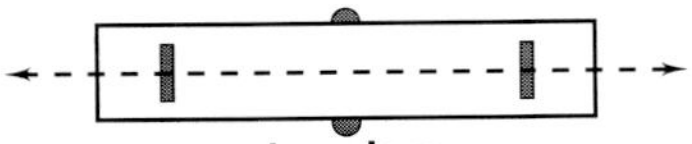

33a.

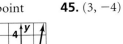

Language	Horizontal Line	Vertical Line	Point
English	B, C, D, E, H, I, K, O, X	A, H, I, M, O, T, U, V, W, X, Y	H, I, N, O, S, X, Z
Greek	B, E, H, Θ, I, K, Ξ, O, Σ, Φ, X	A, Δ, H, Θ, I, Λ, M, Ξ, O, Π, T, Y, Φ, X, Ψ, Ω	Z, H, Θ, I, N, Ξ, O, Φ, X

35. reflectional in *y*-axis **37.** point **45.** (3, −4)

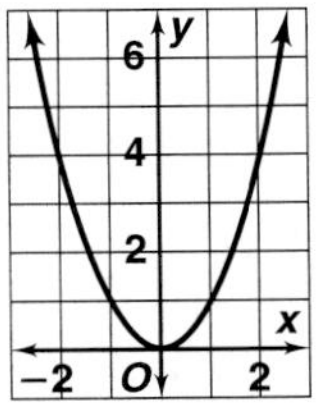

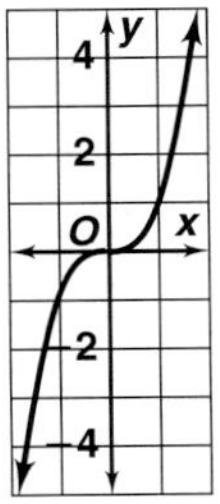

MIXED REVIEW **47.** 1 **49.** $\frac{7}{3}$ **51.** 0; 3

CHECKPOINT **1.** rotational **2.** point **3.** reflectional, point **4.** $C'(-4, 3)$, $A'(-1, 6)$, $L'(-3, 2)$

Lesson 3-6 pages 163–165

ON YOUR OWN

1.

3.

5.
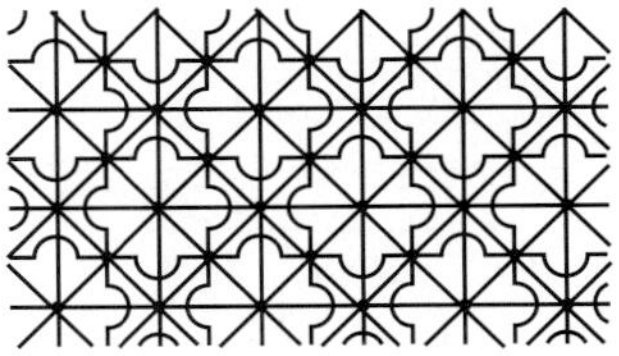

rotational, point, reflectional, glide reflectional, and translational

9.
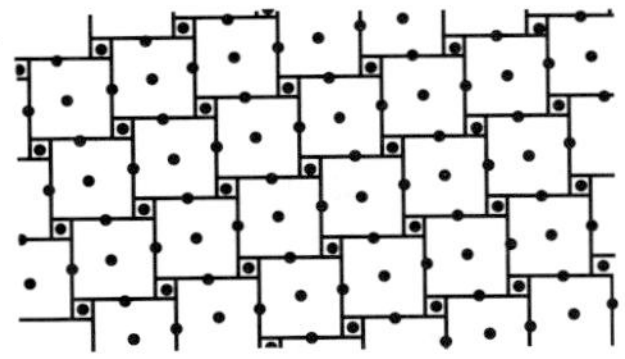

rotational, point, and translational

13.
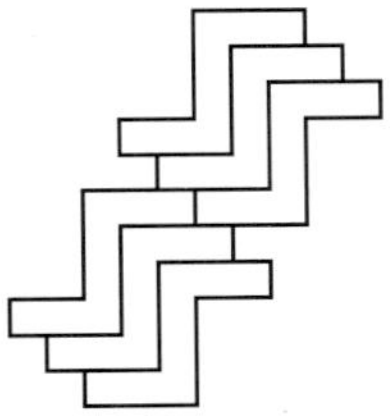

15.
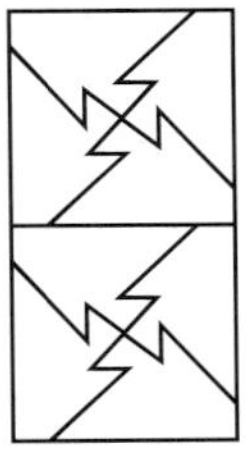

17.

19.

MIXED REVIEW **21.** equilateral, equiangular **23.** scalene, right **25a.** $\overline{BD}$ **25b.** Sample: $\overset{\frown}{BCE}$ **25c.** Sample: $\overset{\frown}{ED}$ **25d.** Sample: $\overline{AE}$ **25e.** Sample: $\angle BAE$ **25f.** Sample: $\overset{\frown}{BE}$ and $\overset{\frown}{ED}$ **27a.** 3; 10 **27b.** 2

Lesson 3-7 pages 168–172

ON YOUR OWN

1.
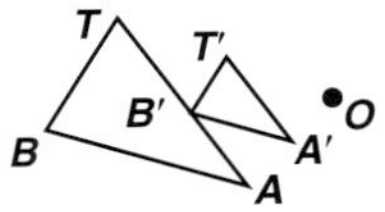

3.
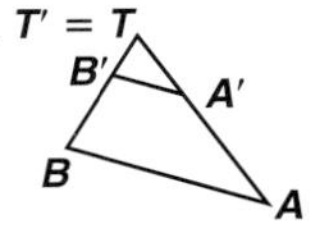

5a. reduction **5b.** $\frac{1}{3}$ **7a.** enlargement **7b.** 3 **9a.** enlargement **9b.** $\frac{3}{2}$ **11a.** reduction **11b.** $\frac{2}{5}$

13. $\begin{matrix} A' & B' & C' \end{matrix}$ $\begin{bmatrix} 3 & 9 & 15 \\ 0 & 6 & 3 \end{bmatrix}$ **15.** $\begin{matrix} A' & B' & C' \end{matrix}$ $\begin{bmatrix} -4 & -8 & -6 \\ 0 & -6 & 0 \end{bmatrix}$ **17.** about 343

Selected Answers

19.

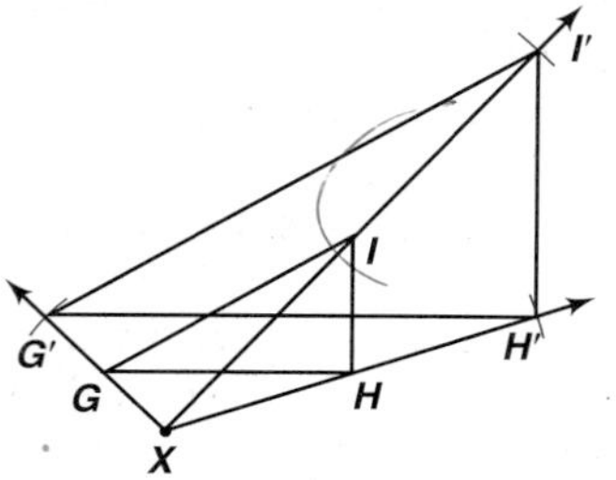

21.

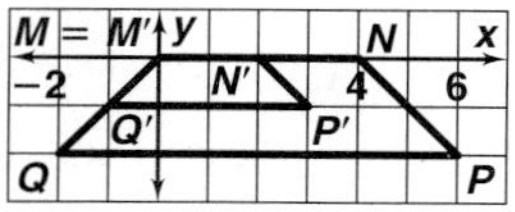

23a. $\begin{matrix} A' & B' & C' & D' \end{matrix}$
$$\begin{bmatrix} -6 & 6 & 6 & -6 \\ -6 & -6 & 6 & 6 \end{bmatrix}$$

23b.

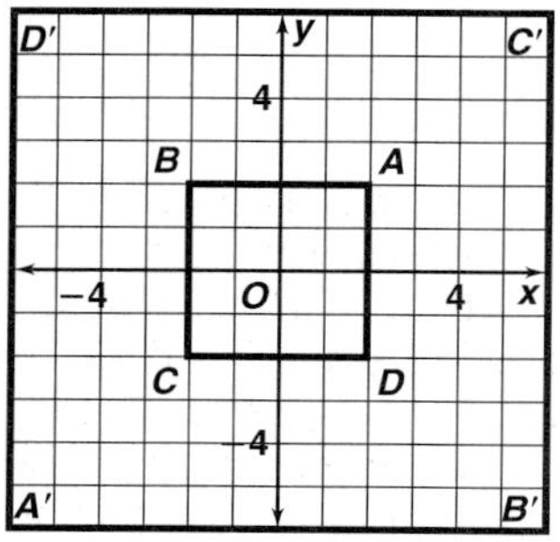

23c. The image of a dilation with a negative factor is the image of a dilation with a positive factor with the same absolute value, rotated 180° about the origin.

25. $\begin{matrix} Q' & R' & T' & W' \end{matrix}$
$$\begin{bmatrix} -9 & -6 & 9 & 9 \\ 12 & -3 & 3 & 15 \end{bmatrix}$$

27. $\begin{matrix} Q' & R' & T' & W' \end{matrix}$
$$\begin{bmatrix} -6 & -4 & 6 & 6 \\ 8 & -2 & 2 & 10 \end{bmatrix}$$

29a. vertex of V **29b.** $\frac{1}{2}$ **33.** 10; 12 **35.** 32; 7.5
37. True; the image and the preimage are similar.
39. False; a dilation does not change orientation.

MIXED REVIEW **41.** $x = 105$; $y = 75$; $z = 35$
43. $x = 85$; $y = 125$ **45.** trapezoid

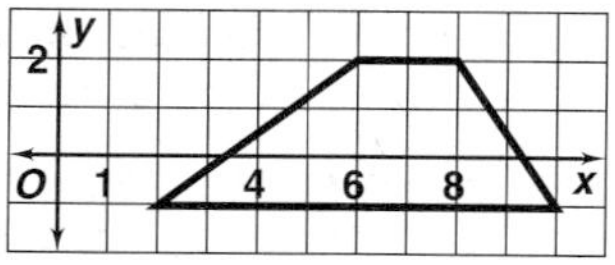

47.

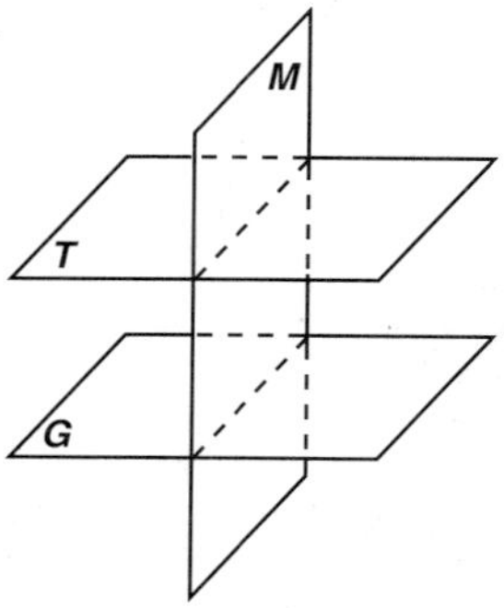

Wrap Up pages 174–177

1.

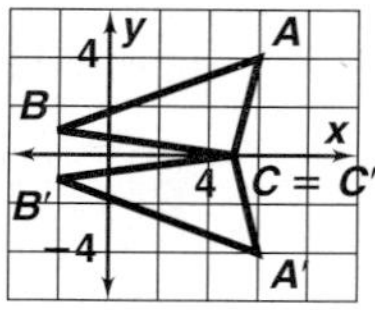

2.

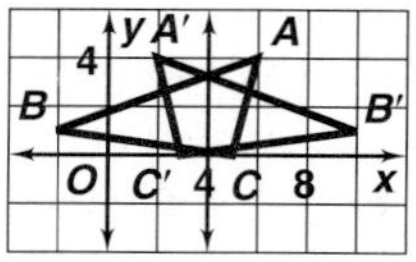

6.

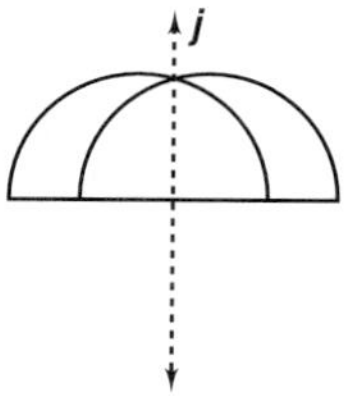

8.

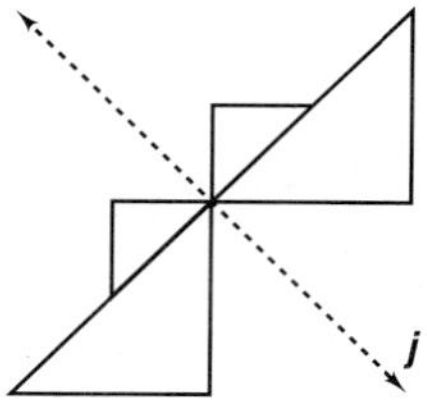

9. E

11. $A'(7, 12)$, $B'(6, 6)$, $C'(3, 5)$ **12.** $R'(-4, 3)$, $S'(-6, 6)$, $T'(-10, 8)$ **13.** $\langle 2, 8 \rangle$ **14.** $\langle -2, -1 \rangle$ **15.** $\langle 11, -4 \rangle$

16.

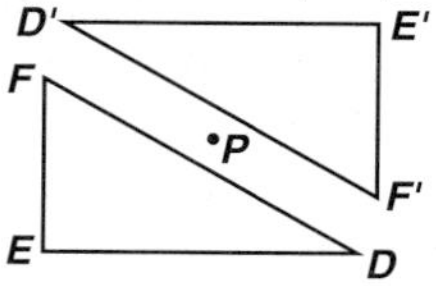

17.

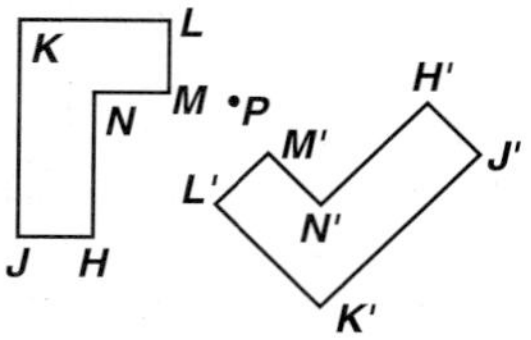

18.

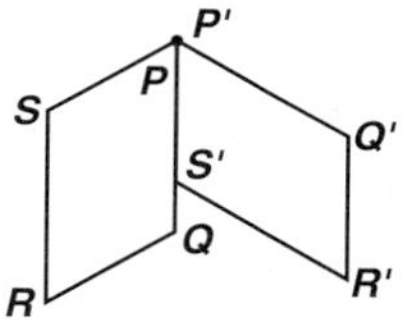

19.

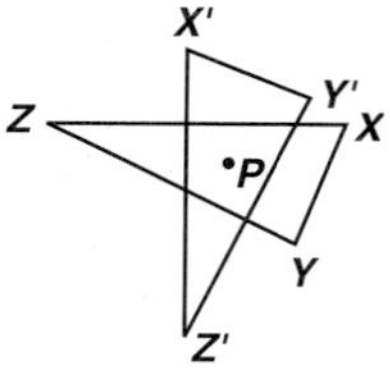

20. $(-2, 5)$ **21.** $(-3, 0)$ **22.** $(-1, -4)$ **23.** $(0, 7)$
24. $(8, -2)$ **25.** $(0, 0)$ **26a.** II **26b.** I **26c.** III
26d. IV **27a.** III **27b.** IV **27c.** II **27d.** I
28. $T'(-4, -9)$, $A'(0, -5)$, $M'(-1, -10)$ **29.** reflectional
30. 72° rotational **31.** 90° rotational, point

32a. 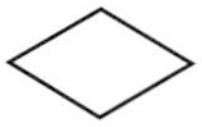

32b. rotational, point, reflectional, translational, glide reflectional

33a.  and 

33b. point, reflectional, translational, glide reflectional

34a.  and 

34b. rotational, point, reflectional, translational, glide reflectional

35. $\begin{matrix} M' & A' & T' & H' \end{matrix}$
$$\begin{bmatrix} -15 & -30 & 0 & 15 \\ 20 & -5 & 0 & 10 \end{bmatrix}$$

36. $\begin{matrix} A' & N' & D' \end{matrix}$
$$\begin{bmatrix} 14 & -8 & 0 \\ -2 & -6 & 4 \end{bmatrix}$$

37. $\begin{matrix} W' & I' & T' & H' \end{matrix}$
$$\begin{bmatrix} 12 & 6 & 9 & 0 \\ 15 & 18 & 24 & 21 \end{bmatrix}$$

38. $\begin{matrix} F' & U' & N' \end{matrix}$
$$\begin{bmatrix} -2 & 2\frac{1}{2} & -1 \\ 0 & 0 & -2\frac{1}{2} \end{bmatrix}$$

39. Rotations, dilations and translations preserve orientation. Reflections and glide reflections reverse orientation. **40.** $(2, 8)$ **41.** $(1, 7)$ **42.** $(5, 4)$
43. isosceles, acute **44.** scalene, right **45.** isosceles,

obtuse **46.** scalene, acute

Preparing for Standardized Tests page 179

1. E **3.** D **5.** A **7.** A **9.** C **11.** rotations, reflections, translations, and glide reflections

CHAPTER 4

Lesson 4-1 pages 185–187

ON YOUR OWN **1.** You send in a proof-of-purchase label; they send you a get-well card. **3.** If <u>$3x - 7 = 14$</u>, then <u>$3x = 21$</u>. **5.** If <u>a triangle is isosceles</u>, then <u>it has two congruent sides</u>. **7.** 1 and 9 are not prime. **9.** Softball and cricket are sports played with a ball and a bat. **11a.** If you grow, then you will eat all of your vegetables. **11b.** If you do not eat all of your vegetables, then you will not grow. **11c.** If you do not grow, then you will not eat all of your vegetables. **15a.** If 2 segments have the same length, then they are ≅. **15b.** If 2 segments are not ≅, then they have different lengths. **15c.** If 2 segments have different lengths, then they are not ≅. **19a.** If you have a passport, then you travel from the U.S. to Kenya. **19b.** true; false **23a.** If the slopes of 2 nonvertical lines are =, then they are ∥. **23b.** true; true **23c.** 2 nonvertical lines are ∥ if and only if their slopes are =. **29.** If the sum of the digits of a number is divisible by 3, then the number is divisible by 3; if a number is divisible by 3, then the sum of the digits of the number is divisible by 3. **33a.** If a polygon is regular, then all its sides are ≅. **33b.** If not all the sides of a polygon are ≅, the polygon is not regular. **35a.** If a transformation is a rotation, then it is an isometry. **35b.** If a transformation is not an isometry, it is not a rotation.

MIXED REVIEW

37.

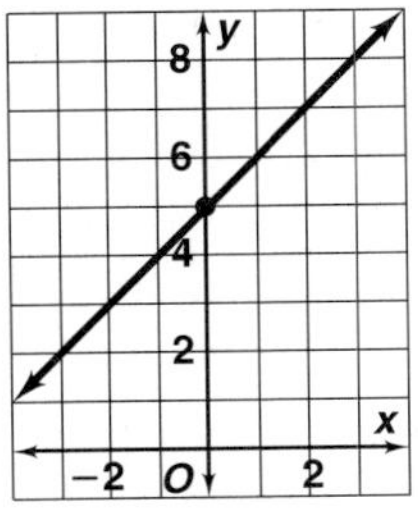

39.

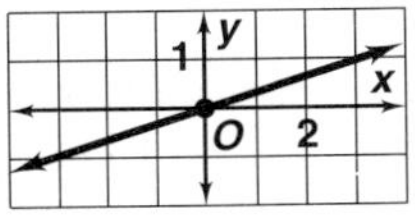

41. $A'(0, -3)$, $B'(4, 6)$, $C'(-6, -1)$ **43.** 35

Lesson 4-2 pages 191–193

ON YOUR OWN **1.** $x = 80$; $y = 40$ **3.** $x = 4.5$; $y = 60$ **5.** $x = 92$; $y = 7$ **7.** $x = 64$; $y = 71$ **9.** The measure of each base angle is 70. **11.** The 3rd vertex must be on the ⊥ bisector of the base, the line $x = 3$. **13.** False; a rectangle need not have 4 ≅ sides. **15.** true **17.** (0, 5), (5, 0), (0, 10), (10, 0), (−5, 5), (5, −5) **19a.** 25 **19b.** 40, 40, 100 **19c.** Isosceles; the △ has 2 ≅ angles. **21a.** isosceles **21b.** 900 ft; 1100 ft **21c.** Answers may vary. Sample: The tower is the ⊥ bisector of the base. **23.** 2.5 **25.** 35 **27.** 60 **29.** 50 **31.** 120 **33.** 70 **35.** No; the base is the side opposite the vertex angle. **37.** $m = 20$; $n = 45$ **39.** $m = 36$; $n = 27$

MIXED REVIEW **41.** (1, 5); $\sqrt{13}$ **43.** (0.5, 6.5); $\frac{1}{2}\sqrt{26}$ **45.** 24 sides **47.** $\overline{DF} \parallel \overline{EG}$ because in a plane, 2 lines ⊥ to a 3rd line are ∥.

Lesson 4-3 pages 196–199

ON YOUR OWN **1.** Answers may vary. Samples: $m\angle V = 45$ because acute angles of a rt. △ are complementary; $UT = TV$ because if 2 ∠s of a △ are ≅, the sides opposite them are ≅. **3.** $ABCD$ is a rectangle; sum of the measures of angles of a quadrilateral = 360 so each angle is 90°. **5.** Answers may vary. Sample: $MP = MN$ and $NO = PO$ because if 2 ∠s of a △ are ≅, the sides opposite them are ≅. **7.** C **9.** E; assume (A) is T. Then (B–F) are all T. But (B) is F if (C–F) are T. So the assumption that (A) is T must be F and (A) is F. Assume (B) is T. Then (C–F) are F. But (C) is T if (B) is T, so the assumption that (B) is T must be F and (B) is F. Since (A) and (B) are F, (C) is F. Since (A–C) are F, (D) is F. Since (A–D) are F, (E) is T. Since (E) is T, (F) is F. **11a.** $\angle ONK$ **11b.** $\angle MNO$ **11c.** $\angle MON$ **13.** d, f, b, a, c, e **15.** By Substitution, $m\angle 1 + m\angle 2 = m\angle 2 + m\angle 3$. Subtracting = quantities from each side yields $m\angle 1 = m\angle 3$. **17.** Given; Isosceles Triangle Thm.; Isosceles Triangle Thm.; Transitive Prop. of ≅

MIXED REVIEW **19.** 147.6, 144, 54, 14.4

21a.

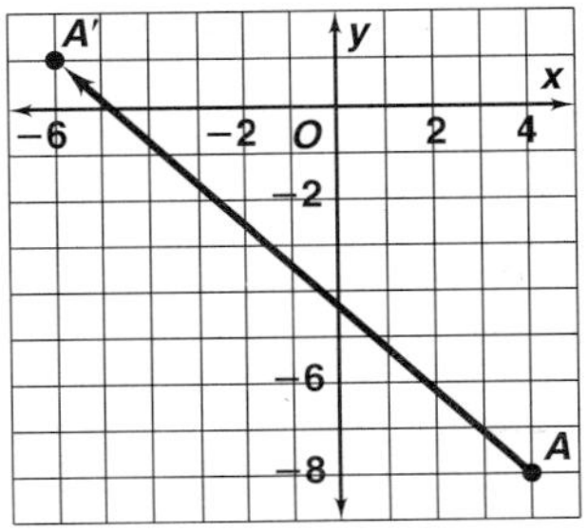

21b. shift left 10 and up 9; ⟨−10, 9⟩

Toolbox page 200

CONJECTURE The midsegment is ∥ to a side of the △ and is half its length.

Selected Answers

EXTEND Opp. angles of the orig. $\triangle$ and the midsegment $\triangle$ have $=$ measures; corr. angles in the 4 smaller $\triangle$s have $=$ measures.

The sides of the midsegment $\triangle$ are each $\frac{1}{2}$ of the corr. side of the original $\triangle$; the 4 $\triangle$s are $\cong$; the 4 $\triangle$s are $\sim$ to the original $\triangle$. The area of each smaller $\triangle$ is $\frac{1}{4}$ the area of the original $\triangle$; the perimeter of each smaller $\triangle$ is $\frac{1}{2}$ the perimeter of the orig. $\triangle$.

Lesson 4-4 pages 203–206

ON YOUR OWN **1.** 9 **3.** 31 **5a.** $H(2, 0)$; $J(4, 2)$ **5b.** slope of $\overline{HJ} = 1$, slope of $\overline{EF} = 1$ ✔ **5c.** $HJ = 2\sqrt{2}$, $EF = 4\sqrt{2}$, $HJ = \frac{1}{2}EF$ **7a.** Answers may vary. Sample:

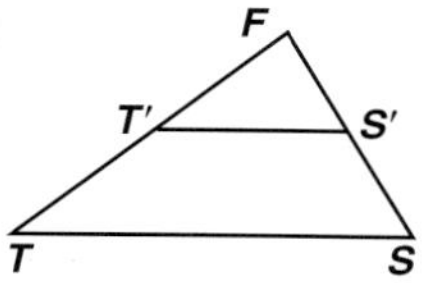

7b. Dilation with center F and scale factor $\frac{1}{2}$ **7c.** The triangles are $\sim$. **9.** 60 **11.** 10 **13.** 45 **15.** 154 cm **17.** 2 cm; the length of a side of the largest square $=$ the length of the diagonal of the middle square. The length of the diagonal of that square is twice the length of the sides of the smallest square.

MIXED REVIEW **19.** 5 **21.** $(-2, 6)$ **23.** Lines m and n do not intersect. **25.** $\triangle ABC$ is isosceles. **27.** 180; it must be a straight angle.

CHECKPOINT **1a.** If the measure of at least one of the angles of a triangle is 60, then the triangle is equilateral. **1b.** true; false **2a.** If all the angles of a polygon are $\cong$, then the polygon is regular. **2b.** true; false **3.** A converse reverses the hypothesis and the conclusion. The truth value of a converse does not depend on the truth value of the original statement. A contrapositive reverses and negates the hypothesis and the conclusion of the original statement. The truth values of a statement and its contrapositive are the same. **4.** C **5a.** 9 **5b.** 54, 54, 72 **5c.** Isosceles; the $\triangle$ has 2 $\cong$ angles. **6.** False; a $\triangle$ with angles 30°, 30°, 120° is an obtuse isosceles $\triangle$. **7.** 36

Lesson 4-5 pages 209–211

ON YOUR OWN **1.** Assume it is not raining outside. **3.** Assume $\triangle PEN$ is scalene. **5.** Assume $\overline{XY} \not\cong \overline{AB}$. **7.** I and II **9.** I and III **11.** This bridge is a bascule. **13.** Sumiko is not an air traffic controller. **15.** Assume $\angle A \cong \angle B$. Then $\triangle ABC$ is isosceles with $BC = AC$. This contradicts the assumption. Therefore, $\angle A \not\cong \angle B$. **17.** E **19.** Assume that the driver had not applied the brakes. Then the wheels would not have locked and there would be no skid marks. There are skid marks. Therefore, the assumption is false. The driver had applied the brakes. **21.** Assume the polygon is a hexagon. Then the sum of measures of its interior angles is 720. But the sum of measures of the polygon's interior angles is 900, not 720. Therefore, the assumption is false. The polygon is not a hexagon. **23.** Assume $m\angle P = 90$. By def., $m\angle Q > 90$. The sum of the angle measures of $\triangle PQR$ is > 180. By the Triangle Angle-Sum Thm., the sum of the angle measures of a $\triangle$ is 180. Therefore, the assumption is false, and an obtuse $\triangle$ cannot contain a rt. angle. **25.** Mr. Pitt was in Maine on April 8 and stayed there at least until 12:05 A.M. Assume Mr. Pitt is the robber. Then he took $<$ 55 min to travel from Maine to Charlotte. It is not possible to travel from Maine to Charlotte in 55 min or less. Therefore, Mr. Pitt is not the robber. **27.** The hole in the roof; of 5 possibilities, 4 have been eliminated.

MIXED REVIEW **29.** Answers may vary. Sample: $ABCD$ and $BCFG$ **31.** Answers may vary. Sample: A, B, C, and D **33.** B **35.**

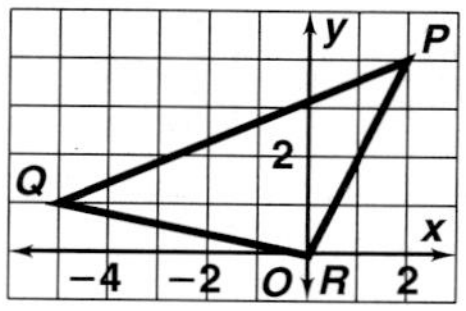

$\overline{PR}$, $\overline{QR}$, $\overline{PQ}$

Toolbox page 212

1. $x \leq -1$ **5.** $a \leq 18$ **9.** $n \leq -3\frac{1}{2}$ **13.** $n \geq -119$ **17.** $b \geq 8$ **21.** $x < -5$ **25.** $a \geq -1$

Lesson 4-6 pages 216–218

ON YOUR OWN **1.** no; $2 + 3 \not> 6$ **3.** Yes; the length of each segment is $<$ the sum of the lengths of the other 2. **5.** Yes; the length of each segment is $<$ the sum of the lengths of the other 2. **7.** Yes; the length of each segment is $<$ the sum of the lengths of the other 2. **9.** $\overline{MN}$, $\overline{ON}$, $\overline{OM}$ **11.** $\overline{TU}$, $\overline{UV}$, $\overline{TV}$ **13.** $\overline{AK}$, $\overline{AR}$, $\overline{KR}$ **15.** $\angle Q$, $\angle R$, $\angle S$ **17.** $\angle G$, $\angle H$, $\angle I$ **19.** $\angle A$, $\angle B$, $\angle C$ **21.** $\angle Z$, $\angle Y$, $\angle X$ **23.** $\frac{1}{2}$ **25b.** The directions of each pair of inequalities match. **25c.** The 3rd side of the 1st triangle is longer than the 3rd side of the 2nd triangle. **27.** The 2 cities may not be both straight ahead. For instance, Topeka might be 110 mi east, and Wichita might be 90 mi south. **29.** $\overline{CD}$ **31.** $4 < z < 20$ **33.** $0 < z < 12$ **35a.** Given **35b.** Isosceles Triangle Thm. **35c.** Angle Addition Post. **35d.** Substitution **35e.** Exterior Angle Thm. **35f.** Transitive Prop. of Inequality

MIXED REVIEW **37.** $S'(-2, -7)$, $Y'(4, -3)$ **39.** 156; 24 **41.** The set of red pts. is 3 units from the set of blue pts.

Lesson 4-7 pages 223–225

ON YOUR OWN

1.

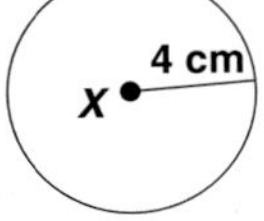

3.

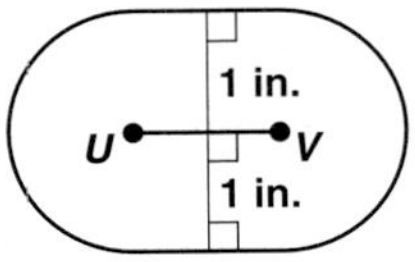

5.

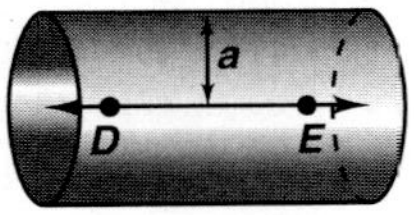

9.

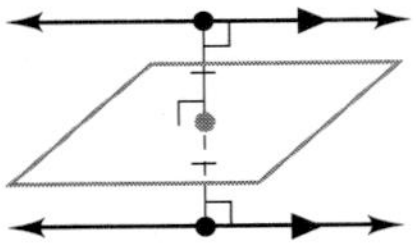

11.

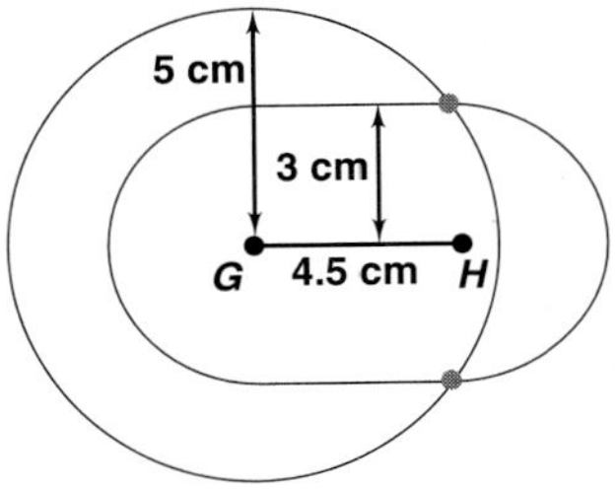

13. The locus is the two points at which the bisector of $\angle JKL$ intersects $\odot C$. **15.** second base **17a.** a circle **17b.** a (smaller) circle

21. The locations that meet Paul's requirement are on the $\perp$ bisector of the segment connecting their offices. The locations that meet Priscilla's requirement are within a circle centered at the downtown with radius 3 mi. The locations that meet both requirements lie along the portion of the $\perp$ bisector that is within the circle.

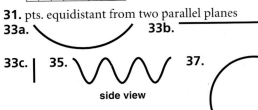

23.

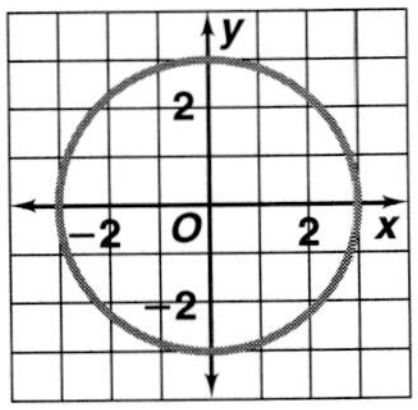

27.

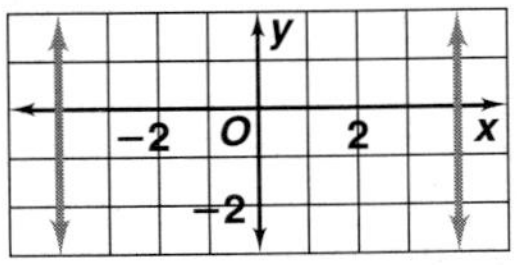

29. pts. equidistant from the sides of $\angle A$

31. pts. equidistant from two parallel planes

33a.

33b.

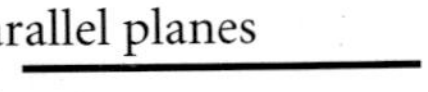

33c.

35.

37.

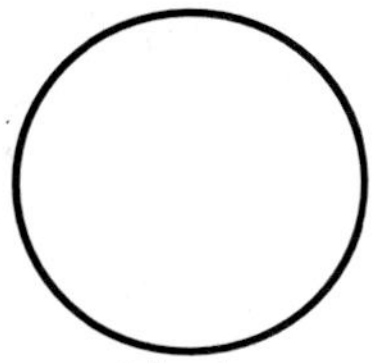

MIXED REVIEW **39.** $\overset{\frown}{CBA}$, $\overset{\frown}{CAB}$

41.

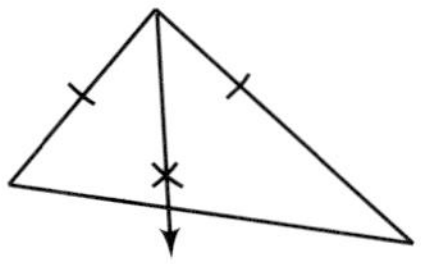

CHECKPOINT **1.** $\overline{AB}$, $\overline{BC}$, $\overline{AC}$ **2.** $\overline{MN}$, $\overline{MO}$, $\overline{NO}$ **3.** $\overline{QR} \cong \overline{RS}$, $\overline{QS}$ **4.** Sample: If $x = |x|$, then $x \geq 0$; assume $x < 0$. **5.** Assume February has at least 30 days. **6.** Assume pentagon has at least 4 rt. angles.

7.

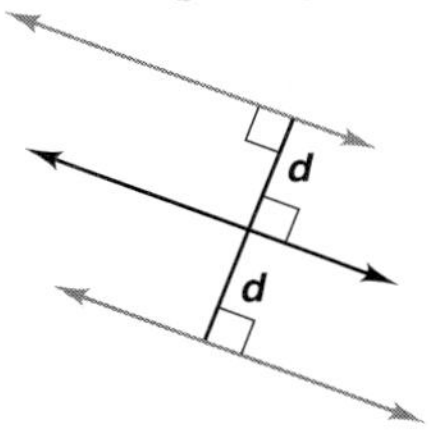

8.

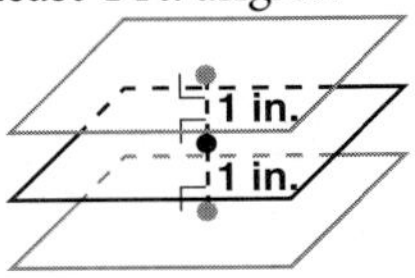

Toolbox page 226

CONJECTURE Each set of lines intersects at a single pt.

EXTEND acute △; right △; obtuse △; The medians and angle bisectors always intersect inside the △. Altitudes intersect inside the △ for acute △s, at a vertex for rt. △s, and outside the △ for obtuse △s.

Lesson 4-8 pages 230–232

ON YOUR OWN **1.** median **3.** none of these **5.** altitude **7a.** $(-2, 0)$ **7b.** $(1, -3)$ **9a.** $(0, 0)$ **9b.** $(-4, 0)$ **11.** IA, IIC, IIIB, IVD **13.** the pt. of concurrency of the $\perp$ bisectors of segments connecting the 3 areas **15a.** $x = 0$, $y = 0$, $y = x$ **15b.** $x = 6$, $y = 6$, $y = x$ **15c.** The line $y = x$; since the altitude from D bisects the base of isosceles $\triangle DEF$, the altitude is also the $\perp$ bisector of the base. **17a.** The triangle balances. **17b.** The triangle balances.

MIXED REVIEW **19.** right **21.** right **23.** Sample: $\overleftrightarrow{AB}$ and $\overleftrightarrow{DE}$ **25.** Sample: $\overleftrightarrow{AB}$, $\overleftrightarrow{BC}$, $\overleftrightarrow{BE}$

Wrap Up pages 234–237

1a. If you are south of the equator, then you are in Australia. **1b.** true; false **2a.** If the measure of an angle is > 90 and < 180, then the angle is obtuse. **2b.** true; true **2c.** An angle is obtuse if and only if its measure is > 90 and < 180. **3a.** If it is cold outside, then it is snowing. **3b.** true; false **4a.** If the sides of a figure are $\cong$, then it is a square. **4b.** true; false **6.** $x = 4$; $y = 65$ **7.** $x = 60$; $y = 60$ **8.** $x = 55$; $y = 62.5$ **9.** $x = 65$; $y = 90$ **10.** d, c, a, b **12.** Answers may vary.

Sample: $\angle ABC$ is a rt. angle by def. of complementary angles and the Angle Addition Post. $\angle A$ and $\angle C$ are complementary because acute angles of a rt. $\triangle$ are complementary. **13.** $\angle E \cong \angle F$ because base $\angle$s of an isosceles $\triangle$ are $\cong$. **14.** Answers may vary. Samples: $\angle HIG \cong \angle JIK$ because they are vert. angles. $\angle HIG \cong \angle G$ and $\angle JIK \cong \angle J$ because base $\angle$s of an isosceles $\triangle$ are $\cong$. **15.** 15 **16.** 12 **17.** 11
18. Assume that the room had been painted with oil-based paint. Then the brushes would be soaking in paint thinner, but they are not. So the assumption is false. The room must not have been painted with oil-based paint.
19. Assume that both numbers are odd. The product of 2 odd numbers is always odd, which contradicts the fact that their product is even. So the assumption that both numbers are odd is false, and at least 1 must be even.
20. Assume that a triangle has 2 or more obtuse angles. By def., the measure of each of these angles is > 90. Then the sum of angle measures of the $\triangle$ is > 180. By the Triangle Angle-Sum Thm., the sum of angle measures of a $\triangle$ is 180. Therefore the assumption was false, and a $\triangle$ has no more than 1 obtuse angle. **21.** Assume 1 angle of an equilateral $\triangle$ is obtuse. Then its measure is > 90 and < 180. By the Isosceles Triangle Thm., the remaining angles are each $\cong$ to the 1st angle. By def. of $\cong$ angles and the Transitive Prop. of $=$, the measures of the 3 angles are $=$, and all the angles are obtuse. Then the sum of the measures of the angles is > 270. But the sum of the measures of angles of a $\triangle$ is 180. Therefore, the assumption is false. An equilateral $\triangle$ cannot have an obtuse angle. **22.** $\angle T, \angle R, \angle S$; $\overline{SR}, \overline{TS}, \overline{TR}$
23. $\angle C, \angle A, \angle B$; $\overline{AB}, \overline{BC}, \overline{AC}$
24. $\angle G, \angle O, \angle F$; $\overline{OF}, \overline{FG}, \overline{OG}$ **25.** B

27.
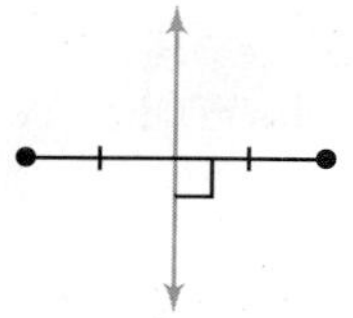

28.
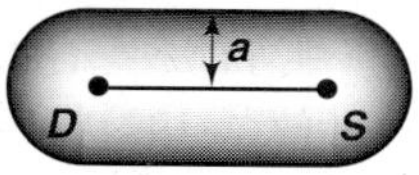

30. $(-1, 0)$
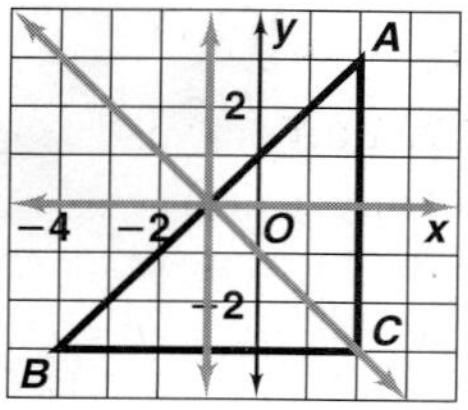

31. $(0, -1)$
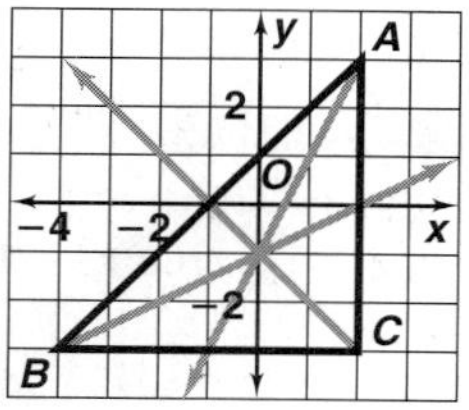

32. $(2, -3)$
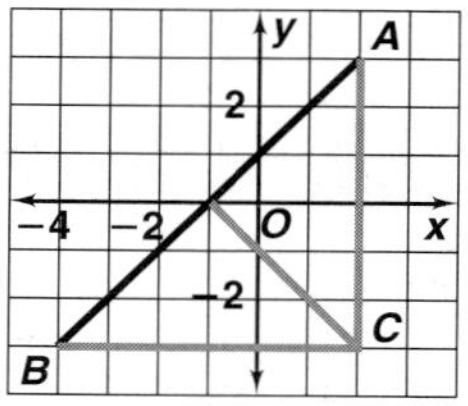

33. 16 cm; 15 cm^2
34. 30.4 ft; 55.8 ft^2
35. 4 m; 0.75 m^2
36. 5 **37.** 13 **38.** 15

Cumulative Review page 239

1. E **3.** A **5.** B **7.** A **9.** C **11.** D
13. Sample:
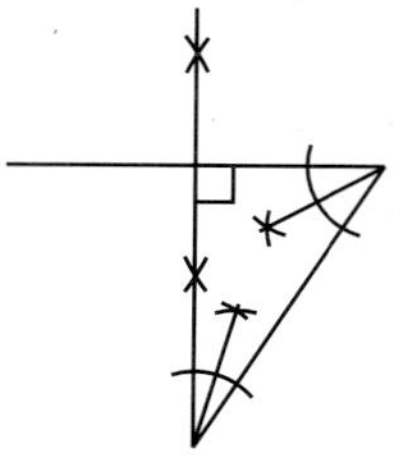

CHAPTER 5

Lesson 5-1 pages 245–248

ON YOUR OWN **1.** about 40 in. **3.** about 20 ft **5.** 36 cm **7.** 24 cm **9.** 78 cm **11.** 74 ft **13.** 15 cm^2 **15.** 14 cm^2 **17.** 12 $units^2$ **19.** 15 $units^2$ **21b.** 38 units **21d.** 54 $units^2$ **23.** 288 cm **25.** 6 yd^2 **27a.** 3492 $in.^2$ or $24\frac{1}{4}$ ft^2 **27b.** $11\frac{3}{4}$ ft^2 **27c.** \$71.80 **27d.** \$9.43 **33a.** Samples: 10 ft by 90 ft, 200 ft; 15 ft by 60 ft, 150 ft **33b.** 30 ft by 30 ft. **35.** 310 cm^2 **37.** 24 cm^2 **39.** 16 **41.** 16

MIXED REVIEW **45.** $(5.5, 5)$ **47.** $(-4, 3)$ **49.** If it is November, then it is Thanksgiving. **51.** between 2 m and 8 m **53.** 2 $units^2$ **55.** 8 $units^2$

Lesson 5-2 pages 252–254

ON YOUR OWN **1.** 15 $units^2$ **3.** 6 $units^2$ **5.** 27 $units^2$ **7.** 4 in. **9.** 240 cm^2 **11.** 20.3 cm^2 **13b.** 24.5 $units^2$ **15b.** 16 $units^2$ **17.** 0.24 **21.** 8 $units^2$ **23.** 8 $units^2$ **25.** 312.5 ft^2 **27.** 12,800 m^2 **29.** The areas of the $\triangle$s are $=$; they have the same bases and $=$ heights. **31.** 126 m^2 **33.** 60 $units^2$ **35.** 4.5 $units^2$

MIXED REVIEW
37. $A'(1, 3), B'(-1, 3), C'(-1, 2)\ D'(1, 0)$
39. $A'(-1, -3), B'(-3, -3), C'(-3, -2)\ D'(-1, 0)$
41. $9 + 16 = 25$ **43.** $36 + 64 = 100$

Toolbox page 255

1. $5\sqrt{2}$ **3.** 8 **5.** $4\sqrt{3}$ **7.** $6\sqrt{2}$ **9.** 6 **11.** $6\sqrt{2}$

Lesson 5-3 pages 259–262

ON YOUR OWN **1.** 10 **3.** $2\sqrt{89}$ **5.** $3\sqrt{2}$ **7.** $2\sqrt{2}$ **9.** 14 ft **11.** obtuse **13.** acute **15.** obtuse **17.** obtuse **19.** acute **21.** right **25.** 7.6 **27.** 19.3 **29.** 11.3 **31.** 4.2 in. **33.** $\frac{9\sqrt{3}}{2}$ m^2 **35.** 10.5 in.^{2}
37a. $|x_2 - x_1|; |y_2 - y_1|$
37b. $PQ^2 = (x_2 - x_1)^2 + (y_2 - y_1)^2$
37c. $PQ = \sqrt{(x_2 - x_1)^2 + (y_2 - y_1)^2}$ **39.** 50 **41.** 15 **43.** 84 **45.** 16 **47.** 12 cm **49.** 17.9 cm **51a.** 5 in.
51b. $\sqrt{29}$ in. or about 5.4 in.
51c. $d_2 = \sqrt{AC^2 + BC^2 + BD^2}$ **51d.** 34 in.
53. $\sqrt{61}$ **55.** $2\sqrt{38}$

MIXED REVIEW
59.

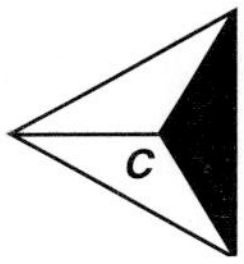

61.

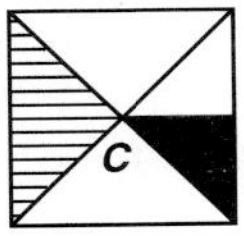

Lesson 5-4 pages 266–268

ON YOUR OWN **1.** $x = 8; y = 8\sqrt{2}$ **3.** $x = 24; y = 12\sqrt{3}$ **5.** $x = \sqrt{2}; y = 2$ **7.** $x = 4; y = 2$ **9a.** 55 ft **9b.** 0.55 min or 33 sec **11.** 9 **13.** $x = 9; y = 18$ **15.** $a = 6; b = 6\sqrt{2}; c = 2\sqrt{3}; d = 6$ **17.** $a = 4; b = 4$ **19.** C **21.** 110.9 cm^2 **23.** 11.3 yd^2 **27a.** $\sqrt{3}$ units **27b.** $2\sqrt{3}$ units **27c.** $s\sqrt{3}$ units

MIXED REVIEW **29.** 1 **31.** 7 units^2; 12.1 units

CHECKPOINT **1.** 84 in.^{2}; 48 in. **2.** 112 cm^2; 48 cm **3.** 72 m^2; 40 m **4.** 12 **5.** $x = 10; y = 10\sqrt{2}$ **6.** $x = 12\sqrt{3}; y = 24$ **7.** B
8. Sample:

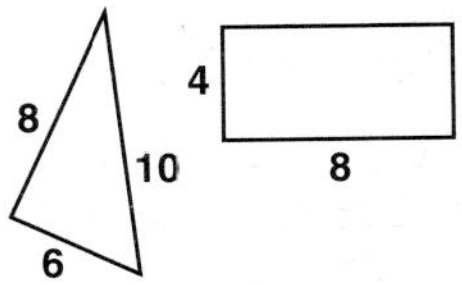

Lesson 5-5 pages 271–273

ON YOUR OWN **1.** 472 in.^{2} **3.** 30 ft^2 **5.** 108,990 mi^2 **7.** 4 ft **9.** $52\sqrt{3}$ ft^3 **11.** 128 m^2 **13a.** $\frac{1}{2}hb_1; \frac{1}{2}hb_2$ **15.** 669 in.^{2} **17.** 18 cm; 12 cm, 24 cm **19.** 49.9 ft^2 **21.** 11.3 cm^2 **23.** 1.5 m^2

MIXED REVIEW **29.** $25\sqrt{3}$ cm^2 **31.** $\frac{100\sqrt{3}}{3}$ m^2

Lesson 5-6 pages 276–278

ON YOUR OWN **1.** 120; 60; 30 **3.** 60; 30; 60 **5.** 12,100 in.^{2} **7.** 128 cm^2 **9.** 841.8 ft^2 **11.** 310.4 ft^2 **15.** 72 cm^2 **17.** $75\sqrt{3}$ ft^2 **21a.** (2.8, 2.8) **21b.** 5.6 units^2 **21c.** 44.8 units^2 **23.** $24\sqrt{3}$ cm^2 **25.** $36\sqrt{3}$ in.^{2}

MIXED REVIEW
27.

29.

31. 25.13 **33.** 31.42

Lesson 5-7 pages 282–284

ON YOUR OWN **1.** 15π cm **3.** 3.7π in. **5.** 56.55 in. **7.** 1.57 yd **9.** 28π cm; 3.5π cm **11.** 36π m; 27π m **13a.** 100 in. **13b.** 50 in. **13c.** $\frac{100}{3}\pi$ in. **15.** 105 ft **17.** 490π mi **19.** 36.13 m **21.** 3.93 m **23.** 2π in. **25.** 12.6

MIXED REVIEW
29.

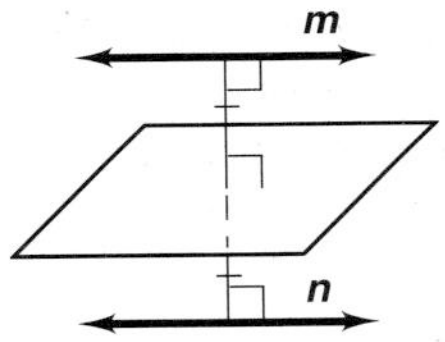

31. about 1 in.^{2}
33. about 1 in.^{2}

CHECKPOINT **1.** 110 cm^2 **2.** $72\sqrt{3}$ in.^{2} **3.** $27\sqrt{3}$ ft^2 **4.** 50.27 in. **5.** 12.57 m **6.** 31.42 ft **7.** 8.80 km **8.** 56.55 mm **9.** 4.5π mm

Lesson 5-8 pages 288–290

ON YOUR OWN **1.** 400π m^2 **3.** $\frac{9}{64}\pi$ in.^{2} **5.** 30 m **7.** 6.25π units^2 **9.** 54.11 m^2 **11.** 11,310 ft^2 **13.** 64π cm^2 **15.** $\frac{169}{6}\pi^2$ m^2 **17.** 1,620,000 m^2 **19.** 18.27 ft^2 **21.** 925.41 ft^2 **23.** 12 in. **25.** $(784 - 196\pi)$ in.^{2} **27.** 4π m^2

29a.

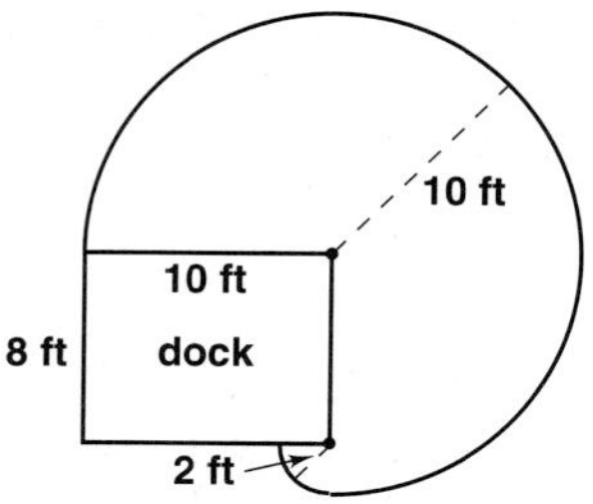

29b. the area of $\frac{3}{4}$ circle with radius 10 ft and $\frac{1}{4}$ circle with radius 2 ft **29c.** about 239 ft^2

MIXED REVIEW **31.** \$90 **33.** $18\sqrt{3}$ cm^2 **35.** 72; isosceles, acute **37.** 68; scalene, acute **39.** 45; isosceles, right

Toolbox page 291

INVESTIGATE The ratios do not change with size.

CONJECTURE As the number of sides increases, each ratio comes closer to 1.

EXTEND Yes; the ratio of the perimeter to the circumference gets closer to 1 faster than the ratio of the areas.

about 63 cm; about 314 cm^2

Wrap Up pages 293–296

1. 32 cm; 64 cm^2 **2.** 38 ft; 78 ft^2 **3.** 32 in.; 40 in^2 **5.** 10 m^2 **6.** 90 $in.^2$ **7.** 33 ft^2 **8.** 16 **9.** $2\sqrt{113}$ **10.** 17 **11.** $x = 9\sqrt{3}$; $y = 18$ **12.** $x = 12\sqrt{2}$ **13.** $x = \frac{20\sqrt{3}}{3}$; $y = \frac{40\sqrt{3}}{3}$ **14.** E **15.** 18 m^2 **16.** 16 ft^2 **17.** $96\sqrt{3}$ mm^2

19.

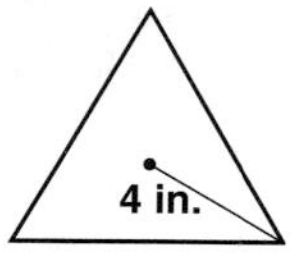

20.8 $in.^2$

20.

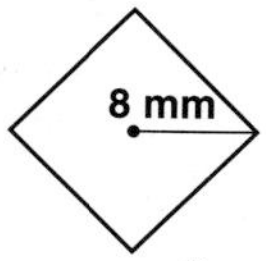

128 mm^2

21.

127.3 cm^2

22. 8π in.; $\frac{22}{9}\pi$ in. **23.** 14π m; $\frac{14}{9}\pi$ m **24.** 6π mm; π mm **25.** 76.97 ft^2 **26.** 18.27 m^2 **27.** 40.96 cm^2 **28.** 24 $in.^2$ **29.** 98.4 cm^2 **30.** 684 ft^2

Preparing for Standardized Tests page 299

1. A **3.** E **5.** B **7.** C **9.** C **11.** No; if an altitude is outside the △, then the △ is obtuse and the altitude is from an acute angle vertex. The altitude from the other acute angle vertex is also outside the △.

CHAPTER 6

Lesson 6-1 pages 304–306

ON YOUR OWN **1.** A, B, D **3.** B **5.** E **7.** A **9.** blue **11.** brown **13a.**

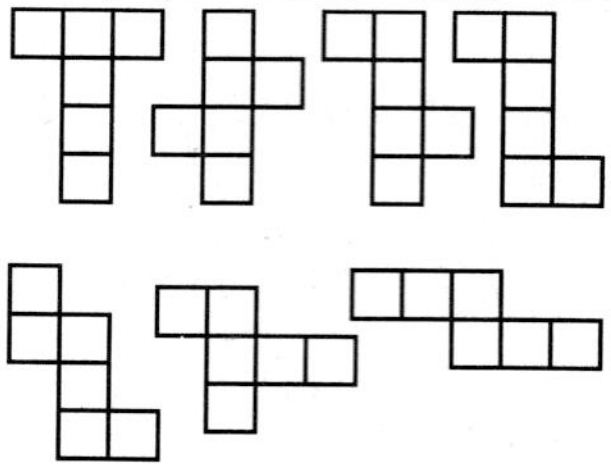

15. 6 in. **17a.** A: icosahedron; B: octahedron; C: hexahedron; D: tetrahedron; E: dodecahedron **17b.** tetrahedron, hexahedron; triangular pyramid, cube **17c.** $12 = 8 + 6 - 2$

MIXED REVIEW **21.** $\angle B$ **23.** 40π cm^2

Toolbox page 307

1. B **3.** B **5.** 10 **7.** 51 **9.** 0.035 **11.** 230 **13.** 5 **15.** 30,000,000,000 **17.** $\frac{5}{18}$ **19.** 108

Lesson 6-2 pages 312–314

ON YOUR OWN **1.** 38 $units^2$ **3.** 38 $units^2$ **5.** 144 ft^2; 216 ft^2 **7.** 288 $in.^2$; 336 $in.^2$ **9a.** right hexagonal prism **9b.** $48\sqrt{3}$ cm^2 **9c.** 240 cm^2 **9d.** $(240 + 48\sqrt{3})$ cm^2 **11.** 880 cm^2 **13.** 36,800 cm^2 **15.** 20 cm **17a.** $A(3, 0, 0)$, $B(3, 5, 0)$, $C(0, 5, 0)$, $D(0, 5, 4)$ **17b.** 5 **17c.** 3 **17d.** 4 **17e.** 94 $units^2$ **21.** 619.1 m^2 **23.** 1726 cm^2 **25.** D **27a.** $r = 0.7$ cm, $h = 4$ cm **27b.** a translation

MIXED REVIEW **29.** 60 cm^2 **31.** $B'(4, 2)$, $I'(0, -3)$, $G'(-1, 0)$ **33.** $B'(12, 2)$, $I'(8, -3)$, $G'(7, 0)$ **35.** 15.3 in. **37.** 17.7 cm

Toolbox page 315

INVESTIGATE about 130 cm^2; arbitrarily large

CONJECTURE either a large side length or a large height; $s = h \approx 4.6$ cm; length of side of base = height; cube

EXTEND 10 cm-by-10 cm-by-10 cm

Lesson 6-3 pages 319–322

ON YOUR OWN **1.** 17 cm **3.** 12.8 m **5.** 264 $in.^2$ **7.** 80 $in.^2$ **13.** 228.1 $in.^2$ **17.** 80.1 m^2 **19.** 43.5 cm^2 **21.** 179.4 $in.^2$ **23.** 62.4 cm^2 **25.** They are =. **29.** 5 cm; 4 cm

MIXED REVIEW **31.** $AB = 5$, $BC = 6$, $AC = 5$; isosceles **33.** $WX = \sqrt{82}$, $XY = 3\sqrt{5}$, $YW = \sqrt{13}$; scalene **35.** If the sum of measures of 2 angles is 90, then they are complementary; true. **37.** 60 cubes **39.** 3 by 3 by 3

CHECKPOINT **1.** 297.6 $in.^2$ **2.** 377.0 cm^2 **3.** 185.6 m^2 **4.** 288.7 ft^2 **5.** 477.5 ft^2

Lesson 6-4 pages 327–329

ON YOUR OWN **1.** 904.8 $in.^3$ **3.** 80 $in.^3$ **5.** 125.7 cm^3 **7.** 280.6 cm^3 **9a.** 28 ft^3 **9b.** 1747 lb **13.** 79 million ft^3 **15.** 5 in. **17.** 6 ft **19a.** 5832 $in.^3$ **19b.** 729 $in.^3$ **21a.** 24 cm **21b.** 3 cm **23.** 140.6 $in.^3$ **25a.** 16π $units^3$ **25b.** 32π $units^3$

MIXED REVIEW **27.** $\overline{AC}, \overline{AB}, \overline{BC}$ **29.** $\overline{BC}, \overline{AB}, \overline{AC}$ **31.** 8 in.

Lesson 6-5 pages 333–336

ON YOUR OWN **1.** 115.5 $in.^3$ **3.** 122.5 $in.^3$ **5a.** 120π ft^3 **5b.** 60π ft^3 **5c.** 240π ft^3 **7.** 300 $in.^3$ **9.** $\frac{16}{3}\pi$ ft^3 **13.** The volumes are =; the volume of the large cone is $\frac{1}{3}$ of the volume of the cylinder; the volume of each small cone is $\frac{1}{3}$ of the volume of half the cylinder. **15.** 15 in. **17a.** 5,920,000 ft^3 **17b.** 267 ft **19.** pyramid **21.** 3 **23.** 9

MIXED REVIEW
25a. $AB = 5$, $BC = 10$, $CD = 5$, $AD = 10$
25b. $AB = \frac{4}{3}$, $BC = -\frac{3}{4}$, $CD = \frac{4}{3}$, $AD = -\frac{3}{4}$
25c. rectangle **27.** 113.1 $in.^2$; 37.7 in. **29.** 19.6 ft^2; 15.7 ft

Lesson 6-6 pages 340–343

ON YOUR OWN **1.** 1794.5 cm^2 **3.** 23.8 $in.^2$ **5.** 64π $units^2$; $\frac{256}{3}\pi$ $units^3$ **7.** 85 lb **9.** 288π cm^3 **11.** $\frac{2048}{3}\pi$ cm^3 **13.** No; the volume of the ice cream is $\frac{4}{3}$ times the volume of the cone. **15a.** 1207 ft^3 **15b.** 10.8 $\frac{lb}{ft^3}$ **15c.** 4191 mi **17a.** 3 in. **17b.** 102.9 $in.^3$ **19.** 2 : 3 **21a.** $6\sqrt{3}$ in.; $3\sqrt{3}$ in. **21b.** 371.7 $in.^3$ **23.** A **25.** The balls weigh 75 lb and 253 lb, respectively.

27a. Cube; the edge of the cube is about 1.61 times as long as the radius r of the sphere. The surface area of the cube, about $15.59r^2$, is > surface area of the sphere, about $12.57r^2$.

MIXED REVIEW **33.** 261.3 m^3

CHECKPOINT **1.** 60.2 ft^2; 22.5 ft^3 **2.** 659.7 cm^2; 1256.6 cm^3 **3.** 332.9 $in.^2$; 377.0 $in.^3$ **4.** 113.1 m^2; 113.1 m^3 **5.** C **6.** The remaining volume is $2\pi r^3$, which is > $\frac{4}{3}\pi r^3$ (volume of one ball).

Lesson 6-7 pages 346–347

ON YOUR OWN **1.** 864π $in.^3$ **3.** 10,368 ft^3 **5.** 501 $in.^3$
7a.
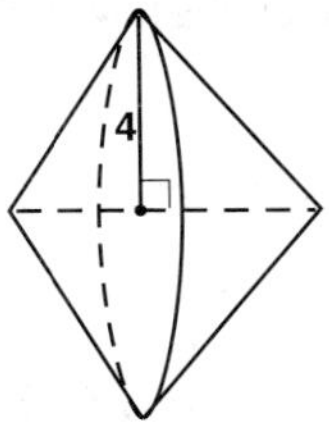

7b. 32π $units^3$ **9.** cone, hemisphere
13a.
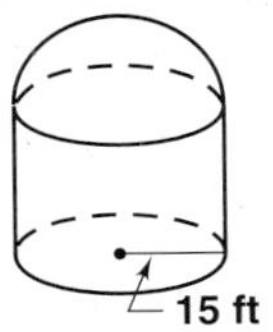

13b. 19 ft; 34 ft **15.** 73 cm^3 **17a.** 237 ft^2 **17b.** 1 can

MIXED REVIEW **19.** Assume $\angle P \cong \angle N$. By the Converse of the Isos. Triangle Thm., $\overline{MN} \cong \overline{MP}$, which contradicts the hypothesis. Therefore, $\angle P \ncong \angle N$. **21.** $\frac{1}{6}$ **23.** $\frac{1}{2}$

Lesson 6-8 pages 350–352

ON YOUR OWN **1.** $\frac{2}{5}$ or 40% **3.** about 61%
5.
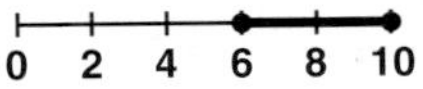

40% **7a.** The erasure must start at least 15 min before the end of the tape.
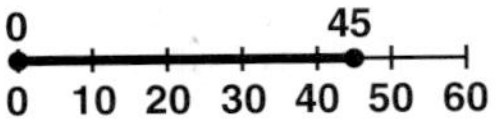

7b.
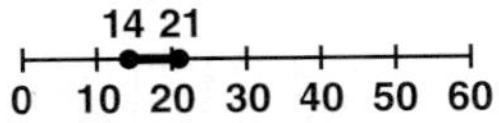

$\frac{7}{45}$ **9.** 4% **11.** about 1.9% **13.** about 26%

MIXED REVIEW **15.** U.S. **17.** 16 cm

Wrap Up pages 354–357

1.
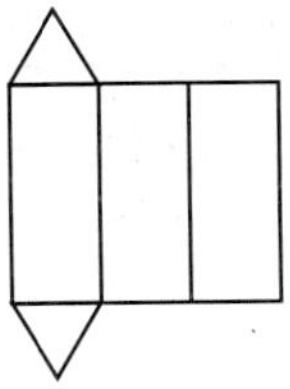
2. D **3.** C **4.** A **5.** B **6.** 36 cm^2 **7.** 66π m^2 **8.** 208 $in.^2$ **9.** 170π ft^2 **10.** 172.8 ft^2; 251.3 ft^2 **11.** 160 cm^2; 224 cm^2 **12.** 37.7 $in.^2$; 50.3 $in.^2$ **13.** 320 m^2; 576 m^2 **14.** 250 ft^3 **15.** 54π cm^3 **16.** 1764 $in.^3$ **17.** 32π m^3 **19.** 235.6 mm^3 **20.** 149.3 ft^3 **21.** 6 m^3 **22.** 301.6 cm^3

23. 314.2 in.^{2}; 523.6 in.^{3} **24.** 153.9 cm^2; 179.6 cm^3 **25.** 50.3 ft^2; 33.5 ft^3 **26.** 8.0 ft^2; 2.1 ft^3 **27.** 8.6 in.^{3} **28.** 263.9 m^3 **29.** 162 cm^3 **30.** 280 in.^{3} **31.** $\frac{1}{2}$ or 50% **32.** $\frac{3}{8}$ or 37.5% **33.** $\frac{1}{6}$ or about 16.7% **34.** No; you model probability by ratio of lengths of segments. **35.** Sample: $\overline{AB}$, $\overline{EF}$, $\overline{CD}$ **36.** Sample: $\overline{AB}$, $\overline{BC}$, $\overline{BF}$

Cumulative Review — page 359

1. E **3.** C **5.** D **7.** C

CHAPTER 7

Toolbox — page 362

INVESTIGATE $m\angle 1 = m\angle 3 = m\angle 5 = m\angle 7$, $m\angle 2 = m\angle 4 = m\angle 6 = m\angle 8$

CONJECTURE Sample: 2 sets of 4 $\cong$ angles are formed.

EXTEND Yes; the angles formed by 3 $\parallel$ lines have a relation similar to the one for the angles formed by 2 $\parallel$ lines; 2

All the angles formed by these lines are rt. angles.

Lesson 7-1 — pages 366–369

ON YOUR OWN **1.** $\angle 1$ and $\angle 2$: corresponding, $\angle 3$ and $\angle 4$: alt. interior, $\angle 5$ and $\angle 6$: corresponding **3.** $\angle 1$ and $\angle 2$: corresponding, $\angle 3$ and $\angle 4$: same-side interior, $\angle 5$ and $\angle 6$: alt. interior **5.** 120 (If $\parallel$ lines, then corres. $\angle$s are $\cong$.); 60 (If $\parallel$ lines, then same-side int. $\angle$s are supplementary.) **7.** 70 (If $\parallel$ lines, then same-side int. $\angle$s are supplementary.); 110 (If $\parallel$ lines, then same-side int. $\angle$s are supplementary.) **9.** If $\parallel$ lines, then alt. int. $\angle$s are $\cong$. **11.** If $\parallel$ lines, then alt. int. $\angle$s are $\cong$. **13.** If $\parallel$ lines, then corres. $\angle$s are $\cong$. **15a.** corresponding **15b.** alt. interior **17a.** $\frac{1}{2}$ **19.** $\angle 1$ and $\angle 9$, $\angle 2$ and $\angle 10$, $\angle 5$ and $\angle 11$, $\angle 6$ and $\angle 12$ **21.** same-side interior angle, ℓ **23.** alt. interior, ℓ **25.** $\angle 3$, $\angle 9$ **27.** $\angle 7$, $\angle 10$ **29a.** *BERT* is a ▱. **29b.** def. of ▱ **29c.** If $\parallel$ lines, then alt. int. $\angle$s are $\cong$. **31.** 70 **33.** 32 **35.** 20 **37.** $v = 42$, $w = 25$, $x = 76$, $y = 37$ **41.** Answers may vary. Sample: Perpendicular parking does not waste any space but makes it harder to park a car. Slanted parking makes it easier to park a car and makes the direction of traffic clear, but there are fewer parking spaces because some space in each corner is wasted.

MIXED REVIEW **43.** (−0.5, 3.5) **45.** (−1.5, −2.5) **47.** 5 **49.** 1790 cm^2; 1937 cm^2 **51.** If there are no clouds in the sky, then the sky is blue.

Toolbox — page 370

1. (−3, −7) **3.** no solution **5.** (8, 17)

Lesson 7-2 — pages 373–376

ON YOUR OWN **1.** $\overleftrightarrow{BE} \parallel \overleftrightarrow{CG}$; if $\cong$ corres. $\angle$s, then lines are $\parallel$. **3.** $\overline{JO} \parallel \overline{LM}$; if supplementary same-side int. $\angle$s, then lines are $\parallel$. **5.** $a \parallel b$; if $\cong$ corres. $\angle$s, then lines are $\parallel$. **7.** $a \parallel b$; if supplementary same-side int. $\angle$s, then lines are $\parallel$. **9.** $\ell \parallel m$; if $\cong$ alt. int. $\angle$s, then lines are $\parallel$. **11.** $a \parallel b$; if supplementary same-side int. $\angle$s, then lines are $\parallel$. **13.** no $\parallel$ lines **15.** Vertical $\angle$s are $\cong$, and if $\cong$ corres. $\angle$s, then lines are $\parallel$.

17.

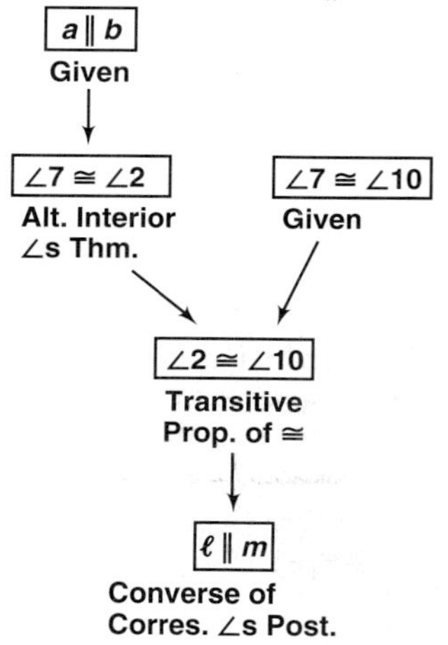

19. If corres. angles are $\cong$, then the lines are $\parallel$. **21.** D **23.** 50 **25.** 31 **27.** 20 **29a.** The $\triangle$s are $\sim$ and isosceles, so the 4 base angles are $\cong$. Then the alt. interior angles are $\cong$, so the lines are $\parallel$. **31.** Each line and the flat surface form a corres. angle that is congruent to the 60° angle of the drawing $\triangle$. Since the corres. angles are $\cong$, the lines are $\parallel$. **33.** Trapezoid; 2 distinct pairs of same-side int. angles are supplementary, so one pair of sides are $\parallel$. **35.** Rectangle; all the angles are rt. angles and 2 pairs of opp. sides are $\parallel$. **37.** 1. $\overline{A'B'}$ is the reflection of $\overline{AB}$ in line m. (Given) 2. $\overline{AA'} \perp m$ and $\overline{BB'} \perp m$ (Def. of reflection) 3. $\overline{AA'} \parallel \overline{BB'}$ (In a plane, 2 lines $\perp$ to 3rd are $\parallel$.)

MIXED REVIEW **39.** 157.1 cm^2 **41.** 2.3 m^2

Lesson 7-3 pages 380–383

ON YOUR OWN

1.

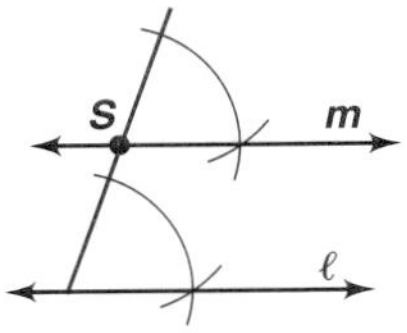

3a.

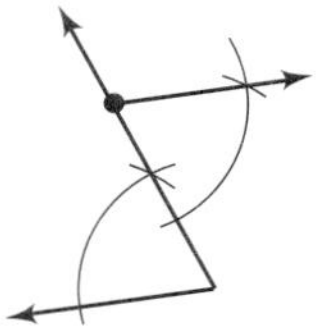

7.

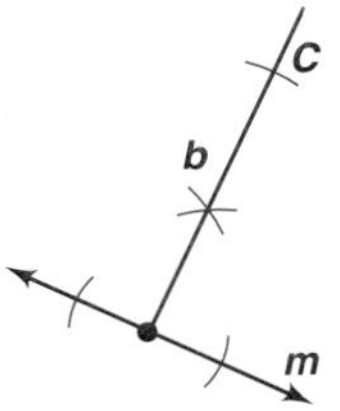

9.

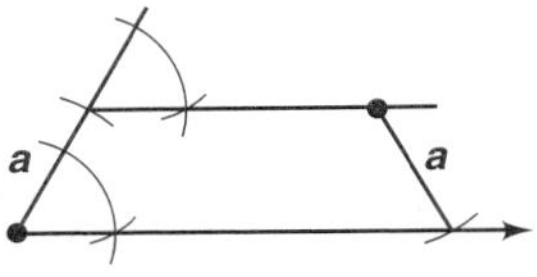

11a–b.

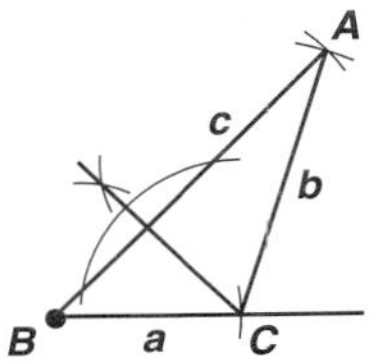

13.

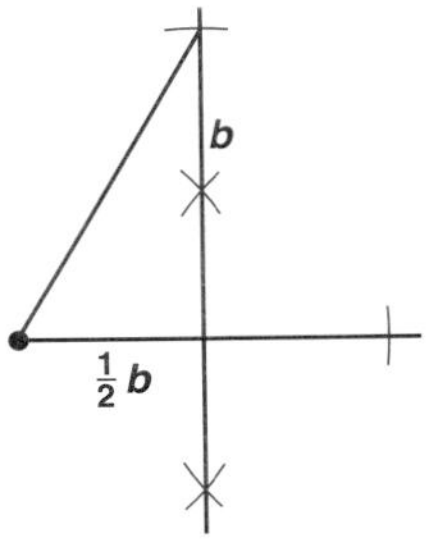

15.

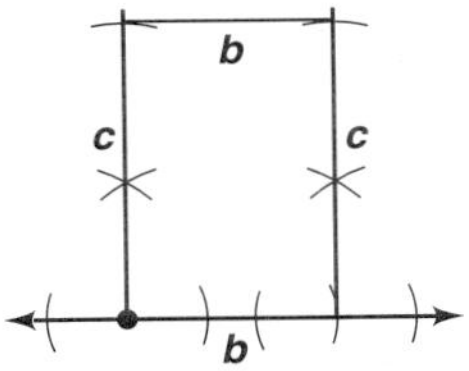

17c. $p \parallel m$; in a plane, 2 lines $\perp$ to 3rd are $\parallel$.

19a. ①-④: construct 2 $\perp$ lines. ⑤: construct a rt. $\triangle$ with a hypotenuse that is twice the length of a leg. **19b.** ①-③: construct an equilateral $\triangle$. ④-⑥: construct an angle bisector. **19c.** ①-③: construct an equilateral $\triangle$. ④-⑤: construct the $\perp$ bisector of a side.

MIXED REVIEW **21.** Cannot form a $\triangle$; $2 + 3 = 5$ **23.** acute; $5^2 < 4^2 + 4^2$ **25a.** $\frac{16}{3}\pi$ cm **25b.** 64π cm^2 **25c.** $\frac{64}{3}\pi$ cm^2

CHECKPOINT **1.** If $\parallel$ lines, then corres. $\angle$s are $\cong$. **2.** If $\cong$ corres. $\angle$s, then lines are $\parallel$. **3.** If $\parallel$ lines, then same-side int. $\angle$s are supplementary. **4.** If $\cong$ alt. int. $\angle$s, then lines are $\parallel$. **5.** Vertical $\angle$s are $\cong$. **6.** If $\parallel$ lines, then alt. int. $\angle$s are $\cong$. **7.** If $\cong$ corres. $\angle$s, then lines are $\parallel$. **8.** If $\parallel$ lines, then corres. $\angle$s are $\cong$. **9.** If supplementary same-side int. $\angle$s, then lines are $\parallel$.

10.

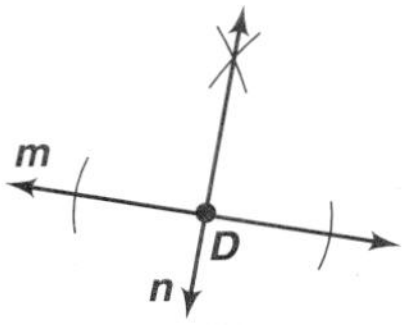

11.

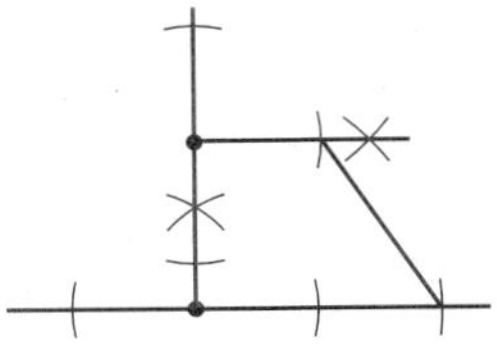

12. C

Toolbox page 384

INVESTIGATE Above and to the left; inside *ABCD;* below and to the right.

Lesson 7-4 pages 388–391

ON YOUR OWN **1.** 2-pt. perspective **3.** 2-pt. perspective **5.** IB, IIC, IIIA

7.

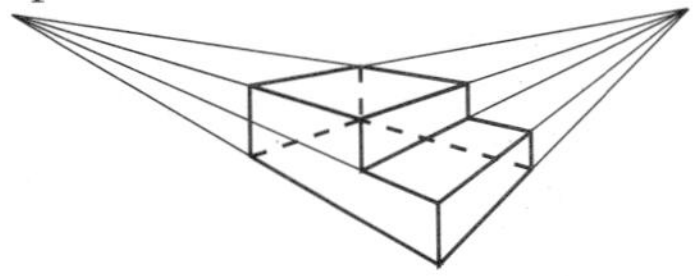

9.

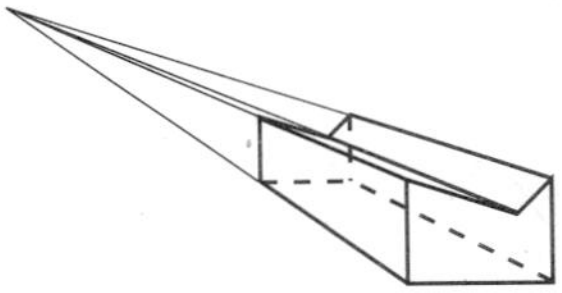

13. The painting is in perspective. The edges of the road converge to a vanishing point.

MIXED REVIEW **21a.** If it is July 4, then it is a national holiday in the United States. **21b.** false; true **23.** a circle with center at $(12, -6)$ and radius 6 **25.** 270

Lesson 7-5 pages 395–397

ON YOUR OWN

1.

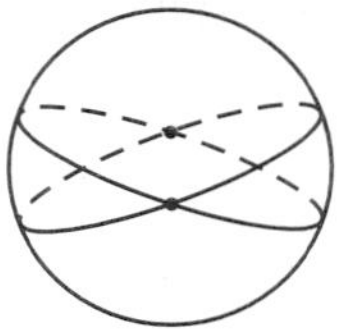

3.

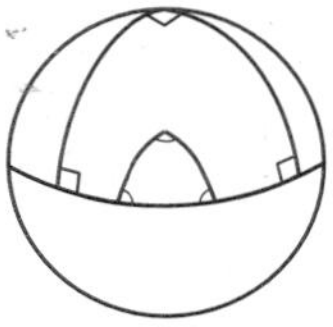

7.

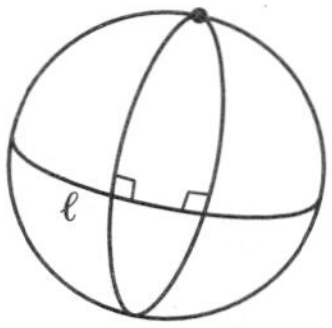

11.

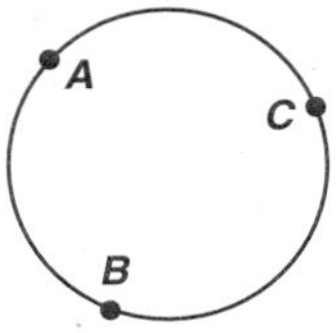

15. true **17.** true **19.** false

MIXED REVIEW

21. yes

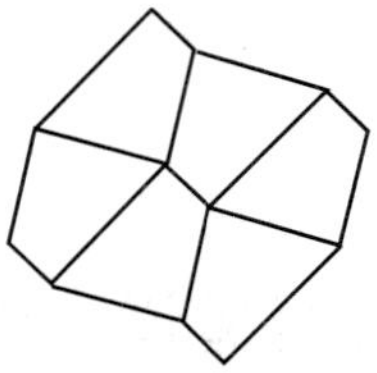

23. yes

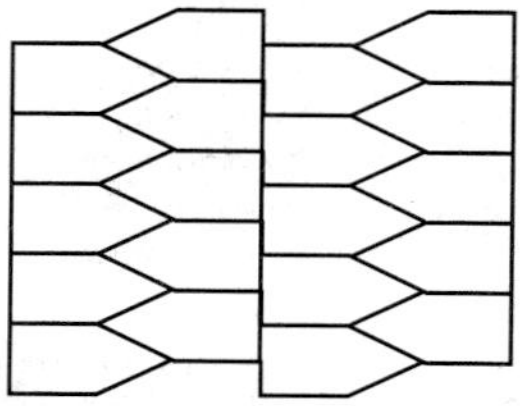

25. $-\frac{2}{3}$ **27.** $\frac{1}{3}$ **29.** 36, 54

Wrap Up — pages 399–401

1. 120, If $\parallel$ lines, then corres. $\angle$s are $\cong$; 120, Vertical $\angle$s are $\cong$. **2.** 75, If $\parallel$ lines, then same-side int. $\angle$s are supplementary; 105, If $\parallel$ lines, then alt. int. $\angle$s are $\cong$. **3.** 55, If $\parallel$ lines, then same-side int. $\angle$s are supplementary; 90, If $\parallel$ lines, then alt. int. $\angle$s are $\cong$. **6.** C **7.** 20 **8.** 20 **9.** 24 **10.**

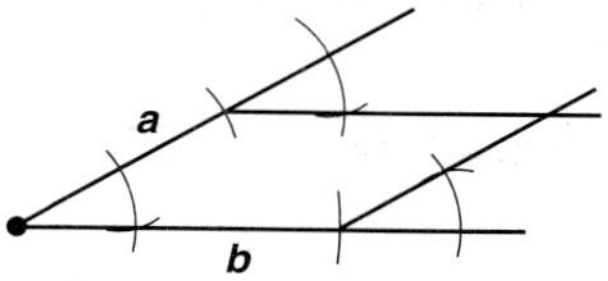

11a.

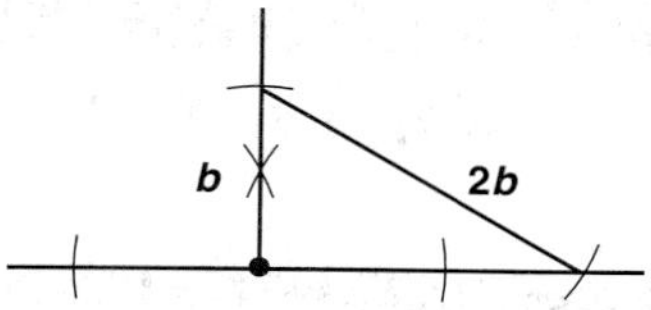

11b. $b\sqrt{3}$
11c. 30, 60, 90

15.

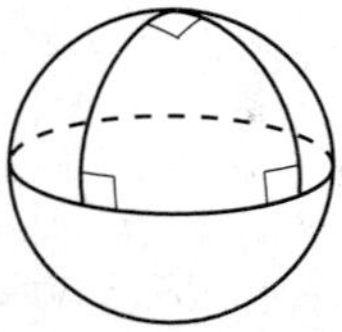

16.

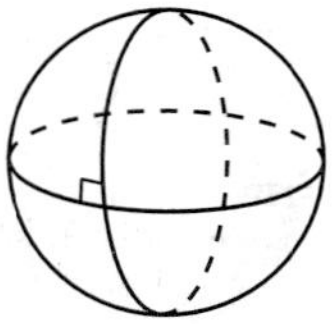

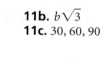

17. See diagram for Exercise 15.
18.

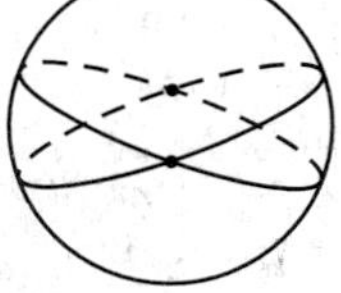

19. $x = 60; y = 30$ **20.** $x = 9; y = 9\sqrt{2}; z = 45$ **21.** $y = 4; z = 2\sqrt{3}$ **22.** $x = 8; y = 55$ **23.** $x = 35; y = 10; z = 10$ **24.** $w = 88; x = 7; y = 52; z = 40$

Preparing for Standardized Tests — page 403

1. C **3.** D **5.** C **7.** B **9.** B

CHAPTER 8

Lesson 8-1 — pages 409–412

ON YOUR OWN **1.** $\overline{XZ}$ **3.** SAS **7.** $\triangle JKL \cong \triangle NMO$; SSS **9.** not possible **11.** $\triangle PQR \cong \triangle NMO$; SAS **13.** $\triangle XYP \cong \triangle QYP$; SAS **17.** yes; SAS Post. **19.** 1. $\overline{AE}$ and $\overline{BD}$ bisect each other. (Given) 2. $\overline{AC} \cong \overline{CE}$; $\overline{BC} \cong \overline{CD}$ (Def. of segment bisector) 3. $\angle ACB \cong \angle ECD$ (Vertical $\angle$s are $\cong$.) 4. $\triangle ACB \cong \triangle ECD$ (SAS) **21a.** $=$ (equality) and $\sim$ (similarity) **23.** 1. $\overline{FG} \parallel \overline{KL}$ (Given)
2. $\angle GFK \cong \angle LKF$ (If $\parallel$ lines, then alt. int. $\angle$s are $\cong$.)
3. $\overline{FG} \cong \overline{KL}$ (Given) 4. $\overline{FK} \cong \overline{FK}$ (Reflexive Prop. of $\cong$)
5. $\triangle FGK \cong \triangle KLF$ (SAS)

MIXED REVIEW **27.** 314.2 mm^3; 282.7 mm^2 **29.** 34, 34

Toolbox — page 413

INVESTIGATE yes; no; yes

INVESTIGATE yes; yes; no; only when the segments coincide

CONJECTURE No; no; the above constructions demonstrate SSA $\triangle$s that are not $\cong$ and AAA $\triangle$s that are not $\cong$.

EXTEND no; no; yes

Lesson 8-2 — pages 416–419

ON YOUR OWN **1.** AAS **3.** AAS **5.** ASA **7.** not possible **9.** $\angle D \cong \angle B$ **11.** $\angle M \cong \angle P$ **15.** yes; SSS **19.** 1. $\overline{FG} \parallel \overline{JH}$ (Given) 2. $\angle FGJ \cong \angle HJG$ (If $\parallel$ lines, then alt. int. $\angle$s are $\cong$.) 3. $\angle F \cong \angle H$ (Given) 4. $\overline{JG} \cong \overline{JG}$ (Reflexive Prop. of $\cong$) 5. $\triangle FGJ \cong \triangle HJG$ (AAS)
21. 1. $\angle MON \cong \angle QOP$ (Vertical $\angle$s are $\cong$.)
2. $\angle N \cong \angle P$ (Given) 3. $\overline{MO} \cong \overline{QO}$ (Given)
4. $\triangle MON \cong \triangle QOP$ (AAS) **23.** The hypotenuse of a rt. $\triangle$ is longer than each leg. So $FE > 5$ cm and $\overline{FE} \not\cong \overline{AC}$.

MIXED REVIEW **25.** (2, 0) **27.** $x = \frac{7\sqrt{3}}{3}; y = \frac{14\sqrt{3}}{3}$ **29.** $x = 4; y = 4\sqrt{2}$

Lesson 8-3 — pages 422–425

ON YOUR OWN **1.** E **3.** $\overline{XV} \cong \overline{TV}$ or $\overline{RX} \cong \overline{RT}$ **5.** $\angle AQC$ and $\angle GJC$ are rt. angles. **7.** $\overline{RT} \cong \overline{NQ}$ **9.** You need to know which side is the hypotenuse. **11b.** $-1; -1; 1$ **11c.** They are rt. $\angle$s. **11d.** $\sqrt{26}; \sqrt{26}$ **11e.** $\triangle EGD$ and $\triangle EGF$ share a common leg $\overline{EG}$. Since the hypotenuses are $\cong$, the right $\triangle$s are $\cong$ by HL Thm. **13.** 1. $\overline{OM}$ and $\overline{ON}$ are radii of $\odot O$. (Given)

2. $\overline{OM} \cong \overline{ON}$ (All radii of a $\odot$ are $\cong$.) 3. $\angle M$ and $\angle N$ are rt. angles. (Given) 4. $\triangle LMO$ and $\angle LNO$ are rt. $\triangle$s. (Def. of rt. $\triangle$s) 5. $\overline{OL} \cong \overline{OL}$ (Reflexive Prop. of $\cong$) 6. $\triangle LMO \cong \triangle LNO$ (HL Thm.) **17.** All sides of a regular polygon are $\cong$. Since the $\triangle$s are rt. $\triangle$s, and the shorter legs are all $\cong$, by HL Thm., all the $\triangle$s are $\cong$.

19.

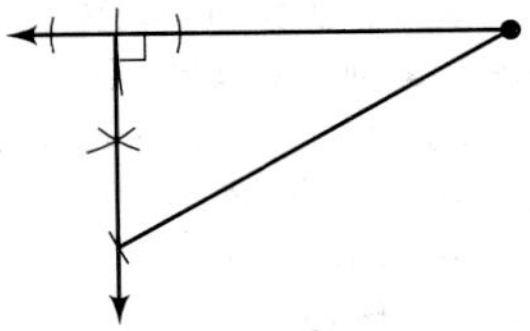

MIXED REVIEW **21.** Angle Addition Post. **23.** If $\parallel$ lines, then same-side int. $\angle$s are supplementary. **25.** If $\parallel$ lines, then corres. $\angle$s are $\cong$. **27.** 7.64 m^2

CHECKPOINT **1.** ASA **2.** SSS **3.** SAS **4.** not possible **5.** AAS **6.** HL **7b.** No; if 2 pairs of angles are $\cong$, the 3rd pair of angles is also $\cong$. Then the $\triangle$s are $\cong$ by AAS or ASA. **8.** E **9.** If the hypotenuse and an acute angle of one right triangle are congruent to the hypotenuse and an acute angle of another right triangle, then the triangles are congruent. The HA Theorem is true. Since the right angles are congruent, the triangles are congruent by AAS.

Lesson 8-4 pages 429–432

ON YOUR OWN **1.** Use ASA to prove $\triangle ABD \cong \triangle CBD$. **3.** Use SAS to prove $\triangle STP \cong \triangle OTP$. **5.** Use HL to prove $\triangle OER \cong \triangle OEP$. **7.** Yes; the $\triangle$s are $\cong$ by SSS, and the angles are $\cong$ by CPCTC. **9a.** Given **9b.** Given **9c.** Def. of angle bisector **9d.** Reflexive Prop. of $\cong$ **9e.** AAS **9f.** CPCTC **11a.** Given **11b.** Def. of $\perp$ lines **11c.** All rt. $\angle$s are $\cong$. **11d.** Given **11e.** Def. of segment bisector **11f.** Reflexive Prop. of $\cong$ **11g.** SAS **11h.** CPCTC **13.** It is given that $\overline{BA} \cong \overline{BC}$ and $\overline{BD}$ bisects $\angle ABC$. $\angle ABD \cong \angle CBD$ by def. of $\angle$ bisector, and $\angle A \cong \angle C$ because base $\angle$s of an isosceles $\triangle$ are $\cong$. Then $\triangle ABD \cong \triangle CBD$ by AAS. $\overline{AD} \cong \overline{CD}$ by CPCTC. Therefore, $\overline{BD}$ bisects $\overline{AC}$ by def. of segment bisector. By the Angle Addition Post., $m\angle ADB + m\angle CDB = 180$. $\angle ADB \cong \angle CDB$ by CPCTC. So, the measure of each angle is $\frac{1}{2}(180) = 90$. Therefore, $\overline{BD} \perp \overline{AC}$ by def. of $\perp$ lines. **15a.** Use SAS on 2 legs and rt. angle. **15b.** $\overline{BF}$; CPCTC **15c.** The $\triangle$s are rt. $\triangle$s with $\cong$ hypotenuses and $\cong$ legs. By HL, $\triangle FAB \cong \triangle BCD$. **17a.** $\overline{PA} \cong \overline{PB}$; $\overline{AC} \cong \overline{BC}$ **17b.** $\overline{CP} \cong \overline{CP}$ by Reflexive Prop. of $\cong$. Then, $\triangle APC \cong \triangle BPC$ by SSS and $\angle APC \cong \angle BPC$ by CPCTC. Therefore, $m\angle APC = m\angle BPC = 90$ by Angle Addition Post., and $\overline{CP} \perp \overline{AB}$ by def. of $\perp$ lines.

MIXED REVIEW **21.** 89% **23.** about 79°

25.

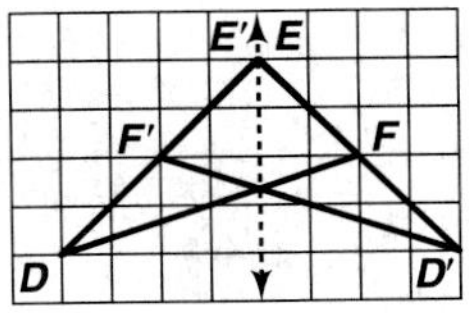

$\angle E$

Lesson 8-5 pages 436–438

ON YOUR OWN **1.** $\angle M$ **3.** $\overline{XY}$ **5.** $\triangle STU$, $\triangle TSR$; SSS **7.** $\triangle PQT$, $\triangle URS$; AAS **9.** $\triangle WYX$, $\triangle ZXY$; HL **11.** Prove $\triangle QTB \cong \triangle QUB$ by ASA. So $\overline{QT} \cong \overline{QU}$. Then use SAS to prove $\triangle QET \cong \triangle QEU$. **13a.** Given **13b.** Reflexive Prop. of $\cong$ **13c.** *ITE* **13d.** SSS **13e.** *TID* **13f.** Given **13g.** All rt. $\angle$s are $\cong$. **13h.** AAS **13i.** $\overline{TD}$; $\overline{RO}$ **13j.** CPCTC **15.** B **17.** Since $\overline{TQ}$ bisects $\overline{PR}$, $\overline{PQ} \cong \overline{RQ}$. Since $\overline{TQ} \perp \overline{PR}$, $\angle PQT$ and $\angle RQT$ are $\cong$ rt. $\angle$s. Also, $\overline{QT} \cong \overline{QT}$, so $\triangle PQT \cong \triangle RQT$ by SAS. $\angle PTQ \cong \angle RTQ$ by CPCTC. Since $\overline{TQ}$ bisects $\angle VQS$, $\angle VQT \cong \angle SQT$. So, $\triangle VQT \cong \triangle SQT$ by ASA, and $\overline{VQ} \cong \overline{SQ}$ by CPCTC.

MIXED REVIEW **19.** isosceles rt. $\triangle$ **21.** 8 units^2; $(8 + 4\sqrt{2})$ units

Toolbox page 439

1. 2, −7 **3.** −0.5, −3 **5.** −0.27, −3.06 **7.** −3, 9 **9.** −0.5, 0.75

Wrap Up pages 441–443

1. SSS; $\triangle DMR \cong \triangle TMR$ **2.** not possible **3.** SAS; $\triangle MNQ \cong \triangle PNO$ **4.** The order of the vertices indicates the corresponding parts. **5.** D **6.** not possible **7.** AAS **8.** ASA **9.** 1. $\overline{PS} \perp \overline{SQ}$ and $\overline{RQ} \perp \overline{QS}$ (Given) 2. $\angle PSQ$ and $\angle RQS$ are rt. $\angle$s. (Def. of $\perp$ lines) 3. $\triangle PSQ$ and $\triangle RQS$ are rt. $\triangle$s. (Def. of rt. $\triangle$s) 4. $\overline{PQ} \cong \overline{RS}$ (Given) 5. $\overline{QS} \cong \overline{QS}$ (Reflexive Prop. of $\cong$) 6. $\triangle PSQ \cong \triangle RQS$ (HL Thm.) **10.** 1. U is the midpt. of $\overline{TV}$. (Given) 2. $\overline{TU} \cong \overline{UV}$ (Def. of midpt.) 3. $\overline{XU} \perp \overline{TV}$ and $\overline{WV} \perp \overline{TV}$ (Given) 4. $\angle XUT$ and $\angle WVU$ are rt. $\angle$s. (Def. of $\perp$ lines) 5. $\triangle TXU$ and $\triangle UWV$ are rt. $\triangle$s. (Def. of rt. $\triangle$s) 6. $\triangle TXU \cong \triangle UWV$ (HL Thm.) **11.** 1. $\overline{LN} \perp \overline{KM}$ (Given) 2. $\angle LNK$ and $\angle LNM$ are rt. $\angle$s. (Def. of $\perp$ lines.) 3. $\triangle KLN$ and $\triangle MLN$ are rt. $\triangle$s. (Def. of rt. $\triangle$s.) 4. $\overline{KL} \cong \overline{ML}$ (Given) 5. $\overline{LN} \cong \overline{LN}$ (Reflexive Prop. of $\cong$) 6. $\triangle KLN \cong \triangle MLN$ (HL Thm.) **13.** Use AAS to show $\triangle TVY \cong \triangle YWX$. $\overline{TV} \cong \overline{YW}$ by CPCTC. **14.** Use ASA to show $\triangle BCE \cong \triangle DCE$. $\angle B \cong \angle D$ by CPCTC. **15.** Use HL to show $\triangle KLM \cong \triangle MNK$. $\overline{KN} \cong \overline{ML}$ by CPCTC. **16.** $\triangle ADB$, $\triangle ACE$; SAS **17.** $\triangle FIH$, $\triangle GHI$; SAS **18.** $\triangle PST$, $\triangle RAT$; ASA **19.** rectangle **20.** square **21.** parallelogram **22.** kite

23. The diagonals are ⊥ bisectors of each other. Each diagonal is an angle bisector of 2 of the angles of the rhombus. The area of the rhombus is half the product of the lengths of the diagonals.

Cumulative Review — page 445

1. E **3.** A **5.** B **7.** A **11.**

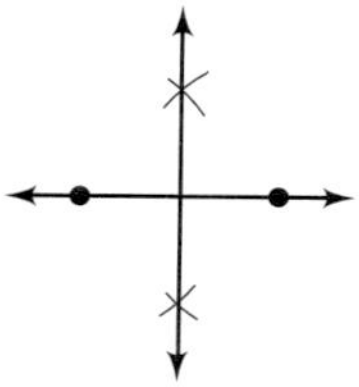

CHAPTER 9

Lesson 9-1 — pages 451–453

ON YOUR OWN **1.** $m\angle 1 = 118$; $m\angle 2 = 62$; $m\angle 3 = 118$ **3.** $m\angle 1 = 81$; $m\angle 2 = 28$; $m\angle 3 = 71$ **5.** 32 **7.** 16 **9.** $ST = 7$ cm; $TW = 17$ cm; $WR = 7$ cm **11.** Pick 6 equally spaced lines on the paper. Place the paper so that the 1st button is on the 1st line and the last button is on the 6th line. Draw a line between the first and last buttons. The remaining buttons should be placed where the drawn line crosses the 4 ∥ lines. **15.** 1. ▱*LENS* (Given) 2. $\overline{LS} \parallel \overline{EH}$ (Def. of ▱) 3. ▱*NGTH* (Given) 4. $\overline{GT} \parallel \overline{EH}$ (Def. of ▱) 5. $\overline{LS} \parallel \overline{GT}$ (2 lines ∥ to a 3rd line are ∥.) **17.** 60 **19.** $x = 109$; $y = 88$; $z = 76$ **21.** $x = y = 6$ **23.** $x = 0$; $y = 5$ **27a.** If a quad. is a ▱, then opp. angles are ≅.

27b.

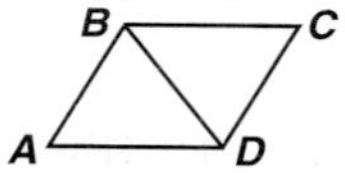

Given: ▱ *ABCD*. Prove: $\angle A \cong \angle C$; $\angle B \cong \angle D$. Draw diagonal $\overline{BD}$. 1. ▱*ABCD* (Given) 2. $\overline{AB} \parallel \overline{CD}$ (Def. of ▱) 3. $\angle ABD \cong \angle BDC$ (If ∥ lines, then alt. int. ∠s are ≅.) 4. $\overline{BC} \parallel \overline{DA}$ (Def. of ▱) 5. $\angle ADB \cong \angle CBD$ (If ∥ lines, then alt. int. ∠s are ≅.) 6. $\overline{BD} \cong \overline{BD}$ (Reflexive Prop. of ≅) 7. $\triangle ABD \cong \triangle CDB$ (ASA) 8. $\angle A \cong \angle C$ (CPCTC) Similarly, use diagonal $\overline{AC}$ to prove $\angle B \cong \angle D$. **29.** 1. ▱*RSTW* and ▱*XYTZ* (Given) 2. $\angle R \cong \angle T$ (Opp. angles of a ▱ are ≅.) 3. $\angle X \cong \angle T$ (Opp. angles of a ▱ are ≅.) 4. $\angle R \cong \angle X$ (Transitive Prop. of ≅) **31.** kite **33.** trapezoid

MIXED REVIEW **41a.** (2.5, 1.5), (2.5, 1.5); $\overline{AC}$ and $\overline{BD}$ bisect each other. **41b.** $\frac{1}{3}$; $\frac{1}{3}$; the slopes are =.
41c. $\overline{AB} \parallel \overline{DC}$ because they are vertical.
41d. parallelogram

Lesson 9-2 — pages 457–460

ON YOUR OWN **1.** Yes; if both pairs of opp. sides of a quad. are ≅, the quad. is a ▱. **3.** Yes; if both pairs of opp. angles of a quad. are ≅, the quad. is a ▱. **5.** Yes; both pairs of opp. sides are ∥ because alt. int. ∠s are ≅. The quad. is a ▱ by def. **7.** Yes; the quad. is a ▱ by def. **9.** yes **11.** yes **13.** yes **15.** 60; yes; both pairs of opp. angles are ≅. **17.** 6; yes; both pairs of opp. sides are ≅. **19.** $x = 3$; $y = 11$ **21.** B **25a.** (4, 0) **25b.** (6, 6) **25c.** (−2, 4) **27.** 1. $\triangle TRS \cong \triangle RTW$ (Given) 2. $\overline{RS} \cong \overline{TW}$; $\overline{ST} \cong \overline{WR}$ (CPCTC) 3. *RSTW* is a ▱. (If both pairs of opp. sides of a quad. are ≅, then the quad. is a ▱.) **29a.** 1. *E* is midpt. of $\overline{BC}$, and *F* is midpt. of $\overline{AD}$. (Given) 2. $BE = \frac{1}{2}BC$; $AF = \frac{1}{2}AD$ (Def. of midpt.) 3. ▱*ABCD* (Given) 4. $AD = BC$ (Opp. sides of a ▱ are ≅.) 5. $\frac{1}{2}AD = \frac{1}{2}BC$ (Mult. Prop. of =) 6. $AF = BE$ (Substitution) 7. $\overline{AD} \parallel \overline{BC}$ (Opp. sides of a ▱ are ≅.) 8. *ABEF* is a ▱. (If 1 pair of opp. sides of a quad. is ≅ and ∥, then the quad. is a ▱.) **29b.** Opp. angles of a ▱ are ≅. **29c.** Opp. sides of a ▱ are ≅. **31a.** ∥ threads remain ∥; the small rectangles are replaced with small nonrectangular ▱s. **31b.** The fabric stretches along one direction of threads and shrinks along the other.

MIXED REVIEW **33.** 50.3 cm^2 **35.** 166.3 ft^2
37. A ▱ is a quad. with 2 pairs of ∥ sides.
39. A rectangle is a ▱ with 4 rt. angles.

CHECKPOINT **1.** $m\angle 1 = 59$; $m\angle 2 = 121$; $m\angle 3 = 59$ **2.** $m\angle 1 = 43$; $m\angle 2 = 62$; $m\angle 3 = 62$ **3.** $m\angle 1 = 106$; $m\angle 2 = 74$; $m\angle 3 = 26$ **4.** $x = 45$; $y = 60$ **5.** $x = 7$; $y = 16$ **6.** $x = 1$; $y = 2$ **7.** D

Toolbox — page 461

INVESTIGATE Sample: The diagonals are ≅. Sample: The diagonals are ⊥.

CONJECTURE Sample: Diags. of a rectangle are ≅; diags. of a rhombus are ⊥ bisectors of each other; diags. of a rhombus bisect the angles of the rhombus.

Sample: All the properties of the diags. of rectangles and rhombuses are true for the diags. of squares.

EXTEND In a ▱, perpendicular diags. determine a rhombus; ≅ diags. determine a rectangle; diags. that bisect the angles of the ▱ determine a rhombus.

Lesson 9-3 pages 465–468

ON YOUR OWN **1a.** rhombus **1b.** $m\angle 1 = 38$; $m\angle 2 = 38$; $m\angle 3 = 38$; $m\angle 4 = 38$ **3a.** rectangle **3b.** $m\angle 1 = 56$; $m\angle 2 = 68$; $m\angle 3 = 112$; $m\angle 4 = 56$ **5.** 26.8 in. **7.** $LB = 10$; $BP = 5$; $LM = 8$ **9.** $LB = 12$; $BP = 12$; $LM = 2\sqrt{11}$ **11.** C **13.** $x = 45$; $y = 45$; $z = 30$ **15.** $x = 7.5$; $y = 3$ **17a.** $\overline{BD}$ **17b.** AE **17c.** EC **17d.** AE; EC **17e.** AE; EC **17f.** AC **19.** 15.9 **21.** Yes; $2.5^2 + 6^2 = 6.5^2$, so the diagonals divide the ▱ into 4 rt. △s. Since the diagonals are ⊥, ▱$JKLM$ is a rhombus. **23.** 24 m^2 **25.** 18 cm^2 **27.** Yes; 4 sides are ≅, so the opp. sides are ≅, and the quad. is a ▱. A ▱ with all sides ≅ is a rhombus. **29.** 1. $\overline{AC} \cong \overline{AC}$ (Reflexive Prop. of ≅) 2. $\overline{AC}$ bisects $\angle BAD$ and $\angle BCD$. (Given) 3. $\angle 1 \cong \angle 2$; $\angle 3 \cong \angle 4$ (Def. of angle bisector) 4. $\triangle ABC \cong \triangle ADC$ (ASA) 5. $\overline{AB} \cong \overline{DA}$; $\overline{BC} \cong \overline{CD}$ (CPCTC) 6. ▱$ABCD$ (Given) 7. $\overline{AB} \cong \overline{CD}$; $\overline{BC} \cong \overline{DA}$ (Opp. sides of a ▱ are ≅.) 8. $\overline{AB} \cong \overline{BC} \cong \overline{CD} \cong \overline{DA}$ (Transitive Prop. of ≅) 9. $ABCD$ is a rhombus. (Def. of rhombus) **31.** In ▱$ABCD$, $\overline{AB} \cong \overline{DC}$ because opp. sides of a ▱ are ≅. $\overline{AD} \cong \overline{AD}$ by the Reflexive Prop. of ≅. Since $\overline{AC} \cong \overline{BD}$, $\triangle BAD \cong \triangle CDA$ by SSS and $\angle BAD \cong \angle CDA$ by CPCTC. $\angle BAD$ and $\angle CDA$ are consecutive angles of a ▱, so they are supp. Therefore, both angles are rt. ∠s and the angles opp. them are rt. ∠s. So ▱$ABCD$ is a rectangle by def. **33a.** Since $ABCD$ is a ▱, opp. ∠s are ≅. So $\angle B \cong \angle D$, therefore $\angle D$ is a rt. angle. In a ▱, consecutive ∠s are supp. Therefore, $\angle A$ and $\angle C$ are each supp. to $\angle B$. The supp. of a rt. angle is a rt. angle, so $\angle A$ and $\angle C$ are rt. ∠s. Therefore, $ABCD$ is a rectangle by def. **33b.** If one ∠ of a ▱ is a rt. angle, then the ▱ is a rectangle.

Toolbox page 469

INVESTIGATE parallelogram

CONJECTURE The quad. whose vertices are midpts. of another quad. is a ▱; yes; no.

EXTEND

- rectangle; rhombus
- The ratio of lengths of sides and perimeters is $\frac{1}{2}$; the ratio of areas is $\frac{1}{4}$; $MNOP \sim FGHE$.

Lesson 9-4 pages 473–476

ON YOUR OWN **1.** $m\angle 1 = 77$; $m\angle 2 = 103$ **3.** $m\angle 1 = 90$; $m\angle 2 = 68$ **5.** $UI = 6$; $IT = 15$; $UT = 21$ **7.** 12 cm, 12 cm, 21 cm, 21 cm **9a.** Isosceles trapezoid; all the large rt. △s appear to be ≅. **9b.** 112; 68; 68 **11a.** No; if one pair of consecutive ∠s is supplementary, then another pair must be also because a pair of opp. ∠s of a kite is ≅. Therefore, the opp. sides are ∥, which means the figure is a ▱ and cannot be a kite. **11b.** Yes; if 2 ≅ angles are rt. angles, they are supplementary. The other 2 angles are also supplementary. **13.** 1.5 **15.** $x = 35$; $y = 30$ **19.** $\overline{AA'} \cong \overline{AA''}$ and $\overline{BA'} \cong \overline{BA''}$, so it is a kite (unless $x = 60$, in which case the 4 sides are ≅ and it is a rhombus). **25a.** isosceles trapezoids **25b.** 45, 135, 135, 45 **27.** 1. Isosceles trapezoid $TRAP$ (Given) 2. $\overline{TA} \cong \overline{PR}$ (Diagonals of an isos. trap. are ≅.) 3. $\overline{TR} \cong \overline{PA}$ (Given) 4. $\overline{RA} \cong \overline{RA}$ (Reflexive Prop. of ≅) 5. $\triangle TRA \cong \triangle PAR$ (SSS) 6. $\angle 1 \cong \angle 2$ (CPCTC)

MIXED REVIEW **29.** $\frac{3}{8}$ **31a.** $x(x + 7)$ or $x^2 + 7x$ **31b.** 78 cm^2 **33.** parallelogram **35.** square

Toolbox page 477

1. 4, $x \neq -3$ **3.** $\frac{x-3}{x+3}$, $x \neq -3$ **5.** $7w + 3$ **7.** $\frac{1}{5t}$, $t \neq 0$ **9.** $4x^2$, $x \neq 3$ **11.** $v + 1$, $v \neq -1$ **13.** $\frac{3}{r}$, $r \neq 0$ **15.** $15c$, $c \neq -2$ **17.** $15x$, $x \neq 0$ **19.** $\frac{3a}{2}$, $a \neq 0$ **21.** $\frac{t+2}{5}$, $t \neq -3$ or 2 **23.** $\frac{1}{2a+10}$, $a \neq -5$ or 5 **25.** $3y^4 + 3y^3$, $y \neq -1$ or 0 **27.** $\frac{3c^2 - 2c - 1}{c^2 + 1}$, $c \neq 1$

Lesson 9-5 pages 480–482

ON YOUR OWN **1.** $W(0, h)$; $Z(b, 0)$ **3.** $W(-b, b)$; $Z(-b, -b)$ **5.** $W(-r, 0)$; $Z(0, -t)$ **7.** $\left(\frac{b}{2}, \frac{h}{2}\right)$, $\sqrt{b^2 + h^2}$ **9.** $(-b, 0)$, $2b$ **11.** $\left(-\frac{r}{2}, -\frac{t}{2}\right)$, $\sqrt{r^2 + t^2}$

15a.

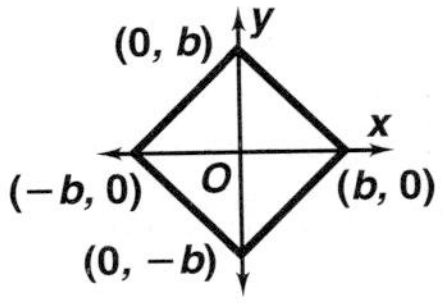

15b. $(-b, 0)$, $(0, b)$, $(b, 0)$, $(0, -b)$ **15c.** $b\sqrt{2}$ **15d.** 1 and −1 **15e.** Yes, because the product of the slopes is −1. **17.** $(c - a, b)$ **19.** $(-b, 0)$ **21a.** $(0, 0, 0)$, $(0, 0, 2a)$, $(0, 2a, 0)$, $(2a, 0, 0)$, $(0, 2a, 2a)$, $(2a, 0, 2a)$, $(2a, 2a, 0)$, $(2a, 2a, 2a)$ **21b.** $(-a, -a, -a)$, $(-a, -a, a)$, $(-a, a, -a)$, $(a, -a, -a)$, $(-a, a, a)$, $(a, -a, a)$, $(a, a, -a)$, (a, a, a)

23.

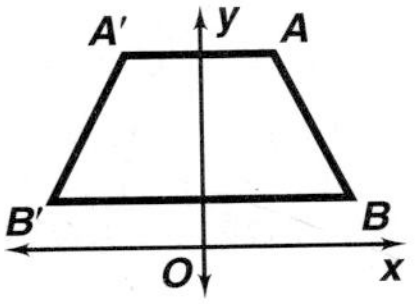

Isosceles trapezoid; the y-axis is the ⊥ bisector of $\overline{AA'}$ and $\overline{BB'}$. In a plane, 2 lines ⊥ to a 3rd line are ∥, so $\overline{AA'} \parallel \overline{BB'}$. Also, $AB = A'B'$.

25a. The diagonals of a rhombus are ⊥.

MIXED REVIEW **27.** $36\sqrt{3}$ in.^{2} **29.** 76 ft^2

Selected Answers

CHECKPOINT **1.** $x = 51; y = 51$ **2.** 3 **3.** $x = 58;$ $y = 32$ **4.** $x = 2; y = 4$ **5.** The diagonals of a rhombus are $\perp$; each diagonal bisects 2 opp. angles of the rhombus.

Lesson 9-6 pages 485–487

ON YOUR OWN **3.** $E(-a, 0), F(-b, c)$; use the Distance Formula to find the lengths of $\overline{EG}$, $\sqrt{(b+a)^2 + c^2}$, and of $\overline{HF}$, $\sqrt{(a+b)^2 + (-c)^2}$. The lengths are =, so the diagonals are $\cong$. **5.** $T(-2a, 0), R(-2b, 2c),$ $D(-a-b, c), E(0, 2c), F(a+b, c), G(0, 0);$ $DE = EF = FG = GD = \sqrt{(a+b)^2 + c^2}$ **7.** $K(-b, a+c), L(b, a+c), M(b, c), N(-b, c)$; slope of $\overline{KL}$ and slope of $\overline{MN}$ are 0. Slope of $\overline{LM}$ and slope of $\overline{NK}$ are undef. Lines with 0 slope and with undef. slope are $\perp$ to each other. Therefore, $KLMN$ is a rectangle. **11a.** $\frac{b}{c}$ **11b.** Let a pt. on line p be (x, y). Then the equation of p is $\frac{y-0}{x-a} = \frac{b}{c}$ or $y = \frac{b}{c}(x-a)$. **11c.** $x = 0$ **11d.** When $x = 0$, $y = \frac{b}{c}(x-a) = \frac{b}{c}(-a) = \frac{-ab}{c}$. So p and q intersect at $\left(0, \frac{-ab}{c}\right)$. **11e.** $\frac{a}{c}$ **11f.** Let a pt. on line r be (x, y). Then the equation of r is $\frac{y-0}{x-b} = \frac{a}{c}$ or $y = \frac{a}{c}(x-b)$. **11g.** $-\frac{ab}{c} = \frac{a}{c}(b-0)$ ✔ $-\frac{ab}{c} = \frac{b}{c}(a-0)$ ✔ **11h.** $\left(0, -\frac{ab}{c}\right)$

MIXED REVIEW **13.** 36% **15.** The percent figure in each category is rounded to the nearest integer.

Wrap Up pages 489–491

1. $m\angle 1 = 101; m\angle 2 = 79; m\angle 3 = 101$ **2.** $m\angle 1 = 71;$ $m\angle 2 = 54; m\angle 3 = 55$ **3.** $m\angle 1 = 38; m\angle 2 = 43;$ $m\angle 3 = 99$ **4.** $m\angle 1 = 52; m\angle 2 = 25; m\angle 3 = 25$ **5.** E **6.** $x = 2; y = 1$ **7.** $x = 29; y = 28$ **8.** $x = 4;$ $y = 5$ **9.** $m\angle 1 = 124; m\angle 2 = 28; m\angle 3 = 62$ **10.** $m\angle 1 = 60; m\angle 2 = 90; m\angle 3 = 30$ **11.** $m\angle 1 = 50;$ $m\angle 2 = 130$ **12.** $m\angle 1 = 90; m\angle 2 = 25$ **13.** 26 in. **14.** 19 ft **15.** 20 cm **16.** 96 ft^2 **18.** (a, b) **19.** $(-a, 0)$ **20.** $(0, c)$ **21.** $(a-b, c)$ **22.** Sample: $A(0, a), B(b, a), C(b, 0), D(0, 0)$; $AC = \sqrt{a^2 + b^2}$, $BD = \sqrt{a^2 + b^2}$, so $AC = BD$, and $\overline{AC} \cong \overline{BD}$. **23.** Sample: $F(0, a), G(a, a), H(a, 0), I(0, 0)$. The slope of $\overline{FH}$ is -1. The slope of $\overline{GI}$ is 1. Since the product of the slopes is -1, $\overline{FH} \perp \overline{GI}$. **31.** 6 **32.** 6 **33.** 3 **34.** 5

Preparing for Standardized Tests page 493

1. E **3.** C **5.** B **7.** C **9.** 140 cm^2

CHAPTER 10

Lesson 10-1 pages 499–502

ON YOUR OWN **1.** true **3.** true **5.** true **7.** true **9.** true **11.** 1 : 360 **13.** no; $\frac{36}{52} \neq \frac{20}{30}$ **15.** $JKLM \sim OPQN$; $\frac{3}{5}$ **17.** No; the corres. angles are not $\cong$. **19.** 6 **21.** 16.5 **23.** 7.5 **25.** -6 or 6 **27.** 3 ft by 2 ft **29.** $(-4, 24), (16, 24)$ **31.** $x = 6; y = 8; z = 10$ **33.** $x = 20; y = 17.5; z = 7.5$ **35.** No; the ratios of the corres. sides are not = ; $\frac{17}{14} \neq \frac{25}{21}$.

39a.

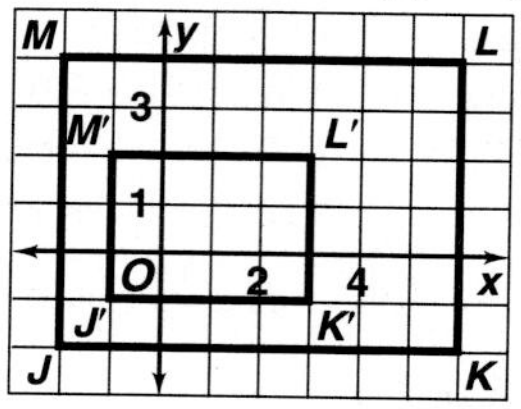

39b. $JKLM \sim J'K'L'M'$ with similarity ratio 2 : 1 **39c.** reduction

MIXED REVIEW **45.** SSS **47.** SAS

Toolbox page 503

1. $t = 0.065c$ **3a.** $t = 1.34n$ **3b.** 14.9 gal

Lesson 10-2 pages 507–510

ON YOUR OWN **1.** no **3.** $\triangle ABC \sim \triangle FED$; SSS $\sim$ Thm. **5.** no **7.** $\triangle MGK \sim \triangle MSP$; SAS $\sim$ Thm. **9.** no **11.** C **13.** 9 **15.** $12\frac{5}{6}$ **17.** 8 **21.** 220 yd **23.** 90 ft **25.** 45 ft **29.** 1. $\overline{BC} \parallel \overline{DF}$ (Given)
2. $\angle YBC \cong \angle YDF$ (If $\parallel$ lines, then corres. $\angle$s are $\cong$.)
3. $\angle DYF \cong \angle DYF$ (Reflexive Prop. of $\cong$)
4. $\triangle BYC \sim \triangle DYF$ (AA $\sim$ Post.) **31a.** 98 m; 98 m **31b.** 420 m^2; 420 m^2 **31c.** No; corres. sides need not be proportional.

MIXED REVIEW **37.** $\pm 2\sqrt{5}$ **39.** $\pm 2\sqrt{10}$

CHECKPOINT **1.** $\triangle ABC \sim \triangle XYZ$; AA $\sim$ Post. **2.** $\triangle LMN \sim \triangle RPQ$; SSS $\sim$ Thm. **3.** $\triangle WST \sim \triangle GJH$; SAS $\sim$ Thm. **4.** $w = 4.5; x = \sqrt{29.25}$ **5.** 12 **6.** 43 ft 9 in.

Lesson 10-3 pages 514–516

ON YOUR OWN **1.** JNK; KNL **3.** $2\sqrt{10}$ **5.** 12 **7.** 121 **9.** b **11.** r, s **13.** c, s **15.** s **17.** 20 **19.** $6\sqrt{3}$ **21.** $(-2, 6), (10, 6)$ **23.** 18 mi; 24 mi **25.** $x = 12; y = 3\sqrt{7}; z = 4\sqrt{7}$ **27.** $x = 4;$ $y = 2\sqrt{13}; z = 3\sqrt{13}$ **29.** 3 **31.** 4.5 **33.** $5\sqrt{3}$ cm

MIXED REVIEW **37.** 1979–1989 **41.** 7.5 mm **43.** 14.5 in.

Lesson 10-4 pages 520–523

ON YOUR OWN **1.** *KS* **3.** *JP* **5.** *KM* **7.** *JP* **9.** 6 **11.** 7.5 **13.** $3\frac{1}{3}$ **15.** 3.6 **17a.** 559 ft **17b.** 671 ft **19.** D **21.** 2.5 **25.** 4.5 cm; 12.5 cm **27a.** If a line that intersects 2 sides of a △ divides them proportionally, then the line is ∥ to the 3rd side.

27b. Sample:

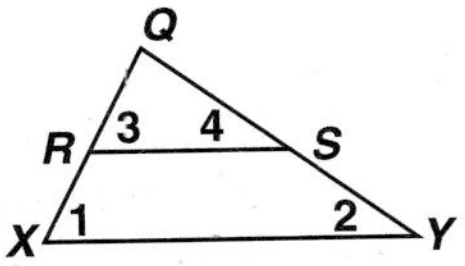

Given: $\frac{QR}{RX} = \frac{QS}{SY}$

Prove: $\overline{RS} \parallel \overline{XY}$

Plan for Proof: To prove that $\overline{RS} \parallel \overline{XY}$, show that $\angle 3 \cong \angle 1$. To prove that $\angle 3 \cong \angle 1$, prove that $\triangle QRS \sim \triangle QXY$ by the SAS ~ Thm. **29.** yes; $\frac{15}{12} = \frac{20}{16}$ **31.** yes; $\frac{45}{63} = \frac{55}{77}$ **33a.** 6 **33b.** 2.5 **33c.** 19.5

MIXED REVIEW **37.** $A'(1, 3)$, $B'(8, 0)$, $C'(-4, -2)$ **39.** $A'(1, -3)$, $B'(8, 0)$, $C'(-4, 2)$ **41.** 49 in.^{2}; 28 in. **43.** 24 cm^2; 24 cm

Lesson 10-5 pages 527–529

ON YOUR OWN **1.** $\frac{1}{2}; \frac{1}{4}$ **3.** $\frac{2}{3}; \frac{4}{9}$ **5.** $\frac{1}{2}$ **7.** $\frac{7}{3}$ **9.** \$384 **11.** 2 : 1 **15.** 12 in.

17a.

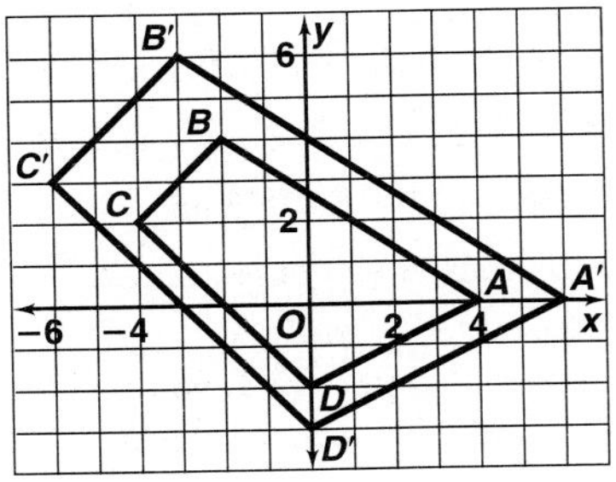

17b. 49.5 units^2 **19.** $\frac{8}{3}; \frac{64}{9}$

MIXED REVIEW **25.** $\sqrt{34}$ in. **29.** 135 m^3; 174 m^2

CHECKPOINT **1.** $x = 15$; $z = 10\sqrt{3}$ **2.** 7.5 **3.** $4\sqrt{5}$ **4.** $x = 1\frac{2}{3}$; $y = 3\frac{3}{5}$ **5.** $\frac{1}{2}$ **6.** 2 **7.** 3 **8.** $\frac{3}{8}$ **9.** D

Toolbox page 530

CONJECTURE The ratio of the surface areas of similar solids is the square of the ratio of their linear dimensions; the ratio of the volumes of similar solids is the cube of the ratio of their linear dimensions.

Lesson 10-6 pages 533–535

ON YOUR OWN **1.** no; $\frac{18}{27} \neq \frac{9}{12}$ **3.** yes; $\frac{4}{6} = \frac{6}{9} = \frac{8}{12}$ **5.** E **7.** $\sqrt[3]{9} : 1$ or $3 : \sqrt[3]{3}$ **9.** about 1000 cm^2 **11.** 9 : 25; 27 : 125 **13.** 5 : 8; 25 : 64 **17.** about 27,000 qt **19a.** 100 times **19b.** 1000 times **19c.** 600 times **19d.** Paul Bunyan's weight is 1000 times the weight of an average person, but his bones can only support 600 times the weight of an average person.

MIXED REVIEW **21.**

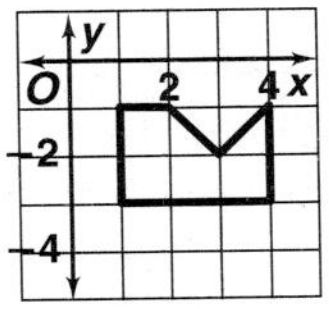

Wrap Up pages 537–539

1. 4 **2.** 9 **3.** $x = 12$; $y = 15$ **4.** $\triangle ABC \sim \triangle FDE$; AA ~ Post. **5.** no **6.** $\triangle XYZ \sim \triangle JKL$; SAS ~ Thm. **7.** C **8.** $x = 15$; $y = 12$; $z = 20$ **9.** $x = 2\sqrt{21}$; $y = 4\sqrt{3}$; $z = 4\sqrt{7}$ **10.** $x = 2\sqrt{3}$; $y = 6$; $z = 4\sqrt{3}$ **12.** 7.5 **13.** 5.5 **14.** 37.5 **15.** 4 : 9 **16.** 9 : 4 **17.** 1 : 4 **18.** 27 : 64 **19.** 64 : 27 **20.** 125 : 343 **22.** $\frac{3}{5}; \frac{4}{5}; \frac{3}{4}$ **23.** $\frac{\sqrt{2}}{2}; \frac{\sqrt{2}}{2}; 1$ **24.** $\frac{\sqrt{3}}{2}; \frac{1}{2}; \sqrt{3}$

Cumulative Review page 541

1. D **3.** B **5.** C **7.** C

CHAPTER 11

Lesson 11-1 pages 547–549

ON YOUR OWN **1.** $\frac{1}{2}$; 2 **3.** 1; 1 **5.** 74.1 **7.** 114.5 **9.** 11.2 **11.** 14.4 **13.** 32 **15.** 48 **17.** 1.6 **19.** 21.4 **21.** about 51° **23.** 52 m

25a.

x	tan $x°$
5	0.1
10	0.2
15	0.3
20	0.4
25	0.5
30	0.6
35	0.7
40	0.8
45	1

x	tan $x°$
50	1.2
55	1.4
60	1.7
65	2.1
70	2.7
75	3.7
80	5.7
85	11.4

25b.

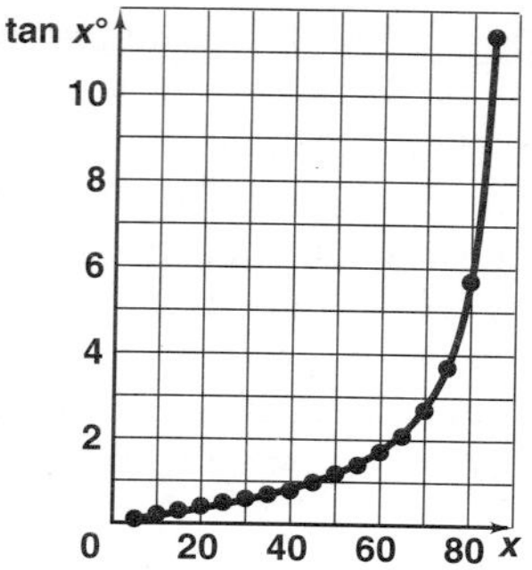

25c. The ratio approaches 0; the ratio increases rapidly. **25d.** Estimates may vary slightly. Sample: 82; 2.5; 74 **27.** $w = 6.7; x = 8.1$ **29.** $w = 59; x = 36$ **31.** 53.1 **33.** 33.7

MIXED REVIEW **37a.** If a flag is a United States flag, then it contains the colors red, white, and blue. **37b.** false; true **39a.** If you live on an island, then you live in Hawaii. **39b.** true; false **43.** $\frac{10}{19}; \frac{3\sqrt{29}}{19}$ **45.** $\frac{9}{15}$ or $\frac{3}{5}$; $\frac{12}{15}$ or $\frac{4}{5}$

Toolbox page 550

INVESTIGATE No; as $m\angle A$ increases, so does the ratio; 0; 1

$m\angle A$	$\frac{DE}{AE}$
10	0.17
20	0.34
30	0.5
40	0.64
50	0.77
60	0.87
70	0.94
80	0.98

CONJECTURE The values match those in the sine column; sine

EXTEND The smaller the measure of the angle, the closer DA is to AE; the values are between 0 and 1.

The greater the measure of the angle, the greater is the ratio; the values are positive and can be arbitrarily large.

$m\angle A$	$\frac{DA}{AE}$
10	0.98
20	0.94
30	0.87
40	0.77
50	0.64
60	0.5
70	0.34
80	0.17

$m\angle A$	$\frac{DE}{DA}$
10	0.18
20	0.36
30	0.58
40	0.84
50	1.19
60	1.73
70	2.75
80	5.67

Lesson 11-2 pages 554–555

ON YOUR OWN **1.** $\frac{7}{25}; \frac{24}{25}$ **3.** $\frac{1}{2}; \frac{\sqrt{3}}{2}$ **7.** 8.3 **9.** 21 **11.** 46 **13.** 106.5 **15.** D **19a.** about 1.51 AU **19b.** about 5.19 AU **21.** $w = 37; x = 7.5$ **23.** $w = 59; x = 20.0$

MIXED REVIEW **25.** $a = 90; b = 40; c = 50$ **27.** $a = 90; b = 12.6; c = 3$ **29.** $\angle 7$ **31.** $\angle 6$ **33.** 180

Lesson 11-3 pages 558–561

ON YOUR OWN **1a.** angle of elevation from the submarine to the boat **1b.** angle of depression from the boat to the submarine **1c.** angle of elevation from the boat to the lighthouse **1d.** angle of depression from the lighthouse to the boat **3.** 4.8° **5.** 560 ft **7.** 986 m **9.** 3.3 km

MIXED REVIEW **15.** about 13.5% **17.** 55–64 **19a.** 35–44 **19b.** 55–64; 55–64 year olds have a higher share of bluegrass sales than their share of population. **21.** (3, −2) **23.** (9, −7)

CHECKPOINT **1.** C **2.** about 17 ft **3.** about 393 m **4.** 15.0 **5.** 61 **6.** 20.8

Toolbox page 562

1. $w = \frac{P - 2\ell}{2}$ **3.** $b = \frac{2A}{h}$ **5.** $b_2 = \frac{2A}{h} - b_1$ **7.** $a = \frac{2A}{p}$ **9.** $r = \sqrt{\frac{A}{\pi}}$ **11.** $w = \frac{3V}{\ell h}$ **13.** $b = c \cos A$ **15.** $\ell = \frac{S - \pi r^2}{\pi r}$

Lesson 11-4 pages 565–567

ON YOUR OWN **1.** $\langle -22, 46 \rangle$ **3.** $\langle -307, -54 \rangle$ **5.** 15° south of west **7.** 20° west of south **9.** 97 mi; 41° south of west **11.** 5300 mi at 26° south of west **15.** The directions of both vectors are the same. The ratio of the magnitude of the image to the magnitude of the orig. vector is k. **17.** No; the description gives the direction but no magnitude. **19.** No; the description gives magnitude but no direction. **21.** about 54 mi/h; 22° north of east **23.** about 0.22 in./h; 32° north of east **25.** Vectors are $\parallel$ if they have the same or opp. directions.

MIXED REVIEW **27.** Neither; you need information about at least 1 acute angle or at least 1 other side. **29.** congruent; AAS **31.** 12.2 m

Lesson 11-5 pages 570–572

ON YOUR OWN **1.** $\langle -2, 1 \rangle$ **3.** $\langle -6, 2 \rangle$ **5.** $\langle -2, -9 \rangle$ **7.** $\langle 1, -1 \rangle$ **9a.** $\frac{2}{3}$ **11a.** 15° south of west **11b.** about 6.7 h or 6 h 40 min

13a.

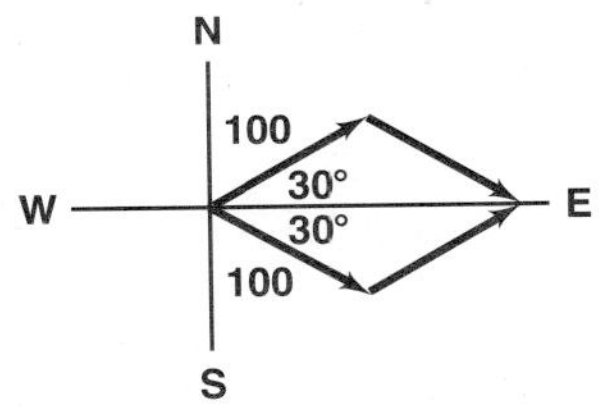

13b. about 173 due east

15a.

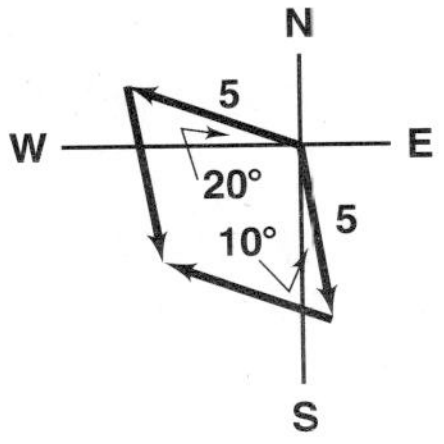

15b. 5 at 40° south of west

17a–b.

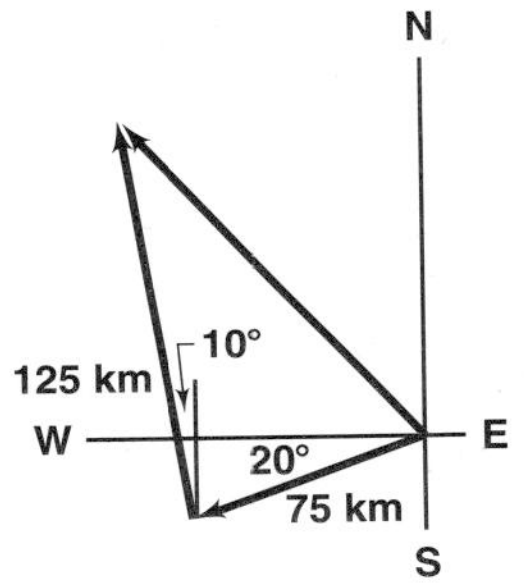

17b. Answers may vary slightly. Sample: 134 km at 43° east of south **19.** white

MIXED REVIEW **21.** 452.4 cm^2 **23.** 5542.6 m^2 **27.** $\overline{GD}$

CHECKPOINT **1.** $\vec{a}$: 6.4; $\vec{c}$: 4.5 **2.** $\langle -2, -9 \rangle$ **3.** 304 mi/h at 9° east of south

Lesson 11-6 pages 575–577

ON YOUR OWN **1.** 83.1 $in.^2$ **3.** 259.8 ft^2 **5.** 47.0 $in.^2$ **7.** 2540.5 yd^2 **9.** 27.7 m^2 **11.** 7554.0 m^2 **13.** 311.3 km^2 **15.** 0.7 ft^2 **17a.** 50 mm^2 **17b.** 116 mm^2 **17c.** 232 mm^3 **19.** 1,459,000 ft^2 **21.** 320 ft

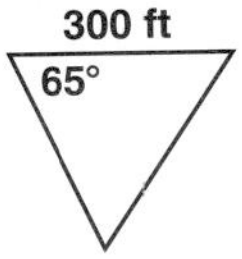

MIXED REVIEW **23.** 5 in. **25.** 8.5 m **27.** 281.5 cm^3; 140.7 cm^2

Wrap Up pages 579–581

1. $\frac{4}{5}; \frac{3}{5}; \frac{4}{3}$ **2.** $\frac{\sqrt{19}}{10}; \frac{9}{10}; \frac{\sqrt{19}}{9}$ **3.** $\frac{\sqrt{3}}{2}; \frac{1}{2}; \frac{\sqrt{3}}{1}$ **4.** $\frac{8}{17}; \frac{15}{17}; \frac{8}{15}$ **5.** 51 **6.** 16.5 **7.** 33 **8.** 13.8 **9.** D **10.** 1410 ft **11.** 280 ft **13.** $\langle 126, 82 \rangle$ **14.** $\langle 37, -93 \rangle$ **15.** $\langle -22, 34 \rangle$ **16.** $\langle -206, -283 \rangle$ **17.** about 168 mi at 27° east of south **18.** about 206 km at 14° west of south **19.** about 503 mi/h at 27° north of west **20.** about 175 m/h at 31° east of north **21.** Multiply each coordinate by the same positive number. **22.** No; the description gives magnitude but no direction. **23.** Yes; the description includes magnitude and direction. **24.** No; the description gives magnitude but no direction. **25.** $\langle 1, 4 \rangle$ **26.** $\langle 4, -6 \rangle$ **27.** $\langle 2, 0 \rangle$ **28.** $\langle 1, -1 \rangle$ **29a.** 5° west of south **29b.** about 3 h **30.** 73.5 ft^2 **31.** 232.5 cm^2 **32.** 80.9 $in.^2$ **33.** 100.8 cm^2 **34.** 88.4 ft^2 **35.** 70.4 m^2 **37.** 25π $in.^2$; 10π in. **38.** 16π cm^2; 8π cm **39.** 121π m^2; 22π m **40.** 4π ft^2; 4π ft

Preparing for Standardized Tests page 583

1. D **3.** C **5.** A **7.** E **9.** C

CHAPTER 12

Lesson 12-1 pages 588–591

ON YOUR OWN **1.** $(x - 2)^2 + (y + 8)^2 = 81$ **3.** $(x - 0.2)^2 + (y - 1.1)^2 = 0.16$ **5.** $(x + 6)^2 + (y - 3)^2 = 64$ **7.** B **9.** C **11.** $(-7, 5)$; 4 **13.** $(0.3, 0)$; 0.2

15. 6

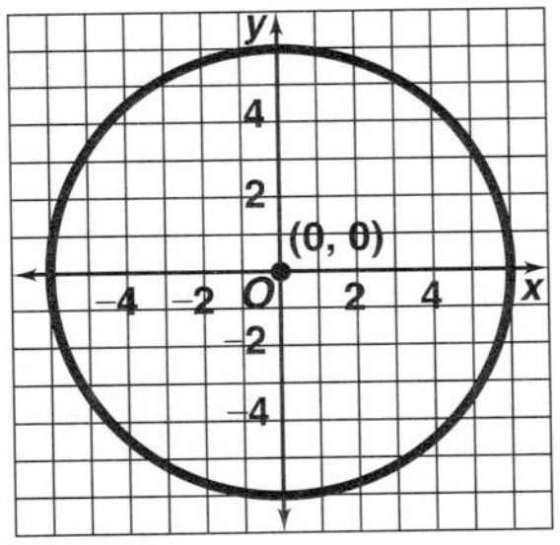

17. 3

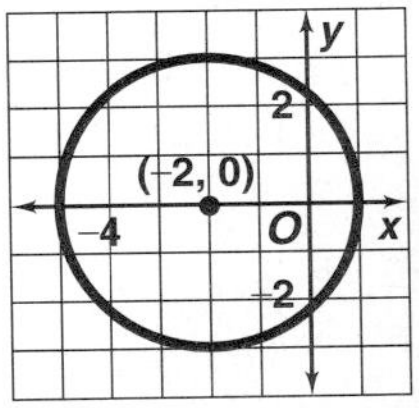

19a. $\langle -5, 3 \rangle$ **19b.** $(0, 0)$; 8 **21.** $x^2 + y^2 = 9$ **23.** $(x - 2)^2 + (y - 2)^2 = 16$ **25.** $(x + 2)^2 + (y - 3)^2 = 9$ **27.** $(x + 2)^2 + (y - 6)^2 = 16$ **29.** $(x - 7)^2 + (y + 2)^2 = 52$ **31.** $(x - 3)^2 + (y + 5)^2 = 100$ **33a.** $x^2 + y^2 = 15{,}681{,}600$ **33b.** 69.1 mi **33c.** 1.2 mi **33d.** about 32 days **35.** $(x - 5)^2 + (y - 3)^2 = 13$ **37.** $(x + 3)^2 + (y + 1.5)^2 = 6.25$ **39.** $(x - 2)^2 + (y + 2)^2 = 41$ **43a.** $\sqrt{6}$ **43b.** $(x + 1)^2 + (y - 3)^2 + (z - 2)^2 = 6$

MIXED REVIEW **45.** $\angle AEB \cong \angle DEC$ because vertical $\angle$s are $\cong$. Then $\triangle ABE \cong \triangle DCE$ by AAS, and $\overline{BE} \cong \overline{CE}$, $\overline{AE} \cong \overline{DE}$ by CPCTC. $\overline{AC} \cong \overline{DB}$ by the Segment Addition Post. $\overline{BC} \cong \overline{BC}$ by Reflexive Prop. of $\cong$. Then $\triangle ABC \cong \triangle DCB$ by SSS. **47.** 512 in.^{3} **49.** 17°; 107° **51.** 3°; 93°

Toolbox — page 592

1. (3, 120°) **3.** (5, 315°) **5.** (2, 240°) **13.** Read the measures of the central $\angle$s for points X and Y from the graph, and subtract the measures.

Lesson 12-2 — pages 596–599

ON YOUR OWN **1.** 63 **3.** 72 **5.** 13 **7.** 9 **9b.** the lines bisecting the angles formed by the axes **9c.** $y = x, y = -x$ **11.** D **13a.** 14.5 in. **13b.** 16.5 in. **15.** 14.2 in. **17.** 1. $\overline{BC}$ is tangent to $\odot A$ at D. (Given) 2. $\overline{BC} \perp \overline{AD}$ (A tangent to a $\odot$ is $\perp$ to the radius drawn to the pt. of tangency.) 3. $\overline{DB} \cong \overline{DC}$ (Given) 4. $\overline{AD}$ is the $\perp$ bisector of $\overline{BC}$. (Def. of $\perp$ bisector) 5. $\overline{AB} \cong \overline{AC}$ (A pt. on the $\perp$ bisector of a segment is equidistant from the endpts. of the segment.) **19.** They are $\cong$; $\overline{CE} \cong \overline{FE}$ and $\overline{FE} \cong \overline{DE}$ because 2 segments tangent to a $\odot$ from a pt. outside the $\odot$ are $\cong$; $\overline{CE} \cong \overline{DE}$ by Transitive Prop. of $\cong$. **21.** Blue segments are common external tangents to the sun and the moon; green segments are common internal tangents to the sun and the moon; red segments are common external tangents to the sun and Earth. **23.** 4 **25.** about 34.6 in.

MIXED REVIEW **27.** $\frac{13}{17}; \frac{2\sqrt{30}}{17}$ **29a.** rectangle **31.** 8.1

Lesson 12-3 — pages 603–606

ON YOUR OWN **1.** 6 **3.** 5 **5.** 7 **7.** 18.8 **9a.** 5.9 cm **11.** 90 **13.** about 123.9 **15.** 6 in. **17.** $8\sqrt{3}$ cm **21.** 1. $\overline{CE} \perp \overline{BD}$ (Given) 2. $\overline{BF} \cong \overline{FD}$ (A diameter $\perp$ to a chord bisects the chord.) 3. $\overline{CE}$ is the $\perp$ bisector of $\overline{BD}$. (Def. of $\perp$ bisector) 4. $\overline{BC} \cong \overline{DC}$ (A point on the $\perp$ bisector of a segment is equidistant from the endpts. of the segment.) 5. $\overset{\frown}{BC} \cong \overset{\frown}{DC}$ ($\cong$ chords have $\cong$ arcs.) **23a.** $\overline{EC}; \overline{ED}; \overline{BC}; \overline{BD}$ **23b.** center of the $\odot$ **23c.** A diameter $\perp$ to a chord bisects the chord and its arc; the $\perp$ bisector of a chord contains the center of the $\odot$. **25.** The thms. are not true for angles, chords, or arcs that are in noncongruent $\odot$s. **27a.** 2 **27b.** 3.5 cm and 15.5 cm

MIXED REVIEW **29.** obtuse **31.** obtuse **33.** acute **35.** 80 **37.** 125

CHECKPOINT **1.** $(x - 1.5)^2 + (y - 0.5)^2 = 2.5$ **2.** $(x - 3.5)^2 + (y - 1)^2 = 46.25$ **3.** $(x + 1.5)^2 + (y + 4)^2 = 22.25$ **4.** 76 cm **5.** 48 in. **6.** 51 m **7.** 24 **8.** 5 **9.** 8 **10b.** 3 **10c.** Kites; the sides inside the $\odot$ are $\cong$ because they are radii of the same $\odot$; the sides outside the $\odot$ are $\cong$ because they are tangent segments to a $\odot$ drawn from a pt. outside the $\odot$.

Lesson 12-4 — pages 610–613

ON YOUR OWN **1.** inscribed in **3.** circumscribed about **5.** 58 **7.** $a = 218; b = 109$ **9.** $a = 85;$ $b = 47.5; c = 90$ **11.** $a = 112; b = 124; c = 42$ **13.** 1. $\angle ABC$ is inscribed in $\odot O$. (Given) 2. $m\angle ABP = \frac{1}{2}m\overset{\frown}{AP}$ and $m\angle PBC = \frac{1}{2}m\overset{\frown}{PC}$ (Inscribed Angle Thm., Case I) 3. $m\angle ABP + m\angle PBC = \frac{1}{2}m\overset{\frown}{AP} + \frac{1}{2}m\overset{\frown}{PC}$ (Addition Prop. of =) 4. $m\angle ABP + m\angle PBC = \frac{1}{2}(m\overset{\frown}{AP} + m\overset{\frown}{PC})$ (Distributive Prop.) 5. $m\angle ABP + m\angle PBC = m\angle ABC$ (Angle Addition Post.) 6. $m\overset{\frown}{AC} = m\overset{\frown}{AP} + m\overset{\frown}{PC}$ (Arc Addition Post.) 7. $m\angle ABC = \frac{1}{2}m\overset{\frown}{AC}$ (Substitution)

15. 1. $m\overset{\frown}{AD} = m\overset{\frown}{BC}$ (Given) 2. $m\angle ABD = \frac{1}{2}m\overset{\frown}{AD}$ and $m\angle BAC = \frac{1}{2}m\overset{\frown}{BC}$ (Measure of an inscribed $\angle$ is half the measure of its intercepted arc.) 3. $m\angle BAC = m\angle ABD$ (Substitution) 4. $\overline{AD} \cong \overline{BC}$ ($\cong$ arcs have $\cong$ chords.) 5. $\angle ACB \cong \angle BDA$ (2 inscribed angles that intercept the same arc are $\cong$.) 6. $\triangle ABD \cong \triangle BAC$ (AAS) **17a.** rectangle **17b.** Opp. angles are $\cong$ and supplementary. So they are rt. angles. **19.** Draw the diameter through the given pt. Construct the line $\perp$ to the diameter through this pt. **21a.** 360

21b. $m\overset{\frown}{DAB} + m\overset{\frown}{BCD} = 360$. $m\angle A = \frac{1}{2}m\overset{\frown}{BCD}$ and $m\angle C = \frac{1}{2}m\overset{\frown}{DAB}$, so $m\angle A + m\angle C = \frac{1}{2} \cdot 360 = 180$. $\angle A$ and $\angle C$ are supplementary. **23.** $\angle ACB$ is inscribed in a semicircle so it is a rt. angle. Then $\overleftrightarrow{BC}$ is $\perp$ to radius $\overline{AC}$ and passes through its endpt. on the $\odot$. Therefore, $\overleftrightarrow{BC}$ is tangent to $\odot A$. **25a.** 30 **25b.** 78 **25c.** 95 **25d.** 105 **25e.** 85 **25f.** 75 **27a.** 64 **27b.** 116 **27c.** 58 **27d.** 58 **27e.** 58 **27f.** 64 **27g.** 32 **31.** 180

MIXED REVIEW **33.** 5 cm **35.** 108 **37.** 130

Lesson 12-5 — pages 617–619

ON YOUR OWN **1.** 46 **3.** $x = 60; y = 70$ **5.** $x = 100;$ $y = 30$ **7.** $x = 50; y = 97.5$ **9a.** 160° **9b.** about 18,800 mi **11.** A **13a.** $360 - x$ **13b.** $180 - x$ **13c.** $180 - y$ **15a.** $m\angle X = 30; 30 < m\angle Y < 180;$ $0 < m\angle Z < 30$ **15b.** The ship is in safe waters if the measure of the angle is ≤ 30. **17.** $x = 80$; $y = 50$; $z = 90$

MIXED REVIEW **19.** 35% **21.** 25–44 **23.** *HF*; *UT*

CHECKPOINT **1.** 58 **2.** $a = 30$; $b = 42$; $c = 80$; $d = 116$ **3.** 30 **4.** $a = 140$; $b = 70$; $c = 47.5$

5. A secant extends outside the ⊙, a chord does not.

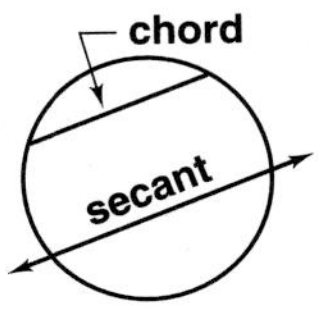

6. D

Toolbox page 620

INVESTIGATE The 2 products are =.

CONJECTURE The products of the lengths of segments formed by the intersection of 2 chords in a ⊙ are =.

INVESTIGATE The 2 products are =.

CONJECTURE When 2 secants, such as $\overleftrightarrow{GF}$ and $\overleftrightarrow{EB}$ above, intersect at a point D outside the circle, $DG \cdot DF = DE \cdot DB$.

EXTEND The product of the lengths of the segments formed by the secant is = to the square of the length of the tangent segment. When $\overleftrightarrow{DF}$ becomes tangent, pts. *F* and *G* coincide. So $DF \cdot DG = DF^2 = DG^2$.

Lesson 12-6 pages 623–626

ON YOUR OWN **1.** 11.5 **3.** 7.8 **5.** $x = 25.8$; $y = 12.4$ **7.** about 271 ft **9.** 1. $\overline{AC}$ and $\overline{BC}$ are secants of ⊙*O*. (Given) 2. $\angle CAE \cong \angle CBD$ (2 inscribed ∠s that intercept the same arc are ≅.) 3. $\angle ACE \cong \angle DCB$ (Reflexive Prop. of ≅) 4. $\triangle AEC \sim \triangle BDC$ (AA~) 5. $\frac{EC}{DC} = \frac{AC}{BC}$ (In similar figures, corres. sides are proportional.) 6. $BC \cdot EC = AC \cdot DC$ (Prop. of Proportions) **11.** 26.7 **13.** 14.1 **15.** 172,000 mi **17c.** The center has the greatest product of the lengths of the segments of chords passing through it. **19.** $x = 8.9$; $y = 2$ **21a.** 1.1 in.

21c.

23a. If a tangent and a secant are drawn from a pt. outside a ⊙, then the product of the lengths of the secant segment and its external segment = the square of the length of the tangent segment.

23b. Substitution **23c.** Distributive Prop.

23d. Addition Prop. of =

MIXED REVIEW **25.** 40.3

Wrap Up pages 628–630

1. $(x - 2)^2 + (y - 5)^2 = 9$
2. $(x + 3)^2 + (y - 1)^2 = 5$
3. $(x - 9)^2 + (y + 4)^2 = 12.25$
4. $x^2 + (y - 1)^2 = 80$ **5.** $(x + 2)^2 + (y - 3)^2 = 85$
6. $(x - 10)^2 + (y - 7)^2 = 468$ **7.** E **8.** 58 in.

9. 84 mm **10.** 9.6 cm **11.** In Ch. 11, a tangent of an angle is the ratio of the length of the opp. leg of a rt. △ to the length of the adjacent leg. In Ch. 12, a tangent line to a ⊙ is a line in the plane of the ⊙ that intersects the ⊙ at exactly 1 pt. **12.** 4.3 **13.** 19.5 **14.** 6.4 **15.** 4.5 **16.** No; the arcs have different radii so they cannot be congruent. **17.** $a = 40$; $b = 140$; $c = 90$ **18.** $a = 118$; $b = 49$; $c = 144$; $d = 98$ **19.** $a = 34$; $b = 68$ **20.** $a = 90$; $b = 90$; $c = 70$; $d = 65$ **21.** $a = 95$; $b = 85$ **22.** 37 **23.** 80 **24.** $x = 57$; $y = 44.5$; $z = 129$; $v = 51$ **25.** 9.5 **26.** 4 **27.** 17.1 **28.** 8.4

Cumulative Review pages 632–633

1. E **3.** A **5.** E **7.** B **9.** D **11.** B **13.** B **15.** C **19.** 4 **21.** 27,475 ft^2

EXTRA PRACTICE

Chapter 1 page 634

1. 37, 42 **3.** $8, \frac{8}{5}$ **5.** 0.0003, 0.000 03 **7.** true **9.** false **11.** false **13.** true **15.** 27 **17.** 6

19. 40

21.

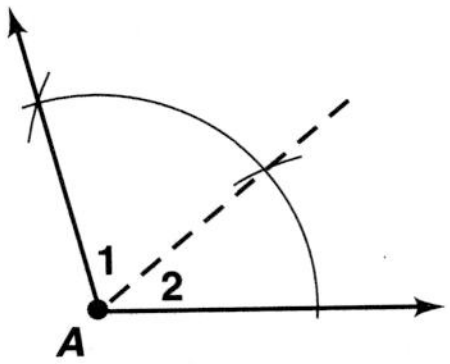

23.

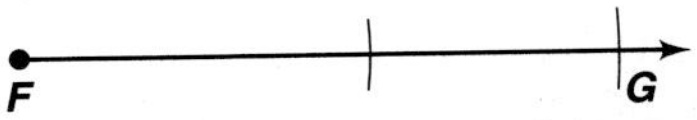

25. 15 **27a.** 1.4 **27b.** (2.5, 0.5) **29a.** 19.1 **29b.** (−0.5, 3) **31a.** 9.5 **31b.** (1.5, −0.5)

Chapter 2 page 635

1. scalene, obtuse **3.** isosceles, acute **5.** 100 **7.** 65 **9.** trapezoid **11.** parallelogram **13.** Sample: $\widehat{ED}$, $\widehat{AE}$, $\widehat{BA}$ **15.** Sample: $\widehat{ED}$, $\widehat{AE}$ **17.** Sample: $\angle APB$ **19.** $\angle G$ **21.** $\angle T$ **23.** $\triangle TAS$

Selected Answers

25.

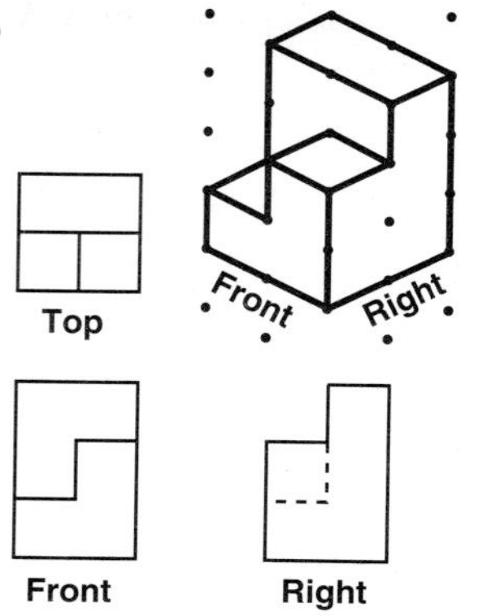

27.

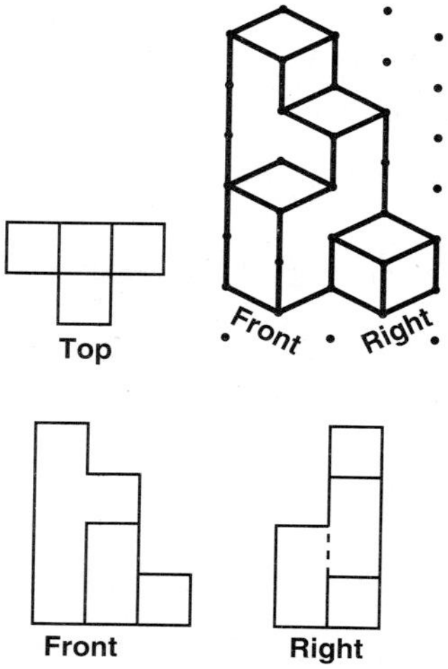

9. E 11. C 13. G 15.

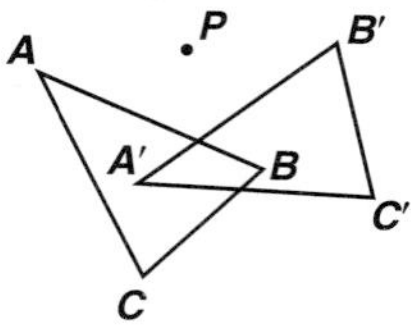

17.

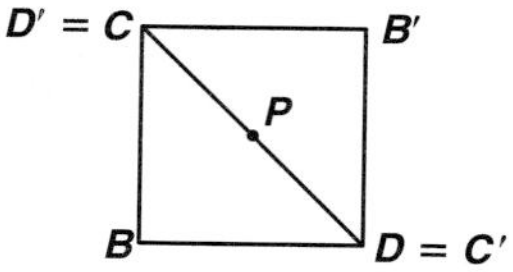

19. rotation
21. translation
23a. line, rotational, point
23b. yes 25a. line
25b. yes

Chapter 4 page 637

1. If 2 angles are ≅, then they are vert.; if 2 angles are not vert., then they are not ≅; if 2 angles are not ≅, then they are not vert. 3. If a car has no doors, then it is blue; if a car is not blue, then it has doors; if a car has doors, then it is not blue. 5. 10 7. 65 9. Assume $\triangle ABC$ is not a rt. △. 11. Assume lines ℓ and m are not ∥. 13. $\overline{JB}, \overline{BP}, \overline{PJ}$ 15. $\overline{CT}, \overline{TA}, \overline{AC}$ 17.

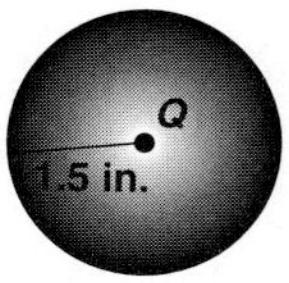

19. altitude
21. angle bisector

Chapter 5 page 638

1. 42 in.; 98 in.^{2} 3. 10 cm; 5 cm^2 5. 15 7. $3\sqrt{5}$ 9. 72 cm^2 11. $\frac{25\sqrt{3}}{4}$ mm^2 13a. 6π cm 13b. 2π cm 15a. 18π cm 15b. $\frac{9\pi}{2}$ cm 17. $\frac{49\pi}{3}$ ft^2 19. $\frac{81\pi}{8}$ cm^2

Chapter 6 page 639

1. cube 3. cylinder 5. 84 ft^2; 108 ft^2 7. 40π in.^{2}; 56π in.^{2} 9. 16 mm^3 11. 15π m^3 13. $\frac{500\pi}{3}$ cm^3; 100π cm^2 15. $\frac{4\pi}{3}$ ft^3; 4π ft^2 17. $\frac{243\pi}{2}$ m^3; 81π m^2 19. 64π ft^3 21. $(60 + 6\pi)$ in.^{3} 23. $\frac{1}{3}$ 25. $\frac{7}{24}$

Chapter 7 page 640

1. $m\angle 1 = 134$; if ∥ lines, then same-side int. ∠s are supplementary; $m\angle 2 = 46$, Angle Addition Post. (or, if ∥ lines, then alt. int. ∠s are ≅). 3. $m\angle 1 = 58$; if ∥ lines, then alt. int. ∠s are ≅; $m\angle 2 = 122$, Angle Addition Post. (or, if ∥ lines, then same-side int. ∠s are supplementary). 5. none 7. $c \parallel d$; if supplementary same-side int. ∠s, then lines are ∥. 9. $r \parallel s$; if ≅ corres. ∠s, then lines are ∥. 11. none

Chapter 3 page 636

1.

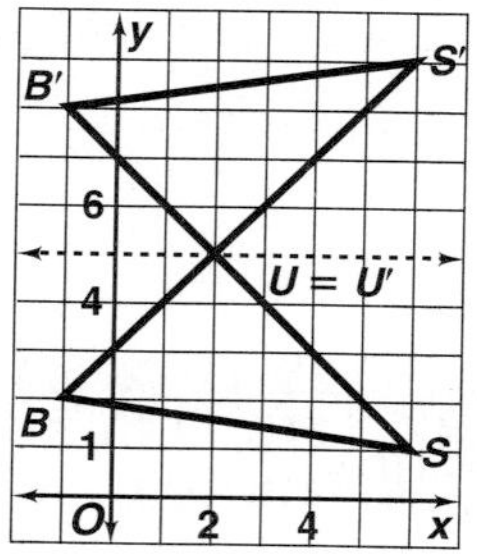

3.

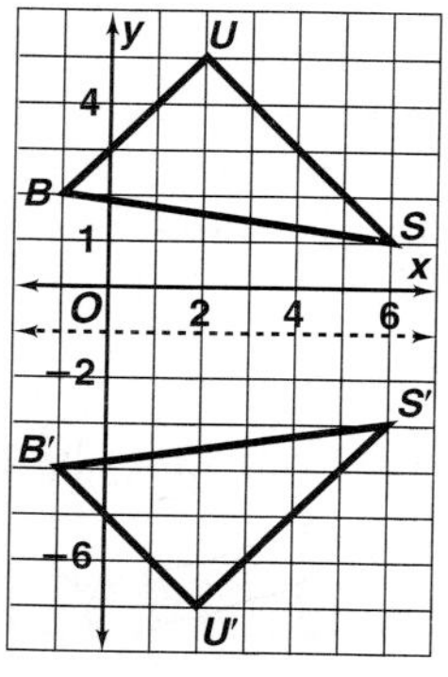

5.

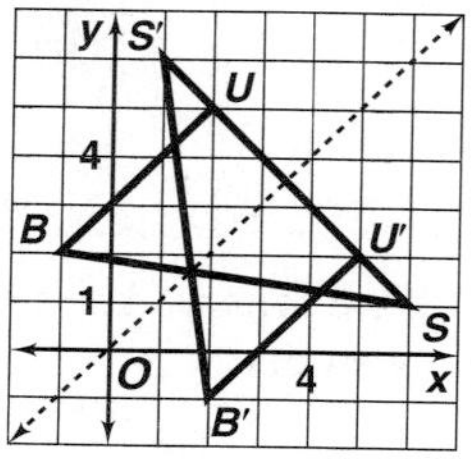

7.

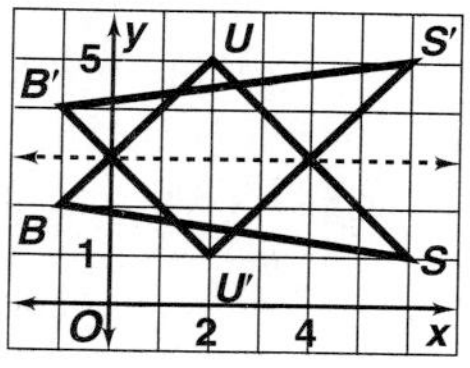

13.

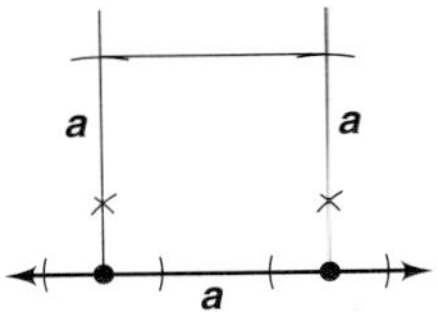

15.

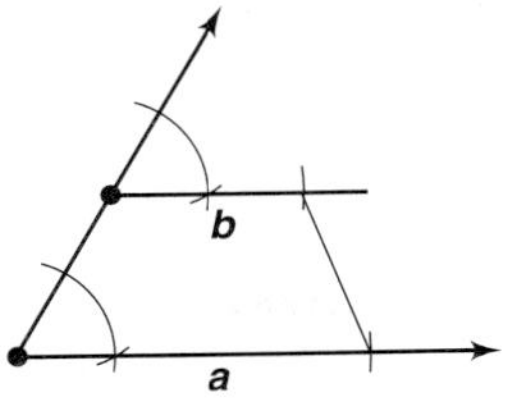

17. 1-pt. **19.** 2-pt.

21.

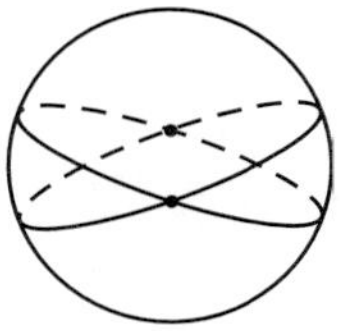

23. Sample: The measure of an exterior angle with vertex *C* is clearly less than $m\angle A + m\angle B$.

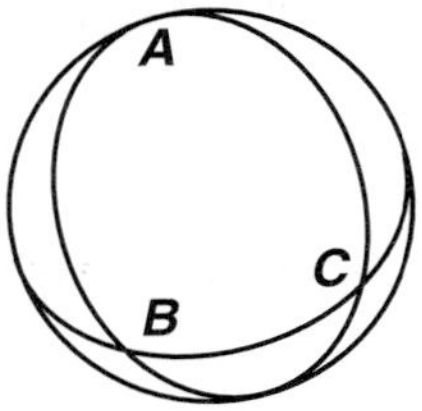

Chapter 8 page 641

1. $\triangle ALE \cong \triangle LAP$; SAS **3.** not possible **5.** $\triangle BOI \cong \triangle TOW$; SAS **7.** $\overline{ED} \cong \overline{RF}$ **9.** $\overline{KS} \cong \overline{PJ}$ **11.** $\overline{OS} \cong \overline{OS}$ by Reflexive Prop. of $\cong$. Since $\angle T \cong \angle E$ and $\angle TSO \cong \angle EOS$, $\triangle TSO \cong \triangle EOS$ by AAS, and $\overline{TO} \cong \overline{ES}$ by CPCTC. **13.** $\triangle QRM \cong \triangle RQS$; SSS **15.** $\triangle FAD \cong \triangle EBC$; AAS

Chapter 9 page 642

1. $x = 12$; $y = 84$ **3.** $x = 30$; $y = 55$ **5.** yes; def. of ▱ **7.** Yes; Triangle Angle-Sum Thm., if both pairs of opp. ∠s of a quad. are $\cong$, then the quad. is a ▱. **9.** square; $m\angle 1 = 45$; $m\angle 2 = 45$ **11.** rectangle; $m\angle 1 = 116$; $m\angle 2 = 64$; $m\angle 3 = 32$; $m\angle 4 = 58$ **13.** $m\angle 1 = 110$; $m\angle 2 = 70$ **15.** $m\angle 1 = 70$; $m\angle 2 = 70$ **17.** $D(0, b)$, $S(a, 0)$ **19.** $D(-c, 0)$, $S(0, -b)$ **21.** Given: Square *DRSQ* with *K*, *L*, *M*, *N* midpts. of $\overline{DR}$, $\overline{RS}$, $\overline{SQ}$, and $\overline{QD}$, respectively. Prove: *KLMN* is a square.

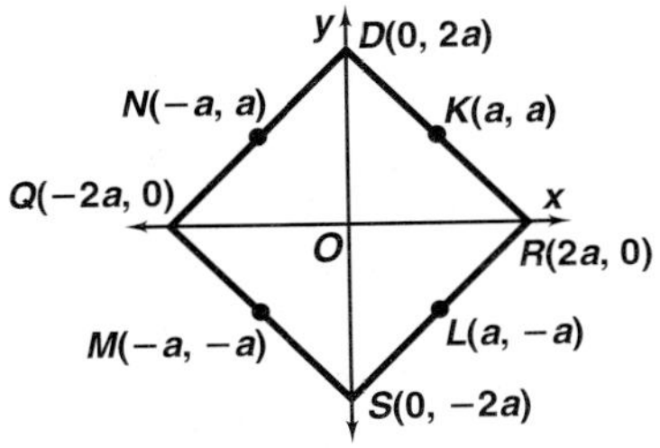

$K(a, a)$, $L(a, -a)$, $M(-a, -a)$, and $N(-a, a)$ are midpts. of the sides of the square. $KL = LM = MN = NK = 2a$. The slopes of $\overline{KL}$ and $\overline{MN}$ are undefined. The slopes of $\overline{LM}$ and $\overline{NK}$ are 0, so adjacent sides are ⊥ to each other. Since all angles are rt. angles, the quad. is a rectangle. A rectangle with all $\cong$ sides is a square.

Chapter 10 page 643

1. 10 **3.** 2 **5.** 4.5 **7.** $\triangle QCT \sim \triangle MCP$; SAS~ **9.** not ~ **11.** 14.4 **13.** $x = \sqrt{70}$; $y = \sqrt{21}$; $z = \sqrt{30}$ **15.** $x = 65$; $y = 60$; $z = 156$ **17.** $18\frac{2}{3}$ **19.** 4 : 3; 16 : 9 **21.** 4 : 9; 8 : 27 **23.** 3 : 4; 9 : 16

Chapter 11 page 644

1. 5.6 **3.** 9.4 **5.** 29 **7.** 50 **9.** 653 ft **11a.** $\langle -49, 142\rangle$, $\langle 38, 47\rangle$ **11b.** $\langle -11, 189\rangle$ **13a.** $\langle -54, 72\rangle$, $\langle -95, -33\rangle$ **13b.** $\langle -149, 39\rangle$ **15.** 30.1 ft^2 **17.** 43.2 cm^2 **19.** 31.2 ft^2

Chapter 12 page 645

1. $x^2 + y^2 = 16$ **3.** $(x - 9)^2 + (y + 3)^2 = 49$ **5.** $(x + 6)^2 + (y + 2)^2 = 13$ **7.** 65 **9.** 6 **11.** 14.8 **13.** 5.3 **15.** $a = 154$; $b = 76$ **17.** $a = 105$; $b = 100$ **19.** $x = 193$; $y = 60.5$ **21.** 10.4 **23.** $x = 112.5$; $y = 67.5$ **25.** 42.5

SKILLS HANDBOOK

page 646

EXERCISES **1.** 10 handshakes **3.** 16 regions

page 647

EXERCISES **1.** 6, 8, 10 **3.** any real number between 0 and 4 **5.** 2 rolls with 36 exposures and 5 rolls with 24 exposures **7.** at least 6

page 648

EXERCISES **1.** 165 toothpicks **3.** 3280 triangles

page 649

EXERCISES **1.** 28 posts **3.** 30 people **5.** 25 trapezoids **7.** 101 s

page 650

EXERCISES **1.** Izzy is the dog; J.T. is the canary; Arf is the goldfish; Blinky is the hamster. **3.** 14 students

page 651

EXERCISES **1.** Route 90 east, Route 128 north, Route 4 north **3.** 23 d **5.** \$2000 **7.** 11.4 mi

page 652

EXERCISES **1.** 39 mm; 51 mm **3.**

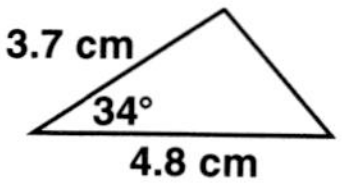

5.

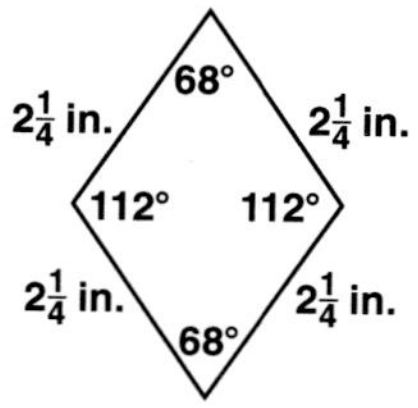

7.

page 653

EXERCISES **1.** 0.4 **3.** 600 **5.** 1008 **7.** $1\frac{5}{16}$ **9.** 34,000 **11.** 4.3 **13.** 15,000 **15.** 30,000 **17.** $2\frac{1}{4}$ **19.** 0.012 **21.** 72 **23.** 1,080,000 **25.** 12.6 **27.** $144\frac{4}{9}$ **29.** about 0.72 **31.** 2419.2 **33.** 13,000,000

page 654

EXERCISES **1.** $23\frac{1}{2}$ ft to $24\frac{1}{2}$ ft **3.** $339\frac{1}{2}$ mL to $340\frac{1}{2}$ mL **5.** 73.15 cm to 73.25 cm **7.** 5.35 mi to 5.45 mi **9.** 8.7 cm to 9.1 cm **11.** 208 m

page 655

EXERCISES **1.** 353.6; 301; no mode **3.** \$40,533; \$28,150; \$18,000 **5.** Median; the mode does not represent most of the data; the mean is significantly affected by the single highest salary. **7.** $67\frac{6}{17}$; 67; 66 **9.** Yes; the mode would be 67.

page 656

EXERCISES **3.** 8 A.M.–2 P.M. **5.** No; you cannot tell how the temp. changed between the measurements.

page 657

EXERCISES **1.** 212; 76; 29; 29; 14

3.

4-yr college	211
2-yr college	75
Tech. college	29
Work	29
Undecided/Other	14

page 658

EXERCISES **1a.** 16.6 **1b.** 15.1 **1c.** 19.4

page 659

EXERCISES **1.** −50 **3.** 15 **5.** 2 **7.** 36 **9.** −2 **11.** 243 **13.** −20 **15.** −4 **17.** B **19.** $2\ell + 2w$ **21.** $-4x - 7$ **23.** $x - 5$ **25.** $r^2 - 2r + 1$ **27.** $y^2 - 2y - 3$ **29.** −1 **31.** $2\pi hr^2 - 4\pi hr + 2\pi h$ **33.** $-x^2 - 8x - 16$

page 660

EXERCISES **1.** $3\sqrt{3}$ **3.** $5\sqrt{6}$ **5.** $2\sqrt{2}$ **7.** $\frac{\sqrt{10}}{5}$ **9.** $\frac{3\sqrt{2}}{4}$ **11.** $5\sqrt{10}$ **13.** $4\sqrt{14}$ **15.** $4\sqrt{10}$ **17.** $2\sqrt{29}$

page 661

EXERCISES **1.** $\frac{5}{3}$ **3.** $\frac{1}{2x}$ **5.** $\frac{6}{7}$ **7.** $\frac{1}{16}$ **9.** $\frac{3}{10}$ **11.** $\frac{r}{24}$ **13.** $\frac{x-3}{3x-2}$ **15.** 9 **17.** $\frac{5}{12}$ **19.** $\frac{5}{13}$ **21.** $\frac{12}{5}$ **23.** $\frac{2}{\pi}$ **25.** $\frac{1}{9}$

page 662

EXERCISES **1.** 17.5 **3.** −4 or 4 **5.** 3 **7.** 16.5 **9.** 78 **11.** 3 **13.** 26.5 **15.** 1 **17.** 12,480 km **19.** 906.25 ft^2

page 663

EXERCISES **1.** 5 **3.** 3 **5.** −2 **7.** −2 **9.** $-1\frac{3}{8}$ **11.** $-\frac{1}{4}$ **13.** 56 **15.** −5

17. $x \geq 2$

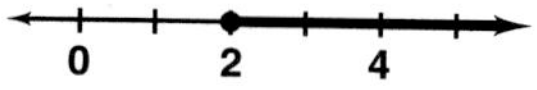

19. $t \geq 4$

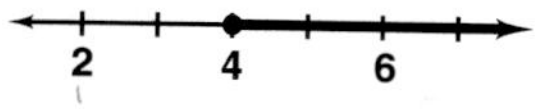

21. $y < -3$

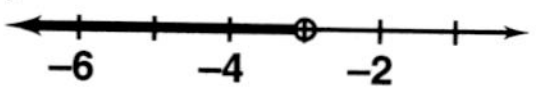

23. $k \leq -2$

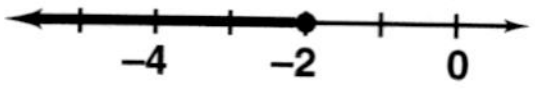

page 664

EXERCISES **1.** −2 **3.** −1 **5.** 0 **7.** undef. **9.** $\frac{1}{3}$; −1 **11.** no intercept; −1 **13.** 1; 1 **15.** −4; no intercept **17.** 1; $-\frac{1}{2}$ **19.** −4; $2\frac{2}{9}$ **21.** undef.; $-\frac{2}{3}$; 0

page 665

EXERCISES **1.**

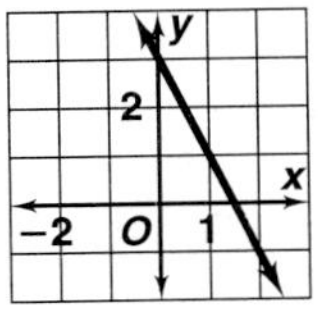

3.

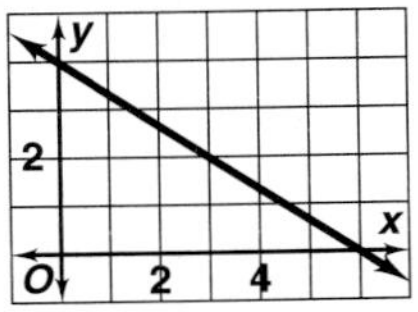

5.

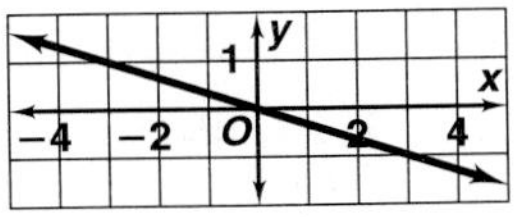

7.

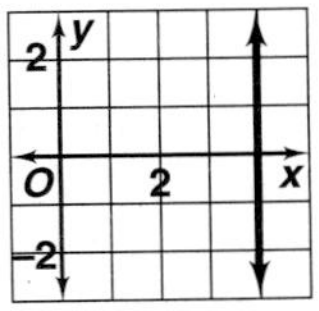

9.

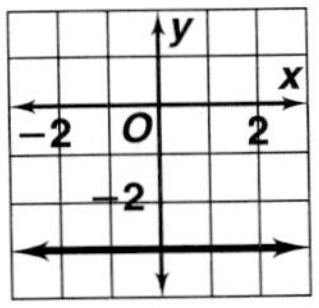

11.

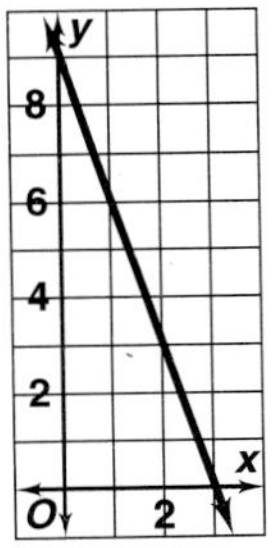

13.

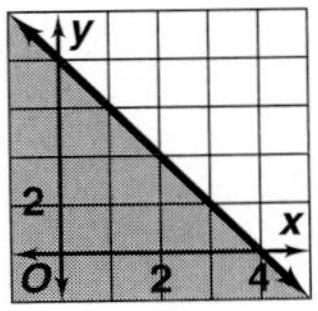

15.

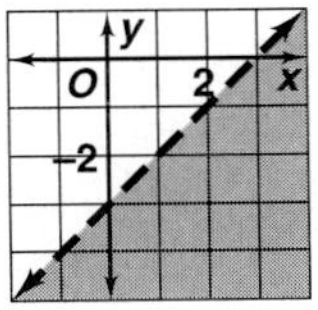

17.

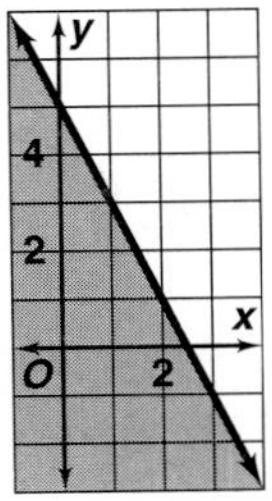

19.

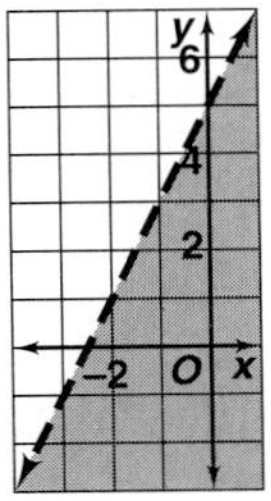

page 666

EXERCISES **1.** $w = \frac{p - 2\ell}{2}$ **3.** $r = \frac{1}{2}\sqrt{\frac{S}{\pi}} = \frac{\sqrt{\pi S}}{2\pi}$ **5.** $d_2 = \frac{2A}{d_1}$ **7.** $a = \frac{2A}{p}$ **9.** $h = \frac{2A}{b}$ **11.** $h = \frac{3V}{\ell w}$ **13.** $b = c \cos A$ **15.** $\ell = \frac{S - \pi r^2}{\pi r} = \frac{S}{\pi r} - r$ **17.** $B = \frac{3V}{h}$ **19.** $\ell = \frac{2S - 2B}{p}$

page 667

EXERCISES **1.** (4, −1) **3.** (4, 1) **5.** $\left(\frac{3}{4}, 1\frac{1}{4}\right)$ **7.** $\left(-\frac{1}{3}, \frac{5}{6}\right)$

Selected Answers

Index

Index

Index

Index

Acknowlegments

Cover Photos

Ge and background, Bill Westheimer; concept car, Ron Kimba

Technical Illustratn

Fran Jarvis, Technical Illustration Manager; AD/Outlook

Illustration

Leo Abbett: 97 ml, 205 t, 3 b
Susan Avishai: 427
Kim Barnes: 24, 199
Tom Barrett: 186, 341 b(it)
Judith Pinkham-Cataldo: 2 inset, 613
Jim DeLapine: 228, 289 b, 9, 340, 390 m
Kathleen Dempsey: 486
Peggy Dressel: 167 b
Howard S. Friedman: 365
Function Thru Form Inc.: , 134, 220 t, 303 all, 312 b, 313 t, 336, 341 t, 345 all, 346 all
Dave Garbot: 247 t, 253 t
GeoSystems Global Corporion: 127 b
Dale Glasgow & Associates: 59, 432
Kelley Hersey: 97, 509
Linda Johnson: 156 m, 157 b
Ellen Korey-Lie: 429, 572
Seymour Levy: 129 m, 150 m 186 inset, 246, 341 b
Andrea G. Maginnis: 17 t, 30, 90, 137 t
Ortelius Design Inc.: 13 t, 24 nset, 270 b, 271, 395
Gary Phillips: 30 m
Lois Leonard Stock: 162 b, 16 b, 508 all, 556 all
Peter Siu: 590
Gary Torrisi: 205 t, 270 t, 352 516, 558 all, 560 m, 563 all, 598 b
Gregg Valley: 224 t
Joe Veno: 4, 32

Feature Design

Alan be Associates

Photography

Photo Research

Sue McDermott

Abbreviations: JC = Jon Chomitz; FPG = Freelance Photographer's Guild; KO = Ken O'Donoghue; PR = Ph Researchers, Inc.; PH = Prentice Hall File Photo; SB = Stock Boston; SM = The Stock Market; MT = ' Thayer; TSI = Tony Stone Images

Front matter: Page vii, David Young-Wolff/P...dit; **viii,** Jerry Jacka; **ix,** Photofest; **x,** Superstock; **xi,** Tony...man/PhotoEdit; ...iv, Stephen **xii,** Tom & Pat Leeson/DRK Photo; **xiii,** ...amp; **xvi,** Alan and Frisch/SB; **xv,** Geoffrey Clifford/Wood...l/SB; **xviii,** Steve Linda Detrick/PR; **xvii,** Bob Daemr Niedorf/The Image Bank.

Chapter 1: Pages 2-3, Anselm...ring/The Image Bank; **2 & 3 insets,** Used with permissi...of Sterling Publishing Co., Inc., 387 Park Ave. Sl, NY, NY...016 from BEST EVER PAPER AIRPLANES by Norm...n Schmidt. ©1994 by Norman Schmidt. A Sterling/Tamos B...ok; **2 & 3 insets,** JC; **5,** David Young-Wolff/PhotoEdit; **8,** Tom Pantages; **9,** JC; **12,** Jack Newton/Masterfile; **16,** Johnny Johnson/DRK Photo; **18,** Superstock; **19,** John Gerlach/DRK Photo; **20,** JC; **21,** Will Ryan/SM; **22,** KO; **26,** Mark Bolster/International Stock Photo; **27,** JC; **29 t,** Glyn Kirk/TSI; **29 b,** Ralph Cowan/FPG; **33,** Michael Hart/FPG; **34,** JC; **36,** KO; **38,** Dan McCoy/SM; **39,** JC; **41,** JC; **44,** KO; **46,** Larry Grant/FPG; **48,** Telegraph Colour Library/FPG; **55,** Mike Agliolo/International Stock Photo; **56,** Paul Chesley/TSI; **57,** Tom Pantages.

Chapter 2: Pages 66-67, Jason Hawkes/TSI; **67 inset,** Russ Lappa **68,** KO; **70,** Sunstar/The Picture Cube; **72,** John Coletti/The Picture Cube; **73 t,** Tom Van Sant/Geosphere Project, Santa Monica/PR; **73 bl & br,** Jerry Jacka; **76,** John Elk/SB; **78,** P Pasley/SB; **79 t,** KO; **79 bl,** Superstock; **79 bml,** Nawroc Photo/Picture Perfect; **79 bmr,** Jerry Jacka; **80,** Barb Adams/FPG; **83,** Gary Brettnacher/TSI; **85,** PhotoN ART, Jacob Albert; **87 t,** B. Busco/The Image Bank; **8** Courtesy Rossotto/SM; **90,** KO; **93,** Michael Nelson; **9** **0,** David copyright ©1996: WHITNEY MUSEUM ASA; **104,** NEW YORK. Photography by Sheldan C SI; **106,** Rod of Paramount's Carowinds; **98,** Dale **109,** ©1996, Microsoft Weintraub/SB; **101,** David Weintr IMAGE (R/3D); **110,** Rosanne Olson/TSI; **105,** Cathl MT; **113,** Reprinted by Planck/Tom Stack & Associat Corporation. Image create es; **115 t,** MT; **115 b,** Rick Barry Durand/Odyssey/C Permission: Tribune MT; **123,** Bob Daemmrich/SB; **124,** Altman/SM.

Chapter 3: Pag ...berts; **129,** Biblioteca Ambrosiana, ...m Art Library; **130,** MAURICE HORN, KO; **126,** Fr...CYCLOPEDIA OF COMICS (PAGE 41), Milan/Th permission of Chelsea House Publishers; **132,** THE W **136,** Chris Michaels/FPG; **140,** NASA; **142,** Jerry repri **5 l,** JC; **145 r,** Alfred Pasieka/SPL/PR; **147 bl,** Viviane Ph SM; **147 bm,** Paul Jablonka/International Stock Photo; **br,** Adam Peirport/SM; **149,** KO; **152 t,** Keystone/Sygma; **152** ...JC; **153,** PH; **154,** Frank Fornier/Contact Press Images; **155 tl,** John Kaprielian/PR; **155 tr,** Floyd Dean/FPG; **155 bl,** Michael Simpson/FPG; **155 br,** Dave Gleiter/FPG; **156,** PH; **159 t,** Tom Pantages; **160,** KO; **161,** KO; **163 l,** Russ Lappa; **163 tr,** M.A. Chappell/Animals Animals; **164 tm,** Guido Alberto Rossi/The Image Bank; **164 tr,** Courtesy of Francois Brisse; **165,** Robert Frerck/Odyssey/Chicago; **167,** Charles Gupton/TSI; **169,** Photofest; **172,** Stephen Frisch/SB; **176 l,** Patti Murray/Animals Animals; **176 m,** Russ Lappa; **178 l,** Don & Pat Valenti/DRK Photo; **178 r,** Jeff Foott/DRK Photo.

Chapter 4: Pages 180-181, Richard T. Nowitz/PR; **181,** Peter Menzel/SB; **182,** Framed. ©William Wegman 1995. 20 x 24 inches. Unique Polacolor ER photograph. Courtesy PaceWildenstein MacGill Gallery. New York; **183,** JC; **185,** ©1977 NEA, Inc.; **188,** Visuals Unlimited; **191,** Fernando Serna/Department of the Air Force; **193,** Russ Lappa, **194,** JC; **198,** John M. Roberts/SM; **199,** Corbis-Bettmann; **204,** Superstock; **207,** MT; **209 t,** Mike Penney/David Frazier Photolibrary; **209 m,** Peter Menzel/SB; **209 b,** Lynn McLaren/The

Picture Cube; **211 t,** Photofest; **211 b,** PH; **213,** JC; **216,** George Holton/PR; **217,** PH; **218,** Larry Lefever/Grant Heilman Photography; **219,** JC; **220,** The Granger Collection; **223,** Mike Shirley/Picture Perfect; **224,** Martin Rogers/TSI; **225,** Globus Brothers/SM; **230,** KO; **232,** KO.

Chapter 5: Pages 240-241, John Madere/SB; **241,** Wolfgang Kaehler/Liason International; **242,** Superstock; **244,** Tony Freeman/PhotoEdit; **246,** Bill Gallery/SB; **249,** MT; **251,** Bill Horsman/SB; **254,** MT; **257,** Tom Dietrich/TSI; **260,** Robert Caputo/SB; **263,** Superstock; **264,** Nathan Bilow/Allsport; **267,** Towers/SM; **270,** Terry Donnelly/TSI; **272,** William R. /Duomo; **273 t,** Doll by Mary Tiger. Photo courtesy of the partment of the Interior, Indian Arts and Crafts Board; Granger Collection; **275,** Photo courtesy of Nida- peri Ross/FPG; **279,** Russ Lappa; **280,** Tony FrencEdit; **283 t,** CALVIN AND HOBBES ©Watterson. McDonaSAL PRESS SYNDICATE. Reprinted with ts reserved; **283 b,** Superstock; **286,** Gerald

Chapter 6: Pay Tobias/Sharpshooters; **288,** Joe Leigh/Stock Ima**290,** America Hurrah Archive, NYC. Triangles, Middle MT; **301,** Russ Lappa; **302,** Purchase Fund and Gveless Tunic (?) with Stepped Arts, Boston; **306,** Tom 800 AD., Textile Income Greig Cranna/SB; **312,** Dav Courtesy of Museum of Fine Scala/Art Resource; **314,** KO; DRK Photo; **308,** KO; **311,** **320,** Richard Clintsman/TSI; **32**lff/PhotoEdit; **313,** KO; **326 t,** David Harp/Folio, Inc.; id Sutherland/TSI; Works; **328,** Mark C. Burnett/PR; **330b,** Uniphoto; **323,** Uniphoto; **337,** KO; **338 l,** Mark C. Burnvoord/The Image Burnett/PR; **340 tml,** Richard Hutchings/Pniphoto; **334,** Freeman/PhotoEdit; **340 tr,** David Young-Wo**tl,** Mark C. **b,** Breck Kent/Earth Scenes; **343,** KO; **344,** Don Tony Tom Stewart/SM; **347,** Robert Brenner/PhotoEdit; lit; **340** Corwin/SB; **351,** Frank Fournier/SM. **345,**

Chapter 7: Pages 360-361, T. Tracy/FPG; **360 tl inset,** John Henley/SM; **360 ml inset,** Bob Daemmrich/SB; **360 bl inset,** Bachman/PR; **361,** John Feingersh/SM; **363,** Charles Feil/SB; **367,** Richard Bryant/ARCAID; **368,** David Frazier/TSI; **373,** JC; **374,** JC; **375,** Paul Yandoli/SB; **376,** AP/Wide World Photos; **379 t,** Photo by Richard Cheek; **379 b,** Jock Reynolds, Installation view from Sol Lewitt: Twenty-five Years of Wall Drawings, 1968-1993, ©1993, Addison Gallery of American Art, Phillips Academy, Andover, MA; **381,** JC; **382 t,** JC; **382 b,** Andrew Brookes/TSI; **385 l,** The Granger Collection; **385 r,** Giraudon/Art Resource; **386,** Artothek; **387 l,** D & J Heaton/SB; **387 r,** Jeff Greenberg/SB; **389,** CALVIN AND HOBBES ©Watterson. Dist. by UNIVERSAL PRESS SYNDICATE. Reprinted with permission. All rights reserved; **390 tl,** Melville McLean/Portland Museum of Art; **390 tr,** Courtesy Allan Stone Gallery, New York; **390 br,** Superstock; **392,** ©AAA used by permission; **392 inset,** JC; **393,** AP/Wide World Photos; **395 tr,** Bob Krist/TSI; **395 bl,** Louis Rosendo/FPG; **395 br,** Glen Allison/TSI; **397,** JC.

Chapter 8: Pages 404-405, David Lissy/The Picture Cube; **405,** KO; **407,** Tom Alexander/TSI; **408,** MT; **409,** PH; **411,** PEANUTS reprinted by permission of United Feature Syndicate, Inc.; **415 l,** Amwell/TSI; **415 r,** Superstock; **417,** Charles Winters/PR; **419,** Charlie Westerman/Liason International; **421 l,** Lori Adamski Peek/TSI; **421 r,** Superstock; **423,** Stepheisch/SB; **424,** John Bechtold/International Stock Photo; **429**if Skoogfors/Woodfin Camp; **434,** Jim Ruck/SM; **435,** Miriam Nathan-Roberts; **436,** Charlie Wrman/Liason International; **438,** MT.

Chapter 9: Pages 446-447, Scott Barrowternational Stock Photo; **447,** Joy Syverson/SB; **451,** Gale ker/SB; **454,** KO; **457 l,** Dave Bartruff/SB; **457 r,** MT; **459,** Sepeitz/Woodfin Camp; **462,** Judith Larzelere; **465,** Jose Carillo/otophile; **467,** Gabe Palmer/SM; **471,** Jan Halaska/Index St **473,** Bernard Van Berg/The Image Bank; **474,** Wallace Gason/Index Stock; **475,** PEANUTS reprinted by permission ofiited Feature Syndicate, Inc.; **476,** Don Smetzer/TSI; **479,** Bachn/Photo Network; **481,** Lawrence Naylor/PR; **482,** Geoffrey Cbrd/Woodfin Camp; **484,** John Eastcott/Woodfin Camp; **48**Kunio Owaki/SM.

Chapter 10: Pages 494-495, Chuck Pl/The Image Bank; **495,** Russ Lappa; **497 l,** Jon Riley/TSI; **497** Boden/Ledingham/Masterfile; **498 l,** ss Lappa; **498 m,** Kathleen Campbell/TSI; **498 r,** Telegrh Colour Library/FPG; **500,** ©AAA reprinted with permissio**501,** KO; **502,** NASA; **505,** Francis Lepine/Earth Scenes; **51**Photofest; **513,** Baron Wolman/TSI; **514 t,** Randy Wells/TS**514 l,** Philip & Karen Smith/TSI; **514 m,** John Chard/TSI; **4 tr,** Mulvehill/The Image Works; **518,** Chuck Kuhn/The Imagbank; **520,** William Helsel/TSI; **522,** Joe Sohm/SM; **523,**lichael Dwyer/SB; **524,** JC; **526,** Alan and Linda Detrick/PR; **52** Bob Daemmrich/The Image Works; **531,** B. Swersey/Gama Liason; **534,** Kevin Schafer/TSI; **535,** The Granger Colltion.

Chapter 11: Pages 542-543, Jim Ole/Uniphoto; **543,** KO; **545,** Richard Steedman/SM; **547,** Curtesy of Katoomba Scenic Railway; **548,** Kaluzny/Thatcher/T **552,** AGE Fotostock/First Light; **554,** Toyohiro Yamada/FPG;**57 tl,** Paul Berger/TSI; **557 tr,** Tom Carroll; **557 bl,** Bob Daemrich/SB; **561,** Hideo Kurihara/TSI; **565,** Gary Buss/FPC**566,** THE FAR SIDE ©1993 FARWORKS, INC/Dist. by UNIVRSAL PRESS SYNDICATE. Reprinted with permission. All rigts reserved; **569,** Superstock; **570,** JC; **572,** NOAA; **574,** Guido Aberto Rossi/The Image Bank; **576,** Philip Jon Bailey/SB; **577,** Rah Cowan/FPG.

hapter 12: Pages 584-585, Supertock; **585,** A Joyful Scene of ng, 1969 by Alma Thomas. Frm the collection of Ruth and 58Kainen. Photo by Gene Youig; **587,** Jose Luis Pelaez/SM; pernhen J. Krasemann/PR; **590,** PEANUTS reprinted by Rosentof United Feature Syndicate, Inc.; **594,** Michael Niedorf/T; **597,** MT **598,** David Parker/SPL/PR; **599,** Steve Richard Quaage Bank; **602 l,** Robert Frerck/TSI; **602 r,** Ehlers/TSI; **61**Folio, Inc.; **603,** Chad Slattery/TSI; **604,** Chad Kobal Collection;s Sorensen/SM; 613, Nicola Goode/The David W. Hamilton Timothy Eagan/Woodfin Camp; **616 r,** Stock; **626,** Telegraph Cmage Bank; **623,** Less Reiss/Index r Library/FPG.

Flowchart information on 158 from *Geometry and Its Applications: Symmetry & Pat* by Nancy Crisler, ©1995, COMAP, Inc. Used with permiss